STRUCTURAL BIOINFORMATICS

Second Edition

STRUCTURAL BIOINFORMATICS

Second Edition

Edited By

Jenny Gu, Ph.D.
Philip E. Bourne, Ph.D.

A JOHN WILEY & SONS, INC., PUBLICATION

Published by John Wiley & Sons, Inc., Hoboken, New Jersey
Published simultaneously in Canada

For general information on our other products and services or for technical support, please contact our
Customer Care Department within the United States at (800) 762-2974, outside the United States at (317) 572-3993
or fax (317) 572-4002.

Wiley also publishes its books in a variety of electronic formats. Some content that appears in print may not be
available in electronic formats. For more information about Wiley products, visit our web site at www.wiley.com.

Library of Congress Cataloging-in-Publication Data:

ISBN 978-0-470-18105-8

Printed in the United States of America

10 9 8 7 6 5 4 3 2 1

CONTENTS

Foreword xi

Preface xv

Acknowledgments xix

Contributors xxi

Section I DATA COLLECTION, ANALYSIS, AND VISUALIZATION   1

1 DEFINING BIOINFORMATICS AND STRUCTURAL BIOINFORMATICS   3
Russ B. Altman and Jonathan M. Dugan

2 FUNDAMENTALS OF PROTEIN STRUCTURE   15
Eric D. Scheeff and J. Lynn Fink

3 FUNDAMENTALS OF DNA AND RNA STRUCTURE   41
Stephen Neidle, Bohdan Schneider, and Helen M. Berman

4 COMPUTATIONAL ASPECTS OF HIGH-THROUGHPUT CRYSTALLOGRAPHIC MACROMOLECULAR STRUCTURE DETERMINATION   77
Paul D. Adams, Ralf W. Grosse-Kunstleve, and Axel T. Brunger

5 MACROMOLECULAR STRUCTURE DETERMINATION BY NMR SPECTROSCOPY   93
John L. Markley, Arash Bahrami, Hamid R. Eghbalnia, Francis C. Peterson, Robert C. Tyler, Eldon L. Ulrich, William M. Westler, and Brian F. Volkman

6   ELECTRON MICROSCOPY IN THE CONTEXT
    OF STRUCTURAL SYSTEMS BIOLOGY   143
    Niels Volkmann and Dorit Hanein

7   STUDY OF PROTEIN THREE-DIMENSIONAL STRUCTURE
    AND DYNAMICS USING PEPTIDE AMIDE HYDROGEN/
    DEUTERIUM EXCHANGE MASS SPECTROMETRY
    (DXMS) AND CHEMICAL CROSS-LINKING WITH
    MASS SPECTROMETRY TO CONSTRAIN MOLECULAR
    MODELING   171
    Sheng Li, Dmitri Mouradov, Gordon King, Tong Liu, Ian Ross,
    Bostjan Kobe, Virgil L. Woods Jr, and Thomas Huber

8   SEARCH AND SAMPLING IN STRUCTURAL
    BIOINFORMATICS   207
    Ilan Samish

9   MOLECULAR VISUALIZATION   237
    Steven Bottomley and Erik Helmerhorst

Section II   DATA REPRESENTATION
             AND DATABASES   269

10 THE PDB FORMAT, mmCIF FORMATS, AND OTHER
    DATA FORMATS   271
    John D. Westbrook and Paula M.D. Fitzgerald

11 THE WORLDWIDE PROTEIN DATA BANK   293
    Helen M. Berman, Kim Henrick, Haruki Nakamura, and John L. Markley

12 THE NUCLEIC ACID DATABASE   305
    Bohdan Schneider, Joanna de la Cruz, Zukang Feng, Li Chen,
    Shuchismita Dutta, Irina Persikova, John D. Westbrook, Huanwang Yang,
    Jasmine Young, Christine Zardecki, and Helen M. Berman

13 OTHER STRUCTURE-BASED DATABASES   321
    J. Lynn Fink, Helge Weissig, and Philip E. Bourne

Section III DATA INTEGRITY AND COMPARATIVE FEATURES   339

14 STRUCTURAL QUALITY ASSURANCE   341
Roman A. Laskowski

15 THE IMPACT OF LOCAL ACCURACY IN PROTEIN AND RNA STRUCTURES: VALIDATION AS AN ACTIVE TOOL   377
Jane S. Richardson and David C. Richardson

16 STRUCTURE COMPARISON AND ALIGNMENT   397
Marc A. Marti-Renom, Emidio Capriotti, Ilya N. Shindyalov, and Philip E. Bourne

17 PROTEIN STRUCTURE EVOLUTION AND THE SCOP DATABASE   419
Raghu P. R. Metpally and Boojala V. B. Reddy

18 THE CATH DOMAIN STRUCTURE DATABASE   433
Frances M. G. Pearl, Alison Cuff, and Christine A. Orengo

Section IV STRUCTURAL AND FUNCTIONAL ASSIGNMENT   457

19 SECONDARY STRUCTURE ASSIGNMENT   459
Claus A. Andersen and Burkhard Rost

20 IDENTIFYING STRUCTURAL DOMAINS IN PROTEINS   485
Stella Veretnik, Jenny Gu, and Shoshana Wodak

21 INFERRING PROTEIN FUNCTION FROM STRUCTURE   515
James D. Watson, Gail J. Bartlett, and Janet M. Thornton

22 STRUCTURAL ANNOTATION OF GENOMES   539
Adam J. Reid, Corin Yeats, Jonathan Lees, and Christine A. Orengo

23 EVOLUTION STUDIED USING PROTEIN STRUCTURE   559
    Song Yang, Ruben Valas, and Philip E. Bourne

Section V MACROMOLECULAR INTERACTIONS   573

24 ELECTROSTATIC INTERACTIONS   575
    Nathan A. Baker and J. Andrew McCammon

25 PREDICTION OF PROTEIN–NUCLEIC ACID
    INTERACTIONS   593
    Timothy Robertson and Gabriele Varani

26 PREDICTION OF PROTEIN–PROTEIN INTERACTIONS
    FROM EVOLUTIONARY INFORMATION   615
    Alfonso Valencia and Florencio Pazos

27 DOCKING METHODS, LIGAND DESIGN, AND
    VALIDATING DATA SETS IN THE STRUCTURAL
    GENOMICS ERA   633
    Natasja Brooijmans

Section VI STRUCTURE PREDICTION   663

28 CASP AND OTHER COMMUNITY-WIDE
    ASSESSMENTS TO ADVANCE THE FIELD
    OF STRUCTURE PREDICTION   665
    Jenny Gu and Philip E. Bourne

29 PREDICTION OF PROTEIN STRUCTURE IN 1D:
    SECONDARY STRUCTURE, MEMBRANE REGIONS,
    AND SOLVENT ACCESSIBILITY   679
    Burkhard Rost

30 HOMOLOGY MODELING   715
    Hanka Venselaar, Elmar Krieger, and Gert Vriend

31 FOLD RECOGNITION METHODS   733
    Adam Godzik

32 *DE NOVO* PROTEIN STRUCTURE PREDICTION: METHODS AND APPLICATION 755
Kevin Drew, Dylan Chivian, and Richard Bonneau

33 RNA STRUCTURAL BIOINFORMATICS 791
Magdalena A. Jonikas, Alain Laederach, and Russ B. Altman

Section VII THERAPEUTIC DISCOVERY 807

34 STRUCTURAL BIOINFORMATICS IN DRUG DISCOVERY 809
William R. Pitt, Alícia Pérez Higueruelo, and Colin R. Groom

35 B-CELL EPITOPE PREDICTION 847
Julia V. Ponomarenko and Marc H.V. van Regenmortel

Section VIII FUTURE CHALLENGES 879

36 METHODS TO CLASSIFY AND PREDICT THE STRUCTURE OF MEMBRANE PROTEINS 881
Marialuisa Pellegrini-Calace and Janet M. Thornton

37 PROTEIN MOTION: SIMULATION 907
Ilan Samish, Jenny Gu, and Michael L. Klein

38 THE SIGNIFICANCE AND IMPACTS OF PROTEIN DISORDER AND CONFORMATIONAL VARIANTS 937
Jenny Gu and Vincent J. Hilser

39 PROTEIN DESIGNABILITY AND ENGINEERING 961
Nikolay V. Dokholyan

40 STRUCTURAL GENOMICS OF PROTEIN SUPERFAMILIES 983
Stephen K. Burley, Steven C. Almo, Jeffrey B. Bonanno, Mark R. Chance, Spencer Emtage, Andras Fiser, Andrej Sali, J. Michael Sauder, and Subramanyam Swaminathan

INDEX 1019

FOREWORD

The quality of the coverage and the timeliness of the first edition of *Structural Bioinformatics* led to its wide usage. In turn, the collection has been adopted by the community as a defining articulation for the utility of bioinformatics and structural biology applied to address a range of functional studies in 21st Century Biology. I personally found the book to be read by students, postdoctoral fellows and more senior researchers around the world. The success reflected the excitement in the growing impact of both domains and also helped to accelerate that impact. As pointed out in the Introduction, bioinformatics is now a mainstream activity within the biological sciences; similarly, the implementation of the structural genomics initiative worldwide as a logical and necessary follow up to the Human Genome Project represents the consensus that structure is indeed important for understanding the mechanisms of molecular function. The increased impact and rapid progress during the 6 years, since the first edition was published, demonstrate the extraordinary significance of the interface between structural biology and informatics. Such significant advances indicate the clear need for an update, which has now been provided through the editorial efforts of Gu and Bourne and the writing team of leading researchers who bring the reader to the edge of the frontier.

The second edition, simultaneously as a textbook and an expert monograph, contains a balanced set of contributions, which include updates and advances over the past 6 years and new innovative domains made possible by the sustained application of structural bioinformatics, which were undoubtedly catalyzed by the first edition. Just as was notable about the first edition, this new comprehensive collection fully captures the spirit of excitement at the "bleeding edge" of biodiscovery. The frontier between computing and biology itself reflects the decades of extraordinary progress and revolutionary advances in both domains. During the past 6 years, the data provided by the completion of the full human and numerous model genomes has accelerated the importance of computing for biology, and brought new funding opportunities and research training needs. At the same time, the deeper insight from complete genome sequencing has been the need to look beyond individual genes to systems and to look across multiple scales of biology, ultimately to establish an integrative view of biology based on experimentation and computation. With funding from within mainstream science programs and the increased recognition by the experimental community, the combined use of information technology and quantitative approaches are central to building an integrative view of biology. The superb collection of articles in this second edition speaks directly and powerfully for the role of structural bioinformatics in this effort for both the basic and the applied life sciences.

Additional academic training programs that include structural bioinformatics have been introduced consistently over the past 6 years. Yet, faculty in top flight research institutions who do research and teach in first class bioinformatics programs remain overwhelmed by the demands; indeed, the challenges of bioinformatics education has been a common theme at the major professional meetings in the field. Thus, textbooks of the highest quality and clarity are essential today, as we all struggle to determine the right curriculum and the right content to train what will be the first generation of students who are truly bioinformaticians. Today, every young aspiring biologist wants to learn about bioinformatics, as do those training in computer science and other quantitative sciences. Understanding the underlying assumptions and intricacies of any bioinformatics algorithm is necessary for proper usage and interpretation of the results obtained with the tools. To teach the large and growing numbers of young scientists who wish to utilize or even contribute to bioinformatics requires authoritative treatments that provide the basis to pull in a new generation of scientists. Such works must also use the best treatments possible to reach a much larger audience, including mature scientists who wish to retrain themselves, and need to set a standard for training everywhere in the world. This collection admirably meets those goals and is a must read for those entering the field and for all of us committed to understanding the interplay of structure and function. By assembling the best thinkers to address systematically all of the challenges at the next stage of the genome effort, Gu and Bourne have created a book that will serve to educate the next generation who will be the future young investigators who will create the tools required to interpret the ever advancing frontier of biology.

At the same time, while the rapid pace of research progress in structural bioinformatics has driven the need for a second edition, the prehistory of modern structural bioinformatics, as is well described in this book, has been retained to remind the readers of some fundamental challenges that should not be neglected and forgotten. This update is not an extensive replicate with minor tweaks of the first edition; instead, the role of historical context and origins of structural bioinformatics as they contribute to present advances are discussed, such as the biannual Critical Assessment of Structure Prediction (CASP) and the initial "exponentiation" of progress in biology resulting from the application of advanced computing technology to structural biology. For more historical content, I refer any reader to the forward for the first edition of *Structural Bioinformatics* or any of my status reports (over the past decade) on computational biology, readily accessible via the internet (or PubMed) in today's world of "e-knowledge."

What is most notable and important for the potential readership of this edition is that the fields of computational biology and bioinformatics have gone from uncertainty and neglect to the buzz words on everyone's lips in less than 10 years. While newly created, contemporary bioinformatics and computational biology training programs are in the process of being incorporated into every biological science domain, and those working at the interface will have increased options even within core disciplines. The reason is obvious: we are already living in the future as biologists; we feel fully the impact of having completely sequenced genomes; we are a part of the transition to high throughput and high information content biological research, and to asking global or systemic questions as the norm, rather than using individual macromolecule-specific probes. Early in this century, the extraordinary, early, and even unanticipated successes of the genome project enabled computer search and modeling techniques to open up new vistas in biology.

The continuing availability of complete genomes coupled with high throughput experimental biological methods, structure determination, and annotated databases will definitely advance our current understanding of protein structures as they relate to biological function, processes, and evolution—a basic research curiosity that has captured central stage in the community. The challenge is to decode the rich information content implicit in genomes and apply the resulting knowledge in service to society. Obtaining a better understanding of biological processes will be achieved through the integration of a variety of methods including the use of structural information, and should be extended in the future to provide improved health care delivery. At the same time, we are only learning how best to exploit computer and information technology to understand biological mechanisms; the educational content of *Structural Bioinformatics* provides a perspective on our progress and educational content to enable the full potential of systematic computational analysis.

Those of us concerned with macromolecular structure, and protein science in particular, have long spoken the mantra: form follows function, a given function requires a specific structure, or, conversely, structure in turn can be seen to determine function. That is, if we know the structure, we can infer many aspects of biochemical and sometimes, even cellular function which can subsequently be experimentally tested. Indeed, given that we now "know" many gene sequences gained implicitly from sequenced genomes, such resources provide the basis for more refined algorithms that either leverage structural information or improve our understanding of protein structures to model them explicitly and more accurately. Subsequently, these improvements facilitate our ability to predict functionality and greatly reduce the search space for experimental efforts by providing a guided focus to test only the most likely functions. Furthermore, computational modeling of these molecules, for both static and dynamic processes, can provide a detailed description of biological processes at the atomic level, an alternative to traditional biological cartoons which have been the descriptive ways in which we biologists think.

The field of structural bioinformatics, to connect the abstract to the practical, continues to push boundaries beyond what was previously thought impossible. The basis for all structural bioinformatics, the central community database for structural biology, is the Protein Data Bank (PDB) and the macromolecule structures it contains. The growth of this resource has been accelerated by structural genomics initiatives, which continue to sustain and enhance the discovery of novel structures and folds paving the way for new opportunities of discovery and insight through structural bioinformatics. Some examples include a more accurate genome annotation and the basis for clarifying evolutionary related questions. The book addresses key points for cutting edge research such as these, beginning with definitions, and conveys the current scope of research and knowledge of protein structures.

Central to this wonderful collection and insightful articles are advances that have been made in building the infrastructure for an integrative approach to understanding biosystems through the power of understanding protein structures and the implications for the mechanics of function. A survey of current resources is provided to highlight the foundation where future developments are needed to integrate experimental data better and provide the basis for abstraction and generalization. Overall, the individual chapters outline the suite of major basic life science questions such as the status of efforts to predict protein structure and how proteins carry out cellular functions, and also the applied life science questions such as how structural bioinformatics can improve health care through accelerating drug discovery. Dictated by the process of uncovering the mechanisms through which macromolecules act, this journey of discovery into the regulation of life's processes will keep biologists entertained for centuries to come. The second edition book is a great guidebook, even more informative than the earlier collection, and represents the basis for this journey. I highly recommend it to all members of our community.

John C. Wooley
Associate Vice Chancellor, Research
University of California, San Diego
La Jolla, CA

PREFACE

Six years have elapsed since the first edition of this book was published. The field of structural bioinformatics has sustained a high level of excitement in that time, leading to innovative developments and considerable progress throughout the topics covered in the first edition and in the extension to many new domains. Through the efforts of the authors of this new edition, we have tried to capture these developments and to provide an accurate, detailed view of the current field. One way of picturing the advances or defining the "structural" change is relatively straightforward; namely, the number of experimental macromolecular structures has doubled since the first edition of this book was published. The Protein Structure Initiative has also led to an increase in the number of novel structures and folds. Overall, the continued growth in experimental structures has created an even richer data source for much of the work described herein. But numbers do not tell the whole story. The complexity of structures, the methods used, the ways structure is represented, our ability to model structures, our understanding of proteomes and their structural coverage, and so on, have also changed.

Describing the advances in "bioinformatics" per se is more difficult. Change in this case reflects both scientific advances and an increase in recognition within the biological sciences for the importance of computational methods. Due in part to the explosion in high-throughput experimental methods, bioinformatics is certainly more mainstream than it was 6 years ago and most experimental (i.e., non-computational) life scientists would acknowledge the role bioinformatics now plays in furthering our understanding of living systems. Some years from now, whether or not bioinformatics will exist as a separate entity, rather than as a core effort in every biological science department, is a subject for debate. What is important here is that there is an active effort to apply computational methods to a rapidly growing corpus of macromolecular structure data. Our primary goal is to provide a comprehensive description of what this field has accomplished to date and to make the reader aware of what we have gained and could gain in the future toward our understanding of living systems through the study of macromolecular structure and the continued, rigorous application of bioinformatics. As such, this edition should provide a fully current, useful reference to those already in the field, and a suitable text for those educating others. The first edition already encouraged new scholars to enter the field and we believe the case for engaging in structural bioinformatics is stronger than ever.

To meet this goal, the second edition includes not only updated chapters, but also new chapters covering mass spectrometry, genome annotation, immunology, protein dynamics and disorder, membrane proteins, protein design capabilities, and evolutionary biology as they relate to macromolecular structure.

Macromolecular structure is often underappreciated and bypassed in practice during the current era of high-throughput biology, especially since researchers can jump directly from genomic sequence to phenotype and conduct biochemical studies on large-scale protein–protein interactions that are often involved in pathways associated with disease states. While a great deal can be learned from such studies, ultimately the devil is in the details which do not arise from traditional functional studies; the field of molecular biophysics, now termed structural biology, came into existence to obtain and use structures to provide those details. We believe structure will play an ever increasingly important role as genome studies seek to explore deeper into the mechanisms of life. As such, computational approaches that analyze structure are essential and we hope this book will be there to guide you.

We begin by describing the scope of this book and the history of the field (Chapter 1). The remainder of the introductory Section I is devoted to the understanding of the data itself, namely protein, DNA and RNA structure, respectively (Chapters 2 and 3). Understanding the nuances (scope, accuracy, completeness, etc.) of structural data is prerequisite to any effective use of that data. Effective data use in turn requires an understanding of the experiments or experimental method that produce the data. The most popular methods for deriving macromolecular structure data are, in order, X-ray crystallography (Chapter 4), NMR spectroscopy (Chapter 5), and electron microscopy (Chapter 6). Constructing structural models of molecules can also be guided with hydrogen–deuterium and cross-linking experiments coupled with mass spectrometry (Chapter 7). The raw data from these methods are most often a set of Cartesian coordinates representing the positions of the atoms in these structures, which are well suited for analysis by computer, but alternative representations of this information rich content are sometimes needed to conduct wide-scale bioinformatics analysis (Chapter 8). That is, structural biology and structural bioinformatics are inherently visual sciences—the tabular output of atomic coordinates can be useful as input for computation, but not for human insight. The visualization of structure has evolved along with the science and many useful tools, mostly free, are available (Chapter 9).

In the early days of structural biology (up to the late 1970s), those in the field could name all the structures that had been solved, some of which had Nobel prizes attached to them. As the field grew this was no longer possible, and databases of structure data began to appear. Consistent use of structural data contained within these databases (and indeed the construction of the databases themselves) requires consistent data representation and Section II is devoted to this topic. Chapter 10 introduces the common data representations used by today's software. The field is very fortunate to have scientists who recognize the importance of having a single source of primary data, the worldwide PDB (wwPDB—Chapter 11), from which a variety of secondary resources are derived. Examples of such resources are provided in Chapters 12 and 13.

As the number of structures has increased, much can be learnt from comparative analysis (Section III), where similarities and differences provide new insights. Chapters 14 and 15 describe structure validation, which is important in understanding the accuracy of the data you are dealing with before a 3D comparison and alignment of structures can be made (Chapter 16). When structure comparisons are made and similarities found, reductionism can be applied to make sense of the vast amount of data. Such reductionism leads to classification in various ways, such as by fold, domain, family, and superfamily (Chapters 17 and 18).

The more we know from comparing structures the more we can learn about structure and functional assignment (Section IV). Secondary structure assignment can now be made consistently and reliably for the majority of structures (Chapter 19). Proteins exist as one or more domains or compact structural and functional units. Hence, automated assignment of domains is important (Chapter 20). Through the structural genomics projects and the NIH

Protein Structure Initiative, structure determination is moving from a functional to a genomic initiative. That is, structures were traditionally determined in an effort to elucidate further details about a known function, and these structural efforts were established based on very extensive prior biological, biochemical and often genetic research, and were done in parallel with continuing biological research on functional properties. In contrast, high-resolution structures with no elucidated or known function are being determined at an accelerated pace, thus making functional assignment critical (Chapter 21). The use of structural information to identify distantly related proteins also serves in annotating genomes (Chapter 22) and clarifying evolutionary relationships (Chapter 23).

Proteins do not act in isolation, that is, most proteins do not function by themselves but act as the result of complex protein–protein, protein–ligand and protein–solvent interactions and are often part of larger macromolecular assemblies. Section V describes these interactions beginning with an introduction to electrostatic forces that have a fundamental impact on recognition between molecules (Chapter 24). The majority of these interactions are not captured in the experimental structure of a complex, but as an apo form of the structure with a signature that can be teased out to predict that interaction. Understanding these signatures when found in protein–DNA and protein–RNA interactions (Chapter 25) and in protein–protein interactions (Chapter 26) aids, for example, in the identification of new transcription sites and reconstruction of protein signaling networks. After the sites of interactions are identified, docking of the molecules is simplified, which is important in drug design (Chapter 27).

While the number of structures is increasing rapidly, the number of protein sequences is increasing much more rapidly; thus, the idea of predicting protein structure from its sequence remains an "obsessive" goal (Section VI). Spurred by an unusual biannual competition, referred to as CASP—the Critical Assessment of Protein Structure Prediction (Chapter 28), progress is being made in subcategories of structure prediction efforts within CASP and the field in general. Structure prediction categories include homology modeling (Chapter 30), fold recognition (Chapter 31) and *ab initio* structure prediction (Chapter 32). Other forms of prediction include secondary structure and membrane components for proteins (Chapter 29). Advances in understanding and predicting RNA structures have also been made and are discussed in Chapter 33.

Structural bioinformatics is playing an increasingly important role in the development of new pharmaceuticals (Section VII). The identification of drug targets, understanding the action of drug binding, and the design of promising leads all involve structural bioinformatics (Chapter 34). In addition to the development of small molecule and peptide-based drugs, contributions are also being made in identifying antigen recognition sites that aid in antibody-based therapeutics (Chapter 35).

Finally, Section VIII identifies challenges at the frontiers of structural bioinformatics. Membrane associated proteins, whose structures are difficult to characterize *in vitro* and thus are underrepresented experimentally, are one example (Chapter 36). Proteins are not static under physiological conditions, yet understanding the dynamics (Chapter 37) and the impact of disorder and conformational variants (Chapter 38), while important to protein function, are all still poorly understood. As our understanding of protein structure improves, so do our design rules and capacity for engineering new proteins for their functions that improve upon nature or provide the potential for novel processes (Chapter 39). The best way to push back these frontiers is with more structures and enhanced generalizations about their roles; structure genomics is doing just that (Chapter 40) and is thus a fitting place to end our tour of structural bioinformatics.

The words that follow are written by many of the leaders in the field and we thank them for their time and energy in sharing what motivates them to unravel the mysteries of nature, which are so beautifully displayed before us in an ever increasing number of macromolecular structures.

Jenny Gu
Philip E. Bourne

Color files of all figures from this book are available for download from the following web address: ftp://ftp.wiley.com/public/sci_tech_med/structural_bioinformatics

ACKNOWLEDGMENTS

"Science does not know its debt to imagination."
Ralph Waldo Emerson

We are grateful to the University of Texas Medical Branch, the University of Münster, and the University of California at San Diego for institutional support in our research and educational endeavors, including this book. Likewise, we thank the Jeanne Kempner Foundation and the US funding agencies, the National Science Foundation, and the National Institutes of Health for their continual support in the advancement of science. This book could not have been completed without the help of all the contributors who have dedicated their expertise, time, and efforts to this extensive update and indispensable resource for the community—we are greatly indebted to them. Additionally, we would like to thank Wolfgang Bluhm, Kristine Briedis, Naomi Cotton, Lynn Fink, Apostol Gramada, Maria Kontoyianni, Kannan Natarajan, Julia Ponomarenko, Peter Rose, Wayne Townsend-Merino, Ruben Valas, Stella Veretnik, Lei Xie, Song Yang, and Zhanyang Zhu for their assistance and support in reviewing the materials for this book. The successful publication of this edition could not have been achieved without the help and patience of Andrea Baier, Tiffany Williams, and Thomas Moore at Wiley-Blackwell.

On a more personal note, JG would like to thank her family, friends, colleagues and mentors for their continued support in taking on new challenging endeavors. PEB would like to sincerely thank his family, Roma, Melanie, and Scott, for their continued understanding of the role science plays in his life.

CONTRIBUTORS

Paul D. Adams, Physical Biosciences Division, Lawrence Berkeley Laboratory, Berkeley, CA and Department of Bioengineering, University of California, Berkeley, CA

Steven C. Almo, Albert Einstein College of Medicine, Bronx, NY

Russ B. Altman, Departments of Bioengineering & Genetics, Stanford University, Stanford, CA

Claus A. Andersen, Siena Biotech Spa., Siena, Italy

Arash Bahrami, Graduate Program in Biophysics and National Magnetic Resonance Facility at Madison, Biochemistry Department, University of Wisconsin-Madison, Madison, WI

Nathan A. Baker, Department of Biochemistry and Molecular Biophysics, Center for Computational Biology, Washington University, St. Louis, MO

Gail J. Bartlett, School of Chemistry, University of Bristol, Cantock's Close, Clifton, Bristol, UK

Helen M. Berman, RCSB Protein Data Bank, Department of Chemistry and Chemical Biology, Rutgers, The State University of New Jersey, Piscataway, NJ

Jeffrey B. Bonanno, Albert Einstein College of Medicine, Bronx, NY

Richard Bonneau, Department of Biology, New York University, New York, NY; and Department of Computer Science, Courant Institute, New York University, New York, NY

Steven Bottomley, School of Biomedical Sciences, Curtin University of Technology, Perth, Western Australia, Australia

Philip E. Bourne, Skaggs School of Pharmacy and Pharmaceutical Sciences, University of California, San Diego, La Jolla, CA

Natasja Brooijmans, Wyeth, Madison, NJ

Axel T. Brunger, The Howard Hughes Medical Institute and Departments of Molecular and Cellular Physiology, Neurology and Neurological Sciences, Structural Biology, and Stanford Synchrotron Radiation Laboratory, Stanford University, Stanford, CA

Stephen K. Burley, Eli Lilly and Company, San Diego, CA

Emidio Capriotti, Structural Genomics Unit, Bioinformatics and Genomics Department, Centro de Investigación Príncipe Felipe, Valencia, Spain

Mark R. Chance, Case Western Reserve University, Cleveland, OH

Li Chen, Department of Chemistry and Chemical Biology, Rutgers, The State University of New Jersey, Piscataway, NJ

Dylan Chivian, Physical Biosciences Division, Lawrence Berkeley National Laboratory, Berkeley, CA

Alison Cuff, Institute of Structural and Molecular Biology, University College London, London, UK

Joanna de la Cruz, Department of Chemistry and Chemical Biology, Rutgers, The State University of New Jersey, Piscataway, NJ

Nikolay V. Dokholyan, Department of Biochemistry and Biophysics, School of Medicine, The University of North Carolina at Chapel Hill, Chapel Hill, NC

Kevin Drew, Department of Biology, New York University, New York, NY; and Department of Computer Science, Courant Institute, New York University, New York, NY

Jonathan M. Dugan, Stanford University, Stanford, CA

Shuchismita Dutta, Department of Chemistry and Chemical Biology, Rutgers, The State University of New Jersey, Piscataway, NJ

Hamid R. Eghbalnia, Department of Molecular and Cellular Physiology, University of Cincinnati, Cincinnati, OH

Spencer Emtage, Eli Lilly and Company, San Diego, CA

Zukang Feng, Department of Chemistry and Chemical Biology, Rutgers, The State University of New Jersey, Piscataway, NJ

J. Lynn Fink, Skaggs School of Pharmacy and Pharmaceutical Sciences, University of California, San Diego, La Jolla, CA

Andras Fiser, Albert Einstein College of Medicine, Bronx, NY

Paula M. D. Fitzgerald, Merck Research Laboratories, Boston, MA

Adam Godzik, Burnham Institute for Medical Research, La Jolla, CA

Ralf W. Grosse-Kunstleve, Physical Biosciences Division, Lawrence Berkeley Laboratory, Berkeley, CA

Colin R. Groom, Biochemistry Department, University of Cambridge, Cambridge, UK

Jenny Gu, Institute for Evolution and Biodiversity, University of Münster, Münster, Germany

Dorit Hanein, Burnham Institute for Medical Research, La Jolla, CA

Erik Helmerhorst, School of Biomedical Sciences, Curtin University of Technology, Perth, Western Australia, Australia

Kim Henrick, Macromolecular Structural Database, European Bioinformatics Institute, Cambridge, UK

Alícia Pérez Higueruelo, Biochemistry Department, University of Cambridge, Cambridge, UK

Vincent J. Hilser, Sealy Center for Structural Biology and Molecular Biophysics, University of Texas Medical Branch, Galveston, TX

Thomas Huber, The University of Queensland, School of Molecular and Microbial Sciences, Brisbane, Queensland, Australia

Magdalena A. Jonikas, Department of Bioengineering, Stanford University, Stanford, CA

Gordon King, The University of Queensland, Institute for Molecular Bioscience and ARC Special Research Centre for Functional and Applied Genomics, Brisbane, Queensland, Australia

Michael L. Klein, Center for Molecular Modeling, Department of Chemistry, University of Pennsylvania, Philadelphia, PA

Bostjan Kobe, The University of Queensland, School of Molecular and Microbial Sciences and Institute for Molecular Bioscience and ARC Special Research Centre for Functional and Applied Genomics, Brisbane, Queensland, Australia

Elmar Krieger, CMBI, NCMLS, Radboud University Medical Centre, Nijmegen, The Netherlands

Alain Laederach, Biomedical Sciences Program, School of Public Health, SUNY, Albany, NY

Roman A. Laskowski, European Bioinformatics Institute, Hinxton, Cambridge, UK

Jonathan Lees, Department of Biochemistry & Molecular Biology, University College London, London, UK

Sheng Li, Department of Medicine and Biomedical Sciences Graduate Program, University of California, San Diego, La Jolla, CA

Tong Liu, Department of Medicine and Biomedical Sciences Graduate Program, University of California, San Diego, La Jolla, CA

John L. Markley, Biochemistry Department, University of Wisconsin-Madison, Madison, WI

Marc A. Marti-Renom, Structural Genomics Unit, Bioinformatics and Genomics Department, Centro de Investigación Príncipe Felipe, Valencia, Spain

J. Andrew McCammon, Departments of Chemistry & Biochemistry and Pharmacology, Howard Hughes Medical Institute, University of California, San Diego, La Jolla, CA

Raghu P. R. Metpally, Bioinformatics and In Silico Drug Design Lab., Computer Science Department, Queens College of City University of New York, Flushing, NY

Dmitri Mouradov, The University of Queensland, School of Molecular and Microbial Sciences, Brisbane, Queensland, Australia

Haruki Nakamura, PDBj, Institute for Protein Research, Osaka University, Osaka, Japan

Stephen Neidle, School of Pharmacy, University of London, London, UK

Christine A. Orengo, Department of Biochemistry & Molecular Biology, University College London, London, UK

Florencio Pazos, Computational Systems Biology Group, National Center for Biotechnology (CNB-CSIC), C/Darwin, Madrid, Spain

Frances M. G. Pearl, EMBL/EBI, The Wellcome Trust Genome Campus, Cambridge, UK

Marialuisa Pellegrini-Calace, EMBL/EBI, The Wellcome Trust Genome Campus, Cambridge, UK

Irina Periskova, Department of Chemistry and Chemical Biology, Rutgers, The State University of New Jersey, Piscataway, NJ

Francis C. Peterson, Department of Biochemistry, Medical College of Wisconsin, Milwaukee, WI

William R. Pitt, UCB Celltech, Slough, UK

Julia V. Ponomarenko, Skaggs School of Pharmacy & Pharmaceutical Science, San Diego Supercomputer Center, University of California, San Diego, CA

Boojala V. B. Reddy, Bioinformatics and In Silico Drug Design Lab., Computer Science Department, Queens College of City University of New York, Flushing, NY

Adam J. Reid, Department of Biochemistry & Molecular Biology, University College London, London, UK

David C. Richardson, Department of Biochemistry, Duke University, Durham, NC

Jane S. Richardson, Department of Biochemistry, Duke University, Durham, NC

Timothy Robertson, Department of Biochemistry, University of Washington, Seattle, WA

Ian Ross, The University of Queensland, Institute for Molecular Bioscience and ARC Special Research Centre for Functional and Applied Genomics, Brisbane, Queensland, Australia

Burkhard Rost, CUBIC, Department of Biochemistry and Molecular Biophysics, Columbia University, New York, NY; Columbia University Center for Computational Biology and Bioinformatics (C2B2), New York, NY; Northeast Structural Genomics Consortium (NESG), Columbia University, New York, NY; and Department of Pharmacology, Columbia University, New York, NY

Andrej Sali, University of California, San Francisco, CA

Ilan Samish, Department of Biochemistry & Biophysics, and Department of Chemistry, University of Pennsylvania, Philadelphia, PA

J. Michael Sauder, Eli Lilly and Company, San Diego, CA

Eric D. Scheeff, Razavi Newman Center for Bioinformatics, The Salk Institute for Biological Studies, La Jolla, CA

Bohdan Schneider, Institute of Biotechnology of the Academy of Sciences of the Czech Republic, Prague, Czech Republic

Ilya N. Shindyalov, San Diego Supercomputer Center, San Diego, La Jolla, CA

Subramanyam Swaminathan, Brookhaven National Laboratory, Upton, NY

Janet M. Thornton, European Bioinformatics Institute, Wellcome Trust Genome Campus, Hinxton, Cambridge, UK

Robert C. Tyler, Department of Biochemistry, Medical College of Wisconsin, Milwaukee, WI

Eldon L. Ulrich, BMRB, Biochemistry Department, University of Wisconsin-Madison, Madison, WI

Ruben Valas, Bioinformatics Graduate Program, University of California, San Diego, La Jolla, CA

Alfonso Valencia, Structural Biology and Biocomputing Programme, Spanish National Cancer Research Centre (CNIO), Madrid, Spain

Marc H. V. van Regenmortel, Biotechnology School of the University of Strasbourg, CNRS, Illkirch, France

Gabriele Varani, Departments of Biochemistry and Chemistry, University of Washington, Seattle, WA

Hanka Venselaar, CMBI, NCMLS, Radboud University Medical Centre, Nijmegen, The Netherlands

Stella Veretnik, San Diego Supercompter Center, University of California, San Diego, La Jolla, CA

Brian F. Volkman, Department of Biochemistry, Medical College of Wisconsin, Milwaukee, WI

Niels Volkmann, Burnham Institute for Medical Research, La Jolla, CA

Gert Vriend, CMBI, NCMLS, Radboud University Medical Centre, Nijmegen, The Netherlands

James D. Watson, European Bioinformatics Institute, Wellcome Trust Genome Campus, Hinxton, Cambridge, UK

Helge Weissig, Activx Biosciences, San Diego, CA

John D. Westbrook, Department of Chemistry and Chemical Biology, Rutgers, The State University of New Jersey, Piscataway, NJ

William M. Westler, National Magnetic Resonance Facility at Madison, Biochemistry Department, University of Wisconsin-Madison, Madison, WI

Shoshana Wodak, The Hospital for Sick Children, Toronto, Ontario, Canada

Virgil L. Woods Jr., Department of Medicine and Biomedical Sciences Graduate Program, University of California, San Diego, La Jolla, CA

Huanwang Yang, Department of Chemistry and Chemical Biology, Rutgers, The State University of New Jersey, Piscataway, NJ

Song Yang, Skaggs School of Pharmacy and Pharmaceutical Sciences, La Jolla, CA

Corin Yeats, Department of Biochemistry & Molecular Biology, University College London, London, UK

Jasmine Young, Department of Chemistry and Chemical Biology, Rutgers, The State University of New Jersey, Piscataway, NJ

Christine Zardecki, Department of Chemistry and Chemical Biology, Rutgers, The State University of New Jersey, Piscataway, NJ

Section I

DATA COLLECTION, ANALYSIS, AND VISUALIZATION

1

DEFINING BIOINFORMATICS AND STRUCTURAL BIOINFORMATICS

Russ B. Altman and Jonathan M. Dugan

WHAT IS BIOINFORMATICS?

The precise definition of bioinformatics is a matter of debate. Some define it narrowly as the development of databases to store and manipulate genomic information. Others define it broadly as encompassing all of computational biology. Based on its current use in the scientific literature, bioinformatics can be defined as the study of two information flows in molecular biology (Altman, 1998). The first information flow is based on the central dogma of molecular biology: DNA sequences are transcribed into mRNA sequences; mRNA sequences are translated into protein sequences; and protein sequences fold into three-dimensional structures that have functions. These functions are selected, in a Darwinian sense, by the environment of the organism, which drives the evolution of the DNA sequence within a population. The first class of bioinformatics applications, then, can address the transfer of information at any stage in the central dogma, including the organization and control of genes in the DNA sequence, the identification of transcriptional units in DNA, the prediction of protein structure from sequence, and the analysis of molecular function. These applications include the emergence of system-wide analyses of biological phenomenon, now called systems biology. Systems biology aims to achieve quantitative understanding not only of the individual players in a biological system but also of the properties of the system itself that emerge from the interaction of all its parts. This field also includes the new field of metagenomics, where we study entire ecosystems of interacting organisms. In the same way that systems biology studies how the molecular entities in a cell combine to make the cell work, metagenomics studies how the individual organisms within an ecological system combine to create that ecology. The initial forays into metagenomics are based on high-throughput sequencing not of individual species (that generally cannot be isolated) but of the mixture of species that create an ecosystem.

The second information flow is based on the scientific method: we create hypotheses regarding biological activity, design experiments to test these hypotheses, evaluate the resulting data for compatibility with the hypotheses, and extend or modify the hypotheses in response to the data. The second class of bioinformatics applications addresses the transfer of information within this protocol, including systems that generate hypotheses, design experiments, store and organize the data from these experiments in databases, test the compatibility of the data with models, and modify hypotheses. The emergence and emphasis on systems-level modeling and interactions in both systems—biology and metagenomics—create major new challenges for our field.

The explosion of interest in bioinformatics has been driven by the emergence of experimental techniques that generate data in a high-throughput fashion—such as high-throughput DNA sequencing, mass spectrometry or microarray expression analysis (Miranker, 2000; Altman and Raychaudhuri, 2001; The Genome International Sequencing Consortium, 2001; Venter et al., 2001). Bioinformatics depends on the availability of large data sets that are too complex to allow manual analysis. The rapid increase in the number of three-dimensional macromolecular structures available in databases such as the Protein Data Bank (PDB,[1] Chapter 11; Berman et al., 2000) has driven the emergence of a subdiscipline of bioinformatics: *structural bioinformatics*. Structural bioinformatics is the subdiscipline of bioinformatics that focuses on the representation, storage, retrieval, analysis, and display of structural information at the atomic and subcellular spatial scales.

Structural bioinformatics, like many other subdisciplines within bioinformatics,[2] is characterized by two goals: the creation of general purpose methods for manipulating information about biological macromolecules and the application of these methods to solve problems in biology and create new knowledge. These two goals are intricately linked because part of the validation of new methods involves their successful use in solving real problems. At the same time, the current challenges in biology demand the development of new methods that can handle the volume of data now available and the complexity of models that scientists must create to explain these data.

Structural Bioinformatics Has Been Catalyzed by Large Amounts of Data

Biology has attracted computational scientists over the past 30 years in two distinct ways. First, the increasing availability of sequence data has been a magnet for those with an interest in string analysis, algorithms, and probabilistic models (Gusfield, 1997; Durbin et al., 1998). The major accomplishments have been the development of algorithms for pair-wise sequence alignment, multiple alignment, the definition and discovery of sequence motifs, and the use of probabilistic models, such as hidden Markov models to find genes (Burge and Karlin, 1997), align sequences (Hughey and Krogh, 1996), and summarize protein families (Bateman et al., 2000). Second, the increasing availability of structural data has been a magnet for those with an interest in computational geometry, computer graphics, and algorithms for analyzing crystallographic data (Chapter 4) and NMR data (Chapter 5) to create credible molecular models. Structural bioinformatics has its roots in this second group. The development of molecular graphics was one of the first applications of computer

---

[1] http://www.rcsb.org.

[2] The International Society for Computational Biology (ISCB, http://www.iscb.org/) is the professional organization for bioinformatics; many developments in structural bioinformatics are reported in the journals and conferences associated with this society.

graphics (Langridge and Gomatos, 1963). The elucidation of the structure of DNA in the mid-1950s and the publication of the first protein crystal structures in the early 1960s created a demand for computerized methods for examining these complex molecules. At the same time, the need for computational algorithms to deconvolute X-ray crystallographic data and fit the resulting electron densities to the more manageable ball-and-stick models created a cadre of structural biologists who were very well versed in computational technologies. The challenges of interpreting NMR-derived distance constraints into three-dimensional structures further introduced computational technologies to biological structure. As the number of three-dimensional structures increased, the need to create methods for storing and disseminating this data led to the creation of the PDB, one of the earliest scientific databases.[1] In the past 10 years, we have seen a third wave of interest in biological problems from a group that was not engaged by the availability of 1D sequence data or 3D structural data. This third wave has arisen in response to the increased availability of RNA expression data and has captured the interest of computational scientists with an interest in statistical analysis and machine learning, particularly in clustering methodologies and classification techniques. The problems posed by these data are different from those seen in both sequence and structural analysis data. The recent introduction of high-throughput DNA sequencing technologies that produce short-length (25–50) snippets of DNA sequence is re-energizing the sequence analysis community with new challenges.

Structural bioinformatics is now in a renaissance with the success of the genome sequencing projects, the emergence of high-throughput methods for expression analysis, and identification of compounds via mass spectrometry. There are now organized efforts in structural genomics (Chapter 40) to collect and analyze macromolecular structures in a high-throughput manner (Teichmann, Chothia, and Gerstein, 1999; Teichmann, Murzin, and Chothia, 2001). These efforts include challenges in the selection of molecules to study, the robotic preparation and manipulation of samples to find crystallization conditions, the analysis of X-ray diffraction data, and the annotation of these structures as they are stored in databases (Section II). In addition, there have been advancements in the capabilities of NMR structure determination, which previously could only study proteins in à limited range of sizes. The solution of the malate synthase G complex from *E. coli* with 731 residues has pushed the frontier for NMR spectroscopy and suggests that NMR is having its own renaissance (Tugarinov et al., 2005). The PDB now has a critical mass of structures that allow (indeed require!) statistical analysis to learn the rules of how active and binding sites are constructed which allow us to develop knowledge-based methods for the prediction of structure and function. Finally, the emergence of this structural information, when linked to the increasing amount of genomic information and expression data, provides opportunities for linking structural information to other data sources to understand how cellular pathways and processes work at a molecular level.

Toward a High-Resolution Understanding of Biology. The great promise of structural bioinformatics is predicated on the belief that the availability of high-resolution structural information about biological systems will allow us to precisely reason about the function of these systems and the effects of modifications or perturbations. The genetic analyses can only associate genetic sequences with their functional consequences, whereas the structural biological analyses offer the additional promise of ultimate insight into the mechanisms of these consequences, and therefore a more profound understanding of how biological function follows from the structure. The promise for structural bioinformatics lies in four areas: (1) creating an infrastructure for building up structural models from component parts, (2) gaining the ability to understand the design principles of proteins, so that new functionalities can be created, (3) learning how to design drugs efficiently based on structural knowledge of their target, and (4) catalyzing the development of simulation models that can give insight into function based on structural simulations. Each of these areas has already seen success, and the structural genomics projects promise to create data sets sufficient to catalyze accelerated progress in all these areas.

With respect to creating an infrastructure for modeling larger structural ensembles, we are already seeing the emergence of a new generation of structures larger by an order of magnitude than the structures submitted to the PDB a few years ago. Some achievements in recent years include (1) the elucidation of the structure of the bacterial ribosome (with more than 250,000 atoms) (Ban et al., 2000; Clemons Jr et al., 2001; Yusupov et al., 2001), (2) the publication of the RNA polymerase structure (with about 500,000 atoms) (Cramer et al., 2000), and (3) the increased ability to solve the structure of membrane proteins (transporters and receptors, in particular) that have proven technically difficult in the past. Each of these allows us to examine the principles of how a large number of component protein and nucleic acid structures can assemble to create macromolecular machines. With these successes, we can now target numerous other cellular ensembles for structural studies.

The design principles of proteins are now in reach both because we have a large "training set" of example proteins to study and because methods for structure prediction are beginning to allow us to identify structures that are unlikely to be stable. There have been preliminary successes in the design of four-helix bundle proteins (DeGrado, Regan, and Ho, 1987) and in the engineering of TIM barrels (Silverman, Balakrishnan, and Harbury, 2001). There has been interesting work in "reverse folding" in which a set of amino acid side chains is collected to stabilize a desired protein backbone conformation (Koehl and Levitt, 1999).

Rational drug design has not been the primary way for discovering major therapeutics (Chapters 27, 34 and 35). However, recent successes in this area give reason to expect that drug discovery projects will increasingly be structure based. One of the most famous examples of rational drug design was the creation of HIV protease inhibitors based on the known three-dimensional crystal structure (Kempf, 1994; Vacca, 1994). Methods for matching combinatorial libraries of chemicals against protein binding sites have matured and are in routine use at most pharmaceutical companies.

The simulation of biological macromolecular dynamics dates almost as far back as the elucidation of the first protein structure (Doniach and Eastman, 1999). These simulations are based on the integration of classical equations of motion and computation of electrostatic forces between atoms in a molecule. Methods for simulation now routinely include water molecules and are able to remain stable (the molecule does not fall apart) and reproduce experimental measurements with some fidelity. The simulation of larger ensembles and structural variants (such as based on known genetic variations in sequence) should lead to a more profound understanding of how structural properties produce functional behavior. The NIH has recognized the importance of simulation and created a national center devoted to physics-based simulation of biological structure (SIMBIOS, http://simbios.stanford.edu/).

Special Challenges in Computing with Structural Data

Structural bioinformatics must overcome some special challenges that are either not present or not dominant in other types of bioinformatics domains (such as the analysis of sequence or microarray data). It is important to remember these challenges when assessing the opportunities in the field. They include the following:

- Structural data are not linear and therefore not easily amenable to algorithms based on strings. In addition to this obvious nonlinearity, there are nonlinear relationships between atoms (the forces are not linear). This means that most computations on structure need to either make approximations or be very expensive.

- The search space for most structural problems is continuous. Structures are represented generally by atomic Cartesian coordinates (or internal angular coordinates) that are continuous variables. Thus, there are infinite search spaces for algorithms attempting to assign atomic coordinate values. Many simplifications can be applied (such as lattice models for 3D structure; Hinds and Levitt, 1994), but these are attempts to manage the inherent continuous nature of these problems.

- There is a fundamental connection between molecular structure and physics. While this statement seems obvious and trivial, it means that when reduced representations, such as pseudoatoms (Wuthrich, Billeter, and Braun, 1983) or lattice models are applied, they become more difficult to relate to the underlying physics that governs the interactions. The need to keep structural calculations physically reasonable is an important constraint.

- Reasoning about structure requires visualization. As mentioned above, the creation of computer graphics was driven, in part, by the need of structural biologists to look at molecules (Chapter 9). This is both a benefit and a detriment; structure is well defined, and well-designed visualizations can provide insight into structural problems. However, graphical displays have a human user as a target and are not easily parsed or understood by computers, and thus represent something of a computational "dead end." The need to have expressive data structures underlying these visualizations allows the information to be understood and analyzed by computer programs and thus opens the possibility of further downstream analysis.

- Structural data, like all biological data, can be noisy and imperfect. Despite some amazing successes in the elucidation of very high-resolution structures, the precision of our knowledge about many structures is likely to be limited by their flexibility, dynamics, or experimental noise (Chapters 14, 15, 37, and 38). Understanding the protein structural disorder may be critical for understanding the protein's function. Thus, we must be comfortable in reasoning about structures for which we only have partial knowledge.

- Protein and nucleic acid structures are generally conserved more than their associated sequence. Thus, sequences will accumulate mutations over time that may make identification of their similarities more difficult, while their structures may remain essentially identical. This is a challenge because sequence information is still much more abundant than structural information, and so for many molecules it is the sequence information that is readily available. The need to identify distant sequential similarities to gain structural insights can be a major challenge.

- Structural genomics will likely produce a large number of structures at the level of the domain—relatively well-defined modules that associate to form larger ensembles. The principles by which these domains associate and cooperatively function pose a major challenge to structural biology (Chapters 17, 18, 20, and 26).

- Finally, we must recognize that there is a major gap in our knowledge of a large fraction of proteins that are not globular and water soluble. In particular, membrane-bound and fibrous proteins are simply not well understood and structures have not been available in the numbers required to allow routine statistical and informatics approaches to their study. The importance of this shortcoming cannot be over emphasized, since these classes of proteins are among the most important ones for understanding a large number of cellular processes of great interest, including signal transduction, cytoskeletal dynamics, and cellular localizations and compartmentalization. Recently, some fascinating structures of membrane-bound transporter proteins, such as a zinc transporter (Lu and Fu, 2007), have improved our understanding of membrane protein structure (Chapter 36).

TECHNICAL CHALLENGES WITHIN STRUCTURAL BIOINFORMATICS

The scientific challenges within structural bioinformatics fall into two rough categories: the creation of methods to support structural biology and structural genomics and the creation of methods to elucidate new biological knowledge. This distinction is not absolute, but is useful for dividing much structural bioinformatics work. The support of experimental structural biology is currently an area of particular interest with the emergence of efforts in high-throughput structural genomics. Informatics approaches are required for many aspects of this enterprise, and can be briefly reviewed here:

- *Target Selection*: Structural genomics efforts with finite resources must select proteins to study carefully. Informatics methods are used to compare the database of existing structures and known sequences with potential targets to identify those that are most likely to add to our structural knowledge base. This selection can be informed by the expected novelty of the structure, and even its importance as reflected in the published literature (Linial and Yona, 2000). A critical part of target selection is the identification of domains within large proteins. Domains are often easier to study initially in isolation, and then in complexes. The definition of domains from sequence data alone is a challenging problem (Chapter 20).
- *Tracking Experimental Crystallization Trials*: One of the major bottlenecks in structural genomics is the discovery of crystallization conditions that work for proteins of interest. In addition to the obvious need of storing and tracking information on proteins, the conditions attempted, and the results, there is also an opportunity to apply machine learning methods to these data to extract rules that may help increase the yield of crystals based on previous experience (Hennessy et al., 2000). Until recently, the results of failed crystallization experiments were not generally available, thus making it difficult to apply automated machine learning methods to these data sets.
- *Analysis of Crystallographic Data*: A long-standing area of computation within structural biology is the algorithm for deconvoluting the X-ray diffraction pattern, which involves computing an inverse Fourier transform with partial information (i.e., with missing phase information). There is interest in *ab initio* methods for automating these computations, and success in this area reduces the number of heavy atom derivatives that must be created for structures of interest

(Gilmore, Dong, and Bricogne, 1998). Multiwavelength anomalous diffraction (MAD) (Hendrickson, 1991) is now the preferred method for solving the crystallographic phase problem. Over one-half of all structures are determined by MAD, a development in keeping with the availability of tunable synchrotron sources. Similarly, once the electron density is computed, there is a challenge in fitting the density to a standard ball-and-stick model of the atoms. While this has been done manually (with graphical computer assistance), there is interest in finding methods for using image processing techniques to automatically identify connected densities and match them to the known shape of protein backbone and side chain elements (Barr and Feigenbaum, 1982). Recent progress has been made on automated electron density map fitting and refinement (Chapter 4).

- *The Analysis of NMR Data*: NMR experiments provide complementary data to the crystallographic analyses. NMR experiments produce two (or higher) dimensional spectra for which each individual peak must be assigned to an atomic interaction. The automated analysis and assignment of atoms in these spectra is a difficult search problem, but the one in which progress has been made to accelerate the analysis of structure (Zimmerman and Montelione, 1995). Given a set of atomic proximities from NMR, we need methods to "embed" these distance measures into three-dimensional structures that satisfy these constraints. Distance geometry (Moré and Wu, 1999), restrained molecular dynamics (Bassolino-Klimas et al., 1996), and other nonlinear optimization methods have been developed for this purpose (Altman, 1993; Williams, Dugan, and Altman, 2001).

- *Assessment and Evaluation of Structures*: Given the results of a crystallographic or NMR structure determination effort, we must check the structures to be sure that they meet certain quality standards. Algorithms have been developed for assessing the basic chemistry of structural models and also for identifying active and binding sites in these structures (Laskowski et al., 1993; Feng, Westbrook, and Berman, 1998; Vaguine, Richelle, and Wodak, 1999). Computational methods are still needed for automatically annotating 3D structures with functional information, based on an understanding of how molecular properties aggregate in three dimensions to produce function (such as binding, catalysis, motion, and signal transduction) (Wei, Huang, and Altman, 1999, Chapter 5).

- *Storing Molecular Structures in Databases*: The storage of the results of structural genomics efforts is an important task, requiring data structures and organizations that facilitate the most common queries. Ideally, databases of structure will store not only the resulting model but also the raw data upon which it is based. The PDB (Chapter 11) is the major repository for three-dimensional structural information of proteins; the Nucleic Acids Database (NDB, Chapter 12) serves this function for nucleic acids. There is also an effort to store the raw data associated with crystallography in the PDB/NDB and the raw data associated with NMR in the BioMagResBank (BMRB).[3]

- *Correlating Molecular Structural Information with Structural and Functional Information Gained from Other Types of Experimentation*: In the end, we perform structural studies in order to get an insight into how the molecules work. Structural studies with crystallography and NMR are two methods that can be used to probe structure–function relationships. The integration of the results of these methods

[3] http://www.bmrb.wisc.edu

with other structural and functional data allows us to build comprehensive models of mechanism, specificity, and dynamics. A major bottleneck for using informatics methods for this integration is the lack of repositories of structural and functional data that can be accessed by computer programs doing systematic analyses. One exception is the noncrystallographic structural data on the 30S and 50S ribosomal subunits stored in the RiboWEB (http://riboweb.stanford.edu/), a knowledge base of ribosomal structural components that stores more than 8000 noncrystallographic structural and functional observations about the bacterial ribosome. It stores its information in structured "information templates" that are easily parsed by computer programs, thus making possible automated comparison and evaluation of structural models. For example, RiboWEB has been used to assess the compatibility of the published ribosomal crystal structures with over 1000 proximity measurements from cross-linking, chemical protection, and labeling experiments (collected during the past 25 years). Incompatibilities between these data and the crystal structures may suggest artifactual data or (more usefully) may suggest areas of important dynamic motion for the ribosome (Whirl-Carrillo et al., 2002).

Understanding the Structural Basis for Biological Phenomenon

Given the structural information created by efforts in X-ray crystallography and NMR, there is a wide range of analytic and scientific challenges to informatics. It is not possible to cover the full scope of activities, but they can be reviewed briefly to show the richness of opportunities in the analysis of structural data.

- *Visualization*: The creation of images of molecular structure remains a primary activity within structural biology (Chapter 9). The complexity of these molecules seems to demand novel display methods that are able to combine structural information with other information sources (such as electrostatic fields, the location of functional sites, and areas of structural or genetic variability). The issues for informatics include the creation of flexible software infrastructures for extending display capabilities and the use of novel methods for rapidly rendering complex molecular structures (Huang et al.,1996; Sanner et al., 1999).
- *Classification*: The database of known structures is already sufficiently large, making it necessary to cluster similar structures together to form families of proteins. These families are often aggregated into superfamilies, and indeed entire structural hierarchies have been created. The structural classification of proteins (SCOP; Chapter 17) is an example of a semiautomated classification of all protein structures (Murzin et al., 1995), and there have been numerous efforts to create automated classification—usually based on the pair-wise comparison of all structures to create a matrix of distances (Chapter 18; Holm and Sander, 1996; Orengo et al., 1997).
- *Prediction*: Despite the growth of the structural databases, the number of known three-dimensional structures has lagged far behind the availability of sequence information. Thus, the prediction of three-dimensional structure remains an area of keen interest. The Critical Assessment for Structure Prediction (CASP;[4] Chapter 28) meetings have provided a biennial forum for the comparison of methods

---

[4] http://predictioncenter.llnl.gov

for structure prediction. The main categories of prediction have been homology modeling (based on high sequence homology to a known structure; Chapter 30; Sánchez and Sali, 1997), threading (based on homology (Chapter 31); Bryant and Altschul, 1995), and *ab initio* prediction (based on no detectable homology; Chapter 32; Osguthorpe, 2000). The diversity of methods invented and evaluated is quite inspiring, and the resulting lessons about how proteins are put together have been significant.

- *Simulation*: The results of crystallographic studies (and to some extent, NMR studies) are primarily static structural models. However, the properties of these molecules that are of the greatest interest are often the results of their dynamic motions. The definition of energy functions that govern the folding of proteins and their subsequent stable dynamics has been an area of great interest since the first structure was determined. Unfortunately, the timescales on which macromolecular dynamics must be sampled (fractions of picoseconds) are much shorter than the timescale on which biologically important phenomena occur (from microseconds to seconds). Nevertheless, the availability of increasingly powerful computers and clever approximation and search methods is enabling molecular simulations of sufficient length and accuracy to emerge, making contributions to our understanding of protein function.[5] The associated computation of electrostatic fields of macromolecular structures (Chapter 24) has emerged as an important component of understanding molecular function (Sheinerman, Norel, and Honig, 1992).

It should be emphasized that although there has been primary focus on protein structures, with respect to the challenges outlined above, there is increasing interest in the same issues for RNA structure. The last decade has shown that the role of RNA molecules in the cell goes far beyond being a passive information carrier as messenger RNA. A large number of structured RNA molecules are involved in gene regulation (through RNA inhibition and other mechanisms), whose 3D structure is critical for understanding their function. The overall challenges for RNA structure are similar to proteins, but the details differ—RNA structure is dominated by electrostatics and not hydrophobic interactions, the secondary structure is easier to predict but offers a more limited repertoire for structural uses, and the molecules are more prone to finding stable misfolded states. Nonetheless, our understanding of structural biology will necessarily include the structure of RNA and RNA–protein complexes (Chapter 3, 12, and 33).

INTEGRATING STRUCTURAL DATA WITH OTHER DATA SOURCES

Structural bioinformatics has existed, in one form or another, since the determination of the first myoglobin structure. One could argue that the roots go back to the time when small molecular structure determination was introduced. In any case, the challenges for the field are clearly abundant and significant. As we look into the coming decade, it appears that a primary challenge in structural bioinformatics will be the integration of structural information with other biological information, to yield a higher resolution understanding

---

[5] The IBM BlueGene project (http://www.research.ibm.com/bluegene) is focused on the creation of a very large supercomputer, with the theoretical capability of simulating the folding of a small protein in about 1 year. The computer is being designed to have 10^{15} floating-point operations per second.

of biological function. The success of genome sequencing projects has created information about all the structures that are present in individual organisms, as well as both shared and unique features of these organisms. Even with the success of structural genomics projects, bioinformatics techniques will most likely be used to create homology models of most of these genomic components. The resulting structures will be studied with respect to how they interact and perform their functions. Similarly, the emergence of high-throughput expression measurements provides an opportunity to understand how the assembly of macromolecular structures is regulated (including the key structural machinery associated with transcription, translation, and degradation). Mass spectroscopic methods that allow the identification of structural modifications and variations (such as genetic mutation or post-translational modifications) will need to be integrated with structural models to understand how they alter functional characteristics. Cross-linking data, particularly *in vivo*, will provide valuable information about the physical association between macromolecules and ligands and the dynamics of molecular ensembles, thus helping us to create a structural portrait of a cell in three dimensions at near-atomic resolution (Tsutsui and Wintrode, 2007). Finally, cellular localization data will allow us to place three-dimensional molecular structures into compartments within the cell, as we build more complex models of how cells are organized structurally to optimize their function. This exciting activity will mark the next phase of structural bioinformatics—when the organization and physical structure of entire cells are understood and represented in computational models that provide insight into how thousands of structures within a cell work together to create the functions associated with life.

REFERENCES

Altman RB (1993): Probabilistic structure calculations: a three-dimensional tRNA structure from sequence correlation data. In: *Proceedings of the First International Conference on Intelligent Systems for Molecular Biology,* July 6–9, 1993, Bethesda, MD. Menlo Park, CA: The AAAI Press.

Altman RB (1998): A curriculum for bioinformatics: the time is ripe. *Bioinformatics* 14:549–550.

Altman RB, Raychaudhuri S (2001): Whole-genome expression analysis: challenges beyond clustering. *Curr Opin Struct Biol* 11:340–347.

Ban N, Nissen P, Hansen J, Moore P, Steitz T (2000): The complete atomic structure of the large ribosomal subunit at 2.4 Å resolution. *Science* 289:878–879.

Barr A, Feigenbaum E (1982): Crysalis. In: *The Handbook of Artificial Intelligence.* Stanford, CA: HeurisTech Press, pp 124–133.

Bassolino-Klimas D, Tejero R, Krystek SR, Metzler WJ, Montelione GT, Bruccoleri RE (1996): Simulated annealing with restrained molecular dynamics using a flexible restraint potential: theory and evaluation with simulated NMR constraints. *Protein Sci* 5:593–603.

Bateman A, Birney E, Durbin R, Eddy SR, Howe KL, Sonnhammer EL (2000): The Pfam protein families database. *Nucleic Acids Res* 28:263–266.

Berman HM, Westbrook J, Feng Z, Gilliland G, Bhat TN, Weissig H, Shindyalov IN, Bourne PE (2000): The protein data bank. *Nucleic Acids Res* 28:235–242.

Bryant SH, Altschul SF (1995): Statistics of sequence–structure threading. *Curr Opin Struct Biol* 5:236–244.

Burge C, Karlin S (1997): Prediction of complete gene structures in human genomic DNA. *J Mol Biol* 268:78–94.

Clemons WM Jr, Brodersen DE, McCutcheon JP, May JLC, Carter AP, Morgan-Warren RJ, Wimberly BT, Ramakrishnan V (2001): Crystal structure of the 30S ribosomal subunit from *Thermus thermophilus*: purification, crystallization and structure determination. *J Mol Biol* 310:827–843.

Cramer P, Bushnell DA, Fu J, Gnatt AL, Maier-Davis B, Thompson NE, Burgess RR, Edwards AE, David PR, Kornberg RD (2000): Architecture of RNA polymerase II and implications for the transcription mechanism. *Science* 288:640–649.

DeGrado W, Regan L, Ho S (1987): The design of a four-helix bundle protein. *Cold Spring Harbor Symp Quant Biol* 52:521–526.

Doniach S, Eastman P (1999): Protein dynamics simulations from nanoseconds to microseconds. *Curr Opin Struct Biol* 9:157–163.

Durbin R, Krogh A, Mitchison G, Eddy S (1998): *Biological Sequence Analysis: Probabilistic Models of Proteins and Nucleic Acids*. Cambridge: Cambridge University Press.

Feng Z, Westbrook J, Berman HM (1998): NUCheck. Rutgers Publication NDB-407. New Brunswick, NJ: Rutgers University.

Gilmore CJ, Dong W, Bricogne G (1998): A multisolution method of phase determination by combined maximisation of entropy and likelihood. VI. The use of error-correcting codes as a source of phase permutation and their application to the phase problem in powder, electron and macromolecular crystallography. *Acta Crystallogr A* 55:70–83.

Gusfield D (1997): *Algorithms on Strings, Trees, and Sequences: Computer Science and Computational Biology*. Cambridge: Cambridge University Press.

Hendrickson WA (1991): Determination of macromolecular structures from anomalous diffraction of synchrotron radiation. *Science* New York, NY 254:51–58.

Hennessy D, Buchanan B, Subramanian D, Wilkosz PA, Rosenberg JM (2000): Statistical methods for the objective design of screening procedures for macromolecular crystallization. *Acta Crystallogr D* 56:817–827.

Hinds D, Levitt M (1994): Exploring conformational space with a simple lattice model for protein structure. *J Mol Biol* 243:668–682.

Holm L, Sander C (1996): Mapping the protein universe. *Science* 273:595–602.

Huang CC, Couch GS, Pettersen EF, Ferrin TE (1996): Chimera: an extensible molecular modeling application constructed using standard components. In: Hunter L, Klein TE, editors. *Pacific Symposium on Biocomputing,* January 3–6, 1996, USA, Hawaii. Singapore: World Scientific Publishing, p 724.

Hughey R, Krogh A (1996): Hidden Markov models for sequence analysis: extension and analysis of the basic method. *CABIOS* 12:95–107.

Kempf D (1994): Design of symmetry-based, peptidomimetic inhibitors of human immunodeficiency virus protease. *Methods Enzymol* 241:334–354.

Koehl P, Levitt M (1999): Structure-based conformational preferences of amino acids. *Proc Natl Acad Sci USA* 96:12524–12529.

Langridge R, Gomatos PJ (1963): The structure of RNA. *Science* 141:694–698.

Laskowski RA, McArthur MW, Moss DS, Thornton JM (1993): PROCHECK: a program to check the stereochemical quality of protein structures. *J Appl Crystallogr* 265:283–291.

Linial M, Yona G (2000): Methodologies for target selection in structural genomics. *Prog Biophys Mol Biol* 73:297–320.

Lu M, Fu D (2007): Structure of the zinc transporter YiiP. *Science* New York, NY 317: 1746–1748.

Miranker AD (2000): Protein complexes and analysis of their assembly by mass spectrometry. *Curr Opin Struct Biol* 10:601–606.

Moré J, Wu Z (1999): Distance geometry optimization for protein structures. *J Global Optim* 15:219–234.

Murzin AG, Brenner SE, Hubbard T, Chothia C (1995): SCOP: a structural classification of proteins database for the investigation of sequences and structures. *J Mol Biol* 247:536–540.

Orengo CA, Michie AD, Jones S, Jones DT, Swindells MB, Thornton JM (1997): CATH: a hierarchic classification of protein domain structures. *Structure* 5(8).1093–1108.

Osguthorpe DJ (2000): *Ab initio* protein folding. *Curr Opin Struct Biol* 10:146–152.

Sánchez R, Sali A (1997): Advances in comparative protein-structure modelling. *Curr Opin Struct Biol* 7:206–214.

Sanner MF, Duncan BS, Carrillo CJ, Olson AJ (1999): Integrating computation and visualization for biomolecular analysis: an example using PYTHON and AVS. In: Altman RB, Lauderdale K, Dunker AK, Hunter L, Klein TE, editors. *Pacific Symposium on Biocomputing*, Hawaii, USA, Singapore: World Scientific Publishing.

Sheinerman FB, Norel R, Honig B (1992): Electrostatic aspects of protein–protein interactions. *Curr Opin Struct Biol* 10:153–159.

Silverman JA, Balakrishnan R, Harbury PB (2001): Reverse engineering the (beta/alpha)8 barrel fold. *Proc Natl Acad Sci USA* 98:3092–3097.

Teichmann SA, Chothia C, Gerstein M (1999): Advances in structural genomics. *Curr Opin Struct Biol* 9:390–399.

Teichmann SA, Murzin AG, Chothia C (2001): Determination of protein function, evolution and interactions by structural genomics. *Curr Opin Struct Biol* 11:354–363.

Tugarinov V, Choy W-Y, Orekhov VY, Kay LE (2005): Solution MR-derived global fold of a monomeric 82-kDa enzyme. *Proc Natl Acad Sci USA* 102(3): 622–627.

Tsutsui Y, Wintrode PL (2007): Hydrogen/deuterium exchange-mass spectrometry: a powerful tool for probing protein structure, dynamics and interactions. Current medicinal chemistry 14: 2344–2358.

The Genome International Sequencing Consortium (2001): Initial sequencing and analysis of the human genome. *Nature* 409:860–921.

Vacca J (1994): Design of tight-binding human immunodeficiency virus type 1 protease inhibitors. *Methods Enzymol* 241:331–334.

Vaguine AA, Richelle J, Wodak J (1999): SFCheck: a unified set of procedures for evaluating the quality of macromolecular structure-factor data and their agreement with the atomic model. *Acta Crystallogr D* 55:191–205.

Venter JC, Adams MD, Myers EW, Li PW, Mural RJ, Sutto GG, Smith HO, Yandell M, et al. (2001): The sequence of the human genome. *Science* 291:1304–1351.

Wei L, Huang ES, Altman RB (1999): Are predicted structures good enough to preserve functional sites. *Structure* 7:643–650.

Whirl-Carrillo M, Gabashvili IS, Bada M, Banatao DR, Altman RB (2002): Mining biochemical information: lessons taught by the ribosome. *RNA* 8:279–289.

Williams GA, Dugan JM, Altman RB (2001): Constrained global optimization for estimating molecular structure from atomic distances. *J Comput Biol* 8:523–547.

Wuthrich K, Billeter M, Braun W (1983): Pseudo-structures for the 20 common amino acids for use in studies of protein conformations by measurements of intramolecular proton–proton distance constraints with nuclear magnetic resonance. *J Mol Biol* 169:949–961.

Yusupov MM, Yusupova GZ, Baucom A, Lieberman K, Earnest TN, Cate JHD, Noller HF (2001): Crystal structure of the ribosome at 5.5 Å resolution. *Science* 292:883–896.

Zimmerman DE, Montelione GT (1995): Automated analysis of nuclear magnetic resonance assignments for proteins. *Curr Opin Struct Biol* 5:664–673.

2

FUNDAMENTALS OF PROTEIN STRUCTURE

Eric D. Scheeff and J. Lynn Fink

THE IMPORTANCE OF PROTEIN STRUCTURE

Most of the essential structures and functions of cells are mediated by proteins. These large, complex molecules exhibit a remarkable versatility that allows them to perform a myriad of activities that are fundamental to life. Indeed, no other type of biological macromolecule could possibly assume all of the functions that proteins have amassed over billions of years of evolution.

Any consideration of protein function must be grounded in an understanding of protein structure. A fundamental principle in all of protein science is that *protein structure leads to protein function*. The distinctive structures of proteins allow the placement of particular chemical groups in specific places in three-dimensional space. It is this precision that allows proteins to act as catalysts (enzymes) for an impressive variety of chemical reactions. Precise placement of chemical groups also allows proteins to play important structural, transport, and regulatory functions in organisms. Since protein structure leads to function and protein functions are diverse, it is no surprise that protein structure is similarly diverse. Further, the functional diversity of proteins is expanded through the interaction of proteins with small molecules, as well as other proteins.

For those who wish to study protein structure, this diversity represents a challenge. Upon their determination of the first three-dimensional globular protein structure (the oxygen-storage protein myoglobin) in 1958, John Kendrew and his coworkers registered their disappointment (Kendrew et al., 1958):

> Perhaps the most remarkable features of the molecule are its complexity and its lack of symmetry. The arrangement seems to be almost totally lacking in the kind of regularities which one instinctively anticipates, and it is more complicated than has been predicated by any theory of protein structure.

Structural Bioinformatics, Second Edition   Edited by Jenny Gu and Philip E. Bourne
Copyright © 2009 John Wiley & Sons, Inc.

Despite these initial frustrations, subsequent studies of the myoglobin structure based on higher quality data (Kendrew, 1961) revealed that the protein did have *some* regularities; these regularities were also observed in other protein structures.

Decades of research has now yielded a coherent set of principles about the nature of protein structure and the way in which this structure is utilized to effect function. These principles have been organized into a four-tier hierarchy that facilitates description and understanding of proteins: primary, secondary, tertiary, and quaternary structure. This hierarchy does not seek to precisely describe the physical laws that produce protein structure but is rather an abstraction to make protein structural studies more tractable.

THE PRIMARY STRUCTURE OF PROTEINS: THE AMINO ACID SEQUENCE

Amino Acids

Proteins are linear polymers of amino acids,[1] and *it is the distinct sequence of component amino acids that determines the ultimate three-dimensional structure of the protein.* The sequence of a protein is often referred to as its primary structure. The concept of proteins as linear amino acid polymers was initially proposed by Fischer and Hofmeister in 1902 (Fruton, 1972). At that time, the prevailing theory in protein science was that proteins lacked a regular structure and consisted of loose associations of small molecules (colloids). This issue was hotly debated for over 20 years, until the linear polymer theory achieved general acceptance in the late 1920s (Fruton, 1972). In 1952, Fred Sanger made the important discovery that proteins could be distinguished by their amino acid sequences (Sanger, 1952). Indeed, he found that *proteins of exactly the same type have identical sequences.* Sanger's work helped to remove remaining doubts about the accuracy of the linear polymer theory.

Amino acids are small molecules that contain an amino group (NH_2), a carboxyl group (COOH), and a hydrogen atom attached to a central alpha (α) carbon (Figure 2.1). In addition, amino acids also have a side chain (or R group) attached to the α-carbon. It is this group, and only this group, that distinguishes one amino acid from another. Furthermore, the side chain confers the specific chemical properties of the amino acid.

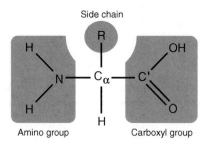

Figure 2.1. The structure of a prototypical amino acid. The chemical groups bound to the central alpha (α) carbon are highlighted in gray. The R-group represents any of the possible 20 amino acid side chains.

---

[1]Specifically, the amino acids used in proteins are *alpha* (α)-amino acids.

Cellular genomes contain coded instructions for the production of multiple proteins, and there are 20 amino acids that can be incorporated into a protein via these instructions. The resulting sequence of a protein can contain any combination and number of the 20 amino acids, in any order. Though amino acids had been known to be the building blocks of proteins prior to the turn of this century, the exact set of amino acids used in proteins was not determined until 1940 (Fruton, 1972). This set of 20 amino acids is considered "standard" in that it is common to all observed organisms. Modified forms of these 20 amino acids do exist in proteins, but these are the product of modifications that occur subsequent to protein synthesis.

There are two recent additions to the 20 standard amino acids that, although they are infrequently observed, merit discussion. In a few proteins, a selenium-containing residue, selenocysteine (Sec), can be incorporated during protein synthesis (Zinoni et al., 1986; Atkins and Gesteland, 2000), and the lysine derivative, pyrrolysine (Pyl), has been found in some proteins in methanogenic bacteria (Srinivasan, James, and Kryzycki, 2002). Both of these amino acids are encoded by a signal that normally serves to stop synthesis of the protein chain. However, if the necessary cellular machinery is available to incorporate these amino acids, the stop signals are co-opted for insertion of what are termed the 21st and 22nd amino acids.

Interestingly, the cellular machinery that incorporates amino acids into a protein chain can be extrinsically altered to incorporate amino acids not included in the standard set (Wang, 2001; Wang, Xie, Schultz, 2006). This allows researchers to genetically encode *unnatural* amino acids that can fluoresce, are glycosylated, can bind metal ions, or are redox-active, into both prokaryotic and eukaryotic organisms. Incorporation of these amino acids serves as an experimental tool for researching aspects of protein structure. To date, over 30 unnatural amino acids have been successfully encoded using this method.

The 20 standard amino acids can be loosely grouped into classes based on the chemical properties conferred by their side chains. Three classes are commonly accepted: hydrophobic, polar, and charged. Within these classes, additional subclassifications are possible, for example, aromatic or aliphatic, large or small, and so on (Taylor, 1986). Figure 2.2 provides one possible amino acid classification.

A few amino acids have distinctive properties that merit closer attention. The side chain of proline forms a bond with its own amino group, causing it to become cyclic.[2] Though proline generally exhibits the properties of an aliphatic nonpolar amino acid, the cyclic construction limits its flexibility, and this impacts the overall structure of proteins that contain it.

Glycine is also of interest because its side chain consists only of a single hydrogen atom. In effect, glycine has no side chain, and this confers a unique property on it amongst the 20 amino acids: glycine is *achiral*. Any carbon bound to four distinct groups (as seen in the other 20 amino acids) is said to be chiral (Figure 2.3). Chiral molecules can exist in two distinct forms, which are in effect mirror images of each other. These two forms have been deemed the D and L forms.[3] In 1952, Fred Sanger discovered that proteins seem to be constructed entirely of L-amino acids (Sanger, 1952). Indeed, for unknown reasons, all known organisms have standardized upon the L form of amino acids for the genetically directed production of proteins. D-Amino acids *are* seen in polypeptides in rare cases, but they are a result of direct enzymatic synthesis (Kreil, 1997).

---

[2]Because of the cyclic bond in proline, this molecule is technically an *imino* acid. However, proline is commonly referred to as one of the 20 "amino" acids.

[3]International chemical convention calls for the designations R and S, respectively, but D and L are the traditional, and currently dominant, terms.

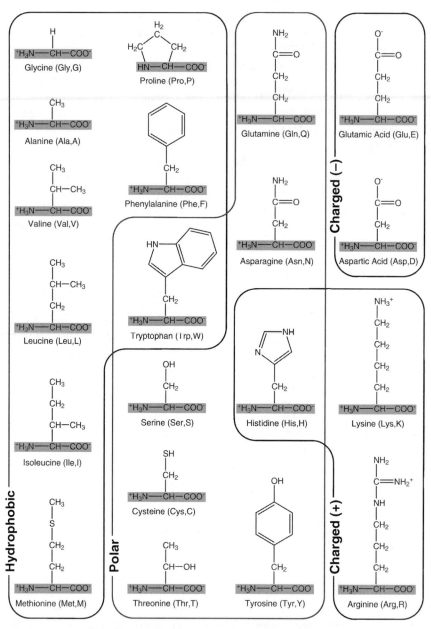

Figure 2.2. The 20 "standard" amino acids used in proteins, grouped on the basis of the properties of their side chains. The shared amino acid structure is shaded gray. Each amino acid is labeled with its full name, followed by its three-letter and one-letter abbreviations. This classification groups amino acids on the basis of the form that predominates at physiological conditions (note that their amino and carboxyl groups are charged under these conditions). Although this classification is a useful guideline, it does not convey the full complexity of side chain properties. For example, tryptophan and histidine do not fall clearly into a single grouping. Tryptophan is somewhat polar due to the nitrogen in its five-membered ring but has a hydrophobic six-membered ring at the end of its side chain. Histidine can be neutral polar and/or positively charged under physiological conditions.

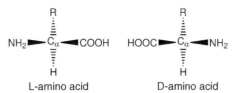

Figure 2.3. The possible stereoisomers of a prototypical amino acid. Note that these structures are mirror images of each other. The L-form is the only type incorporated into proteins via the genetic machinery.

The Peptide Bond

Amino acids can form bonds with each other through a reaction of their respective carboxyl and amino groups. The resulting bond is called the peptide bond, and two or more amino acids linked by such a bond are referred to as a peptide (Figure 2.4). A protein is synthesized by the formation of a linear succession of peptide bonds between *many* amino acids (as directed by the genetic code) and can thus be referred to as a polypeptide. Once an amino acid is incorporated into a peptide, it is referred to as an amino acid residue, and the atoms involved in the peptide bond are referred to as the peptide backbone.

The specific characteristics of the peptide bond have important implications for the three-dimensional structures that can be formed by polypeptides. The peptide bond is planar and quite rigid. Therefore, the polypeptide chain has only a rotational freedom about the bonds formed by the α-carbons. These bonds have been termed the Phi (Φ) and Psi (Ψ) angles (Figure 2.5). However, rotational freedom about the Φ ($C_\alpha-N$) and Ψ ($C_\alpha-C'$) angles is limited by steric hindrance between the side chains of the residues and the peptide backbone. Consequently, the possible conformations of a given polypeptide chain are quite limited. A *Ramachandran plot* (a plot of Φ versus Ψ angles) maps the entire conformational space of a polypeptide and illuminates the allowed and disallowed conformations (Ramachandran and Sasisekharan, 1968) (Figure 2.6). These plots were developed by G.N. Ramachandran in the late 1960s based on studies of sterically allowed Φ and Ψ angle combinations. See Chapter 14 for a more detailed discussion of these plots.

Some key exceptions to these conformational limitations can be attributed to glycine and proline. As noted previously, the side chain of glycine (a single hydrogen atom) is very small. There is markedly reduced steric hindrance about the Φ and Ψ angles of this residue, thus expanding the possible conformational space. Conversely, the cyclic bond

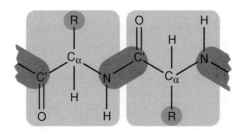

Figure 2.4. The peptide bond. Two peptide units (amino acid residues) are shown shaded in light gray. The peptide bond between them is shaded in dark gray. The R-group represents any of the possible 20 amino acid side chains.

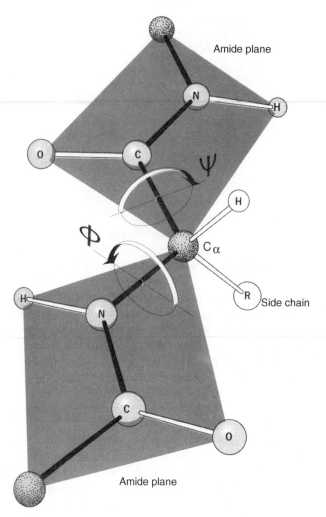

Figure 2.5. The planar characteristics of the peptide bond, and rotation of the peptide backbone about the C_α atom. Note the two planar peptide bonds about a central alpha carbon, shown here as a ball-and-stick model. Rotation is only possible about the Φ (C_α–N) and Ψ (C_α–C') angles. Arrows about the two angles show the direction that is considered positive rotation. In this figure, both angles are approximately 180°. From R.E. Dickerson and I. Geis. *The Structure and Action of Proteins*. New York: Harper & Row, 1969.

present in proline residues reduces the conformational freedom beyond the limitations observed with other amino acids.

THE SECONDARY STRUCTURE OF PROTEINS: THE LOCAL THREE-DIMENSIONAL STRUCTURE

The secondary structure of a protein can be thought of as the local conformation of the polypeptide chain, independent of the rest of the protein. The limitations imposed upon the primary structure of a protein by the peptide bond and hydrogen bonding considerations

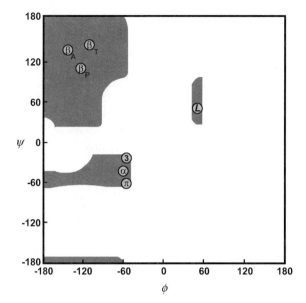

Figure 2.6. A schematic representation of a Ramachandran plot (a plot of Φ versus Ψ angles). Gray regions denote the allowed conformations of the polypeptide backbone. Circles indicate the paired angle values of the repetitive secondary structures. Definitions of symbols: β_A, antiparallel β–sheet; β_P, parallel β–sheet; β_T, twisted β–sheet (parallel or antiparallel); α, right-handed α-helix; L, left-handed helix; 3, 3_{10} helix; π, π helix.

dictate the secondary structure that is possible. During the course of protein structure research, two types of secondary structure have emerged as the dominant local conformations of polypeptide chains: alpha (α) helix and beta (β) sheets. Interestingly, these structures were actually predicted by Linus Pauling, Robert Corey, and H.R. Branson, based on the known physical limitations of polypeptide chains, prior to the experimental determination of protein structures (Pauling and Corey, 1951a; Pauling, Corey, and Branson, 1951). Indeed, if the Ramachandran plot is examined, helices and sheets contain Φ and Ψ angles that fall within the two largest regions of allowed conformation (Figure 2.6). These structures exhibit a high degree of regularity: the particular Φ and Ψ angle combinations in the polypeptide chain are approximately repeated for the duration of the secondary structure.

Although helices and sheets satisfy the peptide bond constraints, this is not the only factor that explains their ubiquity. Both of these structural elements are stabilized by hydrogen bond interactions between the backbone atoms of the participating residues, making them a highly favorable conformation for the polypeptide chain. Helices and sheets are the only *regular* secondary structural elements present in proteins. However, *irregular* secondary structural elements are also observed in proteins and are vital to both structure and function.

α-Helices

A helix is created by a curving of the polypeptide backbone such that a regular coil shape is produced. Because the polypeptide backbone can be coiled in two directions (left or right), helices exhibit handedness. A helix with a rightward coil is known as a right-handed helix. Almost all helices observed in proteins are right-handed, as steric restrictions limit the ability of left-handed helices to form. Among the right-handed helices, the α helix is by far the most prevalent.

An α-helix is distinguished by having a period of 3.6 residues per turn of the backbone coil. The structure of this helix is stabilized by hydrogen bonding interactions between the carbonyl oxygen of each residue and the amide proton of the residue four positions ahead in the helix (Figure 2.7). Consequently, all possible backbone hydrogen bonds are satisfied within the α-helix, with the exception of a few at each end of the helix, where a partner is not available.

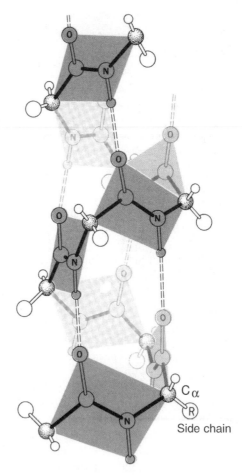

Figure 2.7. Diagram of an α-helix using a ball-and-stick model. The bonds forming the backbone of the polypeptide are shaded dark. The α-helix is stabilized by internal hydrogen bonds formed between the carbonyl oxygen of each residue and the amide proton of the residue four positions ahead in the helix, shown here as dashed lines. Note that the polypeptide backbone curves toward the right, and as such the α-helix is a right-handed helix. From R.E. Dickerson and I. Geis. *The Structure and Action of Proteins*. New York: Harper & Row, 1969.

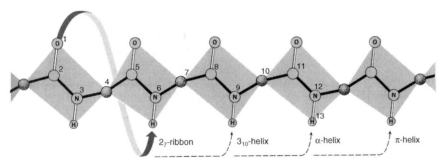

Figure 2.8. The hydrogen bonding patterns of different helical secondary structures. The peptide backbone is shown in an extended conformation, with an arrow denoting the hydrogen bonding pairings that would occur in each type of helix. The common α-helix, depicted in Figure 2.7, forms hydrogen bonds between the carbonyl oxygen of each residue and the amide proton of the residue four positions ahead in the helix. The 3_{10} helix forms hydrogen bonds between the carbonyl oxygen of each residue and the amide proton of the residue three positions ahead, forming a more narrow and elongated helix. The π-helix forms hydrogen bonds between the carbonyl oxygen of each residue and the amide proton of the residue five positions ahead, forming a wider helix. The 2_7 ribbon is not a regular secondary structure, but is shown here to demonstrate all possible hydrogen bond pairings. From R.E. Dickerson and I. Geis. *The Structure and Action of Proteins.* New York: Harper & Row, 1969.

Other helices have also been observed in proteins, though much less frequently due to their less favorable geometry. The 3_{10} helix has a period of three residues per turn, with hydrogen bonds between each residue and the residue three positions ahead (Taylor, 1941; Huggins, 1943) (Figure 2.8). This type of helix is usually seen only in short segments, often at the ends of an α-helix. The very rare pi (π) helix has a period of 4.4 residues per turn, with hydrogen bonds between each residue and the residue five positions ahead, and has only been seen at the ends of α-helices (Low and Baybutt, 1952).

β-Sheets

Unlike helices, β-sheets are formed by hydrogen bonds between adjacent polypeptide chains rather than within a single chain. Sections of the polypeptide chain participating in the sheet are known as β-strands. β-strands represent an extended conformation of the polypeptide chain, where Φ and Ψ angles are rotated approximately 180° with respect to each other. This arrangement produces a sheet that is pleated, with the residue side chains alternating positions on opposite sides of the sheet (Figure 2.9).

Two configurations of β-sheet are possible: parallel and antiparallel. In parallel sheets, the strands are arranged in the same direction with respect to their amino terminal (N) and carboxy terminal (C) ends. In antiparallel sheets, the strands alternate their amino and carboxy terminal ends, such that a given strand interacts with strands in the opposite orientation. β-sheets can also form in a mixed configuration, with both parallel and antiparallel sections, but this configuration is less common than the uniform types mentioned above. Almost all β-sheets exhibit some degree of twist when the sheet is viewed edge on, along an axis perpendicular to the direction of the polypeptide chains. This twist is generally right-handed.

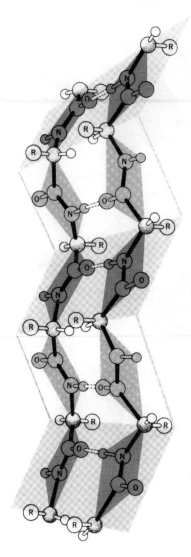

Figure 2.9. Diagram of an antiparallel β-sheet using a ball-and-stick model. The bonds forming the backbone of the polypeptide are shaded dark. The β-sheet is stabilized by hydrogen bonds (shown here as dashed lines) formed between the carbonyl oxygen of a residue on one strand and the amide proton of a residue on the adjacent strand. Note that this arrangement produces a sheet that is pleated, with the residue side chains alternating positions on opposite sides of the sheet. From R.E. Dickerson and I. Geis. *Hemoglobin: Structure, Function, Evolution, and Pathology*. Menlo Park, CA: The Benjamin/Cummings Publishing Company, 1983.

In order for β-strands to interact with each and form a sheet, the amino acid backbone must form a turn. The simplest and tightest turn is the β-hairpin turn (also called type I and type II turns, depending on the conformation of the backbone). Other stereochemically allowed turns include the mirror images of the type I and type II turns (denoted type I′ and type II′) as well as type III and type VI turns (Rotondi, 2006). All are observed in naturally occurring proteins and the type I turn appears to be most commonly used, frequently occurring between antiparallel strands.

An important variant of the classical β-sheet structure is the β-bulge. The β-bulge, most often observed in antiparallel β-sheets, is a hydrogen bond between two residues on one β-strand with one residue on the adjacent strand (Richardson, Getzoff, and Richardson, 1978; Chan et al., 1993). This structure can alter the direction of the polypeptide chain and augment the right-handed twist of the sheet (Richardson, 1981).

Other Secondary Structure

α-helices and β-sheets account for the majority of secondary structure seen in proteins. However, these regular structures are interspersed with regions of irregular structure that are referred to as loop or coil. Loop regions are usually present at the surface of the protein. These regions are often simply transitions between regular structures, but they can also possess structural significance and can be the location of the functional portion, or active site, of the protein.

Most of these structures were predicted and observed in the 1960s and 1970s (Venkatachalam, 1968; Chandrasekaran et al., 1973; Lewis et al., 1973; Chou and Fasman, 1977). Because of their irregularity, these elements are difficult to classify. However, some types of structures are ubiquitous enough to have been loosely categorized. β-Hairpin turns (described above) involve the minimum number of residues (4–5); transitions involving more residues (6–16) are often referred to as omega (Ω) loops because they resemble the shape of the capital Greek letter omega (Richardson, 1981). These loops can involve complex interactions that include the side chains, in addition to the polypeptide backbone. Extended loop regions involving more than 16 residues have also been observed.

Although loop regions are irregular, these structures are generally as well ordered as the regular secondary structures. However, experimental determination of protein structures has shown that some loop regions are disordered and thus do not achieve a stable structure. This type of loop region is referred to as *random* coil.

THE TERTIARY STRUCTURE OF PROTEINS: THE GLOBAL THREE-DIMENSIONAL STRUCTURE

The tertiary structure of a protein is defined as the global three-dimensional structure of its polypeptide chain. In a striking display of foresight, Alfred Mirsky and Linus Pauling correctly described several important aspects of tertiary structure in 1936 when they presented the following hypothesis on the nature of proteins (Mirsky and Pauling, 1936):

> ... one polypeptide chain which continues without interruption throughout the molecule ... folded into a uniquely defined configuration, in which it is held by hydrogen bonds between the peptide nitrogen and oxygen atoms and also between the free amino and carboxyl groups of the diamino and dicarboxyl amino acid residues.

The prediction of Mirsky and Pauling is especially striking considering that very little structural data were available and the linear polymer theory was still unproven at that time. As described previously, hydrogen bonds are important in the stabilization of secondary structure, but Mirsky and Pauling correctly divined their importance in tertiary structure stabilization as well.

While Mirsky and Pauling correctly predicted the role of hydrogen bonds in protein structure, it subsequently became apparent that other forces would also be important. With the determination of the structure of myoglobin by John Kendrew (Kendrew et al., 1958), protein scientists at last could begin to confirm many of their assumptions about various aspects of tertiary structure. Subsequent determination of the myoglobin structure using more accurate data (Kendrew et al., 1960), and the determination of the structure of hemoglobin by Max Perutz (Perutz et al., 1960), allowed scientists to begin to directly catalog aspects of the tertiary structure protein for the first time. Years of experimentation has now made it possible to understand how secondary structural elements combine in three-dimensional space to yield the tertiary structure of a protein.

Side Chains and Tertiary Structure

There are a wide variety of ways in which the various helix, sheet, and loop elements can combine to produce a complete structure. These combinations are brought about largely through interactions between the side chains of the constituent amino acid residues of the protein. Thus, *at the level of tertiary structure the side chains play a much more active role in creating the final structure.* In contrast, backbone interactions are primarily responsible for the generation of secondary structure (particularly in the case of helices and sheets). Most proteins each have a distinctive tertiary structure; the secondary structural elements of these proteins will always form the same tertiary structure. This consistency is vital to proper function of the protein.

Protein Folding

The final three-dimensional tertiary structure of a protein is commonly referred to as its fold. Appropriately, the process by which a linear polypeptide chain achieves its distinctive fold is known as protein folding. Protein folding is a complex process that is not yet completely understood, but many of the important principles are known. In a series of experiments in the early 1960s, Christian Anfinsen and his colleagues demonstrated that a small protein (ribonuclease) could be completely unfolded with chemicals, but when the protein was placed back into normal physiological conditions, it would spontaneously refold and regain full activity, as if it had never been unfolded in the first place (Anfinsen et al., 1961; Haber and Anfinsen, 1962). Anfinsen's work demonstrated a key concept in protein folding: the primary structure of the protein contains all of the information required for it to acquire the correct fold. In other words, *proteins are capable of self-assembly.* Anfinsen also proposed that the native fold of a protein corresponded to its lowest potential energy state and that protein folding was thus a process by which a protein chain spontaneously "falls" from the higher energy unfolded state to the lower energy folded state. Though this description of folding explained *why* proteins fold, it did not describe exactly *how* the process takes place.

Cyrus Levinthal demonstrated the remaining difficulties in protein folding with a simple thought experiment that measured the time a protein chain would take to find its lowest energy conformation through exhaustive search of all possible conformations (Levinthal, 1968). He reported that the universe would likely come to an end before the protein would fold![4] However, most proteins are known to fold in fractions of a second. This conflict between theory and experiment became known as Levinthal's paradox, demonstrating that the search for a protein's native state must not be random, but rather directed in some way such that the protein can fold rapidly.

---

[4]Levinthal's original calculations assumed more flexibility in the protein chain than is actually present and therefore overstated the time a protein would take to find its native state (Zwanzig, Szabo, and Bagchi, 1992). However, the essential point of the argument, that folding must be directed, remains sound.

The need for directed folding led to the current leading model for protein folding, initially proposed by Peter Wolynes and colleagues, in which the folding "energy landscape" is funnel-shaped (Frauenfelder, Sligar, and Wolynes, 1991). The edges of the funnel correspond to high potential energy unfolded states, and as the protein folds it is directed down the funnel into lower energy states until it reaches its native fold. The difficulty with this model is how the protein chain avoids getting stuck in "dips" in the funnel that are low energy, but not the native (lowest energy) fold. The answer appears to be found in the evolutionary process itself. It appears the protein sequences have evolved not only to form their native structures but also to have smooth energy landscapes that lack significant dips. Indeed, small naturally evolved proteins have been observed to fold in a highly "cooperative" fashion. That is, the whole protein chain participates in the folding process, and once the protein begins to fold the transition from unfolded to folded state is extremely rapid, and discrete intermediate structures cannot be isolated. This cooperative folding capability appears to be a specific quality shared by natural sequences and is not necessarily seen in proteins that have been designed in the laboratory (Watters et al., 2007). Further, it should be noted that random polypeptide sequences will almost never fold into an ordered structure; an important property of natural protein sequences is that they have been selected by the evolutionary process to achieve a reproducible, stable structure (Richardson, 1992).

Larger proteins (> 100 residues) appear to fold based on the same principles, but their complexity leads to slower folding, often with clear transition states on the way to the native state, and a more modular folding pattern (Vendruscolo et al., 2003). For these proteins, the more complex folding landscape introduces a new risk: the partly folded protein could start to interact improperly with other proteins, leading to the formation of harmful protein aggregates (Dobson, 2003). To deal with this challenge, cells employ "molecular chaperones," which are proteins that help other proteins to fold correctly and avoid aggregation (Hartl and Hayer-Hartl, 2002).

Thus, protein sequences are subject to several disparate evolutionary pressures. They must fold smoothly, avoiding incorrect alternate structures on the way to the correct one, resist the formation of aggregates, and finally, form a stable native structure once folding is complete. The sequences of natural proteins encode all of these characteristics, allowing them to fold reliably. However, despite the deterministic nature of protein folding, it is not yet possible to accurately predict the final structure of a protein given only its sequence (see Section VII for a discussion of the methods currently used to tackle this problem).

Domains and Motifs

Within the overall protein fold, distinct domains and motifs can be recognized. Domains are compact sections of the protein that represent structurally (and usually functionally) independent regions. In other words, a domain is a subsection of the protein that would maintain its characteristic structure, even if separated from the overall protein. Motifs (also referred to as *supersecondary* structure) are small substructures that are not necessarily structurally independent; generally, they consist of only a few secondary structural elements. Specific motifs are seen repeatedly in many different protein structures; they are integral elements of protein folds. Further, motifs often have a functional significance and in these cases represent a minimal functional unit within a protein. Several motifs can combine to form specific domains. Please see Figure 2.10 for a brief history of protein structure representation and visualization of these domains and motifs.

Figure 2.10. Schematic diagrams. One of the earliest pioneers in protein visualization was Irving Geis (Kendrew et al., 1961; Dickerson and Geis, 1969), who created some of the most definitive representations of protein structure. His hand-drawn depictions were so enlightening that they appear in textbooks to this day (and appear in this chapter). While hand drawings are extremely valuable, they are ultimately impractical because of the large number of protein structures that must be depicted (and the unusual level of talent required). Arthur Lesk and Karl Hardman were the first to popularize the use of computers to automatically generate schematic diagrams, given the (experimentally determined) spatial coordinates of the atoms in the protein (Lesk and Hardman, 1982). For more on molecular visualization, see Chapter 9.

The N-terminal domain of the catalytic core of eukaryotic protein kinase A (PKA, PDB id 1APM: residues 35–123) is depicted here using four different representations. This section of PKA contains a five-stranded antiparallel β-sheet and three helices. Figure 2.10a depicts this domain using an all-atom line representation. As can be seen, it is difficult to determine the overall structural characteristics of this protein using such a representation. Because proteins are often large and complex structures, views at the atomic level tend to obfuscate the important features. For this reason, a variety of schematic diagrams have been developed for the visual representation of protein structure. These diagrams replace the individual residues with shapes that represent the secondary structure they belong to and facilitate recognition of motifs and domains.

Simple topology diagrams are two-dimensional projections of the protein structure that are particularly useful for comparing the tertiary structures of different proteins (Holbrook et al., 1970). In these diagrams, β-strands are represented by arrows that point from the N-terminus to the C-terminus; α-helices are represented by cylinders. Connections between the secondary structural elements (loops) are simply represented as lines. These diagrams clearly illustrate the topology (connectivity) between the secondary structural elements and parallel or antiparallel nature of β-sheets (Figure 2.10b).

Cartoon diagrams illustrate the topology of the protein, as well as the spatial relationship between the structural components. These diagrams represent the three-dimensional structure of the protein as it actually occurs, with the atoms replaced by the same elements used in topology diagrams. Initially conceived of by Jane Richardson, and presented as hand drawings (Richardson, 1981), this representation is now very commonly used in protein visualization software packages (Figure 2.10c). Figure 2.10a and c was generated with the MolScript package (Kraulis, 1991), and is shown in an identical spatial orientation. The cartoon images in Tables 2.1 and 2.2 were also generated with the MolScript package.

TOPS diagrams, developed by Michael Sternberg and Janet Thornton, allow both the topology and the interaction of structural elements to be represented in a two-dimensional format (Sternberg and Thornton, 1977). Here, the secondary structural elements are viewed edge-on, as if they were projecting outward in a direction perpendicular to the plane of the page. β-strands are represented by triangles; an upward-pointing triangle portrays a strand pointing out of the page while a downward-pointing triangle portrays a strand pointing into the page. Helices are represented by circles. Lines represent the loops connecting these elements and also help portray chain direction (Figure 2.10d). Software is available to automatically generate these figures (Flores, Moss, and Thornton, 1994).

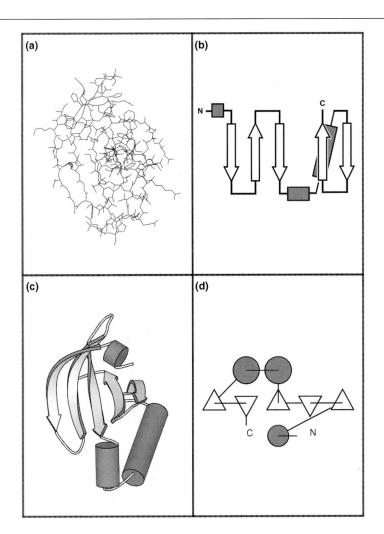

Molecular Interactions in Tertiary Structure

As with secondary structure, intramolecular forces are integral to defining and stabilizing the tertiary structure. The molecular interactions between structural elements, and individual side chains within these elements, help determine the protein fold. Because of the variety of chemical characteristics of the 20 standard amino acids, many types of interactions between them are possible. Furthermore, it is important to note that many proteins exist in an aqueous environment; intermolecular interactions with water (solvent) must be considered in addition to interactions within the protein itself.

A dominant molecular interaction in the tertiary structure of proteins is the hydrophobic effect (Tanford, 1978). Residues with hydrophobic side chains are packed into the *core* of the protein, away from the solvent, while charged and polar residues form the surface of the protein and are able to interact with polar water molecules and solvated ions. Although residues with hydrophobic side chains fit naturally into the core of the protein, their *polar* polypeptide backbone does not. For the backbone to participate in the hydrophobic core, its hydrogen bonding groups must be satisfied such that their polarity is, in effect, neutralized. Ordered secondary structural elements (helices and sheets) provide this neutralization through their regular hydrogen bonding patterns. Thus, secondary structural elements are critical to the formation of the hydrophobic core.

Residues with polar side chains can also participate in the hydrophobic core. They are, however, subject to same restrictions as the polypeptide backbone: they must be involved in an interaction that neutralizes their polarity. Buried polar residues can form hydrogen bonds with other polar residues, or with sections of the polypeptide backbone not participating in a regular secondary structure. Alternately, in some proteins small pockets exist in which buried polar residues satisfy their hydrogen bonds with water molecules. These water molecules are completely isolated from the solvent and are integral to the protein structure.

Charged residues can also occur within the hydrophobic core. This arrangement is possible only if the charged residue is paired with another residue of opposite charge such that the net charge is zero. This interaction, initially proposed by Henry Eyring and Allen Stearn in 1939, is known as an ion pair or salt bridge (Eyring and Stearn, 1939).

Eyring and Stearn also surmised that covalent connections between residue side chains were important in maintaining tertiary structure (Eyring and Stearn, 1939). Indeed, covalent interactions have been observed in some proteins. However, only one of the standard 20 amino acids is capable of participating in a covalent linkage: cysteine. The disulfide bond ($-S-S-$) can occur between the thiol ($-SH$) groups of two cysteine residues.[5] This interaction exists in proteins to further stabilize the protein fold. Cysteine residues do not always participate in disulfide bonds; in proteins, the majority of these residues are not part of a disulfide linkage.

Protein Modifications

Even though amino acids have a wide variety of chemical characteristics, chemical modifications and interaction with nonpolypeptide molecules can further extend the capabilities of proteins. At the level of tertiary structure, it is important to note these phenomena, as they can be critical to both the structure and the function of proteins.

---

[5]When cysteine is involved in a disulfide bond it is referred to as *cystine*.

The chemical properties of the 20 standard amino acids can be extended through side chain modification. In some cases, such modification can be indispensable to the proper formation of tertiary structures. For example, the protein collagen contains a modified proline residue, hydroxyproline, which greatly stabilizes the protein fold (Pauling and Corey, 1951b). In other cases, residue modification has no effect on fold, but rather extends the functional repertoire of the protein.

Molecules, such as carbohydrates and lipids, can be attached to the protein via covalent bonds with specific residues. Carbohydrate modifications are often seen on proteins that function extracellularly and are known to play a role in intracellular protein localization. Lipid modifications can help anchor a protein in the cell membrane.

Proteins can also associate with small molecules or metal atoms (covalently or noncovalently) to diversify their functional and structural capabilities. Many enzymes employ such molecules as cofactors, which assist them in chemical catalysis (Karlin, 1993). Other proteins require these molecules or atoms for proper tertiary structure formation. For example, the zinc-finger motif, a structural motif important for the interaction of some proteins with DNA, cannot form without the covalent interaction of a zinc atom with specific cysteine, or cysteine and histidine, residues (Lee, 1989).

Fold Space and Protein Evolution

One might imagine that the ways in which secondary structures can be combined to form a complete protein fold are almost limitless. Indeed, the variety of protein folds and the chemical diversity of their residues lead to a wide array of functions. This universe of extant folds is often called "fold space." Interestingly, currently available protein structure data suggest that fold space is in fact quite limited, relative to the range of folds that would seem possible *a priori*. Early estimates suggested that there might be about 1000 unique protein folds (Chothia, 1992; Govindarajan, Recabarren, and Goldstein, 1999; Wolf, Grishin, and Koonin, 2000). However, this number is now being approached by the available protein structure data and it appears likely it will be surpassed (Andreeva et al., 2004). More recent estimates suggest that the total number of natural folds lies somewhere between 1,000 and 10,000 (Koonin, Wolf, and Karev, 2002; Grant, Lee, and Orengo, 2004). It is likely that two forces have played a role in the observed limitation of fold space: divergent evolution of protein function and convergent evolution of protein structure.

In the case of divergent evolution, *the number of extant protein folds is limited because they are derived from a relatively small group of shared common ancestor proteins*. These early ancestor proteins would have "discovered" a stable fold, which has then been duplicated and re-used by organisms for many other functions over the course of evolution. Presumably, modification of an existing fold is more likely to occur than the spontaneous generation of a novel fold. There is clear evidence that this sort of modification has occurred over and over; it is possible because the link between protein structure and function is not direct (Todd, Orengo, and Thornton, 2001). Though protein structure leads to protein function, *similar protein structures will not always have similar functions*. There are many cases where two proteins have similar sequences and structures but differ by a few key amino acid residues in an active site and hence have very different functions. Thus, it is important to consider a protein's overall tertiary structure as a *guide* to the function of that protein, rather than a *definition* of the function (see Chapter 21). This functional versatility suggests the possibility that many protein folds will never be seen because organisms have simply not required or developed them.

In the case of convergent evolution, *the number of extant folds is limited because certain folds are biophysically much more favored, and so have been created independently in multiple cases*. Certain folds are clearly overrepresented in the set of proteins of known structure, even when efforts are made to reduce the representational bias inherent in the Protein Data Bank (PDB) (Berman et al., 2000; see Chapter 11). In some cases, there is no discernable sequence similarity between proteins of similar fold, suggesting that they have converged upon a similar structure independently and do not share a common ancestor (Holm and Sander, 1996). Further, some protein folding models have suggested that a small subset of folds will be biophysically favored over all others (Govindarajan and Goldstein, 1996; Li et al., 1996). Hence, it is possible that many folds will never be seen because they are not structurally favorable, and other more favorable folds can be adapted to the needed functions.

The divergent and convergent protein evolution scenarios are not mutually exclusive, and both appear to have had a part in limiting the range of folds in fold space. Even though fold space appears limited, it is still complex enough to make classification and comprehension of protein folds difficult (see Chapters 17 and 18).

Biochemical Classification of Folds

One method of protein classification partially sidesteps the issue of structural organization in favor of biochemical properties. Here, proteins are classified into three major groups: globular, membrane, and fibrous.

Globular proteins exist in an aqueous environment and thus fold as compact structures with hydrophobic cores and polar surfaces. They exhibit the "typical" structural elements that have been discussed above. This class of proteins is well represented in the PDB, partly because these structures are the easiest to determine experimentally (see Chapters 4–6). Because of the variety of determined structures available, globular proteins have been used as the basis for most protein structure studies (Berman et al., 2000). Indeed, the first two experimentally determined structures, myoglobin and hemoglobin, are globular (Kendrew et al., 1958; Perutz et al., 1958) (Table 2.1).

Membrane proteins exhibit many of the same characteristics as globular proteins, but they are distinguished in two important ways. First, they exist in an environment completely different from typical globular proteins: the cell membrane. In the interior of the cell membrane, the protein is surrounded by a hydrophobic environment. Thus, the regions of the protein within the cell membrane must have a hydrophobic *surface* in order to be stable. Some proteins exist almost entirely within the cell membrane, while others have membrane-spanning or membrane-interacting domains. Second, experimental structure determination (Chapters 4–6) of membrane proteins is more difficult than the determination of the structure of globular proteins. These proteins are very poorly represented in the PDB and thus are poorly understood. However, the structures that have been determined suggest that they use the same secondary structural elements and follow the same general folding principles as globular proteins (Table 2.1).

Fibrous proteins differ markedly from both membrane and globular proteins. These proteins are often constructed of repetitive amino acid sequences that form simple, elongated fibers. Their repetitive design is well suited to the structural roles they often play in organisms. Some fibrous proteins consist of a single type of regular secondary structure that is repeated over very long sequences. Others are formed from repetitive atypical secondary structures, or have no discernable secondary structure whatsoever (Table 2.1).

T A B L E 2 . 1 . Biochemical Folds

Fold	Protein Example	Protein Schematic
Globular proteins	Myoglobin (PDB id 1A6M)	
Membrane proteins	Rhodopsin (PDB id 1AT9)	
Fibrous proteins	Collagen (PDB id 1QSU)	

Structural Classification of Folds

As more and more protein structures have been determined, development of increasingly specific fold classifications has become possible. Cyrus Chothia and Michael Levitt derived one of the first such classifications, which grouped proteins on the basis of their predominant secondary structural element (Levitt and Chothia, 1976). This classification consisted of four groups: all α, all β, α/β, and α + β (Figure 2.11). All α-proteins, as the name suggests,

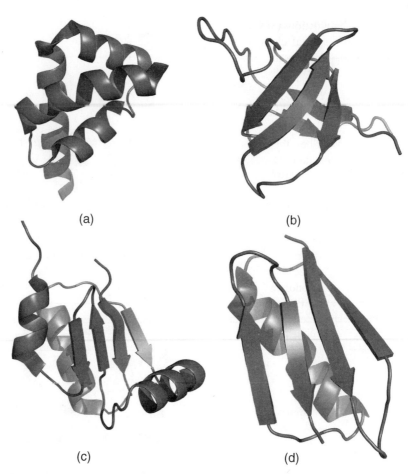

(a)

(b)

(c)

(d)

Figure 2.11. The four structural protein classes. (a) all α, PDB ID 1I2T; (b) all β, PDB ID 1K76; (c) α/β, PDB ID 1H75; (d) α + β, PDB ID 1EM7. All α-proteins contain almost entirely α-helices. All β-proteins contain almost entirely β-sheets. α/β proteins contain both α-helices and β-sheets, often organized as parallel β-strands connected by α-helices. α + β proteins consist of discrete α-helix and β-sheet motifs that are not interwoven (as they are in α/β structure). This figure was generated with PyMol (http://www.pymol.org).

are based almost entirely on an α-helical structure, and all β-structures are based almost entirely on β-sheet. α/β structure is based on a mixture of α-helix and β-sheet, often organized as parallel β-strands connected by α-helices. Finally, α + β structures consist of discrete α-helix and β-sheet motifs that are not interwoven (as they are in α/β structure). As known fold space has become more complex, these types of classifications have been adjusted and extended such that a complete hierarchy is created that places almost all known protein structure into specific subclassifications. Two approaches to this sort of classification (SCOP and CATH) are described in Chapters 17 and 18.

While most proteins achieve a stable structure, it has now become clear that a large number of proteins are intrinsically unstructured, in whole or part. Unstructured proteins appear to be particularly important in cell signaling and regulation. Often, the function of these proteins depends on achieving an ordered structure only under certain conditions or while engaged in specific interactions (Wright and Dyson, 1999; Dunker et al. 2002). It is therefore appropriate to add a special class of "intrinsically unstructured" to the available classifications of protein folds. For more on unstructured proteins, see Section VII, Chapter 38.

THE QUATERNARY STRUCTURE OF PROTEINS: ASSOCIATIONS OF MULTIPLE POLYPEPTIDE CHAINS

Tertiary structure describes the structural organization of a single polypeptide chain. However, many proteins do not function as a single chain, or monomer. Rather, they exist as a noncovalent association of two or more independently folded polypeptides. These proteins are referred to as multisubunit, or multimeric, proteins and are said to have a quaternary structure. The subunits, or protomers, may be identical, resulting in a homomeric protein, or they can be comprised of different subunits resulting in a heteromeric protein (Figure 2.12).

The first observation of quaternary structure has been attributed to Theodor Svedberg. His use of the ultracentrifuge to determine the molecular weights of proteins in 1926 resulted in the separation of multisubunit proteins into their constituent protomers (Klotz, 1970). The concept of quaternary structure was not of general biochemical interest until experiments in the early 1960s on enzyme regulation showed that protein subunits were crucial to understanding higher levels of cellular function (Gerhart and Pardee, 1962; Monod, Changeux, and Jacob, 1963; Klotz, 1970).

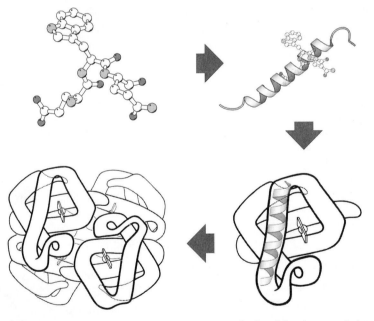

Figure 2.12. The four-tier hierarchy of protein structure, depicted for the protein hemoglobin. Hemoglobin is a multisubunit, heteromeric protein consisting of four all-helical subunits. The figure begins with the depiction of primary structure in the upper left corner and proceeds to quaternary structure in a clockwise direction. The primary and secondary structure depictions were generated with the MolScript package (Kraulis, 1991). Tertiary and quaternary structure depictions from R.E. Dickerson and I. Geis. *Hemoglobin: Structure, Function, Evolution, and Pathology.* Menlo Park, CA: The Benjamin/Cummings Publishing Company, 1983.

TABLE 2.2. Functional Relevance of Quaternary Structure

Functional Relevance	Protein Example	Protein Schematic
Cooperativity The association of subunits that bind the same substrate is often able to enhance binding capabilities of the multimer beyond what is possible with individual subunits. This cooperativity is realized through the ability of the subunits to influence each other based on their close proximity.	Hemoglobin (PDB id 1A3N)	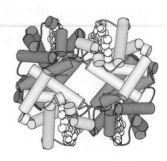
Colocalization of function Different subunits can associate to confer multiple functions on a single protein. Often these functions involve distinct steps in the processing of a single substrate. Thus, the colocalization of function provided by a multisubunit complex can further enhance the abilities of a protein.	Tryptophan synthase (PDB id 1QOP)	
Combinations of subunits Combinatorial shifts in quaternary structure are able to bestow impressive versatility on protein function and regulation. Protein function can be altered by subunit swapping, and protein regulation can be achieved via interactions with different subunits.	Immunoglobulin (PDB id 12E8)	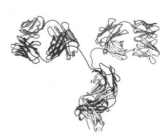
Structural assembly Very large structural proteins are made possible by the association of a large number of small subunits. This component-based assembly simplifies the construction of such structures and allows the information required to code these proteins to be more concise.	Actin (PDB id 1ALM)	

Interestingly, most proteins are folded such that aggregation with other polypeptides is avoided (Richardson, 1992); the formation of multisubunit proteins is therefore a very specific interaction. Quaternary structures are stabilized by the same types of interactions employed in tertiary and secondary structure stabilization. The surface regions involved in subunit interactions generally resemble the cores of globular proteins; they consist of residues with nonpolar side chains, residues that can form hydrogen bonds, or residues that can participate in disulfide bonds. Table 2.2 explains some of the functional advantages that are bestowed upon proteins by their quaternary structure.

CONCLUSION

Protein structure is complex, but it should be noted that by the 1970s protein scientists had determined most of the basic principles. These findings were confirmed in later years, as the pace of protein structure determination increased. These basic principles of protein structure now form the stable foundation needed for researching many of the remaining questions in protein science.

Computational tools now available, and described throughout this book, make management of the complexity of protein structure information more tractable. Protein structures can be more easily determined (see Chapter 40), and hypotheses as to the nature of protein structure can be more easily tested. As a result, it is now possible to pose more sophisticated questions, and in the coming years protein scientists can look forward to an understanding of protein structure and function on a much deeper level.

MORE INFORMATION ON THE INTERNET

The RCSB Protein Data Bank (PDB): http://www.rcsb.org/. [The sole worldwide public repository for macromolecular structure data.]

Education Resources Listing at the PDB. [A collection of links to educational resources covering protein structure.]

TOPS homepage and server: http://www.tops.leeds.ac.uk. [Used to create one of the figures in this chapter.]

MolScript homepage: http://www.avatar.se/molscript/. [A popular package for generating images of protein structure, used to create many of the figures in this chapter.]

PyMol homepage: http://pymol.sourceforge.net. [A popular package for generating images of protein structure, used to create figures in this chapter.]

REFERENCES

General References

Branden C, Tooze J (1999): *Introduction to Protein Structure*. 2nd ed. New York: Garland Publishing, Inc. [Perhaps the finest book on protein structure. It features a well-written text and excellent hand-drawn illustrations.]

Mathews CK, van Holde KE (2000): *Biochemistry*. Redwood City, CA: The Benjamin/Cummings Publishing Company.

Voet D, Voet JG (2005): *Biochemistry,* 3rd ed. New York: John Wiley & Sons, Inc.

Cited References

Andreeva A, Howorth D, Brenner SE, Hubbard TJP, Chothia C, Murzin AG (2004): SCOP database in 2004: refinements integrate structure and sequence family data. *Nucleic Acids Res* 32: D226–D229.

Anfinsen CB, Haber E, Sela M, White FH Jr (1961): The kinetics of formation of native ribonuclease during oxidation of the reduced polypeptide chain. *Proc Natl Acad Sci USA* 47:1309–1314.

Atkins JF, Gesteland RF (2000): The twenty-first amino acid. *Nature* 407:463–465.

Berman HM, Westbrook J, Feng Z, Gilliland G, Bhat TN, Weissig H, Shindyalov IN, Bourne PE (2000): The protein data bank. *Nucleic Acids Res* 28:235–242.

Chan AW, Hutchinson EG, Harris D, Thornton JM (1993): Identification, classification, and analysis of beta-bulges in proteins. *Protein Sci* 2:1574–1590.

Chandrasekaran R, Lakshminarayanan AV, Pandya UV, Ramachandran GN (1973): Conformation of the LL and LD hairpin bends with internal hydrogen bonds in proteins and peptides. *Biochim Biophys Acta* 303:14–27.

Chothia C (1992): One thousand families for the molecular biologist. *Nature* 357:543–544.

Chou PY, Fasman GD (1977): β turns in proteins. *J Mol Biol* 115:135–175.

Dickerson RE, Geis I (1969): *The Structure and Action of Proteins.* New York: Harper and Rowe. [A seminal book on protein structure, featuring the lucid illustrations of Irving Geis.]

Dobson CM (2003): Protein folding and misfolding. *Nature* 426:884–890.

Dunker AK, Brown CJ, Lawson JD, Iakoucheva LM, Obradovic Z (2002): Intrinsic disorder and protein function. *Biochemistry* 41:6573–6582.

Eyring H, Stearn AE (1939): The application of the theory of absolute reaction rates to proteins. *Chem Rev* 24:253–270.

Flores TP, Moss DM, Thornton JM (1994): An algorithm for automatically generating protein topology cartoons. *Protein Eng* 7:31–37.

Frauenfelder H, Sligar SG, Wolynes PG (1991): The energy landscapes and motions of proteins. *Science* 254:1598–1603.

Fruton JS (1972): *Molecules and Life: Historical Essays on the Interplay of Chemistry and Biology.* New York: Wiley–Interscience. [An informative perspective on the early histoxry of protein science.]

Gerhart JC, Pardee AB (1962): The enzymology of control by feedback inhibition. *J Biol Chem* 237:891–896.

Govindarajan S, Goldstein RA (1996): Why are some protein structures so common? *Proc Natl Acad Sci USA* 93:3341–3345.

Govindarajan S, Recabarren R, Goldstein RA (1999): Estimating the total number of protein folds. *Proteins* 35:408–414.

Grant A, Lee D, Orengo C (2004): Progress towards mapping the universe of protein folds. *Genome Biol* 5:107.

Haber E, Anfinsen CB (1962): Side-chain interactions governing the pairing of half-cystine residues in ribonuclease. *J Biol Chem* 237:1839–1844.

Hartl FU, Hayer-Hartl M (2002): Molecular chaperones in the cytosol: from nascent chain to folded protein. *Science* 295:1852–1858.

Holbrook JJ, Liljas A, Steindel SJ, Rossmann MG (1970): Oxidation–reduction. Part A. Dehydrogenases (I), Electron transfer (I). In: Boyer PD, editor. *The Enzymes* 3rd ed., Vol. 11. New York: Academic Press. p. 210.

Holm L, Sander C (1996): Mapping the protein universe. *Science* 273:595–603.

Huggins ML (1943): The structure of fibrous proteins. *Chem Rev* 32:195–218.

Karlin KD (1993): Metalloenzymes, structural motifs, and inorganic models. *Science* 261:701–708.

Kendrew JC, Bodo G, Dintzis HM, Parrish RG, Wyckoff H, Phillips DC (1958): A three-dimensional model of the myoglobin molecule obtained by X-ray analysis. *Nature* 181:662–666. [The publication announcing the experimental determination of the first protein structure. Worth reading for a historical perspective.]

Kendrew JC, Dickerson RE, Strandberg BE, Hart RG, Davies DR, Phillips DC, Shore VC (1960): Structure of myoglobin. *Nature* 185:422–427.

Kendrew JC (1961): The three-dimensional structure of a protein molecule. *Sci Am* 205:96–111. [An interesting follow-up to the determination of the myoglobin structure, it also introduces the illustrations of Irving Geis.]

Klotz IM, Langerman NR, Darnall DW (1970): Quaternary structure of proteins. *Annu Rev Biochem* 39:25–62.

Koonin EV, Wolf YI, Karev GP (2002): The structure the protein universe and genome evolution. *Nature* 420:218–223.

Kraulis PJ (1991): MoLScript: a program to produce both detailed and schematic plots of protein structures. *J Appl Crystallogr* 24:946–950.

Kreil G (1997): D-Amino acids in animal peptides. *Annu Rev Biochem* 66:337–345.

Lee MS (1989): 3D structure of single zinc finger. *Science* 245:635–637.

Lesk AM, Hardman KD (1982): Computer-generated schematic diagrams of protein structures. *Science* 216:539–540.

Levinthal C (1968): Are there pathways for protein folding? *J Chim Phys* 65:44–45.

Levitt M, Chothia C (1976): Structural patterns in globular proteins. *Nature* 261:552–558.

Lewis PN, Momany FA, Scheraga HA (1973): Chain reversals in proteins. *Biochim Biophys Acta* 303:211–229.

Li H, Helling R, Tang C, Wingreen N (1996): Emergence of preferred structures in a simple model of protein folding. *Science* 271:666–669.

Low BW, Baybutt RB (1952): The π-helix: a hydrogen bonded configuration of the polypeptide chain. *J Am Chem Soc* 74:5806–5810.

Mirksy AE, Pauling L (1936): On the structure of native, denatured and coagulated proteins. *Proc Natl Acad Sci USA* 22:439–447. [A highly insightful and well-presented discussion of protein structure from two key pioneers in the field. Worth reading for a historical perspective.]

Monod J, Changeux JP, Jacob F (1963): Allosteric proteins and cellular control systems. *J Mol Biol* 6:306–329.

Pauling L, Corey RB (1951a): Configurations of polypeptide chains with favored orientations around single bonds: two new pleated sheets. *Proc Natl Acad Sci USA* 37:729–740. [This paper and the two that follow below are part of a series of publications in which Linus Pauling and his colleagues discuss different aspects of secondary structure.]

Pauling L, Corey RB (1951b): The structure of fibrous proteins of the collagen–gelatin group. *Proc Natl Acad Sci USA* 37:272–281.

Pauling L, Corey RB, Branson HR (1951): The structure of proteins: two hydrogen bonded helical configurations of the polypeptide chain. *Proc Natl Acad Sci USA* 37:205–211.

Perutz MF, Rossmann MG, Cullis AF, Muirhead H, Will G, North ACT (1960): Structure of hemoglobin. *Nature* 185:416–422. [The publication announcing the experimental determination of the first multisubunit protein structure. Worth reading for a historical perspective.]

Ramachandran GN, Sasisekharan V (1968): Conformation of polypeptides and proteins. *Adv Protein Chem* 23:283–437.

Richardson JS, Getzoff ED, Richardson DC (1978): The β bulge: a common small unit of nonrepetitive protein structure. *Proc Natl Acad Sci USA* 75:2574–2578.

Richardson JS (1981): The anatomy and taxonomy of protein structure. *Adv Protein Chem* 34:167–339. [An exhaustive compendium of known protein structure, still useful to this day.]

Richardson JS (1992): Looking at proteins: representations, folding, packing, and design. *Biophys J* 63:1186–1209. [An examination of protein structure and a discussion of efforts to engineer protein structures from designed amino acid sequences.]

Rotondi KS, Gierasch LM (2006): Natural polypeptide scaffolds: beta-sheets, beta-turns, and beta-hairpins. *Biopolymers* 84:13–22.

Sanger F (1952): The arrangement of amino acids in proteins. *Adv Protein Chem* 7:1–67. [A presentation of the techniques and experimental data used to explore the peptide theory of proteins.]

Srinivasan G, James CM, Kryzycki JA (2002): Pyrrolysine: encoded by UAG in Archaea: charging of a UAG-decoding specialized tRNA. *Science* 296:1459–1462.

Sternberg MJE, Thornton JM (1977): On the conformation of proteins: the handedness of the connection between parallel β strands. *J Mol Biol* 110:269–283.

Tanford C (1978): The hydrophobic effect and the organization of living matter. *Science* 200: 1012–1018.

Taylor HS (1941): Large molecules through atomic spectacles. *Proc Am Phil Soc* 85:1–12.

Taylor WR (1986): The classification of amino acid conservation. *J Theor Biol* 119:205–218.

Todd AE, Orengo CA, Thornton JM (2001): Evolution of function in protein superfamilies, from a structural perspective. *J Mol Biol* 307:1113–1143. [A comprehensive look at the relationship between protein structure and function in enzymes of known structure.]

Vendruscolo M, Paci E, Karplus M, Dobson CM (2003): Structures and relative free energies of partially folded states of proteins. *Proc Natl Acad Sci USA* 100:14817–14821.

Venkatachalam CM (1968): Stereochemical criteria for polypeptides and proteins. V. Conformation of a system of three linked peptide units. *Biopolymers* 6:1425–1436.

Wang L, Brock A, Herberich B, Schultz PG (2001): Expanding the genetic code of *Escherichia coli. Science* 292:498–500.

Wang L, Xie J, Schultz PG (2006): Expanding the genetic code. *Annu Rev Biophys Biomol Struct* 35:225–249.

Watters AL, Deka P, Corrent C, Callender D, Varani G, Sosnick T, Baker D (2007): The highly cooperative folding of small naturally occurring proteins is likely the result of natural selection. *Cell* 128:613–624.

Wolf YI, Grishin NV, Koonin EV (2000): Estimating the number of protein folds and families from complete genome data. *J Mol Biol* 299:897–905.

Wright PE, Dyson HJ (1999): Intrinsically unstructured proteins: re-assessing the protein structure–function paradigm. *J Mol Biol* 293:321–331. [A review discussing the current knowledge of proteins lacking an ordered structure, or achieving an ordered structure only under certain conditions or while engaged in specific interactions.]

Zinoni F, Birkmann A, Stadtman TC, Böck A (1986): Nucleotide sequence and expression of the selenocysteine-containing polypeptide of formate dehydrogenase (formate-hydrogen-lyase-linked) from Escherichia coli. *Proc Natl Acad Sci USA* 83:4650–4654.

Zwanzig R, Szabo A, Bagchi B (1992): Levinthal's paradox. *Proc Natl Acad Sci USA* 89:20–22.

3

FUNDAMENTALS OF DNA AND RNA STRUCTURE

Stephen Neidle, Bohdan Schneider, and Helen M. Berman

INTRODUCTION

In 1946, Avery provided concrete experimental evidence that DNA was the main constituent of genes (Avery et al., 1944); universal acceptance of this idea came with the publication of the Hershey–Chase experiments (Hershey and Chase, 1951). After the seminal discovery of the double helical nature of DNA in 1953 (Watson and Crick, 1953), the focus of nucleic acid structural research turned to fiber diffraction of natural and defined sequences (Arnott 1970; Arnott, Campbell Smith, and Chandrasekaran, 1976a). Through these studies, we gained many insights into nucleic acid structure. We learned that hydration, ionic strength, and sequence affect conformation type and that nucleic acids can adopt a wide variety of structures including single-stranded helices (Arnott, Chandrasekaran, and Leslie, 1976b) and parallel helices (Rich et al., 1961), as well as triple and quadruple helices (Arnott, Chandrasekharan, and Marttila, 1974).

Once it was possible to synthesize and purify oligonucleotides (Khorana et al., 1956), crystallography and later NMR could be used to determine nucleic acid structures. The first crystal structure that was identified as containing all the components that could have allowed us to see a double helix was a very small piece of RNA—the dinucleoside monophosphate UpA (Seeman et al., 1971). But rather than forming a double helix, it displayed unusual conformations, anticipating some of the many conformations that we now know exist in nucleic acids. In 1973, the double helix was visualized at atomic resolution with the determination of the crystal structures of two self-complementary RNA fragments, ApU and GpC (Rosenberg et al., 1973). The determination of the structures of dinucleoside phosphates complexed with drugs followed and laid the foundation for nucleic acid recognition (Tsai et al., 1975).

The publication in 1980 of a structure of more than a full turn of B-DNA (Wing et al., 1980) laid aside the doubts of even the most skeptical researchers (Rodley et al., 1976) that DNA was a right-handed double helix. The structure was also a milestone in our understanding of the fine structure of DNA, whereby it was possible to determine the effects of sequence on structure. Interestingly, it was at the same time that the structure of an unusual left-handed form of DNA Z DNA was solved (Wang et al., 1979).

In parallel with the studies of these synthetic oligonucleotides, researchers were successful in purifying transfer RNA. The publication of the structure of yeast Phe tRNA in 1974 (Kim et al., 1974; Robertus et al., 1974) represented the first and until relatively recently, the only example of a natural intact nucleic acid structure. Now we are seeing an ever-increasing number of structures of RNA that are giving us insights into the RNA world.

In this chapter, we present the principles of nucleic acid structure. We then present a brief overview of the current state of our knowledge of nucleic acid structure determined using X-ray crystallographic methods. Further details on individual nucleic acid structures are provided in the recent book by one of us (Neidle, 2007). Discussion of nucleic acids in complex with proteins is dealt in Chapter 25.

CHEMICAL STRUCTURE OF NUCLEIC ACIDS

In the early years of the twentieth century, chemical degradation studies on material extracted from cell nuclei established that the high molecular weight "nucleic acid" was actually composed of individual acid units termed nucleotides. Four distinct types were isolated—guanylic, adenylic, cytidylic, and thymidylic acids. These could be further cleaved to phosphate groups and four distinct nucleosides. The latter were subsequently identified as consisting of a deoxypentose sugar and one of the four nitrogen-containing heterocyclic bases. Thus, each repeating unit in a nucleic acid polymer constitutes these three units linked together—a phosphate group, a sugar, and one of the four bases.

The bases are planar aromatic heterocyclic molecules and are divided into two groups—the pyrimidine bases thymine and cytosine, and the purine bases adenine and guanine. Their major tautomeric forms are shown in Figure 3.1. Thymine is replaced by uracil in ribonucleic acids. RNA also has an extra hydroxyl group at the $2'$ position of its pentose sugar groups. The sugar present in RNA is ribose; in DNA, it is deoxyribose. The standard nomenclature for the atoms in nucleic acids is shown in Figure 3.1. Accurate bond length and angle geometries for all bases, nucleosides, and nucleotides have been well established by X-ray crystallographic analyses. The most recent surveys (Clowney et al., 1996; Gelbin et al., 1996) have calculated mean values for these parameters (at equilibrium) from the most reliable structures in the Cambridge Structural Database (Allen et al., 1979) and the Nucleic Acid Database (Berman et al., 1992). These have been incorporated in several implementations of the AMBER (Weiner and Kollman, 1981; Case et al., 2005) and CHARMM (Brooks et al., 1983) force fields, widely used in molecular mechanics and dynamics modeling and in a number of computer packages for both crystallographic and NMR structural analyses (Parkinson et al., 1996). Accurate crystallographic analyses, at very high resolution, can also directly yield quantitative information on the electron density distribution in a molecule, and hence on individual partial atomic charges. These charges for nucleosides have hitherto been obtained by *ab initio* quantum mechanical calculations, they but are now available experimentally for all four DNA nucleosides (Pearlman and Kim, 1990).

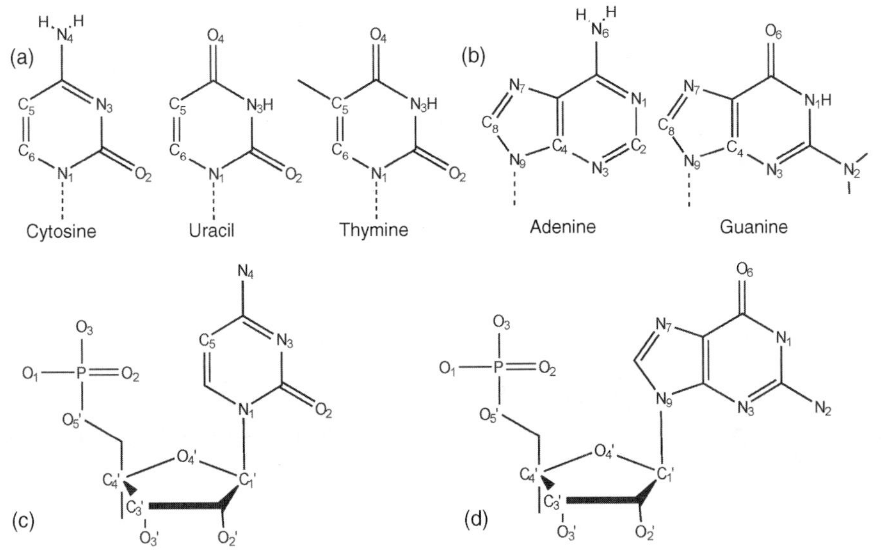

Figure 3.1. Chemical composition and nomenclature of nucleic acid components. (a) Pyrimidines. Uracil occurs in RNA, and DNA base thymine has a methyl group attached to C5. (b) Purines. (c) A pyrimidine nucleotide, shown is cytidine-5'-phosphate. (d) A purine nucleotide, shown is guanosine-5'-phosphate.

Individual nucleoside units are joined together in a nucleic acid in a linear manner through phosphate groups that are attached to the 3' and 5' positions of the sugars (Figure 3.1c and d). Hence, the full repeating unit in a nucleic acid is a 3',5'-nucleotide.

For nucleic acid and oligonucleotide sequences, single-letter codes for the five unit nucleotides are used—A, T, G, C, and U. The two classes of bases can be abbreviated as Y (pyrimidine) and R (purine). Phosphate groups are usually designated as "p". A single oligonucleotide chain is conventionally numbered from the 5'-end; for example, ApGpCpTpTpG has the 5' terminal adenosine nucleoside, with a free hydroxyl at its 5' position, and thus the 3'-end guanosine has a free 3' terminal hydroxyl group. Intervening phosphate groups are sometimes omitted when a sequence is written down. Chain direction is sometimes emphasized with 5' and 3' labels. Thus, an antiparallel double helical sequence can be written as

5'CpGpCpGpApApTpTpCpGpCpG
3'GpCpGpCpTpTpApApGpCpGpC

or simply as (CGCGAATTCGCG)$_2$. In structural publications, the prefix "d" is usually applied to a DNA sequence, as in d(CGAT) to emphasize that the oligonucleotide is a deoxyribose rather than an oligoribonucleotide.

The bond between sugar and base is known as the glycosidic bond. Its stereochemistry is important. In natural nucleic acids, the glycosidic bond is always in the β stereochemistry, which is to say that the base is above the plane of the sugar when viewed onto the plane and therefore on the same face of the plane as the 5' hydroxyl substituent (Figure 3.2d). The absolute stereochemistry of other substituent groups on the deoxyribose sugar ring of DNA is defined such that when viewed end on with the sugar ring oxygen atom O4' at the rear, the

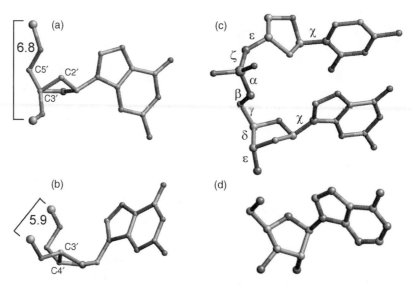

Figure 3.2. Structural features and stereochemistry of nucleic acid components. (a) C2'-*endo* sugar pucker, typical for B-DNA. Atom C3', labeled, is in front and conceals atom C4'. Atoms C2' and C5' point both "up". Two consecutive P atoms have typical distance about 6.4–6.9 Å. (b) C3'-*endo* sugar pucker, typical for A-form nucleic acids (shown for deoxyribose, as in (a)). In C3'-*endo*, atom C3' is "above" the plane of the remaining four sugar ring atoms. P–P distances are shorter than in the case of the C2'-*endo* pucker, between 5.4 and 6.2 Å. (c) Definition of the backbone torsion angles: $\alpha = O3'-P-O5'-C5$, $\beta = P-O5'-C5'-C4'$, $\gamma = O5'-C5'-C4'-C3'$, $\delta = C5'-C4'-C3'-O3'$, $\varepsilon = C4'-C3'-O3'-P$, and $\zeta = C3'-O3'-P-O5'$. Torsions around the glycosidic bonds are defined as $\chi = O4'-C1'-N1-C2$ for pyrimidines (shown is cytosine) and $\chi = O4'-C1'-N9-C4$ for purines (shown is guanine). (d) Stereochemistry of a natural β-nucleoside.

hydroxyl group at the 3' position is below the ring and the hydroxymethyl group at the 4' position is above.

A unit nucleotide can have its phosphate group attached at the either 3'- or 5'-ends and is thus termed either a 3' or a 5' nucleotide. It is chemically possible to construct α-nucleosides and from them α-oligonucleosides, which have their bases in the "below" configuration relative to the sugar rings and their other substituents. These are much more resistant to nuclease attack than standard natural β-oligomers and have been used as antisense oligomers to mRNAs because of their superior intracellular stability.

BASE-PAIR GEOMETRY

The realization that the planar bases can associate in particular ways by means of hydrogen bonding was a crucial step in the elucidation of the structure of DNA. The important early experimental data of Chargaff (Zamenhof, Brawermann, and Chargaff, 1952) showed that the molar ratios of both adenine : thymine and cytosine : guanine in DNA were unity. This led to the proposal by Crick and Watson that in each of these pairs, the purine and pyrimidine bases are held together by specific hydrogen bonds to form planar base pairs. In native double helical DNA, the two bases in a base pair necessarily arise from two separate strands of DNA with intermolecular hydrogen bonds that hold the DNA double helix together (Watson and Crick, 1953).

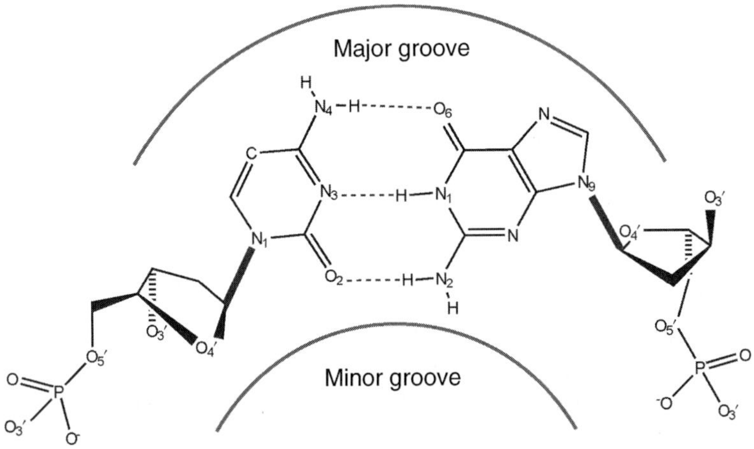

Figure 3.3. The canonical Watson–Crick base pair, shown as the G-C pair. Positions of the minor and major grooves are indicated. The glycosidic sugar–base bond is shown by the bold line; hydrogen bonding between the two bases is shown in dashed lines.

The adenine:thymine (AT) base pair has two hydrogen bonds compared to the three in a guanine : cytosine (GC) one (Figure 3.3). Fundamental to the Watson–Crick arrangement is that the sugar groups are both attached to the bases asymmetrically on the same side of the base pair. This defines the mutual positions of the two sugar–phosphate strands in DNA itself. Atoms at the surface of the sugar phosphate backbone define two indentations with different dimensions called minor and major grooves. By convention, the major groove is faced by C6/N7/C8 purine atoms and their substituents and by C4/C5/C6 pyrimidine atoms and their substituents, and the minor groove by C2/N3 purine and C2 pyrimidine atoms and their substituents.

The two base pairs are required to be almost identical in dimensions by the Watson–Crick model. High-resolution (0.8–0.9 Å) X-ray crystallographic analyses of the ribodinucleoside monophosphate duplexes $(GpC)_2$ and $(ApU)_2$ by A. Rich and colleagues in the early 1970s has established accurate geometries for these AT and GC base pairs (Rosenberg et al., 1976; Seeman et al., 1976). These structure determinations showed that there are only small differences in size between the two types of base pairings, as indicated by the distance between glycosidic carbon atoms in a base pair. The C1′...C1′ distance in the GC base-pair structure is 10.7 Å and 10.5 Å in the AU-containing dinucleoside.

The individual bases in a nucleic acid are flat aromatic rings, but base pairs bound together only by nonrigid hydrogen bonds can show considerable flexibility. The vertical arrangement of bases and base pairs is flexible and restrained mainly by stacking interactions of bases. This flexibility to some extent depends on the nature of the bases and base pairs themselves, but is more related to their base-stacking environments. Thus, descriptions of base morphology have become important in describing and understanding many sequence-dependent features and deformations of nucleic acids. The sequence-dependent features are often considered primarily at the dinucleoside local level, whereas longer range effects, such as helix bending, can also be analyzed at a more global level.

A number of rotational and translational parameters have been devised to describe these geometric relations between bases and base pairs, which were originally defined in the 1989 "Cambridge Accord" (Dickerson et al., 1989). These definitions, together with the Cambridge Accord sign conventions, are given for some key base parameters.

Propeller twist (ω) between bases is the dihedral angle between the normal vectors to the bases, when viewed along the long axis of the base pair. The angle has a negative sign under normal circumstances, with a clockwise rotation of the nearer base when viewed down the long axis. The long axis for a purine–pyrimidine base pair is defined as the vector between the C8 atom of the purine and the C6 of a pyrimidine in a Watson–Crick base pair. Analogous definitions can be applied to other nonstandard base pairings in a duplex including purine–purine and pyrimidine–pyrimidine ones.

Buckle (κ) is the dihedral angle between bases, along their short axis, after propeller twist has been set to $0°$. The sign of buckle is defined as positive if the distortion is convex in the direction $5' \rightarrow 3'$ of strand 1. The change in buckle for succeeding steps, termed *cup*, has been found to be a useful measure of changes along a sequence. Cup is defined as the difference between the buckle at a given step and that of the preceding one.

Inclination (η) is the angle between the long axis of a base pair and a plane perpendicular to the helix axis. This angle is defined as positive for right-handed rotation about a vector from the helix axis toward the major groove.

X and **Y** *displacements* define translations of a base pair within its mean plane in terms of the distance of the midpoint of the base pair long axis from the helix axis. *X* displacement is toward the major groove direction, when it has a positive value. *Y* displacement is orthogonal to this, and it is positive if toward the first nucleic acid strand of the duplex.

The key parameters for base-pair steps are as follows:

Helical twist (Ω) is the angle between successive base pairs, measured as the change in orientation of the $C1'$–$C1'$ vectors on going from one base pair to the next, projected down the helix axis. For an exactly repetitious double helix, helical twist is $360°/n$, where n is the unit repeat defined above.

Roll (ρ) is the dihedral angle for rotation of one base pair with respect its neighbor about the long axis of the base pair. A positive roll angle opens up a base-pair step toward the minor groove. *Tilt* (τ) is the corresponding dihedral angle along the short (i.e., *x*-axis) of the base pair.

Slide is the relative displacement of one base pair compared to another, in the direction of nucleic acid strand 1 (i.e., the *Y* displacement), measured between the midpoints of each C6–C8 base pair long axis.

Unfortunately, there is now some confusion in the literature regarding these parameters, in part because the Cambridge Accord did not define a single unambiguous convention for their calculation, and two distinct types of approaches have been developed to calculate them (Lu et al., 1999). In one approach, the parameters are defined with respect to a global helical axis that need not be linear. The other uses a set of local axes, one per dinucleotide step. Another ambiguity is that a variety of definitions of local and global axes have been used. Fortunately, the overall effect for most undistorted structures is that only a minority of parameters appear to have distinctly different values depending on which method of calculation is used by the widely available programs: CEHS/SCHNAaP (El Hassan and Calladine, 1995; Lu, El Hassan, and Hunter, 1997), CompDNA (Gorin, Zhurkin, and Olson, 1995; Kosikov et al., 1999), Curves (Lavery and Sklenar, 1988; Lavery and Sklenar, 1989), FREEHELIX (Dickerson, 1998), NGEOM (Soumpasis and Tung, 1988; Tung et al., 1994), NUPARM (Bhattacharyya and Bansal, 1989; Bansal et al., 1995), and RNA (Babcock et al., 1993; Babcock and Olson, 1994; Babcock et al., 1994).

To resolve the ambiguities in description of base morphology parameters (Figure 3.4), a standard coordinate reference frame for the calculation of these parameters has been proposed (Lu and Olson, 1999) and endorsed by the successor to the Cambridge Accord, the 1999 Tsukuba Accord (Olson et al., 2001). The right-handed reference frame is shown in Figure 3.5. It has the *x*-axis directed toward the major groove along the pseudo-twofold axis

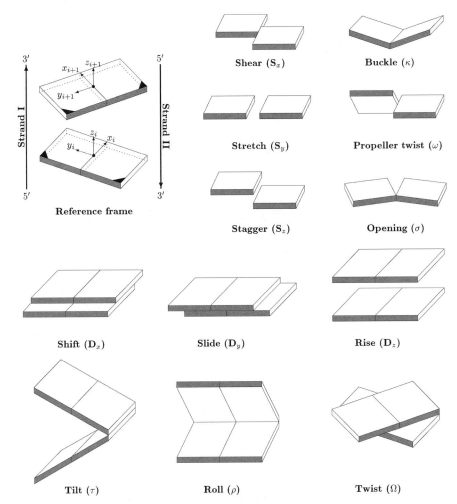

Figure 3.4. Pictorial definitions of some parameters that relate complementary base pairs and sequential base-pair steps. The base-pair reference frame is constructed such that the x-axis points away from the (shaded) minor groove edge. Images illustrate positive values of the designated parameters. Reprinted with permission from Adenine Press from Lu, Babcock, and Olson (1999).

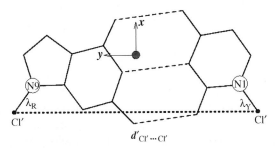

Figure 3.5. Illustration of idealized base-pair parameters, $d_{C1' \cdots C1'}$ and λ, used, respectively, to displace and pivot complementary bases in the optimization of the standard reference frame for right-handed A- and B-DNA, with the origin at • and the x- and y-axes pointing in the designated directions. Reprinted with permission from Olson et al. (2001).

(shown as •) of an idealized Watson–Crick base pair. The *y*-axis is along the long axis of the base pair, parallel to the C1′ ⋯ C1′ vector. The position of the origin clearly depends on the geometry of the bases and the base pair. These have been taken from the published compilations (Clowney et al., 1996; Gelbin et al., 1996). The Tsukuba reference frame is unambiguous and has the advantage of being able to produce values for the majority of local base pair and base step parameters that are independent of the algorithm used.

Although Watson–Crick base pairs are the most prevalent type of pairs in nucleic acids, in single-stranded RNA, 40% of the base pairs form other types. A systematic classification has been adopted (Figure 3.6) (Leontis and Westhof, 2001). In this system,

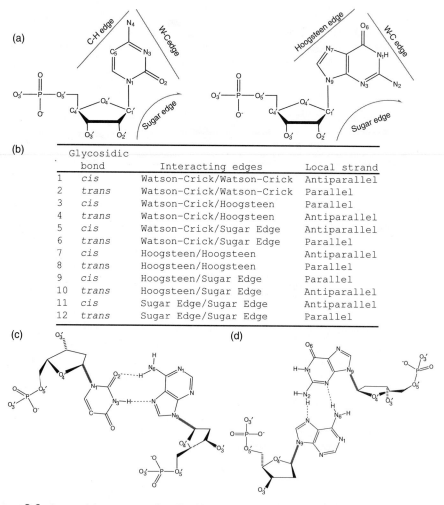

(b)

	Glycosidic bond	Interacting edges	Local strand
1	cis	Watson–Crick/Watson–Crick	Antiparallel
2	trans	Watson–Crick/Watson–Crick	Parallel
3	cis	Watson–Crick/Hoogsteen	Parallel
4	trans	Watson–Crick/Hoogsteen	Antiparallel
5	cis	Watson–Crick/Sugar Edge	Antiparallel
6	trans	Watson–Crick/Sugar Edge	Parallel
7	cis	Hoogsteen/Hoogsteen	Antiparallel
8	trans	Hoogsteen/Hoogsteen	Parallel
9	cis	Hoogsteen/Sugar Edge	Parallel
10	trans	Hoogsteen/Sugar Edge	Antiparallel
11	cis	Sugar Edge/Sugar Edge	Antiparallel
12	trans	Sugar Edge/Sugar Edge	Parallel

Figure 3.6. Base-pairing patterns described by Leontis–Westhof (L-W) nomenclature (Leontis and Westhof, 2001). (a) Definition of the "edges" that can form base pairs with the other base. (b) The 12 base-pair classes in the L-W schema. (c) Noncanonical base pair interacting by Watson–Crick and Hoogsteen edges with *trans* orientation of glycosidic bonds. This L-W class 4 is important in RNA architecture. (d) L-W class 10 base pair formed by the Hoogsteen and Sugar edges with *trans* orientation of the glycosidic bonds.

base pairing is defined as hydrogen bonding between edges of the interacting bases. Each base has three edges: Watson–Crick, sugar, and Hoogsteen (in purines) or C–H (in pyrimidines) as shown in Figure 3.6a; the classes are defined in Figure 3.6b. The canonical Watson–Crick pair is a CG or AT pair interacting through the Watson–Crick edges with *cis* mutual orientation of the glycosidic bonds (Figure 3.3) and antiparallel local orientation of the opposing nucleotide strands. Two examples of non-Watson–Crick bonding are shown in Figures Figure 3.6c and d.

CONFORMATION OF THE SUGAR PHOSPHATE BACKBONE

The five-member deoxyribose sugar ring in nucleic acids is inherently nonplanar. This nonplanarity is termed puckering. The precise conformation of a deoxyribose ring can be completely specified by the five endocyclic torsion angles within it (Figure 3.7). The ring puckering arises from the effect of nonbonded interactions between substituents at the four-ring carbon atoms; energetically, the most stable conformation for the ring is that in which all substituents are as far apart as possible. Thus, different substituent atoms would be expected to produce differing types of puckering. The puckering can be described by either a simple qualitative description of the conformation in terms of atoms deviating from ring coplanarity, or precise descriptions in terms of the ring internal torsion angles.

In principle, there is a continuum of interconvertible puckers, separated by energy barriers. These various puckers are produced by systematic changes in the ring torsion angles. The puckers can be succinctly defined by the parameters P and τ_m (Altona and Sundaralingam, 1972). The value of P, the phase angle of pseudorotation, indicates the type of pucker since P is defined in terms of the five torsion angles τ_0–τ_4:

$$\tan P = \frac{(\tau_4 + \tau_1) - (\tau_3 + \tau_0)}{2\tau_2 (\sin 36° + \sin 72°)}$$

and the maximum degree of pucker, τ_m, by $\tau_m = \tau_2/(\cos P)$.

The pseudorotation phase angle can take any value between 0° and 360°. If τ_2 has a negative value, then 180° is added to the value of P. The pseudorotation phase angle is

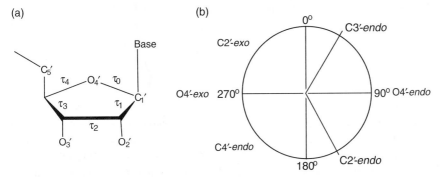

Figure 3.7. (a) The five internal torsion angles in a ribose ring. (b) The pseudorotation wheel for a deoxyribose sugar with indication of the most important ribose conformers.

commonly represented by the pseudorotation wheel, which indicates the continuum of ring puckers (Figure 3.7b). Values of τ_m indicate the degree of puckering of the ring; typical experimental values from crystallographic studies on mononucleosides are in the range of 25–45°. The five internal torsion angles are not independent of each other, and so to a good approximation, any one angle τ_j can be represented in terms of just two variables:

$$\tau_j = \tau_m \cos[P + 0.8p(j-2)]$$

Pseudorotation parameters can be calculated interactively using the PROSIT facility (Sun et al., 2004) at http://cactus.nci.nih.gov/prosit/. Phase angle P correlates quite tightly with the backbone torsion δ, C5′–C4′–C3′–O3′, so that the sugar pucker is effectively connected to the backbone conformation; the $\delta \sim P$ relation can be described by equation $\delta = 40° \cos(P° + 144°) + 120°$ (Privé, Yanagi, and Dickerson, 1991).

A large number of distinct deoxyribose ring pucker geometries have been observed experimentally by X-ray crystallography and NMR techniques. When one ring atom is out of the plane of the other four, the pucker type is an envelope one. More commonly, two atoms deviate from the plane of the other three, with these two on either side of the plane. It is usual for one of the two atoms to have a larger deviation from the plane than the other, in which case the result is a twist conformation. The direction of atomic displacement from the plane is important. If the major deviation is on the same side as the base and C4′–C5′ bond, then the atom involved is termed *endo*. If it is on the opposite side, it is called *exo*. The most commonly observed puckers in crystal structures of isolated nucleosides and nucleotides are either close to C2′-*endo* or C3′-*endo* types (Figure 3.2a and b). The C2′-*endo* family of puckers has P values in the range 140–185°; in view of their position on the pseudorotation wheel, they are sometimes termed S (south) conformations. The C3′-*endo* domain has P values in the range -10 to $+40°$, and its conformation is termed N (north). In practice, these pure envelope forms are rarely observed, largely because of the differing substituents on the ring. Consequently, the puckers are then best described in terms of twist conformations. When the major out-of-plane deviation is on the *endo* side, there is a minor deviation on the opposite, *exo* side. The convention used for describing a twist deoxyribose conformation is that the major out-of-plane deviation is followed by the minor one, for example, C2′-*endo*, C3′-*exo*.

The pseudorotation wheel implies that deoxyribose puckers are free to interconvert. In practice, there are energy barriers between major forms. The exact size of these barriers has been the subject of considerable study (Olson, 1982; Olson and Sussman, 1982). The consensus is that the barrier height depends on the route around the pseudorotation wheel. For interconversion of C2′-*endo* to C3′-*endo*, the preferred pathway is via the O4′-*endo* state, with a barrier of 2–5 kcal/mol found from an analysis of a large body of experimental data (Olson and Sussman, 1982) and a somewhat smaller (potential energy) value of 1.5 kcal/mol from a molecular dynamics study (Harvey and Prabhakarn, 1986). The former value, being an experimental one, represents the total free energy available for interconversion.

Relative populations of puckers can be monitored directly by NMR measurements of the ratio of coupling constants between H1′–H2′ and H3′–H4′ protons. These show that in contrast to the "frozen-out" puckers found in the solid-state structures of nucleosides and nucleotides, there is rapid interconversion in solution. Nonetheless, the relative populations of the major puckers depend on the type of base attached. Purines show a preference for the C2′-*endo* pucker conformational type, whereas pyrimidines favor C3′-*endo*. Deoxyribose nucleosides are primarily at greater than 60% in the C2′-*endo* form, and ribonucleos- ides favor C3′-*endo*. Sugar pucker preferences have their origin in the nonbonded inter- actions between substituents on the sugar ring and to some extent on their electronic characteristics. The C3′-*endo* pucker of ribose would have hydroxyl substituents at the 2′ and 3′ positions further apart than with the C2′-*endo* pucker. Ribonucleosides are therefore significantly more restricted in their mobility than deoxyribonucleotides; this has signifi- cance for the structures of oligoribonucleotides. These differences in puckering equilibrium, and hence in their relative populations in solution and in molecular dynamics simulations, are reflected in the patterns of puckers found in surveys of crystal structures (Murray-Rust and Motherwell, 1978).

Correlations have been found from numerous crystallographic and NMR studies, between sugar pucker and several backbone conformational variables, both in isolated nucleosides/nucleotides and in oligonucleotide structures. These are discussed later in this chapter. Sugar pucker is thus an important determinant of oligo- and polynucleotide conformation because it can alter the orientation of C1′, C3′, and C4′ substituents, resulting in major changes in backbone conformation and overall structure, as indeed is found.

The glycosidic bond links a deoxyribose sugar and a base, being the C1′−N9 bond for purines and the C1′−N1 bond for pyrimidines. The torsion angle χ around this single bond can, in principle, adopt a wide range of values, although as will be seen, structural constraints result in marked preferences being observed. Glycosidic torsion angles are defined in terms of the four atoms: O4′−C1′−N9−C4 for purines and O4′−C1′−N1−C2 for pyrimidines.

Theory has predicted two principal low-energy domains for the glycosidic angle, in accord with experimental findings for a large number of nucleosides and nucleotides. The *anti* conformation has the N1, C2 face of purines and the C2, N3 face of pyrimidines directed away from the sugar ring (Figure 3.8a), so that the hydrogen atoms attached to C8 of purines and C6 of pyrimidines lie over the sugar ring. Thus, the Watson–Crick hydrogen bonding groups of the bases are directed away from the sugar ring. These orientations are reversed for the *syn* conformation, with these hydrogen bonding groups now oriented toward the sugar and especially its O5′ atom (Figure 3.8b). A number of crystal structures of *syn* purine nucleosides have shown hydrogen bonding between the O5′ atom and the N3 base atom, which would stabilize this conformation. Otherwise, the *syn* conformation is slightly less preferred than the *anti* for purines, because of fewer nonbonded steric clashes in the latter case. The principal exceptions to this rule are guanosine-containing nucleotides, which have a slight preference for the *syn* form because of favorable electrostatic interactions between the exocyclic N2 amino group of guanine and the 5′-phosphate group. For pyrimidine nucleotides, the *anti* conformation is preferred over the *syn* because of unfavorable contacts between the O2 oxygen atom of the base and the 5′-phosphate group. The calculated results of molecular mechanics energy minimizations on all four DNA nucleotides in both *syn* and *anti* forms, using the AMBER all-atom force field, are fully in accord with these observations.

The sterically preferred ranges for the two domains of glycosidic angles are

$$anti: -60 > \chi > 180°$$

$$syn: 0 < \chi < 90°$$

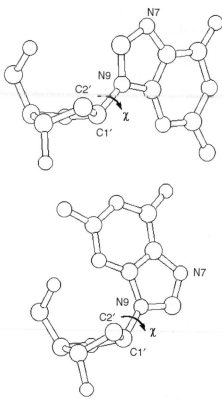

Figure 3.8. (a) A guanosine nucleoside with the glycosidic angle χ set in an *anti* conformation. (b) Guanosine now in the *syn* conformation.

Values of χ in the region of about −90° are often described as "high *anti*." There are pronounced correlations between sugar pucker and glycosidic angle, which reflect the changes in nonbonded clashes produced by C2′-*endo* versus C3′-*endo* puckers. Thus, *syn* glycosidic angles are not found with C3′-*endo* puckers due to steric clashes between the base and the H3′ atom, which points toward the base in this pucker mode.

The phosphodiester backbone of an oligonucleotide has six variable torsion angles (Figure 3.2c), designated as α, β, γ, δ, ε, and ζ, in addition to the five internal sugar torsions $\tau_0 \cdots \tau_4$ and the glycosidic angle χ. As will be seen, a number of these have highly correlated values and therefore correlated motions in a solution environment. Steric considerations alone dictate that the backbone angles are restricted to discrete ranges (Sundaralingam, 1969; Olson, 1982) and are accordingly not free to adopt any value between 0° and 360°. The fact that angles α, γ, and ζ each have three populated ranges, together with the broad range for angles ε and β that includes two staggered regions, leads to a large number of possible low-energy conformations for the unit nucleotide, especially when glycosidic angle and sugar pucker flexibility are taken into account. In reality, only a small number of DNA oligonucleotide and polynucleotide structural classes have actually been observed out of this large range of possibilities; this is doubtless in large part due to the restraints imposed by Watson–Crick base pairing on the backbone conformations when two DNA strands are intertwined together. In contrast, crystallographic and NMR studies on a large number of standard and modified mononucleosides and nucleotides have shown considerably greater conformational diversity. For these, backbone conformations in the solid state and in solution are not always in agreement; the requirements for efficient packing in the crystal can often overcome the modest energy barriers between different torsion angle values. A large range of base–base interactions characterizes many larger RNA molecules, which can therefore adopt a variety of backbone conformations.

A common convention for describing these backbone angles is to term values of $\sim$60° as *gauche+* (g$^+$), $-$60° as *gauche–* (g$^-$), and $\sim$180° as *trans* (t). Thus, for example, angles α (about the P-O5′ bond) and γ (the exocyclic angle about the C4′–C5′ bond) can be in the g$^+$, g$^-$, or t conformations. The two torsion angles around the phosphate group itself, ζ and α, have been found to show a high degree of flexibility in various dinucleoside crystal structures, with the tg$^-$, g$^-$g$^-$, and g$^+$g$^+$ conformations all having been observed (Kim et al., 1973). A- and B-DNA adopt to g$^-$g$^-$ and tg$^-$ conformations; Z-DNA adopts the g$^+$g$^+$, g$^+$t, and g$^-$t conformations. The torsion angle β, about the O5′–C5′ bond, is usually near *trans*. All three possibilities for the γ angle have been observed in nucleoside crystal structures, although the g$^+$ conformation predominates in right-handed oligo- and polynucleotide double helices. The torsion angle δ around the C4′–C3′ bond adopts values that relate to the pucker of the sugar ring, since the internal ring torsion angle τ_3 (also around this bond) has a value of about 35° for C2′-*endo* and about 40° for C3′-*endo* puckers. δ is about 81° for C3′-*endo* and about 140° for C2′-*endo* puckers.

There are a number of correlations involving these backbone torsion angles, as well as sugar pucker and glycosidic angle, that have been observed in mononucleosides and nucleotides and, more recently, in oligonucleotides (Schneider et al., 1997; Packer and Hunter, 1998; Murthy et al., 1999; Varnai et al., 2002; Murray et al., 2003; Sims and Kim, 2003; Hershkovitz et al., 2003; Schneider et al., 2004). Nucleotides are inherently more flexible in solution as well as being more subject to packing forces in the crystal. Some of the more significant correlations are as follows:

1. Between ζ and α, there are clear distinctions for the A-, B-, and Z-conformational classes (Figure 3.9a) that can be considered an "imprint" characterizing the most important features of local nucleic acid conformation. The ζ/α scattergram shows

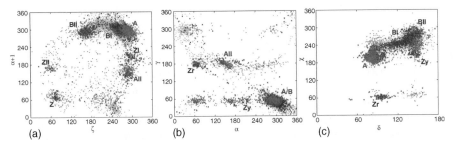

Figure 3.9. Backbone conformations in DNA and RNA. 2D scattergrams show conformations of 1861 nucleotides from 186 structures of uncomplexed DNA (dark blue), 5878 nucleotides from 261 structures of DNA in complexes with proteins or drugs (yellow), and 3752 nucleotides from 132 structures of RNA (red). A, AII, BI, BII, ZI, and ZII are respective double helical forms, Zr/Zy are purine and pyrimidine steps, respectively, in the Z-form. (a) Conformations at the phosphodiester linkage O3′-P-O5′, ζ versus α; (b) torsions α versus γ; (c) relationship between the sugar pucker and the base orientation, δ versus χ. Torsion angles are defined in Figure 3.2. Figure also appears in the Color Figure section.

significant broadening of the torsion distributions in protein/DNA complexes (yellow) and in RNA (red) structures compared to structures of uncomplexed DNA (blue).

2. Between torsions α and γ (Figure 3.9b), the distributions are trimodal, and there is an overall preference for values of α near 300° and γ near 60° that are observed in the majority of both B and A forms. The combination of α near 60° and γ near 300° is observed almost exclusively in DNA/protein complexes. This again demonstrates the broadening of distributions and larger structural diversity of the complexes.

3. Between sugar pucker and glycosidic angle χ, especially for pyrimidine nucleosides (Figure 3.9c), C3'-*endo* pucker is usually associated with median-value *anti* glycosidic angles near 200°. This is a typical arrangement observed in the A-form. C2'-*endo* puckers are commonly found with higher *anti* χ angles, in the range 240–280°, which is typical for the B-form. *Syn* glycosidic angle conformations show a marked preference for C2'-*endo* sugar puckers.

Further analysis of nucleic acid conformations in RNA structures has led to the identification of about 50 discrete RNA dinucleotide conformers (Richardson et al., 2008). These have been compiled in a library of consensus conformations that can be useful for finding motifs in RNA and for the modeling of experimental electron densities.

STRUCTURES OF NUCLEIC ACIDS

DNA Duplexes

The first known structures of nucleic acids were of DNA duplexes derived using fiber diffraction methods. B-DNA, the classic structure first described by Watson and Crick (Watson et al., 1953), has been refined using the linked-atom least-squares procedure developed by Arnott and his group (Kim et al., 1973). In canonical B-DNA (Figure 3.10), the backbone conformation has C2'-*endo* sugar puckers and high *anti* glycosidic angles (−100°). The right-handed double helix has 10 base pairs per complete turn, with the two polynucleotide chains antiparallel to each other and linked by Watson–Crick AT and GC base pairs. The paired bases are almost exactly perpendicular to the helix axis, and they are stacked over the axis itself. Consequently, the base-pair separation is the same as the helical rise, that is, −3.4 Å. An important consequence of the Watson–Crick base-pairing arrangement is that the two deoxyribose sugars linked to the bases of an individual pair are asymmetrically on the same side of it. So, when successive base pairs are stacked on each other in the helix, the gap between these sugars forms two continuous indentations with different dimensions in the surface that wind along, parallel to the sugar–phosphodiester chains. These indentations are termed grooves. The asymmetry in the base pairs results in two parallel types of groove, whose dimensions (especially their depths) are related to the distances of base pairs from the axis of the helix and their orientation with respect to the axis. The wide major groove is almost identical in depth to the much narrower minor groove that has the hydrophobic hydrogen atoms of the sugar groups forming its walls. In general, the major groove is richer in base substituents O6, N6 of purines and N4, O4 of pyrimidines compared to the minor groove (Figure 3.3). This, together with the steric differences between the two, has important consequences for interaction with other molecules.

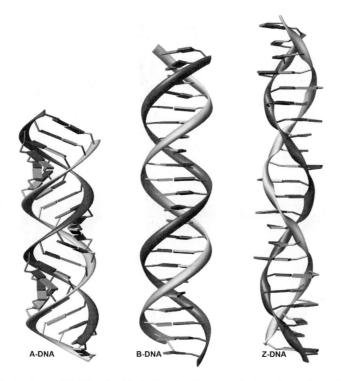

A-DNA   B-DNA   Z-DNA

Figure 3.10. A-, B-, and Z-DNA double helices in the canonical conformations. All helices are formed by 22 base pairs.

Over 200 single-crystal structures with B-DNA conformations have been determined since the first such structure was published (Figure 3.11a) (Drew et al., 1981). The average values of the base morphology parameters are close to the canonical values derived from fiber studies. However, the individual structures have diverse shapes. Many uncomplexed DNA structures are bent by as much as 15° (Dickerson, Goodsell, and Kopka, 1996), and the groove widths are variable (Heinemann, Alings, and Hahn, 1994). The bending is a function of some of the base morphology parameters, in particular twist and roll. Attempts to relate the base morphology parameters to the sequence of the bases have shown some trends. Twist and roll appear to be correlated with regressions that depend on the nature of the bases in the step; purine–pyrimidine, purine–purine, and pyrimidine–purine steps show distinctive differences (Quintana et al., 1992; Gorin, Zhurkin, and Olson, 1995). AT base pairs in AA steps have high propeller twist and bifurcated hydrogen bonds (Yanagi, Prive, and Dickerson, 1991). Stretches of A in sequences appear to be straight (Young et al., 1995). On the contrary, certain steps such as TA and CA show a very large variability due to crystal packing, sequence context, or both (Lipanov et al., 1993). This is consistent with the findings of a combined NMR and low-angle X-ray analysis (Schwieters and Clore, 2007), which has also provided reliable solution data on the amplitudes of motion for B_I/B_{II} phosphate exchanges, sugar pucker, and base morphology.

A novel host–guest approach has been used to analyze DNA sequences that are unable to crystallize on their own (Coté, Yohannan, and Georgiadis, 2000). This uses the N-terminal

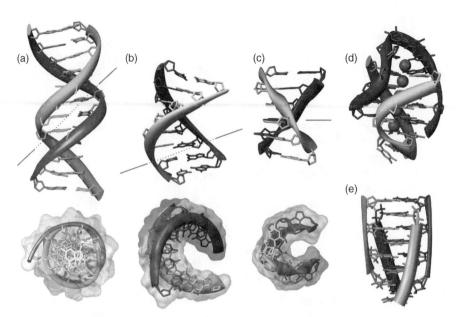

Figure 3.11. Important structural forms of DNA. Molecules are colored by strand. (a) B-DNA dodecamer (Drew et al., 1981); (b) A-DNA octamer (McCall et al., 1985); (c) Z-DNA hexamer (Gessner et al., 1989 ; Frederick et al., 1990); (d) DNA guanine tetraplex. Four guanine tetramers are flanked by two T-T-T-T loops, potassium cations are shown in magenta (Haider, Parkinson, and Neidle, 2002); (e) cytosine tetraplex, or i-motif (Chen et al., 1994). In (a), (b), and (c), two perpendicular views are shown. The minor groove is shown in the red line. Figure also appears in the Color Figure section.

fragment of the Moloney murine leukemia virus reverse transcriptase as a sequence-neutral "host" to bind to a wide range of 16-mer DNA sequences, including ones with mismatches and bound drug molecules. The resulting structures have a dumbbell-like appearance with a protein molecule at each end connected by the oligonucleotide duplex. There is sufficient flexibility in the DNA such that sequence-dependent effects are apparent. For example, a study of the sequence d(CTTTTTAAAAGAAAAG) with its complementary strand has shown that in spite of there being three A tracts in the sequence, the structure overall is not significantly bent (Coté, Pflomm, and Georgiadas, 2003).

Included among the many B-DNA structures are numbers with base mispairs. Interestingly, these mutations cause only local perturbations to the structures, and the overall conformations of these structures remain the same as the parent structures.

Hydration also plays an important role in the structure of B-DNA. The spine of hydration first seen in the dodecamer structure and described in some detail (Dickerson et al., 1983) has proven to be an enduring feature of these molecules. A detailed analysis of the hydration around the bases in DNA has demonstrated that the hydration is local (Schneider and Berman, 1995). Hydrated building blocks from decameric B-DNA were used to construct the dodecamer structure and the spine was faithfully reproduced. That same analysis strongly suggests that the patterns of hydration, be they spines or ribbons, are a function of the base morphology and that there is strong synergy between the hydration and base conformations.

Canonical A-DNA (Figure 3.10) has C3'-endo sugar puckers, which bring consecutive phosphate groups on the nucleotide chains closer together at 5.9 Å, compared to the 7.0 Å in

B-form DNA (Figure 3.2). This alters the glycosidic angle from high *anti* to *anti*. As a consequence, the base pairs are twisted and tilted with respect to the helix axis and are displaced nearly 5 Å from it, in striking contrast to the B-helix. As a consequence, the helical rise is much reduced to 2.56 Å, compared to 3.4 Å for canonical B-DNA. The helix is wider than the B form and has an 11 base-pair helical repeat. The combination of base-pair tilt with respect to the helix axis and base-pair displacement from the axis results in very different groove characteristics for the A double helix compared to the B form. This also results in the center of the A double helix being a hollow cylinder. The major groove is now deep and narrow, and the minor one is wide and very shallow.

Analysis of the A-DNA crystal structures has shown that there are variations in the helical parameters that appear to be related to crystal packing (Ramakrishnan and Sundaralingam, 1993). Two sequences have been found to crystallize in two different space groups—GTGTACAC (Jain, Zon, and Sundaralingam, 1989; Jain, Zon, and Sundaralingam, 1991; Thota et al., 1993) and GGGCGCCC (Shakked et al., 1989). In both cases, the helical parameters are different in different crystal forms. Another analysis of a series of A-DNA duplexes shows that there is an inverse linear relationship between the crystal packing density and the depth of the major groove (Heinemann, 1991).

In the earlier days of nucleic acid crystallography, it had been thought by some that the A conformation was simply a crystalline artifact and not likely to bear any relationship to biology. Now it has been demonstrated that in the TATA binding protein–DNA complex (Kim, Nikolov, and Burley, 1993; Kim et al., 1993), the conformation of the DNA is an A–B chimera (Guzikevich-Guerstein and Shakked, 1996). The apparent deformability of the base geometry seen in A-DNA oligonucleotide crystal structures may prove to be an advantage in forming protein–DNA complexes. All of this shows the innate plasticity of the DNA duplex, a view reinforced by studies that show DNA in transition from A to B helices. Thus, the crystal structures of six d(GGCGCC) duplex hexanucleotides with varying degrees of cytosine bromination or methylation (Vargason, Henderson, and Ho, 2001; Ng and Dickerson, 2001) show features that progressively span A, B, and plausible intermediate states.

One of the earliest determined A-type structures demonstrated a unique arrangement of water molecules consisting of edge-linked pentagons (Shakked et al., 1983). This pattern, like the spine seen in B-DNA, provoked the continued pursuit of the role of hydration in the stabilization of the different forms of DNA. One theory is that the "economy of hydration" seen in A-type structures may provide a structural explanation for the B to A transition as the humidity is lowered (Saenger et al., 1986). In a study of the hydration of a series of A-DNA structures, it was demonstrated that the bases have specific patterns and that both the direct and the water-mediated interactions could be related to recognition properties of DNA and proteins (Eisenstein and Shakked, 1995).

Around the same time that the single-crystal structure of B-DNA was determined, a left-handed conformation called Z-DNA was discovered (Wang et al., 1979). The zigzag phosphate backbone defines a convex outer surface of the major groove and a deep central minor groove (Figures 3.10 and 3.11c). Z-type structures have alternation of cytosine and guanine with a cytosine at the first position. Of the unmodified Z-DNA structures, there are very few examples in which there have been substitutions of A for G and T for C. Modifications of the cytosines with methyl groups at the five positions have allowed for more drastic sequence changes as exemplified by a structure with tandem G's in the center (Schroth, Kagawa, and Ho, 1993) and another with an AT at that position (Wang et al., 1985). A B–Z junction has been observed in the crystal structure (Ha et al., 2005) of a 15-base-pair sequence containing within it a CG hexanucleotide Z-DNA motif. This was tethered by being bound to domains of a Z-DNA binding protein, so that eight base pairs are in the Z-form and six are in a B-DNA conformation. There is a sharp B–Z backbone turn at the junction and the resulting B-helix is of a standard regular conformation.

The basic building block of Z-DNA is a dinucleotide with the twist of the CG steps 9° and of the GC steps 50°. The backbone conformation of Z-DNA is characterized by the alternation of the χ angle of the guanine and the cytosine, where the former is *syn* and the latter *anti*. The values for backbone torsion angles do not resemble those found in either A- or B-DNA. In many Z-DNA crystals, the central step of one chain has a different conformation than the other steps. This conformational polymorphism is called ZI and ZII (Gessner et al., 1989). Analysis of the packing patterns in Z crystals has shown that it is possible to correlate the pattern of water bridges with the presence of the ZI–ZII pattern (Schneider et al., 1992).

The hydration characteristics of Z-DNA crystals are very uniform with a spine of hydration in the central minor groove and tightly bound water on the exterior major groove (Gessner et al., 1994). The effects of solvent reorganization as a result of subtle sequence changes have been offered as a very plausible explanation for the different stabilities of these sequences (Wang et al., 1984; Kagawa et al., 1989).

Drug Complexes in Double Helical DNA

Three major types of drug interactions have been observed. The intercalation mode was first observed in a cocrystal between UpA and ethidium bromide (Tsai et al., 1975) and in a subsequent series of dinucleoside drug complexes (Berman and Young, 1981). The first known intercalation complexes (Wang et al., 1987) with longer stretches contained daunorubicin sandwiched between the terminal CG base pair. Now more than 200 related structures have been determined, thus shedding light on the effects of changes on the drug and the sequence. The determination of the structure of a complex between actinomycin D (Kamitori and Takusagawa, 1994) and an octamer in which the drug is bound to the central GC base pair showed how the double helix can accommodate the drug without itself fraying. Among the large number of intercalation complex crystal structures are several with rhodium-containing molecules. These have been remarkable in showing the consequences of intercalation in longer DNA sequences. The crystal structure (Kielkopf et al., 2000) of a rhodium complex intercalated into an octanucleotide duplex has five independent complex molecules in the crystallographic asymmetric unit. They all show consistent structural features and conformations, with the planar group inserted from the major groove direction and embedded at the center of a significant length of DNA sequence, thus showing the molecular features of intercalation that are relevant to the situation in bulk DNA. A rhodium complex has also been cocrystallized (Pierre, Kaiser, and Barton, 2007) with a mismatched dodecamer d(CGGAAATTCCCG).

The second class of drug complexes involves a drug bound in the minor groove of a B-type helix. Starting with the determination of the structure of netropsin bound to the Dickerson dodecamer (Kopka et al., 1985), there have been over 50 determined structures in which the drugs bind to the minor groove (Figure 3.12a). Several of these contain the Hoechst benzimidazole derivative, which is typical for these structures in having a planar hetero-aromatic cross section whose dimensions complement those of the A/T narrow minor groove. In general, these groove binders show strong sequence specificity that is moderated by a combination of hydrophobic and hydrogen bonding interactions (Tabernero, Bella, and Alemán, 1996). There are examples in this class of complexes in which two drugs are bound

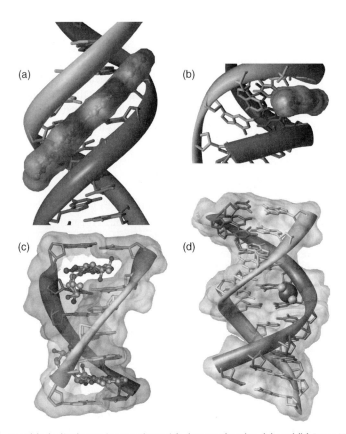

Figure 3.12. Double-helical DNA interacting with drug molecules. (a) and (b) Drug Hoechst 33258 binding to the minor groove of B-DNA (Quintana et al., 1992). (b) Shows the fit of flat drug molecule into the minor groove; (c) drug daunomycin intercalated between base pairs of B-DNA octamer CGATCG (Moore et al., 1989); (d) *cis* platinum covalently bound to guanine bases. Pt is shown as a large sphere with the NH2 groups pointing to the viewer. Two Pt$-$N7(G) bonds (point into the paper plane) bend DNA double helix (Takahara et al., 1995).

in a side-by-side manner in the minor groove of an octameric fragment of DNA (Chen et al., 1994). Many of these molecules have a crescent-like shape that complements the curvature of the minor groove. However, this is not a universal requirement, as has been shown by several crystal structures and associated biophysical studies of linear or near-linear molecules that bind strongly in the minor groove. The crystal structure of the almost linear *meta*-substituted diamidine molecule CGP40215A with d(CGCGAATTCGCG)$_2$ shows it binding in the minor groove A/T region (Nguyen et al., 2002) and has short water-mediated contacts with base edges.

Although covalent adducts are thought to be key in carcinogenesis as well as antitumor activity, in general, it has been very difficult to obtain crystalline samples. There are some examples, such as an anthramycin molecule bound to a dodecamer (Kopka et al., 1994) and several complexes of platinum-containing drugs (see, for example, Silverman et al., 2002). Most of the carcinogenic adducts that have been crystallized are simple modifications to the O6 methyl group of G. Two structures containing this type of modification are crystallized in the same approximate unit cell as the parent dodecamer but are not isomorphous (Gao et al., 1993; Vojtechovsky et al., 1995). This may demonstrate how a local lesion may actually affect interaction and recognition properties. A pharmaceutically important structure is that of the antitumor agent *cis* platinum covalently bound to tandem guanine residues in a DNA duplex. The DNA is strongly bent as a result of this interaction (Takahara et al., 1995) (Figure 3.12d). In addition, several of the complexes containing daunomycin are actually covalent adducts that were formed from the presence of formaldehyde in the crystallization drop (Wang et al., 1991).

DNA Quadruplexes

The existence of a tetrameric arrangement of DNA and RNA helices was first shown in fiber of polyG and polyI (Gellert, Lipsett, and Davies, 1962; Arnott, Chandrasekharan, and Marttila, 1974). Their biological relevance was discovered later when it was hypothesized that these types of conformations may occur in telomeres (Sen and Gilbert, 1988; Rhodes and Giraldo, 1995). Guanine-rich DNA sequences can form multistranded structures as a consequence of the ability of guanine bases to form planar hydrogen-bonded arrays simultaneously involving two of its faces, notably the guanine (G) quartet formed with four guanine bases. The resulting structures can be formed from one, two, or four strands to form fold-back intramolecular, bimolecular or tetramolecular structures, respectively. All require the presence of centrally coordinated potassium, sodium, or similarly sized cations for stability. These structures, termed quadruplexes, have considerable diversity that depends on the number of separate strands involved, on the relative polarities of the strands, and on the intervening nonguanine loop sequences. There are now several examples of crystal structures of quadruplex DNA: d(TGGGGT) forms a tetramolecular quadruplex (Phillips et al., 1997), d(GGGGTTTTGGGG) forms a bimolecular one (Haider, Parkinson, and Neidle, 2003), and d[AGGG(TTAGGG)₃] forms a parallel-stranded intramolecular quadruplex (Parkinson, Lee, and Neidle, 2002). Figure 3.11d shows an example of one such crystal structure.

G-quadruplexes may provide target locations for selective therapeutic action, for example, at the ends of telomeres in human cancer cells, where telomere maintenance is fundamentally distinct from that in normal cells. A few crystal structures of drug complexes are available that are useful in structure-based design studies. All show planar drug molecules stacked onto the terminal G-quartets of quadruplexes and not intercalated within the structure (Haider, Parkinson, and Neidle, 2003; Parkinson, Ghosh, and Neidle, 2007), in accordance with observations that quadruplexes are stable structures that cannot be as readily disrupted by small molecules such as duplex DNA.

RNA Duplexes

In the last few years, there has been a pronounced increase in the number of RNA structures determined. This is due in part to the improved ability to obtain pure material and crystallize it (Wyatt, Christain, and Puglisi, 1991; Wahl et al., 1996). Although RNA is generally single stranded, double-stranded RNA can be readily formed, analogous to duplex DNA. Uracil participates in UA base pairs that are fully isomorphous with A·T pairs in duplex DNA. Duplex RNA is conformationally rather rigid, and its behavior contrasts remarkably with the polymorphism of duplex DNA in that only one major polymorph of the RNA double helix has been observed. This has many features in common with A-DNA and, accordingly, is known as A-RNA. The conformational features of canonical RNA helices have been obtained from fiber diffraction studies on duplex RNA polynucleotides from both viral and synthetic origins. A-RNA is an 11-fold helix, with a narrow and deep major groove and a wide, shallow minor groove, and base pairs inclined to and displaced from the helix axis. A-RNA helices have the C3'-*endo* sugar pucker. Another way it differs from duplex DNA is that RNA helices, though capable of a small degree of bending (up to ca. 15°), do not undergo the large-scale bending seen, for example, in A-tract DNA.

A number of structures of base-paired duplex RNA have been reported, the first being the structures of r(AU) and r(GC) (Rosenberg et al., 1976; Seeman et al., 1976). Crystallographic analyses of sequences, such as the octanucleotide r(CCCCGGGG), have shown helices of length sufficient for a full set of helical parameters to be extracted. This sequence crystallizes in two distinct crystal lattices, enabling the effects of crystal packing factors on structure to be assessed (Portmann, Usman, and Egli, 1995). In each instance, rhombohedral and hexagonal, the RNA helices are very similar, and their features closely resemble those in fiber diffraction canonical A-RNA. The structure of r(UUAUAUAUAUAUAA) was the first to show a full turn of an RNA helix (Dock-Bregeon et al., 1989). Again, the helix is essentially classical A-RNA.

The increasing availability of high-resolution crystal structures has enabled RNA hydration in oligonucleotides to be defined. Some general rules are beginning to emerge for the water arrangements beyond the obvious finding of an inherently greater degree of hydration compared to DNA oligonucleotides, because the 2'-OH group is an active hydrogen bond participant. There is no analogue of the minor groove spine of hydration seen in B-DNA. Again, this is unsurprising since the minor groove in A-RNA is too wide for such an arrangement.

Mismatched and Bulged RNA

RNAs can readily form stable base pair and triplet mismatches and extrahelical regions within the context of "normal" RNA double helices. Such features, notably extrahelical loops, have been the subject of intense structural study, since they are present in large RNAs (tRNA, mRNA, rRNA) and, together with various types of base stacking, are responsible for maintaining their tertiary structures. It is paradoxically common for crystal structures of short sequences containing potential loop regions, such as the UUCG "tetraloop" (an especially stable extrahelical loop structure in solution), not to show such features. This is probably a consequence of the high ionic strength of many crystallization conditions, together with the preference of some sequences to pack in the crystal as helical arrays. So, instead of loops, these crystal structures tend to have runs of non-Watson–Crick mismatched base pairs where the loop would have formed. For example, an A-RNA double helix, albeit with GU and UU base pairs, is formed in the crystal by the sequence r (GCUUCGGC) d (^{Br}U) (Cruse et al., 1994). There is evidence of some deviations from the exact canonical A-RNA duplexes formed by fully Watson–Crick base pairs, since this helix has <10 base pairs per turn. The dodecamer sequence r(GGACUUCGGUCC) (Figure 3.13a) similarly forms a base-paired duplex (Holbrook et al., 1991), with U·C and G·U base pairs. The A-RNA helix here has a significant increase in the width of its major groove, possibly on account of the water molecules that are strongly associated with the mismatch base pairs. Much greater perturbations are apparent in the structure of the dodecamer r(GGCCGAAAGGCC) (Baeyens et al., 1996), where the four non-Watson–Crick base pairs in the center of the sequence form an internal loop with sheared GA and AA base pairs. The resulting structure is

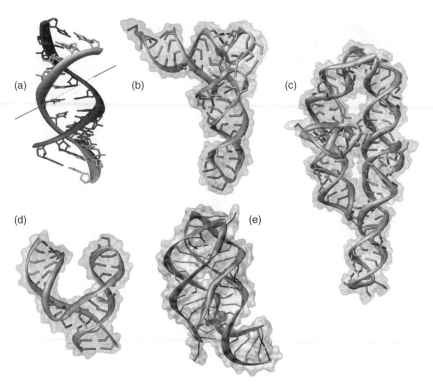

Figure 3.13. Examples of RNA molecules: (a) A-RNA dodecamer with non-Watson–Crick base pairs in the double helix (Holbrook et al., 1991); (b) phenylalanine transfer RNA (Sussman et al., 1978; Brown et al., 1985); (c) group I intron ribozyme (Cate et al., 1996); (d) Hammerhead ribozyme. The shorter strand is the DNA inhibiting the catalyzed reaction (Pley, Flaherty, and McKay, 1994); (e) guanine riboswitch (Batey, Gilbert, and Montange, 2004). Hypoxanthine bound in the active site is shown in yellow (Please see figure at supplementary weblink).

very distorted from A-RNA ideality, with a compression of the major groove, an enlargement of the minor groove width to 13.5 Å, and a pronounced curvature of the resulting helix. This sequence forms a tetraloop in solution (Baeyens et al., 1996). The GU base pair is a prominent and very important element of large RNA structures since it is especially stable (Varani and McClain, 2000) because of its two hydrogen bonds. It is known as the "wobble" pair, since a G in the first position of a codon can accept either a C or a U in the third anticodon (wobble) position.

This tendency of short RNA sequences to maximize helical features is also apparent in the crystal structure of a 29-nucleotide fragment from the signal recognition particle (Wild et al., 1999) that forms 28-mer heteroduplexes rather than a hairpin structure. Even so, this duplex has features of wider interest, since it has a number of non-Watson–Crick base pairs such as a 5′-GGAG/3′-GGAG purine bulge. Their overall effect is to produce backbone distortions so that the helix has non-A-RNA features such as a widening of the major groove by ∼9 Å and local undertwisting of base pairs adjacent to A·C and G·U mismatches.

In this respect, the structural studies of Dumas and Ennifar of the HIV-1 retrovirus dimerization initiation site (DIS) are of great interest. DIS has features that act as signals for RNA packaging and subsequent virulence and are known to contain nonhelical motifs, some of which have been studied by structural methods. In their early studies, the expected secondary structure with two loops in a "kissing-loop" (loop–loop) arrangement was not observed. Instead, the crystal contains duplex with two AG base pairs, each adjacent to an extrahelical, bulged-out adenosine (Ennifar et al., 1999). However, more recent studies with crystals grown under different conditions revealed the expected arrangement of the "kissing loops" between recognition loops (Ennifar and Dumas, 2006), in agreement with an NMR study (Baba et al., 2005).

Transfer RNA

The crystal structure of yeast phenylalanine tRNA (Figure 3.13b) was determined in the early 1970s simultaneously by two groups (Kim et al., 1974; Robertus et al., 1974). Together with a few other tRNAs, these structures remained the sole complex RNA molecular structures available for 20 years until the first ribozyme structure was determined. The structures—one monoclinic (Robertus et al., 1974) and the other orthorhombic (Kim et al., 1974)—are similar. They showed that the molecule is folded into an overall L-shape, with two arms at right angles to each other. The original cloverleaf is still apparent, but with additional interactions between distant parts of the structure. The arms consist of short A-RNA helices with extensive base–base interactions that hold the two arms together. The longer double helical anticodon arm has the short helix of the D stem stacked upon it. The other arm is formed by the helix of the acceptor stem, on which is stacked the four base-pair T arm helix. This key feature of helix–helix stacking has turned out to be of central importance for other complex RNAs. The nine additional base · · · base interactions that maintain the structural fold all tend to be in the elbow region, where the two arms are joined together. Almost all of these are of non-Watson–Crick type, and several are triplet interactions. Other subsequent crystal structures of tRNAs have shown that the overall L shape is invariant, as are many of the tertiary interactions.

Ribozymes

The ability of certain RNA molecules to catalyze chemical reactions, in most cases self-cleavage or cleavage of other RNA molecules, is one of the most convincing pieces of evidence suggesting the existence of an "RNA world" (Gilbert, 1986;Gesteland, Cech, and Atkins, 1999). Several distinct categories of RNA enzymes, called ribozymes, are known to date, with undoubtedly more remaining to be discovered:

- The RNA of self-splicing group I introns from Tetrahymena was the first ribozyme to be discovered (Cech, Zaug, and Grabowski, 1981). To date, there are over a thousand in both eukaryotic and prokaryotic organisms. These ribozymes contain four conserved sequence elements and form a characteristic secondary structure (Cate et al., 1996) (Figure 3.13c). The initial step of the cleavage involves a nucleophilic attack by a conserved guanosine.
- The RNA of self-splicing group II introns also have conserved sequence elements, but have a very different secondary and tertiary structure (Zhang and Doudna, 2002). These structures have a distinct mechanism of cleavage that involves the nucleophilic attack of a conserved adenosine, contained within the intron sequence.
- The RNA subunit of the enzyme, ribonuclease P, is a protein-assisted ribozyme that is conserved in all kingdoms of life and exists in at least two forms (Krasilnikov et al., 2003, 2004).

- Self-cleaving RNAs from viral and plant satellite are smaller than the group I or II intron ribozymes and include the hammerhead ribozyme. The crystal structures of hammerhead ribozymes were the first to be determined (Figure 3.13d). One is of a complex with a DNA strand containing the putative cleavage site (Pley et al., 1994); since ribozymes do not cleave DNA, this is effectively an inhibitor complex. The structure shows three A-RNA stems connected to a central two-domain region encompassing the catalytic core that contains the conserved residues.

- Hairpin ribozymes are another group of self-cleaving RNAs from viral and plant satellite RNAs. Their structures consist of two parallel double helices, one with a bulge forming the catalytic active site (Salter et al., 2006). The glmS ribozyme is currently the only natural ribozyme requiring a small-molecule cofactor for catalysis and can therefore be considered a riboswitch (Klein and Ferre-D'Amare, 2006).

- Structures of other types of ribozymes, such as the hepatitis delta virus (HDV) ribozyme (Ferre-D'Amare, Zhou, and Doudna, 1998), or ribozyme fragments, such as the five catalytic helical segments from the Varkud satellite (VS) ribozyme (Campbell and Legault, 2005), have also been determined.

- Ribosomal RNA is responsible for synthesis of peptide bonds in newly synthesized proteins (discussed below).

RNA Involved in Posttranscriptional Gene Control

RNA molecules have been shown to actively participate in posttranslational control of gene expression. Their involvement includes splicing of pre-mRNA, gene silencing by RNA interference, and control of RNA turnover by riboswitches.

RNA interference (RNAi) is a mechanism of the posttranscriptional inhibition of gene expression called "gene silencing." It involves microRNA (miRNA), which are single-stranded RNA of about 20–25 nucleotides in length that bind to their target mRNA to form double-stranded RNA that are then recognized and cut by a specific RNase III, dicer (Macrae et al., 2006) into short double-stranded RNA segments with dinucleotide overhangs (siRNAs).

The mechanism of RNAi highlights a functional role for double-stranded RNA. The PAZ domain, for example, is an RNA binding domain from the human Argonaute protein that in complex with a nona-nucleotide siRNA-like double helix (Ma, Ye, and Patel, 2004) shows sequence-independent protein binding to the heptameric A-RNA double helix with a 3'-end dinucleotide overhang on both strands.

Riboswitches are noncoding (pre-)mRNA sequences found at their 5' untranslated end that interact with small-molecule ligands and interfere with expression of the genes involved in metabolism of these ligands. These single-stranded RNA molecules act without protein assistance. Riboswitches include oligonucleotide domains that are capable of conformational changes induced by binding of their ligand and downstream regions that contain expression-controlling elements.

One example is seen in the crystal structure of the adenine binding riboswitch that distinguishes between bound adenine and guanine with high specificity. It has two parallel double helices wrapped around each other and connected by three hairpin loops; the 5'- and 3'-end sequences form the third double helix (Serganov et al., 2004). The internal bulges zipper up to form the purine binding pocket. The structure of the guanine riboswitch from a different organism (Figure 3.13e) (Batey, Gilbert, and Montange, 2004) shows very similar architecture. Also, other known structures of riboswitches show similar architecture suggesting its general relevance.

The Ribosome

The ribosome is responsible for protein synthesis in all prokaryotic and eukaryotic cells. It consists of two subunits, each a complex of proteins and ribosomal RNA. The complete 70S prokaryotic ribosome has a total molecular weight of about 2.5 million daltons. In prokaryotes, the 30S subunit contains about 20 proteins and a single RNA molecule of around 1500 nucleotides in length. The larger 50S subunit contains half as many more proteins and a large RNA of about 3000 nucleotides, together with the small (120 nucleotide) 5S RNA. The primary function of the small subunit is to control tRNA interactions with mRNAs. The large subunit controls peptide transfer and undertakes the catalytic function of peptide bond formation.

Structural studies on bacterial ribosomes have been underway for almost 40 years, with the ultimate goal of achieving atomic resolution to understand the mechanics of ribosome function. In the last few years, several groups have successfully determined the structures of ribosomal subunits as well as the whole ribosome (Ban et al., 2000; Schluenzen et al., 2000; Wimberly et al., 2000; Yusupov et al., 2001).

The 3.0 Å crystal structure of the 30S subunit has the complete 16S ribosomal RNA of 1511 nucleotides together with the ordered regions of 20 ribosomal proteins, organized into four well-defined domains. The implication is that there is considerable flexibility between them, which is needed to ensure the movement of messenger and transfer RNAs. The overall shape of the 30S particle is dominated by the structure of the folded RNA. The secondary structure of the RNA shows over 50 helical regions. The numerous loops are mostly small and do not disrupt the runs of helix in which they are embedded. There are extensive interactions between helices, mostly involving coaxial stacking via the minor grooves. In one type of helix–helix interaction, two minor grooves abut each other with consequent distortions from A-type geometry. These distortions, which tend to involve runs of adenines, are facilitated by both extrahelical bulges and noncanonical base pairs as similarly observed in simple RNA structures. Less commonly, the perpendicular packing of one helix against another, also via the minor groove, is mediated by an unpaired purine base. This mode is of special importance since it involves the functionally significant helices in the 30S subunit. The motifs of RNA tertiary structure such as non-Watson–Crick base pairs, base triplets and tetraloops, all contribute to the overall structure.

The majority of the 20 proteins in the ribosomal structure each consist of a globular region and a long flexible arm. The latter has been too flexible to be observed in structural studies on the individual proteins but has been located in the 30S subunit, where it plays an important role in helping to stabilize the RNA folding by essentially filling the numerous spaces in the RNA folds.

The structure identifies the three sites where tRNA molecules bind and function and where the essential proofreading checks for fidelity of code reading and translation occur. These sites are the P (peptidylation) site, where a tRNA anticodon base pairs with the appropriate codon in mRNA and where the peptide chain is covalently linked to a tRNA; the A (acceptor) site, where peptide bonds are eventually formed (the actual peptidyl transferase steps occur in the 50S subunit); and the E (exit) site, where tRNAs are released from the subunit as part of the protein synthesis cycle.

tRNA itself is not present in this crystal structure, but the RNA from a symmetry-related 30S subunit effectively serves to mimic it as the anticodon stem loop. Its interactions with the 30S RNA are also mediated via (1) minor groove surfaces, helped by some contacts with ribosomal proteins and (2) backbone contacts. Interestingly, the exit site of the tRNA is almost exclusively protein associated, whereas the other functional sites are composed of RNA and not protein. It is thus the ribosomal RNA that mediates the functions of the 30S subunit and not the ribosomal proteins.

In the 2.4 Å crystal structure of the 50S subunit, 2711 out of a total of 2923 ribosomal RNA nucleotides have been observed, together with 27 ribosomal proteins, and 122 nucleotides of 5S RNA, with the subunit being about 250 Å in each dimension. Although the RNA itself can be arranged into six domains on the basis of its secondary structure, overall the 50S subunit is remarkably globular, reflecting its greater conformational rigidity compared to the 30S subunit, consistent with its functional need to be more flexible.

The first determined structure of the complete bacterial 70S ribosome (Yusupov et al., 2001) revealed much about the interactions between the subunits that are an essential element to the protein synthesis cycle. More recently, higher resolution structures (Berk et al., 2006; Korostelev et al., 2006; Selmer et al., 2006) (Figure 3.14) have provided more detailed information about the mechanisms of recognition between tRNA and its three ribosomal recognition sites, A, P, and E, and about the assembly between the 30S and 50S subunits.

Functionally variable RNA molecules are an important target for drugs; bacterial ribosomes are targeted by many clinically important antibiotics. With the determination of ribosomal crystal structures, information about the structural details of their interactions

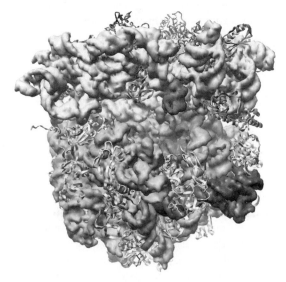

Figure 3.14. The structure of the complete 70s ribosome (Selmer et al., 2006). The 30S subunit is on top, the 50S is on the bottom. RNA molecules are rendered as surfaces: yellow for 16S rRNA, light blue for 23S rRNA, blue for 5S rRNA, and red for the three tRNA molecules. Proteins are shown as helix/turn cartoons in orange and pink for the small and large subunits, respectively. Figure also appears in the Color Figure section.

with different drugs has become abundant (Hermann, 2005), thus shedding light on structural details of translational machinery as well as giving hints for the discovery of new antibiotics.

The ribosome can be targeted at different subsites by chemically diverse molecules. For example, aminoglycoside antibiotics binding at the decoding site of the 16S rRNA (A-site) in the small ribosomal subunit compromises the fidelity of protein synthesis of the targeted bacteria. Structures of complexes between the small subunit and clinically important drugs such as tetracyclines (Brodersen et al., 2000) or paromomycin (Carter et al., 2000) are known. Quite significantly, key structural features of drug/RNA interactions in these large ribosome–drug complexes are preserved at the oligonucleotide level in both solution (Fourmy et al., 1996) and crystal structures (Vicens and Westhof, 2001), opening a way to use these smaller systems for drug design.

Some of the most clinically important antibiotics, for example, macrolides, stepto-gramins, chloramphenicol, and oxazolidones block the peptidyl-transferase catalytic site and/or peptide exit tunnel in the 23S rRNA of the large ribosomal subunit (Harms et al., 2003). Extensive structural studies of ribosome inhibition by these drugs (Hansen et al., 2002; Schlunzen et al., 2003; Tu et al., 2005) have revealed structural details about their binding inside the peptide exit tunnel of the 23S rRNA. Molecular details of the inhibition will be extremely useful in the design of new potent and specific drugs.

CONCLUSION

The past 20 years has seen an exponential growth in the determinations of nucleic acid structures. DNA crystallography has provided information about sequence effects, fine structures, and hydration. RNA crystallography has revealed a rich assortment of structural motifs whose characteristics are now under study.

Continued efforts to uncover the underlying principles of nucleic acid structure will result in much greater insights into the complex functions of these molecules.

ACKNOWLEDGMENTS

The 3D molecular images were created using program Chimera (Pettersen et al., 2004).

BS is supported by a grant LC512 from the Ministry of Education of the Czech Republic. Funds from the NSF for the NDB is gratefully acknowledged.

REFERENCES

Allen FH, Bellard S, Brice MD, Cartright BA, Doubleday A, Higgs H, Hummelink T, Hummelink-Peters BG, Kennard O, Motherwell WDS, Rodgers JR, Watson DG (1979): The Cambridge Crystallographic Data Centre: computer-based search, retrieval, analysis and display of information. *Acta Crystallogr B* 35:2331–2339.

Altona C, Sundaralingam M (1972): Conformational analysis of the sugar ring in nucleosides and nucleotides. A new description using the concept of pseudorotation. *J Am Chem Soc* 94:8205–8212.

Arnott S (1970): The geometry of nucleic acids. *Prog Biophys Mol Biol* 21:267–319.

Arnott S, Chandrasekharan R, Marttila CM (1974): Structures of polyinosinic acid and polyguanylic acid. *Biochem J* 141:537–538.

Arnott S, Campbell Smith PJ, Chandrasekaran R (1976a) Atomic coordinates and molecular conformations for DNA–DNA, RNA–RNA, and DNA–RNA helices. In: Fasman GD, editor. *CRC Handbook Of Biochemistry And Molecular Biology, Nucleic Acids*, Vol. 2, Cleveland: CRC Press, pp. 411–422.

Arnott S, Chandrasekaran R, Leslie AGW (1976b) Structure of the single-stranded polyribonucleotide polycytidylic acid. *J Mol Biol* 106:735–738.

Avery OT, MacLeod CM, McCarthy M (1944): Studies on the chemical nature of the substance inducing transformation of pneumococcal types: induction of transformation by a deoxyribonucleic acid fraction isolated from Pneumococcus Type III. *J Exp Med* 79:137–158.

Baba S, Takahashi K, Noguchi S, Takaku H, Koyanagi Y, Yamamoto N, Kawai G (2005): Solution RNA structures of the HIV-1 dimerization initiation site in the kissing-loop and extended-duplex dimers. *J Biochem* 138:583–592.

Babcock MS, Olson WK (1994): The effect of mathematics and coordinate system on comparability and "dependencies" of nucleic acid structure parameters. *J Mol Biol* 237:98–124.

Babcock MS, Pednault EPD, Olson WK (1993): Nucleic acid structure analysis: a users guide to a collection of new analysis programs. *J Biomol Struct Dyn* 11:597–628.

Babcock MS, Pednault EPD, Olson WK (1994): Nucleic acid structure analysis. Mathematics for local cartesian and helical structure parameters that are truly comparable between structures. *J Mol Biol* 237:125–156.

Baeyens KJ, De Bondt HL, Pardi A, Holbrook SR (1996): A curved RNA helix incorporating an internal loop with G-A and A-A non-Watson–Crick base pairing. *Proc Natl Acad Sci USA* 93:12851–12855.

Ban N, Nissen P, Hansen J, Moore PB, Steitz TA (2000): The complete atomic structure of the large ribosomal subunit at a 2.4 Å resolution. *Science* 289:905–920. [This paper describes a key macromolecular complex involved in protein synthesis.]

Bansal M, Bhattacharyya D, Ravi B (1995): NUPARM and NUCGEN: software for analysis and generation of sequence dependent nucleic acid structures. *CABIOS* 11:281–287.

Batey RT, Gilbert SD, Montange RK (2004): Structure of a natural guanine-responsive riboswitch complexed with the metabolite hypoxanthine. *Nature* 432:411–415.

Berk V, Zhang W, Pai RD, Cate JH (2006): Structural basis for mRNA and tRNA positioning on the ribosome. *Proc Natl Acad Sci USA* 103:15830–15834.

Berman HM, Young P (1981): The interaction of intercalating drugs with nucleic acids. *Annu Rev Biophys Bioeng* 10:87–114.

Berman HM, Olson WK, Beveridge DL, Westbrook J, Gelbin A, Demeny T, Hsieh SH, Srinivasan AR, Schneider B (1992): The Nucleic Acid Database—a comprehensive relational database of three-dimensional structures of nucleic acids. *Biophys J* 63:751–759. [The architecture and functionality of the Nucleic Acid Database (NDB) is presented.]

Bhattacharyya D, Bansal M (1989): A self-consistent formulation for analysis and generation of non-uniform DNA structures. *J Biomol Struct Dyn* 6:635–653.

Brodersen DE, Clemons WM Jr, Carter AP, Morgan-Warren R, Wimberly BT, Ramakrishnan V (2000): The structural basis for the action of the antibiotics Tetracycline, Pactamycin, and Hygromycin B on the 30S ribosomal subunit. *Cell* 103:1143–1154.

Brooks BR, Bruccoleri RE, Olafson BD, States DJ, Swaminathan S, Karplus M (1983): CHARMM: a program for macromolecular energy, minimization, and dynamics calculations. *J Comput Chem* 4:187–217.

Brown, RS, Dewan, JC, Klug, A (1985): Crystallographic and biochemical investigation of the lead(II)-catalyzed hydrolysis of yeast phenylalanine T-RNA. *Biochemistry* 24: 4785–4801.

Campbell DO, Legault P (2005): Nuclear magnetic resonance structure of the Varkud satellite ribozyme stem-loop V RNA and magnesium-ion binding from chemical-shift mapping. *Biochemistry* 44:4157–4170.

Carter AP, Clemons WM Jr, Brodersen DE, Wimberly BT, Morgan-Warren R, Ramakrishnan V (2000): Functional insights from the structure of the 30S ribosomal subunit and its interactions with antibiotics. *Nature* 407:340–348.

Case DA, Cheatham TE III, Darden T, Gohlke H, Luo R, Merz KM Jr, Onufriev A, Simmerling C, Wang B, Woods RJ (2005): The AMBER biomolecular simulation programs. *J Comput Chem* 26:1668–1688.

Cate JH, Gooding AR, Podell E, Zhou KH, Golden BL, Kundrot CE, Cech TR, Doudna JA (1996): Crystal structure of a group I ribozyme domain: principles of RNA packing. *Science* 273:1678–1685. [One of the early structure determinations of a large ribozyme is described.]

Cech T, Zaug A, Grabowski P (1981): In vitro splicing of the ribosomal RNA precursor of Tetrahymena: involvement of a guanosine nucleotide in the excision of the intervening sequence. *Cell* 27:487–496.

Chen X, Ramakrishnan B, Rao ST, Sundaralingam M Side by side binding of two distamycin A drugs in the minor groove of an alternating B-DNA duplex. (1994): *Nat Struct Biol* 1:169–175.

Clowney L, Jain SC, Srinivasan AR, Westbrook J, Olson WK, Berman HM (1996): Geometric parameters in nucleic acids: nitrogenous bases. *J Am Chem Soc* 118:509–518.

Coté ML, Yohannan SJ, Georgiadis MM (2000): Use of an N-terminal fragment from moloney murine leukemia virus reverse transcriptase to facilitate crystallization and analysis of a pseudo-16-mer DNA molecule containing G-A mispairs. *Acta Crystallogr D* 56:1120–1131.

Coté ML, Pflomm M, Georgiadas MM (2003): Staying straight with A-tracts: a DNA analog of the HIV-1 polypurine tract. *J Mol Biol* 330:57–74.

Cruse WBT, Saludjian P, Biala E, Strazewski P, Prange T, Kennard O (1994): Structure of a mispaired RNA double helix at 1.6 Å resolution and implications for the prediction of RNA secondary structure. *Proc Natl Acad Sci USA* 91:4160–4164.

Dickerson RE, Drew HR, Conner BN, Kopka ML, Pjura PE (1983): Helix geometry and hydration in A-DNA, B-DNA and Z-DNA. *Cold Spring Harb Symp Quant Biol* 47:13–24.

Dickerson RE, Bansal M, Calladine CR, Diekmann S, Hunter WN, Kennard O, von Kitzing E, Lavery R, Nelson HCM, Olson W, Saenger W, Shakked Z, Sklenar H, Soumpasis DM, Tung C-S, Wang AH-J, Zhurkin VB (1989): Definitions and nomenclature of nucleic acid structure parameters. *EMBO J* 8:1–4.

Dickerson RE, Goodsell D, Kopka ML (1996): MPD and DNA bending in crystals and in solution. *J. Mol. Biol.* 256:108–125.

Dickerson RE (1998): DNA bending: the prevalence of kinkiness and the virtues of normality. *Nucleic Acids Res* 26:1906–1926.

Dock-Bregeon AC, Chevrier B, Podjarny A, Johnson J, de Bear JS, Gough GR, Gilham PT, Moras D (1989): Crystallographic structure of an RNA helix: [U(UA)6A]2. *J. Mol. Biol.* 209:459–474.

Drew HR, Wing RM, Takano T, Broka C, Tanaka S, Itakura K, Dickerson RE (1981): Structure of a B-DNA dodecamer: conformation and dynamics. *Proc Natl Acad Sci USA* 78:2179–2183. [The first structure determination of a full turn of B-DNA is presented.]

Eisenstein M, Shakked Z (1995): Hydration patterns and intermolecular interactions in A-DNA crystal structures. Implications for DNA recognition. *J Mol Biol* 248:662–678.

El Hassan MA, Calladine CR (1995): The assessment of the geometry of dinucleotide steps in double-helical DNA: a new local calculation scheme with an appendix. *J Mol Biol* 251:648–664.

Ennifar E, Yusupov M, Walter P, Marquet R, Ehresmann B, Ehresmann C, Dumas P (1999): The crystal structure of the dimerization initiation site of genomic HIV-1 RNA reveals an extended duplex with two adenine bulges. *Structure* 7:1439–1449.

Ennifar E, Dumas P (2006): Polymorphism of bulged-out residues in HIV-1 RNA DIS kissing complex and structure comparison with solution studies. *J Mol Biol* 356:771–782.

Ferre-D'Amare AR, Zhou K, Doudna JA (1998): Crystal structure of a hepatitis delta virus ribozyme. *Nature* 395:567–674.

Frederick CA, Williams LD, Ughetto G, van der Marel GA, van Boom JH, Rich A, Wang AH-J (1990): Structural comparison of anticancer drug–DNA complexes: adriamycin and daunomycin. *Biochemistry* 29:2538–2549.

Fourmy D, Recht MI, Blanchard SC, Puglisi JD (1996): Structure of the A site of Escherichia coli 16S ribosomal RNA complexed with an aminoglycoside antibiotic. *Science* 274:1367–1371.

Gao Y-G, Sriram M, Denny WA, Wang AH-J (1993): Minor groove binding of SN6999 to an alkylated DNA: molecular structure of d(CGC[e^{6}G]AATTCGCG)-SN6999 complex. *Biochemistry* 32: 9639–9648.

Gelbin A, Schneider B, Clowney L, Hsieh S-H, Olson WK, Berman HM (1996): Geometric parameters in nucleic acids: sugar and phosphate constituents. *J Am Chem Soc* 118:519–528.

Gellert M, Lipsett MN, Davies DR (1962): Helix formation by guanylic acid. *Proc Natl Acad Sci USA* 48:2013–2018.

Gessner RV, Frederick CA, Quigley GJ, Rich A, Wang AH-J (1989): The molecular structure of the left-handed Z-DNA double helix at 1.0-Å atomic resolution. *J Biol Chem* 264:7921–7935.

Gessner RV, Quigley GJ, Egli M (1994): Comparative studies of high resolution Z-DNA crystal structures.1. Common hydration patterns of alternating DC-DG. *J Mol Biol* 236:1154–1168.

Gesteland RF, Cech TR, Atkins JF (1999): *The RNA World*, 2nd ed, Vol. 37, Cold Spring Harbor monograph series. Cold Spring Harbor, NY: Cold Spring Harbor Laboratory Press.

Gilbert W (1986): Origin of life - the RNA world. *Nature* 319:618.

Gorin AA, Zhurkin VB, Olson WK (1995): B-DNA twisting correlates with base-pair morphology. *J Mol Biol* 247:34–48.

Guzikevich-Guerstein G, Shakked Z (1996): A novel form of the DNA double helix imposed on the TATA-box by the TATA-binding protein. *Nat Struct Biol* 3:32–37.

Ha SC, Lowenhaupt K, Rich A, Kim YG, Kim KK (2005): Crystal structure of a junction between B-DNA and Z-DNA reveals two extruded bases. *Nature* 437:1183–1186.

Haider SM, Parkinson GN, Neidle S (2002): Crystal structure of the potassium form of an Oxytricha nova G-quadruplex. *J Mol Biol* 320:189–200.

Haider SM, Parkinson GN, Neidle S (2003): Structure of a G-quadruplex–ligand complex. *J Mol Biol* 326:117–125.

Hansen JL, Ippolito JA, Ban N, Nissen P, Moore PB, Steitz TA (2002): The structures of four macrolide antibiotics bound to the large ribosomal subunit. *Mol Cell.* 10:117–128.

Harms, JM, Bartels, H, Schlunzen, F, Yonath, A (2003): Antibiotics acting on the translational machinery. *J Cell Sci* 116: 1391–1393.

Harvey S, Prabhakarn M (1986): Ribose puckering: structure, dynamics, energetics, and the pseudorotation cycle. *J Am Chem Soc* 108:6128–6136.

Heinemann U (1991): A note on crystal packing and global helix structure in short A-DNA duplexes. *J Biomol Struct Dyn* 8:801–811.

Heinemann U, Alings C, Hahn M (1994): Crystallographic studies of DNA helix structure. *Biophys Chem* 50:157–167.

Hermann T (2005): Drugs targeting the ribosome. *Curr Opin Struct Biol* 15:355–366.

Hershey AD, Chase M (1951): Genetic recombination and heterozygosis in bacteriophage. *Cold Spring Harb Symp Quant Biol* 16:471–479.

Hershkovitz, E, Tannenbaum, E, Howerton, SB, Sheth, A, Tannenbaum, A, Williams, LD (2003): Automated identification of RNA conformational motifs: theory and application to the HM LSU 23S rRNA. *Nucleic Acids Res* 31(21): 6249–6257.

Holbrook SR, Cheong C, Tinoco I Jr, Kim S-H (1991): Crystal structure of an RNA double helix incorporating a track of non-Watson–Crick base pairs. *Nature* 353:579–581.

Jain SC, Zon G, Sundaralingam M (1989): Base only binding of spermine in the deep groove of the A-DNA octamer d(GTGTACAC). *Biochemistry* 28:2360–2364.

Jain S, Zon G, Sundaralingam M (1991): The hexagonal crystal structure of the A-DNA octamer d (GTGTACAC) and its comparison with the tetragonal structure-correlated variations in helical parameters. *Biochemistry* 30:3567–3576.

Kagawa TF, Stoddard D, Zhou G, Ho PS (1989): Quantitative analysis of DNA secondary structure from solvent accessible surfaces: the B- to Z-DNA transition as a model. *Biochemistry* 28:6642–6651.

Kamitori S, Takusagawa F (1994): Multiple binding modes of anticancer drug actinomycin D: x-ray, molecular modeling, and spectroscopic studies of d(GAAGCTTC)(2)-actinomycine D complexes and its host DNA. *J Am Chem Soc* 116:4154–4165.

Khorana HG, Tener GM, Moffatt JG, Pol EH (1956): A new approach to the synthesis of poly-nucleotides. *Chem Ind* 1523.

Kielkopf CL, Erkkila KE, Hudson BP, Barton JK, Rees DC (2000): Structure of a photoactive rhodium complex intercalated into DNA. *Nat Struct Biol* 7:117–121.

Kim S-H, Berman HM, Newton MD, Seeman NC (1973): Seven basic conformations of nucleic acid structural units. *Acta Crystallogr B* 29:703–710.

Kim S-H, Suddath FL, Quigley GJ, McPherson A, Sussman JL, Wang AH-J, Seeman NC, Rich A (1974): Three-dimensional tertiary structure of yeast phenylalanine transfer RNA. *Science* 185:435–440.

Kim JL, Nikolov DB, Burley SK (1993): Co-crystal structure of TBP recognizing the minor groove of a TATA element. *Nature* 365:520–527. [One of the first structural determinations of tRNA is described.]

Kim Y, Geiger JH, Hahn S, Sigler PB (1993): Crystal structure of a yeast TBP/TATA-box complex. *Nature* 365:512–520.

Klein DJ, Ferre-D'Amare AR (2006): Structural basis of glmS ribozyme activation by glucosamine-6-phosphate. *Science* 313:1752–1756.

Kopka ML, Yoon C, Goodsell D, Pjura P, Dickerson RE (1985): The molecular origin of DNA-drug specificity in netropsin and distamycin. *Proc Natl Acad Sci USA* 82:1376–1380.

Kopka ML, Goodsell DS, Baikalov I, Grzeskowiak K, Cascio D, Dickerson RE (1994): Crystal structure of a covalent DNA-drug adduct: anthramycin bound to C-C-A-A-C-G-T-T-G-G and a molecular explanation of specificity. *Biochemistry* 33:13593–13610.

Korostelev A, Trakhanov S, Laurberg M, Noller HF (2006): Crystal structure of a 70S ribosome–tRNA complex reveals functional interactions and rearrangements. *Cell* 126:1065–1077.

Kosikov KM, Gorin AA, Zhurkin VB, Olson WK (1999): DNA stretching and compression: large-scale simulations of double helical structures. *J Mol Biol* 289:1301–1326.

Krasilnikov AS, Yang X, Pan T, Mondragon A (2003): Crystal structure of the specificity domain of ribonuclease P. *Nature* 421:760–764.

Krasilnikov AS, Xiao Y, Pan T, Mondragon A (2004): Basis for structural diversity in homologous RNAs. *Science* 306:104–107.

Lavery R, Sklenar H (1988): The definition of generalized helicoidal parameters and of axis curvature for irregular nucleic acids. *J Biomol Struct Dyn* 6:63–91.

Lavery R, Sklenar H (1989): Defining the structure of irregular nucleic acids: conventions and principles. *J Biomol Struct Dyn* 6:655–667.

Leontis NB, Westhof E (2001): Geometric nomenclature and classification of RNA base pairs. *RNA* 7:499–512.

Lipanov A, Kopka ML, Kaczor-Grzeskowiak M, Quintana J, Dickerson RE (1993): Structure of the B-DNA decamer C-C-A-A-C-9-T-T-G-G in two different space groups: conformational flexibility of B-DNA. *Biochemistry* 32:1373–1380.

Lu X-J, El Hassan MA, Hunter CA (1997): Structure and conformation of helical nucleic acids: analysis program (SCHNAaP). *J Mol Biol* 273:668–680.

Lu X-J, Babcock MS, Olson WK (1999): Mathematical overview of nucleic acid analysis programs. *J Biomol Struct Dyn* 16:833–843.

Lu X-J, Olson WK (1999): Resolving the discrepancies among nucleic acid conformation analyses. *J Mol Biol* 285:1563–1575.

Ma JB, Ye K, Patel DJ (2004): Structural basis for overhang-specific small interfering RNA recognition by the PAZ domain. *Nature* 429:318–322.

Macrae IJ, Zhou K, Li F, Repic A, Brooks AN, Cande WZ, Adams PD, Doudna JA (2006): Structural basis for double-stranded RNA processing by Dicer. *Science* 311:195–198.

McCall, M, Brown, T Kennard, O (1985): The crystal structure of d(G-G-G-G-C-C-C-C). A model for poly(dG)•poly(dC). *J Mol Biol* 183:385–396.

Moore, MH, Hunter, WN, d'Estaintot, BL Kennard, O (1989): DNA-Drug interactions. The crystal structure of d(CGATCG) complexed with daunomycin. *J Mol Biol* 206: 693–670.

Murray LJ, Arendall WB III, Richardson DC Richardson JS (2003): RNA backbone is rotameric. *Proc Natl Acad Sci USA* 100:13904–13909.

Murray-Rust P, Motherwell S (1978): Computer retrieval and analysis of molecular geometry. III. Geometry of the beta-1'-aminofuranoside fragment. *Acta Crystallogr B* 34:2534–2546.

Murthy VL, Srinivasan R, Draper DE, Rose GD (1999): A complete conformational map for RNA. *J Mol Biol* 291 (2):313–327.

Neidle S (2007) *Principles of Nucleic Acid Structure*. London: Elsevier/Academic Press.

Ng H-L, Dickerson RE (2001): DNA structure from A to B. *Proc Natl Acad Sci USA* 98:6986–6989.

Nguyen B, Lee MPH, Hamelberg D, Joubert A, Bailly C, Brun R, Neidle S, Wilson WD (2002): Strong binding in the DNA minor groove by an aromatic diamidine with a shape that does not match the curvature of the groove. *J Am Chem Soc* 124:13680–13681.

Olson WK (1982): How flexible is the furanose ring? II. An updated potential energy estimate. *J Am Chem Soc* 104:278–284.

Olson WK, Sussman JL (1982): How flexible is the furanose ring? I. A comparison of experimental and theoretical studies. *J Am Chem Soc* 104:270–278.

Olson WK, Bansal M, Burley SK, Dickerson RE, Gerstein M, Harvey SC, Heinemann U, Lu X-J, Neidle S, Shakked Z, Sklenar H, Suzuki M, Tung C-S, Westhof E, Wolberger C, Berman HM (2001): A standard reference frame for the description of nucleic acid base-pair geometry. *J Mol Biol* 313:229–237.

Packer M, Hunter C (1998): Sequence-dependent DNA structure: the role of the sugar-phosphate backbone. *J Mol Biol* 280:407–420.

Parkinson G, Vojtechovsky J, Clowney L, Brünger AT, Berman HM (1996): New parameters for the refinement of nucleic acid containing structures. *Acta Crystallogr D* 52:57–64.

Parkinson GN, Lee MPH, Neidle S (2002): Crystal structure of parallel quadruplexes from human telomeric DNA. *Nature* 417:876–880.

Parkinson GN, Ghosh R, Neidle S (2007): Structural basis for binding of porphyrin to human telomeres. *Biochemistry* 46:2390–2397.

Pearlman D, Kim S-H (1990): Atomic charges for DNA constituents derived from single-crystal X-ray diffraction data. *J Mol Biol* 211:171–187.

Pettersen EF, Goddard TD, Huang CC, Couch GS, Greenblatt DM, Meng EC, Ferrin TE (2004): UCSF Chimera–a visualization system for exploratory research and analysis. *J Comput Chem* 25:1605–1612.

Phillips K, Dauter Z, Murchie AIH, Lilley DMJ, Luisi B (1997): The crystal structure of a parallel-stranded guanine tetraplex at 0.95 Å resolution. *J Mol Biol* 273:171–182.

Pierre VC, Kaiser JT, Barton JK (2007): Insights into finding a mismatch through the structure of a mispaired DNA bound by a rhodium intercalator. *Proc Natl Acad Sci USA* 104:429–434.

Pley HW, Flaherty KM, McKay DB (1994): Three-dimensional structure of a hammerhead ribozyme. *Nature* 372:68–74.

Portmann S, Usman N, Egli M (1995): The crystal structure of r(CCCCGGGG) in two distinct lattices. *Biochemistry* 34:7569–7575.

Privé GG, Yanagi K, Dickerson RE (1991): Structure of the B-DNA decamer C-C-A-A-C-G-T-T-G-G and comparison with isomorphous decamers C-C-A-A-G-A-T-T-G-G and C-C-A-G-G-C-C-G-T-G-G. *J Mol Biol* 217:177–199.

Quintana JR, Grzeskowiak K, Yanagi K, Dickerson RE (1992): The structure of a B-DNA decamer with a central T-A step: C-G-A-T-T-A-A-T-C-G. *J Mol Biol* 225:379–395.

Ramakrishnan B, Sundaralingam M (1993): Crystal packing effects on A-DNA helix parameters—a comparative study of the isoforms of the tetragonal and hexagonal family of octamers with differing base sequences. *J Biomol Struct Dyn* 11:11–26.

Rhodes D, Giraldo R (1995): Telomere structure and function. *Curr Opin Struct Biol* 5:311–322.

Rich A, Davies DR, Crick FHC, Watson JD (1961): The molecular structure of polyadenlyic acid. *J Mol Biol* 3:71–86.

Richardson, JS, Schneider, B, Murray, LW, Kapral, GJ, Immormino, RM, Headd, JJ, Richardson, DC, Ham, D, Hershkovits, E, Williams, LD, Keating, KS, Pyle, AM, Micallef, D, Westbrook, J, Berman, HM (2008): RNA backbone: consensus all-angle conformers and modular string nomenclature (an RNA Ontology Consortium contribution). *RNA* 14(3): 465–481.

Robertus JD, Ladner JE, Finch JT, Rhodes D, Brown RS, Clark BFC, Klug A (1974): Structure of yeast phenylalanine tRNA at 3 Å resolution. *Nature* 250:546–551. [One of the first structural determinations of tRNA is described.]

Rodley GA, Scobie RS, Bates RHT, Lewitt RM (1976): A possible conformation for double-stranded polynucleotides. *Proc Natl Acad Sci USA* 73:2959–2963.

Rosenberg JM, Seeman NC, Kim JJP, Suddath FL, Nicholas HB, Rich A (1973): Double helix at atomic resolution. *Nature* 243:150–154.

Rosenberg JM, Seeman NC, Day RO, Rich A (1976): RNA double helical fragments at atomic resolution: II. The structure of sodium guanylyl-3′,5′-cytidine nonhydrate. *J Mol Biol* 104:145–167.

Saenger W, Hunter WN, Kennard O (1986): DNA conformation is determined by economics in the hydration of phosphate groups. *Nature* 324:385–388.

Salter J, Krucinska J, Alam S, Grum-Tokars V, Wedekind JE (2006): Water in the active site of an all-RNA hairpin ribozyme and effects of Gua8 base variants on the geometry of phosphoryl transfer. *Biochemistry* 45:686–700.

Schluenzen F, Tocilj A, Zarivach R, Harms J, Gluehmann M, Janell D, Bashan A, Bartels H, Agmon I, Franceschi F, Yonath A (2000): Structure of functionally activated small ribosomal subunit at 3.3 Å resolution. *Cell* 102:615–623.

Schlunzen F, Harms JM, Franceschi F, Hansen HA, Bartels H, Zarivach R, Yonath A (2003): Structural basis for the antibiotic activity of ketolides and azalides. *Structure* 11:329–338.

Schneider B, Ginell SL, Jones R, Gaffney B, Berman HM (1992): Crystal and molecular structure of a DNA fragment containing a 2-aminoadenine modification: the relationship between conformation, packing, and hydration in Z-DNA hexamers. *Biochemistry* 31:9622–9628.

Schneider B, Berman HM (1995): Hydration of the DNA bases is local. *Biophys J* 69:2661–2669.

Schneider B, Neidle S, Berman HM (1997): Conformations of the sugar-phosphate backbone in helical DNA crystal structures. *Biopolymers* 42:113–124.

Schneider B, Moravek Z, Berman HM (2004): RNA conformational classes. *Nucleic Acids Res* 32:1666–1677.

Schroth GP, Kagawa TF, Ho PS (1993): Structure and thermodynamics of nonalternating C-G base pairs in Z-DNA: the 1.3 Å crystal structure of the asymmetric hexanucleotide d(m5CGGGm5CG). d(m5CGCCm5CG). *Biochemistry* 32:13381–13392.

Schwieters CD, Clore GM (2007): A physical picture of atomic motions within the Dickerson DNA dodecamer in solution derived from joint ensemble refinement against NMR and large-angle X-ray scattering data. *Biochemistry* 46:1152–1166.

Seeman NC, Sussman JL, Berman HM, Kim S-H (1971): Nucleic acid conformation: crystal structure of a naturally occurring dinucleoside phosphate (UpA). *Nat New Biol* 233:90–92.

Seeman NC, Rosenberg JM, Suddath FL, Kim JJP, Rich A (1976): RNA double helical fragments at atomic resolution: I. The crystal and molecular structure of sodium adenylyl-3′-5′-uridine hexahydrate. *J Mol Biol* 104:109–144. [The first determination of the structure of the double helix is presented.]

Selmer M, Dunham CM, Murphy FV, Weixlbaumer A, Petry S, Kelley AC, Weir JR, Ramakrishnan V (2006): Structure of the 70S ribosome complexed with mRNA and tRNA. *Science* 313:1935–1942.

Sen D, Gilbert W (1988): Formation of parallel four-stranded complexes by guanine-rich motifs in DNA and its implications for meiosis. *Nature* 334:364–366.

Serganov A, Yuan YR, Pikovskaya O, Polonskaia A, Malinina L, Phan AT, Hobartner C, Micura R, Breaker RR, Patel DJ (2004): Structural basis for discriminative regulation of gene expression by adenine- and guanine-sensing mRNAs. *Chem Biol* 11:1729–1741.

Shakked Z, Rabinovich D, Kennard O, Cruse WBT, Salisbury SA, Viswamitra MA (1983): Sequence-dependent conformation of an A-DNA double helix. The crystal structure of the octamer d(G-G-T-A-T-A-C-C). *J Mol Biol* 166:183–201.

Shakked Z, Guerstein-Guzikevich G, Eisenstein M, Frolow F, Rabinovich D (1989): The conformation of the DNA double helix in the crystal is dependent on its environment. *Nature* 342:456–460.

Silverman AP, Bu W, Cohen SM, Lippard SJ (2002): 2.4-Å crystal structure of the asymmetric platinum complex [Pt(ammine)(cyclohexylamine)]$^{2+}$ bound to a dodecamer DNA duplex. *J Biol Chem* 277:49743–49749.

Sims GE, Kim S-H (2003): Global mapping of nucleic acid conformational space: dinucleoside monophosphate conformations and transition pathways among conformational classes. *Nucleic Acids Res* 31:5607–5616.

Soumpasis DM, Tung CS (1988): A rigorous base pair oriented description of DNA structures. *J Biomol Struct Dyn* 6:397–420.

Sun G, Voigt JH, Filippov IV, Marquez VE, Nicklaus MC (2004): PROSIT: pseudo-rotational online service and interactive tool, applied to a conformational survey of nucleosides and nucleotides. *J Chem Inf Comput Sci* 44:1752–1762.

Sundaralingam M (1969): Stereochemistry of nucleic acids and their constituents. IV. Allowed and preferred conformations of nucleosides, nucleoside mono-, di-, tri-, tetraphosphates, nucleic acids and polynucleotides. *Biopolymers* 7:821–860.

Sussman JL, Holbrook SR, Warrant RW, Church GM, Kim S-H (1978): Crystal structure of yeast phenylalanine transfer RNA. I. Crystallographic refinement. *J Mol Biol* 123:607–630.

Tabernero L, Bella J, Alemán C (1996): Hydrogen bond geometry in DNA–minor groove binding drug complexes. *Nucleic Acids Res* 24:3458–3466.

Takahara PM, Rosenzweig AC, Frederick CA, Lippard SJ (1995): Crystal structure of double-stranded DNA containing the major adduct of the anticancer drug cisplatin. *Nature* 377:649–652.

Thota N, Li XH, Bingman C, Sundaralingam M (1993): High resolution refinement of the hexagonal A-DNA octamer d(GTGTACAC) at 1.4 Å resolution. *Acta Crystallogr D* 49:282–291.

Tsai C-C, Jain SC, Sobell HM (1975): X-ray crystallographic visualizations of drug-nucleic acid intercalative binding: structure of an ethidium–dinucleoside monophosphate crystalline complex, ethidium: 5-iodouridylyl(3′-5′) adenosine. *Proc Natl Acad Sci USA* 72:628–632.

Tu D, Blaha G, Moore PB, Steitz TA (2005): Structures of MLSBK antibiotics bound to mutated large ribosomal subunits provide a structural explanation for resistance. *Cell* 121:257–270.

Tung C-S, Soumpasis DM, Hummer G (1994): An extension of the rigorous base-unit oriented description of nucleic acid structures. *J Biomol. Struct Dyn* 11:1327–1344.

Varani G, McClain W (2000): The G × U wobble base pair. A fundamental building block of RNA structure crucial to RNA function in diverse biological systems. *EMBO Rep* 1:18–23.

Vargason JM, Henderson K, Ho PS (2001): A crystallographic map of the transition from B-DNA to A-DNA. *Proc Natl Acad Sci USA* 98:7265–7270.

Varnai P, Djuranovic D, Lavery R, Hartmann B (2002): Alpha/gamma transitions in the B-DNA backbone. *Nucleic Acids Res* 30:5398–5406.

Vicens Q, Westhof E (2001): Crystal structure of paromomycin docked into the eubacterial ribosomal decoding A site. *Structure* 9:647–658.

Vojtechovsky J, Eaton M, Gaffney B, Jones R, Berman H (1995): Structure of a new crystal form of a DNA dodecamer containing $T\sqrt{(O^6Me)G}$ base pairs. *Biochemistry* 34:16632–16640.

Wahl MC, Ramakrishnan B, Ban CG, Chen X, Sundaralingam M (1996): RNA—synthesis, purification, and crystallization. *Acta Crystallogr D* 52:668–675.

Wang AH-J, Quigley GJ, Kolpak FJ, Crawford JL, van Boom JH, van der Marel GA, Rich A (1979): Molecular structure of a left-handed double helical DNA fragment at atomic resolution. *Nature* 282:680–686.

Wang AH-J, Hakoshima T, van der Marel GA, van Boom JH, Rich A (1984): AT base pairs are less stable than GC base pairs in Z-DNA: the crystal structure of d(m⁵CGTAm⁵CG). *Cell* 37:321–331.

Wang AH-J, Gessner RV, van der Marel GA, van Boom JH, Rich A (1985): Crystal structure of Z-DNA without an alternating purine–pyrimidine sequence. *Proc Natl Acad Sci USA* 82:3611–3615.

Wang AH-J, Ughetto G, Quigley GJ, Rich A (1987): Interactions between an anthracycline antibiotic and DNA: molecular structure of daunomycin complexed to d(CpGpTpApCpG) at 1.2-Å resolution. *Biochemistry* 26:1152–1163.

Wang AH-J, Gao YG, Liaw Y-C, Li Y-K (1991): Formaldehyde cross-links daunorubicin and DNA efficiently: HPLC and x-ray diffraction studies. *Biochemistry* 30:3812–3815.

Watson JD, Crick FHC (1953): A structure for deoxyribose nucleic acid. *Nature* 171:737–738.

Weiner PK, Kollman PA (1981): AMBER: assisted model building with energy refinement. A general program for modeling molecules and their interactions. *J Comput Chem* 2:287–303.

Wild K, Weichenrieder O, Leonard G, Cusack S (1999): The 2A structure of helix 6 of the human signal recognition particle RNA. *Structure* 7:1345–1352.

Wimberly BT, Brodersen DE, Clemons WM Jr, Morgan-Warren R, Carter AP, Vonrhein C, Hartsch T, Ramakrishnan V (2000): Structure of the 30S ribosomal subunit. *Nature* 407:327–339. [This paper describes the structure of the 30S subunit of the ribosome.]

Wing R, Drew HR, Takano T, Broka C, Tanaka S, Itakura K, Dickerson RE (1980): Crystal structure analysis of a complete turn of B-DNA. *Nature* 287:755–758.

Wyatt J, Christain M, Puglisi J (1991): Synthesis and purification of large amounts of RNA olignucleotides. *BioTechniques* 11:764–769.

Yanagi K, Prive GG, Dickerson RE (1991): Analysis of local helix geometry in three B-DNA decamers and eight dodecamers. *J Mol Biol* 217:201–214.

Young MA, Ravishanker G, Beveridge DL, Berman HM (1995): Analysis of local helix bending in crystal structures of DNA oligonucleotides and DNA-protein complexes. *Biophys J* 68:2454–2468.

Yusupov MM, Yusupova GZ, Baucom A, Lieberman K, Earnest TN, Cate JHD, Noller HF (2001): Crystal structure of the ribosome at 5.5 Å resolution. *Science* 282:883–896.

Zamenhof S, Brawermann G, Chargaff E (1952): On the desoxypentose nucleic acids from several microorganisms. *Biochim Biophys Acta* 9:402–405.

Zhang L, Doudna JA (2002): Structural insights into group II Intron catalysis and branch-site selection. *Science* 295:2084–2088.

4

COMPUTATIONAL ASPECTS OF HIGH-THROUGHPUT CRYSTALLOGRAPHIC MACROMOLECULAR STRUCTURE DETERMINATION

Paul D. Adams, Ralf W. Grosse-Kunstleve, and Axel T. Brunger

INTRODUCTION

The desire to understand biological processes at a molecular level has led to the routine application of X-ray crystallography. However, significant time and effort are usually required to solve and complete a macromolecular crystal structure. Much of this effort is in the form of manual interpretation of complex numerical data using a diverse array of software packages and the repeated use of interactive three-dimensional graphics. The need for extensive manual intervention leads to two major problems: significant bottlenecks that impede rapid structure solution (Burley et al., 1999) and the introduction of errors due to subjective interpretation of the data (Mowbray et al., 1999). These problems present a major impediment to the success of structural genomics efforts (Burley et al., 1999; Montelione and Anderson, 1999) that require the whole process of structure solution to be as streamlined as possible. See Chapter 40 for a detailed description of structural genomics. The automation of structure solution is thus necessary as it has the opportunity to produce minimally biased models in a short time. Recent technical advances are fundamental to achieving this automation and make high-throughput structure determination an obtainable goal.

HIGH-THROUGHPUT STRUCTURE DETERMINATION

Automation in macromolecular X-ray crystallography has been a goal for many researchers. The field of small-molecule crystallography, where atomic resolution data are routinely

Structural Bioinformatics, Second Edition   Edited by Jenny Gu and Philip E. Bourne
Copyright © 2009 John Wiley & Sons, Inc.

collected, is already highly automated. As a result, the current growth rate of the Cambridge Structural Database (CCSD) (Allen, Kennard, and Taylor, 1983) is more than 15,000 new structures per year. This is approximately 10 times the growth rate of the Protein Data Bank (PDB) (Berman et al., 2000). See Chapters 11–13 for further details of structural databases. Automation of macromolecular crystallography can significantly improve the rate at which new structures are determined. The goal of automation moved to a position of prime importance with the development of the concept of structural genomics (Burley et al., 1999; Montelione and Anderson, 1999) and the routine application of high-resolution macromolecular crystallography to study protein–ligand complexes for drug discovery (Nienaber et al., 2000). To exploit the information present in the rapidly expanding sequence databases, it has been proposed that the structural database must also grow. Increased knowledge about the relationship between sequence, structure, and function will allow sequence information to be used to its full extent. The success of structural genomics requires macromolecular structures to be solved at a rate significantly faster than that at present. This high-throughput structure determination depends on automation to reduce the bottlenecks related to human intervention throughout the whole crystallographic process. Automation of structure solution from the experimental data relies on the development of algorithms that minimize or eliminate subjective input, the development of algorithms that automate procedures traditionally performed by manual intervention, and finally, the development of software packages that allow a tight integration between these algorithms. Truly automated structure solution requires the computer to make decisions about how best to proceed in the light of the available data.

The automation of macromolecular structure solution applies to all the procedures involved beginning with data collection to structure refinement. There have been many technological advances that make macromolecular X-ray crystallography easier. In particular, cryoprotection to extend crystal life (Garman, 1999), the availability of tunable synchrotron sources (Walsh et al., 1999a), high-speed CCD data collection devices (Walsh et al., 1999b), and the ability to incorporate anomalously scattering selenium atoms into proteins have all made structure solution much more efficient (Walsh et al., 1999b). The desire to make structure solution more efficient has led to investigations into the optimal data collection strategies for multiwavelength anomalous diffraction (Gonzalez et al., 1999; Gonzalez, 2007) and phasing using single anomalous diffraction with sulfur or ions (Dauter and Dauter, 1999; Dauter et al., 1999). It has been shown that MAD phasing using only two wavelengths can be successful (Gonzalez et al., 1999). The optimum wavelengths for such an experiment are those that give a large contrast in the real part of the anomalous scattering factor (e.g., the inflection point and high-energy remote). However, it has also been shown that, in general, a single wavelength collected at the anomalous peak is sufficient to solve a macromolecular structure (Rice, Earnest, and Brunger, 2000). Such an approach minimizes the amount of data to be collected and increases the efficiency of synchrotron beamlines, and is becoming a more widely used technique.

DATA ANALYSIS

The first step of structure solution, once the raw images have been processed, is assessment of data quality. The intrinsic quality of the data must be quantified and the appropriate signal extracted. Observations that are in error must be rejected as outliers. Some observations will be rejected at the data-processing stage, where multiple observations are available. However, if redundancy is low then probabilistic methods can be used (Read, 1999). The prior expectation, given either by a Wilson distribution of intensities or model-based structure— factor probability distributions, is used to detect outliers. This method is able to reject strong observations that are in error, which tend to dominate the features of electron density and Patterson maps. This method could also be extended to the rejection of outliers during the model refinement process.

When using isomorphous substitution or anomalous diffraction methods for experimental phasing, the relevant information lies in the differences between the multiple observations. In the case of anomalous diffraction, these differences are often very small, being of the same order as the noise in the data. In general, the anomalous differences at the peak wavelength are sufficient to locate the heavy atoms, provided that a large enough anomalous signal is observed (Grosse-Kunstleve and Brunger, 1999; Weeks et al., 2003). However, in less routine cases it can be very important to extract the maximum information from the data. One approach used in MAD phasing is to analyze the data sets to calculate F_A structure factors, which correspond to the anomalously scattering substructure (Terwilliger, 1994). Several programs are available to estimate the F_A structure factors: XPREP (Bruker, 2001), MADSYS (Hendrickson, 1991), and SOLVE (Terwilliger and Berendzen, 1999a). In another approach, a specialized procedure for the normalization of structure factor differences arising from either isomorphous or anomalous differences has been developed to facilitate the use of direct methods for heavy atom location (Blessing and Smith, 1999).

Merohedral twinning of the diffraction data can make structure solution difficult and in some cases impossible. The twinning occurs when a crystal contains multiple diffracting domains that are related by a simple transformation such as a twofold rotation about a crystallographic axis, a phenomenon that occur in certain space groups or under certain combinations of cell dimensions and space group symmetry (Parsons, 2003). As a result, the observed diffraction intensities are the sum of the intensities from the two distinctly oriented domains. Fortunately, the presence of twinning can be detected at an early stage by the statistical analysis of structure–factor distributions (Yeates, 1997). If the twinning is only partial, it is possible to detwin the data. Perfect twinning typically makes structure solution using experimental phasing methods difficult, but the molecular replacement method and refinement (see later) still can be successfully used.

HEAVY ATOM LOCATION AND COMPUTATION OF EXPERIMENTAL PHASES

The location of heavy atoms in isomorphous replacement or the location of anomalous scatterers was traditionally performed by manual inspection of Patterson maps. However, in recent years labeling techniques such as selenomethionyl incorporation have become widely used. This leads to an increase in the number of atoms to be located, rendering manual interpretation of Patterson maps extremely difficult. As a result, automated heavy atom location methods have proliferated. The programs SOLVE (Terwilliger and Berendzen, 1999a) and CNS (Brunger et al., 1998; Grosse-Kunstleve and Brunger, 1999) use Patterson-based techniques to find a starting heavy atom configuration that is then completed using difference Fourier analyses. Shake-and-Bake (SnB) (Weeks and Miller, 1999), SHELX-D (Sheldrick and Gould, 1995), and HySS (Grosse-Kunstleve and Adams, 2003) use direct methods reciprocal space phase refinement combined with modifications in real space. SnB refines phases derived from randomly positioned atoms, while SHELX-D derives starting phases by automatic inspection of the Patterson map. All methods have been used with great success to solve substructures with more than 60 selenium sites. SHELX-D and SnB have been used to find up to 150 and 160 selenium sites, respectively. The HySS program from PHENIX provides a high degree of automation, terminating the search once a successful solution has been found.

After the heavy atom or anomalously scattering substructure has been located, experimental phases can be calculated and the parameters of the substructure refined. A number of modern maximum likelihood based methods for heavy atom refinement and phasing are readily available: MLPHARE (Otwinowski, 1991), CNS (Brunger et al., 1998), SHARP (de La Fortelle and Bricogne, 1997), SOLVE (Terwilliger and Berendzen, 1999a), and Phaser (McCoy, Storoni, and Read, 2004). Programs such as PHENIX (Adams et al., 2002) have the advantage of fully integrating heavy atom location (using HySS), site refinement/phasing (using SOLVE or Phaser), and automated choice of heavy atom hand.

DENSITY MODIFICATION

Often the raw phases obtained from the experiment are not of sufficient quality to proceed with structure determination. However, there are many real space constraints, such as solvent flatness, that can be applied to electron density maps in an iterative fashion to improve initial phase estimates. This process of density modification is now routinely used to improve experimental phases prior to map interpretation and model building. However, due to the cyclic nature of the density modification process, where the original phases are combined with new phase estimates, introduction of bias is a serious problem. The γ correction was developed to reduce the bias inherent in the process and has been applied successfully in the method of solvent flipping (Abrahams, 1997), and also implemented in the CNS package. The γ correction has been generalized to the γ perturbation method in the DM program, part of the CCP4 suite (Collaborative Computational Project 4, 1994), and can be applied to any arbitrary density modification procedure including noncrystallographic symmetry (NCS) averaging and histogram matching (Cowtan, 1999). After bias removal, histogram matching is significantly more powerful than solvent flattening for comparable volumes of protein and solvent (Cowtan, 1999). More recently, a reciprocal space maximum likelihood formulation of the density modification process has been devised and implemented in the program RESOLVE (Terwilliger, 2000; Terwilliger, 2002a; Terwilliger, 2003a). This method has the advantage that a likelihood function can be directly optimized with respect to the available parameters (phases and amplitudes), rather than indirectly through a weighted combination of starting parameters with those derived from flattened maps. In this way, the problem of choices of weights for phase combination is avoided. The concept of statistical density modification has been developed further in the program PIRATE (Cowtan, 2004), where many different probability distributions are used to classify the density.

MOLECULAR REPLACEMENT

The method of molecular replacement is commonly used to solve structures for which a homologous structure is already known. As the database of known structures expands as a result of structural genomics efforts, this technique will become more and more important.

The method attempts to locate a molecule or fragments of a molecule, whose structure is known, in the unit cell of an unknown structure for which experimental data are available. To make the problem tractable, it has traditionally been broken down into two consecutive three-dimensional search problems: a search to determine the rotational orientation of the model followed by a search to determine the translational orientation for the rotated model (Rossmann and Blow, 1962). The method of Patterson correlation (PC) refinement is often used to optimize the rotational orientation prior to the translation search, thus increasing the likelihood of finding the correct solution (Brunger, 1997). With currently available programs, structure solution by molecular replacement usually involves significant manual input. Recently, however, methods have been developed to automate molecular replacement. One approach has used the exhaustive application of traditional rotation and translation methods to perform a complete six-dimensional search (Sheriff, Klei, and Davis, 1999). More recently, less time-consuming methods have been developed. The EPMR program implements an evolutionary algorithm to perform a very efficient six-dimensional search (Kissinger, Gehlhaar, and Fogel, 1999). A Monte Carlo simulated annealing scheme is used in the program Queen of Spades to locate the positions of molecules in the asymmetric unit (Glykos and Kokkinidis, 2000). To improve the sensitivity of any molecular replacement search algorithm, maximum likelihood methods have been developed in the Phaser program (Read, 2001; Storoni, McCoy, and Read, 2004; McCoy et al., 2005). The traditional scoring function of the search is replaced by a function that takes into account the errors in the model and the uncertainties at each stage. This approach is seen to greatly improve the chances of finding a correct solution using the traditional approach of rotation (Storoni, McCoy, and Read, 2004) and translation searches (McCoy et al., 2005). In addition, the method performs anisotropic correction of the experimental data and a statistically correct treatment of simultaneous information from multiple search models using multivariate statistical analysis (Read, 2001). This allows information from different structures to be used in highly automated procedures, while minimizing the risk of introducing bias. In the future, molecular replacement algorithms may permit experimental data to be exhaustively tested against all known structures to determine whether a homologous structure is already present in a database, which could then be used as an aid in structure determination.

MAP INTERPRETATION

The interpretation of the initial electron density map, calculated using either experimental phasing or molecular replacement methods, is often performed in multiple stages (described later), the final goal being the construction of an atomic model. If the interpretation cannot proceed to an atomic model, it is often an indication that the diffraction data collection must be repeated with improved crystals. Alternatively, repeating previous computational steps in data analysis or phasing may generate revised hypotheses about the crystal, such as a different space group symmetry or estimate of unit cell contents. Clearly, completely automating the process of structure solution will require that these eventualities are taken into consideration and dealt with in a rigorous manner.

The first stage of electron density map interpretation is an overall assessment of the information in a given map. The standard deviation of the local root-mean-square electron density can be calculated from the map. This variation is high when the electron density map has well-defined protein and solvent regions and is low for maps calculated with random phases (Terwilliger, 1999; Terwilliger and Berendzen, 1999b). A similar, more discriminating, analysis can be performed by the calculation of the skewness of the histogram of electron density values in the unit cell (Podjarny, 1976). It has also been shown that the correlation of the local root-mean-square density in adjacent regions in the unit cell can be used as a measure of the presence of distinct, contiguous solvent and macromolecular regions in an electron density map (Terwilliger and Berendzen, 1999c).

Currently, the process of analyzing an experimental electron density map to build the atomic model is a time-consuming, subjective process and almost entirely graphics based. Sophisticated programs such as COOT (Emsley and Cowtan, 2004), O (Jones et al., 1991), XtalView (McRee, 1999), QUANTA (Oldfield, 2000), TurboFrodo (Jones, 1978), and MAIN (Turk, 2000) are commonly used for manual rebuilding. These greatly reduce the effort required to rebuild models by providing libraries of side-chain rotamers and peptide fragments (Kleywegt and Jones, 1998), map interpretation tools, and real space refinement of rebuilt fragments (Jones et al., 1991). However, it has been shown that there are substantial differences among the models built manually by different people when presented with the same experimental data (Mowbray et al., 1999). The majority of time spent in completing a crystal structure is in the use of interactive graphics to manually modify the model. This manual modification is required either to correct parts of the model that are incorrectly placed or to add parts of the model that are currently missing. This process is prone to human error because of the large number of degrees of freedom of the model and the possible poor quality of regions of the electron density map.

Although interactive graphics systems for manual model building have made the process dramatically simpler, there have also been significant advances in making the process of map interpretation and model building truly automated. One route to automated analysis of the electron density map is the recognition of larger structural elements, such as α-helices and β-strands. Location of these features can often be achieved even in electron density maps of low quality using exhaustive searches in either real space (Kleywegt and Jones, 1997) or reciprocal space (Cowtan, 1998; Cowtan, 2001), the latter having a significant advantage in speed because the translation search for each orientation can be calculated using a fast Fourier transform. The automatic location of secondary structure elements from skeletonized electron density maps can be combined with sequence information and databases of known structures to build an initial atomic model with little or no manual intervention from the user (Oldfield, 2000). This method has been seen to work even at relatively low resolution ($d_{min} \sim 3.0$ Å). However, the implementation is still graphics based and requires user input. A related approach in the program MAID also uses a skeleton generated from the electron density map as the starting point for locating secondary structure elements (Levitt, 2001). Trial points are extended in space by searching for connected electron density at C_α distance (approximately 3.7 Å) with standard α-helical or β-strand geometry. Real space refinement of the fragments generated is used to improve the model. Both of these methods suffer from the limitation that they do not combine the model building process with the generation of improved electron density maps derived from the starting phases and the partial models.

To completely automate the model building process, methods have been developed that combine automated identification of potential atomic sites in the map with model refinement. In the ARP/warp system, an iterative procedure is used that describes the electron density map as a set of unconnected atoms from which protein-like patterns, primarily the main-chain trace from peptide units, are extracted. From this information and knowledge of the protein sequence, a model can be automatically constructed (Perrakis, Morris, and Lamzin, 1999). This powerful procedure, known as warpNtrace in ARP/wARP, can gradually build a more complete model from the initial electron density map and in many cases is capable of building the majority of the protein structure in a completely automated way. Unfortunately, this method currently has the limitation of a need for relatively high-resolution data ($d_{min} < 2.3$ Å). Data that extend to this resolution are available for less than 60% of the ~16,500 X-ray structures in the PDB. Therefore, other approaches have been developed to automatically interpret maps at lower resolution (Holton et al., 2000; Terwilliger, 2002b; Terwilliger, 2003b; Terwilliger, 2003c; Terwilliger, 2003d). In the PHENIX system (Adams et al., 2002), the combination of secondary structure fragment location and fragment extension by RESOLVE (Terwilliger, 2003d) with iterated structure refinement by *phenix.refine* (Afonine, Grosse-Kunstleve, and Adams, 2005) for map improvement provides an automated model building method that is relatively insensitive to resolution and is capable of typically building 70% or more of a structure even at 3.0 Å. With this technology, it is has now been possible to investigate the variability of models by building many models against the same data (DePristo, de Bakker, and Blundell, 2004; DePristo et al., 2005; Terwilliger et al., 2007).

Methods have recently been developed for the automated location and fitting of small molecules into difference electron density maps, a process critical to the crystallographic screening of potential therapeutic compounds bound to their target molecules. These methods have used reduction of the difference electron density to a simpler representation (Zwart, Langer, and Lamzin, 2004; Aishima et al., 2005) or systematic searching against the density map with rigid fragments of the small molecule (Terwilliger et al., 2006). This is still an active area of research, where problems of small molecule disorder and partial occupancy present significant challenges to robust automation.

REFINEMENT

In general, the atomic model obtained by automatic or manual methods contains some errors and must be optimized to best fit the experimental diffraction data and prior chemical information. In addition, the initial model is often incomplete and refinement is carried out to generate improved phases that can then be used to compute a more accurate electron density map. However, the refinement of macromolecular structures is often difficult for several reasons. First, the data to parameter ratio is low, creating the danger of overfitting the diffraction data. This results in a good agreement of the model to the experimental data even when it contains significant errors. Therefore, the apparent ratio of data to parameters is often increased by incorporation of chemical information, that is, bond length and bond angle restraints obtained from ideal values seen in high-resolution structures (Hendrickson, 1985). Second, the initial model often has significant errors often due to the limited quality of the experimental data or a low level of homology between the search model and the true structure in molecular replacement. Third, local (false) minima exist in the target function. The more local minima and the deeper they are will more likely lead to a failed refinement. Fourth, model bias in the electron density maps complicates the process of manual rebuilding between cycles of automated refinement.

Methods have been devised to address these difficulties. Cross-validation, in the form of the free R-value, can be used to detect overfitting (Brunger, 1992). The radius of convergence of refinement can be increased by the use of stochastic optimization methods such as molecular dynamics-based simulated annealing (Brunger, Kuriyan, and Karplus, 1987). Most recently, improved targets for refinement of incomplete, error-containing models have been obtained using the more general maximum likelihood formulation (Murshudov, Vagin, and Dodson, 1997; Pannu et al., 1998). The resulting maximum likelihood refinement targets have been successfully combined with the powerful optimization method of simulated annealing to provide a very robust and efficient refinement scheme (Adams et al., 1999). For many structures, some initial experimental phase information is available from either isomorphous heavy atom replacement or anomalous diffraction methods. These phases represent additional observations that can be incorporated in the refinement target. Tests have shown that the addition of experimental phase information greatly improves the results of refinement (Pannu et al., 1998; Adams et al., 1999). It is anticipated that the maximum likelihood refinement method will be extended further to incorporate multivariate statistical analysis, thus allowing multiple models to be refined simultaneously against the experimental data without introducing bias (Read, 2001).

The refinement methods used in macromolecular structure determination work almost exclusively in reciprocal space. However, there has been renewed interest in the use of real space refinement algorithms that can take advantage of high-quality experimental phases from anomalous diffraction experiments or NCS averaging. Tests have shown that the method can be successfully combined with the technique of simulated annealing (Chen, Blanc, and Chapman, 1999).

The parameterization of the atomic model in refinement is of great importance. When the resolution of the experimental data is limited, then it is appropriate to use chemical constraints on bond lengths and angles. This torsion angle representation is seen to decrease overfitting and improve the radius of convergence of refinement (Rice and Brunger, 1994). If data are available to a high enough resolution, additional atomic displacement parameters can be used. Macromolecular structures often show anisotropic motion, which can be resolved at a broad spectrum of levels ranging from whole domains down to individual atoms. The use of the fast Fourier transform to refine atomic anisotropic displacement parameters in the program REFMAC has greatly improved the speed with which such models can be generated and tested (Murshudov et al., 1999). The method has been shown to improve the crystallographic R-value and free R-value as well as the fit to geometric targets for data with resolution higher than 2 Å. New programs, such as *phenix.refine* (Afonine, Grosse-Kunstleve, and Adams, 2005), are being developed with the explicit goal of increasing the automation of structure refinement, which still remains a significant bottleneck in structure completion.

VALIDATION

Validation of macromolecular models and their experimental data (Vaguine, Richelle, and Wodak, 1999) is an essential part of structure determination (Kleywegt, 2000). This is important during both the structure solution process and coordinate and data deposition at the Protein Data Bank, where extensive validation criteria are also applied (Berman et al., 2000). More recently, the MolProbity structure validation suite has been developed (Lovell et al., 2003; Davis et al., 2004). This applies numerous geometric validation criteria to assess both global and local correctness of the model. This information can be readily used to correct errors in the model. See Chapters 14 and 15 for more descriptions of validation methods based on stereochemistry and atomic packing. In the future, the repeated application of validation criteria in automated structure solution will help avoid errors that may still occur as a result of subjective manual interpretation of data and models.

CHALLENGES TO AUTOMATION

Noncrystallographic Symmetry

It is not uncommon for macromolecules to crystallize with more than one copy in the asymmetric unit. This leads to relationships between atoms in real space and diffraction intensities in reciprocal space. These relationships can be exploited in the structure solution process. However, the identification of NCS is generally a manual process. A method for automatic location of proper NCS (i.e., a rotation axis) has been shown to be successful even at low resolution (Vonrhein and Schulz, 1999). A more general approach to finding NCS relationships uses skeletonization of electron density maps (Spraggon, 1999). A monomer envelope is calculated from the solvent mask generated by solvent flattening. The NCS relationships between monomer envelopes can then be determined using standard molecular replacement methods. When a model is being built automatically, it has been shown that the NCS relationships can be extracted from the local features of the electron density map (Pai, Sacchettini, and Ioerger, 2006).

These methods could be used in the future to automate the location of NCS operators and determine molecular masks. In the case of experimental phasing using heavy atoms or anomalous scatterers, it is possible to locate the NCS from the sites (Lu, 1999; Terwilliger, 2002c). The RESOLVE program automates this process such that NCS averaging can be automatically performed as part of the phase improvement procedure. NCS information can also be used in structure refinement (Kleywegt, 1996) to either decrease the number of refined parameters (NCS constraints) or increase the number of restraints (NCS restraints). The CNS program implements both these methods, and most other refinement programs implement NCS restraints. It should be noted that NCS sometimes could be very close to crystallographic symmetry. For example, a translational relationship between the molecules in the asymmetric unit can lead to very weak reflections that may be interpreted as systematic absences resulting from crystallographic centering or rotational symmetry in a molecular complex. As a result, this may lead to an assignment of a higher symmetry space group than is the case of the true crystal symmetry. These possible complications should always be considered during structure solution as they can lead to stalled R-factors in structure refinement.

Disorder

Except in the rare case of very well-ordered crystals of extremely rigid molecules, disorder of one form or another is a component of macromolecular structures. This disorder may take the form of discrete conformational substates for side chains (Wilson and Brunger, 2000), surface loops, or small changes in the orientation of entire molecules throughout the crystal. The degree to which this disorder can be identified and interpreted typically depends on the quality of the diffraction data. With low to medium resolution data, dual side-chain conformations are occasionally observed. With high-resolution data (1.5 Å or better), multiple side-chain and main-chain conformations are often seen. The challenge for automated structure solution is the identification of the disorder and its incorporation into the atomic model without the introduction of errors due to misinterpretation of the data. Disorder of whole molecules within the crystal, as a result of small differences in packing between neighboring unit cells, cannot be visualized in electron density maps. However, the effect on refinement statistics such as the R and free-R values can be significant because no single atomic model can fit the observed diffraction data well. One approach to the problem is to simultaneously refine multiple models against the data (Burling and Brunger, 1994). An alternative approach is the refinement of translation–libration–screw (TLS) parameters for whole molecules or subdomains of molecules (Winn, Isupov, and Murshudov, 2001). This introduces only a few additional parameters to be refined while still accounting for the majority of the disorder. However, it still remains a challenge to automatically identify subdomains. The use of normal modes as an alternative parameterization for the molecular flexibility has the potential for refinement of the structures at much lower resolution (Delarue and Dumas, 2004; Poon et al., 2007), while also avoiding the need to identify subdomains.

CONCLUSIONS

Over the last decade, there have been many significant advances toward automated structure determination. Programs such as PHENIX (Adams et al., 2002) and AutoSHARP (Vonrhein et al., 2006) combine large functional blocks in an automated fashion. The program CNS (Brunger et al., 1998) provides a framework in which different algorithms can be combined and tested, using a powerful scripting language. The CCP4 suite (Collaborative Computational Project 4, 1994) provides a large number of separate programs that can be easily run from a graphic user interface.

Some progress toward full automation has been made by linking together existing programs, which is typically achieved using scripting languages and/or the World Wide Web. However, long-term robust solutions, such as the PHENIX system (Adams et al., 2002), are fully integrating the latest crystallographic algorithms within a modern computer software environment. Eventually, complete automation will need structure solution to be intimately associated with data collection and processing. When automated software permits the heavy atom location and phasing steps of structure solution to be performed in a few minutes, it will enable real-time assessment of diffraction data, as it is collected at synchrotron beamlines. Map interpretation will need to be significantly faster than the present situation, with initial analysis of the electron density taking minutes rather than the hours or days required currently.

REFERENCES

Abrahams JP (1997): Bias reduction in phase refinement by modified interference functions: introducing the gamma correction. *Acta Crystallogr D* 53:371–376.

Adams PD, Pannu NS, Read RJ, Brunger AT (1999): Extending the limits of molecular replacement through combined simulated annealing and maximum likelihood refinement. *Acta Crystallogr D* 55:181–190.

Adams PD, Grosse-Kunstleve RW, Hung L-W, Ioerger TR, McCoy AJ, Moriarty NW, Read RJ, Sacchettini JC, Sauter NK, Terwilliger TC (2002): PHENIX: building new software for automated crystallographic structure determination. *Acta Crystallogr D* 58:1948–1954.

Afonine PV, Grosse-Kunstleve RW, Adams PD (2005): A robust bulk-solvent correction and anisotropic scaling procedure. *Acta Crystallogr D* 61:850–855.

Aishima J, Russel DS, Guibas LJ, Adams PD, Brunger AT (2005): Automated crystallographic ligand building using the medial axis transform of an electron-density isosurface. *Acta Crystallogr D* 61:1354–1363.

Allen FH, Kennard O, Taylor R (1983): Systematic analysis of structural data as a research technique in organic chemistry. *Acc Chem Res* 16:146–153.

Berman HM, Westbrook J, Feng Z, Gilliland G, Bhat TN, Weissig H, Shindyalov IN, Bourne PE (2000): The protein data bank. *Nucleic Acids Res* 28:235–242.

Blessing RH, Smith GD (1999): Difference structure–factor normalization for heavy-atom or anomalous-scattering substructure determinations. *J Appl Crystallogr* 32:664–670.

Bruker (2001): *Analytical X-Ray Solutions.* Madison, WI.

Brunger AT, Kuriyan J, Karplus M (1987): Crystallographic *R* factor refinement by molecular dynamics. *Science* 235:458–460.

Brunger AT (1992): The free *R* value: a novel statistical quantity for assessing the accuracy of crystal structures. *Nature* 355:472–474.

Brunger AT (1997): Patterson correlation searches and refinement. *Methods Enzymol* 276:558–580.

Brunger AT, Adams PD, Clore GM, Gros P, Grosse-Kunstleve RW, Jiang J-S, Kuszewski J, Nilges M, Pannu NS, Read RJ, Rice LM, Simonson T, Warren GL (1998): Crystallography & NMR system (CNS): a new software system for macromolecular structure determination. *Acta Crystallogr D* 54:905–921. [The design and implementation of the widely used CNS program is described. The use of a scripting language to develop, test, and implement new features is a powerful feature of the software.].

Burley SK, Almo SC, Bonanno JB, Capel M, Chance MR, Gaasterland T, Lin D, Sali A, Studier FW, Swaminathan S (1999): Structural genomics: beyond the human genome project. *Nat Genet* 23:151–157.

Burling FT, Brunger AT (1994): Thermal motion and conformational disorder in protein crystal structures: comparison of multi-conformer and time-averaging models. *Isr J Chem* 34:165–175.

Chen Z, Blanc E, Chapman MS (1999): Real-space molecular-dynamics structure refinement. *Acta Crystallogr D* 55:464–468.

Collaborative Computational Project 4 (1994): The CCP4 suite: programs for protein crystallography. *Acta Crystallogr. D* 50:760–763.

Cowtan K (1998): Modified phased translation functions and their application to molecular-fragment location. *Acta Crystallogr D* 54:750–756.

Cowtan K (1999): Error estimation and bias correction in phase-improvement calculations. *Acta Crystallogr D* 55:1555–1567.

Cowtan K (2001): Fast Fourier feature recognition. *Acta Crystallogr D* 57:1435–1444.

Cowtan K (2004): Statistical phase improvement without a solvent boundary. *Acta Crystallogr A* 60: S14.

Dauter Z, Dauter M (1999): Anomalous signal of solvent bromides used for phasing of lysozyme. *J Mol Biol* 289:93–101.

Dauter Z, Dauter M, de La Fortelle E, Bricogne G, Sheldrick GM (1999): Can anomalous signal of sulfur become a tool for solving protein crystal structures? *J Mol Biol* 289:83–92.

Davis IW, Murray LW, Richardson JS, Richardson DC (2004): MolProbity: structure validation and all-atom contact analysis for nucleic acids and their complexes. *Nucleic Acids Res* 32: W615–W619.

de La Fortelle E, Bricogne G (1997): Maximum-likelihood heavy-atom parameter refinement in the MIR and MAD methods. *Methods Enzymol* 276:472–494.

Delarue M, Dumas P (2004): On the use of low-frequency normal modes to enforce collective movements in refining macromolecular structural models. *Proc Natl Acad Sci USA* 101:6957–6962.

DePristo MA, de Bakker PIW, Blundell TL (2004): Heterogeneity and inaccuracy in protein structures solved by X-ray crystallography. *Structure* 12:831–838.

DePristo MA, de Bakker PIW, Johnson RJK, Blundell TL (2005): Crystallographic refinement by knowledge-based exploration of complex energy landscapes. *Structure* 13:1311–1319.

Emsley P, Cowtan K (2004): Coot: model-building tools for molecular graphics. *Acta Crystallogr D* 60:2126–2132.

Garman E (1999): Cool data: quantity AND quality. *Acta Crystallogr. D* 55:1641–1653.

Glykos NM, Kokkinidis M (2000): A stochastic approach to molecular replacement. *Acta Crystallogr D* 56:169–174.

Gonzalez A, Pedelacq J-D, Sola M, Gomis-Rueth FX, Coll M, Samama J-P, Benini S (1999): Two-wavelength MAD phasing: in search of the optimal choice of wavelengths. *Acta Crystallogr D* 55:1449–1458.

Gonzalez A (2007): A comparison of SAD and two-wavelength MAD phasing for radiation-damaged Se-MET crystals. *J Synchrotron Radiat* 14:43–50.

Grosse-Kunstleve RW, Brunger AT (1999): A highly automated heavy-atom search procedure for macromolecular structures. *Acta Crystallogr D* 55:1568–1577.

Grosse-Kunstleve RW, Adams PD (2003): Substructure search procedures for macromolecular structures. *Acta Crystallogr D* 59:1966–1973.

Hendrickson WA (1985): Stereochemically restrained refinement of macromolecular structures. *Methods Enzymol* 115:252–270.

Hendrickson WA (1991): Determination of macromolecular structures from anomalous diffraction of synchrotron radiation. *Science* 254:51–58.

Holton T, Ioerger TR, Christopher JA, Sacchettini JC (2000): Determining protein structure from electron-density maps using pattern matching. *Acta Crystallogr D* 56:722–734.

Jones TA (1978): A graphics model building and refinement system for macromolecules. *J Appl Crystallogr* 11:268–272.

Jones TA, Zou J-Y, Cowan SW, Kjeldgaard M (1991): Improved methods for the building of protein models in electron density maps and the location of errors in these models. *Acta Crystallogr A* 47:110–119.

Kissinger CR, Gehlhaar DK, Fogel DB (1999): Rapid automated molecular replacement by evolutionary search. *Acta Crystallogr D* 55:484–491.

Kleywegt GJ (1996): Use of non-crystallographic symmetry in protein structure refinement. *Acta Crystallogr D* 52:842–857.

Kleywegt GJ, Jones TA (1997): Template convolution to enhance or detect structural features in macromolecular electron-density maps. *Acta Crystallogr D* 53:179–185.

Kleywegt GJ, Jones TA (1998): Databases in protein crystallography. *Acta Crystallogr D* 54:1119–1131.

Kleywegt GJ (2000): Validation of protein crystal structures. *Acta Crystallogr D* 56:249–265.

Levitt DG (2001): A new software routine that automates the fitting of protein X-ray crystallographic electron-density maps. *Acta Crystallogr D* 57:1013–1019.

Lovell SC, Davis IW, Arendall WB III, de Bakker PIW, Word JM, Prisant MG, Richardson JS, Richardson DC (2003): Structure validation by Cα geometry: φ, ψ and C$_\beta$ deviation. *Proteins* 50:437–450.

Lu G (1999): FINDNCS: A program to detect non-crystallographic symmetries in protein crystals from heavy atoms sites. *J Appl Crystallogr* 32:365–368.

McCoy AJ, Storoni LC, Read RJ (2004): Simple algorithm for a maximum-likelihood SAD function. *Acta Crystallogr D* 60:1220–1228.

McCoy AJ, Grosse-Kunstleve RW, Storoni LC, Read RJ (2005): Likelihood-enhanced fast translation functions. *Acta Crystallogr D* 61:458–464.

McRee DE (1999): XtalView/Xfit—a versatile program for manipulating atomic coordinates and electron density. *J Struct Biol* 125:156–165.

Montelione GT, Anderson S (1999): Structural genomics: keystone for a Human Proteome Project. *Nat Struct Biol* 6:11–12.

Mowbray SL, Helgstrand C, Sigrell JA, Cameron AD, Jones TA (1999): Errors and reproducibility in electron-density map interpretation. *Acta Crystallogr D* 55:1309–1319.

Murshudov GN, Vagin AA, Dodson EJ (1997): Refinement of macromolecular structures by the maximum-likelihood method. *Acta Crystallogr D* 53:240–255.

Murshudov GN, Vagin AA, Lebedev A, Wilson KS, Dodson EJ (1999): Efficient anisotropic refinement of macromolecular structures using FFT. *Acta Crystallogr D* 55:247–255.

Nienaber VL, Richardson PL, Klighofer V, Bouska JJ, Giranda VL, Greer J (2000): Discovering novel ligands for macromolecules using X-ray crystallographic screening. *Nat Biotechnol* 18:1005–1108.

Oldfield T, (2000): A semi-automated map fitting procedure. In: Bourne PE, Watenpaugh K, editors. *Crystallographic Computing 7: Macromolecular Crystallographic Data (Crystallographic Computing)*. Oxford University Press.

Otwinowski Z (1991): Maximum likelihood refinement of heavy atom parameters. In: Wolf W, Evans PR, Leslie AGW, editors. *Isomorphous Replacement and Anomalous Scattering, Proceedings of Daresbury Study Weekend*. Warrington:SERC Daresbury Laboratory, pp 80–85.

Pai R, Sacchettini J, Ioerger T (2006): Identifying non-crystallographic symmetry in protein electron-density maps: a feature-based approach. *Acta Crystallogr D* 62:1012–1021.

Pannu NS, Murshudov GM, Dodson EJ, Read RJ (1998): Incorporation of prior phase information strengthens maximum-likelihood structure refinement. *Acta Crystallogr D* 54:1285–1294.

Parsons S (2003): Introduction to twinning. *Acta Crystallogr. D* 59:1995–2003.

Perrakis A, Morris R, Lamzin VS (1999): Automated protein model building combined with iterative structure refinement. *Nat Struct Biol* 6:458–463. [An automated method for building and refining a protein model is described. An iterative procedure is used that describes the electron density map as a set of unconnected atoms from which protein-like patterns are extracted. This method is currently used by crystallographers to automate model building when high-resolution data are available (approximately 2.3 Å or better).].

Podjarny AD, (1976): Thesis, Weizmann Institute of Science, Rehovot.

Poon BK, Chen X, Lu M, Vyas NK, Quiocho FA, Wang Q, Ma J (2007): Normal mode refinement of anisotropic thermal parameters for a supramolecular complex at 3.42-Å crystallographic resolution. *Proc Natl Acad Sci USA* 104:7869–7874.

Read RJ (1999): Detecting outliers in non-redundant diffraction data. *Acta Crystallogr D* 55:1759–1764.

Read RJ, (2001): Pushing the boundaries of molecular replacement with maximum likelihood. *Acta Crystallogr D* 57:1373–1382. [This paper describes the basis for the application of maximum likelihood scoring methods to the problem of molecular replacement. Subsequent implementation of these methods in the program Phaser has made a major contribution to the success of difficult molecular replacement problems.].

Rice LM, Brunger AT (1994): Torsion angle dynamics: reduced variable conformational sampling enhances crystallographic structure refinement. *Proteins* 19:277–290.

Rice LM, Earnest TN, Brunger AT (2000): Single wavelength anomalous diffraction phasing revisited: a general phasing method? *Acta Crystallogr D* 56:1413–1420.

Rossmann MG, Blow DM (1962): The detection of sub-units within the crystallographic asymmetric unit. *Acta Crystallogr* 15:24–31.

Sheldrick GM, Gould RO (1995): Structure solution by iterative peaklist optimization and tangent expansion in space group P1. *Acta Crystallogr B* 51:423–431.

Sheriff S, Klei HE, Davis ME (1999): Implementation of a six-dimensional search using the AMoRe translation function for difficult molecular-replacement problems. *J Appl Crystallogr* 32:98–101.

Spraggon G (1999): Envelope skeletonization as a means to determine monomer masks and non-crystallographic symmetry relationships: application in the solution of the structure of fibrinogen fragment D. *Acta Crystallogr D* 55:458–463.

Storoni LC, McCoy AJ, Read RJ (2004): Likelihood-enhanced fast rotation functions. *Acta Crystallogr D* 60:432–438.

Terwilliger TC (1994): MAD phasing: Bayesian estimates of F_A. *Acta Crystallogr D* 50:11–16.

Terwilliger TC (1999): σ_R^2, a reciprocal-space measure of the quality of macromolecular electron-density maps *Acta Crystallogr D* 55:1174–1178.

Terwilliger TC, Berendzen J (1999a): Automated MAD and MIR structure solution. *Acta Crystallogr D* 55:849–861.

Terwilliger TC, Berendzen J (1999b): Discrimination of solvent from protein regions in native Fouriers as a means of evaluating heavy-atom solutions in the MIR and MAD methods. *Acta Crystallogr D* 55:501–505.

Terwilliger TC, Berendzen J (1999c): Evaluation of macromolecular electron-density map quality using the correlation of local r.m.s. density. *Acta Crystallogr D* 55:1872–1877.

Terwilliger TC (2000): Maximum-likelihood density modification. *Acta Crystallogr D* 56:965–972. [A procedure is described for reciprocal space maximization of a likelihood function based on experimental phases and characteristics of the electron density map. This powerful approach to phase improvement is able to generate minimally biased phase estimates and will be a valuable tool in the future for all aspects of phase improvement and phase combination.].

Terwilliger TC (2002a): Statistical density modification with non-crystallographic symmetry. *Acta Crystallogr D* 58:2082–2086.

Terwilliger TC (2002b): Automated structure solution, density modification, and model building. *Acta Crystallogr D* 58:1937–1940.

Terwilliger TC (2002c): Rapid automatic NCS identification using heavy-atom substructures. *Acta Crystallogr D* 58:2213–2215.

Terwilliger TC (2003a): Statistical density modification using local pattern matching. *Acta Crystallogr D* 59:1688–1701.

Terwilliger TC (2003b): Automated main-chain model building by template-matching and iterative fragment extension. *Acta Crystallogr D* 59:38–44. [This paper describes a method for automated model building using secondary structure elements and small fragments. This method is readily applied to medium to low resolution data (3.2 Å and better).].

Terwilliger TC (2003c): Automated side-chain model building and sequence assignment by template matching. *Acta Crystallogr D* 59:45–49.

Terwilliger TC (2003d): Improving macromolecular atomic models at moderate resolution by automated iterative model building, statistical density modification and refinement. *Acta Crystallogr D* 59:1174–1182.

Terwilliger TC, Klei H, Adams PD, Moriarty NW, Cohn JD (2006): Automated ligand fitting by core-fragment fitting and extension into density. *Acta Crystallogr D* 62:915–922.

Terwilliger TC, Grosse-Kunstleve RW, Afonine PV, Adams PD, Moriarty NW, Zwart P, Read RJ, Turk D, Hung LW (2007): Interpretation of ensembles created by multiple iterative rebuilding of macromolecular models. *Acta Crystallogr D* 63:597–610. [This paper describes the automated building of multiple molecular models against synthetic and real diffraction data sets to better understand what information can be extracted from the multiple models. The results indicate that the variation between the models is a reflection of the uncertainty in the data, rather than different physical conformations of the molecules in the crystal.].

Turk D (2000): MAIN 96: an interactive software for density modifications, model building, structure refinement and analysis. In: Bourne PE, Watenpaugh K, editors. *Crystallographic Computing 7: Macromolecular Crystallographic Data (Crystallographic Computing)*. Oxford University Press.

Vaguine AA, Richelle J, Wodak SJ (1999): SFCHECK: a unified set of procedures for evaluating the quality of macromolecular structure–factor data and their agreement with the atomic model. *Acta Crystallogr D* 55:191–205.

Vonrhein C, Schulz GE (1999): Locating proper non-crystallographic symmetry in low-resolution electron-density maps with the program GETAX. *Acta Crystallogr D* 55:225–229.

Vonrhein C, Blanc E, Roversi P, Bricogne G (2006): Automated structure solution with autoSHARP. *Methods Mol Biol* 364:215–230.

Walsh MA, Evans G, Sanishvili R, Dementieva I, Joachimiak A (1999a): MAD data collection—current trends. *Acta Crystallogr D* 55:1726–1732.

Walsh MA, Dementieva I, Evans G, Sanishvili R, Joachimiak A (1999b): Taking MAD to the extreme: ultrafast protein structure determination. *Acta Crystallogr D* 55:1168–1173.

Weeks CM, Miller R (1999): The design and implementation of SnB v2.0. *J Appl Crystallogr* 32:120–124.

Weeks CM, Adams PD, Berendzen J, Brunger AT, Dodson EJ, Grosse-Kunstleve RW, Schneider TR, Sheldrick GM, Terwilliger TC, Turkenburg MG, Uson I (2003): Automatic solution of heavy-atom substructures. *Methods Enzymol* 374:37–83.

Wilson MA, Brunger AT (2000): The 1.0 Å crystal structure of Ca^{2+} bound calmodulin: an analysis of disorder and implications for functionally relevant plasticity. *J Mol Biol* 301:1237–1256.

Winn MD, Isupov MN, Murshudov GN (2001): Use of TLS parameters to model anisotropic displacements in macromolecular refinement. *Acta Crystallogr D* 57:122–133.

Yeates TO (1997): Detecting and overcoming crystal twinning. *Methods Enzymol* 276:344–358.

Zwart PH, Langer GG, Lamzin VS (2004): Modelling bound ligands in protein crystal structures. *Acta Crystallogr D* 60:2230–2239.

5

MACROMOLECULAR STRUCTURE DETERMINATION BY NMR SPECTROSCOPY

John L. Markley, Arash Bahrami, Hamid R. Eghbalnia,
Francis C. Peterson, Robert C. Tyler, Eldon L. Ulrich,
William M. Westler, and Brian F. Volkman

Bioinformatics plays an important role in biomolecular structure determination by NMR spectroscopy. Increasingly, structural biology is being driven by information gained from gene sequencing and hypotheses derived from bioinformatics approaches. This is particularly so for structural proteomics efforts. A major rationale for structural proteomics is that a more complete mapping of peptide sequence space onto conformational space will lead to efficiencies in determining structure–function relationships (Burley, 2000; Heinemann, 2000; Terwilliger, 2000; Yokoyama, 2000). Longer range scientific goals are the prediction of structure and function from sequence and simulations of the functions of a living cell. Structural proteomics is part of a wider functional genomics effort, which promises to assign functions to proteins within complex biological pathways and to enlarge the understanding and appreciation of complex biological phenomena (Thornton et al., 2000). Because of its potential to greatly broaden the targets for new pharmaceuticals, structural proteomics is expected to join combinatorial chemistry and screening as an integral approach to modern drug discovery (Dry, McCarthy, and Harris, 2000). It is clear that much larger databases of structures, dynamic properties of biomolecules, biochemical mechanisms, and biological functions are needed to approach these goals. Because of its ability to provide atomic-resolution structural and chemical information about proteins, NMR spectroscopy is positioned to play an important role in this endeavor. Already, NMR spectroscopy contributes about 15% of the protein structures deposited at the Protein Data Bank (PDB). In addition, NMR spectroscopy is used routinely in high-throughput screens to determine protein–ligand interactions (Mercier et al., 2006; Hajduk and Greer, 2007). NMR

Structural Bioinformatics, Second Edition   Edited by Jenny Gu and Philip E. Bourne
Copyright © 2009 John Wiley & Sons, Inc.

also is a key tool in mechanistic enzymology and in studies of protein folding and stability. As discussed here, advances in key technologies promise rapid increases in the efficiency and scope of NMR applications to structural and functional genomics. Bioinformatics figures prominently in these advances.

This Chapter discusses the current status and prospects of macromolecular structure determination by solution state NMR spectroscopy. Although the focus is on proteins, the approaches can be generalized to other classes of biological macromolecules. Solid-state NMR spectroscopy shows great promise for structural studies of proteins that may not be amenable to investigation in solution, such as membrane proteins, and for functional investigations of protein in the solid state. Solid-state NMR strategies for biomolecular sample preparation, data collection, and analysis are developing rapidly and are expected to assume prominence in the next few years. Further discussion of this highly specialized field is beyond the scope of this chapter, and interested readers are directed to recent reviews (McDermott, 2004; Baldus, 2006; Hong, 2006).

This chapter starts with an overview of the physical basis for NMR spectroscopy and a discussion of NMR experiments and instrumentation. Then, the basic approaches used in determining structures and dynamic properties of proteins are presented. Sample requirements and sample preparation, including labeling methods, are discussed. Next a protocol for NMR structure determination covering all steps from protein characterization to data deposition is detailed. This is followed by a section on evolving bioinformatics technology for achieving a more automated, probabilistic approach to protein NMR structure determination. The final section summarizes various bioinformatics resources that are available to support biomolecular NMR spectroscopy.

INTRODUCTION TO PROTEIN STRUCTURE DETERMINATION BY NMR

Comparison of NMR Spectroscopy and X-ray Crystallography

Current methods for high-throughput protein structure determination are single-crystal diffraction and solution-state NMR spectroscopy. The two methods have complementary features, as X-ray crystallography represents a mature and rapid approach for proteins that form suitable crystals while NMR has advantages for structural studies of small proteins that are partially disordered, exist in multiple stable conformations in solution, show weak interactions with important cofactors or ligands, or do not crystallize readily. NMR spectroscopy is an incremental method that can rapidly provide useful information concerning overall protein folding, local dynamics, existence of multiple folded conformations, or protein–ligand or protein–protein interactions in advance of a three-dimensional structure. This information can be useful in designing strategies for 3D structure determinations by either NMR or X-ray crystallography. Several ongoing structural proteomics pilot projects are employing a combination of X-ray crystallography and NMR spectroscopy.

Physical Basis for Biomolecular NMR Spectroscopy

As summarized schematically in Figure 5.1, NMR spectroscopy investigates transitions between spin states of magnetically active nuclei in a magnetic field. The most important magnetically active nuclei for proteins are the proton (^1H), carbon-13 (^{13}C), nitrogen-15

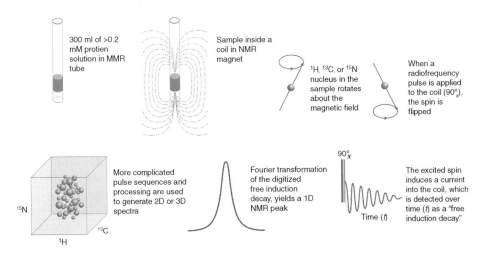

Figure 5.1. Schematic illustrating several basic principles of biomolecular NMR spectroscopy. (Clockwise from upper left) The NMR sample solution is placed in an NMR tube, which is then inserted into the static magnetic field of the NMR spectrometer. NMR-active nuclei (e.g., ^{1}H, ^{13}C, and ^{15}N) within the sample precess (like a spinning gyroscope) with their net magnetization aligned with the direction of the magnetic field; the precession frequency is governed by the type of nucleus but is modulated by the chemical environment of the atom in the macromolecule. Application of a radiofrequency pulse excites the nucleus by flipping its spin to the excited energy level opposing the external magnetic field. A "90° pulse" flips the rotating magnetization into the plane perpendicular to the axis of the magnetic field (the "x–y plane"), and the rotating magnetization induces an oscillating current (the "free induction decay") into the coil that detects the signal. Fourier transformation of the time-domain signal (free induction decay) converts it to a frequency-domain signal (the "NMR peak"). More complicated pulse sequences are used in experiments that create two- or three-dimensional NMR spectra; the type of 3D spectrum illustrated is the result of a "triple-resonance" experiment in which the three orthogonal frequency axes display the peak positions of combinations of ^{1}H, ^{13}C, and ^{15}N nuclei that are related by spectral connectivities created by the pulse sequence.

(^{15}N), and phosphorus-31 (^{31}P). Each of these stable (nonradioactive) isotopes has a nuclear spin of one half and, as a consequence, two spin states: one at lower energy, with the magnetic spin paired with the external field, and one at higher energy, with the magnetic spin opposing the external field. The magnetic moment of each nucleus precesses about the external magnetic field and is influenced by other fields. Influences on a given spin are generated by neighboring spins in the molecule, giving rise to intrinsic NMR parameters, or by radio-frequency pulses and/or pulsed field gradients as programmed by the NMR spectroscopist. In an NMR experiment, a time sequence of radiofrequency pulses and pulsed field gradients are applied to the spins present in the molecule studied. The excited spins are allowed to interact with one another and with the external magnetic field. Then, the state of the system is read out by detection of the current induced by the nuclear spins of the sample in the receiver coil of the NMR spectrometer. The physical basis for NMR is well understood, and the spectroscopic consequences of a given pulse sequence applied to a particular molecule can be simulated to a good level of precision (Cavanagh et al., 2006; Ernst, Bodenhausen, and Wokaun, 1987).

NMR Experiments

The NMR spectrometer can be programmed to operate on sets of nuclei of single or multiple atom types and to accommodate particular spectral features (chemical shift range, spin–spin coupling range, etc.). NMR spectroscopy is unique in that the Hamiltonian of the system under study can be manipulated easily by the application of radiofrequency pulses and/or pulsed field gradients. The versatility in creating sequences of complex "spin gymnastics" makes possible a myriad of NMR experiments by which particular parameters can be investigated. A huge variety of NMR experiments are at one's disposal when collecting data for a structure determination or functional investigation. The NMR field is still young and dynamic, and these experiments continue to evolve; thus the optimal set of experiments for a structure determination or structure–function study is a matter of exploration and individual taste. The approaches described here are ones that we have found to be effective.

Structure from NMR

The classical approach to structure determination is to first use multidimensional and multinuclear NMR methods to determine "sequence-specific assignments," that is, resolve signals from the ^{1}H, ^{15}N, and ^{13}C nuclei of a protein and to assign them to specific nuclei in the covalent structure of the molecule. The assigned chemical shifts (cs) themselves provide reliable information about the secondary structure of the protein (Wishart, Sykes, and Richards, 1992; Wishart and Sykes, 1994a; Wishart and Case, 2001; Eghbalnia et al., 2005c), and the oxidation states of cysteine residues (Sharma and Rajarathnam, 2000), and can be used to test or validate structural models. Additional structural restraints are obtained from an interpretation of data from one or more different classes of NMR experiments: (1) nuclear overhauser effect (NOE) spectra, which provide ^{1}H $-$ ^{1}H distance constraints; (2) empirical torsion angle constraints based on chemical shifts (Cornilescu, Delaglio, and Bax, 1999); and (3) residual dipolar couplings from partially oriented proteins (Bax, Kontaxis, and Tjandra, 2001; Lipsitz and Tjandra, 2004; Prestegard, Bougault, and Kishore, 2004; Bax and Grishaev, 2005), which provide spatial constraints (with respect to the orientation axes) for pairs of coupled nuclei. Additional hydrogen bond constraints are determined from hydrogen exchange experiments, chemical shifts, and/or *trans*-hydrogen bond couplings (Cordier and Grzesiek, 1999; Cordier et al., 1999).

Molecular Size Limitations

Protein NMR methods have advanced to the point where small to medium sized protein domain structures can be determined in a routine manner. Key milestones in the development of this technique include NOE-based resonance assignment and structure refinement (Wüthrich, 1986), efficient stable isotope labeling methods (Markley and Kainosho, 1993), detection of multinuclear correlations (Westler, Ortiz-Polo, and Markley, 1984; Ortiz-Polo et al., 1986; Oh et al., 1988; Westler et al., 1988a, 1988b; Oh, Westler, and Markley, 1989) coupled with indirect detection and multinuclear NMR spectroscopy (Kay, Marion, and Bax, 1989; Kay et al., 1990). More recent advances have included the use of pulsed field gradients for coherence selection and line-narrowing by transverse relaxation-optimized spectroscopy (TROSY) (Czisch and Boelens, 1998; Pervushin et al., 1998; Salzmann et al., 1998; Zhu, Kong, and Sze, 1999; Zhu et al., 1999; Fernandez and Wider, 2003; Hu,

Eletsky, and Pervushin 2003). The improvements in commercially available NMR hardware and remarkable advances in desktop computing capabilities have also been important.

Protein domains of up to roughly 25 kDa generally are amenable to high-throughput NMR structure determination, provided that soluble, stable, $U - {}^{13}C$- and $U - {}^{15}N$-labeled samples can be obtained. This isotope enrichment enables the use of sensitive multinuclear correlation experiments for the task of matching all chemical shifts in the protein with specific amino acid residues and atoms. Once all ${}^1H$, ${}^{15}N$, and ${}^{13}C$ resonances have been assigned, the correlations observed in nuclear Overhauser enhancement experiments can be interpreted in terms of short (< 5 Å) inter-proton distances. NMR structures are obtained from constrained molecular dynamics calculations, with these NOE-derived ${}^1H$–${}^1H$ distances as the primary experimental constraints. As a consequence of chemical shift degeneracy, many NOE correlations may have multiple assignment possibilities, and the results of preliminary structure calculations are used to eliminate unlikely candidates on the basis of inter-proton distances. Refinement continues in an iterative manner until a self-consistent set of experimental constraints produces an ensemble of structures that also satisfies standard covalent geometry and steric overlap considerations. Recent advances in software have automated many of the steps in structure determination (Lopez-Mendez and Güntert, 2006). Structures are validated and reported in the Protein Data Bank in a manner analogous to those obtained from X-ray crystallographic methods (see Chapters 14 and 15).

Stable Isotope Labeling

Structures of smaller proteins can be determined with samples at natural abundance. If the protein is reasonably soluble (≥ 1 mM), and particularly if a high-sensitivity cryogenic probe is available, it is possible to obtain ${}^1H$–${}^{13}C$ and ${}^1H$–${}^{15}N$ correlations at natural abundance. This additional information can be useful for signal assignments and to provide additional chemical shifts for secondary structure analysis. These structures are generally of lower quality than those determined from labeled proteins because of the greater difficulty in obtaining complete assignments. High-throughput strategies for protein structure determination generally are designed for proteins 'double-labeled' with ${}^{13}C$ and ${}^{15}N$ ($M_r > {\sim}25,000$ Da) or "triple-labeled" with ${}^{13}C$, ${}^{15}N$, and ${}^2H$ ($M_r < {\sim}25,000$ Da). NMR spectra of larger molecules are characterized by increasingly rapid decay of the signal to be detected leading to increased line widths and lowered sensitivity (signal-to-noise ratio) and by greater overlap and signal degeneracy. Even with triple-labeled proteins, these problems make it increasingly difficult to solve NMR structures of proteins larger than 40 kDa with current methodology. Using higher magnetic fields and TROSY methodology suggests that these limitations can be overcome to some extent. Kay and coworkers have developed a labeling strategy for large molecules by ${}^{15}N$, ${}^{13}C$, and ${}^2H$ labeling while the methyl groups of isoleucine, leucine, and valine are protonated (Gardner and Kay, 1998; Tugarinov, Kanelis, and Kay, 2006).

Selective labeling approaches make it possible to obtain information about selected regions of proteins as large as 150 kDa. These methods include labeling by single residue type (Markley and Kainosho, 1993) and segmental labeling by peptide semisynthesis, including intein technology (Yamazaki et al., 1998; Xu et al., 1999). In the designer labeling scheme, known as stereo-array isotope labeling (SAIL) (Kainosho et al., 2006), a complete stereospecific and regiospecific pattern of stable isotopes is incorporated into the protein using cell-free protein synthesis. SAIL, by significantly reducing the proton density and thus reducing the resonance overlap and line width, has great potential in extending the molecular weight range for structure determination by NMR.

Data Collection and Instrumentation

Even with protein sample concentrations in the millimolar range (>10 mg/mL), coaddition of 8 or 16 transients may be required at each time point in some two- or three-dimensional experiments to achieve sufficient signal-to-noise (s/n). Cryogenic probe technology can reduce the data acquisition period by up to a factor of 10. This results from the three- to fourfold enhancement in sensitivity achieved by cooling the probe and preamplifier circuitry to very low temperatures (Styles et al., 1984). Because of the square-root relationship between time and sensitivity in signal averaging, an inherent sensitivity improvement of threefold is equivalent to a ninefold increase in signal averaging. In practical terms, the s/n of an experiment acquired with one or two transients at each time increment on a cryogenic probe should be equivalent (if not superior) to one acquired with 16 transients using a conventional (noncooled) probe. For this example, the cryogenic probe results in an eightfold reduction in experiment time with no negative consequences. This is especially important for NOE experiments. There are some caveats associated with the use of cryogenic probes. First, the three- to fourfold sensitivity increases generally are realized only for simple 1D experiments in nonionic solutions. The RF field inhomogeneity of the ^{13}C and ^{15}N coils decreases the efficiency of the probes in multipulse triple-resonance experiments. Practically, for a protein solution at low ionic strength, the sensitivity increase is closer to twofold. However, this increase yields either a fourfold saving in time or, more importantly for systems of poor solubility, the ability to lower the sample concentration by a factor of two. The most severe limitation of the cryogenic probe is that they lose their sensitivity advantage in solutions having high ionic strength. Recent studies have shown that by changing the shape of the sample tube to limit the penetration of the electric field into the salty solution, the sensitivity can be recovered even in solutions of high ionic strength (de Swiet, 2005).

Approaches for decreasing the overall experimental time by collecting sparse data sets have been developed, such as HIFI (Eghbalnia et al., 2005a), GFT (Kim and Szyperski, 2003), and reconstruction of 3D from 2D planes (Kupce and Freeman, 2003). These methods decrease the collection time simply by collecting projected 2D planes out of a 3D spectrum and can reduce the data collection time by nearly an order of magnitude. The ultimate limit for rapid data collection is the longitudinal relaxation rate (R_1) of the nucleus to be detected. Data collection schemes, such as "SOFAST," that use optimal recycle rates have produced 2D spectra in less than 1 min (Schanda and Brutscher, 2005), and these strategies have been developed into a suite of pulse sequences for triple-resonance data collection (Lescop, Schanda, and Brutscher, 2007).

Protein Dynamics from NMR

NMR signals report on the chemical properties of nuclei including their relative motions with respect to the magnet and molecular frames. Consequently, NMR spectroscopy is uniquely suited for investigations of the kinetics and thermodynamics of molecular motions, segmental motions, dynamic conformational equilibria, and ligand binding equilibria— as reviewed in greater detail elsewhere (Palmer III, Kroenke, and Loria, 2001; Wand, 2001; Kern and Zuiderweg, 2003; Mittermaier and Kay, 2006). Whereas regions of proteins that are statically or dynamically disordered may fail to produce resolvable electron density, NMR signals from such regions generally are observed and can be investigated (Dyson and Wright, 2005).

Dynamic processes observable by NMR spectroscopy cover a vast timescale: from picoseconds to seconds, days, and even months. Nuclear spin relaxation rates are sensitive to protein motions ranging from picosecond to milliseconds. Overall molecular tumbling on the nanosecond timescale is the dominant factor in relaxation of most nuclei in folded proteins. Internal protein motions on the picosecond timescale cause a slight averaging of the dipolar interactions that in turn reduces the overall spectral density. These variations can be measured for the backbone of each residue in a protein by determining the time constants for ^{15}N longitudinal (T_1) and transverse (T_2) relaxation rates as well as the $^1H(^{15}N)$ heteronuclear NOE (Palmer, 1997). Analogous measurements of ^2H or ^{13}C relaxation parameters can yield information on side chain dynamics (Igumenova, Frederick, and Wand, 2006).

The motional effect on relaxation rates can be expressed as an order parameter (S^2) in a model-free formalism as originally expounded by Lipari and Szabo (1982a, 1982b). Order parameters range from 0 to 1, with smaller values reflecting increased internal motion. Often the simple single exponential model is insufficient to fit the relaxation data, and additional exponentials describing other rapid motions must be added to extend the basic model-free approach (Clore et al., 1990; Mandel, Akke, and Palmer III, 1995). However, it should be noted that as parameters are added to the analysis, the available experimental data points may not suffice to constrain the system to a unique neighborhood of solutions (Andrec, Montelione, and Levy, 2000). Outliers in the transverse relaxation rate often arise from motions in the microsecond-to-millisecond time regime that report on chemical or conformational exchange phenomena (Volkman et al., 2001). Relaxation dispersion methods detect these slower motions directly (Akke and Palmer, 1996), and exchange rates have been correlated with protein binding and folding (Sugase, Dyson, and Wright, 2007), rate-limiting steps in enzyme catalysis (Eisenmesser et al., 2005), and even conformational dynamics of the 20S proteasome— a 670 kDa complex (Sprangers and Kay, 2007). For conformational dynamics involving millisecond-to-second interconversion between two or more states giving rise to distinct NMR signals, traditional 2D exchange spectroscopy provides another method for quantification of exchange rates, and ultrafast 2D NMR pulse schemes enable real-time monitoring of kinetic processes on the 1–10 s timescale (Schanda, Forge, and Brutscher, 2007).

Dynamics on longer timescales can be interrogated through hydrogen exchange kinetics (Huyghues-Despointes et al., 2001). In this experiment, a protein that has the natural abundance of ^1H atoms on the amide groups is dissolved in a solution containing ∼100% ^2H_2O and then quickly transferred to the spectrometer. By observing the loss of amide ^1H signals due to exchange with the ^2H atoms, the kinetics can be followed. Protons with larger protection factors are interpreted to be in more highly structured regions with more stable hydrogen bonds. Highly labile protons may exchange completely in the interval between the addition of ^2H_2O and acquisition of the first NMR spectrum. A complementary method (CLEANEX) for rapidly exchanging amides uses solvent saturation transfer to quantify exchange rates (Hwang, van Zijl, and Mori, 1998). In combination, saturation transfer and H/D exchange measurements can monitor local backbone fluctuations from milliseconds to months. This broad repertoire of NMR methods enables quantitative, site-specific measurement of fast motions and slower conformational exchange process associated with catalysis, allostery, folding, and signaling—central features of most biological processes.

PREPARATION OF PROTEIN SAMPLES FOR NMR

Approaches to Protein Production and Labeling

A variety of protein expression systems have been investigated in support of structural proteomics (Yokoyama, 2003). By far the most widely used and most successful platform is heterologous overexpression of proteins in *Escherichia coli* cells. It is a general experience, however, that a certain percentage of proteins does not express abundantly in soluble form in bacteria. Insolubility arises either from an intrinsic property of the protein (for example, aggregation due to an exposed hydrophobic surface) or because inappropriate folding mechanisms in the expression host permit aggregation of folding intermediates (Henrich, Lubitz, and Plapp, 1982; Goff and Goldberg, 1987; Chrunyk et al., 1993). Expression can be particularly difficult for eukaryotic proteins that consist of multiple domains, that require cofactors or protein partners for proper folding, or that normally undergo extensive post-translational modification. In industry, the next most widely used approach is protein production from insect cells by a baculovirus system. However, this approach is not widely used for NMR spectroscopy, because labeled media for insect cells are highly expensive. Recently, a more economical medium for Sf9 insect cells has been described, which promises to improve the practicality of this approach (Walton et al., 2006). Yeast cells offer another alternative, and approaches for large-scale production of labeled proteins from the yeast *Pichia pastoris* have been described (Wood and Komives, 1999; Pickford and O'Leary, 2004).

Cell-free biological systems are capable of synthesizing proteins with high speed and accuracy approaching *in vivo* rates (Kurland, 1982; Pavlov and Ehrenberg, 1996). Commercial implementations of this technology incorporate recent advances as described below (Spirin et al., 1988; Baranov et al., 1989; Endo et al., 1992; Sawasaki et al., 2002a, 2002b). *E. coli* cell-free systems give high yields (Kigawa et al., 1999), and active extracts are easily prepared (Torizawa et al., 2004; Matsuda et al., 2007). Membrane proteins can be produced in soluble form by incorporating detergents into cell-free systems (Klammt et al., 2004; Klammt et al., 2006; Liguori et al., 2007). Two groups have made extensive use of cell-free (*in vitro*) protein production for NMR spectroscopy: the RIKEN Structural Genomics Group (Kigawa, Muto, and Yokoyama, 1995) and the Center for Eukaryotic Structural Genomics (Vinarov, Loushin Newman, and Markley, 2006).

In our experience, a combination of protein production in *E. coli* cells and in a wheat germ cell-free system provides an economical approach for producing eukaryotic proteins for NMR structural analysis. The two methods are complementary in the sense that many proteins that fail to be produced in soluble, folded state from *E. coli* can be made from the wheat germ cell-free system, whereas, when production is successful from *E. coli*, the yields can be very high (Tyler et al., 2005a).

Cell-Based Methods

As noted above, protein production in *E. coli* has an established record of being the most successful approach for providing protein targets for structural proteomics. Suitable expression vectors are readily available, and the method is more economical than production in eukaryotic cells. Furthermore, promising proteins can be subsequently labeled metabolically with heavy atom-labeled amino acids for X-ray or stable isotopes for NMR (Edwards et al., 2000). ^{13}C–, ^{15}N–, and ^{2}H-metabolic labeling for NMR work was pioneered with

E. coli expression systems (Markley and Kainosho, 1993). A recent important advance has been the development of self-induction media containing glucose, glycerol, and lactose for protein production from *E. coli*, which support very high cell densities while removing the need to follow cell growth to initiate induction and associated centrifugation steps. This approach has been fine-tuned recently to improve yields (Blommel et al., 2007). The approach can be used to produce stable isotope labeled samples for NMR spectroscopy (Tyler et al., 2005b), although the costs for isotopes are higher, because ^{13}C-glycerol must be added in addition to ^{13}C-glucose and ^{15}N-ammonia. Additional cost saving can be achieved with *E. coli* production systems through the use of disposable polyethylene terephthalate (PET) bottles as growth vessels, which do not require cleaning or sterilization (Sanville et al., 2003).

Cell-Free Methods

We have published detailed descriptions of the protocols, which have been developed for the production of labeled proteins for NMR spectroscopy from a wheat germ cell-free system (Vinarov et al., 2004; Vinarov and Markley, 2005; Vinarov et al., 2006). Cell-free protein production methods have several advantages: (1) volumes can be kept manageable—in favorable cases, sufficient protein for a structural investigation can be produced from a reaction volume of under 6–12 mL; (2) proteins can be produced which would be toxic to cells; (3) the ratio of desired protein to unwanted protein is low, simplifying purification; (4) in a cell-free system, one has the potential for introducing metal ions, cofactors, or other agents that promote correct folding; (5) enzymes and substrates can be added to promote post-translational modifications; and (6) labeled amino acids can be introduced without scrambling of the label— this enables clean selective labeling and newer labeling strategies, such as stereo-array isotope labeling (Kainosho et al., 2006).

The production of labeled proteins from cell-free systems requires that labeled amino acids be supplied in the reaction mixture. The optimal concentration of each amino acid in the cell-free reaction mixture is about 1 mM. At currently achievable protein yields (0.4–1.1 mg protein/mL of reaction mixture), a reaction volume of 6–12 mL is needed to produce enough protein for an NMR sample. $[U-^{15}N]$- and $[U-^{15}N, U-^{13}C]$-amino acids are commercially available at prices that make cell-free protein production of labeled proteins competitive with production from *E. coli* cells when relative labor costs are taken into account. Figure 5.2 illustrates the automated cell-free production and purification of three protein targets incorporating ^{15}N-amino acids for screening for suitability as structural targets.

Sample Concentration, Volume, and Stability

The higher the protein concentration the faster NMR data can be collected, provided that the protein does not aggregate. Practical lower limit protein concentrations are about 200 μM with ordinary probes and about 60 μM with cryogenic probes. The decreased concentration requirement of the cryogenic probe may make it possible to collect data from proteins that aggregate at higher concentrations. With ordinary NMR tubes, a sample volume of 300–500 μL is required, depending on the length of the detection coil in the probe. Susceptibility-matched cells (Shigemi Inc.) enable the use of smaller volumes: 200 μL in a 5 mm OD tube or 60 μL in a 3 mm OD tube. In cases where the protein is sufficiently soluble (≥ 1.5 mM), it is possible to use a 1 mm microcoil NMR probe to greatly reduce the amount of protein required (Peti et al., 2005; Aramini et al., 2007).

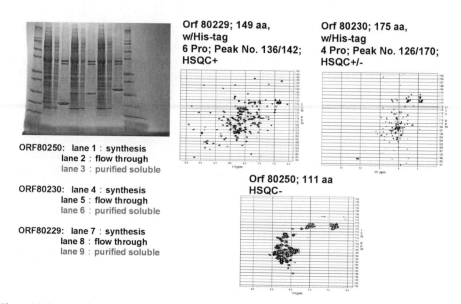

Figure 5.2. Examples of ^{15}N-labeled His-tagged proteins (two *Arabidopsis* and one human target) produced by a wheat germ cell-free robotic system (CellFree Sciences DT-II). The robot starts with a DNA plasmid and carries out automated transcription and translation and metal ion affinity protein purification. The SDS gel scans show molecular weight markers on either side, and for each target, the total reaction mixture, column flow-through, and purified soluble fraction seem incomplete. The two bands, at higher molecular weight, correspond to proteins present in the wheat germ extract that bind to the affinity column. They do not show up in the NMR spectra because they are not labeled with ^{15}N. The only manual steps required in preparing the NMR samples were buffer exchange and concentration of the purified protein. The three targets illustrate how ^{1}H–^{15}N HSQC NMR is used to screen targets for their suitability for structure determination. HSQC + indicates a spectrum with the expected number of well-resolved peaks. HSQC +/− indicates a protein that is partially unstructured and may be salvageable as a structural target by either solvent optimization or by trimming the protein expressed. HSQC indicates a natively unfolded or aggregated protein unsuitable for NMR structure determination (Loushin Newman, Markley, Song, and Vinarov, unpublished results).

The sample must remain stable over the data collection period. If the stability of the sample is a limiting factor, it is possible to use separate samples for different NMR experiments. Ideally, all data sets are collected with the same sample, because slight differences in solution conditions can lead to chemical shift differences that make it difficult to compare results from different experiments. One of the advantages of cryogenic probes and higher fields is that the overall data collection time for each experiment is shortened. This makes it possible to investigate systems that are less stable over time.

PROTOCOL FOR PROTEIN STRUCTURE DETERMINATION BY NMR

As noted above, NMR methodology is evolving rapidly and competing approaches are being explored in various structural proteomics centers. The following sections outline a protocol for semiautomated protein structure determination by NMR spectroscopy. Proteins ($\leq$ 25 kDa) labeled with ^{15}N emerging from the cell-based or cell-free protein production pipeline are subjected to 1D and 2D NMR screening to determine their folding and aggregation states under a variety of solution conditions (Figure 5.2). Ideally, an NMR spectrometer equipped with a sample preparation robot, sample changer, and cryogenic probe is used for these screens (a 600 MHz system is adequate for this purpose). NMR screens are used to determine aggregation state (through measurement of transverse diffusion by gradient NMR), assess folding state (from analysis of the amide region of 1D ^1H NMR spectra and the pattern of ^1H–^{15}N correlation spectra), monitor ligand binding (by detecting chemical shift perturbations), and survey proteins for the presence of unstructured regions. The results of the NMR screens are used in deciding whether to proceed with NMR structure determination, transfer the protein to the crystallization trials pipeline, or re-engineer the protein to improve its suitability for NMR analysis. This may be accomplished by removing unstructured regions to simplify the 2D and 3D NMR spectra or by isolating individual domains to achieve greater stability or solubility and discourage aggregation (Lytle et al., 2006; Peterson et al., 2007b). Proteins that pass these screens as candidates for NMR structure determination are double-labeled with ^{15}N and ^{13}C as described above.

A 600 MHz NMR spectrometer equipped with a cryogenic probe is well suited to the collection of protein triple-resonance spectra. If available, spectrometers with cryogenic probes operating at or above 750 MHz may be employed in collecting multidimensional NOE and isotope-filtered/selected NOE spectra with enhanced resolution. Measurement of residual dipolar couplings (RDCs) for structure refinement and validation (Tolman et al., 1995; Tjandra and Bax, 1997; Tjandra et al., 1997; Prestegard, 1998; Tian, Valafar, and Prestegard, 2001) requires that samples be partially oriented in the magnetic field. The orientation can be achieved by dissolving the protein in an anisotropic medium, such as phospholipid bicelles (Sanders, Schaff, and Prestegard, 1993; Bax and Tjandra, 1997; Ottiger and Bax, 1999), phage particles (Hansen, Mueller, and Pardi, 1998), or stretched polyacrylamide gels (Chou et al., 2001). Identification of a suitable alignment medium may require additional screening, based on the ranges of pH and temperature that are compatible with the target protein and the degree of association between the protein and medium. In the absence of a robust, "universal" alignment medium, RDC analysis is treated as a supplemental approach to this protocol, and structures are solved using primarily distance constraints derived from NOESY spectra and dihedral angle constraints predicted from chemical shifts. In the absence of RDCs, validation is performed by using a variety of software tools that analyze the coordinates and constraint data.

Standard Data Collection Protocols

The improved sensitivity from cryogenic probes permits data collection sufficient for a protein structure in 7–10 days, rather than 1 or 2 months. If the protein concentration is sufficiently high that data can be recorded without signal averaging, the total data collection period can be as short as 30 h. The 2D and 3D pulse schemes listed in Table 5.1 have been in routine use in many laboratories for several years.

Updated versions of these pulse sequences that are compatible with cryogenic probes are provided as part of the standard pulse sequence libraries by NMR instrument manufacturers and are available from the NMRFAM web site (<http://www.nmrfam.wisc.edu>) and the BioMagResBank (BMRB) web site (http://www.bmrb.wisc.edu/tools/choose_pulse_info.php). Reduced dimensionality versions of these pulse sequences have

T A B L E 5 . 1 . Typical Series of NMR Experiments
Collected for a Full Structure Determination

2D ^{1}H–^{15}N HSQC
2D ^{1}H–^{15}N heteronuclear NOE
3D ^{1}H–^{15}N NOESY–HSQC
3D ^{1}H–^{13}C NOESY–HSQC (aliphatic)
3D ^{1}H–^{13}C NOESY–HSQC (aromatic)
2D ^{1}H–^{13}C HSQC (aromatic)
2D ^{1}H–^{13}C CT-HSQC (aliphatic)
2D ^{1}H–^{13}C CT-HSQC (aromatic)
3D HNCO
3D HN(CO)CA
3D HNCA
3D HN(CA)CO
3D HNCACB
3D CBCA(CO)NH
3D C(CO)NH
3D H(CCO)NH
3D HBHA(CO)NH
3D HN(CO)CACB
3D HCCH-TOCSY

also been developed and should yield further reductions in the time required for data acquisition (Montelione et al., 2000; Szyperski et al., 2002; Eghbalnia et al., 2005a). Acquisition of all spectra on a single sample of doubly labeled [U–^{15}N, U–^{13}C] protein is preferred, since it minimizes sample-dependent variations between NMR experiments.

In cases where the protein of interest is a homo-/heterodimer or protein–ligand complex, we prepare a differentially labeled sample by mixing equal amounts of unlabeled protein and uniformly ^{15}N/^{13}C labeled protein. This sample allows us to obtain 3D NOESY data that are F1-filtered and F3-edited for either ^{15}N or ^{13}C. The resulting spectra will only contain NOEs that arise between a proton directly bound to an NMR active heteronucleus and a proton bound to any NMR inactive nucleus while suppressing all other cross-peaks by isotope filtering (Stuart, 1999). These experiments are essential for correctly defining the dimer interfaces.

Processing and Analyzing NMR Datasets: The Stepwise Approach

The usual approach in a biomolecular NMR study is to first convert time-domain data to frequency-domain spectra by Fourier transformation. Then peaks are picked from each spectrum and analyzed. Methods have been developed for automated peak picking or global analysis of spectra to yield models consisting of peaks with known intensity, frequency, phase, and decay rate or linewidth (in each dimension) (Chylla and Markley, 1995; Chylla, Volkman, and Markley, 1998). Our current protocols for processing, peak picking, and assignment of NMR spectra primarily use the programs NMRPipe (Fourier transformation) (Delaglio et al., 1995), XEASY (peak picking and semiautomated assignment) (Bartels et al., 1995), SPSCAN (automated peak picking; available at http://www.personal.uni-jena. de/~b1glra/spscan/manual/index.html), and GARANT (fully automated assignment) (Bartels et al., 1996) on Linux and Mac OS X computing platforms. More recently, for automated backbone and side chain assignments, we have been using PINE (http://miranda. nmrfam.wisc.edu/PINE/). The iterative process of NOE assignment and structure calculations relies primarily on XEASY and CYANA (automated NOE assignment and torsion angle dynamics calculations) (Güntert, Mumenthaler, and Wüthrich, 1997; Herrmann, Güntert, and Wüthrich, 2002b).

Convert and Process Raw Data. Before processing, time-domain data files acquired on Bruker spectrometers must be converted to the proper format for NMRPipe (Delaglio et al., 1995). This is accomplished using the bruk2pipe program supplied with NMRPipe. The corresponding bruk2pipe and NMRPipe scripts for each experiment type are stored in the database at the time of the initial setup. Postconversion to XEASY format is also easily performed with the pipe2xeasy program, available for download from http://www. personal.uni-jena.de/~b1glra/spscan/download/.

Peak Picking of All Spectra. Standard methods for obtaining chemical shift assignments begin with triple-resonance experiments that correlate various combinations of backbone and side chain ^{13}C signals with the amide ^{15}N and ^{1}H signals. A 2D ^{1}H–^{15}N HSQC serves as the initial reference spectrum for directing the identification of signals in the 3D HNCO spectrum, which has the highest resolution and sensitivity of all the triple-resonance experiments. 2D HSQC peaks are picked and then transferred to the 3D HNCO using the strip plot features of XEASY. This array of HNCO strips is manually inspected for completeness and to eliminate spurious noise or artifact peaks. Peak picking of all other $^{1}H^{N}$ – correlated 3D spectra (HN(CA)CO, HNCA, HN(CO)CA, C(CO)NH, CBCA (CO)NH, etc.) is performed by SPSCAN in an automated fashion with the HNCO peaklist as a reference by execution of a single input script (available from BMRB). After inspection in XEASY and acceptance of the final peak lists, the data tabulated from 3D triple-resonance spectra are ready for analysis by automated methods for obtaining sequence-specific chemical shift assignments.

Sequence-Specific Assignments

A number of automated assignment strategies have been described in the literature. These methods have been implemented in programs like AUTOASSIGN (Zimmerman et al., 1997), CONTRAST (Olson and Markley, 1994), GARANT (Bartels et al., 1996), and others. GARANT, which has been used extensively at CESG (Center for Eukaryotic Structural Genomics), shares the same file formats with XEASY, SPSCAN, and CYANA, simplifying the data pathway by eliminating file conversion problems, and it has the ability to provide partial side chain assignments, along with backbone assignments. PINE, a new system developed at NMRFAM (Bahrami et al., manuscript in preparation) (discussed below) accepts a flexible range of experimental peaklists in XEASY and other formats and provides backbone assignments via a web-based server (http://pine.nmrfam.wisc.edu/).

Depending on the quality of the NMR spectra and the resulting peaklists, chemical shift assignments generated by automated software are typically ~75–90% accurate and complete. Manual correction and verification of chemical shift assignments to a level of at least 90% completeness is required before proceeding to automated structure determination (Jee and Güntert, 2003). Side chain ^{1}H and ^{13}C chemical shift assignments are completed using the 3D HBHA(CO)NH, H(CCO)NH, C(CO)NH, and HCCH–TOCSY spectra (for aliphatic side chains) and the ^{13}C – edited 3D (aromatic) spectrum for aromatic side chains. A list of verified chemical shift assignments serves as the input for TALOS, a program that generates backbone φ and ψ dihedral angle constraints from a combination of secondary ^{1}H, ^{15}N, and ^{13}C shifts and pattern matching to a database of tripeptide conformations from experimental structures for which chemical shift values are also known (Cornilescu, Delaglio, and Bax, 1999). Tabular chemical shift data in the TALOS input format is easily generated using the CYANA macro "TALOSLIST." We have streamlined this process with two scripts. One converts chemical shift assignments into the proper format and executes TALOS. After inspection of φ/ψ predictions in the graphical interface provided with TALOS, the table of results is converted into a CYANA-readable list of dihedral angle constraints with a second script (available from BMRB).

NOE Assignment and Structure Calculation

Semiautomated assignment of NOEs using simple chemical shift filters to suggest possible assignment combinations is an error-prone iterative process incompatible with a high-throughput structural proteomics pipeline. The NOEASSIGN module of the torsion angle dynamics (TAD) program CYANA provides a fully automated approach for the iterative assignment of NOEs (Güntert, Mumenthaler, and Wüthrich, 1997; Herrmann, Güntert, and Wüthrich, 2002b; Güntert, 2004) that has performed robustly in our experience. Over the past 3 years, we have used NOEASSIGN to solve 14 unique structures by providing unassigned NOESY peak lists, dihedral angle constraints generated by TALOS (Cornilescu, Delaglio, and Bax, 1999), and a chemical shift list as input (Lytle et al., 2004a; Lytle et al., 2004b; Vinarov et al., 2004; Peterson et al., 2005; Waltner et al., 2005; Lytle et al., 2006; Peterson et al., 2006a; Peterson et al., 2006b; Peterson et al., 2007a; Tuinstra et al., 2007). NOEASSIGN uses the concepts of network anchoring and constraint combination (Herrmann, Güntert, and Wüthrich, 2002a; Güntert, Mumenthaler, and Wüthrich, 1997; Güntert, 2004) in performing seven rounds of iterative NOE assignment with the result of the previous round used as a structural model (Figure 5.3). This automated routine enables the user to go from unassigned peak lists to folded structure in as little as 10–20 min, depending on the size of the protein and the available computational power (a 16 node/32 processor cluster was used here). NOESY peak assignments from NOEASSIGN are then verified or corrected manually using XEASY, and refined with additional CYANA calculations to generate a final structural model that meets basic acceptance criteria for agreement with experimental constraints, Ramachandran statistics, and coordinate precision.

Resolving Symmetric Dimers. A significant proportion of structural proteomics targets form stable oligomers in solution. Dimers and other oligomers with total molecular weight less than 30 kDa are amenable to high-throughput NMR methods, and this highlights the need for an automated approach to solving structures of symmetric assemblies. CYANA supports automated structure calculations on symmetric homodimers and higher order oligomers. Dimeric proteins do not typically present a special problem for X-ray crystallography, but they offer a challenge for NMR methods because inter- and intramolecular NOEs are indistinguishable in uniformly ^{15}N/^{13}C-labeled samples due to chemical shift degeneracy between symmetry-related subunits. Thus, NOEs that define the dimer interface are not readily apparent in standard isotope-edited NOESY spectra. To identify intermolecular NOEs at the dimer interface, we prepare a differentially labeled sample by mixing equal amounts of unlabeled protein and uniformly ^{15}N/^{13}C-labeled protein and acquire a 3D F1-^{13}C/^{15}N-filtered, F3-^{13}C-edited NOESY spectrum

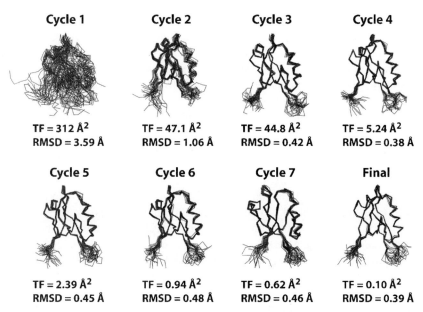

Figure 5.3. Automated structure calculations using the NOEASSIGN module of CYANA. The structure of the second PDZ domain (residues 450–558) from Par-3 (PDZ2) was determined using the NOEASSIGN module of CYANA. Six files, including the sequence file, a chemical shift list, three unassigned NOESY peak lists with intensity values, and a list of dihedral angle constraints from TALOS were provided as input. The structural model in cycle 1 is generated based on chemical shift mapping, network-anchoring, and constraint combination. Continual improvements in the structural model are observed in the progression from cycle 1 to the final calculation as NOE assignments are verified and filtered against the previous structural model starting in cycle 2. The final structural model was defined by a total of 1,487 NOE (759 short range, 199 medium range, and 487 long range) and 117 dihedral angle constraints, and was completed in ∼10 min using a 32 node cluster.

(Stuart, 1999). This spectrum contains only NOEs arising between a proton directly bound to ^{13}C nucleus and a proton bound to any NMR inactive nucleus while suppressing all other cross-peaks by isotope filtering. Constraints generated from the filtered NOESY spectrum can therefore be unambiguously assigned to protons on opposing faces of the dimer interface.

In theory, CYANA should be able to resolve the correct monomeric fold and dimer interface from a combination of isotope edited and filtered NOE peaks when the MOLECULES DEFINE command is utilized in an NOEASSIGN run. MOLECULES DEFINE specifies the residue range for each monomer and works in conjunction with the MOLECULES IDENTITY and MOLECULES SYMDIST commands to minimize differences between torsion angles and distances, respectively, between corresponding atoms in the two monomers. Calculation of a symmetric homodimer requires the user to provide a sequence and chemical shift lists that contain explicit values for both subunits but only a single set of (nonduplicated) peak lists. Duplication of the symmetry-related NOESY distance constraints is performed automatically in NOEASSIGN when the MOLECULES DEFINE command is implemented. In our experience, NOEASSIGN fails to correctly define a homodimer unless ∼20–30 intermolecular NOEs are manually assigned and maintained during the course of the iterative structure refinement. For example, 26 preassigned intermolecular NOEs were required for NOEASSIGN to converge to the correct structure for MZF1 (Peterson et al., 2006a). Once a structural model is obtained from NOEASSIGN, it can be used to seed a second NOEAS-SIGN run with all peak assignments reset at the beginning of the calculation. The second NOEASSIGN calculation typically generates more NOE constraints and a structural ensemble with lower root mean squared deviation (RMSD) and target function values when compared with the initial calculation. This ensemble is usually a suitable starting point for manual refinement in CYANA, final refinement in explicit water, or deposition in the PDB.

NMR Molecular Replacement Using CYANA/NOEASSIGN. As structural proteomics efforts increase the number of known folds and their representation in sequence space, the likelihood of finding a structural homologue for a particular protein increases. A 3D model of the target protein of interest may be generated from the structure of a likely homologue using MODELLER (Sali and Blundell, 1993) or other homology modeling programs. Inclusion of a homology model as a reference structure for the NOEASSIGN algorithm often improves the quality of the final model when compared with the results of a *de novo* calculation. In the early stages of iterative NOE assignment, knowledge of the global fold can provide a valuable distance filter that allows the selection of many unique distance constraints that would otherwise be treated ambiguously due to chemical shift degeneracy. We have applied this method to monomeric proteins with outstanding results, but the impact is even greater when applied to symmetric homodimers.

To solve the NMR structure of the homodimeric SCAN domain from ZNF24, we constructed a ZNF24 model (Figure 5.4a) based on the structure of tumor suppressor protein myeloid zinc finger 1 (MZF1; PDB ID 2FI2) (Peterson et al., 2006a). The model was included as a reference structure in an NOEASSIGN calculation performed on completely unassigned NOESY peaklists. The ensembles generated by NOEASSIGN converged to a domain swapped dimeric assembly (Figure 5.4b) similar to that observed for other members of the SCAN family (Ivanov et al., 2005; Peterson et al., 2006a). Consequently, the ZNF24 dimer structure was solved in much less time than the MZF1 structure.

Protein crystal structures solved by molecular replacement methods are susceptible to systematic error due to model bias, and it is important to assess the extent to which NMR refinement may be biased by a reference structure. In contrast to crystallography, NOE-based refinement appears relatively insensitive to and unconstrained by the orientation of different structural elements in a seed model. Model-based filtering of NOE assignments relies on coarse estimates of interatomic distances for the initial round of refinement only, and is insensitive to long-range differences between the reference model and the true structure of the target protein. Consequently, the NMR analogue of molecular replacement may be more robust than the crystallographic version. To illustrate this point, we compared the NMR structure of the ZNF24 homodimer calculated using NOEASSIGN with the reference model created with MODELLER from the MZF1 structure (Figure 5.4d) and with an X-ray crystal structure of ZNF24 (Figure 5.4e). The reference model and the structure produced by NOEASSIGN exhibited clear differences in the overall arrangement of structural elements, suggesting that this method does not introduce a significant structural bias. Moreover, the NMR structure agrees very closely with the X-ray structure (Figure 5.4c), which was solved independently using phases obtained from single-wavelength anomalous dispersion, providing additional evidence that the automated NOEASSIGN algorithm converged on the correct structure for ZNF24.

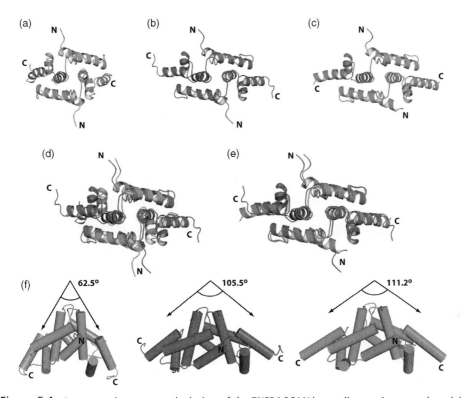

Figure 5.4. Automated structure calculation of the ZNF24 SCAN homodimer using a seed model. (a) Seed model (green) of ZNF24 constructed using MODELLER software and the MZF1 SCAN domain structure (PDB 2FI2). (b) NMR structure (magenta) of ZNF24 generated using the NOEASSIGN module of CYANA and the ZNF24 seed model shown in panel a. The NMR structure is defined by 2595 NOEs and 134 dihedral angles per monomer. (c) X-ray structure (cyan) of ZNF24 solved by single-wavelength anomalous diffraction using a mercury derivative. (d) Overlay of the ZNF24 model (green) and the NMR structure determined using the NOEASSIGN module of CYANA. (e) Comparison of the NMR and the X-ray structures. (f) Side view of all three models showing the change in the angle between helix 5 from each monomer as the seed model of ZNF24 is refined by NOEASSIGN to fit the experimental data. The more than 40° difference in the helix 5 angle between the seed model and the final NMR structure of ZNF24 suggests that the seed model does not bias the structure calculation. This is supported by the close agreement in helix angle when the NMR and X-ray structures are compared. The figure also appears in the Color Figure section.

Final Refinement

The final stages of refinement often require inspection of NOE assignments that result in consistently violated distance constraints. NOEs that have overestimated intensities due to peak overlap can produce too restrictive constraints and require manual adjustment. A fully automated refinement scheme may be able to resolve many such conflicts, but manual intervention is typically required at some point to achieve the desired goals for precision (backbone RMSD below ~0.6 Å), agreement with experimental constraints (none violated by more than 0.5 or 5 Å), and good torsion angle geometry (no residues in disallowed regions of φ/ψ space). CYANA provides diagnostic output with concise summaries of all the necessary data for each ensemble of structures calculated, often simplifying the search for problematic restraints or assignments.

Average constraint density is an informative parameter that typically correlates with coordinate precision. NMR structures defined by more than 15–20 nontrivial constraints/ residue typically exhibit high coordinate precision, with RMSD values of 0.5–0.7 Å for backbone atoms and 1.0–1.2 for all nonhydrogen atoms. However, it is also important to assess the local mobility of the polypeptide backbone using ^{15}N relaxation measurements, since flexible regions will often be devoid of long-range NOEs. For example, ^{1}H–^{15}N heteronuclear NOE values show that most residues of a PDZ domain are ordered but an internal loop and both termini are flexible on picosecond–nanosecond timescales (Figure 5.5a). Consequently, the local NOE constraint density is expected to be low (Figure 5.5b), and no extra effort should be made to identify long-range NOE constraints for those residues. Automated NOE assignment algorithms may incorrectly identify long-range distance constraints for flexible residues, and these assignments can be rejected based on the evidence of local disorder provided by ^{15}N relaxation measurements. Atomic coordinate precision, as reflected in local RMSD values (Figure 5.5c), mirrors the local constraint density, and this local flexibility is illustrated by the structural ensemble (Figure 5.5d).

When a complete and consistent set of experimental constraints has been obtained, and the resulting structural ensemble meets the basic acceptance criteria, a final refinement calculation is performed in explicit water with physical force field parameters. A method for Cartesian space molecular dynamics in explicit water using XPLOR–NIH was shown previously to improve the stereochemical quality of NMR structural ensembles generated with simple force fields consisting only of terms for covalent geometry, experimental constraints, and a simple repulsive potential (Linge et al., 2003). This conclusion was reinforced through the recalculation of structures for over 500 proteins from standardized sets of restraints (Nederveen et al., 2005). Improvements in a variety of validation criteria were observed upon water refinement of NMR ensembles, particularly for the percentage of residues in the most favored region of the Ramachandran (ϕ/ψ) plot. Systematically poor Ramachandran statistics for NMR structures calculated with simplistic force fields can be attributed to the unrealistic constraining of peptide bond planarity (ω angle). In CYANA, for example, ω is fixed at 0° or 180°, corresponding to the *cis* and *trans* peptide bond isomers. However, examination of high-resolution crystal structures in the PDB revealed a strong correlation between the standard deviation of ω and the proportion of residues in the most favored ϕ/ψ regions. Consequently, relaxation of peptide bond rigidity during water refinement in Cartesian space is accompanied by changes in ϕ and ψ, which improve overall quality of the backbone conformation.

To simplify the water refinement routine, we have developed a script (available from BMRB) that converts the output of a CYANA calculation into input for water refinement and initiates the water refinement routine that utilizes a variety of software packages including XPLOR–NIH, CNS–SOLVE, PROFIT, PROCHECK–NMR, WHAT IF, and AQUA. While this routine is not fully integrated into CYANA calculations, the user can initiate a water refinement run with this single script once a complete and consistent set of experimental constraints has been achieved. Starting from the CYANA calculation directory, the script first creates a validation directory with a standardized structure and copies the required output files from CYANA to the validation directory. The CYANA files are then modified to ensure that residue naming conventions and numbering are compatible with the water refinement routine. Specialized input files that specify disulfide bonds, metal centers or *cis*-peptide

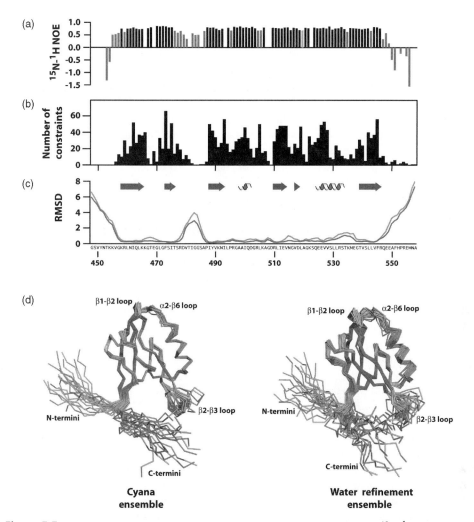

Figure 5.5. Refinement and validation of structural models. (a) Heteronuclear ^{15}N–^{1}H NOE values for the second PDZ domain from Par-3. NOE values below 0.7 are shown in light gray. (b) NOE distance constraint density plotted by residue number. Distance constraints used in the final structure calculation were derived from 3000 assigned peaks observed in three 3D NOESY spectra. Elimination of redundant NOEs and meaningless constraints (distances that impose no conformational restriction or are held fixed by covalent geometry) yielded a final set of 1222 NOE distance constraints. Including dihedral angle constraints from TALOS, a total of 1339 NOE and dihedral angle constraints were used in the structure calculation corresponding to ~12 restraints per residue. However, if residues with ^{15}N–^{1}H NOE values below 0.7 are excluded from consideration, the restraint density increases to ~17 restraints per residue. Structures defined by <10, 10–15, and >15 constraints per residue typically correspond to low, medium, and high-quality structures, respectively. (c) The root mean squared deviation for the structural ensembles before (green line) and after (orange line) water refinement in explicit water plotted as a function of residue number. (d) Structural ensembles produced in CYANA using NOEASSIGN before (left panel) and after water refinement (right panel) show differences in conformational sampling for flexible loop regions as reflected by increased RMSD values in panel c. The helix, strand and loop elements in each ensemble are colored cyan, magenta, and salmon, respectively. The figure also appears in the Color Figure section.

bonds are also copied to the appropriate location during the initial setup. The script then initiates the validation run by parsing the input data and converting the distance and dihedral angle constraints into the XPLOR format. The initial structural model from CYANA is analyzed against the distance and dihedral angle constraints using WHAT IF and PRO-CHECK–NMR. The structural model is then refined using restrained molecular dynamics in explicit water using XPLOR–NIH as described above. Finally, the refined structural model is verified against the distance and dihedral angle constraints as previously described. A comparison of the input and refinement statistics shows that the Ramachandran statistics of the torsion angle dynamics ensemble from CYANA and the WHAT IF RMS Z scores are systematically improved by refinement in explicit water (Table 5.2).

Validation and Assessment of Structural Models

The optimal validation approach to be taken depends on the experimental data collected and how they were used in deriving the family of conformers that represents the structure. Two approaches are used for the validation of biomolecular NMR structures. In the first, the agreement between experimental data and coordinates is assessed. The second measures the match between many aspects of the geometry and the expected range of permissible values for each geometric parameter. Ranges of expected values can be derived theoretically or can

TABLE 5.2. Structure Statistics for PDZ 2 of Mouse Par-3 (Residues 450–558)

Average Atomic RMSD to the Mean Structure (Å)	Cyana	Water Refinement
Residues 457–478, 487–531,538–546		
Backbone (C^α, C′, N)	0.45 ± 0.10	0.65 ± 0.14
Heavy atoms	1.01 ± 0.15	1.22 ± 0.16
Z-score		
First-generation packing quality	-4.69 ± 0.19	-2.36 ± 0.24
Second-generation packing quality	-2.98 ± 0.23	-1.97 ± 0.26
Ramachandran plot appearance	-3.91 ± 0.36	-2.33 ± 0.35
Chi-1/chi-2 roamer normality	-6.09 ± 0.24	-1.70 ± 0.58
Backbone conformation	-6.89 ± 0.77	-7.41 ± 0.77
Overall quality	-34.9 ± 3.1	-32.4 ± 2.65
RMS Z-score		
Bond lengths	0.12 ± 0.00	0.76 ± 0.02
Bond angles	0.27 ± 0.00	0.75 ± 0.03
Omega angle restraints	0.01 ± 0.01	0.66 ± 0.05
Side chain planarity	0.03 ± 0.00	0.64 ± 0.08
Improper dihedral distribution	0.93 ± 0.02	5.21 ± 0.59
Inside/Outside distribution	1.05 ± 0.02	1.04 ± 0.02
Ramachandran statistics (% of all residues)		
Most favored	77.0 ± 3.0	83.3 ± 2.5
Additionally allowed	20.4 ± 3.3	13.7 ± 2.7
Generously allowed	2.2 ± 1.1	1.4 ± 1.2
Disallowed	0.4 ± 0.6	1.1 ± 0.8

be obtained from databases of protein structures (such as PDB), nucleic acid structures (such as NDB), and small molecules (such as Cambridge Structural Database, CSD). These types of structure validation provide complementary checks on the quality of structures and are described below along with various existing software packages.

As with X-ray structures, the stereochemical quality of protein models can be checked with the software tools PROCHECK (Laskowski et al., 1993) and WHAT IF (Vriend, 1990). NMR specific tools include AQUA and PROCHECK–NMR (Laskowski et al., 1996). Recently, WHAT IF has been extended to include a check on hydrogen geometry (Doreleijers et al., 1999b). Structural studies on peptides (Engh and Huber, 1991) and nucleic acids (Parkinson et al., 1996) solved at high resolution and present in the CSD resulted in a set of reference values and standard deviations of bond lengths and angles for heavy atoms that are taken as reference values for larger biomolecules. In most common force fields, the force constants for geometry constraints are chosen to reflect the variation in these geometric parameters. The check on bond lengths and angles is one among the various checks that should be performed before accepting a refined model. Severe deviations from the commonly accepted standard geometry signal the problematic regions in a structure.

Software packages used to solve and refine NMR structures (X-PLOR, CNS, DYANA, AMBER, and others) provide a rich set of diagnostic tools and a summary of the violations between the experimental data and the coordinates. Structures uploaded at deposition through the ADIT-NMR system can be checked using PROCHECK (Laskowski et al., 1993) and MAXit (Berman et al., 2000a). A number of additional software tools are available for evaluating the quality of derived NMR structures. Many of these can be downloaded from the World Wide Web or are available as web servers. These include Verify3D (Bowie, Luthy, and Eisenberg, 1991; Lüthy, Bowie, and Eisenberg, 1992), ProSA (Sippl, 1993; Wiederstein and Sippl, 2007), MolProbity (Lovell et al., 2003; Davis et al., 2004; Davis et al., 2007), VADAR (Willard et al., 2003). WHAT IF (Vriend,) and AQUA (Laskowski et al., 1996; Doreleijers et al., 1999b). In general, NMR structures of acceptable quality meet the following validation criteria: (1) an RMSD for NOE violations less than $\sim$0.05 Å and no persistent NOE violation greater than 0.5 Å across the ensemble of structures (Doreleijers, Rullmann, and Kaptein, 1998); (2) an NOE completeness of at least 50% for all NMR observable proton contacts within 4 Å (Doreleijers et al., 1999a); (3) all torsion angles within the range of the restraints; and (4) chemical shift values within acceptable ranges, unless verified independently. At the time the structures are released, full documentation of the structure and the experimental data, as described in the IUPAC publication "Recommendations for the Presentation of NMR Structures of Proteins and Nucleic Acids" (Markley et al., 1998), should be provided, as well as the reports generated by the software packages mentioned above.

A number of approaches to validate NMR structures against NOE restraints or peak volumes and against residual dipolar couplings have been proposed (Gronwald and Kalbitzer, 2004). Kalbitzer and coworkers have developed an improved software package for back-calculating NOESY spectra from structures that take into account relaxation and scalar coupling effects; this approach and an associated "R" factor offer a promising way of validating NMR structures (Gronwald et al., 2000; 2007). Brünger and coworkers have developed a complete cross-validation technique, in analogy to the free R-factor used in X-ray crystallography, which provides an independent quality assessment of NMR structures based on the NOE violations (Brünger, 1992). The free R-factors of different NMR structures are comparable only if the NOE intensities have been translated into distance restraints in exactly the same way. Montelione and coworkers have proposed a combined scoring system for assessing the quality of NMR structures (Huang et al., 2005). The groups of Clore (Clore, Gronenborn, and Tjandra, 1998) and Bax (Ottiger and Bax, 1999) have developed a quality factor for residual dipolar couplings. In cases where dipolar coupling data can be obtained easily, this provides an attractive approach. A chemical shift validation quality factor also has been proposed (Cornilescu et al., 1998).

Validation of structural models against assigned chemical shifts is another promising technique for checking structures. The measurement of chemical shifts is easy and precise but the back-calculation of the expected chemical shift from a structure is less trivial (Case, 1998). Williamson and coworkers have related measured proton chemical shifts of proteins to values calculated on the basis of NMR and X-ray structures (Williamson, Kikuchi, and Asakura, 1995). The chemical shifts derived from NMR structures showed the same degree of agreement with the measured proton chemical shifts as X-ray structures with a resolution between 2 and 3 Å (σ of 0.35 ppm). Methods for back-calculating chemical shifts from structure are improving rapidly (Case, 2000; Xu and Case, 2001; Neal et al., 2003; Zhang, Neal, and Wishart, 2003). In addition to the AVS software tools (Moseley, Sahota, and Montelione, 2004), BMRB uses the linear analysis of chemical shift (LACS) (Wang et al., 2005) software application and chemical shift back-calculation routines in NMRPipe (F. Delaglio, unpublished) to validate assigned chemical shift depositions.

Deposition of Completed NMR Structures

Deposition of coordinates, chemical shifts, and structural constraints in public databases requires extensive data formatting and the entry of multiple kinds of information. In an effort to accelerate the deposition of new structures and improve the informational content of the entries, the PDB developed the AutoDep Input Tool (ADIT), a tool that accepts formatted input files using an expanded form of the macromolecular Crystallographic Information File (mmCIF) dictionary (Bourne et al., 1997; Berman et al., 2000b) or files in the PDB format. An mmCIF contains all the required structural data and experimental information needed to complete the deposition and automatically populates the query fields when parsed using ADIT. The BMRB in collaboration with the PDB used this system in developing a one-stop deposition system based on the NMR-STAR dictionary (ADIT-NMR; http://deposit.bmrb.wisc.edu/bmrb-adit/). This system is now available for submitting NMR-derived atomic coordinates, restraints, and all other experimental data to the PDB and BMRB. This interface, which can be accessed from the Research Collaboratory for Structural Bioinformatics (RCSB), Protein Data Bank (PDB), PDBj, or BMRB, minimizes the time required to complete the deposition to both data banks by allowing the user to upload and answer all queries once. The PDB takes responsibility for annotating the coordinates, and the BMRB takes responsibility for annotating the NMR data and restraints. The EBI is developing a new deposition system, but currently uses the older AutoDep system for structure depositions.

Generation of the mmCIF file can be conducted in either a semiautomated or manual fashion using the PDB_extract software available from the PDB website (Yang et al., 2004). This software extracts critical structural and experimental information from the output and log files generated by the programs commonly used in NMR and X-ray structure determination and combines it with the structural coordinates to generate a single file for upload. The resulting file can then be verified and/or edited manually to ensure that all information has been captured correctly prior to upload through the ADIT–NMR system. Alternatively, a template file generated by PDB_extract for NMR structures can be edited manually to include all the relevant information for the deposition of a particular target and passed to

PDB_extract with the structural coordinates to assemble the mmCIF required for deposition. Our protocol calls for deposition of completed structures using a manually edited template file in conjunction with a script (available from BMRB) that automates the preparation of the final structural ensemble, distance and dihedral angle constraints, and chemical shift list for deposition. The script generates a single file containing both the distance and the dihedral angle restraints and runs PDB_extract to generate an mmCIF that can be uploaded to the PDB and BMRB data banks using the ADIT–NMR interface after a final review for accuracy. After the mmCIF, constraints file, and chemical shift list are uploaded to ADIT-NMR, the depositor must answer only few queries to complete the deposition to both databases. In our experience, this approach is considerably faster and more reliable than performing each structure deposition by completing all fields of the ADIT-NMR form via the web interface.

PROBABILISTIC APPROACHES AND AUTOMATION

Since the 1980s, the field of automated NMR analysis has been making strides toward a sound algorithmic foundation. A great deal of thought has gone into modeling the "analysis pipeline" (Figure 5.6). The principal aim of automated analysis is to ensure that the performance is satisfactory, or nearly optimal, even when there is uncertainty about the underlying data. The general approach has been to assume that the protein structure can be described as a bundle of conformers that satisfy an empirical force field and structural constraints. Variation in the conformers arises from uncertainty in the data, with large

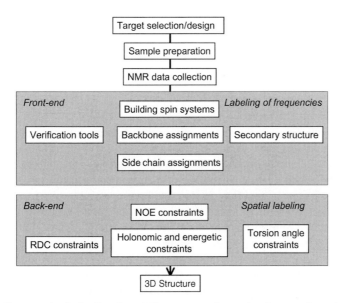

Figure 5.6. The standard pipeline for NMR structure determination consists of a number of discrete tools that, in reality, are intertwined by iterative cycles involving the detection and correction of errors. The front-end of NMR analysis is mostly concerned with "labeling of frequencies" derived from spectra while the back-end focuses on "spatial labeling." Normally, the front-end and back-end are part of separate analyses; but ideally they should become part of a single analytic process.

variations possibly indicating errors in prior steps in the pipeline (such as peak assignments or ambiguity in NOE assignments) that could be "corrected" so as to lead to a tighter bundle of conformers. Over time, it became clear that such an approach can lead to overconstrained bundles. The reason is that in this model of uncertainty, every resonance assignment, experimentally derived restraint, bundle of structures, or protein residue in the sphere of uncertainty is deemed to be equally likely at each stage, and the decision is reduced to a binary accept/reject of observed data for use in the subsequent stage. Under such conditions, and combined with the incompleteness of the NMR experimental data, the algorithm is therefore obliged to guarantee equally satisfactory performance for every residue of every bundle within this sphere of uncertainty, irrespective of the level of uncertainty that may exist within the context of a specific NMR experiment and the specific protein under consideration.

Inherent uncertainties and incompleteness in experimental data as well as errors in modeling approximations (e.g., force field definitions) can lead to errors in attaining the correct fold or bundle of conformers. In short, NMR structure determination is a noisy, inverse, and ill-posed problem. In automating protein structure determination, it is particularly essential to carefully consider issues arising from uncertainties, including ways to explicitly measure and report them, because their impact can be remarkably difficult to detect otherwise. Errors can be broadly classified as either random or systematic. Random errors, often referred to as noise, lead to increased variance in results. In the language of robust analysis, noise is generally thought to have its largest impact on the reliability of results, but not necessarily on their accuracy. Systematic errors are usually more difficult to detect by computational methods, and they typically affect the accuracy of the results. To deal with sources of error and to address challenges faced in the traditional design of the pipeline, a recent approach has been adopted to include some notion of weight to the uncertainty in the data by assigning a probability to results from each discrete step and to each subset of data. Thus, instead of assuring equally satisfactory performance at every possible residue or for every nucleus, the aim becomes to maximize the expected value of the performance of the algorithm. With this reformulation, there is reason to believe that the resulting model structures will better reflect the uncertainties in data than those based on deterministic uncertainty models.

The prospects for carrying out a fully probabilistic, automated program for NMR structure determination have been bolstered by advances in theoretical areas, such as algorithms that exploit relationships between local statistical potentials and global characteristics. It is now understood that global phenomena that obey a certain (large) family of probability distributions can often be described, or reasonably approximated, in terms of local potentials. Conversely, a set of local potentials could lead to a global description given by the same (large family) of probability distributions. These linkages are based on the Hammersly–Clifford theorem (Hammersley and Clifford, 1971; Spitzer, 1971; Besag, 1974), which implies that the efficacy of approximations given by statistical potentials is intimately tied to the accuracy of the description of the local potential. To understand the behavior of a certain global model, its sensitivity to estimates, and the reliability and accuracy of solutions, we can use an approach that combines elements of statistical mechanics and statistical decision theory while relying on local potential descriptions (Chentsov, 1982; Janyszek and Mrugala, 1989; Ruppeiner, 1995), which seems promising.

In the following sections, we first discuss how probabilistic approaches to discrete steps in the NMR structure determination pipeline (data collection, peak assignments, secondary structure determination, and outlier detection) can achieve remarkably effective outcomes.

Then, we demonstrate how iterative combination of information from these discrete steps can lead to increased reliability and improved accuracy. The final section discusses how probabilistic tools can be used to address the ill-posed, inverse problem of NMR structure determination.

Adaptive, Probabilistic Data Collection

One of the challenges in protein NMR spectroscopy is to minimize the time required for multidimensional data collection. An innovative approach to this problem, "high-resolution iterative frequency identification (HIFI) for NMR" (HIFI-NMR) (Eghbalnia et al., 2005a), combines the reduced dimensionality approach for collecting n-dimensional data as a set of tilted, two-dimensional (2D) planes with features that take advantage of the probabilistic paradigm. For 3D triple-resonance experiments of the kind used to assign protein backbone and side chain resonances, the probabilistic algorithm used by HIFI-NMR automatically extracts the positions (chemical shifts) of peaks with considerable time saving compared to conventional approaches in data collection, processing, and peak picking.

The HIFI-NMR algorithm starts with the collection of two-orthogonal 2D planes. In cases where $^1H-^{15}N$ and $^1H-^{13}C$ 2D planes are common to multiple experiments, they need only be collected once. The algorithm then moves to the list of experiments specified for data collection and starts with the first experiment. For this experiment, tilted planes are selected adaptively, one-by-one, by use of probabilistic predictions. A robust statistical algorithm extracts information from each plane, and results are incorporated into the online algorithm that maintains a probabilistic representation of peak positions in 3D space. The updated probabilistic representation of the peak positions is used to derive the optimal angle for the next plane to be collected. The online software predicts whether the collection of this additional data plane will improve the model of peak positions. If the answer is "yes," the next plane is collected and the results incorporated into the evolving model; if the answer is "no," data collection is terminated for the experiment. The algorithm then proceeds to next NMR experiment in the list to be done. Once all experiments have been run, the peak lists with associated probabilities generated directly for each experiment, without total reconstruction of the three-dimensional (3D) spectrum, are ready for use in subsequent assignment or structure determination steps.

In achieving the steps outlined above, the probabilistic algorithm in HIFI-NMR divides the 3D spectral space into discrete voxels on the basis of prior information: the number of expected peaks; the nuclei involved and their expected chemical shift ranges; and the data collection parameters. Next, an empirical estimate for the prior probability distribution of peak locations in the multidimensional spectral space is obtained. The resulting joint probability distribution estimated in this step signifies the probability of a peak's existence in a specific voxel of the spectrum. Let $P_{pr}(X)$ denote the prior probability for the presence of a peak at position X, where the elements of the vector $X = (x_1, x_2, x_3)$ represent the voxels in three-dimensional space. The initial step in the HIFI approach is to collect data for the two orthogonal planes ($0°$ and $90°$ planes). For experiments involving ^1H, ^{15}N, and ^{13}C dimensions (e.g., HNCO, HNCACB), the $0°$ ($^{13}C - ^1H$), and the $90°$ ($^{15}N - ^1H$) planes are collected and peak-picked. Because the ^1H dimension is common to the two planes, the possible candidates for peaks in 3D space can be generated by considering all combinatorial possibilities for peaks in the two planes that have the same ^1H chemical shift within a given tolerance. This generates many more candidates than actual peaks, but will include most, not all, real peaks. These candidate peaks restrain the prior probability space generated in the earlier step. At the end of this stage, $P_0(x)$ designates the new probability distribution that includes information gained about the probability space consisting of candidate peaks.

The next best tilted plane is determined by assuming that the candidate peaks have been observed on a plane at angle θ and by evaluating the information theoretic impact of this observation on the probability of the peaks. The measure of this influence is the divergence S_θ between the initial probability $P_0(x)$ (probability at the initial or the 0th step) and the predicted probability after the impact of the projection at angle θ has been taken into account $Q_\theta(x)$:

$$S_\theta = - \sum Q_\theta \ln \frac{Q_\theta}{P_0}, \tag{5.1}$$

where the sum is taken over all voxels in the spectral space.

Next, each candidate peak x is examined in turn in the light of the data from the new tilted plane and prior data, and Bayes' rule is used to update its probability $P_0(x)$ of being a "real peak,"

$$P_1(x) \equiv P(x|x\_observed\_in\_\theta) = \frac{P(x\_observed\_in\_\theta|x)P_0(x)}{P(x\_observed\_in\_\theta)} \tag{5.2}$$

or

$$P_1(x) \equiv P(x|x\_not\_observed\_in\_\theta) = \frac{P(x\_not\_observed\_in\_\theta|x)P_0(x)}{P(x\_not\_observed\_in\_\theta)} \tag{5.3}$$

in which $x\_not\_observed\_in\_\theta$ is the complement event for $x\_observed\_in\_\theta$. The term in the denominator, $P(x\_observed\_in\_\theta)$ or $P(x\_not\_observed\_in\_\theta)$ represents the probability of a voxel x in plane θ to be considered a peak (or not a peak), regardless of any other consideration. The probabilities $P(x\_observed\_in\_\theta|x)$ and $P(x\_not\_observed\_in\_\theta|x)$ do not experience much angular variation and can be reasonably approximated by the empirical data. This process is continued until no significant information is obtained further. At this stage, n peaks with probabilities above a threshold are reported as the peak list. The value of n is estimated automatically from the number of amino acid residues in the protein, the type of experiment, and the probability distribution $P(x)$. The value of n can be adjusted in a subsequent postprocessing step to include additional expert information regarding the data observed.

Probabilistic approaches are useful in a variety of situations where the robustness of analysis results needs to be validated. This includes the experimental determination of NMR coupling constants. HIFI-C (Cornilescu et al., 2007) represents the extension HIFI-NMR's adaptive and intelligent data collection approach to the rapid and robust collection of coupling data. In this case, data collected from one or more optimally tilted 2D planes are used to determine couplings with high resolution and precision in less time than required by a 3D experiment. A further benefit of the approach is that data from independent planes provide a statistical measure of reliability for each measured coupling. Fast data collection can enable measurements in cases where sample stability is a limiting factor, as frequently happens with proteins in the orienting media used for determining reduced dipolar coupling data.

Probabilistic Approach to NMR Peak Assignment

Resonance assignment programs can be categorized by the methods they use in the mapping steps. These methods include stochastic approaches, such as simulated annealing/Monte Carlo algorithms (Buchler et al., 1997; Lukin et al., 1997; Leutner et al., 1998), genetic algorithms (Bartels et al., 1997), exhaustive search algorithms (Atreya et al., 2000; Andrec and Levy, 2002; Coggins and Zhou, 2003; Jung and Zweckstetter, 2004), heuristic comparison-based algorithms that predict chemical shifts derived from homologous proteins (Gronwald et al., 1998), and heuristic best-first algorithms (Zimmerman et al., 1994; Li and Sanctuary, 1997; Hyberts and Wagner, 2003). For example, AutoAssign (Moseley, Monleon, and Montelione, 2001) is a constraint-based expert system that uses a heuristic best-first mapping algorithm. Among these algorithms, those that rely on stochastic approaches have the best potential for satisfactorily addressing issues arising from noise.

In the PISTACHIO assignment algorithm (Eghbalnia et al., 2005b), the cost function is recast as a measure of "system energy" and the optimization problem is restated in physical terms as that of finding a "ground state." This restatement enables one to ask questions that turn out to have tractable solutions with practical applications. For example, the approach makes it possible to ask for a "typical low energy configuration" and to find a computationally feasible (polynomial time) solution that gives a great deal of information about the "ground state." This is pertinent to the NMR assignment problem because the available data may not support a unique set of assignments. An additional advantage to this approach is that it yields a set of assignment configurations with associated probabilities that can be explored and refined further as additional data becomes available. The local state of the statistical system is constructed in three steps. First, the peak list data are parsed into the set of all possible tripeptide spin systems and their associated scores. Next, tripeptides are assembled to match the sequence and to achieve the optimal probabilities for correct assignments. Each of the three steps serves to restrict the size of the combinatorial search space and to minimize the impact of noise on the analysis. The impact of noise is further ameliorated because the global ground state of the model corresponds to a state that is "statistically closest" to reported observations.

The starting point is to define a "local cost model" $J(Y)$ in terms of a set of cost functions J_N and their configurations Y_N that represent the spin system for a given tripeptide:

$$J(Y) = \sum_{Y_N \subseteq Y} J_N(Y_N).$$

To minimize the overall cost J, one can maximize the following expression:

$$e^{-J(Y)} = \exp\left(\sum_{Y_N \subseteq Y} -J_N(Y_N)\right)$$
$$= \prod_{Y_N \subseteq Y} \exp(-J_N(Y_N)).$$

To treat this as a probability, a normalization factor is introduced,

$$P(Y) = \frac{1}{Z} \prod_{Y_N \subseteq Y} \exp(-J_N(Y_N)),$$

where the cost J is stated on the basis of pair-wise matching of tripeptides:

$$J = \sum_{i=1}^{n}\sum_{i=1}^{n} l_{ij} B(\sigma(i), \sigma(j)) + \sum_{i=1}^{n} C(\sigma(i), i) = \sum \Psi + \sum \Phi$$

$$e^{-J} = e^{-\Sigma\Psi} e^{-\Sigma\Phi}.$$

(5.4)

Upon dividing by the normalization factor to obtain probabilities, and by denoting the set of all permutations as Σ, the following expression is obtained:

$$P(\Sigma) = \frac{1}{Z} \prod_{i,j} \exp(-\Psi(\sigma(i), \sigma(j))) \prod_{i} \exp(-\Phi(\sigma(i), i))$$

$$i \neq j.$$

(5.5)

In Eq. 5.5, the second term measures the merit of assigning a tripeptide spin system Φ to a given tripeptide, whereas in the first term, 'I', provides a measure of the compatibility of the spin systems of adjacent, overlapped tripeptides. To achieve the goal of deriving probabilities for pairing spin systems with residues given by the Φ term, the marginal probability rather than the joint probability given by Eq. 5.5 must be calculated. One major practical advantage of this formulation is its generality because it does not rely on a specific experiment or set of experiments; rather, it enables an easy and efficient incorporation of data from multiple experiments into the framework. PISTACHIO is available as part of the PINE web server described below.

Probabilistic Secondary Structure Determination

Protein secondary structure plays an important role in classifying proteins (Lesk and Rose, 1981) and in analyzing their functional properties (Przytycka, Aurora, and Rose, 1999). Several algorithms have been developed for identifying secondary structure from NMR chemical shifts alone; these include the $\Delta\delta$ method (Reily, Thanabal, and Omecinsky, 1992), the chemical shift index method (Wishart and Sykes, 1994b), the database approach by TALOS (Cornilescu, Delaglio, Bax, 1999), and the probability-based method (Wang and Jardetzky, 2002). A supervised machine-learning approach has been used to combine data on sequence chemical shifts (Hung and Samudrala, 2003).

PECAN (protein energetic conformational analysis from NMR chemical shifts) is a probabilistic approach that utilizes a local statistical potential to combine the separate predictive potential of the sequence and chemical shifts in determining protein secondary structure (Eghbalnia et al., 2005c). Probabilistic and energetic models share some common ground through the Boltzman–Gibbs distribution. PECAN derives a residue-specific statistical energy function that is used to derive probabilistic secondary structure determinations. Suppose that w_n is the number of observations of state n and E_n is the energy of state n. Further, assume that w_n is approximately proportional to the probability for state n. Then, for a system in equilibrium in one of N possible states, the energy and probability are related as

$$p(E_n) = \left(\frac{1}{Z}\right) e^{-\beta E_n} \qquad Z = \sum_{n=1}^{N} e^{-\beta E_n} \qquad \beta = 1/kT,$$

(5.6)

where k is the Boltzmann constant, T is the temperature, Z is the normalization factor called the partition function, and $p(E_n)$ is the probability of state E_n. The constant β sets a monotonic scale for units of energy and can be set equal to a convenient constant. The energy term E is given by

$$E_i = - \sum_{j=i-l}^{i+l} (\Lambda_{j,j\pm1}(V_j) + c_j V_j), \qquad \text{where } \Lambda_{j,j\pm1}(V_j) = (H_{j,j+1} + H_{j-1,j})V_j. \quad (5.7)$$

Eq. 5.7 represents a finite model of length l, where the term V_i represents the bias potential vector, which is determined by constructing optimized residue-specific density estimates that are related to energy through Eq. 5.6. $H_{j,j+1}$ is a transition matrix for each state and represents the propensities of residues j and $j + 1$ to be in each of the possible state combinations. The values for the term $H_{j,j+1}$ are derived from a database of experimental values. The two products $H_{j,j+1}V_j$ and $H_{j-1,j}V_j$ represent two different stochastic mixings of the initial vector V_j. The parameter c_j, which can be replaced by a constant c, represents the proportional influence of the two terms and is optimized by reference to the database of experimental data.

Each site can be viewed as being in one of the three geometrically defined states: helix (H), strand (extended) (E), or "random coil" (R) (Figure 5.7). R is defined simply as neither

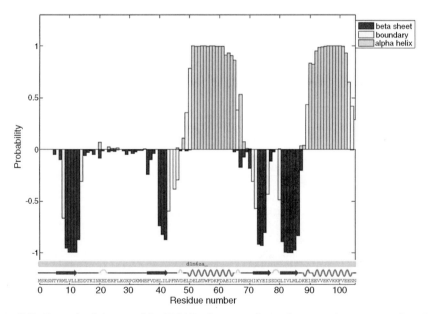

Figure 5.7. Example of the use of the PECAN software to determine secondary structure from NMR chemical shifts and protein sequence (Eghbalnia et al., 2005c). The data are from BMRB entry 4083. The horizontal axis represents the sequence number. The vertical axis represents the probability of helix (gray), represented by positive values, and extended structure (black) represented by negative values. Values near zero indicate random coil residues. The white bars indicate the identification of regions without a discrete structural designation. Because a 3D structure of this protein is known (PDB 1CEY), the PECAN results can be compared with those derived from analysis of the protein conformation by DSSP (Kabsch and Sander, 1983), as shown at the bottom of the plot in cartoon representation where strands are arrows and helices are waves.

H nor E. Each site has an associated energy that is coherent in the finite neighborhood l. For NMR chemical shifts of proteins collected under varying conditions of pH and temperature, PECAN yielded an average of 90% accuracy (93.5% helix, 83.5% strand, and 89.5% random coil) on a large data set (more than 85,000 residues). PECAN achieved a similar rate of success with a second set of data not previously considered (97 proteins; 12,000 residues). For 99% of the proteins, the identification accuracy exceeded 75%, and for 96% of the proteins, the accuracy exceeded 80%. PECAN is available from a web server at http://bija.nmrfam.wisc.edu/PECAN/ and also is incorporated into the PINE server described below.

Validation

One validation approach that we have found valuable takes advantage of the finding that, for a correctly referenced protein data set, linear regression plots of $\Delta\delta^{13}C^{\alpha}_{i}$, $\Delta\delta^{13}C^{\beta}_{i}$, or $\Delta\delta^{1}H^{\alpha}_{i}$ versus $(\Delta\delta^{13}C^{\alpha}_{i} - \Delta\delta^{13}C^{\beta}_{i})$ pass through the origin from two directions, the helix-to-coil and strand-to-coil directions (Figure 5.8). Thus, linear analysis of chemical shifts (Wang et al., 2005) can be used to detect referencing errors and to recalibrate the ^{1}H and ^{13}C

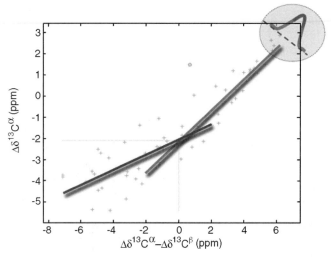

Figure 5.8. Example showing the analysis of assigned backbone $^{13}C^{\alpha}$ and $^{13}C^{\beta}$ chemical shifts by the LACS algorithm (Wang et al., 2005). Data shown are for a single protein (BMRB entry 4998). The left-hand line represents the statistically linear trend of chemical shift values in α-helical regions and the right-hand line represents those for residues in β-sheet (extended) regions of the protein. Residues in nonhelical, nonextended regions are represented by points near the intersection of the two lines. This probabilistic analysis is useful for detecting possible misreferencing (as indicated by the deviation of the crossing point of the two lines from (0,0) expected for properly referenced ^{13}C data) and for correcting the data so that the chemical shifts better represent secondary structure. The analysis can also detect outlier values that may represent misassignments or, as in the case shown (histidine, residue 56, indicated by arrow), a residue whose backbone carbon chemical shifts are pH dependent and thus not a reliable indicator of secondary structure. The probabilistic detection of outliers is depicted in the inset, which shows the normal distribution of points about the right-hand line.

chemical shift scales, if needed, and to recognize outliers that may indicate a possible misassignment or abnormal backbone chemical shift. In the context of statistical potentials, the model of LACS assumes a trimodal Gaussian distribution. LACS is available from a web server at http://bija.nmrfam.wisc.edu/MANI-LACS/.

Integration of Probabilistic Steps within the Pipeline

The integration of individual probabilistic steps into a more global iterative analysis can be considered as the process of assigning a finite set of tags or labels to a collection of observations, subject to side conditions. This challenge is notable for its computational complexity and has theoretical and practical significance to a range of applications in bioinformatics. The combinatorial labeling paradigm is particularly problematic in modeling the early stages of NMR structure determination in which there is the need to associate amino acid labels, secondary structure labels, and other discrete attributes, to the observed data (NMR peaks with associated intensities and positions). The observed data often are incomplete and contain false positives and negatives. These characteristics render the robust assignment of labels remarkably challenging. The front-end labeling process associates one or more NMR parameters with a physical entity (e.g., nucleus, residue, tripeptide, helix, and chain); the back-end labeling process associates NMR parameters with constraints that define conformational states (e.g., dihedral angle, distance, and dipole orientation) (Figure 5.6).

One approach to carry out labeling in the presence of noise and incomplete data is cost function. For example, in the most straightforward case of using a chemical shift index to assign secondary structure, an entirely local three-state cost function was utilized (Wishart and Sykes, 1994a). Numerous cost functions of empirical or statistical nature have been proposed for problems ranging from assignment, secondary structure determination, dihedral angle restraints, to molecular dynamics force fields. Other approaches have used training algorithms, neural networks, or genetic algorithms. However, none of these approaches provides a principled answer to the questions of how many residues or resonances should be assigned, or how reliable and consistent such assignments need to be. To assign the best labels, the algorithmic process often optimizes additional parameters or introduces auxiliary criteria. As a result, it becomes necessary to introduce a variety of parameter setting rules, stopping rules, or other inspection methods. Often, validation tools are not utilized to test for systematic errors or outliers in the data, and no decision is tested against subsequent downstream decisions derived from experimental data (to ensure, for example, that the label from one step of data analysis is not in conflict with a label from another step).

A different view of labeling can be formulated within the context of information theory that explicitly implements the idea that NMR data analysis is motivated primarily by interest in some derived quantity (e.g., the protein structure) and that the solution strategy should focus on preserving information relevant to this goal, rather than trying to derive an optimal similarity metric in the local space of data. We imagine that each frequency label occurs together with a corresponding set of labels for the nuclei of a residue, along with additional labels indicating other information such as secondary structure state. In this formulation, there is no need to define a similarity measure; this measure arises from the iterative optimization principle itself. This approach is attractive because it frees us from the need to specify in advance what it means for the labels to be optimal. In addition, the basic idea can be extended with similar benefits to the entire NMR structure determination pipeline so as to identify in advance what it means for an NMR-derived structure to be most consistent with the data. Importantly, the approach enables the unbiased comparison of alternative algorithms for structure determination and analysis.

We have developed an algorithm called PINE that implements this approach to labeling and takes into account the interconnectedness of different stages of analysis (Figure 5.9). PINE begins with a set of local statistical potentials. It then proceeds iteratively until a stationary state for a consistent global similarity measure is achieved. The resulting software enables a seamless and robust integration of multiple steps in the NMR structure determination pipeline. PINE accepts combinations of NMR data specified by the user from an extensive list of NMR experiments, and provides as output a probabilistic assignment of backbone and side chain signals and the secondary structure of the protein. It also identifies, verifies, and rectifies problems related to referencing, assignment, or outlying data. PINE can make use of prior information supplied from selective labeling or spin system assignments derived independently by other means. PINE currently incorporates, and improves upon, the

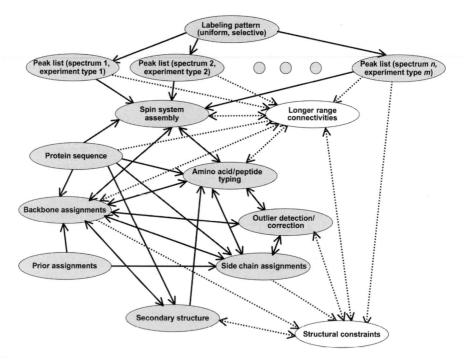

Figure 5.9. Modeled relationships linking independent stages of evidence in protein NMR spectroscopy. The shaded ovals and solid lines indicate parts of the scheme currently implemented in the software package PINE (Bahrami, Markley, Assadi, and Eghbalnia, manuscript in preparation). PINE takes, as input, peak lists from different experiments, the protein sequence, information about labeling, and any prior assignments (used to override those determined). The analysis starts with an initial generation of an "almost exhaustive" list of spin systems in relation to the input data. The output currently gives backbone and side chain assignments and secondary structure. Future versions of PINE will work with longer range connectivities (from TOCSY and NOESY data) and, in addition, provide structural constraints (unshaded ovals and dotted lines).

capabilities of PISTACIO, PECAN, and LACS. Extension of this core of algorithms can, in principle, support the additional statistical potentials involved in the back-end process of NMR structure determination.

The input to PINE consists of the amino acid sequence and peak lists from one or more specified NMR experiments; the peak lists can be either probabilistic (e.g., peak lists generated by HIFI–NMR) or traditional types generated by popular NMR analysis tools. Starting with the spectral alignment step, which consists of a global and a local alignment, a probabilistic model is invoked. This can be viewed as assigning a distance probability for a peak i to belong to a spin system (SS_j) as follows:

$$P_{\text{distance}}{}^i(SS_j) = \frac{f(\text{peak}_i, SS_j)}{\sum_k f(\text{peak}_i, SS_k)} \tag{5.8}$$

$$f(\text{peak}_i, SS_j) = \begin{cases} e^{-d(i,j)/C}, & \text{if} \quad d(i,j) \leq 2C \\ 0, & \text{if} \quad d(i,j) > 2C, \end{cases} \tag{5.9}$$

where $d(i,j)$ is the distance of the peak i and the reference peak in spin system j, and C is a constant that can be set to the resolution of the spectra (Lukin et al., 1997). In the spin system generation, nuclei of the spin system may have multiple chemical shift candidates with associated probabilities. Those probabilities can be derived from the distance probabilities described above along with the probability of existence of the peak itself (e.g., from HIFI-NMR-generated peak lists).

$$P_{SS_j}(i) = \frac{P_{\text{hifi}}(i)P_{\text{distance}}{}^i(SS_j)}{\sum_{k=1}^{N} P_{\text{hifi}}(k)P_{\text{distance}}{}^k(SS_j)}. \tag{5.10}$$

The total number of states of the spin system includes the combinatorial set of all nuclei choices, plus an extra state called *null state*. For the case of nonprobabilistic peak picking (traditional NMR experiments), the volume of the peak may be used to derive an estimate for significance of the peak to be used instead of P_{hifi}. The preservation of all information at this step is particularly important for larger proteins where noise peaks and real peaks are closely interspersed in the spectral data.

The amino acid typing (scoring) step is the key in the integration of subnetworks. This process makes use of chemical shift probability density functions calculated from BMRB-PDB data for each atom of every amino acid in three states: alpha-helix, beta strand, and random coil. Amino acid typing is achieved by computing the probability of a chemical shift belonging to each residue i by

$$p_{\text{cs}}(i) = \frac{p_{\text{helix}}(i)f_{\text{helix}}^i(\text{cs}) + p_{\text{strand}}(i)f_{\text{strand}}^i(\text{cs}) + p_{\text{coil}}(i)f_{\text{coil}}^i(\text{cs})}{\sum_j p_{\text{helix}}(j)f_{\text{helix}}^j(\text{cs}) + p_{\text{strand}}(j)f_{\text{strand}}^j(\text{cs}) + p_{\text{coil}}(j)f_{\text{coil}}^j(\text{cs})}, \tag{5.11}$$

where f_{helix}^i, f_{strand}^I, and f_{coil}^I are the chemical shift probabilities of the related atom in amino acid i and $p_{\text{helix}}(i)$, $p_{\text{strand}}(i)$, and $p_{\text{coil}}(i)$ are the secondary structure probability of the amino acid i with the condition

$$p\,\text{helix}(i) + p\,\text{strand}(i) + p\,\text{coil}(i) = 1. \tag{5.12}$$

The above method connects amino acid typing and secondary structure states through a conditional dependency model. Notice that for the first iteration, where no assignment is available, secondary structure information follows a uniform distribution.

Deriving the backbone and side chain assignments from amino acid typing and other experimental data (connectivity experiments) is a computationally challenging element of the whole network. The approach in this step follows the model described in subsection, "Sequence-Specific Assignments." First, the probabilistic assignments are analyzed by LACS for possible outlier detection. For each nucleus i and its assignment candidates j, one assigns a probability of being an outlier $P_{i,j}$(outlier). Outliers need to be re-evaluated at each iteration. LACS also detects and corrects any possible referencing error in the spectra at this stage. Next, the PECAN algorithm (Eghbalnia et al., 2005c) is used to derive a probabilistic secondary structure determination. As the assignments are probabilistic (not unique), the combinatorial set of possible assignments and their corresponding secondary structures must be considered. After computing the probability of each residue i to be in each of three states ($x = $ H, S, C) for different assignment configurations ($c_1, c_2, \ldots, c_n$), the overall probability can be calculated by

$$P_i(x) = \sum_{j=1}^{n} p(c_j)p(x|c_j), \qquad (5.13)$$

in which $p(x|c_j)$ is the probability of residue i to be in state x with the assignment configuration c_j, and $p(c_j)$ is the probability of the configuration c_j.

Posterior probabilities derived in the assignment process are used as prior probabilities in the next round of assignment, provided that (1) LACS has not flagged the assignment as an outlier, (2) the assignment is fully consistent with the predicted secondary structure, and (3) the assignment is consistent with established connectivities. If any of the above conditions does not hold, scores are re-evaluated in the next iteration. The iteration process continues until a stationary or quasistationary state is reached. The iteration process leads to "self-correction" through appropriate adjustments to the similarity metric to preserve maximum information. The integration of individual steps within PINE leads to results that surpass those achieved by stepwise application of the separate steps in the pipeline (Figure 5.10). Results from multiple automated analyses by PINE for a range of proteins demonstrate average assignment accuracies well above the level of 90% needed for subsequent stages of automation. PINE is available from a web server at http://pine.nmrfam.wisc.edu/.

Inferential Structure Determination

The back-end stages of structure determination involve the specification of three-dimensional coordinates for atoms of the macromolecule as derived from indirect observations in the earlier stages. In application domains where the forward model is very accurate and the numerical predictions of the model are in good agreement with the measurements, the procedure for dealing with noise and the sparsity of data is well understood. However, NMR data are noisy and sparse, and the forward model is insufficiently accurate to determine a unique structure. Thus, errors resulting from model misspecification can be significant. A common approach is to model these errors by including some "model-noise" and by regularizing the forward model to include restraint terms that enforce agreement with the observed data.

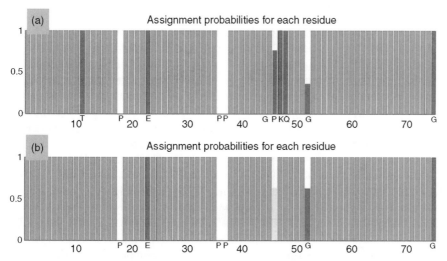

Figure 5.10. Representation of automated assignments determined for the protein ubiquitin from a set of input data generated by HIFI-NMR. The heights of the vertical bars indicate assignment probabilities. Correct assignments are indicated by green (high probability) and yellow (low probability) bars; incorrect assignments, as determined from the known structure, are indicated by red bars. (a) Results achieved by the PISTACHIO software package alone. (b) Results from the PINE software package. The comparison demonstrates the improvement in performance achieved by the multistep iterative probabilistic approach embodied in PINE (A. Bahrami, A. H. Assadi, J. L. Markley, and H. R. Eghbalnia, manuscript in preparation). Figure also appears in the Color Figure section.

The current paradigm in NMR structure calculation is analogous to the process involved in inverting a noisy forward model. A typical scenario involves the minimization of a hybrid energy composed of an energy term representing the physical energy and a "regularization" term that measures the agreement of the model with the data through empirical measures. The components of the physical energy term describe the physical properties of the macromolecule, such as interactions between the atoms, and comprise the "force field." The specification of the force field, however, does not sufficiently remove the degeneracy, and a regularization term is necessary to control the noise through agreement with the data. The degree to which this strategy works crucially depends on the quality of data. Specifically, when NMR structure determination is viewed as an ill-posed inverse problem, it becomes clear that the inversion strategy is successful if and only if the hybrid energy function does not describe a very rough energy surface. In cases where the hybrid energy represents a rough landscape, the determination of the weights for the regularization term becomes an expert's task. In the absence of guiding principles, it is difficult to objectively evaluate the quality and reliability of determined structures. When structure ensembles depend on user-specific parameter settings and on the minimization protocol, statistically meaningful error estimates for atomic coordinates become practically infeasible.

Bayesian and maximum likelihood approaches have already proven useful for data analysis in NMR spectroscopy and X-ray crystallography. Intuitively, the uncertainties can be modeled by associating distributions to the parameter values that define the physical model of the protein. By replacing parameter values with distributions of values, and subsequently solving for a distribution of structure coordinates, the sensitivity of the structure coordinates to the parameters, and therefore the reliability of the resulting structure bundle, can be examined. This can allow the time (and space) evolution of the physical system to depart from deterministic evolution, thereby accommodating the effect of insufficiently modeled background events. For example, it is common practice to include the variance of the model-noise process in the collection of the parameters we wish to estimate from the data. It has been pointed out that allowing "model-noise" results in a good fit to the data, even if the deterministic part of the model is qualitatively wrong (Judd, 2003).

One approach to address the information insufficiency in structure determination is by amending the experimental data with some prior knowledge. Inferential structure determination views structure determination as an inference problem that requires reasoning from noisy and incomplete data (Cox, 1946). This approach does not lead to a unique structure; rather, it provides weights P_i for each conformation X_i —indicating the degree to which each conformation is supported by the experimental data. This inferential approach provides a principled way for handling the calculus of the inference problem (Cox, 1946; Jaynes, 2003). In this setting, the distribution of the probabilities P_i is interpreted as the information content of the data. A very sharp distribution of P_i values indicates a practically unique structure, while a flat distribution indicates lack of sufficient information for deciding among the conformational alternatives. The discrete setting can be readily generalized to the continuous model for the Cartesian coordinates of a macromolecule where probability densities now play the role of the probability mass. This Bayesian inferential approach demands that the probabilities be objective in the sense that they only depend on data D, and on relevant prior information I. The formal construction, through Bayes' theorem, sets the posterior density to be proportional to the product of the likelihood of data given the model and the prior probability of a given macromolecular structure. This dependence on the prior is reflected in, for example, the use of the forward model; although other priors could be included. Other parameters that are required but cannot be measured, often referred to as nuisance parameters, are added to the list of parameters to be estimated. Therefore, the estimated density is formally a conditional probability that depends on the data D and the prior information I. This is a departure from the frequency view of probability.

In practice, the approach involves simulations of the posterior and sampling from areas of high probability mass. Once the terms in the Bayesian model have been established, the simulation provides an objective measure of agreement between data and the resulting structural model within numerical and simulation precision. Increased sampling of the conformational space and the suppression of entropically unfavorable states can provide a more realistic model of the macromolecule. The influence of parameters on the resulting model is transparent. For example, if the forward model does not incorporate any explicit information about protein dynamics, the formulation clearly shows that motion and imprecision due to experimental factors are indistinguishable. In addition, the sensitivity of the atomic positions to the parameters and the functional form of the force field can be explored in an objective manner.

It is often argued that the plausible specification of a prior is difficult or impossible. This difficulty can be addressed. Methods under development can achieve excellent results with little or no prior information. The robustified versions of these methods can guard against unexpected inferences resulting from prior selections. The current inferential approach focuses on the final stage of structure determination and does not rely on the propagation of uncertainties along the structure determination pipeline (Habeck, Nilges, and Rieping, 2005; Rieping, Habeck, and Nilges, 2005; Nilges et al., 2006). Although it is conceptually straightforward to include additional nuisance parameters, the increasing computational complexity and the presence of additional uncertainty models may lead to solutions that must be independently tested for stability and reliability. In the future, a comprehensive NMR pipeline based on inferential principles and capable of modeling other macromolecular features, for example, dynamics, should be capable of providing an objective, detailed, and feature-rich insight regarding the internal consistency, robustness, and accuracy of the resulting structural model.

DATABASES FOR BIOMOLECULAR NMR

The major international repository for biomolecular NMR data is Ulrich, et al., (2008), which is affiliated with the Research Collaboratory for Structural Bioinformatics (http://www.rcsb.org/) and is a member of the Worldwide Protein Data Bank (wwPDB, http://wwpdb.org/) (Berman et al., 2003). Coordinates for structural models derived from NMR data, as from crystallography, are archived at the Protein Data Bank (PDB, http://www.rcsb.org/pdb) (Berman et al., 2000).

Table 5.3 summarizes NMR data classes that are specific to biomolecular structure–function relationships. The primary information to be archived in a protein NMR investigation includes; (1) a complete description of the protein system studied (specification of each constituent and the stoichiometry of interacting constituents); (2) the solution conditions for each protein sample investigated (solvent, pH, temperature, pressure, and the concentration and isotopic labeling of each constituent); (3) a full description of each NMR experiment used in collecting data for the sample; (4) the NMR parameters and derived information obtained in the investigation; and (5) the unprocessed, time-domain NMR spectra collected. The present goal of the databases (PDB and BMRB) is to accommodate the inclusion of additional information on the preparation of the sample and the methods for structure determination at the level of detail provided in the methods section of a research journal. BMRB also accepts full data sets (time-domain spectra plus intermediate results) contributed by authors associated with a structure determination and now has archived more than 90 such sets to date.

In the interest of maintaining a common chemical shift standard for biomolecules, the international community has adopted the methyl signal of internal 2,2-dimethylsilapentane-5-sulfonic acid (DSS) at low concentration as the ^1H chemical shift standard. Chemical shifts of nuclei other than ^1H are referenced indirectly to the ^1H standard through the application of a conversion factor for each nucleus derived from ratios of NMR frequencies (Markley et al., 1998). The BMRB (Ulrich, Markley, and Kyogoku, 1989; Seavey et al., 1991) contains over 3.1 million experimental chemical shifts for proteins, nucleic acids, and small molecules as well as more than 14,000 experimental coupling constants. As of 2008, there were 2357 BMRB entries with associated PDB entries; of these, 1404 BMRB were available with chemical shifts referenced consistently using the IUPAC recommendations.

NMR structures in the PDB are represented by a description of the input data used for structure determination and refinement, the methods used to determine the structure, a statistical analysis of the results, and the structure itself, which usually is represented by a family of conformers that best satisfy the input constraints along with geometric and energetic criteria. One of these conformers is specified as being the single representative structure.

BMRB archives additional information about the chemical properties and dynamics of proteins derived from NMR spectroscopy, including hydrogen exchange rates at specified sites and pH titration parameters (pK_a values, Hill coefficients, and pH-dependent spectro-

TABLE 5.3. Summary of the NMR Data Classes That are Specific to Structural and Functional Genomics

Raw NMR Data

Time-domain data: free induction decays from a particular experiment with a particular sample under defined conditions. The multiplicity of the spectrum (1D, 2D, 3D, 4D) results from the number of time domains sampled

Processed NMR Data

Frequency-domain data: spectra derived from time-domain data by Fourier transformation and/or other signal processing methods

NMR Parameters Extracted from NMR Data

Peak lists derived from individual data sets
Chemical shifts
$^1H-^1H$ NOE
J-couplings
Residual dipolar couplings
NMR relaxation rates

Derived Information

Percentage of expected peaks observed in a data set
NMR peak assignments
Percentage of theoretical peaks assigned
Covalent structure
Bond hybridizations
Secondary structural elements
Interatomic distances
Torsion angles
Hydrogen bonds
Order parameters and other dynamic information
Solvent exposure
Three-dimensional structure (coordinates of the family of conformers that best correspond to the experimental data; coordinates of the conformer that is designated as "representative")
Binding constants
pH titration parameters (pK_a values, Hill coefficients, titration shifts)
Hydrogen exchange rates
Delocalization of unpaired electrons
Thermodynamics and kinetics of structural interconversions
Disordered regions

scopic shifts) for titratable groups and relaxation parameters (heteronuclear NOE, T_1, T_{1rho}, T_2, and order parameters). Figure 5.11 illustrates the full range of information at BMRB.

NMR-STAR is the data model and tag-value data format for biomolecular NMR developed by BMRB in collaboration with the PDB, the Collaborative Computing Project for NMR (CCPN) (Fogh et al., 2002), and a number of contributors from the NMR community. NMR-STAR is an implementation of the STAR format developed for the communication of information in a compact ASCII format (Hall, 1991; Hall and Spadaccini, 1994). The mmCIF data dictionary and data format are STAR implementations used in biomolecular

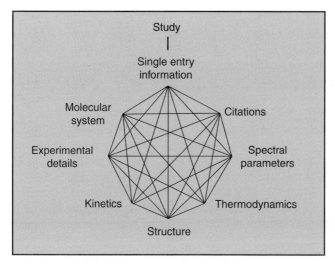

Figure 5.11. Potential information content of a BMRB entry for a particular target (study) showing the interrelatedness of the data to be collected.

crystallography (see Chapter 10). BMRB has developed software tools for interconverting NMR-STAR and mmCIF. NMR-STAR is used by BMRB as its data input format and by many biomolecular NMR software packages as a data exchange format. Specification of the NMR-STAR format and software tools for operating on NMR-STAR files are available from the BMRB web site (http://www.bmrb.wisc.edu). BMRB data are exported in the ASCII NMR-STAR format and in a format that can be loaded easily into a relational database.

BMRB established two new databases of structural restraints used to calculate NMR structures: DOCR (Database of Converted Restraints) and FRED (Filtered Restraints Database) (Doreleijers et al., 2005). Both databases were constructed in collaboration with EBI and CCPN groups from restraint data available from the PDB. The DOCR database contains restraint files with atom nomenclature consistent with the nomenclature in the corresponding atomic coordinate files. The restraints data in the FRED database have been filtered to remove redundant values and other restraints that do not impact the structure calculations. Over 500 protein structures were recalculated using restraints from these databases and the resulting structures analyzed for their quality (Nederveen et al., 2005). The resulting database of recalculated structures (RECOORD) is available on the web (http://www.ebi.ac.uk/msd-srv/docs/NMR/recoord/main.html). The DOCR and FRED databases now contain over 4000 PDB NMR entries. The data are available in NMR-STAR format, X-PLOR, or an XML format.

As BMRB expands, the database of publicly accessible biomolecular NMR data also expands, creating a highly valuable resource for data mining and the development of statistical correlations between NMR observables and the structural and chemical properties of biomolecules. Examples include SHIFTY (Wishart et al., 1997), RefDB (Zhang, Neal, and Wishart, 2003), TALOS (Cornilescu, Delaglio, and Bax, 1999), AVS (Moseley, Sahota, and Montelione, 2004), PISTACHIO (Eghbalnia et al., 2005b), PECAN (Eghbalnia et al., 2005c), and other probabilistic studies (Wang, Eghbalnia, and Markley, 2006). BMRB routinely uses the AVS software package to inform depositors of suspicious or anomalous assigned chemical shifts within their depositions.

ACKNOWLEDGMENTS

We thank the members of the CESG, NMRFAM, and BMRB teams for their contributions. Support from NIH grants P41 LM05799, 1U54 GM074901, and P41 RR02301 is gratefully acknowledged.

REFERENCES

Akke M, Palmer III, AG (1996): Monitoring macromolecular motions on microsecond–millisecond time scales by R1R-R1 constant relaxation time NMR spectroscopy. *J Am Chem Soc* 118:911–912.

Andrec M, Montelione GT, Levy RM (2000): Lipari-Szabo mapping: A graphical approach to Lipari-Szabo analysis of NMR relaxation data using reduced spectral density mapping. *J Biomol NMR* 18:83–100.

Andrec M, Levy RM (2002): Protein sequential resonance assignments by combinatorial enumeration using ^{13}C alpha chemical shifts and their (i, i-1) sequential connectivities. *J Biomol NMR* 23:263–270.

Aramini JM, Rossi P, Anklin C, Xiao R, Montelione GT (2007): Microgram-scale protein structure determination by NMR. *Nat Methods* 4:491–493.

Atreya HS, Sahu SC, Chary KV, Govil G (2000): A tracked approach for automated NMR assignments in proteins (TATAPRO). *J Biomol NMR* 17:125–136.

Baldus M (2006): Molecular interactions investigated by multidimensional solid-state NMR. *Curr Opin Struct Biol* 16:618–623.

Baranov VI, Morozov IY, Ortlepp SA, Spirin AS (1989): Gene-expression in a cell-free system on the preparative scale. *Gene* 84:463–466.

Bartels C, Xia TH, Billeter M, Güntert P, Wüthrich K (1995): The program XEASY for computer-supported NMR spectral-analysis of biological macromolecules. *J Biomol NMR* 5:1–10.

Bartels C, Billeter M, Güntert P, Wüthrich K (1996): Automated sequence-specific NMR assignment of homologous proteins using the program GARANT. *J Biomol NMR* 7:207–213.

Bartels C, Güntert P, Billeter M, Wüthrich K (1997): GARANT: a general algorithm for resonance assignment of multidimensional nuclear magnetic resonance spectra. *J Comput Chem* 18:139–149.

Bax A, Tjandra N (1997): High-resolution heteronuclear NMR of human ubiquitin in an aqueous liquid crystalline medium. *J Biomol NMR* 10:289–292.

Bax A, Kontaxis G, Tjandra N (2001): Dipolar couplings in macromolecular structure determination. *Methods Enzymol* 339:127–174.

Bax A, Grishaev A (2005): Weak alignment NMR: a hawk-eyed view of biomolecular structure. *Curr Opin Struct Biol* 15:563–570.

Berman HM, Westbrook J, Feng Z, Gilliland G, Bhat TN, Weissig H, Shindyalov IN, Bourne PE (2000): The protein data bank. *Nucleic Acids Res* 28:235–242.

Berman HM, Bhat TN, Bourne PE, Feng Z, Gilliland G, Weissig H, Westbrook J (2000a): The Protein Data Bank and the challenge of structural genomics. *Nat Struct Biol* 7(Suppl):957–959.

Berman HM, Westbrook J, Feng Z, Gilliland G, Bhat TN, Weissig H, Shindyalov IN, Bourne PE (2000b): The Protein Data Bank. *Nucl Acids Res* 28:235–242.

Berman HM, Henrick K, Nakamura H (2003): Announcing the worldwide protein data bank. *Nature Structural Biology* 10:980.

Besag J (1974): Spatial interaction and statistical-analysis of lattice systems. *J Royal Stat Soc Ser B* 36:192–236.

Blommel PG, Becker KJ, Duvnjak P, Fox BG (2007): Enhanced bacterial protein expression during auto-induction obtained by alteration of lac repressor dosage and medium composition. *Biotechnol Prog* 23:585–598.

Bourne PE, Berman HM, McMahon B, Watenpaugh KD, Westbrook JD, Fitzgerald PMD (1997): The macromolecular crystallographic information file (mmCIF). *Methods Enzymol* 277: 571–590.

Bowie JU, Luthy R, Eisenberg D (1991): A method to identify protein sequences that fold into a known three-dimensional structure. *Science* 253:164–170.

Brünger AT (1992): Free R-value: a novel statistical quanity for assessing the accuracy of crystal-structures. *Nature* 355:472–475.

Buchler NE, Zuiderweg ER, Wang H, Goldstein RA (1997): Protein heteronuclear NMR assignments using mean-field simulated annealing. *J Magn Reson* 125:34–42.

Burley SK (2000): An overview of structural genomics. *Nat Struct Biol* 7:932–934.

Case DA (1998): The use of chemical shifts and their anisotropies in biomolecular structure determination. *Curr Opin Struct Biol* 8:624–630.

Case DA (2000): Interpretation of chemical shifts and coupling constants in macromolecules. *Curr Opin Struct Biol* 10:197–203.

Cavanagh J, Fairbrother W, Palmer III AG, Skelton NJ Rance M (2006): *Protein NMR Spectroscopy: Principles and Practice*, 2nd ed. San Diego: Academic Press.

Chentsov NN (1982): *Statistical Decision Rules and Optimal Inference*. Providence, RI: American Mathematical Society.

Chou JJ, Gaemers S, Howder B, Louis JM, Bax A (2001): A simple apparatus for generating stretched polyacrylamide gels, yielding uniform alignment of proteins and detergent micelles. *J Biomol NMR* 21:377–382.

Chrunyk BA, Evans J, Lillquist J, Young P, Wetzel R (1993): Inclusion-body formation and protein stability in sequence variants of interleukin-1-beta. *J Biol Chem* 268:18053–18061.

Chylla RA, Markley JL (1995): Theory and application of the maximum likelihood principle to NMR parameter estimation of multidimensional NMR data. *J Biomol NMR* 5:245–258.

Chylla RA, Volkman BF, Markley JL (1998): Practical model fitting approaches to the direct extraction of NMR parameters simultaneously from all dimensions of multidimensional NMR spectra. *J Biomol NMR* 12:277–297.

Clore GM, Gronenborn AM, Tjandra N (1998): Direct structure refinement against residual dipolar couplings in the presence of rhombicity of unknown magnitude. *J Magn Reson* 131:159–162.

Clore GM, Szabo A, Bax A, Kay LE, Driscoll PC, Gronenborn AM (1990): Deviations from the simple two-parameter model-free approach to the interpretation of nitrogen-15 nuclear magnetic relaxation of proteins. *J Am Chem Soc* 112:4989–4991.

Coggins BE, Zhou P (2003): PACES: Protein sequential assignment by computer-assisted exhaustive search. *J Biomol NMR* 26:93–111.

Cordier F, Grzesiek S (1999): Direct observation of hydrogenbonds in proteins by interresidue 3H J NC′ scalar couplings. *J Am Chem Soc* 121:1601–1602.

Cordier F, Rogowski M, Grzesiek S, Bax A (1999): Observation of through-hydrogen-bond (2h)J(HC′) in a perdeuterated protein. *J Magn Reson* 140:510–512.

Cornilescu G, Marquardt JL, Ottinger M, Bax A (1998): Validation of protein structure from anisotropic carbonyl chemical shifts in a dilute liquid crystalline phase. *J Am Chem Soc* 120:6836–6837.

Cornilescu G, Delaglio F, Bax A (1999): Protein backbone angle restraints from searching a database for chemical shift and sequence homology. *J Biomol NMR* 13:289–302.

Cornilescu G, Bahrami A, Tonelli M, Markley JL, Eghbalnia HR (2007): HIFI-C: a robust and fast method for determining NMR couplings from adaptive 3D to 2D projections. *J Biomol NMR* 38:341–351.

Cox RT (1946): Probability, frequency and reasonable expectation. *Am J Phys* 14:1–13.

Czisch M, Boelens R (1998): Sensitivity enhancement in the TROSY experiment. *J Magn Reson* 134:158–160.

Davis IW, Murray LW, Richardson JS, Richardson DC (2004): MOLPROBITY: structure validation and all-atom contact analysis for nucleic acids and their complexes. *Nucl Acids Res* 32: W615–W619.

Davis IW, Leaver-Fay A, Chen VB, Block JN, Kapral GJ, Wang X, Murray LW, Arendall III WB Snoeyink J, Richardson JS, Richardson DC (2007): MolProbity: all-atom contacts and structure validation for proteins and nucleic acids. *Nucl Acids Res* 35:W375–383.

de Swiet TM (2005): Optimal electric fields for different sample shapes in high resolution NMR spectroscopy. *J Magn Reson* 174:331–334.

Delaglio F, Grzesiek S, Vuister GW, Zhu G, Pfeifer J, Bax A (1995): NMRPIPE: a multidimensional spectral processing system based on UNIX pipes. *J Biomol NMR* 6:277–293.

Doreleijers JF, Rullmann J.A.C, Kaptein R (1998): Quality assessment of NMR structures: a statistical survey. *J Mol Biol* 281:149–164.

Doreleijers JF, Raves ML, Rullmann T, Kaptein R (1999a): Completeness of NOEs in protein structure: a statistical analysis of NMR data. *J Biomol NMR* 14:123–132.

Doreleijers JF, Vriend G, Raves ML, Kaptein R (1999b): Validation of nuclear magnetic resonance structures of proteins and nucleic acids: hydrogen geometry and nomenclature. *Proteins* 37:404–416.

Doreleijers JF, Nederveen AJ, Vranken W, Lin J, Bonvin AM, Kaptein R, Markley JL, Ulrich EL (2005): BioMagResBank databases DOCR and FRED containing converted and filtered sets of experimental NMR restraints and coordinates from over 500 protein PDB structures. *J Biomol NMR* 32:1–12.

Dry S, McCarthy S, Harris T (2000): Structural genomics in the biotechnology sector. *Nat Struct Biol* 7:946–949.

Dyson HJ, Wright PE (2005): Intrinsically unstructured proteins and their functions. *Nat Rev Mol Cell Biol* 6:197–208.

Edwards AE, Arrowsmith CH, Christendat D, Dharamsi A, Friesen JD, Greenblatt JF, Vedadi M (2000): Protein production: feeding the crystallographers and NMR spectroscopists. *Nat Struct Biol* 7:970–972.

Eghbalnia HR, Bahrami A, Tonelli M, Hallenga K, Markley JL (2005a): High-resolution iterative frequency identification for NMR as a general strategy for multidimensional data collection. *J Am Chem Soc* 127:12528–12536.

Eghbalnia HR, Bahrami A, Wang L, Assadi A, Markley JL (2005b): Probabilistic identification of spin systems and their assignments including coil–helix inference as output (PISTACHIO). *J Biomol NMR* 32:219–233.

Eghbalnia HR, Wang L, Bahrami A, Assadi A, Markley JL (2005c): Protein energetic conformational analysis from NMR chemical shifts (PECAN) and its use in determining secondary structural elements. *J Biomol NMR* 32:71–81.

Eisenmesser EZ, Millet O, Labeikovsky W, Korzhnev DM, Wolf-Watz M, Bosco DA, Skalicky JJ, Kay LE, Kern D (2005): Intrinsic dynamics of an enzyme underlies catalysis. *Nature* 438:117–121.

Endo Y, Otsuzuki S, Ito K, Miura K (1992): Production of an enzymatic active protein using a continuous-flow cell-free translation system. *J Biotechnol* 25:221–230.

Engh RA, Huber R (1991): Accurate bond and angle parameters for X-ray protein structure refinement. *Acta Cryst* A47:392–400.

Ernst RR, Bodenhausen G, Wokaun A (1987): *Principles of Nuclear Magnetic Resonance in One and Two Dimensions.* Oxford: Oxford University Press.

Fernandez C, Wider G (2003): TROSY in NMR studies of the structure and function of large biological macromolecules. *Curr Opin Struct Biol* 13:570–580.

Fogh R, Ionides J, Ulrich E, Boucher W, Vranken W, Linge JP, Habeck M, Rieping W, Bhat TN, Westbrook J, et al., (2002): The CCPN project: an interim report on a data model for the NMR community. *Nat Struct Biol* 9:416–418.

Gardner KH, Kay LE (1998): The use of ^{2}H, ^{13}C, ^{15}N multidimensional NMR to study the structure and dynamics of proteins. *Annu Rev Biophys Biomol Struct* 27:357–406.

Goff SA, Goldberg AL (1987): An increased content of protease la, the Ion gene-product, increases protein-degradation and blocks growth in *Escherichia coli*. *J Biol Chem* 262: 4508–4515.

Gronwald W, Willard L, Jellard T, Boyko RF, Rajarathnam K, Wishart DS, Sonnichsen FD, Sykes BD (1998): CAMRA: chemical shift based computer aided protein NMR assignments. *J Biomol NMR* 12:395–405.

Gronwald W, Kirchhofer R, Gorler A, Kremer W, Ganslmeier B, Neidig KP, Kalbitzer HR (2000): RFAC, a program for automated NMR R-factor estimation. *J Biomol NMR* 17:137–151.

Gronwald W, Kalbitzer HR (2004): Automated structure determination of proteins by NMR spectroscopy. *Prog Nucl Magn Reson Spectrosc* 44:33–96.

Gronwald W, Brunner K, Kirchhofer R, Trenner J, Neidig KP, Kalbitzer HR (2007): AUREMOL-RFAC-3D, combination of R-factors and their use for automated quality assessment of protein solution structures. *J Biomol NMR* 37:15–30.

Güntert P, Mumenthaler C, Wüthrich K (1997): Torsion angle dynamics for NMR structure calculation with the new program DYANA. *J Mol Biol* 273:283–298.

Güntert P (2004): Automated NMR structure calculation with CYANA. *Methods Mol Biol* 278:353–378.

Habeck M, Nilges M, Rieping W (2005): Bayesian inference applied to macromolecular structure determination. *Phys Rev E* 72:031912.

Hajduk PJ, Greer J (2007): A decade of fragment-based drug design: strategic advances and lessons learned. *Nature Rev* 6:211–219.

Hall SR (1991): The STAR file: a new format for electronic data transfer and archiving. *J Chem Inform Comput Sci* 31:326–333.

Hall SR, Spadaccini N (1994): The STAR file: detailed specifications. *J Chem Inform Comput Sci* 34:505–508.

Hammersley JM, Clifford P (1971): Markov field on finite graphs and lattices. (Unpublished manuscript.).

Hansen MR, Mueller L, Pardi A (1998): Tunable alignment of macromolecules by filamentous phage yields dipolar coupling interactions. *Nat Struct Biol* 5:1065–1074.

Heinemann U (2000): Structural genomics in Europe: slow start, strong finish?. *Nat Struct Biol* 7:940–942.

Henrich B, Lubitz W, Plapp R (1982): Lysis of *Escherichia coli* by induction of cloned Phi-X174 genes. *Mol Gen Genet* 185:493–497.

Herrmann T, Guntert P, Wüthrich K (2002a): Protein NMR structure determination with automated NOE-identification in the NOESY spectra using the new software ATNOS. *J Biomol NMR* 24:171–189.

Herrmann T, Guntert P, Wüthrich K (2002b): Protein NMR structure determination with automated NOE assignment using the new software CANDID and the torsion angle dynamics algorithm DYANA. *J Mol Biol* 319:209–227.

Hong M (2006): Oligomeric structure, dynamics, and orientation of membrane proteins from solid-state NMR. *Structure* 14:1731–1740.

Hu K, Eletsky A, Pervushin K (2003): Backbone resonance assignment in large protonated proteins using a combination of new 3D TROSY-HN(CA)HA, 4D TROSY-HACANH and ^{13}C-detected HACACO experiments. *J Biomol NMR* 26:69–77.

Huang YJ, Moseley HN, Baran MC, Arrowsmith C, Powers R, Tejero R, Szyperski T, Montelione GT (2005): An integrated platform for automated analysis of protein NMR structures. *Methods Enzymol* 394:111–141.

Hung LH, Samudrala R (2003): Accurate and automated classification of protein secondary structure with PsiCSI. *Prot Sci* 12:288–295.

Huyghues-Despointes BM, Pace CN, Englander SW, Scholtz JM (2001): Measuring the conformational stability of a protein by hydrogen exchange. *Methods Mol Biol* 168:69–92.

Hwang TL, van Zijl PCM, Mori S (1998): Accurate quantitation of water–amide proton exchange rates using the phase-modulated CLEAN chemical EXchange (CLEANEX-PM) approach with a Fast-HSQC (FHSQC) detection scheme. *J Biomol NMR* 11:221–226.

Hyberts SG, Wagner G (2003): IBIS: a tool for automated sequential assignment of protein spectra from triple resonance experiments. *J Biomol NMR* 26:335–344.

Igumenova TI, Frederick KK, Wand AJ (2006): Characterization of the fast dynamics of protein amino acid side chains using NMR relaxation in solution. *Chem Rev* 106:1672–1699.

Ivanov D, Stone JR, Maki JL, Collins T, Wagner G (2005): Mammalian SCAN domain dimer is a domain-swapped homolog of the HIV capsid C-terminal domain. *Mol Cell* 17:137–143.

Janyszek H, Mrugala R (1989): Riemannian geometry and the thermodynamics of model magnetic systems. *Phys Rev A* 39:6515–6523.

Jaynes ET (2003): *Probability Theory: The Logic of Science.* Cambridge: Cambridge University Press.

Jee J, Güntert P (2003): Influence of the completeness of chemical shift assignments on NMR structures obtained with automated NOE assignment. *J Struct Funct Genomics* 4:179–189.

Judd K (2003): Chaotic-time-series reconstruction by the Bayesian paradigm: right results by wrong methods. *Phys Rev E* 67:026212.

Jung Y.-S, Zweckstetter M (2004): Mars: robust automatic backbone assignment of proteins. *J Biomol NMR* 30:11–23.

Kabsch W, Sander C (1983): Dictionary of protein secondary structure: pattern recognition of hydrogen-bonded and geometrical features. *Biopolymers* 22:2577–2637.

Kainosho M, Torizawa T, Iwashita Y, Terauchi T, Mei Ono A, Guntert P (2006): Optimal isotope labelling for NMR protein structure determinations. *Nature* 440:52–57.

Kay LE, Marion D, Bax A (1989): Practical aspects of 3D heteronuclear NMR of proteins. *J Magn Reson* 84:72–84.

Kay LE, Ikura M, Tschudin R, Bax A (1990): 3-Dimensional triple-resonance NMR-spectroscopy of isotopically enriched proteins. *J Magn Reson* 89:496–514.

Kern D, Zuiderweg ER (2003): The role of dynamics in allosteric regulation. *Curr Opin Struct Biol* 13:748–757.

Kigawa T, Muto Y, Yokoyama S (1995): Cell-free synthesis and amino acid-selective stable isotope labeling of proteins for NMR analysis. *J Biomol NMR* 6:129–134.

Kigawa T, Yabuki T, Yoshida Y, Tsutsui M, Ito Y, Shibata T, Yokoyama S (1999): Cell-free production and stable-isotope labeling of milligram quantities of proteins. *FEBS Lett* 442:15–19.

Kim S, Szyperski T (2003): GFT NMR, a new approach to rapidly obtain precise high-dimensional NMR spectral information. *J Am Chem Soc* 125:1385–1393.

Klammt C, Lohr F, Schafer B, Haase W, Dotsch V, Ruterjans H, Glaubitz C, Bernhard F (2004): High level cell-free expression and specific labeling of integral membrane proteins. *Eur J Biochem* 271:568–580.

Klammt C, Schwarz D, Lohr F, Schneider B, Dotsch V, Bernhard F (2006): Cell-free expression as an emerging technique for the large scale production of integral membrane protein. *FEBS J* 273:4141–4153.

Kupce E, Freeman R (2003): Reconstruction of the three-dimensional NMR spectrum of a protein from a set of plane projections. *J Biomol NMR* 27:383–387.

Kurland CG (1982): Translational accuracy *in vitro*. *Cell* 28:201–202.

Laskowski RA, MacArthur MW, Moss DS, Thornton JM (1993): PROCHECK: a program to check the stereochemical quality of protein structures. *J Appl Crystallogr* 26:283–291.

Laskowski RA, Rullmann J.A.C, MacArthur MW, Kaptein R, Thornton JM (1996): AQUA and PROCHECK-NMR: programs for checking the quality of protein structures solved by NMR. *J Biomol NMR* 8:477–486.

Lescop E, Schanda P, Brutscher B (2007): A set of BEST triple-resonance experiments for time-optimized protein resonance assignment. *J Magn Reson* 187:163–169.

Lesk AM, Rose GD (1981): Folding units in globular proteins. *Proc Natl Acad Sci USA* 78:4304–4308.

Leutner M, Gschwind RM, Liermann J, Schwarz C, Gemmecker G, Kessler H (1998): Automated backbone assignment of labeled proteins using the threshold accepting algorithm. *J Biomol NMR* 11:31–43.

Li KB, Sanctuary BC (1997): Automated resonance assignment of proteins using heteronuclear 3D NMR. 2. Side chain and sequence-specific assignment. *J Chem Inf Comput Sci* 37:467–477.

Liguori L, Marques B, Villegas-Mendez A, Rothe R, Lenormand JL (2007): Production of membrane proteins using cell-free expression systems. *Expert Rev Proteomics* 4:79–90.

Linge JP, Williams MA, Spronk CA, Bonvin AM, Nilges M (2003): Refinement of protein structures in explicit solvent. *Proteins* 50:496–506.

Lipari G, Szabo A (1982a): Model-free approach to the interpretation of nuclear magnetic resonance relaxation in macromolecules. 1. Theory and range of validity. *J Am Chem Soc* 104:4546–4559.

Lipari G, Szabo A (1982b): Model-free approach to the interpretation of nuclear magnetic resonance relaxation in macromolecules. 2. Analysis of experimental results. *J Am Chem Soc* 104:4559–4570.

Lipsitz RS, Tjandra N (2004): Residual dipolar couplings in NMR structure analysis. *Annu Rev Biophys Biomol Struct* 33:387–413.

Lopez-Mendez B, Güntert P (2006): Automated protein structure determination from NMR spectra. *J Am Chem Soc* 128:13112–13122.

Lovell SC, Davis IW, Arendall III WB de Bakker PI, Word JM, Prisant MG, Richardson JS, Richardson DC (2003): Structure validation by C-alpha geometry: phi, psi and C-beta deviation. *Proteins* 50:437–450.

Lukin JA, Gove AP, Talukdar SN, Ho C (1997): Automated probabilistic method for assigning backbone resonances of (13C, 15N)-labeled proteins. *J Biomol NMR* 9:151–166.

Lüthy R, Bowie JU, Eisenberg D (1992): Assessment of protein models with three-dimensional profiles. *Nature* 356:83–85.

Lytle BL, Peterson FC, Kjer KL, Frederick RO, Zhao Q, Thao S, Bingman C, Johnson KA, Phillips GN Jr, Volkman BF (2004a): Structure of the hypothetical protein At3g17210 from Arabidopsis thaliana. *J Biomol NMR* 28:397–400.

Lytle BL, Peterson FC, Qiu SH, Luo M, Zhao Q, Markley JL, Volkman BF (2004b): Solution structure of a ubiquitin-like domain from tubulin-binding cofactor B. *J Biol Chem* 279:46787–46793.

Lytle BL, Peterson FC, Tyler EM, Newman CL, Vinarov DA, Markley JL, Volkman BF (2006): Solution structure of *Arabidopsis thaliana* protein At5g39720.1, a member of the AIG2-like protein family. *Acta Crystallogr F* 62:490–493.

Mandel AM, Akke M, Palmer III AG (1995): Backbone dynamics of *Escherichia coli* ribonuclease HI: correlations with structure and function in an active enzyme. *J Mol Biol* 246:144–163.

Markley JL, Kainosho M (1993): Stable isotope labeling and resonance assignments in larger proteins. In: Roberts G.C.K. editor. *NMR of Biological Macromolecules: A Practical Approach.* Oxford, England: Oxford University Press, pp 101–152.

Markley JL, Bax A, Arata Y, Hilbers CW, Kaptein R, Sykes BD, Wright PE, Wüthrich K (1998): Recommendations for the presentation of NMR structures of proteins and nucleic acids. *J Mol Biol* 280:933–952.

Matsuda T, Koshiba S, Tochio N, Seki E, Iwasaki N, Yabuki T, Inoue M, Yokoyama S, Kigawa T (2007): Improving cell-free protein synthesis for stable-isotope labeling. *J Biomol NMR* 37:225–229.

McDermott AE (2004): Structural and dynamic studies of proteins by solid-state NMR spectroscopy: rapid movement forward. *Curr Opin Struct Biol* 14:554–561.

Mercier KA, Baran M, Ramanathan V, Revesz P, Xiao R, Montelione GT, Powers R (2006): FAST-NMR: functional annotation screening technology using NMR spectroscopy. *J Am Chem Soc* 128:15292–15299.

Mittermaier A, Kay LE (2006): New tools provide new insights in NMR studies of protein dynamics. *Science* 312:224–228.

Montelione GT, Zheng D, Huang YJ, Gunsalus KC, Szyperski T (2000): Protein NMR spectroscopy in structural genomics. *Nat Struct Biol* 7:982–985.

Moseley HN, Monleon D, Montelione GT (2001): Automatic determination of protein backbone resonance assignments from triple resonance nuclear magnetic resonance data. *Methods Enzymol* 339:91–108.

Moseley HN, Sahota G, Montelione GT (2004): Assignment validation software suite for the evaluation and presentation of protein resonance assignment data. *J Biomol NMR* 28:341–355.

Neal S, Nip AM, Zhang H, Wishart DS (2003): Rapid and accurate calculation of protein 1H, 13C and 15N chemical shifts. *J Biomol NMR* 26:215–240.

Nederveen AJ, Doreleijers JF, Vranken W, Miller Z, Spronk CA, Nabuurs SB, Guntert P, Livny M, Markley JL, Nilges M, et al, (2005): RECOORD: a recalculated coordinate database of 500 + proteins from the PDB using restraints from the BioMagResBank. *Proteins* 59:662–672.

Nilges M, Habeck M, O'Donoghue SI, Rieping W (2006): Error distribution derived NOE distance restraints. *Proteins* 64:652–664.

Oh BH, Westler WM, Darba P, Markley JL Protein carbon-13 spin systems by a single two-dimensional nuclear magnetic resonance experiment. *Science* 240:(1988): 908–911.

Oh BH, Westler WM, Markley JL (1989): Carbon-13 spin system directed strategy for assigning cross peaks in the COSY fingerprint region of a protein. *J Am Chem Soc* 111:3083–3085.

Olson Jr JB , Markley JL (1994): Evaluation of an algorithm for the automated sequential assignment of protein backbone resonances: a demonstration of the connectivity tracing assignment tools (CONTRAST) software package. *J Biomol NMR* 4:385–410.

Ortiz-Polo G, Krishnamoorthi R, Markley JL, Live DH, Davis DG, Cowburn D (1986): Natural-abundance ^{15}N NMR studies of Turkey ovomucoid third domain. Assignment of peptide ^{15}N resonances to the residues at the reactive site region via proton-detected multiple-quantum coherence. *J Magn Reson* 68:303–310.

Ottiger M, Bax A (1999): Bicelle-based liquid crystals for NMR-measurement of dipolar couplings at acidic and basic pH values. *J Biomol NMR* 13:187–191.

Palmer III AG (1997): Probing molecular motion by NMR. *Curr Opin Struct Biol* 7:732–737.

Palmer III AG Kroenke CD, Loria JP (2001): Nuclear magnetic resonance methods for quantifying microsecond-to-millisecond motions in biological macromolecules. *Methods Enzymol* 339: 204–238.

Parkinson G, Voitechovsky J, Clowney L, Brünger AT, Berman HM (1996): Bond lengths and angles, DNA/RNA. *Acta Cryst D* 52:57–64.

Pavlov MY, Ehrenberg M (1996): Rate of translation of natural mRNAs in an optimized in vitro system. *Arch Biochem Biophys* 328:9–16.

Pervushin KV, Riek R, Wider G, Wüthrich K (1998): Transverse relaxation-optimized spectroscopy (TROSY) for NMR studies of aromatic spin systems in 13 C-labeled proteins. *J Am Chem Soc* 120:6394–6400.

Peterson FC, Lytle BL, Sampath S, Vinarov D, Tyler E, Shahan M, Markley JL, Volkman BF (2005): Solution structure of thioredoxin h1 from *Arabidopsis thaliana*. *Prot Sci* 14:2195–2200.

Peterson FC, Hayes PL, Waltner JK, Heisner AK, Jensen DR, Sander TL, Volkman BF (2006a): Structure of the SCAN domain from the tumor suppressor protein MZF1. *J Mol Biol* 363:137–147.

Peterson FC, Thorpe JA, Harder AG, Volkman BF, Schwarze SR (2006b): Structural determinants involved in the regulation of CXCL14/BRAK expression by the 26 S proteasome. *J Mol Biol* 363:813–822.

Peterson FC, Deng Q, Zettl M, Prehoda KE, Lim WA, Way M, Volkman BF (2007a): Multiple WASP-interacting protein recognition motifs are required for a functional interaction with N-WASP. *J Biol Chem* 282:8446–8453.

Peterson FC, Hayes PL, Waltner JK, Heisner AK, Jensen DR, Sander TL, Volkman BF (2007b): Erratum to "Structure of the SCAN domain from the tumor suppressor protein MZF1" [J Mol Biol 363 (2006) 137-147]. *J Mol Biol* 366:346.

Peti W, Page R, Moy K, O'Neil-Johnson M, Wilson IA, Stevens RC, Wüthrich K (2005): Towards miniaturization of a structural genomics pipeline using micro-expression and microcoil NMR. *J Struct Funct Genomics* 6:259–267.

Pickford AR, O'Leary JM (2004): Isotopic labeling of recombinant proteins from the methylotrophic yeast *Pichia pastoris*. *Methods Mol Biol* 278:17–33.

Prestegard JH (1998): New techniques in structural NMR: anisotropic interactions. *Nat Struct Biol* 5:517–522.

Prestegard JH, Bougault CM, Kishore AI (2004): Residual dipolar couplings in structure determination of biomolecules. *Chem Rev* 104:3519–3540.

Przytycka T, Aurora R, Rose GD (1999): A protein taxonomy based on secondary structure. *Nat Struct Biol* 6:672–682.

Reily MD, Thanabal V, Omecinsky DO (1992): Structure-induced carbon-13 chemical shifts: a sensitive measure of transient localized secondary structure in peptides. *J Am Chem Soc* 114:6251–6252.

Rieping W, Habeck M, Nilges M (2005): Inferential structure determination. *Science* 309:303–306.

Ruppeiner G (1995): Riemannian geometry in thermodynamic fluctuation theory. *Rev Modern Phys* 67:605–659.

Sali A, Blundell TL (1993): Comparative protein modelling by satisfaction of spatial restraints. *J Mol Biol* 234:779–815.

Salzmann M, Pervushin KV, Wider G, Senn H, Wüthrich K (1998): TROSY in triple-resonance experiments: new perspectives for sequential NMR assignment of large proteins. *Proc Natl Acad Sci USA* 95:13585–13590.

Sanders II CR Schaff JE, Prestegard JH (1993): Orientational behavior of phosphatidylcholine bilayers in the presence of aromatic amphiphiles and a magnetic field. *Biophys J* 64:1069–1080.

Sanville MC, Stols L, Quartey P, Kim Y, Dementieva I, Donnelly MI (2003): A less laborious approach to the high-throughput production of recombinant proteins in *Escherichia coli* using 2-liter plastic bottles. *Prot Expr Purif* 29:311–320.

Sawasaki T, Hasegawa Y, Tsuchimochi M, Kamura N, Ogasawara T, Kuroita T, Endo Y (2002a): A bilayer cell-free protein synthesis system for high-throughput screening of gene products. *FEBS Lett* 514:102–105.

Sawasaki T, Ogasawara T, Morishita R, Endo Y (2002b): A cell-free protein synthesis system for high-throughput proteomics. *Proc Natl Acad Sci USA* 99:14652–14657.

Schanda P, Brutscher B (2005): Very fast two-dimensional NMR spectroscopy for real-time investigation of dynamic events in proteins on the time scale of seconds. *J Am Chem Soc* 127:8014–8015.

Schanda P, Forge V, Brutscher B (2007): Protein folding and unfolding studied at atomic resolution by fast two-dimensional NMR spectroscopy. *Proc Natl Acad Sci USA* 104:11257–11262.

Seavey BR, Farr EA, Westler WM, Markley JL (1991): A relational database for sequence-specific protein NMR data. *J Biomol NMR* 1:217–236.

Sharma D, Rajarathnam K (2000): ^{13}C NMR chemical shifts can predict disulfide bond formation. *J Biomol NMR* 18:165–171.

Sippl MJ (1993): Recognition of errors in three-dimensional structures of proteins. *Proteins* 17: 355–362.

Spirin AS, Baranov VI, Ryabova LA, Ovodov SY, Alakhov YB (1988): A continuous cell-free translation system capable of producing polypeptides in high-yield. *Science* 242:1162–1164.

Spitzer F (1971): Markov random fields and Gibbs ensembles. *Amer Math Monthly* 78:142–154.

Sprangers R, Kay LE (2007): Quantitative dynamics and binding studies of the 20S proteasome by NMR. *Nature* 445:618–622.

Stuart AC, Borzilleri KA, Withka JM, Palmer III AG (1999): Compensating for variations in 1H-13C scalar coupling constants in isotope-filtered NMR experiments. *J Am Chem Soc* 121:5346–5347.

Styles P, Soffe NF, Scott CA, Cragg DA, Row F, White DJ, White PCJ (1984): A high-resolution NMR probe in which the coil and preamplifier are cooled with liquid helium. *J Magn Reson* 60:397–404.

Sugase K, Dyson HJ, Wright PE (2007): Mechanism of coupled folding and binding of an intrinsically disordered protein. *Nature* 447:1021–1025.

Szyperski T, Yeh DC, Sukumaran DK, Moseley HN, Montelione GT (2002): Reduced-dimensionality NMR spectroscopy for high-throughput protein resonance assignment. *Proc Natl Acad Sci USA* 99:8009–8014.

Terwilliger TC (2000): Structural genomics in North America. *Nat Struct Biol* 7:935–939.

Thornton JM, Todd AE, Milburn D, Borkakoti N, Orengo CA (2000): From structure to function: approaches and limitations. *Nat Struct Biol*, 7:(Suppl.): 991–994. [In process citation.]

Tian J, Valafar H, Prestegard JH (2001): A dipolar coupling based strategy for simultaneous resonance assignment and structure determination of protein backbones. *J Am Chem Soc* 123:11791–11796.

Tjandra N, Bax A (1997): Direct measurement of distances and angles in biomolecules by NMR in a dilute liquid crystalline medium. *Science* 278:1111–1114.

Tjandra N, Omichinski JG, Gronenborn AM, Clore GM, Bax A (1997): Use of dipolar 1 H-15 N and 1 H-13 C couplings in the structure determination of magnetically oriented macromolecules in solution. *Nat Struct Biol* 4:732–738.

Tolman JR, Flanagan JM, Kennedy MA, Prestegard JH (1995): Nuclear magnetic dipole interactions in field-oriented proteins: information for structure determination in solution. *Proc Natl Acad Sci USA* 92:9279–9283.

Torizawa T, Shimizu M, Taoka M, Miyano H, Kainosho M (2004): Efficient production of isotopically labeled proteins by cell-free synthesis: a practical protocol. *J Biomol NMR* 30:311–325.

Tugarinov V, Kanelis V, Kay LE (2006): Isotope labeling strategies for the study of high-molecular-weight proteins by solution NMR spectroscopy. *Nat Protoc* 1:749–754.

Tuinstra RL, Peterson FC, Elgin ES, Pelzek AJ, Volkman BF (2007): An engineered second disulfide bond restricts lymphotactin/XCL1 to a chemokine-like conformation with XCR1 agonist activity. *Biochemistry* 46:2564–2573.

Tyler RC, Aceti DJ, Bingman CA, Cornilescu CC, Fox BG, Frederick RO, Jeon WB, Lee MS, Newman CS, Peterson FC, et al, 2005a Comparison of cell-based and cell-free protocols for producing target proteins from the *Arabidopsis thaliana* genome for structural studies. *Proteins* 59:633–643.

Tyler RC, Sreenath HK, Singh S, Aceti DJ, Bingman CA, Markley JL, Fox BG (2005b): Auto-induction medium for the production of [U-15N]- and [U-13C, U-15N]-labeled proteins for NMR screening and structure determination. *Prot Expr Purif* 40:268–278.

Ulrich EL, Markley JL, Kyogoku Y (1989): Creation of a nuclear magnetic resonance data repository and literature database. *Prot Seq Data Anal* 2:23–37.

Ulrich EL, Akutsu H, Doreleijers JF, Harano Y, Ioannidis YE, Lin J, Livny M, Mading S, Maziuk D, Miller Z, Nakatani E, Schulte CF, Tolmie DE, Wenger RK, Yao H, Markley J (2008): BioMagResBank. *Nucleic Acids Res* 36:D402–D408.

Vinarov DA, Lytle BL, Peterson FC, Tyler EM, Volkman BF, Markley JL (2004): Cell-free protein production and labeling protocol for NMR-based structural proteomics. *Nat Methods* 1:149–153.

Vinarov DA, Markley JL (2005): High-throughput automated platform for nuclear magnetic resonance-based structural proteomics. *Expert Rev Proteomics* 2:49–55.

Vinarov DA, Loushin Newman CL, Markley JL (2006): Wheat germ cell-free platform for eukaryotic protein production. *FEBS J* 273:4160–4169.

Vinarov DA, Loushin Newman CL, Tyler EM, Markley JL (2006): Protein production using the wheat germ cell-free expression system. In: Wingfield PT, editor. *Current Protocols in Protein Science*. John Wiley & Sons, Chapter 5.

Volkman BF, Lipson D, Wemmer DE, Kern D (2001): Two-state allosteric behavior in a single-domain signaling protein. *Science* 291:2429–2433.

Vriend G (1990): WHAT IF: a molecular modeling and drug design program. *J Mol Graph* 8:52–56.

Waltner JK, Peterson FC, Lytle BL, Volkman BF (2005): Structure of the B3 domain from *Arabidopsis thaliana* protein At1g16640. *Prot Sci* 14:2478–2483.

Walton WJ, Kasprzak AJ, Hare JT, Logan TM (2006): An economic approach to isotopic enrichment of glycoproteins expressed from Sf9 insect cells. *J Biomol NMR* 36:225–233.

Wand AJ (2001): Dynamic activation of protein function: a view emerging from NMR spectroscopy. *Nat Struct Biol* 8:926–931.

Wang L, Eghbalnia HR, Bahrami A, Markley JL (2005): Linear analysis of carbon-13 chemical shift differences and its application to the detection and correction of errors in referencing and spin system identifications. *J Biomol NMR* 32:13–22.

Wang L, Eghbalnia HR, Markley JL (2006): Probabilistic approach to determining unbiased random-coil carbon-13 chemical shift values from the protein chemical shift database. *J Biomol NMR* 35:155–165.

Wang Y, Jardetzky O (2002): Probability-based protein secondary structure identification using combined NMR chemical-shift data. *Prot Sci* 11:852–861.

Westler WM, Ortiz-Polo G, Markley JL (1984): Two-dimensional 1 H-13C chemical-shift correlated spectroscopy of a protein at natural abundance. *J Magn Reson* 58:354–357.

Westler WM, Kainosho M, Nagao H, Tomonaga N, Markley JL (1988a): Two-dimensional NMR strategies for carbon–carbon correlations and sequence-specific assignments in carbon-13 labeled proteins. *J Am Chem Soc* 110:4093–4095.

Westler WM, Stockman BJ, Hosoya Y, Miyake Y, Kainosho M, Markley JL (1988b): Correlation of carbon-13 and Nitrogen-15 chemical shifts in uniformly labeled proteins by heteronuclear two-dimensional NMR spectroscopy. *J Am Chem Soc* 110:6265–6258.

Wiederstein M, Sippl MJ (2007): ProSA-web: interactive web service for the recognition of errors in three-dimensional structures of proteins. *Nucl Acids Res* 35:W407–410.

Willard L, Ranjan A, Zhang H, Monzavi H, Boyko RF, Sykes BD, Wishart DS (2003): VADAR: a web server for quantitative evaluation of protein structure quality. *Nucl Acids Res* 31:3316–3319.

Williamson MP, Kikuchi J, Asakura T (1995): Application of 1 H NMR chemical shifts to measure the quality of protein structures. *J Mol Biol* 247:541–546.

Wishart DS, Sykes BD, Richards FM (1992): The chemical shift index: a fast and simple method for the assignment of protein secondary structure through NMR spectroscopy. *Biochemistry* 31:1647–1651.

Wishart DS, Sykes BD (1994a): The ^{13}C chemical shift index: a simple method for the identification of protein secondary structure using ^{13}C chemical shifts. *J Biomol NMR* 4:171–180.

Wishart DS, Sykes BD (1994b): The ^{13}C chemical-shift index: a simple method for the identification of protein secondary structure using ^{13}C chemical-shift data. *J Biomol NMR* 4:171–180.

Wishart DS, Watson MS, Boyko RF, Sykes BD (1997): Automated 1H and ^{13}C chemical shift prediction using the BioMagResBank. *J Biomol NMR* 10:329–336.

Wishart DS, Case DA (2001): Use of chemical shifts in macromolecular structure determination. *Methods Enzymol* 338:3–34.

Wood MJ, Komives EA (1999): Production of large quantities of isotopically labeled protein in *Pichia pastoris* by fermentation. *J Biomol NMR* 13:149–159.

Wüthrich K (1986): *NMR of Proteins and Nucleic Acids.* New York, NY: Wiley-Interscience.

Xu R, Ayers B, Cowburn D, Muir TW (1999): Chemical ligation of folded recombinant proteins: segmental isotopic labeling of domains for NMR studies. *Proc Natl Acad Sci USA* 96:388–393.

Xu XP, Case DA (2001): Automated prediction of N-15, C-13(alpha), C-13(beta) and C-13' chemical shifts in proteins using a density functional database. *J Biomol NMR* 21:321–333.

Yamazaki T, Otomo T, Oda N, Kyogoku Y, Uegaki K, Ito N, Ishino Y, Nakamura H (1998): Segmental isotope labeling for protein NMR using peptide splicing. *J Am Chem Soc* 120:5591–5592.

Yang H, Guranovic V, Dutta S, Feng Z, Berman HM, Westbrook JD (2004): Automated and accurate deposition of structures solved by X-ray diffraction to the Protein Data Bank. *Acta Crystallogr D* 60:1833–1839.

Yokoyama S (2000): RIKEN Structural Genomics Initiative. International Conference on Structural Genomics 2000, November 2–5, Yokohama, Japan. p 40.

Yokoyama S (2003): Protein expression systems for structural genomics and proteomics. *Curr Opin Chem Biol* 7:39–43.

Zhang H, Neal S, Wishart DS (2003): RefDB: a database of uniformly referenced protein chemical shifts. *J Biomol NMR* 25:173–195.

Zhu G, Kong XM, Sze KH (1999): Gradient and sensitivity enhancement of 2D TROSY with water flip-back, 3D NOESY-TROSY and TOCSY-TROSY experiments. *J Biomol NMR* 13:77–81.

Zhu G, Xia YL, Sze KH, Yan XZ (1999): 2D and 3D TROSY-enhanced NOESY of N-15 labeled proteins. *J Biomol NMR* 14:377–381.

Zimmerman DE, Kulikowski CA, Wang L, Lyons BA, Montelione GT (1994): Automated sequencing of amino acid spin systems in proteins using multidimensional HCC(CO)NH-TOCSY spectroscopy and constraint propagation methods from artificial intelligence. *J Biomol NMR* 4:241–256.

Zimmerman DE, Kulikowski CA, Huang Y, Feng W, Tashiro M, Shimotakahara S, Chien C, Powers R, Montelione GT (1997): Automated analysis of protein NMR assignments using methods from artificial intelligence. *J Mol Biol* 269:592–610.

6

ELECTRON MICROSCOPY IN THE CONTEXT OF STRUCTURAL SYSTEMS BIOLOGY

Niels Volkmann and Dorit Hanein

INTRODUCTION

As modern molecular biology moves from single molecules toward more complex multi-molecular machines, the need for structural information about these assemblies grows. Nuclear magnetic resonance (NMR) spectroscopy (see Chapter 5) and X-ray crystallography (see Chapter 4) are well-established approaches for obtaining atomic structures of biological macromolecules, but it has become increasingly clear that the structures of individual components of assemblies can only be a first step to understanding a biological phenomenon.

Based on the analysis of genome sequences, it was suggested that life depends on about 200–300 core biological processes (Martin and Drubin, 2003). Each of these processes involves multiple proteins, often organized into large heterogeneous assemblies, with a wide range of morphologies and complexity. The "parts list" that has emerged from the genome project and from structural genomics is far from a "wiring diagram" that we need to understand these processes. Detailed knowledge of the parts of a system usually provides only limited insight into the dynamics and function of the system as a whole. A complete understanding requires not only knowledge of the transient and steady-state structures at near-atomic details; it also requires knowledge of structural pathways and ligand interactions in a cellular context.

Due to dramatic improvements in experimental methods and computational techniques, electron microscopy has matured into a powerful and diverse collection of methods that allow the visualization of the structure and the dynamics of an extraordinary range of biological assemblies at resolution spanning from molecular (about 2–3 nm) to near atomic

Structural Bioinformatics, Second Edition   Edited by Jenny Gu and Philip E. Bourne
Copyright © 2009 John Wiley & Sons, Inc.

(0.3 nm). Many of the restrictions of X-ray crystallography or NMR spectroscopy do not apply to electron microscopy. Crystalline order is helpful but not necessary, there is no upper size limit for the structures studied, the quantities of sample needed are relatively small, and cryo-methods enable the observation of molecules in their native aqueous environment (Dubochet et al., 1988). All in all, imaging of large and multicomponent cellular machinery close to physiological conditions is possible using electron microscopy and image analysis.

In the early years of electron microscopy, electron micrographs of molecules in a thin film of heavy atom stain were used to produce structures that were interpreted directly. Later, the interpretation of the two-dimensional images as projected density summed along the direction of the electron beam led to the ability to reconstruct the three-dimensional object that was imaged (DeRosier and Klug, 1968). The 1970s marked the development of electron cryomicroscopy in which macromolecules are examined without the use of heavy atom stains by embedding the specimens in a thin film of rapidly frozen water (Dubochet et al., 1988). This use of unstained specimens led to structure determination of the molecules themselves rather than the structure of a stain-excluding volume (negative stain). The staining procedures greatly enhance the signal-to-noise ratio for imaging of biological macromolecules but are severely limited by preservation artifacts. The signal-to-noise ratio in electron cryomicroscopy is much lower, but it allows imaging of biological specimens close to their native, fully hydrated state.

Up to July 2007, seven atomic resolution structures have been obtained by electron cryomicroscopy of thin two-dimensional crystalline arrays (Henderson et al., 1990; Kühlbrandt et al., 1994; Nogales, Wolf, and Downing, 1998; Murata et al., 2000; Gonen et al., 2004; Hiroaki et al., 2006; Holm et al., 2006). For most biological macromolecules and assemblies it has not yet been possible to determine their structure beyond 0.7–3 nm resolution by using electron microscopy and image analysis. While this resolution precludes atomic modeling directly from the data, near-atomic models can often be generated by combining high-resolution structures of individual components in a macromolecular complex with a low-resolution structure of the entire assembly.

In recent years, electron microscopy has emerged as the primary tool for correlating structural and dynamical information ranging from atomic resolution structures obtained by NMR or X-ray crystallography all the way up to whole microorganisms and tissue sections. Constant improvements in computational approaches and the wide applicability of relatively low-cost computing clusters have aided the automation of experiment and data analysis, thus allowing the field to efficiently address questions that were out of reach only a few years ago. A combination of electron microscopy with bioinformatics-based technologies such as pattern recognition, database searches, or homology modeling is used to generate molecular models of large assemblies as well as to have increasing applications for interpreting structures of much higher complexity including whole microorganisms (see, for example, Nickell et al., 2003; Kurner, Frangakis, and Baumeister, 2005; Komeili et al., 2006). The hope for the future is that electron microscopy, together with other structural bioinformatics tools, will enable a comprehensive, high-resolution structural mapping of entire cells, thus laying the structural foundation for understanding biological systems at unprecedented detail and providing the spatial context for systems biology (Aloy and Russell, 2005; Bork and Serrano, 2005; Aloy and Russell, 2006). This chapter gives an overview of key aspects of electron microscopy and puts them into context with structural bioinformatics and systems biology. More information on various aspects of electron microscopy can be obtained from several recent review articles (Chiu et al., 2005; Jiang and Ludtke, 2005; Lucic et al., 2005;

Chiu et al., 2006; Frey et al., 2006; Leis et al., 2006; Mitra and Frank, 2006; Renault et al., 2006).

ELECTRON OPTICS AND IMAGE FORMATION

Electron cryomicroscopy provides three-dimensional electron density maps of macromolecules very similar to the electron density maps determined by X-ray crystallography. In the imaging process of electron microscopy, the incident electron beam passes through the specimen and individual electrons are either unscattered or scattered by the specimen. Scattering occurs either elastically, with no loss of energy, or inelastically, with energy transfer from scattering electrons to electrons in the specimen, thus leading to radiation damage. The electrons emerging from the specimen are collected and focused by the imaging optics of the microscope (Figure 6.1). In the viewing area, either the electron diffraction pattern or the image can be seen directly by eye on the phosphor screen, detected by a CCD camera, or recorded on photographic film or imaging plate.

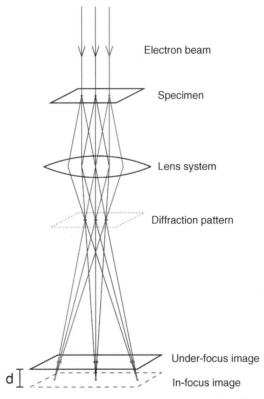

Figure 6.1. Simplified schematic diagram showing the principle of image formation in the electron microscope. The incident electron beam illuminates the specimen. Scattered and unscattered electrons are collected by the objective lens system and focused back to form first an electron diffraction pattern and then an image. In practice, an in-focus image has no contrast, so images are recorded with the objective lens system slightly defocused (d), taking advantage of the out-of-focus phase contrast mechanism.

Structural information can only be obtained from coherent, elastic scattering of electrons. The amplitudes and phases of the scattered electron beam are directly related to the Fourier components of the atomic distribution in the specimen. When the scattered beams are recombined with the unscattered beam in the image, they create an interference pattern that, for thin specimens, is directly related to density variations in the specimen. Thin samples of biological molecules fulfill the weak phase approximation, a theory of image formation that is used to describe the phase-contrast images of weak scattering specimens. Although there is practically no contrast when the image is in focus, spherical aberration and defocus combine to give a phase-contrast image. The imaging characteristics are described by the contrast transfer function (CTF), which can be derived from the weak phase approximation. The CTF describes the contrast transfer as a function of spatial frequency. It has alternating bands of positive and negative contrast (Figure 6.2), appearing in diffraction images as Thon rings. In order to restore the correct structural information, the images must be corrected for the CTF. For high-resolution studies, images must be collected at a range of defocus values to fill in missing data caused by zeros in the CTF whose positions vary with the actual defocus.

The most important consequence of inelastic scattering is the deposition of energy in the specimen, leading to radiation damage. Scattering events with X-rays are about

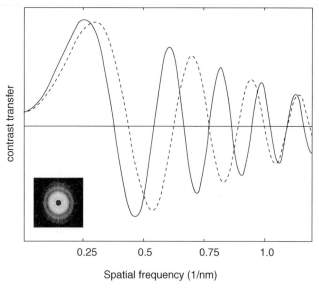

Spatial frequency (1/nm)

Figure 6.2. Graph of contrast transfer function plotted against spatial frequency. The lines represent underfocus levels of 2 μm (dashed line) and 3 μm (solid line). The inset shows the typical Thon rings that are caused by the modulation effect of the CTF and are visible in diffraction patterns of electron micrographs recorded with defocused lens systems. The finite source size and non-monochromatic electron beam both cause Gaussian decay of the CTF. In practice, contrast is much weaker at high spatial frequencies as a result of further decay caused by other factors like inelastic scattering or specimen drift. When the CTF crosses the zero line (the straight line in the plot), the phases need to be flipped to correct the contrast reversal at the corresponding spatial frequencies. At the spatial frequencies where the CTF actually approaches zero, no information is present. By collecting various data sets with different defocus, these information gaps can be filled.

1000 times more damaging than those with electrons, but the cross section for electron scattering is 10^5 times greater. Therefore, radiation damage is a much more serious problem for electron microscopy, and cooling is essential for imaging of high-resolution detail. Still, radiation damage is limiting even at low temperature. Therefore, the image exposures are chosen to be the weakest possible to obtain a measurable signal. Consequently, the signal-to-noise ratio of the recorded images is extremely low. This low signal-to-noise limits the amount of information that can be obtained from an image of a single biological macromolecule. The high-resolution structure cannot be determined from a single molecule alone but requires the averaging of the information from at least 10,000 molecules in theory and even more in practice (Henderson, 1995). For example, 5 million molecules were used to determine the atomic structure for bacteriorhodopsin (Henderson et al., 1990).

The electron optical resolution of electron microscopy is in the order of 0.1 nm, much coarser than the diffraction limit imposed by the electron wavelength. The resolution is restricted by the small aperture size needed because of aberrations in the electromagnetic lenses. However, additional resolution restrictions for macromolecules come from radiation sensitivity, specimen movement in the electron beam, and low contrast. These effects have so far limited the resolution of macromolecular imaging to 0.3–0.4 nm in the best cases. Particularly for single particles, the loss of contrast beyond 2 nm resolution is a major limitation. This limit can be extended by the use of a field emission gun (FEG) electron source. The small apparent source size gives a highly coherent illumination that provides much better phase contrast at high resolution.

A microscope can be used either as an imaging or as a diffraction instrument. Unlike diffraction experiments, in which phase information is lost, the image recorded by microscopy contains both amplitude and phase information. However, a much higher electron dose is needed to record the image than for the electron-diffraction pattern. Mechanical stability is particularly critical for obtaining phases. Movement does not affect the amplitudes in electron diffraction provided the crystalline area stays in the beam, but any movement during image recording distorts the phases and may easily make the image unusable.

THREE-DIMENSIONAL RECONSTRUCTION

The determination of three-dimensional structure by electron cryomicroscopy follows a common scheme for all macromolecules. Briefly, each sample must be prepared in a relatively homogeneous, aqueous form. This specimen is then rapidly frozen (vitrified) as a thin film, transferred to the electron microscope and imaged under low-dose conditions (less than 5000 electrons/nm^2). Before image analysis, the best micrographs are selected in which the electron exposure is optimal; there is no specimen movement, there is minimal astigmatism, and there is a reasonable amount of defocus. Interesting areas are boxed out for further use.

An electron micrograph consists of two-dimensional projections of a three-dimensional object. To retrieve its three-dimensional structure, sufficiently sampled angular views of the object need to be aligned and combined. An object might possess crystalline, helical, icosahedral, or rotational symmetry; or no symmetry at all. The presence of symmetry means that redundant motifs are provided in the specimen, thereby enhancing the signal-to-noise ratio of the image, providing geometric constraints for the alignment of the objects, and

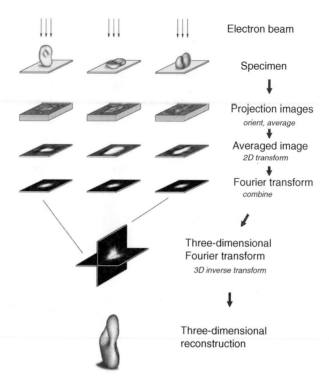

Figure 6.3. The principle of 3D reconstruction from 2D projections using the Arp2/3 complex (Volkmann et al., 2001a) as an example. The electron beam illuminates the molecule in different orientations. This gives rise to sets of 2D projection images that need to be classified and oriented. To enhance the signal, as many projection images of the same molecule orientation as possible are averaged. The 2D Fourier transform of each projection (shown below the averaged projection images) is a section through the 3D Fourier transform of the underlying structure. The 3D Fourier transform is represented by two intersecting transform sections derived from the top and front view of the structure. Once enough sections are available, the full 3D Fourier transform can be interpolated and inverse transformed into a 3D density reconstruction.

reducing the number of images required to obtain a reconstruction. The exact steps of image analysis and image acquisition vary according to the symmetry and nature of the specimen. There are three basic tasks that are common to all samples (Figure 6.3). First, images of the object must be obtained in a sufficient number of orientations. This can be achieved using the natural design of the object (helical filaments or icosahedral viruses), using experimental design (by tilting the sample holder in the microscope to specified angles), or through a random distribution of orientations (single particles). Second, the orientation and center of the object needs to be determined. Iterative refinement of these parameters is usually carried out by cross-comparison between different images or by projection images of preliminary models. Third, image shifts must be applied computationally, either in real space or Fourier space, to bring all views of the object to a common origin. Only then a three-dimensional reconstruction can be calculated. The different sample geometries require different data-collection schemes and image-processing approaches. These are described in the following sections.

Crystalline Arrays

Ordered protein arrays, often only one-molecule thick, are too insubstantial to be analyzed by X-ray crystallography. Due to the comparatively large cross section of electron scattering and the resulting increase in scattering power, structure determination using electron beams is feasible for these samples. The strategy followed is to build up a three-dimensional Fourier transform of the repeating unit by recording data from arrays tilted through various angles. The calculated Fourier transform of each image provides phases and amplitudes of a central plane through the three-dimensional Fourier transform. For well-ordered arrays, electron diffraction yields amplitudes superior to those calculated from the image and are consequently incorporated instead. In addition to the direct measurement of phases that distinguishes electron crystallography from X-ray crystallography, the images can be corrected for short-range lattice disorder such as bending or wrinkling of the arrays (Henderson et al., 1990; Kunji et al., 2000). This extends the resolution of the calculated transform from each image. There is a missing cone of data because the maximum tilt angle possible is 60–70°. This results in anisotropic resolution and distortions: features parallel to the plane of the array are better resolved than features perpendicular to the plane. Other possible complications with obtaining high-resolution structures from crystalline arrays include difficulties in finding proper crystallization conditions and lattice defects. Once images of sufficient quality are acquired, lattice reflections calculated from each image are assembled into the three-dimensional Fourier transform, which is interpolated and inversely transformed to give the three-dimensional density map.

Electron crystallography is well suited for intrinsic membrane proteins that are visualized in their natural environment embedded in a lipid bilayer. Bacteriorhodopsin, which naturally exists as planar arrays in the cell, was the first molecule with an atomic model provided by electron crystallography (Henderson et al., 1990). The plant light-harvesting complex is another example of an atomic model entirely based on electron crystallographic analysis (Kühlbrandt et al., 1994). More recently, electron-crystallography-based models of the aquaporin membrane protein family (Murata et al., 2000; Gonen et al., 2004; Hiroaki et al., 2006) and of the microsomal glutathione transferase 1 (Holm et al., 2006) also became available. In addition, the three-dimensional structures of about a dozen or so membrane proteins were determined at a resolution of 0.5–1 nm (for recent review see Renault et al., 2006). Other examples include an atomic model obtained for tubulin, a major component of the cytoskeleton (Nogales, Wolf, and Downing, 1998), and several other proteins that are not integral membrane components have also been studied at somewhat lower resolution (reviewed in Stahlberg et al., 2001). Finally, a crystallographic approach was also used to obtain a subnanometer resolution reconstruction of the quasicrystalline acrosomal bundle (Schmid et al., 2004).

Helical Assemblies

Many biological assemblies occur naturally in helical form, particularly cytoskeleton filaments. These filamentous structures are particularly attractive targets for electron microscopy as they are not usually amenable to crystallization due to their natural tendency to polymerize. To resolve their structures, these filamentous structures were traditionally targeted by specifically developed helical reconstruction approaches that make explicit use of the helical symmetry in Fourier space (DeRosier and Moore, 1970). Actin filaments studied using electron microscopy, for example, include the following: actin filaments with bound bacterial proteins (Galkin et al., 2002b) and with bound capping proteins (McGough et al., 1997; Galkin et al., 2003), actin complexed with domains of cytoskeletal proteins such as α-actinin (McGough, Way, and DeRosier, 1994), fimbrin (Hanein, Matsudaira, and DeRosier, 1997; Hanein et al., 1998), utrophin (Moores, Keep, and Kendrick-Jones, 2000; Galkin et al., 2002a), talin (Lee et al., 2004), and vinculin (Janssen et al., 2006). These structures have all been studied using electron microscopy and helical reconstruction techniques (see, for example, De La Cruz et al., 2000; Orlova and Egelman, 2000; Steinmetz et al., 2000; Galkin et al., 2002c).

Conformational changes in motor proteins (myosin, kinesin, and NCD) are also being investigated by using electron cryomicroscopy and helical reconstruction techniques with actin- or microtubule-bound motors (for reviews see Vale and Milligan, 2000; Volkmann and Hanein, 2000). Many other helical structures are being investigated using these techniques including myosin thick filaments (Woodhead et al., 2005), nucleoprotein filaments (Galkin et al., 2005; Esashi et al., 2007), filamentous bacteriophage (Wang et al., 2006), and bacterial pili (Mu, Egelman, and Bullitt, 2002; Craig et al., 2006).

Helical crystallization has also been used for structure determination, particularly for membrane proteins, which can be induced to form tubular crystals. Diffraction from a helix occurs on a set of layer lines related to the pitch repeat. An advantage of helical diffraction analysis over crystalline arrays is that the repeating unit is naturally presented over a range of angular views and tilting is not usually necessary for three-dimensional reconstruction (DeRosier and Moore, 1970). However, it is more difficult to collect a large enough number of the repeating units and to achieve high resolution. Possible complications that arise during helical analysis include partial decoration of the helix under study, bending of the helix in the plane of the image and perpendicular to the plane, or imperfect helical symmetry. As a consequence, the majority of the helical reconstructions are in the 1.5–3 nm range, but 0.4–0.5 nm resolution has also been achieved using tubular crystals of the nicotinic acetylcholine receptor (Miyazawa et al., 1999) and for bacterial flagellar filaments (Yonekura, Maki-Yonekura, and Namba, 2003).

Hybrid techniques that are primarily based on addition of aspects from single-particle analysis to the reconstruction process (Egelman, 2000; Holmes et al., 2003; Egelman, 2007; Yonekura and Toyoshima, 2007) have recently gained popularity. These techniques address some of the potential problems of traditional helical reconstruction techniques such as partial decoration and mixed populations. Helical structures that were previously inaccessible to image analysis for these reasons can now be processed (for a recent overview, see Egelman, 2007). However, there still appears to be an advantage in using the more traditional helical reconstruction techniques for well-ordered helical arrays (Pomfret, Rice, and Stokes, 2007).

Single-Particle Analysis

Isolated single particles (macromolecules) offer certain practical advantages for electron cryomicroscopy. Because there is no requirement for crystallization, virtually any particle is eligible. For more detailed information on single-particle reconstruction see the textbook by Frank (Frank, 2006b). In summary, single-particle reconstruction takes advantage of the fact that molecular complexes often exist as many copies in the specimen, visible as isolated particles, distinguished only by their orientations. Thus, a snapshot of a sufficiently large number of particles covers the complete angular range of possible orientations. A small number of micrographs, each of a different field of view, often contain enough particles to reconstruct the molecule in three dimensions. The averaging of many copies of the structure by this approach reduces the noise and carries potential for reaching high resolution, even for particles without symmetry. Computationally, the most challenging task is to determine and refine the particle orientations.

Orientation analysis is much more straightforward if the sample is biochemically homogeneous and is also facilitated by internal particle symmetry. In order to generate a high-resolution reconstruction, an initial starting model at moderate resolution (~3 nm) is acquired as a first step. Methods for doing that include the "random conical tilt" method (Radermacher, 1988), "angular reconstitution" (van Heel, 1987), the use of models of related structures, or electron tomography (see below), and, most recently, the "orthogonal tilt" reconstruction method (Leschziner and Nogales, 2006). The second stage involves cyclic model-based refinement. At each stage, more accurate values for the viewing angles of each particle are obtained by matching it against projections of the current model, each of which represents a particular view. Translational refinement is also fine-tuned. Then, a refined reconstruction is calculated and the procedure is iterated exhaustively until convergence.

The resolution attained depends on several factors, including the number of particles in the data set, the accuracy of the orientation parameters and the quality of the original data. Even under the most favorable conditions macromolecules yield noisy, low-contrast images. For the determination of a successful orientation, a sufficient signal for discrimination among projections must be generated. In practice, this requirement places a lower size limit on macromolecules that can be analyzed by this technique with the current limit for unstained cryosamples being at 200 kDa for particles without internal symmetry (Volkmann et al., 2001a; Adair and Yeager, 2002). The use of staining significantly increases the signal-to-noise ratio and therefore allows reconstruction of even smaller particles but staining also caries the risk of introducing artifacts (Cheng et al., 2006). Particles of 0.5–10 MDa are considered optimal for high-resolution single-particle analysis.

Because 1 nm is the spacing typical of close-packed α-helices, density maps with resolution higher than 1 nm are particularly informative, especially for proteins with high α-helical contents. The first single-particle analyses at resolutions below 1 nm were of the icosahedral capsids of hepatitis B virus (Böttcher, Wynne, and Crowther, 1997; Conway et al., 1997) and papillomavirus (Trus et al., 1997). Icosahedral particles have the advantage of a 60-fold symmetry, which reduces the amount of particles required for averaging. As a consequence, the number of icosahedral structures at resolution better than 1 nm has been steadily increasing. Cytoplasmic polyhedrosis virus (Zhou et al., 2003), rice dwarf virus (Zhou et al., 2001), reovirus (Zhang et al., 2003b), herpes simplex virus (Zhou et al., 2000), adenovirus (Fabry et al., 2005; Saban et al., 2005), Semliki forest virus (Mancini et al., 2000), dengue virus (Zhang et al., 2003a), the icosahedral core of pyruvate dehydrogenase (Borgnia et al., 2004), and phages PM2 (Huiskonen et al., 2004), P22 (Jiang et al., 2003), and φ29 (Morais et al., 2005) were all solved at subnanometer resolution.

The number of lower symmetry particles solved below 1 nm resolution is much more limited. These include the structures of the D6 hexagonal barrel form of clathrin (Fotin et al., 2004), the isolated connector from phage T7 (Agirrezabala et al., 2005), the skeletal muscle ryanodine receptor (Ludtke et al., 2005), and three reconstructions of ribosomes or ribosomal subunits (Matadeen et al., 1999; Valle et al., 2003; Menetret et al., 2005). Curiously, the highest resolution (0.6 nm) single-particle reconstruction published to date is GroEL, which possesses only 14-fold symmetry (Ludtke et al., 2004). Although these results are very encouraging, the vast majority of single-particle analyses today yield reconstructions with resolutions considerably lower than 1 nm, thus precluding direct analysis and modeling of α-helical arrangements.

Electron Tomography

The most general method for obtaining three-dimensional information by electron microscopy is tomography. The method is not only applicable to isolated particles but also to pleomorphous structures such as mitochondoria, other organelles, or even whole cells (for recent reviews see Lucic, Forster, and Baumeister, 2005; McIntosh, Nicastro, and Mastronarde, 2005; Frey et al., 2006; Leis et al., 2006). Special issues of the *Journal of Structural Biology* (Koster and Agard, 1997; McEwen and Koster, 2002), as well as textbooks edited by Frank (Frank, 2006a) and McIntosh (McIntosh, 2007) are all devoted to electron tomography. In this technique, a series of images is taken of a single specimen as the specimen is tilted over a wide range of angles. Sometimes, for better angular coverage, another tilt series is taken with the specimen rotated by 90°.

Tomography is the only method available for reconstruction of specimen with unique structure (no multiple copies). An entire cell for example would fall into this category, as it would be impossible to find two cells that are exactly identical. Today, with the use of computer-controlled microscopes and the availability of charge-coupled device (CCD) cameras, it has become possible to image large-scale structures at a resolution of better than 5 nm, with data sets comprising up to 150 projections with a cumulative dose as low as 5000 electrons/nm^2. Tomography is undergoing considerable growth at present due to the realization that molecular information can be obtained from unstained, frozen-hydrated whole cells (Grimm et al., 1998; Medalia et al., 2002), potentially allowing to map entire proteomes at high resolution and in their native environment (Nickell et al., 2006). The main disadvantage of the tomographic approach is that radiation damage builds up during the multiple exposures as the specimen is being tilted. Although data collection with extremely low doses of radiation is under development, the experimental realization and image processing under such conditions still poses great challenges. Similar to electron crystallography, there is a missing wedge of data because of the maximum tilt angle, resulting in anisotropic resolution and distortions. Features perpendicular to the electron beam are better resolved than features parallel to the beam. Owing to the low electron dose, the signal-to-noise ratio is well below one for frozen-hydrated samples.

While the main area of application for electron tomography is large, multicomponent objects, structures of purified single macromolecules have also been determined in negative stain (Rockel et al., 1999) as well as in vitrified ice (Nitsch et al., 1998; Liu et al., 2006; Nickell et al., 2007). In these cases, the three-dimensional, noisy tomograms of the single macromolecules were aligned and averaged in three dimensions. To generate an averaged, high-quality three-dimensional structure, only a few hundred particles are required using tomographic techniques (Koster et al., 1997), an advantage of using electron tomography instead of single-particle analysis that requires several thousand particles for structure determination.

Averaging of structures imaged in their native cellular environment has also been done in some cases, primarily for specimen with considerable symmetry to facilitate alignment. These include microtubule-related structures (Nicastro et al., 2006; Sui and Downing, 2006), nuclear pore complex (Beck et al., 2004), and envelope proteins of viruses (Förster et al., 2005; Zanetti et al., 2006; Zhu et al., 2006). Two studies used averaging of asymmetric particles, namely, ribosomes (Ortiz et al., 2006) and flagellar motors (Murphy, Leadbetter, and Jensen, 2006). The resolution currently achievable by this tomographic single-particle approach can approach 2 nm for purified molecules and can be better than 3 nm for symmetrical particles in their cellular environment. For the asymmetrical particles the resolution was ~5 nm for ribosomes and 7 nm for the flagellar motor respectively.

The three-dimensional alignment and averaging of tomographic data is highly nontrivial owing to defocus gradients in the tilt series, exceedingly low signal-to-noise ratio, and the missing data caused by the tilt geometry. As a consequence, in spite of high self-consistency for both data sets (better than 3 nm resolution after averaging), significantly different averaged structures were observed for the AIDS virus envelope spike when different image processing procedures were used (Zanetti et al., 2006; Zhu et al., 2006).

Because samples for electron microscopy need to be thin enough for electrons to penetrate, only organisms or cellular protrusions with a maximum thickness of ~1 μm can be investigated without additional sample preparation steps. Sectioning of frozen-hydrated specimen is possible but still faces severe experimental difficulties (Al-Amoudi, Norlen, and Dubochet, 2004). As a consequence, electron tomography of frozen-hydrated samples is currently restricted to small prokaryotic cells (reviewed in Jensen and Briegel, 2007), viruses (reviewed in Grünewald and Cyrklaff, 2006), or thin protrusions of *Dictyostelium* cells (Medalia et al., 2002; Medalia et al., 2007). Larger cells or tissue are still largely the domain of heavily processed, stained sections (see, for example, Marsh, 2005; Briggman and Denk, 2006; Frey et al., 2006).

A recent promising development toward electron cryotomography of eukaryotic cell types that are normally too thick for cryotomography is the technology of "ventral membrane preparations" (Anderson et al., in preparation) where the top portion of the cell is removed and only the ventral portion of the cell stays attached to the substrate. The resulting samples are substantially less than 1 μm in thickness independent of cell type, the ventral portion is completely intact, and the samples are amenable to cryotomography (Anderson et al., in preparation).

Mixing and Matching

A recent trend is the innovative combination of the more traditional methodologies mentioned above to push the limits of electron microscopy even further. For example, single-particle image-processing methods can be exploited for disorder correction of subunits or groups of subunits in ordered assemblies such as helical filaments (Egelman, 2000; Egelman, 2007; Yonekura and Toyoshima, 2007) or crystalline arrays (Sherman et al., 1998; Verschoor, Tivol, and Mannella, 2001). On the other hand, helical orientation parameters (that tend to be more accurate than single-particle orientation parameters) can be used to aid single-particle reconstructions of asymmetric particles attached to the end of helices (Yonekura et al., 2000; Narita et al., 2006; Narita and Maeda, 2007). A combination of helical analysis and electron crystallographic techniques is being used to interpret two-dimensional paracrystalline arrays of filamentous structures (Sukow and DeRosier, 1998; Volkmann et al., 2001b), and electron tomography is being used to generate starting models for single-particle analysis (Walz et al., 1999).

Imaging Protein Dynamics

In electron cryomicroscopy, molecules can be imaged in their native, fully hydrated environment, unrestricted by a crystal lattice. This gives the opportunity to study

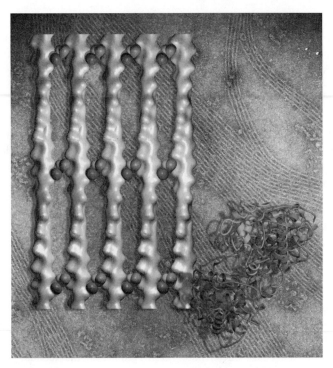

Figure 6.4. Example of a hybrid study that combines elements of electron crystallography and helical reconstruction with homology modeling and molecular docking approaches to elucidate the structure of an actin–fimbrin cross-link (Volkmann et al., 2001b). Fimbrin is a member of a large superfamily of actin-binding proteins and is responsible for cross-linking of actin filaments into ordered, tightly packed networks such as actin bundles in microvilli or stereocilia of the inner ear. The diffraction patterns of ordered paracrystalline actin–fimbrin arrays (background) were used to deduce the spatial relationship between the actin filaments (white surface representation) and the various domains of the cross-linker (the two actin-binding domains of fimbrin are pink and blue, the regulatory domain cyan). Combining this data with homology modeling and data from docking the crystal structure of fimbrin's N-terminal actin-binding domain into helical reconstructions (Hanein et al., 1998) allowed us to build a complete atomic model of the cross-linking molecule (foreground, color scheme as in surface representation of the array). Figure also appears in the Color Figure section.

macromolecules in their various functional states and to determine the associated conformational changes. Actin- and microtubule-based motor proteins are good examples where function-related motions can be detected by electron cryomicroscopy. Changes of the nucleotide state of several actin-bound myosin isoforms, for example, induce large conformational changes that are readily identified by eye (see, for example, Whittaker et al., 1995; Jontes and Milligan, 1997; Carragher et al., 1998). In addition, recent advances in computational methodology and image processing allow also more subtle changes to be identified and quantified in these systems (Volkmann et al., 2000; Volkmann et al., 2005; Trybus et al., 2007). Another example for glimpses of a molecular machine in motion is provided by recent electron microscopy studies of the ribosome, which began to unravel the structural changes associated with different functional states in the translation process

(Frank and Agrawal, 2000; VanLoock et al., 2000; Gabashvili et al., 2001; Spahn et al., 2001). Electron microscopy of the chaperonin GroEL has revealed a large repertoire of hinge rotations and allosteric movements in the chaperonin ATPase cycle (Ranson et al., 2001). The structures of herpes simplex virus (Heymann et al., 2003) and the phages HK97 (Lata et al., 2000) and P22 (Jiang et al., 2003) have also been studied in various functional states.

Because the freezing step in electron cryomicroscopy is very rapid (considerably less than 1 ms), it is possible to capture very short-lived structural states by spraying a ligand onto the specimen just before freezing. The spray method allowed capturing the acetylcholine-activated state of the nicotinic acetylcholine receptor, which has a lifetime of only 10 ms (Berriman and Unwin, 1994). A stopped-flow mixer system combined with an atomizer spray has been developed for actomyosin kinetic studies (Walker et al., 1999). However, particle states may not be fully synchronized if the half life of the structural state investigated is in the same order as the time needed for cryofixation. Particles at the leading edge of the sample holder will be frozen slightly earlier than particles at the trailing edge (Kasas et al., 2003).

A possibility to detect dynamics of mixed conformations directly in electron microscopic reconstructions is the analysis of the local variability using statistics on the single molecular units that contribute to the final average. Variants of this idea were developed for single-particle analysis (Liu, Boisset, Frank, 1995b; Penczek et al., 2006b) as well as for helical reconstructions (Rost, Hanein, and DeRosier, 1998). Reconstruction methods for dealing with mixed conformations are based on multivariate statistical analysis of the images using either multiple references, sometimes obtained by normal mode analysis of the initial reconstruction (Brink et al., 2004), or by using areas in the initial reconstruction identified as being highly variable in order to sort the data into different groups (Penczek, Frank, and Spahn, 2006a).

COMBINATION WITH OTHER APPROACHES: HYBRID METHODS

In the three-dimensional structure determination of macromolecules, X-ray crystallography covers the full range from small molecules to very large assemblies such as viruses with molecular masses of megadaltons. The limiting factors are expression, crystallization, stability, and homogeneity of the structure, and obtaining the correct phasing information for deciphering the diffraction pattern. In the case of NMR, structures can be determined from molecules in solution, but the size limit, although increasing, is presently of the order of 100 kDa. Dynamic aspects can be quantified, but again the structures of mixed conformational states cannot be determined. Electron microscopy provides complementary information to these other methods, being able to tackle very large assemblies as well as transient or mixed states, and usually requires small amounts of material. In addition, electron microscopy is capable of imaging macromolecules in their native cellular environment. However, for the majority of the biological specimens studied, it has not yet been possible to determine their structures beyond 1–3 nm. By combining electron microscopy and image analysis with other sources of information, the gap between high-resolution information, as obtained by X-ray crystallography and NMR, and lower-resolution information, such as that coming from light microscopy, can be bridged (Figure 6.4). Studies using this "hybrid methods" approach are becoming more and more common. A recent dedicated issue of the *Journal of Structural Biology* (Hanein, 2007) gives a good overview of current trends in this area.

Electron Microscopy and X-Ray Crystallography

Combining electron microscopy and crystallographic data can take two forms. When the whole specimen can be crystallized, the electron microscopy results can be used as an initial molecular replacement-phasing model. The principal difficulty here is the frequently poor overlap between the data from electron microscopy (highest resolution frequently 1–3 nm) and the X-ray data (spots with resolution lower than 3 nm are often hidden in the beam stop). If successful phasing can be initiated with the electron microscopic reconstruction, it can be used to find heavy atoms, as was the case in the analysis of a ribosomal subunit (Ban et al., 1998). Electron microscopic reconstructions can also be used to improve and extend crystallographic phases by noncrystallographic symmetry averaging (see, for example, Prasad et al., 1999) or as constraints in maximum-likelihood phasing procedures (Volkmann et al., 1995).

When the complete assembly of interest can be imaged by electron microscopy but cannot be crystallized, lower resolution information from electron microscopy can often be combined with the high-resolution information from atomic models of the assembly components. Manual fitting is widely used for this purpose (for a review see Baker and Johnson, 1996). In this method, the fit of the atomic model into isosurface envelopes calculated from the electron microscopic reconstruction is judged by eye and corrected manually until the fit "looks best." Objective scoring functions have been used occasionally to assess the quality and to refine the initial manual fit. If the components of the assembly under study are large molecules with distinctive shapes at the resolution of the reconstruction, manual fitting can often be performed with relatively little ambiguity (see, for example, Rayment et al., 1993; Smith et al., 1993). However, divergent models of the same complex docked by eye have also been reported (Hoenger et al., 1998; Kozielski, Arnal, and Wade, 1998).

Approaches aiming at automated, quantitative docking of atomic structures into lower resolution reconstructions from electron microscopy have been developed. Most of these methods employ global, exhaustive searches for the best fit, using various density correlation measures (Volkmann and Hanein, 1999; Rossmann, 2000), sometimes combined with density filtering operations (Roseman, 2000; Chacon and Wriggers, 2002), or by matching vector distributions derived by vector quantization of the atomic model and the reconstruction (Wriggers, Milligan, and McCammon, 1999). Conformational flexibility can be handled to some extent by using modular fitting of structural domains (Volkmann and Hanein, 2003) or by using refinement approaches that explicitly allow deformation of the structure (Wriggers and Birmanns, 2001; Chen et al., 2003; Tama et al., 2004; Velazquez-Muriel et al., 2006).

Open issues in the development of these automated docking approaches include estimation of fitting quality, validation of results, estimation of fitting errors, and detection of ambiguities. A promising concept in this regard is that of solution sets (Volkmann and Hanein, 1999; Volkmann and Hanein, 2003). In this approach, the global search is followed by a statistical analysis of the distribution of the fitting criterion. The analysis results in the definition of confidence intervals that lead eventually to solution sets. These sets contain all fits that satisfy the data within the error margin defined by the chosen confidence level. Structural parameters of interest can then be evaluated as properties of these sets. For example, the uncertainty of each atom position of the fitted structure can be approximated by calculating the root-mean-square deviation for each atom using all members of the solution set. Ambiguities in the fitting are clearly reflected in the shape of the solution set (Volkmann and Hanein, 2003). The size of the solution set can serve as a normalized goodness-of-fit criterion. The smaller the set, the better the data determines the position of the fitted atomic structure.

The statistical nature of the approach allows the use of standard statistical tests, such as Student's t-test, to evaluate differences between models in different functional states and to help model conformational changes. It also allows estimating the probability that a certain residue is involved in the interaction between two components (Figure 6.5). This probabilistic ranking of residues in terms of their involvement in binding gives a better starting point for the design of mutagenesis experiments.

Difference mapping between the density calculated from the fitted model and the reconstruction from electron microscopy (similar to F_c–F_o maps in crystallography) is a powerful tool in locating portions of the structure that are not present in the crystal structure. For example, recent fitting of myosin crystal structures into reconstructions of actin-bound

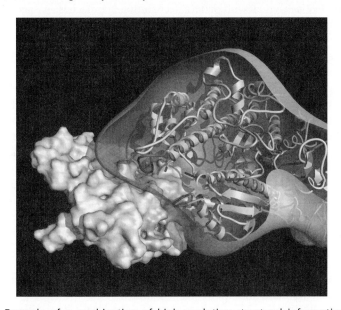

Figure 6.5. Example of a combination of high-resolution structural information from X-ray crystallography and medium-resolution information from electron cryomicroscopy (here 2.1 nm). Actin and myosin were docked into helical reconstructions of actin decorated with smooth-muscle myosin (Volkmann et al., 2000). Interaction of myosin with filamentous actin has been implicated in a variety of biological activities including muscle contraction, cytokinesis, cell movement, membrane transport, and certain signal transduction pathways. Attempts to crystallize actomyosin failed due to the tendency of actin to polymerize. Docking was performed using a global search with a density correlation measure (Volkmann and Hanein, 1999). The estimated accuracy of the fit is 0.22 nm in the myosin portion and 0.18 nm in the actin portion. One actin molecule is shown on the left as a molecular surface representation. The yellow area denotes the largest hydrophobic patch on the exposed surface of the filament, a region expected to participate in actomyosin interactions. The fitted atomic model of myosin is shown on the right. The transparent envelope represents the density corresponding to myosin in the 3D reconstruction. The solution set concept (see text) was used to evaluate the results and to assign probabilities for residues to take part in the interaction. The tone of red on the myosin model is proportional to this statistically evaluated probability (the more red, the higher the probability). Figure also appears in the Color Figure section.

myosin revealed the location of a functionally important myosin loop of about 10 residues in reconstructions of about 2 nm (Volkmann et al., 2000; Volkmann et al., 2003; Volkmann et al., 2005). This loop was not resolved in any of the crystal structures due to structural flexibility. Presumably, actin binding stabilizes the loop so it becomes detectable in the reconstructions.

Pattern Recognition

In the absence of atomic models, structural interpretation of large macromolecules at intermediate resolution is a difficult task, although structural information is still present. Individual domains, components, and structural elements must be identified in the reconstruction to yield a tentative atomic model. If the resolution of the reconstruction under study is better than 1 nm, long α-helices appear as cylindrical density and become identifiable by eye as well as with pattern-recognition approaches. A specialized method for helix recognition in subnanometer resolution reconstructions was developed (Jiang et al., 2001a). This approach incorporates a multistep process including cross-correlation, density segmentation, segment quantification, and explicit description of the identified helices. The final helices are represented as cylinders, each specified by its length and six orientation parameters. The information encoded by these parameters, as well as the relative position and orientation of the helices to each other, can then be used to identify homologous structures based on spatial arrangement of secondary structural elements of proteins using a library based on protein structures in the PDB or to match helical regions from secondary structure prediction with the derived helical fragments in order to map sequence to structure. The latter method was used to derive the folds of rice dwarf virus (Zhou et al., 2001), bacteriophage P22 (Jiang et al., 2003), and herpes simplex virus (Baker et al., 2003). The fold of rice dwarf virus was later largely confirmed by X-ray crystallography (Nakagawa et al., 2003).

In addition to α-helices, it is sometimes feasible to detect β-sheets in subnanometer resolution reconstructions (Kong and Ma, 2003; Baker, Ju, and Chiu, 2007). At this resolution β-sheets appear as roughly planar density segments but tend to adopt a variety of twists and bends with potentially vastly different shapes and sizes. These properties make β-sheets significantly more difficult to detect than α-helices.

Pattern recognition approaches are also being developed for the interpretation of noisy, low-resolution tomograms of cells and large subcellular structures. The main experimental difficulty in interpreting these types of tomograms is the assignment of density to a particular molecular component. In this context, the pattern recognition can be divided in a feature extraction and a template-matching step. First, features of interest are extracted from the tomogram using segmentation algorithms (Frangakis and Hegerl, 2002; Volkmann, 2002). The second step consists of classification of the extracted features by template matching using a database of known atomic structures. A feasibility study of pattern recognition algorithms was conducted using calculated volumes and tomographic data sets containing isolated particles (Böhm et al., 2000; Frangakis et al., 2002). The tests demonstrate the feasibility of this strategy by showing that a distinction between the proteasome and the thermosome, molecules with similar shape but slightly different dimensions, can be made with reasonable confidence even in experimental tomograms at a resolution between 4 and 8 nm. These studies were later followed up with a proof of concept for detection and mapping of ribosome distributions in two prokaryotic systems (Ortiz et al., 2006; Seybert, Herrmann, and Frangakis, 2006).

Structure Prediction

Recently, a substantial synergy is developing between electron cryomicroscopy and structure prediction methodology. Secondary structure prediction is being used together with helix and sheet recognition (see above) to determine connectivity between the detected structural elements and to predict the overall fold of the underlying atomic structure (Zhou et al., 2001; Baker et al., 2003; Jiang et al., 2003). Electron crystallographic reconstructions of transmembrane proteins at better than 1 nm resolution are being used to assist in constructing *ab initio* models with the aim of accurately predicting side-chain interactions between transmembrane helices (Baldwin, Schertler, Unger, 1997; Fleishman et al., 2004b). The underlying idea is to restrict the conformational search of the transmembrane-helices to the positions obtained from the reconstruction. To construct the model, the spatial constraints from the reconstructions are combined with general energetic considerations as well as biochemical data and bioinformatics techniques such as phylogenetic conservation analysis (Fleishman et al., 2004a).

If the resolution of the reconstruction is below 1 nm, helices or sheets cannot be reliably detected. If structural components are known to atomic resolution, density-based docking of these structures into the reconstruction can yield tentative atomic models. In the absence of such information, one possible way for generating high-resolution information on a structure is the combination of homology modeling and density-based docking (Topf and Sali, 2005). In fact, this combination is becoming increasingly common (see, for example, Volkmann et al., 2001b; Baker et al., 2002; Gao et al., 2003; Fotin et al., 2004; Liu, Taylor, and Taylor, 2004; Sengupta et al., 2004; Topf et al., 2005; Topf et al., 2006). Fitting a homology model based on a remotely related template is generally better than fitting the template itself, and the most accurate models can often be identified by the density score at resolutions as low as 1.5 nm (Topf et al., 2005).

If no homologous structures are available, structural information can still be extracted from electron microscopic reconstructions. A structural model for the herpes virus VP26 core domain using *ab initio* modeling and a subnanometer resolution reconstruction has been proposed (Baker et al., 2006a), and a recent study shows that scoring of low-resolution reconstructions can be used efficiently for selecting near-native structures from a large pool of not only homology models but also any type of theoretical models, even if the resolution of the reconstruction is below 2 nm (Shacham, Sheehan, and Volkmann, 2007).

FUTURE DIRECTIONS

Technical advances in data acquisition and computational methods have made it possible to reconstruct biological macromolecular complexes at resolutions ranging from 3 to 0.35 nm. The development of more powerful computational methods coupled with the availability of faster computers with large storage capabilities will continue to have a major impact on the field. Parallel computing should further speed up the image analysis and three-dimensional reconstruction process. Automation of data collection (see, for example, Ziese et al., 2002; Carragher et al., 2004; Mastronarde, 2005; Typke et al., 2005) and image analysis (see, for example, Ludtke, Baldwin, and Chiu, 1999; Jiang et al., 2001b; Zhu et al., 2004; Baker et al., 2006b; van der Heide et al., 2007) continues to be an important factor toward higher throughput and integration into the mainstream of structural biology. Electron microscopy provides complementary information to that from atomic-resolution techniques such as

X-ray crystallography or NMR. Further development of methods for combining these data sources by docking is likely to play a major role in the future. A closer integration of these docking methods with structural database searches and structure prediction is also an attractive possibility for the future.

Electron tomography in conjunction with pattern recognition and other structural bioinformatics tools will make it possible to provide a bridge between cell biological function and molecular mechanism. Together with emerging technologies such as double labeling for electron and light microscopy (Giepmans et al., 2005), these methods will not only allow high-resolution mapping of entire cells but also direct correlation with dynamics information. Thus, electron microscopy is likely to provide major contributions for defining the detailed spatiotemporal framework that is necessary for pushing systems biology to the next level (Bork and Serrano, 2005; Lemerle, Di Ventura, and Serrano, 2005; Ridgway, Broderick, and Ellison, 2006). In summary, further technical progress combined with systematic integration of bioinformatics and systems biology tools should allow electron microscopy to be a major player in the future of structural bioinformatics.

ACKNOWLEDGMENTS

This work was supported by NIH research grant GM 076503 to NV.

REFERENCES

Adair BD, Yeager M (2002): Three-dimensional model of the human platelet integrin $\alpha_{IIb}\beta_3$ based on electron cryomicroscopy and X-ray crystallography. *Proc Natl Acad Sci USA* 99:14059–14064.

Agirrezabala X, Martin-Benito J, Valle M, Gonzalez JM, Valencia A, Valpuesta JM, Carrascosa JL (2005): Structure of the connector of bacteriophage T7 at 8 Å resolution: structural homologies of a basic component of a DNA translocating machinery. *J Mol Biol* 347:895–902.

Al-Amoudi A, Norlen LP, Dubochet J (2004): Cryo-electron microscopy of vitreous sections of native biological cells and tissues. *J Struct Biol* 148:131–135.

Aloy P, Russell RB (2005): Structure-based systems biology: a zoom lens for the cell. *FEBS Lett* 579:1854–1858.

Aloy P, Russell RB (2006): Structural systems biology: modelling protein interactions. *Nat Rev Mol Cell Biol* 7:188–197.

Anderson K, Page C, Evans JG, Luxenburg C, Addadi L, Geiger B, Matsudaira P, Volkmann N, Hanein, D (in preparation): The three-dimensional structure of podosomes.

Baker TS, Johnson JE (1996): Low resolution meets high: towards a resolution continuum from cells to atoms. *Curr Opin Struct Biol* 6:585–594.

Baker ML, Serysheva II, Sencer S, Wu Y, Ludtke SJ, Jiang W, Hamilton SL, Chiu W (2002): The skeletal muscle Ca^{2+} release channel has an oxidoreductase-like domain. *Proc Natl Acad Sci USA* 99:12155–12160. [Epub 12002 September 12156.]

Baker ML, Jiang W, Bowman BR, Zhou ZH, Quiocho FA, Rixon FJ, Chiu W (2003): Architecture of the herpes simplex virus major capsid protein derived from structural bioinformatics. *J Mol Biol* 331:447–456.

Baker ML, Jiang W, Wedemeyer WJ, Rixon FJ, Baker D, Chiu W (2006a): *Ab initio* modeling of the herpesvirus VP26 core domain assessed by CryoEM density. *PLoS Comput Biol* 2:e146.

Baker ML, Yu Z, Chiu W, Bajaj C (2006b): Automated segmentation of molecular subunits in electron cryomicroscopy density maps. *J Struct Biol* 156:3 432–441.

Baker ML, Ju T, Chiu W (2007): Identification of secondary structure elements in intermediate-resolution density maps. *Structure* 15:7–19.

Baldwin JM, Schertler GF, Unger VM (1997): An alpha-carbon template for the transmembrane helices in the rhodopsin family of G-protein-coupled receptors. *J Mol Biol* 272:144–164.

Ban N, Freeborn B, Nissen P, Penczek P, Grassucci RA, Sweet R, Frank J, Moore PB, Steitz TA (1998): A 9 Å resolution X-ray crystallographic map of the large ribosomal subunit. *Cell* 93:1105–1115.

Beck M, Forster F, Ecke M, Plitzko JM, Melchior F, Gerisch G, Baumeister W, Medalia O (2004): Nuclear pore complex structure and dynamics revealed by cryoelectron tomography. *Science* 306:1387–1390.

Berriman J, Unwin N (1994): Analysis of transient structures by cryo-microscopy combined with rapid mixing of spray droplets. *Ultramicroscopy* 56:241–252.

Böhm J, Frangakis AS, Hegerl R, Nickell S, Typke D, Baumeister W (2000): From the cover: toward detecting and identifying macromolecules in a cellular context: template matching applied to electron tomograms. *Proc Natl Acad Sci USA* 97:14245–14250.

Borgnia MJ, Shi D, Zhang P, Milne JL (2004): Visualization of alpha-helical features in a density map constructed using 9 molecular images of the 1.8 MDa icosahedral core of pyruvate dehydrogenase. *J Struct Biol* 147:136–145.

Bork P, Serrano L (2005): Towards cellular systems in 4D. *Cell* 121:507–509.

Böttcher B, Wynne SA, Crowther RA (1997): Determination of the fold of the core protein of hepatitis B virus by electron cryomicroscopy. *Nature* 386:88–91.

Briggman KL, Denk W (2006): Towards neural circuit reconstruction with volume electron microscopy techniques. *Curr Opin Neurobiol* 16:562–570.

Brink J, Ludtke SJ, Kong Y, Wakil SJ, Ma J, Chiu W (2004): Experimental verification of conformational variation of human fatty acid synthase as predicted by normal mode analysis. *Structure* 12:185–191.

Carragher BO, Cheng N, Wang ZY, Korn ED, Reilein A, Belnap DM, Hammer JA, Steven AC (1998): Structural invariance of constitutively active and inactive mutants of Acanthamoeba myosin IC bound to F-actin in the rigor and ADP-bound states. *Proc Natl Acad Sci USA* 95:15206–15211.

Carragher B, Fellmann D, Guerra F, Milligan RA, Mouche F, Pulokas J, Sheehan B, Quispe J, Suloway C, Zhu YX, Potter CS (2004): Rapid, routine structure determination of macromolecular assemblies using electron microscopy: current progress and further challenges. *J Synchrotron Radiat* 11:83–85.

Chacon P, Wriggers W (2002): Multi-resolution contour-based fitting of macromolecular structures. *J Mol Biol* 317:375–384.

Chen JZ, Furst J, Chapman MS, Grigorieff N (2003): Low-resolution structure refinement in electron microscopy. *J Struct Biol* 144:144–151.

Cheng Y, Wolf E, Larvie M, Zak O, Aisen P, Grigorieff N, Harrison SC, Walz T (2006): Single particle reconstructions of the transferrin–transferrin receptor complex obtained with different specimen preparation techniques. *J Mol Biol* 355:1048–1065.

Chiu W, Baker ML, Jiang W, Dougherty M, Schmid MF (2005): Electron cryomicroscopy of biological machines at subnanometer resolution. *Structure* 13:363–372.

Chiu W, Baker ML, Almo SC (2006): Structural biology of cellular machines. *Trends Cell Biol* 16:144–150.

Conway JF, Cheng N, Zlotnick A, Wingfield PT, Stahl SJ, Steven AC (1997): Visualization of a 4-helix bundle in the hepatitis B virus capsid by cryo-electron microscopy. *Nature* 386:91–94.

Craig L, Volkmann N, Arvai AS, Pique ME, Yeager M, Egelman EH, Tainer JA (2006): Type IV pilus structure by cryo-electron microscopy and crystallography: implications for pilus assembly and functions. *Mol Cell* 23:651–662.

De La Cruz EM, Mandinova A, Steinmetz MO, Stoffler D, Aebi U, Pollard TD (2000): Polymerization and structure of nucleotide-free actin filaments. *J Mol Biol* 295:517–526.

DeRosier DJ, Klug A (1968): Reconstruction of three-dimensional structures from electron micrographs. *Nature* 217:130–134.

DeRosier DJ, Moore PB (1970): Reconstruction of three-dimensional images from electron micrographs of structures with helical symmetry. *J Mol Biol* 52:355–369.

Dubochet J, Adrian M, Chang J-J, Homo J-C, Lepault J, McDowall AW, Schultz P (1988): Cryo-electron microscopy of vitrified specimens. *Q Rev Biophys* 21:129–228.

Egelman EH (2000): A robust algorithm for the reconstruction of helical filaments using single-particle methods. *Ultramicroscopy* 85:225–234.

Egelman EH (2007): The iterative helical real space reconstruction method: surmounting the problems posed by real polymers. *J Struct Biol* 157:83–94.

Esashi F, Galkin VE, Yu X, Egelman EH, West SC (2007): Stabilization of RAD51 nucleoprotein filaments by the C-terminal region of BRCA2. *Nat Struct Mol Biol* 14:468–474.

Fabry CM, Rosa-Calatrava M, Conway JF, Zubieta C, Cusack S, Ruigrok RW, Schoehn G (2005): A quasi-atomic model of human adenovirus type 5 capsid. *EMBO J* 24:1645–1654.

Fleishman SJ, Harrington S, Friesner RA, Honig B, Ben-Tal N (2004a): An automatic method for predicting transmembrane protein structures using cryo-EM and evolutionary data. *Biophys J* 87:3448–3459.

Fleishman SJ, Unger VM, Yeager M, Ben-Tal N (2004b): A Calpha model for the transmembrane alpha helices of gap junction intercellular channels. *Mol Cell* 15:879–888.

Förster F, Medalia O, Zauberman N, Baumeister W, Fass D (2005): Retrovirus envelope protein complex structure *in situ* studied by cryo-electron tomography. *Proc Natl Acad Sci USA* 102:4729–4734.

Fotin A, Cheng Y, Sliz P, Grigorieff N, Harrison SC, Kirchhausen T, Walz T (2004): Molecular model for a complete clathrin lattice from electron cryomicroscopy. *Nature* 432:573–579. [Epub 2004 Oct 2024.]

Frangakis AS, Hegerl R (2002): Segmentation of two- and three-dimensional data from electron microscopy using eigenvector analysis. *J Struct Biol* 138:105–113.

Frangakis AS, Bohm J, Forster F, Nickell S, Nicastro D, Typke D, Hegerl R, Baumeister W (2002): Identification of macromolecular complexes in cryoelectron tomograms of phantom cells. *Proc Natl Acad Sci USA* 99:14153–14158.

Frank J, Agrawal RK (2000): A ratchet-like inter-subunit reorganization of the ribosome during translocation. *Nature* 406:318–322.

Frank J (2006a): *Electron Tomography: Methods for Three-Dimensional Visualization of Structures in the Cell.* New York: Springer.

Frank J (2006b): *Three-Dimensional Electron Microscopy of Macromolecular Assemblies: Visualization of Biological Molecules in Their Native State.* Oxford: Oxford University Press.

Frey TG, Perkins GA, Ellisman MH (2006): Electron tomography of membrane-bound cellular organelles. *Annu Rev Biophys Biomol Struct* 35:199–224.

Gabashvili IS, Gregory ST, Valle M, Grassucci R, Worbs M, Wahl MC, Dahlberg AE, Frank J (2001): The polypeptide tunnel system in the ribosome and its gating in erythromycin resistance mutants of L4 and L22. *Mol Cell* 8:181–188.

Galkin VE, Orlova A, VanLoock MS, Rybakova IN, Ervasti JM, Egelman EH (2002a): The utrophin actin-binding domain binds F-actin in two different modes: implications for the spectrin super-family of proteins. *J Cell Biol* 157:243–251.

Galkin VE, Orlova A, VanLoock MS, Zhou D, Galan JE, Egelman EH (2002b): The bacterial protein SipA polymerizes G-actin and mimics muscle nebulin. *Nat Struct Biol* 9:518–521.

Galkin VE, VanLoock MS, Orlova A, Egelman EH (2002c): A new internal mode in F-actin helps explain the remarkable evolutionary conservation of actin's sequence and structure. *Curr Biol* 12:570–575.

Galkin VE, Orlova A, VanLoock MS, Shvetsov A, Reisler E, Egelman EH (2003): ADF/cofilin use an intrinsic mode of F-actin instability to disrupt actin filaments. *J Cell Biol* 163:1057–1066.

Galkin VE, Esashi F, Yu X, Yang S, West SC, Egelman EH (2005): BRCA2 BRC motifs bind RAD51-DNA filaments. *Proc Natl Acad Sci USA* 102:8537–8542.

Gao H, Sengupta J, Valle M, Korostelev A, Eswar N, Stagg SM, Van Roey P, Agrawal RK, Harvey SC, Sali A, Chapman MS, Frank J (2003): Study of the structural dynamics of the *E. coli* 70S ribosome using real-space refinement. *Cell* 113:789–801.

Giepmans BN, Deerinck TJ, Smarr BL, Jones YZ, Ellisman MH (2005): Correlated light and electron microscopic imaging of multiple endogenous proteins using quantum dots. *Nat Methods* 2:743–749.

Gonen T, Sliz P, Kistler J, Cheng Y, Walz T (2004): Aquaporin-0 membrane junctions reveal the structure of a closed water pore. *Nature* 429:193–197.

Grimm R, Singh H, Rachel R, Typke D, Zillig W, Baumeister W (1998): Electron tomography of ice-embedded prokaryotic cells. *Biophys J* 74:1031–1042.

Grünewald K, Cyrklaff M (2006): Structure of complex viruses and virus-infected cells by electron cryotomography. *Curr Opin Microbiol* 9:437–442.

Hanein D, Matsudaira P, DeRosier DJ (1997): Evidence for a conformational change in actin induced by fimbrin (N375) binding. *J Cell Biol* 139:387–396.

Hanein D, Volkmann N, Goldsmith S, Michon AM, Lehman W, Craig R, DeRosier D, Almo S, Matsudaira P (1998): An atomic model of fimbrin binding to F-actin and its implications for filament crosslinking and regulation. *Nat Struct Biol* 5:787–792.

Hanein D (2007): Structural analysis of supramolecular assemblies by hybrid methods. *J Struct Biol* 158:135–136.

Henderson R, Baldwin JM, Ceska TA, Zemlin F, Beckmann E, Downing KH (1990): An atomic model for the structure of bacteriorhodopsin. *Biochem Soc Trans* 18:844.

Henderson R (1995): The potential and limitations of neutrons, electrons and X-ray for atomic resolution microscopy of unstained biological molecules. *Q Rev Biophys* 28:171–194.

Heymann JB, Cheng N, Newcomb WW, Trus BL, Brown JC, Steven AC (2003): Dynamics of herpes simplex virus capsid maturation visualized by time-lapse cryo-electron microscopy. *Nat Struct Biol* 10:334–341.

Hiroaki Y, Tani K, Kamegawa A, Gyobu N, Nishikawa K, Suzuki H, Walz T, Sasaki S, Mitsuoka K, Kimura K, Mizoguchi A, Fujiyoshi Y (2006): Implications of the aquaporin-4 structure on array formation and cell adhesion. *J Mol Biol* 355:628–639.

Hoenger A, Sack S, Thormahlen M, Marx A, Muller J, Gross H, Mandelkow E (1998): Image reconstructions of microtubules decorated with monomeric and dimeric kinesins: comparison with X-ray structure and implications for motility. *J Cell Biol* 141:419–430.

Holm PJ, Bhakat P, Jegerschold C, Gyobu N, Mitsuoka K, Fujiyoshi Y, Morgenstern R, Hebert H (2006): Structural basis for detoxification and oxidative stress protection in membranes. *J Mol Biol* 360:934–945.

Holmes KC, Angert I, Kull FJ, Jahn W, Schroder RR (2003): Electron cryo-microscopy shows how strong binding of myosin to actin releases nucleotide. *Nature* 425:423–427.

Huiskonen JT, Kivela HM, Bamford DH, Butcher SJ (2004): The PM2 virion has a novel organization with an internal membrane and pentameric receptor binding spikes. *Nat Struct Mol Biol* 11:850–856.

Janssen ME, Kim E, Liu H, Fujimoto LM, Bobkov A, Volkmann N, Hanein D (2006): Three-dimensional structure of vinculin bound to actin filaments. *Mol Cell* 21:271–281.

Jensen GJ, Briegel A (2007): How electron cryotomography is opening a new window onto prokaryotic ultrastructure. *Curr Opin Struct Biol* 17:260–267.

Jiang W, Baker ML, Ludtke SJ, Chiu W (2001a): Bridging the information gap: computational tools for intermediate resolution structure interpretation. *J Mol Biol* 308:1033–1044.

Jiang W, Li Z, Zhang Z, Booth CR, Baker ML, Chiu W (2001b): Semi-automated icosahedral particle reconstruction at sub-nanometer resolution. *J Struct Biol* 136:214–225.

Jiang W, Li Z, Zhang Z, Baker ML, Prevelige PE Jr, Chiu W (2003): Coat protein fold and maturation transition of bacteriophage P22 seen at subnanometer resolutions. *Nat Struct Biol* 10:131–135.

Jiang W, Ludtke SJ (2005): Electron cryomicroscopy of single particles at subnanometer resolution. *Curr Opin Struct Biol* 15:571–577.

Jontes JD, Milligan RA (1997): Brush border myosin-I structure and ADP-dependent conformational changes revealed by cryoelectron microscopy and image analysis. *J Cell Biol* 139:683–693.

Kasas S, Dumas G, Dietler G, Catsicas S, Adrian M (2003): Vitrification of cryoelectron microscopy specimens revealed by high-speed photographic imaging. *J Microsc* 211:48–53.

Komeili A, Li Z, Newman DK, Jensen GJ (2006): Magnetosomes are cell membrane invaginations organized by the actin-like protein MamK. *Science* 311:242–245.

Kong Y, Ma J (2003): A structural-informatics approach for mining beta-sheets: locating sheets in intermediate-resolution density maps. *J Mol Biol* 332:399–413.

Koster AJ, Agard DA (1997): Editorial. *J Struct Biol* 120:207–209.

Koster AJ, Grimm R, Typke D, Hegerl R, Stoschek A, Walz J, Baumeister W (1997): Perspectives of molecular and cellular electron tomography. *J Struct Biol* 120:276–308.

Kozielski F, Arnal I, Wade R (1998): A model of the microtuble–kinesin complex based on electron cryomicroscopy and X-ray crystallography. *Curr Biol* 8:191–198.

Kühlbrandt W, Wang DN, Fujiyoshi Y (1994): Atomic model of plant light-harvesting complex by electron crystallography. *Nature* 367:614–621.

Kunji ER, von Gronau S, Oesterhelt D, Henderson R (2000): The three-dimensional structure of halorhodopsin to 5 Å by electron crystallography: a new unbending procedure for two-dimensional crystals by using a global reference structure. *Proc Natl Acad Sci USA* 97:4637–4642.

Kurner J, Frangakis AS, Baumeister W (2005): Cryo-electron tomography reveals the cytoskeletal structure of *Spiroplasma melliferum*. *Science* 307:436–438.

Lata R, Conway JF, Cheng N, Duda RL, Hendrix RW, Wikoff WR, Johnson JE, Tsuruta H, Steven AC (2000): Maturation dynamics of a viral capsid: visualization of transitional intermediate states. *Cell* 100:253–263.

Lee HS, Bellin RM, Walker DL, Patel B, Powers P, Liu H, Garcia-Alvarez B, de Pereda JM, Liddington RC, Volkmann N, Hanein D, Critchley DR, Robson RM (2004): Characterization of an actin-binding site within the talin FERM domain. *J Mol Biol* 343:771–784.

Leis AP, Beck M, Gruska M, Best C, Hegerl R, Baumeister W, Leis JW (2006): Cryo-electron tomography of biological specimens. *IEEE Signal Process Mag* 23:95–103.

Lemerle C, Di Ventura B, Serrano L (2005): Space as the final frontier in stochastic simulations of biological systems. *FEBS Lett* 579:1789–1794.

Leschziner AE, Nogales E (2006): The orthogonal tilt reconstruction method: an approach to generating single-class volumes with no missing cone for *ab initio* reconstruction of asymmetric particles. *J Struct Biol* 153:284–299.

Liu W, Boisset N, Frank J (1995): Estimation of variance distribution in three-dimensional reconstruction. II. Applications. *J Opt Soc Am A* 12:2628–2635.

Liu J, Taylor DW, Taylor KA (2004): A 3-D reconstruction of smooth muscle alpha-actinin by CryoEm reveals two different conformations at the actin-binding region. *J Mol Biol* 338:115–125.

Liu J, Taylor DW, Krementsova EB, Trybus KM, Taylor KA (2006): Three-dimensional structure of the myosin V inhibited state by cryoelectron tomography. *Nature* 442:208–211.

Lucic V, Forster F, Baumeister W (2005): Structural studies by electron tomography: from cells to molecules. *Annu Rev Biochem* 74:833–865.

Ludtke SJ, Baldwin PR, Chiu W (1999): EMAN: semiautomated software for high-resolution single-particle reconstructions. *J Struct Biol* 128:82–97.

Ludtke SJ, Chen DH, Song JL, Chuang DT, Chiu W (2004): Seeing GroEL at 6 Å resolution by single particle electron cryomicroscopy. *Structure* 12:1129–1136.

Ludtke SJ, Serysheva II, Hamilton SL, Chiu W (2005): The pore structure of the closed RyR1 channel. *Structure* 13:1203–1211.

Mancini EJ, Clarke M, Gowen BE, Rutten T, Fuller SD (2000): Cryo-electron microscopy reveals the functional organization of an enveloped virus, Semliki Forest virus. *Mol Cell* 5:255–266.

Marsh BJ (2005): Lessons from tomographic studies of the mammalian Golgi. *Biochim Biophys Acta* 1744:273–292.

Martin AC, Drubin DG (2003): Impact of genome-wide functional analyses on cell biology research. *Curr Opin Cell Biol* 15:6–13.

Mastronarde DN (2005): Automated electron microscope tomography using robust prediction of specimen movements. *J Struct Biol* 152:36–51.

Matadeen R, Patwardhan A, Gowen B, Orlova EV, Pape T, Cuff M, Mueller F, Brimacombe R, van Heel M (1999): The *Escherichia coli* large ribosomal subunit at 7.5 Å resolution. *Struct Fold Des* 7:1575–1583.

McGough A, Way M, DeRosier D (1994): Determination of the alpha-actinin-binding site on actin filaments by cryoelectron microscopy and image analysis. *J Cell Biol* 126:433–443.

McEwen B, Koster A (2002): Proceedings of the Academic Colloquium on electron Tomography, October 17–20, 2001, Amsterdam, the Netherlands. *J Struct Biol* 138:1.

McGough A, Pope B, Chiu W, Weeds A (1997): Cofilin changes the twist of F-actin: implications for actin filament dynamics and cellular function. *J Cell Biol* 138:771–781.

McIntosh JR (2007): *Cellular Electron Microscopy*, Vol. 79. San Diego, CA: Academic Press.

McIntosh R, Nicastro D, Mastronarde D (2005): New views of cells in 3D: an introduction to electron tomography. *Trends Cell Biol* 15:43–51.

Medalia O, Weber I, Frangakis AS, Nicastro D, Gerisch G, Baumeister W (2002): Macromolecular architecture in eukaryotic cells visualized by cryoelectron tomography. *Science* 298:1209–1213.

Medalia O, Beck M, Ecke M, Weber I, Neujahr R, Baumeister W, Gerisch G (2007): Organization of actin networks in intact filopodia. *Curr Biol* 17:79–84.

Menetret JF, Hegde RS, Heinrich SU, Chandramouli P, Ludtke SJ, Rapoport TA, Akey CW (2005): Architecture of the ribosome-channel complex derived from native membranes. *J Mol Biol* 348:445–457.

Mitra K, Frank J (2006): Ribosome dynamics: insights from atomic structure modeling into cryo-electron microscopy maps. *Annu Rev Biophys Biomol Struct* 35:299–317.

Miyazawa A, Fujiyoshi Y, Stowell M, Unwin N (1999): Nicotinic acetylcholine receptor at 4.6 Å resolution: transverse tunnels in the channel wall. *J Mol Biol* 288:765–786.

Moores CA, Keep NH, Kendrick-Jones J (2000): Structure of the utrophin actin-binding domain bound to F-actin reveals binding by an induced fit mechanism. *J Mol Biol* 297:465–480.

Morais MC, Choi KH, Koti JS, Chipman PR, Anderson DL, Rossmann MG (2005): Conservation of the capsid structure in tailed dsDNA bacteriophages: the pseudoatomic structure of phi29. *Mol Cell* 18:149–159.

Mu XQ, Egelman EH, Bullitt E (2002): Structure and function of Hib pili from *Haemophilus influenzae* type b. *J Bacteriol* 184:4868–4874.

Murata K, Mitsuoka K, Hirai T, Walz T, Agre P, Heymann JB, Engel A, Fujiyoshi Y (2000): Structural determinants of water permeation through aquaporin-1. *Nature* 407:599–605.

Murphy GE, Leadbetter JR, Jensen GJ (2006): *In situ* structure of the complete *Treponema primitia* flagellar motor. *Nature* 442:1062–1064.

Nakagawa A, Miyazaki N, Taka J, Naitow H, Ogawa A, Fujimoto Z, Mizuno H, Higashi T, Watanabe Y, Omura T, Cheng RH, Tsukihara T (2003): The atomic structure of rice dwarf virus reveals the self-assembly mechanism of component proteins. *Structure* 11:1227–1238.

Narita A, Takeda S, Yamashita A, Maeda Y (2006): Structural basis of actin filament capping at the barbed-end: a cryo-electron microscopy study. *EMBO J* 25:5626–5633.

Narita A, Maeda Y (2007): Molecular determination by electron microscopy of the actin filament end structure. *J Mol Biol* 365:480–501.

Nicastro D, Schwartz C, Pierson J, Gaudette R, Porter ME, McIntosh JR (2006): The molecular architecture of axonemes revealed by cryoelectron tomography. *Science* 313:944–948.

Nickell S, Hegerl R, Baumeister W, Rachel R (2003): *Pyrodictium* cannulae enter the periplasmic space but do not enter the cytoplasm, as revealed by cryo-electron tomography. *J Struct Biol* 141:34–42.

Nickell S, Kofler C, Leis AP, Baumeister W (2006): A visual approach to proteomics. *Nat Rev Mol Cell Biol* 7:225–230.

Nickell S, Mihalache O, Beck F, Hegerl R, Korinek A, Baumeister W (2007): Structural analysis of the 26S proteasome by cryoelectron tomography. *Biochem Biophys Res Commun* 353:115–120.

Nitsch M, Walz J, Typke D, Klumpp M, Essen LO, Baumeister W (1998): Group II chaperonin in an open conformation examined by electron tomography. *Nat Struct Biol* 5:855–857.

Nogales E, Wolf SG, Downing KH (1998): Structure of the alpha beta tubulin dimer by electron crystallography. *Nature* 391:199–203.

Orlova A, Egelman EH (2000): F-actin retains a memory of angular order. *Biophys J* 78:2180–2185.

Ortiz JO, Forster F, Kurner J, Linaroudis AA, Baumeister W (2006): Mapping 70S ribosomes in intact cells by cryoelectron tomography and pattern recognition. *J Struct Biol* 156:334–341.

Penczek PA, Frank J, Spahn CM (2006a): A method of focused classification, based on the bootstrap 3D variance analysis, and its application to EF-G-dependent translocation. *J Struct Biol* 154:184–194.

Penczek PA, Yang C, Frank J, Spahn CM (2006b): Estimation of variance in single-particle reconstruction using the bootstrap technique. *J Struct Biol* 154:168–183.

Pomfret AJ, Rice WJ, Stokes DL (2007): Application of the iterative helical real-space reconstruction method to large membranous tubular crystals of P-type ATPases. *J Struct Biol* 157:106–116.

Prasad BV, Hardy ME, Dokland T, Bella J, Rossmann MG, Estes MK (1999): X-ray crystallographic structure of the Norwalk virus capsid. *Science* 286:287–290.

Radermacher M (1988): Three-dimensional reconstruction of single particles from random and nonrandom tilt series. *J Electron Microsc Tech* 9:359–394.

Ranson NA, Farr GW, Roseman AM, Gowen B, Fenton WA, Horwich AL, Saibil HR (2001): ATP-bound states of GroEL captured by cryo-electron microscopy. *Cell* 107:869–879.

Rayment I, Holden HM, Whittaker M, Yohn CB, Lorenz M, Holmes KC, Milligan RA (1993): Structure of the actin–myosin complex and its implications for muscle contraction. *Science* 261:58–65.

Renault L, Chou HT, Chiu PL, Hill RM, Zeng X, Gipson B, Zhang ZY, Cheng A, Unger V, Stahlberg H (2006): Milestones in electron crystallography. *J Comput Aided Mol Des* 20:519–527.

Ridgway D, Broderick G, Ellison MJ (2006): Accommodating space, time and randomness in network simulation. *Curr Opin Biotechnol* 17:493–498.

Rockel B, Walz J, Hegerl R, Peters J, Typke D, Baumeister W (1999): Structure of VAT, a CDC48/p97 ATPase homologue from the archaeon *Thermoplasma acidophilum* as studied by electron tomography. *FEBS Lett* 451:27–32.

Roseman AM (2000): Docking structures of domains into maps from cryo-electron microscopy using local correlation. *Acta Crystallogr D* 56:1332–1340.

Rossmann MG (2000): Fitting atomic models into electron-microscopy maps. *Acta Crystallogr D* 56:1341–1349.

Rost LE, Hanein D, DeRosier DJ (1998): Reconstruction of symmetry deviations: a procedure to analyze partially decorated F-actin and other incomplete structures. *Ultramicroscopy* 72:187–197.

Saban SD, Nepomuceno RR, Gritton LD, Nemerow GR, Stewart PL (2005): CryoEM structure at 9 Å resolution of an adenovirus vector targeted to hematopoietic cells. *J Mol Biol* 349:526–537.

Schmid MF, Sherman MB, Matsudaira P, Chiu W (2004): Structure of the acrosomal bundle. *Nature* 431:104–107.

Sengupta J, Nilsson J, Gursky R, Spahn CM, Nissen P, Frank J (2004): Identification of the versatile scaffold protein RACK1 on the eukaryotic ribosome by cryo-EM. *Nat Struct Mol Biol* 11:957–962. [Epub 2004 August 2029.]

Seybert A, Herrmann R, Frangakis AS (2006): Structural analysis of *Mycoplasma pneumoniae* by cryo-electron tomography. *J Struct Biol* 156:342–354.

Shacham E, Sheehan B, Volkmann N (2007): Density-based score for selecting near-native atomic models of unknown structures. *J Struct Biol* 158:188–195.

Sherman MB, Soejima T, Chiu W, van Heel M (1998): Multivariate analysis of single unit cells in electron crystallography. *Ultramicroscopy* 74:179–199.

Smith TJ, Olson NH, Cheng RH, Liu H, Chase ES, Lee WM, Leippe DM, Mosser AG, Rueckert RR, Baker TS (1993): Structure of human rhinovirus complexed with Fab fragments from a neutralizing antibody. *J Virol* 67:1148–1158.

Spahn CM, Beckmann R, Eswar N, Penczek PA, Sali A, Blobel G, Frank J (2001): Structure of the 80S ribosome from *Saccharomyces cerevisiae*—tRNA–ribosome and subunit–subunit interactions. *Cell* 107:373–386.

Stahlberg H, Fotiadis D, Scheuring S, Remigy H, Braun T, Mitsuoka K, Fujiyoshi Y, Engel A (2001): Two-dimensional crystals: a powerful approach to assess structure, function and dynamics of membrane proteins. *FEBS Lett* 504:166–172.

Steinmetz MO, Hoenger A, Stoffler D, Noegel AA, Aebi U, Schoenenberger CA (2000): Polymerization, three-dimensional structure and mechanical properties of *Dictyostelium* versus rabbit muscle actin filaments. *J Mol Biol* 303:171–184.

Sui H, Downing KH (2006): Molecular architecture of axonemal microtubule doublets revealed by cryo-electron tomography. *Nature* 442:475–478.

Sukow C, DeRosier D (1998): How to analyze electron micrographs of rafts of actin filaments crosslinked by actin-binding proteins. *J Mol Biol* 284:1039–1050.

Tama F, Miyashita O, Brooks CL 3rd (2004): Flexible multi-scale fitting of atomic structures into low-resolution electron density maps with elastic network normal mode analysis. *J Mol Biol* 337:985–999.

Topf M, Baker ML, John B, Chiu W, Sali A (2005): Structural characterization of components of protein assemblies by comparative modeling and electron cryo-microscopy. *J Struct Biol* 149:191–203.

Topf M, Sali A (2005): Combining electron microscopy and comparative protein structure modeling. *Curr Opin Struct Biol* 15:578–585.

Topf M, Baker ML, Marti-Renom MA, Chiu W, Sali A (2006): Refinement of protein structures by iterative comparative modeling and CryoEM density fitting. *J Mol Biol* 357:1655–1668. [Epub 2006 February 1652.]

Trus BL, Roden RB, Greenstone HL, Vrhel M, Schiller JT, Booy FP (1997): Novel structural features of bovine papillomavirus capsid revealed by a three-dimensional reconstruction to 9 Å resolution. *Nat Struct Biol* 4:413–420.

Trybus KM, Guschin M, Liu H, Hazelwood L, Krementsova E, Volkmann N, Hanein D (2007): Effect of calcium on calmodulin bound to the IQ motifs of myosin V. *J Biol Chem* 282:32 23316–23325.

Typke D, Nordmeyer RA, Jones A, Lee J, Avila-Sakar A, Downing KH, Glaeser RM (2005): High-throughput film-densitometry: an efficient approach to generate large data sets. *J Struct Biol* 149:17–29.

Vale RD, Milligan RA (2000): The way things move: looking under the hood of molecular motor proteins. *Science* 288:88–95.

Valle M, Zavialov A, Li W, Stagg SM, Sengupta J, Nielsen RC, Nissen P, Harvey SC, Ehrenberg M, Frank J (2003): Incorporation of aminoacyl-tRNA into the ribosome as seen by cryo-electron microscopy. *Nat Struct Biol* 10:899–906.

van der Heide P, Xu XP, Marsh BJ, Hanein D, Volkmann N (2007): Efficient automatic noise reduction of electron tomographic reconstructions based on iterative median filtering. *J Struct Biol* 158:196–204.

van Heel M (1987): Angular reconstitution: a posteriori assignment of projection directions for 3D reconstruction. *Ultramicroscopy* 21:111–124.

VanLoock MS, Agrawal RK, Gabashvili IS, Qi L, Frank J, Harvey SC (2000): Movement of the decoding region of the 16S ribosomal RNA accompanies tRNA translocation. *J Mol Biol* 304:507–515.

Velazquez-Muriel JA, Valle M, Santamaria-Pang A, Kakadiaris IA, Carazo JM (2006): Flexible fitting in 3D-EM guided by the structural variability of protein superfamilies. *Structure* 14:1115–1126.

Verschoor A, Tivol WF, Mannella CA (2001): Single-particle approaches in the analysis of small 2d crystals of the mitochondrial channel VDAC. *J Struct Biol* 133:254–265.

Volkmann N, Schlünzen F, Vernoslava EA, Urzhumstev AG, Podjarny AD, Roth M, Pebay-Peyroula E, Berkovitch-Yellin Z, Zaytzev-Bashan A, Yonath A (1995): On *ab initio* phasing of ribosomal particles at very low resolution. *Joint CCP4 and ESF-EACBM Newsletters* 31.

Volkmann N, Hanein D (1999): Quantitative fitting of atomic models into observed densities derived by electron microscopy. *J Struct Biol* 125:176–184.

Volkmann N, Hanein D (2000): Actomyosin: law and order in motility. *Curr Opin Cell Biol* 12:26–34.

Volkmann N, Hanein D, Ouyang G, Trybus KM, DeRosier DJ, Lowey S (2000): Evidence for cleft closure in actomyosin upon ADP release. *Nat Struct Biol* 7:1147–1155.

Volkmann N, Amann KJ, Stoilova-McPhie S, Egile C, Winter DC, Hazelwood L, Heuser JE, Li R, Pollard TD, Hanein D (2001a): Structure of Arp2/3 complex in its activated state and in actin filament branch junctions. *Science* 293:2456–2459.

Volkmann N, DeRosier D, Matsudaira P, Hanein D (2001b): An atomic model of actin filaments cross-linked by fimbrin and its implications for bundle assembly and function. *J Cell Biol* 153:947–956.

Volkmann N (2002): A novel three-dimensional variant of the watershed transform for segmentation of electron density maps. *J Struct Biol* 138:123.

Volkmann N, Hanein D (2003): Docking of atomic models into reconstructions from electron microscopy. *Methods Enzymol* 374:204–225.

Volkmann N, Ouyang G, Trybus KM, DeRosier DJ, Lowey S, Hanein D (2003): Myosin isoforms show unique conformations in the actin-bound state. *Proc Natl Acad Sci USA* 100:3227–3232.

Volkmann N, Liu H, Hazelwood L, Krementsova EB, Lowey S, Trybus KM, Hanein D (2005): The structural basis of myosin V processive movement as revealed by electron cryomicroscopy. *Mol Cell* 19:595–605.

Walker M, Zhang XZ, Jiang W, Trinick J, White HD (1999): Observation of transient disorder during myosin subfragment-1 binding to actin by stopped-flow fluorescence and millisecond time resolution electron cryomicroscopy: evidence that the start of the crossbridge power stroke in muscle has variable geometry. *Proc Natl Acad Sci USA* 96:465–470.

Walz J, Koster AJ, Tamura T, Baumeister W (1999): Capsids of tricorn protease studied by electron cryomicroscopy. *J Struct Biol* 128:65–68.

Wang YA, Yu X, Overman S, Tsuboi M, Thomas GJ Jr, Egelman EH (2006): The structure of a filamentous bacteriophage. *J Mol Biol* 361:209–215.

Whittaker M, Wilson-Kubalek EM, Smith JE, Faust L, Milligan RA, Sweeney HL (1995): A 35-A movement of smooth muscle myosin on ADP release. *Nature* 378:748–751.

Woodhead JL, Zhao FQ, Craig R, Egelman EH, Alamo L, Padron R (2005): Atomic model of a myosin filament in the relaxed state. *Nature* 436:1195–1199.

Wriggers W, Milligan RA, McCammon JA (1999): Situs: a package for docking crystal structures into low-resolution maps from electron microscopy. *J Struct Biol* 125:185–195.

Wriggers W, Birmanns S (2001): Using situs for flexible and rigid-body fitting of multiresolution single-molecule data. *J Struct Biol* 133:193–202.

Yonekura K, Maki S, Morgan DG, DeRosier DJ, Vonderviszt F, Imada K, Namba K (2000): The bacterial flagellar cap as the rotary promoter of flagellin self-assembly. *Science* 290:2148–2152.

Yonekura K, Maki-Yonekura S, Namba K (2003): Complete atomic model of the bacterial flagellar filament by electron cryomicroscopy. *Nature* 424:643–650.

Yonekura K, Toyoshima C (2007): Structure determination of tubular crystals of membrane proteins. IV. Distortion correction and its combined application with real-space averaging and solvent flattening. *Ultramicroscopy* 107:12 1141–1158.

Zanetti G, Briggs JA, Grunewald K, Sattentau QJ, Fuller SD (2006): Cryo-electron tomographic structure of an immunodeficiency virus envelope complex *in situ*. *PLoS Pathog* 2:e83.

Zhang W, Chipman PR, Corver J, Johnson PR, Zhang Y, Mukhopadhyay S, Baker TS, Strauss JH, Rossmann MG, Kuhn RJ (2003a): Visualization of membrane protein domains by cryo-electron microscopy of dengue virus. *Nat Struct Biol* 10:907–912.

Zhang X, Walker SB, Chipman PR, Nibert ML, Baker TS (2003b): Reovirus polymerase lambda 3 localized by cryo-electron microscopy of virions at a resolution of 7.6 Å. *Nat Struct Biol* 10:1011–1018.

Zhou ZH, Dougherty M, Jakana J, He J, Rixon FJ, Chiu W (2000): Seeing the herpes virus capsid at 8.5 Å. *Science* 288:877–880.

Zhou ZH, Baker ML, Jiang W, Dougherty M, Jakana J, Dong G, Lu G, Chiu W (2001): Electron cryomicroscopy and bioinformatics suggest protein fold models for rice dwarf virus. *Nat Struct Biol* 8:868–873.

Zhou ZH, Zhang H, Jakana J, Lu XY, Zhang JQ (2003): Cytoplasmic polyhedrosis virus structure at 8 Å by electron cryomicroscopy: structural basis of capsid stability and mRNA processing regulation. *Structure* 11:651–663.

Zhu Y, Carragher B, Glaeser RM, Fellmann D, Bajaj C, Bern M, Mouche F, de Haas F, Hall RJ, Kriegman DJ, Ludtke SJ, Mallick SP, Penczek PA, Roseman AM, Sigworth FJ, Volkmann N, Potter CS (2004): Automatic particle selection: results of a comparative study. *J Struct Biol* 145:3–14.

Zhu P, Liu J, Bess J Jr, Chertova E, Lifson JD, Grise H, Ofek GA, Taylor KA, Roux KH (2006): Distribution and three-dimensional structure of AIDS virus envelope spikes. *Nature* 441:847–852.

Ziese U, Janssen AH, Murk JL, Geerts WJ, Van der Krift T, Verkleij AJ, Koster AJ (2002): Automated high-throughput electron tomography by pre-calibration of image shifts. *J Microsc* 205:187–200.

7

STUDY OF PROTEIN THREE-DIMENSIONAL STRUCTURE AND DYNAMICS USING PEPTIDE AMIDE HYDROGEN/DEUTERIUM EXCHANGE MASS SPECTROMETRY (DXMS) AND CHEMICAL CROSS-LINKING WITH MASS SPECTROMETRY TO CONSTRAIN MOLECULAR MODELING

Sheng Li, Dmitri Mouradov, Gordon King, Tong Liu, Ian Ross, Bostjan Kobe, Virgil L. Woods Jr, and Thomas Huber

INTRODUCTION

As more genomes are fully sequenced, there is a shift in interest to explore the intricate interplay of proteins in complex pathways and to study how protein structures vary in the course of these myriad interactions. While many high-throughput structural genomics programs seek to address these problems, it is apparent that producing structural models of protein complexes and large multidomain proteins using conventional techniques can be a slow process taking many months to years. Size limitations in NMR and the ability of a protein or protein complex to crystallize are the greatest hurdles. These bottlenecks have encouraged research into alternate approaches for structure determination such as mass spectrometry based techniques and examination of chemically cross-linked proteins.

Structural Bioinformatics, Second Edition   Edited by Jenny Gu and Philip E. Bourne
Copyright © 2009 John Wiley & Sons, Inc.

While mass spectrometry has not been traditionally viewed as a structure determination tool, advancements in MS techniques in combination with molecular modeling have opened up an entire new field in high-throughput structure determination. One such approach is enhanced peptide amide hydrogen/deuterium exchange mass spectrometry (DXMS) that focuses on liquid chromatography–mass spectrometry (LC–MS) based determination of these exquisitely informative exchange rates. When the technique is taken to its limit, single amide level resolution of exchange rates is possible. These values, one for each amide (amino acid) in the protein, reflect the accessibility of solvent water to each peptide amide hydrogen in the protein, with access provided by native state fluctuations in the structure of the protein. Thus, these measurements can directly report the thermodynamic stability of the entire protein at the resolution of single amino acids. With appropriate methodologies, these measurements can be achieved with small amounts of large proteins, without the need for protein crystallization or undue concentration.

Three examples of probing protein structure/dynamics via DXMS are presented here. We first describe DXMS investigation of conformational changes of the manganese transport regulator (MntR) in solution upon binding divalent transition metal ions. Metal binding rigidifies the protein and therefore reduces the entropic cost of DNA binding. Second, we describe how PAS (Per-Arnt-Sim) domain allostery and light-induced conformational changes in photoactive yellow protein (PYP) upon I_2 intermediate formation were probed with DXMS. Finally, we show the structural basis for the binding of anthrax lethal factor to oligomeric protective antigen is elucidated by DXMS analysis.

Another emerging technique takes advantage of the ability of chemical cross-linking reagents to provide distance constraints for pairs of functional groups within a protein or between interacting proteins. This hybrid biochemical/bioinformatics approach derives distance constraints from chemical cross-linking experiments to be used in molecular modeling. The technique uses high-accuracy mass spectrometers to locate the exact insertion site of chemical cross-links that can be used as sparse distance constraints for quick and inexpensive low-resolution structure prediction. While cross-linking has been used for many years identifying interacting proteins, it has been the advancement in mass spectrometry techniques (Mann and Talbo, 1996) that has made this technique appealing for structural determination studies. The accuracy of current mass spectrometers and the fragmentation capability allow the identification of the exact insertion point (Chen, Chen, and Anderson, 1999; Pearson, Pannell, and Fales, 2002; Dihazi and Sinz, 2003; Kruppa, Schoeniger, and Young, 2003). The cross-linking technique is able to remove size limitations imposed by both NMR and crystallography (for large complexes), as only proteolytic fragments are analyzed.

In this chapter, we also introduce the techniques used for structure determination using chemical cross-linking and molecular modeling. We highlight current available cross-linking reagents and mass spectrometry techniques. While it has been proven difficult to identify cross-linked peptides from a forest of native species, we introduce innovations in cross-linker design and new analysis techniques that have greatly improved their identification. Finally, examples that effectively use chemical cross-links for structure prediction when combined with molecular modeling/docking (Young et al., 2000; Mouradov et al., 2006) are demonstrated. This hybrid biochemical and computational technique can be used to quickly and inexpensively obtain low-resolution structure prediction.

Principles of Mass Spectrometry

Mass spectrometers have been an integral part of analytical science since first invented in 1918, with applications ranging from identifying unknown compounds to determining structure of compounds based on fragmentation. This is carried out by measurement of the mass-to-charge (m/z) ratio of the ionized forms of the molecules of interest. While many types of mass spectrometers have been developed in recent years, they all consist of three basic components: an ion source, a mass analyzer, and an ion detector (Figure 7.1).

The analyte is often subjected to extensive processing before ion generation. Protein or peptide samples can be proteolyzed, and then subjected to liquid chromatography, typically employing reverse-phase (hydrophobic) supports, with proteins/peptides applied to the column in aqueous solution, and acetonitrile gradient-eluted molecules continuously directed to the ion source.

The ion source ionizes (either positively or negatively) molecules in the sample. The two most common ion sources are electrospray ionization (ESI) and matrix-assisted laser desorption/ionization (MALDI). Electrospray ionization sprays the analyte (dissolved in a volatile solvent) from a small, highly charged capillary into gas (typically nitrogen) at approximately atmospheric pressure. The charge density on the aerosol microdroplets greatly increases as the solvent evaporates, driving a process called Coulombic fission, which eventuates in rapid and repeated droplet fission and evaporation, with the charge efficiently transferred to analyte ions in the gas phase. The aerosol is generated in close proximity to a pinhole-sized "leak" in the mass spectrometer, through which ions are drawn into a relatively high-vacuum regime and then directed to the mass analyzer.

In MALDI ionization, the analyte is first placed on a plate within the high-vacuum regime, and then ionized by a pulsed laser beam. "Matrix molecules" are typically mixed with the analyte, before it is dried onto the MALDI plate. The matrix facilitates ionization of the analyte and protects the biomolecule from being fragmented by the laser beam. It is hypothesized that the matrix is first ionized, with the transfer of a part of their energy to the analyte, in turn ionizing it. It should be noted that ESI typically produces multiply charged ions, while MALDI produces primarily single charged ions.

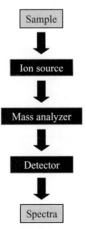

Figure 7.1. Schematic showing the three basic components that all mass spectrometers contain (black boxes). The sample is first ionized and then separated according to m/z. Finally, all m/z ratios are recorded and combined into a spectrum.

Mass analyzers separate the ions according to their m/z ratio using either static or dynamic fields with a combination of either magnetic or electric fields. One of the most common analyzers is time-of-flight (TOF) that measures time taken for a charged particle to reach the detector (after acceleration from an electric field). Lighter particles with the same charge will reach the detector before heavier particles.

A quadrupole mass analyzer works as a mass filter, transmitting one ion at a time based on its m/z ratio. Over time, the analyzer goes through a specified mass range that in turn creates a mass spectrum. A quadrupole ion trap works in the same manner, but it has the capability to trap ions and then eject them sequentially according to their m/z ratio.

In recent years, a high-resolution and precision mass analyzer was developed in the form of Fourier transform ion cyclotron resonance (FTICR) mass spectrometry. It uses the cyclotron frequency of the ions in a fixed magnetic field to measure the m/z ratios. The ions are injected into an electric/magnetic ion trap (called a Penning trap) where they are bent by the magnetic field into a circular motion in a plane perpendicular to the field and excited to a larger cyclotron radius by an oscillating electric field, as the original radius of the motion is too small to be detected. The resulting signal is called a free induction decay (FID) and consists of a complex frequency versus a time spectrum containing all the signals. By deconvoluting this signal using a Fourier transform allows the conversion of the composite signal to m/z values for each constituent ion contributing to the spectrum.

Multiple steps of mass analysis can be carried out using tandem mass spectrometry. The most common use of this is for peptide fragmentation. In this process, the first mass analyzer selects ions to be fragmented (precursor masses), while the second mass analyzer allows for fragmentation by collision-induced dissociation (CID). CID fragments molecular ions by allowing them to collide with neutral gas molecules (helium, nitrogen, or argon). Finally, the third analyzer separates the fragmented ions to allow the measurement of their m/z. An example of this is a MALDI-TOF/TOF mass spectrometer, where a CID is separated by two TOF analyzers.

The mass spectrometer's final element detects the arrival of an ion at a precise location and time, and from this information the m/z of the ion can be calculated. Combining all recorded m/z for all ions detected in the sample produces an m/z spectrum for the sample.

Modern mass spectrometers are powerful enough to resolve masses to within several parts per million. Differences in mass of less than 1 Da are readily resolved. The constituent atoms of peptides naturally exist as mixtures of isotopes of varying weights, and thus a spectrum derived from a specific peptide does not have a single m/z, but consists of a closely clustered group of ions that differ in mass by the equivalent of 1 Da. This cluster of masses for the peptide is termed the isotopic envelope.

H/D Exchange Chemistry

Amide hydrogens can be exchanged with solvent hydrogen through either acid-, base-, or water-catalyzed reactions (Bai et al., 1993):

$$k_{ch} = k_H [H^+] + k_{OH} [OH^-] + k_{H_2O}. \tag{7.1}$$

At low pH, the acid-catalyzed reaction dominates, while the rates of the base-catalyzed reactions increase at higher pH values. The water-catalyzed reaction is independent of pH. Given the temperature dependence of exchange rates, the slowest exchange rates at room temperature are observed at about pH 2.7. Much higher exchange rates are observed near neutral pH, where the amide hydrogen exchange reaction is mostly base-catalyzed.

The exchange rate of peptide amide hydrogens reflects its precise and unique environment within the protein's three-dimensional structure, and there is one such hydrogen for each amino acid in the protein, except for proline. A backbone amide hydrogen can exhibit highly variable exchange rates with solvent hydrogen, with rates ranging over eight orders of magnitude in folded proteins (Engen and Smith, 2001). In contrast, amide hydrogen exchange rates in peptides lacking secondary and tertiary structure vary only about 100-fold, depending primarily on neighboring amino acid side chains (Bai et al., 1993).

The exchange kinetics of amide hydrogens can be followed by deuterium isotope labeling with exchange times ranging from seconds to days. The exchange rates of hydrogens on $-OH$, $-SH$, $-NH_2$, $-COOH$, and $-CONH_2$ groups and the amino and carboxy termini are much faster. Carbon-centered hydrogens do not exchange under normal conditions and undergo isotope substitution only following activation by chemical treatment, such as reaction with hydroxyl radicals (Goshe and Anderson, 1999).

Several features affect the rate of amide hydrogen exchange, including participation in hydrogen bonding (Hilser and Freire, 1996), distance from the protein surface (Resing, Hoofnagle, and Ahn, 1999), and the flexibility of the peptide chain (Zhang, Post, and Smith, 1996). The degree of retardation in amide hydrogen exchange rate that results from the amide's physical environment is termed its "protection factor" (pf):

$$pf = k_{ch}/k_{ex}, \tag{7.2}$$

where k_{ex} is the observed exchange rate and k_{ch} is the "intrinsic" exchange rate calculated at a given pH and temperature in unstructured peptide chain (Bai et al., 1993).

H/D Exchange Thermodynamics

Formalisms to relate the observed rates of amide hydrogen exchange to thermodynamic stabilization of proteins have been developed (Englander and Kallenbach, 1983). Amide hydrogens of proteins in the native, folded state are proposed to exchange according to the following equation:

$$closed \underset{k_{cl}}{\overset{k_{op}}{\rightleftharpoons}} open \overset{k_{ch}}{\longrightarrow} exchanged, \tag{7.3}$$

$$k_{ex} = k_{op} k_{ch}/(k_{cl} + k_{ch}). \tag{7.4}$$

where k_{op} is the rate at which amide hydrogen converts from closed state to open state and k_{cl} is the rate amide hydrogen converts from open state to closed state. For most proteins at or below neutral pH, amide H/D exchange occurs by an EX2 mechanism (Sivaraman, Arrington, and Robertson, 2001), where $k_{cl} \gg k_{ch}$. In EX2 condition, Eq. 7.4 can be simplified as

$$k_{ex} = k_{op} k_{ch}/k_{cl} = k_{ch}/K_{cl}. \tag{7.5}$$

The closing equilibrium constant at each amide ($K_{cl} = k_{cl}/k_{op}$) is equal to the protection factor (pf) and can be translated into the stabilization free energy of closed state (ΔG_{cl}) by Eq. 7.6:

$$\Delta G_{cl} = -RT \ln(K_{cl}) = -RT \ln(pf) = -RT \ln(k_{ch}/k_{ex}). \tag{7.6}$$

Therefore, the ratio of measured H/D exchange rates in the folded protein (k_{ex}) and the calculated "intrinsic" rates (k_{ch}) can be converted into the free energy of amide hydrogen at a given condition. Thus, the measurement of exchange rates of backbone amide hydrogen serves as a precise thermodynamic sensor of the local environment.

H/D Exchange for Protein Structure and Dynamics

Frequently, the hydrogen exchange rates of two or more physical states of a protein, such as with and without protein binding partner (here represented by k_{ex+} and k_{ex-}), are measured to locate stabilization free energy changes upon the perturbation ($\Delta G_{-\to+}$):

$$\Delta G_{-\to+} = \Delta G_+ - \Delta G_- = -RT \ln(k_{ex-}/k_{ex+}). \tag{7.7}$$

The change in free energy upon protein binding ($\Delta G_{-\to+}$) can be monitored by H/D exchange rates.

OVERVIEW OF DXMS METHODOLOGY

Dramatic advances in mass spectrometry and improvements in the various steps within the experimental hydrogen exchange procedures have resulted in the development of automated systems for high-throughput, high-resolution H/D exchange analysis (Woods Jr, 1997; Woods Jr, 2001a; Woods Jr, 2001b; Woods Jr and Hamuro, 2001; Pantazatos et al., 2004). The system (Figure 7.2), described in this section, incorporates the latest of these enhancements, including solid-phase proteolysis, automated liquid handling, and streamlined data reduction software (Hamuro et al., 2003a ; Hamuro et al., 2003b; Black et al., 2004; Hamuro et al., 2004)

H/D Exchange Reaction

To initiate an H/D exchange reaction, a protein sample, initially in nondeuterated buffer, is incubated in a buffer with 50–90% mole fraction deuterated water. There are almost no restrictions on reaction conditions that allow exchange behavior to be studied as a function of protein and buffer composition, solution pH, and in the presence and absence of ligands. To

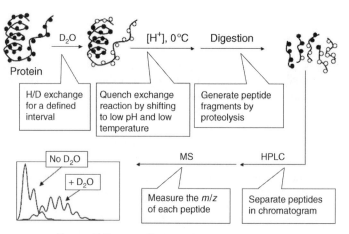

Figure 7.2. Overall H/D exchange experiment.

follow the deuterium buildup of individual amide hydrogen or sets of hydrogens, several on-exchange time points are sampled for each condition.

Quench of Exchange Reaction

Following incubation in a deuterated environment for a defined interval, the exchange reaction is "exchange quenched" by diluting the protein sample with a cold, acidic solution (pH ~ 2.5 and $0\,^{\circ}\mathrm{C}$). The quench conditions significantly slow the amide exchange reaction and limit undesirable back exchange. Subsequent experimental procedures are conducted near the quench conditions to minimize the loss of incorporated deuterium.

Protein Fragmentation by Proteolysis

To localize the rate of deuterium buildup to specific amides, the analyte protein is fragmented into a collection of peptides using combinations of endo- and exoproteases. Due to the low pH of the quench conditions in which the protein and peptide samples are maintained after deuterium labeling, the acid-reactive proteases such as pepsin must be employed. Studies with combinations of acid-reactive endoproteinases and carboxypeptidases have been employed to achieve greater sequence coverage and higher amide resolution (Englander et al., 2003; Woods Jr and Hamuro, 2001).

Digestion Optimization

The digestion conditions are optimized prior to conducting multiple H/D exchange experiments, to ensure high sequence coverage. Calculation of the differences in deuterium content between overlapping peptides is the preferred method to localize incorporated deuterium atoms (Hamuro et al., 2002a). Variable digestion parameters include the type and bed volume of the protease columns, the transit time of the protein over the protease column, the type and concentration of denaturant (Hamuro et al., 2002a), and the inclusion of reducing reagents such as tris(2-carboxyethyl)phosphine hydrochloride (TCEP) (Yan et al., 2002).

HPLC Separation

The peptides generated by proteolysis are separated using reverse-phase HPLC to minimize mass overlap and ionization suppression caused by ion competition in the electrospray source (Woods Jr, 1997). The optimized LC gradient parameters efficiently separate peptides while minimizing loss of deuterium through back exchange with solvent. Increased sensitivity can be achieved by using capillary HPLC columns and nanoelectrospray methods (Wang and Smith, 2003).

Mass Analysis

The majority of reported H/D studies employ quadrupole ion-trap (QIT) instruments due to their ease of use, excellent sensitivity, ability to perform MS/MS experiments, compact size, and low cost. Other reports discuss the use of instruments with higher mass resolving power such as the hybrid QTOF instruments (Wang and Smith, 2003). A few groups have utilized FTICR mass spectrometry, which offers ultrahigh mass-resolving power and improved mass accuracy (Akashi and Takio, 2001; Lanman et al., 2003).

Automation of H/D Exchange by MS

A fully automated system for performing detailed studies has been developed to improve the reproducibility and throughput (Figure 7.2) (Hamuro et al., 2003a). It consists of two functional components: a sample deuteration device and a protein processing unit. The preparation operations (shown at the top of Figure 7.2) are performed by two robotic arms equipped with low-volume syringes and two temperature-controlled chambers, one held at 25 °C and the other at 1 °C. To initiate the exchange experiment, a small amount of protein solution is mixed with a deuterated buffer and the mixture is then incubated for a programmed period of time in the temperature-controlled chamber. This on-exchanged sample is immediately transferred to the cold chamber where a quench solution is added to the mixture. The exchange-quenched solution is then injected onto the protein processing system that includes injection loops, protease column(s), a trap column, an analytical column, electronically controlled valves, and isocratic and gradient pumps. The injector, columns, and valves reside in a low-temperature chamber to minimize the loss of deuterium by back exchange (Figure 7.3). The quenched protein solution is pumped in series through a

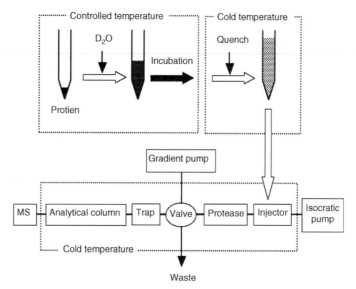

Figure 7.3. Diagram of a fully automated system for acquiring H/D exchange MS data starting with a stock solution of the nondeuterated protein. In this system (Hamuro et al., 2003a), the liquid handler mixes a small amount of concentrated protein solution with a selected deuterated buffer and the mixture is incubated for a programmed period of time. The exchange reaction is conducted in a temperature-controlled chamber held at 25 °C. The mixture is then transferred to an acidic quench solution held at 1 °C. After quenching the exchange reaction, the entire sample is injected onto an LC–MS system that includes injection loops, protease column(s), a trap, an analytical column, and isocratic and gradient pumps. The injector, columns and electronically controlled valves reside in a low-temperature chamber to minimize the loss of deuterium by back exchange. The quenched protein solution is pumped in series over a column containing the immobilized protease and a reverse-phase trap to capture the peptide fragments. The gradient pump is activated following the digestion and the peptides captured on the trap are eluted into the mass spectrometer after separation in the analytical column.

column containing an immobilized protease and a trap column to capture the peptide fragments. The gradient pump is activated following digestion and the peptides captured on the trap column are eluted and separated over an analytical reverse-phase HPLC column directly into the mass spectrometer.

Automated Data Analysis

A software system capable of extracting and cataloging the large number of data points obtained during each experiment has been developed (Hamuro et al., 2003a). The automated system streamlines most data handling steps and reduces the potential for errors associated with manual manipulation of large data sets. In the first processing step, the centroid mass value is obtained for the isotopic envelope of each peptide ion observed in every LC–MS data file associated with the experiment. This step includes peak detection, selection of retention time window, selection of m/z range, and calculation of envelope centroid. The second step involves correction for deuterium losses subsequent to sample quench with reference to measured peptide-specific losses (Hamuro et al., 2003a; Zhang and Smith, 1993). After calculating the percent deuterium incorporation for each peptide at each time point, H/D exchange data are displayed in a number of formats, often as a stacked bar chart that is aligned with the protein sequence.

EXAMPLES OF APPLICATIONS OF DXMS

DXMS Analysis of Conformational Changes of the Manganese Transport Regulator

The Manganese Transport Regulator of Bacillus Subtilis. Manganese transport regulator (MntR) is a metalloregulatory protein found in *Bacillus subtilis*, a member of the iron-responsive diphtheria toxin repressor (DtxR) family (Que and Helmann, 2000). MntR responds to Mn^{2+} and Cd^{2+} by binding to cognate DNA recognition sequences and repressing the *mntABCD* and *mntH* operons, and forms a stable, metal-independent homodimer in solution. Crystallographic studies (Glasfeld et al., 2003; Kliegman et al., 2006; DeWitt et al., 2007) show that MntR consists of two domains: the N-terminal domain containing a winged helix–turn–helix (HTH) motif involved in DNA binding, and the C-terminal domain that contains the dimerization interface. The overall conformations of MntR in the various metal-bound forms are almost identical, with only minor structural variations between the mobile region and the metal binding sites, which exhibits mononuclear (Zn^{2+} and Co^{2+}) or dinuclear (Mn^{2+}, Cd^{2+}, and Ca^{2+}) metal active sites for metal-loaded structures.

DXMS was used to compare the solution structure of apo-MntR and holo-MntR to identify regions affected by metal binding and to elucidate how the formation of the binuclear metal site contributes to the mechanism of metal-mediated activation for DNA binding of this protein. In addition, comparison of exchange rates between apo-MntR, Mn^{2+}-bound, Cd^{2+}-bound, and Co^{2+}-bound forms of MntR is used to reveal the solution dynamics of MntR.

DXMS Experimental Methods Employed with MntR. Deuterated samples were prepared at 25 °C by diluting 20 μL of protein stock solution with 60 μL of deuterated buffer (Hamuro et al., 2002a; Hamuro et al., 2002b; Hamuro et al., 2003b; Woods Jr, 2003), followed by "on-exchange" incubation for various times (10, 30, 60, 150, 600, 3600, 14,400, and 86,400 s) prior to quenching in 120 μL quench buffer (0.8% formic acid, 13.3% glycerol, and 6.2 M guanidine hydrochloride (GuHCl)) at 0 °C, and were stored frozen at −80 °C. The frozen samples were then thawed at 0 °C before passing over an immobilized pepsin column at 100 μL/min with collection of proteolytic products by a reverse-phase C18 column. Subsequently, the C18 column was eluted with a linear gradient of 4–36% acetonitrile over 30 min. The eluate was electrosprayed directly into the mass spectrometer and MS data were analyzed using specialized DXMS software to determine the number of deuterons incorporated with each fragmented peptide.

DXMS Analysis of MntR. The N-terminal DNA binding domain (1–62) of MntR showed less protection than the C-terminal (83–142) dimerization domain, while the linker region between N- and C-terminal domain (65–80) showed intermediate protection. However, in regions affected by metal binding (11–15, 65–80, 93–111), holo-MntR shows different H/D exchange pattern when compared with apo-MntR. Peptides 11–15 and 65–80 show more protection in Co-MntR, and even more protection in Cd-MntR and Mn-MntR. In the dimer interface region (93–111), Cd-MntR and Mn-MntR also show more protection than either apo-MntR or Co-MntR.

Dynamics of MntR in Solution. The DXMS behavior of apo-MntR suggests that the N-terminal domain, containing the HTH DNA binding motif, shows more conformational flexibility in solution than the C-terminal dimerization domain, which is consistent with crystallography studies (Glasfeld et al., 2003; Kliegman et al., 2006). Well-protected regions (11–15 and 30–32) in DNA binding domain are involved in hydrophobic interactions and further in metal-mediated DNA binding. We also saw protection of helix 5 (86–91) and helix 6 (114–116) in the dimer interface region, suggesting that a hydrophobic core was formed to initiate dimer formation. Biophysical studies have confirmed that apo-MntR exists as homodimer in solution (Lieser et al., 2003). The interhelix interaction of helix 7 (118–141) implies that this region is also involved in dimer formation.

The comparison between the DXMS behavior of apo-MntR and holo-MntR can map the conformational changes associated with metal binding. The deuteration levels of three segments in MntR (11–15, 65–80, and 93–111) were significantly lower in the presence of metal ions, suggesting that these regions were involved in the metal binding and dramatically rigidified upon metal binding. In addition, these regions showed even more protection in Cd-MntR or Mn-MntR than that in Co-MntR, consistent with their coordination geometries. Cd-MntR and Mn-MntR formed a binuclear metal site within MntR, while in the case of Co-MntR, a mononuclear metal site was formed.

Mechanism of Metal Activation. Based on the result that strong protection was observed on helices 1, 4, and 5 in response to metal binding, a mechanism was proposed that the metal binding event solidifies the MntR structure to constrain the high mobility between the DNA binding and dimerization domains. The lack of deuteration changes between apo-MntR and holo-MntR for the DNA binding domain implied that the mechanism of activation does not involve large conformational changes, but rather a structural rigidification upon metal binding. Since the DNA binding motif in apo-MntR is preformed, the protein can readily bind DNA once the metal binding event restricts the interdomain motion. In such a context, the mechanism of DNA binding activation of MntR relies on reducing the mobility of a preorganized protein via metal binding.

DXMS Analysis of PAS Domain Allostery and Light-Induced Conformational Changes in Photoactive Yellow Protein Upon I₂ Intermediate Formation

Photoactive Yellow Protein I₂ Intermediate. Photoactive yellow protein, a small bacterial photoreceptor protein that exemplifies the PAS domain superfamily of sensors and signal transducers, is a 125-amino acid residue cytosolic blue light receptor from the purple bacterium *Halorhodospira halophila* and presumably serves as the sensor for negative phototaxis (Cusanovich and Meyer, 2003; Hellingwerf, Hendriks, and Gensch, 2003). PYP adopts an α/β fold (Figure 7.4) with a central six-stranded antiparallel β-sheet flanked by α-helices on both sides and covalently binds a negatively charged *p*-hydroxycinnamic acid chromophore with Cys69 via a thioester linkage. Upon absorption of blue light ($l_{max} = 446\,nm$), PYP undergoes a light-activated reaction cycle that starts with early intermediates I_0 and I_1, via long-lived intermediate I_2 that represents the biologically important signaling state, and returns to the ground state or dark state.

The PYP fold represents the structural prototype of the PAS domain (Pellequer et al., 1998) found in diverse multidomain proteins, mediating protein–protein interaction

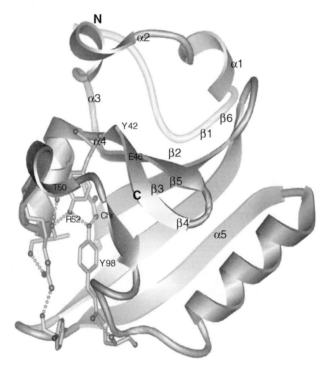

Figure 7.4. PYP fold and active-site hydrogen bonding network. The chromophore (Chr) and active-site residues that participate in a hydrogen bonding network (dotted lines) are shown with bonds. The chromophore forms hydrogen bonds with its thioester oxygen to the backbone amide group and with its phenolate oxygen to the side chains of Tyr42 and Glu46. The side chain of Thr50 is in hydrogen bonding distance to the hydroxyl group of Tyr42 and the main- chain carbonyl oxygen of Glu46. Arg52 forms hydrogen bonds to the backbone carbonyl oxygens of Thr50 and Tyr98 by its guanidinium group and shields the chromophore from solvent. N- and C-termini are labeled (PDB ID: 2PHY).

and functioning as sensors and signal transducers. The PAS/PYP fold can be divided into four segments: (i) the N-terminal cap (1–28), including α-helices 1 and 2; (ii) the PAS core (29–69), including β-strands 1–3, intervening α-helices 3 and 4, and a short π-helix, (iii) the helical connector (70–87), consisting predominantly of the long helix α5; and (iv) the β-scaffold (88–125), containing β-strands 4–6.

In PYP crystallographic structures, light induced conformational changes are localized near the chromophore binding site. During I_2 formation, the phenolic ring of the chromophore rotates out of the binding pockets via the disruption of dark-state hydrogen binding network and becomes protonated and solvent accessible. Additional structural and dynamic changes have also been observed with various biophysical techniques (NMR, FTIR, CD, etc.).

DXMS was used to reveal the detailed protein conformational and dynamic changes upon I_2 formation. As a result, an allosteric tense (T) to relaxed (R) state transition observed in PYP was proposed to provide a general mechanism and pathway for signal transduction directed by the conserved structures of the PYP/PAS module.

DXMS Experimental Methods Employed with PYP. Deuterated samples were prepared as above, except that during deuterium exchange, the samples were either kept in the dark (for dark-state samples) or illuminated with >400 nm light for accumulation of the I_2 intermediate. The subsequent sample processing was as mentioned earlier.

DXMS Analysis of PYP Dark State and I_2 Intermediate. Fifty-seven high-quality identical peptides were obtained for both PYP dark state and I_2 intermediate that cover the entire PYP sequence without any gaps. The H/D exchange map for the PYP dark state (Figure 7.5a) shows that adjacent segments with similar H/D exchange rates delimit the four regions defining the PYP/PAS module. Residues 1–27 show high H/D exchange rates that correspond with the solvent-exposed PYP/PAS N-terminal cap. Residues 28–73 have very low H/D exchange rates that map to the conserved PAS core. Residues 78–88 in the helical connector show intermediate H/D exchanges rates corresponding to the long helix α5. The remaining C-terminal segments (88–125), corresponding to β-scaffold of PYP/PAS

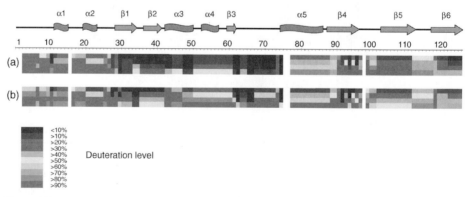

Figure 7.5. Amide H/D exchange for the PYP ground state (a) and the I_2 intermediate (b). Each block represents protein sequence segments defined by overlapping peptides and consists of four time points (from top: 30, 300, 3000, and 30,000 s). The deuteration levels of each such segment at each time point are shown by different colors from blue (<10% deuteration) through red (>90% deuteration), as indicated in the figure. Residues 1–2, 16, 76–77, and 98 are not covered by any sequence segment. The secondary structure elements are shown on top of the protein sequence; α stands for α-helix, β for β-sheet. Figure also appears in the Color Figure section.

module, exhibit exchange rates reflecting their secondary structural elements. In strand β4 (91–96), single-residue resolution highlights a pattern of alternative fast and slow H/D exchange, showing strong protection provided by hydrogen bonding between β4 and β5.

For I_2 intermediate, although PYP remains stable, overall deuterium incorporation is ~17% higher than for the dark state. The largest increases in H/D exchange for I_2 are focused on the chromophore binding site (43–73), where less protection is observed from strand β2 to the connector helix (Figure 7.6). The most significant decreases in H/D exchange are found in the PYP N-terminus: residues 3–5, 6, 7, and 10 become more protected in I_2. Significant changes are also found at the junction of the N-terminal cap and strand β1 (less protection), residue 34–35 (more protection), and strand β4 (less distinct alternating protection pattern). The C-terminal end and N-terminal cap show less changes between the dark state and I_2 intermediate. Changes in H/D exchange kinetics between the dark state and I_2 state were also examined (Figure 7.6).

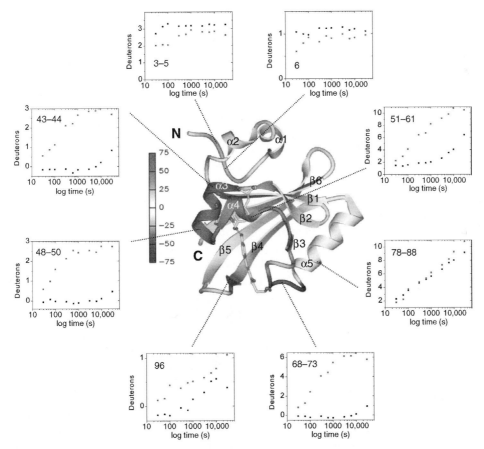

Figure 7.6. Average deuteration changes upon I_2 state formation mapped on PYP structure. Average deuterium incorporation changes upon I_2 formation from Figure 7.4 are mapped on the PYP ground-state ribbon diagram colored from −75% to +75% change according to the color bar. α-Helices 1–5, β-strands I–VI, and N- and C-termini are labeled. The chromophore and active-site residues Tyr42 and Glu46, which form hydrogen bonds (green dotted lines) to the phenolate oxygen of the chromophore in the ground state, are shown for better orientation as well. Figure also appears in the Color Figure section.

Integrating Structural Changes in PYP Upon I_2 State Formation. The DXMS results reveal conformational and dynamic changes in PYP upon light-activated I_2 intermediate formation in solution. During I_2 formation, the largest deuteration increases are found in the continuous stretch of residues 43–73 (Figure 7.6), which spans from α-helix 3, contributing to the hydrogen binding network, short β-strand 3 and the overlapping π-helix, to chromophore-bearing Cys69. The increased H/D exchange rates for segments around the chromophore binding site (43–44, 48–50, 51–61, 68–73) illustrate the relaxation of PAS core via *trans–cis* isomerization of the chromophore, disruption of the active-site hydrogen binding network, and loss of the dark state. The overall pattern of H/D exchange rates in this region suggests a breathing mode, in which the π-helix remains attached to the β-sheet, while preceding helices α3 and α4 become less hydrogen bonded and more solvent exposed. Overall, the connections between PAS core region and the rest of PYP dark state structure are weakened upon I_2 formation. The light-activated conformational and dynamic changes at the chromophore binding site subsequently propagate through the PYP protein to the β4–β5 hairpin, the β-sheet, and the *N*-terminus.

Mapping the Signaling Pathway. By combining our DXMS results with crystallographic structures, we can map the following signal transduction pathways for PYP:

PAS Core α-*Helices:* The pathway for destabilization of the PAS core started from α-helix 3, which exhibits highest increase in deuteration for PYP during I_2 formation. The increased flexibility and unfolding of helix α3 propagate to neighboring segments β2, α4, and β3, as seen from deuteration increases encompassing residue 36–62. DXMS analyses also show the release of the central PAS core from its connections to the rest of the PYP structure, as well as the release of α3 from the N-terminal cap.

β-*Hairpin:* In the β4–β5 hairpin, the connecting loop (97–102) and the C-terminal end of edge strand β4 (89–96) are affected by the movements of the neighboring chromophore. Significantly altered deuterium incorporation shown in residues 92, 94, 96, and 99 upon I_2 formation reflects the disruption of the dark-state hydrogen bonding network between β4 and the chromophore binding site, where large increases in deuteration were also observed.

β-*Sheet Interface:* DXMS results show that the β-sheet relaxes and becomes less distorted and more regular in the I_2 intermediate on one side of the β-sheet. However, the first seven N-terminal residues, opposite end of the β-sheet from the chromophore, show a significant decrease in deuteration upon I_2 formation. This is in good agreement with the N-terminal truncation experiments (Harigai et al., 2001; Harigai et al., 2003), where the N-terminus is important for tuning the lifetime of the I_2 state and catalyzing PYP recovery.

Allosteric Transition: Based on the above results and analyses, we propose that PYP/PAS module undergoes an allosteric transition, switching between two conformations (tense and relaxed states) for signaling. The versatile PYP/PAS module can propagate a conformational signal in three directions: through the PAS core helices, the β-hairpin, or the β-sheet interface.

Structural Determinants for the Binding of Anthrax Lethal Factor to Oligomeric Protective Antigen

Anthrax Lethal Toxin. Three toxin components, that is, edema factor (EF), lethal factor (LF), and protective antigen (PA), secreted by *Bacillus anthracis*, the causative agent of anthrax, assemble at the host cell surface to form a series of toxic, noncovalent complexes. Toxin assembly begins when PA binds to a cell surface receptor (Bradley et al., 2001; Scobie et al., 2003) and is enzymatically cleaved into N-terminal fragment (PA_{20}) that dissociates into the media and a C-terminal fragment (PA_{63}) that heptamerizes to form the prepore (($PA_{63})_7$). LF binds to oligomeric forms of PA_{63} via its homologous N-terminal $\sim$250-residue domain (LF_N) to generate the toxin complex (LF-PA) that is then internalized into acidified compartments. This internalization converts an existing prepore conformation into a membrane-spanning pore, allowing LF to be translocated across the endosomal membrane to the cytosol.

Previous mutagenesis studies (Cunningham et al., 2002; Lacy et al., 2002) showed that a small convoluted patch of seven residues formed on the surface of LF_N constituted the binding site for PA, while the site on PA was found to span the interface between two adjacent subunits of oligomeric PA. This result suggested that additional residues on LF_N may also be involved in binding to PA. DXMS is applied to identify regions within LF_N that are involved in binding to PA. Findings show that four regions of the primary structure displayed protection from solvent in the presence of PA. These regions form a single continuous surface on LF_N that is in contact with PA. Further mutagenesis study revealed the mechanism and binding interface of LF to oligomeric forms of PA.

DXMS Analysis of LF_N Alone and LF_N–PA_{63} Complexes. One hundred fifty-five fragmented peptides were obtained and collectively covered most of the LF_N sequence, except for the first $\sim$60 residues that produce highly positively charged species that are difficult to identify in the mass spectrometer. The results of H/D exchange of LF_N in the absence of PA are mapped onto the crystal structure of LF_N (Figure 7.7). Most of the peptides derived from the surface of LF_N showed increased deuteration at various stages of

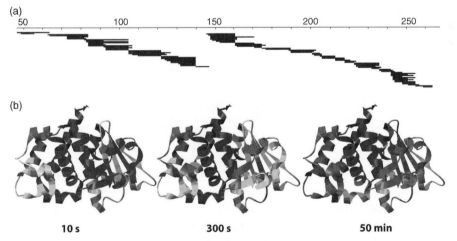

Figure 7.7. Pepsin fragmentation and H/D exchange behavior for LF_N. (a) Pepsin digest coverage map of LF_N. Each black line represents one of the 155 peptide fragments that were identified, analyzed, and used in this study. (b) Ribbon diagrams of LF_N (residues 27–263 of LF_N; PDB entry 1J7N) showing the degree of deuterium content incorporated at three different time points. The level of deuteration for each peptide is colored according to the following scheme: blue, <33% deuterated; yellow, 33–67% deuterated; red, >67% deuterated. This figure was generated using Chimera (Pettersen et al., 2004). Figure also appears in the Color Figure section.

2

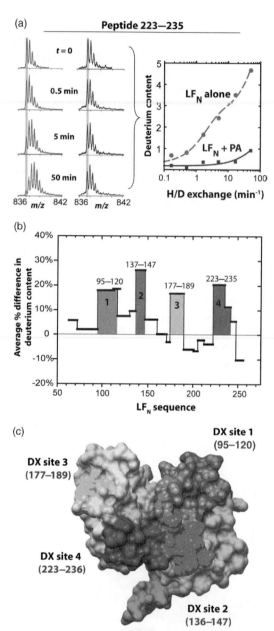

Figure 7.8. Deuterium uptake for LF$_N$ in the absence and presence of PA. (a) The mass spectra of a peptide corresponding to residues 223–235 in the absence (left) and presence (right) of PA. In the absence of PA, the centroid of the peaks increases steadily, whereas in the presence of PA, the centroid of peaks does not shift significantly. Gray lines are arbitrary reference points used to visualize the peak shift. Data from spectra were then used to calculate the deuterium content for LF$_N$ peptides in the absence (●) and presence of PA (■) over six time points. Data are fit to the sum of double exponentials and shown for LF$_N$ alone as a dashed red line, and for LF$_N$ + PA as a solid blue line. (b) The average percent difference in deuterium content over the six time points was plotted for representative peptides that spanned the LF$_A$ sequence. Positive values are

exchange, whereas the core regions (148–61, 213–22) showed little changes. The lack of deuteration within the central core helices indicates that LF_N has a rigidly folded nucleus under native conditions.

Because of the inherent tendency of LF_N–PA heptamer complexes to aggregate at high concentrations, the PA dimer is selected as an alternative binding partner for LF_N. When two mutated forms of PA that are oligomerization deficient are combined in the presence of LF_N, a ternary complex is formed. DXMS results of the ternary complex showed that four discrete sites in LF_N (i.e., DX site 1, 95–120; DX site 2, 137–147, DX site3, 177–189, and DX site 4, 223–235) became significantly more protected from solvent exchange in the presence of PA (Figure 7.8). When mapped onto the surface of LF_N, these four sites converged to form a continuous surface that is larger than that previously implied by mutagenesis studies (Lacy et al., 2002) that only covered DX sites 3 and 4. To address the role of the two additional sites identified here, we constructed a series of mutants within and near these regions and tested their effects on PA binding.

Mutaganesis Analysis of LF_N–PA_{63} Complexes. Twenty-seven mutants were generated at surface-exposed residues within and near DX sites 1 and 2. Mutated proteins were ^{35}S-labeled using *in vitro* incorporation of [^{35}S]Met at the five methionine sites in LF_N, and the proteins were then tested for ability to bind cells in the presence and absence of PA. The majority of the binding defective mutants identified here localized to DX site 2–4. These results suggested evidence that two residues on the PA binding surface of LF_N (Glu-135 and Asp-182), which are $\sim$40 Å apart, interact with the same basic residue (Lys-197) on adjacent subunits of the precore. This result supports earlier findings that oligomerization of PA_{63} is required for LF or EF binding, that the footprint of LF_N and EF_N is large enough to block the binding sites on two PA subunits, and that the heptameric prepore binds only three molecules of EF and/or LF at saturation. The Glu-135–Lys-197 and Asp-182–Lys-197 interaction limit to the two possible orientations of LF_N docked onto the prepore. The fact that mutations of Asp-182 and Lys-197 have greater effects on binding than subunit B mutations, Glu-135 and Lys-197, favors the orientation shown in Figure 7.9c.

These results demonstrate how the four sites within LF_N identified by DX participated in the surface binding interactions and also provide an explanation for how a single LF_N bridges two PA subunits. Most of the residues implicated in binding are expected to be charged at neutral pH values. The affinity of such an interface is expected to decrease with decreasing pH. Thus, the LF–PA interface appears to have been designed to have an inherent pH release trigger, which provides an elegant solution for dissociation of this nanomolar-affinity complex in the endosome. Our study demonstrates the power of H/D exchange mass spectrometry approaches for mapping complex protein–protein interaction surfaces.

---

taken to represent decreases in deuterium uptake in the presence of PA, relative to the absence of PA. The four sites with significant decreases in deuterium uptake in the presence of PA, denoted as DX sites 1–4, are colored blue, green, yellow, and red, respectively. (c) Surface representation of the proposed binding surface of LF_N with the four DX sites (1–4) identified by H/D exchange colored in blue, green, yellow, and red, respectively. Figure also appears in the Color Figure section.

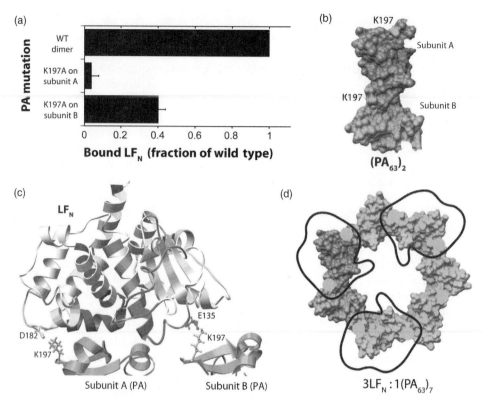

Figure 7.9. H/D exchange and mutagenesis define a model for the LF_N–PA interaction. (a) Data represent the fraction of mutant ^{35}S-LFN bound to PA and mutants of PA on cells relative to that seen for the WT ternary complex. Error bars represent SE of the mean. (b) Surface representation of dimeric PA with K197 shown in yellow on subunits A and B. (c) LF_N–PA_2 model adapted from Lacy et al. with the two subunits (A and B) of PA shown in pink and cyan. DX sites 1–4 identified using H/D exchange are colored blue, green, yellow, and red, respectively. D182 and E135 side chains are shown for LF_N and K197 side chains on both PA subunits are shown. (d) Top view of LF_N binding to domain 1 of heptameric PA. Black trace represents outline of LF_N. Yellow circles indicate K197 positions. Figures were generated using Chimera. Figure also appears in the Color Figure section.

DXMS ANALYSIS: CONCLUDING REMARKS

We have reviewed examples of how hydrogen/deuterium exchange mass spectrometry can be used to study protein structure, conformation, dynamics, and binding interactions. Our examples include the conformational changes of MntR, conformation and dynamics of PYP, and structural dynamics of LF_N–PA complex. In the conformational studies of MntR, the metal-mediated DNA binding mechanism of MntR was elucidated, indicating that metal binding acts to rigidify MntR, therefore limiting the mobility of the protein and reducing the energy lost from DNA binding. In the conformational studies of PYP, wild-type PYP undergoes a light-induced tense to relaxed state transition upon formation of I_2 signaling intermediate, yielding a more open, flexible, and relaxed protein fold. In the study of LF_N–PA complex, DXMS combined with directed mutational analysis defined the surface on LFN that interacts with PA and further reveals the structural basis for anthrax lethal toxin assembly.

At present, the structural inferences drawn from DXMS analysis are of low resolution, even when the data acquired are of high quality and information density. Limitations of the presently employed bioinformatics approaches to exploit and decipher the data are at the heart of the problem and there is a pressing need for improvement in this area.

HYBRID BIOCHEMICAL/BIOINFORMATICS APPROACH TO LOW-RESOLUTION STRUCTURE DETERMINATION USING CHEMICAL CROSS-LINKERS AND MASS SPECTROMETRY

Basic Principles of Using Chemical Cross-Linking for Structure Determination

Two general approaches have been developed for analyzing cross-linked samples: bottom-up and top-down. The bottom-up process can be broken down into four distinct steps (Figure 7.10a). Firstly, the introduction of a cross-linking reagent into the protein(s) of interest by a one- or two-step chemical reaction. Secondly, proteolysis of the sample, followed by mass analysis of the fragments using a variety of available MS techniques. Finally, identified cross-links are used as distance constraints in molecular docking and/or modeling.

The top-down approach (Figure 7.10b) is a more direct approach to analyzing cross-linked species (Kelleher et al., 1999; McLafferty et al., 1999). It uses Fourier transform ion cyclotron resonance mass spectrometers (FTICR-MS) for both the purification and fragmentation steps. The cross-linked products are isolated in the ion cyclotron resonance (ICR) cell before being fragmented via one of several available fragmentation techniques. This technique has been mostly applied to intramolecular cross-linked products. The biggest disadvantage of the top-down approach is that large assemblies of proteins are difficult to analyze and may require the protein to be proteolysed into smaller fragments before FTICR analysis.

Chemical Cross-Linkers Overview

General Reagent Architecture. Chemical cross-linking has been used for decades for a wide variety of applications; hence, many variations of chemical cross-linkers are now commercially available. Based on their chemical functionalities, they can be organized into one of the four general classes: homobifunctional, heterobifunctional, trifunctional, and zero-length cross-linkers (Figure 7.11) (Sinz, 2006).

Homobifunctional cross-linkers are designed for simple one-step reactions. They contain two identical functional groups connected by a spacer of various lengths. Heterobifunctional cross-linkers contain two different functional groups and allow each to be incorporated in separate chemical reactions. This is especially useful when probing a complex of two proteins, where one side of the cross-linker can be reacted to one of the binding partners, while the second one is only activated when the other binding partner is added.

The next logical step was the introduction of another functional group to produce trifunctional cross-linkers. The third functional group is usually some kind of affinity tag (Trester-Zedlitz et al., 2003). Affinity tags allow for separation of cross-linkers from native

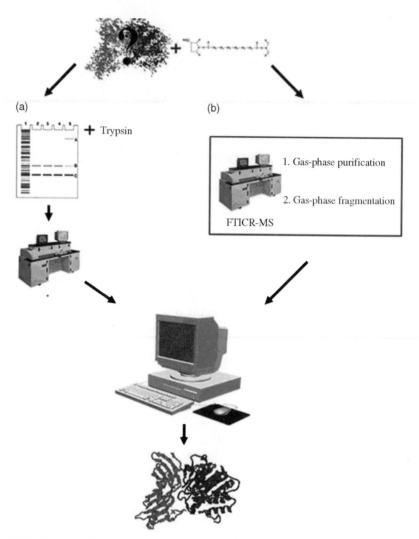

Figure 7.10. The overall process for determining protein complex determination using chemical cross-linking and mass spectrometry in a top-down and bottom-up approach. After introducing the chemical cross-linker of choice in target protein/complex, the bottom-up approach (a) involves separation on an SDS-PAGE gel, in-gel digestion, and extraction of peptides. The extracted peptides are further analyzed by mass spectrometry. In the top-down approach (b) the separation and digestion is done in-source in an FTICR-MS. Both the approaches then use the recorded peptide masses to identify insertion points of cross-link by using a number of available online tools. The distance constraints can then be used in docking or modeling tools.

peptides. Finally, zero-length cross-linkers covalently link amino acids without the incorporation of a linker (Duan and Sheardown, 2006).

Cross-Linker Reactive Groups. Even though many cross-linkers are commercially available, their reactions with amino acids are based on only a handful of organic reactions (Figure 7.12). Mixing and matching a limited number of reactive groups with various spacers creates a large variety of cross-linkers for numerous applications. Most reactive groups can

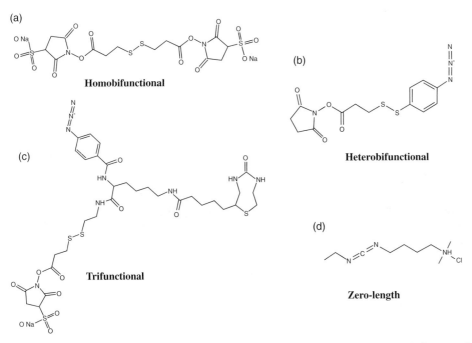

Figure 7.11. Examples of commercially available cross-linkers. (a) DTSSP is a homobifunctional amine reactive cross-linker. (b) SADP is a heterobifunctional cross-linker with amine and sulfhydryl-reactive groups. (c) Sulfo-SBED is a trifunctional cross-linker containing an amine-reactive and photoreactive group as well as a biotin tag. (d) EDC is a zero-length cross-linker, which reacts with amine and carboxylic acid groups.

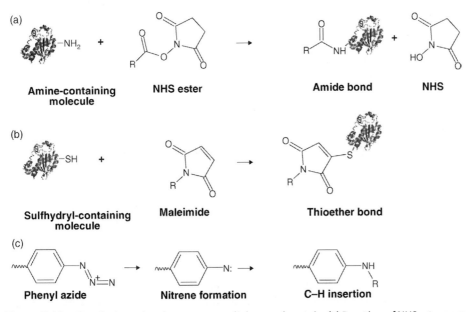

Figure 7.12. Chemical reactions between cross-linkers and protein. (a) Reaction of NHS ester cross-linker and an amine forming an amide bond. (b) Reaction of maleimide and a sulfhydryl forming a thioether bond. (c) One of many possible radical reactions of phenyl azide.

be categorized into one of the following reagents: amine-reactive, sulfhydryl-reactive, and photoreactive.

The two main types of amine reactive cross-linkers are NHS esters and imidoesters. The chemistry of both involves the reaction with primary amines. NHS esters form amide bonds by reacting with ε-amines on lysine residues and free α-amines on N-termini of proteins (Bragg and Hou, 1975; Lomant and Fairbanks, 1976). Since reacted lysines are cleaved at extremely low frequencies by trypsin, larger fragments are formed when using this reagent. This may cause an increased difficulty in mass spectrometric identification. Similar to NHS esters, imidoesters also react with ε-amines and free α-amines of N-termini, however, they form amidine linkages instead of amide bonds.

Sulfhydryl-reactive reagents such as maleimides target thiol groups of cysteines to form thioether bonds. Unlike in amine reactive cross-linkers, lysine residues remain untouched allowing for efficient cleavage by trypsin. If the number of free thiols in a protein is low, introduction of new sulfhydryls can be performed by reacting primary amines with 2-iminothiolane.

Finally, photoreactive cross-linkers require UV or visible light to become reactive. The radical reaction in the presence of light forms a nitrene group that can be inserted into C−H or N−H and allows for a controlled two-step reaction where one end of a cross-linker is inserted in the absence of light. Unlike other reactive groups, cross-linkers containing a photoreactive functional group eliminate amino acid specificity and allow multiple residues to become cross-linking candidates; however, this makes finding the exact point of insertion difficult, as well as causing low yield of specific insertions.

The latest development in this approach has made it feasible to study interactions in living systems, which is extremely useful in understanding cellular processes. Two main strategies exist for this kind of approach, the first one being the incorporation of amino acids that can be photoactivated into proteins by the cell's own translation machinery (Suchanek, Radzikowska, and Thiele, 2005). Such "man-made" amino acids as "photoleucine" are structurally similar to leucine and allow for straightforward incorporation. The second strategy involves expressing a tagged protein, then introducing formaldehyde that can pass through membranes and form cross-links between interacting proteins (Vasilescu, Guo, and Kast, 2004). A purification step can extract the protein of interest along with any proteins cross-linked to it for identification.

Types of Cross-Linker Insertions

An important part of cross-linking is to understand what kind of products can be formed by reacting cross-linkers with proteins. Three types of products can be observed after proteolysis; these are assigned to type 0, 1, or 2 (Figure 7.13) (Schilling et al., 2003).

Type 0 product results when one end of the cross-linker is reacted with the protein while the other end is hydrolyzed, prohibiting it to further react. While this product does not provide distance constraints, it may be used as information regarding the surface accessibility of the amino acid. A type 1 product results when both reactive groups react within the same peptide. While these products do provide some distance constraints, in most cases they provide limited information since they are closely located in sequence.

When a cross-link connects two peptides together, a type 2 product is formed. The peptides must have either a proteolytic cleavage between them or originated from separate proteins. In the hybrid approach described here, this type of product usually provides the most useful information.

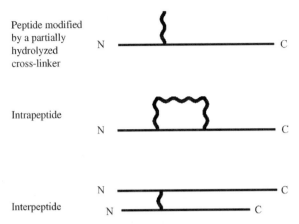

Figure 7.13. The three types of cross-linking products formed by bifunctional cross-linking reagents.

Mass Spectrometry Techniques and Limitations

Over the years, innovations in mass spectrometry techniques have allowed approaches such as the one described in this chapter to become a reality. With various types of mass spectrometers available, it is important to note the advantages and disadvantages of each.

MALDI-TOF Mass Spectrometers. Matrix-assisted laser desorption/ionization time of flight (MALDI-TOF) mass spectrometers have been utilized in numerous cross-linking studies (Itoh, Cai, and Khorana, 2001; Sinz and Wang, 2001; Pearson, Pannell, and Fales, 2002; Trester-Zedlitz et al., 2003; Sinz, Kalkhof, and Ihling, 2005). MALDI-TOF requires minimal amounts of material needed for analysis. A quick overlay of spectra from modified and nonmodified digested samples can lead to assignment of potential cross-linked peptides by mass alone (assuming a well-calibrated instrument).

However, a major drawback of MALDI-MS is its ionization preference for arginine-containing sequences as opposed to lysine-containing sequences (Krause, Wenschuh, and Jungblut, 1999). Arginines with a more basic side chain have a greater ionization efficiency than lysine and these are more easily detectable by MALDI-MS (Krause, Wenschuh, and Jungblut, 1999). As trypsin is the most commonly used enzyme for proteolytic digestion, this becomes a sizable weakness. Conversion of lysines to homoarginines using O-methylisourea can be used to overcome this problem (Beardsley and Reilly, 2002). MALDI-TOF/TOF mass spectrometer can also boost its effectiveness for identifying cross-links as a MS/MS sequencing step can be added, as well as a preceding liquid chromatography separation step. Recently, a unique fragmentation pattern was observed when using MALDI-TOF/TOF to analyze samples cross-linked with a disulfide-based cleavable cross-linker, making it an invaluable tool.

Electrospray Ionization Mass Spectrometers. Another widely used mass spectrometry for cross-linking analysis is electrospray ionization (Mouradov et al., 2006). Unlike MALDI-TOF, ESI does not display preference for arginine-containing peptides over those containing lysine. As with MALDI-TOF, a liquid chromatography step can be used prior to MS analysis. The formation of multiple ionization states by ESI-MS makes for a more time-consuming analysis.

Fourier Transform Ion Cyclotron Resonance Mass Spectrometers. Fourier transform ion cyclotron resonance mass spectrometers have ultrahigh resolution. Coupled with gas-phase purification and fragmentation, it replaces SDS-PAGE separation and proteolytic digestion that are required for analysis using other mass spectrometers (Dihazi and Sinz, 2003; Schulz et al., 2004). However, high cost and lack of availability limit its use in cross-linking studies.

Techniques for Cross-Link Identification

The biggest bottleneck with using chemical cross-linking for structural studies of proteins is the correct identification of cross-linked fragments within mass spectra that are mainly dominated by native tryptic fragments. While the cross-linking and mass spectrometry analysis steps can be done in a matter of days, the assignment of cross-links can be a difficult and often time-consuming exercise. Analysis of MS/MS fragmentation can help in identification; however, the spectra created will include simultaneous fragmentation of two peptides, greatly complicating the fragmentation pattern interpretation. Various approaches and specifically designed cross-linkers have been created to facilitate rapid and accurate identification of cross-linked fragments.

Heavy Cross-Linking Reagents. One of the first innovations to facilitate rapid identification of cross-linkers has been the use of isotope labeling. This approach commonly consists of replacing nonpolar hydrogen atoms by deuterium in the cross-link reagent, creating a chemically identical but heavier reagent. If this "heavy" reagent is used in a 1:1 ratio with the nondeuterated analogue cross-linker, doublet peaks separated by a defined mass (based on the number of deuterium atoms) will be observed in mass spectrometry analysis (Muller et al., 2001). Isotope-labeled cross-linkers, as well as their nonlabeled analogues, are commercially available and allow rapid identification of potential cross-linked fragments.

Cleavable Cross-Linking Reagents. Cleavable cross-linkers contain a disulfide bond or carboxylic ester in the spacer. The comparison of cleaved and noncleaved samples can pinpoint to cross-links within a spectrum as masses corresponding to type 2 cross-links will disappear after the cleavage, creating instead two masses corresponding to each arm of the cross-linker (Bennett et al., 2000; Davidson and Hilliard, 2003). Cleavage of cross-linked peptides also allows for a simplified MS/MS fragmentation pattern on the separate arms.

Affinity Tagged Cross-Linking Reagents. The addition of affinity tags (such as biotin) to cross-links allows for effective separation of modified peptides from native peptides via a purification step. Trifunctional cross-linkers containing affinity tag enrich low abundant cross-linked fragments for mass spectrometry analysis (Sinz, Kalkhof, and Ihling, 2005).

Fragmentation of DTSSP Cross-Linker. The latest advancement in identification of cross-linkers does not require any specially labeled cross-linkers or purifications steps. Instead, it uses the fragmentation properties of a commercially available cleavable cross-linker (DTSSP) for rapid identification (King et al., 2008). This technique takes advantage of a unique asymmetric breakage of disulfide bonds, which forms easily recognizable (66 amu) doublet signals during MALDI-TOF/TOF analysis. Two duplets will

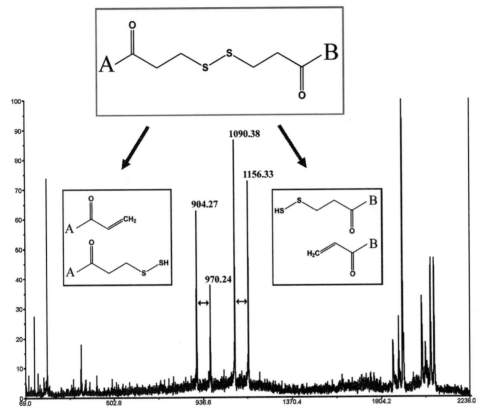

Figure 7.14. Unique fragmentation of DTSSP cross-linked fragments using MALDI TOF/TOF-MS. The two sets of peaks, which are 66 amu apart, represent the asymmetric fragmentation of each half of the cross-linked species.

be formed for a type 2 cross-link, one for each of the two peptides (Figure 7.14). The doublet fragmentation appears analogous to the β-elimination of cysteine that produces thiocysteine and dehydroalanine.

Structure Determination Using Sparse Constraints

With an ever-growing list of tools and techniques for identification of chemically inserted cross-links in proteins, the computational challenge is to use these sparse distance constraints for the purpose of structure prediction. Sparse distance constraints have been shown to improve success of threading algorithms and to predict the orientation of proteins in the complex. Early examples focused on postfiltering of threading results based on validation of the experimental constraints. Newer techniques allow the use of sparse constraints directly in building better models in the first place.

In Silico Hypothetic Cross-Linking Analysis. Albrecht et al. (2002) conducted an *in silico* analysis of how much distance constraints from chemical cross-linking can improve success with fold recognition by threading. The analysis was carried out on 81 non-redundant single-domain proteins (Hobohm96-25 database) whose pair-wise sequence identity does not exceed 25%. Cross-linking constraints were selected for aspartate, glutamate, and lysine residues between 8 and 12 Å apart based on the known structures of the proteins. Constraints were used as a postfiltering method on alignments based on the 123D threading program. Various scoring functions on the validity of distance constraints were used including simply counting the number of satisfied constraints and a more complex scoring function that gives higher scores for satisfied constraints conserved among members of a fold class. The results show that even sparse constraints derived from cross-links can improve recognition rates from about 54–65% up to 58–73% depending on the quality of the initial alignment.

Postfiltering of Threading Using Cross-Link Derived Constraints. While the previous study used hypothetical cross-links, Young et al. (2000) used a similar postfiltering approach on a fibroblast growth factor-2 protein where 15 cross-links were experimentally identified using mass spectrometry techniques. Again the 123D threading tool was utilized and the top 20 structures were considered (when searched against a 635 protein database sharing less than 30% identity). Using just the 123D tool, the protein is incorrectly identified as belonging to the beta-clip fold family. However, a simple scoring function based on the number of satisfied constraints (within the 24 Å maximum cross-linking distance of the BS^3 cross-linker used) was then used to sort through the 20 top models. The first-, second-, and fourth-ranked structures all correctly identify the FGF-2 structure as part of the beta-trefoil family.

Distance Constraints in Threading. The next evolutionary step would be to incorporate the sparse distance constraints into the actual threading algorithm. Xu et al. (2000) detail an approach that uses NMR nuclear Overhauser effect (NOE) distance restraints to improve threading performance. Even though NOEs provide less sparse data, a similar approach can also be applied to more sparse constraints derived from chemical cross-linking and mass spectrometry. Xu et al. employ a "divide and conquer" strategy that divides the structures into substructure sets comprising only of one secondary structure, and then optimally aligns substructures with subsequences. Two conditions must be always met to incorporate the distance constraints: (1) a link must be present for a constraint to be aligned to two cores, and (2) linked cores must not be aligned to sequence positions that violate constraints. Results show that even a small number of NOEs were sufficient to improve threading success in difficult to predict proteins from 70% to 92.7%.

Distance Constraints in Docking Algorithms. The most promising application for distance constraints derived from cross-linking and mass spectrometry may be the structure determination of multidomain proteins and protein complexes, specifically where the structures of the separate components are known but the orientation with respect to each other, however, is not. Docking tools such as "HADDOCK" (Dominguez, Boelens, and Bonvin, 2003) have been developed to use distance constraints as an input. Docking algorithms specifically optimized to incorporate sparse constraints from cross-linking data have also been successfully employed.

Studies by Mouradov et al. (2006) and Forwood et al. (2007) have shown how docking algorithms, specifically designed to incorporate sparse distance constraints from cross-links, can be used to determine how the separate units in a protein complex and multidomain protein interact (Figure 7.15). The docking algorithm used a systematic six-dimensional search over all rotations in steps of five degrees and all Cartesian translations of 1.0 Å up to

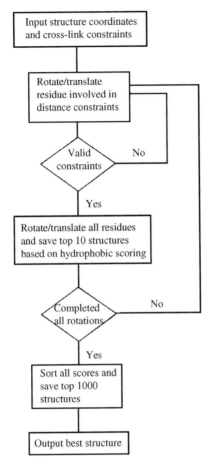

Figure 7.15. Flow diagram of docking algorithm used in studies by Mouradov et al. and Forwood et al. to compute models based on constraints from chemical cross-linking. The optimization step, of only rotating/translating amino acids involved in the constraint to check for validity before rotating/ translating the rest of the molecule, allows for a large number of configurations to be analyzed.

±66 Å along each coordinate axis. Given the cross-linker reagent used in the two studies, the maximal Cα–Cα distance between cross-linked residues was estimated at 25 Å; models with distances larger than 25 Å were therefore immediately excluded from further analysis. An optimized prescreening test was added, where only the coordinates of those residues involved in the constraints were rotated and translated. Only when the constraints were met, were the rest of the coordinates rotated and translated and a full analysis was carried out. This allowed screening for a large set of docking positions that would have been computationally costly using a conventional method. Models were scored by a simple hydrophobic energy score that counts the number of contacts (<8 Å) between hydrophobic amino acids (A, V, L, I, F, C, M, W). For each rotation, the 10 best scoring models were retained. Final models were sorted according to their hydrophobic score and the 1000 best models considered. This approach has been applied to predict the structure of a protein complex to within 4 Å root-mean-square deviation (RMSD) with information from only three interprotein cross-links as well as to propose structure of the multidomain mouse protein acytel-CoA thioesterase 7 (Acot7).

RECENT APPLICATIONS

In the recent years, the hybrid approach described here has been implemented in proof-of-concept experiments as well as in structure prediction. Much research has been put into simplifying identification of cross-linked fragments.

Back et al. (2002) presented a novel approach to screen proteolytic mass spectra maps of cross-linked protein complexes for the presence of cross-linked peptides. The approach was based on the incorporation of ^{18}O heavy water into the C-termini of proteolytic peptides. Cross-links linking two peptides are readily distinguished in mass spectra by a characteristic 8 amu shift when compared to normal water digests due to the incorporation of two ^{18}O atoms in each C-terminus (Figure 7.16). As proof of principle, Back et al. successfully applied the method to a complex of two DNA repair proteins (Rad18–Rad6) and identified the interaction site.

A similar approach was undertaken by Muller at al. (2001); however, instead of heavy and light water, labeled cross-linkers were used. The deuterium-labeled cross-linking reagents bis(sulfosuccinimidyl)-glutarate-d(4), -pimelate-d(4), and -sebacate-d(4) were synthesized (labeled cross-linkers are now commercially available) as well as their undeuterated counterparts. In these cross-links, four nonpolar hydrogens present on the linker were replaced with deuterium. The cross-linking reaction is carried out in a 1:1 mixture of nonlabeled and deuterium-labeled isomers. When compared to native peptides, those that have been modified will be observed as mass signals with 4 Da separation

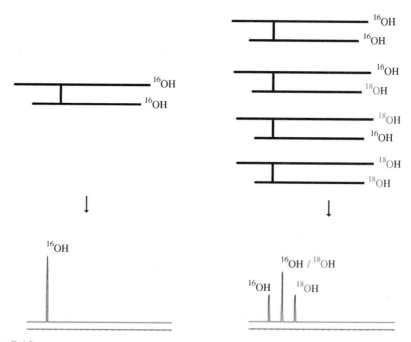

Figure 7.16. The formation of a type 2 cross-link in the presence of normal water (left) and in the presence of a 1:1 mixture of normal and ^{18}O heavy water (right). Each $+2$ amu peak in the 1:1 mixture shows the insertion of an extra ^{18}O heavy water.

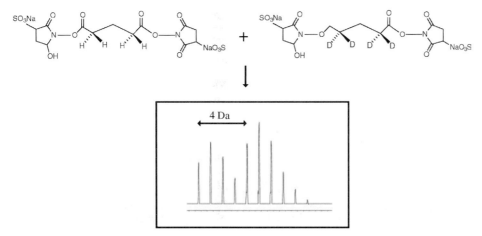

Figure 7.17. Cross-linking using a 1:1 mixture of nonlabeled and deuterium-labeled cross-linker isomers results in a distinct pattern, with doublet peaks 4 amu apart resulting from the four deuterium atoms present in the deuterium-labeled cross-linker.

(Figure 7.17), allowing for a straightforward detection. This methodology facilitated a refined view of the complex between Op18 and tubulin corroborating the tubulin "capping" activity of the N-terminal domain of Op18.

Alternative approaches to deriving distance constraints from cross-linking experiments have been proposed. While traditionally a bottom-up approach is used for cross-linking experiments where the sample is digested in gel before analyzing the peptide fragments mass, a top-down approach initially proposed by McLafferty et al. (1999) is becoming popular. This approach is used to address some possible limitations of the bottom-up approach, including time-consuming separation steps, such as separating cross-linked complexes, as well as a susceptibility to artifacts due to man-made peptide modifications (formylation, oxidation, and sodiation). Kruppa, Schoeniger, and Young (2003) demonstrate that cross-link insertion points can be identified by the use of electrospray ionization Fourier transform mass spectrometry (FTMS) on a crude cross-linked sample. The need for chromatographic separation is replaced with a "gas-phase purification" in the FTMS analyzer cell, while proteolytic digestion is replaced by a gas-phase fragmentation and MS/MS analysis. Using a lysine-specific homobifunctional cross-linking reagent disuccinimidyl suberate (DSS) and ubiquitin, two cross-links were identified and confirmed to be consistent with the known protein structure. This approach lends itself to a true high-throughput methodology where various cross-linkers and reactions can be analyzed quickly and inexpensively.

Various cross-linking studies show exactly how the derived distance can be used to propose distance constraint. Young et al. (2000) used BS^3 cross-linker to correctly identify the fold of a fibroblast growth factor 2 (FGF2) protein. BS^3 cross-links were identified by TOF-MS and confirmed by postsource decay MS. The 18 identified cross-links correctly identified the FGF2 as a member of the IL-1β family even though the sequence identity was less than 13%. While the cross-links were not directly used in model building, they were used to help sort through models obtained by threading. Without the distance constraints, the FGF2 protein would have been incorrectly identified as a member of the β-clip fold family. Furthermore, Young et al. were able to build a structure based on homology IL-1β (without

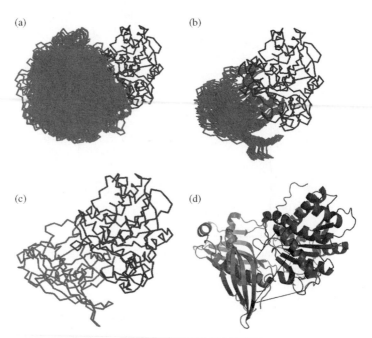

Figure 7.18. Docking of carboxypeptidase A (dark gray) to latexin (black) using in-house rigid body docking software. (a) The top 1000 structures are shown. (b) The top-scoring cluster based on average hydrophobic interaction scoring. (c) The top-scoring docked structure. (d) The positions of the three identified cross-links (black lines) are also shown with respect to the 3D structure.

the use of cross-links) to within 4.8 Å root-mean-square deviation to the backbone. This work was one of the first to show how cross-links can be used to filter out incorrect threading matches.

Mouradov et al. (2006) derived a structure of a complex between carboxypeptidase A and latexin based on chemical cross-linking and molecular docking. While the crystal structures of the separate molecules were known, the complex between the two proteins had yet to be solved. The BS^3 cross-linker was used in combination with LC–ESI-MS to identify cross-linked peptides. The three identified cross-linked fragments between the two molecules were used in a simple rigid body docking algorithm that implemented a hydrophobic scoring function. Shortly after this work, a crystal structure of homologous proteins was solved between carboxypeptidase A and latexin, allowing to test the success/accuracy of the approach. The comparison of the backbone revealed that the predicted structure of the complex was only ~4 Å RMSD (C$_\alpha$) from the crystal structure (Figure 7.18).

This approach was also carried out on a multidomain protein, long-chain acyl-CoA thioesterase 7 (Forwood et al., 2007). The thioesterase is composed of two hot-dog fold domains and converts acyl-coenzyme A to coenzyme A and free fatty acids. The crystal structures of the two domains were solved separately; however, the full-length protein could not be crystallized. The Use of BS^3 and DTSSP (3,3′-dithiobis[sulfosuccinimidylpropionate]) cross-linking reagents, coupled with MALDI-TOF, allowed for the identification of eight interdomain cross-links. These cross-links were used for both homology modeling and molecular docking (Figure 7.19). For modeling, the distance constraints were used in conjunction with homology to the structure of bacterial thioesterases. The model was

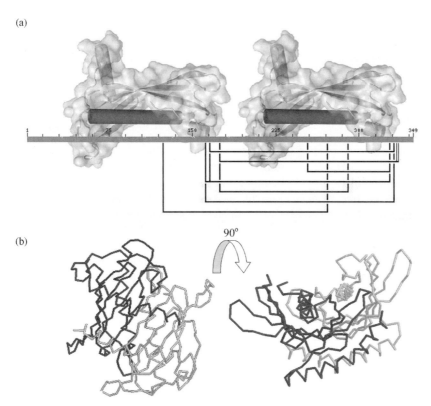

Figure 7.19. (a) 2D view of identified cross-links connecting the two hot-dog fold domains of the acyl-CoA thioesterase. (b) Orthogonal view of the derived orientation of the two hot-dog fold domains based on cross-linking data.

confirmed using rigid body docking with the two domains as separate molecules (using no prior assumptions on structural arrangement).

Until recently, the top-down approach has been limited to analysis of protein structures from intramolecular cross-linking experiments. Schmidt et al.(2005)) combined the advantages of Fourier transform ion cyclotron resonance, such as excellent mass resolution, ultrahigh mass measurement accuracy, and high sensitivity, with enzymatic digestion that allows the analysis of large protein complexes. To further facilitate the identification of cross-linked fragments, deuterium-labeled cross-linkers were also used. The study involved the complex between calmodulin and a 25-amino acid peptide from the C-terminal region of adenylyl cyclase. Five separate lysine residues in calmodulin linked to one single lysine in the 25-amino acid peptide were identified. While no structure was proposed, the study proposes a convenient approach to obtain protein structural data in limited time.

One of the most promising applications of cross-linking is in membrane proteins that play a central role in signaling, transport, and energy. Despite their importance, relatively few membrane protein structures have been solved using X-ray crystallography or NMR when compared to soluble proteins. Jacobsen et al. (2006) have demonstrated that structural information can be derived from proteins purified in their native membrane. The G-protein-coupled receptor rhodopsin was used as the model system. Seven different cross-linking reagents in combination with FTICR mass spectrometry were used to identify a total of 28 intramolecular cross-links. All but one cross-link was theoretically viable based on the crystal structure. The one inexplicable cross-link is in the C-terminal region for which biochemical data from other studies suggest that it is highly mobile in solution.

CONCLUSION WITH RESPECT TO CHEMICAL CROSS-LINKAGE METHODS

The techniques described in this chapter have been shown to provide distance data on proteins that are useful to determine 3D structures at low to medium resolution. While chemical cross-linking provides a limited number of constraints, they can be enough to filter correct from incorrect models or even predict orientation of interacting domains or proteins. While only a limited number of studies have used this technique to propose a structure, much research has gone into developing more efficient ways to identify exact insertion points of cross-links. These innovations, along with mass spectrometry becoming a more commonly used research tool in many departments, demonstrate the potential for chemical cross-linking to become a powerful and widely used approach for swift and inexpensive structure determination.

ACKNOWLEDGMENTS

This work was supported by NIH Grants CA099835, CA118595, GM037684, AI0220221, AI022160, GM20501 and a Discovery Grant (UC10591) from the University of California Industry–University Cooperative Research Program, Biogen Idec Inc. corporate sponsor (V.L.W). B.K. is an ARC Federation Fellow and a National Health and Medical Research Council (NHMRC) Honorary Research Fellow.

REFERENCES

Akashi S, Takio K (2001): Structure of melittin bound to phospholipid micelles studied using hydrogen-deuterium exchange and electrospray ionization Fourier transform ion cyclotron resonance mass spectrometry. *J Am Soc Mass Spectrom* 12(12):1247–1253.

Albrecht M, Hanisch D, Zimmer R, Lengauer T (2002): Improving fold recognition of protein threading by experimental distance constraints. *In Silico Biol* 2(3):325–337.

Back JW, Notenboom V, de Koning LJ, Muijsers AO, Sixma TK, de Koster CG, de Jong LZ (2002): Identification of cross-linked peptides for protein interaction studies using mass spectrometry and O-18 labeling. *Anal Chem* 74(17):4417–4422.

Bai Y, Milne JS, Mayne L, Englander SW (1993): Primary structure effects on peptide group hydrogen exchange. *Proteins* 17(1):75–86.

Beardsley RL, Reilly JP (2002): Optimization of guanidination procedures for MALDI mass mapping. *Anal Chem* 74(8):1884–1890.

Bennett KL, Kussmann M, Bjork P, Godzwon M, Mikkelsen M, Sorensen P, Roepstorff P (2000): Chemical cross-linking with thiol-cleavable reagents combined with differential mass spectrometric peptide mapping—a novel approach to assess intermolecular protein contacts. *Protein Sci* 9(8):1503–1518.

Black BE, Foltz DR, Chakravarthy S, Luger K, Woods VL Jr, Cleveland DW (2004): Structural determinants for generating centromeric chromatin. *Nature* 430(6999):578–582.

Bradley KA, Mogridge J, Mourez M, Collier RJ, Young JA (2001): Identification of the cellular receptor for anthrax toxin. *Nature* 414(6860):225–229.

Bragg PD, Hou C (1975): Subunit composition, function, and spatial arrangement in Ca^{2+}-activated and Mg^{2+}-activated adenosine triphosphatases of *Escherichia coli* and *Salmonella typhimurium*. *Arch Biochem Biophys* 167(1):311–321.

Chen XH, Chen YH, Anderson VE (1999): Protein cross-links: universal isolation and characterization by isotopic derivatization and electrospray ionization mass spectrometry. *Anal Biochem* 273(2):192–203.

Cunningham K, Lacy DB, Mogridge J, Collier RJ (2002): Mapping the lethal factor and edema factor binding sites on oligomeric anthrax protective antigen. *Proc Natl Acad Sci USA* 99(10):7049–7053.

Cusanovich MA, Meyer TE (2003): Photoactive yellow protein: a prototypic PAS domain sensory protein and development of a common signaling mechanism. *Biochemistry* 42(17):4759–4770.

Davidson WS, Hilliard GM (2003): The spatial organization of apolipoprotein A-I on the edge of discoidal high density lipoprotein particles—a mass spectrometry study. *J Biol Chem* 278(29): 27199–27207.

DeWitt MA, Kliegman JI, Helmann JD, Brennan RG, Farrens DL, Glasfeld A (2007): The conformations of the manganese transport regulator of *Bacillus subtilis* in its metal-free state. *J Mol Biol* 365(5):1257–1265.

Dihazi GH, Sinz A (2003): Mapping low-resolution three-dimensional protein structures using chemical cross-linking and Fourier transform ion-cyclotron resonance mass spectrometry. *Rapid Commun Mass Spectrom* 17(17):2005–2014.

Dominguez C, Boelens R, Bonvin A (2003): HADDOCK: a protein–protein docking approach based on biochemical or biophysical information. *J Am Chem Soc* 125(7):1731–1737.

Duan X, Sheardown H (2006): Dendrimer crosslinked collagen as a corneal tissue engineering scaffold: mechanical properties and corneal epithelial cell interactions. *Biomaterials* 27(26): 4608–4617.

Engen JR, Smith DL (2001): Investigating protein structure and dynamics by hydrogen exchange MS. *Anal Chem* 73(9):256A–265A.

Englander SW, Kallenbach NR (1983): Hydrogen exchange and structural dynamics of proteins and nucleic acids. *Q Rev Biophys* 16(4):521–655.

Englander JJ, Del Mar C, Li W, Englander SW, Kim JS, Stranz DD, Hamuro Y, Woods VL Jr (2003): Protein structure change studied by hydrogen-deuterium exchange, functional labeling, and mass spectrometry. *Proc Natl Acad Sci USA* 100(12):7057–7062.

Forwood JK, Thakur AS, Guncar G, Marfori M, Mouradov D, Meng W, Robinson J, Huber T, Kellie S, Martin JL, Hume DA, Kobe B (2007): Structural basis for recruitment of tandem hotdog domains in acyl-CoA thioesterase 7 and its role in inflammation. *Proc Natl Acad Sci USA* 104(25): 10382–10387.

Glasfeld A, Guedon E, Helmann JD, Brennan RG (2003): Structure of the manganese-bound manganese transport regulator of *Bacillus subtilis*. *Nat Struct Biol* 10(8):652–657.

Goshe MB, Anderson VE (1999): Hydroxyl radical-induced hydrogen/deuterium exchange in amino acid carbon–hydrogen bonds. *Radiat Res* 151(1):50–58.

Hamuro Y, Burns L, Canaves J, Hoffman R, Taylor S, Woods V (2002a): Domain organization of D-AKAP2 revealed by enhanced deuterium exchange-mass spectrometry (DXMS). *J Mol Biol* 321(4):703–714.

Hamuro Y, Wong L, Shaffer J, Kim JS, Stranz DD, Jennings PA, Woods VL Jr, Adams JA (2002b): Phosphorylation driven motions in the COOH-terminal Src kinase, CSK, revealed through enhanced hydrogen-deuterium exchange and mass spectrometry (DXMS). *J Mol Biol* 323(5): 871–881.

Hamuro Y, Coales SJ, Southern MR, Nemeth-Cawley JF, Stranz DD, Griffin PR (2003a): Rapid analysis of protein structure and dynamics by hydrogen/deuterium exchange mass spectrometry. *J Biomol Tech* 14(3):171–182.

Hamuro Y, Zawadzki KM, Kim JS, Stranz DD, Taylor SS, Woods VL Jr (2003b): Dynamics of cAPK type IIbeta activation revealed by enhanced amide H/^{2}H exchange mass spectrometry (DXMS). *J Mol Biol* 327(5):1065–1076.

Hamuro Y, Anand GS, Kim JS, Juliano C, Stranz DD, Taylor SS, Woods VL Jr (2004): Mapping intersubunit interactions of the regulatory subunit (RIalpha) in the type I holoenzyme of protein kinase A by amide hydrogen/deuterium exchange mass spectrometry (DXMS). *J Mol Biol* 340(5): 1185–1196.

Harigai M, Yasuda S, Imamoto Y, Yoshihara K, Tokunaga F, Kataoka M (2001): Amino acids in the N-terminal region regulate the photocycle of photoactive yellow protein. *J Biochem* 130(1):51–56.

Harigai M, Imamoto Y, Kamikubo H, Yamazaki Y, Kataoka M (2003): Role of an N-terminal loop in the secondary structural change of photoactive yellow protein. *Biochemistry* 42(47):13893–13900.

Hellingwerf KJ, Hendriks J, Gensch T (2003): Photoactive yellow protein, a new type of photoreceptor protein: will this "yellow lab" bring us where we want to go?. *J Phys Chem A* 107(8):1082–1094.

Hilser VJ, Freire E (1996): Structure-based calculation of the equilibrium folding pathway of proteins. Correlation with hydrogen exchange protection factors. *J Mol Biol* 262(5):756–772.

Itoh Y, Cai K, Khorana HG (2001): Mapping of contact sites in complex formation between light-activated rhodopsin and transducin by covalent crosslinking: use of a chemically preactivated reagent. *Proc Natl Acad Sci USA* 98(9):4883–4887.

Jacobsen RB, Sale KL, Ayson MJ, Novak P, Hong JH, Lane P, Wood NL, Kruppa GH, Young MM, Schoeniger JS (2006): Structure and dynamics of dark-state bovine rhodopsin revealed by chemical cross-linking and high-resolution mass spectrometry. *Protein Sci* 15 (6):1303–1317.

Kelleher NL, Lin HY, Valaskovic GA, Aaserud DJ, Fridriksson EK, McLafferty FW (1999): Top down versus bottom up protein characterization by tandem high-resolution mass spectrometry. *J Am Chem Soc* 121(4):806–812.

King GJ, Jones A, Kobe K, Huber T, Mouradov D, Hume DA, Ross IL (2008): Identification of disulfide-containing chemical cross-links in proteins using MALDI-TOF/TOF-mass spectrometry. *Anal Chem* 80(13): 5036–5043.

Kliegman JI, Griner SL, Helmann JD, Brennan RG, Glasfeld A (2006): Structural basis for the metal-selective activation of the manganese transport regulator of *Bacillus subtilis*. *Biochemistry* 45(11):3493–3505.

Krause E, Wenschuh H, Jungblut PR (1999): The dominance of arginine-containing peptides in MALDI-derived tryptic mass fingerprints of proteins. *Anal Chem* 71(19):4160–4165.

Kruppa GH, Schoeniger J, Young MM (2003): A top down approach to protein structural studies using chemical cross-linking and Fourier transform mass spectrometry. *Rapid Commun Mass Spectrom* 17(2):155–162.

Lacy DB, Mourez M, Fouassier A, Collier RJ (2002): Mapping the anthrax protective antigen binding site on the lethal and edema factors. *J Biol Chem* 277(4):3006–3010.

Lanman J, Lam TT, Barnes S, Sakalian M, Emmett MR, Marshall AG, Prevelige PE Jr (2003): Identification of novel interactions in HIV-1 capsid protein assembly by high-resolution mass spectrometry. *J Mol Biol* 325(4):759–772.

Lieser SA, Davis TC, Helmann JD, Cohen SM (2003): DNA-binding and oligomerization studies of the manganese(II) metalloregulatory protein MntR from *Bacillus subtilis*. *Biochemistry* 42(43): 12634–12642.

Lomant AJ, Fairbanks G (1976): Chemical probes of extended biological structures—synthesis and properties of cleavable protein cross-linking reagent [dithiobis(succinimidyl-S-35 propionate). *J Mol Biol* 104(1):243–261.

Mann M, Talbo G (1996): Developments in matrix-assisted laser desorption ionization peptide mass spectrometry. *Curr Opin Biotechnol* 7(1):11–19.

McLafferty FW, Fridriksson EK, Horn DM, Lewis MA, Zubarev RA (1999): Biochemistry—biomolecule mass spectrometry. *Science* 284(5418):1289–1290.

Mouradov D, Craven A, Forwood JK, Flanagan JU, Garcia-Castellanos R, Gomis-Ruth FX, Hume DA, Martin JL, Kobe B, Huber T (2006): Modelling the structure of latexin–carboxypeptidase A complex based on chemical cross-linking and molecular docking. *Protein Eng Des Sel* 19(1):9–16.

Muller DR, Schindler P, Towbin H, Wirth U, Voshol H, Hoving S, Steinmetz MO (2001): Isotope tagged cross linking reagents. A new tool in mass spectrometric protein interaction analysis. *Anal Chem* 73(9):1927–1934.

Pantazatos D, Kim JS, Klock HE, Stevens RC, Wilson IA, Lesley SA, Woods VL Jr (2004): Rapid refinement of crystallographic protein construct definition employing enhanced hydrogen/deuterium exchange MS. *Proc Natl Acad Sci USA* 101(3):751–756.

Pearson KM, Pannell LK, Fales HM (2002): Intramolecular cross-linking experiments on cytochrome *c* and ribonuclease A using an isotope multiplet method. *Rapid Commun Mass Spectrom* 16(3): 149–159.

Pellequer JL, Wager-Smith KA, Kay SA, Getzoff ED (1998): Photoactive yellow protein: a structural prototype for the three-dimensional fold of the PAS domain superfamily. *Proc Natl Acad Sci USA* 95(11):5884–5890.

Pettersen EF, Goddard TD, Huang CC, Couch GS, Greenblatt DM, Meng EC, Ferrin TE (2004): UCSF chimera—a visualization system for exploratory research and analysis. *J Comput Chem* 25(13): 1605–1612.

Que Q, Helmann JD (2000): Manganese homeostasis in *Bacillus subtilis* is regulated by MntR, a bifunctional regulator related to the diphtheria toxin repressor family of proteins. *Mol Microbiol* 35(6):1454–1468.

Resing KA, Hoofnagle AN, Ahn NG (1999): Modeling deuterium exchange behavior of ERK2 using pepsin mapping to probe secondary structure. *J Am Soc Mass Spectrom* 10(8):685–702.

Schilling B, Row RH, Gibson BW, Guo X, Young MM (2003): MS2Assign, automated assignment and nomenclature of tandem mass spectra of chemically crosslinked peptides. *J Am Soc Mass Spectrom* 14(8):834–850.

Schmidt A, Kalkhof S, Ihling C, Cooper DMF, Sinz A (2005): Mapping protein interfaces by chemical cross-linking and Fourier transform ion cyclotron resonance mass spectrometry: application to a calmodulin/adenylyl cyclase 8 peptide complex. *Eur J Mass Spectrom* 11(5): 525–534.

Schulz DM, Ihling C, Clore GM, Sinz A (2004): Mapping the topology and determination of a low-resolution three-dimensional structure of the calmodulin–melittin complex by chemical cross-linking and high-resolution FTICRMS: direct demonstration of multiple binding modes. *Biochemistry* 43(16):4703–4715.

Scobie HM, Rainey GJ, Bradley KA, Young JA (2003): Human capillary morphogenesis protein 2 functions as an anthrax toxin receptor. *Proc Natl Acad Sci USA* 100(9):5170–5174.

Sinz A, Wang K (2001): Mapping protein interfaces with a fluorogenic cross-linker and mass spectrometry: application to nebulin–calmodulin complexes. *Biochemistry* 40 (26):7903–7913.

Sinz A, Kalkhof S, Ihling C (2005): Mapping protein interfaces by a trifunctional cross-linker combined with MALDI-TOF and ESI-FTICR mass spectrometry. *J Am Soc Mass Spectrom* 16(12): 1921–1931.

Sinz A (2006): Chemical cross-linking and mass spectrometry to map three-dimensional protein structures and protein–protein interactions. *Mass Spectrom Rev* 25(4):663–682.

Sivaraman T, Arrington CB, Robertson AD (2001): Kinetics of unfolding and folding from amide hydrogen exchange in native ubiquitin. *Nat Struct Biol* 8(4):331–333.

Suchanek M, Radzikowska A, Thiele C (2005): Photo-leucine and photo-methionine allow identification of protein–protein interactions in living cells. *Nat Methods* 2(4):261–267.

Trester-Zedlitz M, Kamada K, Burley SK, Fenyo D, Chait BT, Muir TW (2003): A modular cross-linking approach for exploring protein interactions. *J Am Chem Soc* 125(9):2416–2425.

Vasilescu J, Guo XC, Kast J (2004): Identification of protein–protein interactions using *in vivo* cross-linking and mass spectrometry. *Proteomics* 4(12):3845–3854.

Wang L, Smith DL (2003): Downsizing improves sensitivity 100-fold for hydrogen exchange-mass spectrometry. *Anal Biochem* 314(1):46–53.

Woods VL Jr (1997) Method for characterization of the fine structure of protein binding sites employing amide hydrogen exchange. U.S. Patent 5,658,739, USA.

Woods V Jr (2001a) Methods for the high-resolution identification of solvent-accessible amide hydrogens in polypeptides or proteins and for the characterization of the fine structure of protein binding sites. U.S. Patent 6,291,189, USA.

Woods V Jr (2001b) Method for characterization of the fine structure of protein binding sites using amide hydrogen exchange. US Patent 6,331,400, USA.

Woods VL Jr, Hamuro Y (2001): High resolution, high-throughput amide deuterium exchange-mass spectrometry (DXMS) determination of protein binding site structure and dynamics: utility in pharmaceutical design. *J Cell Biochem* 37:89–98.

Woods V Jr (2003) Methods for identifying hot-spot residues of binding proteins and small compounds that bind to the same. U.S. Patent 6,599,707, USA.

Xu Y, Xu D, Crawford OH, Einstein JR, Serpersu E (2000): Protein structure determination using protein threading and sparse NMR data (extended abstract). Paper presented at *Proceedings of the Fourth Annual International Conference on Computational Molecular Biology*, Tokyo, Japan.

Yan X, Zhang H, Watson J, Schimerlik MI, Deinzer ML (2002): Hydrogen/deuterium exchange and mass spectrometric analysis of a protein containing multiple disulfide bonds: solution structure of recombinant macrophage colony stimulating factor-beta (rhM-CSFbeta). *Protein Sci* 11(9): 2113–2124.

Young MM, Tang N, Hempel JC, Oshiro CM, Taylor EW, Kuntz ID, Gibson BW, Dollinger G (2000): High throughput protein fold identification by using experimental constraints derived from intramolecular cross-links and mass spectrometry. *Proc Natl Acad Sci USA* 97(11): 5802–5806.

Zhang Z, Smith DL (1993): Determination of amide hydrogen exchange by mass spectrometry: a new tool for protein structure elucidation. *Protein Sci* 2(4):522–531.

Zhang Z, Post CB, Smith DL (1996): Amide hydrogen exchange determined by mass spectrometry: application to rabbit muscle aldolase. *Biochemistry* 35(3):779–791.

8

SEARCH AND SAMPLING IN STRUCTURAL BIOINFORMATICS

Ilan Samish

The road was still paved with yellow brick, but these were much covered by dried branches and dead leaves from the trees, and the walking was not all good.
L. Frank Baum, in *The Wonderful Wizard of Oz*, 1900.

INTRODUCTION

Search and Sampling—Tools of Structural Bioinformaticians

Structural bioinformatics aims to gain knowledge about biological macromolecules utilizing computational methods applied to structural and structure-related data (see definition in Chapter 1). Search and sampling methods aim at providing a focused "yellow brick road" toward understanding the properties of complex biomolecular structures. Such methods must often be tailored for this unique field. First, intrinsic to the "bio" nature of the data is high complexity in the number and the type of variables. Structural data is derived experimentally from various sources and is presented in different resolutions (Chapters 2–6). Moreover, other types of data often need to be integrated in the processing of structural data. These include (1) the exponentially growing sequence data, for example, in the form of evolutionary conservation (Chapters 17, 23, 26), as well as (2) biophysical data, for example, stability of proteins, distance restraints between residues as derived from mass spectroscopy (Chapter 7) or analysis of mutational data. Second, the required representative space (defined later in the chapter) one wishes to study is often biased, not fully covered or cannot be fully sampled. Scientific research tends to produce focused data on specific regions, rather than, unveiling the full space. Third, often the available data types or the integration of data from different sources requires specialized search and sampling. Since the physically based rules governing the behavior of biological macromolecules are still not fully understood, or may be not practical to compute, phenomenological data are often utilized. In the resulting

Structural Bioinformatics, Second Edition   Edited by Jenny Gu and Philip E. Bourne
Copyright © 2009 John Wiley & Sons, Inc.

"knowledge-based" parameterization, the interplay between the constituting variables need not be fully understood to be useful. Yet, the generality of such parameterization and the correlation between resulting parameters have to be addressed. Finally, with the diminishing boundaries between structural bioinformatics and computational biophysics, the former may expand in scope by adding more data as well as new complex variables, for example, the fourth dimension of time. Notably, these required factors for search and sampling are evident whether studying a single macromolecule, an ensemble of related macromolecules, or a (sub) space of structural data.

The process of structural bioinformatics search and sampling methods can be divided into several general groups or steps. First, the *sampling stage* seeks to ensure that the "answer space" of the assessed question is properly covered. Second, the *search stage* sets to select the important and relevant information within the vast available data. As shown in Figure 8.1, these two stages are intimately entwined. Third, the *analysis stage* deals with inferring the biological meaning of the data, and as a result the data can often be classified into few or numerous divisions. Finally, the *optimization and quality assurance stage* examines the analyzed data with its related quality and chooses the most correct answer. Alternatively, this stage may result in a decision to conduct further fine-tuning analysis, reduce the data derived from the search space, or even go back and revise the sampling process. While the latter two stages are an integral part of the search and sampling process, here we will focus on the first two stages.

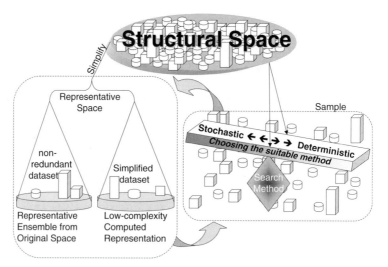

Figure 8.1. Search, sampling, and the representation of a structural space. A redundant structural space, containing different numbers (or clusters) of each constituent structure can be simplified into a representative space. The latter contains a representative of each of the original constituents (or constituent clusters) with each member described in a full or a simplified form. Proper sampling is required to cover the structural space in a sufficient manner that will not filter out important features, and yet will reduce the biasing redundancy and achieve a computationally manageable space. When choosing the deterministic or stochastic search method, these considerations should be weighted against the computational complexity and ease of applying the chosen method. Search methods that are fit to the complex structural space and the questions asked by the method assess the structural space and provide a representative space, which in turn can be further searched.

The methods of search and sampling structural space comprise a broad group of tools essential to numerous structural bioinformatics applications. In this brief chapter, we cannot exhaustively cover the mathematical foundations associated with these methods, nor their numerous structural bioinformatics applications. A short introduction of terms and philosophy is introduced in the following section. Further description of such techniques can be found in other sources relating to (1) general methods, for example "Monte Carlo applications in science" (Liu, 2001) or "Markov chain Monte Carlo in practice" (Gilks et al., 1996), (2) bioinformatics from a machine learning approach (Baldi and Brunak, 2001), and (3) structural bioinformatics and computational biophysics books (Becker MacKerell et al., 2001; Leach, 2001; Schlick, 2002). Here, we aim to present the rationale behind some of the different algorithms along with their advantages, limitations, and examples of applications to structural bioinformatics research. The chapter opens with a brief introduction to Bayesian statistics, which is the underlying formalism and rationale to most methods developed in this field. Indeed, some of the studies cited in this chapter utilize the Bayesian formalism explicitly (e.g., Simons et al., 1997; Dunbrack, ; Dunbrack, 2002), while others follow the Bayesian rationale despite not being presented through this formalism. Next, some of the major methods and applications for representative sampling and search of conformational space are described. As a concrete applicative example and without losing generality, we focus on the classic challenge of exploring conformational space of proteins, mainly at the side-chain level.

Underlying Terms: From Frequentist (or Sampling) Statistics to Bayesian Statistics

A very basic introduction to the underlying terms and a way of thinking as it relates to statistical fundamentals of search and sampling is essential. Statistics was presented in the 18th century for the analysis of data about the state (*statisticum* in Latin is "of the state"). Within the broad quantitative examination "of the state," more than one approach was developed. Frequentist statisticians focus on having methods guaranteed to work most of the time, given minimal assumptions. The probability is given only when dealing with well-defined multiple random experiments. Each experiment can have only one of two outcomes; it can occur or it does not occur. As a random experiment, there is no *a priori* knowledge that will bias our knowledge on the outcome of the experiment. Note, as the set of all possible occurrences is often called "sample space," frequentist statistics is often referred to as "sampling statistics." However, here we will not use this term to avoid confusion. Contrary to frequentist statistics, Bayesian statistics (also known as Bayesian inference) attempts to make inferences that take into account all (or all "relevant") available information and answer the question of interest given the particular data set. The observations are used to update, or newly infer, the probability that a hypothesis may be true. The degree of belief in the hypothesis changes as the evidence is accumulated. A debate still exists between the two methods, often resulting from misleading interpretations of common statistical terms (MacKay, 2005). In most but not all structural bioinformatics studies, Bayesian statistics is the method of choice (Dunbrack, 2001).

Formally, frequentist statistics deals with a function, also called statistic, S, of the data, y. The statistic is a summary of multiple occurrences, N, of a specified experiment obtained from repeated trials. A hypothesis, H, about the data is tested as true or false. Thus, the hypothesis cannot have a measured likelihood or probability but rather can be accepted or rejected. Most often the null hypothesis that is at odds with $S(y)$ is utilized with the overall objective of rejecting it. The null hypothesis is rejected if the probability of the data is small (e.g., less than 5%) given the truth of the hypothesis. This requires a comparison of the statistic of the observed data (also called actual data) to the distribution of all possible (and usually simulated) data $p(S(y_{simulated})|H)$. Thus, if $\int^{\infty} S(y_{actual}) p(S(y_{simulated})|H) \, dS(y_{simulated}) < 0.05$, then the null hypothesis can be rejected. The probability distribution and its associated quantities such as the variance σ^2 are defined in terms of parameters θ that may be inferred from the data. Confidence intervals can be defined as the range of parameter values for which the statistic S, such as the, average, $\bar{y}$, would arise 95% of the time. For example, a 95% confidence interval for the mean θ (for normally distributed data) is obtained as follows:

$$\Pr\left(\bar{y} - \sigma/\sqrt{N} < \theta < \bar{y} + \sigma/\sqrt{N}\right) = 0.95. \tag{8.1}$$

Unlike frequentist statistics, probability in Bayesian inference is the degree of belief in the truth of the statement. For example, in confidence interval terms, this would mean the probability of the parameter given the data, rather than the probability of the data given the parameter; that is,

$$\Pr\left(\theta - \sigma/\sqrt{N} < \bar{y} < \theta + \sigma/\sqrt{N}\right) = 0.95. \tag{8.2}$$

The probability of any hypothesis, H, is conditional upon the information, I, we have on the system, namely, $p(H|I)$. The sum rule and the product rule form two basic rules upon which Bayesian inference relies. The sum rule states that given the information, I, we have on a system, the cumulative probabilities for a hypothesis, H, and its inverse, denoted $\bar{H}$ (note, this is the same common annotation used to denote average), to be true is equal to one, that is, $p(H|I) + p(\bar{H}|I) = 1$. The product rule states that the probability of occurrence of two hypotheses, or events, is equal to the probability of one event multiplied by the conditional probability of the other event given the occurrence of the other event, namely, $p(A, B|I) = p(A|B, I)p(B|I) = p(B|A, I)p(A|I)$. As in all these probabilities the same information on the system is assumed, so "I" can be dropped from the equation resulting with $p(A, B) = p(A|B)p(B) = p(B|A)p(A)$. In Bayesian inference, the probability of a hypothesis given the data is pursued. In the sum rule, event A can be replaced by hypothesis, H, and event B by the data, D, resulting in $p(H|D)p(D) = p(D|H)p(H)$. Dividing both sides by the marginal distribution of the data provides a way to measure the posterior distribution—the probability of the hypothesis after examining the data. This distribution is expressed by Bayes' rule (Bayes, 1763):

$$p(H|D) = \frac{p(D|H)p(H)}{p(D)}. \tag{8.3}$$

The prior probability, $p(H)$, is of the hypothesis before considering any data. If the prior distribution may remain uniform over the entire distribution of the data, then the data are said to be uninformative, or if the distribution may be higher for some ranges of the data, then it is said to be informative data. In either case, it should be proper, that is, normalized to integrate to one over the entire data range. The likelihood of the data given that the hypothesis is true is expressed by $p(D|H)$. Uncertainties in measurement and physical theory connecting the data to the hypothesis are expressed by this term. Finally, $p(D)$

is the marginal distribution of the data—a quantity that should be constant with respect to the parameters and act only as a normalization factor.

Bayesian statistics is often applied to specific questions using Bayesian models in which prior distributions are computed for all parameters and a likelihood function formulized for the data given the parameters. Then the posterior distribution of the parameters is computed (analytically or via simulation) based on the data. Finally, the model fitness to the data is evaluated by utilizing posterior predictive simulations. Data predicted from posterior distribution should resemble the data that generated the distribution and be physically realistic.

While frequentist theory has an analytical power in quantifying error as well as differences between groups of data, Bayesian statistics is often more intuitive to scientific research. The very basic scientific way of thinking includes the use of the prior distribution and the focus on parameter value ranges for the data. The possible subjectivity of the Bayesian prior distributions was not accepted by early frequentist statisticians. Yet, the latter use test statistics calculated on hypothetical and often unintuitive data. Some see the subjective need for prior distribution in Bayesian statistics as an advantage. Importantly, the choice of the prior distribution requires considering the full range of possibilities for the hypothesis. Further, it is possible to consider varying degrees of informativity, including the possibility of coupling parameters from unrelated data. As a result, Bayesian statistics is often the underlying rationale for structural bioinformatics.

SAMPLING STRUCTURAL SPACE

Sampling of a space can be achieved by multiple random accesses to the space or by a continuous walk within the space. The ergodic hypothesis, described in the next section, presents the equality between both approaches. Whatever approach is utilized, accurate sampling should ensure proper coverage of the sampled space. A proper coverage includes a full coverage of the space under the given covering resolution and, complementary, the lack of focusing on one area within the space. Often, a representative set is built from the original space to achieve proper sampling. The next sections will describe the definition of the representative spaces in the full protein and in the side-chain hierarchy levels. Notably, when dealing with structural space with multiple local minima, for example, an energy landscape, often the best search is conducted by sampling the full space at low resolution and areas of interest at higher resolution. This important topic will be discussed under the "search" segment of this chapter.

Basic Terms: The Ergodic Hypothesis, Providing the (Detailed) Balance Between Thermodynamic and Dynamic Averages

The ergodic hypothesis is a basic underlying principle in structural bioinformatics and, more generally, in statistical thermodynamics. Historically, the word "ergodic" is an amalgamation of the Greek words *ergon* (work) and *odos* (path). The ergodic hypothesis claims that over a sufficiently long period, the time spent in a region of phase space of microstates with the same energy is proportional to the size of this region. In the context of structural bioinformatics, the phase space is often the energy landscape, that is, the energy of the investigated macromolecule(s) for each possible conformation. The hypothesis was first suggested by Ludwig Boltzmann in the 1870s as part of statistical mechanics. Neumann proved the *quasiergodic hypothesis* (Neumann, 1932), and Sinai gave a stronger version for a particular system. For a historical background as well as yet to be resolved aspects, see Szasz (1994).

For sampling algorithms, the ergodic hypothesis implies that after a sufficiently long time, the sampling algorithm makes an equal number of steps in different equal volumes or regions of phase space. In other words, the ensemble, or *thermodynamic average*, for a phase space region is equivalent to the time average, also termed the *dynamic average*. The thermodynamic average is defined as the average over all points in phase space at a single time point, while the dynamic average averages over a single point in phase space at all times. In this context, the assumption of equilibrium means that moving from state X to state Y is as likely as moving from state Y to state X —that is, the system obeys the *detailed balance* criterion. Utilizing biased techniques, the detailed balance may be violated but only if properly corrected to approach the requested description of equilibrium.

In the field of structural bioinformatics, detailed balance sets the stage for (1) studying structures using sampling of microstates, for example, via Monte Carlo (MC) simulations, (2) studying a dynamic analysis of a single trajectory, for example, molecular dynamics (MD), and even more importantly, (3) enabling the interchange between the two. Of course, any interchange between the two approaches depends on the accuracy of each method. This accuracy includes complementary aspects such as the size of phase space, the density of sampling the space, and the completeness of the sampling within the defined space. The completeness and the density of the sampling define the representativeness of the assessed phase space. For example, representative sampling is necessary to establish accurate thermodynamic averages as well as the phase space *probability density*, the probability of finding the system at each state in phase space. In MD, the spatial precision often required for accurate solution of the equations of motion necessitates a small time step for the trajectory propagation coupled with a simulation time that fits the assessed phase space. This is to ensure sampling of all the required phase space taking into account the need to overcome, directly or indirectly, local barriers in the rough energy landscape of the macromolecule. Often, while the local explored region, for example energy landscape, is fully ergodic, high-energy barriers prevent the sampling of distant low-energy regions. The *quasiergodic* sampling requires designated methods to overcome the challenging energy landscape architecture. Approaches to avoid this *local minimum trap* include raising the energy of the local low-energy basins, for example, accelerated molecular dynamics (Hamelberg et al., 2004), or the occasional high-energy jumping above barriers (presented below). Therefore, structural bioinformatics research must constantly address the aspect of ergodicity when drawing general conclusions from the limited available data.

Defining a Nonredundant Data Set for Representative Sampling of Available Data

To understand the structural (sub)space of proteins, one needs a representative description of this space. Even if the size of the space did not pose a limitation, which it often does, the large amount of redundant information will bias any requested underlying characterization of this space. In constructing a nonredundant representative structural data set, several aspects should be addressed. First, the multidimensional phase space should be defined. Second, low-quality noninformative data should be excluded. Third, and more difficult to implement, the data should be representatively distributed in phase space. Following the ergodic hypothesis, any general conclusions about a structural (sub)space should aim at utilizing data that equally samples the different parts of phase space rather than averaging all the available and intrinsically biased data. Thus, the derivation of a nonredundant structural space is a key step in the computational analysis of structural data.

Defining a Nonredundant Data Set in the Protein Structure Level. One possibility of characterizing structural (sub)space is to utilize available structural data and attempt to overcome the known limitations in the representativeness of the data. Here, we will use the Protein Data Bank (Berman et al., 2002) (Chapter 11) as an example of the challenges in creating a representative set. As shown in Figure 8.2, this database is naturally biased toward the focus of the research that generated it as well as the ease of generating the data. Well-represented proteins include medically relevant small proteins, historical "model proteins" that were crystallized under many conditions and modifications, and proteins from interesting creatures, for example, *Homo sapiens*. In contrast, some structures are underrepresented as they are technically hard to obtain. These include very large proteins, intrinsically mobile proteins, and membrane proteins.

How does one define, sample, and generate a representative nonredundant set? Prior to answering this challenge, the representative nature should be defined. The topic is especially relevant in the current age of structural genomics (Chapter 40), where the focus is to fill gaps and achieve a fully covered structural space. Typically, geometric and sequence redundancy are addressed. Other definitions for representativeness such as functional differentiation (Friedberg and Godzik, 2007) or a combination of classifiers such as secondary structure content, chain topology, and domain size (Figure 8.2) (Hou et al., 2003) were also suggested. Several "nonredundant" databases exclude redundant structures based on sequence identity. For example, the PDBSELECT database provides structures with low sequence similarity (Hobohm et al., 1992; Hobohm and Sander, 1994). The PDB search engine (Deshpande et al., 2005), PISCES (Wang and Dunbrack, 2005) and Astral (Brenner et al., 2000) allow for the creation of user-defined nonredundant sets utilizing user-determined sequence redundancy, resolution and other structural quality cutoffs.

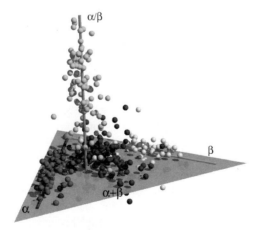

Figure 8.2. PDB fold space—in this representative description of the PDB, each of the 498 fold groups that were available in the four-class seconday-structure-based division of SCOP (2003). The four classes are depicted in different shades (or colors - see color figure online). Notably, even after deleting the redundancy within the folds, intrinsic bias of the current data has to be taken into account, for example, the abundance of all-α folds. Reprinted with permission from Hou et al. (2003).

A widely accepted automatically generated nonredundant database based on representativeness of fold space is still missing. This is partially due to the subjective definition of folds. Topics such as complexity, definition of redundancy, and representativeness of a subset have been addressed since the early days of the PDB. Fischer et al. built a sequence-independent nonredundant PDB resulting in 220 representative structures (Fischer et al., 1995). A less subjective definition of fold representativeness can by obtained by selecting representative structures from the different subgroups of the fold, as obtained from classification databases such as SCOP (Murzin et al., 1995), CATH (Greene et al., 2007), or DALI (Holm and Sander, 1996) (Chapters 17 and 18).

Protein structures can be represented by smaller units of structure (Chapter 2). The building blocks often comprise subunits of sequences or, in the case of structures, domains. The domain (Chapter 20) is classically defined as a group of amino acids within the protein structure that is stable to the point of folding independently from the rest of the protein. Domains often consist of all or part of one chain, but may consist of parts of different chains. Both types of building blocks are currently utilized, but consideration should be taken when deciding which strategy to use. On the one hand, it is easier to define and work with sequence information. On the other hand, the structural domain definition may be more accurate for some applications where structural folding units are addressed. Finally, the correlation between structures has to be addressed. For example, an independent distribution of structures must sample structure space without a bias toward a specific common ancestor of a subgroup within a given common "superfold." The CATH-based Nh3D database addresses this important aspect (Thiruv et al., 2005).

Hence, while the importance of nonredundant structural databases is well established and while there are databases of nonredundant structures or means to compute such sets on-the-fly, the field is still evolving. With the increasing number and accuracy of available structures and unique domains/folds contributed by the structural genomics efforts, topics such as a proper definition of a structural phase space and the proper sampling of the defined space continue to be addressed.

Defining a Nonredundant Data Set at the Protein Residue Level. At the backbone level (Chapter 2), Φ and Ψ are sufficient to describe the conformation of each residue in the protein chain. The Ramachandran plot that describes these angles exhibits distinct clustering that corresponds to the different secondary structures. Similarly, at the side-chain level, clustering the dihedral angles is an efficient method to describe conformational space. Figure 8.3 introduces the notion of *rotamers*. The dihedral angle between the C_α and C_β (namely, $N–C_\alpha–C_\beta–X_{\gamma(1)}$, where X denotes atoms of type C, N, O, or S) is termed χ_1, the dihedral angle between C_β and C_γ (namely, $C_\alpha–C_\beta–X_{\gamma(1)}–X_{\delta(1)}$) is termed χ_2, and so on. The set of side-chain dihedral angles determines a single side-chain conformation termed rotational isomer, or rotamer (Dunbrack, 2002). Rotamers are not evenly distributed and represent local energetic preferences. Beyond the general preference for staggered conformation ($60°$, $180°$, $-60°$), the backbone conformation limits the side chains to specific conformations.

Computational efficiency may be achieved through better data representation. Working with a *rotamer library* that defines the average dihedral angles for the most common rotamer clusters for each residue helps to reduce the dimensionality of the data. This has a dramatic effect as often such data have to be accessed multiple times. Other efficient structural data processing schemes, for example, geometric hashing (Nussinov and Wolfson, 1991), also make use of preprocessed data that can later be efficiently accessed. In the case of rotamer

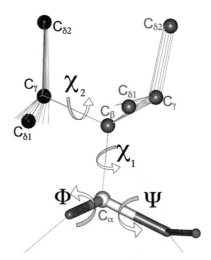

Figure 8.3. A superposition of several leucine residues depicts two clusters (differing in this case, solely in the χ_1 dihedral angle). A conformer description of these clusters will provide two coordinate systems, each including a conformation of a central residue within each cluster (depicted by balls). A rotamer description will provide for each cluster the average χ_1 and χ_2 angles. In either case, additional information can be provided regarding the relative frequency of each cluster, the variation of the parameters within the cluster, and the quality of the structures from which these coordinates were derived. For backbone-dependent libraries, the Φ and ψ angles corresponding to the assessed residues are provided.

libraries, the transformation to dihedral space forms an additional measure of data reduction. The transformation assumes an idealized (within error) covalent geometry. The resulting rotamer library can be viewed as a look-up table that allows the user to process structural data within a more manageable structural space.

Early studies utilized as few as 84 rotamers to represent the 19 side-chain conformations (Ponder and Richards, 1987). More recently, others constructed very large libraries with as many as 50,000 rotamers (Peterson et al., 2004). Holm and Sander have shown that in most cases increasing the library size does not improve our ability to identify the global low-energy solution (Holm and Sander, 1992). Rotamer libraries provide the relative frequency of each rotamer, providing an additional level of information. Rotamers that are not described in the library are denoted as "non-rotameric" conformations. In general, rotamers fall within well-defined statistical bounds. Yet, in specific regions, for example functional sites, rotamers may define unique conformations. These rotamers are often stabilized by a specific set of interactions to other protein or cofactor moieties. Therefore, by simply examining the distribution of rotamers in the protein, functional insight can be deciphered.

Importantly, the accurate representation of a rotamer library highly depends on the accuracy of the raw data applied. For example, Richardson and coworkers (Lovell et al., 2000) filtered the data in several ways: Beyond the selection of high-resolution proteins (resolution better than 1.7 Å) other structural quality assurance (Chapters 14–15) filters were applied: Side chains with high intrinsic thermal motion (as marked by B-values greater than 40Å) were removed. Likewise, side chains that exhibit steric clashes after properly adding hydrogens to the heavy atoms were not included and side-chain flipping was applied to residues (Asn, Gln, and His) that were not modeled correctly in the original PDB file. This led to improvement in the resulting rotamer library and underscores the need to examine raw data very carefully prior to drawing conclusions. An alternative approach is to move from the coordinate data to the source electron density data and eliminate residues with poor density (Shapovalov and Dunbrack, 2007).

Rotamer libraries can also depend on the backbone conformation, secondary structure, or protein fragment in which the specific side chain is located. Backbone-dependent libraries provide a different rank order of rotamers (or probabilities) for a representative set of their dependent variables. Dunbrack and Karplus utilized Bayesian statistics to derive a backbone-dependent library by dividing the Ramachandran plot into $20° \times 20°$ bins for the Φ and ψ dihedral angles (Dunbrack and Karplus, 1994). Bayesian statistics were applied to improve the sparsely populated regions and offer complete coverage of both backbone-independent and backbone-dependent data (Dunbrack and Cohen, 1997). Recent versions of this library (Dunbrack, 2002) incorporate various filtering methods including those offered by Lovell et al. (2000).

Using a discrete set of rotamers to model geometry-sensitive side-chain interactions may result in clashes. The use of reduced atom radii or the addition of local minimization are common methods to tackle this challenge. For example, short MD runs were utilized to define rotamers for modeling protein–protein interactions (Camacho, 2005). Alternatively, the original rotamer library can be expanded to include *subrotamers* by adding representative structures in standard deviation increments around the original rotamers (De Maeyer et al., 1997; Simons et al., 1997; Mendes et al., 1999). The completeness of the rotamer library was found to be specifically important for modeling buried side chains (Eyal et al., 2004). Not unrelated, many search and sampling procedures may have different results for buried and solvent-exposed residues (Voigt et al., 2000). The difference is due to the different physical properties each group has to optimize, namely, hydrophobic packing for the buried residues and the solvent interface for the solvent-exposed residues.

An alternative to the use of rotamer libraries is the use of *conformer libraries*. Here, a single central side-chain coordinate is extracted for each cluster of side-chain conformations, thus maintaining the real covalent geometry and all the natural degrees of freedom. Such data contain more information than a rotamer that is represented by a cluster's average information. Moreover, the size of conformer libraries can be easily fine-tuned. Each conformer can be defined using angle similarities (Xiang and Honig, 2001) or root mean square deviation (RMSD) (Shetty et al., 2003). In both cases, accuracy was shown to correlate with the number of conformers in the library, indicating that several thousand conformers were required. Mayo and colleagues have shown that conformer libraries are superior to rotamer libraries in the computational design of enzymes that include small molecules (Lassila et al., 2006). The use of a conformer library is tailored for a detailed study of the varying natural flexibility around rotamers, for example, around functional sites or interactions with other molecules. In such locations, the usage of a rotamer that depicts an average cluster may limit the sampling of the high-variation and nonideal geometries that are typical to these sites. Yet, with very large conformer libraries that include rare conformers, erroneous raw data may easily enter the library making the filtering of mistakes critical. Finally, as an evolving field, objective benchmarks, as found say in the CASP competition (Chapter 28), are required to assess the relative success and weakness of the different libraries.

To have better sampling of conformational space in real proteins, rotamer libraries can be expanded beyond the 20 amino acid side chains to include additional structural elements. Posttranslational modifications as well as common cofactors can be modeled using a similar approach. Moreover, with the increasing study of non-amino acid foldamers (Goodman et al., 2007), rotamers can be constructed for other types of building blocks, for example, β-amino acids (Shandler and DeGrado, unpublished work). Finally, different regions within the protein may be modeled by different rotamer types. Focusing on the scoring function rather than the search algorithm, Edelman and coworkers showed that the local solvent-accessible surface of each residue is an important consideration in choosing the correct rotamer (Eyal et al., 2004). Indeed, Baker and colleagues have shown that water can be incorporated into a *solvated rotamer library* (Jiang et al., 2005). Water molecules are tightly clustered around side-chain functional groups in positions based on hydrogen bond geometry, enabling the use of a "centroid" water atom as a natural extension to the original side-chain rotamer. This solvated rotamer library enhances the ability to model and design protein–protein interactions as well as the interior of proteins where specific water molecules may play an important role.

Another way to cope with the large search space is to move to a coarse-grained description of the structure (Chapter 37). A coarse-grained simplification may be achieved by moving from a full description of the side-chain conformation to the use of structurally representative pseudoatoms or centroids. Suggested over 30 years ago by Levitt (1976), this method can be expanded to use more than one pseudoatom for each side chain corresponding to the side chain's complexity and functional moiety (Herzyk and Hubbard, 1993). Such methods are useful when extensive sampling is required. Examples include *ab initio* membrane protein structure prediction (Becker et al., 2003) as well as the Baker's lab *ab initio* modeling (Chapter 32) using fragments stitched with the help of a Bayesian scoring function (Simons et al., 1997). More generally, lower resolution models resulting from coarse-grained data can be used as input to a more detailed sampling of the phase space in the proximity of the coarse-grained model.

Finally, one can further reduce the complexity of structures by searching sequence space using only a representative small set of amino acids. As Hecht and coworkers have shown, a binary hydrophobic and hydrophilic code is sufficient for the design of complex structural motifs such as four helix bundles (Kamtekar et al., 1993; Bradley et al., 2007). Moreover, even when working with all the residues, not all conformations should be part of the set. Known structural motifs with specific functions, for example, a catalytic triad of a requested functional site or other highly conserved residues, can be constrained to the required conformation. Similarly, unlike buried residues, solvent-exposed residues need not be scored for packing. Hence, the residue-level search space should be tailored to the questions being investigated.

SEARCH METHODS

Introduction—Stochastic Versus Deterministic Search Methods

The complexity of search problems within structural bioinformatics is exemplified by the Levinthal paradox for protein folding (Levinthal, 1969). According to this paradox, even for a simplified representation of a protein, the number of potential backbone conformations increases exponentially with the size of the protein to numbers that are beyond computation capabilities even for small proteins. The problem is further amplified in protein design (Kang and Saven, 2007) where multiple amino acids and their related environment need to be assessed for each variable residue position. Searching such an immense structural space to find a single structure requires specialized search methods. These methods can be classified as *deterministic* or *stochastic*. Deterministic methods have access to the complete data, and if they converge (or are run for a sufficiently long time for time-dependent methods), such methods will find the lowest energy configuration. In contrast, stochastic approaches are not guaranteed to find the lowest energy configuration. A fully deterministic approach is an *exhaustive search* in which all of the investigated structural space is assessed in the same level of detail. Such a computationally intensive search is possible only for a small search space. Examples include highly simplified representations of structural space such as two-dimensional lattice models where questions such as sequence designability can be approached analytically (Kussell and Shakhnovich, 1999), or small but detailed structural space problems, such as combinatorial mutagenesis of a few amino acids within a large protein, for example (Shlyk-Kerner et al., 2006).

The appropriate search method that should be applied depends on the particular model, the energy function, and the type of the information to be derived. Search methods classified as deterministic include *dead-end elimination* (DEE), *self-consistent field methods* (SCMF), *belief propagation*, *MD*, *branching methods* (e.g., *branch and bound* or *branch and terminate*), *graph decomposition*, and *linear programming*. In contrast, stochastic methods include different versions of MC *simulations*, *simulated annealing* (SA), *graph search algorithms* such as A*, *neural networks* (*NN*), *genetic algorithms* (GA), and *iterative stochastic elimination* (ISE). For an introduction to applications of search methods used in conformational analysis, see Becker (2001). For a brief overview of search methods in the context of macromolecular interactions (Chapters 25–27), see the review by the Wolfson–Nussinov group (Halperin et al., 2002). For search methods in the context of protein design, see the relevant sections in Saven (2001), Park et al. (2005), and Rosenberg and Goldblum (2006). The protein design community has invested in developing search techniques as this is one of the main challenges of the field. In the following sections, several search methods will be outlined and structural bioinformatics applications of these methods will be demonstrated.

Deterministic Search Methods

Dead-End Elimination. DEE is a useful pruning method that iteratively trims dead-end branches from the search tree (De Maeyer et al., 2000). In essence, branches within a search tree that stem from a nonoptimized criterion, for example, a high-energy conformation that can be proven to lie outside the solution, should be eliminated. Note, hereafter "energy" will be used to exemplify DEE though any other criteria can be utilized. This method of identifying solutions that are absolutely incompatible with the solution is in contrast to methods such as GA where the focus is on maintaining solutions that may be part of the global optimum.

The DEE scheme requires several components: First, the set of discrete independent variables should be well defined. Second, a precomputed value, for example, the energy, should be associated with each variable. Third, the relationships between the different variables should be defined, often using an energy function. Complementary to this requirement, the DEE scheme is limited to problems in which the score or energy may be expressed as pair interactions. Finally, a threshold criterion should define when a potential solution cannot be part of the final solution as it cannot be included in the *global minimum energy conformation* (GMEC). As conceptually demonstrated in Figure 8.4, several elimination criteria were developed within the DEE scheme. The main strength of the method is in the iterative use of this distilling procedure until no dead end, that is, GMEC

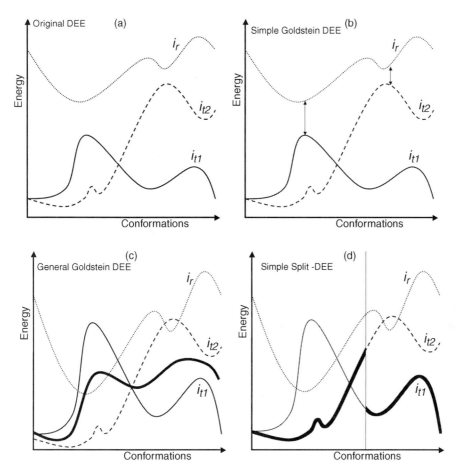

Figure 8.4. Examples of DEE criteria. The x-axis represents all possible protein conformations and the y-axis the net energy contributions produced by the interactions with a specific rotamers at position i. The assessed rotamer, i_r, and two reference rotamers, i_{t1} and i_{t2}, are depicted by dotted, full, and broken lines, respectively. (a) Original DEE: i_r is eliminated by i_{t1} but not by i_{t2}. (b) Simple Golstein DEE: i_r is eliminated by either i_{t1} or i_{t2}. (c) General Goldstein DEE: i_r cannot be eliminated by either i_{t1} or i_{t2} but can be eliminated by a weighted average (thick line) of the two. (d) Simple split DEE: i_r is eliminated by i_{t1} and by i_{t2} in the partitions (thick line) corresponding to splitting rotamers k_{v1} and k_{v2}, respectively. Modified after Pierce et al., (2000).

incompatible member, is found in the assessed space. The algorithm ends when no conditions can be established to further eliminate dead-end branches resulting in a residual set of rotamers. The resulting rotamer set contains a unique conformation in which the GMEC is identified directly. If the resulting set does not correspond to a unique solution but is sufficiently small, exhaustive enumeration or tree search can identify the GMEC result.

As reviewed by the authors who introduced the DEE theorem in 1992 (Desmet et al., 1992), the method became popular in structural bioinformatics applications within a very few years (De Maeyer et al., 2000). Variations of the four main components of DEE were tailored to fit specific studies. For example, additional screening assays may quickly reduce the immense search space. Alternatively, rotamers with a clear steric clash with the protein backbone atoms should not be further considered. Accurate homology modeling (Chapter 30) depends on the exact positioning of side chains on the selected backbone template. De Maeyer et al. have coupled their DEE algorithm to a detailed rotamer library that was pruned also by a knowledge-based criterion of 70% overlap between side chains (De Maeyer et al., 1997). The criterion was derived from comparing the deviation between modeled and crystallographic structures ensuring that the chosen threshold will not include possible GMEC solutions.

In all the threshold criteria, the most important component of DEE, elimination relies on energetic data from a precalculated matrix describing the pair-wise interaction energies between two rotamers as well as the interactions of all rotamers with the protein scaffold. The common Goldstein criterion (Goldstein, 1994) claims that a rotamer r at some residue position i will be dead-ending if, when compared with some other rotamer t that is at the same residue position, the following inequality is satisfied:

$$E(i_r)-E(i_t) + \sum_j \min_s[E(i_r j_s)-E(i_t j_s)] > 0 \qquad (8.4)$$

where $E(i_r)$ and $E(i_t)$ are rotamer-template energies, $E(i_r j_s)$ and $E(i_t j_s)$ are rotamer–rotamer energies for rotamers on residues i and j, and the function minimums selects the rotamer s on residue j that minimizes the argument of the function (Figure 8.4b). According to the original DEE criterion (Figure 8.4a), only when the best energy of a specific rotamer is worse than the worst energy of another rotamer at the same position, it can be eliminated, that is, the minimization part of the above equation is decomposed to a minimization of $E(i_r j_s)$ and maximization of $E(t_r j_s)$, and each rotamer is compared to the rotamers of any protein conformation. As shown in Eq. 8.4, the elimination criterion was relaxed by Goldstein to include rotamers for which other rotamers can be found that have lower energy against an identical conformational background (Goldstein, 1994). In other words, the Goldstein criterion DEE claims that a rotamer can be eliminated if the rotamer's contribution to the total energy is always reduced by using an alternative rotamer. A conformational background is obtained by fixing each residue in the conformation that most favors the assessed rotamer. A further relaxation of the simple Golstein criterion, termed "general Goldstein criteria" (Figure 8.4c), compares the rotamer to the average of other rotamers taken from the same conformational background (Goldstein, 1994). A rotamer can be eliminated if the average of the other rotamers has a lower energy compared to the assessed rotamer, as then there will always be at least one rotamer that is better. Additional relaxing modifications to the DEE pruning criteria include the use of "split-DEE" (Figure 8.4d). Here, the conformational space is split into partitions within which each of the candidate rotamers can be compared singly (Pierce et al., 2000).

An alternative modification is to use dead-ending pairs (also termed "doubles") in which a specific combination is dead-ending. In such a case, each one of the rotamers building the pair may not be dead-ending if applied in a different combination. The idea can be generalized into grouping several side chains into *super rotamers* (Goldstein, 1994; De Maeyer et al., 2000). The super rotamer contains all the possible combinations of the individual side-chain rotamers. While the pruning power of a single rotamer's interaction with a super rotamer is more intuitive, the implementation of super rotamers is more complex. "Generalized DEE" compares clusters of rotamers instead of individual rotamers and adds a split-DEE scheme for the clusters, thus improving the convergence properties enabling efficient redesign of large proteins (Looger and Hellinga, 2001). The choice of weights to achieve the "average" adds additional complexity. The "enhanced-DEE" algorithm finds these weights via linear programming (Lasters et al., 1995).

The first fully automated design and subsequent experimental validation of a novel sequence for an entire protein utilized the DEE pruning approach (Dahiyat and Mayo, 1997). The fully automated protein design scheme uses a combination of several search methods, each tailored for a different segment of the design scheme (Dahiyat and Mayo, 1996): Following the use of the generalized DEE algorithm, a MC simulation (described later) was executed for the top sequences selected by the DEE. The MC was carried out as the theoretical and actual potential surfaces may be different. In other words, a GMEC selected by the DEE procedure from the theoretical energy landscape may differ from the actual GMEC that is part of the experimental "real" energy landscape. Interestingly, the MC scoring function was derived by comparing structural properties of the designed proteins to experimental stability (melting temperature) in the frame of a quantitative structure–activity relationship (QSAR) (Chapter 34). The QSAR feedback selected polar and nonpolar surface burial parameters, as well as the relative weight of each parameter within the scoring function, thus improving the design algorithm.

Very recently, DEE was used to investigate multistate and flexible structures. The X-DEE (extended-DEE) application searches within discrete subspaces for low-energy states of the protein, for example, the photocycle intermediates of the bacteriorhodopsin light-driven proton pump (Kloppmann et al., 2007). Following the many DEE runs for a large number of subspaces, gap-free lists of low-energy states are generated. Such lists can be utilized for a kinetic study of the transitions between states. To overcome the difficulty of standard DEE in tackling multistate design problems, a type-dependent DEE was recently suggested (Yanover et al., 2007). Standard DEE allows for a rotamer at a certain position to be a potential candidate for elimination by any other rotamer at that position. In contrast, in type-dependent DEE a rotamer may be eliminated based only on comparisons with other rotamers of the same amino acid type. Adding ranges of spatial locations to each rotamer is another approach to intrinsically take into account varying states or natural flexibility. Flexible backbone DEE was applied by adding a constraining box around each residue, thus limiting the measured rotameric interactions by upper and lower bounds. (Georgiev and Donald, 2007). As demonstrated, the pruning power of DEE can be utilized for many cases as long as the rotamer interaction constraints are formulated properly.

Finally, one of the major limitations of DEE is that it does not scale well with the size of the protein. The limitation is especially relevant for large proteins and for the larger structural space that needs to be scanned in protein design. As briefly discussed, one possibility is to hierarchically split a given problem and use the DEE method only for the part of the challenge for which it is optimized.

Self-Consistent Mean Field. SCMF, also known as mean field theory, aims at reducing the cost of combinatorics generated by the interaction terms when summing over all states of a multibody problem. The main idea of the mean field theory is to replace all interactions to any one body with an average or effective interaction. This reduces any multibody problem into an effective one-body problem. Thus, the method is far faster, compared to methods such as DEE, and usually scales polynomially (rather than exponentially) as a function of the size of the assessed variables (e.g., number of residues). Yet, due to the simplified parameterization, SCMF is less accurate and unlike DEE, may fail in finding the optimal solution (Cootes et al., 2000; Voigt et al., 2000). The accuracy is not easy to compare to other methods and depends on the *mean field approximation* (MFA). The MFA is the minimizing reference system that transforms the multibody interaction to a one-body (zero order) approximation, preferably using noncorrelated degrees of freedom. Thus, SCMF should be applied only in cases where the MFA is valid within the targeted accuracy. In SCMF, the average field, such as energy, is determined self-consistently as the overall cutoff threshold is lowered. SCMF will be demonstrated here via examples that are carefully tailored to fit the MFA and the consistency requirements.

Some of the first applications of SCMF for biological molecules were in the field of protein structure prediction (Section VI in this book). For example, a mean field approach for assigning a unique fold, known as fold recognition (Chapter 31), was applied by Friedrrichs and Wolynes (1989). Finkelstein and Reva (1991) used SCMF to predict β-sandwich structures. While short-range interactions were modeled explicitly, the large number of atoms contributing to the long-range interaction component utilized a mean field approach. Using only the strand directionality (but not strand length) and sequence, their SCMF algorithm identified the folds correctly. As their mean field equation for inter- and intrastrand interactions converged quickly and independently of the starting "field," this early work concludes that for fold recognition it is the local hydrophobicity that determines the fold.

SCMF was applied for side-chain modeling (Mendes et al., 1999) and expanded to include loop modeling (also known as "gap closure") (Koehl and Delarue, 1994; Koehl and Delarue, 1995). Loops connecting secondary structures are the least accurate region in homology modeling (Chapter 30) as well as in other protein structure prediction schemes. Consequently, the mean field approach is especially relevant for such modeling. Koehl and Delarue sampled several potential fragments as loop candidates to connect the modeled secondary structures. Each such position requires rotamer library-based modeling of the best side-chain conformation. A similar multicopy approach was used for side-chain modeling. The conformational matrix was weighted according to the relative probabilities of the different backbone fragments. For loop modeling, rather than modeling each fragment discretely, a MFA was built from the conformational matrix and utilized to assess the rotamer–rotamer energies when modeling each side chain sequentially. The iterative procedure was carried out till convergence, confirming the self-consistency of this search method. This loop and side-chain modeling was applied in the 3D-JIGSAW comparative modeling program (Bates et al., 2001). To overcome the lower resolution of the SCMF approach, the online server (http://www.bmm.icnet.uk/servers/3djigsaw) refines the results by trimming the loops for good geometry by adjusting torsion angles as well as adding a short minimization step.

Stochastic Search Methods

Unlike deterministic methods, stochastic methods have a random component and may give a different answer for each run. The "random" character is often given by a random number generator. Notably, it is important to choose a generator that is indeed random, as not all are. One can also test this component to see whether over a large number of runs the scattering is indeed random. For more on commonly used random number generators and pitfalls therein, see Schlick (2002).

Monte Carlo Simulations. MC simulations, a collective name for the usage of an iterative application of random perturbation steps that can be accepted or rejected, has become one of the most common stochastic search procedures. MC is applied in numerous disciplines having a large search space, a large uncertainty of specific parameters, or a combination of both, for example, structural bioinformatics. This simple and efficient technique was largely established in Los Alamos by Metropolis et al. setting the method of generation of trial states along with criteria for accepting or rejecting the MC step (Metropolis et al., 1953). MC can simulate a time-dependent process, such as Brownian dynamics or a static state, such as a rotamer library-based conformational search. For an equilibrium simulation, the probability of residing in a particular state should be that of an equilibrium distribution. As introduced earlier in the chapter in the frame of ergodicity, this implied that the detailed balance criterion must be maintained - a prime concern when setting the MC simulation conditions.

The very name of MC hints at the gambling aspect of this simulation—a random move is attempted using information derived from the latest state. Some criteria for the move are set in advance. These include the amount and type of perturbation applied at each step, namely, the relative segment of the investigated system that undergoes change and the magnitude of the change. More important, in classical MC the direction of the move is random. Next, and most important, the move has to be accepted or rejected. In order to get out of local minima moves that are less favorable should be occasionally accepted. Classic MC applications simulate an equilibrium state that follows a Boltzmann-type distribution—the distribution originally suggested by Metropolis. As such, moves that have a lower minimum will always be accepted, while moves that are less favorable will be accepted with a probability corresponding to the Boltzmann distribution. This simple acceptance criterion not only channels the moves toward the local minima but also allows occasional jumps to bypass local energy barriers, hence exploring other minima. Finally, convergence of the MC application should be addressed. As a stochastic procedure, finding the GMEC is not guaranteed. Consequently, overcoming local minima traps is often approached by applying multiple independent MC runs, each with a predefined number of steps.

Here, MC will be demonstrated for the side-chain conformational search challenge hopefully without losing generality. The conformation of a protein sequence is composed as a discrete set of residue conformations, that is, $\alpha = \{\alpha_1, \ldots, \alpha_N\}$. Such a set can be modeled as a *Markov Chain*; a discrete-time stochastic process with a Markov property. Having the Markov property means the next state depends solely on the present state and does not directly depend on the other previous states. A single component, for example, α_i, of this Markov Chain MC (often termed MCMC) can be altered. Alternatively, several, or all, of the components can be altered simultaneously. In either case, a new trial rotamer sequence α' is generated. Following the Metropolis criterion (Metropolis et al., 1953), detailed balance can be achieved by selecting the general form of:

$$\alpha = \min\left(1, \frac{p(\alpha')q(\alpha|\alpha')}{p(\alpha)q(\alpha'|\alpha)}\right) \tag{8.5}$$

where $p(\alpha)$ is the equilibrium (long time) probability of configuration α, and $q(\alpha|\alpha')$ is the trial probability associated with generating configuration α' given that we started with configuration α. The equilibrium probability has the form $p(\alpha) \propto \exp(-\beta F(\alpha))/Z(\beta)$, where $F(\alpha)$ is an energy-based scoring function that quantifies how well the sequence matches a particular target structure, and $\beta \propto 1/T$, where T is an effective temperature, and the partition function $Z(\beta)$ ensures that $p(\alpha)$ is normalized. In classic MC, the trial distribution is independent of the initial distribution, that is, $q(\alpha|\alpha') = q(\alpha'|\alpha)$. Thus, the equation is reduced to the Metropolis acceptance criteria (Metropolis et al., 1953):

$$\alpha = \min(1, \exp(-\beta(F(\alpha') - F(\alpha)))). \tag{8.6}$$

Convergence of the MC protocol can be achieved using a number of methods. For example, a predetermined number of MC steps can be assigned or the simulation can be terminated if a predetermined number of steps does not improve the result. To sample the required landscape comprehensively while focusing on the most probable region in greater detail, a SA scheme may be applied. SA involves a temperature that is initially high allowing for large jumps between local minima and then gradually decreased to carefully investigate the region of interest. The exact cooling rate, which is proportional to the number of MC iterations, is often decided using a trial-and-error procedure in which convergence is assessed on case studies. Notably, the general method of SA can be applied with acceptance probabilities other than the Boltzmann one.

The details of the degree of change between one state (α) and another (α') depend on the specific MC simulation conducted. In conformational search challenges, changes may include specific residue locations, identities, and rotamers. For example, DeGrado and coworkers applied a SA-MC scheme of 10,000 iterations to design the side chains of transmembrane helices that specifically target natural transmembrane proteins (Yin et al., 2007). Using an exponential cooling schedule, each SA-MC iteration included a single mutation of a protein–protein interface residue. Then, rotamer optimization was applied to the entire ensemble of residues in the helical pair. A further decrease in search size was implemented by including only residues prevalent in transmembrane proteins. The rotamer search space was minimized by including only highly probable rotamers and optimized using the DEE method. Hence, a combination of search and sampling methods was tailored for this protein design task.

OVERCOMING MC LIMITATIONS.   MC requires a large number of iterations to achieve convergence as it is challenged by the large phase space to be explored as well as by the roughness of the energy landscape. Thus, one of the main challenges in MC simulations is how to focus the search, namely, minimize the sampled space to the minimum required for convergence. While SA will change the probability to accept a less favorable trial as the simulation proceeds, it will still give each trial configuration an equal likelihood, that is, $q(\alpha|\alpha') = q(\alpha'|\alpha)$. Changing this relationship may enable a biased search toward areas of increased likelihood, hopefully without missing regions bordered by high-energy barriers. Indeed, to avoid local minima traps a very slow cooling rate is often required. Next, we will briefly present examples of MC modifications to overcome these limitations including MC-quench, biased MC, SCMF-MC, and replica exchange-biased MC.

A *MC-quench* procedure recognizes that each trial configuration had an equal likelihood and was assessed sequentially. In other words, each position is not optimized to the final structure but to a trial predecessor of this structure.Consequently, when the end result is achieved, there may be specific locations that may be further improved as their environment has changed after they were assessed. Assessing side-chain position, Mayo and coworkers found that by taking the resulting structure from an MC-SA procedure and minimizing the energy of all possible rotamers of each location the solution is improved dramatically (Voigt et al., 2000).

A *biased MC* procedure can focus the simulation toward more probable attempted trials. Similar to accepting less preferred moves in classical MC, biased MC will not throw away moves toward less preferred regions but rather provide such moves with a lower probability. Thus, the stochastic movement is guided by the local energy facilitating large movements and increased acceptance rates. Mathematically, we have

$$\alpha = \min\left(1, \frac{W(\alpha')}{W(\alpha)}\right) = \min\left(1, \frac{\prod_k z_{s_k}(\alpha')}{\prod_j z_{s_j}(\alpha)}\right), \tag{8.7}$$

where W, the Rosenbluth weight (Rosenbluth and Rosenbluth, 1955), can be defined as the relative portion of available sites multiplied by the weight of the previous state. This can be described by

$$z_{s_k}(\alpha') = \sum \exp(-\beta F_{s_i}(\alpha|\alpha'_{s_1} \cdots \alpha'_{s_{i-1}})). \tag{8.8}$$

Such an approach requires a computational overhead in the need to calculate the relative probabilities in each biased MC step. This overhead should be weighted against the added efficiency resulting from the biasing. Therefore, the cost-effectiveness of biased MC is best valued in specific applications. For example, growing a molecule toward acceptable structures is often done via the *configurational-biased MC* method (Siepmann, 1990). McCammon and coworkers recently applied the method for the continuous exploration of side-chain placement (Jain et al., 2006). This interesting application combines a molecular mechanics potential function with a rotamer library that can be extended around the discrete rotamers. Such fine-tuning of flexibility is especially important in the most interesting areas of the protein, namely, functional sites and protein–protein interfaces. Notably, for specific cases a very efficient end-transfer configurational-biased MC can be applied, for example, deleting one end of a polymer and growing it on the other side (Arya and Schlick, 2007). Such a technique is best suited for sampling linear combinations of biopolymers such as nucleic acids (Chapter 3) as applied by Arya and Schlick, in combination with other MC methods, to study chromatic folding (Arya and Schlick, 2006).

Biased MC was applied, for example, in an early attempt at homology modeling. Holm and Sander (1992) attempted to bias the rotamer weights used for side-chain optimization. Statistics on rotamer searches in the optimized configurations from a large number of short MC runs were collected and the new "learned" rotamer frequencies were applied to bias new moves toward high-probability regions. Baker and coworkers applied the RosettaDesign algorithm (itself an extension of their *ab initio* structure prediction algorithm—see Chapter 32) in an on-the-fly biased-MC way (Dantas et al., 2003). Here, the first MC runs tested all possible 20 residues for each position in the designed protein. The list of acceptable mutations was utilized to bias a second MC procedure enabling the sampling of a much larger rotamer library, thus refining the results. In the same line of thought, one can conduct MC procedures iteratively with each iteration hierarchically focusing on a different search space, for example, cycling between side-chain design and backbone optimization (Kuhlman et al., 2003).

Synthesizing the sampling power of MC and the speed of SCMF methods, Zou and Saven (2003) offered a *SCMF-biased MC* method for designing and sampling protein sequences. Following a SCMF approach (described earlier in the chapter), the sequence identity probabilities for each location were precalculated prior to the MC sampling. These weights are utilized to bias the MC sampling resulting in efficient sampling of low-energy sequences and enabling fast SA cooling rates. Following the above general equation, as q $(\alpha|\alpha')/q(\alpha'|\alpha) = p(\alpha)/p(\alpha')$, the acceptance probability is reduced to

$$\alpha = \min\left(1, \frac{p(\alpha')}{p(\alpha)} \exp(-\beta(F(\alpha')-F(\alpha)))\right). \tag{8.9}$$

An alternative approach to more efficient scanning of the target landscape focuses on the temperature protocol applied to the simulation, that is, protocols other than SA. For example, the low-temperature simulation may periodically be allowed to jump (jump walking or *J-walking*) to a high-temperature simulation (Frantz et al., 1990). A variation of J-walking is the *replica exchange method* (REM) (Geyer and Thompson, 1995). In this method, noninteracting copies (replicas) of the original system at different temperatures are simulated independently and simultaneously. Pairs of replicas are exchanged with a specific transition probability that maintains each temperature's equilibrium ensemble distribution, thus realizing a random walk in energy space and facilitating the passage over energy barriers. Such a method is most suitable for sampling the conformation of a rough energy landscape, as shown for a pentapeptide conformation search (Hansmann, 1997). Yang and Saven applied REM-MC and REM-biased MC to redesign the calcium binding protein calbindin (Yang and Saven, 2005). The output energy, the redesigned sequence, was lowest when applying REM-biased MC compared to REM-MC that in turn was lower relative to SA-MC. Possibly even more important, an analysis of specific pivotal residues showed that the REM-biased MC method overcome local minima traps showing an improved sampling of the low-energy sequences.

In conclusion, despite the two limiting aspects of MC, large sampling space and lack of guarantee to convergence, MC became widely used due to a high success rate and many incorporated tricks applied to overcome limitations. There is no one answer to the best way to tailor an MC approach to a specific problem. Unfortunately, there have been few benchmarks that objectively compare sampling methods, let alone MC variations. The success of MC and other stochastic methods is not only due to our understanding of the investigated search problems and the availability of fast computers, but also due to the fact that most often multiple answers can satisfy the investigated question. Finally, there continues to be a wide variety of MC variations, and the above examples are only used to introduce this powerful method.

Stochastic Population Search and Sampling Methods. Most search and sampling methods focus on finding a single best solution. However, often it would be beneficial to aim at finding a (relatively) small top-scoring population that can be either further assessed by higher resolution methods or fully checked experimentally. One option to achieve such a scheme is by hierarchically combining methods of increasing resolution. Thus, every method does not necessarily provide a population but a representative solution of smaller subpopulations that are subject to further higher resolution search and sampling. Examples include side-chain and main-chain modeling, where each part of the problem should be addressed in a different resolution (Kuhlman et al., 2003; Yin et al., 2007). In other cases, for example, energy minimization, a protocol of different minimization methods with growing resolution results is hierarchically devised, for example, grid search, followed by conjugated gradient and adapted basis Newton Raphson (Chapter 37).

A hierarchical population approach may be devised by deterministic approaches that do not converge but do result in a manageable top-scoring population. For example, in interesting work by Singh and coworkers on side-chain positioning, *integer linear programming* was combined with simple DEE to reduce search space (Kingsford et al., 2005). In brief, linear programming aims to minimize a linear equation subject to constraints. In this specific case, this is the energy function of rotamers subject to choosing a single rotamer per residue and calculating interresidue contact energy only for the chosen rotamers. By relaxing the integrality of the latter constraint to include only rotamers that interact, the algorithm is reduced to a much faster polynomial-time linear programming heuristic. However, this heuristic will not find the GMEC requiring full integer linear programming in some cases. In such cases, DEE is applied first to prune the search space considerably. Next, the solutions not pruned by DEE are subject to the integer linear programming. Despite the fact that the DEE pruning mechanism is not tailored to find the optimal subpopulation of solutions, this application yields high-accuracy results similar to those obtained by far less efficient algorithms. In this example, DEE is specifically targeted not to find the GMEC but to enable a manageable data set for an algorithm that could not run practically on the full data set. In this context, it should be noted that others have also combined the power of DEE with complementary search techniques to produces low-energy populations. For example, Leach and Lemon (1998) combined DEE with the A* algorithm that finds the "least cost" path of a graph. With the added computational cost, they assess the side-chain contribution to the conformational entropy of the folded protein.

Iterative Stochastic Elimination (ISE), developed by the Goldblum group, aims at producing a manageable ensemble of a best population that can be further searched by more detailed or even exhaustive search methods (Glick et al., 2002). Rather than focusing on finding one GMEC solution, this algorithm puts emphasis on forming a high-scoring ensemble. After a quick elimination of backbone clashing rotamers and side-chain to side-chain clashing rotamer pairs, the algorithm introduces an interesting stochastic stage. Following sampling at random of a large population of conformations, a distribution function of the energies in all sampled conformations is constructed. Then, the rotamers of the best and worst results are compared. The stochastic stage is repeated with every cycle excluding rotamers that were present in the set of the worst results but not in the set of best results of the previous iteration. The method was applied successfully on a number of other search challenges, for example, predicting the conformation of protein–protein interfaces using data from structures in the unbound state (Noy et al., 2007).

"*Biological*" approaches such as GA or NN have been applied to a number of search and sampling challenges in structural bioinformatics. For example, GA was utilized for protein structure prediction (Pedersen and Moult, 1996) and protein design (Jones, 1994). Briefly, the investigated degrees of freedom, for example, translational, rotational, and internal degrees of freedom, are encoded into "genes" that are represented by the real values of these degrees of freedom. Mutations at a fixed rate, often corresponding to an average of one mutation per sequence, are applied to a population of random sequences. A scoring function, such as energy, is calculated for each mutant. The best scoring mutant sequences are kept as well as subjected to crossovers among themselves. These "most fit" surviving sequences, namely, the new generation, is subjected again to a cycle of mutations and crossovers. The cycles are fixed to a certain number or until the energy of the new generations does not improve. The basic algorithm has many additional possible modifications, for example, a migration operator responsible for moving individual sequences from one subpopulation to another. In the case of conformational searching, the genes are usually a subset of coordinates with a potential energy defining the fitness.

The population dynamics of GAs makes larger moves than the typical MC, and thus can sample a large space rapidly overcoming local energy barriers. While the overall number of required steps is usually small, the generation of each GA cycle is more costly than other methods. In addition, while keeping only the top-scoring members of the population, the nature of this error-embedded and inbred population may result in slower, and sometimes lack of, convergence. This is especially true of highly coupled systems and cases where the coupling is not adjacent in location, for example, structural interactions of locations that are not close in sequence. Mayo and coworkers compared the GA to other protein design algorithms (Voigt et al., 2000). They found that GA performs better than the often locally trapped MC and SCMF methods in spite of the possible intrinsic convergence problems. Yet, the simple fine-tuning of MC to MC-quench (section "Overcoming MC limitations") outperformed GA in the tested examples. Brooks and coworkers compared MD-SA, GA, and MC as search procedures for flexible docking (Vieth et al., 1998). All three methods showed overall similar efficiency with GA scoring the best when searching a small space but not for the larger space, where MD-SA was the best.

A neural network, another method to extract patterns embedded in high-dimensional data, is often applied in conjunction with other methods. Hwang and Liao (1995) applied NN along with a MC-SA scheme for side-chain modeling. Kono and Doi (1994) applied a NN algorithm to design protein cores using a reduced set of residue types. Schneider et al. applied NN for peptide design by processing QSAR information from an experimental training set of peptides and computationally designed low-energy scoring antagonist peptides against the β_1-adenoreceptor (Schneider et al., 1998).

Probabilistic approaches to protein design are complementary to those that search for particular sequences via optimization of a scoring function (Kono and Saven, 2001; Saven, 2002; Fu et al., 2003; Park et al., 2005). Such methods estimate the site-specific probabilities of the amino acids within sequences that are structurally consistent with a target structure. These structural interactions are quantified by atomistic energy functions that address rotamer–backbone, rotamer–rotamer, and rotamer–solvent interactions. In addition, a variety of constraints may be introduced to specify function, such as the placement of side-chain ligands to form a metal binding site. The site-specific amino acid probabilities are determined by maximizing an effective entropy function subject to constraints on the sequences. The method has been extended so as to bias SA methods (Zou and Saven, 2003) and include symmetric quaternary structures (Fu et al., 2003). Such a statistical computationally assisted design strategy (SCADS) has been applied to design a variety of proteins, including water-soluble analogues of the membrane spanning potassium channel KcsA (Slovic et al., 2004).

Other Methods to Search the Complex Phase Space—Random Walk, Umbrella Sampling, and Replica Exchange. Local minima within a phase space, such as an energy landscape, pose a challenge to detailed searching of the space. Sampling such a surface must specifically look for the small but potentially deep wells distributed along the rough energy landscape. Several designated methods for better sampling of a complex structural space have been developed in the context of molecular dynamics. Consequently, complementary additional sampling topics will be discussed in

Chapter 37. For an example of replica exchange sampling, see section "Overcoming MC Limitations."

DATA ANALYSIS AND REDUCTION

A topic intimately connected to the search and sampling challenge is the way data are analyzed and reduced. While this requires a chapter of its own, some basic terms are briefly introduced here. Also, some topics in this field that are common in the study of protein motion (Chapter 37) are discussed therein. Notably, the search and sampling strategy should be targeted to the available data analysis methods, which in turn may provide results requiring further search and sampling of the target space.

Similarity measures, for example, via support vector machines (SVM) and designated cluster and pattern analysis, for example, principal component analysis (PCA) and normal mode analysis (NMA) are powerful tools to decipher high-dimensionality data. NMA is rapidly developing into bringing quick biophysical insight that was not long ago not in reach—readers are referred to the excellent recent text-book on NMA theory and applications edited by Cui and Bahar (2006) and to Chapter 37. PCA can be viewed as a generalizing extension of the NMA concept (Cui and Bahar, 2006). PCA aims to reduce the dimensionality of a data set consisting of a large number of interrelated variables, while retaining as much as possible of the variation present in the data set. This reduction is achieved by transforming to a new set of variables, the principal components, which are uncorrelated and ordered so that the first few retain most of the variation present in all the original variables (Jolliffe, 2002). PCA involves calculation of the eigenvalue decomposition of a data covariance matrix or can utilize singular value decomposition (SVD) of the data matrix. SVD is a factorization of rectangular matrices, that is, a generalization of the eigenvalue decomposition for square matrices. The SVD approach is more efficient compared to the classic principal component calculation, as formulation of the covariance matrix is not required. Beyond the straightforward use in protein motion (Chapter 37), for example, as shown by Ichiye and Karplus (1991), this general method can be utilized to compare complex data such as protein structures (Zhou et al., 2006) or NMR spectra (Ross and Senn, 2003).

SVM is a machine learning approach for classifying data (Noble, 2006). Briefly, a classifying hyperplane between groups of objects is defined using a known training set. The hyperplane is constructed so that it will optimize a "maximum margin," namely, a maximum distance from each of the objects within the groups it separates. It is built using a kernel function of higher dimensionality compared to the original data. Notably, the dimensionality must not be too high to avoid overfitting as well as increased complexity. Finally, accounting for the intrinsic noise of the data, a soft margin, extending from the hyperplane to the nearest object in each group, is assigned. The high prediction power of SVM and suitability to the intrinsic nature of structural bioinformatics data resulted in growing popularity of this method. Just to name some examples, recently SVMs were used to predict (1) functionally flexible regions from primary sequence (Gu et al., 2006), (2) protein binding sites using sequence and structure homolog information (Chung et al., 2006), (3) protein folding kinetic order and rate using kinetic and mechanistic information from known structure (Capriotti and Casadio, 2007), (4) folds within multidomain protein structures (Redfern et al., 2007), (5) membrane binding proteins from primary sequence using information from known such structures (Bhardwaj et al., 2006), and (6) local ordered regions within the "random coil" secondary structure region using sequence information (Zimmermann and Hansmann, 2006).

CONCLUDING REMARKS

In this introductory chapter, a glimpse into the vast topic of search and sampling in structural bioinformatics is provided. As evident in the many references to other chapters in this book, search and sampling methods are pivotal in most applications of structural bioinformatics research. As a broad and evolving field, many search and sampling methods and specific applications were not included. With increasing computational power, the use of search and sampling methods producing the single solution or low-energy populations of the requested target is expected to increase. Beyond the possibility of using more detailed computational methods to further assess such a population, high-throughput experimental techniques can also be applied to such a manageable set of candidate results. However, despite the growing computer power, many questions now and into the future in structural bioinformatics lie in an expanding space far too large to exhaustively scan, so search and sampling techniques will continue to be critical.

Contrary to the success stories described here, some structural bioinformatics results are incorrect as a result of the lack of sufficient understanding of limitations and errors within the different search and sampling methods. Thus, even more important than understanding the success of each method is fully recognizing the limitations and optimal resolution achieved by each one of these methods. We hope this chapter is a step in this direction.

ACKNOWLEDGMENTS

Thanks to Jason E. Donald, Roland L. Dunbrack, Scott Shandler, Joanna S. Slusky, and Cinque Soto for critical reading of the manuscript. IS was funded by the European Molecular Biology Organization (EMBO) and by the Human Frontier Science Program Organization (HFSPO).

REFERENCES (Including Remarks on Recommended Further Reading)

Arya G, Schlick T (2006): Role of histone tails in chromatin folding revealed by a mesoscopic oligo-nucleosome model. *Proc Natl Acad Sci USA* 103(44):16236–16241.

Arya G, Schlick T (2007): Efficient global biopolymer sampling with end-transfer configurational bias Monte Carlo. *J Chem Phys* 126(4):044107.

Baldi P, Brunak S (2001): *Bioinformatics: The Machine Learning Approach*. MIT Press.

Bates PA, Kelley LA, MacCallum RM, Sternberg MJ (2001): Enhancement of protein modeling by human intervention in applying the automatic programs 3D-JIGSAW and 3D-PSSM. *Proteins* (Suppl. 5): 39–46.

Bayes T (1763): An essay towards solving a problem in the doctrine of chances. *Philos Trans Roy Soc* 53:370.

Becker OM (2001): Conformational analysis. In: Becker OM, MacKerell AD, Roux B, Watanabe M, editors. *Computational Biochemistry and Biophysics*. Marcel Dekker, pp 69–90.

Becker OM, MacKerell AD, et al. , editors (2001): *Computational Biochemistry and Biophysics*. New York: Marcel Dekker, Inc. [An excellent text on computational biochemistry and biophysics that is focused largely on molecular motion. The book includes several important introductory chapters along with advanced applications.]

Becker OM, Shacham S, et al. (2003): Modeling the 3D structure of GPCRs: advances and application to drug discovery. *Curr Opin Drug Discov Devel* 6(3):353–361.

Berman HM, Battistuz T, et al. (2002): The Protein Data Bank. *Acta Crystallogr D Biol Crystallogr* 58 (Pt 6 No 1):899–907.

Bhardwaj N, Stahelin RV, et al. (2006): Structural bioinformatics prediction of membrane-binding proteins. *J Mol Biol* 359(2):486–495.

Bradley LH, Wei Y, et al. (2007): Protein design by binary patterning of polar and nonpolar amino acids. *Methods Mol Biol* 352:155–166.

Brenner SE, Koehl P, et al. (2000): The ASTRAL compendium for protein structure and sequence analysis. *Nucleic Acids Res* 28(1):254–256.

Camacho CJ (2005): Modeling side-chains using molecular dynamics improve recognition of binding region in CAPRI targets. *Proteins* 60(2):245–251.

Capriotti E, Casadio R (2007): K-Fold: a tool for the prediction of the protein folding kinetic order and rate. *Bioinformatics* 23(3):385–386.

Chung JL, Wang W, et al. (2006): Exploiting sequence and structure homologs to identify protein–protein binding sites. *Proteins* 62(3):630–640.

Cootes AP, Curmi PM, et al. (2000): Automated protein design and sequence optimisation: scoring functions and the search problem. *Curr Protein Pept Sci* 1(3):255–271.

Cui Q, Bahar I, editors. (2006): *Normal Mode Analysis: Theory and Applications to Biological and Chemical Systems. Mathematical and Computational Biology Series.* Chapman & Hall/CRC. [A new and highly recommended book on theory and applications of normal mode analysis. The book is complementary to the book edited by Becker et al., (2001) and presents the theory behind NMA as well as applications such as structural refinement, large motions, conformational sampling, and diffusion in liquids.]

Dahiyat BI, Mayo SL (1996): Protein design automation. *Protein Sci* 5(5):895–903.

Dahiyat BI, Mayo SL (1997): De novo protein design: fully automated sequence selection. *Science* 278(5335):82–87.

Dantas G, Kuhlman B, et al. (2003): A large scale test of computational protein design: folding and stability of nine completely redesigned globular proteins. *J Mol Biol* 332(2):449–460.

De Maeyer M, Desmet J, et al. (1997): All in one: a highly detailed rotamer library improves both accuracy and speed in the modelling of side chains by dead-end elimination. *Fold Des* 2 (1):53–66.

De Maeyer M, Desmet J, et al. (2000): The dead-end elimination theorem: mathematical aspects, implementation, optimizations, evaluation, and performance. *Methods Mol Biol* 143:265–304.

Deshpande N, Addess KJ, et al. (2005): The RCSB Protein Data Bank: a redesigned query system and relational database based on the mmCIF schema. *Nucleic Acids Res* 33(Database issue): D233–D237.

Desmet J, Demaeyer M, et al. (1992): The dead-end elimination theorem and its use in protein side-chain positioning. *Nature* 356(6369):539–542.

Dunbrack RL Jr, Karplus M (1994): Conformational analysis of the backbone-dependent rotamer preferences of protein side chains. *Nat Struct Biol* 1(5):334–340.

Dunbrack RL Jr, Cohen FE (1997): Bayesian statistical analysis of protein side-chain rotamer preferences. *Protein Sci* 6(8):1661–1681.

Dunbrack RL Jr (2001): Bayesian statistics in molecular and structural biology. *Computational Biochemistry And Biophysics.* In: Becker OM, MacKerell AD, Roux B, Watanabe M,editors. New York: Marcel Dekker, p. 512.

Dunbrack RL Jr (2002): Rotamer libraries in the 21st century. *Curr Opin Struct Biol* 12(4):431–440.

Dwyer MA, Looger LL, et al. (2004): Computational design of a biologically active enzyme. *Science* 304(5679):1967–1971.

Eyal E, Najmanovich R, et al. (2004): Importance of solvent accessibility and contact surfaces in modeling side-chain conformations in proteins. *J Comput Chem* 25(5):712–724.

Finkelstein AV, Reva BA (1991): A search for the most stable folds of protein chains. *Nature* 351(6326):497–499.

Fischer D, Tsai CJ, et al. (1995): A 3D sequence-independent representation of the protein data bank. *Protein Eng* 8(10):981–997.

Frantz DD, Freeman DL, et al. (1990): Reducing quasi-ergodic behavior in Monte Carlo simulations by J-walking—applications to atomic clusters. *J Chem Phys* 93(4):2769–2784.

Friedberg I, Godzik A (2007): Functional differentiation of proteins: implications for structural genomics. *Structure* 15(4):405–415.

Friedrichs MS, Wolynes PG (1989): Toward protein tertiary structure recognition by means of associative memory Hamiltonians. *Science* 246(4928):371–373.

Fu X, Kono H, et al. (2003): Probabilistic approach to the design of symmetric protein quaternary structures. *Protein Eng* 16(12):971–977.

Georgiev I, Donald BR (2007): Dead-end elimination with backbone flexibility. *Bioinformatics* 23 (13):i185–i194.

Geyer CJ, Thompson EA (1995): Annealing Markov-Chain Monte-Carlo with applications to ancestral inference. *J Am Stat Assoc* 90(431):909–920.

Gilks WR, Richardson S et al., editors (1996): *Markov Chain Monte Carlo in Practice.* Chapman & Hall/CRC.

Glick M, Rayan A, et al. (2002): A stochastic algorithm for global optimization and for best populations: a test case of side chains in proteins. *Proc Natl Acad Sci USA* 99(2):703–708.

Goldstein RF, (1994): Efficient rotamer elimination applied to protein side-chains and related spin glasses. *Biophys J* 66(5):1335–1340.

Goodman CM, Choi S, et al. (2007): Foldamers as versatile frameworks for the design and evolution of function. *Nat Chem Biol* 3(5):252–262.

Greene LH, Lewis TE, et al. (2007): The CATH domain structure database: new protocols and classification levels give a more comprehensive resource for exploring evolution. *Nucleic Acids Res* 35(Database issue):D291–D297.

Gu J, Gribskov M, et al. (2006): Wiggle-predicting functionally flexible regions from primary sequence. *PLoS Comput Biol* 2(7):e90.

Halperin I, Ma B, et al. (2002): Principles of docking: an overview of search algorithms and a guide to scoring functions. *Proteins* 47(4):409–443.

Hamelberg D, Mongan J, et al. (2004): Accelerated molecular dynamics: a promising and efficient simulation method for biomolecules. *J Chem Phys* 120(24):11919–11929.

Hansmann UHE (1997): Parallel tempering algorithm for conformational studies of biological molecules. *Chem Phys Lett* 281(1–3):140–150.

Herzyk P, Hubbard RE (1993): A reduced representation of proteins for use in restraint satisfaction calculations. *Proteins* 17(3):310–324.

Hobohm U, Scharf M, et al. (1992): Selection of representative protein data sets. *Protein Sci* 1(3):409–417.

Hobohm U, Sander C (1994): Enlarged representative set of protein structures. *Protein Sci* 3(3):522–524.

Holm L, Sander C (1992): Fast and simple Monte Carlo algorithm for side chain optimization in proteins: application to model building by homology. *Proteins* 14(2):213–223.

Holm L, Sander C (1996): Mapping the protein universe. *Science* 273(5275):595–603.

Hou J, Sims GE, et al. (2003): A global representation of the protein fold space. *Proc Natl Acad Sci USA* 100(5):2386–2390.

Hwang JK, Liao WF (1995): Side-chain prediction by neural networks and simulated annealing optimization. *Protein Eng* 8(4):363–370.

Ichiye T, Karplus M (1991): Collective motions in proteins: a covariance analysis of atomic fluctuations in molecular dynamics and normal mode simulations. *Proteins* 11(3):205–217.

Jain T, Cerutti DS, et al. (2006): Configurational-bias sampling technique for predicting side-chain conformations in proteins. *Protein Sci* 15(9):2029–2039.

Jiang L, Kuhlman B, et al. (2005): A "solvated rotamer" approach to modeling water-mediated hydrogen bonds at protein–protein interfaces. *Proteins* 58(4):893–904.

Jolliffe IT, (2002): *Principal Component Analysis.* Springer.

Jones DT (1994): De novo protein design using pairwise potentials and a genetic algorithm. *Protein Sci* 3(4):567–574.

Kamtekar S, Schiffer JM, et al. (1993): Protein design by binary patterning of polar and nonpolar amino acids. *Science* 262(5140):1680–1685.

Kang SG, Saven JG (2007): Computational protein design: structure, function and combinatorial diversity. *Curr Opin Chem Biol* 11(3):329–334.

Kingsford CL, Chazelle B, et al. (2005): Solving and analyzing side-chain positioning problems using linear and integer programming. *Bioinformatics* 21(7):1028–1036.

Kloppmann E, Ullmann GM, et al. (2007): An extended dead-end elimination algorithm to determine gap-free lists of low energy states. *J Comput Chem* 28(14):2325–2335.

Koehl P, Delarue M (1994): Application of a self-consistent mean field theory to predict protein side-chains conformation and estimate their conformational entropy. *J Mol Biol* 239(2):249–275.

Koehl P, Delarue M (1995): A self consistent mean field approach to simultaneous gap closure and side-chain positioning in homology modelling. *Nat Struct Biol* 2(2):163–170.

Kono H, Doi J (1994): Energy minimization method using automata network for sequence and side-chain conformation prediction from given backbone geometry. *Proteins* 19(3):244–255.

Kono H, Saven JG (2001): Statistical theory for protein combinatorial libraries. Packing interactions, backbone flexibility, and the sequence variability of a main-chain structure. *J Mol Biol* 306(3):607–628.

Kuhlman B, Dantas G, et al. (2003): Design of a novel globular protein fold with atomic-level accuracy. *Science* 302(5649):1364–1368.

Kussell EL, Shakhnovich EI (1999): Analytical approach to the protein design problem. *Phys Rev Lett* 83(21):4437–4440.

Lassila JK, Privett HK, et al. (2006): Combinatorial methods for small-molecule placement in computational enzyme design. *Proc Natl Acad Sci USA* 103(45):16710–16715.

Lasters I, De Maeyer M, et al. (1995): Enhanced dead-end elimination in the search for the global minimum energy conformation of a collection of protein side chains. *Protein Eng* 8(8):815–822.

Leach AR, Lemon AP (1998): Exploring the conformational space of protein side chains using dead-end elimination and the A* algorithm. *Proteins* 33(2):227–239.

Leach AR (2001): *Molecular Modeling: Principles and Applications.* Prentice Hall. [A practical guide to macromolecular modeling. This 2nd edition goes deep into the scientific and the practical challenges facing molecular modelers.]

Levinthal C (1969): How to fold graciously. In: DeBrunner P, Tsibris J, Munck E. *Mossbauer Spectroscopy in Biological Systems.* Proceedings of a Meeting Held at Allerton House, Monticello, IL. Urbana, IL: University of Illinois Press, pp 22–24.

Levitt M (1976): A simplified representation of protein conformations for rapid simulation of protein folding. *J Mol Biol* 104(1):59–107.

Liu JS (2001): *Monte Carlo Strategies in Scientific Computing.* Springer.

Looger LL, Hellinga HW (2001): Generalized dead-end elimination algorithms make large-scale protein side-chain structure prediction tractable: implications for protein design and structural genomics. *J Mol Biol* 307(1):429–445.

Looger LL, Dwyer MA, et al. (2003): Computational design of receptor and sensor proteins with novel functions. *Nature* 423(6936):185–190.

Lovell SC, Word JM, et al. (2000): The penultimate rotamer library. *Proteins* 40(3):389–408.

MacKay DJC, (2005): *Information Theory, Inference and Learning Algorithms.* Cambridge: Cambridge University Press.

Mendes J, Baptista AM, et al. (1999): Improved modeling of side-chains in proteins with rotamer-based methods: a flexible rotamer model. *Proteins* 37(4):530–543.

Mendes J, Soares CM, et al. (1999): Improvement of side-chain modeling in proteins with the self-consistent mean field theory method based on an analysis of the factors influencing prediction. *Biopolymers* 50(2):111–131.

Metropolis N, Rosenbluth AW, et al. (1953): Equation of state calculations by fast computing machines. *J Chem Phys* 21(6):1087–1092.

Murzin AG, Brenner SE, et al. (1995): SCOP: a structural classification of proteins database for the investigation of sequences and structures. *J Mol Biol* 247(4):536–540.

Neumann JV (1932): Proof of the Quasi-Ergodic Hypothesis. *Proc Natl Acad Sci USA* 18(1):70–82.

Noble WS (2006): What is a support vector machine? *Nat Biotechnol* 24(12):1565–1567.

Noy E, Tabakman T, et al. (2007): Constructing ensembles of flexible fragments in native proteins by iterative stochastic elimination is relevant to protein–protein interfaces. *Proteins* 68(3):702–711.

Nussinov R, Wolfson HJ (1991): Efficient detection of three-dimensional structural motifs in biological macromolecules by computer vision techniques. *Proc Natl Acad Sci USA* 88(23): 10495–10499.

Park S, Kono H, et al. (2005): Progress in the development and application of computational methods for probabilistic protein design. *Comp Chem Eng*, 29:407–421.

Park S, Yang X, et al. (2004): Advances in computational protein design. *Curr Opin Struct Biol* 14(4):487–494.

Pedersen JT, Moult J (1996): Genetic algorithms for protein structure prediction. *Curr Opin Struct Biol* 6(2):227–231.

Peterson RW, Dutton PL, et al. (2004): Improved side-chain prediction accuracy using an ab initio potential energy function and a very large rotamer library. *Protein Sci* 13(3):735–751.

Pierce NA, Spriet JA, et al. (2000): Conformational splitting: a more powerful criterion for dead-ending elimination. *J Comput Chem* 21(11):999–1009.

Ponder JW, Richards FM (1987): Internal packing and protein structural classes. *Cold Spring Harb Symp Quant Biol* 52:421–428.

Redfern OC, Harrison A, et al. (2007): CATHEDRAL: a fast and effective algorithm to predict folds and domain boundaries from multidomain protein structures. *PLoS Comput Biol* 3(11):e232.

Rosenberg M, Goldblum A (2006): Computational protein design: a novel path to future protein drugs. *Curr Pharm Des* 12(31):3973–3997.

Rosenbluth MN, Rosenbluth AW (1955): Monte-Carlo calculation of the average extension of molecular chains. *J Chem Phys* 23(2):356–359.

Ross A, Senn H (2003): Automation of biomolecular NMR screening. *Curr Top Med Chem* 3(1):55–67.

Saven JG (2001): Designing protein energy landscapes. *Chem Rev* 101(10):3113–3130.

Saven JG (2002): Combinatorial protein design. *Curr Opin Struct Biol* 12(4):453–458.

Schlick T (2002): *Molecular Modeling and Simulation: An Interdisciplinary Guide.* Springer [A comprehensive guide on molecular modeling and simulation. The book is aimed for newcomers and novices alike. It assumes little background yet focuses on the mathematical formulation.]

Schneider G, Schrodl W, et al. (1998): Peptide design by artificial neural networks and computer-based evolutionary search. *Proc Natl Acad Sci USA* 95(21):12179–12184.

Shapovalov MV, Dunbrack RL Jr (2007): Statistical and conformational analysis of the electron density of protein side chains. *Proteins* 66(2):279–303.

Shetty RP, De Bakker PI, et al. (2003): Advantages of fine-grained side chain conformer libraries. *Protein Eng* 16(12):963–969.

Shlyk-Kerner O, Samish I, et al. (2006): Protein flexibility acclimatizes photosynthetic energy conversion to the ambient temperature. *Nature* 442(7104):827–830.

Siepmann JI (1990): A method for the direct calculation of chemical potentials for dense chain systems. *Mol Phys* 70(6):1145–1158.

Simons KT, Kooperberg C, et al. (1997): Assembly of protein tertiary structures from fragments with similar local sequences using simulated annealing and Bayesian scoring functions. *J Mol Biol* 268(1):209–225.

Slovic AM, Kono H, et al. (2004): Computational design of water-soluble analogues of the potassium channel KcsA. *Proc Natl Acad Sci USA* 101(7):1828–1833.

Szasz D (1994): Boltzmann's Ergodic Hypothesis, a Conjecture for Centuries? Presented at International symposium in honor of Boltzmann's 150th birthday, Vienna.

Thiruv B, Quon G, et al. (2005): Nh3D: a reference dataset of non-homologous protein structures. *BMC Struct Biol* 5:12.

Vieth M, Hirst JD, et al. (1998): Assessing search strategies for flexible docking. *J Comput Chem* 19 (14):1623–1631.

Voigt CA, Gordon DB, et al. (2000): Trading accuracy for speed: a quantitative comparison of search algorithms in protein sequence design. *J Mol Biol* 299(3):789–803.

Wang G, Dunbrack RL Jr (2005): PISCES: recent improvements to a PDB sequence culling server. *Nucleic Acids Res* 33(Web Server issue):W94–W98.

Xiang Z, Honig B (2001): Extending the accuracy limits of prediction for side-chain conformations. *J Mol Biol* 311(2):421–430.

Yang X, Saven JG (2005): Computational methods for protein design and protein sequence variability: biased Monte Carlo and replica exchange. *Chem Phys Lett* 401(1–3):205–210.

Yanover C, Fromer M, et al. (2007): Dead-end elimination for multistate protein design. *J Comput Chem* 28(13):2122–2129.

Yin H, Slusky JS, et al. (2007): Computational design of peptides that target transmembrane helices. *Science* 315(5820):1817–1822.

Zhou X, Chou J, et al. (2006): Protein structure similarity from principle component correlation analysis. *BMC Bioinformatics* 7:40.

Zimmermann O, Hansmann UH (2006): Support vector machines for prediction of dihedral angle regions. *Bioinformatics* 22(24):3009–3015.

Zou JM, Saven JG (2003): Using self-consistent fields to bias Monte Carlo methods with applications to designing and sampling protein sequences. *J Chem Phys* 118(8):3843–3854.

9

MOLECULAR VISUALIZATION

Steven Bottomley and Erik Helmerhorst

INTRODUCTION

Visual images are powerful. Images can turn something that may be unseen, abstract, theoretical, or complex into something real and purposeful. Visual images are not just decorations. In a glance, they can add meaning to a message, help people learn, and provide insight that is not possible by looking at the raw data alone. Scientific visualization generates visual images from data generated by scientific experiment and theory. It aims to discover new knowledge by distilling mountains of data into something immediately useful. Traditionally, line graphs, two-dimensional (2D) plots, three-dimensional (3D) plots, mathematical models, physical models, specimens, photographs, drawings, movies, and symbols are used to visualize scientific data. The computer has now become a convenient, and powerful, tool not only for traditional scientific visualization, but also for more challenging visualization tasks involving computer graphics, animations, modeling, and simulations. The importance of computer visualization to science is showcased in annual 'Visualization Challenge' competitions (Nesbit and Bradford, 2006) and IEEE Visualization Conferences (Table 9.1).

Molecular Visualization

Molecular visualization is a way of seeing the abstract and unseen atomic and molecular world. It uses graphics to study structure, properties, and functions of molecules (Breithaupt, 2006). Consequently, it is also called molecular graphics (Henkel and Clarke, 1985). Small molecules (Figure 9.1a) and macromolecules (Figure 9.1b) can be represented in various ways such as a formula or character format in one dimension (1D), 2D drawings, 2D tables, 3D physical models, or 3D computer-generated models. The SMILES string format (Weininger, 1988) is useful for representing atom connectivity and bond order of small molecules, but awkward for representing macromolecules. Consequently, the

Structural Bioinformatics, Second Edition   Edited by Jenny Gu and Philip E. Bourne
Copyright © 2009 John Wiley & Sons, Inc.

TABLE 9.1. Web sites on Various Topics of Molecular Visualization[a]

History of computer molecular graphics, molecular modeling, and physical models
Dalton, J. 1808 "A New System of Chemical Philosophy" at web.lemoyne.edu/~GIUNTA/dalton.html
Bernstein H "The Historical Context of RasMol Development" at www.openrasmol.org/history.html
Martz E and Francoeur E "History of visualization of biological macromolecules" at www.umass.
    edu/microbio/rasmol/history.htm
Connolly ML "Molecular Surfaces: A Review" at www.netsci.org/Science/Compchem/feature14.html
O'Donnell TJ "The Scientific and Artistic Uses of Molecular Surfaces" at www.netsci.org/Science/
    Compchem/feature15.html

Recent Physical Molecular Models
Centre for BioMoleclular Modeling at www.rpc.msoe.edu/cbm/
The Scripps Molecular Graphics Laboratory at mgl.scripps.edu/
Richard Garratt: www.hwi.buffalo.edu/ACA/ACA04/abstracts/text/W0278.pdf

Programs or Program Resources
Pov-Ray. Open source ray tracing software at www.povray.org
Molecular visualization program list maintained by Erik Martz and Trevor Kramer at molvisindex.org
OpenGL source and software at www.opengl.org/

Examples of Macromolecule Animations
wishart.biology.ualberta.ca/moviemaker/gallery/index.html
www.hms.harvard.edu/news/releases/699clathrin.html
www.umass.edu/microbio/chime/pe_beta/pe/protexpl/morfdoc.htm
www.moleculesinmotion.com/
www.bmb.psu.edu/faculty/tan/lab/gallery_protdna.html

Examples of Molecular Graphics or Molecular Models as Art
Molecular Graphics Art Show: mgl.scripps.edu/people/goodsell/mgs_art
Artn: www.artn.com/portfolio.cfm
Biografx: www.biografx.com/virtualgallery/galleryintro.html
IEEE Visualization Conferences: vis.computer.org

Virtual Reality
Molecular Visualization by immersive virtual reality at the University of Groningen
www.rug.nl/cit/hpcv/vr_visualisation/molecular_visualisation/mol_visualisation

[a]Note that the "http://"part of the URL is omitted for clarity but is implied.

primary structure of protein, DNA, and RNA is commonly represented by their one-letter codes. The peptide bond between residues of proteins, or the phosphodiester bond between nucleotides in DNA and RNA, is implicit in the 1D strings. Recently, the primary structure of DNA has also been represented as a 2D graph (Dai, Liu, and Wang, 2006). Character strings, using standard abbreviations for monosaccharides (and other components in a heteropolysaccharide), are used to represent polysaccharides, but the anomer configuration and bond direction (e.g., $\beta1-4$) also must be included in the string. The common 2D line drawings for chemical structure (also less commonly called chemical graphs) effectively represent atom connectivity and bond order in small molecules. However, this representation is awkward for macromolecules, because of the enormity of the data and the resulting visual clutter. The 2D drawings of proteins, DNA, or RNA simplify the representation by using visual abstractions and models to emphasize gross structural elements such as secondary structures. Proteins, in particular, are drawn or painted using cartoons (Richardson, 1981; Richardson, 1985b) and other representations (Dickerson and Geis, 1969; Goodsell, 1991; Goodsell, 1992;

Goodsell, 2000; Chapter 2) to show secondary structures. In addition, the structure and topology of proteins are visualized by star-like graphs (Randic, Zupan, and Vikic-Topic, 2007) or computer-generated 2D drawings using programs such as TOPS (Westhead et al., 1999), TopDraw (Bond, 2003), and Polyview (Porollo and Meller, 2007). The secondary structure of RNA can also be visualized using computer-generated 2D drawings (Wiese and Glen, 2006).

Tables represent molecules in abstract form listing atoms, their connectivity, or coordinates in particular formats such as MDL molfile, mmCIF, or PDB (Chapter 10). In addition, molecules can be represented in table-like formats such as extensible markup language (XML) or chemical markup language (CML) for web display.

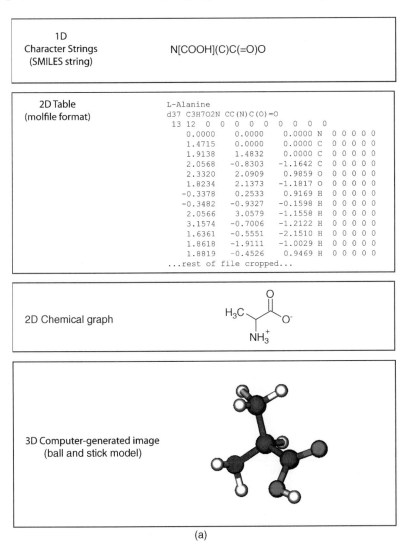

(a)

Figure 9.1. (a) Representations of small molecules. Small molecules can be represented as character strings, 2D drawings, 2D tables, or 3D images. ʟ-Alanine is used as an example. (b) Representation of proteins. Macromolecules can be represented as character strings, 2D drawings, 2D tables, or 3D images. Carboxypeptidase A is used as an example.

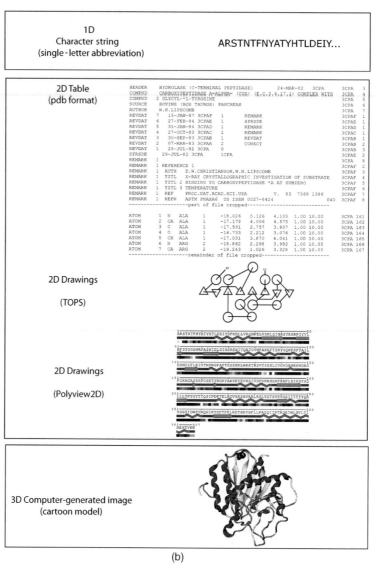

(b)

Figure 9.1. (*continued*)

Both 1D and 2D representations of molecules are useful, but limited in their application to large complex molecules such as proteins, DNA, and RNA. The inclusion of an additional dimension provides the information needed to effectively visualize these macromolecules. Molecules are 3D entities and their function intimately depends upon their structures. Consequently, the challenges in computer molecular visualization involve devising suitable models of 3D structure that can be projected onto a 2D computer screen. Molecular visualization is multidisciplinary and involves aspects of physics, chemistry, mathematics, computer science, and computer graphics. Any discussion of molecular visualization necessarily touches on one, or more, of these subjects. Molecular visualization also requires the integration of several different models including atomic, molecular, and protein models, as well as mathematical and geometrical models. Molecular models, also known as molecular representations, are a very important way for us to clarify, and understand, what is happening at the molecular level. We will see later how some of these models are used in molecular visualization.

A broader view of molecular visualization could encompass any chemical, biochemical, or physical technique used to visualize molecules. These include the common dye staining, fluorescent, radioactive, and bioluminescence techniques used in biochemistry, genomics, and proteomics (Hu et al., 2004). The field of molecular imaging (Weissleder, 1999; Weissleder and Mahmood, 2001; Blasberg and Gelovani-Tjuvajev, 2002; Rollo, 2003) uses radioactive, or fluorescent, molecular probes in techniques as diverse as *in vivo* magnetic resonance imaging (MRI) (Artemov, 2003), positron emission tomography (PET) (Phelps, 2000), and cytogenetics (de Jong, 2003). Molecular imaging also includes techniques such as scanning tunneling microscopy (STM) and atomic force microscopy (Silva, 2005). While some of these techniques may resolve single molecules (Zlatanova and van Holde, 2006; Cornish and Ha, 2007), most involve gross visualization of molecules, or molecular aggregates, and do not resolve atomic detail. In this chapter, we will concentrate on the interactive computer graphic 3D visualization of biomolecules (with an emphasis on proteins) at the atomic level.

THE PROCESS OF MOLECULAR VISUALIZATION

Raw Data to 3D Coordinates

The structures in the atomic and molecular world can be resolved using techniques such as X-ray crystallography, nuclear magnetic resonance (NMR), cryoelectron microscopy, and electron tomography (ET). The data generated by these experiments need to be processed by molecular visualization and modeling software to generate a model of the structure (Lamzin and Perrakis, 2000). Some of the commonly used software in X-ray crystallography includes the CCP4 suite of programs (Collaborative Computational Project, 1994), MAID (Levitt, 2001), ARP/wARP (Cohen et al., 2004), O (Jones et al., 1991), CNS (Brunger et al., 1998), XtalView (McRee, 1999), and X-PLOR (Brunger, 1992). These programs use X-ray diffraction data obtained from a purified protein crystal and generate electron density maps (e.g., CCP4), build the molecular model (e.g., O), and refine the model (e.g., X-PLOR). Many of these programs not only are fully automated, but also allow the user to interact at various points in the analysis. Figure 9.2 shows an outline of the process involved in visualizing X-ray crystallography data. The result of this process is a file containing the 3D orthogonal coordinates (x, y, z) of the refined molecule. These coordinates specify the position of each atom in the molecule within three-dimensional Cartesian space. The coordinates are saved as a file in PDB format, although other formats such as mmCIF are also used (Chapter 10). The 3D coordinates can be used as the input for visualization of the molecule either as a physical or computer-generated model.

3D Coordinates to Molecular Visualization by Physical Models

Despite significant technological improvements in computer graphics, tangible physical models that are not confined to the computer screen are still very useful. Physical models offer an impressionable, persistent, and high impact way to "see" and "feel" the molecule giving the investigator a perspective that promotes understanding and insight that is not

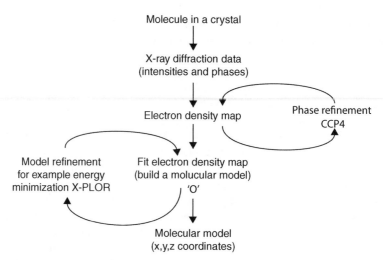

Figure 9.2. Visualizing X-ray crystallography data. X-ray diffraction data is processed by software to model the protein structure and produce the final 3D orthogonal coordinates in a pdb file.

always possible with computer graphic models. The building materials used to model molecules include wax, clay, wire, plastic, and wood. Physical models are still very useful in teaching and learning because they provide a multisensory learning experience that also enhances collaboration (Gillet et al., 2005).

Some of the early physical models include Pauling's space-filled wood molecular models (Corey and Pauling, 1953), Watson and Crick's wire model of the DNA double helix (Watson and Crick, 1953), and Kendrew's plasticine "sausage" model of myoglobin (Kendrew, 1961). Eric Martz and Eric Francoeur give a brief, but excellent, history of molecular visualization with pictures of early physical, and computer graphic, molecular models on their history of molecular visualization Web site (Table 9.1).

Recent physical models have been developed by Timothy Herman at the Centre for BioMolecular Modeling, Arthur Olson's group at the Scripps Molecular Graphics Laboratory, and Richard Garratt at the Instituto de Fisica Sao Carlos (see Table 9.1 for Web sites). The Centre for BioMolecular Modeling uses rapid prototyping techniques such as stereolithography, laminated object manufacturing, selective laser sintering, fused deposition modeling, and Z-corp 3D printing to rapidly build high-fidelity molecular models. The Olson group is also investigating the concept of "augmented reality," which is an interesting approach that combines physical models, computer graphics, and virtual reality techniques (Sankaranarayanan et al., 2003; Gillet et al., 2005). Garratt and Pereira (2006) developed an innovative kit to build plastic cartoon models of proteins suitable for teaching.

The problems associated with most physical models, however, are that they take a relatively long time to build; are not always easy to store, edit, manipulate, or change; have limited interactivity; and quantitative information (such as bond distances and angles) is not readily available. These issues are particularly evident when trying to model large molecules such as proteins. For example, molecule kits are still available to build your own insulin protein (the antidiabetic hormone), but the construction time for this relatively small protein containing 51 residues is in the order of weeks. In contrast, visualizing insulin using 3D interactive computer graphics takes seconds. Computer-generated 3D molecular models readily address the disadvantages of tangible physical models.

3D Coordinates to Molecular Visualization by Computer-Generated Models

Molecular visualization uses interactive computer graphics to generate a virtual 3D image of a molecule on a 2D computer screen from the 3D coordinates of the molecule's structure. Computer graphics play a central role in producing the image, and the user interacts with the image shown on the computer display. Interactive computer graphics is a large, growing, and actively researched area. It is multidisciplinary and encompasses subjects such as mathematics (particularly geometry and trigonometry), optics (including illumination), psychology (particularly human perception), physics, engineering, computer hardware and software, algorithms, solid modeling, rendering, and animation (Foley et al., 1990). Consequently, we will only briefly discuss the basic aspects of computer hardware, software, and computer graphics techniques to help us understand the process involved in generating the molecule we see on the computer screen.

Computer Hardware

Computer hardware can be constituted of one, or more, of the following components: central processing unit (CPU); CPU-associated memory such as random access memory (RAM) and read-only memory (ROM); storage memory devices (e.g., hard drive, tape drive, and thumb drive); input devices (e.g., keyboard, mouse, tablet, pen), output devices (e.g., printer, plotter, display screen); and communications devices (e.g., modem and network interface).

Three-dimensional computer graphics is complex and involves many calculations and computer operations. Intuitively, we know that it takes more steps to draw a triangular prism (a simple 3D image with four points and six lines) than to draw a triangle (a simple 2D image with three points and lines). If we want to make the prism look "solid," then we need to provide additional information to specify the surfaces (i.e., the geometrical planes of the prism) and which surfaces are visible from a particular view of the prism. If we also want to add colors, textures, lighting, and other such effects to make the pyramid look more "realistic," then the number of steps and computer operations increases. Extending the reasoning of this simple example to a protein containing thousands of atoms, it is easy to see how considerable computational power is needed to view the final 3D image. The more information we include in the 3D image the greater the computer power we need to process, visualize, and interact with the image within a reasonable time.

A computer's "power" is related to the increased speed with which the CPU (variously called a chip, microchip, microprocessor, or processor) can perform basic computer operations and calculations. This processing power also depends upon the corresponding advances in other components of the computer, such as higher capacity memory (RAM, ROM, and disk memory), increased speed of internal electronic pathways (data buses), computer architecture, computer input and output devices, and software. In particular, parallel processing architectures (where multiple CPUs and memories are used to increase the number of computer operations that can be performed simultaneously) and dedicated graphics processors have enabled computer graphics to produce stunning photorealistic images (Kerlow, 2004). The relevance of this increased computer power to computer graphics, and to molecular visualization in particular, is that we can generate images with increased clarity, greater interaction, and in less time (so we can get our work done faster!).

The computer display is obviously important, because it enables us to see the computer-generated molecules. During the 1960s, molecular visualization computer systems used vector graphics displays (Francoeur, 2002). Vector graphics is a term used variously to describe the hardware for drawing a 2D shape, a particular image data format, and the mathematics used to represent graphics in software or dedicated graphics processors. The original type of display used with a computer was a cathode ray tube (CRT). A CRT uses an electron beam to irradiate and excite a phosphor to glow on the screen. The line has to be repeatedly drawn or refreshed (at about 30 times per second) to achieve persistent vision, because the phosphor rapidly decays (Foley et al., 1990). The data for this refresh are stored in memory and constituted of the coordinate data and other commands. This is also called a random scan, because the electron beam can scan the complete X–Y dimensions of the available screen, and a point-to-point drawing can occur at any position directed by the order of commands in the vector graphics software. Vector displays are fast and do not use much memory, but can only trace relatively simple lines, planes, and a few colors.

Most current computer molecular visualization systems use raster graphic displays rather than vector displays. Raster graphics is a term used to describe the hardware for drawing a 2D shape, a particular image data format, and particular computer graphic techniques. The image to be displayed on the screen is initially stored in memory (called a frame or refresh buffer) as a two-dimensional array (the raster) of picture elements (pixels) called a bitmap. A pixel is a defined area on the display that can be activated and attenuated by the electronic signal generated from the computer (e.g., an electron beam of a CRT). Each pixel can also be activated to show any of thousands or millions of colors depending upon the number of bits (bit depth) used for color in each pixel. A bit depth of 8 bits per pixel would allow 256 colors, while a 24 bits per pixel would allow millions of colors to be displayed. A raster display traces out one horizontal row of pixels at a time (called a scan line) to eventually map the stored bitmap to the screen. The screen has to be repeatedly drawn or refreshed (about 60 times per second) to maintain persistent vision. A computer screen with a resolution of 1280×1024 has 1280 pixels per horizontal row and 1024 pixels per vertical row and is considered high resolution. Raster graphics displays use a CRT although newer technologies such as plasma and liquid crystal display (LCD) are also used. Raster displays can be used to show color, complex shapes, volumes, textures, surfaces, and photorealistic images. RasMol is one of the most commonly used molecular visualization programs for raster graphics displays and its name is purported to be a mnemonic for Raster Molecules (see Martz and Francoeur, Table 9.1).

A graphics processing unit (GPU) is a microprocessor dedicated to manipulating and displaying computer graphics. The GPU takes over from the CPU in handling various 2D and 3D graphics tasks. The GPU is responsible for implementing most of the visualization, or graphics, pipeline (see below). Consequently, they help increase the speed of graphics processing and hence molecular visualization. Modern GPUs are also programmable, which means there is more flexibility in manipulating, and presenting, the image. Some molecular visualization applications take advantage of the GPU (e.g., Chimera and VMD), some still depend on the CPU for processing power (e.g., RasMol), and some bypass the GPU altogether to maintain their "platform independence" (e.g., Jmol).

Computer Software

Computer software consists minimally of application programs (also commonly known as "programs" or "software"), such as a molecular visualization program, and an operating system. We will discuss molecular visualization programs in a little more detail later. The operating system provides the basic operating environment of the computer and manages (among other things) the memory (RAM and disk), internal communication with specialized coprocessors such as graphic processing units (GPU), memory storage, and input/output devices. Table 9.2 shows that molecular visualization is carried out on various operating systems and the choice of a particular operating system may depend upon personal preference, reliability, stability, support for extended architectures or modularization, support from developers, graphics libraries or other graphics resources within the system, a suitable user interface such as a graphical user interface (GUI), or commercial considerations. The application software is developed to perform a particular task or range of tasks and works in conjunction with the operating system, the hardware resources of the computer system (e.g., GPU), and other software (e.g., OpenGL).

The Visualization Pipeline

Computer visualization is a complex process that is dealt with by breaking it up into several conceptual stages called a visualization pipeline. The aim of the visualization pipeline is to transform the 3D coordinates of an object (which is a molecule in a molecular visualization program) into a virtual 3D image on a 2D computer display. It is called a pipeline to reflect the fact that it involves an interconnected series of processes where the result of one process usually feeds into the subsequent process. The processes deal with both the shape and the appearance of the molecule. The shape of a molecule is dealt with by the 3D geometric processes of the pipeline and its appearance is dealt with by the rendering processes of the pipeline. Figure 9.3 shows that the pipeline has three general stages: modeling, geometry, and rendering. There are also processes within each stage. The actual sequence of the stages and the type or number of processes within each stage may vary between the application software, the GPU, and the application programming interface (API). An API, such as OpenGL (Angel, 2002), is an interface between hardware and application software. It is a set of rules, and graphics commands, that standardizes the use of hardware GPUs by application software. An API is essentially written into application software and also a GPU. In this way, programmers can write their software to an API, and it should run on any GPU using the same API.

The perception of depth, and hence 3D, of objects presented on a 2D display can be enhanced by many techniques including lighting, motion, color, perspective shading, textures, and shadows. Each of these techniques is applied at particular stages of the visualization pipeline. The modeling and geometry stages of the pipeline deal with the processes that create and define the 3D geometry of an object and are performed by the application program and the GPU. The application defines the particular model, or molecular representation (see the section "Molecular Models"), while the geometry stage involves transformations through various coordinate systems. A computer graphics system capable of producing 3D images generally uses five different coordinate systems: a global, or world, coordinate system; a local coordinate system; a view (eye) coordinate system; a picture (projection) plane coordinate system; and the display screen coordinate system (Mortenson, 1989). We will not go into the details of these coordinate systems here except to say that the initial 3D coordinates of the molecule are progressively transformed into a 2D coordinate system suitable for the computer display and for effective interaction by the user. Other processes (usually applied to a particular coordinate system) determine colors, lighting effects, how the object is clipped to fit on screen (especially if zooming the molecule), and the perspective of the object in geometry stage. Perspective is a particularly useful depth cue and helps create the illusion of 3D from a 2D image. Depth is also dealt with by z-buffering at the next stage of the pipeline (discussed later).

TABLE 9.2. Examples of Recent Molecular Visualization Programs

Program	Platform[a]	Type[b]	Reference or Web Site[c]
Free Molecular Visualization Programs			
Astex-Viewer	M,W,L	JA	Oldfield (2004)
Biodesigner	W	SA	www.pirx.com/biodesigner/index.shtml
Carbohydra	W	SA	Kuttel et al. (2006)
CheVi	L	SA	www.simbiosys.ca/chevi/
Chime	M,W	PI	molviz.org
Chimera	W,L,U	SA	Pettersen et al. (2004)
			www.cgl.ucsf.edu/chimera/
Cn3D	M,W,L,U	SA	Wang et al. (2000)
CrystalMaker	M,W	SA	www.crystalmaker.com/
Deep View	M,W,L,U	SA	Guex and Peitsch (1997)
			www.expasy.org/spdbv/
3DNA	W,L,U	SA	Lu and Olson (2003)
			rutchem.rutgers.edu/~olson/3DNA/
Dino	M,L,U	SA	cobra.mih.unibas.ch/dino/intro.php
DSVisualizer	W,L	SA	www.accelrys.com/products/dstudio/
eMovie	M,W,L	PI	Hodis et al. (2007)
			www.weizmann.ac.il/ISPC/eMovie.html
Epitope Viewer	M,W,L,U	JSA	Beaver, Bourne, and Ponomarenko (2007)
			www.immuneepitope.org
Facio	W	SA	www1.bbiq.jp/zzzfelis/Facio.html
FirstGlance	M,W,L	JA	molvis.sdsc.edu/fgij/
FPV	W,L,U	J3DSA	Can et al. (2003)
GoCav	M,W,L	J3D	Smith et al. (2006)
			www.bindingmoad.org/
ICM Browser	M,W,L,U	SA	www.molsoft.com/
iMol	M	SA	www.pirx.com/iMol/
Jmol	M,W,L,U	JSA,JA	jmol.sourceforge.net
jAMVLE	M,W,L,U	JSA	Bottomley et al., (2006)
			wabri.org.au/jamvle
JavaMAGE	M,W,L,U	JA	kinemage.biochem.duke.edu/software/javamage. php
JMV	M,W,L,U	J3DSA	www.ks.uiuc.edu/Research/jmv/
KiNG	M,W,L	JSA	kinemage.biochem.duke.edu/software/king.php
MAGE	M,W,L,U	SA	Richardson and Richardson (1992)
			kinemage.biochem.duke.edu/software/mage.php
MICE	W,L,U	J3DSA	Tate, Moreland, and Bourne (2001)
			mice.sdsc.edu/
Molekel	M,W,L,U		Portmann and Lüthi (2000)
			www.cscs.ch/a-display.php?id=138
MOLMOL	W,U	SA	Koradi, Billeter, and Wuthrich (1996)
			hugin.ethz.ch/wuthrich/software/molmol/index.html
Molscript	U	SA	www.avatar.se/molscript/
MOLVIE	M,W,L,U	JA	Sun, Li, and Xu (2002)
MolViewX	M	SA	Smith (1995, 2004)
			www.danforthcenter.org/smith/MolView/molview. html
MovieMol	U	SA	www.ifm.liu.se/compchem/moviemol/

TABLE 9.2. (*Continued*)

Program	Platform[a]	Type[b]	Reference or Web Site[c]
MSV	U	SA	mgl.scripps.edu/people/sanner/html/msv.html
Protein Explorer	M,W,L,U	PI	Martz (2002)
			proteinexplorer.org
ProteinShop	M,L	SA	Crivelli et al. (2004)
PyMol	M,W,L,U	SA	pymol.sourceforge.net/
Qmol	W,U	SA	Gans and Shalloway (2001)
			www.mbg.cornell.edu/Shalloway_Lab_QMOL. cfm
QuickPDB	M,W,L,U	JA	cl.sdsc.edu/QuickPDB.html
RasMol	M,W,L,U	SA	Sayle and Milner-White (1995)
			openrasmol.org
Ribbons 2.0	U	SA	Carson (1994)
			www.msg.ucsf.edu/local/programs/ribbons/ribbons. html
Sirius	M,W,L	SA	sirius.sdsc.edu/index.php
STM4	M,W,L,U	SA	Valle (2005). Free but requires AVS/Express below.
			www.cscs.ch/~mvalle/STM4/index.html
Sweet 2	M,W,L,U	WS	Bohne, Lang, and von der Lieth (1999)
			www.glycosciences.de/modeling/sweet2/doc/index. php
TexMol	W,L	SA	Bajaj et al. (2004)
VIBE	U(VR)	SA	Cruz-Neira, Langley, and Bash (1996)
VMD	L,U	SA	Humphrey, Dalke, and Schulten (1996)
			www.ks.uiuc.edu/Research/vmd/
WebMol	M,W,L,U	JA	Walther (1997)
			www.cmpharm.ucsf.edu/~walther/webmol.html
YasaraView	W,U	SA	www.yasara.org/yasaradl.htm

Server-Based Molecular Visualization Programs

AISMIG	M,W,L,U	WS	Bohne-Lang, Groch, and Ranzinger (2005)
			www.dkfz-heidelberg.de/spec/aismig/
iMolTalk	M,W,L,U	WS	Diemand and Scheib (2004)
			i.moltalk.org/
MovieMaker	M,W,L,U	WS	wishart.biology.ualberta.ca/moviemaker
PDB2multiGIF	M,W,L,U	WS	Bohne (1998)
			www.glycosciences.de/modeling/pdb2mgif/
Polyview 3D/2D	M,W,L,U	WS	Porollo and Meller (2007)
			polyview.cchmc.org/
PPG	M,W,L,U	WS	Binisti, Salim, and Tuffery (2005)

Commercial Molecular Visualization Programs

Amira	M,W,L,U	SA	www.tgs.com/products/amira_features.asp
AVS/Express	M,W,L,U	SA	www.avs.com
Bio3D	W	SA	www.cambridgesoft.com
Chem3D	M,W,L	SA	www.cambridgesoft.com
Sculpt	W	SA	www.mdl.com/products/predictive/sculpt/index.jsp
Spartan	M,W,L,U	SA	www.wavefun.com

(*Continued*)

TABLE 9.2. (*Continued*)

Program	Platform[a]	Type[b]	Reference or Web Site[c]
Molecular Visualization as a Component of X-Ray Crystallography and NMR Programs			
ARP/warp	U	SA	Cohen et al. (2004) www.embl-hamburg.de/ARP/
CNS	U,L	SA	Brunger et al. (1998)
CCP4 suite	U	SA	Winn (2003); Collaborative Computational Project, 1994 www.ccp4.ac.uk/
Frodo			now superseded by O
MAID	U	SA	Levitt (2000) www.msi.umn.edu/~levitt/
O	M,W,L,U	SA	Jones et al. (1991) xray.bmc.uu.se/alwyn/
X-PLOR	U	SA	Brunger (1992) atb.csb.yale.edu/xplor/
XtalView	U	SA	McRee (1999) www.sdsc.edu/CCMS/
Molecular Visualization as a Component of Molecular Modeling Programs			
Catalyst	L,U	SA	www.accelrys.com/
Sybyl	L,U	SA	www.tripos.com/
Molecular Visualization as a Component of Bioinformatics Programs			
CLC Bio	M,W,L	SA	www.clcbio.com/
Geneious	M,W,L	SA	www.geneious.com/
VectorNTI	M,W,L	SA	www.invitrogen.com/

[a]M, Macintosh; W, Windows; L, Linux; U, Unix. No distinction is made between versions, or distributions, of the operating system. Versions may also be run cross platform with suitable simulators, for example, Parallels Windows simulator for Macintosh.
[b]SA, "stand-alone"; PI, "plug in" for a web browser or another program; JA, java applet; J3D, java 3D; VR, virtual reality environment; WS, web server.
[c]A Web site URL, reference, or both is given wherever possible. The "http://" part of the URL is omitted for clarity but is implied. Please note that Web site URLs were checked, and operating, before publication, but unfortunately they do tend to change or expire over time.

The rendering and rastorization stage of the visualization pipeline comprise processes that determine the final 3D appearance of the molecule. The aim is to make a realistic image from a geometric model. Many of the effects applied at the rendering stage produce photorealistic images. This is an unfortunate term when applied to viewing molecules since what we are viewing is far from realistic and certainly not photographic. The molecule itself is a model derived from physical data, theory, and prefiltering from X-ray crystallography or NMR spectroscopy. Furthermore, atoms can be depicted as physical objects, and rendering processes can make them look like particular materials (e.g., plastic or metal) with textures, shadows, and colors. These physical attributes obviously do not exist in the real atom. It is all an abstraction and an illusion. However, it is an abstraction that is useful as it can translate into descriptions or predictions of structural and functional properties in the real molecule.

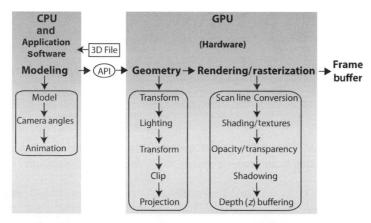

Figure 9.3. Graphics Visualization Pipeline. The graphics visualization pipeline consists of model-ing, geometry, and rendering stages that all contribute to transforming the 3D coordinates of an object into a virtual 3D image on a 2D computer display. API = application programming interface.

Although screen space is an x–y (i.e., 2D) mapping of a 3D object to the screen, the z-values have been carried through all of the operations in the pipeline. There are various ways to use the depth information implicit in the z-values. One of the more common ways is to store the values in memory called a z-buffer. An algorithm then uses the z-values in the buffer to determine the visibility of parts of the object from a particular coordinate view. A near part of the object might obscure, to some extent, a far section of the object, and the z-buffer algorithm would remove the hidden part of the object from view. Consequently, this helps create the illusion of depth (3D) on the 2D screen.

The final stage of the visualization pipeline takes the rendered image and converts it to a bitmap (a raster graphic) for presentation on a raster graphics display. The bitmap is first stored in a frame buffer (a memory of the bitmap) and then transmitted to the display. Vector graphics can be used in drawing and the geometry of an object, but must ultimately be converted to a bitmap for raster display. Any interaction with the image sends new instructions flowing through the visualization pipeline to produce the modified image. This continual "image–interact–image" loop is an active feedback process that helps the user to interpret, and understand, molecular structure, properties, and function.

Graphics Objects

Creating the 2D, and 3D, geometry of a molecule (or other object) involves vector graphics. Vector graphics uses mathematically derived 2D and 3D geometrical shapes called primi-tives. This is in contrast to raster graphics, which is the representation of images as a collection of pixels and is the type of graphic seen on the computer screen. Vector graphics are eventually converted, by the visualization pipeline, to raster graphics for display on the computer screen. The basic geometrical shapes are described as primitives, because they can be used to construct more complex shapes. The type and number of primitives available in any computer graphics system will depend upon the application software and the GPU. The basic 2D primitives include points, lines, circle, curves, and polygons. The basic 3D primitives are regular polyhedra that include sphere, cylinder, tetrahedron, and cube. All 3D geometric shapes can be created either as polygonal structures or as curved surfaces. A polygon consists of vertices and edges that connect the vertices. Figure 9.4 shows that polygons can be

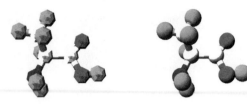

Few polygons per atom sphere   Many polygons per atom sphere
(low resolution)   (higher resolution)

Figure 9.4. Building Atoms with Polygons. Polygons can be combined in a mesh to approximate any shape and are used to build images of atoms, molecules, and macromolecules.

combined in a "mesh" to approximate any shape. A mesh is constructed such that two polygons share an edge. The combination of triangles, in particular, is very useful strategy for forming meshes and building shapes. The greater the number of polygons used to describe a shape, the smoother the shape and the greater the computational power needed.

Curves can also be combined to form a mesh, or curved surface, to model 3D shapes. There are different ways to mathematically determine, and draw, curves with the three most common being: Hermite curves, Bèzier curves, and various splines (Foley et al., 1990). Curved surfaces avoid the artefacts that can occur when approximating a shape with polygons. Curves are useful for generating smooth shapes and have been used in the ribbon model for representing protein secondary structures (see section "Molecular Models" below). The general process of combining polygons, or curved surfaces, to create a 3D object is called tessellation. All molecular visualization applications will have these geometric and rendering capabilities, but will vary in how many of these features they offer.

MOLECULAR MODELS

Visualization of molecules using models began as early as 1808 with the introduction of Dalton's chemical symbols (Figure 9.1) and has progressed to models generated with modern computer graphics. Brief but excellent histories on the development of molecular visualization, and molecular models, can be found elsewhere in articles (Francoeur, 1997; Del Re, 2000; Olson, 2001; Francoeur, 2002; Tate, 2003; Goodsell, 2005), or at Web sites (Table 9.1).

Molecules can be modeled or represented in various ways (Leach, 2001; Goodsell, 2005), and we will discuss some common molecular models in the next section. The models are approximations to reality and are designed to help clarify particular aspects of a molecule's structure, function, or properties (Laskowski, Watson, Thornton, 2003). Each model gives a different perspective and insight into molecular structure. They also serve as a metaphor for the atomic world we cannot see (Goodsell, 2005) and allow us to compare size and structure of molecules. Structural models of proteins are used to represent the primary, secondary, supersecondary, tertiary, or quaternary structural hierarchy of a protein molecule (Chapter 2, Branden and Tooze, 1999; Banaszak, 2000; Lesk, 2001). They are also used to represent functional domains, binding interactions, solvent-accessible areas, solvent-excluded areas, and as a framework for other physical and chemical properties (e.g., electrostatics). No single model can represent everything. Consequently, the choice of model depends on the particular structure, property, or function that needs to be illustrated.

Most molecular models are represented as physical objects because we can render physical objects in various ways that help clarify, and simplify, structure or properties.

However, implicit in this representation are the physics of atomic and molecular structure. While some molecular models are based on quantum mechanics, most structural models are based upon molecular mechanics assumptions (Leach, 2001). That is, the atom is modeled as a hard sphere with the radius of the sphere equal to, or some multiple of, the atom's van der Waals radius. Some molecular models such as the ribbon, backbone, and trace models may not represent atoms as spheres, but molecular mechanics assumptions are usually inherent in the model or in the data used by the model. Covalent bonds are represented (as they are in the usual 2D chemical graphs) as line segments, or 3D cylinders, joining each of the atoms. We will discuss aspects of the more common molecular models later.

Figure 9.5 shows that a molecular representation starts out as a fundamental mathematical, chemical, or physical chemistry model that is converted into an algorithm and written in a particular programming language. This algorithm is then either interpreted or compiled into machine code read by the computer. The efficiency of the application software, and the representation of a particular molecular model, is intimately related to the efficiency of the algorithms used in the software (Harel, 1998).

An algorithm is a precisely defined procedure for accomplishing a particular task. An algorithm can be likened to a recipe for baking a cake. However, in computer graphics the recipe is coded in a programming language, rather than a written document, and is implemented by a particular application program (e.g., molecular visualization program) on a computer system. The programming language can change, but the algorithm always remains the same. The efficiency of an algorithm is measured both by the time and the amount of storage space (memory) needed to perform a particular process (e.g., draw a CPK atom). Memory space is determined by the amount of data and the number of storage locations needed by the algorithm to perform the particular process. Time is determined by

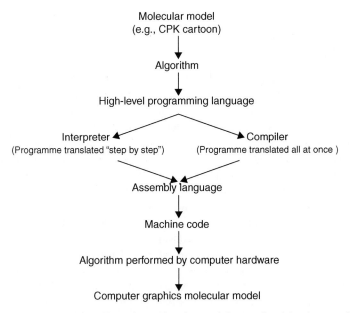

Figure 9.5. Implementing algorithms for molecular models. An algorithm is a precisely defined procedure for accomplishing a particular task. Algorithms are implemented in computer software programs to visualize various molecular models.

the number of elementary steps needed by the algorithm and the speed with which the CPU can perform those steps. The "big-O" notation (for "order of") describes the dependence of an algorithm on both time and space. For example, an algorithm that is described as $O(N)$ performs N elementary instructions, and its running time, or memory requirement, is linear with N. The "big-O" notation can give some idea of the speed and the efficiency of the algorithm, but the actual run time will also vary with the computer architecture and on the size of input data (Harel, 1998). The importance of this to molecular visualization is that some molecular models run faster, or slower, than others depending upon the complexity of their algorithms and their implementation on the graphics hardware. We will now discuss some of the common molecular model algorithms used in molecular visualization.

Wireframe, Stick, CPK, and Ball and Stick

Figure 9.6 shows some of the common molecular models used to represent protein structure. The wireframe model is the simplest molecular model where line segments represent bonds and the atoms are at the vertices of the line segments. It is often used as the default model when a molecule is first opened by, or loaded into, a molecular visualization application. It is faster to render because it needs only the coordinates of the atoms to draw a line segment (one of the graphics primitives) between them. Consequently, it takes much less time and memory to process than a more complex model and leaves the CPU more time to respond to other user commands. The disadvantage of a wireframe model is that it is difficult to discern color and 3D structure especially when viewing larger molecules such as proteins.

The stick model (derived from physical models developed by Dreiding, 1959) can be seen as essentially a thicker wireframe where the bond is represented as a cylinder. This is also called a "licorice" model in some molecular visualization applications (e.g., VMD). The cylinder can be drawn by increasing the diameter of the line segment or by rendering a cylinder using the geometric primitives available with the application or the GPU. Color and other effects can be added to the model. The issues with this model are the same as for the wireframe molecular model.

The CPK model takes its name after its originators and is also known as the space-fill model (it is also a color scheme, see below). It was derived from wood and plastic physical models developed by Corey and Pauling (1953) and later Koltun (1965). Each atom is represented as an opaque sphere with a radius equal to its van der Waals radius. The spheres can be rendered as a 3D primitive, as tessellated polygons or curved surfaces, depending upon the molecular visualization program and the GPU. Color, lighting, transparency, textures, shading, and other effects can also be added to the spheres. Consequently, greater computation is needed to render the model especially in proteins where there are thousands of atoms. Algorithms have been developed to increase the speed of rendering CPK models (Schafmeister, 1990). If the molecule is moved, then the CPK model may be rendered with open meshes (small number of polygons), dots, or the CPK model is removed altogether. Thus, processing time is not unnecessarily consumed by the need to continuously render and redraw the screen with each movement of the molecule. The CPK model implies volume and is also one of the first surface representations of a molecule. It is useful for showing the size and shape of a molecule, for showing when atoms may collide in certain conformations due to steric hindrance, and for showing cavities. However, it can also obscure details and cause visual clutter when used to model larger molecules and proteins (Francoeur, 1997).

The ball and stick model is a hybrid between a stick model and a CPK model. Bonds are represented as cylinders and atoms as spheres or sometimes ellipsoids (Burnett and

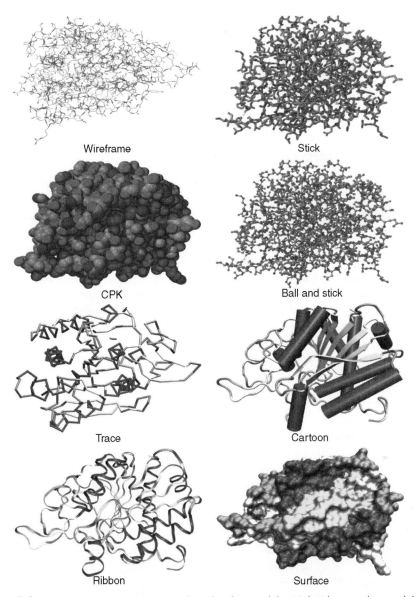

Figure 9.6. Common computer-generated molecular models. Molecules can be modeled or represented in various ways. Each model gives a different perspective and insight into molecular structure and function. Carboxypeptidase A (3CPA.pdb) is used as an example. The CPK, stick, ball and stick were drawn by jAMVLE. The remainder were drawn by VMD. All files rendered by Pov-Ray. Figure also appears in the Color Figure section.

Johnson, 1996). Unfortunately, perhaps, some molecular visualization applications have also called this a CPK model (e.g., VMD). Color, lighting, textures, shading, and other effects can be added to the both bond cylinders and atom spheres. However, the radii of the atom spheres are usually much less than the true van der Waals radii to reduce visual clutter and enable faster rendering. The ball and stick model is useful for information on topology and for measuring bond angles, torsional angles, and bonds lengths. However, Figure 9.6 shows that the ball and stick model still creates visual clutter when used to model proteins.

Wireframe, stick, CPK, and ball and stick molecular models may be considered "all-atom" models in that each atom in the protein is explicitly included in the model. The following molecular models involve algorithms that use a particular subset of atoms. They were designed to simplify and clarify protein structures and show elements of structure obscured by all-atom representations.

Backbone, Trace, Cartoons, and Ribbons

The backbone model (possibly based on physical models of Rubin and Richardson (1972) and Keefe and Howe (1973)) was introduced for proteins and is conceptually similar to the stick model for smaller molecules. The backbone of a polypeptide is minimally constituted of the carbonyl carbon, the alpha carbon, and the amide nitrogen of each residue (Chapter 2). However, the backbone model's algorithm uses line segments to join each consecutive alpha carbon in the polypeptide. The advantages of this model are that it simplifies the structure; it avoids visual clutter; it is faster to render; it can reveal elements of secondary structure such as helices, strands, and turns; it can show the conformation and folding of the polypeptide chain; it is useful in tracing electron density maps; and can be used to determine protein dynamics (Higo and Umeyama, 1997). A disadvantage is that it is visually less impressive than cartoons or ribbons.

The trace model is similar to the backbone model except that spline curves are used to generate the trace through the alpha carbon atoms. The mathematics of the curve may mean that the curve does not exactly intersect the coordinates of each alpha carbon, but this makes the model look smoother. The backbone and trace models are also useful for protein structure comparison techniques (used in threading and homology modeling), where usually only the coordinates of alpha carbons are considered (Johnson and Lehtonen, 2000).

The cartoon models of proteins were derived from hand drawn models of protein secondary structures by Richardson (1981, 1985a and 1985b). The computer graphics for cartoon models were further developed by Kraulis (1991) in the program Molscript. The advantages of cartoon models are as for the backbone model above. In particular, it simplifies protein structure to reveal the direction of the protein chain, protein folding, and elements of secondary, supersecondary, and domain structure. Secondary structure determination is important in categorizing proteins in both the CATH and SCOP databases (Chapters 17 and 18). The cartoons can be rendered as solid graphic objects (e.g., helices as cylinders and β-strands as broad arrows) and have various lighting, shading, and other effects applied.

Ribbon models were derived from Richardson's hand drawn models (Richardson, 1981; Richardson, 1985b) and later developed by Carson (Carson and Bugg, 1986; Carson, 1987). They could also be labeled a "cartoon," but ribbons are usually listed as a separate rendering option in molecular visualization programs. The justification is based on the mathematical model, and algorithm, used to graphically create the ribbon that is based on a B-spline curve. The peptide planes of the peptide bond are used as the coordinates for the spline. Initially, the ribbons were generated as separate spline line segments winding their way down the polypeptide chain (Carson and Bugg, 1986) but later the splines were used to form curve patches that could be rendered as solid ribbons (Carson, 1987). Later development of the ribbons algorithm allowed ribbons to take full advantage of raster computer graphics capabilities including color, transparency, shading, and lighting that produce ribbons of varying styles (Carson, 1994).

Surface Models

That part of a protein exposed to the surrounding environment of solvent, and other molecules, is generally described as its solvent-accessible surface. It is this surface and its constituent atoms that are important to the protein's function (Via et al., 2000). The volume or bulk of a protein is also implicit with some surface models, and this helps to define the shape and size of the molecule as well as the extent of a particular molecular property (see below). One of the first models to show explicit surface boundaries and volume was the CPK model discussed above. However, this is an all-atom model, and it does not clearly discriminate between those atoms that are on the surface available to interact with the surrounding solvent and those that are buried. Algorithms are used to smooth the "bumpy" all-atom representation, reduce the clutter inherent in the CPK model, and produce an overall shape for the protein that represents its surface topography.

Various methods have been developed to describe a protein's surface (Blinn, 1982; Connolly, ; Max, 1984; Goodford, 1985; Richards, 1985; Goodsell, Mian, and Olson, 1989; Goodsell and Olson, 1990; Agishtein, 1992; Laskowski, 1995; Sanner et al., 1996;Cai, Zhan, and Maigret, 1998). Michael Connolly has written an excellent review on the development of surface models on the NetScience Web site (Table 7.1). We will only briefly discuss some of the main concepts and algorithms.

Lee and Richards (1971) described a protein's solvent-accessible surface by using the "rolling ball" algorithm. The algorithm uses a sphere of a particular radius to roll over the surface atoms of the protein. The radius of the sphere can be altered, but it is usually set to the 1.5 Å radius of a water molecule. Each atom is represented by its van der Waals radius, and the center of the sphere traces out the solvent-accessible surface. Connolly (1983a 1983b) further developed this algorithm to run on computer graphics systems and describe the related concept of the solvent-excluded surface (otherwise known as the molecular surface). The solvent-excluded surface is where the atom's van der Waals radius comes into contact with the probe and forms the boundary for the excluded volume. An alternative approach (called a "volume-based" approach) to determine the accessible surface of a protein uses a probe to scan the complete 3D internal, and external, space of the protein, rather than just rolling it over the surface (Goodford, 1985; Goodsell and Olson, 1990).

Surfaces are complex graphical objects and can be rendered as dots, meshes, or solids (where the mesh and solid can be constituted of tessellated polygons or curves). Color, lighting, textures (Teschner et al., 1994), shading, and other effects can be added to the surface. Surfaces take longer to render, and research is directed to develop algorithms to reduce the rendering time (Sanner, Olson, and Spehner, 1996; Bajaj et al., 2004).

Properties of a molecule such as electrostatic potential, hydrophobicity, and some noncovalent interactions can be modeled as a surface plane or as a surface that defines a volume in or around the molecule. The surface plane approach maps the property to a curved plane with proximity to the residue or atom comprising that property. The plane can be rendered with different colors or textures. The surface volume approach maps the property to an existing surface (such as a Connolly surface), either as a contiguous "density cloud" (also known as volume rendering), or as 3D isovalue contour (Goodsell and Olson, 1989; Henn et al., 1996; Olson, 2001; Holtje et al., 2003; Goodsell, 2005; see also O'Donnell Web site listed in Table 7.1). Some field-like properties, such as electrostatics, vary in size and direction both in and around the molecule. Consequently, it usually is more informative to display properties as density clouds, or 3D isovalue contours, with varying volumes in and around the molecule. The extra information provided by molecular properties can be visualized by varying the opacity, texture, or color of the rendered volume or surface.

Colors, Covalent and Noncovalent Interactions, and Hydrogen Atoms

The CPK color scheme was based upon the colors of the space-filling models developed by Corey and Pauling (1953). In this scheme, carbon is light gray, nitrogen is light blue, oxygen is red, hydrogen is white, and sulfur is yellow. Many other color schemes can be used to highlight other features, such as atom types, a structural part of the molecule, or molecular properties (e.g., charge and polarity). Most molecular visualization programs enable the user to define colors.

Single covalent bonds are usually represented as line segments or cylinders in models and can either be calculated by the molecular visualization program or generated directly from the "CONECT" records of the PDB file. The molecular visualization program calculates the van der Waals radii of two adjacent atoms and if the distance is less than the sum of the van der Waals radii of the two atoms, then a bond is drawn between them. Double bonds are not usually represented in the PDB file, so the molecular visualization program also has to calculate these using the distances between atoms where the length of a single bond is greater than a double bond that in turn is greater than a triple bond.

Noncovalent (nonbonded) interactions occur within protein structures, between proteins, and between proteins and ligands and include hydrogen bonds, hydrophobic interactions, ionic interactions, aromatic interactions, and metal complexation (Bohm and Schneider, 2003). Hydrogen bonds (Pimentel and McClelan, 1960) are usually modeled, because they are a common interaction and have important roles in protein structure and function (Jeffrey, 1991). The geometry of the hydrogen bond is also well known, and as such, it is relatively easy to design algorithms to detect and display them. Hydrogen bonds are calculated using bond angles, bond distances, and sometimes with reference to atom types. If two atoms are within particular bond distances and bond angles, then a dotted line is usually drawn to signify the bond. Molecular visualization applications may have different ways of determining and drawing the hydrogen bond. The hydrogen atom locations (see below) may also have to be calculated before hydrogen bond placement.

Hydrophobic interactions are also important in protein–protein interactions (Mueller and Feigon, 2002), protein folding, and protein stability (Hendsch and Tidor, 1994; Wimley et al., 1996). Various criteria can be used to determine the hydrophobic and aromatic (Kolomeisky and Widom, 1999; Sacile and Ruggiero, 2002) propensities of proteins, and these can be mapped as an isovalue surface or mapped on to an existing surface (see Surfaces above).

Aromatic interactions, such as pi–pi (McGaughey, Gagne, and Rappe, 1998) and pi–cation (Ma and Dougherty, 1997), have important roles in enzyme–substrate interactions, ion channels, and membrane receptors (Kumpf and Dougherty, 1993; Dougherty, 1996; Ma and Dougherty, 1997; Gallivan and Dougherty, 1999; Pletneva et al., 2001). However, the geometry and conditions for these interactions is not always straightforward, and it is difficult to design appropriate algorithms to detect them in protein structures. Consequently, they may not always be available as models in molecular visualization applications.

Many protein structures solved by X-ray crystallography do not have implicit hydrogens because the resolution of the technique does not allow for definitive identification of the small hydrogen atom's location. Consequently, the molecular visualization program has to calculate the position of the hydrogen atom in the structure. The calculation is usually based on the atom type. The atom type is a classification of atoms based upon their chemical environments and bonding propensities. For example, carbon can be classified as sp^3 hybridized, sp^2 hybridized, sp hybridized, aromatic, or carboxylate carbon. Each atom type will have a defined geometry and will bond to a different number of hydrogen atoms.

Animations

Animation can be a useful and powerful molecular visualization tool (Max, 1990; McClean et al., 2005; Breithaupt, 2006; Hodis et al., 2007). The addition of movement adds another dimension to help clarify and communicate information about molecular structure and function. Molecular animations are not purely artistic representations but are based upon actual 3D coordinates and molecular dynamics. The Database of Macromolecular Movements (Echols, Milburn, and Gerstein, 2003), which is a collection of data on the flexibility of protein and RNA structures, is an example of the utility of animation. RasMol, one of the earliest molecular visualization programs, can produce simple animations using scripts or animated gifs (Bohne, 1998). MovieMol (Table 7.2), MovieMaker (Maiti, Van Domselaar, and Wishart, 2005) and eMovie (Hodis et al., 2007) are some of the programs specifically designed to produce molecular animations. Alternatively, generic animation programs such as, Macromedia's Flash, Apple's QuickTime, or Autodesk's Maya, can be used to produce animations based on actual 3D coordinate data. We will not go into the details of making animations, but Table 9.1 lists a few of the many Web sites that feature animations for both research and education.

MOLECULAR VISUALIZATION PROGRAMS

Molecular visualization programs enable molecules to be viewed as different models or molecular representations that reflect underlying physical, chemical, or structural principles. The user can also manipulate, and interact with, the molecular image to control both its content and appearance. Molecular visualization applications vary considerably in the number, and range, of features available for interacting with, and viewing, the molecule. However, most programs have the ability to import, export, and save molecules in PDB and other molecular file formats; display the molecule in standard molecular model types (see above); move the molecule by rotation and translation; scale (zoom in or zoom out); select or "pick" parts of the molecule; and color the molecule according to various standard or nonstandard color schemes. Additional features may include other types of molecular model types, molecular surfaces, rendering effects (such as lighting, shading, depth cues, and perspective), stereo view, structural alignment, and molecular modeling tools (e.g., the ability to mutate residues and build molecules). Most programs process the input files by using internal algorithms to analyze the structure and check such things as atom distances, bonds, and connectivity. Any problems would either be automatically fixed or reported to the user.

Most molecular visualization applications have a graphical user interface and the user can directly interact with the program through pull-down, pop-up, or contextual menus, and mouse actions. Other applications (e.g., Molscript) will have a command line interface that accepts specific commands, or a command language, to control the function of the program and the molecule display. Command line interfaces allow the user some flexibility in manipulating and displaying the molecule that would otherwise not be available as a menu item. Some applications will be able to accept a series of commands (called scripts or macros) stored in a text file that can be loaded by the program to automatically execute the commands. These command line interfaces, with scripting, allow the user to automatically perform complex or repetitive operations. Finally, some applications have the best of all worlds and have a combined GUI and command line interface (e.g., RasMol, jAMVLE).

Most molecular visualization programs will display a range of small molecules, proteins, DNA, and RNA. However, there are also programs (see Table 9.2) that specialize in visualizing particular structures such as 3D immune epitopes (Epitope Viewer), large macromolecular assemblies (Chimera or TexMol), DNA (3DNA), and carbohydrates (Carbohydra or Sweet II).

Molecular Visualization versus Molecular Modeling

What is the difference between a molecular visualization program and a molecular modeling program? A molecular modeling program has the ability to construct, simulate, and predict molecular structures, properties, or functions. They can measure and manipulate the geometry of the molecule's structure, calculate various molecular properties according to particular force fields, test various molecular conformations by performing energy minimizations, and compare structures (e.g., protein–protein, protein–ligand, or protein–drug). We have already seen that various molecular models are used by molecular visualization applications. Some molecular visualization programs also have the ability to construct molecules and perform energy minimizations. Thus, we could say that some molecular visualization programs also support molecular modeling. However, the emphasis of a molecular visualization program is the visual representation of molecular structure and properties. This could also be envisaged as the graphical "front end" of a molecular modeling program where the "back end" is the underlying computational ability of the molecular modeling program.

Types of Molecular Visualization Programs

Table 9.2 shows that there are many software programs available for 3D molecular visualization that operate on a variety of computer platforms. These are used as stand-alone programs (e.g., Deep View, Mage, RasMol, and VMD), integrated within a web browser (e.g., Chime and Protein Explorer), as java applets integrated within the web browser (e.g., Jmol, WebMol, and QuickPDB), as stand-alone Java-based programs (e.g., Jmol, jAMVLE, KiNG, and Cn3D), as interactive server-based programs (e.g., AISMIG and iMolTalk), as an integral component in a suite of other bioinformatics (e.g., Vector NTI, CICbio, and Geneious) and molecular modeling (e.g., Catalyst and Sybil) programs, or as a distributed system (Zhu et al., 2004).

Molecular visualization programs can export molecule files in various formats that can then be used by other programs or online servers. For example, more fully featured rendering programs, such as Pov-Ray (Table 9.1) and Render3D (Merritt, 1994), are often used to produce publication quality molecular images from molecular files. Molecules can also be visualized by online generation of pictures or animations (Binisit et al., 2005), by generic programs such as QuickTime, or by viewers for web-based markup languages such as virtual reality modeling (or markup) language (VRML) and X3D (Liu and Sourin, 2006;

Willis, 2006). Molecular visualization programs can be implemented in virtual reality environments (Luo et al., 2004; Ferreira, Sharma, and Mavroidis, 2005; Gillet et al., 2005), collaborative virtual environments (Bourne et al., 1998; Tate, Moreland, and Bourne, 2001; Marchese, Mercado, and Pan, 2003; Chastine et al., 2005), and with haptic technology (Sankaranarayanan et al., 2003; Wollacott and Merz, 2007). In addition, software development toolkits such as the MBT toolkit (Moreland et al., 2005), and component-based software (Sanner, 2005), offer a flexible and modular approach to quickly develop molecular visualization software. These can be used to cater for new visualization methods, new data types (such as large macromolecular assemblies), and to offer the software through the Internet.

There are advantages and disadvantages for each type of molecular visualization program. The stand-alone applications take full advantage of the computer system's hardware and software configuration, and this leads to more options available for visualization. This includes a greater range of molecular models, various interactive tools for editing and modifying molecules, and powerful rendering capabilities. However, the stand-alone applications may not be available for all computer platforms (i.e., limited "portability"); need continuous development to keep up with changes in operating software, computer hardware, data types, and molecular visualization concepts; can be awkward to use in conjunction with web-based educational materials (Bottomley et al., 2006); have limited ability to be customized; and are usually offered "as is" to the user.

Web browser integration programs (e.g., Chime), and some Java applets, offer good integration with web-based educational materials. However, they can have compatibility issues with different web browsers or computer platforms. They also do not take advantage of the operating system software or hardware and, consequently, may offer relatively limited options for visualization or interactivity.

Java applications offer independence from either the web browser or the operating system. Java is an interpreted language, and its programs are compiled into a "virtual computer" called the Java Virtual Machine. The Java Virtual Machine, and Java interpreter, needs to be installed on the computer platform to run Java code. Java's independence from the operating system means that software can be developed once and used with minimal problems on most computer platforms (i.e., Java offers "portability"). It also provides some protection from system crashes, because it operates through the Java Virtual Machine that essentially acts as an interface to the operating system. Furthermore, Java applications can be flexible, modularized, and offered as open-source software (e.g., Jmol). This allows them to be updated, and modified, by the scientific community to cater for new developments in molecular visualization. The lead time for implementing new developments in open-source applications depends upon the enthusiasm of the scientists working on the software, but it can be relatively fast compared to updates to stand-alone applications. Jmol, in particular, has a growing and enthusiastic community of developers (e.g., www.jmol.org) that bodes well for its future use in molecular visualization. Jmol does not need a GPU, and this can be an advantage for those computer systems that do not have a 3D graphics card. However, this can also prevent it from taking full advantage of the operating system software and graphics accelerating hardware.

General VRML, or X3D, viewers may have compatibility issues with computer platform or browsers, offer limited interactivity, and may not always use the computer systems GPUs. A recent development is ProteinX3D (Willis, 2006; Willis, 2007), which is a Java program that uses XML data from the PDB database to visualize proteins.

Applications of Molecular Visualization

Molecular visualization is a powerful concept and tool to simplify, clarify, model, analyze, illustrate, and communicate molecular structure, properties, and function. A molecular visualization arises not only as a result of a dedicated interactive display program (e.g., RasMol), but also as a result of computational analyses generated by many other programs (e.g., from molecular modeling or bioinformatics programs). Consequently, molecular visualization can be applied in myriad situations (too many to list here), and some examples are art (Gaber and Goodsell, 2000; Goodsell, 1991; also see Table 9.1 for the Web sites of the Molecular Graphics Art Show, "Artn," and Biografx), education (Gilbert, 2005; Bottomley et al., 2006), sequence or structure alignment (Chapter 16; Catherinot and Labesse, 2004), protein docking (Chapter 27), structure superposition (Chapter 16; Sumathi et al., 2006), structure-based drug design and development (Meyer, Swanson, and Williams, 2000; Chapter 34), viewing binding cavities (Smith et al., 2006), viewing multiple protein–ligand complexes (O'Brien et al., 2005), viewing protein interfaces (Teyra et al., 2006), mapping motifs to structure (Bennett, Lu and Brutlag, 2003), homology and homology modeling (Chapter 30), protein classification (Chapters 17 and 18), viewing ligand binding modes (Chapter 27; Ivanov, Palyulin, and Zefirov, 2007), structure prediction (Chapters 28, 29, 30, 31, 32), determining drug efflux mechanisms (Kiralj and Ferreira, 2006), and even used as a "toy" (Laszlo, 2000).

Issues with Molecular Visualization

Most molecular visualization applications are robust. This means that they are usually reliable, do the job required of them, and are not prone to crashing your computer. Consequently, users tend to trust the software and the result produced. However, there are philosophical (Hendry, 1999; Francoeur, 2000; Zeidler, 2000) and practical issues with molecular visualization. In particular, there is a danger in accepting any biological software, including molecular visualization applications, at face value. Molecular visualization applications can contain errors and may interpret data in different ways. For example, RasMol often incorrectly lists the number of polypeptide chains in its summary window, because it interprets nonpolypeptide hetero groups as a chain.

There are also issues in the design of the software. The user interface varies considerably between different molecular visualization applications (see above). The interface can sometimes be rudimentary (e.g., a command line), complex (e.g., contain various options and parameters), and may contain language (e.g., computer graphics terminology) not immediately understandable by the biologist. This can cause confusion especially if the user lacks the relevant knowledge. Consequently, users may only choose the "default" parameters supplied by the software. They may not know how to adjust parameters to refine their analysis, nor the effect that such changes may have on their results (Bottomley, 2004; Goodsell, 2006). This could lead to inappropriate application of the software and questionable relevance and validity of the results. Default parameters also differ between molecular visualization programs, and choosing the default space-filling representation in one program (e.g., jAMVLE) can result in a completely different representation in another program (e.g., CrystalMaker).

As Goodsell (2006) has correctly pointed out, there is the danger of confusing the "image" with science. Optical illusions demonstrate that our visual acuity is so persistent that we sometimes "see" things in images that are not really there. This can be an asset when we want to create the illusion of volume in an atom on a computer screen, but it can be a liability when trying to make valid interpretations of images generated from scientific data. Moreover, molecular visualization software is built from assumptions inherent in its algorithms, mathematics, engineering, and programming. Consequently, software is not always as objective as we may like to believe and the results are necessarily an approximation to reality. Visualization is both objective and subjective. It is objective in that it uses raw data, but it is subjective depending upon how we transform, and interpret, the raw data. There will always be a compromise between the two. That is, we can try to be as objective as possible in generating our data, but as soon as we start transforming the data, we necessarily presuppose some model or theory. Our interpretation of that transformation will be subjective based on the underlying model or theory. The user can also choose a particular molecular model, or rendering technique, based only on personal preference. This will bias the image in some way and may be chosen to satisfy a particular audience (Goodsell, 2006). For example, an austere, precisely defined, molecular representation would be appropriate for a knowledgeable scientific audience, but a more colorful, and artistic, representation would be appropriate for a nontechnical audience. Thus, it is possible to use molecular visualization to inadvertently, or even deliberately, mislead the intended audience.

There is also a temptation to assume that we do not have to know how the software works as long as it gives the results we are seeking. However, we should carefully consider many factors that may affect the efficacy, the robustness, and the biological relevance of the software as failure to do so may result in misuse (Bottomley, 2004). The image on the computer display is the final stage of the molecular graphics visualization pipeline, but it is the penultimate stage in interpretation. Visual images are powerful, but it still needs an experienced, knowledgeable, biologist to interpret the result and give it biological relevance and validity.

REFERENCES

Agishtein ME (1992): Fuzzy molecular surfaces. *J Biomol Struct Dyn* 9:759–768.

Angel ES (2002): *Interactive Computer Graphics with OpenGL*. 3rd ed. Addison-Wesley.

Artemov D (2003): Molecular magnetic resonance imaging with targeted contrast agents. *J Cell Biochem* 90:518–524.

Bajaj C, Djeu P, Siddavanahalli V, Thane A (2004): *TexMol: Interactive Visual Exploration of Large Flexible Multi-component Molecular Complexes*. Proceeding of the Annual IEEE Visualization, Austin, TX, pp. 243–250.

Banaszak LJ (2000): *Foundations of Structural Biology*. Academic Press.

Beaver JE, Bourne PE, Ponomarenko JV (2007): EpitopeViewer: a Java application for the visualization and analysis of immune epitopes in the Immune Epitope Database and Analysis Resource (IEDB). *Immunome Res* 3:3.

Bennett SP, Lu L, Brutlag DL (2003): 3MATRIX and 3MOTIF: a protein structure visualization system for conserved sequence motifs. *Nucleic Acids Res* 31:3328–3332.

Binisti C, Salim AA, Tuffery P (2005): PPG: online generation of protein pictures and animations. *Nucleic Acids Res* 33:W320–W323.

Blasberg RG, Gelovani-Tjuvajev J (2002): In vivo molecular-genetic imaging. *J Cell Biochem* Suppl 39:172–183.

Blinn JF (1982): A generalization of algebraic surface drawing. *ACM Trans Graph* 1:235–256.

Bohm HJ, Schneider G (2003): *Protein–Ligand Interactions*. Weinheim: Wiley-VCH.

Bohne A (1998): PDB2multiGIF: a web tool to create animated images of molecules. *J Mol Model* 4:344–346.

Bohne A, Lang E, von der Lieth CW (1999): SWEET—WWW-based rapid 3D construction of oligo- and polysaccharides. *Bioinformatics* 15:767–768.

Bohne-Lang A, Groch WD, Ranzinger R (2005). AISMIG—an interactive server-side molecule image generator. *Nucleic Acids Res* 33:W705–W709.

Bond CS (2003): TopDraw: a sketchpad for protein structure topology cartoons. *Bioinformatics* 19:311–312.

Bottomley S (2004): Bioinformatics: smartest software is still just a tool. *Nature* 429: 241.

Bottomley S, Chandler D, Morgan E, Helmerhorst E (2006): jAMVLE, a new integrated molecular visualization learning environment. *Biochem Mol Biol Educ* 34:343–349.

Bourne PE, Gribskov M, Johnson G, Moreland J, Weissig H (1998): A prototype molecular interactive collaborative environment (MICE). In: Altman R, Dunker K, Hunter L, and Klein T, editors. *Pac Symp Biocomput:* 118–129.

Branden C, Tooze J (1999): *Introduction to Protein Structure*. 2nd ed. Garland Publishing.

Breithaupt H (2006): Seeing is understanding. Improvements in computer software and hardware are revolutionizing three-dimensional imaging in biology. *EMBO Rep* 7:467–470.

Brunger AT (1992): *X-PLOR, Version 3.1. A System for X-Ray Crystallography and NMR.* New Haven, CT: Yale University Press.

Brunger AT, Adams PD, Clore GM, DeLano WL, Gros P, Grosse-Kunstleve RW, Jiang JS, Kuszewski J, Nilges M, Pannu NS, et al. (1998): Crystallography & NMR system: a new software suite for macromolecular structure determination. *Acta Crystallogr D* 54:905–921.

Burnett MN, Johnson CK (1996): *ORTEP-III: Oak Ridge Thermal Ellipsoid Plot Program for Crystal Structure Illustrations*.

Cai W, Zhan M, Maigret B (1998): New approach for representation of molecular surface. *J Comput Chem* 19:1805–1815.

Can T, Wang Y, Wang YF, Su J (2003): FPV: fast protein visualization using Java 3D. *Bioinformatics* 19:913–922.

Carson M (1987): Ribbon models of macromolecules. *J Mol Graph* 5:103–106.

Carson M (1994): Ribbons 2.0. *J Appl Crystallogr* 24:958–961.

Carson M (1997): Ribbons. *Methods Enzymol* 277:493–505.

Carson M, Bugg CE (1986): Algorithm for ribbon models of proteins. *J Mol Graph* 4: 121–122.

Catherinot V, Labesse G (2004): ViTO: tool for refinement of protein sequence–structure alignments. *Bioinformatics* 20:3694–3696.

Chastine JW, Zhu Y, Brooks JC, Owen GS, Harrison RW, Weber IT (2005): *A Collaborative Multi-View Virtual Environment for Molecular Visualization and Modeling*. Proceedings of the Third International Conference on Coordinated & Multiple Views in Exploratory Visualization (CMV'05). London, UK: IEEE Computer Society.

Cohen SX, Morris RJ, Fernandez FJ, Jelloul MB, Kakaris M, Parthasarathy V, Lamzin VS, Kleywegt GJ, Perrakis A (2004): Towards complete validated models in the next generation of ARP/wARP. *Acta Crystallogr D* 60:2222–2229.

Collaborative Computational Project (1994): The CCP4 suite: programs for protein crystallography. *Acta Crystallogr D* 50:760–763.

Connolly ML (1983a): Solvent-accessible surfaces of proteins and nucleic acids. *Science* 221:709–713.

Connolly ML (1983b): Analytical molecular surface calculation. *J Appl Crystallogr* 16:548–558.

Corey RB, Pauling L (1953): Molecular models of amino acids, peptides, and proteins. *Rev Sci Instrum* 24:621–627.

Cornish PV, Ha T (2007): A survey of single-molecule techniques in chemical biology. *ACS Chem Biol* 2:53–61.

Crivelli S, Kreylos O, Hamann B, Max N, Bethel W (2004): ProteinShop: a tool for interactive protein manipulation and steering. *J Comput Aided Mol Des* 18:271–285.

Cruz-Neira C, Langley R, Bash P (1996): Vibe: a virtual biomolecular environment for interactive molecular modeling. *Comput Chem* 20:469–475.

Dai Q, Liu X, Wang T (2006): A novel 2D graphical representation of DNA sequences and its application. *J Mol Graph Model* 25:340–344.

de Jong H (2003): Visualizing DNA domains and sequences by microscopy: a fifty-year history of molecular cytogenetics. *Genome* 46:943–946.

Del Re, G. (2000): Models and analogies in science. *HYLE Int J Philos Chem* 6:5–15.

Dickerson RE, Geis I (1969): *The Structure and Action of Proteins.* Benjamin/Cummings.

Diemand AV, Scheib H (2004): iMolTalk: an interactive, internet-based protein structure analysis server. *Nucleic Acids Res* 32:W512–W516.

Dougherty DA (1996): Cation–pi interactions in chemistry and biology: a new view of benzene, Phe, Tyr, and Trp. *Science* 271:163–168.

Dreiding AS (1959): Einfache Molekularmodelle. *Helv Chim Acta* 17:1339–1344.

Echols N, Milburn D, Gerstein M (2003): MolMovDB: analysis and visualization of conformational change and structural flexibility. *Nucleic Acids Res* 31:478–482.

Ferreira A, Sharma G, Mavroidis C (2005): New trends in bio-nanorobotics using virtual reality technologies. Proceedings of the 2005 IEEE International Conference on Robotics and Biomimetics, Hong Kong.

Foley JD, van Dam A, Feiner SK, Hughes JF (1990): *Computer Graphics: Principles and Practice.* 2nd ed. Addison-Wesley.

Francoeur E (1997): The forgotten tool: the design and use of molecular models. *Soc Stud Sci* 27:7–40.

Francoeur E (2000): Beyond dematerialization and inscription. Does the materiality of molecular models really matter? *HYLE Int J Philos Chem* 6:63–84.

Francoeur E (2002): Cyrus Levinthal, the Kluge and the origins of interactive molecular graphics. *Endeavour* 26:127–131.

Gaber BP, Goodsell DS (2000): The art of molecular graphics. Escape from flatland. *J Mol Graph Model* 18:71–72.

Gallivan JP, Dougherty DA (1999): Cation–pi interactions in structural biology. *Proc Natl Acad Sci USA* 96:9459–9464.

Gans JD, Shalloway D (2001): Qmol: a program for molecular visualization on Windows-based PCs. *J Mol Graph Model* 19:557–559, 609.

Garratt RC, Pereira HM (2006): *A Guide to the use of 3D Cartoon Models for Teaching Protein Structure in the Post-Genomic Era.* In Silico Analysis of Proteins. 20th Anniversary of Swiss-Prot, Fortaleza, Brazil p. RP214.

Gilbert JK (2005): *Visualization in Science Education.* Springer.

Gillet A, Sanner M, Stoffler D, Olson A (2005): Tangible interfaces for structural molecular biology. *Structure* 13:483–491.

Goodford PJ (1985): A computational procedure for determining energetically favorable binding sites on biologically important macromolecules. *J Med Chem* 28:849–857.

Goodsell DS (1991): Inside a living cell. *Trends Biochem Sci* 16:203–206.

Goodsell DS (1992): A look inside the living cell. *Am Sci* 80:457–465.

Goodsell DS (2000): Illustrating the machinery of life. *J Biocommun* 27:12–18.

Goodsell DS (2005): Visual methods from atoms to cells. *Structure* 13:347–354.

Goodsell DS (2006): Seeing the Nanoscale. *Nanotoday* 1:44–49.

Goodsell DS, Mian IS, Olson AJ (1989): Rendering volumetric data in molecular systems. *J Mol Graph* 7:41–47.

Goodsell DS, Olson AJ (1989): Molecular applications of volume rendering and 3-D texture maps. Proceedings of the 1989 Chapel Hill workshop on Volume visualization, Chapel Hill, NC, pp. 27–31.

Goodsell DS, Olson AJ (1990): Automated docking of substrates to proteins by simulated annealing. *Proteins* 8:195–202.

Guex N, Peitsch MC (1997): SWISS-MODEL and the Swiss-PdbViewer: an environment for comparative protein modeling. *Electrophoresis* 18:2714–2723.

Harel D (1998): *Algorithmics: The Spirit of Computing*. 2nd ed. Addison-Wesley.

Hendry RF (1999): Molecular models and the question of physicalism. *HYLE Int J Philos Chem* 5:143–160.

Hendsch AS, Tidor B (1994): Do salt bridges stabilise proteins? A continuum electrostatic analysis. *Protein Sci* 3:211–226.

Henkel JG, Clarke FH (1985): *Molecular Graphics on the IBM PC Microcomputer*. Orlando, FL: Academic Press.

Henn C, Teschner M, Engel A, Aebi U (1996): Real-time isocontouring and texture mapping meet new challenges in interactive molecular graphics applications. *J Struct Biol* 116:86–92.

Higo J, Umeyama H (1997): Protein dynamics determined by backbone conformation and atom packing. *Protein Eng* 10:373–380.

Hodis E, Schreiber G, Rother K, Sussman JL (2007): eMovie: a storyboard-based tool for making molecular movies. *Trends Biochem Sci* 32:199–204.

Holtje H-D, Sippl W, Rognan D, Folkers G (2003): *Molecular Modeling: Basic Principles and Applications*. 2nd ed. Wiley-VCH.

Hu Z, Mellor J, Wu J, DeLisi C (2004): VisANT: an online visualization and analysis tool for biological interaction data. *BMC Bioinformatics* 5:17.

Humphrey W, Dalke A, Schulten K (1996): VMD: visual molecular dynamics. *J Mol Graph* 14:33–38.

Ivanov AA, Palyulin VA, Zefirov NS (2007): Computer aided comparative analysis of the binding modes of the adenosine receptor agonists for all known subtypes of adenosine receptors. *J Mol Graph Model* 25:740–754.

Jeffrey GA, Saenger W (1991): *Hydrogen Bonding in Biological Structures*. Berlin: Springer-Verlag.

Johnson MS, Lehtonen JV (2000): Comparison of protein three-dimensional structures. In: Higgins D, Taylor W, editors. *Bioinformatics: Sequence, Structure, and Databanks: A Practical Approach*. Oxford University Press, pp. 15–50.

Jones TA, Zou JY, Cowan SW, Kjeldgaard M (1991): Improved methods for building protein models in electron density maps and the location of errors in these models. *Acta Crystallogr A* 47:110–119.

Keefe WE, Howe D-B (1973): A Method of Construction of Protein Backbone Models. *Rev Sci Instrum* 44:1414–1415.

Kendrew JC (1961): The three-dimensional structure of a protein molecule. *Sci Am* 205:96–110.

Kerlow IV (2004): *The Art of 3D Computer Animation and Effects*. 3rd ed. John Wiley & Sons.

Kiralj R, Ferreira MM (2006): Molecular graphics approach to bacterial AcrB protein-beta-lactam antibiotic molecular recognition in drug efflux mechanism. *J Mol Graph Model* 25:126–145.

Kolomeisky AB, Widom B (1999): Model of the hydrophobic interaction. *Faraday Discuss* 112:81–89.

Koltun WL (1965): Precision space-filling atomic models. *Biopolymers* 3:665–679.

Koradi R, Billeter M, Wuthrich K (1996): MOLMOL: a program for display and analysis of macromolecular structures. *J Mol Graph* 14:51–55.

Kraulis P (1991): MOLSCRIPT: a program to produce both detailed and schematic plots of protein structures. *J Appl Crystallogr* 24:946–950.

Kumpf RA, Dougherty DA (1993): A mechanism for ion selectivity in potassium channels: computational studies of cation–pi interactions. *Science* 261:1708–1710.

Kuttel M, Gain J, Burger A, Eborn I (2006): Techniques for visualization of carbohydrate molecules. *J Mol Graph Model* 25:380–388.

Lamzin VS, Perrakis A (2000): Current state of automated crystallographic data analysis. *Nat Struct Biol*. 7 (Suppl):978–981.

Laskowski RA (1995): SURFNET: a program for visualizing molecular surfaces, cavities, and intermolecular interactions. *J Mol Graph* 13:323–330.

Laskowski RA, Watson JD, Thornton JM (2003): From protein structure to biochemical function? *J Struct Funct Genomics* 4:167–177.

Laszlo P (2000): Playing with molecular models. *HYLE Int J Philos Chem* 6:85–97.

Leach AR (2001): *Molecular Modelling: Principles and Applications*. 2nd ed. Pearson Education Limited.

Lee B, Richards FM (1971): The interpretation of protein structures: estimation of static accessibility. *J Mol Biol* 55:379–400.

Lesk AM (2001): *Introduction to Protein Architecture*. Oxford University Press.

Levitt DG (2001): A new software routine that automates the fitting of protein X-ray crystallographic electron density maps. *Acta Crystallogr D* 57:1013–1019.

Liu Q, Sourin A (2006): *Function-Based Shape Modeling And Visualization In X3d*. Proceedings of the eleventh international conference on 3D web technology. Columbia, MD: ACM Press. pp. 131–141.

Lu XJ, Olson WK (2003): 3DNA: a software package for the analysis, rebuilding and visualization of three-dimensional nucleic acid structures. *Nucleic Acids Res* 31:5108–5121.

Luo Y, Guo P, Hasegawa S, Sato M (2004): *An Interactive Molecular Visualization System for Education in Immersive Multi-Projection Virtual Environment*. Proceedings of the Third International Conference on Image and Graphics (ICIG04), IEEE Computer Society, Hong Kong, China, pp. 485–488.

Ma JC, Dougherty DA (1997): The cation–pi interaction. *Chem Rev* 97:1303–1324.

Maiti R, Van Domselaar GH, Wishart DS (2005): MovieMaker: a web server for rapid rendering of protein motions and interactions. *Nucleic Acids Res* 33:W358–W362.

Marchese FT, Mercado J, Pan Y (2003): *Adapting Single-User Visualization Software for Collaborative Use*. Proceedings of the Seventh International Conference on Information Visualization (IV03), IEEE Computer Society.

Martz E (2002): Protein Explorer: easy yet powerful macromolecular visualization. *Trends Biochem Sci* 27:107–109.

Max N (1984): Computer representation of molecular surfaces. *J Mol Graph* 2:8–13.

Max N (1990): Computer animation in mathematics, science, and art. In: Chudnovksy D, Jenks R, editors. *Computers in Mathematics*. New York: Marcel Dekker, Inc., 321–333.

McClean P, Johnson C, Rogers R, Daniels L, Reber J, Slator BM, Terpstra J, White A (2005): Molecular and cellular biology animations: development and impact on student learning. *Cell Biol Educ* 4:169–179.

McGaughey GB, Gagne M, Rappe AK (1998): pi-Stacking interactions. Alive and well in proteins. *J Biol Chem* 273:15458–15463.

McRee DE (1999): XtalView/Xfit–a versatile program for manipulating atomic coordinates and electron density. *J Struct Biol* 125:156–165.

Merritt EA (1994): Raster3D Version 2.0: a program for photorealistic molecular graphics. *Acta Crystallogr D*, 50:869–873.

Meyer EF, Swanson SM, Williams JA (2000): Molecular modelling and drug design. *Pharmacol Ther* 85:113–121.

Moreland JL, Gramada A, Buzko OV, Zhang Q, Bourne PE (2005): The Molecular Biology Toolkit (MBT): a modular platform for developing molecular visualization applications. *BMC Bioinformatics* 6:21.

Mortenson ME, (1989): *Computer Graphics: An Introduction to the Mathematics and Geometry*. Oxford: Heinemann Newnes.

Mueller TD, Feigon J (2002): Solution structures of UBA domains reveal a conserved hydrophobic surface for protein–protein interactions. *J Mol Biol* 319:1243–1255.

Nesbit J, Bradford M (2006): 2006 Visualization Challenge. *Science* 313:1729.

O'Brien SE, Brown DG, Mills JE, Phillips C, Morris G (2005): Computational tools for the analysis and visualization of multiple protein–ligand complexes. *J Mol Graph Model* 24:186–194.

Oldfield TJ (2004): A Java applet for multiple linked visualization of protein structure and sequence. *J Comput Aided Mol Des* 18:225–234.

Olson AJ, (2001): *Molecular graphics and animation*. In: Rossmann M, Arnold E, editors *International Tables for Crystallography*. The Netherlands: Kluwer International Publishers. 357–368.

Pettersen EF, Goddard TD, Huang CC, Couch GS, Greenblatt DM, Meng EC, Ferrin TE (2004): UCSF Chimera—a visualization system for exploratory research and analysis. *J Comput Chem* 25:1605–1612.

Phelps ME (2000): PET: the merging of biology and imaging into molecular imaging. *J Nucl Med* 41:661–681.

Pimentel GC, McClelan AL (1960): *The Hydrogen Bond*. San Fransisco: Freeman.

Pletneva EV, Laederach AT, Fulton DB, Kostic NM (2001): The role of cation–pi interactions in biomolecular association. Design of peptides favoring interactions between cationic and aromatic amino acid side chains. *J Am Chem Soc* 123:6232–6245.

Porollo A, Meller J (2007): Versatile annotation and publication quality visualization of protein complexes using POLYVIEW-3D. *BMC Bioinformatics* 8:316.

Portmann S, Lüthi HP (2000): MOLEKEL: an interactive molecular graphics tool. *CHIMIA* 54:766–770.

Randic M, Zupan J, Vikic-Topic D (2007): On representation of proteins by star-like graphs. *J Mol Graph Model* 26:290–305.

Richards FM (1985): Calculation of molecular volumes and areas for structures of known geometry. *Methods Enzymol* 115:440–464.

Richardson DC, Richardson JS (1992): The kinemage: a tool for scientific communication. *Protein Sci* 1:3–9.

Richardson JS (1981): The anatomy and taxonomy of protein structure. *Adv Protein Chem* 34:167–339.

Richardson JS (1985a): Describing patterns of protein tertiary structure. *Methods Enzymol* 115:341–358.

Richardson JS (1985b): Schematic drawings of protein structures. *Methods Enzymol* 115:359–380.

Rollo FD (2003): Molecular imaging: an overview and clinical applications. *Radiol Manage* 25:28–32.

Rubin B, Richardson JS (1972): The simple construction of protein alpha-carbon models. *Biopolymers* 11:2381–2385.

Sacile R, Ruggiero C (2002): Modelling hydrophobicity to identify secondary structure signals in globular proteins. Proceedings of the IEEE-EMBS Special Topic Conference on Molecular, Cellular and Tissue Engineering, pp. 134–135.

Sankaranarayanan G, Weghorst S, Sanner M, Gillet A, Olson A (2003): Role of haptics in teaching structural molecular biology. Proceedings of the 11th Symposium on Haptic Interfaces for Virtual Environment and Teleoperator Systems, pp 363–366.

Sanner MF (2005): A component-based software environment for visualizing large macromolecular assemblies. *Structure* 13:447–462.

Sanner MF, Olson AJ, Spehner JC (1996): Reduced surface: an efficient way to compute molecular surfaces. *Biopolymers* 38:305–320.

Sayle RA, Milner-White EJ (1995): RASMOL: biomolecular graphics for all. *Trends Biochem Sci* 20:374.

Schafmeister C (1990): Fast algorithm for generating CPK images on graphics workstations. *J Mol Graph 8: 201–206* 211.

Silva LP (2005): Imaging proteins with atomic force microscopy: an overview. *Curr Protein Pept Sci* 6:387–395.

Smith RD, Hu L, Falkner JA, Benson ML, Nerothin JP, Carlson HA (2006): Exploring protein–ligand recognition with Binding MOAD. *J Mol Graph Model* 24:414–425.

Smith TJ (1995): MolView: a program for analyzing and displaying atomic structures on the Macintosh personal computer. *J Mol Graph* 13:122–125.

Smith TJ (2004): MolViewX: a molecular visualization program for the Macintosh OS X system. *J Appl Crystallogr* 37:654–657.

Sumathi K, Ananthalakshmi P, Roshan MN, Sekar K (2006): 3dSS: 3D structural superposition. *Nucleic Acids Res* 34:W128–W132.

Sun HD, Li M, Xu Y (2002): MOLVIE: an interactive visualization environment for molecular structures. *Comput Methods Programs Biomed* 71:85–90.

Tate JG (2003): Molecular Visualization. In: Bourne PE, Weissig H, editors.Structural Bioinformatics. Methods of Biochemical Analysis, Vol. 44. Wiley. p 135.

Tate JG, Moreland JL, Bourne PE (2001): Design and implementation of a collaborative molecular graphics environment. *J Mol Graph Model* 19:280–287.

Teschner M, Henn C, Vollhardt H, Reilling S, Brickmann J (1994): Texture mapping: a new tool for molecular graphics. *J Mol Graph* 12:98–105.

Teyra J, Doms A, Schroeder M, Pisabarro MT (2006): SCOWLP: a web-based database for detailed characterization and visualization of protein interfaces. *BMC Bioinformatics* 7:104.

Valle M (2005): STM3: a chemistry visualization platform. *Zeitschrift für Kristallographie* 220:585–588.

Via A, Ferre F, Brannetti B, Helmer-Citterich M (2000): Protein surface similarities: a survey of methods to describe and compare protein surfaces. *Cell Mol Life Sci* 57:1970–1977.

Walther D (1997): WebMol—a Java-based PDB viewer. *Trends Biochem Sci* 22:274–275.

Wang Y, Geer LY, Chappey C, Kans JA, Bryant SH (2000): Cn3D: sequence and structure views for Entrez. *Trends Biochem Sci* 25:300–302.

Weininger D (1988): SMILES, a chemical language and information system. 1. Introduction to methodology and encoding rules. *J Chem Inf Comput Sci* 28:31–36.

Weissleder R (1999): Molecular imaging: exploring the next frontier. *Radiology* 212:609–614.

Weissleder R, Mahmood U (2001): Molecular imaging. *Radiology* 219:316–333.

Westhead DR, Slidel TW, Flores TP, Thornton JM (1999): Protein structural topology: automated analysis and diagrammatic representation. *Protein Sci* 8:897–904.

Wiese KC, Glen E (2006): *jViz, Rna–An Interactive Graphical Tool for Visualizing RNA Secondary Structure Including Pseudoknots.* 19th IEEE Symposium on Computer-Based Medical Systems (CBMS06), pp. 659–664.

Willis S (2006): X3D protein ribbon models for the visual analysis of co-evolving amino acids structural relationship, Web3D consortium. X3D Scenarios and Case Studies: Real World Applications—Real World Solutions.

Willis S (2007): Protein CorreLogo: an X3D representation of co-evolving pairs, tertiary structure, ligand binding pockets and protein-protein interactions in protein families. 12th International Conference on 3D Web Technology, Umbria, Italy, University of Perugia.

Wimley WC, Gawrisch K, Creamer TP, White SH (1996): Direct measurement of salt-bridge solvation energies using a peptide model system: implications for protein stability. *Proc Natl Acad Sci* 93:2985–2990.

Winn MD (2003): An overview of the CCP4 project in protein crystallography: an example of a collaborative project. *J Synchrotron Radiat* 10:23–25.

Wollacott AM, Merz KM Jr (2007): Haptic applications for molecular structure manipulation. *J Mol Graph Model* 25:801–805.

Zeidler P (2000): The epistemological status of theoretical models of molecular structure. *HYLE Int J Philos Chem* 6:17–34.

Zhu H, Chana T-Y, Wang L, Cai W, See S (2004): A prototype of distributed molecular visualization on computational grids. *Future Generation Computer Systems* 20:727–737.

Zlatanova J, van Holde K (2006): Single-molecule biology: what is it and how does it work?. *Mol Cell* 24:317–329.

Section II

DATA REPRESENTATION
AND DATABASES

10

THE PDB FORMAT, mmCIF FORMATS, AND OTHER DATA FORMATS

John D. Westbrook and Paula M.D. Fitzgerald

INTRODUCTION

In this chapter, the data formats and protocols used to represent primary macromolecular structure data are presented. The historical format used by the Protein Data Bank (PDB) is described first. Dictionary-based representations such as the macromolecular Crystallographic Information File (mmCIF) and extensions such as the PDB Exchange dictionary are presented. The translation of the PDB Exchange dictionary into XML or PDB Markup Language (PDBML) is also described. Finally, protocols that provide data access through application program interfaces are described.

THE PDB FORMAT

The Protein Data Bank (http://www.pdb.org/; see also Chapter 11) (Bernstein et al., 1977; Berman et al., 2000) was established in 1971 by Walter Hamilton at Brookhaven National Laboratory, in response to community requests for a central repository of information on biological macromolecular structures. Seven structures were included in the PDB at its inception. The essential elements of the format used to encode these first entries are still the core of the PDB format used today. Because of the simplicity of the format and its consistency in representing three-dimensional structures, the PDB format remains the most widely supported means of exchanging macromolecular structure data.

The PDB format consists of a collection of fixed format records that describe the atomic coordinates, chemical and biochemical features, experimental details of the structure determination, and some structural features such as secondary structure assignments, hydrogen bonding, and biological assemblies and active sites. The details of the format

Structural Bioinformatics, Second Edition   Edited by Jenny Gu and Philip E. Bourne
Copyright © 2009 John Wiley & Sons, Inc.

```
             1         2         3         4         5         6         7
    1234567890123456789012345678901234567890123456789012345678901234567890123456789
    ATOM    145  N   VAL A  25      32.433  16.336  57.540  1.00 11.92       A1  N
    ATOM    146  CA  VAL A  25      31.132  16.439  58.160  1.00 11.85       A1  C
    ATOM    147  C   VAL A  25      30.447  15.105  58.363  1.00 12.34       A1  C
    ATOM    148  O   VAL A  25      29.520  15.059  59.174  1.00 15.65       A1  O
    ATOM    149  CD AVAL A  25      30.385  17.437  57.230  0.28 13.88       A1  C
    ATOM    150  CB BVAL A  25      30.166  17.399  57.373  0.72 15.41       A1  C
    ATOM    151  CG1AVAL A  25      28.870  17.401  57.336  0.28 12.64       A1  C
    ATOM    152  CG1BVAL A  25      30.805  18.788  57.449  0.72 15.11       A1  C
    ATOM    153  CG2AVAL A  25      30.835  18.826  57.661  0.28 13.58       A1  C
    ATOM    154  CG2BVAL A  25      29.909  16.996  55.922  0.72 13.25       A1  C
```

Figure 10.1. An abbreviated example of the column oriented data format for PDB ATOM records. The ATOM records in this example contain fields for the record name, atom serial number, an atom name, a residue name, a polymer chain identifier, a residue number, the x, y, z Cartesian coordinates, the isotropic thermal parameter and the occupancy. Atoms with serial numbers 149–154 also contain a label for alternative conformation in column 17.

are described in the PDB Contents Guide (Callaway et al., 1996) updated and available at http://www.wwpdb.org. This document enumerates the field formats for each PDB record and remark and describes the PDB conventions for naming atoms, peptides, and nucleotides.

Each item of the data in the PDB format is assigned to a range of character positions in one of many PDB record types (HEADER, SOURCE, REMARK, etc.). The ATOM records shown in Figure 10.1 encode the atomic coordinate data. ATOM records are among the more than 45 named data records in the PDB format. These named data records have strict column formatting rules.

During its early history, the PDB served as a simple repository and a point of dissemination for structure data and was used primarily by crystallographers and NMR spectroscopists. PDB entries during this early period resemble journal publications and contain lengthy descriptive text sections encoded in the REMARK records. An example of how refinement information was coded in pre-1994 PDB entries is shown in Figure 10.2a.

As the number of structures in the archive increased and the user base broadened, the PDB format required elaboration to enable comparative analysis of the data in the archive. At a minimum, such comparative studies require a consistent representation of the data and an increase in the types of the data included in each entry. Accordingly, extensions in the PDB format were advanced in 1992 (Protein Data Bank, 1992), 1996 (Callaway et al., 1996), 2006 and 2008 (http://www.wwpdb.org/docs.html). Figure 10.2b is an example of how refinement information is coded in the current PDB format. Prior to 2007, changes in the PDB format were not backwardly propagated to earlier entries. In early 2007, the wwPDB (http://www. wwpdb.org and see Chapter 11) produced a version of the PDB archive in a single consistent PDB format, (Henrick et al., 2008; Lawson et al., 2008). Although the evolution of the PDB format has significantly increased the encoding precision and level of detail for both the biochemical and the experimental descriptions, the format of the PDB coordinate records has remained largely unchanged.[1]

While the PDB format has served as the standard for representing macromolecular structure data for over three decades, both the underlying data and the user requirements for

---

[1] Coordinate records such as ATOM records were extended in column positions beyond 72 to include a segment identifier, element symbol, and atomic charge. In the earliest PDB format, column positions beyond 72 were reserved for punch card sequence numbers.

```
(a) REMARK   3
    REMARK   3 REFINEMENT. HENDRICKSON AND KONNERT RESTRAINED PARAMETERS
    REMARK   3 LEAST SQUARES.   THE FINAL R VALUE IS 0.136 FOR 21962
    REMARK   3 REFLECTIONS WITH I .GT. SIGMA(I) AND R IS 0.126 FOR
    REMARK   3 18168 REFLECTIONS WITH I .GE. 3*SIGMA(I) IN THE RANGE
    REMARK   3 8.0 TO 1.8 ANGSTROMS.


(b) REMARK   3
    REMARK   3 REFINEMENT.
    REMARK   3   PROGRAM   : REFMAC 5.2.0013
    REMARK   3   AUTHORS   : MURSHUDOV,VAGIN,DODSON
    REMARK   3
    REMARK   3    REFINEMENT TARGET : ENGH & HUBER
    REMARK   3
    REMARK   3  DATA USED IN REFINEMENT.
    REMARK   3   RESOLUTION RANGE HIGH (ANGSTROMS) :  2.10
    REMARK   3   RESOLUTION RANGE LOW  (ANGSTROMS) : 41.89
    REMARK   3   DATA CUTOFF            (SIGMA(F)) :  0.000
    REMARK   3   COMPLETENESS FOR RANGE        (%) : 93.8
    REMARK   3   NUMBER OF REFLECTIONS             : 118496
    REMARK   3
    REMARK   3  FIT TO DATA USED IN REFINEMENT.
    REMARK   3   CROSS-VALIDATION METHOD          : THROUGHOUT
    REMARK   3   FREE R VALUE TEST SET SELECTION  : RANDOM
    REMARK   3   R VALUE     (WORKING + TEST SET) : 0.243
    REMARK   3   R VALUE            (WORKING SET) : 0.199
    REMARK   3   FREE R VALUE                     : 0.248
    REMARK   3   FREE R VALUE TEST SET SIZE   (%) : 2.500
    REMARK   3   FREE R VALUE TEST SET COUNT      : 2967
```

Figure 10.2.   (a) An example portion of PDB REMARK 3 given in the data format used prior to 1992. In this example, information about the refinement is given as free text. (b) An example portion of PDB REMARK 3 given in the current data format. In this example, the text is more structured.

this data have changed dramatically. Together these considerations have posed informatics challenges that the current PDB format cannot fully address.

The macromolecular structure data represented in a PDB entry has increased in both type and complexity. In addressing changes in experimental methodology, the PDB format has been extended with new REMARK records. For example, the organization and information content of REMARK 3 that encodes refinement information has been modified and extended for each new refinement program and program version. Although extending REMARK records in this way captures information in a manner that is easy for a human to read, the diversity of organization of these data makes it very difficult to design software that can automatically and reliably extract information from these records. Data in these records are also defined only in terms of the program that computed the information. Information between programs may not be directly comparable.

The PDB format uses fixed width fields to represent data, and this places absolute limits on the size of certain items of data. For instance, the maximum number of atom records that can be represented in a single-structure model is limited to 99,999, and the field width of the identifier for polymer chains is limited to a single character. Although these restrictions were certainly reasonable when the format was first defined, this is no longer the case. Many large molecular systems, such as the ribosomal subunit structures, cannot be represented in a single PDB entry. These entries must be divided into multiple PDB files, seriously complicating their use.

As the size and diversity of structure data in the PDB archive has grown, it has become an increasingly important resource in structural biology. User requirements for PDB data have

```
SEQRES    1    396  MET ASP GLU ASN ILE THR ALA ALA PRO ALA ASP PRO ILE
SEQRES    2    396  LEU GLY LEU ALA ASP LEU PHE ARG ALA ASP GLU ARG PRO

. . .

. . .

ATOM      1   N   MET      5        41.402  11.897  15.262  1.00 48.61
ATOM      2   CA  MET      5        40.919  13.262  15.600  1.00 47.70
ATOM      9   N   PHE      6        39.627  14.840  14.228  1.00 48.66
ATOM     10   CA  PHE      6        39.199  15.440  12.964  1.00 45.33

. . .
```

Figure 10.3. An abbreviated example illustrating sequence inconsistency between PDB records. The PDB SEQRES records describe the sequence of the polymer that was crystallized. The sequence labels in the PDB ATOM records report the coordinates and the residue sequence observed in the refined structure. These should be consistent; as shown in this example, there have often been exceptions. This example highlights the sequence conflict between ASP in the chemical sequence (SEQRES) with PHE, residue number 6 in the ATOM records.

grown from accessing individual entries to analysis and comparison of experimental and structure data across the entire archive. The latter has been facilitated by the increased accessibility of database technologies. The support of comparative analysis and database applications requires data uniformity and internal consistency that are typically beyond the needs of the software accessing individual entries, such as molecular graphics applications.

The difficulty of reliably extracting experimental information from each entry has already been discussed in terms of the format variation of REMARK records used to encode experimental details (e.g., refinement information in REMARK 3). Internal consistency problems within PDB entries arise in cases in which portions of the structure or individual structural elements are referenced in different PDB records in a noncorresponding manner. Because the PDB format does not specify the precise relationship between the polymer sequence given in the SEQRES records and the observed residue sequence within the ATOM records, the sequence information that is presented in the PDB entry cannot be used directly. To use these data, additional sequence alignment must be performed to resolve gaps and possible conflicts. An example of this is shown in Figure 10.3.

Consistency problems can also arise in other records that reference structural features such as those records describing secondary structure, active sites, and biological assemblies. Although the relationships between these records and the coordinate data that they reference are obvious to the experienced user, they can only be understood by a careful reading of the format description document. Because these relationships are not electronically accessible, each such relationship must be coded as a special case by any software that needs to validate interrecord consistency.

mmCIF: A DICTIONARY-BASED APPROACH TO DATA DESCRIPTION

The Crystallographic Information File (Hall, Allen, and Brown, 1991) was created to archive information about crystallographic experiments and results (Hall, 1991) and is the format in which all structures described in articles sent to *Acta Crystallographia C* are submitted. In 1990, the International Union of Crystallography (IUCr) formed a working group (Fitzgerald et al., 1992) to expand this dictionary so that it would be able to do the same for macromolecules.

The original short-term goal of the working group was to fulfill the mandate set by the IUCr: to define mmCIF data names that needed to be included in the CIF dictionary in order to adequately describe the macromolecular crystallographic experiment and its results. This implied the need to describe all of the data items included in a PDB entry. Long-term goals were also established: to provide sufficient data names so that the experimental section of a structure paper could be written automatically and to facilitate the development of tools so that computer programs could easily access and validate the mmCIF data files.

In order to describe the progress of this project and to solicit community feedback, several informal and formal meetings were held. The first meeting, hosted by Eleanor Dodson, convened in April 1993 at the University of York. The attendees included the mmCIF working group, structural biologists, and computer scientists. A major focus of the discussion was whether the formal structure of the dictionary that was implemented using the then-current Dictionary Definition Language (DDL 1.0) (Westbrook et al., 2005a) was adequate to deal with the complexity of the macromolecular data items. Criticisms included the idea that the data typing was not strong enough and that there were no formal links among the data items. A working group was formed to address these issues. The second workshop was hosted by Phil Bourne in Tarrytown, NY, in October 1993. The meeting focused on the development of software tools and the requirements of an enhanced DDL. In October 1994, a workshop hosted by Shoshana Wodak at the Free University of Brussels resulted in the adoption of a new DDL that addressed the various problems that had been identified at the preceding workshops. The evolving mmCIF dictionary was cast in this new DDL 2 (Westbrook et al., 2005a) and was presented at the ACA meeting in Montreal in July 1995. This dictionary was open for further community review. The dictionary was placed on a world wide web site and community comments were solicited via a list server. Lively discussions via this mmCIF list server ensued, resulting in the continuous correction and updating of the dictionary. Software was developed and was also presented on this www site. A workshop held at Rutgers University in 1997 provided tutorials for using both the dictionary and the software tools that had been developed at that time.

In January 1997, an mmCIF dictionary containing 1700 definitions was completed and submitted to the IUCr committee that oversees dictionary development (COMCIFS) for review, and in June 1997, Version 1.0 was released (Fitzgerald et al., 1996; Bourne et al., 1997). The method adopted for managing dictionary extensions involves the use of a scientific journal as a model. Proposed extensions are sent to the editors of the mmCIF dictionary (Fitzgerald et al., 1996; Editorial Board: Paula Fitzgerald, editor; Helen Berman, associate editor) who send the new definitions to a member of the board of editors for scientific review. These editors have expertise in various areas covered by the dictionary. Once the definitions are reviewed for their scientific content, they are sent to technical editors. More than 100 new definitions were proposed after the dictionary was released in 1997 and these have been reviewed using the procedures outlined. Version 2 of the mmCIF dictionary was released in the fall of 2000 and this new version contains many of these new definitions.

Software libraries to parse and access data in CIF and mmCIF have been produced for a number of popular languages including C/C++, JAVA, FORTRAN, PERL, and Python (http://sw-tools.pdb.org or http://www.iucr.org for lists of programs).

mmCIF Dictionary and Data File Syntax

The syntax used in both mmCIF data files and dictionaries derived from the STAR (Self-defining Text Archive and Retrieval) (Hall, 1991) grammar and is similar in most respects to the syntax used by core CIF for describing small molecule crystallography.

```
_refine.ls_d_res_high                          2.10
_refine.ls_d_res_low                           41.89
_refine.pdbx_ls_sigma_F                        0.00
_refine.ls_percent_reflns_obs                  93.850
_refine.ls_number_reflns_obs                   118496
_refine.pdbx_ls_cross_valid_method             THROUGHOUT
_refine.pdbx_R_Free_selection_details          RANDOM
_refine.ls_R_factor_obs                        0.243
_refine.ls_R_factor_R_work                     0.199
_refine.ls_R_factor_R_free                     0.248
_refine.ls_percent_reflns_R_free               2.500
_refine.ls_number_reflns_R_free                2967
```

Figure 10.4. An example portion of the mmCIF REFINE category containing the same information as in the PDB REMARKs in Figure 10.2a–b. This example illustrates keyword–value pair formatting that is characteristic of the mmCIF syntax.

In its simplest form, an mmCIF data file looks like a paired collection of data item names and values. Figure 10.4 illustrates the assignments of values to selected refinement parameters analogous to the PDB format data in Figure 10.2b. The syntax is described here.

The leading underscore character identifies data item names. The underscore character is followed by a text string interpreted as containing both a category name and an item name separated by a period. The keyword portion of the name is the unique identifier of the data item within the category. In the examples shown in Figure 10.4, all of the data items belong to the REFINE category. This example also illustrates the one-to-one correspondence required between item names and item values. Data category and data item names are not case sensitive.

Figure 10.5 illustrates how text strings are expressed. Short text strings may be enclosed in single or double quotation marks. Text strings that span multiple lines are enclosed by semicolons that are placed at the first character position of the line. There are two special characters used as placeholders for item values that for some reason cannot be explicitly assigned. The question mark (?) is used to mark an item value as missing. A period (.) may be

```
_diffrn_measurement.diffrn_id              'Data set 1'
_diffrn_measurement.device                 '3-circle camera'
_diffrn_measurement.method                 'omega scan'
_diffrn_measurement.details
;440 frames, 0.20 degrees, 150 sec, detector distance 12 cm,
detector angle 22.5 degrees
;
_diffrn_measurement.specimen_support       ?
```

Figure 10.5. An example illustrating the encoding of text strings using mmCIF. Short strings, such as *Data set 1* and *omega scan*, are surrounded by either single or double quotation marks. Multiline strings such as the value of _diffrn_measurement.details are encapsulated by semicolons in the first column of the beginning and ending lines of the string.

```
loop_
_atom_site.group_PDB
_atom_site.type_symbol
_atom_site.label_atom_id
_atom_site.label_comp_id
_atom_site.label_asym_id
_atom_site.label_seq_id
_atom_site.label_alt_id
_atom_site.cartn_x
_atom_site.cartn_y
_atom_site.cartn_z
_atom_site.occupancy
_atom_site.B_iso_or_equiv
_atom_site.footnote_id
_atom_site.auth_seq_id
_atom_site.id
ATOM N N   VAL A 11 . 25.369 30.691 11.795 1.00 17.93 . 11 1
ATOM C CA  VAL A 11 . 25.970 31.965 12.332 1.00 17.75 . 11 2
ATOM C C   VAL A 11 . 25.569 32.010 13.881 1.00 17.83 .  11 3
# [data omitted]
```

Figure 10.6. An abbreviated example of the mmCIF category ATOM_SITE. This category is organized as a table and illustrates the use of the *loop_* directive followed by the list of data item names as the simple means of declaring an mmCIF table. The data values that follow the list of data item names are assigned to each data item (column) in turn. Here, the value, ATOM, is assigned to the column for _atom_site.group_PDB in each of the three rows in this example.

used to identify that there is no appropriate value for the item or that a value has been intentionally omitted.

Vectors and tables of data may be encoded using a *loop_* directive. To build a table, the data item names corresponding to the table columns are preceded by the *loop_* directive and followed by the corresponding rows of data. The mmCIF example in Figure 10.6 builds a table of atomic coordinates.

The use of the *loop_ directive* has a few restrictions. First, it is required that all of the data items within the loop belong to the same data category. Second, the number of data values following the loop must be an exact multiple of the number of data item names. Finally, mmCIF does not support the nesting of *loop_* directives.

Data blocks are used to organize related information and data. A data block is a logical partition of a data file or dictionary created using a *data_* directive. A data block may be named by appending a text string after the *data_* directive, and a data block is terminated by either another *data_* directive or by the end of the file. Figure 10.7 shows a very simple example of a pair of abbreviated data blocks.

Figure 10.7 illustrates how data blocks can be used to separate similar information pertaining to different structures. This separation is required because the mmCIF syntax prohibits the repetition of the same category at multiple places within the same data block. As a result, the simple concatenation of the contents of the above two data blocks into a single data block would be syntactically incorrect.

```
#
# --- Lines beginning with # are treated as comments
#
data_X987A
_entry.id                        X987A
_exptl_crystal.id                'Crystal A'
_exptl_crystal.colour            'pale yellow'
_exptl_crystal.density_diffrn    1.113
_exptl_crystal.density_Matthews  1.01

_cell.entry_id                       X987A
_cell.length_a                       95.39
_cell.length_a_esd                    0.05
_cell.length_b                       48.80
_cell.length_b_esd                    0.12
_cell.length_c                       56.27
_cell.length_c_esd                    0.06

# Second data block
data_T100A

_entry.id                        T100A
_exptl_crystal.id                'Crystal B'
_exptl_crystal.colour            'orange'
_exptl_crystal.density_diffrn    1.156
_exptl_crystal.density_Matthews  1.06

_cell.entry_id                       T100A
_cell.length_a                       68.39
_cell.length_a_esd                    0.05
_cell.length_b                       88.70
_cell.length_b_esd                    0.12
_cell.length_c                       76.27
_cell.length_c_esd                    0.06
```

Figure 10.7. An abbreviated example illustrating the organization of mmCIF files in data blocks. Data blocks are declared using the *data_* directive and optionally followed by a data block name. In this example, data blocks X987A and T100A are declared. The information in these data blocks is treated as logically distinct even if the data block exists within the same data file.

Definitions in mmCIF data dictionaries are encapsulated in named save frames (Figure 10.8). A save frame is a syntactical element that begins with the *save_* directive and is terminated by another *save_* directive. Save frames are named by appending a text string to the *save_* directive. In the mmCIF dictionary, save frames are used to encapsulate item and category definitions. The mmCIF dictionary is composed of a data block containing thousands of save frames, where each save frame contains a different definition. Save frames appear in data dictionaries but they are not used in data files. Save frames may not be nested.

The content of this dictionary definition has the same item–value pair organization as in the previous data file examples. DDL2 dictionary definitions typically contain a small number of items that specify the essential features of the item. The example definition shown in Figure 10.8 includes a description or text definition, the name and the category of the item, a code indicating that the item is optional (not mandatory), the name of a related definition in the core CIF dictionary, and a code specifying that the data type is *text*. A further description of the elements of the dictionary definitions is presented in the next section.

```
save__exptl.details
    _item_description.description
;   Any special information about the experimental work prior to
    the intensity measurement. See also _exptl_crystal.preparation.
;
_item.name                      '_exptl.details'
_item.category_id                exptl
_item.mandatory_code             no
_item_aliases.alias_name        '_exptl_special_details'
 item_aliases_dictionary         cif_cor_dic
_item_aliases.version            2.0.1
_item_type.code                  text
save_
```

Figure 10.8. An example of the mmCIF data definition for data item `_exptl.details`. This definition contains a textual definition, name and category identity, a code indicating the item is optional, an alias name to a previous dictionary, and a data type.

Semantic Elements of the mmCIF Data Dictionary

The elements of DDL provide the organizational framework for building data dictionaries such as mmCIF. The role of the DDL is to define which data items may be used to construct the definitions in the data dictionary and to define also the relationships between these defining data items.

The dictionary language contains no information about a particular discipline such as macromolecular crystallography; rather, it defines the data items that can be used to describe a discipline. The contents of the mmCIF dictionary are metadata, or data about data. The contents of the DDL are meta-metadata, the data defining the metadata. DDL defines data items that describe the general features of a data item, such as a textual description, a data type, a set of examples, a range of permissible values, or perhaps a discrete set of permitted values. Consequently, data modeling using DDL can be applied in many application areas, not just macromolecular structure description.

The lowest level of organization provided by the DDL is the description of an individual data item. Collections of related data items are organized in categories. Categories are essentially tables in which each repetition of the group of related items adds a row. The terms *category* and *data item* are used here in order to conform with the previous use of these terms by STAR and CIF applications; these terms could be replaced by *relation* and *attribute* (or table and column), terms which are commonly used to describe the relational model that underlies the DDL.

Within a category, the set of data items determining the uniqueness of their group are designated as *key* items in the category. No data item group in a category is allowed to have a set of duplicate values of its key items. Each data item is assigned membership in one or more categories. Parent–child relationships may be specified for items belonging to multiple categories. These relationships permit the specification of the very complicated hierarchical data structures required to describe macromolecular structure.

Other levels of organization in addition to category are also supported. Related categories may be collected together in category groups, and parent relationships may be specified for these groups. This higher level of association provides a method for organizing a large and complicated collection of categories into smaller, more relevant, and potentially interrelated groups. This effectively provides a chaptering mechanism for large and complicated dictionaries, such as mmCIF. Within the level of a category, subcategories of data items may be defined among groups of related data items. The subcategory provides a mechanism to identify, for example, that the data items month, day, and year collectively define a date.

For categories, subcategories, and items, *methods* may be specified. Methods are computational procedures that are defined and expressed in a programming language (e.g., C/C++, PERL, Python, or JAVA) and stored within a dictionary. Among other things, these dictionary methods may be used to calculate a missing value or to check the validity of a particular value.

The highest levels of data organization provided by DDL2 are the data block and the dictionary. The dictionary level collects a set of related definitions into a single unit, and provides the attributes for a detailed revision history on the collection.

The detailed features of the DDL used to build the mmCIF data dictionary are described elsewhere (Westbrook and Bourne, 2000; Westbrook et al., 2005a and b).

mmCIF Dictionary Content

Version 1.0 of the mmCIF dictionary contains approximately 1,700 definitions describing the macromolecular experiment and its structural results. This dictionary includes definitions describing all aspects of macromolecular structure; experimental details about crystallization, data collection, data processing, phasing, and refinement; and other supporting data categories describing citation and software. A complete discussion of the contents of the mmCIF dictionary has been previously provided (Bourne et al., 1997; Fitzgerald et al., 2005).

The following section provides a summary of a portion of mmCIF data categories describing chemical structure. In this discussion, the data categories are presented in the form of a schematic diagram, a brief description, and a set of examples. In the diagrams, boxes enclose the data items within each mmCIF category, and arrows indicate the correspondence between data items that are common to multiple categories, with the arrowheads pointing in the direction of the parent data item. The category key data items are indicated with black dots.

mmCIF Molecular Entities. An entity is a chemically distinct part of an mmCIF entry. There are three types of entities: polymer, nonpolymer, and water. A common name, systematic name, source information, and keyword description can be assigned to each mmCIF entity. The relationships between categories that describe these entity features are illustrated in Figures 10.9 and 10.10 is an example of an entity description taken from an HIV protease structure (PDB 5HVP; Fitzgerald et al., 1990).

mmCIF Polymer and Nonpolymer Entities. Additional data categories are provided to describe polymeric entities. Polymer type, sequence length, and information about nonstandard linkages and chirality may be specified. The monomer sequence for each polymer entity is listed in category ENTITY_POLY_SEQ. This sequence information is directly linked to the sequence specified in the coordinate list. It is also linked to the full chemical description of each monomer or nonstandard monomer in the CHEM_COMP category group. The relationships between categories describing polymer entities are illustrated in Figure 10.11 and Figure 10.12 shows an example of the description of a polymeric entity for a simple protein.

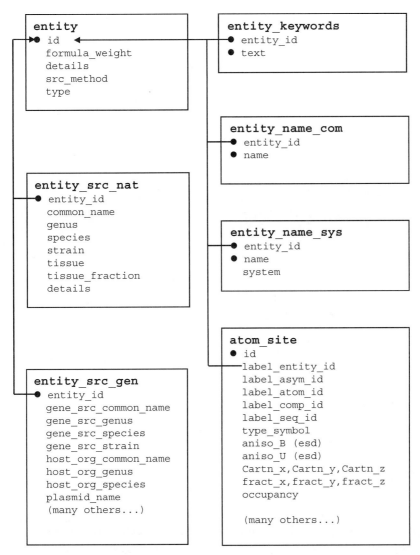

Figure 10.9. A schematic diagram illustrating the content and relationships among a portion of the mmCIF categories describing chemical entities, entity names, entity source organism, and the relationship between the entity and atomic level of description. In this figure, boxes enclose the data items within each mmCIF category, and arrows indicate the correspondence between data items that are common to multiple categories, with the arrowheads pointing in the direction of the parent data item. The category key data items are indicated with black dots.

Nonpolymeric entities are treated as individual chemical components. These entities may be fully described in the CHEM_COMP group of categories in the same manner as monomers within a polymeric entity. Like polymeric entities, each nonpolymeric entity carries both an entity identifier and a component identifier. These identifiers form part of the label used to identify each atom (_atom_site.label_entity_id and _atom_-site.label_comp_id). For polymeric entities, the monomer identifier and the component identifier are the same; however, the atom label also includes an additional field

```
loop_
_entity.id
_entity.type
_entity.formula_weight
_entity.details
1   polymer        10916
;   The enzymatically competent form of HIV protease is a dimer. This
    entity corresponds to one monomer of an active dimer.
;
2   non-polymer    '647.2'   '.'
3   water          18        '.'
#
loop_
_entity_name_com.entity_id
_entity_name_com.name
1   'HIV-1 protease monomer'
1   'HIV-1 PR monomer'
2   'acetyl-pepstatin'
2   'acetyl-Ile-Val-Asp-Statine-Ala-Ile-Statine'
3   'water'
#
loop_
_entity_src_gen.entity_id
_entity_src_gen.gene_src_common_name
_entity_src_gen.gene_src_strain
_entity_src_gen.host_org_common_name
_entity_src_gen.pdbx_host_org_scientific_name
_entity_src_gen.plasmid_name
1 'HIV-1' 'NY-5' 'bacteria'  'Escherichia coli'   'pB322'
#
```

Figure 10.10. An abbreviated example of the mmCIF description of the chemical features of HIV protease. In this example, three entities are defined. One entity is a monomer of HIV protease, the second is the peptidic inhibitor, and the third is solvent. Common names for these entities are specified. The source organism from which the protein sequence was obtained is also specified in this example.

for the sequence position (`_atom_site.label_seq_id`). An example for a drug–DNA complex illustrating both polymer and nonpolymer entity descriptions is shown in Figure 10.13a.

Atomic Positions. The refined coordinates are stored in the ATOM_SITE category. Atomic positions and their associated uncertainties may be stored in either Cartesian or fractional coordinates along with the option to store temperature factors and occupancies for each position.

Each atomic position must be uniquely identified by the data item _atom_site.id. Each position must also include a reference (`_atom_site.type_symbol`) to the table of elemental symbols in category ATOM_TYPE. All other data items in ATOM_SITE category are optional.

A typical atomic position for a macromolecule includes a variety of label information, as illustrated in Figure 10.13b. The data items that label atomic positions can be divided into two groups: those that are integrated into higher level structural descriptions and those that are provided to hold alternative nomenclatures. The data items in the former group are prefixed by `label_` and the latter carry an `auth_` prefix.

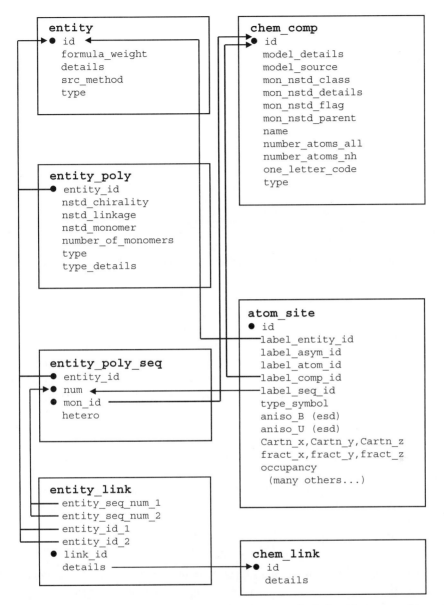

Figure 10.11. A schematic diagram illustrating the content and relationships among the mmCIF categories describing polymer entities. In this figure, boxes enclose the data items within each mmCIF category, and arrows indicate the correspondence between data items that are common to multiple categories, with the arrowheads pointing in the direction of the parent data item. The category key data items are indicated with black dots.

THE PDB EXCHANGE AND OTHER DATA DICTIONARIES

The wwPDB (Berman et al., 2003 and see Chapter 11) uses technology based on mmCIF for content description, data exchange, and data archiving. The PDB Exchange dictionary (Westbrook et al., 2005a) incorporates the content of the mmCIF dictionary, as well as

```
loop_
_entity_poly.entity_id
_entity_poly.type
_entity_poly.nstd_chirality
_entity_poly.nstd_linkage
_entity_poly.nstd_monomer
1  polypeptide(L)   no   no   no
#
loop_
_entity_poly_seq.entity_id
_entity_poly_seq.num
_entity_poly_seq.mon_id
    1    1   PRO   1    2   GLN   1    3   ILE   1    4   THR   1    5   LEU
    1    6   TRP   1    7   GLN   1    8   ARG   1    9   PRO   1   10   LEU
    1   11   VAL   1   12   THR   1   13   ILE   1   14   LYS   1   15   ILE
    1   16   GLY   1   17   GLY   1   18   GLN   1   19   LEU   1   20   LYS
    1   21   GLU   1   22   ALA   1   23   LEU   1   24   LEU   1   25   ASP
#  -  -  -  abbreviated  -  -  -
```

Figure 10.12. An abbreviated mmCIF description of polymer entity, including the enumeration of the three-letter residue codes of the entity polymer sequence.

extensions for high-throughput Structural Genomics (ISGO, 2001); noncrystallographic methods such as NMR and cryoelectron microscopy; protein production; and internal data management and status tracking. Within this dictionary framework the wwPDB provides a consistent representation of experiment and structure of macromolecular systems spanning a scale from small oligomers to subcellular assemblies and molecular machines.

In addition to the PDB Exchange dictionary, the mmCIF methodology has been used to describe a number of other content areas. All of these dictionaries have been developed to be consistent with the mmCIF data model. These dictionaries, which are all available from the mmCIF resource site (http://mmcif.pdb.org), cover the following content areas:

- The imgCIF dictionary (Hammersley et al., 2005) provides details of crystallographic data collection including data from image detectors in ASCII and binary data format.
- The BIOSYNC (http://biosync.pdb.org) dictionary describes the features and facilities provided by synchrotron beamlines.
- The MDB dictionary is an extension of the mmCIF dictionary for homology models.
- The Symmetry extension (Brown, 2005) supplements the mmCIF dictionary with detailed aspects of crystallographic symmetry.
- The NMRSTAR dictionary supplements the structural description of NMR in the PDB Exchange dictionary with the experimental details collected and archived by the Biological Magnetic Resonance Bank (BMRB; Seavey et al., 1991).

SUPPORTING OTHER FORMATS

Much attention in the area of data format has recently been focused on technologies related to the extensible markup language (XML). XML provides a framework for structuring complex information and documents. XML is used as a data exchange vehicle in a variety

```
(a)  #
     loop_
     _entity.id
     _entity.type
     _entity.src_method
     _entity.pdbx_description
     _entity.formula_weight
     1 polymer     syn "5'-D(*CP*GP*AP*TP*CP*G)-3'" 1809.232
     2 non-polymer syn ADRIAMYCIN                      543.526

     3 water       nat water              18.015
     #
     loop_
     _entity_keywords.entity_id
     _entity_keywords.text
         1   'NUCLEIC ACID'
         2   'DRUG'
     #
     loop_
     _entity_name_com.entity_id
     _entity_name_com.name
         2   DOXORUBICIN
         3   WATER

     #
     _entity_poly.entity_id                      1
     _entity_poly.type                           polydeoxyribonucleotide
     _entity_poly.nstd_linkage                   no
     _entity_poly.nstd_monomer                   no
     _entity_poly.pdbx_seq_one_letter_code_can   CGATCG
     #
     loop_
     _entity_poly_seq.entity_id
     _entity_poly_seq.num
     _entity_poly_seq.mon_id
     1 1 DC
     1 2 DG
     1 3 DA
     1 4 DT
     1 5 DC
     1 6 DG
     #
```

Figure 10.13. An abbreviated example of an mmCIF description of a drug-DNA complex. This example includes the description of both the DNA strand and the drug adriamycin. The mmCIF categories depicted in Figures 10.9 and 10.10 are populated in this example. In particular, this example illustrates how mmCIF describes the chemistry of both polymer and nonpolymer molecules. (a) The chemical entities in this drug complex are defined. The first entity is the DNA polymer strand and the second entity is the drug adriamycin. The nucleotide sequence of the DNA is enumerated. (b) The list of chemical components in the complex is specified. This includes the nucleotide monomers, the drug, and the solvent. The atomic coordinates of a portion of the first nucleotide (cytosine) are given. This coordinate data is explicitly related to the chemical description through identifiers for entity (_atom_site.label_entity_id = 1), component (_atom_site.label_comp_id = DC), and sequence order (_atom_site.label_seq_id =1).

```
(b) loop_
    _chem_comp.id
    _chem_comp.type
    _chem_comp.mon_nstd_flag
    _chem_comp.name
    _chem_comp.formula
    _chem_comp.formula_weight
    DC  'DNA linking' N "2'-DEOXYCYTIDINE-5'-MONOPHOSPHATE"   'C9 H14 N3 O7 P'   307.199
    DG  'DNA linking' N "2'-DEOXYGUANOSINE-5'-MONOPHOSPHATE"  'C10 H14 N5 O7 P'  347.224
    DA  'DNA linking' N "2'-DEOXYADENOSINE-5'-MONOPHOSPHATE"  'C10 H14 N5 O6 P'  331.224
    DT  'DNA linking' N "THYMIDINE-5'-MONOPHOSPHATE"          'C10 H15 N2 O8 P'  322.211
    DM2 non-polymer   . ADRIAMYCIN                            'C27 H29 N O11'    543.526
    HOH non-polymer   . WATER                                 'H2 O'              18.015
    #
    #
    loop_
    _atom_site.group_PDB
    _atom_site.id
    _atom_site.type_symbol
    _atom_site.label_atom_id
    _atom_site.label_alt_id
    _atom_site.label_comp_id
    _atom_site.label_asym_id
    _atom_site.label_entity_id
    _atom_site.label_seq_id
    _atom_site.pdbx_PDB_ins_code
    _atom_site.Cartn_x
    _atom_site.Cartn_y
    _atom_site.Cartn_z
    _atom_site.occupancy
    _atom_site.B_iso_or_equiv
    _atom_site.auth_seq_id
    _atom_site.auth_comp_id
    _atom_site.auth_asym_id
    _atom_site.auth_atom_id
    _atom_site.pdbx_PDB_model_num
    ATOM 1 O  "O5'" . DC A 1 1 ? 9.462  20.142 22.444 1.00 8.67 1  DC  A "O5'" 1
    ATOM 2 C  "C5'" . DC A 1 1 ? 9.675  21.129 23.376 1.00 8.13 1  DC  A "C5'" 1
    ATOM 3 C  "C4'" . DC A 1 1 ? 11.117 21.115 23.806 1.00 7.93 1  DC  A "C4'" 1
    ATOM 4 O  "O4'" . DC A 1 1 ? 11.336 19.954 24.661 1.00 7.29 1  DC  A "O4'" 1
    ATOM 5 C  "C3'" . DC A 1 1 ? 12.140 21.021 22.673 1.00 7.51 1  DC  A "C3'" 1
    ATOM 6 O  "O3'" . DC A 1 1 ? 13.292 21.776 23.049 1.00 7.06 1  DC  A "O3'" 1
    ATOM 7 C  "C2'" . DC A 1 1 ? 12.473 19.529 22.650 1.00 8.43 1  DC  A "C2'" 1
    ATOM 8 C  "C1'" . DC A 1 1 ? 12.371 19.133 24.132 1.00 6.43 1  DC  A "C1'" 1
    #[data abbreviated]
```

Figure 10.13. (*Continued*)

of software platforms. A growing number of commercial and public domain software applications provide support for XML. Many of these applications provide sophisticated display and query features that can be applied to the XML data. Owing to the availability of off-the-shelf access and query tools, this format would appear to be a good choice for storing structure data.

The representation of PDB structure data in XML is based on the mapping of content from the PDB Exchange dictionary to an XML schema. XML schemas (http://www.w3c.org/XML/Schema) provide a means of defining the structure, content, and semantics of XML documents analogous to the role the mmCIF dictionaries play for mmCIF data files. Like mmCIF dictionaries, XML schemas provide support for strong data typing, complex data types, enumerations, range restrictions, and parent–child and key relationships.

In creating the PDBML schema, the naming conventions and logical organization used within PDB exchange dictionary have been preserved. The content mapping between data dictionary and schema is fully software driven (Westbrook et al., 2005c). A consequence of maintaining the logical organization of the mmCIF dictionary mapping is that PDBML data files lack the embedded hierarchical organization characteristic to many XML files. Instead, the hierarchical structure in PDBML files is represented in XML `key-keyref` relationships just as with the mmCIF parent–child relationships. An important advantage of representing data in regular table-like containers is that a custom parser is not required to navigate a complex hierarchy of data elements, and that extensions to the data model require little if any supporting software changes.

An example is given in Figures 10.14a–b; the former is a PDBML encoding of the data shown in an mmCIF representation in the latter. This example illustrates the essential mapping between mmCIF and PDBML. Each mmCIF category is encapsulated with an XML element that pairs the category name with the appended token *Category*. Each row of the mmCIF category is enclosed in an element named after the category and including any key mmCIF data items as XML attributes. Nonkey mmCIF data items are represented as XML elements. All data XML element names are prefixed with the XML namespace, *PDB*x. This example also illustrates the large overhead of XML tags. For a medium-sized protein structure of 200 residues, this results in a 10-fold increase in data file size. Figure 10.14c shows an alternative style in which the atom coordinate records are stored as a string without detailed markup. In the latter example, the string of atomic coordinate information is matched against a regular expression in the PDBML schema defining the record organization. While the latter approach gains storage efficiency, this efficiency is achieved with a loss of the detailed access and search functionality of the fully marked-up form.

SUPPORTING APPLICATION PROGRAM INTERFACES

No single file format is satisfactory for all users and applications. One way to avoid file format issues entirely is to provide access to data through an application program interface (API). Depending on the language implementation the API provides access to data through a collection of functions, procedures, or methods. While a wide range of API technologies are available, two approaches have been commonly applied for exchanging macromolecular structure data.

The Object Management Group (OMG) standardizes the developed API by using its Common Object Request Broker Architecture (CORBA). CORBA supports an interface definition language (IDL) for defining programmable interfaces that are both language and platform independent. CORBA has also been developed to support distributed cross-platform access. The CORBA IDL for macromolecular structure (http://cgi.omg.org/cgibin/http://www://cgi.omg.org/cgi-bin/doc?lifesci/00-02-02) is based on the mmCIF data representation and provides efficient program access to all of the data in PDB entries.

Access to data and query functionality for a wide range of biological resources is provided by web services based on the Web Services Description Language (WSDL; http://www.w3.org/TR/wsdl) and the Simple Object Access Protocol (SOAP) (http://www.w3.org/TR/2000/NOTE-SOAP-20000508/). This approach leverages XML data encoding and widely used network transport protocols such as HTTP to provide distributed cross-platform access to data and query services. WSDL provides XML-encoded definitions of data and query services along with the content of messages exchanged between servers and their clients. SOAP provides the protocol by which messages are transmitted over the network. At

```
(a)  <PDBx:atom_siteCategory>
        <PDBx:atom_site id="1">
            <PDBx:group_PDB>ATOM</PDBx:group_PDB>
            <PDBx:type_symbol>O</PDBx:type_symbol>
            <PDBx:label_atom_id>O5&apos;</PDBx:label_atom_id>
            <PDBx:label_alt_id xsi:nil="true" />
            <PDBx:label_comp_id>DC</PDBx:label_comp_id>
            <PDBx:label_asym_id>A</PDBx:label_asym_id>
            <PDBx:label_entity_id>1</PDBx:label_entity_id>
            <PDBx:label_seq_id>1</PDBx:label_seq_id>
            <PDBx:Cartn_x>9.462</PDBx:Cartn_x>
            <PDBx:Cartn_y>20.142</PDBx:Cartn_y>
            <PDBx:Cartn_z>22.444</PDBx:Cartn_z>
            <PDBx:occupancy>1.00</PDBx:occupancy>
            <PDBx:B_iso_or_equiv>8.67</PDBx:B_iso_or_equiv>
            <PDBx:auth_seq_id>1</PDBx:auth_seq_id>
            <PDBx:auth_comp_id>DC</PDBx:auth_comp_id>
            <PDBx:auth_asym_id>A</PDBx:auth_asym_id>
            <PDBx:auth_atom_id>O5&apos;</PDBx:auth_atom_id>
            <PDBx:pdbx_PDB_model_num>1</PDBx:pdbx_PDB_model_num>
        </PDBx:atom_site>
        <PDBx:atom_site id="2">
            <PDBx:group_PDB>ATOM</PDBx:group_PDB>
            <PDBx:type_symbol>C</PDBx:type_symbol>
            <PDBx:label_atom_id>C5&apos;</PDBx:label_atom_id>
            <PDBx:label_alt_id xsi:nil="true" />
            <PDBx:label_comp_id>DC</PDBx:label_comp_id>
            <PDBx:label_asym_id>A</PDBx:label_asym_id>
            <PDBx:label_entity_id>1</PDBx:label_entity_id>
            <PDBx:label_seq_id>1</PDBx:label_seq_id>
            <PDBx:Cartn_x>9.675</PDBx:Cartn_x>
            <PDBx:Cartn_y>21.129</PDBx:Cartn_y>
            <PDBx:Cartn_z>23.376</PDBx:Cartn_z>
            <PDBx:occupancy>1.00</PDBx:occupancy>
            <PDBx:B_iso_or_equiv>8.13</PDBx:B_iso_or_equiv>
            <PDBx:auth_seq_id>1</PDBx:auth_seq_id>
            <PDBx:auth_comp_id>DC</PDBx:auth_comp_id>
            <PDBx:auth_asym_id>A</PDBx:auth_asym_id>
            <PDBx:auth_atom_id>C5&apos;</PDBx:auth_atom_id>
            <PDBx:pdbx_PDB_model_num>1</PDBx:pdbx_PDB_model_num>
        </PDBx:atom_site>
        <!--- data abbreviated   -->
    </PDBx:atom_siteCategory>
```

Figure 10.14. (a) An abbreviated example of PDBML encoding of data from the atom_site category. In this example, the coordinates of two atoms are included. Each coordinate record is enclosed within <PDBx:atom_site> tags. The other item names are taken directly from the item names defined in the mmCIF dictionary less the category name. For instance, the value of _atom_site.id the unique identifier or category key for this category is presented as an XML attribute (i.e., <id="1">). Other data items are encoded as XML data elements (e.g., <PDBx:type_symbol></PDBx:type_symbol>). All of the PDBML element names are qualified with the XML namespace, PDBx. (b) The data corresponding to Figure 10.14a expressed using mmCIF. In the mmCIF representation, the keywords defining the data items are specified once as part of the *loop_* declaration. (c) The PDBML presentation for the data in Figure 10.14a using the more efficient alternative atom record markup.

```
(b)  #
     loop_
     _atom_site.group_PDB
     _atom_site.id
     _atom_site.type_symbol
     _atom_site.label_atom_id
     _atom_site.label_alt_id
     _atom_site.label_comp_id
     _atom_site.label_asym_id
     _atom_site.label_entity_id
     _atom_site.label_seq_id
     _atom_site.pdbx_PDB_ins_code
     _atom_site.Cartn_x
     _atom_site.Cartn_y
     _atom_site.Cartn_z
     _atom_site.occupancy
     _atom_site.B_iso_or_equiv
     _atom_site.auth_seq_id
     _atom_site.auth_comp_id
     _atom_site.auth_asym_id
     _atom_site.auth_atom_id
     _atom_site.pdbx_PDB_model_num
     ATOM 1 O  "O5'" . DC  A 1 1 ? 9.462   20.142 22.444 1.00 8.67 1   DC  A "O5'" 1
     ATOM 2 C  "C5'" . DC  A 1 1 ? 9.675   21.129 23.376 1.00 8.13 1   DC  A "C5'" 1
     # --  data abbreviated --


(c)  <PDBx:category_atom_record>
         <PDBx:atom_record id="1">
          ATOM 1 A A 1 1 ? . DC DC O O5' O5'   9.462 20.142 22.444 1.00 8.67
         </PDBx:atom_record>
         <PDBx:atom_record id="2">
          ATOM 1 A A 1 1 ? . DC DC C C5  C5'   9.675 21.129 23.376 1.00 8.13
         </PDBx:atom_record>
         <!--- data abbreviateda -->
     </PDBx:category_atom_record>
```

Figure 10.14. (*Continued*)

the time of this writing, web services were deployed for testing at each of the wwPDB distribution sites.

CONCLUSION

In this chapter, the syntactical features of a number of ways to represent macromolecular structure data have been discussed. Each of these has its strengths and weaknesses. The PDB format is simple and accessible with very simple software tools. PDBML provides great flexibility and is well supported by commercial and public domain software. Far more important than the details of syntax is the ability of a particular data representation to precisely define data in a manner that is completely electronically accessible. The particular strength of the mmCIF approach is that it is based on a data dictionary. This data dictionary provides the detailed semantics, including precise definitions, and examples combined with a robust metadata model, which can be exploited by software to perform detailed checks on individual data items as well as checks of the internal consistency between data items. The dictionary also provides the necessary semantic information to support lossless translation into alternative data formats and application program interfaces.

ACKNOWLEDGMENTS

The development of the mmCIF dictionary and the associated DDL was an enormous community task, and the authors realize that any list of contributors to the effort will certainly be incomplete. Much of the mmCIF dictionary development was done by the original working group, including Enrique Abola, Helen Berman, Phil Bourne, Eleanor Dodson, Art Olson, Wolfgang Steigemann, Lynn Ten Eyck, and Keith Watenpaugh. Evaluation and critique of the dictionary development process was greatly aided by the input from COMCIFS, the IUCr committee with oversight over this process (I. David Brown and Brian McMahon). Many members of the community provided valuable input during the public review of the mmCIF dictionary. Frances Bernstein, Herbert Bernstein, Dale Tronrud, Kim Henrick, and Peter Keller were particularly active in this review process. Sydney Hall, Michael Scharf, Peter Grey, Peter Murray-Rust, Dave Stampf, and Jan Zelinka contributed to defining the requirements for the mmCIF DDL. The development of the PDB Exchange dictionary has been a collaborative effort within the wwPDB with significant contributions from Helen Berman, Kim Henrick, John Markley, Eldon Ulrich and Cathy Lawson. PDBML was also developed as a wwPDB collaboration with significant participation from Nobutoshi Ito, Haruki Nakamura, Helen Berman, and Kim Henrick.

REFERENCES

Berman HM, Westbrook J, Feng Z, Gilliland G, Bhat TN, Weissig H, Shindyalov IN, Bourne PE (2000): The Protein Data Bank. *Nucleic Acids Res* 28:235–242.

Berman HM, Henrick K, Nakamura H (2003): Announcing the worldwide Protein Data Bank. *Nat Struct Biol* 10(12):980.

Bernstein FC, Koetzle TF, Williams GJ, Meyer EE, Brice MD, Rodgers JR, Kennard O, Shimanouchi T, Tasumi M (1977): Protein Data Bank: a computer-based archival file for macromolecular structures. *J Mol Biol* 112:535–542.

Bourne PE, Berman HM, Watenpaugh K, Westbrook JD, Fitzgerald PMD (1997): The macromolecular Crystallographic Information File (mmCIF). *Methods Enzymol* 277:571–590.

Brown ID, (2005): Symmetry Dictionary (symCIF): In: Hall SR, McMahon B, editors. *International Tables for Crystallography*, Vol. G. Definition and Exchange of Crystallographic Data. Dordrecht, The Netherlands: Springer. pp. 459–466.

Callaway J, Cummings M, Deroski B, Esposito P, Forman A, Langdon P, Libeson M, McCarthy J, Sikora J, Xue D, Abola E, Bernstein F, Manning N, Shea R, Stampf D, Sussman J (1996): *Protein Data Bank Contents Guide: Atomic Coordinate Entry Format Description*. Upton, NY: Brookhaven National Laboratory.

Fitzgerald PMD, McKeever BM, VanMiddlesworth JF, Springer JP, Heimbach JC, Leu C-T, Kerber WK, Dixon RAF, Darke PL (1990): Crystallographic analysis of a complex between Human Immunodeficiency Virus Type 1 protease and acetyl-pepstatin at 2.0 Å resolution. *J Biol Chem* 265:14209–14219.

Fitzgerald PMD, Berman HM, Bourne PE, Watenpaugh K (1992): Macromolecular CIF working group, International Union of Crystallography.

Fitzgerald PMD, Berman HM, Bourne PE, McMahon B, Watenpaugh K, Westbrook J (1996): The mmCIF dictionary: community review and final approval. IUCr Congress and General Assembly, August 8–17, Seattle, WA. *Acta Cryst* A52 (Suppl):MSWK. CF. 06.

Fitzgerald PMD, Westbrook JD, Bourne PE, McMahon B, Watenpaugh KD, Berman HM (2005): The macromolecular Crystallographic Information File (mmCIF): *International Tables for*

Crystallography, Vol. G. Definition and Exchange of Crystallographic Data. Dordrecht, The Netherlands: Springer. pp. 295–443.

Hall SR, (1991): The STAR File: a new format for electronic data transfer and archiving. *J Chem Inf Comput Sci* 31:326–331.

Hall SR, Allen AH, Brown ID (1991): The Crystallographic Information File (CIF): a new standard archive file for crystallography. *Acta Cryst* A47:655–685.

Hammersley AP, Bernstein HJ, Westbrook JD (2005): Image Dictionary (imgCIF): In: Hall SR, McMahon B, editors. *International Tables for Crystallography*, Vol. G. Definition and Exchange of Crystallographic Data. Dordrecht, The Netherlands: Springer. pp. 444–458.

Henrick, K, Feng, Z, Bluhm, W, Dimitropoulos, D, Doreleijers, JF, Dutta, S, Flippen-Anderson, JL, Ionides, J, Kamada, C, Krissinel, E, Lawson, CL, Markley, JL, Nakamura, H, Newman, R, Shimizu, Y, Swaminathan, J, Velankar, S, Ory, J, Ulrich, EL, Vranken, W, Westbrook, J, Yamashita, R, Yang, H, Young, J, Yousufuddin, M, Berman, H (2008): Remediation of the Protein Data Bank Archive. *Nucleic Acids Res* 36:D426–D33.

International Structural Genomics Organization Task Force on Deposition, Annotation and Curation of Primary Information (2001): Task Force Reports from the Second International Structural Genomics Meeting, Arilie, Va. http://www.nigms.nih.gov/news/reports/airlie_tasks.html.

Lawson, CL, Dutta, S, Westbrook, JD, Henrick, K, Berman, HM (2008): Representation of viruses in the remediated PDB archive. *Acta Cryst* D64:874–882.

Protein Data Bank (1992): Protein Data Bank atomic coordinate and bibliographic entry format description. Upton, NY: Brookhaven National Laboratory.

Seavey BR, Farr EA, Westler WM, Markley JL (1991): A relational database for sequence-specific protein NMR data. *J Biomolecular NMR* 1:217–236.

Westbrook J, Bourne PE (2000): STAR/mmCIF: an extensive ontology for macromolecular structure and beyond. *Bioinformatics* 16:159–168.

Westbrook JD, Berman HM, Hall SR (2005a): Specification of a relational Dictionary Definition Language (DDL2). In: Hall SR, McMahon B, editors. *International Tables for Crystallography*, Vol. G. Definition and Exchange of Crystallographic Data. Dordrecht, The Netherlands: Springer. pp. 61–70.

Westbrook JD, Henrick K, Ulrich EL, Berman HM (2005b): The Protein Data Bank exchange data dictionary. In: Hall SR, McMahon B, editors. *International Tables for Crystallography*, Vol. G. Definition and Exchange of Crystallographic Data. Dordrecht, The Netherlands: Springer. pp. 195–198.

Westbrook J, Ito N, Nakamura H, Henrick K, Berman HM (2005c): PDBML: The representation of archival macromolecular structure data in XML. *Bioinformatics* 21:988–992.

11

THE WORLDWIDE PROTEIN DATA BANK

Helen M. Berman, Kim Henrick, Haruki Nakamura, and John L. Markley

INTRODUCTION

The Protein Data Bank was established at Brookhaven National Laboratory (BNL) (Bernstein et al., 1977) in 1971 as an archive for biological macromolecular crystal structures. It represents one of the earliest community-driven molecular biology data collections. In the beginning, the archive held seven structures, and with each passing year a handful more were deposited. In the 1980s, the number of deposited structures began to increase dramatically. This was due to the improvements in technology for all aspects of the crystallographic process, the addition of structures determined by other methods, and changes in community views about data sharing. By the early 1990s, many journals required a PDB ID for publication; now, virtually all journals require not only coordinates but also the primary experimental data be deposited with the PDB. As of December 2, 2008, the archive contained 54559 structures.

The initial goal of the PDB was to archive author-submitted structures determined by X-ray crystallography. Today, structures are deposited in the PDB that have been determined using X-ray crystallography, nuclear magnetic resonance (NMR), and, most recently, cryo-electron microscopy (cryoEM). In addition to the structural biologists who deposit data, the majority of PDB users are a diverse group of researchers in the biomedical sciences (biologists, chemists, physicists, etc.), as well as educators and students at all levels. To ensure that the PDB would remain a single uniform archive freely accessible via a collection of FTP tools, the worldwide PDB (wwPDB; www.wwpdb.org/) was formed in 2003 (Berman, Henrick, and Nakamura, 2003) with an agreement among three depositions centers in the United States, Europe, and Japan: the Research Collaboratory for Structural Bioinformatics (RCSB PDB), the Macromolecular Structure Database at the European Bioinformatics Institute (now called Protein Data Bank Europe (PDBe)), and the PDB Japan (PDBj). The BioMagResBank (BMRB), which archives biomolecular NMR data, joined the wwPDB in 2006. In this chapter, we describe the collection, validation, annotation, and distribution of data by the wwPDB. We additionally provide a brief description of the services offered by the member sites.

Structural Bioinformatics, Second Edition   Edited by Jenny Gu and Philip E. Bourne
Copyright © 2009 John Wiley & Sons, Inc.

DATA ACQUISITION AND PROCESSING

Content of the Data Collected by the wwPDB

The PDB archive consists of entries containing three-dimensional Cartesian coordinates, information specific to the method of structure determination, and more recently experimental data. These include structure factors for X-ray experiments and constraints generated by NMR experiments. NMR depositions in the PDB contain links to the BMRB archive of web-accessible constraints, chemical shifts, and other data relevant to NMR structures. In the near future, volumes for electron microscopy will become a part of the PDB archive. Table 11.1 contains the general information that the wwPDB collects for all structures as well as the information specific to X-ray, NMR, and cryoEM experiments.

The definitions of the data items collected are in the PDB Exchange Dictionary (PDBx) (Westbrook et al., 2005a) that is based on the mmCIF syntax. The mmCIF dictionary contains 1,700 terms that define the macromolecular structure and the crystallographic experiment (Fitzgerald et al., 2005). PDBx contains additional terms for NMR and cryoEM. The BMRB-developed NMR-STAR Dictionary specifies additional data items linked to NMR structures. These terms were developed in collaboration with the depositors who are experts in these methods. Terms needed for tracking and other information management purposes are also contained in the PDBx.

Data Deposition Sites

Data are deposited to the PDB at one of the wwPDB member sites. Because it is critical that the final archive is kept uniform, the content and format of the final files, as well as the methods used to check them, are the same.

Each deposition to the PDB is represented by a PDBid—a four-character code. The PDBid is assigned arbitrarily and is an immutable reference to the structure. PDBids are never reused and remain the link between the structure and the literature reference that describes that structure. Experimental NMR data processed and archived at the BMRB are identified by a unique and immutable integer tag. Hyperlinks in the related PDB and BMRB files provide seamless access to all information.

The RCSB PDB (deposit.pdb.org/adit/) and PDBj (pdbdep.protein.osaka-u.ac.jp/adit/) use the program ADIT for data deposition and validation. PDBe processes data that are submitted to the site via AutoDep (autodep.ebi.ac.uk/) (Tagari et al., 2006).

BMRB sites at Madison (batfish.bmrb.wisc.edu/bmrb-adit/) and Osaka (nmradit. protein.osaka-u.ac.jp/bmrb-adit/) use ADIT-NMR to collect both coordinate and experimental data. The experimental data are processed by BMRB and the coordinate data by RCSB PDB.

After processing is complete, coordinates, structure factor files, and constraints from all the sites are sent to the RCSB PDB for inclusion in the archive.

Validation and Annotation

Validation refers to the procedure for assessing the quality of deposited atomic models (structure validation) and for assessing how well these models fit the experimental data (experimental validation). Annotation refers to the process of adding information resulting from the validation to the entry. The wwPDB validates structures and associated data by using accepted community standards.

TABLE 11.1. Content of Data in the PDB

Content of All Depositions

Source: specifications such as genus, species, strain, or variant of gene (cloned or synthetic);
    expression vector and host, or description of method of chemical synthesis
Sequence: Full sequence of all macromolecular components
Chemical structure of cofactors and prosthetic groups
Names of all components of the structure
Broad description of the function and/or the composition of the structure
Literature citations for the structure submitted
Three-dimensional coordinates
Keywords and experimental method

Additional Items for X-ray Structure Determinations

Temperature factors and occupancies assigned to each atom
Crystallization conditions, including pH, temperature, solvents, salts, and methods
Crystal data, including the unit cell dimensions and space group
Presence of noncrystallographic symmetry
Data collection information describing the methods used to collect the diffraction data including
    instrument, wavelength, temperature, and processing programs
Data collection statistics including data coverage, Rsym, I/sigma I, data above 1, 2, 3 sigma levels and
    resolution limits
Refinement information including R factor, resolution limits, number of reflections, method of
    refinement, sigma cutoff, geometry rmsd, sigma
Structure factors: h, k, l, Fobs, sigma Fobs, Intensity, sigma intensity, Flag for free R test

Additional Items for NMR Structure Determinations

For an ensemble, the model number for each coordinate set that is deposited and an indication if one
    should be designated as a representative
Data collection information describing the types of methods used, instrumentation, magnetic field
    strength, console, probe head, and sample tube
Sample conditions, including solvent, macromolecule concentration ranges, concentration ranges of
    buffers, salts, antibacterial agents, other components, isotopic composition
Experimental conditions, including temperature, pH, pressure, and oxidation state of structure
    determination and estimates of uncertainties in these values
Noncovalent heterogeneity of sample, including self-aggregation, partial isotope exchange,
    conformational heterogeneity resulting in slow chemical exchange
Chemical heterogeneity of the sample (e.g., evidence for deamidation or minor covalent species)
A list of NMR experiments used to determine the structure including those used to determine
    resonance assignments, NOE/ROE data, dynamical data, scalar coupling constants, and those used
    to infer hydrogen bonds and bound ligands. The relationship of these experiments to the constraint
    files is given explicitly.
Constraint files used to derive the structure as described in the Task Force recommendations
Links (where available) to associated files at BMRB containing chemical shift, filtered constraint, and
    other NMR data

Additional Data Items for Cryoelectron Microscopy

Sample preparation, including aggregation state, concentration, buffer, pH, sample support, and
    description of vitrification procedure

(*continued*)

TABLE 11.1. (*Continued*)

3D reconstruction procedure, including method, nominal and actual pixel size, resolution, CTF correction method, magnification calibration, and text description of the reconstruction procedure

3D model fitting procedure, including fitting procedure type and program used, IDs of models used

3D model refinement, including whether refinement is done in real space or reciprocal space, type of protocol, and refinement target criteria. If sample is a 2D or 3D array, symmetry and repeat parameters.

If the entry contains a virus, details of the virus host including type, species, and growth cell, as well as details of the virus including type, isolate, and International Committee on Taxonomy of Viruses ID.

The following checks are run:

Covalent Bond Distances and Angles: Proteins are compared against standard values from Engh and Huber (1991); nucleic acid bases are compared against standard values from Clowney et al. (1996); and sugar and phosphates are compared against standard values from Gelbin et al. (1996).

Stereochemical Validation: All chiral centers of proteins and nucleic acids are checked for correct stereochemistry.

Atom Nomenclature: The nomenclature of all atoms is checked for compliance with IUPAC standards (IUPAC-IUB Joint Commission on Biochemical Nomenclature, 1983; Markley et al. 1998) and adjusted if necessary.

Close Contacts: The distances between all atoms within the asymmetric unit of crystal structures and the unique molecule of NMR structures are calculated. For crystal structures, contacts between symmetry-related molecules are checked as well.

Ligand and Atom Nomenclature: Residue and atom nomenclature is compared against a standard dictionary (www.wwpdb.org/ccd.html) for all ligands as well as standard residues and bases. Unrecognized ligand groups are flagged and any discrepancy in known ligands is listed as extra or missing atom. New ligands are added to the dictionary as they are deposited.

Sequence Comparison: The sequence provided by the depositor is compared against the sequence derived from the coordinate records. This information is displayed in a table where any differences or missing residues are annotated. During the annotation process the sequence database references provided by the author are checked for accuracy. If no reference is given, a BLAST (Altschul et al., 1990) search is used to find the best match. Any conflict between the depositor's sequence and the sequence derived from the coordinate records is further resolved and annotated by comparison with other sequence databases as needed.

Distant Waters: The distances between all water oxygen atoms and all polar atoms (oxygen and nitrogen) of the macromolecules, ligands, and solvent in the asymmetric unit are calculated. Distant solvent atoms are repositioned using crystallographic symmetry such that they fall within the solvation sphere of the macromolecule.

Geometry: The torsion angle distributions and peptide bond deviations from *cis* and *trans* conformation are also checked. In addition, derived data such as site records, helix and sheet records for secondary structure are also annotated.

Checks of NMR Data by the BMRB: NMR constraints are checked for consistency with the three-dimensional structure and atom nomenclature, chemical shifts are checked for possible referencing errors and outliers.

In almost all cases, serious errors detected by these checks have been corrected through annotation and correspondence with the authors.

The wwPDB continuously reviews the validation methods used and will continue to integrate new procedures as they become available and are accepted as community standards.

Data Processing Statistics

Figure 11.1a shows the growth of PDB data, indicating how the complexity of structures released into the archive has increased over time.

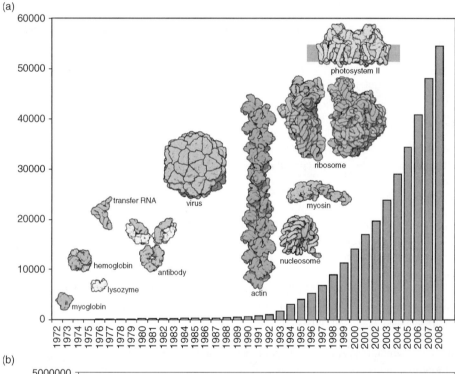

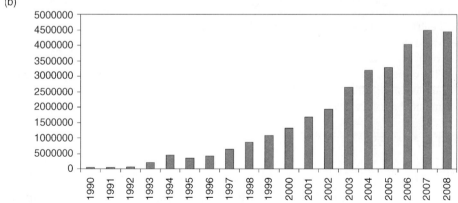

Figure 11.1. Growth in the PDB archive as of December 2, 2008. (a) Growth chart of the PDB showing the total number of structures available in the PDB archive per year and highlighting example structures from different time periods. (b) Number of residues released in the PDB per year.

TABLE 11.2. Demographics of Data Released in the PDB (as of December 2, 2008)

Experimental Technique	Molecule Type				
	Proteins, Peptides, and Viruses	Nucleic Acids	Protein–Nucleic Acid Complexes	Carbohydrates and Other	Total
X-ray diffraction and other	43,520	1,101	2,011	24	46,656
NMR	6,624	827	140	7	7,598
Electron microscopy	142	12	50	0	204
Other	91	4	4	2	101
Total	50,377	1,944	2,205	33	54,559

As of December 2, 2008, the PDB contained more than 54000 publicly accessible structures; of these entries, 46656 (86%) were determined by X-ray methods, 7598 (14%) were determined by NMR, and 204 were determined by cryoEM. Overall, 73% of the entries have released experimental data. As of this writing, in 2008 more than 99% of X-ray structures had structure factors deposited. For NMR structures, 98% were accompanied by restraint files and 44% by chemical shifts and other supporting data (Table 11.2).

Data Uniformity

A key goal of the wwPDB is to make the archive as consistent and error-free as possible. Before the files are released, all new depositions are reviewed by annotators. Errors found subsequent to release by authors and PDB users are addressed as rapidly as possible and the entries are revised.

The member sites of wwPDB have collaborated to remediate all the data in the PDB archive (Henrick et al., 2008; Lawson et al., 2008). In particular, the monomer components and the ligands have been examined and a new dictionary has been constructed that includes the stereochemistry, nomenclature, ideal coordinates, model coordinates as well as SMILES, and InChI representations. A Chemical Component Dictionary (www.wwpdb.org/ccd.html) contains the definitions for all components. All the instances of these components contained in the PDB entries have been matched against this dictionary and where necessary corrections have been made to the PDB files. The sequences, citations, and some experimental data items have also been corrected.

The corrected files are available at the wwPDB site (ftp://ftp.wwpdb.org). Coupled with this remediation effort has been a careful review of all annotation practices that are documented at (www.wwpdb.org/docs.html).

DATA ACCESS

FTP

The "PDB archive" is the collection of flat files that are maintained in three different formats: the legacy PDB file format (Bernstein et al., 1977); the PDB exchange format (PDBx) that follows the mmCIF syntax (Fitzgerald et al., 2005; Westbrook et al., 2005a) (mmcif.pdb.org); and the PDBML/XML format (Westbrook et al., 2005b) that is a direct translation of the PDB exchange format. Each wwPDB site distributes the same PDB archive via FTP. The archive is updated weekly.

Time-stamped snapshots of the PDB archive are added each year to ftp://snapshots. rcsb.org. They provide a frozen copy of the archive as it appeared at that time for research and historical purposes. Scripts are available to download all, or part, of a snapshot automatically.

Web Sites

In addition to providing access to the PDB archive, each wwPDB site provides services and resources that provide different views and analyses of the structural data contained within the PDB archive.

RCSB PDB

The RCSB PDB (www.pdb.org) (Berman et al., 2000) provides a web site with several functionalities. In addition to being able to download files in PDB, PDBx, and PDBML formats, there are different types of search capabilities. These include search by PDB ID, author, or keyword; search within a relational database that is populated by the data in the PDBx/PDBML files; or browsing characteristics that have been integrated from external resources (Deshpande et al., 2005). These include biological process, cellular component, and molecular function as defined by Gene Ontology (The Gene Ontology Consortium, 2000); enzyme classification, Medical Subject Headings (MeSH terms), source organism as defined by the NCBI Taxonomy, gene location from Entrez (Wheeler et al., 2006), and folds as classified by SCOP (Murzin et al., 1995) and CATH (Orengo et al., 1997). Many data items on the Structure Summary page for each structure are links to searches for other structures with similar properties. A variety of tabular reports can be made for groups of structures with information about the experiment, the chemistry and biology, and citations. Static pictures and interactive graphics are provided for each structure and protein–ligand interaction from the web site. Some of these interactive views use the molecular biology toolkit (MBT; mbt.sdsc.edu; Moreland et al., 2005). A variety of summary statistics are available as histograms and in tabular form. Advanced searches can be performed on the structure data in combination with the data integrated from over 30 different external sources. Multimedia tutorials are available describing these features.

Also contained at the RCSB PDB site is the *Molecule of the Month* authored by David S. Goodsell. This feature describes a molecule in such a way that nonspecialists can learn about a particular molecular system. The BioSync (biosync.pdb.org) resource provides information about the characteristics of synchrotron beamlines (Kuller et al., 2002).

For application developers, a variety of Web Services are available to directly extract subsets of data.

PDBj

PDBj (www.pdbj.org) maintains a variety of services for querying, displaying, and analyzing PDB data. xPSSS, in addition to providing PDBid and keyword searches, also allows more sophisticated text-based searches. xPSSS takes advantage of PDBj's native extensible markup language (XML) database by using XPath and XQuery to perform even very complicated queries efficiently. Structure Navigator can be used to locate PDB entries that are similar to a query structure or PDB ID. Structural similarity is defined by the newly developed score used in the stand-alone program ASHs (Alignment of Structural Homologues), which can be run or downloaded from www.pdbj.org/ASH/. ASH shows high sensitivity and selectivity in recognizing CATH or SCOP domains, without requiring large amounts of CPU (Standley, Toh, and Nakamura, 2004; Standley, Toh, and Nakamura, 2005; Standley, Toh, and Nakamura, 2007). Sequence Navigator is a BLAST interface to PDBj. The input to Sequence Navigator can be a PDB ID or an amino acid sequence. Most of the query services described above can be executed as Web Services using the Simple Object Application Protocol (SOAP) or through web browsers. jV version 3 (www.pdbj.org/jV, formerly known as PDBjViewer) is a program to display molecular graphics of proteins and nucleic acids (Kinoshita and Nakamura, 2004). The eProtS, Encyclopedia of Protein Structures, is an educational resource for learning biological functions and structural characteristics of protein molecules. The eProtS now contains explanations for a few hundred protein molecules that are of particular interest. eProtS is implemented as a Wiki so that motivated structural biologists can contribute articles for the proteins they have solved. Two derived databases are also available from PDBj. ProMode is a database of normal mode analyses (NMA) of proteins. eF-site (electrostatic surface of functional site) is a database for molecular surfaces of proteins' functional sites, displaying the electrostatic potentials and hydrophobic properties together on the Connolly surfaces of the active sites, for analyses of the molecular recognition mechanisms. eF-seek is a service for searching the similar molecular surface of the known active sites (Standley et al. 2008).

PDBe

PDBe (www.ebi.ac.uk/msd) (Velankar et al., 2005) has developed a variety of search and retrieval systems, some of which are summarized here. MSDChem (www.ebi.ac.uk/msd-srv/chempdb) and MSDsite (Golovin et al., 2004) (www.ebi.ac.uk/msd-srv/msdsite) extract the ligand binding information and characterize their chemical and geometrical environments using the MSD relational database thus allowing a user to obtain detailed information about a specific ligand or macromolecule–ligand interaction. MSDanalysis (www.ebi.ac.uk/msd-as/MSDvalidate) allows statistical analysis of macromolecular structure. These include the following: *MSDvalidate*: the validation of structures within the PDB and uploaded files; *MSDstatistics*: the view of statistical information from the molecules in the PDB; *MSDselection*: the creation of selections of PDB molecules based on statistical distributions of molecular data; *MSDresidue statistics*: the analysis of the residue-based data from macromolecular structure; *MSDdatabase*: a web interface that allows SQL queries to be made against the MSD database; and *MSDmine*: a fully featured analysis of the MSD database based on an SQL like interface.

PISA is an interactive tool for the exploration of macromolecular interfaces, prediction of probable quaternary structures (assemblies) and database searches of structurally similar interfaces and assemblies (Krissinel and Henrick, 2005). MSDMOTIF is a database that allows to perform fast Phi/Psi search, secondary structure patterns search, and small 3D structural motifs' complexes search (www.ebi.ac.uk/msd-srv/msdmotif/). It also allows visualization as well as sequence multiple alignment and 3D multiple alignment.

MSDtemplate provides matching of a small number of amino acids by geometry to a protein (www.ebi.ac.uk/msd-srv/MSDtemplate). A template can be an active site, ligand environment, or any set of amino acids that are of interest. The MSD has a library of templates generated from data mining the PDB looking for statistically significant collections of residues. These templates from data mining include all the known active sites and metal binding sites.

BMRB

The BMRB (www.bmrb.wisc.edu) (Ulrich, Markley, and Kyogoku, 1989) archive can be searched by database ID, keywords, author, molecule name, sequence, sample conditions, or other characteristics. The website contains statistics and resources for the NMR community. BMRB archives data connected with an NMR structure as recommended by the wwPDB NMR Task Force for structural biology depositions: structural constraints, assigned chemical shifts, and NOESY spectra and/or peak lists. BMRB also supports more comprehensive NMR data depositions by members of structural genomics community, which may include additional primary (time domain) spectra and intermediate results.

The BMRB provides software and tools for validating NMR data and associated structures prior to deposition and tools for visualizing various types of NMR data with or without an associated three-dimensional structure. The BMRB is responsible for processing constraint files and for providing constraints to the PDB archive that are consistent with structural models and converted to machine-readable NMR-STAR format. Filtered restraints, further processed for use by test calculation protocols and longitudinal analyses and validations, are available online.

FUTURE

Structural biology is a fast evolving field that poses challenges to the collection, curation, and distribution of macromolecular structure data. Since 1998, the rate of deposition has tripled, and the total holdings have increased by a factor of five. The structures have increased in complexity with many molecular machines now part of the archive. In addition, the use of data harvesting has made it possible to collect more information about each structure. The challenge is to be able to continue to collect data rapidly and at the same time maintain quality.

The maintenance and further development of the PDB archive is a worldwide effort. The willingness of the global community to share ideas, software, and data provides a unique resource for biological research.

ACKNOWLEDGMENTS

The RCSB PDB is operated by Rutgers, the State University of New Jersey and the San Diego Supercomputer Center and the Skaggs School of Pharmacy and Pharmaceutical Sciences at the University of California, San Diego. It is supported by funds from the National Science Foundation, the National Institute of General Medical Sciences, the Office of Science, Department of Energy, the National Library of Medicine, the National Cancer Institute, the National Center for Research Resources, the National Institute of Biomedical Imaging and Bioengineering, the National Institute of Neurological Disorders and Stroke, and the National Institute of Diabetes and Digestive and Kidney Diseases.

EBI-MSD gratefully acknowledges the support of the Wellcome Trust (GR062025MA), the EU (TEMBLOR, NMRQUAL, and IIMS), CCP4, the BBSRC, the MRC and EMBL.

PDBj is supported by grant-in-aid from the Institute for Bioinformatics Research and Development, Japan Science and Technology Agency (BIRD-JST), and the Ministry of Education, Culture, Sports, Science, and Technology (MEXT).

The BMRB is supported by NIH grant LM05799 from the National Library of Medicine.

REFERENCES

Altschul SF, Gish W, Miller W, Myers EW, Lipman DJ (1990): Basic local alignment search tool. *J Mol Biol* 215:403–410.

Berman HM, Henrick K, Nakamura H (2003): Announcing the worldwide Protein Data Bank. *Nat Struct Biol* 10(12):980.

Berman HM, Westbrook J, Feng Z, Gilliland G, Bhat TN, Weissig H, Shindyalov IN, Bourne PE (2000): The Protein Data Bank. *Nucleic Acids Res* 28:235–342.

Bernstein FC, Koetzle TF, Williams GJB, Meyer EF Jr, Brice MD, Rodgers JR, Kennard O, Shimanouchi T, Tasumi M (1977): Protein Data Bank: a computer-based archival file for macro-molecular structures. *J Mol Biol* 112:535–542.

Clowney L, Jain SC, Srinivasan AR, Westbrook J, Olson WK, Berman HM (1996): Geometric parameters in nucleic acids: nitrogenous bases. *J Am Chem Soc* 118:509–518.

Deshpande N, Addess KJ, Bluhm WF, Merino-Ott JC, Townsend-Merino W, Zhang Q, Knezevich C, Xie L, Chen L, Feng Z, Kramer Green R, Flippen-Anderson JL, Westbrook J, Berman HM, Bourne PE (2005): The RCSB Protein Data Bank: a redesigned query system and relational database based on the mmCIF schema. *Nucleic Acids Res* 33:D233–D237.

Engh RA, Huber R (1991): Accurate bond and angle parameters for X-ray protein structure refinement. *Acta Crystallogr* A47:392–400.

Fitzgerald PMD, Westbrook JD, Bourne PE, McMahon B, Watenpaugh KD, Berman HM (2005): 4.5 Macromolecular dictionary (mmCIF). In: Hall SR, McMahon B, editors. *International Tables for Crystallography*, Vol. G. Definition and Exchange of Crystallographic Data. Dordrecht, The Netherlands: Springer. pp. 295–443.

Gelbin A, Schneider B, Clowney L, Hsieh S-H, Olson WK, Berman HM (1996): Geometric parame-ters in nucleic acids: sugar and phosphate constituents. *J Am Chem Soc* 118(3):519–528.

Golovin A, Oldfield TJ, Tate JG, Velankar S, Barton GJ, Boutselakis H, Dimitropoulos D, Fillon J, Hussain A, Ionides JM, John M, Keller PA, Krissinel E, McNeil P, Naim A, Newman R, Pajon A, Pineda J, Rachedi A, Copeland J, Sitnov A, Sobhany S, Suarez-Uruena A, Swaminathan GJ, Tagari M, Tromm S, Vranken W, Henrick K (2004): E-MSD: an integrated data resource for bioinformatics. *Nucleic Acids Res* 32 (Database issue):D211–D216.

Henrick, K, Feng, Z, Bluhm, W, Dimitropoulos, D, Doreleijers, JF, Dutta, S, Flippen-Anderson, JL, Ionides, J, Kamada, C, Krissinel, E, Lawson, CL, Markley, JL, Nakamura, H, Newman, R, Shimizu, Y, Swaminathan, J, Velankar, S, Ory, J, Ulrich, EL, Vranken, W, Westbrook, J, Yamashita, R, Yang, H, Young, J, Yousufuddin, M, Berman, H (2008): Remediation of the Protein Data Bank Archive, *Nucleic Acids Res* 36:D426–D33.

IUPAC-IUB Joint Commission on Biochemical Nomenclature (1983) Abbreviations and symbols for the description of conformations of polynucleotide chains. *Eur J Biochem* 131:9–15.

Kinoshita K, Nakamura H (2004): eF-site and PDBjViewer: database and viewer for protein functional sites. *Bioinformatics* 20:1329–1330.

Krissinel E, Henrick K (2005): Detection of protein assemblies in crystals. In: Berthold MR, Glen R, Diederichs K, Kohlbacher O, Fischer I, editors. *Computational Life Sciences. First International Symposium: CompLife 2005, Konstanz, Germany, September 25–27, 2005, Proceedings*, Vol. 3695. Berlin: Springer-Verlag. pp. 163–174.

Kuller A, Fleri W, Bluhm WF, Smith JL, Westbrook J, Bourne PE (2002): A biologist's guide to synchrotron facilities: the BioSync web resource. *TIBS* 27:213–215.

Lawson, CL, Dutta, S, Westbrook, JD, Henrick, K, Berman, HM (2008): Representation of viruses in the remediated PDB archive, *Acta Cryst* D64:874–882.

Markley JL, Bax A, Arata Y, Hilbers CW, Kaptein R, Sykes BD, Wright PE, Wüthrich K (1998): Recommendations for the presentation of NMR structures of proteins and nucleic acids. IUPAC-IUBMB-IUPAB Inter-Union Task Group on the standardization of data bases of protein and nucleic acid structures determined by NMR spectroscopy. *J Biomol NMR* 12:1–23.

Moreland JL, Gramada A, Buzko OV, Zhang Q, Bourne PE (2005): The molecular biology toolkit (MBT) a modular platform for developing molecular visualization applications. *BMC Bioinformatics* 6(1):21.

Murzin AG, Brenner SE, Hubbard T, Chothia C (1995): SCOP: a structural classification of proteins database for the investigation of sequences and structures. *J Mol Biol* 247:536–540.

Orengo CA, Michie AD, Jones S, Jones DT, Swindells MB, Thornton JM (1997): CATH: a hierarchic classification of protein domain structures. *Structure* 5:1093–1108.

Standley DM, Toh H, Nakamura H (2004): Detecting local structural similarity in proteins by maximizing number of equivalent residues. *Proteins* 57:381–391.

Standley DM, Toh H, Nakamura H (2005): GASH: an improved algorithm for maximizing the number of equivalent residues between two protein structures. *BMC Bioinformatics* 6:221.

Standley DM, Toh H, Nakamura H (2007): ASH structure alignment package: sensitivity and selectivity in domain classification. *BMC Bioinformatics* 8:116.

Standley, DM, Kinjo, AR, Kinoshita, K, Nakamura, H (2008): Protein structure databases with new web services for structural biology and biomedical research, *Brief Bioinform* 9(4):276–285.

Tagari M, Tate J, Swaminathan GJ, Newman R, Naim A, Vranken W, Kapopoulou A, Hussain A, Fillon J, Henrick K, Velankar S (2006): E-MSD: improving data deposition and structure quality. *Nucleic Acids Res* 34(Database issue):D287–D290.

The Gene Ontology Consortium (2000) Gene ontology: tool for the unification of biology. *Nat Genet* 25:25–29.

Ulrich EL, Markley JL, Kyogoku Y (1989): Creation of a nuclear magnetic resonance data repository and literature database. *Protein Seq Data Ana* 2:23–37.

Velankar S, McNeil P, Mittard-Runte V, Suarez A, Barrell D, Apweiler R, Henrick K (2005): E-MSD: an integrated data resource for bioinformatics. *Nucleic Acids Res* 33(Database issue): D262–D265.

Westbrook J, Henrick K, Ulrich EL, Berman HM (2005a): 3.6.2 The Protein Data Bank exchange data dictionary. In: Hall SR, McMahon B, editors. *International Tables for Crystallography*, Vol. G. Definition and Exchange of Crystallographic Data. Dordrecht, The Netherlands: Springer. pp. 195–198.

Westbrook J, Ito N, Nakamura H, Henrick K, Berman HM (2005b): PDBML: The representation of archival macromolecular structure data in XML. *Bioinformatics* 21:988–992.

Wheeler DL, Barrett T, Benson DA, Bryant SH, Canese K, Chetvernin V, Church DM, DiCuccio M, Edgar R, Federhen S, Geer LY, Helmberg W, Kapustin Y, Kenton DL, Khovayko O, Lipman DJ, Madden TL, Maglott DR, Ostell J, Pruitt KD, Schuler GD, Schriml LM, Sequeira E, Sherry ST, Sirotkin K, Souvorov A, Starchenko G, Suzek TO, Tatusov R, Tatusova TA, Wagner L, Yaschenko E (2006): Database resources of the National Center for Biotechnology Information. *Nucleic Acids Res* 34s(Database issue):D173–D180.

12

THE NUCLEIC ACID DATABASE

Bohdan Schneider, Joanna de la Cruz, Zukang Feng, Li Chen,
Shuchismita Dutta, Irina Persikova, John D. Westbrook, Huanwang
Yang, Jasmine Young, Christine Zardecki, and Helen M. Berman

INTRODUCTION

The Nucleic Acid Database (NDB) (Berman et al., 1992) was established in 1991 as a resource for specialists in the field of nucleic acid structure. Over the years, the NDB has developed generalized software for processing, archiving, querying, and distributing structural data for nucleic acid containing structures. The core resource of the NDB has been its relational database of nucleic acid containing crystal structures; now the database has incorporated structures derived by both X-ray crystallography and methods of NMR spectroscopy. Recognizing the importance of standard data representation in building a database, the NDB became an active participant in the mmCIF project and provided the test bed for this format. From a foundation of well-curated data, the NDB created a searchable relational database of primary and derivative data with very rich query and reporting capabilities. This robust database was unique in that it allowed researchers to make comparative analyses of nucleic acid containing structures selected from the NDB by means of the many attributes stored in the database.

In 1992, the NDB assumed responsibility for processing all nucleic acid crystal structures that were deposited into the PDB, and in 1996, it became a direct deposit site. To meet data processing requirements, the NDB created the first validation software for nucleic acids (Feng, Westbrook, and Berman, 1998). The NDB continues to provide a high level of information about nucleic acids and serves as a specialty database for its community of researchers. When the NDB was established, the known world of nucleic acid structures consisted of DNA and RNA oligonucleotides, a few protein–DNA complexes, and some tRNA structures. Annotation of structural features was performed manually, and structures were easily classified into a few known molecular architectures by visual inspection. However, in the last 10 years investigators have uncovered a whole new universe of nucleic

Structural Bioinformatics, Second Edition   Edited by Jenny Gu and Philip E. Bourne
Copyright © 2009 John Wiley & Sons, Inc.

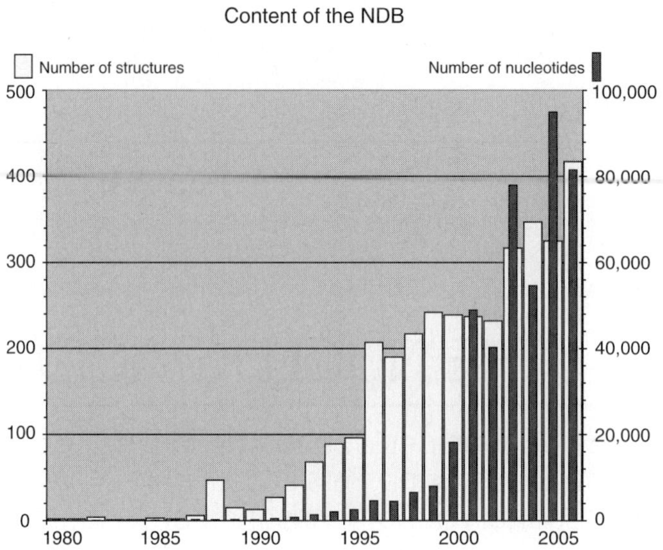

Content of the NDB

Figure 12.1. Available nucleic acids and nucleotides per year at the time of writing (1980–2006). The total number of nucleic acid-only structures added per year is indicated by the light-colored bars (scale on the left); dark bars show the number of nucleotides in each year's structures (scale on the right).

acid structures (Figure 12.1; see also Chapter 3). There are many ribozyme structures and many different types of protein–nucleic acid complexes represented in almost 2000 structures. Since information about ribosomal structures began to emerge in 2000, the number of nucleotide residues resident in the NDB has increased several fold (Moore, 2001).

In this chapter, we describe the architecture and capabilities of the NDB and then present some of the research that has been enabled by this resource.

DATA PROCESSING AND VALIDATION

The NDB created a robust data processing system that produces high-quality data that are readily loaded into a database. The full capability of this system was recently demonstrated by the successful processing of ribosomal subunits, which are very large and complex structures.

Early on, the NDB adopted the Macromolecular Crystallographic Information File (mmCIF) (Chapter 10; Fitzgerald et al., 2006) as its data standard. This format has several advantages from the point of view of building a database: (1) the definitions for the data items are based on a comprehensive dictionary of crystallographic terminology and molecular structure descriptions; (2) it is self-defining; and (3) the syntax contains explicit rules that further define the characteristics of the data items, particularly the relationships between individual data items (Westbrook and Bourne, 2000). The latter feature is important because it allows for rigorous checking of the data.

The tools first developed by the NDB project are now used by the RCSB PDB and PDBj (both are members of the wwPDB; www.wwpdb.org; Berman, Henrick, and Nakamura, 2003) for processing both proteins and nucleic acids (Berman et al., 2000). A validation tool, NUCheck (Feng, Westbrook, and Berman, 1998) was developed especially for nucleic acids. It verifies valence geometry, torsion angles, intermolecular contacts, and the chiral centers of the sugars and phosphates. The dictionaries used for checking the structures were developed by the NDB Project from analyses of high-resolution, small-molecule structures (Clowney et al., 1996; Gelbin et al., 1996) from the Cambridge Structural Database (CSD) (Allen et al., 1979). The torsion angle ranges for double helical DNA forms were derived from an analysis of well-resolved nucleic acid structures (Schneider, Neidle, and Berman, 1997). One important outgrowth of these validation projects was the creation of the force constants and restraints that are now in common use for crystallographic refinement of nucleic acid structures (Parkinson et al., 1996). In addition to geometry checks, the molecular model is checked against the experimental data using SFCheck (Vaguine, Richelle, and Wodak, 1999).

Once primary annotation and validation are complete, nucleic acid-specific structural and functional annotations specific to nucleic acids are added. These vary from broadly characterizing nucleic acid conformations as double helical A, B, or Z type to identifying simple structural features that are present, such as a bulge or helical loop. Proteins from protein–nucleic acid complexes are annotated for their function in the complexes as (1) structural proteins (e.g., ribosomal and histone proteins), (2) regulatory proteins (e.g., different types of transcription and translation cofactors), or (3) enzymes (e.g., polymerases, topoisomerases, and endonucleases).

THE DATABASE

Information Content of the NDB

Structures available in the NDB include RNA and DNA oligonucleotides with two or more bases either alone or complexed with ligands, natural nucleic acids such as tRNA, and protein–nucleic acid complexes. The archive stores both primary and derived information about the structures (Table 12.1). The primary data include the crystallographic coordinate data, structure factors, and information about the experiments used to determine the structures, such as crystallization information, data collection, and refinement statistics. Derived structural information, such as valence geometry, torsion angles, and intermolecular contacts, is calculated and stored in the database. Structural and functional features annotated specifically for nucleic acid containing structures are also loaded into the database.

Some structural features of nucleic acids have historically been derived using different algorithms, and therefore, it can be difficult to provide the most reliable values. Whenever possible, the NDB has promoted standards that allow for structural comparisons. For example, different values for base morphology parameters are produced by different programs (Lavery and Sklenar, 1989; Babcock, Pednault, and Olson, 1994; Bansal, Bhattacharyya, and Ravi, 1995; Lu, El Hassan, and Hunter, 1997; Dickerson, 1998). This meant that it was not possible to compare any two structures using published data, making it necessary to recalculate these values for any analysis. To help resolve this problem, the NDB cosponsored the *Tsukuba Workshop on Nucleic Acid Structure and Interactions* to which all the key software developers in this field were invited. They decided that a single reference frame would be used to calculate these values and agreed upon the definition of that reference frame (Olson et al., 2001). This work fully quantitates the proposal for base morphology made previously at a meeting in Cambridge (Dickerson et al., 1989). As a result, these programs have been amended so that they produce very similar values for the base morphology parameters. The NDB has recalculated these values for all the structures in the

TABLE 12.1. The Information Content of the NDB

Identification, processing information	
Unique codes	NDB, PDB codes
Processing information	Deposition, release dates
Primary experimental information stored in the NDB	
Chemical description	The chemical composition of the deposition (content of all chemical constituents of the deposition), detailed description of chemical modifications of the nucleotides
Sequence description	Sequences of polymers, nucleic acids and proteins, reported
Citation	Authors, title, journal, volume, pages, year
For NMR structures:	
Coordinate information	Atomic coordinates for all models submitted
Experimental data	Spectrometer, types of NMR experiments, solution content, derivatization
For X-ray structures:	
Coordinate information	Atomic (orthogonal or fractional) coordinates, occupancies, and temperature factors for the asymmetric unit
Experimental data	Cell dimensions; space group for the x-ray structures, spectrometer, solution content, etc. for the NMR depositions
Data collection description	Radiation source and wavelength; data collection device; temperature; resolution range; total and unique number of reflections
Crystallization description	Crystallization method; temperature; pH value; solution composition
Refinement information	Program used; number of reflections used for refinement; data cutoff; resolution range; different R-factors; refinement of temperature factors and occupancies
Derived information stored in the NDB	
Structure summary	Descriptor briefly characterizing the structure
Distances	Chemical bond lengths; virtual bonds between phosphorus atoms
Angles	Valence angles, virtual angles involving phosphorus atoms
Torsion angles	Backbone and side chain torsion angles for nucleic acids and proteins; pseudo-rotational parameters for the sugar rings
Base pairing	Detailed description of base pairing, classification by Leontis and Westhof (2001) and Saenger (1984) with full identification of all pairs.
Base morphology	Base morphology parameters calculated by 3DNA (Lu and Olson, 2003) using the standard reference frame (Olson et al., 2001).

TABLE 12.1. (*Continued*)

Special geometrical	Nonbonded contacts, crystal-related geometries (for X-ray structures), symmetry-related coordinates, coordinates for symmetry-related strands, root mean square deviations from small molecule standards for valence geometries
Sequence pattern statistics	
Manually annotated structural features	Type of helix, classification of double helix, presence of loops, bulges, three/four way junctions
Program assigned structural features	Presence of A/B/Z type of helix, presence of triple, quadruple helices, loops, bulges, three/four way junctions

This table shows the types of data found in the NDB organized by content type. The database can be searched and reports can be generated using these characteristics.

repository using program 3DNA (Lu and Olson, 2003) and made them available as output from NDB searches and prepared reports.

User Web Access

The core of the NDB project is its relational database that can be used to query and report the stored structures. This resource is supplemented with other services. To secure wide and easy distribution of the NDB, all these services are accessible from an integrated web interface (http://ndbserver.rutgers.edu). The NDB home page links to the search and Atlas pages, coordinate and structure data, standards for nucleic acid geometry, prepared database reports, and tools (mainly programs) directly related to calculation or analysis of nucleic acid structural properties.

The Query Capabilities

The NDB relational database has all the data organized into more than 90 tables, with each table containing 5–20 data items. These tables include both experimental and derived information. For instance, the citation table contains all the items relevant for the relevant literature reference(s), the entity table contains critical items related to chemical entities observed in the structure, and the "entity_poly" table contains the sequences of polymer(s) observed in the structure that are indexed to the entity table.

A web interface was designed to make the query capabilities of the NDB as widely accessible as possible. There are two ways to query the NDB over the web: by simple "Search" and by "Integrated Search". The simple search option is a web form in which the user enters text in a box or selects an option from the pull-down menu. Possible queries include NDB and PDB identification codes, author names, year of publication or release, experimental method, and chemical and structural characteristics, such as the presence of non-Watson–Crick base pairing, different types of double helices (B/A/Z/right-handed),

TABLE 12.2. Reports Available for the NDB

Report Name	Contains
NDB status	NDB and PDB IDs, processing status, release date, descriptor, authors of the entries
Primary citation	Authors, title, journal, volume, pages, year
Descriptor	Descriptor with NDB and PDB IDs
NA backbone torsions	Backbone torsions with unique residue identifier
Base pair parameters	Buckle, propeller, opening, shear, stretch, stagger with unique pair identifier
Base step parameters	Twist, rise, helical rise, tilt, roll, inclination, shift, slide, unique step identifier
Refinement information	Resolution limits and reflections used in refinement, R factor
Cell dimensions	Crystallographic cell constants, space group

These reports can be automatically generated for any database query results.

tetraplex, or three- or four-way junctions. If no selection is made, the entire database is selected. After the query is executed, a list of structure IDs and descriptors that match the desired conditions is listed and the user can proceed to view the reported search results.

The "Integrated Search" provides the user with a wide range of options for structure selection and the use of Boolean operators to make complex queries. The interface consists of a series of pull-down menus that allow the selection of key components of database entries: citation information, experiment type and details, types of molecules present (DNA, RNA, protein, and ligand), presence of nucleic acid modifications, and nucleic acid conformation type (single/double/multiple helix, type of loop, and conformation type). After the user selects the desired defining criteria, the query is saved and executed; it can also be further modified for subsequent searches.

Several different reports or Atlas pages (see later) can be retrieved using either the Simple or Integrated Search tools (Table 12.2). Multiple reports can be generated for the same group of selected structures; this capability is particularly convenient for quickly producing reports based on derived features, such as torsion angles and base morphology.

Atlas Pages

An NDB Atlas report page summarizes the most important characteristics of a structure and provides information about the authors, citation, and sequence(s) of nucleic acids and proteins. These pages offer experimental information, a molecular view of the biologically relevant assembly of the reported structure, and links to tables of derived quantities, such as torsion angles, base morphology, and base pairing. Structural coordinates can be downloaded in either the PDB or mmCIF format, and a link is provided to the RCSB PDB's Structure Explorer page (Figure 12.2).

Atlas pages are created directly from the NDB database for all entries and are organized by experimental method (X-ray or NMR), and by structure type into broad groups of structures (with possible overlaps). This permits effective browsing of the structures in the database. Atlas pages are accessible from database query results.

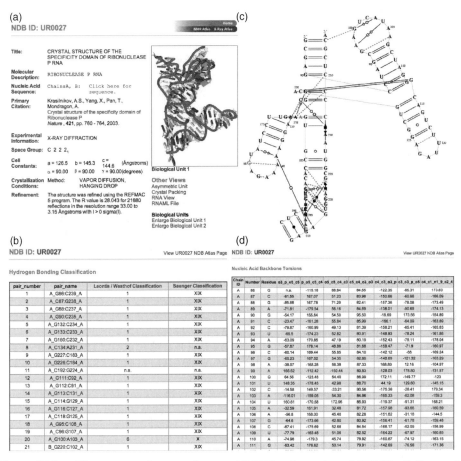

Figure 12.2. Different NDB Atlas pages for entry UR0027, a ribonuclease P RNA (Krasilnikov et al., 2003). (a) The main Atlas page provides summary information and links to coordinates, images, and tables of derived features of the entry; (b) classification of hydrogen bonding in base pairs of UR0027; (c) schematic 2D drawing of base pairing calculated and drawn by RNAView (Yang et al., 2003); and (d) torsion angles for UR0027.

DISTRIBUTION OF INFORMATION

In addition to these services, the NDB provides structural coordinate files, experimental data files, database reports, software programs, and other resources. They are available from the web (http://ndbserver.rutgers.edu/) and the ftp server (ftp://ndbserver.rutgers.edu).

Structural Coordinates and Experimental Data

Structural coordinate files can be downloaded in two formats, mmCIF and PDB. Coordinates for the complete biological assembly are distributed for crystal structures that do not have the asymmetric unit identical to the biological assembly. When available, experimental files, structure factors for X-ray structures, and distance restraints for NMR structures are made accessible to the community users.

Database Reports

The "NDB Reports" section of the website contains a large variety of precomputed information and tables useful to researchers looking for structure identifiers, citations, cell dimensions, and structure summaries. Prepared reports can be downloaded in the ASCII or PostScript formats for all structures currently in the NDB or for groups of structures similar to groups in the Atlas pages. Results from the program SFCheck (Vaguine, Richelle, and Wodak, 1999), which calculates agreement between coordinate model and structure factors, can be downloaded for all crystals structures with structure factors.

Standards and Software Tools

The "Standards" section provides dictionaries of standard ("ideal") geometries of nucleic acid components as well as their parameter files for X-PLOR (Brünger, 1992). It also provides a full explanation of the Standard reference frame (Olson et al., 2001). The "Tools" section links to downloadable and web-based programs, a few of which are described here.

Several programs are available for examining RNA-containing structures. For any NDB structure or in a structure uploaded by the user, RNAViewer (Yang et al., 2003) generates a 2D representation of the base pairing using annotations according to Leontis and Westhof (2001). The Base Pair Viewer (BPView) can be used to select and interactively visualize base pairs or triples. RNAMLView (http://fjossinet.u-strasbg.fr/rnamlview) can display and edit 2D diagrams of RNA secondary structures with tertiary interactions created by RNAView. These programs use the RNAML syntax for exchanging RNA information (Waugh et al., 2002).

Two programs, predictdnaht and PDNA-pred, analyze protein DNA-binding structural motifs, namely, the helix–turn–helix (HTH) motif. Both can be used to predict if a protein target, either from the NDB archive or uploaded by the user, is a DNA-binding protein with the HTH motif (McLaughlin and Berman, 2003).

APPLICATIONS OF THE NDB

The NDB has been used to analyze characteristics of nucleic acids alone and complexed with proteins. The ability to select structures according to many different criteria has made it possible to create appropriate data sets for study. A few examples are given here.

The conformational characteristics of A-, B-, and Z-DNA were examined by using carefully selected examples of well-resolved structures in these classes (Schneider, Neidle, and Berman, 1997). Conformation wheels (Figure 12.3a) for each conformation as well as scattergrams of selected torsion angles (Figure 12.3b) were created. These diagrams can now be used to assess and classify new structures. Studies of B-DNA helices have shown that the base steps have characteristic values that depend on their sequence (Gorin, Zhurkin, and Olson, 1995). Plots of twist versus roll are different for purine–purine, purine–pyrimidine, and pyrimidine–purine steps. This particular analysis was used to derive energy parameters for B-DNA sequences (Olson et al., 1998).

The interfaces in protein–nucleic acid complexes have been systematically studied. In an analysis of protein–DNA complexes, 26 complexes were selected in which the proteins were nonhomologous (Jones et al., 1999) (Figure 12.4). The results showed that there are amino acid propensities at these interfaces that are markedly different from those in protein–protein complexes. Generally, while protein–protein complexes have larger and

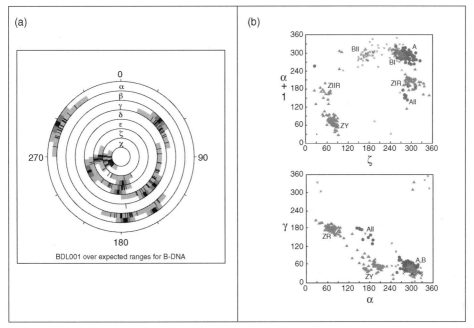

Figure 12.3. Conformations for the double helical DNA. (a) Conformation wheel showing the torsion angles for a B-DNA dodecamer BDL001 (Drew et al., 1981). Black lines show actual values of torsion angles, and cyan background their allowed range in the B-type DNA conformation (Schneider, Neidle, and Berman, 1997). The gray shades in the outer rings show the average value(s) of the torsions in dark gray flanked by values of one and two estimated standard deviations in lighter gray; (b) Two scattergrams between the backbone torsions ζ (C3′-O3′-P-O5′), α (O3′-P-O5′-C5′), and γ (O5′-C5′-C4′-C3′) that show clustering of the most typical DNA forms (see Chapter 3 for more information on nucleic acid conformations). Figure also appears in the Color Figure section.

more hydrophobic interfaces, the protein–DNA interface is dominated by hydrophilic and positively charged residues Arg, Lys, Asn, Gln, Ser, and Thr. It was also possible to place complexes into three classes: double-headed, single-headed, and enveloping. A similar analysis has also been done for protein–RNA complexes (Jones et al., 2001). Finally, detailed analyses of the hydrogen bonding patterns at the protein–DNA interface have found that CH ··· O bonds are surprisingly common (Mandel-Gutfreund et al., 1998).

Some analyses have been done to investigate the potential impact of crystal packing on conformation. Although there are more than 30 different crystal forms of B-DNA in the NDB, the actual number of packing motifs (Figure 12.5) remains relatively small, with the most common motifs being minor groove–minor groove, stacking–lateral backbone, and major groove–backbone (Timsit, Youri, and Moras, 1992).

Minor groove–minor groove interactions, in which the guanine of one duplex form hydrogen bonds with the guanines of a neighboring duplex, are seen not only in dodecamer structures but also in an octamer sequence that has three duplexes in the asymmetric unit (Urpi et al., 1996). The second motif contains duplexes stacked one above another with the adjoining phosphates forming lateral interactions. A large number of variations of this motif have been observed in decamer (Grzeskowiak et al., 1991) and hexamer structures (Cruse et al., 1986; Tari and Secco, 1995). The third type of packing involves the major groove of one helix

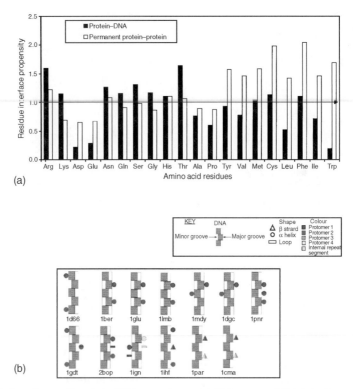

(a)

(b)

Figure 12.4. Interfaces of protein–DNA complexes. (a) Histogram of the interface residue propensities calculated for 26 protein–DNA complexes and compared to those for permanent protein–protein complex (Jones and Thornton, 1996). A propensity of >1 indicates that a residue occurs more frequently at the interface than on the protein surface. (b) Schematic drawings that model protein–DNA complexes for double-headed binding proteins. The diagrams give an indication of the predominant secondary structure of the binding motif, protein symmetry, and the type and relative position of the DNA groove bound. The secondary structure of the predominant binding motifs are indicated using different symbols analogous to those used in TOPS diagrams (Westhead et al., 1998). The symmetry of each protein is indicated by using a different color for each (pseudo)symmetry-related element. A single symbol shaded in two colors indicates that there are secondary structures of this type contributed by more than one symmetry-related element. Reprinted with permission from Jones et al. (1999). Figure also appears in the Color Figure section.

interacting with the phosphate backbone of another (Timsit et al., 1989). Sequence appears to be a large factor in determining these motifs, but it is not the only one. For example, the first structures known to exhibit the major groove–phosphate interactions contained a cytosine that formed a hydrogen bond to the phosphate. However, not all structures containing this motif have this hydrogen bond (Wood et al., 1997). The particular sequence in this crystal is even more intriguing because it also crystallizes into another form in which the terminal flips out to form a minor groove interaction with another duplex (Spink et al., 1995).

The task of trying to determine the relative effects of base sequence and crystal packing on the values of the base morphology parameters is hampered to some degree by the uneven distribution of the 16 different base steps among the different crystal types. Some steps like CG are very well represented in B structures, whereas others such as AC have very few

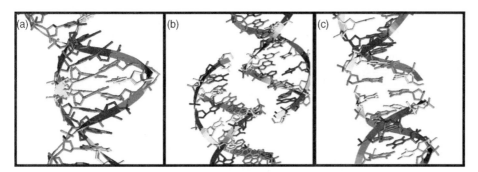

Figure 12.5. Examples of packing motifs in B-DNA duplexes. From left to right: (a) minor groove–minor groove interactions in BDL042 (Leonard and Hunter, 1993); (b) major groove–backbone interactions in BDJ060 (Goodsell, Grzeskowiak, and Dickerson, 1995); (c) stacking inter-actions in BDJ025 (Grzeskowiak et al., 1991). The bases are colored green for guanine, yellow for cytosine, red for adenine, and blue for thymine. Reprinted from Berman, Gelbin, and Westbrook (1996) with permission from Elsevier Science. Figure also appears in the Color Figure section.

representatives in the data set. Nonetheless, there are a few steps that occur in crystals with different packing motifs. An analysis of the CG steps across all crystal types show that its conformation is relatively insensitive to crystal packing and the distribution is similar to that found for all steps (see Berman, Gelbin, and Westbrook, 1996). On the other hand, the variability of the CA step appears to depend not simply on crystal type but on the packing motif. The values of twist for CA steps in minor groove–minor groove motifs are smaller than those for the major groove–backbone motif. Very high values are displayed for CA steps in the stacking–lateral backbone motif. Plots of twist versus roll for CG steps show the distribution noted by others (Gorin, Zhurkin, and Olson, 1995); clustering is not observed to be dependent on crystal type. On the other hand, the same plots for CA steps show very distinctive differences that appear to depend on the packing motifs. It is important to note here that these motifs encompass several crystal types so that the structural variability observed is a function of a particular type of structural interaction rather than a particular crystal form. Before any definitive statements can be made about all the steps, it will be necessary to have much more data.

In a series of systematic studies of the hydration patterns of DNA double helices, it was found that the hydration patterns around the bases are well defined and are local (Schneider and Berman, 1995). That is, small changes in the conformation of the backbone do not affect the hydration around the bases. It was also found that there are more diffuse patterns around the phosphate backbone that are dependent on the conformational class of the DNA (Schneider, Patel, and Berman, 1998). These analyses were used in an attempt to predict the binding sites of protein side chains on the DNA. The hydration sites of the DNA were predicted for a series of protein–nucleic acid complexes, and then compared with the known location of the amino acid side chains in these structures. The results were surprisingly good; the side chain sites and the predicted hydration sites in most cases were very close (Woda et al., 1998) (Figure 12.6).

Variable conformations of nucleic acids and complicated folds of large RNA molecules, such as ribosomal RNA, present a challenge to structural scientists. Several groups have analyzed torsional distributions of RNA backbone approximately at the nucleotide level (Murthy et al., 1999; Hershkovitz et al., 2003; Murray et al., 2003; Sims and Kim, 2003; Schneider, Moravek, and Berman, 2004) and have demonstrated that RNA prefers relatively small number of conformations. These studies used traditional small-molecule descriptors—torsion angles.

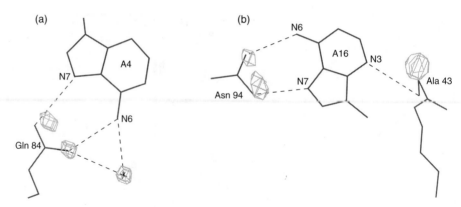

Figure 12.6. Hydration at the protein–DNA interface. Shown are amino acids and bases inter-acting in the (a) Lambda repressor-operator/DNA structure PDR010 (Beamer and Pabo, 1992) and (b) MAT α2 homeodomain-operator/DNA complex PDT005 (Wolberger et al., 1991). The density cages represent the predicted hydration sites around the bases and are calculated without the knowledge of the amino acid positions. The agreement between the predicted hydration sites and the positions of the amino acid hydrogen bonding atoms is very good. In figure (a), the hydration site and the actual crystallographic water (labeled with a cross) also overlap well.

These and other atomic scale descriptors, such as base morphology, are too detailed to describe the architecture of large RNA molecules. They need to be supplemented by descriptors that provide more robust definitions of larger structural motifs that scale between several and dozens of nucleotides. New algorithms need to be developed to allow automatic identification of these motifs in structures across the archive as well as in new depositions and represent them in the database schema in such a way that they can be queried and reported.

A first step in creating a more robust definition for categorizing RNA structures has been conducted for a database containing RNA motifs (SCOR) (Klosterman et al., 2002; Klosterman et al., 2004). The goal of the RNA Ontology Consortium (ROC) is to create a set of better definitions and systematic methods for searching structural motifs by combining base pairing, base stacking, and the backbone conformations (Leontis et al., 2006). However, systematizing a set of robust definitions for these larger structural motifs remains a challenging task ahead.

ACKNOWLEDGMENT

Funding from the National Science Foundation and the Department of Energy is gratefully acknowledged.

REFERENCES

Allen FH, Bellard S, Brice MD, Cartright BA, Doubleday A, Higgs H, Hummelink T, Hummelink-Peters BG, Kennard O, Motherwell WDS, Rodgers JR, Watson DG (1979): The Cambridge Crystallographic Data Centre: computer-based search, retrieval, analysis and display of informa-tion. *Acta Crystallogr B* 35:2331–2339.

Babcock MS, Pednault EPD, Olson WK (1994): Nucleic acid structure analysis, Mathematics for local Cartesian and helical structure parameters that are truly comparable between structures. *J Mol Biol* 237:125–156.

Bansal M, Bhattacharyya D, Ravi B (1995): NUPARM and NUCGEN: software for analysis and generation of sequence dependent nucleic acid structures. *CABIOS* 11:281–287.

Berman HM, Olson WK, Beveridge DL, Westbrook J, Gelbin A, Demeny T, Hsieh SH, Srinivasan AR, Schneider B (1992): The Nucleic Acid Database—a comprehensive relational database of three-dimensional structures of nucleic acids. *Biophys J* 63:751–759.

Berman HM, Gelbin A, Westbrook J (1996): Nucleic acid crystallography: a view from the Nucleic Acid Database. *Prog Biophys Mol Biol* 66:255–288.

Berman HM, Westbrook J, Feng Z, Gilliland G, Bhat TN, Weissig H, Shindyalov IN, Bourne PE (2000): The Protein Data Bank. *Nucleic Acids Res* 28 (1):235–242.

Berman HM, Henrick K, Nakamura H (2003): Announcing the worldwide Protein Data Bank. *Nat Struct Biol* 10(12):980.

Brünger AT (1992): *X-PLOR, Version 3.1, A System for X-Ray Crystallography and NMR.* New Haven, CT: Yale University Press.

Clowney L, Jain SC, Srinivasan AR, Westbrook J, Olson WK, Berman HM (1996): Geometric parameters in nucleic acids: nitrogenous bases. *J Am Chem Soc* 118:509–518.

Cruse WBT, Salisbury SA, Brown T, Cosstick R, Eckstein F, Kennard O (1986): Chiral phosphor-othioate analogues of B-DNA. The crystal structure of Rp-d|Gp(S)CpGp(S)CpGp(S)C|. *J Mol Biol* 192:891–905.

Dickerson RE, Bansal M, Calladine CR, Diekmann S, Hunter WN, Kennard O, von Kitzing E, Lavery R, Nelson HCM, Olson W, Saenger W, Shakked Z, Sklenar H, Soumpasis DM, Tung C-S, Wang AH-J, Zhurkin VB (1989): Definitions and nomenclature of nucleic acid structure parameters. *EMBO J* 8:1–4.

Dickerson RE (1998): DNA bending: the prevalence of kinkiness and the virtues of normality. *Nucleic Acids Res* 26:1906–1926.

Feng Z, Westbrook JD, Berman HM (1998): *NUCheck,* Rutgers Publication NBD-407. Rutgers University, New Brunswick, New Jersey.

Fitzgerald PMD, Westbrook JD, Bourne PE, McMahon B, Watenpaugh KD, Berman HM (2006): Macromolecular dictionary (mmCIF). In: Hall SR, McMahon B, editors. *International Tables for Crystallography. Definition and Exchange of Crystallographic Data.* vol. G. Dordrecht: Springer. pp. 295–443.

Gorin AA, Zhurkin VB, Olson WK (1995): B-DNA twisting correlates with base-pair morphology. *J Mol Biol* 247 (1):34–48.

Gelbin A, Schneider B, Clowney L, Hsieh S-H, Olson WK, Berman HM (1996): Geometric parameters in nucleic acids: sugar and phosphate constituents. *J Am Chem Soc* 118 (3): 519–528.

Grzeskowiak K, Yanagi K, Privé GG, Dickerson RE (1991): The structure of B-helical C-G-A-T-C-G-A-T-C-G and comparison with C-C-A-A-C-G-T-T-G-G: the effect of base pair reversal. *J Biol Chem* 266:8861–8883.

Hershkovitz E, Tannenbaum E, Howerton SB, Sheth A, Tannenbaum A, Williams LD (2003): Automated identification of RNA conformational motifs: theory and application to the HM LSU 23S rRNA. *Nucleic Acids Res* 31(21):6249–6257.

Jones S, van Heyningen P, Berman HM, Thornton JM (1999): Protein–DNA interactions: a structural analysis. *J Mol Biol* 287:877–896.

Jones S, Daley DTA, Luscombe NM, Berman HM, Thornton JM (2001): Protein–RNA interactions: a structural analysis. *Nucleic Acids Res* 29:934–954.

Klosterman PS, Tamura M, Holbrook SR, Brenner SE (2002): SCOR: a structural classification of RNA database. *Nucleic Acids Res* 30:392–394.

Klosterman PS, Hendrix DK, Tamura M, Holbrook SR, Brenner SE (2004): Three-dimensional motifs from the SCOR, structural classification of RNA database: extruded strands, base triples, tetraloops and U-turns. *Nucleic Acids Res* 32 (8):2342–2352.

Lavery R, Sklenar H (1989): Defining the structure of irregular nucleic acids: conventions and principles. *J Biomol Struct Dyn* 6:655–667.

Leontis NB, Westhof E (2001): Geometric nomenclature and classification of RNA base pairs. *RNA* 7:499–512.

Leontis NB, Altman RB, Berman HM, Brenner SE, Brown JW, Engelke DR, Harvey SC, Holbrook SR, Jossinet F, Lewis SE, Major F, Mathews DH, Richardson JS, Williamson JR, Westhof E (2006): The RNA Ontology Consortium: an open invitation to the RNA community. *RNA* 12(4):533–541.

Lu X-J, El Hassan MA, Hunter CA (1997): Structure and conformation of helical nucleic acids: analysis program (SCHNAaP). *J Mol Biol* 273:668–680.

Lu XJ, Olson WK (2003): 3DNA: a software package for the analysis, rebuilding and visualization of three-dimensional nucleic acid structures. *Nucleic Acids Res* 31 (17):5108–5121.

Mandel-Gutfreund Y, Margalit H, Jernigan RL, Zhurkin VB (1998): A role for CH $\cdots$ O interactions in protein–DNA recognition. *J Mol Biol* 277:1129–1140.

McLaughlin WA, Berman HM (2003): Statistical models for discerning protein structures containing the DNA-binding helix–turn–helix motif. *J Mol Biol* 330:43–55.

Moore P (2001): The ribosome at atomic resolution. *Biochemistry* 40:3243–3250.

Murray LJ, Arendall WB III, Richardson DC, Richardson JS (2003): RNA backbone is rotameric. *Proc Natl Acad Sci USA* 100 (24):13904–13909.

Murthy VL, Srinivasan R, Draper DE, Rose GD (1999): A complete conformational map for RNA. *J Mol Biol* 291 (2):313–327.

Olson WK, Gorin AA, Lu X-J, Hock LM, Zhurkin VB (1998): DNA sequence-dependent deformability deduced from protein–DNA crystal complexes. *Proc Natl Acad Sci USA* 95 (19): 11163–11168.

Olson WK, Bansal M, Burley SK, Dickerson RE, Gerstein M, Harvey SC, Heinemann U, Lu X-J, Neidle S, Shakked Z, Sklenar H, Suzuki M, Tung C-S, Westhof E, Wolberger C, Berman HM (2001): A standard reference frame for the description of nucleic acid base-pair geometry. *J Mol Biol* 313:229–237.

Parkinson G, Vojtechovsky J, Clowney L, Brunger AT, Berman HM (1996): New parameters for the refinement of nucleic acid-containing structures. *Acta Crystallogr D* 52:57–64.

Saenger W (1984): *Principles of Nucleic Acid Structure, Springer Advanced Texts in Chemistry.* New York: Springer-Verlag.

Schneider B, Berman HM (1995): Hydration of the DNA bases is local. *Biophys J* 69:2661–2669.

Schneider B, Neidle S, Berman HM (1997): Conformations of the sugar–phosphate backbone in helical DNA crystal structures. *Biopolymers* 42:113–124.

Schneider B, Patel K, Berman HM (1998): Hydration of the phosphate group in double helical DNA. *Biophys J* 75:2422–2434.

Schneider B, Moravek Z, Berman HM (2004): RNA conformational classes. *Nucleic Acids Res* 32(5): 1666–1677.

Sims GE, Kim S-H (2003): Global mapping of nucleic acid conformational space: dinucleoside monophosphate conformations and transition pathways among conformational classes. *Nucleic Acids Res* 31(19):5607–5616.

Spink N, Nunn C, Vojetchovsky J, Berman H, Neidle S (1995): Crystal structure of a DNA decamer showing a novel pseudo four-way helix–helix junction. *Proc Natl Acad Sci USA* 92(23): 10767–10771.

Tari LW, Secco AS (1995): Base-pair opening and spermine binding—B-DNA features displayed in the crystal structure of a gal operon fragment: implications for protein–DNA recognition. *Nucleic Acids Res* 23(11):2065–2073.

Timsit Y, Westhof E, Fuchs RPP, Moras D (1989): Unusual helical packing in crystals of DNA bearing a mutation hot spot. *Nature* 341:459–462.

Timsit Y, Moras D (1992): Crystallization of DNA. *Methods Enzymol* 211:409–429.

Urpi L, Tereshko V, Malinina L, Huynh-Dinh T, Subirana JA (1996): Structural comparison between the d(CTAG) sequence in oligonucleotides and trp and met repressor–operator complexes. *Nat Struct Biol* 3:325–328.

Vaguine AA, Richelle J, Wodak SJ (1999): SFCHECK: a unified set of procedures for evaluating the quality of macromolecular structure–factor data and their agreement with the atomic model. *Acta Crystallogr D* 55:191–205.

Waugh A, Gendron P, Altman R, Brown JW, Case D, Gautheret D, Harvey SC, Leontis N, Westbrook J, Westhof E, Zuker M, Major F (2002): A standard syntax for exchanging RNA information. *RNA* 8:707–717.

Westbrook J, Bourne PE (2000): STAR/mmCIF: An extensive ontology for macromolecular structure and beyond. *Bioinformatics* 16:159–168.

Woda J, Schneider B, Patel K, Mistry K, Berman HM (1998): An analysis of the relationship between hydration and protein–DNA interactions. *Biophys J* 75:2170–2177.

Wood AA, Nunn CM, Trent JO, Neidle S (1997): Sequence-dependent crossed helix packing in the crystal structure of a B-DNA decamer yields a detailed model for the Holliday junction. *J Mol Biol* 269:827–841.

Yang H, Jossinet F, Leontis N, Chen L, Westbrook J, Berman HM, Westhof E (2003): Tools for the automatic identification and classification of RNA base pairs. *Nucleic Acids Res* 31:3450–3460.

13

OTHER STRUCTURE-BASED DATABASES

J. Lynn Fink, Helge Weissig, and Philip E. Bourne

INTRODUCTION

The single repository for experimentally derived macromolecular structures is the Protein Data Bank (PDB) (Bernstein et al., 1977; Berman et al., 2000; Berman et al., 2007) described in Chapter 11. The *primary* data provided by the PDB are the Cartesian coordinates, occupancies, and temperature factors for the atoms in these structures. Additional information given includes literature references, author names, experimental details, links to the sequence in the sequence databases, and some limited annotation of the biological function (Chapter 10). Collated into a single entry, due to the restrictions of the PDB format, or into multiple entries for very large X-ray structures and large NMR ensembles, these data constitute a concise description of the three-dimensional form of a molecule. The PDB currently releases the primary structure data once per week as requested by the depositor, whereupon a number of sites worldwide acquire these data via the Internet, derive additional information, and constitute a set of *secondary* resources. Secondary resources cover features such as stereochemical quality (Table 13.1), protein structure classification (Table 13.2), protein–protein interaction data (Table 13.3), structure visualization (Table 13.4), and data on specific protein families. The secondary resources described in this chapter can be viewed as downstream of the PDB in an information flow diagram (Figure 13.1). The number of these secondary resources is growing every year and no attempt is made at a complete overview, but rather to give a synopsis from several classes of resource (Figure 13.1) of what is available. A current compendium of secondary resources is maintained by the PDB at http://www.pdb.org/pdb/static.do?p=general_information/web_links/index.html. More details on popular, well-established, structure-based databases are available in other chapters. Chapter 5 includes a description of the NMR-specific BioMagResBank resource; the Nucleic Acid Database (NDB) is described in Chapter 12; the comparative fold classification databases SCOP and CATH are described in Chapters 17 and 18, respectively; Chapter 14 includes brief descriptions of stereochemical- quality-oriented resources and

Structural Bioinformatics, Second Edition   Edited by Jenny Gu and Philip E. Bourne
Copyright © 2009 John Wiley & Sons, Inc.

TABLE 13.1. Popular Software and Resources for Protein Structure Validation

Resource	Details
PDBSum	Summaries for all protein structures including validation checks. http://www.biochem.ucl.ac.uk/bsm/pdbsum/
Procheck	Structure validation suite. http://www.biochem.ucl.ac.uk/~roman/procheck/procheck.html (Laskowski et al., 1996)
What_Check	Detailed stereochemical quality summaries for all protein structures. Part of the Whatif package. http://www.cmbi.kun.nl/gv/whatcheck/
SFCheck	Validate the experimental structure factors associated with an X-ray diffraction experiment (Vaguine et al., 1999).
PDB validation server	Validate the format and content of a PDB entry using the same software procedures as used by the PDB. Includes those listed above in this table. http://pdb.rutgers.edu/validate/
Protein–protein interaction server	http://www.biochem.ucl.ac.uk/bsm/PP/server/server_help.html (Jones and Thornton, 1996)
Protein–DNA interaction server	http://www.biochem.ucl.ac.uk/bsm/DNA/server/ (Jones et al., 1999)

TABLE 13.2. Resources Classifying Protein Structure

Resource	Details
SCOP	The Structure Classification of Proteins. http://scop.mrc-lmb.cam.ac.uk/scop/ (Murzin et al., 1995; Andreeva et al., 2004).
CATH	Class (C), Architecture (A), Topology (T), and Homologous superfamily (H). http://www.biochem.ucl.ac.uk/bsm/cath_new/index.html (Orengo et al., 1997; Greene et al., 2007)
DALI	DALI Domain Dictionary. http://www.embl-ebi.ac.uk/dali/domain/ (Dietmann et al., 2001)
VAST	Vector Alignment Search Tool. http://www.ncbi.nlm.nih.gov/Structure/VAST/vast.shtml (Gibrat et al., 1996)
CE	Polypeptide chain comparison. http://cl.sdsc.edu/ce.html (Shindyalov and Bourne, 1998)
3Dee	Protein Domain Definitions. http://jura.ebi.ac.uk:8080/3Dee/help/help_intro.html (Siddiqui and Barton, 1995)
CAMPASS	Cambridge database of protein alignments organized as structural superfamilies http://www-cryst.bioc.cam.ac.uk/~campass/ (Sowdhamini et al., 1998).

TABLE 13.3. Popular Resources of Protein Interactions

Resource	Details
DIP	Database of Interacting Proteins. http://dip.doe-mbi.ucla.edu/ (Xenarios et al., 2002)
BIND	The Biomolecular Interaction Network Database. http://www.bind.ca/ (Bader et al., 2001; Bader et al., 2003)
MINT	Molecular Interactions Database. http://tweety.elm.eu.org/mint/index.html

TABLE 13.4. Popular Resources Visualizing Macromolecular Structures

Resource	Details
Jena Image Library	Images depicting biological function and useful links to other resources. http://www.imb-jena.de/IMAGE.html (Reichert and Suhnel, 2002)
PDBSum	Summaries for all protein structures including protein–ligand interaction. http://www.biochem.ucl.ac.uk/bsm/pdbsum/
NDB Atlas	Protein–DNA complexes. http://ndbserver.rutgers.edu/NDB/NDBATLAS/
STING	Sequence and property browser. http://mirrors.rcsb.org/SMS/
GRASS	Static GRASP images of electrostatic and surface properties. http://trantor.bioc.columbia.edu/GRASS/surfserv_enter.cgi
General	World Index of Molecular Visualization Resources. http://molvis.sdsc.edu/visres/

additional resources are referenced throughout. The reader is also referred to the annual edition of *Nucleic Acids Research* dedicated to molecular biology databases, which appears in January and includes descriptions of many of the resources outlined here.

THE ADDED VALUE PHILOSOPHY

At the time of the inception of the PDB in the early 1970s, the database had only a few entries available and information technology to manage these data was in its infancy. However, as the number of entries in the PDB grew slowly during the 1980s, comparative analysis of these entries became possible with the support of new algorithms and improved computational technology on this growing body of data and the availability of databases to efficiently access these data.

Comparative analysis of proteins within the PDB revealed deficiencies in both the content and the format of the data. This is discussed in Chapter 10 and not considered further here. Today the PDB is committed to provide consistent and complete information on the macromolecular structure and the experiment used to determine that structure. These rich and complex biological data provide many with the opportunity to add value to these data. Consequently, researchers are faced with a large array of resources from which to choose to interpret structure-based data. This chapter introduces a subset of these resources that we consider to be important to a large audience.

OTHER PRIMARY INFORMATION RESOURCES

Not all primary information on macromolecular structure is located in the PDB (Figure 13.1). We consider three additional sources of primary information in this section. First, there is information on crystallization conditions that have been extracted from the literature. Second, there is information on small organic molecules, a number of which are covalently or noncovalently bound to large biological macromolecules. Third, there is the growing body of information derived from structural genomics projects.

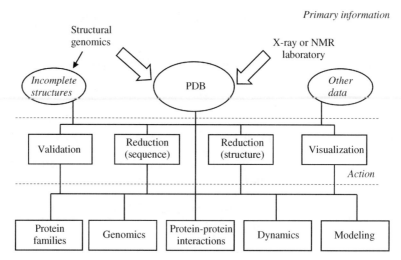

Figure 13.1. The flow of macromolecular structure data. *Primary information* is derived directly from experiment. All completed macromolecular structures in the public domain are deposited with the PDB. It is anticipated that in the future incomplete structures will also be available from the structural genomics projects. Additional primary information, such as sequences, crystallization conditions, and the structure of small-molecule ligands are available in primary resources other than the PDB. A variety of *actions* are performed on these primary data and a set of *secondary resources* result.

Biological Macromolecule Crystallization Database—http://xpdb.nist.gov:8060/BMCD4/

The Biological Macromolecule Crystallization Database (BMCD) contains crystal data and the crystallization conditions, which have been compiled by human annotators from the literature (Gilliland et al., 1994). Currently, BMCD includes 5247 crystal entries. These entries include proteins, protein–protein complexes, nucleic acids, nucleic acid–nucleic acid complexes, protein–nucleic acid complexes, and viruses. In addition to the reported crystallization data collected from literature, the BMCD holds data from the NASA Protein Crystal Growth Archive, a compilation of data generated from microgravity experiments conducted by NASA and other international space agencies.

BMCD addresses what is often the most difficult and time-consuming step in the determination of a macromolecular structure by X-ray crystallography—the crystallization of the macromolecule (see Chapter 4). The late Max Perutz once said "crystallization is a little like hunting, requiring knowledge of your prey and a certain cunning." While some structure models deposited in the PDB report the conditions used for the successful crystallization, many do not. Hennessy et al. (2000) have recently documented the usefulness of the information stored in BMCD for predicting crystallization conditions. The structural genomics projects (Chapter 40) are also collecting and storing information on failed crystallization experiments, signaling a new era. These negative results can be considered as useful as those that led to successful diffraction experiments.

Cambridge Structural Database: Small-Molecule Organic Structures—http://www.ccdc.cam.ac.uk/products/csd/

The Cambridge Structural Database (CSD) is the small molecule equivalent of the PDB serving as a primary resource for crystal structure information of nearly a quarter million organic and metallorganic compounds (Allen and Kennard, 1993; Allen and Taylor, 2004). Crystal structures, deposited directly to the CSD or manually annotated from the literature, are derived from both X-ray and neutron diffraction. The CSD contains three distinct types of information for each entry that can be categorized according to its dimensionality:

1. *One-Dimensional Information*: This includes all of the bibliographic information for the particular entry and a summary of the structural and experimental data. The text and numerical information include the names of the authors, compound names, and full journal references, as well as the crystallographic cell dimensions and space group. Where applicable, descriptions of absolute configuration, polymorphism form, and any drug or biological activity are also included.

2. *Two-Dimensional Information*: Data encoded as a chemical connection table including atom and bond properties and a chemical diagram of the molecule. Atom properties include the element symbol, the number of connected non-hydrogen atoms, the number of connected hydrogen atoms, and the net charge.

3. *Three-Dimensional Information*: Data used to generate a 3D representation of the molecule. These data include the atomic Cartesian coordinates, the space group symmetry, the covalent radii, and the crystallographic connectivity established by using those radii.

The data format used by the CSD, CIF, or Crystallographic Information File (Hall et al., 1991; Brown and McMahon, 2002) is a small molecule version of the macromolecular CIF (mmCIF, Chapter 10). Both CIF and mmCIF are endorsed and maintained by the International Union of Crystallography (IUCr).

CSD is distributed by the Cambridge Crystallographic Data Center (CCDC) as a commercial product for local installation. Network access to CSD information is currently made available free of charge to academic users in the United Kingdom and Europe. In addition, individual entries can be retrieved from the CSD using a simple form that requires at least a name, an e-mail address and, for location of the entry, the CSD accession number, and a complete journal reference of the CSD entry. Upon submission of the form, results are returned within three business days via e-mail.

Structures Not Yet Available

It is useful to know what macromolecular structures will likely be available at some point in the future. Two resources provide this information and both are maintained by the PDB. The first are those structures already solved and deposited in the PDB, but not yet available. These can be reviewed at http://www.rcsb.org/pdb/statistics/statisticsPieChart. do?content=status-pie&seqid=100. These structures are either on hold pending publication of the associated paper, or on hold for a longer period to permit the depositor to fully exploit their data. This period usually does not exceed one year. The second are those structures being determined by the structural genomic projects worldwide. These data range from a description of the target sequence under consideration, to a status of the structure determination, to a final 3D model. Details can be found at http://targetdb.pdb.org. A brief discussion of structural genomics concludes this chapter.

SECONDARY RESOURCES

The resources described in this section are presented in no particular order and represent a cross section of what is available worldwide. Additional resources are listed in Tables 13.1–13.4. Where available, information is provided on current update frequencies, data formats, and the underlying technology used. In most cases, users of these secondary resources can expect a delay between the release of a structure by the PDB and the availability of derivative information on the structure through a secondary resource. As the rate of deposition of structures increases (Chapter 40), resources that rely on semiautomated update requiring human annotation lag behind. As indicated throughout this book, the future will likely see a concerted effort in new and improved algorithms to automatically generate and statistically validate secondary information.

Sequence and Structure Relationships to Provide Nonredundant Data: The ASTRAL Compendium for Sequence and Structure Analysis—http://astral.berkeley.edu/

The PDB file format does not always provide an explicit relationship between the SEQRES records of biological sequence information and the ATOM/HETATM records that contain the Cartesian coordinates for each amino acid or nucleotide. While this shortcoming has been fully addressed in the mmCIF format, most structural bioinformatics software currently uses PDB files. The ASTRAL compendium (Brenner et al., 2000; Chandonia et al., 2004) is a collection of data files and tools, providing a partially curated mapping of these records as produced by the program pdb2cif (Bernstein et al., 1998). The mapping is distributed in a text format named Rapid Access Format (RAF) that can easily be parsed by computer programs. The RAF file includes mappings for all PDB chains represented in the first seven classes of SCOP (see Chapter 17). It is used as the definitive sequence mapping resource for ASTRAL and SCOP but is also intended as a useful resource for any PDB user.

Using RAF data the primary role of ASTRAL is to maintain nonredundant sequence sets corresponding to unique protein domains as defined by SCOP (Chapter 17). This information is helpful for the analysis of evolutionary relationships between domains based on sequence alignments. It also serves to reduce the redundancy in the PDB by filtering out protein sequences with varying degrees of sequence identity, leaving a representative conforming to the most accurate structure determination. The PDB has also recently begun to offer a similar capability.

Redundancy arises in the following ways, given a repository that accepts all structures solved by the worldwide community. First, different groups can determine the same or very similar structures independently. Second, a point mutation, occurring naturally or introduced posttranslationally to analyze biological function or structure folding, leads to a very similar structure. For example, approximately 1000 lysozyme structures can be found in the PDB where almost every position in the structure, and combinations thereof, has been modified. Third, structures are often determined multiple times with different ligands bound to them (e.g., HIV1 proteases with different inhibitors bound), without significant change in the protein itself.

PDBselect (Hobohm and Sander, 1994) was the first widely used reduced set of protein structure data. When performing such reduction, an important question that arises is as to how does one choose the representative? All approaches employ an initial ranking of structures based on the widely used quality parameters for X-ray structures: resolution and R-factor. ASTRAL uses, in addition, a Stereochemical Check Score (SCS) combining scores from Procheck (Laskowski et al., 1996) and What_Check (Hooft et al., 1996), two well-known stereochemical quality assessment programs. Structures are chosen as representatives for others at sequence identity cutoffs at set percentages based on the AEROSPACI (Aberrant Entry Re-Ordered Summary PDB ASTRAL Check Index)[1] score. ASTRAL also provides access to nonredundant sets filtered by E-value for data classified by SCOP, that is, a representative set for each class, fold, superfamily, and family.

Providing Links to Literature, Sequence, and Genome Information: The Molecular Modeling Database (MMDB)—http://www.ncbi.nlm.nih.gov/Structure/MMDB/mmdb.shtml

The Macromolecular Database (MMDB) maintained by the National Center for Biotechnology Information (NCBI) contains all experimentally determined structures from the PDB (Ohkawa et al., 1995; Chen et al., 2003). It is updated on a monthly basis and provides linkage to structural information from NCBI's integrated query interface Entrez. Entries in MMDB are specified using an Abstract Syntax Notation One (ASN.1; http://asn1.elibel.tm.fr/). MMDB provides access to coordinates, sequences, all bibliographic information, and taxonomy data, as well as the authors and deposition dates together with the PDB-assigned classification and compound information of a PDB entry.

The assignment of the correct species of origin of a specific PDB chain is based on a semiautomated procedure in which a human expert validates the automatically assigned taxonomy annotation based on sequence comparisons with GenBank and SwissProt. A set of rules ensures the consistency of this approach. Missing annotations are generated from literature information or using BLAST searches. Artificially generated protein and nucleic acid chains (excluding trivial modifications like single amino acid substitutions or His-tags) are labeled as "synthetic."

Beyond enabling the query for structures based on the textual information described, MMDB also provides structural neighbor assignments produced by the Vector Alignment Search Tool (VAST) (Gibrat et al., 1996). Each chain of each entry in MMDB is compared with every other chain to compile a list of structural neighbors. These are made available for individual chains as well as for domains. In addition, MMDB can be queried with user-supplied coordinate sets to find entries based on structural similarity.

The information stored in MMDB and Entrez allows the seamless exploration and query of literature references, sequence information, and taxonomical and genomic data associated with macromolecular structures. While other resources, including the PDB, provide links to some of these data, MMDB uniquely combines them into a single resource. NCBI also provides several graphical tools including the application CND3 for 3D structure visualization and a WWW-based chromosome browser.

---

[1] Previously, structural representatives were selected based on the "Summary PDB ASTRAL Check Index" (SPACI) score. The AEROSPACI score is the SPACI score adjusted with penalties for aberrant structures.

Derived Secondary Structure of Proteins—http://www.sander.ebi.ac.uk/dssp/

The Derived Secondary Structure of Proteins (DSSP) resource provides secondary structure assignments computed from structure using an algorithm developed in the early 1980s by Kabsch and Sander (1983). The DSSP resource consists of the DSSP program itself (licensed at no cost to academic users and available for commercial licensing) and the DSSP-generated flat files, one per PDB entry. Using a standardized representation, the DSSP file contains the secondary structure assignments, geometric structure, and solvent exposure for each residue. These data are also available from a variety of Web sites. In contrast, the PDB files provide annotator validated secondary structure assignments based on the PROMOTIF program (Hutchinson and Thornton, 1996).

Protein Quaternary Structure—http://pqs.ebi.ac.uk/

Structure determination does not always provide the functional form of a biological macromolecule. Rather, it provides the tertiary structure as found in the asymmetric unit of the crystal that is not necessarily a biologically functional unit. Proteins often form quaternary structures—the macromolecular assembly of two or more copies of tertiary structure elements that form homo- or heteromultimers that confer biological function. Viral protein coats are beautiful examples of biological function inferred by the organization of tertiary structure into a quaternary biologically active assembly (see for examples specific to viruses VIrus Particle ExploreR, VIPER; http://mmtsb.scripps.edu/viper/viper.html) (Reddy et al., 2001). The Protein Quaternary Structure (PQS) resource maintained by the Macromolecular Structure Database (MSD) group at the European Bioinformatics Institute (EBI) provides an automatically derived assessment of the biological unit of a PDB entry determined by X-ray crystallography (Henrick and Thornton, 1998).

The Cartesian coordinates found in a PDB entry generally correspond to the asymmetric unit of the molecule as found in the crystal and represent the unique atomic positions that are refined against the experimental data. However, these coordinates do not necessarily correspond to the biologically active molecule. The necessary crystallographic symmetry operations as defined by the space group and possibly noncrystallographic symmetry[2] must be applied to generate the biologically active quaternary structure. This is done through the application of rotation and translation to the individual chemical components listed in the PDB entry.

Automating this procedure is nontrivial. The process must distinguish between an assembly that is a truly biologically active molecule and an assembly that is a number of discreet biologically active components associated through crystal packing, but having no physiological relevance. It should be noted that the PDB now seeks to capture the interpretation of the biologically active molecule from the structural biologists depositing the structure data rather than attempt to only determine it automatically.

The PQS procedure is well documented on the Web site and only a synopsis is given here. For nonvirus structure, PQS performs two steps—generate the assembly and assess the assembly for the likelihood it is a quaternary structure. The first step involves applying

---

[2] If the molecule exhibits its own symmetry, then refinement of the structure may be undertaken only on the part considered unique; the tertiary structure is then generated from the coordinates present in the PDB and the application of noncrystallographic symmetry. See PDBids 3HHB and 4HHB for contrasting examples from the same molecule, deoxy hemoglobin.

any noncrystallographic symmetry and then recursively adding symmetry-related contents to the asymmetric unit. If close contacts are found, this is considered a candidate quaternary structure. The second step determines the nature of the contacts using the solvent accessible surface. The premise is that components forming a quaternary complex will have a lower solvent accessible surface than those existing as discreet globular proteins.

PQS provides its results in the form of PDB formatted files that include the list of all symmetry operators and calculated coordinates. In addition, PQS provides a description of the quaternary structure, for example, "homodimer" or "heterotetradecamer." Virus entries are treated differently in that several files are provided to include the complete virion and separate, symmetry-related files as well as a file containing all chains needed to describe the unique protein–protein interfaces.

Comparisons between literature-derived information as well as information provided by individual researchers were used to determine a rough measure of accuracy for the PQS procedure (Henrick and Thornton, 1998). Using 6739 entries available from the PDB in December 1997, 1398 were determined to be potential homodimers. Of these, 244 were assigned to have nonspecific (crystal packing) contacts. This could be confirmed for 31 entries based on the available textual information. The remaining 1154 entries were assigned true homodimer status. This could be confirmed for 385 entries, could not be confirmed for 386 entries, and was found to be false positives for 383 entries. Of those 383, 190 were lysozymes, which exhibit very strongly associated crystallographic packing, underscoring the difficulty in automatically determining the difference between specific and nonspecific macromolecular associations. Other examples of seemingly incorrect predictions include a prediction of a 24 meric assembly of the transcription repressor protein rop (PDB identifier 1GTO). Although the biologically active molecule in fact is a DNA-associated dimer, the authors of the crystal structure describe a "hyperstable helical bundle" in the crystal structure, possibly due to artificial solid-state interactions.

In summary, while caution needs to be exercised in using PQS-generated quaternary structure predictions, the resource nevertheless provides a valuable starting point to the determination of the biologically active molecules represented by the asymmetric units given in the PDB entries.

Protein–Ligand Interactions—ReliBase—http://relibase.ccdc.cam.ac.uk/

The biological function and regulation of proteins oftentimes involves the binding of smaller organic or inorganic molecules that commonly are grouped together under the term ligands. Metal ions, anions, solvate molecules (except water), cofactors, and inhibitors are generally all regarded as ligands. ReliBase, developed primarily by Dr. Manfred Hendlich and now maintained at the Cambridge Crystallographic Data Center, contains experimental PDB structures with ligands and structures where only the ligand-binding partner was modeled into the structure (Hendlich, 1998; Hendlich et al., 2003). DNA and RNA strands are visualized in result sets as ligands but cannot be searched. ReliBase provides access to its entries via text queries over the header, compound, and source records of the PDB files, as well as the names of authors, chemical names of ligands, and their PDB-assigned three letter codes. In addition, ReliBase can be queried using a protein sequence or SMILES strings (string representations of 2D structural fragments or molecules) (Siani et al., 1994). Finally, it is possible to search ReliBase using 3D diagrams drawn using a Java applet.

ReliBase results are easy to browse and include 2D diagrams of ligands, bibliographic and some additional textual information from the PDB entries as well as convenient links to searches for similar ligands, binding sites, or protein chains. Query results can be stored as hit lists, which can be used in SMILES or 2D/3D searches. In addition, binding sites can be superimposed and visualized in different ways using static images, graphical applets, or client side visualization tools such as Rasmol (Sayle and Milner-White, 1995). Third-party tools integrated into ReliBase include the sequence search package FASTA (Pearson, 1990; Pearson, 1994) and the computational chemistry tool kit CACTVS (Ihlenfeldt et al., 2002) that is used to generate 2D diagrams in ReliBase.

ReliBase is the product of several industrial/academic partnerships and is written in C^{++} with a Perl CGI WWW front end. Stand-alone distributions for several platforms are available from the CCDC upon request.

Protein Families

Often macromolecular structure information is only a part of a larger study on a particular family of proteins that are functionally related. Resources capturing such comprehensive information are usually developed by individual research laboratories with interest in specific protein families. The general notion is to be narrow but deep versus resources like the PDB, which are broad but shallow with respect to their information content. Stated in another way, the PDB contains a limited amount of information on all macromolecular structures; resources such as those described in this section integrate structure as part of additional information on a specific protein family. A couple of these resources are highlighted to indicate the kind of content that is available. Similar resources to those discussed here exist for chaperonins, the P450 family, cytokines, esterases, G-protein-coupled receptors, glucoamylases, kinesins, thyroid hormone receptors, topoisomerases, and viruses. A more complete list and associated Web links can be found at the CMS Molecular Biology Resource at http://restools.sdsc.edu/biotools/biotools25.html.

HIV Proteases—http://mcl1.ncifcrf.gov/hivdb/index.html

The HIV Protease Database (HIVdb) archives experimentally determined structures of human immunodeficiency virus 1 (HIV1), human immunodeficiency virus 2 (HIV2), and simian immunodeficiency virus (SIV) proteases and their complexes (Vondrasek et al., 1997; Vondrasek and Wlodawer, 2002). The structures contained in HIVdb include 124 structures not currently available through the PDB and were made available by several pharmaceutical companies for exclusive use by the resource. An additional 148 structures are taken from the PDB.

The information provided by HIVdb includes tabular listings of ligand/enzyme complexes, enzyme inhibitors, and proteinase mutants. In addition, analytical information on volume analysis, interaction energy, surface analysis, subsite occupation, and structural superpositions is made available in graphical form. The resource is searchable through a simple text field and results are presented in tabular form including bibliographical information, PDB accession numbers, if applicable, and inhibitor information including graphical representations.

HIVdb was developed in the group of Dr. Alex Wlodawer and is maintained at the National Cancer Institute.

Metalloprotein—http://metallo.scripps.edu/

The Metalloprotein Database and Browser (MDB) is part of the Metalloprotein Structure, Bioinformatics and Design Program at The Scripps Research Institute (TSRI). MDB provides quantitative information and tools to visualize protein metal-binding sites from structures taken from the PDB (Castagnetto et al., 2004). Approximately, one third of all structures in the PDB contain a metal ion.

Entries are extracted from the PDB and added to MDB with a set of automatic tools that periodically scan newly released PDB structures for the occurrence of metal ions. An indexing tool extracts first- and second-shell data, recognizes multinuclear and cluster-containing sites, and classifies metal-binding sites according to criteria such as the number of metal ions in the site, the types of ions, and metal coordination. Noncovalent interactions are also determined within and among indexed shells.

MDB can be queried with a variety of methods ranging from simple text-based queries to fairly complex SQL queries that fully realize the power of the underlying, fully documented relational database schema. Real-time three-dimensional viewing of binding sites is provided through a Java applet that enables the user to inspect interatomic distances, bond angles, and torsion angles. Structure superpositions, stereoviewing, and selection of atoms based on distance are also possible.

In addition to the interactive query and analysis interfaces provided to users, MDB offers noninteractive gateways for incorporation of MDB data into stand-alone programs. Most notably, MDB supports an XML-RPC-based interface, a remote procedure calling protocol that uses the Hypertext Transfer Protocol (http) and Extensible Markup Language (XML) for the exchange of data. XML-RPC is simple protocol, which allows complex data structures to be transmitted, processed, and returned. The protocol would, for example, allow a metal-site design program to obtain an up-to-date list of observed ranges for a certain geometric feature (e.g., torsion angle) to compare a suggested model value with those found in known metalloproteins.

MDB is build on top of the relational database system MySQL and uses the powerful Web scripting language PHP as a front end. The Java applet is also used by other sites such as the IMB Jena Image Library of Biological Macromolecules (Reichert et al., 2000; Reichert and Suhnel, 2002) as a gateway to MDB.

Macromolecular Motions Database—http://molmovdb.org/

The Macromolecular Motions Database (MolMovDB) describes and systematizes known motions that occur in proteins and other macromolecules. Associated with MolMovDB are a set of free software tools and servers for structural motion analysis (Gerstein and Krebs, 1998; Krebs and Gerstein, 2000; Flores et al., 2006).

MolMovDB addresses an important phenomenon in biochemistry, the precise movement of many atoms within a macromolecule that often plays a crucial role in its function. Macromolecular motions are essential in, for example, enzymatic reactions, allosteric regulation of activity, transporter functionality, and locomotion. Due to the involved timescales, which range from subnanosecond loop closures to refolding spanning several seconds, it is near impossible to study these motions with a single computational approach like molecular dynamics due to the computational intractability.

MolMovDB currently contains more than 20,000 entries. Of these, 19,600 have been automatically extracted from the PDB, 230 have been manually curated, and 200 have been submitted by users. Protein motions are categorized first by the information available on the motion, its size (distinguished are fragment, domain, and chain motions), and lastly by type of motion. Motions of proteins involving fragment or domain motions are primarily characterized as consisting either a "shear" motion (sliding of a continuously maintained and tightly packed interface) or a "hinge" motion (movement of two domains connected by a flexible linker without a continuously maintained interface). Motions of subunits are predominantly classified as "allosteric," "nonallosteric," or "complex." Each individual motion in the database is assigned a mnemonic accession code and a classification code. For example, the motion in calmodulin is accessible under the identifier "cm" and is classified as a "known domain motion, hinge mechanism" (D-h-2). A total of 29 such classifiers were established and are documented.

MolMovDB is searchable by keyword and/or by PDB identifier. Curated entries are also listed for easy access. Each entry is accompanied by its classification, links to PDB structures (via their PartsList entries, see Section 4.10), a description of the motion, and particular values describing the motion. Movies are associated with each entry and available in several formats. The Morph Server software automatically generates these movies and produces 2D and 3D animations of plausible pathways between two end points of a particular motion. A typical morph takes a few minutes to compute and results are stored for later access. Morphing involves an adiabatic mapping algorithm to interpolate two PDB input files. A particular pathway is broken up into several equal length steps, at each step interpolated coordinates are subjected to an energy minimization "refinement" to correct bond length, bond angle, and torsion angle aberrations. The Morph Server is accessible as a stand-alone tool for users wishing to generate their own movies based on two given structures.

MolMovDB exists as a combination of XML files and MYSQL with a Perl-based CGI front end; some computationally intensive components of the site (Morph Server) are partially implemented in C/C^{++}, FORTRAN, and Python/MMTK. The WWW front end is easy to navigate for any user but SQL dumps are also available for advanced users upon request from the maintainers.

PartsList: Dynamic Fold Comparisons—http://bioinfo.mbb.yale.edu/partslist/

The number of structures in the PDB is expected to increase significantly in the next few years, specifically with the advent of structural genomics (see also Chapter 40 and a short perspective at the end of this chapter). However, the number of protein folds is quite limited and analyses and re-analyses of this finite PartsList from an expanding number of perspectives will probably become more and more informative as the list reaches completeness. The resource described in this section, PartsList, allows users to dynamically compare this emerging and linked set of protein folds.

PartsList is based on the SCOP (see Chapter 17) fold classification and functions as supplemental annotation to SCOP. Folds in PartsList (represented by domains corresponding to specific folds and/or superfamilies in SCOP) are ranked on a growing number of currently more than 180 attributes. These attributes include the occurrence in completely sequenced genomes, the number of occurrences of a fold in the PDB, participation in protein–protein interactions, the number of known functions associated with a fold, the amino acid composition, participation in protein motions, and the level of similarity based on a comprehensive set of structural alignments using the Gerstein/Levitt algorithm (Gerstein and Levitt, 1998; Qian et al., 2001).

Three ways of visualizing the fold rankings are provided by PartsList: first, a profiler emphasizing the progression of high and low ranks across many preselected attributes, next a dynamic comparer for custom comparisons, and finally a numerical rankings correlator. Traditional single-structure reports are provided to summarize information related to genome occurrence, expression level, motion, function, and interaction with additional links to many other resources.

The ranking provided by PartsList allows a comparison of folds using a unified approach. The numerical values associated with each rank can be used to compare the very different attributes of a fold, for example, expression levels and participation in protein–protein interaction. Access to tabular comparisons is made available for all individual fold rankings according to individual attributes. For example, users can readily switch between occurrence, interaction, motion, or alignment information for a fold identified with the Profiler, Comparer, or Correlator tool. In addition, PartsList is searchable by PDB or SCOP accession number and text files (summary tables and structural alignments) are made available for download.

PartsList is maintained in Prof. Mark Gerstein's group at Yale University. The resource provides "extrinsic" information on protein folds, that is, putting a fold into the context of all other folds according to specific criteria.

Automated Comparative Modeling: Swiss-Model—
http://swissmodel.expasy.org/

Protein modeling involves the generation of a theoretical model of a protein structure based on its sequence and one or more known structures with more or less similar sequences. In recent years, many automated approaches have been reported in the literature and several servers are available for users to generate their own structural models (see Chapters 29–32). The Swiss-Model server (Guex and Peitsch, 1997; Schwede et al., 2003) is one example of many structure prediction and modeling resources and the reader is referred to a more comprehensive listing available at http://restools.sdsc.edu/biotools/biotools9.html.

Swiss-Model offers several modes in which users can generate and refine their models. In addition, the structure viewing program Swiss-PDBViewer has been tightly integrated with the modeling resource. Swiss-PDBViewer enables the analysis of several proteins at the same time. Proteins can be superimposed to generate structural alignments to compare relevant parts, for example, their active sites. Amino acid mutations, hydrogen bonds, bond angles, and distances between atoms are displayed via a graphic and menu interface. Swiss-PDBViewer can also read electron density maps for detailed interpretation of structures, various modeling tools are integrated and command files for use in popular energy minimization packages can be generated. While both Swiss-Model and Swiss-PDBViewer can be used independently, the combination of both can be used to generate structural models.

Swiss-Model uses structure templates extracted from the PDB, their sequences, and the ProModII modeling package to generate the actual models. Users are able to submit their own templates in PDB format for use in ProModII. The automatic template selection step involves a BLAST query of the Swiss-Model template database given user definable threshold values. The subsequent modeling procedure employed by ProModII involves the following steps: (1) superposition of related 3D structures, (2) generation of a multiple alignment with the sequence to be modeled, (3) generation of a framework for the new sequence, (4) a rebuild lacking loops, (5) completion and correction of the structural backbone, (6) correction and rebuilding of side chains, (7) verification of the model structure's quality and a check of its packing, and (8) refinement of the structure by energy minimization and molecular dynamics. Generated models are sent to users by email and can be imported, analyzed, and manipulated in Swiss-PDBViewer.

Swiss-Model and Swiss-PDBViewer were developed in the group of Dr. Manuel Peitsch and are maintained at part of the Expert Protein Analysis System (ExPASy) server of the Swiss Institute of Bioinformatics.

Other Sources of Targets and Prediction Methods

Although the cost of structure determination is decreasing rapidly, it will probably never become as cheap as the cost of sequencing. Hence, the ratio of the number of structures to the number of sequences will remain at several orders of magnitude. Yet, as the number of structures continues to rise, they provide a rich source of template information for structure prediction using techniques such as homology modeling and threading. Progress in these areas is monitored by the Critical Assessment of Structure Prediction (CASP) experiments that are conducted every two years (Chapter 28). At the CASP meetings, prediction methods are compared, rated, and hotly debated (Venclovas et al., 2001). Predictions can be performed in 1D (secondary structure, solvent accessibility), 2D (interresidue distances), and 3D (*ab initio* prediction, homology modeling such as implemented by Swiss-Model and threading). Resources even exist to evaluate prediction servers (e.g., EVA, http://cubic.bioc.columbia.edu/eva/ and LiveBench, http://meta.bioinfo.pl/LiveBench/). To facilitate these prediction efforts, if the depositor permits, sequences of solved protein structures are now released ahead of the structures by the PDB to permit unbiased experiments from a continuous source of new target. Another source of targets are the sequences registered by the structural genomics projects in a target database maintained by the PDB at http://targetdb.pdb.org/.

STRUCTURAL DATABASES OF THE FUTURE

Integration Over Multiple Resources

The world of online information available to structural biologists has become extremely balkanized as the number of resources available as well as the information content provided by these resources has increased exponentially in the last decade (Williams, 1997). Most databases available today on the Web provide a good number of cross-links to other resources with relevant information. However, in almost all nontrivial cases (i.e., those cases where the link is not simply based on an obvious identifier in the remote resource), these cross-links have to be added and maintained by human curators. To create such links automatically, database maintainers have to first agree on a common nomenclature or provide a comprehensive ontology of the information available through their resources for interconnection with other ontologies. Much progress has been made in the last few years in this area and the PDB curation efforts of the RCSB are a notable example.

A new effort, the BioLit project, aims to obviate the need for this human curation step by integrating the open access literature and biological data. These tools are being implemented using the entire corpus of the Public Library of Science (PLoS), which is leading the open access movement, and the Protein Data Bank as testing platforms. The tools are being designed, however, to be generally applicable to all open access literature and other biological data.

The BioLit tools capture metadata from an article or manuscript by identifying relevant terms and identifiers and adding markup to the original NLM DTD-based XML document containing the open access article. Terms relating to the life sciences are identified using ontologies and controlled vocabularies specific to this field such as the Gene Ontology (Ashburner et al., 2000; Harris et al., 2004) and Medical Subject Headings (MeSH). These metadata are captured in different ways depending on the status of the article. One tool will allow this information to be captured while the manuscript is being written. This strategy gives the author full and fine control over the exact metadata that are captured. The tool will prompt the author with choices or will allow the author to customize the metadata if no appropriate matches are found in the resources that the tool has knowledge of. Cross-references to biological databases will also be detected and added to the metadata, allowing the manuscript content to be more easily integrated with the database. Articles that have already been published can be postprocessed through a related tool that identifies the same types of metadata and generates similar XML markup. The metadata may not be as rich using this approach since the author has not had direct input, but the capture of any information is a significant advance. Effective use of these tools will make the integration between data, resources, and literature nearly seamless.

The Impact of Structural Genomics

Structural genomics (Burley et al., 1999) (see also Chapter 40) is an effort to develop and employ high-throughput structure determination for purposes including the filling in of protein fold space to facilitating comparative modeling, the determining of as many protein structures from a given genome as possible, or the furthering of our understanding of specific diseases or biochemical pathways. Although the goals may differ, the process is the same and it was initially estimated that a large number of structures (over 35,000) (Weissig and Bourne, 1999) would have been generated by now. Many of these structures will be incomplete, having been discarded in a partially completed state, since they were not deemed useful for the goals of a given project. Others will be complete, but for the first time functionally unclassified. Although efforts are under way to ensure the central deposition of all structural genomics results, many of these data might not be available centrally from the PDB given the expected lack of annotation or their level of incompleteness. While this situation will likely change, it may be that the user will need to visit multiple sources of structure information for a complete coverage of all available macromolecular structures.

Many structural genomics centers report their results to TargetDB, the structural genomics target registration database maintained by the RCSB (http://targetdb.rcsb.org/). This database currently contains over 140,000 entries (10 times the number available as reported in the first edition of this book), some of which will be solved and further enrich the large variety of databases of derived information described in this chapter. While resource maintainers are faced with new challenges to judge and automatically handle the quality of the shear amount of structure information available, users will shortly have an even richer collection of resources available from which to study structure–function relationships. The fact that these resources already greatly enhance our understanding of biological systems is a testament not only to those individuals who produce the primary structure data, but also to all those who have developed and maintained the resources described herein.

Now that the structural genomics initiative has been operational for several years, we can evaluate how well the actual progress has matched our estimations. According to a recent review, the cost of solving a structure at a structural genomics center is indeed less compared to traditional methods (Chandonia and Brenner, 2006). In addition, structural genomics has made an appreciable advance in exploring the fold space even though it has fallen short of expectations. However, structures solved using traditional methods appear to be cited more often in the literature. This may be because scientists using traditional methods tend to focus on structures that are of particular interest to the community, although a recent editorial suggests (Anon, 2007) that the intended consumers of structural genomics output are unfortunately largely unaware of these efforts and thus may not be accessing these structures. It seems clear that structural genomics has made a significant contribution and, now, it is a matter of tapping into this underutilized resource to make better use of the data.

REFERENCES

Allen FH, Kennard O (1993): 3D search and research using the Cambridge Structural Database. *Chem Des Autom News* 8:31–37.

Allen FH, Taylor R (2004): Research applications of the Cambridge Structural Database (CSD). *Chem Soc Rev* 33(8):463–475.

Andreeva A, Howorth D, et al. (2004): SCOP database in 2004: refinements integrate structure and sequence family data. *Nucleic Acids Res* 32(Database issue):D226–D229.

Anon (2007): Looking ahead with structural genomics. *Nat Struct Mol Biol* 14(1):1.

Ashburner M, Ball CA, et al. (2000): Gene ontology: tool for the unification of biology. The Gene Ontology Consortium. *Nat Genet* 25(1):25–29.

Bader GD, Donaldson I, et al. (2001): BIND: The Biomolecular Interaction Network Database. *Nucleic Acids Res* 29(1):242–245.

Bader GD, Betel D, et al. (2003): BIND: The Biomolecular Interaction Network Database. *Nucleic Acids Res* 31(1):248–250.

Berman HM, Bhat TN, et al. (2000): The Protein Data Bank and the challenge of structural genomics. *Nat Struct Biol* 7(Suppl):957–959.

Berman H, Henrick K, et al. (2007): The worldwide Protein Data Bank (wwPDB): ensuring a single, uniform archive of PDB data. *Nucleic Acids Res* 35(Database issue):D301–D303.

[The standard reference for the PDB since it has been managed by the Research Collaboratory for Structural Bioinformatics.] Bernstein FC, Koetzle TF, et al. (1977): The Protein Data Bank: a computer-based archival file for macromolecular structures. *J Mol Biol* 112(3):535–542.

[The original PDB reference.] Bernstein H, Bernstein F, et al. (1998): CIF Applications VIII. pdb2cif: translating PDB entries into mmCIF format. *J Appl Cryst* 31:282–295.

Brenner SE, Koehl P, et al. (2000): The ASTRAL compendium for protein structure and sequence analysis. *Nucleic Acids Res* 28(1):254–256.

Brown ID, McMahon B, (2002): CIF: the computer language of crystallography. *Acta Crystallogr B* 58(Part 3 Part 1):317–324.

Burley SK, Almo SC, et al. (1999): Structural genomics: beyond the human genome project. *Nat Genet* 23(2):151–157.

[Original and highly cited article defining structural genomics.] Castagnetto JM, Hennessy SW, et al. (2004): MDB: the Metalloprotein Database and Browser at The Scripps Research Institute. *Nucleic Acids Res* 30(1):(2002) 379–382.

Chandonia JM, Hon G, et al. The ASTRAL Compendium in 2004. *Nucleic Acids Res* 32(Database issue):D189–D192.

Chandonia JM, Brenner SE (2006): The impact of structural genomics: expectations and outcomes. *Science* 311(5759):347–351.

Chen J, Anderson JB, et al. (2003): MMDB: Entrez's 3D-structure database. *Nucleic Acids Res* 31(1):474–477.

Dietmann S, Park J, et al. (2001): A fully automatic evolutionary classification of protein folds: Dali Domain Dictionary version 3. *Nucleic Acids Res* 29(1):55–57.

Flores S, Echols N, et al. (2006): The Database of Macromolecular Motions: new features added at the decade mark. *Nucleic Acids Res* 34(Database issue):D296–D301.

Gerstein M, Krebs W (1998): A database of macromolecular motions. *Nucleic Acids Res* 26(18): 4280–4290.

Gerstein M, Levitt M (1998): Comprehensive assessment of automatic structural alignment against a manual standard, the scop classification of proteins. *Protein Sci* 7(2):445–456.

Gibrat JF, Madej T, et al. (1996): Surprising similarities in structure comparison. *Curr Opin Struct Biol* 6(3):377–385.

Gilliland GL, Tung M, et al. (1994): Biological Macromolecule Crystallization Database, Version 3.0: new features, data and the NASA archive for protein crystal growth data. *Acta Crystallogr D Biol Crystallogr* 50(Part 4):408–413.

Greene LH, Lewis TE, et al. (2007): The CATH domain structure database: new protocols and classification levels give a more comprehensive resource for exploring evolution. *Nucleic Acids Res* 35(Database issue):D291–D297.

Guex N, Peitsch MC (1997): SWISS-MODEL and the Swiss-PdbViewer: an environment for comparative protein modeling. *Electrophoresis* 18(15):2714–2723.

Hall SR, Allen FH, et al. (1991): The crystallographic information file (CIF): a new standard archive file for crystallography. *Acta Crystallogr D Biol Crystallogr* A47:655–685.

Harris MA, Clark J, et al. (2004): The Gene Ontology (GO) database and informatics resource. *Nucleic Acids Res* 32(Database issue):D258–D261.

Hendlich M (1998): Databases for protein–ligand complexes. *Acta Crystallogr D Biol Crystallogr* 54 (Part 6 Part 1):1178–1182.

Hendlich M, Bergner A, et al. (2003): Relibase: design and development of a database for comprehensive analysis of protein–ligand interactions. *J Mol Biol* 326(2):607–620.

Henrick K, Thornton JM (1998): PQS: a protein quaternary structure file server. *Trends Biochem Sci* 23(9):358–361.

Hennessy D, Buchanan B, Subramanian D, Wilkosz PA, Rosenberg JM (2000): Statistical methods for the objective design of screening procedures for macromolecular crystallization. *Acta Cryst* D56:817–827.

Hobohm U, Sander C (1994): Enlarged representative set of protein structures. *Protein Sci* 3(3):522–524.

[A widely used set of structures in bioinformatics based on non-redundancy of sequence.]Hooft RW, Vriend G, et al. (1996): Errors in protein structures. *Nature* 381(6580):272.

Hutchinson EG, Thornton JM (1996): PROMOTIF: a program to identify and analyze structural motifs in proteins. *Protein Sci* 5(2):212–220.

Ihlenfeldt WD, Voigt JH, et al. (2002): Enhanced CACTVS browser of the Open NCI Database. *J Chem Inf Comput Sci* 42(1):46–57.

Jones S, van Heyningen P, Berman HM, Thornton JM. (1999): Protein-DNA interactions: A structural analysis *J Mol Biol* 287(5): 877–896.

Jones S, Thornton, JM (1996): Principles of protein-protein interactions. *Proc Natl Acad Sci U S A* 93:13–20.

Kabsch W, Sander C (1983): Dictionary of protein secondary structure: pattern recognition of hydrogen-bonded and geometrical features. *Biopolymers* 22(12):2577–2637.

Krebs WG, Gerstein M, (2000): The morph server: a standardized system for analyzing and visualizing macromolecular motions in a database framework. *Nucleic Acids Res* 28(8): 1665–1675.

Laskowski RA, Rullmannn JA, et al. (1996): AQUA and PROCHECK-NMR: programs for checking the quality of protein structures solved by NMR. *J Biomol NMR* 8(4):477–486.

Murzin AG, Brenner SE, et al. (1995): SCOP: a structural classification of proteins database for the investigation of sequences and structures. *J Mol Biol* 247(4):536–540.

[Original paper describing the most widely used protein structure classification scheme.]Ohkawa H, Ostell J, et al. (1995). MMDB: an ASN.1 specification for macromolecular structure. *Proc Int Conf Intell Syst Mol Biol* 3:259–267.

Orengo CA, Michie AD, et al. (1997): CATH: a hierarchic classification of protein domain structures. *Structure* 5(8):1093–1108.

Pearson WR (1990): Rapid and sensitive sequence comparison with FASTP and FASTA. *Methods Enzymol* 183:63–98.

Pearson WR (1994): Using the FASTA program to search protein and DNA sequence databases. *Methods Mol Biol* 25:365–389.

Qian J, Stenger B, et al. (2001): PartsList: a web-based system for dynamically ranking protein folds based on disparate attributes, including whole-genome expression and interaction information. *Nucleic Acids Res* 29(8):1750–1764.

Reddy VS, Natarajan P, et al. (2001): Virus Particle Explorer (VIPER), a website for virus capsid structures and their computational analyses. *J Virol* 75(24):11943–11947.

Reichert J, Jabs A, et al. (2000): The IMB Jena Image Library of biological macromolecules. *Nucleic Acids Res* 28(1):246–249.

Reichert J, Suhnel J, (2002): The IMB Jena Image Library of Biological Macromolecules: 2002 update. *Nucleic Acids Res* 30(1):253–254.

Sayle RA, Milner-White EJ (1995): RASMOL: biomolecular graphics for all. *Trends Biochem Sci* 20(9):374.

Schwede T, Kopp J, et al. (2003): SWISS-MODEL: an automated protein homology-modeling server. *Nucleic Acids Res* 31(13):3381–3385.

Shindyalov IN, Bourne PE (1998): Protein structure alignment by incremental combinatorial extension (CE) of the optimal path. *Protein Eng* 11(9):739–747.

Siani MA, Weininger D, et al. (1994): CHUCKLES: a method for representing and searching peptide and peptoid sequences on both monomer and atomic levels. *J Chem Inf Comput Sci* 34(3):588–593.

Siddiqui AS, Barton GJ (1995): Continuous and discontinuous domains: an algorithm for the automatic generation of reliable protein domain definitions. *Protein Sci* 4(5):872–884.

Sowdhamini R, Burke DF, et al. (1998): CAMPASS: a database of structurally aligned protein superfamilies. *Structure* 6(9):1087–1094.

Vaguine AA, Richelle J, Wodak SJ (1999) SFCHECK: A unified set of procedures for evaluating the quality of macromolecular structure factor data and their agreement with the atomic model. *Acta Cryst* D55:191–205.

Venclovas C, Zemla A, et al. (2001): Comparison of performance in successive CASP experiments. *Proteins* Suppl 5:163–170.

Vondrasek J, van Buskirk CP, et al. (1997): Database of three-dimensional structures of HIV proteinases. *Nat Struct Biol* 4(1):8.

Vondrasek J, Wlodawer A (2002): HIVdb: a database of the structures of human immunodeficiency virus protease. *Proteins* 49(4):429–431.

Weissig H, Bourne PE, (1999): An analysis of the Protein Data Bank in search of temporal and global trends. *Bioinformatics* 15(10):807–831.

Williams N (1997): How to get databases talking the same language. *Science* 275(5298):301–302.

Xenarios I, Salwinski L, et al. (2002): DIP, the Database of Interacting Proteins: a research tool for studying cellular networks of protein interactions. *Nucleic Acids Res* 30(1):303–305.

Section III

DATA INTEGRITY AND COMPARATIVE FEATURES

14

STRUCTURAL QUALITY ASSURANCE

Roman A. Laskowski

INTRODUCTION

The experimentally determined 3D structures of proteins and nucleic acids represent a key knowledge base from which a vast understanding of biological processes has been derived over the past half century. Individual structures have provided explanations of specific biochemical functions and mechanisms, while comparisons of structures have given insights into general principles governing these complex molecules, the interactions they make, their biological roles, and their evolutionary relationships. The 3D structures are held and looked after by the Worldwide Protein Data Bank (wwPDB) that, as of August 2007, held over 45,000 entries.

These structures form the foundation of Structural Bioinformatics; all structural analyses depend on them and would be impossible without them. Therefore, it is crucial to bear in mind two important truths about them, both of which result from the fact that they have been determined experimentally. The first is that the result of any experiment is merely a "model" that aims to give as good an explanation for the experimental data as possible. The term "structure" is commonly used, but you should realize that this should be correctly read as "model." As such, the model may be an accurate and meaningful representation of the molecule, or it may be a poor one. The quality of the data and the care with which the experiment has been performed will determine which it is. Independently performed experiments can arrive at very similar models of the same molecule; this suggests that both are accurate representations, that they are good models.

The second important truth is that any experiment, however carefully performed, will have errors associated with it. These come in two distinct varieties: systematic errors and random errors. Systematic errors relate to the *accuracy* of the model—how well it corresponds to the "true" structure of the molecule in question. These often include errors of interpretation. In X-ray crystallography, for example, the molecule(s) need to be fitted to the electron density computed from the diffraction data. If the data are poor, and the quality of

Structural Bioinformatics, Second Edition   Edited by Jenny Gu and Philip E. Bourne
Copyright © 2009 John Wiley & Sons, Inc.

the electron density map is low, it can be difficult to find the correct tracing of the molecule(s) through it. A degree of subjectivity is involved and errors of mistracing and "frame-shift" errors, described later, are not uncommon, and in some cases the chain has been traced so badly as to render the final model completely wrong. In NMR spectroscopy, judgments must be made at the stage of spectral interpretation where the individual NMR signals are assigned to the atoms in the structure most likely to be responsible for them.

Random errors, on the other hand, depend on how precisely a given measurement can be made. All measurements contain errors at some degree of precision. If a model is essentially correct, the sizes of the random errors will determine how *precise* the model is. The distinction between accuracy and precision is an important one. It is of little use having a very precisely defined model if it is completely inaccurate.

The sizes of the systematic and random errors may limit the types of questions a given model can answer about the given biomolecule. If the model is essentially correct, but the data was of such poor quality that its level of precision is low, then it may be of use for studies of large scale properties, such as protein folds, but worthless for detailed studies requiring the atomic position to be precisely known; for example, to help understand a catalytic mechanism.

STRUCTURES AS MODELS

To make the point about 3D structures being merely models, it is instructive to consider the subtly different types of model obtained by the two principal experimental techniques: X-ray crystallography and NMR spectroscopy. Figure 14.1 shows the two different interpretations of the same protein given by the two methods, as explained below. The models are of the protein rubredoxin with a bound metal ion held in place by four cysteines. Both can be found in the PDB: the X-ray model (PDB code 1IRN) has a zinc as its metal ion while in the NMR model (1BFY), the metal ion is iron.

Models from X-Ray Crystallography

Figure 14.1a is a representation of the protein model as obtained by X-ray crystallography. It is not a standard depiction of a protein structure; rather, its aim is to illustrate some of the components that go into the model. The components are: the x-, y-, z-coordinates, B-factors and occupancies of all the individual atoms in the structure. These parameters, together with the theory that explains how X-rays are scattered by the electron clouds of atoms, aim to account for the observed diffraction pattern. The x-, y-, z-coordinates define the mean position of each atom while its B-factor and occupancy aim to model its apparent disorder about that mean. This disorder may be the result of variations in the atom's position in time, due to the dynamic motions of the molecule, or variations in space, corresponding to differences in conformation from one location in the crystal to another, or both. The higher the atom's disorder, the more "smeared out" its electron density. B-factors model this apparent smearing around the atom's mean location; at high resolution a better fit to the "observations" can often be obtained by assuming the B-factors to be anisotropic, as represented by the ellipsoids in Figure 14.1a. Occasionally, the data can be explained better by assuming that certain atoms can be in more than one place—say, due to alternative conformations of a particular side chain (indicated by the arrows showing the two alternative positions of the glutamate side chain in Figure 14.1a). The atom's occupancy defines how

(a)   (b)

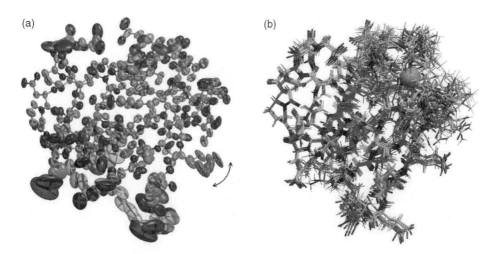

Figure 14.1. The different types of model generated by X-ray crystallography and NMR spectroscopy. Both are representations of the same protein: rubredoxin. (a) In X-ray crystallography the model of a protein structure is given in terms of atomic coordinates, occupancies and B-factors. The side chain of Glu50 has two alternative conformations, with the change from one conformation to the other identified by the double-headed arrow. The B-factors on all the atoms are illustrated by "thermal ellipsoids," which give an idea of each atom's anisotropic displacement about its mean position. The larger the ellipsoid, the more disordered the atom. Note that the main chain atoms tend to be better defined than the side chain atoms, some of which exhibit particularly large uncertainty of position. The coordinates and B-factors come from PDB entry 1IRN, which has a zinc ion bound and was solved at 1.2 Å resolution and refined with anisotropic B-factors. (b) The result of an NMR structure determination is a whole ensemble of model structures, each of which is consistent with the experimental data. The ensemble shown here corresponds to 10 of the 20 structures deposited as PDB code 1BFY. In this case, the metal ion is iron. The more disordered regions represent either regions that are more mobile, or regions with a paucity of experimental data, or a combination of both. The region around the iron-binding site appears particularly disordered, in stark contrast with the same region in the X-ray model in (a), where the B-factors are low and the model well-ordered. Both diagrams were generated with the help of the Raster3D program (Merritt and Bacon, 1997). Figure also appears in the Color Figure section.

often it is found in one conformation and how often in another (for example, in Figure 14.1a the occupancy of one of the conformations of the Glu is 56% and that of the other is 44%).

Models from NMR Spectroscopy

The data obtained from NMR experiments are very different, so the models obtained differ in their nature, too. Generally, the sample of protein or nucleic acid is in solution, rather than in crystal form, which means that molecules that are difficult to crystallize, and hence impossible to solve by crystallography, can often be solved by NMR instead. The spectra measured by NMR provide a diversity of information about the molecule's average structure in solution together with its dynamics. The most numerous, but often least precise, data are from Nuclear Overhauser Effect SpectroscopY (NOESY) experiments where the intensities of particular signals correspond to the separations between the spatially close protons (≤ 6 Å) in the structure. The spectra from COrrelated SpectroscopY (COSY)-type experiments give more precise information on the separations of protons up to three covalent bonds apart, and in some cases on the presence, or even length, of specific hydrogen bonds. Dipolar coupling experiments give information on the relative orientation of particular backbone covalent bonds (Clore and Gronenborn, 1998).

The proton separations obtained from the experimental data are converted into distance and angular restraints that are used to generate models of the structure that are consistent with these restraints. Various techniques are used to obtain these models; the most commonly used are molecular dynamics based simulated annealing procedures similar to those used in X-ray structure refinement. The end result is not a single model, but rather an ensemble of models that are all consistent with the given restraints, as illustrated in Figure 14.1b.

The reasons for generating an ensemble of structures from NMR data are twofold. Firstly, the NMR data are relatively less precise and less numerous than experimental restraints from X-rays, so a diversity of structures are consistent with them. Secondly, the molecules may genuinely possess heterogeneity in solution. In fact, it has even been suggested that low-resolution X-ray models would benefit from being generated as ensembles, much like their NMR counterparts, to better represent the experimental data (DePristo, de Bakker, and Blundell, 2004).

For general use, an ensemble of models is rather more difficult to handle than a single model. Ensembles deposited in the PDB can typically comprise 20 models, with the largest ensemble in the PDB, as of August 2007, having 184 models (PDB code 2HYN; Potluri et al., 2006). One of the models may be designated the "representative" of the whole ensemble, or, alternatively, a separate averaged model may be computed from all the ensemble members. This "average structure" has to be energy minimized to counteract the unphysical bond lengths and angles that the averaging process introduces. Such a structure tends to have a separate PDB identifier from that of the ensemble—so the same structure, or rather the outcome of the same experiment, appears as two separate entries in the PDB. This is clearly potentially confusing and the use of separate files is now discouraged. The representative member of an ensemble is usually taken to be the structure that differs least from all other structures in the ensemble. An algorithmic Web-based tool called OLDERADO (Kelley, Gardner, and Sutcliffe, 1996) allows to select such a representative from an ensemble (http://www.ebi.ac.uk/msd-srv/olderado), but no single algorithm is universally agreed upon.

Models from Electron Microscopy (EM)

X-ray and NMR models are by far the most common of the experimentally determined models of macromolecular structures. However, there are a small number of models in the PDB (153 models as of August 2007) that have been derived from electron microscopy (EM) experiments, and these require special caution. The experimental data underlying these models is of especially low resolution, typically in the range of 6–15 Å—well below the limits where individual side chains, let alone atoms, can be discerned—and the structures tend to be of large multiprotein and/or RNA complexes. Yet the models in the PDB entries do contain atomic coordinates; usually, just the C^{α} backbone atoms, but in some cases all main chain and side chain atoms! The latter are derived by fitting the coordinates of one or more crystal structures, solved at higher resolution, into the EM maps. The resultant coordinates of the final model are, at best, an approximation of the atomic positions of the various components making up the sample and will not have been subjected to any rigorous positional or geometric refinement procedures. Consequently, they should not be relied on to any great extent.

Other Experimentally Determined Models

The PDB also contains models solved by powder, fiber, electron and neutron diffraction studies as well as electron tomography, infrared spectroscopy and solution scattering. Each method has its strengths, weaknesses, and peculiarities, so structures solved by these methods need to be carefully assessed on a case-specific basis, with reference to the literature describing them.

Theoretical Models

Particular skepticism should be reserved for models that are not directly based on any experimental measurement. These are the so-called "theoretical models" and are obtained either by homology modeling or "threading" techniques. Homology models rely on the fact that proteins sharing high-sequence identity tend to have very similar overall 3D structures; proteins sharing a sequence identity greater than around 35% will have essentially the same fold. This makes it possible to generate a "homology model" of the 3D structure of a given protein based on the 3D model of a closely related protein. The modeling can be automated and there are several servers that generate models from user-submitted sequences, the best known being SWISS-MODEL (Schwede et al., 2003). Even where the model is a good one, that is, obtained from a very similar protein, one needs to be cautious about taking the model's atomic positions at face value as they do not represent positions based on any experimental data. In fact, it is likely that any experimental errors in the template model will not only be propagated into the homology model but even exacerbated in the process.

The PDB, which used to treat theoretical models much like experimentally derived ones, has, since July 2002, kept them apart in a separate ftp site: ftp://ftp.wwpdb.org/pub/pdb/data/ structures/models/current. New models are no longer accepted (Berman et al., 2006). As of August 2007, there were 1358 theoretical models on the models site.

The least reliable models of all are predicted models such as those generated by *ab initio* and fold recognition or "threading" methods. These methods can be described as straw-clutching measures of last resort, employed when a given protein sequence has no relatives in the PDB and so is not amenable even to homology modeling. Occasionally, these methods approximate the right answer—usually for small, single-domain proteins where they may produce topologically near correct models (Moult, 2005)—but generally, they are wildly wrong. So, it is safe to assume that the level of error in any such model will be high.

AIMS

The primary take-home message of this chapter is that not all structural models, whether in the PDB or not, are equally reliable and each should be critically assessed before being used for one's own work. It may seem slightly churlish to reject any structure given the amount of time, care, and hard work, the experimentalists or modelers may have put into creating it; but if you put unsound data into your analysis, you will get unsound conclusions out of it. Some models will be unreliable simply because the experimental data from which they were derived were of poor (or nonexistent) quality, while others may be unreliable as a result of human error. Whatever the reason, it is important to weed out any uncertain models as early as possible.

This chapter aims to explain the problems of relying on 3D models uncritically and provides some rules of thumb for filtering out the defective ones: what are the symptoms to look for, how to detect them, and at what point should you reject a model?

ERROR ESTIMATION AND PRECISION

All scientific measurements contain errors. No measurement can be made infinitely precisely; so, at some point, say after so many decimal places, the value quoted becomes unreliable. Scientists acknowledge this by estimating and quoting standard uncertainties on their results. For example, the latest value for Boltzmann's constant is $1.3806503(24) \times 10^{-23}$ J/K, where the two digits in brackets represent the standard uncertainty (s.u.) in the last two digits quoted for the constant.

Compare this with the situation we have in relation to the 3D structures of biological macromolecules. Figure 14.2 shows a typical extract from the atom details section of a PDB file. It relates to a single amino acid residue (a leucine) and shows the information deposited about each atom in the protein's structure.

Looking at only the columns representing the x-, y-, z-coordinates you'll notice that each value is quoted to three decimal places. This suggests a precision of about 1 in 10^5 in the given example. Similarly, the B-factors (in the final column) are each quoted to two decimal places. Is it possible that the atomic positions and B-factors were really so precisely defined? What are the error bounds on these values? Are the values accurate to the first place of decimals? The second? The third?

In fact, with the exception of a very few PDB structures, no error bounds are given. As of August 2007, there were 10 such exceptions, out of over 45,000 structures, the largest being a protein chain of 313 amino acid residues and four bound ligands (human aldose reductase, PDB code 2I17). All the 10 had been solved at atomic resolution (ranging from 0.89 to 1.3 Å) and refined by the full-matrix least squares method that will be discussed later.

	Atom number	Atom name	Residue name	Residue number	Atomic coordinates x	y	z	Occup-ancy	B-factor
Atom	1	N	Leu	1	-15.159	11.595	27.068	1.00	18.46
Atom	2	CA	Leu	1	-14.294	10.672	26.323	1.00	9.92
Atom	3	C	Leu	1	-14.694	9.210	26.499	1.00	12.20
Atom	4	O	Leu	1	-14.350	8.577	27.502	1.00	13.43
Atom	5	CB	Leu	1	-12.829	10.836	26.772	1.00	13.48
Atom	6	CG	Leu	1	-11.745	10.348	25.834	1.00	15.93
Atom	7	CD1	Leu	1	-11.895	11.027	24.495	1.00	13.12
Atom	8	CD2	Leu	1	-10.378	10.636	26.402	1.00	15.12

Figure 14.2. An extract from a PDB file of a protein structure showing how the atomic coordinates and other information on each atom are deposited. The atoms are of a single leucine residue in the protein. The contents of each column are labeled above the column. It can be seen that the x-, y-, z-coordinates of each atom are given to three places of decimals.

Thus, in the overwhelming majority of cases one cannot tell how precisely defined the values are. Why is this so? What kind of scientific measurement is this? And how are we to judge how much reliance should be placed on the data given?

ERROR ESTIMATES IN X-RAY CRYSTALLOGRAPHY

Estimation of Standard Uncertainties

In X-ray crystallography, it is actually possible, in theory, to calculate the standard uncertainties of the atomic coordinates and B-factors. In fact, it is routinely done for the X-ray crystal structures of small molecules such as those deposited in the Cambridge Structural Database (CSD) (Allen et al., 1979). The calculations of the s.u.s are performed during the refinement stage of the structure determination. As you learnt in Chapter 4, refinement involves modifying the initial model to improve the match between the experimentally determined structure factors—as obtained from the observed X-ray diffraction pattern—and the calculated structure factors—as obtained from applying scattering theory to the current model of the structure. Figure 14.3 illustrates this principle.

In practice, refinement is usually a long drawn-out procedure requiring many cycles of computation interspersed here and there with manual adjustments of the model using molecular graphics programs to nudge the refinement process out of any local minimum that it may have become trapped in. Furthermore, because in protein crystallography, the data-to-parameter ratio is poor (the data being the reflections observed in the diffraction pattern and the parameters being those defining the model of the protein structure: the atomic x-, y-, z-coordinates, B-factors, and occupancies) the data needs to be supplemented by additional information. This extra information is applied by way of geometrical restraints. These are "target" values for geometrical properties such as bond lengths and bond angles and are typically obtained from crystallographic studies of small molecules. The refinement process aims to prevent the bond lengths and angles in the model drifting too far from these target values. This is achieved by applying additional terms to the function being minimized of the form:

$$\sum_{k=1}^{\text{distance}} \omega_k (d_{k0} - d_k)^2,$$

where d_k and d_{k0} are the actual and target distance, and ω_k is the weight applied to each restraint.

If the structure is refined using full-matrix least-squares refinement, a by-product of this method is that the s.u.s of the refined parameters, such as the atomic coordinates and B-factors, can be obtained. However, their calculation involves inverting a matrix the size of which depends on the number of parameters being refined. The larger the structure, the more atomic coordinates and B-factors, the larger the matrix. As matrix inversion is an order n^3 process, it has tended to be unfeasible for molecules of the size of proteins and nucleic acids; these have several thousand parameters and consequently a matrix whose elements number several millions or tens of millions. This is why s.u.s have been routinely published for small-molecule crystal structures, but not for structures of biological macromolecules. It is purely a matter of size.

However, as faster workstations with larger memories have become available, the situation has started to change, and calculation of atomic errors has become more practicable (Tickle

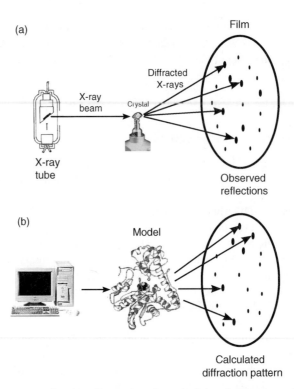

Figure 14.3. A schematic diagram illustrating the principle of structure refinement in X-ray crystallography. (a) X-rays are passed through a crystal of the molecule(s) of interest, generating a diffraction pattern from which, by one method or another (see Chapter 4), an initial model of the molecular structure is calculated. (b) Using the model, it is possible to apply scattering theory to calculate the diffraction pattern we would expect to observe. Usually this will differ from the experimental pattern. The process of structure refinement involves iteratively modifying the model of the structure until a better and better fit between the observed and calculated patterns is obtained. The goodness of fit of the two sets of data is measured by the reliability index, or *R*-factor.

et al., 1998). Indeed, s.u.s are now frequently calculated for small proteins using SHELX (Sheldrick and Schneider, 1997), the refinement package originally developed for small molecules, but sadly, the s.u. data are still not commonly deposited in the PDB file. So this makes us none-the-wiser about the precision with which any given atom's location has been determined.

So what is to be done? What other information is there on the reliability of an X-ray crystal structure? What should one look for?

First of all, one can get an idea of the global quality of the structure from certain parameters that are commonly cited in the literature and quoted in the header records of the PDB file itself, as described next.

Global Parameters for X-Ray Structures

Figure 14.4 shows an extract from the header records of a PDB file showing some of the commonly cited global parameters.

```
...
Remark    2 Resolution. 2.20Å

...
Remark    3   R value              (Working set) : 0.198
Remark    3   FREE R value                       : 0.255
Remark    3   FREE R value test set size   (%)   : 10.2

...
Remark    3   Estimated coordinate error
Remark    3   ESD From luzzati plot         (A) : 0.23
Remark    3   ESD From sigmaa               (A) : 0.23
Remark    3   LOW Resolution cutoff         (A) : 5.00
Remark    3
Remark    3   Cross-validated estimated coordinate error.
Remark    3   ESD FROM C-V LUZZATI PLOT     (A) : 0.30
Remark    3   ESD FROM C-V SIGMAA           (A) : 0.27

...
```

Figure 14.4. Extracts from the header records of a PDB file (1ydv) showing some of the statistics pertaining to the quality of the structure as a whole. These include the resolution, R-factor, R_{free}, and various estimates of average positional errors that range from 0.23 to 0.30 Å. The R_{free} has been calculated on the basis of 10.2% of the reflections removed at the start of refinement and not used during it.

Resolution. The resolution at which a structure is determined provides a measure of the amount of detail that can be discerned in the computed electron density map. The reflections at larger scattering angles, θ, in the diffraction pattern correspond to higher resolution information coming as they do from crystal planes with a smaller interplanar spacing. The high-angle reflections tend to be of a lower intensity and more difficult to measure and the greater the disorder in the crystal, the more of these high-angle reflections will be lost. Resolution relates to how many of these high-angle reflections can be observed, although the value actually quoted can vary from crystallographer to crystallographer, as there is no clear definition of how it should be calculated. The higher resolution shells will tend to be less complete and some crystallographers will quote the highest resolution shell giving a 100% complete data set, whereas others may simply cite the resolution corresponding to the highest angle of scatter observed.

The higher the resolution, the greater the level of detail, and hence the greater the accuracy of the final model. The resolution attainable for a given crystal depends on how well ordered the crystal is, that is, how close the unit cells throughout the crystal are to being identical copies of one another. A simple rule of thumb is that the larger the molecule, the lower will be the resolution of the data collected.

Figure 14.5 shows an example of how the electron density for a single side chain improves, as resolution increases. In general, side chains are difficult to make out at very low resolution (4 Å or lower), and the best that can be obtained is the overall shape of the molecule and the general locations of the regions of regular secondary structure. Models at such low resolution are clearly of no use for investigating side chain conformations or interactions! At 3 Å resolution, the path of a protein's chain can be traced through the density and at 2 Å, the side chains can be confidently fitted.

The most precise structures are the atomic resolution ones (from around 1.2 Å resolution up to around 0.9 Å). Here the electron density is so clear that many of the hydrogen atoms become visible, and alternative occupancies become more easily distinguishable. These structures require fewer geometrical constraints during refinement and hence give a better indication of the true geometry of protein structures.

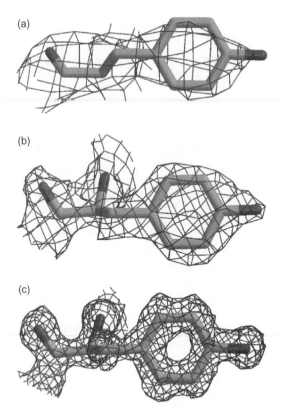

Figure 14.5. The effect of resolution on the quality of the electron density. The three plots show the electron density, as the wire-frame cage, surrounding a single tyrosine residue. The residue is Tyr100 from concanavalin A as found in three PDB structures solved at (a) 3.0 Å resolution (PDB code 1VAL), (b) 2.0 Å (1CON), and (c) 1.2 Å (1JBC). At the lowest resolution the electron density is merely a shapeless blob, but as the resolution improves so the individual atoms come into clear focus. The electron density maps were taken from the Uppsala Electron Density Server (http://eds.bmc.uu.se) and rendered using BobScript (Esnouf, 1997) and Raster3D (Merritt and Bacon, 1997).

The resolution of X-ray structures in the PDB varies from atomic resolutions structures to very low-resolution structures at around 4.0 Å, with a definite peak at around 2.0 Å. Structures solved at resolutions below 4 Å tend to be solved by electron microscopy. The lowest quoted resolution, as at August 2007, was 70.0 Å for a whole series of structural models of skeletal muscle crossbridges in insect flight muscle, solved by electron tomography; for example, PDB entry 1M8Q (Chen et al., 2002).

Resolution is probably the clearest measure of the likely quality of the given model. However, bear in mind that, because there is no single, consistent definition of resolution its value can be overstated (Weissig and Bourne, 1999). What is more, in X-ray crystallography poor resolution data can sometimes result in models of quite respectable quality if the crystals exhibit a high degree of noncrystallographic symmetry. By averaging the data at symmetry related positions one can build models that would otherwise be impossible. This is the trick that allows very large complexes such as the protein coats of viruses to be solved crystallographically despite apparently poor data.

R-Factor. The R-factor is a measure of the difference between the structure factors calculated from the model and those obtained from the experimental data. In essence, it is a measure of the differences in the observed and computed diffraction patterns schematically illustrated in Figure 14.3. Higher values correspond to poorer agreement with the data, while lower values correspond to better agreement. Typically, for protein and nucleic acid structures, values quoted for the R-factor tend to be around 0.20 (or equivalently, 20%). Values in the range 0.40–0.60 can be obtained from a totally random structure, so structures with such values are unreliable and probably would never be published. Indeed, 0.20 seems to be something of a magical figure and many structures are deemed finished once the refinement process has taken the R-factor to this mystical value.

As a reliability measure, however, the R-factor is in itself somewhat unreliable. It is quite easily susceptible to manipulation, either deliberate or unwary, during the refinement process, and so models with major errors can still have reasonable-looking R-factors. For example, one of the early incorrect structures, cited by Brändén and Jones (1990), was that of ferredoxin I, an electron transport protein. The fully refined structure was deposited in 1981 as PDB code 2FD1, with a quoted resolution of 2.0 Å and an R-factor of 0.262. Due to the incorrect assignment of the crystal space group during the analysis of the X-ray diffraction data, this structure turned out to be completely wrong. The replacement structure, reanalyzed by the original authors and having the correct fold, was deposited as PDB entry 3DF1 in 1988. Its resolution was given as 2.7 Å and its R-factor as 0.35. On the face of it, therefore, mere comparison of the resolution and R-factor parameters would lead one to believe the first of the two structures to be the more reliable! The reason why an R-factor as low as 0.262 was achieved for a totally incorrect structure was that the coordinates included 344 water molecules, many extending far out from the protein molecule itself. This is a large number of waters for a protein containing only 107 residues. A well-known rule of thumb suggests that a reasonable number of waters in a given structure is roughly one water molecule for each protein residue, and waters should only be added to the structure if they make plausible hydrogen bonds. This "one water per residue" rule has been shown to be true for structures solved at around 2 Å resolution, although at higher resolution more waters can be detected; at 1.0 Å resolution one can reasonably expect to detect 1.6–1.7 waters per residue (Carugo and Bordo, 1999).

Incidentally, the 3DF1 model of ferredoxin has itself been twice superseded: first by entry 4df1 in mid-1988 and then by entry 5DF1 in 1993. The last of these had a quoted resolution of 1.9 Å and R-factor of 0.215.

The ferredoxin case is one of overfitting, that is, having too many parameters for the available experimental data. In the original 2FD1 model, the discrepancy between the observed and computed diffraction patterns was literally washed out by the addition of so many water molecules. One can always fit a model, however wrong, to the experimental data and obtain a reasonable R-factor if there is an excess of parameters over observations.

R_{free}. A more reliable reliability factor is Brünger's free R-factor, or R_{free} (Brünger, 1992). This is less susceptible than the standard R-factor to manipulation during refinement. It is calculated in the same way, but using only a small fraction of the experimental data, typically 5–10%. Crucially, this fraction, known as the test set, is excluded from the structure refinement procedure that uses the remaining 90–95% of the data, or working set. Thus, unless there are correlations between the data in the test and working sets, the R_{free} provides an independent measure of the fit of the model to the data and cannot be biased by the refinement process.

The value of R_{free} will tend to be larger than the R-factor, although it is not clear what a good value might be. Brünger has suggested that any value above 0.40 should be treated with caution (Brünger, 1997). There were 51 structures in the PDB, as of August 2007, in this category. Not surprisingly, most are fairly low-resolution structures, solved at 3.0 Å or lower, but a few, rather worryingly, are not.

Average Positional Error. Even though atomic coordinate s.u.s are not commonly given, it is quite usual for an estimate of the *average* positional error of a structure's coordinates to be cited. There are two principal methods for estimating the average positional errors: the Luzzati plot (Luzzati, 1952) and the σ_A plot (Read, 1986).

The Luzzati plot is obtained by partitioning the reflections from the diffraction pattern into bins according to their value of sin θ, where θ is the reflection's scattering angle, and then calculating the R-factor for each bin. The value calculated for each bin is plotted as a function of sin θ/λ, where λ is the wavelength of the radiation used. The resulting plot is compared against the theoretical curves of Luzzati (1952) to obtain an estimate of the average positional error. One problem with this method is that the actual curves do not usually resemble the theoretical ones well at all, and so the error estimate is somewhat crude and often merely provides an upper limit on the error. Better results are obtained if the R_{free} is used instead of the traditional R-factor.

The σ_A plot provides a better estimate. It involves plotting $\ln \sigma_A$ against $(\sin \theta/\lambda)^2$, where σ_A is a complicated function that has to be estimated for each $(\sin \theta/\lambda)^2$ bin, as described in Read (1986). The resultant plot should give a straight line the slope of which provides an estimate of the average positional error.

Most refinement programs compute both the Luzzati and Read error estimates, so these values are commonly cited in the PDB file. You will find them in the file's header records under the now unfashionable term "estimated standard deviation" (ESD) (see Figure 14.4).

Bear in mind that an average s.u. is exactly what it says: an average over the whole structure. The s.u.s of the atoms in the core of the molecule, which tends to be more ordered, will be lower than the average, while those of the atoms in the more mobile and less well-determined surface—and often more biologically interesting—regions will be higher than the average.

Atomic B-Factors. A more direct, albeit merely qualitative, way of determining the precision of a given atom's coordinates is to look at its associated B-factor. B-factors are closely related to the positional errors of the atoms, although the relationship is not a simple one that can be easily formulated (Tickle et al., 1998). It is safe to say, however, that atoms in a structure with the largest B-values will also be those having the largest positional uncertainty. So if high levels of precision are required in your analysis, leave out the atoms having the highest B-factors. As a rule of thumb, atoms with B-values in excess of 40.0 are often excluded as being too unreliable. Similarly, if atoms in your region of interest, such as an active site, are all cursed with high B-factors then your region of interest is not well determined and you will either need to be careful about the conclusions you draw from it, or else seek out a different structural model of the same protein where the region is better defined.

Other Parameters. The above parameters are those that are commonly given in the header records of the PDB entry but are by no means the only information that can be used for assessing the usefulness of a given structural model. Other checks, such as on the geometry of the model or its agreement with the experimental data can be made by running the appropriate software or accessing the appropriate Web server, as will be discussed later.

Rules of Thumb for Selecting X-Ray Crystal Structures

Many analyses in Structural Bioinformatics require the selection of a dataset of 3D structures on which analysis can be performed. A commonly used rule of thumb for selecting reliable structures for such analyses, where reasonably accurate models are required, is to choose those models that have a quoted resolution of 2.0 Å or better, and an R-factor of 0.20 or lower. These criteria will give structures that are likely to be reasonably reliable down to the conformations of the side chains and local atom–atom interactions. One example that uses such a dataset is the Atlas of Protein Side Chain Interactions (http://www.ebi.ac.uk/thornton-srv/databases/sidechains) that depicts how amino acid side chains pack against one another within the known protein structures.

Of course, the selection criteria depend on the type of analysis required. For some analyses, only atomic resolution structures (i.e., 1.2 Å or better) will do, as in the accurate derivation of geometrical properties of proteins; for example, side chain torsional conformers and their standard deviations (EU 3D Validation Network, 1998), or fine details of the peptide geometry in proteins that can reveal subtle information about their local electronic features (Esposito et al., 2000). For other types of analysis, structures solved down to 3 Å may be good enough, as in any comparison of protein folds. One interesting example is that of the lactose operon repressor. Three structures of this protein were solved to 4.8 Å resolution, giving accurate position for only the protein's C^{α} atoms (Lewis et al., 1996). However, because the three structures were of the protein on its own, of the protein complexed with its inducer, and of the protein complexed with DNA, the global differences between the three structures showed how the protein's conformation changed between its induced and repressed states. Thus, even these very low-resolution structures were able to help explain how this particular protein achieves its biological function (Lewis et al., 1996).

Often the above rule of thumb (resolution ≤ 2.0 Å, and R-factor ≤ 0.20) is supplemented by a check on the year when the structure was determined. Structures are more likely to be less accurate, the older they are simply because experimental techniques have improved markedly since the early pioneering days of the 1960s and 1970s (Weissig and Bourne, 1999; Kleywegt and Jones, 2002). Indeed, many of the early structures have been replaced by more recent and accurate determinations.

ERROR ESTIMATES IN NMR SPECTROSCOPY

The theory of NMR spectroscopy does not provide a means of obtaining s.u.s for atomic coordinates directly from the experimental data, so estimates of a given structure's accuracy and precision have to be obtained by more indirect means.

Global Parameters for NMR Structures

As mentioned above, a number of models can be derived that are compatible with the NMR experimental data. It is difficult to distinguish whether this multiplicity of models reflects real motion within the molecules or simply results from insufficient experimentally derived restraints. (Compare how the most poorly defined regions of the X-ray model of rubredoxin in Figure 14.1a do not necessarily correspond to the most poorly defined regions of the NMR model in Figure 14.1b, although remembering that one structure was in crystal form, and the other in solution.) Generally, the agreement of NMR models with the NMR data is measured by the agreement between the distance and angular restraints applied during the refinement of the models and the corresponding distances and angles in the final models. Large numbers of severe violations indicate a serious problem of data interpretation and model building.

However, the errors associated with the original experimental data are sufficiently large that it is almost always possible to generate models that do not violate the restraints, or do so only slightly. Consequently, it is not possible to distinguish a merely adequate model from an excellent one by looking for restraint violations alone.

Traditionally, the "quality" of a structure solved by NMR has been measured by the root-mean-squared deviation (RMSD) across the ensemble of solutions. Regions with high RMSD values are those that are less well defined by the data. In principle, such RMSD measures could provide a good indicator of uncertainty in the atomic coordinates; however, the values obtained are rather dependent on the procedure used to generate and select models for deposition. An experimentalist choosing the "best" few structures for deposition from a much larger draft ensemble can obtain very misleading statistics for the PDB entry. For example, the "best" few structures may, in effect, be the same solution with minor variations, so the RMSD values will be small. Structures further down the original list may provide alternative solutions, which are slightly less consistent with the data, but which are radically different.

The number of experimentally derived restraints per residue gives another measure of quality. This provides an indication of how effectively the NMR data define the structure in a manner analogous to the resolution of X-ray structures. Indeed, the number of restraints per residue correlates with the stereochemical quality of the structures to an extent, but some restraints may be completely redundant and no consistent method of counting is used by depositors. A better guide is the completeness of NOEs (Doreleijers et al., 1999) that compares the numbers of NOEs that one would expect to get, given the final model, against the number that were actually observed. The completeness gives a slightly better correlation with stereochemical parameters. However, the calculation of expected NOEs is complicated by the fact that it is made from a static model, so some expected NOEs may not in fact be detectable due to the internal dynamics of the molecule in question.

None of the above measures gives a true indication of the accuracy of the models—that is, how well they represent the true structure—and few of them are reported in the PDB file.

In recent years, NMR equivalents of the crystallographic R-factor have been introduced. One method involves the use of dipolar couplings. These provide long-range structural restraints that are independent of other NMR observables such as the NOEs, chemical shifts, and couplings constants that result from close spatial proximity of atoms. Because the expected dipolar couplings can be computed for a given model, they provide a means of comparing the observed with the expected, and obtaining an R-factor that is a measure of the difference between the two (Clore and Garrett, 1999). What is more, it is also possible to obtain a cross-validated R-factor, equivalent to the crystallographic R_{free}, wherein a subset of dipolar couplings are removed prior to the start of structure refinement and used only for computing the R-factor. This gives an unbiased measure of the quality of the fit to the experimental data. However, in the case of NMR, one cannot use a single test set of data; one has to perform a complete cross-validation. The reason for this is that, whereas in crystallography each reflection contains information about the whole molecule, in NMR each dipolar coupling does not. So a complete cross-validation is required, which means that a number of calculations have to be performed, each using a different selection of test and working data sets; the test set, which usually comprises 10% of the whole data set, being selected at random each time.

Another technique for calculating an NMR R-factor uses the NOEs and involves back calculation of the NMR intensities from the models obtained and comparison with those observed in the experiment. This technique is implemented in the program RFAC (Gronwald et al., 2000) that calculates not only an overall R-factor for the entire structure, but also local R-factors, including residue-by-residue R-factors and individual R-factors for different groups of NOEs (e.g., medium range NOEs, long range NOEs, inter-residue NOEs, etc.).

An additional back-calculation method for checking structure quality is to calculate the expected frequencies (positions) of spectral peaks from the structure and compare them with those observed. This comparison has the advantage that the frequencies are not usually a target of the structure refinement procedure (Williamson, Kikuchi, and Asakura, 1995).

However, once again, the measures described here are not generally included in the deposited PDB files and so have to be laboriously obtained from the relevant literature reference.

Rules of Thumb for Selecting NMR Structures

Historically, the rule of thumb for selecting NMR structures for inclusion in structural analyses has been the simple one of excluding them altogether! This early prejudice stems from the fact that they were viewed as being of generally lower quality than X-ray structures as a number of studies have been keen to point out (see Spronk et al., 2004, and references therein). There has never been an easy way of distinguishing the good from the bad NMR models using a simple rule of thumb as in the case of X-ray structures. However, NMR structures provide much valuable information about protein and DNA structures not available from X-ray studies. For example, just under 15% of PDB structures come from NMR experiments (as at August 2007), although, as mentioned previously, this includes a lot of double counting with many structures being deposited as two separate PDB entries: one for the whole ensemble and the other for an energy minimized average structure. So in terms of models derived from separate NMR experiments, this percentage is far lower. However, when one selects data sets of nonhomologous sequences in the PDB at the 30% sequence identity level, one finds that just over 16% are NMR structures. So to exclude all NMR models involves discarding many unique and important proteins that perhaps can only ever be solved by NMR.

Because there is no standard information provided in the header information of NMR PDB files relating to either the quality of the experimental data or the resultant models, one has to either read and interpret the original paper describing the structure or, more practically, rely on measures of the stereochemical quality of the structure, as will be described later.

There are two very useful Web sites that provide "improved" versions of several hundred old NMR models. The improved models have been generated by re-refining the original experimental data using more up-to-date force fields and refinement protocols. The first is DRESS, a Database of Refined Solution NMR Structures (Nabuurs et al., 2004), which contains 100 re-refined NMR models (http://www.cmbi.kun.nl/dress) and the second, more recent, database is RECOORD (Nederveen et al., 2005) that contains 545 models (http://www.ebi.ac.uk/msd/NMR/recoord). Along with giving the new coordinates, the databases provide indicators of how the newer models are an improvement over the old in terms of stereochemical quality measures. So, if the structure you need is an early NMR model, it is worth checking whether it is included in either of these two databases.

ERRORS IN DEPOSITED STRUCTURES

Serious Errors

At the end of 2006, a letter to the journal "Science" caused something of a stir in the structural biology community. The authors wrote to say that they were withdrawing three of their Science papers, and five corresponding PDB entries, due to serious errors in their models (Chang et al., 2006). The errors were the result of a trivial mistake made during initial processing of the X-ray data. As a consequence, the final models had serious errors in their handedness and topology as was revealed when another group solved a similar structure.

Such a high profile retraction is fairly rare but the concern about serious errors in structural models has been around for a long time (Bränden and Jones, 1990) and there have been a number of serious errors in both X-ray and NMR structures documented in the literature (Kleywegt, 2000). As in the example above, many of the erroneous models have been retracted by their original authors, or replaced by improved versions. It is common for structures to be re-refined, or solved with better data, and the models in the PDB replaced by the improved versions. So it is instructive to bear in mind that, at any time, there are likely to be models in the PDB that are in need of replacement!

The old models do not all disappear. There is a growing graveyard of "obsolete" structures—some very much incorrect, others merely slightly mistaken—available from the wwPDB (ftp://ftp.wwpdb.org/pub/pdb/data/structures/obsolete).

Of all the errors, the most serious are those where the model is, essentially, completely wrong such as when the trace of the protein chain follows the wrong path through the electron density resulting in a completely incorrect fold. One example, that of ferredoxin I, has already been mentioned in the discussion of the crystallographic R-factor above. Another example is depicted in Figures 14.6a and b that show an incorrect and correct model of photoactive yellow protein. There is practically no similarity between the two.

The next most serious errors are where all, or most, of the secondary structural elements have been correctly traced, but the chain connectivity between them is wrong. An example is given in Figures 14.6c and d. Here the erroneous model has most of the correct secondary structure elements, and has them arranged in the correct architecture, but the protein sequence has been incorrectly traced through them (in one case going the wrong way down a β-strand). Consequently, most of the protein's residues have ended up in the wrong place in the 3D structure. Errors such as this arise because the loop regions that connect the secondary structure elements tend to be more flexible, and more disordered, so their electron density tends to be more poorly defined and difficult to interpret correctly. In the case shown in Figure 14.6c, the interpretation of the electron density was made especially difficult by the simple fact that the *primary sequence* of the protein was unknown at the time. Thus, the sequence of the protein had to be inferred from the blobs of electron density as the chain was being fitted into it; usually, the sequence is a crucial guide to the tracing of the protein chain. Needless to say, it was subsequently found that the inferred primary sequence was just as much in error as the 3D structure.

Less serious are frame shift errors, although they can often result in a significant part of the model being incorrect. These errors occur where a residue is fitted into the electron

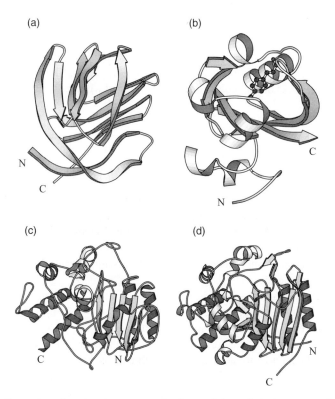

Figure 14.6. Examples of seriously wrong protein models and their corrected counterparts. (a) Incorrect model of photoactive yellow protein (PDB code, 1PHY, an all-C^α atom model), and (b) the corrected model (2PHY, all atoms plus bound ligand). Superposition of the two models gives an RMSD of 15 Å between equivalent C^α atoms. Such a high value is hardly surprising given that the folds of the two models are so completely different. (c) Incorrect model of D-alanyl-D-alanine peptidase (1PTE, an all-C^α atom model), and (d) corrected model (3PTE, all atoms). The initial model had been solved at low resolution (2.8 Å) at a time when the protein's sequence was unknown, so tracing the chain had been much more difficult than usual. Many of the secondary structure elements were correctly detected, but incorrectly connected. The matching secondary structures are shown as the darker shaded helices and strands. The connectivity between them is completely different in the two models, with the earlier model having completely wrong parts of the sequence threaded through the secondary structure elements. Indeed, you can see that the central strand of the β-sheet runs in the opposite direction in the two models. The N- and C-termini of all models are indicated. All plots were generated using the Molscript program (Kraulis, 1991).

density that belongs to the next residue. The frame shift persists until a compensating error is made when two residues are fitted into the density belonging to a single residue (Jones and Kjeldgaard, 1997). These mistakes often occur in loop regions, and almost exclusively at very low resolution (3 Å or lower). Frame shift errors were recently investigated in a data set of 842 protein chains belonging to the oligosaccharide binding (OB)-fold (Venclovas, Ginalski, and Kang, 2004). Of the 842 chains, 12 (1.4%) were found to have frame shift errors. The 12 chains were contained in five separate PDB entries (some of which were multimers) deposited between 1983 and 2000 and ranging in resolution from 2.2 to 3.2 Å.

The authors of the study concluded that around 1% of all protein sequences in the PDB could contain such errors.

The least serious model-building errors involve the fitting of incorrect main chain or side chain conformations into the density. Of course, even such errors, depending on where they occur, can have an effect on the biological interpretation of what the structure does and how it does it.

Typical Errors

Typically, the majority of the models deposited in the PDB will be essentially correct. The remaining errors will be the random errors associated with any experimental measurement. As mentioned above, for X-ray structures, the average s.u.s—estimated on the basis of the Luzzati and σ_A plots—can provide an idea of the magnitude of these errors. The values range from around 0.01 to 1.27 Å. Note that the latter value approaches the length of some covalent bonds! The median of the quoted s.u.s corresponds to estimated average coordinate errors of around 0.28 Å. It has to be remembered that these values are estimates, and are applied as an average over the whole model.

Figure 14.7 gives a feel of some typical uncertainties in atomic positions, showing positional uncertainties of 0.2, 0.3, and 0.39 Å.

A surprising result from a recent study was that some types of errors that are relatively easy to detect and fix were still turning up in newly released PDB entries (Badger and Hendle, 2002). The authors found that, in the PDB entries they looked at (all released on a single day) 3.6% of the amino acids contained an error of some kind, with the worst model having errors in 10.6% of its amino acids. The most common errors were His/Asn/Gln side chain flips. These side chains are symmetrical in terms of shape and so will fit their electron density equally well when rotated by 180°. However, their chemical properties are not symmetrical and, in general, one orientation will result in a better hydrogen-bonding network than the other. This can be critical when the side chain in question is part of a catalytic or binding site and all such errors should really have been fixed prior to the deposition of the models in the PDB.

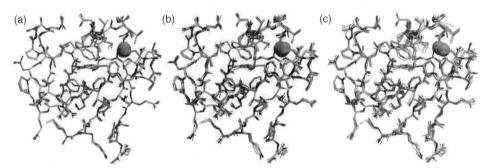

Figure 14.7. Examples of typical uncertainties in atomic positions for (a) an s.u. of 0.2 Å (b) 0.3 Å, and (c) 0.39 Å. The protein is the same rubredoxin from Figure 14.1a. Of course, as shown in Figure 14.1a, the distribution of uncertainties would not normally be so uniform, with higher variability in the surface side chain atoms than, say, the buried main chain atoms. Figure also appears in the Color Figure section.

STEREOCHEMICAL PARAMETERS

Given that all structural models contain some degree of error, what else, other than the global parameters described above, can help assess the quality of each one? One very powerful class of checks are those that examine a model's geometry, stereochemistry and other structural properties. These checks compare a given protein or nucleic acid structure against what is already known about these molecules. The "knowledge" comes from high-resolution structures of small molecules plus systematic analyses of the existing protein and nucleic acid structures in the PDB. The latter analyses rely on the vast body of structures solved to date providing a knowledge base of what is "normal" for these biomolecules.

The advantage of such tests of "normality" is that they do not require access to the original experimental data. Although it is possible to obtain the experimental data for many PDB entries—structure factors in the case of X-ray structures, and distance restraints for NMR ones—this is still the minority of entries, and deposition of these data is still at the discretion of the depositors. Furthermore, to make use of the data requires appropriate software packages and expert know-how. The stereochemical tests, on the contrary, require no experimental data and are easy to run and understand. So, in fact, any 3D structure, whether experimentally determined, or the result of homology modeling, molecular dynamics, threading or blind guesswork can be checked. The software for checking structures is freely available and there are now many Web servers that provide these checks for you, as will be mentioned later.

Before describing the checks, one crucial point needs to be stressed at the start. The majority of the checks compare a given structure's properties against what is the "norm." Yet this norm has been derived from existing structures and could be the result of biases introduced by different refinement practices. Furthermore, outliers, such as an excessively long bond length, or an unusual torsion angle, should not be construed as errors. They may be genuine, for example, as a result of strain in the conformation, say, at the active site. The only way of verifying whether oddities are errors or merely oddities is by referring back to the original experimental data. Indeed, the experimenters who solved the structure may already have done this, found the apparent oddity to be correct and commented to that effect in the literature.

Having said that, if a single structure exhibits a large number of outliers and oddities, then it probably does have problems and perhaps should be excluded from any analyses.

Proteins

The Ramachandran Plot.   For proteins, the best known, and most powerful, check of stereochemical quality is the Ramachandran plot (Ramachandran, Ramakrishnan, and Sasisekharan, 1963). This is a plot of the ψ main chain torsion angle versus the ϕ main chain torsion angle (see Chapter 2) for every amino acid residue in the protein (except the two terminal residues because the N-terminal residue has no ϕ and the C-terminus has no ψ). In the resulting ϕ–ψ scatterplot, the points tend to be excluded from certain "disallowed" regions and have a tendency to cluster in certain favorable regions (Figure 14.8). The disallowed regions are where steric hindrance between side chain atoms makes certain ϕ–ψ combinations difficult or even impossible to achieve. Glycine, which has no side chain to speak of, has a much greater freedom of movement in terms of its ϕ–ψ combinations, although it still has regions from which it is excluded (Figure 14.8b). Conversely, proline, which has two covalent connections to the backbone, is more restricted than other amino

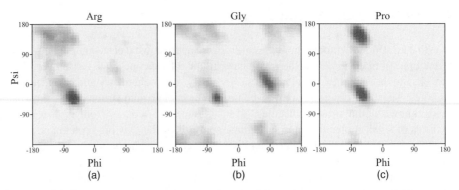

Figure 14.8. Differences in Ramachandran plots for (a) arginine, representing a fairly standard amino acid residue, (b) glycine that, due to its lack of a side chain, is able to reach the parts of the plot that other residues cannot reach, and (c) proline that, due to its restraints on the movement of the main chain, has a restricted range of φ values. The darker regions correspond to the more densely populated regions as observed in a representative sample of protein structures.

acids and its accessible regions of the φ–ψ plot are confined to a narrow range of φ values (Figure 14.8c).

The favorable regions on the plot correspond to the regular secondary structures: right-handed helices, extended conformation (as found in β-strands), and left-handed helices. These are the dark regions in Figure 14.9 labeled A, B, and L, respectively. Even residues in loops tend to lie within these favored regions. Figure 14.9a shows a typical Ramachandran plot with the residues showing a tight clustering in the most favored regions with few or none in the disallowed regions. The favorable and disallowed regions are determined from analyses of existing structures in the PDB (Morris et al., 1992; Kleywegt and Jones, 1996).

Figure 14.9b shows a somewhat pathological Ramachandran plot. It comes from a structure that shall remain nameless. Here, the majority of the residues lie in the disallowed regions and it can be confidently concluded that the model has serious problems.

One caveat concerns proteins containing D-amino acids rather than the more common L-amino acids. These residues have the opposite chirality so their φ–ψ values will be negative with respect to their L-amino cousins. The Ramachandran plot for D-amino acids is the same as for L-amino acids, but with every point reflected through the origin. Thus, proteins such as gramicidin A (e.g., PDB code 1GRM), which have many D-amino acids, give Ramachandran plots that look particularly troubling, but which may be perfectly correct.

Few models are as extreme as the one in Figure 14.9b. The tightness of clustering in the favorable regions varies as a function of resolution, with atomic resolution structures exhibiting very tight clustering (EU 3D Validation Network, 1998). At lower resolutions, as the data quality declines and the model of the protein structure becomes less accurate, the points on the Ramachandran plot tend to disperse and more of them are likely to be found in the disallowed regions.

One feature that makes the Ramachandran plot such a powerful indicator of protein structure quality is that it is difficult to fool (unless one does so intentionally by, say, restraining φ–ψ values during structure refinement as is sometimes done for NMR structures). This reliability was demonstrated by Gerard Kleywegt in Uppsala who once attempted to deliberately trace a protein chain *backward* through its electron density to see whether it

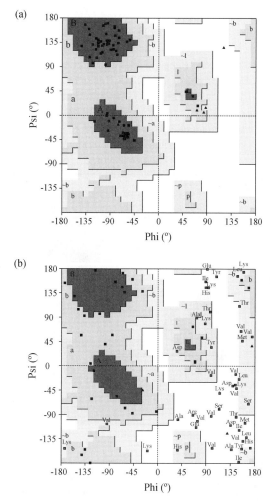

Figure 14.9. Ramachandran plots for (a) a typical protein structure, and (b) a poorly defined protein structure. Each residue's φ–ψ combination is represented by a black box, except for glycine residues which are shown as black triangles. The most darkly shaded regions of the plot correspond to the most favorable, or "core," regions (labeled A for α-helix, B for β-sheet, and L for left-handed helix) where the majority of residues should be found. The progressively lighter regions are the less favored zones, with the white region corresponding to "disallowed" φ–ψ combinations for all but glycine residues. Residues falling within these disallowed regions are shown by the labeled boxes. The plot in (a) is for PDB code 1ubi, which is of the chromosomal protein ubiquitin. All but one of the protein's 66 nonglycine and nonproline residues are in the core regions of the Ramachandran plot (giving a "core percentage" of 98.5%). What is more, the points cluster reasonably well in the core regions. The structure was solved by X-ray crystallography at a resolution of 1.8 Å. The plot in (b) exhibits many deviations from the core regions. The structure was solved by NMR, in the early days of the technique, and has a core percentage of 6.8% while over a third of its residues lie in the disallowed regions. The plots were obtained using the PROCHECK program.

would refine and give the sorts of quality indicators that could fool people into believing it to be a reasonable model (Kleywegt and Jones, 1995). Of the parameters that he tried to fool, the two that seemed least gullible were the R_{free} factor mentioned above, and the Ramachandran plot. Figure 14.10 shows this backward-traced model, together with its Ramachandran plot and secondary structure diagram (referred to later). The Ramachandran plot looks most unhealthy, with several residues in disallowed regions and no significant clustering in the most highly favored regions. The structure itself can be found in the PDBsum database under its fake PDB code of "fake": http://www.ebi.ac.uk/pdbsum/fake.

A simple measure of the overall "quality" of a model can be derived from the plot by computing the percentage of residues in the most favorable, or "core" regions. (Glycines and prolines are excluded from this percentage because of their unique distributions of available ϕ–ψ combinations.) Using the core regions shown in Figure 14.9, which are defined by the

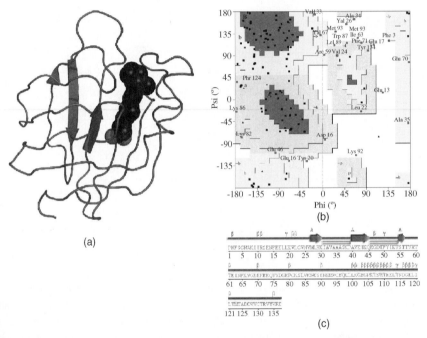

Figure 14.10. A deliberately erroneous model created by tracing the protein of cellular retinoic acid-binding protein type II (CRABP) *backward* through its electron density. While careful and cunning refinement enabled some of the structural quality measures to suggest this to be a reasonable model, other measures show it for what it is: a fake. (a) A cartoon of the model shows that it has little in the way of regular secondary structure. The dark spheres represent a small molecule—all-trans-retinoic acid—bound to the protein (and happily inserted into the electron density the right way round!). (b) The protein's Ramachandran plot reveals many of its residues falling in the disallowed regions, and few clustering in the most favored regions. (c) A schematic diagram of the secondary structure underlines the paucity of regular secondary structure. The only detectable secondary structure is a single strand indicated by the arrows. The large number of residues that appear to be involved in β- and γ-turns (identified by the β and γ symbols) reflect the parts of the sequence that have been threaded through the α-helical regions of the electron density but which do not form the hydrogen bonding patterns that characterize α-helices.

PROCHECK program (see below), one generally finds that the atomic resolution structures have well over 90% of their residues in these most favorable regions. For lower and lower resolution structures, this percentage drops, with the structures solved to 3.0–4.0 Å tending to have a core percentage around 70%. NMR structures also show increasing core percentage with increasing experimental information. However, NMR structures can have relatively good side chain positions even with a poor core percentage as NMR data restrain side chains more strongly than the backbone because of the large number of side chain protons.

Side Chain Torsion Angles. Like its main chain, a protein's side chains also have preferred conformations for their torsion angles. These conformations are known as rotamers and again are a result of steric hindrance making some conformations inaccessible. The side chain torsion angles are labeled χ_1, χ_2, χ_3, and so on. The first of these, χ_1, is defined as the torsion angle about $N–C^\alpha–C^\beta–A^\gamma$, where A^γ is the next atom along the side chain (for example, in lysine the A^γ atom is C^γ). The next, χ_2, is defined as $C^\alpha–C^\beta–A^\gamma–A^\delta$, and so on. The χ_1 and χ_2 distributions are both trimodal with the preferred torsion angle values being termed gauche-minus ($+60°$), trans ($+180°$), and gauche-plus ($-60°$). A plot of χ_2 against χ_1 for each residue has 3×3 preferred combinations, although the strength of each depends very much on the residue type. Figure 14.11 shows some examples of the χ_1–χ_2 distributions for different amino acid types.

Like the Ramachandran plot, a plot of the χ_1–χ_2 torsion angles can indicate problems with a protein model as these, like the ϕ and ψ torsion angles, tends not to be restrained during refinement. Also, like the Ramachandran plot, the clustering in the most favorable regions on the χ_1–χ_2 plot becomes tighter as resolution improves (EU 3D Validation Network, 1998). For example, the standard deviation of the χ_1 torsion angles about their ideal position tends to be around 8° for atomic resolution structures and can go as high as 25° for structures solved at 3.0 Å. Similarly, the corresponding standard deviations for χ_2 tend to be 10° and 30°, respectively.

Bad Contacts. Another good check for structures to be wary of is the count of bad and unfavorable atom–atom contacts that they contain. Too many and the model is likely to be a poor one.

The simplest checks are those that merely count "bad" contacts. A bad contact is where two nonbonded atoms have a center-to-center distance that is smaller than the sum of their van der Waals radii. The check should be applied not only to intraprotein contacts within the given protein structure, but, for X-ray crystal structures, also to atoms from molecules related by crystallographic, and noncrystallographic symmetry.

More sophisticated checks consider each atom's environment and determine how happy that atom is likely to be in that environment. For example, the ERRAT program (Colovos and Yeates, 1993) counts the numbers of nonbonded contacts, within a cutoff distance of 3.5 Å, between different pairs of atom types. The atoms are classified as carbon (C), nitrogen (N), and oxygen (O)/sulfur, so there are six distinct interaction types: CC, CN, CO, NN, NO, and OO. If the frequencies of these interaction types differ significantly from the norms (as obtained from well-refined high-resolution structures), the protein model may be somewhat suspect. Using a 9-residue sliding window and obtaining the interaction frequencies at each window position can locate local problem regions.

One level up in sophistication is the DACA method (Vriend and Sander, 1993) that is implemented in the WHAT IF (Vriend, 1990) and WHAT_CHECK (Hooft et al., 1996) programs. DACA stands for Directional Atomic Contact Analysis and compares the 3D environment surrounding each residue fragment in the protein with normal environments

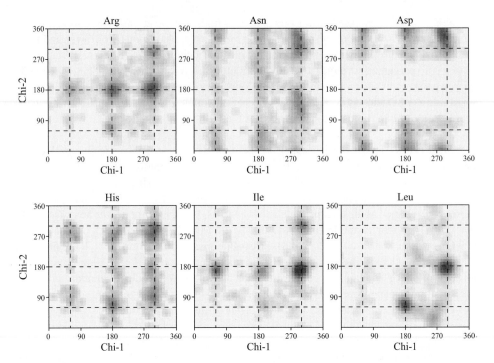

Figure 14.11. Examples of χ_1–χ_2 distributions for six different amino acid residue types: Arg, Asn, Asp, His, Ile and Leu. The darker regions correspond to the more densely populated regions as observed in a representative sample of protein structures. The dotted lines represent "idealized" rotameric torsion angles at 60°, 180°, and 300° (equivalent to −60°). It can be seen that the true rotameric conformations differ slightly from these values and that the different side chain types have very different χ_1–χ_2 distribution preferences.

computed from a high-quality data set of protein structures. There are 80 different main chain and side chain fragment types. The environment of each is essentially the count of different nonbonded atoms in each $1\,\text{Å} \times 1\,\text{Å} \times 1\,\text{Å}$ cell of a $16\,\text{Å} \times 16\,\text{Å} \times 16\,\text{Å}$ cube surrounding the fragment.

A similar approach is used in the ANOLEA program (Atomic Non-Local Environment Assessment) that calculates a "nonlocal energy" for atom–atom contacts based on an atomic mean force potential (Melo and Feytmans, 1998). A relative newcomer to the field, which has become quite popular, is MolProbity (Lovell et al., 2003) that performs a detailed all-atom contact analysis within the given structure. The structure must include hydrogens so, where it doesn't, a first step in the analysis is the addition and full optimization of all hydrogen atoms, both polar and nonpolar. Chapter 15 provides a more detailed description of the methodology of this program.

Secondary Structure. Occasionally, the model of a protein structure is so bad that one can tell immediately from its secondary structure. Most proteins have around 50–60% of their residues in regions of regular secondary structure, that is, in α-helices and β-strands. However, if a model is really poor, the main chain oxygen and nitrogen atoms responsible for the hydrogen bonding that maintains the regular secondary structures can lie beyond normal hydrogen bonding distances; so the algorithms that assign secondary structure (Chapter 19)

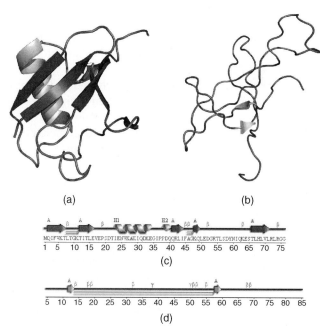

Figure 14.12. Cartoons and secondary structure plots of two protein models in the PDB. (a) and (c) A typical protein showing an expected 50–60% of its residues in α-helices (shown schematically by the helical regions) and β-strands (arrows). (b) and (d) A poorly defined model that has hardly any regions of secondary structure at all. The labels and symbols on the secondary structure plots correspond to various secondary structure motifs. The β and γ symbols identify β- and γ-turns, while the hairpinlike symbols correspond to β-hairpins. The helices are labeled H1–H3 in (c), and strands are labeled A for β-sheet A. The Ramachandran plots for both models are shown in Figure 14.9. The sequence of the protein in (d) has been removed to hinder identification. The above plots were obtained from the PDBsum database.

may fail to detect some of the α-helices and β-strands that the correct protein structure contains. Figure 14.12 gives an example of the secondary structure contents for a typical protein and for the protein that had the poor Ramachandran plot in Figure 14.9b. The backward-traced protein at http://www.ebi.ac.uk/pdbsum/fake, mentioned above, also has very low secondary structure content (Figure 14.10c).

Other Parameters. Other parameters that can be used to validate protein structures include counts of unsatisfied hydrogen bond donors and hydrogen bonding energies as computed by the WHAT_CHECK program mentioned below (Hooft et al., 1996). Some programs—Prosa, Verify3D, and HARMONY—use knowledge-based potentials of some sort to assess how "happy" each residue is to be in the local environment that it finds itself in the current model. Very many unhappy residues make for a sad structure overall.

C-Alpha Only Structures. As of August 2007, there were around 274 structures in the PDB (out of over 45,000) containing one or more protein chains for which only the C^α coordinates had been deposited. The deposition of C^α-only coordinate sets is usually done where the data quality has been too poor to resolve more of the structure. This was common in the early days of protein crystallography but nowadays only tends to be done for very large structures, such as the recently determined structure of a 50s ribosomal subunit at 3.5 Å (PDB code 1XBP; Schlünzen et al., 2004).

The standard validation checks are of no use for such models, lacking as they are in so much of their substance. However, there is an equivalent to the Ramachandran plot for these structures (Kleywegt, 1997). The parameters plotted are the C^α–C^α–C^α–C^α torsion angle as a function of the C^α–C^α–C^α angle for every residue in the protein. As with the Ramachandran plot, there are regions of this plot that tend to be highly populated, and others that appear to be "forbidden." So a structure with many outliers in the forbidden zones should be treated with caution. The checks are incorporated in the STAN Web server (see later).

Nucleic Acids

Finding validation tools for DNA and RNA is trickier than for proteins. The PDB's validation tool, AutoDep Input Tool (ADIT), incorporates a program called NuCheck (Feng, Westbrook, and Berman, 1998) for validating the geometry of DNA and RNA. Binary versions of the ADIT package are included in the PDB Validation Suite (see Table 14.1).

A program specifically developed for checking the geometry of RNA structures, but which can also be used for DNA structures, is MC-Annotate (Gendron, Lemieux, and Major, 2001). It computes a number of "peculiarity factors," based on various metrics including torsion angles and root-mean-squared deviations from standard conformations,

T A B L E 1 4 . 1 . Freely Available Programs for Checking Structure Coordinates

Program name	Reference	URL
ERRAT	Colovos and Yeates (1993)	www.doe-mbi.ucla.edu/People/Yeates/Gallery/Errat.html
HETZE[a]	Kleywegt and Jones (1998)	alpha2.bmc.uu.se/usf/xutil.html
MC-Annotate[b]	Gendron et al. (2001)	www-lbit.iro.umontreal.ca/mcannotate-simple
MolProbity	Lovell et al. (2003)	molprobity.biochem.duke.edu
NUCheck[b]	Feng, Westbrook, and Berman (1998)	See PDB Validation Suite below
PDB Validation Suite	Westbrook et al. (2003)	sw-tools.pdb.org/apps/VAL
PROCHECK	Laskowski et al. (1993)	www.biochem.ucl.ac.uk/~roman/procheck/procheck.html
PROCHECK-NMR	Laskowski et al. (1996) www.biochem.ucl.ac.uk/~roman/procheck_nmr/procheck_nmr.html	
Prosa 2003	Sippl (1993)	prosa.services.came.sbg.ac.at/download/download.php
WHAT_CHECK	Hooft et al. (1996)	swift.cmbi.ru.nl/gv/whatcheck

[a] For hetero compounds only.
[b] For RNA and DNA only.

which can highlight irregular regions in the structure that may be in error or merely under strain.

Hetero Groups

The geometry of hetero compounds, as deposited in the structures in the PDB tends to be of widely varying quality (Kleywegt et al., 2003). This is because, while there are well-established libraries giving standard bond lengths and bond angles for use in refinement of proteins and nucleic acids, there can be no equivalent for the wide variety of small ligand molecules that can bind to them. For any new ligand molecule, the target bond lengths and angles need to be individually compiled from scratch, based on its atom types and bond orders. This requires a thorough understanding of the molecule's chemistry, so errors in the target values do occur. The PRODRG server (http://davapc1.bioch.dundee.ac.uk/programs/prodrg) can help reduce the effort involved and reduce the risk of errors (Schuttelkopf and van Aalten, 2004).

The HETZE program (Kleywegt and Jones, 1998) is one of the few validation methods that checks various geometrical parameters of the hetero compounds associated with PDB structures. These include bond lengths, torsion angles, and some virtual torsion angles, the information principally coming from the small-molecule structures in the Cambridge Structural Database (Allen et al., 1979).

SOFTWARE FOR QUALITY CHECKS

There are a large number of freely available programs that can perform the sorts of quality checks described above on proteins, nucleic acids, and hetero compounds. Table 14.1 lists a selection of these, with URLs giving links to their download information. The most commonly used programs are described below.

PROCHECK

The PROCHECK program (Laskowski et al., 1993), although now quite long in the tooth, is still commonly used for checking the quality of protein structures. It produces a number of plots in PostScript format, including Ramachandran plots, both for the protein as a whole and for each type of amino acid; χ_1–χ_2 plots for each amino acid type; main chain bond lengths and bond angles; secondary structure plot; deviations from planarity of planar side chains; and so on. Parameters that deviate from the norm are highlighted. The program has been incorporated into the PDBsum database (see later) and so its analyses are available for all PDB entries, including the obsolete ones. A version specifically for NMR, and other, ensembles is called PROCHECK-NMR (Laskowski et al., 1996) and provides plots that compare the various stereochemical parameters across all the models of the ensemble.

WHAT_CHECK

The WHAT_CHECK program (Hooft et al., 1996) is a subset of Gert Vriend's WHAT IF package (Vriend, 1990). It contains an enormous number of checks and produces a long and very detailed output of discrepancies of the given protein structure from the norms. It also includes the DACA method, mentioned above, for analyzing nonbonded contacts.

Precomputed WHAT_CHECK results for all PDB entries can be found in the PDBREPORT database (see later).

PDB Validation Suite

The PDB's own internal suite of validation programs is available for download from the RCSB Web site as executable binaries (see Table 14.1). Although principally aimed at experimentalists preparing their models for deposition, it can also be used for validating other structural models. Where the corresponding experimental data are available, these will be included in the validation checks.

Other Programs

There are a number of other programs available, which perform a variety of different checks on a given structure (see Table 14.1). Prosa 2003, the successor to Prosa II (Sippl, 1993), scores a protein structure using knowledge-based potentials representing atom-pair and protein–solvent interactions, as determined from a data set of known protein structures. The Verify3D program (Bowie et al., 1991) considers the environment of each residue in the 3D structure in terms of secondary structure type, buried surface area, and fraction of side chain area covered by polar atoms. A similar approach is given by HARMONY (Pugalenthi et al., 2006). The TAP program (Tosatto and Battistutta, 2007) validates the local torsion angles from propensities computed from known protein structure models. PROVE compares atomic volumes against a set of precalculated standard values (Pontius, Richelle, and Wodak, 1996). Volumes are calculated using Voronoi polyhedra to define the space that each atom occupies by placing dividing planes between it and its neighbors. The ERRAT and MolProbity programs have already been mentioned. Some programs are also available as Web servers (see later) and some are only available via Web servers (i.e., cannot be downloaded for local use).

QUALITY INFORMATION ON THE WEB

Much of the information given by the above packages is readily available on the Web, so one can often get it without installing and running any of the software itself. Several sites provide precomputed quality criteria for all existing structures in the PDB. Other sites allow you upload your own PDB file, via your Web browser, and will run their validation programs on it and provide you with the results of their checks.

PROCHECK Analyses in PDBsum

The PDBsum Web site at http://www.ebi.ac.uk/pdbsum (Laskowski, Chistyakov, and Thornton, 2005) provides numerous structural analyses of all PDB structures. This includes the full PROCHECK output and a mini-Ramachandran plot for all entries containing one or more protein chains. Moving the mouse over the mini-Ramachandran pops up a full-sized version that can give an at-a-glance idea of the likely quality of the structural model in question. It is possible to obtain the PDBsum analyses, including full PROCHECK results, for structures not in the PDB using the Generate button on the PDBsum home page that returns the results on a password-protected set of pages.

PDBREPORT–WHAT_CHECK Results

The PDBREPORT database (http://swift.cmbi.ru.nl/gv/pdbreport) gives a precomputed WHAT_CHECK report for any structure in the PDB. The report is a detailed listing (supplemented by an even more detailed listing called the Full report) of the numerous analyses that have been precomputed using the WHAT_CHECK program. These include space group and symmetry checks, geometrical checks on bond lengths, bond angles, torsion angles, proline puckers, bad contacts, planarity checks, checks on hydrogen bonds, and more, including an overall summary report intended for users of the model.

RCSB Geometry Analyses

The RCSB Web site (http://www.pdb.org) also has geometrical analyses on each entry, consisting of tables of average, minimum and maximum values for the protein's bond lengths, and bond angles and dihedral angles. Unusual values are highlighted. It is also possible to view a backbone representation of the structure in RasMol colored according to the "Fold Deviation Score"—the redder the coloring, the more unusual are the residue's conformational parameters.

The Uppsala Electron Density Server (EDS). For X-ray crystal structures that

have been deposited in the PDB together with their experimental data, the Uppsala Electron Density Server (EDS) at http://eds.bmc.uu.se (Kleywegt et al., 2004) provides useful information on the local quality of the structure. For example, the plot of the real space R-factor (RSR) shows how well each residue fits its electron density. Other useful plots include the occupancy-weighted average temperature factor and a Z-score associated with the residue's RSR for the given resolution. The tall spikes in these plots identify residues that are either particularly disordered or do not sit well within their electron density and so may need to be treated with caution. A particularly nice feature is that clicking on any residue in these plots pops up a window running a Java-based 3D viewer showing the model and electron density in the region of the residue.

Validation Servers on the Web

The servers described above provide quality information on structural models that are already in the PDB. To check the quality of a model that is not in the PDB, one can submit it to one of the validation servers listed in Table 14.2. Most of these validate protein structures and use programs that are freely available for in-house use (see Table 14.1). However, the servers can often be easier and more convenient to use, and of course save you from having to download and install the programs.

Some servers run individual programs while others run compilations of two or more programs. Of the individual programs, ANOLEA, ERRAT, HARMONY, MolProbity, Prosa 2003, TAP, and Verify3D have already been mentioned. The STAN (Structure Analysis) server generates a standard Ramachandran plot, a "C^α-Ramachandran" plot, a plot analyzing any nucleic acid chains, a check for possible misidentification of small cations as waters, and check for incorrect *cis*-peptide bonds. Volume, Area, Dihedral Angle Reporter (VADAR) is a compilation of over 15 different programs (Willard et al., 2003), including PROCHECK and WHAT_CHECK. The COLORADO3D server (Sasin and Bujnicki, 2004) includes ANOLEA, PROSA, PROVE, and Verify3D, while the JCSG server includes

TABLE 14.2. WWW Servers for Checking Uploaded Structure Coordinates Online

Server Name	Reference	URL
ANOLEA	Melo and Feytmans (1998)	protein.bio.puc.cl/cardex/ servers/anolea
COLORADO3D	Sasin and Bujnicki (2004)	asia.genesilico.pl/colorado3d
EDS	Kleywegt et al. (2004)	eds.bmc.uu.se
ERRAT	Colovos and Yeates (1993)	nihserver.mbi.ucla.edu/ ERRATv2
HARMONY	Pugalenthi et al. (2006)	caps.ncbs.res.in/harmony
JCSG Validation Server	Joint Center for Structural Genomics	www.jcsg.org/scripts/prod/ validation/sv_final.cgi
MolProbity	Lovell et al. (2003)	molprobity.biochem.duke.edu
PDBsum Generate	Laskowski et al. (2005)	www.ebi.ac.uk/thornton-srv/ databases/pdbsum/upload. html
PDB Validation Server	Westbrook et al. (2003)	deposit.rcsb.org/validate
ProSA-Web	Wiederstein and Sippl (2007)	prosa.services.came.sbg.ac.at/ prosa.php
STAN	Kleywegt and Jones (1996), Kleywegt (1997)	xray.bmc.uu.se/cgi-bin/gerard/ rama_server.pl
TAP	Tosatto and Battistutta (2007)	protein.cribi.unipd.it/tap
Verify3D	Bowie et al. (1991)	nihserver.mbi.ucla.edu/ Verify_3D
VADAR	Willard et al. (2003)	redpoll.pharmacy.ualberta.ca/ vadar

PROCHECK, WHAT_CHECK, ERRAT, and PROVE. Finally, the PDBsum Generate server, which is not strictly speaking a validation server, generates a complete password-protected PDBsum analysis of any uploaded PDB file, including a full PROCHECK output.

CONCLUSIONS

The main aim of this chapter has been to impress upon you that the macromolecular structures that form the very foundation of Structural Bioinformatics are not all of the same quality and can undermine those foundations if not carefully selected. All structures are just models devised to satisfy data obtained experimentally. As such, they will contain errors, both systematic and random. Some structures have been found to be seriously incorrect—that is, they are inaccurate models of the molecules they represent and in many cases have been replaced by more accurate models. Most structures are reasonably accurate but inevitably contain random errors, as is symptomatic of any experimental measurement. The quality of structures as a whole has improved over the past few years and this trend is expected to continue. However, determining which is a good structure and which is not is still not clear. Even traditional measures such as the resolution and R-factor for X-ray structures, and number of restraints for NMR structures do not always separate the good from the bad. Most often, other quality measures need to be taken into account when selecting a good data set.

The chapter has surveyed the information available, and some of the additional tests that can be performed to ensure that the reliability of any of the structures used is consistent with the conclusions to be drawn from them.

ACKNOWLEDGMENTS

Thanks to Drs Mark Williams and Gert Vriend for their valuable comments on the text of this chapter.

REFERENCES

Allen FH, Bellard S, Brice MD, Cartwright BA, Doubleday A, Higgs H, Hummelink T, Hummelink-Peters BG, Kennard O, Motherwell WDS, Rodgers JR, Watson DG (1979): The Cambridge Crystallographic Data Centre: computer-based search, retrieval, analysis and display of information. *Acta Crystallogr* B35:2331–2339.

Badger J, Hendle J (2002): Reliable quality-control methods for protein crystal structures. *Acta Cryst* D58:284–291.

Berman HM, Burley SK, Chiu W, Sali A, Adzhubei A, Bourne PE, Bryant SH, Dunbrack RL Jr, Fidelis K, Frank J, Godzik A, Henrick K, Joachimiak A, Heymann B, Jones D, Markley JL, Moult J, Montelione GT, Orengo C, Rossmann MG, Rost B, Saibil H, Schwede T, Standley DM, Westbrook JD (2006): Outcome of a workshop on archiving structural models of biological macromolecules. *Structure* 14:1211–1217.

Bowie JU, Lüthy R, Eisenberg D (1991): A method to identify protein sequences that fold into a known three-dimensional structure. *Science* 253:164–170.

Brändén C-I, Jones TA (1990): Between objectivity and subjectivity. *Nature* 343:687–689. [One of the first papers to raise the issue of errors in protein structures, citing five examples of published protein structures that had been found to contain serious errors and had been replaced by more accurate models. The paper outlined the sources of errors that can arise during crystallographic structure determination and listed various means of limiting them.]

Brünger AT (1992): Free R value: a novel statistical quantity for assessing the accuracy of crystal structures. *Nature* 355:472–475. [This paper represents an important milestone in accurate protein structure determination. It describes the application of cross-validation, a standard statistical technique, to the calculation of the R-factor, giving the R_{free}, which is unbiased and not vulnerable to artificial reduction. The R_{free} is now a standard measure of a structure's goodness of fit to the data from which it was determined.]

Brünger AT (1997): Free R value: cross-validation in crystallography. *Methods Enzymol* 277:366–396. [A more detailed exposition of the above.]

Carugo O, Bordo D (1999): How many water molecules can be detected by protein crystallography? *Acta Crystallogr* D55:479–483.

Chang G, Roth C.B, Reyes CL, Pornillos O, Chen Y-J, Chen AP (2006): Retraction of Pornillos et al. *Science* 310 (5756):1950–1953. Retraction of Reyes and Chang, *Science* 308 (5724) 1028–1031. Retraction of Chang and Roth, *Science* 293 (5536) 1793–1800. *Science* 314, 1875. [Brief letter to Science retracting three Science papers and withdrawing 5 structural models from the PDB due to serious errors in the models.]

Chen LF, Winkler H, Reedy MK, Reedy MC, Taylor KA (2002): Molecular modeling of averaged rigor crossbridges from tomograms of insect flight muscle. *J Struct Biol* 138:92–104.

Clore GM, Gronenborn AM (1998): New methods of structure refinement for macromolecular structure determination by NMR. *Proc Natl Acad Sci* 95:5891–5898.

Clore GM, Garrett DS (1999): *R*-factor, free *R*, and complete cross-validation for dipolar coupling refinement of NMR structures. *J Am Chem Soc* 121:9008–9012. [Description of measures equivalent to the crystallographic *R*-factor and R_{free} for structures solved by solution NMR.]

Colovos C, Yeates TO (1993): Verification of protein structures: patterns of nonbonded atomic interactions. *Protein Sci* 2:1511–1519.

DePristo MA, de Bakker PI, Blundell TL (2004): Heterogeneity and inaccuracy in protein structures solved by X-ray crystallography. *Structure* 12:831–838.

Doreleijers JF, Raves ML, Rullmann T, Kaptein R (1999): Completeness of NOEs in protein structure: a statistical analysis of NMR. *J Biomol NMR* 14:123–132.

Esnouf RM (1997): An extensively modified version of MolScript that includes greatly enhanced coloring capabilities. *J Mol Graph Model* 15:132–134.

Esposito L, Vitagliano L, Zagari A, Mazzarella L (2000): Experimental evidence for the correlation of bond distances in peptide groups detected in ultrahigh-resolution protein structures. *Protein Eng* 13:825–828.

EU 3D Validation Network (1998): Who checks the checkers? Four validation tools applied to eight atomic resolution structures. *J Mol Biol* 276:417–436. [An examination of the standard software tools used for validating protein structures by testing their performance on structures solved to atomic resolution which, by definition, are as good as can be obtained. For the most part the validation parameters were found to hold for these high-resolution structures, although some modifications were called for, particularly in the tightening of the values. Surprisingly, the analysis also suggested certain modification to the refinement protocols used for such high-quality structures.]

Feng Z, Westbrook J, Berman HM (1998): *NUCheck. Computer program.* Rutgers publication NDB-407. Rutgers University, New Brunswick, New Jersey.

Gendron P, Lemieux S, Major F (2001): Quantitative analysis of nucleic acid three-dimensional structures. *J Mol Biol* 308:919–936.

Gronwald W, Kirchhöfer R, Görler A, Kremer W, Ganslmeier B, Neidig K-P, Kalbitzer HR (2000): RFAC, a program for automated NMR *R*-factor estimation. *J Biomol NMR* 17:137–151.

Hooft RWW, Vriend G, Sander C, Abola EE (1996): Errors in protein structures. *Nature* 381:272. [This paper is the reference to be cited for the WHATCHECK program, although it hardly describes that program at all. Rather, it describes a detailed analysis, using the program, of "errors" in the 3442 entries in the PDB at the time; an analysis that stirred up not a little controversy amongst the crystallographic community at the time.]

Jones TA, Kjeldgaard M (1997): Electron density map interpretation. *Methods Enzymol* 277:173–208.

Kelley LA, Gardner SA, Sutcliffe MJ (1996): An automated approach for clustering an ensemble of NMR-derived protein structures into conformationally related subfamilies. *Protein Eng* 9:1063–1065.

Kleywegt GJ, Jones TA (1995): Where freedom is given, liberties are taken. *Structure* 3:535–540.

Kleywegt GJ, Jones TA (1996): Phi/psi-chology: Ramachandran revisited. *Structure* 4:1395–1400. [A detailed study of the Ramachandran plot: one of many, but a good one.]

Kleywegt GJ (1997): Validation of protein models from C-alpha coordinates alone. *J Mol Biol* 273:371–376. [Derivation of the equivalent of the Ramachandran plot for protein structures for which only the coordinates of the C^{α} atoms have been determined.]

Kleywegt GJ, Jones TA (1998): Databases in protein crystallography. *Acta Crystallogr* D54: 1119–1131.

Kleywegt GJ (2000): Validation of protein crystal structures. *Acta Crystallogr* D56:249–265. [An excellent and detailed overview of the types and causes of errors in X-ray crystal structures and the measures that crystallographers need to take to reduce them as far as possible.]

Kleywegt GJ, Jones TA (2002): Homo crystallographicus—Quo Vadis? *Structure* 10:465–472.

Kleywegt GJ, Henrick K, Dodson EJ, van Aalten DMF (2003): Pound-wise but penny-foolish: how well do micromolecules fare in macromolecular refinement? *Structure* 11:1051–1059.

Kleywegt GJ, Harris MR, Zou J-Y, Taylor TC, Wählby A, Jones TA (2004): The Uppsala Electron Density Server. *Acta Crystallogr* D60:2240–2249.

Kraulis PJ (1991): MOLSCRIPT: a program to produce both detailed and schematic plots of protein structures. *J Appl Crystallogr* 24:946–950.

Laskowski RA, MacArthur MW, Moss DS, Thornton JM (1993): PROCHECK: a program to check the stereochemical quality of protein structures. *J Appl Crystallogr* 26:283–291.

Laskowski RA, Rullmann JAC, MacArthur MW, Kaptein R, Thornton JM (1996): AQUA and PROCHECK-NMR: programs for checking the quality of protein structures solved by NMR. *J Biomol NMR* 8:477–486.

Laskowski RA, Chistyakov VV, Thornton JM (2005): PDBsum more: new summaries and analyses of the known 3D structures of proteins and nucleic acids. *Nucleic Acids Res* 33:D266–D268.

Lewis M, Chang G, Horton NC, Kercher MA, Pace HC, Schumacher MA, Brennan RG, Lu P (1996): Crystal structure of the lactose operon repressor and its complexes with DNA and inducer. *Science* 271:1247–1254.

Lovell SC, Davis IW, Arendall WB III, de Bakker PIW, Word JM, Prisant MG, Richardson JS, Richardson DC (2003): Structure validation by C-alpha geometry: phi, psi, and C-beta deviation. *Proteins Struct Funct Genet* 50:437–450.

Luzzati PV (1952): Traitement statistique des erreurs dans la determination des structures cristallines. *Acta Crystallogr* 5:802–810. [A description, in French, of the Luzzati plot that has come to be used for estimating the average positional errors in crystal structures, despite this not having been the paper's original purpose; its aim had been to estimate the positional changes required to reach an *R*-factor of zero.]

Melo F, Feytmans E (1998): Assessing protein structures with a nonlocal atomic interaction energy. *J Mol Biol* 277:1141–1152.

Merritt EA, Bacon DJ (1997): Raster3D: photorealistic molecular graphics. *Methods Enzymol* 277: 505–524.

Morris AL, MacArthur MW, Hutchinson EG, Thornton JM (1992): Stereochemical quality of protein structure coordinates. *Proteins Struct Funct Genet* 12:345–364. [Derivation of a number of stereochemical parameters that appeared to be good indicators of protein structure quality in that they were well correlated with resolution. As most of the parameters are not among those restrained during structure refinement, they provide a useful independent measure of how well a structure agrees with what appears to be the norm for proteins. A few discrepancies from the norm are to be expected for any structure, but many suggest there may be something seriously wrong with it.]

Moult J (2005): A decade of CASP: progress, bottlenecks and prognosis in protein structure prediction. *Curr Opin Struct Biol* 15:285–289.

Nabuurs SB, Nederveen AJ, Vranken W, Doreleijers JF, Bonvin AMJJ, Vuister GW, Vriend G, Spronk CAEM (2004): DRESS: a database of re-refined solution NMR structures. *Proteins* 55:483–486.

Nederveen AJ, Doreleijers JF, Vranken W, Miller Z, Spronk CAEM, Nabuurs SB, Guntert P, Livny M, Markley JL, Nilges M, Ulrich EL, Kaptein R, Bonvin AMJJ (2005): RECOORD: a recalculated coordinate database of 500+ proteins from the PDB using restraints from the BioMagResBank. *Proteins* 59:662–672.

Pontius J, Richelle J, Wodak SJ (1996): Deviations from standard atomic volumes as a quality measure for protein crystal structures. *J Mol Biol* 264:121–136.

Potluri S, Yan AK, Chou JJ, Donald BR, Bailey-Kellogg C (2006): Structure determination of symmetric homo-oligomers by a complete search of symmetry configuration space, using NMR restraints and van der Waals packing. *Proteins* 65:203–219.

Pugalenthi G, Shameer K, Srinivasan N, Sowdhamini R (2006): HARMONY: a server for the assessment of protein structures. *Nucleic Acids Res* 34:W231–W234.

Ramachandran GN, Ramakrishnan C, Sasisekharan V (1963): Stereochemistry of polypeptide chain configurations. *J Mol Biol* 7:95–99. [The classic analysis of the distribution of φ–ψ torsion angles in protein main chains, which has given us the Ramachandran plot: one of the most powerful methods for checking whether a protein structure seems reasonable or contains severe errors or strained conformations.]

Read RJ (1986): Improved Fourier coefficients for maps using phases from partial structures with errors. *Acta Crystallogr* A42:140–149. [Derivation of the σ_A plot, which is used to estimate the average positional errors in crystal structures.]

Sasin JM, Bujnicki JM (2004): COLORADO3D, a Web server for the visual analysis of protein structures. *Nucleic Acids Res* 32:W586–W589.

Schlünzen F, Pyetan E, Fucini P, Yonath A, Harms JM (2004): Inhibition of peptide bond formation by pleuromutilins: the structure of the 50S ribosomal subunit from *Deinococcus radiodurans* in complex with tiamulin. *Mol Microbiol* 54:1287–1294.

Schuttelkopf AW, van Aalten DM (2004): PRODRG: a tool for high-throughput crystallography of protein–ligand complexes. *Acta Crystallogr* D60:1355–1363.

Schwede T, Kopp J, Guex N, Peitsch MC (2003): SWISS-MODEL: an automated protein homology-modeling server. *Nucleic Acids Res* 31:3381–3385.

Sheldrick GM, Schneider TR (1997): SHELXL: high-resolution refinement. *Methods Enzymol* 277:319–343. [SHELX was a least-squares structure refinement method originally developed for small-molecule crystallography, but which has been adapted and developed over the years for handling macromolecular structures. In protein crystallography, it is most commonly used for refining atomic resolution structures and can be used to calculate standard uncertainties in the atomic positions by full-matrix refinement in the final cycle.]

Sippl MJ (1993): Recognition of errors in three-dimensional structures of proteins. *Proteins Struct Funct Genet* 17:355–362.

Spronk CAEM, Nabuurs SB, Krieger E, Vriend G, Vuister GW (2004): Validation of protein structures derived by NMR spectroscopy. *Prog NMR Spectrosc* 45:315–337. [A clear exposition of NMR structure determination and the methods used for validating both experimental data and the final structural models.]

Tickle IJ, Laskowski RA, Moss DS (1998): Error estimates of protein structure coordinates and deviations from standard geometry by full-matrix refinement of γB- and βB2-crystallin. *Acta Crystallogr* D54:243–252.

Tosatto SCE, Battistutta R (2007): TAP score: torsion angle propensity normalization applied to local protein structure evaluation. *BMC Bioinform*, 8:155.

Venclovas Č, Ginalski K, Kang C (2004): Sequence-structure mapping errors in the PDB: OB-fold domains. *Prot Sci* 13:1594–1602.

Vriend G (1990): WHAT IF: a molecular modeling and drug design program. *J Mol Graph* 8:52–56.

Vriend G, Sander C (1993): Quality control of protein models: directional atomic contact analysis. *J Appl Crystallogr* 26:47–60.

Weissig H, Bourne PE (1999): An analysis of the Protein Data Bank in search of temporal and global trends. *Bioinformatics* 15:807–831. [An interesting overview of the quality of structures in the PDB and how this has improved with time. It includes an analysis of obsolete entries—entries that have since been superseded—and the various reasons for their replacement.]

Westbrook J, Feng Z, Burkhardt K, Berman HM (2003): Validation of protein structures for Protein Data Bank. *Methods Enzymol* 374:370–385.

Wiederstein M, Sippl MJ (2007): ProSA-Web: interactive Web service for the recognition of errors in three-dimensional structures of proteins. *Nucleic Acids Res*, 35:W407–W410.

Willard L, Ranjan A, Zhang H, Monzavi H, Boyko RF, Sykes BD, Wishart DS (2003): VADAR: a Web server for quantitative evaluation of protein structure quality. *Nucleic Acids Res* 31:3316–3319.

Williamson MP, Kikuchi J, Asakura Y (1995): Application of ^{1}H NMR chemical shifts to measure the quality of protein structures. *J Mol Biol* 247:541–546.

15

THE IMPACT OF LOCAL ACCURACY IN PROTEIN AND RNA STRUCTURES: VALIDATION AS AN ACTIVE TOOL

Jane S. Richardson and David C. Richardson

INTRODUCTION

The enormous wealth of macromolecular structure data already available and the even greater wealth soon to come—from structural genomics, from the push for atomic-resolution structures, and from the push to solve much larger biological complexes, often including nucleic acids as well as proteins—provides a treasure trove of functional, interactional, and evolutionary data that change the idea how one can do biology. To make an effective use of this great resource, however, it is important, among other things, to take into account the very large spread of accuracy in those data. Relatively low-resolution structures can be among the most valuable ones if they are of critical molecules or of large and complex cellular machinery. These structures show overall fold and relative positioning of their interacting parts and they often illuminate function in surprising ways, but one should not expect to learn from them fine details in an active or allosteric site or the critical local differences that determine specificity for molecular interactions. Cryoelectron microscopy techniques are developing rapidly and are especially good at characterizing movements in large complexes. In combination with crystal structures of the components, models with full atomic coordinates can be built; those show overall positioning well, but the interface details are only approximate. At the other extreme, increasing numbers of structures are being solved at better than 1 Å resolution, where one can reliably detect minute changes at catalytic sites or disentangle multiple conformations of side chains, loops, ions, and waters. Neutron diffraction can add further critical details of H-bonding and protonation.

Within an individual structure, however, there can be even wider variability in local accuracy. Regardless of resolution, most structures have a few parts disordered enough that

Structural Bioinformatics, Second Edition   Edited by Jenny Gu and Philip E. Bourne
Copyright © 2009 John Wiley & Sons, Inc.

they are not visible at all in a crystallographic electron-density map (or have no observable NMR restraints), and other parts only weakly tied to data. In some cases (especially at chain termini), the coordinates of those atoms will be omitted in the data bank file, but more often disordered areas are indicated by a high crystallographic B-factor or by highly divergent conformations in an NMR ensemble. If a particular part of a structure is important to the question being asked, these telltale signs should always be heeded—they indicate that the local conformation is highly uncertain, either because of molecular motions or because of problems with the data. Regions of locally ambiguous data unfortunately not only produce a higher level of random errors but also quite often result in systematic errors of interpretation that move atoms by several Ångstroms.

Global evaluations of structure quality are valuable to a bioinformatics end user in choosing which PDB file for a given molecule to include in a broad analysis. Local evaluation scores are even more valuable, however, since no level of global quality can protect against a large local error in the region of specific interest. This chapter will concentrate on explaining how it is feasible for an end user to evaluate a local region of interest within a structure, and determine either that it belongs to the reliable majority or that it is one of the rare (but not rare enough!) cases poorly determined by the data, or even is physically impossible.

Many of the basic quality indicators such as resolution, B-factor, R, and free R residuals (measures of how well the model accounts for the observed data) for crystallography, or model root-mean-square deviations (RMSD) and restraint violations for NMR, are directly reported in the Protein Data Bank (PDB) coordinate file (Chapter 10). Beyond those indicators, the subject known as structure validation (Chapter 14) provides further tools for assessing both overall and local accuracy of structures. Traditional validation programs such as ProCheck (Laskowski et al., 1993; available under the PDBSum link at the PDB site), WhatCheck (Hooft et al., 1996; http://swift.cmbi.ru.nl/gv/whatcheck/), or the Adit deposition process at the PDB (Westbrook et al., 2003; http://deposit.rcsb.org/validate/) provide a broad set of widely used tools, centering especially on ideality of molecular geometry (bond lengths and angles) and on whether protein backbone and side chain dihedral angle combinations occur outside the preferred "core" regions. The largest outliers on some of those criteria are reported in remark 500 of the PDB file header. There also are validation programs that evaluate the experimental data (if available), others that provide an evaluation of the empirical favorability of each residue's surroundings (especially useful for detecting mistakes in chain tracing), and web servers that run several of the above programs for you on an uploaded file, as described in Chapter 14.

Of special importance in validation are independent criteria, not explicitly part of the target function optimized by the structure refinement process, because the new information makes their deviations more sensitive and robust indicators of problems. The two such classic indicators are (1) the backbone ϕ, ψ, or "Ramachandran" plot (Ramachandran, Ramakrishnan, and Sasisekharan, 1963; Laskowski et al., 1993; Kleywegt and Jones, 1996; Lovell et al., 2003), since joint ϕ, ψ values are not in usual target functions, and (2) the free R-factor (Brunger, 1992), which measures agreement between the model and a designated 5–10% of the data that are deliberately kept out of refinement to provide an unbiased indicator of progress in model quality.

Recently, we have discovered, in a surprisingly simple place, a plentiful new source of information for an unbiased and sensitive validation criterion: the hydrogen atoms. They constitute about half of the atoms, but for expediency or technical reasons, they are almost always either left out altogether or not treated fully. H atoms are, of course, important and present in NMR structures, although often not given their full atomic radius.

In macromolecular crystallography, polar H atoms are often added to better define H-bonds but with no vander Waals terms, while nonpolar H atoms are added and refined against the data only at ultrahigh (near 1 Å) resolution. The main reason for this is that hydrogens diffract X-rays very poorly, so that they can be directly detected only under the best of conditions.[1] The second reason is that including hydrogens doubles the number of refinement parameters if their coordinates are treated as fully independent variables, which is acceptable only when there is a large enough number of experimental observations. Finally, only recently has computer speed allowed the extra cost in time, either for structure refinements or for theoretical calculations. The volume of the H atom is accounted for in a standard way by using larger "united atom" radii for the other atoms, but the directionality and specificity of H interactions are not represented. The net result of all this is that the crystallographers have obligingly ignored half their atoms in refinement, managing to do quite well without them, but opening up the opportunity for us to use the correctness of the hydrogen's tight and specific packing interactions as both a global and especially a local validation criterion. This relatively new method (Word et al., 1999a) is called all-atom contact analysis and is most often accessed on the MolProbity web server (http://molprobity.biochem.duke.edu; Davis et al., 2007).

As an active tool applied to the structural database, all-atom contact analysis has two different goals. The first, long-term goal is to actually improve the accuracy of the database entries, by having structural biologists apply the criteria themselves and fix many errors before coordinates are deposited (a similar process occurred several years ago with routine application of free R and Ramachandran plot criteria). The second goal is to give users of the database an easy and effective way to assess local structural accuracy. The first goal would produce higher grade ore for data mining, whereas the second improves the extraction process.

METHODOLOGY OF ALL-ATOM CONTACT ANALYSIS

The all-atom method must start off with a reliable way to add hydrogen atoms and optimize their positions, which is done by the program Reduce (Word et al., 1999; Davis et al., 2007); run either separately or as the first step of the MolProbity service. A great many of the hydrogen positions are completely determined by the heavier atoms: methylene H, backbone NH, aromatic H, and so on. The placement of hydrogens involved in OH rotations and His protonation, on the contrary, must be optimized relative to the surrounding structure. Less obviously, the 180° "flip" orientations of Asn and Gln side chain amides (as well as flips of His rings) also need to be optimized; they are fairly often incorrect as deposited, because the N and O atoms of amides or the N and C of histidine rings are not easily distinguished by the experimental X-ray data. However, the choice can reliably be made if both H-bonding and potential clashes of the NH_2 are considered (Word et al., 1999b). This process can be done automatically for the user in MolProbity, including kinemage displays that animate between the two alternatives to show the evidence for each change. Figure 15.1a and b shows the two contact displays for a doubly interacting Asn–Gln pair whose H-bonds are equally strong in either flip state. Here, the original choice has an impossibly bad clash with the Gln CαH whereas the flipped state fits well. The flip of a side chain amide is a small

---

[1] The invisibility of hydrogen atoms is actually very fortunate, because it produces the beautifully clear separation between hydrophobic side chains in protein interiors at moderate resolution.

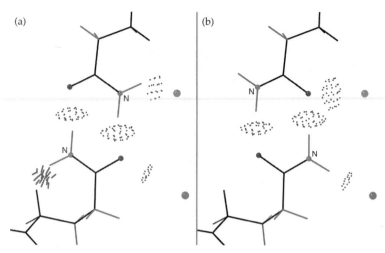

Figure 15.1. Resolving the ambiguity in a pair of doubly H-bonded side chain amides that have four equivalent H-bonds to each other and to waters, in the two best of four possible flip states. Color Figure version shows: H-bonds as pillows of pale green dots), (a) The second-best, but incorrect, flip state (pink side chains), with a large, physically impossible clash of the Gln Nε H with Hα (red spikes at lower left). (b) The correct flip orientation (green side chains), free of bad clashes after exchange of N and O atoms in both amides. From the 1.6 Å peroxidase of 1ARU (Fukuyama et al., 1995).

change but can be crucial if it affects an H-bond at an active, allosteric, or binding site. The Asn/Gln/His flip corrections are entirely automatic, are highly reliable (Word et al., 1999b; Higman et al., 2004), and do not alter agreement with the diffraction data. Any protein crystal structure could benefit from using this functionality, and a bioinformatics study would be justified in adopting a flipped state with a clear score benefit.

Surprisingly, we have found that most methyl rotations do not actually need to be optimized because they are remarkably relaxed in protein structures, with departures from staggered orientation seldom much above 10°. NH_3 groups of lysine or N-termini and side chain methyls of methionine do, however, need rotational optimization. The Reduce program handles nucleic acids and small-molecule ligands as well as proteins, and interactions with individual bound waters are treated by a simplified model. The complexity of the hydrogen addition process is due to the fact that the movable H atoms often occur in interacting H-bond networks and must be optimized as a group rather than individually. In practice, such H-bonding "cliques" are small enough, given our simplified model for water molecules, that exhaustive evaluation of all possible hydrogen positions is computationally tractable. Recent implementation of branch-pruning and other algorithm improvements has sped up the process by 50-fold on average (Davis et al., 2007), usually running in a few seconds, and in less than a minute even on ribosome structures.

All-atom contacts are calculated by the program Probe (Word et al., 1999a) from a Reduce-modified PDB file that now includes hydrogens. The usual output is contact surfaces as color-coded dots in the "kinemage" format for display in the Mage or KiNG graphics programs (Richardson and Richardson, 1992; Davis et al., 2004) as shown in the figures of this chapter, but other display formats, numerical scores, or lists of serious clashes can also be produced. Typically, Probe is run on an entire PDB file, but it can also calculate the

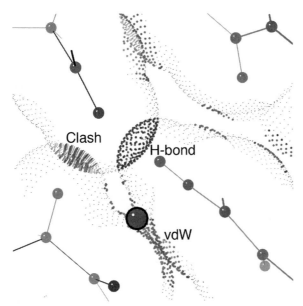

Figure 15.2. Slice through a small section of protein structure showing the relation of all-atom contact surfaces (larger dots) to the atomic van der Waals surfaces (little gray dots) and to the 0.25 Å-radius probe sphere (gray ball) used in the calculation. [Note: This figure should be viewed in the color version, since red versus green color is the primary carrier of information in the all-atom contact representation; color figures also distinguish backbone in white from side chains in cyan, please see supplementary color figures at ftp://ftp.wiley.com/public/sci_tech_med/structural_bioinformatics. The small probe sphere is rolled over the van der Waals surface of each atom, leaving a contact dot only when the probe touches another noncovalently bonded atom. The dots are colored by the local gap width between the two atoms: blue when nearly maximum 0.5 Å separation, shading to bright green near perfect van der Waals contact (0 Å gap). When suitable H-bond donor and acceptor atoms overlap, the dots are shown in pale green, forming the lens or pillow shapes of hydrogen bonds. When incompatible atoms interpenetrate, their overlap is emphasized with "spikes" instead of dots, and with colors ranging from yellow for negligible overlaps to bright reds and hot pinks for serious clash overlaps ≥ 0.4 Å. Kinemage-format contact dots also carry color information about their source atom (they can be shown with O's in red, S's in yellow, and so on); in KiNG or Mage, one can toggle between the two color schemes.

internal contacts for a small region or just the contacts between two pieces (i.e., a ligand and a protein), using a flexible command-line syntax of atom selection or a set of check-boxes in the relevant section of the MolProbity site.

Figure 15.2 illustrates a simple example of all-atom contact surfaces for a small region to show the appearance of favorable van der Waals contacts, favorable H-bond overlaps, and unfavorable atomic overlap, color-coded by the local gap distance between the two contacting atoms. The all-atom contact algorithm rolls a small spherical probe on the surface of each atom, drawing a colored dot only when the probe intersects another noncovalently bonded atom. This is a bit like the inverse of solvent-exposed surface (Lee and Richards, 1971; Connolly, 1983), where only occluded surface is shown; however, our much smaller probe means that only atom pairs within 0.5 Å of touching will count as contacts. These contacts are extremely sensitive to fine details of how well the structure fits together. If a local conformation is in the right energy well but not quite correct, it will usually produce just yellow and orange overlap dots. However, it is very difficult to fit anything in a completely wrong conformation without producing red "clash" overlaps, even after refinement has done its best at adjustments. Therefore, the primary way of interpreting the all-atom contact results is simply that lots of soothing green (such as seen in Figure 15.3a and b) means the structure is correct, while an area of red spikes has some sort of problem. In fact, for an all-atom kinemage displayed in Mage or KiNG one can turn off everything but the bad clashes and quickly spot all problem areas even in a large structure, as shown for the 324-residue dimer in Figure 15.3c.

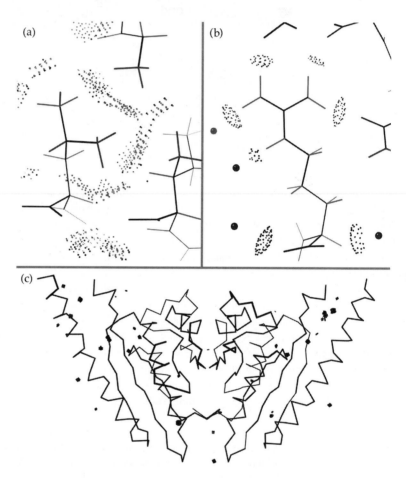

Figure 15.3. All-atom contact examples from the dimer of 1MJH (Zarembinski et al., 1998), a well-determined structural-genomics protein at 1.7 Å resolution. (a) All contacts for one of the typically well-packed and well-fit regions of aliphatic side chains, with the green of close van der Waals contacts predominant. (b) All contacts for an ARG side chain, with the five planar H-bonds (lens-shaped groups of pale green dots) of its guanidinium NHs formed either to protein O atoms or to waters (pink balls). (c) An overview of the dimer, with only the Cα backbone and the serious clashes ≥0.4 Å (red spikes) shown. When interactively displayed in KiNG or MAGE, it is easy to locate and fix the small number of isolated problems, including two flipped-over HIS rings at the putative active site and a Lys with a high B value factor squeezed into insufficient space between two hydrophobic side chains. This figure also appears in Color Figure section.

In addition to graphical display, several scoring schemes suitable for different purposes produce numerical evaluations of the contact, H-bond, and clash terms. These scores are not energies, however, because the serious clash overlaps represent model errors, not real strains in the structure. When used to understand features of molecular architecture, such as side chain packing, overlaps are treated simply as tight contacts, but for structure validation and error correction purposes, the clash overlaps are very much the dominant issue. We consider a serious clash (one that usually indicates some sort of misfitting) to occur where two incompatible atoms overlap by 0.4 Å or more. The overall "clash score" of a structure is the number of serious clashes per 1000 atoms. For protein structures, the MolProbity web site also produces an overall "MolProbity score," which is a weighted combination of the clash score with dihedral-angle scores on updated Ramachandran and side chain rotamer criteria. For choosing the best data set example among closely related structures, we currently use the average of the resolution and the MolProbity score as our primary criterion. Other considerations are availability of structure factors, absence of modifications affecting the study, and so on. For NMR structures we have not yet devised a weighted equivalence between density of NMR data and crystallographic resolution; therefore, within a group of similar NMR structures, currently we would use the single model with the best overall MolProbity score.

COMPLEMENTARY RELATIONSHIP WITH MORE TRADITIONAL CRITERIA

The well-ordered parts of the very best X-ray and NMR structures fit the all-atom contact criteria almost perfectly, with extensive contacts throughout the interior, an absence of even modest clashes, and most atoms showing the green dot patches of ideal van der Waals contact as in Figure 15.3a and b (and, at even higher resolution, in Figure 15.6 below). Such agreement is a strong confirmation that our algorithms and parameters have been chosen reasonably and that the changes recommended by this method go in the right direction. All-atom clash score is strongly correlated with other indicators of structure quality: overall parameters such as resolution or number of NMR restraints correlate with overall clash or MolProbity scores, and both resolution and local crystallographic B-factor correlate strongly with locally measured, per-residue clash score (Figure 15.4a and b).

Another relationship is that different categories of local validation criteria are best used in concert, because a given problem usually shows up in only a subset of them. For instance, if the X-ray refinement terms for geometry were heavily weighted relative to agreement with the experimental data, then the bond angles will not be distorted, but clashes will show; if clashes are between non-H atoms, then refinement may relieve them at the expense of geometry. Model-to-data local accuracy measures (such as real-space correlation) are a highly valuable category for those structures where they are available. The common model-based criteria applicable to all structures fall into three broad categories (with the most powerful measures in each category underlined):

- Geometry: bond lengths, bond angles, planarity, chirality.
- Dihedrals: single-angle preferences, ϕ,ψ angles, side chain rotamers.[2]
- Sterics: van der Waals, H-bonds, all-atom clashes.

---

[2] For proteins; nucleic acid measures are discussed below.

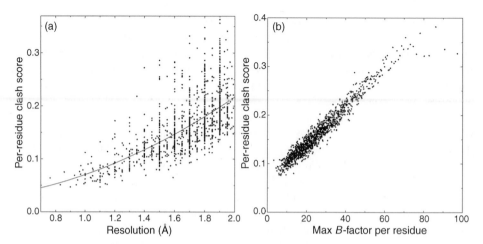

Figure 15.4. Correlation of all-atom clash scores with other indicators of structure quality. (a) Per-residue clash score (fraction of residues with serious overlaps ≥ 0.4 Å, averaged in 1000-residue bins) as a function of resolution, for 5000 representative protein structures between 0.8 and 2.0 Å resolution. The relationship is highly significant and is still improving down near 1 Å. (b) Per-residue clash score per 1000 residues, as a function of maximum crystallographic B-factor in that residue, for the above 5000 proteins (over a million residues). At low B, less than 10% of residues have a bad clash, while more than 30% do at B >60. Clashes level off for the highest B ranges, because those atoms are often exposed at the surface with few neighbors.

We therefore recommend that local structure validation include all four categories when structure factors have been deposited, and all of the last three categories otherwise.

To implement the above recommendation, and to update and extend earlier tools, MolProbity includes new versions of geometrical and dihedral angle quality measures as well as all-atom contact analysis. No early rotamer libraries were quality-filtered at the residue level, and all included at least a few physically impossible rotamers; users should turn either to the Penultimate rotamer library in MolProbity (Lovell et al., 2000) or to the Bayesian library of Dunbrack (Dunbrack, 2002) and their updates.

Similarly, we are all indebted to the original ProCheck Ramachandran criteria that created the foundations of structure validation, but they were based on the entire unfiltered PDB of 1991 and have never been updated. Therefore those criteria are not correct in detail, allowing many conformations that are actually not possible, and flagging as outliers a region with somewhat strained conformation but adopted by validly fit and functionally important residues at many active sites (see Lovell et al., 2003). In current PDB file headers, these original ProCheck criteria are used to flag Ramachandran outliers; it is certainly a very bad sign if there are many of them, but if there is only one and it is at a functionally important site, then it is most probably correct. The latter Ramachandran analyses on the EDS site (Kleywegt and Jones, 1996) or in WhatCheck (Hooft et al., 1996) use larger, more accurate data sets to define accurate core ϕ, ψ regions that encompass 98% of the high-quality data and are very reliable as global evaluations; however, they do not attempt to distinguish the truly worrisome errors within the remaining 2%. The MolProbity ϕ, ψ scores and distributions delineate both favored core regions (98% of good data) and disallowed regions (outside 99.95% of good data). MolProbity also explicitly treats the distributions for Gly, Pro, and pre-Pro as separate cases (Lovell et al., 2003).

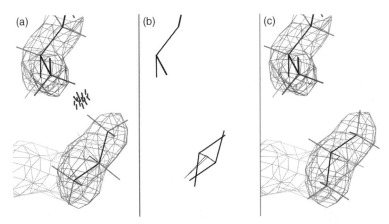

Figure 15.5. Diagnosis and correction of a backward-fit leucine side chain. (a) All-atom contacts for the original side chain, with a substantial clash and an eclipsed χ^2 angle. The model fits in the electron density fairly well, but the direction of the tetrahedral branch at $C\gamma$ is unclear. (b) Original and refit side chains, showing how both occupy approximately the same space but in opposite orientations. (c) Good all-atom contacts and slightly better density match for the Leu, now refit with ideal geometry and a highly favored side chain rotamer conformation. One can be confident that the conformation shown in (c) and not in (a) is in the correct local energy well. Leu 473 from the 1XWL DNA polymerase at 1.7 Å resolution (Keifer et al., 1997).

In MolProbity, the rotamer, Ramachandran, and geometry evaluations are used to supplement the novel all-atom clash information to make a robust diagnosis of local problem areas. If structure factors are available, examination of the electron density map in the area of a serious clash usually reveals that the density is either weak or its shape is somewhat ambiguous, making a misfitting more likely than in clearer areas. For example, in a side chain that branches at the $C\beta$ or $C\gamma$ (Thr, Val, Ile, or Leu) the electron density fairly often shows a straight bar shape rather than bending at the tetrahedral junction, making it possible to misfit the χ_1 angle by 180°. When that happens (as for the Leu in Figure 15.5a), there are nearly always all-atom clashes, the side chain rotamer is an outlier, and for $C\beta$ branches the bond-angle geometry around the $C\alpha$ is almost always badly distorted through forcing the $C\gamma$ atoms to fit into the bar-shaped electron density, although connected to a $C\beta$ that has been fit on the wrong side of the bar. Figure 15.5b shows both the original and the refit leucine side chains, emphasizing the difference in their conformations, though they occupy nearly the same space; Figure 15.5c shows the excellent fit obtainable in a good rotamer with ideal geometry. When this sort of local problem has been clearly identified, a structural bioinformaticist can avoid taking the (probably incorrect) local conformation too seriously, while a structural biologist should refit and re-refine to correct the problem.

Traditionally, electron-density difference maps are used to identify this kind of problem (for instance, sometimes showing a pair of positive and negative peaks at the real and the misfit $C\beta$). The Uppsala Electron Density Server (Kleywegt et al., 2004; http://eds.bmc.uu.se/eds/) is a valuable source of viewable electron density and difference density maps for those PDB files with available structure factors. For those same files, the EDS also supplies several forms of linear sequence plots of the numerical agreement between the model and the local map (e.g., the real-space correlation function); this gives information similar to the B-factor but more reliably comparable between structures, and it is much easier for a noncrystallographer to interpret than the maps themselves. In contrast, we do not find the estimated uncertainties produced by the Cruickshank, Luzzatti, or SigmaA methods (see Chapter 14) to be a reliable guide, because they only report the uncertainty of an atom position within the density where it was fit and do not reflect the possibility that it was fit in the wrong place. Empirical studies of the coordinate differences between multiple determinations of the same structure typically find actual RMSD values to be four–five times higher than those estimated uncertainties (Kleywegt, 1999).

Local problems in polypeptide chain tracing, such as a sequence out of register by 2 within a β strand, are often flanked at each end by clusters of all-atom clashes and bad bond angles. Neither all-atom contacts nor geometrical ideality is suitable in general; however, for identifying incorrect chain folds, both are too sensitive and too local. This task is probably best done by the sort of "threading" methods used in fold recognition (Chapter 31) and homology modeling (Chapter 30).

In drawing conclusions from a structure or comparison, it seldom matters if one or two parameters are slightly off (e.g., a torsion angle by 15°), but it is often critical if the backbone or side chain is actually in the wrong conformation (e.g., a torsion off by 90–180°) as this will change which atoms are in position to interact, say, with a ligand. Diagnosing such errors, or at least locating places where they are likely, is an important part of structural analysis, and it can be done fairly easily with currently available tools.

USING MOLPROBITY AND RELATED FACILITIES

All-atom contacts, along with the related validation criteria of updated dihedral-angle and geometry evaluations in both global and per-residue forms, can be used in two ways: either online at the MolProbity web site (for user-friendly analysis of one or a few structures) or by scripted command-line runs of the separate programs (for bulk evaluations on large numbers of files). The whole software is available from the kinemage web site free, open source, and multiplatform.

The MolProbity service (Davis et al., 2007) works on coordinate files in either old or new (v3.0) PDB format and on most of the variant formats produced by refinement or modeling programs; the file can either be fetched from the PDB or NDB (Chapters 11 and 12) or uploaded from the user's computer. For example, on a small, very high-resolution structure such as the 1BRF thermophilic rubredoxin in Figure 15.6, the multicriterion kinemage shows that the all-atom contacts are excellent throughout (green, with some yellow and blue); there are no Asn/Gln/His flips, no rotamer or Ramachandran outliers, and no large geometry deviations. If everything is turned off except the bad overlaps, it is immediately obvious that there is a single serious clash between two surface side chains (a Glu and a Lys). In the electron density, it is clear that both side chains occupy more than one conformation. If the crystallographer is looking at this clash report, he should investigate that region to see if it can be corrected; if a bioinformaticist is doing the evaluation, he now knows all of this structure is of extremely high quality except for the two clashing side chains, whose detailed conformation cannot be trusted. This example is a small protein for clarity of presentation in static two-dimensional form, but in an interactive display it is easy to locate the problem regions even on a large, lower-resolution structure and to zoom in and examine them, or else to center on the area of special interest and check for nearby problems.

The two areas in which such clash and dihedral reports have had the greatest impact are both for crystal structures: detecting and fixing protein side chains fit in the wrong rotamer (Arendall et al., 2005) and finding places where nucleic acid backbone conformations are incorrect (Murray et al., 2003; Wang et al., 2008). The most common side chain misfittings are for the Asn/Gln/His flips described above, Thr/Val/Ile/Leu tetrahedral branches (as in

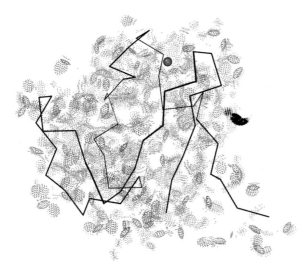

Figure 15.6. Cα backbone, bound Fe (gray ball), and all-atom contacts for the entire structure of 1BRF rubredoxin (Bau et al., 1998), a small protein structure at 0.95 Å resolution. The dense green dot patches (gray in the black-and-white version), signifying well-packed contacts in the molecule and a well-fit model, are seen consistently throughout the structure except for a single, large red (or black) clash between two surface side chains at right. 1BRF thus illustrates both how precisely the all-atom contact criteria are satisfied in atomic-resolution protein structures and also how occasional local errors can be found even in such extremely high-quality structures. This figure also appears in the Color Figure section.

Figure 15.5), Met conformations, and Arg guanidinium groups flipped over in their planar density. The reasons for problems with Thr/Val and with Leu are discussed in detail in ref. Lovell et al. (2000); for validation purposes, it suffices to know that these problems occur fairly often and that they almost always produce bad clashes and usually distort Cα–Cβ geometry (Lovell et al., 2003). We expect that quite soon MolProbity will do automatic correction of misfit Leu side chains when structure factors are available; automated diagnosis and correction of this sort of fitting problem is also being built into the PHENIX integrated crystallographic software system (Adams et al., 2002). Met can be difficult because the heavy Sδ atom produces diffraction ripples in the electron density that weakens the information for the nearby Cγ and Cε; all-atom clash and rotamer information can usually make the correct choice clear. Arg side chains have four χ angles and are not easy to fit; therefore, once the guanidinium has been maneuvered into its flat, triangular density, it will usually not be changed either by refinement or by manual rebuilding. However, the guanidinium is asymmetrical, so that its H-bonding and steric clashes with nearby atoms will always be wrong if it happens to fit upside-down. This rather common problem is a real concern for analysis of protein/nucleic acid complexes, because arginines are the most important side chains in such interactions, either for DNA or for RNA.

RNA: VALIDATION, STRUCTURE IMPROVEMENT, AND CONFORMER STRINGS

The protein/DNA structures at the basis of genetic control are a very important part of structural biology and bioinformatics (see Chapter 25), and RNA and ribonucleoprotein

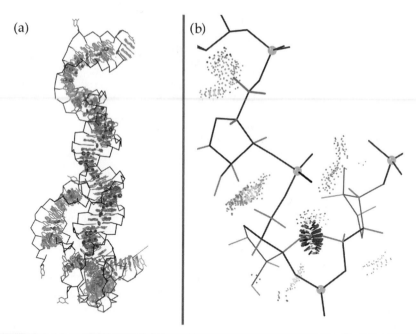

Figure 15.7. Base and backbone all-atom contacts in the 5S RNA from the 2.4 Å ribosome structure of 1S72 (Klein, Moore, and Steitz, 2004). (a) Base–base contacts, showing the long columns of well-fit base stacking (flat layers of green and blue dots). (b) A section of the backbone–backbone contacts, mostly very nicely packed but with one impossible overlap of C3$'$ and C5$'$ hydrogens (red spikes). This figure also appears in the Color Figure section.

complex structures are one of the fastest growing areas of novel biological import, including ribosomes, spliceosomes, RNAi, self-splicing introns, and riboswitches (Chapter 33). In nucleic acid crystal structures, the bases are large, rigid, and well determined (Figure 15.7a), and the phosphate density is generally unambiguous, even at the moderate-to-low resolution (2.5–3.5 Å) typical of the most biologically interesting structures. In those same structures, however, the rest of the sugar–phosphate backbone has too many rotatable bonds per observable atom and is quite prone to errors when in conformations less well understood than standard B-DNA or A-RNA. H atom clashes, however, mark the incorrect conformations extremely clearly. Figure 15.7b shows all-atom contacts for just the backbone of part of a 5S ribosomal RNA; most areas show excellent contacts, but one residue is in a physically impossible conformation. When analyzing nucleic acid structures, all-atom contacts also provide a quick and pleasing way to visualize the stacking between base pairs (see Figure 15.7a) that are almost always well fit and favorably packed. In RNA, the base pairing, both Watson–Crick and noncanonical, dominates the energetics and specificity of the 3D structure, but the detailed backbone structure is central to RNA catalysis, RNA aptamers, and the specificity of RNA binding to proteins or antibiotics. It is therefore important both to diagnose and correct errors in RNA backbone conformation, and also to help structure analysts evaluate the local accuracy of RNA backbone conformation and of RNA/protein interactions, including arginine/RNA contacts (see above).

Most geometrical validation programs can handle nucleic acid as well as protein bond lengths and angles, and some include other checks such as the noncanonical base-pair types and sugar stereochemistry in MCAnnotate (Gendron, Lemieux, and Major, 2001). Mol-Probity's geometrical analysis of whether ribose pucker is likely to be incorrect is a powerful diagnostic tool for misfittings that significantly perturb local backbone and are often accompanied by all-atom clashes and deviant bond angles. Structural biologists can use such diagnoses to guide manual rebuilding, but the multiple variables and constraints in RNA make that process quite difficult. Automated tools are being developed for model building and refitting of RNA, the first of which is RNABC (Wang et al., 2008); such facilities should help improve the level of backbone accuracy in future RNA structures.

Dihedral-angle analysis for RNA is also a rapidly growing area of interest. As for protein ϕ, ψ or side chain rotamers, the RNA backbone dihedral angles have fairly weak constraints individually but constitute quite powerful validation criteria when analyzed in combination. The RNA Ontology Consortium has developed a consensus nomenclature and a list of about 50 distinct RNA backbone conformers found to occur frequently in the well-ordered parts of high-quality RNA structures (Richardson et al., 2008). These conformers are described for the suite unit (sugar to sugar) rather than for the nucleotide unit (phosphate to phosphate), because correlations among the backbone dihedrals are stronger within the suite and it relates the positions of successive bases. Each suite conformer is given a 2-character name, such as 1a for A-form or 5z for the start of an S-motif. For an input RNA coordinate file, MolProbity will report the name (or outlier status) for each suite, along with a conformer-match parameter called "suiteness."

MolProbity also produces a linear string of the RNA conformer names that describe the specific backbone conformation (suite names alternate with base sequence). For instance, the string for the primary strand of an S-motif is N1aN5zA4sG#aU1aA1a. This nomenclature is illustrated in Figure 15.8 for each suite in three superimposed examples of GNRA

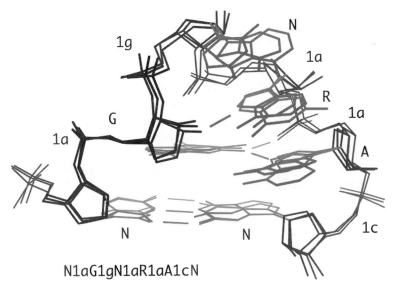

N1aG1gN1aR1aA1cN

Figure 15.8. Three superimposed GNRA tetraloops from RNA structures, with sugar-phosphate backbone shown dark and bases lighter. The closing Watson–Crick base pair is at the bottom, and lines indicate the three conserved tetraloop H-bonds: one in the non-canonical G–A pair and two base–backbone H-bonds. The two-character suite conformer name is labeled next to each backbone suite, and the consensus linear suite string of the motif is shown below: N1aG1gN1aR1aA1cN.

tetraloops, with the suite string given below. Often, as in this case, the backbone conformation of an RNA structural motif is somewhat better conserved than the base sequence. These suite strings open up new possibilities in structural bioinformatics for adapting sequence-comparison tools to search for RNA structural motifs and to compare structure between RNA molecules at many different scales.

USING LOCAL ACCURACY IN BIOINFORMATIC ANALYSES

There are at least four different ways in which all-atom contacts and other measures of local accuracy can be used to promote stronger and clearer conclusions in structural bioinformatic studies. The first, and simplest, is to filter the reference data by omitting instances with poor local quality measures. Larger samples produce statistically more significant conclusions, other things being equal. However, increasing sample size with instances known to be likely in error will degrade rather than improve the results. For example, high crystallographic B-factors for a protein side chain mean that the electron density was low, smeared out, or otherwise ambiguous and therefore the atomic positions and conformational parameters are poorly determined. Recent side chain rotamer libraries have therefore improved reliability and avoided artifacts by omitting side chains with high B-factors (>40, usually) from their reference data sets (Kuszewski et al., 1997; Lovell et al., 2000; Dunbrack, 2002).

The second strategy is the diagnosis of systematic errors by plotting or otherwise analyzing feature occurrence as a function of quality parameter (e.g., resolution, B-factor, NMR restraints per residue, steric clash score, and so on). True features should maintain or increase relative occurrence frequency as each quality criterion improves, while features that tend to disappear as accuracy improves are highly suspect. Such plots were used to distinguish backward-fit "decoy" Leu rotamers from genuine Leu rotamers (Figure 15.5a of Lovell et al., 2000) and to deprecate the plus α, *trans* β, minus γ RNA backbone conformer at C3'-endoriboses (Murray et al., 2003). This method is of quite general utility and should be considered for other types of bioinformatic data as well.

The third strategy is consulting local quality measures to determine whether an outlier in some observed pattern represents a genuine and perhaps informative exception or is simply an error. For example, Videau et al. (2004) examine the conservation within and between protein families for a new *cis*-Pro turn motif found at a dozen quite different functional sites and very rare elsewhere. In the case of DNA polymerase structures, there is one outlier *trans*-Pro example (see Figure 15.9a) that breaks the otherwise strong and simple pattern of complete conservation within the bacterial and T7 type I DNA polymerases, where it helps bind the template DNA strand, contrasted with complete absence in all other DNA or RNA polymerases. (Note that this *cis*-Pro motif is thought to be somewhat energetically strained, so that it is rapidly lost once an alternative way of filling its functional role has evolved.) As is often true, global quality parameters are not very helpful here; the outlier example (1QSS, boxed in Figure 15.9a) is in a structure with mid-range or only slightly worse values of resolution, R free, RMSD bond-angle deviation, and even overall MolProbity score compared to the nine other related polymerase structures. No structure factors were deposited for 1QSS, so one cannot evaluate the local fit to electron density in this case, and the local backbone B-factors are not especially high. Fortunately, the model-based local accuracy measures from the MolProbity multichart or multikin give an unambiguous answer, as shown in Figure 15.9b. In 1QSS, the four residues forming the *trans*-peptide turn at Pro 579 show 10 local validation flags, four of which are very serious: an all-atom

(a)

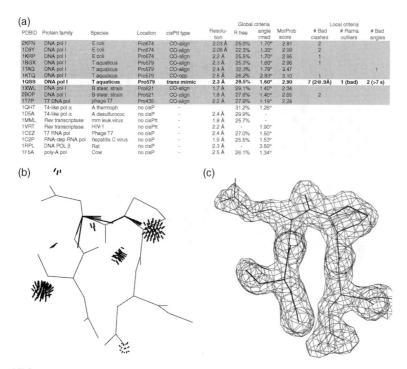

PDBID	Protein family	Species	Location	cisPtt type	Resolu-tion	Global criteria R free	angle rmsd	MolProb score	Local criteria # Bad clashes	# Rama outliers	# Bad angles
2KFN	DNA pol I	E coli	Pro674	CO-align	2.03 Å	25.0%	1.70°	2.81	2		
1D8Y	DNA pol I	E coli	Pro674	CO-align	2.08 Å	22.3%	1.33°	2.99	2		
1KRP	DNA pol I	E coli	Pro674	CO-align	2.2 Å	25.5%	1.70°	2.95	1		
1BGX	DNA pol I	T aquaticus	Pro579	CO-align	2.3 Å	25.3%	1.60°	2.96	1		
1TAQ	DNA pol I	T aquaticus	Pro579	CO-align	2.4 Å	32.3%	1.79°	3.47			1
1KTQ	DNA pol I	T aquaticus	Pro579	CO-opp	2.5 Å	26.2%	2.83°	3.10	1		
1QSS	**DNA pol I**	**T aquaticus**	**Pro579**	**trans mimic**	**2.3 Å**	**28.5%**	**1.60°**	**2.90**	**7 (2@.9Å)**	**1 (bad)**	**2 (>7 s)**
1XWL	DNA pol I	B stear. strain	Pro621	CO-align	1.7 Å	29.1%	1.40°	2.34			
2BDP	DNA pol I	B stear. strain	Pro621	CO-align	1.8 Å	27.6%	1.40°	2.85	2		
1T7P	T7 DNA pol	phage T7	Pro435	CO-align	2.2 Å	27.9%	1.19°	2.28			
1QHT	T4-like pol α	A thermoph	no cisP	-		31.2%	1.26°				
1D5A	T4-like pol α	A desulfurococ	no cisP	-	2.4 Å	29.9%	-				
1MML	Rev transcriptase	mm leuk virus	no cisPtt	-	1.8 Å	25.7%	-				
1VRT	Rev transcriptase	HIV-1	no cisPtt	-	2.2 Å	-	1.90°				
1CEZ	T7 RNA pol	Phage T7	no cisP	-	2.4 Å	27.0%	1.50°				
1C2P	RNA-dep RNA pol	hepatitis C virus	no cisP	-	1.9 Å	25.5%	1.53°				
1RPL	DNA POL β	Rat	no cisP	-	2.3 Å	-	3.50°				
1F5A	poly-A pol	Cow	no cisP	-	2.5 Å	26.1%	1.34°				

(b)

(c)

Figure 15.9. Assessing the validity of a pattern-breaking instance in a structural bioinformatic comparison. (a) A new motif called the *cis*-Pro touch-turn is apparently conserved at a DNA-binding site within the type I bacterial and T7 DNA polymerases (shaded), but entirely absent in the more distantly related families. However, the 1QSS Taq polymerase example (bold, unshaded) breaks the pattern with a *trans*-Pro turn. 1QSS is near average in global quality parameters, but severe local problems stand out unmistakably in the chart. (b) Local steric and angle validation outliers (in black) around the contested peptide in the multicriteria kinemage for 1QSS. (c) In contrast, the *cis*-Pro touch-turn in 1XWL (Keifer et al., 1997) has no local validation outliers at all, shows the *cis* peptide CO orientation clearly in the electron density contours (2Fo-Fc at 1.5 σ), and adopts the characteristically close, flat contact between the two flanking peptides. Overall, then, we can conclude that the pattern of conservation holds, because the anomalous 1QSS example is almost certainly fit incorrectly (Videau et al., 2004).

backbone–backbone steric clash with overlap of 0.9 Å, a bad Ramachandran outlier, and two bond-angle deviations $>7\sigma$, all involving the *trans*-Pro peptide in question. The equivalent *cis*-Pro regions in the nine other models show between zero and two validation flags, none of which is strong indicator of backbone problems. Therefore, one can conclude that the *trans*-Pro turn in 1QSS is an error and the *cis*-Pro touch-turn is almost certainly conserved across these type I DNA polymerases.

The fourth strategy for utilizing local validation criteria, especially all-atom contacts, is in interactive evaluation of whether an individual-residue species difference would be compatible with the known structure of a related comparison molecule. This methodology has been tested out in practice in two different ways: one is to test the suitability of proposed single-site mutations (Ghaemmaghami et al., 1998; Word et al., 2000); the second is to enhance the process of crystallographic model building or model improvement (Richardson, Arendall, and Richardson, 2003; Arendall et al., 2005). This capability is available in

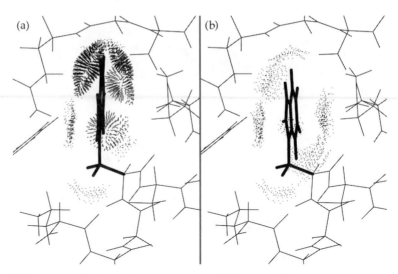

Figure 15.10. A test of alternative sequence possibilities to substitute Trp for Tyr at a buried position in the N-terminal domain of λ repressor, using the interactive side chain mutation and refitting functions in MAGE or KiNG. (a) One of the initial rotamer trials (side chain pointing away), with impossibly bad clashes on both sides of the Trp ring. (b) The best of the exact rotamers (side chain toward you), with excellent van der Waals packing (blue, green, and yellow dots), indicating that the Trp side chain can indeed fit without perturbing the structure significantly. Starting coordinates from 1LMB (Beamer and Pabo, 1992). This figure also appears in the Color Figure section.

a number of different graphics and modeling programs. Two of the most powerful and user-friendly programs for this purpose are KiNG (Davis et al., 2004) and Coot (Emsley and Cowtan, 2004). Both offer side chain mutation, rotamer-guided refitting, interactive update of all-atom contact dots, and display of electron density if available. Coot has a very complete and convenient set of crystallographic tools, while KiNG has interactive report of rotamer quality and a tool called the backrub (Davis et al., 2006) for controlled small backbone adjustments known to occur in mutations or rotamer changes. Either program rapidly solves the problem posed in Figure 15.10a of providing a favorable buried Trp mutation in a domain of λ repressor. If all rotamers of the proposed new side chain showed clashes as bad as Figure 15.10a (even when χ angles and neighboring side chains are adjusted slightly), one would conclude that the replacement was not possible without significant rearrangement of the surrounding structure. However, Figure 15.10b shows excellent contacts around a good Trp rotamer, with optimal fit only 3° away in χ^2 and no backrub needed. This Trp mutant was produced and found to have a stability and folding rate at least as good as, and an NMR spectrum very close to, that of the parent λ repressor domain (Ghaemmaghami et al., 1998). In general, if a satisfactory conformation can be found for the mutated side chain, which means the new amino acid can be accommodated without changing anything else, and therefore new properties of the mutant can be ascribed directly to the change of side chain. If no acceptable conformation can be found, the mutation might still turn out to be stable and functional, but it could not be so without the structure rearranging. Predicting the nature of such rearrangements and their functional consequences is currently still beyond the capabilities of even the most sophisticated modeling tools, and so this simple method has given you an answer nearly as good as can be done.

The all-atom contact tools are also valuable for NMR structures, but that use is less powerful and less straightforward. H atoms are explicitly included already in NMR refinement, and NMR structures are solved in terms of local distances, not absolute Cartesian coordinates. Injudicious application of contact criteria can expand the structure undesirably, but if some models in an ensemble score well and others score badly, the latter are not apt to be correct. One general conclusion from all-atom contact analysis of NMR structures is that for the best-determined cases the interiors excellently fit all-atom criterion and the surface regions would then benefit from a final refinement step with all atoms at 100% radius (rather than the lower percentages often used). All-atom contact analysis is also of considerable use in evaluating theoretical model structures (e.g., Bhattacharya et al., 2007), but again its interpretation is much less robust than when applied to crystallographic structures. A serious clash still means that something must be wrong, but the model with the fewest clashes is not necessarily the most nearly correct.

RELEVANT WEB SITES

http://molprobity.biochem.duke.edu. The MolProbity web site offers global and local quality validation for protein and nucleic acid X-ray or NMR structures: all-atom contact analysis, updated dihedral-angle criteria, and geometrical ideality. Some diagnosed problems can be automatically corrected online.

http://kinemage.biochem.duke.edu. The Richardson lab web site provides multiplatform, open-source software, selected data sets and libraries, and other resources for MolProbity use, structure improvement, and kinemage graphics.

http://eds.bmc.uu.se/eds/. The Electron Density Server provides electron density and difference density maps, real-space correlation plots, and other crystallographic validation tools that complement the MolProbity criteria.

http://www.ysbl.york.ac.uk/~emsley/coot/. Coot (part of ccp4) is a capable and user-friendly graphics system for crystallographic model building. It utilizes information from MolProbity and can refit with interactive all-atom contacts.

http://phenix-online.org. PHENIX is an integrated package for macromolecular crystallography: phasing, model building, refinement, and validation. The MolProbity criteria are utilized in the automated decision steps.

http://ndbserver.rutgers.edu/. The Nucleic Acid DataBase.

http://www.rcsb.org/pdb/. The Protein Data Bank.

REFERENCES

Adams PD, Grosse-Kunstleve RW, Hung L-W, Ioerger TR, McCoy AJ, Moriarty NW, Read RJ, Sacchettini JC, Sauter NK, Terwilliger TC (2002): PHENIX: building new software for automated crystallographic structure determination. *Acta Cryst D* 58:1948–1954.

Arendall WB III, Tempel W, Richardson JS, Zhou W, Wang S, Davis IW, Liu Z-J, Rose JP, Carson WM, Luo M, Richardson DC, Wang B-C (2005): A test of enhancing model accuracy in high-throughput crystallography. *J Struct Func Genomics* 6:1–11. [Documents the effectiveness of MolProbity structure improvement.]

Bau R, Rees DC, Kurtz DM, Scott RA, Huang HS, Adams MWW, Eidsness MK (1998): Crystal structure of rubredoxin from *Pyrococcus furiosus* at 0.95 Å resolution, and the structures of

N-terminal methionine and formylmethionine variants of Pf Rd. Contributions of N-terminal interactions to thermostability. *J Biol Inorg Chem* 3:484–493. [1BRF]

Beamer LJ, Pabo CO (1992): Refined 1.8 Å crystal structure of the lambda repressor–operator complex. *J Mol Biol* 227:177–196. [1LMB]

Bhattacharya A, Wunderlich Z, Monleon D, Tejero R, Montelione GT (2007): Assessing model accuracy using the homology modeling automatically software. *Proteins Struct Func Bioinf*, published online July 19.

Brunger AT (1992): Free R-value: a novel statistical quantity for assessing the accuracy of crystal structures. *Nature* 355:472–475.

Connolly ML (1983): Solvent-accessible surfaces of proteins and nucleic acids. *Science* 221:709–713.

Davis IW, Murray LW, Richardson JS, Richardson DC (2004): MolProbity: structure validation and all-atom contact analysis for nucleic acids and their complexes. *Nucl Acids Res* 32: W615–W619.

Davis IW, Arendall WB III, Richardson DC, Richardson JS (2006): The Backrub motion: how protein backbone shrugs when a side chain dances. *Structure* 14:265–274.

Davis IW, Leaver-Fay A, Chen VB, Block JN, Kapral GJ, Wang X, Murray LW, Arendall WB III, Snoeyink J, Richardson JS, Richardson DC (2007): MolProbity: all-atom contacts and structure validation for proteins and nucleic acids. *Nucl Acid Res* 35:W375–W383. [Recent description of the MolProbity web site functions.]

Dunbrack RL Jr (2002): Rotamer libraries in the 21st century. *Curr Opin Struct Biol* 12:431–440.

Emsley P, Cowtan K (2004): Coot: model-building tools for molecular graphics. *Acta Cryst D* 60: 2126–2132.

Fukuyama K, Kunishima N, Amada F, Kubota T, Matsubara H (1995): Crystal structures of cyanide- and triiodide-bound forms of *Arthromyces ramosus* peroxidase at different pH values. Perturbations of active site residues and their implication in enzyme catalysis. *J Biol Chem* 270: 21884–21892. [1ARU]

Gendron P, Lemieux S, Major F (2001): Quantitative analysis of nucleic acid three-dimensional structures. *J Mol Biol* 308:919–936.

Ghaemmaghami, S, Word, JM, Burton, RE, Richardson, JS, Oas, TG (1998): Folding kinetics of a fluorescent variant of monomeric λ repressor. *Biochemistry* 37: 9179–9185.

Higman VA, Boyd J, Smith LJ, Redfield C (2004): Asparagine and glutamine side-chain conformation in solution and crystal: a comparison for hen-egg-white lysozyme using residual dipolar couplings. *J Biomol NMR* 30:327–346. [An NMR study that confirms MolProbity's assignment of amide flips.]

Hooft RWW, Vriend G, Sander C, Abola EE (1996): Errors in protein structures. *Nature* 381:272–272.

Keifer JR, Mao C, Hansen CJ, Basehore SL, Hogrefe HH, Braman JC, Beese LS (1997): Crystal structure of a thermostable bacillus DNA polymerase I large fragment at 2.1 Å resolution. *Structure* 5:95–108.

Klein DJ, Moore PB, Steitz TA (2004): The roles of ribosomal proteins in the structure, assembly and evolution of the large ribosomal subunit. *J Mol Biol* 340:141–177.

Kleywegt GJ, Jones TA (1996): Phi/psi-chology: Ramachandran revisited. *Structure* 4:1395–1400.

Kleywegt GJ (1999): Experimental assessment of differences between related protein crystal structures. *Acta Crystallogr D* 55:1878–1884.

Kleywegt GJ, Harris MR, Zou JY, Taylor TC, Wahlby A, Jones TA (2004): The Uppsala electron-density server. *Acta Cryst D* 60:2240–2249.

Kuszewski J, Gronenborn AM, Clore GM (1997): Improvements and extensions in the conformational database potential for the refinement of NMR and X-ray structures of proteins and nucleic acids. *J Magn Reson* 125:171–177.

Laskowski RA, Macarthur MW, Moss DS, Thornton JM (1993): ProCheck: a program to check the stereochemical quality of protein structures. *J Appl Crystallogr* 26:283–291.

Lee BK, Richards FM (1971): The interpretation of protein structures: estimation of static accessibility. *J Mol Biol* 55:379–400.

Lovell SC, Word JM, Richardson JS, Richardson DC (2000): The Penultimate rotamer library. *Prot Struct Func Genet* 40:389–408. [An improved side chain rotamer library, from the Top240 data set filtered by all-atom clashes and *B*-factors; analyzes systematic side-chain misfittings prevalent in the structural database.]

Lovell SC, Davis IW, Arendall WB III, de Bakker PIW, Word JM, Prisant MG, Richardson JS, Richardson DC (2003): Structure validation by Cα geometry: φ, ψ and Cβ deviation. *Prot Struct Func Genet* 50:437–450. [Updated Ramachandran criteria from the Top500 data set *B*-filtered and smoothed, for the general, Gly, Pro, and pre-Pro distributions.]

Murray LJW, Arendall WB III, Richardson JS, Richardson DC (2003): RNA backbone is Rotameric. *Proc Natl Acad Sci USA* 100:13904–13909.

Ramachandran GN, Ramakrishnan C, Sasisekharan V (1963): Stereochemistry of polypeptide chain configurations. *J Mol Biol* 7:95–99.

Richardson DC, Richardson JS (1992): The Kinemage: a tool for scientific illustration. *Prot Sci* 1:3–9.

Richardson JS, Arendall WB III, Richardson DC (2003): New tools and data for improving structures, using all-atom contacts. In: Carter CW Jr, Sweet RM, editors. *Methods in Enzymology*, Vol. 374. New York: Academic Press. pp. 385–412.

Richardson JS, Schneider B, Murray LW, Kapral GJ, Immormino RM, Headd JJ, Richardson DC, Ham D, Hershkovits E, Williams LD, Keating KS, Pyle AM, Micallef D, Westbrook J, Berman HM (2008): RNA backbone: consensus all-angle conformers and modular string nomenclature (an RNA Ontology Consortium contribution). *RNA* 14 (3):465–481. [Describes the basis for most MolProbity RNA validation and introduces string analysis of RNA structural motifs.]

Videau LL, Arendall WB III, Richardson JS, Richardson DC (2004): The *cis*-ProTouch-Turn: A rare motif preferred at functional sites. *Prot Struct Func Bioinf* 56:298–309. [A structural bioinformatics study that uses MolProbity validation to assess rare cases.]

Wang X, Kapral G, Murray L, Richardson D, Richardson J, Snoeyink J (2008): RNABC: forward kinematics to reduce all-atom steric clashes in RNA backbone. *J Math Biol* 56 (1–2): 253–278.

Westbrook J, Feng ZK, Burkhardt K, Berman HM (2003): Validation of protein structures for the Protein Data Bank. *Methods Enzymol* 374:370–385.

Word JM, Lovell SC, LaBean TH, Taylor HC, Zalis ME, Presley BK, Richardson JS, Richardson DC (1999a): Visualizing and quantifying molecular goodness-of-fit: small-probe contact dots with explicit hydrogens. *J Mol Biol* 285:1711–1733. [Original description, test cases, and discussion of the all-atom contact method; explains parameter and algorithm choices in the Probe program and analyzes the contacts in a database of 100 high-resolution structures, the "Top100."]

Word JM, Lovell SC, Richardson JS, Richardson DC (1999b): Asparagine and glutamine: using hydrogen atom contacts in the choice of side-chain amide orientation. *J Mol Biol* 285:1735–1747. [Primary reference for the program Reduce; explains H addition and optimization for protein, nucleic acid, and small-molecule "heterogens," including the analysis of H-bond networks and the correction of Asn/Gln/His flips].

Word JM, Bateman RC Jr, Presley BK, Lovell SC, Richardson DC (2000): Exploring steric constraints on protein mutations using MAGE/PROBE. *Prot Sci* 9:2251–2259. [Use of all-atom contacts for evaluating whether a specific sequence change is compatible with a given protein conformation.]

Zarembinski TI, Hung LW, Mueller-Dieckmann HJ, Kim KK, Yokota H, Kim R, Kim S-H (1998): Structure-based assignment of the biochemical function of a hypothetical protein: a test case of structural genomics. *Proc Natl Acad Sci USA* 95:15189–15193. [1MJI]

16

STRUCTURE COMPARISON AND ALIGNMENT

Marc A. Marti-Renom, Emidio Capriotti,
Ilya N. Shindyalov, and Philip E. Bourne

INTRODUCTION

Since the 1970s, after the seminal work of Rossmann and Argos (1978) comparing binding sites of known enzyme structures, the comparison and alignment of protein structures has come to be a fundamental and widely used task in computational structure biology. Three main steps are needed for comparing two protein structures: first, the detection of their common similarities; second, the alignment of the structures based on such similarities; and third, a statistical measure of the similarity. Considering the first two steps, structure comparison refers to the analysis of the similarities and differences between two or more structures, and structure alignment refers to establishing which amino acid residues are equivalent between them. The majority of commonly used methods do a reasonably good job in recognizing obvious similarities between protein structures. However, the alignment of two or more structures is a more difficult task, and its accuracy may depend on the method or program used as well as what the user is trying to accomplish, which will be discussed subsequently. All programs that are briefly described in this chapter perform both steps and are commonly known as protein structure alignment methods.

It is also important to immediately clear up any confusion between structure alignment and structure superposition since such terms are often interchanged in the literature. As mentioned above, structure alignment tries to identify the equivalences between pairs of amino acid residues from the structures to superpose, while structure superposition requires the previous knowledge of such equivalences. Thus, structure superposition tries to solve the simpler geometrical task of minimizing the distance between already known equivalent residues of the superimposed structures by finding a transformation that produces either the lowest root-mean-square deviation (RMSD) or the maximal equivalences within an

Structural Bioinformatics, Second Edition   Edited by Jenny Gu and Philip E. Bourne
Copyright © 2009 John Wiley & Sons, Inc.

RMSD cutoff. Structure superposition methods have been around for some time (Diamond, 1976; Kabsch, 1976; Hendrickson, 1979; Kearsley, 1989). However, structure comparison and alignment methods, recently reviewed (Carugo, 2007; Mayr, Domingues, and Lackner, 2007), were developed later (Usha and Murthy, 1986; Sali and Blundell, 1990; Boberg, Salakoski, and Vihinen, 1992; Kikuchi, 1992; Shapiro et al., 1992; Holm and Sander, 1993b; Johnson, Overington, and Blundell, 1993; Orengo et al., 1993; Overington et al., 1993; Holm and Sander, 1994a; Lessel and Schomburg, 1994).

We begin this chapter by introducing the use of protein structure comparison and alignment for characterizing a fundamental principle in biology. Then we describe the general approach to structure comparison by outlining some of the most widely used methods. Next, we introduce two particular scenarios involving protein structure comparison, multiple structure alignment and flexible structure alignment. Finally, the large-scale application of methods for protein structure comparison and their impact on characterizing structure space is introduced in the context of structure genomics. As a quick guide, a list of common Internet resources for protein structure comparison and alignment is provided in Table 16.1.

Impact of Protein Structure Comparison and Alignment

Similarly to sequence-based alignment methods, structure-based alignment methods have been widely used for characterizing biological processes. In fact, this book includes a broad overview of several approaches that rely on protein structure comparison and alignment:

- Chapters 17 and 18 introduce two widely accepted structure classification systems, the SCOP (Andreeva et al., 2004) and CATH (Greene et al., 2007) databases. Both efforts result in a hierarchical classification of the known structure space of protein domains.
- Chapter 21 focuses on methods for inferring protein function from structure (Godzik, Jambon, and Friedberg, 2007). In such approaches, structure alignments usually play an important role. Functional inference is relevant to structure genomics, which results in a rapid increase in the number of experimentally determined protein structures of unknown function (Chapter 40).
- Chapters 30 to 32 introduce protein structure prediction and model evaluation, which rely heavily on structure alignment methods for classifying the structure space, assessing the likely accuracy of a model, and/or evaluating its actual accuracy.

Chothia and Lesk (1986) first observed, when the number of structures was limited, that protein structure was more conserved than protein sequence. As such, protein structures can provide protein sequence alignments of an accuracy that would not be achievable from sequence alignments alone. This ability is becoming a major contribution to the field of structural bioinformatics and is best illustrated in the consideration of evolution studied through protein structure (Chapters 17, 18, and 23).

On the Relationship Between Sequence and Structure

Since evolution conserves protein structure more than protein sequence, it follows that the number of possible structure folds is less than the number of sequence families. How much is

TABLE 16.1. Popular Internet Resources for Structure Comparison and Alignment

Name	T[a]	Reference	Root URL
CATH	D	Greene et al. (2007)	http://www.cathdb.info
CE	P	Shindyalov and Bourne (1998)	http://www.sdsc.edu
CE-MC	S	Guda et al. (2004)	http://bioinformatics.albany.edu
DALI	D	Holm and Sander (1996)	http://www.ebi.ac.uk
DBAli	D	Marti-Renom et al. (2007)	http://www.dbali.org
FATCAT	S	Ye and Godzik (2004)	http://fatcat.burnham.org
EXPRESSO	S	Armougom et al. (2006)	http://www.tcoffee.org
GANGSTA	S	Kolbeck et al. (2006)	http://gangsta.chemie.fu-berlin.de
KENOBI/K2	S	Szustakowski and Weng (2000)	http://zlab.bu.edu
MAMMOTH	S	Ortiz, Strauss and Olmea (2002)	http://ub.cbm.uam.es
MAMMOTH-Mult	S	Lupyan, Leo-Macias, and Ortiz (2005)	http://ub.cbm.uam.es
MultiProt	S	Shatsky, Nussinov, and Wolfson (2004)	http://bioinfo3d.cs.tau.ac.il
MUSTANG	S	Konagurthu et al. (2006)	http://www.cs.mu.oz.au
LGA	S	Zemla (2003)	http://as2ts.llnl.gov
lovoAlign	S	Martinez, Andreani, and Martinez (2007)	http://www.ime.unicamp.br
SARF2	S	Alexandrov (1996)	http://123d.ncifcrf.gov
SCOP	D	Andreeva et al. (2004)	http://scop.mrc-lmb.cam.ac.uk/scop
SSAP	S	Orengo and Taylor (1996)	http://www.cathdb.info
STAMP	S	Russell, Copley, and Barton (1996)	http://www.compbio.dundee.ac.uk
POSA	S	Ye and Godzik (2005)	http://fatcat.burnham.org
ProFit	S	Not published	http://www.bioinf.org.uk
SALIGN	P	Not published, MODELLER manual	http://www.salilab.org
TM-Align	S	Zhang and Skolnick (2005b)	http://zhang.bioinformatics.ku.edu
TOPOFIT	S	Ilyin, Abyzov, and Leslin (2004)	http://mozart.bio.neu.edu
VAST	S	Madej, Gibrat, and Bryant (1995)	http://www.ncbi.nlm.nih.gov/Structure/VAST/

[a] Type: Program (P); Server (S); Database (D).

implied by "less than" is remarkable. There are a total of 20^{300} possible sequences of 300 residues, which is more than the number of atoms in the universe. However, evolution has selected a very small subset of those protein sequences (less than 30,000 in human) and an even smaller number of protein folds (1000–5000) (Reeves et al., 2006) (Chapter 2). As stated above, such a reduction from sequence space to structure space was first quantified by Chothia and Lesk in the 1980s (Lesk and Chothia, 1980; Lesk and Chothia, 1982; Chothia and Lesk, 1986; Chothia and Lesk, 1987), later confirmed by Sander and Schneider (1991), and recently updated by Rost (1999). To illustrate this relationship here, we have taken a set

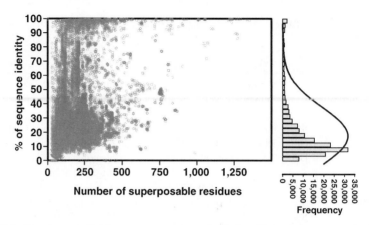

Figure 16.1. Structure similarity versus sequence similarity. Plotted data was obtained from 159,777 pair-wise structural alignments by MAMMOTH comparing 1000 randomly selected protein chains against the complete structural space deposited in the PDB as of March 2007($\sim$87,000 chains). The 1,000 chains set was obtained from a nonredundant set of structures where alignments between any two chains in the list fails at least one of the following four cutoffs: a minimum of 20% sequence identity, a minimum of 75% of Cα atoms aligned within 4 Å, a maximum of 3 Å Cα RMSD, and a maximum of 50 residues difference in length. Sequence identity is plotted against the number of superposed residues (left). The frequency distribution plotted against sequence identity with the 159,777 pair-wise structural alignments are shown on the right.

of 1000 randomly selected chains from 11,900 nonredundant chains in DBAli (Marti-Renom, Ilyin, and Sali, 2001; Marti-Renom et al., 2007) to generate the plots in Figure 16.1. The data for the plots were obtained from 154,777 pair-wise alignments stored in DBAli, using as a query each of the 1000 selected chains that were aligned by MAMMOTH (Ortiz, Strauss, and Olmea, 2002) against all $\sim$87,000 chains in the Protein Data Bank (PDB) (Berman et al., 2000). Only pair-wise alignments that aligned at least 75% of the Cα atoms of a query structure are plotted. As already observed by Rost (1997), a substantial number of the similar pairs of structures have sequence identities near the values of randomly selected pairs of sequences ($\sim$10%). Moreover, Rost assessed that the symmetric shape of the distribution at low sequence identity (<40%) indicates that most sequences are in evolutionary equilibrium making it very difficult to differentiate between convergent and divergent evolution for those sequences. These sequence–structure relationships include the so-called midnight zone (i.e., 0–20% sequence identity) and the so-called twilight zone (i.e., 20–40% sequence identity). The homology between two sequences in the midnight zone is very difficult to determine from sequence methods, and similarity can only be detected using structure alignment methods. Homology can be detected using sequence and/or structure alignment methods in the well-populated twilight zone. In summary, for a large number of evolutionarily related proteins, structure alignments provide valuable insights not achievable from sequence alone.

As discussed elsewhere in this book, it is dangerous to consider these findings as absolute—they most certainly are not. The relationship between primary protein sequence, structure, and biological function is complex and still partially uncharacterized. As George Bernard Shaw once said, "the golden rule is that there are no golden rules." Such a statement clearly applies to what we know from comparing protein sequence and structure spaces. For example, there are cases of structures containing regions of high sequence similarity, and yet sharing no or little structure similarity. Consider that the viral capsid protein (1PIV:1) shares an 80-residue stretch with glycosyltransferase (1HMP:A) where there is >40% sequence identity, yet the structures within those regions are completely different (i.e., mostly β versus mostly α, respectively). In short, structure alignment methods and their results, although only a guideline, have been essential for characterizing the relationship between sequence and structure and its implications in protein evolution.

GENERAL APPROACH TO STRUCTURE COMPARISON AND ALIGNMENT

Structure comparison and alignment is an NP-hard problem, which is only computationally tractable by using heuristics. As a result, the variety of solutions for aligning two protein structures reflect the differences in the heuristics used by each method. Moreover, even if the NP-hard problem were computationally tractable, it is very likely that for pairs of distantly related protein structures, the final alignment would not reveal new biological insights. In fact, for such protein pairs, different structure comparison methods may produce alignments that differ in every position (Godzik, 1996). Thus, it is important to experiment with a different algorithms and to assess the strengths and weaknesses of each method when the quality of the alignment is important to your research. There is a significant body of literature on protein structure comparison and alignment methods to help you in this regard. The reviews listed here are not exhaustive but do provide a historical perspective of the field. In 1994, Orengo comprehensively reviewed a series of methods used in protein structure domain classification (Orengo, 1994). Later on, Gibrat and coworkers reviewed the first structure comparison methods that were fast enough to be effectively used on large numbers of protein structures (Gibrat, Madej, and Bryant, 1996). Lemmen and Lengauer (2000) reviewed the more general field of molecular superposition within drug discovery. The authors provided a perspective on how superposition methods may effectively be used for database screening. In 2001, Koehl highlighted that although protein structure methods were mature enough to be applied in large-scale experiments, it was clear that a reliable score was still needed for assessing the significance of remote structure similarities (Koehl, 2001). Finally, a recent review by Carugo (2007) outlined most of the available methods for pair-wise, multiple, and flexible structure alignment.

In the next section, we outline a few of the most widely used methods for protein structure alignment. However, it is beyond the scope of this chapter to deal with each method in detail, nor to compare them. The intent is to give the reader a sense of the similarities and differences between such approaches. We refer the reader to the original papers for a full description of the methods and the resulting web resources listed in Table 16.1. All methods need to address three problems:

1. *Representation*: How to represent the input structures in a coordinate-independent space suitable for alignment.
2. *Optimization*: How to sample the space of possible alignment solutions between the structures.
3. *Scoring*: How to score a given alignment and determine its statistical significance.

We discuss seven widely used and cited protein structure alignment methods: DALI (Holm and Sander, 1993b), SSAP (Orengo and Taylor, 1996), VAST (Madej, Gibrat, and Bryant, 1995; Gibrat, Madej, and Bryant, 1996), SARF2 (Alexandrov, 1996),

CE (Shindyalov and Bourne, 1998), MAMMOTH (Ortiz, Strauss, and Olmea, 2002), and SALIGN (Sali and Blundell, 1990) with regard to how they address the three problems of structure alignment. A broader list of available methods can be found in Table 16.1.

Briefly, the DALI algorithm, which is used in the FSSP database (Holm and Sander, 1994b), aligns two structures by generating a comparison matrix of intramolecular distances and optimizes that matrix using a Monte Carlo procedure. The SSAP algorithm, which is used in the CATH database (Greene et al., 2007), is a method for automatically comparing 3D structures using a double dynamic programming optimizer. The VAST algorithm, which is part of the NCBI's structure computational services, is a fast similarity search method based on a vector representation of protein structures. The SARF2 algorithm, designed as a similarity search method, compares the spatial arrangements of secondary structure elements computed from the coordinates of proteins. The CE algorithm, also implemented in a multiple structure alignment method CE–MC (Guda, Pal, and Shindyalov, 2006), uses a combinatorial extension (CE) method to extend highly similar pairs of residues, optimizing the best path using dynamic programming (CE) and Monte Carlo optimization (CE–MC). The MAMMOTH algorithm, also implemented in a multiple structure alignment method MAMMOTH-Mult (Lupyan, Leo-Macias, and Ortiz, 2005) and used in the DBAli database (Marti-Renom, Ilyin, and Sali, 2001; Marti-Renom et al., 2007), is a fast method for aligning two structures based on a vector representation of intramolecular distances compared by a dynamic programming optimizer. Finally, the SALIGN command of the MODELLER package (Sali and Blundell, 1993), which is also used in the DBAli database, compares structure properties calculated from the 3D coordinates of two or more proteins that are then aligned by a dynamic programming optimizer.

We now look at each of these methods in more detail according to the three issues associated with comparison.

Protein Structure Representation

DALI uses a distance matrix to represent each structure (Phillips, 1970). Thus, proteins are effectively transformed into 2D arrays of distances between all their Cα atoms. This has the advantage of placing all structures in a simplified common frame of reference. Conceptually, the problem is then straightforward, as if one is imagining each structure's contact map transparently overlaid. Overlap along the diagonal then represents similar backbone conformations (secondary structure) and off-diagonal similarity in tertiary structure. Moving one sheet of paper horizontally or vertically relative to the other to achieve overlap represents gap insertion into one or other of the structures. A later version of DALI introduced an initial quick lookup of common secondary structure elements (SSEs) between the two proteins.

SSAP (Sequence Structure Alignment Program) uses the Cβ atoms to generate a set of vectors connecting residues (in the case of glycine, a dummy Cβ is used). Such vectors effectively represent the structure in two dimensions providing both position and directionality.

VAST (Vector Alignment Search Tool), as the name suggests, represents structures as a set of vectors. In this case, the vectors are calculated from the secondary structure elements whose type, directionality, and connectivity infer the structure topology of the protein.

SARF2 transforms the coordinate representation in a set of SSEs using the Cα atom of each residue to calculate the deviation of α-helices and β-sheets from typical SSEs conformations.

CE (Combinatorial Extension) represents proteins as a set of Cα distances for octamers (i.e., between eight consecutive residues in the structure). Each pair of octameric fragments that can be aligned within a given threshold is considered an aligned fragment pair (AFP).

MAMMOTH (MAtching Molecular Models Obtained from THeory) transforms the original coordinates of the protein structure into a set of six unit-vectors calculated from the Cα trace of consecutive heptamers (Chew et al., 1999).

SALIGN represents proteins by a set of properties or features either calculated from their sequences and structures or arbitrarily defined by the user. Such properties are sequence residue type, interresidue distance, fractional side-chain accessibility, secondary structure, local structure conformation, and a user-specified feature.

COMPARISON ALGORITHM AND OPTIMIZATION

DALI creates a set of submatrices of fixed size by collapsing the original distance matrices into regions of overlap. Submatrices are then joined if there is an overlap between adjacent fragments. The optimal superposition of the final matrices is then obtained using the branch and bound algorithm (Holm and Sander, 1996).

SSAP searches for the optimal structure alignment by using a double dynamic programming algorithm. First, a set of selected matching positions is defined by applying a dynamic programming algorithm to the matrix of differences between Cβ vectors of positions i and k (i is the residue index in the first protein and, k is the residue index in the second protein) and all other positions in their respective proteins. Second, the final S_{ik} matrix is obtained by comparing vectors between Cβ atoms at pairs of positions i and j of the same protein to the Cβ atoms from the selected matching positions. The final alignment is then computed over the matrix of scores S_{ik} by a second dynamic programming step.

VAST uses a Gibbs sampling algorithm from seed SSE pairs to find alternative alignments of SSEs and scoring them by comparing the matches with randomly generated sets of SSE pairs. The final alignment is further refined using a Monte Carlo optimization procedure.

SARF2 evaluates pairs of similar SSEs between two structures by comparing the angle between them, the shortest distance between their axes, the closest point on the axes, and the minimum and maximum distances from each SSE to their medium line. SARF2 implements a graph-based optimizer used to solve the maximum clique problem for searching the largest ensembles of the mutually compatible pairs of SSEs. Finally, an extension and refinement of the alignment is computed by adding additional residues to the alignment until a user-defined RMSD threshold is reached.

CE uses a combinatorial extension algorithm to identify and combine the most similar AFPs between the compared structures. Three similarity thresholds guide the heuristic procedure for finding the optimal alignment between two proteins. First, a threshold is used to define a set of AFPs between the two structures and to select the AFP that will seed the structure alignment. Second, an iterative process is used to identify new AFPs to be added to the seed alignment (i.e., with a single AFP in the first iteration). The alignment will be then extended if the addition of a new AFP maintains the alignment score within the second threshold. Finally, a third threshold will be used to identify the best possible alignments within a set of solutions. To speed up the process, new extensions of the alignment will be performed only with AFPs within 30 residues distance to the current alignment ends. Further optimization is performed on selected alignments by using a dynamic programming algorithm over an interprotein distance matrix.

MAMMOTH obtains a similarity matrix between any two heptamers by calculating a URMS (unit-vector root mean square) through optimally superposing their unit-vectors. Then, a dynamic programming algorithm computes the optimal path over the similarity matrix. Finally, a variant of the heuristic implemented in MaxSub (Siew et al., 2000) is used to identify the largest local structure alignment within a given RMSD threshold.

SALIGN computes a dissimilarity matrix between equivalent properties from two or more structures. The dissimilarity score is computed by comparing a weighted sum of the six properties representing the proteins. Then, the final alignment will be obtained by finding the optimal path in the matrix by a local or global dynamic programming algorithm using either an affine gap penalty or an environment-dependent gap penalty function.

Statistical Analysis of Results

DALI computes the statistical significance of an alignment score by using as a background the distribution of scores from an all-against-all comparison of 225 representative structures with less than 30% sequence identity (Hobohm et al., 1992). Such a statistic is expressed as the number of standard deviations from the average score derived from the database background distribution (i.e., a Z-score).

SSAP does not explicitly calculate the statistical significance of the SSAP score. However, the scores are empirically calibrated against known structure alignments from the CATH database. Thus, a SSAP score higher than 70 is indicative of topological similarities between the compared structures.

VAST computes a *p*-value to assess the statistical significance of an alignment score. Such a statistic is calculated in a similar manner to its sequence counterpart, BLAST (Altschul et al., 1990). Thus, the *p*-value for an alignment by VAST is proportional to the probability that its score can be obtained by randomly aligning SSE pairs. As is also true of BLAST, the considered population of possible solutions weights the final *p*-value.

SARF2 final alignment score is calculated as a function of the RMSD and the number of matched Cα atoms between the compared structures. The statistical significance of the final score is then obtained by comparing it to the background distributing of scores from aligning the leghemoglobin protein against a set of 426 nonredundant structures (Fischer et al., 1995).

CE computes a Z-score for the final alignment using a set of alignments between representative structures with less than 25% sequence identity (Hobohm et al., 1992). The RMSD and gap score for such alignments are then used to generate normal distributions to calculate the final Z-score of the computed alignment. This normal distribution was later updated with a more realistic extreme value distribution (Jia et al., 2004).

MAMMOTH calculates a *p*-value statistic to assess the significance of a pair-wise alignment. The *p*-value estimation is based on an extreme value fitting of the scores resulting from a set of random structure alignments (Abagyan and Batalov, 1997).

SALIGN does not explicitly calculate a statistical significance of the score from the final alignment. Thus, the user is simply presented with the final dissimilarity score obtained by the optimizer. However, when comparing structures, SALIGN returns a quality score, which corresponds to the average percentage of equivalent Cα atoms within 3.5 Å between all pairs of structures in the alignment.

HOW WELL ARE WE DOING?

Most structure comparison methods will detect global structural similarity between two proteins. However, diverse methods may identify different structure similarity for local alignmnets. Even when local or global similarity can be detected, the details of the sequence alignment derived from structure comparison may differ. Godzik showed that different methods for structure comparison could result in very different alignments for pairs of proteins with low sequence identity (Godzik, 1996). Differences could be so extreme that two methods may result in alignments different at every position. Similar conclusions were obtained comparing several structure classification systems (Hadley and Jones, 1999) or structurally aligning the catalytic core of several protein kinases (Scheeff and Bourne, 2006). Given the heuristics used in protein structure alignment methods, such differences are not surprising. Any method for protein structure alignment needs to balance coverage versus accuracy. In other words, a method may align the core of a protein at very high accuracy (i.e., very low RMSD) and very low coverage (i.e., omitting loop regions), while a second method may prefer to increase the coverage (i.e., include the loop regions in the alignment) to the detriment of accuracy (i.e., increasing the RMSD). How best to address this problem? In part, the answer lies in the question that you wish to address. Certainly, maximizing the biological relevance of a result is going to be the most desirable outcome in the majority of cases. We will come back to this issue at the end, and for now consider the implications of not achieving the optimum biological alignment.

Consider the case of comparing expert hand-generated alignments of protein kinases against those produced by the CE algorithm (Scheeff and Bourne, 2005). CE was unable to reproduce an optimal, manually curated alignment of 18 protein kinase structures of low sequence similarity (<40%; available from http://www.sdsc.edu/pb/kinases). The structures showed significant diversity from the hand-curated set in loop regions as well as in some of their secondary structure elements. A different set of parameters optimizing the alignment of highly conserved regions of the structures might have resulted in more biologically relevant results. In a general application, including large-scale computations, such parameters would have to be optimized for typical families of globular proteins. This requirement makes the production of highly accurate alignments for all protein families in the PDB almost impossible. However, better scoring functions that incorporate structure and functional information about a particular family may help the development of more accurate methods.

SAMPLE RESULTS FROM STRUCTURE COMPARISON AND ALIGNMENT

Consider three examples that illustrate the importance of protein structure comparison and alignment for characterizing and quantifying structural and functional similarities between apparently unrelated proteins.

The first example, shown in Figure 16.2, corresponds to the alignment between a membrane protein (colicin A; 1COL:A) and an accessory pigment to chlorophyll (c-phycocianin; 1CPC:A). On first glance, the function of these proteins is very different. Colicin A forms voltage-gated channels in the lipid bilayers of membranes, whereas phycocianin is a pigment from the light harvesting phycobiliprotein family. Holm and Sander (1993a) detected a surprising similarity between these two folds with six α-helices sequentially aligned (Figure 16.2). Such a discovery implies that both sequences had

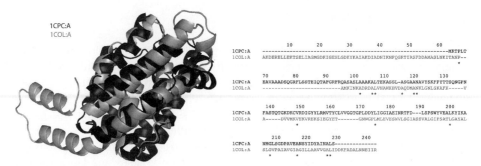

Figure 16.2. Structure alignment for c-phycocyanin (1CPC:A) (black) and colicin A (1COL:A) (gray) as computed by SALIGN. The alignment extended over 86 residues with a 0.97 Å RMSD. The sequence identity of the superposed residues with respect to the shorter of the two structures was 11.9%.

undergone convergent evolution to form a stable 3-on-3 α-helical sandwich fold. Interestingly, it was subsequently discovered that phycocianins can aggregate forming clusters that then adhere to the membrane forming the so-called phycobilisomes. Such a functional relationship may indeed point to convergent evolution from a distant common ancestor.

The second example, which is extracted from the work of one of our groups (Tsigelny et al., 2000), illustrated how the combination and integration of different sources of information, including structural alignments, could help to functionally characterize a protein. In our work, two new EF-hand motifs were identified in acetylcholinesterase (AChE) and related proteins by combining the results from a hidden Markov model sequence search, Prosite pattern extraction, and protein structure alignments by CE. It was also found that the α–β hydrolase fold family, including acetylcholinesterases, contains putative Ca^{2+} binding sites, indicative of an EF-hand motif, and which in some family members may be critical for heterologous cell associations. This putative finding represented the second characterization of an EF-hand motif within an extracellular protein, which previously had only been found in osteonectins. Thus, structure alignment had contributed to our understanding of an important family of proteins.

Finally, the third example, also from a previous work of one of our groups (McMahon et al., 2005), combined information from structural alignments deposited in the DBAli database and experiments to analyze the sequence and fold diversity of a C-type lectin domain. We demonstrated that the C-type lectin fold adopted by a major tropism determinant sequence, a retroelement-encoded receptor binding protein, provides a highly static structural scaffold in support of a diverse array of sequences. Immunoglobulins are known to fulfill the same role of a scaffold supporting a large variety of sequences necessary for an antigenic response. C-type lectins were shown to represent a different evolutionary solution taken by retroelements to balance diversity against stability.

MULTIPLE STRUCTURE ALIGNMENT

Our discussions thus far have involved only pair-wise structure comparison and alignment, or at best, alignment of multiple structures to a single representative in a pair-wise fashion (i.e., progressive pair-wise structure alignment). Most of the available methods for multiple structure alignment start by computing all pair-wise alignments between a set of structures but then use them to generate the optimal consensus alignment between all the structures.

A characteristic of all multiple structure alignment methods is that not all the pair-wise alignments used in the multiple alignment may be optimal. Once computed, multiple structure alignments, which encode weak yet definitive sequence relationships, can be used as seed alignments for iterative searches of the nonredundant sequence databases to derive hidden Markov models (HMMs) or profiles for use in fold, family, and superfamily annotation.

Several methods have been developed for the alignment of multiple structures (Table 16.1). Here, we briefly outline three different methods: CE–MC (Guda et al., 2001; Guda et al., 2004), MAMMOTH-Mult (Lupyan, Leo-Macias, and Ortiz, 2005), and SALIGN (Madhusudhan et al., unpublished).

CE-MC (combinatorial extension–Monte Carlo) refines a set of pair-wise structure alignments using a Monte Carlo optimization technique. That is, the algorithm iteratively modifies the multiple alignment, initialized with pair-wise alignments of a master structure against all other structures, by a random set of moves, which are then accepted with a probability proportional to the gain in the alignment score. The iterative process is stopped when the optimal alignment, which is based on a defined distance score for each block of aligned residues across the alignment, cannot be improved by random moves. Initially two specific families, protein kinases and aspartic proteinases, were tested and compared against manually curated alignments and those from the HOMSTRAD database (Stebbings and Mizuguchi, 2004). CE–MC improved the overall number of aligned residues while preserving key catalytic residues for those families. Using a larger benchmark of 66 protein families, on average, an additional 12% of residues was observed to be aligned.

MAMMOTH-Mult is an extension of the MAMMOTH algorithm used for pair-wise structure alignment. The alignment starts by generating all pair-wise alignments. An average linkage procedure is then used to cluster all structures based on their pair-wise structural similarity, resulting in a dendrogram tree. MAMMOTH-Mult then generates a multiple structure alignment by iteratively aligning, in a pair-wise fashion, the branches of the dendrogram that are closest to each other. The authors demonstrated that the method, which produces a typical multiple alignment every 5 s of a single CPU, produced more accurate alignments compared to other previously existing multiple structure alignment methods (Lupyan, Leo-Macias, and Ortiz, 2005).

SALIGN can be applied to align three or more protein structures using two different approaches, a tree-based or a progressive alignment. Similar to MAMMOTH-Mult, during the tree-based alignment, a dendrogram representing all pair-wise structural relationships is calculated to guide the multiple structure alignment. SALIGN first aligns the two most similar structures in the dendogram, which are then treated as a unit and aligned to the next closest structure or group of structures. This iterative process is finished when SALIGN reaches the root of the tree. Alternatively, if progressive alignment is chosen, the structures are gradually aligned in the order they are input to MODELLER. The progressive alignment method is computationally less intensive than using a tree-based approach.

FLEXIBLE STRUCTURE ALIGNMENT

Flexible structure alignments are becoming increasingly important given our increased understanding of protein fold space, which moves away from the notion of discrete folds to more of a densely populated continuum (next section). A logical outcome of the continuum model is that protein domains are difficult to delineate and that it is better that structure be considered as sets of suprasecondary structures (i.e., continuous or discontinuous small numbers of SSEs). In this model, the differences between related protein structures may lie in the relative orientation of such subdomain protein fragments.

Current methods for protein structure comparison and alignment cannot address subtle changes in the angle between those protein fragments. Ye and Godzik have developed a method for flexible structure alignment called FATCAT (Ye and Godzik, 2003; Ye and Godzik, 2004) and applied it to identify structural similarities in database searches. Briefly, the FATCAT algorithm adds a limited number of "twists" between AFPs (i.e., aligned fragment pairs between the two structures), which are treated as rigid bodies. Thus, the final score is proportional to the alignment score for having a number of AFPs in the alignment and the "penalty" of including twist to join the AFPs. In addition, FATCAT will allow twists that result in a decrease of the RMSD. Dynamic programming is used to refine the final alignment based on the similarity matrix upon superposition of the AFPs resulting from the first step. The authors demonstrated that FATCAT produced more accurate alignments when using a test set of multidomain proteins. The FATCAT algorithm has been applied broadly to produce the Flexible Structural Neighborhood database (Li, Ye, and Godzik, 2006).

MAPPING PROTEIN FOLD SPACE

Ever since the first protein structures were experimentally determined, researchers have attempted to divide and classify them. The most recent view of the protein structure space introduced proteins as combinations of subdomain fragments, which in turn result in a structurally dense and continuous description of the fold space (Haspel et al. 2003; Kihara and Skolnick 2003; Tendulkar et al. 2004; Friedberg and Godzik, 2005a; Friedberg and Godzik, 2005b; Zhang and Skolnick 2005a). It has even been suggested that these fragments may be evolutionary linked to ancestral peptides in an RNA-based world (Lupas, Ponting, and Russell 2001; Soding and Lupas 2003). However, the most accepted view of protein structures divides them into domains (Chapter 20). Domains are considered evolutionary units to the extent that they can be excised from the chain and yet continue to fold correctly with a well-defined hydrophobic core, often still exhibiting biological activity (Rossman, 1981; Holm and Sander, 1996). Given this view, domains can then be considered a particular representation of recurrent and independent protein fragments that may be observed in different folds or environments. Thus, proteins with similar folds could be described as proteins sharing similar arrangement of protein domains or fragments (Ye et al., 2003).

Independently of how domains (or structural units) are defined (Holland et al., 2006), what seems clear today is that the protein fold space is quite dense and continuous. With the exception of nonglobular proteins, such as membrane and disordered proteins, the PDB may already contain most of the recurrent structural units (Kihara and Skolnick, 2003). However, the sequence diversity possible using those recurring structural units is by no means represented in the PDB. The gap between known sequence and structural space is one of the main driving forces behind structural genomics (Chapter 40) as well as a major limitation for complete coverage of large-scale comparative structure prediction methods.

The SCOP (Murzin et al., 1995; Andreeva et al., 2004), DALI (Holm and Sander, 1996; Holm and Sander, 1999), and CATH (Orengo et al., 1997; Greene et al., 2007) databases made the first comprehensive attempts to map protein structure space at the domain level.

These three classification systems use a somewhat different definition of a domain and hence differences in classification result (Day et al., 2003). This clearly reflects the difficulty of uniquely defining a domain (Chapter 20). Given the notion of a more continuous protein fold space than previously suspected, it may be more profitable to characterize structures at the level of the sub-domain. Indeed, we can find plenty of references to recurrent subdomain structures in the literature such as *greek-key*, *jelly-roll*, β-*propeller*, α-*solenoid*, and so on, which are well accepted yet not systematically defined. The challenge then becomes identifying the proper resolution of a map that is needed to solve the problem at hand. One of our groups, using the structural alignments stored in the DBAli database, has made an attempt to visually map the continuity and density of structural space at subdomain resolution. As of August 2007, the DBAli database contained ~1.67 billion pair-wise structure alignments calculated by MAMMOTH. Using these comparisons, we have created a map of the structural relationships between all members of our nonredundant set of 11,900 PDB chains (see Figure 16.1 for details about this set). In this map, two protein chains (vertices) are linked (edges) if at least 40% of their Cα atoms can be superimposed within 4 Å and result in an alignment of at least 40% or 20% sequence identity (Figure 16.3a and b, respectively). Effectively, the resolution of such maps corresponds to aligning fragments of approximately 50 residues. At such resolution, it is the sequence discontinuity that separates protein fold space. More specifically, when the sequence identity threshold is as low as 20%, 4679 of all nonredundant chains are joined into a single largest cluster. This means that for ~40% of the vertices in the map, a path can be found by linking superposed fragments of at least 50 residues. However, such continuity completely breaks by increasing the sequence identity threshold to 40% resulting in 10,121 clusters and only ~1% of chains forming part of

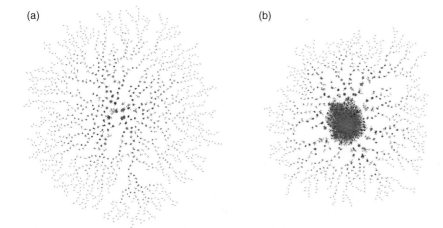

Figure 16.3. Map of protein fold space. Two protein chains (vertices) are linked (edges) if at least 40% of their Cα atoms can be superposed within 4 Å and result in an alignment of at least 40% or 20% sequence identity for panels a and b, respectively. (a) Map at 40% sequence identity of the 11,900 nonredundant set. A total of 2530 clusters with two or more structures are shown. The largest cluster in the map contains 115 chains. The map does not show the 7591 singletons remaining after clustering. (b) Map at 20% sequence identity of the 11,900 nonredundant set. A total of 1521 clusters with two or more structures are shown. The largest cluster in the map contains 4679 chains. The map does not show the 2924 singletons remaining after clustering. Both maps were produced with the LGL program (Adai et al., 2004) and rendered by the lgl2ps script (Fred P. Davis).

the largest cluster. Similar observations were already reached by one of our groups (Shindyalov and Bourne, 2000). In that work, we clustered groups of structures based on the results of an all-against-all comparison with the CE program. Such clustering resulted in a highly repetitive set of superposed substructures not detectable by sequence similarity alone. We observed as well that for some chains, different substructures constituted all or parts of well-annotated folds. We concluded that mapping protein fold space using such substructures could be useful for finding remote sequence homology and predicting the structure and function of proteins (Friedberg and Godzik, 2005a; Friedberg and Godzik, 2005b).

In summary, if the protein fold space is a continuous and dense territory, then many different and systematic definitions of recurrent fragments are possible. It is important to stress the need to carefully identify and define the appropriate level of resolution for answering the biological question to be addressed. In short, there are great opportunities for original thinking in defining the structural units relevant for characterizing protein structure evolution. With the advent of structural genomics, this opportunity is even more pronounced.

THE IMPACT OF STRUCTURAL GENOMICS

The motivation for improved protein structure comparison, alignment, and characterization is currently defined quite simply by quantity—the rate of increase in the number of experimentally determined new folds and the number of structures adopting each fold. A recent review by Carugo (2007) already highlights the sudden increase in automatic methods for structure comparison and alignment. Such methods require not only high accuracy and coverage but also fast execution to cope with the increasing number of structures. Automation should seek to reach the quality of human annotation, since no expert is able to keep up with rate of growth of the PDB. The increase in the number of structures is, in part, due to a move toward high-throughput structure determination. Despite some recent controversy about the cost and impact of structural genomics (SG) (Chandonia and Brenner, 2006; Levitt, 2007; Liu, Montelione, and Rost, 2007; Petsko, 2007), it seems fair (at least to us) to say that the technological advances accomplished by the SG consortia have increased the success rate of structure determination while decreasing the cost per structure. At the time of writing (August 2007), the PDB contained 5129 deposited structures from the SG consortia with a rate of more than 800 per year being added over the past 4 years. About half of those depositions were for structures with sequences that shared, at the date of submission, less than 30% sequence identity to any other known structure in the PDB. Such structures are then used for computationally predicting the structure of all their known homologous sequences. For example, the New York Structural GenomiX Research Consortium has deposited 426 structures in the PDB from which the ModPipe computational package (Eswar et al., 2003) was able to predict 575,035 fold assignments and 56,302 reliable 3D models (i.e., ~130 protein structure predictions per new released structure). The reader can visit http://targetdb.pdb.org for up-to-date detailed statistics about the SG deposited structures in the PDB (Chen et al., 2004).

It is important to note that an increase in deposited structures does not just imply quantity, but also variety, complexity, and singularity. In recent years, the rate of deposition of complex and nonglobular structures deposited in the PDB has also increased. This trend is likely to continue. The Protein Structure Initiative (PSI) responsible for structural genomics in the United States has recently approved its second phase of funding to four of its original pilot centers as well as six new specialized centers, including two centers dedicated to membrane proteins. Therefore, traditional laboratories as well as those within the SG consortia will likely continue to increase the number and variety of predicted and determined structures deposited in the PDB or modeling databases such as ModBase (Pieper et al., 2006). One of the major challenges will be to characterize the functions of a growing number of deposited structures with unknown function. In particular, biologists will be faced with the problem of characterizing the intricate network of interactions between single and multiple domain proteins. Diverse sources of information will be needed for tackling such challenges. For example, high-resolution protein–protein interactions can be characterized by fitting known or modeled structure into a lower resolution structure map of a complex, which could be determined by cryoelectron microscopy (Topf and Sali, 2005).

THE FUTURE

Protein structure comparison and alignment is a well-studied area as the Wikipedia page http://en.wikipedia.org/wiki/Structural_alignment_software will attest. The field is plagued by problems that afflict other areas of bioinformatics—lack of use of benchmark datasets, papers without available software, and minor advances over previous efforts. Notwithstanding, a significant understanding of the problem has been reached and methods continue to improve. We point to some of what we believe are still open challenges for protein structure comparison and alignment:

- *Accurate and Fast Methods for Multiple Structure Alignment*: Existing methods for multiple structure alignment are reaching unprecedented levels of coverage and accuracy. However, some of the most accurate methods are still computationally prohibitive to be applied in large-scale and continuous experiments.
- *Flexible Structure Alignment*: While significant progress has been made, biological features that depend on flexibility have yet to be considered as part of the alignment procedure; for example, difference between open and closed states of an enzyme induced by cofactors present in the structure and known allosteric interactions.
- *Biologically Relevant Alignments*: Current methods for structure comparison and alignment usually focus on optimizing geometrical similarities between two or more structures. However, function is not always related to geometrical features and chemical, physical, or evolutionary information can help in finding the most relevant alignment between structures. Thus, methods that are able to account for additional biological information might lead to more accurate alignments. Rather than align a set of amino acids based solely on their 3D atomic coordinates, alignments will include a variety of parameters that reflect secondary, tertiary, possibly quaternary features, and functional features of the structures under study. This will require significantly better annotation of structures than exist today if this is to be done in a high-throughput mode.
- *Automatic Optimization of Parameters*: Related to the points raised above, new methods that identify optimal parameters for each protein family would likely result in a larger number of accurate alignments.
- *Clustering and Classification*: Currently, the PDB contains more than 90,000 protein chains. A full set of comparisons requires that approximately 4×10^9 comparisons be

computed and stored. This is overwhelming for all but the fastest algorithms and extensive computer clusters. As a result, compromises are made by introducing various types of redundancy to reduce the number of computations to be made, thereby losing important information. Faster and more biologically meaningful clustering and classification algorithms are needed.

- *Biologically Relevant Division of the Structural Space*: Defining and identifying unique structural units that are recurrent between protein structures remains an unresolved issue. Domains and subdomains are currently used, but more fine-grained features may be needed.

- *Leverage of Structure Alignments*: During the last decade, we have seen the development of several methods for fast and reliable pair-wise structure alignment. Some of such methods have been applied in large-scale comparison of all-against-all structures in the PDB. However, such alignments, which are then normally deposited in databases, are barely used outside of the groups that generated them. A double effort is needed to make the data easily accessible to other developers as well as to develop new computational methods that leverage the benefits of such databases.

The ever-increasing number of structures and the scientific insights that structure comparison and alignment can bring to classification, functional understanding, provision of powerful search tools (e.g., HMMs seeded by structure alignments) and so on will see continued efforts to meet these challenges and guarantee that structure comparison and alignment will remain an active area of research in years to come.

ACKNOWLEDGMENTS

We thank Dr. Fred P. Davis for the Perl script to render LGL output graphs. MAMR and EC research is supported by the European Union (Marie Curie Program FP6-039722), the Spanish Ministerio de Educación y Ciencia (BIO2007/66670), and the Generalitat Valenciana (GV/2007/065).

REFERENCES

Abagyan RA, Batalov S (1997): Do aligned sequences share the same fold? *J Mol Biol* 273(1): 355–368.

Adai AT, Date SV, Wieland S, Marcotte EM (2004): LGL: creating a map of protein function with an algorithm for visualizing very large biological networks. *J Mol Biol* 340(1):179–190.

Alexandrov NN (1996): SARFing the PDB. *Protein Eng* 9(9):727–732.

Altschul SF, Gish W, Miller W, Myers EW, Lipman DJ (1990): Basic local alignment search tool. *J Mol Biol* 215(3):403–410.

Andreeva A, Howorth D, Brenner SE, Hubbard TJ, Chothia C, Murzin AG (2004): SCOP database in 2004: refinements integrate structure and sequence family data. *Nucleic Acids Res* 32 (Database issue):D226–D229.

Armougom F, Moretti S, Poirot O, Audic S, Dumas P, Schaeli B, Keduas V, Notredame C (2006): Expresso: automatic incorporation of structural information in multiple sequence alignments using 3D-Coffee. *Nucleic Acids Res* 34 (Web Server issue):W604–W608.

Berman HM, Westbrook J, Feng Z, Gilliland G, Bhat TN, Weissig H, Shindyalov IN, Bourne PE (2000): The Protein Data Bank. *Nucleic Acids Res* 28(1):235–242.

Boberg J, Salakoski T, Vihinen M (1992): Selection of a representative set of structures from Brookhaven Protein Data Bank. *Proteins* 14(2):265–276.

Carugo O (2007): Recent progress in measuring structural similarity between proteins. *Curr Protein Pept Sci* 8(3):219–241.

Chandonia J-M, Brenner S (2006): The impact of structural genomics: expectations and outcomes. *Science* 311(5759):347–351.

Chen L, Oughtred R, Berman HM, Westbrook J (2004): TargetDB: a target registration database for structural genomics projects. *Bioinformatics* 20(16):2860–2862.

Chew L, Huttlenlocher D, Kedem K, Kleinberg J (1999): Fast detection of common geometric substructure in proteins. *J Comput Biol* 6:313–325.

Chothia C, Lesk AM (1986): The relation between the divergence of sequence and structure in proteins. *EMBO J* 5(4):823–826.

Chothia C, Lesk AM (1987): The evolution of protein structures. *Cold Spring Harb Symp Quant Biol*, 52(399–405).

Day R, Beck DA, Armen RS, Daggett V (2003): A consensus view of fold space: combining SCOP, CATH, and the Dali Domain Dictionary. *Protein Sci* 12(10):2150–2160.

Diamond R (1976): Comparison of conformations using linear quadratic transformations. *Acta Crystallogr A* 32:1–10.

Eswar N, John B, Mirkovic N, Fiser A, Ilyin VA, Pieper U, Stuart AC, Marti-Renom MA, Madhusudhan MS, Yerkovich B, Sali A (2003): Tools for comparative protein structure modeling and analysis. *Nucleic Acids Res* 31(13):3375–3380.

Fischer D, Tsai CJ, Nussinov R, Wolfson H (1995): A 3D sequence-independent representation of the protein data bank. *Protein Eng* 8(10):981–997.

Friedberg I, Godzik A (2005a): Connecting the protein structure universe by using sparse recurring fragments. *Structure* 13(8):1213–1224.

Friedberg I, Godzik A (2005b): Fragnostic: walking through protein structure space. *Nucleic Acids Res* 33 (Web Server issue):W249–W251.

Gibrat JF, Madej T, Bryant SH (1996): Surprising similarities in structure comparison. *Curr Opin Struct Biol* 6(3):377–385.

Godzik A (1996): The structural alignment between two proteins: is there a unique answer? *Protein Sci* 5:1325–1338.

Godzik A, Jambon M, Friedberg I (2007): Computational protein function prediction: are we making progress? *Cell Mol Life Sci* 64(19–20) 2505–2511.

Greene LH, Lewis TE, Addou S, Cuff A, Dallman T, Dibley M, Redfern O, Pearl F, Nambudiry R, Reid A, Sillitoe I, Yeats C, Thornton JM, Orengo CA (2007): The CATH domain structure database: new protocols and classification levels give a more comprehensive resource for exploring evolution. *Nucleic Acids Res* 35 (Database issue):D291–D297.

Guda C, Scheeff ED, Bourne PE, Shindyalov IN (2001): A new algorithm for the alignment of multiple protein structures using Monte Carlo optimization. *Pac Symp Biocomput* 275–286.

Guda C, Lu S, Scheeff ED, Bourne PE, Shindyalov IN (2004): CE-MC: a multiple protein structure alignment server. *Nucleic Acids Res* 32 (Web Server issue):W100–W103.

Guda C, Pal LR, Shindyalov IN (2006): DMAPS: a database of multiple alignments for protein structures. *Nucleic Acids Res* 34 (Database issue):D273–D276.

Hadley C, Jones DT (1999): A systematic comparison of protein structure classifications: SCOP, CATH and FSSP. *Structure* 7:1099–1112.

Haspel N, Tsai CJ, Wolfson H, Nussinov R (2003): Hierarchical protein folding pathways: a computational study of protein fragments. *Proteins* 51(2):203–215.

Hendrickson WA (1979): Transformations to optimize the superposition of two similar structures. *Acta Crystallogr A* 35:158–163.

Hobohm U, Scharf M, Schneider R, Sander C (1992): Selection of representative protein data sets. *Protein Sci* 1(3):409–417.

Holland T, Veretnik S, Shindyalov I, Bourne P (2006): Partitioning protein structures into domains: why is it so difficult? *J Mol Biol* 361(3):562–590.

Holm L, Sander C (1993a): Globin fold in a bacterial toxin. *Nature* 361(6410):309.

Holm L, Sander C (1993b): Protein structure comparison by alignment of distance matrices. *J Mol Biol* 233(1):123–138.

Holm L, Sander C (1994a): Searching protein structure databases has come of age. *Proteins* 19(3): 165–173.

Holm L, Sander C (1994b): The FSSP database of structurally aligned protein fold families. *Nucleic Acids Res* 22(17):3600–3609.

Holm L, Sander C (1996): Mapping the protein universe. *Science* 273(5275):595–603.

Holm L, Sander C (1999): Protein folds and families: sequence and structure alignments. *Nucleic Acids Res* 27(1):244–247.

Ilyin VA, Abyzov A, Leslin CM (2004): Structural alignment of proteins by a novel TOPOFIT method, as a superimposition of common volumes at a topomax point. *Protein Sci* 13(7): 1865–1874.

Jia Y, Dewey TG, Shindyalov IN, Bourne PE (2004): A new scoring function and associated statistical significance for structure alignment by CE. *J Comput Biol* 11(5):787–799.

Johnson MS, Overington JP, Blundell TL (1993): Alignment and searching for common protein folds using a data bank of structural templates. *J Mol Biol* 231(3):735–752.

Kabsch W (1976): Solution for best rotation to relate two sets of vectors. *Acta Crystallogr A* 45:922–923.

Kearsley SK (1989): Structural comparison using restraint inhomogeneous transformations. *Acta Crystallogr A* 45:628–635.

Kihara D, Skolnick J (2003): The PDB is a covering set of small protein structures. *J Mol Biol* 334(4): 793–802.

Kikuchi T (1992): Similarity between average distance maps of structurally homologous proteins. *J Protein Chem* 11(3):305–320.

Koehl P (2001): Protein structure similarities. *Curr Opin Struct Biol* 11(3):348–353.

Kolbeck B, May P, Schmidt-Goenner T, Steinke T, Knapp EW (2006): Connectivity independent protein-structure alignment: a hierarchical approach. *BMC Bioinformatics* 7:510.

Konagurthu AS, Whisstock JC, Stuckey PJ, Lesk AM (2006): MUSTANG: a multiple structural alignment algorithm. *Proteins* 64(3):559–574.

Lemmen C, Lengauer T (2000): Computational methods for the structural alignment of molecules. *J Comput Aided Mol Des* 14(3):215–232.

Lesk AM, Chothia C (1980): How different amino acid sequences determine similar protein structures: the structure and evolutionary dynamics of the globins. *J Mol Biol* 136(3):225–270.

Lesk AM, Chothia C (1982): Evolution of proteins formed by beta-sheets. II. The core of the immunoglobulin domains. *J Mol Biol* 160(2):325–342.

Lessel U, Schomburg D (1994): Similarities between protein 3-D structures. *Protein Eng* 7(10): 1175–1187.

Levitt M (2007): Growth of novel protein structural data. *Proc Natl Acad Sci USA* 104(9):3183–3188.

Li Z, Ye Y, Godzik A (2006): Flexible Structural Neighborhood—a database of protein structural similarities and alignments. *Nucleic Acids Res* 34 (Database issue):D277–D280.

Liu J, Montelione GT, Rost B (2007): Novel leverage of structural genomics. *Nat Biotechnol* 25(8): 849–851.

Lupas AN, Ponting CP, Russell RB (2001): On the evolution of protein folds: are similar motifs in different protein folds the result of convergence, insertion, or relics of an ancient peptide world? *J Struct Biol* 134(2–3):191–203.

Lupyan D, Leo-Macias A, Ortiz AR (2005): A new progressive-iterative algorithm for multiple structure alignment. *Bioinformatics* 21(15):3255–3263.

Madej T, Gibrat JF, Bryant SH (1995): Threading a database of protein cores. *Proteins* 23(3): 356–369.

Marti-Renom MA, Ilyin VA, Sali A (2001): DBAli: a database of protein structure alignments. *Bioinformatics* 17(8):746–747.

Marti-Renom MA, Pieper U, Madhusudhan MS, Rossi A, Eswar N, Davis FP, Al-Shahrour F, Dopazo J, Sali A (2007): DBAli tools: mining the protein structure space. *Nucleic Acids Res* 35(Web Server issue):W393–W397.

Martinez L, Andreani R, Martinez JM (2007): Convergent algorithms for protein structural alignment. *BMC Bioinformatics* 8(1):306.

Mayr G, Domingues FS, Lackner P (2007): Comparative analysis of protein structure alignments. *BMC Struct Biol* 7:50.

McMahon SA, Miller JL, Lawton JA, Kerkow DE, Hodes A, Marti-Renom MA, Doulatov S, Narayanan E, Sali A, Miller JF, Ghosh P (2005): The C-type lectin fold as an evolutionary solution for massive sequence variation. *Nat Struct Mol Biol* 12(10):886–892.

Murzin AG, Brenner SE, Hubbard T, Chothia C (1995): SCOP: a structural classification of proteins database for the investigation of sequences and structures. *J Mol Biol* 247:536–540.

Orengo CA, Flores TP, Taylor WR, Thornton JM (1993): Identification and classification of protein fold families. *Protein Eng* 6(5):485–500.

Orengo C (1994): Classification of protein folds. *Curr Biol* 4(429–440).

Orengo CA, Taylor WR (1996): SSAP: sequential structure alignment program for protein structure comparison. *Methods Enzymol* 266:617–635.

Orengo CA, Michie AD, Jones S, Jones DT, Swindells MB, Thornton JM (1997): CATH—a hierarchic classification of protein domain structures. *Structure* 5:1093–1108.

Ortiz AR, Strauss CE, Olmea O (2002): MAMMOTH (matching molecular models obtained from theory) an automated method for model comparison. *Protein Sci* 11(11):2606–2621.

Overington JP, Zhu ZY, Sali A, Johnson MS, Sowdhamini R, Louie GV, Blundell TL (1993): Molecular recognition in protein families: a database of aligned three-dimensional structures of related proteins. *Biochem Soc Trans* 21(Pt 3)(3):597–604.

Petsko GA (2007): An idea whose time has gone. *Genome Biol* 8(6):107.

Phillips DC (1970): The development of crystallographic enzymology. *Biochem Soc Symp* 30:11–28.

Pieper U, Eswar N, Davis FP, Braberg H, Madhusudhan MS, Rossi A, Marti-Renom M, Karchin R, Webb BM, Eramian D, Shen MY, Kelly L, Melo F, Sali A (2006): MODBASE: a database of annotated comparative protein structure models and associated resources. *Nucleic Acids Res* 34 (Database issue):D291–D295.

Reeves GA, Dallman TJ, Redfern OC, Akpor A, Orengo CA (2006): Structural diversity of domain superfamilies in the CATH database. *J Mol Biol* 360(3):725–741.

Rossmann MG, Argos P (1978): The taxonomy of binding sites in proteins. *Mol Cell Biochem* 21(3): 161–182.

Rossman MG (1981): Evolution of glycolytic enzymes. *Philos Trans R Soc Lond B Biol Sci* 293(1063): 191–203.

Rost B (1997): Protein structures sustain evolutionary drift. *Fold Des* 2:S19–S24.

Rost B (1999): Twilight zone of protein sequence alignments. *Protein Eng* 12(2):85–94.

Russell RB, Copley RR, Barton GJ (1996): Protein fold recognition by mapping predicted secondary structures. *J Mol Biol* 259(3):349–365.

Sali A, Blundell TL (1990): Definition of general topological equivalence in protein structures. A procedure involving comparison of properties and relationships through simulated annealing and dynamic programming. *J Mol Biol* 212(2):403–428.

Sali A, Blundell TL (1993): Comparative protein modelling by satisfaction of spatial restraints. *J Mol Biol* 234(3):779–815.

Sander C, Schneider R (1991): Database of homology-derived protein structures and the structural meaning of sequence alignment. *Proteins* 9:56–68.

Scheeff ED, Bourne PE (2005): Structural evolution of the protein kinase-like superfamily. *PLoS Comput Biol* 1(5):e49.

Scheeff ED, Bourne PE (2006): Application of protein structure alignments to iterated hidden Markov model protocols for structure prediction. *BMC Bioinformatics* 7:410.

Shapiro A, Botha JD, Pastore A, Lesk AM (1992): A method for multiple superposition of structures. *Acta Crystallogr A* 48(Pt 1):11–14.

Shatsky M, Nussinov R, Wolfson HJ (2004): A method for simultaneous alignment of multiple protein structures. *Proteins* 56(1):143–156.

Shindyalov IN, Bourne PE (1998): Protein structure alignment by incremental combinatorial extension (CE) of the optimal path. *Protein Eng* 11(9):739–747.

Shindyalov IN, Bourne PE (2000): An alternative view of protein fold space. *Proteins* 38(3): 247–260.

Siew N, Elofsson A, Rychlewski L, Fischer D (2000): MaxSub: an automated measure for the assessment of protein structure prediction quality. *Bioinformatics* 16(9):776–785.

Soding J, Lupas AN (2003): More than the sum of their parts: on the evolution of proteins from peptides. *Bioessays* 25(9):837–846.

Stebbings LA, Mizuguchi K (2004): HOMSTRAD: recent developments of the Homologous Protein Structure Alignment Database. *Nucleic Acids Res* 32(Database issue):D203–D207.

Szustakowski JD, Weng Z (2000): Protein structure alignment using a genetic algorithm. *Proteins* 38(4):428–440.

Tendulkar AV, Joshi AA, Sohoni MA, Wangikar PP (2004): Clustering of protein structural fragments reveals modular building block approach of nature. *J Mol Biol* 338(3):611–629.

Topf M, Sali A (2005): Combining electron microscopy and comparative protein structure modeling. *Curr Opin Struct Biol* 15(5):578–585.

Tsigelny I, Shindyalov IN, Bourne PE, Sudhof TC, Taylor P (2000): Common EF-hand motifs in cholinesterases and neuroligins suggest a role for Ca2 + binding in cell surface associations. *Protein Sci* 9(1):180–185.

Usha R, Murthy MR (1986): Protein structural homology: a metric approach. *Int J Pept Protein Res* 28(4):364–369.

Ye Y, Godzik A (2003): Flexible structure alignment by chaining aligned fragment pairs allowing twists. *Bioinformatics* 19(2):II246–II255.

Ye Y, Jaroszewski L, Li W, Godzik A (2003): A segment alignment approach to protein comparison. *Bioinformatics* 19(6):742–749.

Ye Y, Godzik A (2004): FATCAT: a web server for flexible structure comparison and structure similarity searching. *Nucleic Acids Res* 32(Web Server issue):W582–W585.

Ye Y, Godzik A (2005): Multiple flexible structure alignment using partial order graphs. *Bioinformatics* 21(10):2362–2369.

Zemla A (2003): LGA: A method for finding 3D similarities in protein structures. *Nucleic Acids Res* 31(13):3370–3374.

Zhang Y, Skolnick J (2005a): The protein structure prediction problem could be solved using the current PDB library. *Proc Natl Acad Sci USA* 102(4):1029–1034.

Zhang Y, Skolnick J (2005b): TM-align: a protein structure alignment algorithm based on the TM-score. *Nucleic Acids Res* 33(7):2302–2309.

17

PROTEIN STRUCTURE EVOLUTION AND THE SCOP DATABASE

Raghu P. R. Metpally and Boojala V. B. Reddy

INTRODUCTION

The structure of a protein can elucidate its function and its evolutionary history (see Chapters 18, 21 and 23). Extracting this information requires knowledge of the structure and its relationships with other proteins. These in turn require a general knowledge of the folds that proteins adopt and detailed information about the structure of many proteins. Nearly all proteins have structural similarities with other proteins, and in many cases, share a common evolutionary origin. The knowledge of these relationships makes important contributions to structural bioinformatics and other related areas of science. Further, these relationships will play an important role in the interpretation of sequences produced by genome projects. To facilitate understanding and access to the information available for known protein structures, Murzin, Brenner, Hubbard, and Chothia (1995) have constructed a structural classification of proteins (SCOP) database. The SCOP database is based on evolutionary relationships and on the principles that govern their three-dimensional structure. It provides for each entry links to coordinates, images of the structure, interactive viewers, sequence data, and literature references. The database is freely accessible on the World Wide Web (http://scop.mrc-lmb.cam.ac.uk/scop). To understand the rationale behind SCOP, we begin with a discussion of protein evolution from a sequence structure and functional perspective.

THE EVOLUTION OF PROTEINS

Proteins that have descended from the same ancestor retain memory of that ancestor through the sequence, structure, and function. Murzin (1998) discussed various aspects of the evolution of sequence, structure, fold, and function with specific examples. Strong

Structural Bioinformatics, Second Edition Edited by Jenny Gu and Philip E. Bourne
Copyright © 2009 John Wiley & Sons, Inc.

sequence similarity alone is considered to be sufficient evidence for common ancestry. Close structural and functional similarity taken together is also accepted as sufficient evidence for distant homology between proteins that lack significant sequence similarity. But neither structural nor functional similarity alone is considered to be strong evidence. Proteins of independent origin may well have similar structures due to physicochemical reasons and they may also evolve similar functions due to functional selection referred to as divergent evolved structures. Although in theory descendents from the same ancestor may have different functions and even different structures, they would be very difficult to detect.

THE EVOLUTION OF FOLD

It is generally accepted that, in distantly related proteins, structure is more conserved than sequence. Proteins that have diverged beyond significant sequence similarity still retain the architecture and topology of their ancestral fold. The reasons for this structural conservation are not completely understood. In principle, a protein chain can have more than one stable fold. Of the competing theories, one theory states that selective pressure originating from the physical demands on the protein chain results in protein evolution towards adopting a stable and fast folding conformation. Another theory states as protein evolution convergences upon a set of small, highly stable protein folds, evolutionary pressures were then dominated by functional constraints resulting in the present state of affairs. Evolution requires that existing proteins continue to function and such processes would be interrupted if the protein's fold changed significantly. Notwithstanding, a protein fold can change in a minor way during evolution by keeping all the secondary structure elements the same, packed and connected exactly the same way, but with different organization of connecting loops. The structural similarity between seemingly unrelated proteins are often explained by "convergence" to a stable fold as opposed to divergence from a common ancestor. Convergence does not only imply that the given proteins are of independent origin, but that they also had different original folds. With no evidence for their difference in original fold, these examples are referred to as "parallel" evolution.

THE EVOLUTION OF ENZYMATIC CATALYSIS

Proven cases of distantly related enzymes with very different functions are very rare and therefore are of great value for the understanding of the origin of enzymatic activity. The precision and complexity of the active-site structures of modern enzymes did not happen by chance but evolved from primitive catalytic features of the ancestral protein. Primitive enzymes were probably less efficient, but they were also likely to have a broader range of activities. If this were the case, the devolution of these activities would allow improvements in both efficiency and specificity through customization of the active-site architectures with additional specificity-determining groups. Within the active site, the original catalytic features are likely to remain conserved and can be revealed by structural comparison. When a protein evolves to a new function, the protein fold can change before becoming fully optimized after developing new functional constraints. A simple way of altering a protein fold without a large destabilization effect is to change its topology while maintaining its architecture. This can be done by the internal swapping of similar α-helices and β-strands or by reversing the direction of some of its secondary structures. There are many proteins with similar secondary structures and architectures but different topologies that could be related in such a way (e.g., an immunoglobulin domain and plastocyanin; an SH3 domain and GroES).

THE COMPARISON OF STRUCTURES

Most present methods for structure comparison measure the significance of structural similarity by a score derived from the number of residues that are common in the structural core and their root mean square deviation. These values mainly reflect the arrangement of regular secondary structures, whose packing and topology is determined by stereochemical rules rather than by evolutionary constraints. The common ancestry may manifest itself in more subtle features, such as the conservation of rare and unusual topological and packing details, and other structural irregularities that are less likely to occur independently. Within the common structural core, there may also be conserved turns, α-helical caps, β-bulges and other small structural elements that, in general, may occur in different orientations independent of the protein fold.

Early work on protein structures showed that there are striking regularities with the secondary structure assembly (Levitt and Chothia, 1976; Chothia, Levitt, and Richardson, 1977) and polypeptide topologies (Richardson, 1976; Sternberg and Thornton, 1976; Richardson, 1977). These regularities arise from the intrinsic physical and chemical properties of proteins (Chothia, 1984; Finkelstein and Ptitsyn, 1987) and provide the basis for the classification of protein folds (Levitt and Chothia, 1976; Richardson, 1981). Structure comparison (Holm and Sander, 1993; Shindyalov and Bourne, 1998) and structure classification (Orengo et al., 1993; Overington et al., 1993; Yee and Dill, 1993) provides further insight into the systematic relationship among protein structures. Resources are now available for recognition of the relationships between protein structures and several are discussed in this book (Chapters 16, 17 and 18). The SCOP database hierarchically organizes proteins according to their structures and evolutionary origin (Murzin et al., 1995; Lo Conte et al., 2000). It forms a resource that allows researchers to study the nature of protein folds, to help focus research investigations, and to rely on expert-defined relationships. SCOP is the most cited resource for classifying proteins (Table 17.1). In the context of structural bioinformatics it provides a reductionism that facilitates many of the studies outlined in this book.

SCOP HIERARCHY

The method used to classify proteins in SCOP is manually achieved by the visual inspection and comparison of structures as the human mind is quite adept at recognizing slightly dissimilar objects as belonging to a common architecture. The assistance of automatic tools is sometimes used only to make more complex tasks manageable and to provide generality. Thus, simplified representations of protein structures can be grouped by expert inspection (Murzin et al., 1995). SCOP differs from other protein classification methods like Class Architecture Topology Homologous (CATH) and DALI, in that the protein domains are classified manually compared to CATH which approaches protein domain classification with a combination of manual and automated procedures (Orengo et al., 1997). The Dali Domain Dictionary uses a fully automated approach of defining and classifying domains (Dietmann et al., 2001).

TABLE 17.1. A Partial List of References Where SCOP Classification was Used as the Basis for Analysis

Apic G, et al. (2001): *J Mol Biol* 310(2):311–325.
Aung Z, Tan KL (2006): *J Bioinform Comput Biol* 4(6):1197–1216.
Babbitt PC, Gerlt JA (2000): *Adv Protein Chem* 55:1–28.
Bertone P, Gerstein M (2001): *IEEE Eng Med Biol Mag* 20(4):33–40.
Bhaduri A, et al. (2004): *BMC Bioinform* 5(35).
Bukhman YV, Skolnick J (2001): *Bioinformatics* 17(5):468–478.
Caetano-Anolles G, Caetano-Anolles D (2005): *J Mol Evol* 60(4):484–498.
Chou KC, Maggiora GM (1998): *Protein Eng* 11(7): 523–538.
D'Alfonso G, et al. (2001): *J Struct Biol* 134(2–3):246–256.
Dengler U, et al. (2001): *Proteins* 42(3): 332–344.
Dokholyan NV, Shakhnovich EI (2001): *J Mol Biol* 312(1):289–307.
Dosztanyi Z, Torda AE (2001): *Bioinformatics* 17(8):686–699.
Douguet D, et al. (2006): *Bioinformatics* 22(21):2612–2618.
Dunbrack RL Jr (2006): *Curr Opin Struct Biol* 16(3):374–384.
Friedberg I, et al. (2004): *Curr Opin Struct Biol* 14(3):307–312.
George RA, et al. (2004): *Bioinformatics* 20(Suppl. 1):I130–I136.
Glasner ME, et al. (2006): *Curr Opin Chem Biol* 10(5):492–497.
Gowri VS, et al. (2003): *Nucleic Acids Res* 31(1):486–488.
Grigoriev IV, et al. (2001): *Protein Eng* 14(7):455–458.
Helling R, et al. (2001): *J Mol Graph Model* 19(1):157–167.
Holland TA, et al. (2006): *J Mol Biol* 361(3):562–590.
Hou Y, et al. (2003): *Bioinformatic* 19(17):2294–2301.
Innis CA, et al. (2004): *J Mol Biol* 337(4):1053–1068.
Jefferson ER, et al. (2007): *Nucleic Acids Res* 35(Database issue):D580–D589.
Kinoshita K, et al. (1999): *Protein Sci* 8(6):1210–1217.
Koonin EV, et al. (1998): *Curr Opin Struct Biol* 8(3):355–363.
Lupas AN, et al. (2001): *J Struct Biol* 134(2–3):191–203.
Melo F, Marti-Renom MA (2006): *Proteins*: 63(4):986–995.
Paoli M (2001): *Prog Biophys Mol Biol* 76(1–2):103–130.
Pasquier C, et al. (2001): *Proteins* 44(3):361–369.
Pawlowski K, Godzik A (2001): *J Mol Biol* 309(3):793–806.
Przytycka T, et al. (1999): *Nat Struct Biol* 6(7):672–682.
Reddy BV, Kaznessis YN (2005): *J Bioinform Comput Biol* 3(5):1137–1150.
Sandhya S, et al. (2005): *J Biomol Struct Dyn* 23(3):283–298.
Shi J, et al. (2001): *J Mol Biol* 310(1):243–257.
Stambuk N, Konjevoda P (2001): *Int J Quantum Chem* 84:13–22.
Stark A, et al. (2004): *Structure* 12(8):1405–1412.
Stebbings LA, Mizuguchi K (2004): *Nucleic Acids Res* 32(Database issue):D203–D207.
Taylor WR, et al. (2001): *Rep Prog Phys* 64(4):517–590.
Teichmann SA, et al. (2001a): *Curr Opin Struct Biol* 11(3):354–363.
Teichmann SA, et al. (2001b): *J Mol Biol* 311(4):693–708.
Thangudu RR, et al. (2007): *Proteins* 67(2):255–261.
Thornton JM (2001): *Science* 292(5524):2095–2097.
Todd AE, et al. (2001): *J Mol Biol* 307(4):1113–1143.
Torshin IY (2001): *Front Biosci* 6:A1–A12.
Veretnik S, et al. (2004): *J Mol Biol* 339(3):647–678.
Wilson D, et al. (2007): *Nucleic Acids Res* 35(Database issue):D308–D313.
Winter C, et al. (2006): *Nucleic Acids Res* 34(Database issue):D310–D314.
Xu Y, et al. (2000): *Bioinformatics* 16(12):1091–1104.
Zhang C, DeLisi C (2001): *Cell Mol Life Sci* 58(1):72–79.

SCOP classes: 1.71 release							
All-α	All-β	α/β	α+β	Multi-domain	Membrane and cell surface	Small proteins	Other proteins
226-Folds	149-Folds	134-Folds	286-Folds	48-Folds	49-Folds	79-Folds	189-Folds
							Coiled coil (7) Low resolution (24) Peptides (116) Designed proteins (42)

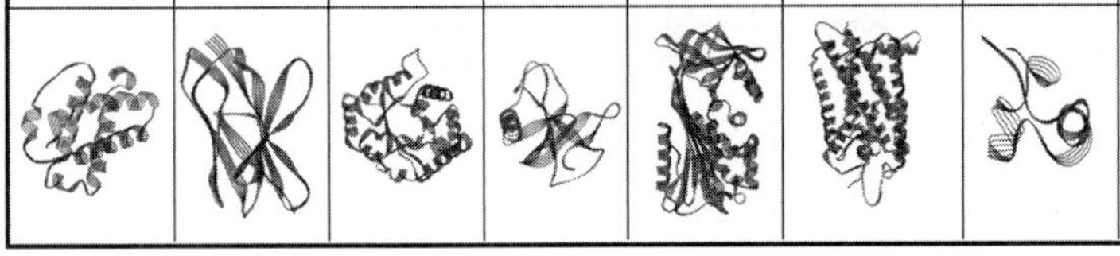

Figure 17.1. SCOP classification of protein folds (release 1.71, October 2006). The classification structure of SCOP is presented here with the number of representative folds in each class. Proteins are largely classified based on the organization of secondary structural elements.

The SCOP database is organized into a number of hierarchical levels, with the principle levels being family, superfamily, fold, and class. Within this hierarchy, the unit of categorization is the protein domain since domains are typically the units of protein evolution, structure, and function. Small and medium sized proteins usually have a single domain and are treated as such. The domains in larger proteins are classified individually. Thus different regions of a single protein may appear in multiple places in the SCOP hierarchy under different folds or, in the case of repeated domains, several times under the same fold.

In SCOP, *families* contain protein domains that share a clear common evolutionary origin, as evidenced by sequence identity or extremely similar structure and function. *Superfamilies* consist of families whose proteins share very common structure and function despite low sequence identities, and therefore there is reason to believe that the different families are evolutionary related. *Folds* consist of one or more superfamilies that share a common core structure (i.e., same secondary structure elements in the same arrangement with the same topological connections). Finally, depending on the type and organization of secondary structural elements, folds are then grouped into four major *classes*. In addition, there are several other classes for proteins that are very atypical and therefore difficult to classify (Figure 17.1).

The following explanation derived from the work of the authors of SCOP database (Brenner et al., 1996; Lo Conte et al., 2002; Andreeva et al., 2004) describes how the protein structures at each of the different levels have been classified.

CLASSES

Initially a protein structure is classified into domains. The basic idea of a domain is a region of the protein that has its own hydrophobic core and has relatively little interaction with the rest of the protein so that it is structurally independent. Identification of domains can be done with absolute certainty using evolutionary information (see Chapter 20). Typically, domains are collinear in sequence, which aids their identification, but occasionally one domain will have another inserted into it. Occasionally two homologous domains will intertwine by swapping some topologically equivalent parts of their chains.

Defining a domain structure and categorizing it in the appropriate class is a straightforward task, although there are several difficult cases. It should be readily apparent whether a domain consists exclusively of α-helices, β-sheet, or some mixture thereof. It is possible that an all-β protein can have small adornments of α- or 3_{10}-helix. Similarly, all-α structures may actually have several regions of 3_{10}-helix, and small β-sheets outside the α-helical core, in rare cases.

Domains with a mixture of helix and sheet structures are divided into two classes, α/β (alpha and beta), and α + β (alpha plus beta). The α/β domains consist principally of a single β-sheet, with α-helices joining the C-terminus of one strand to the N-terminus of the next. Commonly in α/β proteins, two subclasses are observed: in one subclass, the β-sheet is wrapped to form a barrel surrounded by α-helices; in the other, the central sheet is more planar and is flanked on either side by the helices. Domains that have the α and β units largely separated in sequence fall into the α + β class. The strands in these structures do not have the intervening helices and are typically joined by hairpins instead. This structural topology results in antiparallel sheets such as are found in all-β class folds. However, α + β structures may have either one or a small cluster of helices packing tightly and integrally against the sheet.

In addition to the four classes of globular protein structures, SCOP contains a few other classes, namely: multi-domain proteins, membrane and cell surface proteins, small proteins, coiled coil proteins, low resolution structures, peptides, and designed proteins (SCOP 1.71 release, October 2006). *Multi-domain proteins* have different domains placed in different classes. However, the different domains of these proteins have never been seen independently of each other. Multi-domain proteins are the result of duplication and combination of domains. They represent the majority of proteins in a proteome. The *small protein class* has structures in which there are very few helices or sheets and are stabilized by disulfide bridges or metal ligands rather than by hydrophobic cores. *Membrane proteins* frequently have unique structures because of their unusual environment and are therefore placed in a separate class. *Coiled coil proteins*, *small peptides and fragments*, and *designed proteins* are placed separately and are not a true *class*. *Coiled coil proteins* consist of two or more α-helices that wind around one another with a slight left-handed superhelical twist and are characterized by a heptad repeat pattern in which residues in the first and fourth position are hydrophobic, and residues in the fifth and seventh position are predominantly charged or polar. *Designed proteins* are experimental structures of proteins that are theoretically designed and are essentially non-natural sequences. *Low-resolution structures* are also grouped as a separate class.

FOLDS

Identification of the fold of the protein is the most difficult stage of classification. If the proteins have the same major secondary structures in the same arrangement with the same topological connections they are classified as onefold. A short description of its major structural features is used as the name of the fold. Different proteins with the same fold usually have peripheral elements of secondary structure and turn regions that differ in size and conformation. In the more divergent cases, these differing regions may form half or more of each structure. For proteins placed together in the same fold category, the structural similarity probably arises from the physics and chemistry of proteins favoring certain packing arrangements and chain topologies. There may be cases where a common evolutionary origin is obscured by the extent of the divergence in sequence, structure and

TABLE 17.2. SCOP Mirror Sites at Different Locations Around the World

Location	Site	URL
Australia	Walter and Eliza Hall Institute	http://scop.wehi.edu.au/scop/
	Australian National Genomic Information Service (ANGIS)	http://molmod.angis.org.au/scop/
China	Institute of Physical Chemistry, Peking University	http://mdl.ipc.pku.edu.cn/scop/
India	Bioinformatics Centre, Madurai Kamaraj University (MKU)	http://gene.tn.nic.in/scop/
	Bioinformatics Centre, Pune University	http://202.41.70.60/scop/
	Centre for DNA Fingerprinting and Diagnostics (CDFD)	http://www.cdfd.org.in:5555/scop/
	Indian Institute of Science (IISc)	http://scop.physics.iisc.ernet.in/scop/
Israel	Weizmann Institute	http://pdb.weizmann.ac.il/SCOP/
Italy	Center for Biomedical Engineering, Politecnico of Turin	http://loki.polito.it/scop/
Japan	Biomolecular Engineering Research Institute (BERI)	http://www.beri.or.jp/scop/
Korea	Korea Advanced Institute of Science and Technology	http://scop.kaist.ac.kr/scop
	Korea Institute of Science and Technology Information	http://scop.ccbb.re.kr/
Russia	Institute of Protein Research RAS	http://scop.protres.ru/
Singapore	National University of Singapore	http://scop.bic.nus.edu.sg/
Taiwan	National Tsing Hua University	http://scop.life.nthu.edu.tw/
UK	SCOP Home Server in Cambridge	http://scop.mrc-lmb.cam.ac.uk/scop/
USA	University of Arkansas for Medical Sciences (UAMS)	http://bioinformatics.uams.edu/scop/
	University of California, Berkeley	http://scop.berkeley.edu/

function. In such cases, it is possible that the discovery of new structures, with folds between those of the previously known structures, will make clear their common evolutionary relationship.

As of SCOP release 1.71 (October 2006) there are 226 classified folds in all α proteins, 149-folds in all β proteins, 134-folds in α/β, 286-folds in $\alpha + \beta$, and 48-folds in multi-domain proteins classes (Figure 17.1, Table 17.2). The best way to characterize fold is to look first at the major architectural features and then identify the more subtle characteristics.

SUPERFAMILIES

Protein structures classified in the same superfamily are probably related evolutionarily and therefore must share a common fold that often perform similar functions. If the functional relationship is sufficiently strong, for example, the conserved interaction with substrate or cofactor molecules, the shared fold can be relatively small, provided it includes the active sites. Proteins from the same superfamily may have low sequence identities but their structural, and in many cases, functional features suggest a common evolutionary origin.

FAMILIES

Proteins are clustered together into families based on one of the following criteria used to establish a common evolutionary origin. First, all proteins have a sequence identity of greater than 30%. Second, in the absence of high sequence identity, proteins are connected through similar functions and structures. A small number of SCOP families show some evolutionary relationship that is above the standard that establishes the family categorization but is below the superfamily level. Proteins that have similar domain organizations and share a common fold in the catalytic domain (e.g., the dihydrodipi-colinate reductase, glyceraldehyde-3-phosphate and glucose-6-phosphate dehydro-genases) are suggested to be more closely related than those sharing a common fold in their coenzyme domain only.

The major limitation of sequence comparison is that it fails to identify many of the structural relationships in SCOP because the sequence relationship has become too weak (for evolutionary related proteins) or never existed (for evolutionarily unrelated proteins with similar folds). Structure–structure comparison programs use various methods to recognize similar arrangements of atomic coordinates and thus identify domains of similar structure. Although these methods lack complete accuracy, they can be used to suggest a shared fold between proteins of interest and others in SCOP. Manual inspection must then be used to verify the choice of fold and to select the appropriate superfamily. The selection of superfamily is the most challenging step of protein classification, for it ascribes a biological interpretation to chemical and physical data. Therefore, the assignment of all proteins of known structure to evolutionary related superfamilies is perhaps the single most powerful and important features of the SCOP database.

ORGANIZATION AND CAPABILITIES OF THE SCOP RESOURCE

The SCOP database was originally created as a tool for understanding protein evolution through sequence–structure relationships and determining if new sequences and new structures are related to previously known protein structures. On a more general level, the highest levels of classification provide an overview of the diversity of protein structures. The specific lower levels are helpful for comparing individual structures with their evolutionary and structurally related counterparts.

Information in the SCOP database is interactively accessible as a set of HTML pages and through a search engine at http://scop.mrc-lmb.cam.ac.uk/scop. Previous SCOP releases, starting with SCOP 1.48, are also available online at the home SCOP site at MRC (http://scop.mrc-lmb.cam.ac.uk/scop-x.xx, where "x.xx" is the release number).

For rapid and effective access to SCOP, a number of mirrors have been established (Table 17.3). The facilities at various sites may differ with some sites providing sequence similarities and some other sites providing sequence and structure based phylogenic relationships (Balaji and Srinivasan, 2001; Gowri et al., 2003; Balaji and Srinivasan, 2007). SCOP can be used for detailed searching of particular families and browsing of the whole

TABLE 17.3. SCOP 1.71 Release has 27,599 PDB Entries, 75,930 Domains and 1 Literature Reference (Excluding Nucleic Acids and Theoretical Models)

Class	Folds	Superfamilies	Families
All alpha proteins	226	392	645
All beta proteins	149	300	594
α/β proteins	134	221	661
$\alpha+\beta$ proteins	286	424	753
Multi-domain proteins	48	48	64
Membrane and cell surface proteins	49	90	101
Small proteins	79	114	186
Total	971	1589	3004

database with a variety of techniques for navigation. Easy access to data and images make SCOP a powerful general-purpose interface, providing a level of classification not present in the Protein Data Bank (PDB).

BROWSING THROUGH THE SCOP HIERARCHY

SCOP is organized as a tree structure. Entering at the top of the hierarchy the user can navigate through the levels of class, fold, superfamily, family, and species to the leaves of the tree, which are the structural domains of individual PDB entries. The sequence similarity search facility allows any sequence of interest to be searched against databases of protein sequences classified in SCOP using the algorithms BLAST (Altschul et al., 1997) and FASTA (Pearson and Lipman, 1988), and structural similarity search using SSM (Krissinel and Henrick, 2004) and identifying noncanonical interactions (NCI) (Babu, 2003). SCOP can be entered from the list of PDB chains found to be similar and the similarities are then displayed visually. The keyword search facility returns a list of SCOP pages containing the word entered or combinations of words separated by Boolean operators. Besides that, the standard keyword search now accepts *sunids* (possibly right-truncated) *sccss* and EC numbers, as well as words that appear in any of the SCOP pages, PDB identifiers and SCOP *sids*. It also accepts ASTRAL identifiers, including those for the new genetic domain sequences. Pages are provided that order folds, superfamilies and families by date of entry into PDB.

In addition to the structural and evolutionary relationships contained within SCOP, each entry has links to images of the structure, interactive molecular viewers, the atomic coordinates, data on functional conformational changes, sequence data, homologues and MEDLINE abstracts.

LINKING TO OTHER STRUCTURE AND SEQUENCE DATABASES

A new set of links to external resources has been added at the level of SCOP domains. For each domain in the first seven classes, there are links to supplementary information related to that domain in Pfam (Finn et al., 2006), SUPERFAMILY (Gough et al., 2001), PartsList (Qian et al., 2001), ASTRAL (Chandonia et al., 2004), and in case there is one or more sequences predicted to have that fold, PRESAGE (Brenner, Barken, and Levitt, 1999). Links to Pfam provide alignments to homologues from sequence databases for most of the SCOP domains. SUPERFAMILY is a collection of Hidden Markov Models (HMMs) for superfamilies in SCOP, and of HMM-based genome assignments to SCOP superfamilies. PartsList adds genomic, functional, and structural information to most of the SCOP entries. ASTRAL compendium provides databases and tools useful for analyzing protein structures and their sequences. It provides genetic domain sequence and coordinates. PRESAGE is a collaborative resource for structural genomics (SG) with a collection of proteins' annotations reflecting current experimental status, structural assignment models, and predicted folds. Linking to SCOP from external sources is also set up to be more straightforward.

SCOP REFINEMENTS TO ACCOMMODATE STRUCTURAL GENOMICS

The explosion in genome sequence data and technological breakthroughs in protein structure determination inspired the launch of structural genomics initiatives (see also Chapter 40), with the goal of high-throughput structural characterization of all possible protein sequence families, with the long-term hope of significantly impacting the life sciences, biotechnology and drug discovery fields (Brenner, 2001; Todd et al., 2005). In the next 10 years, the daunting perspective of 10,000 new proteins expected to come out of the various SG projects (Brenner, 2001). To accommodate the growing data, SCOP has made new features and enhancements to make its growth, usage and integration in the larger context of structured biological information available on the web starting with release 1.55.

NEW FEATURES IN SCOP

In regards to the SCOP infrastructure, although SCOP is essentially a hierarchy, a superposed mechanism for crosslinking between nodes of the tree makes it a more general form of graph. This allows for more complex representation of biological relationships than the parentchild relationships in a tree. The original design and implementation of SCOP is based on a *description* of the underlying data structure rather than the data structure itself, it would be easy to introduce new classification levels everywhere, such as a *suprafamily* between *family* and *superfamily* for example. Once the description is modified accordingly, the rest of the database will fall into place automatically. The possibility of extensions for SCOP include a new set of identifiers, unique integers (*sunid*) associated to each node of the hierarchy, and a new set of concise classification strings (*sccs*) that were introduced (from release 1.55) and will be kept stable across SCOP releases. A *sunid* is simply a number that uniquely identifies each entry in the SCOP hierarchy, from root to leaves, including entries corresponding to the protein level for which there was no explicit reference before. A *sccs* is a compact representation of a SCOP domain classification, including only the most relevant levels; for class (alphabetical), fold, superfamily, and family (all numerical) to which each domain belongs to. For example, the *sccs* for the ribosome anti-association factor domain (PDB entry 1g61, chain A) is d.126.1.1, where "d" represents the class, "126" the fold, "1" the superfamily, and the last "1" the family. Also, the associated *sunids* are 53,931 for class; 55,908 for fold; 55,909 for superfamily; 55,910 for family; 55,911 for protein; 55,912 for species; and 41,126 for domain. The new identifiers unambiguously link to a SCOP entry and refer to related research work and in the literature. The old SCOP domain identifier are still retained and are valid (Lo Conte et al., 2002).

SCOP introduced three easy-to-parse files to obtain all the information retained in SCOP, with the exclusion of comments. Together, they replace and extend the now obsolete old parseables files (dir.dom.scop.txt and dir.lin.scop.txt). The first file, dir.hie.scop.txt, has no precursor in releases before 1.55. It represents the SCOP hierarchy in terms of *sunid*. Each entry corresponds to a node in the tree and has two additional fields: the *sunid* of the parent of that node (i.e., the node one step up in the tree), and the list of *sunids* for the children of that node (i.e., the nodes one step down in the tree). A second file, dir.cla.scop.txt, contains a description of all domains, their definition and their classification, both in terms of *sunid* and *sccs*. The third file, dir.des.scop.txt, contains a description of each node in the hierarchy, including English names for proteins, families, superfamilies, folds and classes.

INTEGRATION WITH OTHER DATABASES

A new project, started from release 1.63, aims to rationalize and integrate the SCOP information with data regarding protein families housed by prominent sequence and structural databases, including Pfam, InterPro, CATH and MSD. A milestone in this ambitious goal is the provision of stricter and more precise definitions behind the different classification schemes used in these different databases. As part of the database integration, technical developments were taken up: implementation of Simple Object Access Protocol (SOAP) technology to link the databases dynamically and modernize interface capabilities of SCOP.

RECLASSIFICATION

Not only do the structure and function of proteins evolve, so do their classification. The need for reclassification becomes apparent when accumulating evidence regarding their evolutionary relationships become available from experimental data, refinements, data classification reorganization, and error corrections. This dynamic nature of SCOP classification is one of its main features and needs to be accounted for in applications that use the SCOP database. If there is new evidence about protein relationships, then this may result in a redefinition of domain boundaries and/or rearrangements of nodes in the SCOP hierarchy. A typical example is when a part of a large novel protein was initially classified as a single multi-domain entry is subsequently observed as a stand-alone protein or a combination of different domain types and is therefore reclassified as a separate domain. Frequently, reclassification also occurs for two separately classified proteins discovered to be evolutionarily related through a recently determined intermediate. The appearance of such proteins in the structural databases often help to identify more distant relationships between protein domains and lead to a classification rearrangement that unifies distinct protein superfamilies. Another factor that also influences reclassification is its integration with other databases. Starting with release 1.63, a refinement of the SCOP classification was initiated to introduce a number of changes mostly at the levels below superfamily to facilitate integration. For example, some of the reclassified superfamilies include membrane all-α proteins, viral capsids and coat proteins, antibody domains, E-set domains, and protein kinases (Andreeva et al., 2004).

SCOP USAGE

SCOP has broad utility with a wide range of users. Experimental structural biologists may wish to explore the region of "structure space" near their protein of current research. Molecular biologists may find the classification helpful because the categorization assists in locating proteins of interest and the links make exploration easy. As such, two important observations are made from this protein structure classification. First, strikingly skewed distributions occur at all levels relative to what exists in Nature, based on analysis of complete genomes (Brenner, Chothia, and Hubbard, 1997). This probably reflects the experimentalists' bias toward particular proteins and protein families, as well as to the bias of nature toward certain protein superfamilies and folds. Second, a retrospective analysis on the growth of structural data gives an estimate of the total numbers of protein folds and superfamilies that exist in Nature and has shown them to be very limited indeed. It seems that we may have already seen the majority of folds and have determined at least one structure for some half of all superfamilies (Chothia, 1992; Brenner, Chothia, and Hubbard,1997).

SCOP FROM A USER'S PERSPECTIVE

We summarize here some areas of protein structural bioinformatics where the SCOP database has been used extensively: (i) SCOP classified groups of proteins were used as a reference set of data to develop several automatic classification methods used in analyzing families, superfamilies, and folds. These classifications were then extensively used for integrative structural data mining to develop predictive methods and structure comparison tools (Chou and Maggiora, 1998; Przytycka et al., 1999; Lackner et al., 2000; Bertone and Gerstein, 2001; Bukhman and Skolnick, 2001; Pasquier et al., 2001; Stambuk and Konjevoda, 2001; Torshin, 2001; Aung and Tan, 2006); (ii) SCOP classified proteins were extensively used in understanding evolution of protein enzymatic functions (Konin et al., 1998; Murzin, 1998; Powlowski and Godzik, 2001; Todd et al., 2001; George et al., 2004; Glasner, et al., 2006), evolutionary change of protein folds (Caetano and Caetano, 2005; Panchenko et al., 2005; Grishin, 2001; Lupas et al., 2001; Zhang and DeLisi, 2001), and hierarchical structural evolution (Dokholyan and Shakhnovich, 2001; Paoli, 2001); (iii) SCOP classification of proteins at superfamily and fold levels were used to study distantly related proteins with the same fold (Grigoriev et al., 2001; Teichmann et al., 2001; Thornton, 2001; Hou et al., 2003; Sandhya et al., 2005; Melo and Marti, 2006); (iv) SCOP is used to study sequence and structure variability and their dependence in homologous proteins (D'Alfonso et al., 2001; Gowri et al., 2003; Panchenko et al., 2005); (v) SCOP families are used to derive amino acid similarity matrices and substitution tables useful for sequence comparison and fold recognition studies (Dosztanyi and Torda, 2001; Shi et al., 2001; Dunbrack, 2006); (vi) SCOP is helpful in studying the structural anatomy of folds and domains, to extract structural principles for use in protein design experiments (Dangler, 2001; Helling et al., 2001; Taylor et al., 2001; Teichmann et al., 2001); (vii) SCOP domains have been used to study domain combinations and their decomposition in multi-domain proteins (Chou and Maggiora, 1998; Kinoshita et al., 1999; Xu et al., 2000; Apic et al., 2001; Veretnik et al., 2004; Holland et al., 2006); (viii) structural genome projects have been using SCOP extensively in identifying new targets and finding functional sites in structural proteins (Friedberg et al., 2004; Innis et al., 2004; Stark et al., 2004; Wilson et al., 2007); (ix) SCOP database used to study protein–protein interactions (Reddy and Kaznessis, 2005;

Douguet et al., 2006; Winter et al., 2006; Jefferson et al., 2007); (x) SCOP database have been used for developing value added and more specialized databases (Bhaduri, et al., 2004; Stebbings and Mizuguchi, 2004; Thangudu et al., 2007). From this brief synopsis it should be apparent that we owe the SCOP authors a debt of gratitude for providing a resource that has had great impact on the field of structural bioinformatics.

REFERENCES

Altschul SF, Madden TL, Schaffer AA, Zhang J, Zhang Z, Miller W, Lipman DJ (1997): Gapped BLAST and PSI-BLAST: a new generation of protein database search programs. *Nucleic Acids Res* 25(17):3389–3402.

Andreeva A, Howorth D, Brenner SE, Hubbard TJ, Chothia C, Murzin AG (2004): SCOP database in 2004: refinements integrate structure and sequence family data. *Nucleic Acids Res* 32(Database issue):D226–D229.

Babu MM (2003): NCI: a server to identify non-canonical interactions in protein structures. *Nucleic Acids Res* 31(13):3345–3348.

Balaji S, Srinivasan N (2001): Use of a database of structural alignments and phylogenetic trees in investigating the relationship between sequence and structural variability among homologous proteins. *Protein Eng* 14(4):219–226.

Balaji S, Srinivasan N (2007): Comparison of sequence-based and structure-based phylogenetic trees of homologous proteins: inferences on protein evolution. *J Biosci* 32(1):83–96.

Brenner SE, Chothia C, Hubbard TJ, Murzin AG (1996): Understanding protein structure: using SCOP for fold interpretation. *Methods Enzymol* 266:635–643.

Brenner SE, Chothia C, Hubbard TJ (1997): Population statistics of protein structures: lessons from structural classifications. *Curr Opin Struct Biol* 7(3):369–376.

Brenner SE, Barken D, Levitt M (1999): The PRESAGE database for structural genomics. *Nucleic Acids Res* 27(1):251–253.

Brenner SE (2001): A tour of structural genomics. *Nat Rev Genet* 2(10):801–809.

Chandonia JM, Hon G, Walker NS, Lo Conte L, Koehl P, Levitt M, Brenner SE (2004): The ASTRAL Compendium in 2004. *Nucleic Acids Res* 32(Database issue):D189–D192.

Chothia C, Levitt M, Richardson D (1977): Structure of proteins: packing of alpha-helices and pleated sheets. *Proc Natl Acad Sci U S A* 74(10):4130–4134.

Chothia C (1984): Principles that determine the structure of proteins. *Annu Rev Biochem* 53:537–572.

Chothia C (1992): Proteins. One thousand families for the molecular biologist. *Nature* 357(6379):543–544.

Dietmann S, Park J, Notredame C, Heger A, Lappe M, Holm L (2001): A fully automatic evolutionary classification of protein folds: Dali Domain Dictionary version 3. *Nucleic Acids Res* 29(1):55–57.

Finkelstein AV, Ptitsyn OB (1987): Why do globular proteins fit the limited set of folding patterns? *Prog Biophys Mol Biol* 50(3):171–190.

Finn RD, Mistry J, Schuster-Bockler B, Griffiths-Jones S, Hollich V, Lassmann T, Moxon S, Marshall M, Khanna A, Durbin R, Eddy SR, Sonnhammer EL, Bateman A (2006): Pfam: clans, web tools and services. *Nucleic Acids Res* 34(Database issue):D247–D251.

Gough J, Karplus K, Hughey R, Chothia C (2001): Assignment of homology to genome sequences using a library of hidden Markov models that represent all proteins of known structure. *J Mol Biol* 313(4):903–919.

Gowri VS, Pandit SB, Karthik PS, Srinivasan N, Balaji S (2003): Integration of related sequences with protein three-dimensional structural families in an updated version of PALI database. *Nucleic Acids Res* 31(1):486–488.

Grishin NV (2001): Fold change in evolution of protein structures. *Journal of structural biology* 134: 167–185.

Holm L, Sander C (1993): Protein structure comparison by alignment of distance matrices. *J Mol Biol* 233(1):123–138.

Krissinel E, Henrick K (2004): Secondary-structure matching (SSM), a new tool for fast protein structure alignment in three dimensions. *Acta Crystallogr D Biol Crystallogr* 60(Pt 12 Pt 1): 2256–2268.

Lackner P, Koppensteiner WA, Sippl MJ, Domingues FS (2000): ProSup: a refined tool for protein structure alignment. *Protein engineering* 13:745–752.

Levitt M, Chothia C (1976): Structural patterns in globular proteins. *Nature* 261(5561):552–558.

Lo Conte L, Ailey B, Hubbard TJ, Brenner SE, Murzin AG, Chothia C (2000): SCOP: a structural classification of proteins database. *Nucleic Acids Res* 28(1):257–259.

Lo Conte L, Brenner SE, Hubbard TJ, Chothia C, Murzin AG (2002): SCOP database in 2002: refinements accommodate structural genomics. *Nucleic Acids Res* 30(1):264–267.

Murzin AG, Brenner SE, Hubbard T, Chothia C (1995): SCOP: a structural classification of proteins database for the investigation of sequences and structures. *J Mol Biol* 247(4):536–540.

Murzin AG (1998): How far divergent evolution goes in proteins. *Current opinion in structural biology* 8:380–387.

Orengo CA, Flores TP, Taylor WR, Thornton JM (1993): Identification and classification of protein fold families. *Protein Eng* 6(5):485–500.

Orengo CA, Michie AD, Jones S, Jones DT, Swindells MB, Thornton JM (1997): CATH—a hierarchic classification of protein domain structures. *Structure* 5(8):1093–1108.

Overington JP, Zhu ZY, Sali A, Johnson MS, Sowdhamini R, Louie GV, Blundell TL (1993): Molecular recognition in protein families: a database of aligned three-dimensional structures of related proteins. *Biochem Soc Trans* 21(Pt 3, no. 3):597–604.

Panchenko AR, Wolf YI, Panchenko LA, Madej T (2005): Evolutionary plasticity of protein families: coupling between sequence and structure variation. *Proteins* 61:535–544.

Pearson WR, Lipman DJ (1988): Improved tools for biological sequence comparison. *Proc Natl Acad Sci USA* 85(8):2444–2448.

Qian J, Stenger B, Wilson CA, Lin J, Jansen R, Teichmann SA, Park J, Krebs WG, Yu H, Alexandrov V, Echols N, Gerstein M (2001): PartsList: a web-based system for dynamically ranking protein folds based on disparate attributes, including whole-genome expression and interaction information. *Nucleic Acids Res* 29(8):1750–1764.

Richardson JS (1976): Handedness of crossover connections in beta sheets. *Proc Natl Acad Sci U S A* 73(8):2619–2623.

Richardson JS (1977): beta-Sheet topology and the relatedness of proteins. *Nature* 268 (5620):495–500.

Richardson JS (1981): The anatomy and taxonomy of protein structure. *Adv Protein Chem* 34:167–339.

Shindyalov IN, Bourne PE (1998): Protein structure alignment by incremental combinatorial extension (CE) of the optimal path. *Protein Eng* 11(9):739–747.

Sternberg MJ, Thornton JM (1976): On the conformation of proteins: the handedness of the beta-strand-alpha-helix-beta-strand unit. *J Mol Biol* 105(3):367–382.

Todd AE, Marsden RL, Thornton JM, Orengo CA (2005): Progress of structural genomics initiatives: an analysis of solved target structures. *J Mol Biol* 348(5):1235–1260.

Yee DP, Dill KA (1993): Families and the structural relatedness among globular proteins. *Protein Sci* 2(6):884–899.

18

THE CATH DOMAIN STRUCTURE DATABASE

Frances M. G. Pearl, Alison Cuff, and Christine A. Orengo

INTRODUCTION

Protein sequences change during evolution due to both mutations in their residues and the insertion and deletion of residues. These changes give rise to families of related proteins. The earliest protein family resources were first established in the 1970s by the pioneering work of Dayhoff and many other sequence databases have been established since then. These resources are derived solely from sequence data and relationships are often detected using alignment methods based on powerful dynamic programming algorithms adapted from the realm of computer science. Such methods very efficiently handle residue insertions and deletions occurring between distant evolutionary relatives.

Structural data have always been sparser than the sequence data due to the technical challenges of structure determination. There is currently over two orders of magnitude discrepancy between the sequence and structure resources. Thus, while the Protein Data Bank (PDB) contains about 42,500 structural entries, the sequence databank at the NCBI (GenBank) contains over 60 million entries.

Although the first crystal structures were solved in the early 1970s, it was not until the mid-1990s that structural classifications began to emerge, primarily with SCOP (Murzin et al., 1995; Andreeva et al., 2004), DALI (Holm and Sander, 1996), and CATH (Orengo et al., 1997; Greene et al., 2007) databases and data resources. Since then, several other structure classifications and nearest-neighbor resources have arisen, including 3Dee (Dengler, Siddani and Barton, 2001), DaliDD(Holm and Sander, 1998; Dietmann and Holm, 2001) and VAST (Madej et al., 1995). These resources use a variety of different algorithms for comparing 3D structures (see Chapter 16). They also differ in methods for measuring similarity between the structures and clustering them into fold groups or protein superfamilies and families. However, comparisons between three of the largest

Structural Bioinformatics, Second Edition   Edited by Jenny Gu and Philip E. Bourne
Copyright © 2009 John Wiley & Sons, Inc.

classifications SCOP, DALI, and CATH (Hadley and Jones, 1999; Day et al., 2003) reveal a reasonable degree of correspon-dence between protein families generated using different protocols.

Since structure is much more highly conserved than sequence during evolution, the discovery of structural alignment algorithms and the development of structural classifications have made a significant contribution to understanding evolutionary mechanisms as they have enabled much more distant evolutionary relatives to be identified. Furthermore, knowledge of a protein structure can provide important clues to the functional mechanism and biological role of the protein, for example, protein–substrate and protein–protein interactions (see Chapters 25 and 26). Because a large proportion of the structural core of the protein (often more than 50%) is conserved even in very distant relatives, structure alignments are much more accurate than sequence alignments and this improves the identification of conserved structural features or sequence motifs that are often associated with protein function.

The largest structure classifications (SCOP, CATH) currently contain 1500–2100 protein superfamilies. These superfamilies, however, only map to approximately half of the nonredundant sequences in the GenBank sequence database (∼46% on the basis of equivalent residues). The ongoing structure genomics initiatives, described in Chapter 40, will significantly increase the number of novel structures determined over the next 10 years to increase this coverage. Current estimates predict that there could be up to 100,000 new structures before the end of the decade. Because of the manner in which proteins are being selected for structure determination, these new structures will predominantly be from currently underrepresented or distantly related to known protein families in the structure classifications. Therefore, we may soon have structural representatives for most of the major protein families and those of particular medical and biological interest, although some classes of structures such as transmembrane proteins may remain difficult to determine. It is also likely that methods for detecting distant relatives will improve in parallel as a consequence of the growth in the sequence and structure databases. Links between the sequence and structure databases are increasing. The InterPro initiative has integrated several sequence databases (Pfam, PRINTS, PROSITE, SWISS-PROT). This has been extended so that structural assignments from SCOP and CATH and the European Macro-molecular Structure Database (EMSD) are now included.

Structural classifications will therefore play an increasingly important role as repre-sentatives from more protein families are structurally determined and the mapping between structural families and genomic sequences improves. These resources will provide key data for understanding function at the molecular level. In this chapter we describe the CATH structural classification, its development, and the methods used to update and search the resource. Analysis of the classification has revealed that some protein families are very highly populated, a finding that has important implications for understanding evolutionary mechanisms, and is discussed later in the chapter. Information regarding the mapping of structural domains in CATH superfamilies onto genome sequences and a summary of some insights into protein evolution that can be gleaned from analyzing these data are presented in Chapters 22 and 23.

HISTORICAL DEVELOPMENT

The CATH domain structure database was established in 1993 when less than 3000 protein structures had been determined. Nearly a decade later, the database has expanded considerably and contains ~30,000 protein structure entries from the PDB comprising over 93,000 structural domains. CATH's related resource Gene3D, contains approximately 1 million domains extracted from GenBank and UniProt entries that are assigned to one of the 2091 CATH homologous superfamilies using profile-based approaches. Since the domain is considered to be an important evolutionary unit and also because structural prediction and homology modeling methods are often more successful on a domain basis, CATH was established as a domain-based database.

Most publicly available structure classifications are derived using sequence- and/or structure-based protocols. These range from the completely automated approaches of DALI and the DALI Domain Database to the largely manual approach used in compiling the SCOP database. In the CATH database, semiautomated protocols are used for clustering structures both phonetically on the basis of structural similarity and phylogenetically on the basis of apparent evolutionary relatedness. Any ambiguities in the assignments from automated protocols are validated manually and major bottlenecks in the classification correspond to the detection of domain boundaries and the verification of homologous relationships.

CATH is a hierarchical classification comprising four major levels (see Figure 18.1). In fact, CATH is an acronym for these levels: Class, Architecture, Topology, and Homology. At the top, the protein (C)lass is determined by the secondary structure composition and packing using an automated approach. (A)rchitecture describes the orientation of the secondary structures in 3D space, regardless of their connectivity. For example, a large number of protein structures adopt alpha–beta barrel architectures, in which a central barrel of beta strands is enclosed within an outer barrel comprising a layer of alpha helices (see Figure 18.2). At the next level in the hierarchy, (T)opology, both secondary structure orientation and connectivity between the secondary structures are taken into account in describing the fold of the protein. For the example shown in Figure 18.1, the three-layer alpha–beta sandwich architecture contains more than 100 different folds or topologies in which the secondary structures adopt a similar shape in 3D but the connectivities between them can differ considerably as shown by the schematic representations in the illustration. At the fourth and perhaps most biologically important level in the classification, (H)omologous superfamily, proteins are grouped according to whether there is sufficient evidence (structural, sequence, and/or functional similarity) to support an evolutionary relationship. Within each homologous superfamily, proteins are clustered into sequence families at different levels of sequence identity (35, 60, 95, and 100%). More recently, protocols have been developed for identifying functional families within each superfamily.

There are currently three major classes within CATH, corresponding to mainly alpha, mainly beta, and alpha–beta domains. Other categories distinguished at the class level are multidomain proteins; domains comprising few secondary structures and three groups corresponding to proteins at different stages in the classification and pending assignment to a particular fold group or superfamily. In release 3.01, January 2007, CATH contained 40 architectures; 1084 fold groups, and 2091 homologous superfamilies. Further statistics on the database and discussion of the population of different levels are given in the following section.

CURRENT METHODOLOGIES FOR IDENTIFYING STRUCTURAL SIMILARITIES AND EVOLUTIONARY RELATIONSHIPS IN CATH

Phylogenetic and phonetic relationships in the CATH database were initially identified using the powerful structure comparison algorithm, SSAP, devised by Taylor and Orengo (1989)

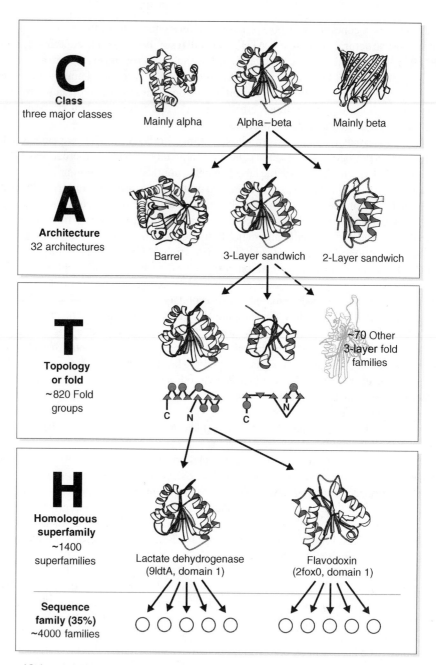

Figure 18.1. Schematic representation of the class, architecture, topology/fold, and homologous superfamily levels in the CATH database.

in 1989, which incorporated a modified dynamic programming algorithm performing at two levels and thus enabling comparison of 3D information. More recently, fast graph-based methods for structure comparisons have been implemented to enable the database to keep pace with the structure genomics initiatives. In fact, sequence-based methods are first used to detect close relatives, as these are much faster than structure comparison and pair-wise methods are reliable for relatives with 35% or more sequence identity. Sensitive profile-based sequence methods are used to detect more distant homologues that can then be verified by structural similarity.

The strategy used in classifying new structures into the database can thus be broken down into the following major steps: (1) Close relatives are identified first using pair-wise sequence methods. Structure comparison is used to extract accurate domain boundaries. (2) Domain boundaries for all other entries are determined by running sequence profiles, structure comparison protocols, and *ab intio* domain boundary prediction algorithms. Domain boundaries are then manually validated. (3) Sequence profiles and structural comparison methods are used to identify distant evolutionary relatives or fold matches, and structures are assigned to a superfamily or fold group. (4) Finally, any new unclassified structures are manually assigned to architectures within CATH or assigned new architectures (see Figure 18.3). Only Stage 1 is fully automated, the other stages include algorithms and manual validation protocols used at different stages of the classification. These methods are described in the following section.

CLASSIFYING CLOSE HOMOLOGUES (CHOPCLOSE)

To maintain the integrity of CATH domain boundaries within the classification, only very close homologues are completely classified automatically. Several studies have shown that when two proteins share more than 30% identity in their sequences, they have similar

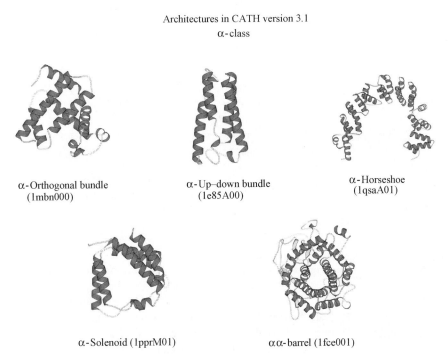

Architectures in CATH version 3.1
α-class

α-Orthogonal bundle
(1mbn000)

α-Up–down bundle
(1e85A00)

α-Horseshoe
(1qsaA01)

α-Solenoid (1pprM01)

αα-barrel (1fce001)

Figure 18.2. Molscript representations of the major architectures in the CATH hierarchy.

Architectures in CATH version 3.1
β-class

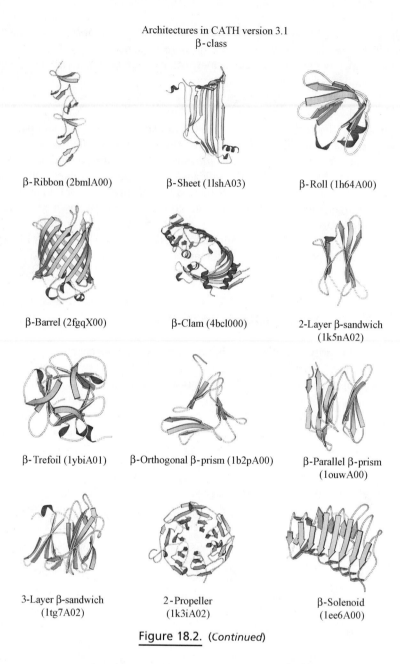

β-Ribbon (2bmlA00)   β-Sheet (1lshA03)   β-Roll (1h64A00)

β-Barrel (2fgqX00)   β-Clam (4bcl000)   2-Layer β-sandwich
                                                      (1k5nA02)

β-Trefoil (1ybiA01)   β-Orthogonal β-prism (1b2pA00)   β-Parallel β-prism
                                                         (1ouwA00)

3-Layer β-sandwich   2-Propeller   β-Solenoid
(1tg7A02)   (1k3iA02)   (1ee6A00)

Figure 18.2. (*Continued*)

structures and can be assigned to the same superfamily. Furthermore, pair-wise sequence alignment methods are reasonably robust at these levels of sequence identity. In CATH, the global alignment method of Needleman and Wunsch (1970) has been implemented. With the demonstration by Rost and Sander showing homologue identification to be dependent on sequence length, with unrelated small proteins of less than 100 residues sometimes exhibiting sequence identities as high as 30%, a more cautious threshold of 80% identity is used for homologue detection.

Architectures in CATH version 3.1
αβ class

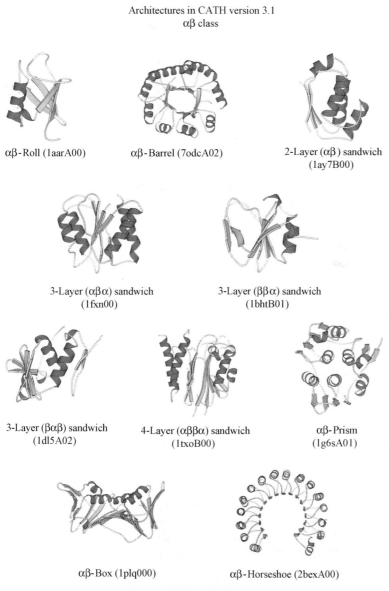

αβ-Roll (1aarA00) αβ-Barrel (7odcA02) 2-Layer (αβ) sandwich
(1ay7B00)

3-Layer (αβα) sandwich 3-Layer (ββα) sandwich
(1fxn00) (1bhtB01)

3-Layer (βαβ) sandwich 4-Layer (αββα) sandwich αβ-Prism
(1dl5A02) (1txoB00) (1g6sA01)

αβ-Box (1plq000) αβ-Horseshoe (2bexA00)

Figure 18.2. (*Continued*)

Having identified a close homologue, the SSAP residue-based structural comparison algorithm (Orengo and Taylor, 1996) is run against the data to obtain the optimal alignment between the structures. Benchmarking has shown that SSAP is more effective than both pairwise sequence alignment and sequence profiles at delineating domain boundaries.

To maintain further consistency within the domain family, we checked that at least 80% of the larger protein is aligned against the smaller protein, and that no more than a 10 residue extension is observed at either end of the alignment. Relatives identified by these pairwise methods are clustered into their respective families using a multilinkage clustering algorithm.

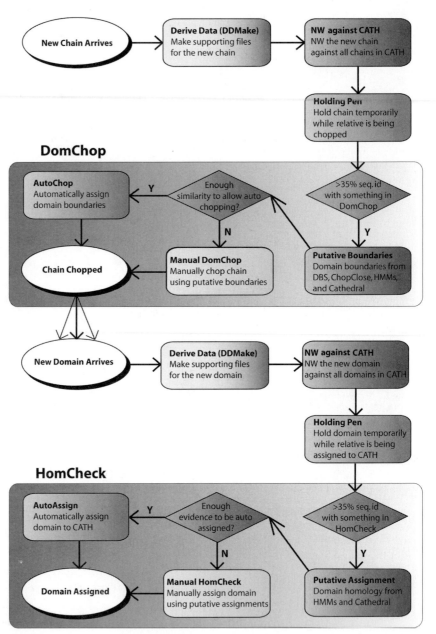

Figure 18.3. A schematic showing the CATH update protocol. The first major step is DomChop, where one or more domains are assigned for each chain. In many cases, these boundaries can be automatically assigned; those that cannot are manually curated. The newly created domains are then classified into CATH families. This part of the protocol is called HomCheck. As with DomChop, any domains that cannot be automatically classified are manually curated. Dark gray boxes denote production of metadata, diamond-shaped boxes workflow decisions, and light gray boxes manual curation. Definitions of abbreviations and terms used are as follows: NW(Needleman–Wunsch sequence alignment algorithm); HMM (hidden Markov model); ChopClose (program that determines domain boundaries based on sequence identity with CATH domains); and CATHEDRAL (structure comparison program).

IDENTIFICATION OF DOMAIN BOUNDARIES

In the January 2007 release of CATH, over 35% of PDB entries were multidomain proteins, of which two thirds comprise only two domains. Furthermore, nearly one third of domains in CATH are discontinuous, that is, the domain is formed from discontinuous segments of the polypeptide chain. It is well established that domains are known to recur in different multidomain contexts and that domain shuffling is a common evolutionary mechanism often responsible for creating new or modified functions within an organism (see Todd et al. (2001) and Teichmann et al. (2001) for reviews). Recent analysis of CATH revealed that at least 70% of the domains from multidomain proteins recurred in different multidomain families or also as single-domain proteins.

Identification of domain boundaries is a difficult process and most have an error rate of between 20% and 30% associated with them, although numerous automatic algorithms have been devised (see Chapter 20, Jones et al. (1998), and Veretnik et al. (2004)). The problem arises from the fact that no simple quantitative definition of a structural domain exists. Qualitatively, domains have been described by various authors as compact semi-independent folding units and many algorithms attempt to locate them by searching for large hydrophobic clusters indicative of domain cores and by dividing structures to maximize internal residue contacts within putative domains while minimizing external contacts between them.

During the CATH classification, any protein unassigned by the automatic sequence-method ChopClose (see above) are analyzed by several algorithms, first to delineate their domains and second for homology identification. CATHEDRAL and HMMscan (see below) are run to detect sequence and structural relatives of domains already in CATH, and the DBS suite runs three independent automatic domain definition programs. These comprise PUU (Holm and Sander, 1994), Domak (Siddiqui and Barton, 1995), and DETECTIVE (Swindells, 1995) (see Jones et al. (1998) for a description of methods). The data derived from these algorithms are displayed on the "domain chopping" page for that chain, along with relevant literature and assignments from other family classifications (e.g., Pfam). The curator is then able to make an informed decision on the most appropriate domain boundaries. Domain boundaries for a chopped chain in the CATH database can be seen on the page for that chain on the CATH Web site (see http://www.cathdb.info/cgi-bin/cath/Chain.pl?chain_id=1o7jA for an example).

Any ambiguous assignments can be manually validated by coloring putative domains in the excellent molecular viewing package, RASMOL (Sayle and Milner-White, 1995). The majority of domain assignment errors can be found in this way, although in difficult cases, boundary identification is a somewhat subjective process as witnessed by the fact that about 20% of boundary definitions in the SCOP and CATH databases disagree. This step is one of the most time consuming but important stages in the classification. Although domain boundary detection from structural data is difficult, it is considerably easier than assignment from sequence data and many sequence databases now incorporate the SCOP and CATH boundaries or use them to validate their own assignments.

METHODS TO DETECT SEQUENCE AND STRUCTURAL RELATIVES

Identifying Distant Homologues

Having determined domain boundaries, homology assignment for the constituent domains is undertaken. Profile-based methods are used as these can detect even very distant homologues. In these approaches, a sequence is scanned against a large sequence database (e.g., GenBank at the NCBI) initially using pair-wise methods to find close homologues, which are then multialigned to derive a sequence profile capturing the most specific residue preferences of the sequence and its relatives. Further iterations can result in highly specific profiles for the family to which the sequence belongs.

A number of protocols have been implemented based on hidden Markov models (HMMs). The SAMT method of Karplus et al. (1998) has been used to build profiles for each nonidentical representative in CATH and also for functionally coherent families within the database. New structures are therefore scanned against both SAMT profiles for representative structures already classified in CATH. For example, there are currently over 7000 SAMT profiles, including those built from individual domains and those generated from multiple alignments of functionally related proteins.

A protocol (Samosa) exploiting models built using multiple structural alignments to improve accuracy gives some improvement in sensitivity (4–5%). However, a protocol exploiting an eightfold expanded HMM library based on sequence relatives of structural domains derived from sequences in GenBank gives an increase of nearly 10% in sensitivity. HMM–HMM, profile–profile-based approaches have also been implemented using the PRC protocol of Madera and coworkers. These allow recognition of extremely remote homologues, some of which are not easily detected by structure comparison methods (discussed later). HMM-based scans developed for the CATH classification protocol are collectively referred to as HMMscan.

Profile-based methods enable nearly 15% of very distant homologues (<35% sequence identity) to be detected. This percentage will increase as the sequence databases, and therefore the intermediate sequence libraries within CATH, continue to grow with the international genome projects.

STRUCTURE-BASED METHODS FOR IDENTIFYING STRUCTURAL HOMOLOGUES AND RELATED FOLDS (SSAP AND CATHEDRAL)

A significant proportion (~20%, currently) of very distant relatives can only be recognized by comparing their structures directly. There are now many examples of evolutionary relatives that have diverged to an extent where no significant sequence similarity can be detected and yet the structural fold adopted by the protein remains highly similar. For example, the globins sequence identities of relatives can fall below 10% but the structures and oxygen binding functions of the proteins remain highly conserved. As with sequence alignment, any method for comparing distant structural relatives must be able to cope with the extensive insertions or deletions occurring during evolution. These are usually restricted to the loops between secondary structures where they are less likely to affect the fold and therefore the stability of the protein.

A recent analysis of CATH families (Reeves et al., 2006) found that on average, at least 50% of the secondary structures in the core of the protein are structurally well conserved across a family. However, in some families secondary structures and sometimes quite large supersecondary motifs can be inserted or deleted. In the most extreme cases, some related domains have a fivefold difference in size. In addition to residue insertions and deletions, shifts in the secondary structures can also occur to modulate the effects of volume changes caused by residue substitutions. Chothia and Lesk (1986) demonstrated quite significant movements of up to 40° in some secondary structure pairs for two large protein families studied (the globins and the immunoglobulins). Although recent analyses of CATH families showed that most secondary structure orientations (>60%) vary by up to 10°, significant numbers of secondary structures (~40%) have been observed to vary more than this and some superfamilies tolerate up to 80° in variation.

Overall the CATH analysis showed that although the core structural motif of protein folds is well conserved, significant structural embellishments to this core and shifts in secondary structure orientation can be found in about 2% of CATH families that are highly recurrent in the genomes. Sometimes these evolutionary changes give rise to modified or diverse functions within the family.

SSAP—SEQUENTIAL STRUCTURE ALIGNMENT PROGRAM

To cope with these structural variations, Taylor and Orengo adapted the dynamic programming methods used so successfully in sequence alignments to compare 3D structures (see Orengo (1994) for a detailed description and review of applications). Instead of comparing residue identities, the method compares the structural environments of residues between two proteins. These can be simply encoded as the set of vectors from the Cβ atom of a particular residue to the Cβ atoms of all other residues within the same protein. Since there is no prior knowledge of which residues are equivalent between the proteins, dynamic programming must be applied at two levels; a lower level in which the structural environments of all residue pairs between the proteins are compared and an upper summary level in which information from putative equivalent pairs are accumulated. The method is therefore sometimes described as double dynamic programming.

SSAP has been benchmarked and optimized using sets of validated structural homologues. The implemented logarithmic scoring scheme was optimized to be largely independent of the size and class of the proteins, although proteins with large proportions of alpha helices tend to give slightly higher scores as the local similarity is so highly conserved. The score is normalized to be in the range of 0–100 for identical proteins, irrespective of size and structures with similar folds tending to give scores above 70, while homologous proteins often give higher scores of 80 and above. Once two structures have been aligned, they are also superposed using the ProFit implementation of rigid body superposition (see Table 18.2) and a normalized RMSD calculated (RMSD × size of the largest domain structure/number of residues aligned). Both scores are used to assess the degree of structural similarity and determine whether the structures are likely to be homologous.

CATHEDRAL—CATHS EXISTING DOMAIN RECOGNITION ALGORITHM

Although SSAP has proven to be reliable and robust, giving extremely accurate structurally based residue alignments, it is computationally expensive and large structures such as TIM barrels with more than 300 residues that can take several days to scan against the database, even with the most powerful machines currently available. Although this has not proved to be a major bottleneck to date, structure genomics initiatives are expected to substantially increase the numbers of structures determined annually. Currently, over 100 new structures are determined weekly and this may double or treble over the next decade. Therefore, to keep pace with structure genomics, a novel algorithm has been designed that combines the robust residue-based structural alignment method of SSAP with a fast secondary and tertiary structure comparison algorithm based on graph theory. Graph theory had previously been implemented in the fast structural comparison algorithm such as POSSUM designed by Artymiuk and coworkers (Grindley et al., 1993).

The method implemented in CATH—CATHEDRAL (CATHs Existing Domain Recognition Algorithm) compares secondary structures between proteins, as there is an order of magnitude fewer of these than residues. These are represented as linear vectors and are associated with nodes in a protein graph. Edges between the nodes are characterized by the orientations of the secondary structures and the distances between their midpoints. Additional angles are used to describe the tilt and rotations of the vectors. Ullmans subgraph isomorphism algorithm is used to detect corresponding structural motifs between proteins, from comparison of their graphs. Parameters for recognizing fold similarities have been optimized using validated relatives from CATH.

A graph for each new entry is scanned against a graph for each of the existing domains in CATH. If a putative domain is identified, the domains are then aligned using the SSAP algorithm. A support vector machine is used to combine different scores that account for similarities in the size, class of the domains being compared, and the structural environments of equivalent residues. The combined scores are then used to select the best fold match for each putative domain and benchmarking experiments show that CATHEDRAL outperforms many equivalent approaches by recognizing the correct fold as the top match in a database scan 94% of the time. CATHEDRAL is also used in the preliminary stages of classifying new structures in CATH to recognize domains within multidomain structures and identify their boundaries. Benchmarking on a data set of nonredundant multidomain proteins has shown that at least 80% of domain boundaries could be recognized within ±15 residues, which compared with a 40% success rate for the most sensitive HMM-based protocols.

METHODS FOR GENERATING MULTIPLE STRUCTURE ALIGNMENTS (CORA) AND PROTOCOLS FOR USING 3D TEMPLATES USED TO IDENTIFY DISTANT STRUCTURAL RELATIONSHIPS

Multiple structure alignments are generated for each superfamily in CATH using a modified version of the SSAP algorithm (CORA—COnsensus Residue Attributes). This is based on progressive alignment of relatives using a single-linkage tree derived from the pair-wise SSAP similarity scores (see Orengo (1994) for a more detailed description). The most similar structures are aligned first and the next relative, most similar to the aligned structures, is then iteratively selected and aligned in the order dictated by the tree. After addition of each relative, a consensus structure is derived consisting of average vectors between residue positions and information on the variability of these vectors. Further relatives are aligned against this consensus structure, weighting the alignment of conserved positions more highly.

Once all relatives have been aligned, information on the conservation of the residue structural environment, including residue contacts, and various other attributes (e.g., accessibility, torsional angles) is compiled and encoded in a 3D template representing the set of structures. In a manner analogous to the improvements obtained using sequence profiles, structural templates have been shown to be more effective at recognizing distant homologues as they capture the most conserved structural characteristics of the family or superfamily. New structures can be scanned against libraries of CORA templates again using the double dynamic programming algorithm and the percentage of highly conserved contacts associated with the superfamily is calculated for the putative relative to determine homology. The pattern of conserved residue contacts has been shown to be a highly characteristic topological fingerprint for a particular superfamily.

CORA templates have been generated for each superfamily containing two or more nonidentical structures and multiple templates have been generated for each cluster of more closely related structures or families within more populated superfamilies. Scanning against the template library increases recognition performance of up to 100 times faster than performing pair-wise searches with SSAP on representative structures from the super-families. The templates also show increased sensitivity and selectivity over the pair-wise SSAP and this can be expected to increase as more structures are determined and coverage within each superfamily increases.

SEQUENCE, STRUCTURAL, AND FUNCTIONAL VALIDATION OF HOMOLOGY

As with domain assignment, homology assignment is a crucial part of the classification process and is manually validated. Results from all the sequence and structure comparison algorithms (HMMscan, SSAP, CATHEDRAL) and functional information from the litera-ture and other family databases (e.g., Pfam) are collated and displayed on an internal webpage to aid the manual curation process. Both fold and homology assignments can then easily be confirmed from these data.

Protocols are currently being devised for automatic comparisons of functional annota-tions between proteins to speed up the validation of homologues during classification. For some very remotely related homologues, confidence in an assignment can be improved by combining information from multiple prediction methods. We have investigated the benefits of machine learning methods to do this automatically. A neural network was trained using a data set of 14,000 diverse homologues (<35% identity) and 14,000 nonhomologous pairs that have been collected using different homologue comparison methods including structure comparison (CATHEDRAL, SSAP), sequence comparison (HMM–HMM), and informa-tion on functional similarity. The latter was obtained by comparing EC classification codes between close relatives of the distant homologues and using a semantic similarity scoring scheme for comparing GO terms, based on a method by Lord et al. (2003). On a separate validation set of 14,000 homologous pairs, 97% of the homologues can be recognized at an error rate of <4%.

We do not expect this approach to be viable for more than about 50% of new relatives, a result suggested by the performance of similar protocols employed by other groups, because this validation stage is time consuming and serves as another major bottleneck in the classification process. However, we expect the new algorithm to significantly improve the frequency of CATH releases.

THE DICTIONARY OF HOMOLOGOUS SUPERFAMILIES (DHS)

Sequence, structural, and functional data for each homologous superfamily in CATH are stored in a *Dictionary of Homologous Superfamilies* (DHS), which is also accessible over the web at the CATH resource. This was originally established in 1999 by Bray and updated in 2006. It now also contains all the sequence relatives for CATH superfamilies identified in

GenBank. Information on pair-wise sequence relationships between all nonidentical structures in each superfamily, namely, sequence identities, and expectation values from PSI-BLAST (Altschul et al., 1997) and SAMT are also stored.

There are also pair-wise structural alignments between all nonidentical structures in the superfamily. Plots of sequence identities versus structural similarity scores can be used to illustrate the structural plasticity observed within a given superfamily and highlight any obvious outliers that may indicate problems in the alignment. Structures within highly populated and structurally diverse superfamilies are also clustered into structural subgroups using multilinkage clustering and a threshold on the pair-wise SSAP structural similarity score of 85. CORA multiple structure alignments are generated for each superfamily and structural subgroup and can be viewed (see http://www.biochem.ucl.ac.uk/bsm/dhs/). These are annotated in various ways: by residue identities or physicochemical properties, by secondary structure, and by PROSITE motifs. RASMOL viewers for multiple superpositions of relatives allow the 3D location of conserved sequence motifs to be easily identified. The multiple structural alignments are also displayed as 2DSEC plots in which secondary structures are drawn as circles (α-helices) and triangles (β-strands) so that common secondary structures can be identified and secondary structure embellishments unique to a particular relative or group of relatives can also be viewed.

THE GENE3D RESOURCE

A related CATH-based web resource, Gene3D (Yeats et al., 2006), uses the assignments of gene sequences from CATH superfamilies to provide structural annotations for completed genomes. The March 2007 release of Gene3D (v5.0) contains 248 complete genomes obtained from UniProt. For each gene within the genome, the location of gene regions that match individual structural domains from CATH superfamilies is shown.

Importantly, Gene3D contains increasing amounts of functional data for each superfamily, including domain data from Pfam, metabolic pathway data from KEGG, functional annotation from COGs and GO, and protein–protein interaction data from MINT, IntAct, and MIPS. Links to the relevant CATH superfamily and DHS entries are also provided. Statistics are given on the distributions of superfamilies and fold groups identified within each genome, enabling comparison of fold and superfamily usage between genomes. Further information on Gene3D and the evolutionary insights that can be derived from the predicted CATH data are presented in Chapter 22.

THE CATH WEB SITE AND SERVER

CATH is available over the internet at the site shown in Table 18.1. The site can be used for browsing the hierarchy and there are representative MOLSCRIPT illustrations at each level together with links to other local resources (DHS, Gene3D, and PDBsum). Each level has its own unique numeric identifier, which is never changed although some numbers may disappear. For example, if new evidence suggests, two superfamilies should be merged.

Previously, CATH data were generated using a group of independent programs and flat files. Over the past few years, we have developed an updated protocol for CATH that is driven by a suite of programs with a central library and a PostgreSQL database system. A classification pipeline has been established that links in a completely automated fashion

TABLE 18.1. Description of Levels in the Classification at the Architecture Level

Primary Classification Number	Description of Level
1	Mainly α
2	Mainly β
3	αβ
4	Few secondary structures
5	Multidomain proteins
6	Single-domain proteins classified by sequence but not structure
7	Ambiguous multidomain proteins whose domain boundary assignment requires manual validation. Protein chains clustered by sequence methods
8	New proteins classified by sequence methods
9	Chains from multichain domains classified by sequence

the different programs that analyze the sequences and structures of protein chains and domains.

A suite of webpages has recently been developed to aid in the manual curation stages of DomChop and HomCheck (see Table 18.2). These pages display all available metadata from the prediction algorithms used in these steps (e.g., ChopClose. CATHEDRAL, HMMscan). Information regarding relevant literature and other classification data such as Pfam are also shown.

As these pages can be universally accessed, they provide biologists with interim data on their protein chain and/or domains of interest before they are fully classified into the CATH database.

THE CATHEDRAL SERVER

Newly determined structures can be submitted to the CATHEDRAL server that scans the query against representatives from the database to determine the putative superfamily or fold group to which the structure belongs or whether the protein comprises one or more novel folds. Domain boundaries can be supplied by the user or will alternatively be determined automatically.

CATHEDRAL identifies the nearest fold group for each domain. Lists of structural neighbors are provided together with links to the appropriate superfamilies in CATH and the DHS. The user can also view superpositions of the structure with other relatives from the superfamily or fold group.

IS FOLD CLASSIFICATION A LEGITIMATE REPRESENTATION OF DOMAIN STRUCTURE SPACE?

The universe of protein structures can be represented as a CATH wheel (Figure 18.4), derived by using one representative from each diverse sequence family (i.e., relatives clustered at 35% sequence identity). Using one representative is important as many

TABLE 18.2. URLs of CATH Domain Structure Database and Related Resources

Resource	URL
CATH Database: Classification of structural domains in the PDB. Domains are grouped by Class, Architecture, Topology (Fold), and Homologous superfamily. There are links to PDB sum	http://cathwww.biochem.ucl.ac.uk/latest/
SSAP Server: The SSAP server allows users to compare the structures of two proteins and view the subsequent structural alignment	http://cathwww.biochem.ucl.ac.uk/cgi-bin/cath/ SsapServer.pl
CATH Server: The CATH server allows users to compare a PDB or novel structure against a representative library of structures in CATH.	http://cathwww.biochem.ucl.ac.uk/cgi-bin/cath/ CathedralServer.pl
Dictionary of Homologous Superfamilies: This resource displays the structural alignments for all members of a homologous superfamily classified in the CATH database. The alignments are augmented with ligand information and SWISS-PROT annotations	http://www.cathdb.info/bsm/dhs/
Gene3D: Database of precalculated structural assignments for genes and whole genomes. The data are derived using PSI-BLAST and IMPALA	http://cathwww.biochem.ucl.ac.uk:8080/Gene3D/
IMPALA server: This server allows the user to screen a sequence against the CATH set of IMPALA sequence profiles for protein structural domains	http://www.biochem.ucl.ac.uk/bsm/cath/Impala/

This includes the HOMCHECK, DOMCHECK, and PROFIT Webpages as well as the Madera PRC resource.

homologous superfamilies have multiple redundant entries in sequence and structure databases, which can introduce considerable bias in the data if not treated correctly.

The distribution of families in the Protein Data Bank can be seen from Figure 18.4 showing that the current diverse sequence families of mainly α (26%), mainly β (22%), and $\alpha\beta$ structures (48%) are distributed within 40 architectures. Twenty-eight of these architectures are well defined, and the remaining groups can be thought of as bins comprising assorted irregular or complex folds or folds containing few secondary structures that are often stabilized by disulfide bonds. Furthermore, among the well-defined architectures, some are much more highly populated than others and it can be seen from Figure 18.4 that about half of the folds adopt one of six regular symmetric architectures: the mainly α-bundles, the two-layer β-sandwiches and β-barrels, and the two- and three-layer $\alpha\beta$-sandwiches and $\alpha\beta$-barrels.

Recent analysis of the structural relationships between fold groups in CATH using the GRATH algorithm has revealed that some fold groups are particularly "gregarious." That is, they have quite large motifs (up to 40% of the structure) in common with many other folds within the database. In particular, those folds containing common supersecondary structural motifs, such as β meanders, Greek keys, α–β plait motifs, or α-hairpins, match similar motifs in many other folds. α–β folds were found to be particularly gregarious, as were other folds

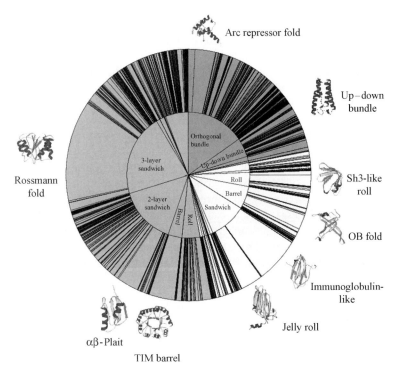

Figure 18.4. CATH wheel plot showing the distribution of nonhomologous structures, that is, a single representative from each homologous superfamily (H-level in CATH) among the different classes (C), architecture (A), and fold families (T) in the CATH database. Protein classes shown are mainly alpha (dark gray), mainly beta (light gray), and alpha–beta (mid-gray). Within each class, the angle subtended for a given segment reflects the proportion of structures within the identified architectures (inner circle) or fold families (outer circle). The superfold families are indicated and are illustrated with a Molscript drawing of a representative from the family.

found in the highly populated architectural regions of fold space (e.g., α–β plaits). This conforms to the idea of a structural continuum where in some regions of fold space there can be considerable overlap between different fold groups. Often it is difficult to distinguish between these folds and the definition of folds becomes somewhat subjective. It tends to be in these regions of fold space that structural classifications such as CATH and SCOP disagree.

Domains that exhibit low gregariousness usually have very distinctive folds, for example, the β trefoil folds, and often comprise unusual motifs or motifs packed in unusual arrangements as in the superhelices. These folds are more representative of discrete islands rather than progressions in a structural continuum.

The idea of a structural continuum is perhaps not surprising as it has long been known that certain structural motifs are favored and often recur within folds for a particular class; for example, the β-hairpins and the beta Greek keys in the mainly β class, the α + β and α/β motifs in the αβ class (Richardson, 1981), the TIM barrel fold in the αβ class (see Figure 18.2 for a representative picture) comprises eight recurring αβ motifs, and these motifs are also found to recur in the αβ Rossmann folds (see also Figure 18.2), though the orientation between two of the motifs differs giving rise to a different topology for the fold.

In addition to considering the structural overlap between different fold groups, it is also interesting to consider the structural deviation or drift within an evolutionary superfamily as this is another mechanism causing overlap between fold groups in hierarchical classifications, such as CATH and SCOP. Recent analyses of CATH reveal that in most superfamilies, the domain relatives are structurally very similar to each other if the global structural similarity between domain structures is considered, that is, at least 60% or more of the larger domain is matched between two domains. However, there are some notable exceptions to this. Considerable deviations in structure can occur between relatives in a small percentage of families (<5%). Many Rossmann fold superfamilies (CATH code 3.40.50) have within them a significant number of protein domains that are very dissimilar to each other in terms of structure.

This phenomenon, which can be described as structural drift, is often caused by the insertions of extensive secondary structures outside the common structural core (see Figures 18.5 and 18.6). These insertions can be described as secondary structure embellishments (Reeves et al., 2006). Many families in which this phenomenon is observed are highly populated in the genomes (Figure 18.6). For example, superfamilies adopting the Rossmann fold are the most highly populated structural families in the genomes and paralogous domains have diverged considerably in structure (Figure 18.5). This structural divergence has been accompanied by extensive modification of the function and a total of 168 GO molecular function terms are currently associated with the Rossmann fold. Interestingly, those super-

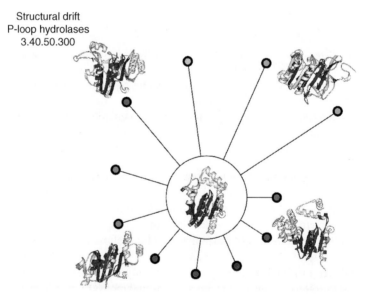

Figure 18.5. Foldspin plot showing an example of structural drift for the Rossmann superfamily 3.40.50.300. Structural comparisons were made between the protein structure in the center of the wheel with other structures in the same superfamily; these are represented by spokes on the wheel. The length of each spoke correlates with the structural similarity score between the two structures; the more similar the structures, the shorter the spoke. The structure for which all comparisons are made and some of these structures it is compared against are shown as molscripts; the dark gray regions highlight the structural features common to both structures and the light gray regions indicate where differences occur. It can clearly be seen that although there is a common structural "core," there are also some significant variations.

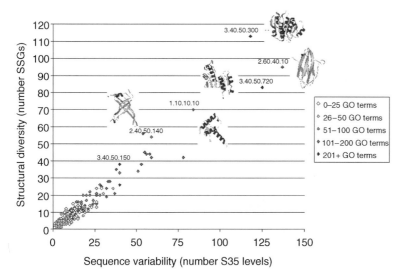

Figure 18.6. Relationship between sequence variability, structural variability, and functional diversity in CATH superfamilies. Structural variation in a CATH superfamily as measured by the number of diverse structural subgroups (SSAP score <80 between groups) is plotted against sequence diversity as measured by the number of sequence diverse subfamilies in the CATH-DHS (<35% sequence identity between groups). The color of each point reflects the number of functions identified in that superfamily using GO as follows: white (0–25), light gray (26–50), mid-gray (51–100), dark gray (101–200), and black (200+).

families most vulnerable to structural drift tend to belong to fold groups that have very regular architectures, for example, two-layer mainly β and two- and three-layer αβ architectures. These architectures appear to be particularly tolerant to secondary structural embellishments that tend to occur at the edges of the β-sheets where they cause little disruption to the overall fold.

Although structural overlaps between fold groups are frequently observed when small motif similarities are analyzed (i.e., comprising <40% of the larger domain), far fewer overlaps are detected when examining structural similarity at a more global level (>60% of larger domain). In fact, fewer than 10% of homologous superfamilies exhibit significant structural overlaps with domains in different fold groups. Although both structural drift and overlap at the global level occur in comparatively few superfamilies, it is important to note that these superfamilies tend to be the most highly populated in terms of the number of available structural domains and their prevalence in the genomes. Therefore, it is likely that the CATH classification will present neighborhood information as well as a strictly hierarchical grouping of structures in the future. For each domain structure, significant structural overlaps with domains in other superfamilies and fold groups are already presented in CATH.

POPULATION OF SUPERFAMILIES AND FAMILIES WITHIN FOLDS

The current population of the different levels in the CATH hierarchy for the January 2007 release is shown in Figures 18.4 and 18.7, and the number of new folds being determined each

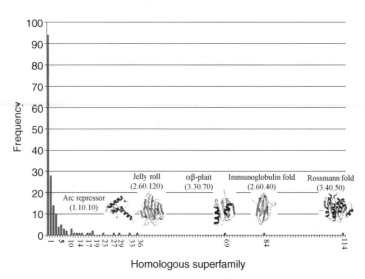

Figure 18.7. Populations of different levels in the CATH hierarchy; homologous superfamilies within fold groups. CATH version 3.1 was used to generate the histograms.

year appears to be slowly decreasing (Figure 18.8). Early analysis of CATH revealed a small number of highly populated fold groups (~10) containing many different homologous superfamilies and families. A recent re-examination of these data shows a greater than 10-fold expansion of the database from 3000 to 35,000 domain structures, a continuing trend as reflected in Figure 18.4 showing eight large fold groups containing nearly one third

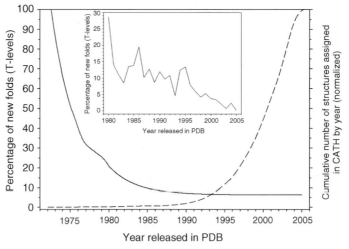

Figure 18.8. Annual decrease in the percentage of new structures classified in CATH that are observed to possess a novel fold. The raw data for years 1972–2005 was fit to a single exponential equation by nonlinear regression using Sigma Plot (SPSS, version 9.0) and the fit is shown as a solid black line. The inset shows a close-up of the raw data for new topologies over the years 1980–2005. For comparison, the numbers of structural domains solved each year and deposited in the PDB and classified in CATH is depicted by the dashed line.

of all the homologous superfamilies in CATH. These highly populated fold groups have been described as superfolds and other classifications (e.g., SCOP) have reported similar observations of frequently occurring domains (FODs) within the database.

The popularity of these folds may be a result of divergent or convergent evolution. Divergent evolution gives rise to families of proteins in which the structure is generally well conserved but sequences may have changed to the extent that no significant similarity remains. In paralogs, which arise from duplication of the gene within an organism, the protein function may also have been modified or changed. Thus, the apparently diverse superfamilies within these superfolds may in fact be extremely distant relatives whose relationships cannot easily be verified from the available sequence or functional data.

Alternatively, Ptitsyn, Finkelstein, and others have suggested that there may be a limited number of folds in nature due to physical constraints on the packing of secondary structures. Chothia has suggested approximately 1000 folds and other similar estimates of a few thousand folds have also been made. Structures sharing the same fold but arising from different ancestral proteins have been described as analogues.

The task of determining homology is often complicated by the lack of evolutionary clues. Murzin and others have shown several interesting examples of relatives possessing similar folds but disparate sequences and functions. In these cases, homology has been suggested by the presence of rare structural characteristics, for example, a conserved β-bulge. Detailed knowledge of the structural family can often provide insights that promote detection of evolutionary fingerprints. However, such practices are not readily amenable to automated protocols. The TIM barrel fold is one of the most highly populated, with 18 superfamilies in a recent CATH release. However, several detailed studies have recently provided evidence of common ligand binding motifs or unusual structural characteristics suggesting that many of these superfamilies should be merged (Nagano, Orengo, and Thornton, 2002). When classifying new relatives in CATH, many types of evidence are manually considered during curation (e.g., matching sequence motifs, similarity in rare structural motifs, and clues about similarity in functional mechanisms) to try to improve the detection of very remote homologues. These data are presented to the curator on the HomCheck webpages (see Table 18.2) and are also available to biologists interested in a particular classification for a domain in CATH.

FOLD CLASSIFICATION IN CATH

When deciding how to classify folds and whether a structural continuum exists, it is important to consider the extent to which related proteins in a superfamily may have residue insertions or deletions that cause significant elaborations or alterations of the protein structure. Although some families such as the plekstrin homology domain and the lipocalins are structurally highly conserved and do not tolerate much deviation, other superfamilies contain structures that have changed dramatically between distant relatives not only in fold but also in architecture.

In these cases where more than one fold group can be identified within the superfamily, the hierarchical nature of CATH classification has broken down. At present, there are relatively few occurrences of these dramatic changes and they are documented on the CATH Web site as cross-hits (see Table 18.2). As more structures are elucidated so that fold space becomes more highly populated and as methods improve for detecting remote homologues, further examples of these extreme changes in fold may be identified.

TABLE 18.3. Names of the Researchers and Advisors Who Have Contributed to the CATH Database Since It Was Established in 1993

List of Contributors (Alphabetical)

Sarah Addou	Andrew Martin
Adrian Akpor	Alex D. Michie
Chris F. Bennett	Rekha Nambudiry
James E. Bray	Christine A. Orengo
Daniel W. A. Buchan	Frances M. G. Pearl
Alison Cuff	Kanchan Phadwal
Tim Dallman	Gabrielle Reeves
Beniot Dessailly	Adam Reid
Mark Dibley	Jane S. Richardson
Lesley Greene	Elisabeth Rideal
Andrew Harrison	Oliver Redfern
Gail Hutchinson	Adrian J. Shepherd
Azara Janmohamed	Stathis Sideris
Susan Jones	Ian Sillitoe
Roman Laskowski	Mark B. Swindells
David Lee	Wille R. Taylor
Jon Lees	Janet M. Thornton
Loredana Lo Conte	Annabel E. Todd
Russell Marsden	

ACKNOWLEDGMENTS

We would like to thank all those who have contributed to the massive undertaking of this resource (Table 18.3)

REFERENCES

Altschul SF, Madden TL, Schaffer AA, Zhang J, Zhang Z, Miller W, Lipman DJ (1997): Gapped Blast and PSI-BLAST: a new generation of protein database search programs. *Nucleic Acids Res* 25:3389–3402.

Andreeva A, Howorth D, Brenner SE, Hubbard TJP, Chothia C, Murzin AG (2004): SCOP database in 2004: refinements integrate structure and sequence family data. *Nucleic Acids Res* 32: D226–D229.

Chothia C, Lesk AM (1986): The relation between divergence of sequence and structure in proteins. *EMBO J* 5:823–826.

Day R, Beck DA, Armen RS, Daggett V (2003): A consensus view of fold space: combining SCOP, CATH, and the Dali Domain Dictionary. *Protein Sci* 12(10):2150–2160.

Dengler U, Siddiqui AS, Barton GJ (2001): Protein structural domains: analysis of the 3Dee domains database. *Proteins* 42(3):322–344.

Dietmann S, Holm L (2001): Identification of homology in protein structure classification. *Nat Struct Biol* 8:953–957.

Greene LH, Lewis TE, Addou S, Cuff A, Dallman T, Dibley M, Redfern O, Pearl F, Nambudiry R, Reid A, Sillitoe I, Yeats C, Thornton JM, Orengo CA (2007): The CATH domain structure database: new protocols and classification levels give a more comprehensive resource for exploring evolution. *Nucleic Acids Res* 35:D291–D297.

Grindley HM, Artymiuk PJ, Rice DW, Willet P (1993): Identification of tertiary structure resemblance in proteins using a maximal common subgraph isomorphism algorithm. *J Mol Biol* 299 (3):707–721.

Hadley C, Jones DT (1999): A systematic comparison of protein classifications: SCOP, CATH and FSSP. *Structure* 7:1099–1112.

Holm L, Sander C (1994): Searching protein structure database has come of age. *Proteins* 19:165–173.

Holm L, Sander C (1996): Mapping the protein universe. *Science* 273:595–603.

Holm L, Sander C (1998): Dictionary of recurrent domains in protein structures. *Proteins* 33:88–96.

Jones S, Stewart M, Michie A, Swindells MB, Orengo CA, Thornton JM (1998): Domain assignment for protein structures using a consensus approach: characterisation and analysis. *Protein Sci* 7:233–242.

Karplus K, Barrett C, Hughey R (1998): Hidden Markov Models (HMMs) for detecting remote homologies. *Bioinformatics* 14:846–856.

Lord PW, Stevens RD, Brass A, Goble CA (2003): Semantic similarity measures as tools for exploring the gene ontology. *Pac Symp Biocomput* pp. 601–612.

Madej T, Gibrat JF, Bryant SH (1995): Threading a database of protein cores. *Proteins* 23(3): 356–369.

Murzin AG, Brenner SE, Hubbard T, Chothia C (1995): SCOP: a structural classification of proteins database for the investigation of sequences and structures. *Nucleic Acids Res* 247 (4):536–540.

Nagano N, Orengo CA, Thornton JM (2002): One fold with many functions: the evolutionary relationships between TIM barrel families based on their sequences, structures and functions. *Mol Biol* 321(5):741–765.

Needleman SB, Wunsch CD (1970): A general method applicable to the search for similarities in the amino acid sequence of two proteins. *J Mol Biol* 48(3):443–453.

Orengo CA (1994): Classification of protein folds. *Curr Opin Struct Biol* 4:429–440.

Orengo CA, Taylor WR (1996): SSAP: sequential structure alignment program for protein structure comparison. *Methods Enzymol* 266:617–635.

Orengo CA, Michie AD, Jones DT, Swindells MB, Thornton JM (1997): CATH: a hierarchic classification of protein domain structures. *Structure* 5:1093–1108.

Reeves GA, Dallman TJ, Redfern OC, Akpor A, Orengo CA (2006): Structural diversity of domain superfamilies in the CATH database. *J Mol Biol* 360:725–741.

Richardson JS (1981): The anatomy and taxonomy of protein structure. *Adv Prot Chem* 34:167–339.

Sayle RA, Milner-White EJ (1995): RASMOL: biolomecular graphics for all. *Trends Biochem Sci* 20:374.

Siddiqui RA, Barton GJ (1995): Continuous and discontinuous domains: an algorithm for the automatic generation of reliable protein domain definitions. *Protein Sci* 4:872–884.

Swindells MB (1995): A procedure for the automatic determination of hydrophobic cores in protein structures. *Protein Sci* 4:93–102.

Taylor WR, Orengo CA (1989): Protein structure alignment. *J Mol Biol* 208:1.

Teichmann SA, Rison SC, Thornton JM, Riley M, Gough J, Chothia C (2001): The evolution and structural anatomy of the small molecule metabolic pathways in *Escherichia coli*. *J Mol Biol* 311:693–708.

Todd AE, Orengo CA, Thornton JM (2001): Evolution of function in protein superfamilies, from a structural perspective. *J Mol Biol* 7:1113–1143.

Veretnik S, Bourne PE, Alexandrov NN, Shindyalov IN (2004): Toward consistent assignment of structural domains in proteins. *J Mol Biol* 339(3) 647–678.

Yeats C, Maibaum M, Marsden R, Dibley M, Lee D, Addou S, Orengo CA (2006): *Nucleic Acids Res* 34:D281–D284.

Section IV

STRUCTURAL AND FUNCTIONAL ASSIGNMENT

19

SECONDARY STRUCTURE ASSIGNMENT

Claus A. Andersen and Burkhard Rost

The Task. When looking at how chains of amino acids in proteins are ordered, we notice regular macro elements in the 3D structure of practically all known structures: helices and strands (see Chapter 2). These structures were coined protein secondary structure by Linderstrøm-Lang (1952) in the context of a protein's primary, tertiary, and quaternary structural definitions. There is no unique inherent physical characteristic to systematically assign secondary structure from 3D coordinates. Instead there are many different assignment schemes, each constructed to reflect one or more aspects of protein structure. Here we will address these aspects, the assignment schemes, and how they have been used to study proteins.

Aspects Reflected in Secondary Structure Assignments. When protein secondary structure is assigned, the complexity of the all atom 3D structure is reduced dramatically. This level of abstraction is obtained by choosing as determinants for the assignment such aspects as physical interaction energies (e.g. H-bonds and van der Waals), geometrical idealization (e.g. idealized cylinder or C_α-distance masks) and/or other structural descriptions optimized to reflect appealing characteristics (e.g. expert assignment, invariance upon thermal fluctuations or predictability). Another generic aspect of all methods is that they assign secondary structure based on residue independent aspects.

Notation. When treating protein secondary structure, we will use the terms: *state*, *class*, and *regular secondary structure*. In the literature, these are not used consistently so we refer to the following notation: *states* are the types of secondary structures defined by a particular method, for example, G refers to 3_{10}-helix in the dictionary of secondary structure of proteins (DSSP); *classes* are the groups of similar states, for example, G, H and I all describing helices in DSSP; *regular secondary structure* are all the states belonging to a direct secondary structure assignment. Note that *nonregular structure* thus refers to the remaining states, which on occasion go by names such as random coil or loop. Super secondary structure refers to the relative spatial distances and orientations between two or more secondary structure elements.

Usage of Secondary Structure. The application of secondary structure assignments is very diverse and covers many areas of protein analyses, most noticeably, within structure

Structural Bioinformatics, Second Edition   Edited by Jenny Gu and Philip E. Bourne
Copyright © 2009 John Wiley & Sons, Inc.

inspection (Chapters 14 and 15), visualization (see Chapter 9) and the related task of structure comparison and classification (see Chapters 16, 17 and 18). Secondary structure has also been employed in protein modeling and structure prediction (see Chapter 29), as well as in studies of protein folding, dynamics, interactions, and function. The one-dimensional nature of the secondary structure description has furthermore been used in sequence alignment. Some applications will be treated in more detail below after the assignment methods have been described.

History: From Expert to Automatic Assignment of Protein Secondary Structure. Pauling and colleagues correctly predicted the idealized protein secondary structures of α-helices (Pauling et al., 1951), π-helices (Pauling et al., 1951), and of β-sheets (Pauling and Corey, 1951) based on intra-backbone hydrogen bonds. Five decades later, we know that on an average about half of the residues in proteins participate in helices or sheets (Berman et al., 2000). Pauling and colleagues incorrectly predicted that 3_{10}-helices would not occur in proteins due to unfavorable bond angles; approximately 4% of the residues are observed in this conformation (Andersen, 2001). Initially, crystallographers assigned secondary structure by eye from the 3D structures. This was the only way to assign secondary structure at that time. However, it lacked consistency since experts occasionally disagreed. This was particularly problematic when comparing secondary structure predictions and was actually the primary objective for Kabsch and Sander (1983a), and Kabsch and Sander (1983b) to automate the assignment in their DSSP program. Originally developed to improve secondary structure prediction, DSSP has remained the standard in the field, most popular for its relatively reliable assignments. Curiously, the prediction method for which Kabsch and Sander created to address this issue was never published.

Experimental Investigations of Protein Secondary Structure. The structure of a protein can be determined at various levels of precision and timescale. X-ray crystallography (see Chapter 4) is widely used and generally provides a static snapshot with all atom resolution, whereas NMR (see Chapter 5) furthermore can measure dynamic motion of proteins in solution, but not below the millisecond regime (Doerr, 2007). Optical spectroscopy is a much faster technique and has been used to inspect H-bond dynamics at a picosecond time scale for a small β-turn peptide (Kolano et al., 2006). In particular circular dichroism (CD) and Raman spectroscopy are used to characterize overall protein secondary structure dynamics in solution, since the helix and sheet structures give strong characteristic spectra which are highly correlated with X-ray data (Tetin et al., 2003; Janes, 2005; Lees et al., 2006). This allows the rapid assessment of conformational changes resulting from ligand binding, macromolecular interactions, and so on, and conformational assessment of natively unfolded proteins (Pelton and McLean, 2000; Maiti et al., 2004). Spectroscopy resolution can be further enhanced with residue-specific isotope labeling, for example, to dissect the conformation of helical peptides at the residue level (Decatur, 2000; Fesinmeyer et al., 2005). Attempts have been made to determine the protein secondary structure and stability by mass spectrometry (Villanueva et al., 2002), but the specific technique presented is not likely to be a valuable conformational probe (Beynon, 2004).

SECONDARY STRUCTURE CONCEPTS

The hydrogen bond is used by many methods to describe and assign protein secondary structures, so we will introduce this concept and some definitions employed. Using the hydrogen bond spurs from the notion of assigning secondary structure based on the local energy gained in stabilizing the polypeptide chain in a given conformation. Following this notion, the energetic calculation can also be extended to the rest of the protein backbone atom interactions by calculating electrostatic and van der Waals interaction energies, described here as the backbone–backbone interaction energy.

Likewise, mathematical concepts are applied to secondary structure assignment. These may be basic geometrical objects, which can be readily comprehended (e.g. a straight line or cylinder) or may require some introduction as done for Voronoï tessellation presented here.

Hydrogen Bond Energy

Pauling (1939) established the hydrogen bond as an important principle in chemistry. The rich network of hydrogen bonds in water creates a very particular environment in which polar molecules participate, while nonpolar molecules disrupt the network of hydrogen bonds. This results in missing water–water hydrogen bonds and therefore a relative energy cost compared to the hydrogen bonded case (4 kcal/mol for isoleucine and leucine when compared to glycine (Creighton, 1993)). This energy cost is in the order of two hydrogen bonds (hydrogen bonds are in the range of -2 kcal/mol) and can be avoided or minimized by packing via agglomerating nonpolar molecules, thereby resulting in the hydrophobic effect.

The packing of nonpolar residues in the core is believed to be the main driving force in tertiary structure formation of proteins, while the specific secondary structures are governed by intramolecular hydrogen bonds (Hvidt and Westh, 1998). Packing the nonpolar residues in the core also means burying the polar backbone atoms and breaking the water–backbone hydrogen bonds. To avoid this heavy energetic cost, the polarities are paired (forming hydrogen bonds) in the protein core, thus fixing the protein conformation. If instead the protein backbone were nonpolar, the protein core elements would then be free to move around, thus leading to a highly dynamic protein structure and thereby preventing the protein from functioning reliably and efficiently.

Approximately 90% of the backbone C=O and NH groups participate in hydrogen bonds (Baker and Hubbard, 1984). Using the Coulomb hydrogen bond definition (see below), we found that approximately 62% of the backbone C=O and NH groups participate in intra-backbone hydrogen bonds (Andersen, 2001). Pauling defined secondary structure by the intra-backbone hydrogen bonds, and this has later become the prevalent means of assigning secondary structure. Thus, for simplicity, we refer to intra-backbone hydrogen bonds when using the term "hydrogen bond".

Angle-Distance Hydrogen Bond Assignment

There are many different angles and distances that can be measured and used to identify the hydrogen bond. Baker and Hubbard (1984) assigned hydrogen bonds according to the inter-atom angle NHO $= q$ and distance r_{HO} in the hydrogen bond. A hydrogen bond is assigned when

$$q > 120° \quad \text{and} \quad r_{HO} < 2.5 \text{Å}$$

This is similar to other rigid distance and angle constraints published (Bordo and Argos, 1994; Jeffrey and Saenger, 1994), and can be simplified further by considering only the donor–acceptor distance, that is, in this case hydrogen bonds are assigned when

$r_{HO} < 3.5$ Å. Although a rather crude way of assigning hydrogen bonds, it has sufficed for several decades and is still being used.

Coulomb Hydrogen Bond Energy Calculation

One way of finding hydrogen bonds is by calculating the Coulomb energy in the bond, as applied in DSSP (see below) focusing on the electrostatic attraction (Figure 19.1). The Coulomb energy for the attraction and repulsion is given by

$$E = f\delta^+ \delta^- \left(\frac{1}{r_{NO}} + \frac{1}{r_{HC'}} - \frac{1}{r_{HO}} - \frac{1}{r_{NC'}} \right) \tag{19.1}$$

where $f = 332$ Å kcal/$(e^2 \text{mol})$ is the dimensional factor, and $\delta^+ = 0.20e$ and $\delta^- = -0.42e$ are the polar charges given in units of the elementary electron charge e. A cut-off level has been set for the weakest acceptable hydrogen bond so that the resulting energy is bound by $E < -0.5$ kcal/mol in DSSP. In practice, the hydrogen atom position is usually not given in periplasmic binding protein (PDB files requiring an extrapolation. For example, the hydrogen atom position that is needed to calculate the two distances r_{OH} and $r_{HC'}$ in Equation 19.1 is usually not given in the PDB files. DSSP circumvents this problem by using an approximate position, assuming that the covalent bond between O=C' is parallel to the covalent N−H bond adjacent to the same polypeptide bond. The direction of the O=C' vector is kept while its length is set to 1 Å, that is, the length of the N−H bond (Creighton, 1993). The position of the H-atom is extrapolated using the direction of the C'=O vector when starting out from the position of the N-atom. These approximations made by DSSP simplify the calculation of the H-atom position and appear to be rather accurate despite the assumptions that were being made. When compared to the original bond angles and distances (Creighton, 1993), we found the DSSP approximation to yield an average error around 0.07 Å (Andersen, 2001). The assumption of the *trans*-peptide bond, giving rise to the rigid peptide plane, was used by DSSP as well as our tests. Partitioning *ab initio* energy calculations of the hydrogen bond into classical components showed that about 75% is electrostatic (Coulombic) and less than 5% comes from polarization and charge transfer, for moderate strength bonds (Jeffrey and Saenger, 1994). Note that the Coulomb energy term does not incorporate atom–atom repulsion to penalize steric clashes and does not give rise to a characteristic hydrogen bond length.

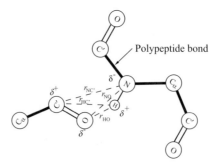

Figure 19.1. Distances used to calculate the Coulomb hydrogen bond energy.

Empirical Hydrogen Bond Energy Calculation

An empirical hydrogen bond energy calculation can be derived from the hydrogen bond geometry in crystal structures or from polypeptides, peptides, amino acids, and small organic compounds (Boobbyer et al., 1989; Wade et al., 1993) as applied in secondary STRuctural IDEntification method (STRIDE) (see below). The total energy E_{hb} depends on the NO distance energy E_r, (reflecting optimal atom distance and atom boundary) and on three bonding angles through the expressions E_p and E_t (reflecting favorable hydrogen bond angles extrapolated from electron orbital interactions):

$$E_{hb} = E_r \cdot E_t \cdot E_p \tag{19.2}$$

The distance dependency energy E_r is similar to the Lennard–Jones potential for the van der Waals interaction, but uses powers of eight and six instead of 12 and six. Thus reducing the slope of the atom–atom superposition term, whereby the penalty for superpositions is more lenient towards experimental inaccuracies in atom position determination.

$$E_r = \left(\frac{4\, r_m^6}{r^6} - \frac{3\, r_m^8}{r^8} \right) E_m \tag{19.3}$$

where r is the NO distance, r_m the optimal distance, and E_m the optimal energy. For intra-backbone hydrogen bonds $r_m = 3.0$ Å and $E_m = -2.8$ kcal/mol is used. The two angular dependent terms are

$$
\begin{aligned}
E_p &= \cos^2(\theta) \\
E_t &= \begin{cases} [0.9 + 0.1 \sin(2t_i)]\cos(t_0) & 0° < t_i \le 90° \\ K_1[K_2 - \cos^2(t_i)]\cos(t_0) & 90° < t_i \le 110° \\ 0 & 110° \le t_i \end{cases}
\end{aligned} \tag{19.4}
$$

where the angles θ, t_i, and t_0 are specified in Figure 19.2.

Backbone–Backbone Interaction Energy

The Coulomb and van der Waals interaction energy calculations can also be applied to all backbone atom interactions (i.e. involving the atoms N, C_α, C′, O, H_N and H_α) as applied in β-spider (see below), thereby covering the two potential backbone hydrogen bonds formed

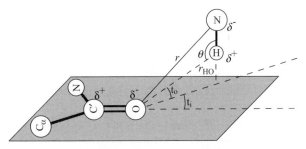

Figure 19.2. Angles and distances defining the empirical hydrogen bond. *Note:* Figure similar to the one in Frishman and Argos (1995).

between two residues (described above), as well as the C_α–$H_\alpha \cdots O{=}C$ hydrogen bond and the $C{=}O \cdots C{=}O$ dipole. This energy evaluation is calculated for each atom pair (E_{ij}^A) and subsequently summed over all pairs between two residues (E_{ij}^R). E_{ij}^A has the same functional form as the Amber force field (Cornell et al., 1995):

$$E_{ij}^A = \varepsilon_{ij}^* \left(\frac{R_{ij}^*}{R_{ij}} \right)^{12} - 2\varepsilon_{ij}^* \left(\frac{R_{ij}^*}{R_{ij}} \right)^6 + 332 \frac{Q_i Q_j}{R_{ij}} \tag{19.5}$$

where E_{ij}^A is the energy between atoms i and j with observed distance R_{ij} (Å), mixing rules ε_{ij}^*, polar charges Q_i, Q_j, and optimal distance R_{ij}^*. The optimal distance, mixing rules, and polar charges were taken from Amber. In the special cases of glycine and proline, the backbone constellation has been adjusted accordingly by adding another H_α and removing the H_N, respectively. In β-spider the hydrogen atom positions were extrapolated geometrically from the N, C_α, C′ and O coordinates using bond lengths, valence, and torsion angles from Amber (Cornell et al., 1995).

$$E_{ij}^R = \sum_{\{N,C_\alpha,C',O,H_N,H_\alpha\}}^{i} \sum_{\{N,C_\alpha,C',O,H_N,H_\alpha\}}^{j} E_{ij}^A \tag{19.6}$$

Voronoï Tessellation for Geometrical Residue Partitioning

By geometrically deriving a polyhedron around the C_α of each residue using Voronoï tessellation a new definition of contact maps is described. This is very informative about the relative packing of residues and has been applied to assign secondary structure by VoTAP (see below). Each C_α is contained within a Voronoï cell, which is determined by Delaunay tetrahedral decomposition. This is a unique decomposition with nonoverlapping cells that only contain one C_α atom each and where all internal space inside the protein is contained within a cell as shown in Figure 19.3.

Converting Secondary Structure States to Three Classes

Several secondary structure assignment methods are presently available, but DSSP continues to be the most widely used method followed by STRIDE. In fact, most prediction methods are based on DSSP assignments. Typically, the eight DSSP states are converted into three classes using the following convention: [GHI] → h, [EB] → e, [TS" "] → c, which reads 3_{10}-, α-, π-helices are grouped into one helix class; extended β-sheets and β-bridges are grouped into one sheet class; and the remaining secondary structure states: turn, bend and "not assigned" are grouped into one coil class.

Usually, 3_{10}-helices and β-bridges constitute short secondary structure segments that have some structural similarity to α-helix and β-strand, respectively. However, they do have different sequence characteristics. Prediction methods, in general, are more precise in the core of regular secondary structure segments than at the termini (Rost, 1995; Cuff and Barton, 1999). Thus, 3_{10}-helices and β-bridges are more difficult to predict than α-helices and β-strands. Therefore an alternative conversion that has been used more recently yields a seemingly higher level of prediction accuracy: [H] → h, [E] → e, [GIBTS" "] → c.

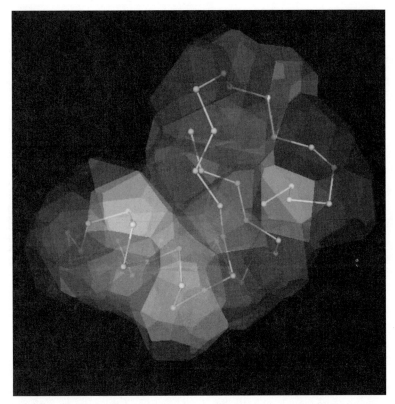

Figure 19.3. Voronoï tessellation partitions the protein space into polyhedra surrounding the C_α atom of each residue (Dupuis et al., 2004). A unique C_α atom contact map can thus be defined as the polyhedra sharing a contact surface. The example shown is Crambin (Teeter, 1984; PDB: 1CRN) visualized from an angle similar to the one employed in Figure 19.9, using the voro3D tool made available to the community by Frank Dupuis.

ASSIGNMENT METHODS

DSSP: H-Bond Pattern-Based Assignment

The so-called DSSP by Kabsch and Sander (1983a) performs its sheet and helix assignments solely based on the backbone–backbone hydrogen bonds. The DSSP method defines a hydrogen bond when the bond energy is below −0.5 kcal/mol from a Coulomb approximation of the hydrogen bond energy (see Section "Secondary Structure Concepts"). The structure assignments are defined such that visually appealing and unbroken structures result. In case of overlaps, α-helix is given first priority, followed by β-sheet. This procedure does not affect the Coulomb approximation, rather the realization of "unbroken structures' addresses the step from individual hydrogen bonds to assigning macrostructures to groups of such bonds (Figure 19.4).

An α-helix assignment (DSSP state "H") starts when two consecutive amino acids have $i \rightarrow i + 4$ hydrogen bonds, and ends likewise with two consecutive $i \rightarrow i - 4$ hydrogen bonds. This definition is also used for 3_{10}-helices (state "G" with $i \rightarrow i + 3$

```
PDB:1CRN
  #   Residue  AA  Structure   BP1 BP2  ACC   N-H-->O    O-->H-N    N-H-->O    O-->H-N
 ....
 15    15      V   H >< S+      0   0    99   -4,-1.7    3,-1.3     2,-0.2    -2,-0.2
 16    16      c   H 3<>S+      0   0    18   -4,-2.5    5,-0.8     1,-0.3    -2,-0.2
 17    17      R   H ><5S+      0   0    94   -4,-2.0    3,-1.6     1,-0.2    -1,-0.3
 18    18      L   T <<5S+      0   0   144   -3,-1.3   -1,-0.2    -4,-0.6    -2,-0.2
 19    19      P   T 3 5S-      0   0   107    0, 0.0   -1,-0.3     0, 0.0    -2,-0.1
 20    20      G   T < 5 +      0   0    53   -3,-1.6   -3,-0.2     1,-0.2    -2,-0.1
 21    21      T       < -      0   0    37   -5,-0.8   -1,-0.2     1,-0.1     5,-0.1
 22    22      P     >>  -      0   0    81    0, 0.0    4,-2.2     0, 0.0     3,-0.7
 23    23      E   H 3> S+      0   0    70    1,-0.2    4,-2.5     2,-0.2     5,-0.1
 24    24      A   H 3> S+      0   0    63    1,-0.2    4,-1.7     2,-0.2    -1,-0.2
 25    25      I   H <> S+      0   0    99   -3,-0.7    4,-1.8     2,-0.2    -1,-0.2
 26    26      c   H X  S+      0   0     0   -4,-2.2    4,-1.9     2,-0.2     6,-0.4
 27    27      A   H X  S+      0   0    12   -4,-2.5    4,-2.7    -5,-0.2     5,-0.5
 28    28      T   H <  S+      0   0   120   -4,-1.7   -1,-0.2     1,-0.2    -2,-0.2
 29    29      Y   H <  S+      0   0   176   -4,-1.8   -1,-0.2    -5,-0.2    -2,-0.2
 30    30      T   H <  S-      0   0    24   -4,-1.9   -2,-0.2    -3,-0.2    -3,-0.2
 31    31      G   S <  S+      0   0    35   -4,-2.7   -3,-0.2     1,-0.4    -4,-0.1
 32    32      b        -       0   0     5   -5,-0.5   -1,-0.4    -6,-0.4     2,-0.3
 33    33      I   E    -A      3  0A    51  -30,-2.8  -30,-2.4    -3,-0.1     2,-0.5
 34    34      I   E    -A      2  0A    78   -2,-0.3  -32,-0.2   -32,-0.2     3, 0.0
 ....
```

β-sheet label
β-bridge 2 label
β-bridge 1 label
Chirality assignments
bend assignments
π-helix hydrogen bonds
α-helix hydrogen bonds
3₁₀-helix hydrogen bonds
Secondary structure synopsis

Figure 19.4. Explanation of DSSP output example segment from Crambin (Teeter, 1984). The first two columns contain the unique DSSP residue number and the corresponding PDB residue number. The third column (here empty) indicates the chain identifier if there are multiple chains. Then follows the amino acid "AA" in one letter codes (*Note*: lower case letters are all cysteines, in order to match up cysteine-bridges, for example, residue 16 has a disulfide bond to residue 26). The "Structure" section starts with the secondary structure synopsis (HBEGITS listed in order of priority in case of overlaps) and is followed by helix hydrogen bond indications for 3_{10}-, α-, and π-helix hydrogen bonds, where ">" indicates an acceptor, "<" a donor and "X" both. The bend and chirality are each given a column followed by the β-bridge label columns (lower case labels are parallel β-bridges and upper case are antiparallel). The DSSP numbers of their partners are written in the "BP1" and "BP2" columns. Each β-sheet is also given a label (independent of the β-bridge labels) indicated in the adjacent column. The "ACC" column contains the solvent accessible surface measured in Å^2 by estimating the number of water molecules in contact with the present residue. The two strongest backbone–backbone hydrogen bonds are then listed, where "N–H → O" are donor hydrogen bonds and "O → H–N" acceptor hydrogen bonds. The format indicates the relative sequence position of the hydrogen bond partner followed by the energy in kcal/mol (e.g. "−5,−0.8" means that the partner residue's DSSP number is 5 less than the present 1 and that the hydrogen bond energy is −0.8 kcal/mol). The DSSP output also contains the C_α coordinates, phi–psi angles and other angles which were left out in this figure due to limitations of space.

hydrogen bonds) and for π-helices (state "I" with $i \rightarrow i + 5$ hydrogen bonds) as well. The helix definition does not assign the edge residues having the initial and final hydrogen bonds in the helix. A minimal size helix is set to have two consecutive hydrogen bonds in the helix, leaving out single helix hydrogen bonds, which are assigned as turns for all three helices (state "T").

β-Sheet residues (state "E") are defined as either having two hydrogen bonds in the sheet, or being surrounded by two hydrogen bonds in the sheet. This implies three sheet residue types: antiparallel and parallel with two hydrogen bonds or surrounded by hydrogen bonds. The minimal sheet consists of two residues at each partner segment. Isolated residues fulfilling this hydrogen bond criterion are labeled as β-bridge (state "B"). The recurring H-bonding patterns connecting the partnering strands in a β-sheet are occasionally interrupted by one or more so-called β-bulge residues. In DSSP these residues are also assigned as β-sheet "E" and may comprise up to four residues on one strand and up to one residue on the partnering strand. These interruptions in the β-sheet H-bonding pattern are only assigned as sheet if they are surrounded by H-bond forming residues of the same type, that is, either parallel or antiparallel. The remaining two DSSP states "S" and " " (space) indicate a bend in the chain and the unassigned/other state, respectively.

DSSPcont: Continuous DSSP Assignment Reflecting Protein Motion

The aim of the continuous assignment was to reflect the structural variability due to thermal motion in a way so that regions which do not vary upon thermal motion have crisp assignments (almost discrete values), while regions which undergo thermal motion should reflect this in their continuous assignment and ideally reflect the occupancy of each secondary structure state. We estimated this by following the energy-based secondary structure assignment of DSSP and letting the strength of the hydrogen bond reflect thermal motion in the assignment (Andersen et al., 2002). This concept led us to develop a continuous extension of DSSP. This continuous assignment is based upon multiple runs of DSSP with different hydrogen bond thresholds. Then, we compiled a weighted average over the individual DSSP assignments to assign secondary structure to each residue. We determined the weights by applying the above criterion for "good" assignments starting with structural homologues from the FSSP (Holm and Sander, 1998) database. Inspecting the structural alignments in detail, we noted a number of possible reasons for observed structural differences:

1. Different solution composition, spatial grouping, and/or environment of the proteins.
2. Uncertainties/errors in the experimental structure determination setup.
3. Minor thermal fluctuations (even though mostly averaged out).
4. Local amino acid substitutions causing the structural change.
5. Insertions/deletions adjacent to the local stretch in question.
6. Nonlocal changes forcing a new local conformation.
7. Other less likely causes, for example, prion-like switching.

Our objective was a secondary structure assignment method de-emphasizing the effects of 1–3 while capturing differences caused by sequence changes. However, for structural alignments of homologues, we cannot separate these effects as illustrated by a comparison between two related structures: periplasmic binding protein (PDB: 4MBP; Quiocho et al., 1997) and putrescine binding protein (PDB: 1POT; Sugiyama et al., 1996). The structural alignment was obtained from FSSP with a Z-score of 23.2 and an RMSD of 3.6 Å over 303 residues. We will focus on a small 10-residue segment (Figure 19.5a) that has spiraling structure (α-helix, 3_{10}-helix or turn) and a β-bridge at the penultimate

PDB : 4MBP

	AA	Structure	BP1	BP2	ACC	N-H-->O	O-->H-N	N-H-->O	O-->H-N
82	D	>> -	0	0	103	-2,-0.4	4,-1.9	1,-0.1	3,-0.8
83	K	H 3> S+	0	0	155	1,-0.2	4,-1.5	2,-0.2	-1,-0.1
84	A	H >4 S+	0	0	63	2,-0.2	3,-0.7	1,-0.2	4,-0.4
85	F	H X> S+	0	0	8	-3,-0.8	3,-2.5	1,-0.3	4,-0.6
86	Q	H >< S+	0	0	65	-4,-1.9	3,-0.9	1,-0.3	-1,-0.3
87	D	T << S+	0	0	95	-4,-1.5	-1,-0.3	-3,-0.7	-2,-0.2
88	K	T <4 S+	0	0	111	-3,-2.5	217,-2.2	-4,-0.4	218,-0.4
89	L	B << S-G	304	0C	2	-3,-0.9	215,-0.2	-4,-0.6	214,-0.1
90	Y	>> -	0	0	75	213,-1.5	3,-1.4	-2,-0.2	4,-0.8

PDB : 1POT

	AA	Structure	BP1	BP2	ACC	N-H-->O	O-->H-N	N-H-->O	O-->H-N
103	K	S < S+	0	0	117	-3,-1.5	2,-0.8	1,-0.2	-1,-0.2
104	L	> +	0	0	2	-3,-0.4	3,-1.5	1,-0.1	-1,-0.2
105	T	T 3 +	0	0	97	-2,-0.8	3,-0.2	1,-0.2	-1,-0.1
106	N	T > S+	0	0	38	1,-0.1	3,-2.6	2,-0.1	-1,-0.2
107	F	G X +	0	0	34	-3,-1.5	3,-2.0	1,-0.3	-1,-0.1
108	S	G 3 S+	0	0	104	1,-0.3	-1,-0.3	-3,-0.2	-2,-0.1
109	N	G < S+	0	0	35	-3,-2.6	189,-1.8	170,-0.1	190,-0.8
110	L	B < S-L	272	0E	8	-3,-2.0	187,-0.2	187,-0.2	5,-0.1
111	D	> –	0	0	35	185,-3.0	3,-1.6	-2,-0.2	-1,-0.1

	PSIpred	SSpro	PROF
4MBP	DKAFQDKLY	DKAFQDKLY	DKAFQDKLY
	hhhhhhccc	chhhhhccc	chhhhhccc
1POT	KLTNFSNLD	KLTNFSNLD	KLTNFSNLD
	ccccccccc	ccccccccc	ccccccccc

Figure 19.5. DSSP assignments for similar structures: 4MBP and 1POT. (a) The DSSP assignment for two segments taken from two structurally similar proteins (periplasmic binding protein 4MBP: Quiocho et al., 1997, and putrescine binding protein 1POT: Sugiyama et al., 1996) illustrates that the observed differences between these segments may originate from sequence differences. The boxed letters shown in the column next to the amino acid sequence give the final DSSP assignment: G = 3_{10}-helix, H = α-helix, T = turn, B = β-bridge, and S = bend. The next column shows the hydrogen bonds (>: hydrogen bond acceptor, <: hydrogen bond donor and X: both), with indications of the hydrogen bond length, that is, i → i + (3,4) for 3_{10} and α-helices, respectively. (b) All the predictions from PSIpred (Jones, 1999a; Jones, 1999b), SSpro (Baldi et al., 1999) and PROFphd (Rost, 1996) (see Chapter 29) correctly spot the α-helix signal in 4MBP, while missing this signal for 1POT. This may indicate that the altered sequence changed the structure significantly in this region. Here, "h" refers to the DSSP class helix (H or G) and "e" to the DSSP nonregular class. *Note:* The predictions are cut out from those for the entire protein.

position (Figure 19.5a). Based on the assignment alone one might characterize the differences as problems in the assignment process, since both segments have 3_{10}-helix hydrogen bonds over the entire stretch. On the contrary, 1POT has no α-helix hydrogen bonds resulting in the assignment of 3_{10}-helix. The results from three high-quality prediction methods (Figure 19.5b) suggest that the structural differences resulted from the sequence divergence. This means that the secondary structure assignments of the two segments should not necessarily be the same. This line of reasoning can be extended from short helices to short sheets and to the N- or C-terminal ends of helices and strands (caps).

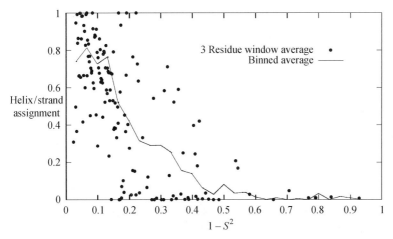

Figure 19.6. Protein motion and secondary structure. Using one set of coordinates from an ensemble of NMR models, the continuous DSSP assignment reproduces the segments in proteins that experimentally had a high degree of motion due to thermal fluctuations in water. Protein motion has been independently measured by the order parameter 1-S2, by the tumbling of the N–H backbone bond-vector. 1-S2 is low when the amino acid is fixed as in the protein core, and it is high when the residue fluctuates. 1-S2 is shown versus the continuous DSSP assignment grouping helices (GHI) and strands (EB). The points are averages over a window segment of three consecutive residues; the line gives an average of helix/strand assignments. *Source*: Figure reproduced from Andersen (2001).

Therefore, we chose to optimize the weights for DSSPcont based on the comparisons between different NMR models for the same protein.

We found that the single residue RMSD between models of high-quality NMR structures correlated well with thermal fluctuations in water as independently measured by the order parameter. The resulting continuous DSSP assignments were constructed to reflect the differences between NMR models of the same protein, so that the assignments reflect segments with thermal fluctuations (Figure 19.6). This means that more the sequence segment fluctuates, the lower the probability for the assigned helix/sheet will become. Information of this type can also be obtained directly from crystal structures. Overall, we found that the continuous assignment of secondary structure reflected the average occupancy of secondary structure assignments. In particular, our continuous assignment for a single NMR structure is similar to the average obtained over all models.

STRIDE: H-Bond and phi/psi Angle-Based Assignment Mimicking Experts

The STRIDE by Frishman and Argos (1995) uses an empirically derived hydrogen bond energy (see Section "Secondary Structure Concepts") and phi–psi torsion angle criteria to assign secondary structure. Torsion angles are given α-helix and β-sheet propensities according to how close they are to their regions in Ramachandran plots (see Chapter 2, Figure 19.6) (Ramachandran and Sasisekharan, 1968). The method fixes five internal parameters for α-helix and four for β-sheets. The parameters are optimized to mirror visual assignments made by crystallographers for a set of proteins. However, crystallographers often disagree in their assignment of secondary structure, which STRIDE aims to even out by averaging over many structures. Since the secondary structure categories have different

```
PDB:1CRN
REM  |---Residue---|    |--Structure--|  |-Phi-|  |-Psi-|  |-Area-|    1CRN
....
ASG  VAL -  15   15   H    AlphaHelix   -69.24   -41.22    93.8     1CRN
ASG  CYS -  16   16   H    AlphaHelix   -56.67   -36.00    18.4     1CRN
ASG  ARG -  17   17   H    AlphaHelix   -77.07   -16.13    94.1     1CRN
ASG  LEU -  18   18   H    AlphaHelix   -53.21   -46.17   143.0     1CRN
ASG  PRO -  19   19   C          Coil   -77.19    -7.60   108.9     1CRN
ASG  GLY -  20   20   C          Coil   106.26     7.31    52.1     1CRN
ASG  THR -  21   21   C          Coil   -52.67   136.34    38.4     1CRN
ASG  PRO -  22   22   C          Coil   -56.98   146.62    81.9     1CRN
ASG  GLU -  23   23   H    AlphaHelix   -56.41   -36.19    68.9     1CRN
ASG  ALA -  24   24   H    AlphaHelix   -63.43   -34.86    61.3     1CRN
ASG  ILE -  25   25   H    AlphaHelix   -74.77   -37.89    98.2     1CRN
ASG  CYS -  26   26   H    AlphaHelix   -64.95   -31.69     0.0     1CRN
ASG  ALA -  27   27   H    AlphaHelix   -62.04   -54.03    11.6     1CRN
ASG  THR -  28   28   H    AlphaHelix   -68.78   -25.49   121.1     1CRN
ASG  TYR -  29   29   H    AlphaHelix   -67.59   -36.30   174.0     1CRN
ASG  THR -  30   30   H    AlphaHelix  -108.96   -18.47    23.4     1CRN
ASG  GLY -  31   31   C          Coil    91.82    -3.07    36.1     1CRN
ASG  CYS -  32   32   C          Coil   -69.52   164.38     4.6     1CRN
ASG  ILE -  33   33   E        Strand  -129.76   157.03    51.0     1CRN
ASG  ILE -  34   34   E        Strand  -111.56   129.59    78.0     1CRN
....
```

Figure 19.7. Explanation of STRIDE output. The STRIDE output for Crambin is shown to explain the format and for comparison to Figure. The format is simple and easily parsed, with "ASG" as the first word in the lines used for assignment. The residue columns comprise the three-letter amino acid code, the chain identifier ("-" for single chains), the PDB residue number and the STRIDE residue number, which starts from one for every new chain. The two structure columns contain the one-letter structure assignments (HGIEBbTC) and its short description. The columns with phi and psi are followed by the column with solvent accessibility (measured in $Å^2$).

parameters, their assignment thresholds are independent for the hydrogen bond and phi–psi torsion angles. By construction, the STRIDE assignments agreed better with the expert assignments than DSSP, at least for the dataset used to optimize the free parameters. In particular, the authors reported that every 11th β-sheet and every 32nd α-helix was more in register with the expert assignments for the data set used (Figure 19.7).

Like DSSP, STRIDE assigns the shortest α-helix ("H") if it contains at least two consecutive $i \rightarrow i + 4$ hydrogen bonds. In contrast to DSSP, helices are elongated to comprise one or both edge residues if they have acceptable phi–psi angles, similarly a short helix can be removed if the phi–psi angles are unfavorable. This implies that hydrogen bond patterns may be ignored if the phi–psi angles are unfavorable. The sheet category does not distinguish between parallel and antiparallel sheets. The minimal sheet ("E") is composed of two residues each in one of five possible hydrogen bond conformations, that is, two more than for DSSP. The dihedral angles are incorporated into the final assignment criterion as was done for the α-helix. β-Sheet bulges are accepted applying the same criterion as DSSP. Single residue sheets, that is, β-bridges are labeled as "B" for the three DSSP hydrogen bond conformations and as "b" for the remaining two. 3_{10}- ("G") and π-helices ("I") are implemented according to the DSSP scheme, but with the empirical hydrogen bond criterion. Turns are assigned according to the phi–psi angles of residue $i + 1$ and $i + 2$ as described in Wilmot and Thornton (1990). The "C" symbol is used whenever none of the above structure requirements is met.

STICK: Continuous Assignment Based on Line Segments

The standard method used to define line segments is to fit an axis through each secondary structure element (e.g. DEFINE). This approach has difficulties, with both inconsistent definitions of secondary structure and the problem of fitting a single straight line to a bent structure. STICK avoids these problems by finding a set of line segments independently of any external secondary structure definition (Taylor, 2001). This allows the segments to be used as a novel basis for secondary structure definition by taking the average rise per residue along each axis to characterize the segment. This practice has the advantage that secondary structures are described by a single (continuous) value that is not restricted to the conventional classes of α-helix, 3_{10}-helix, and β-strand. This latter property allows structures without "classic" secondary structures to be encoded as line segments that can be used in comparison algorithms. When compared over a large number of pairs of homologous proteins, the current method was found to be slightly more consistent than a widely used method based on hydrogen bonds.

Beta-Spider: Packing Energy Assignment

The stabilizing factors maintaining the secondary structure were used in β-spider (Parisien and Major, 2005) as the primary parameters for assignment. This was done by calculating the packing energy of the backbone interactions in the form of Coulomb electrostatic and van der Waals forces (see Section "Secondary Structure Concepts"). For possible β-sheet interacting strands, the packing energy was calculated for tri-peptide pairs and defined as "favored" if at least -5 kcal/mol (-1.67 kcal/mol per residue). This may seem as more than three times the interaction energy required by DSSP H-bonds (-0.5 kcal/mol per residue), but the number of possible polar atom interactions per residue has also tripled (two H-bond donors/acceptors and one dipole), and furthermore the van der Waals energy is included. Two geometrical considerations are also used, which require that the sheet residue pairs are maximally 6 Å apart and; that the torsion $\langle \vec{C}_{\beta i}, \vec{C}_{\alpha i}, \vec{C}_{\alpha j}, \vec{C}_{\beta j} \rangle \leq 90°$. So, if $C_{\alpha i}$ and $C_{\alpha j}$ are aligned, the projected angle spanned between the two C_β must be less than $90°$, that is, point either towards each other or away from each other (glycine is exempted). β-bulges up to three residues in length are also allowed within the sheet. In summary, a β-spider sheet must contain at least one energetically "favored" tri-peptide pair and start/end with a residue pair following the geometrical requirements and be at least two residues long. This setup increases the set of tri-peptide H-bonding motifs allowed within a β-sheet, resulting in approximately 11% and 6% increase in number of motif matches for parallel and antiparallel sheets as compared to DSSP, respectively. At present only a β-sheet assignment scheme has been published.

XTLsstr: Circular Dichroism Driven Assignment

The circular dichroism of a protein in the far ultraviolet range is determined principally by amide–amide interactions of the backbone (King and Johnson, 1999), which has driven the authors to develop a protein secondary structure assignment scheme: XTLsstr. The program calculates two backbone dihedral angles and three distances (two of which are simple two atom H-bond distances), which are used for the assignment. By visual inspection the authors have developed range definitions for each secondary structure (3_{10}-, α-helices, β-sheets and two turn types) that would be consistent with the amide–amide interactions observed in CD.

VoTAP: Residue Contact Assignment Based on Geometry

By first dividing the protein volume into residue specific polyhedra using Voronoï tessellation, a residue contact map is automatically defined (Dupuis et al., 2004) (see

Section "Secondary Structure Concepts"). This contact map is unique and does not depend on a distance threshold, here two residues are in contact if they share one face of their polyhedra. The contacts were divided into strong and normal contacts based on the contact surface area size in a residue specific manner, thus yielding three residue–residue contact designations 0, 1, and 2 for no, normal, and strong contact, respectively. VoTAP assigns residues into the three overall classes helix (3_{10}, α and π), sheet (antiparallel, parallel, and β-bridge), and coil. The assignment was performed by fitting contact pattern statistics to the consensus assignment of DSSP, PSEA, DEFINE and STRIDE in two steps. The first step focuses on the diagonal of the contact matrix by comparing residue contact quintuplets. A lookup table for each quintuplet was created for the consensus assignment and a probability for each secondary structure class stored. A secondary structure class probability is subsequently assigned to each residue using a sliding window. In the second step the off-diagonal contacts are used to assign sheet residues if they follow the parallel or antiparallel β-plated sheet pattern. By VoTAP helices and sheets are constrained to be at least three residues in length.

DEFINE: Idealized C_α-Distance Mask-Based Assignment

The algorithm DEFINE (Richards and Kundrot, 1988) assigns secondary structures by matching C_α-coordinates with a linear distance mask of the ideal secondary structures. First, strict matches are found, which subsequently are elongated and/or joined allowing moderate irregularities or curvature. The algorithm locates the starts and ends of α- and 3_{10}-helices, β-sheets, sharp turns and omega-loops. With these classifications, the authors are able to assign 90–95% of all residues to at least one of the given secondary structure classes.

To assign α-helices the linear mask is matched with each row in the distance matrix of the query protein and the root mean square difference between the distances in the mask and the ones observed in the query protein is calculated as a measure of cumulative discrepancy. If a segment longer than four residues matches the mask within the allowed cumulative discrepancy limit (default $\varepsilon = 1$ Å), and α-helix is assigned to the segment. Assigned α-helices are checked whether they start or end with a 3_{10}-helix, but individual 3_{10}-helices and π-helices are not investigated.

In order to assign β-sheets as a single category, the authors have applied a linear distance mask taken from ideal antiparallel sheets. The problems associated with backbone bend-ability inside sheets and curvatures in larger sheets have been "solved" by excluding nonrigid sheets from the definition. The minimum length of sheets is set to be four residues. According to Pauling's definition of a β-sheet, each strand must pair to another strand to form a sheet. In contrast, DEFINE may assign unpaired strands.

P-Curve: Idealized Protein Curvature-Based Assignment

Sklenar et al. (1989) based their assignment scheme P-Curve on a mathematical analysis of protein curvature. Using differential geometry, they calculated a helicoidal axis on the basis of the fixed axis systems of a series of peptide planes. The secondary structure assignments are performed by motif matching, where the parameters in the motif are the radius of the helicoidal system along with a series of tilting, rolling, and twisting measures describing geometrical differences between two peptide planes. This parameter analysis is achieved mainly by the use of the C_α-coordinates. The P-Curve assignment differs significantly from those performed from phi–psi angles or hydrogen bonds, since different parameters are used (e.g. helicoidal radius, tilting, rolling, and twisting). Furthermore, the degrees of freedom allowed when matching a P-Curve motif are quite different from those allowed when matching a DEFINE linear distance mask. For example, while the linear distance mask of DEFINE fits poorly to a curved β-strand, the local P-Curve parameters are likely to fit better. The assigned secondary structures are recognized by matching known structural motifs. These motifs are based on idealized values for helicoidal parameters. The following motifs are used: right- and left-handed α-helix, 3_{10}-, and π-helix, parallel and antiparallel β-sheets and some other structures of little interest here. Note that like DEFINE, P-Curve may assign the category sheet to unpaired strands.

PALSSE: Linear Element Assignment for Structure Comparison

With the objective to describe, in a vector form, the two major classes of secondary structure for structure comparison, PALSSE performs a three-class assignment based on C_α-coordinates (Majumdar et al., 2005). The helix class assigned includes α-helices, 3_{10}-helices, π-helices, and turns that show a helical propensity in view of the observation that many α-helices start and end with tighter (3_{10}), looser (π) or nonbackbone H-bonded turns/helices. Similarly the β-strand class used includes parallel-sheets, antiparallel-sheets, β-bridges, β-bends, and β-hairpins. This results in many regular structure assignments where approximately 80% of residues are reported in the helix or sheet classes. The high coverage is important for structure comparison and similarity searches where the secondary structure elements are represented as vectors, thus allowing a higher degree of differentiation between proteins.

KAKSI: C_α and phi/psi-Based Assignment Mimicking Experts

The KAKSI assignment has been designed to best fit the secondary structure assignments done by experts in the PDB file header (Martin et al., 2005). This is done by defining allowed C_α distance measures and phi–psi angle values using a single sliding window for helices and two sliding windows for β-sheets to ensure partnering strands in the β-sheet. First helices are assigned followed by β-sheet assignment on the remaining residues, with minimal lengths of five and three residues, respectively. When comparing the KAKSI helix class to DSSP and STRIDE the authors map 3_{10}-, α-, and π-helices into one helix class, and β-sheets and β-bridges into one sheet class.

Other Secondary Structure Assignment Methods

P-SEA (Labesse et al., 1997) assigns helices and β-strands using only the C_α-coordinates. This is primarily done using a short range C_α distance mask ($i \rightarrow (i + 2, i + 3, i + 4)$) and two angle criteria for each secondary structure. The helix class assigned covers α-, 3_{10}-, and π-helices, and the strand class covers parallel and antiparallel β-strands, with minimal lengths of five and three residues, respectively.

SEGNO (Cubellis et al., 2005) performs its assignments based on C_α-coordinates, phi–psi angles, and an angle-distance hydrogen bond. For helix assignment the C_α-atoms must primarily reside within an imaginary cylinder helix, inspired from Richardson and Richardson (1988). The axis is defined by the mean of a sliding window of four C_α-atoms and

TABLE 19.1. Availability of Secondary Structure Assignment Programs

Program	Internet
DSSP	http://www.cmbi.kun.nl/gv/dssp
STRIDE	http://webclu.bio.wzw.tum.de/stride/
DSSPcont	http://cubic.bioc.columbia.edu/services/DSSPcont
VoTAP	http://www.lmcp.jussieu.fr/%7Emornon/voronoi.html
Beta-spider	http://www-lbit.iro.umontreal.ca/bSpider/
XTLsstr	http://oregonstate.edu/dept/biochem/faculty/johnson.html
KAKSI	http://migale.jouy.inra.fr/mig/mig_fr/servlog/kaksi/
PALSSE	http://prodata.swmed.edu/palsse/palsse.php
SEGNO	http://www.bioinf.man.ac.uk/~lovell/segno.shtml
SecStr	http://www.mbfys.lu.se/Services/SecStr/

the cylinder radius is 1.7–3 Å. The β-strand assignment is based on favorable phi–psi angles of at least three residues, and strands are associated into sheets using the angle–distance hydrogen bond. The authors report that this gives a stronger amino acid trend at the helix caps and also improves secondary structure guided sequence alignments. At present the method is not available so it wasn't possible to compare it directly to the other methods presented, but its website is reported in Table 19.1.

SECSTR (Fodje and Al-Karadaghi, 2002) was developed to identify and study the rare π-helices. It uses a DSSP like hydrogen bond definition and a Pauling $i \rightarrow i + 5$ hydrogen bond π-helix assignment scheme requiring at least two consecutive bonds. Approximately 10 times more π-helices were found compared to DSSP by giving priority to the strongest hydrogen bond instead of giving priority to α-helices and thus $i \rightarrow i + 4$ bonds in the assignment. This amounted to 104 overlooked π-helices extracted from a nonhomologous set of high-quality X-ray structures, which were verified by manual inspection.

Local protein structure analyses investigating frequently reoccurring small segments of the polypeptide chain also describe protein structure at the same scale as secondary structure and is being used for structure prediction by *ab initio* methods (see Chapter 32), covered nicely in a recent review (Offmann, Tyagi, and de Brevern, 2007).

SECONDARY STRUCTURE STATISTICS AND COMPARISON

Secondary Structure Frequency and Length

The regular secondary structures (helices and sheets), as defined by DSSP, comprise a bit more than half the protein residues (see Table 19.2), where the α-helix is most abundant with 31.3% followed by antiparallel β-sheets with 15.7% and parallel β-sheets with 5.7%. The remaining residues thus comprise a bit less than 50% (depending on the definition of regular structure used (see below)), which introduces a natural skewness of relevance within secondary structure prediction. The average lengths of the α-helix and β-sheet are 11.2 and 4.4 residues, respectively, which in terms of physical length interestingly enough are quite similar 16.8 and 14–15 Å, respectively (using physical distances for an α-helix of 1.5 Å per residue (Branden and Tooze, 1991) and residue span in fully extended strands is 3.2 Å residue in parallel β-sheets and 3.4 Å per residue in antiparallel β-sheets (Creighton, 1993)).

TABLE 19.2. DSSP Secondary Structure Statistics from a Set of 707
Nonhomologous Protein Chains

	Secondary Structure Statistics Using DSSP Assignments			
	α-Helix: "H"	β-Sheet: "E"	Parallel β-Sheet: "E"[a]	Antiparallel β-Sheet: "E"[a]
Frequency (%)	31.3	20.4[b]	5.7	15.7
Average length	11.2 residues	4.4 residues[c]	4.0 residues	4.6 residues
	3_{10}-helix: "G"	π-helix: "I"	β-bridge: "B"	other: " ","T","S"
Frequency (%)	3.7	0.04	1.3	43.3
Average length	3.4 residues	5.2 residues	1 residue per definition	6.8 residues

[a] The following DSSP residues were counted: sheet H-bonded, middle and single bulge.
[b] The sum of the parallel and antiparallel sheet frequencies is higher than the total β-sheet residues, since a residue may be in two sheets of different type.
[c] The average length of sheets reported is different from the average length of connected "E" stretches (5.2 residues), since overlapping sheets are counted as one.

Residue Distributions for Secondary Structure

The amino acids typically found in α-helices differ considerably from those found in β-sheets (Figure 19.8). Alanine and leucine often occur in α-helices, while proline and glycine are rare. In β-sheets valine and isoleucine are over-represented, while glycine, aspartic acid and proline are under-represented. Shorter structures such as 3_{10}-helices and β-bridges have distinct residue distributions. For 3_{10}-helices, the alanine and leucine signal has disappeared, instead the sequences are dominated by proline which often is observed as a helix initiator and breaker. For β-bridges, we no longer find a preference for valine and isoleucine. This indicates the role of the side chain in defining secondary and tertiary structure. In general, these preferences have long been the basis of secondary structure prediction methods.

Comparison

The secondary structure assignment schemes delineated above have each been designed using one or more aspects of protein structure and with one or more applications in mind. Good quality assignments applied within CD spectroscopy analyses may not necessarily be optimal when applied within structure comparison or vice versa. It is therefore inherently difficult to perform a qualitative comparison among protein secondary structure assignments, so the main focus of the comparison presented will be quantitative. In essence one should choose the secondary structure assignment scheme that is most consistent with the investigation where it is applied.

We used the simple structure of Crambin as an example to point out differences in the assignment schemes (Figure 19.9, note that the P-Curve assignment was taken from the original publication, Sklenar et al., 1989). When comparing the three-class assignments (α-helix, β-sheet, and other), STRIDE and DSSP are identical except for one residue at the C-cap of an α-helix, whose last H-bond is weak as reported by DSSPcont (reported in the figure as a grey letter). This assignment was confirmed by XTLsstr and KAKSI with

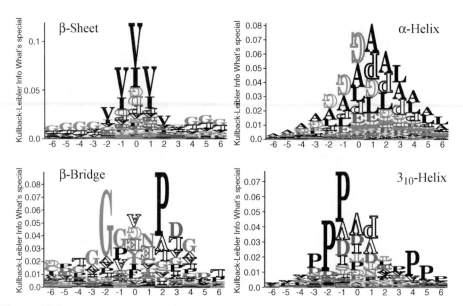

Figure 19.8. Sequence distributions for secondary structure. The four graphs show alignment statistics for β-sheets, α-helices, β-bridges and 3_{10}-helices, by the Kullback–Leibler information at positions surrounding the one assigned (position 0). The number of aligned segments are: (α-helix) 41803, (3_{10}-helix) 4952, (β-sheets) 27320, (β-bridges) 1851. These segments were retrieved from a data set of 707 nonhomologous protein chains using the DSSP assignment. At a given position, we therefore observed the 20 amino acids with a certain frequency; the Kullback–Leibler information calculates the information content of the observed frequencies with respect to the background frequencies (irrespectively of the structure). The more an observed set of frequencies differs from the background, the higher the respective letter. If an amino acid at a given position is observed less frequently than in the background, it is drawn upside-down and hollow.

some variation on cap assignments. P-Curve, STICK and PALSSE also assigned the same elements with capping variations, but reported an additional helix or sheet towards the protein's C-term. Looking at the sheet region in detail (Figure 19.9b), we see that the residues 39 and 40, assigned sheet by P-Curve only, are distant from any residue on a putatively pairing strand. According to Pauling, such an assignment would not be valid. The two sheets have been extended by beta-spider showing that high backbone–backbone packing energy not caught by standard H-bonds is keeping the strands together. Longer sheets were also identified by XTLsstr, KAKSI, PALSSE and partially by STICK and P-Curve indicating that the backbone conformation is extended also for these amino acids, so the extension of the two DSSP/STRIDE sheets appears reasonable. VoTAP identifies the two primary α-helices, but is the only method which, for this particular protein, does not assign the two sheets observed by other methods.

Are the discrepancies observed for Crambin representative? Secondary structure capping assignment do indeed constitute the major differences between methods as the exact specification of where a regular secondary structure ends is not well defined as reported by Colloc'h et al. (1993) for DSSP, P-Curve, and DEFINE. Investigating DSSP in particular, we also found variability of where the helix or sheet starts and ends between high-quality

NMR models for the same protein, where the NMR model variability was found to correlate with inherent protein motion (Andersen et al., 2002). Helix and sheet core segments are also the assignments which correlate best with circular dichroism spectra (Sreerama et al., 1999).

Martin et al. (2005) have performed a comparison of some of the methods described above on a high-quality X-ray dataset (resolution < 1.7Å and R-factor < 0.19) containing 689 protein structures. They used a comparison score called C_3 which compares two assignments as the number of identical regular secondary structure residue assignments relative to the total number of residues assigned to a regular structure, where the helix and sheet class assignments were compared. They found that DSSP and STRIDE are very similar ($C_3 = 95\%$) followed by the expert assignments in PDB ($C_3 > 87\%$). KAKSI is then the nearest neighbor to the group with C_3 in the 81–84% range followed by PSEA and XTLsstr in the 78–81% range. XTLsstr differs the most from the other assignment methods with a C_3 score down to 76% when compared to PSEA. Similar to what Colloc'h et al. (1993) has reported for DSSP; Martin et al. (2005) found that STRIDE assigns many short helices.

Another observation from the Crambin example is that the methods do not mix helix and sheet assignments on the same residue. This has generally been observed to hold true for DSSP, P-Curve, and DEFINE, where only 0.32% of the residues showed conflicts between helix and sheet assignments (Colloc'h et al., 1993).

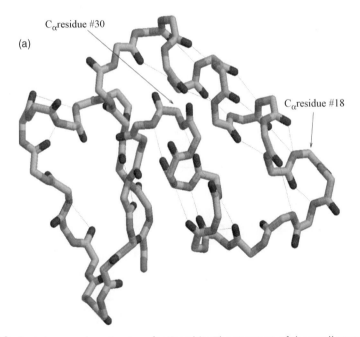

Figure 19.9. Protein secondary structure for Crambin. The structure of the small protein Crambin (Teeter, 1984, PDB: 1CRN) is shown from two angles: (a) the image of the two helices and (b) the central short sheet. (c) The overall assignment of secondary structure is shown for the different assignment methods, where the overall elements were found to agree between most methods shown. Please note that beta-spider only assigns sheets and the DSSPcont assignments have been discretized to crisp (100%: bold or space), high (>90%: normal), mixt (>50%: grey) variants of the DSSP assignment. Each assignment method was used with its default/standard settings supplied.

(b)

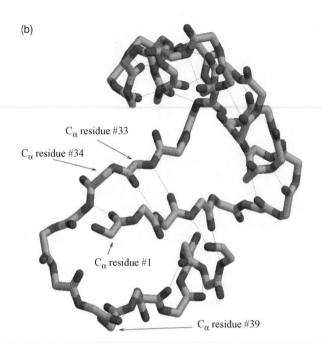

C_α residue #33

C_α residue #34

C_α residue #1

C_α residue #39

(c)   **Crambin: Abyssinian cabbage seed protein**

```
              ....:....1....:....2....:....3....:....4....:.
Sequence:     TTCCPSIVARSNFNVCRLPGTPEAICATYTGCIIIPGATCPGDYAN
DSSP:         EE SSHHHHHHHHHHHTTT  HHHHHHHS EE SSS   GGG
DSSPcont:     EE SSHHHHHHHHHH TTT  HHHHHHHT EE SSS   GGG
STRIDE:       EE  HHHHHHHHHHH     HHHHHHH  EE        TTTTT
P-Curve:      EEEE HHHHHHHHHH     HHHH   EE   EE
STICK:        EEE  HHHHHHHHHHH    HHHHH  EE        HH
VoTAP:             HHHHHH         HHHHH
Beta-Spider:  EEEE                       EEEE
XTLsstr:      EEE  HHHHHHHHGGG   PHHHHHHHH EEE      PTTT
KAKSI:         EEE HHHHHHHHHHHHH HHHHHHHHH EEE       Tt
PALSSE:       EEEE HHHHHHHHHHHHHHH HHHHHHHHHHEEE     HHHHH
```

Figure 19.9. (*continued*)

APPLICATIONS OF SECONDARY STRUCTURE

Secondary structure is being used in many areas of structural bioinformatics, from structure visualization, classification, and comparison to predictions, all covered in this book. Here we will delineate some areas where secondary structure is applied in investigations within different contexts, for example, in understanding biological processes and disease.

Secondary Structure and Protein Function

Circular dichroism spectroscopy is often used to measure differences in protein secondary structure contents under different conditions such as mutated residues or changing environments. An example is the thimet metallo endo-peptidase (THOP1) which is known to be modulated by changes in calcium concentration. To study how this change comes about, key aspartic acid residues believed to bind calcium have been mutated, and the calcium induced change in α-helix content studied by CD (Oliviera, 2005). See also Chapter 21 describing functional inference from structure.

Secondary Structure and Disease

Protein aggregation plays an important part in several well-known diseases such as Alzheimer's, Parkinson's, Huntington's, and prion disease (Kajava et al., 2006). The internal structure of amyloid fibrils in Alzheimer's disease (AD) and type 2 diabetes is a ladder of β-sheet structure arranged in a cross-β conformation (Stromer and Serpell, 2005). Whether, for example, the amyloid-β plaques are causing Alzheimer's disease or are mere agglomerates of excess amyloid-β is debated (Watson et al., 2005), but antiaggregates breaking the β-sheet formation are investigated to prevent AD (Chacon et al., 2004; Rzepecki et al., 2004). The formation of amyloid aggregation has been studied in close detail by Cerda-Costa et al. (2007), who found that the C-terminus of the molecule (comprising the last and edge β-strand) is the major contributor to amyloid fibril formation. 3D structure analysis revealed the stability of amyloid fibrils, their self-seeding characteristic and their tendency to form polymorphic structures (Nelson et al., 2005).

A conserved N-capping box has been found to be important for the structural autonomy of the prion α-helix, where the disease associated D202N mutation destabilizes the helical conformation (Gallo et al., 2005).

Secondary Structure Mimicking Compounds

To develop a drug that can inhibit protein–protein interactions, compounds are being specifically synthesized to mimic helices and strands (Song et al., 2001; Kutzki et al., 2002; Antuch et al., 2006). Antuch et al. reported α-helix mimetic compounds disrupting the Bcl-w/Bak protein–protein interaction, which is important in cancer (Wagner, 2005). Song et al. described nonpeptidic β-strand mimetic compounds that inhibit the HIV-1 protease dimerization necessary for its enzymatic activity. Helices represent one of the most common recognition motifs in proteins (Che et al., 2007), therefore characterization and analysis of the secondary structures involved in protein–protein interactions becomes important. See also Chapter 27 for a description of protein–ligand design.

Secondary Structure and Alignment

Taking local structural characteristics into account improves the detection of remote homologues and the alignment quality (Wallqvist et al., 2000; Shi et al., 2001; Qiu and Elber, 2006). In FUGUE, Shi et al. took the secondary structure, solvent accessibility, and hydrogen bonding status into account when aligning structures and sequences, where the alignment gap penalties are dependent on secondary structure and its conservation. SSALN, a similar method by Qiu and Elber (2006), was found to outperform CLUSTALW and GenThreader using the Fisher's benchmark.

Secondary Structure Capping

Helix capping has recently been studied in greater detail, for example, the capping dynamics of a glycine α_L C-capping motif has been studied in detail (Bang et al., 2006). Bang et al., using chemical synthesis, X-ray crystallography, and thermodynamic data, determined that local conformational strain is responsible for most of the energy penalty in an α_L C-capping motif (Rose, 2006). Helix caps are often stabilized by H-bonds other than the ones used for assignment (described above), as shown by Manikandan and Ramakumar (2004) for the $C-H \cdots O$ H-bond. The donor carbon studied was C_α, C_β, C_γ, C_δ or C_ε, depending on the residue, and the acceptor was the backbone $C'{=}O$ oxygen. The α-helices were assigned using PROMOTIF (Hutchinson and Thornton, 1996) and their ends defined using the Aurora–Rose nomenclature (Aurora and Rose, 1998) with N-terminal residues as N-cap to N5 and C-terminal residues as C5 to C'''. They found on average 1.5 and 3.9 $C-H \cdots O$ H-bonds per N- and C-term, respectively, thus indicating their relevance in stabilizing α-helix caps.

CONCLUSION

Assigning secondary structure from 3D coordinates is an important problem. Many successful solutions have been proposed over the past 25 years. One of the first automated assignment schemes was DSSP, which has become the standard in the field, followed by STRIDE. In fact, secondary structure assignment may be one of the exceptional examples of tools in structural biology and bioinformatics that have not been revolutionized by the explosion of data. For most residues, most of the available methods agree in their assignment. Methods tend to differ mainly in locating the caps of regular secondary structure segments and in distinguishing among more subtle differences (e.g. α-, 3_{10}-, or π-helix). These differences reflect the aspects used for the assignment and the application(s) for which the method was designed. Furthermore, there are indications that some fuzziness in the caps is inherent to protein motion (observed by NMR and CD) and that structural homologues also tend to differ more frequently there. Since the applications of secondary structure vary, their assignment schemes will as well, due to the differences in objectives. Optimization for structure comparison will aim at a high-regular structure content, while optimization for prediction will aim at clear capping signals and optimization for spectroscopists will aim at reflecting the experimental readout, so the story will continue.

ABBREVIATIONS

3D	three-dimensional
CD	circular dichroism
DEFINE	method assigning secondary structure from 3D coordinates based on linear distance masks of ideal secondary structure (Richards and Kundrot, 1988)
DSSP	program and database assigning secondary structure and solvent accessibility for proteins of known 3D structure from hydrogen bonding patterns
DSSPcont	continuous assignment of secondary structure for proteins of known 3D structure (Andersen et al., 2002)
H-bond	hydrogen bond
NMR	nuclear magnetic resonance

P-Curve   curvature-based assignment of secondary structure from 3D (Sklenar et al., 1989)
PDB   Protein Data Bank of experimentally determined 3D structures of proteins (Berman et al., 2000)
RMSD   root-mean square deviation
STRIDE   secondary STRuctural IDEntification method to assign secondary structure from 3D using hydrogen bonds and torsion angles (Frishman and Argos, 1995)
VoTAP   Voronoï tessellation based protein secondary structure assignment (Dupuis et al., 2004)

ACKNOWLEDGMENTS

Thanks to Jinfeng Liu (CUBIC, Columbia) for computer assistance and Jenny Gu for helpful comments on the manuscript. The work of BR was supported by grants 1-P50-GM62413-01 and RO1-GM63029-01 from the National Institutes of Health. Last, not the least, thanks to all those who deposit their experimental data in public databases, and to those who maintain these databases.

REFERENCES

Andersen CA (2001): Protein structure and the diversity of hydrogen bonds. *The Technical University of Denmark, Ph.D. Thesis.*

Andersen CA, Palmer AG, et al. (2002): Continuum secondary structure captures protein flexibility. *Structure* 10(2):175–184.

Antuch W, Menon S, et al. (2006): Design and modular parallel synthesis of a MCR derived alpha-helix mimetic protein–protein interaction inhibitor scaffold. *Bioorg Med Chem Lett* 16(6): 1740–1743.

Aurora R, Rose GD (1998): Helix capping. *Protein Sci* 7(1):21–38.

Baker EN, Hubbard RE (1984): Hydrogen bonding in globular proteins. *Prog Biophys Mol Biol* 44 (2): 97–179.

Baldi P, Brunak S, et al. (1999): Exploiting the past and the future in protein secondary structure prediction. *Bioinformatics* 15(11):937–946.

Bang D, Gribenko AV, et al. (2006): Dissecting the energetics of protein alpha-helix C-cap termination through chemical protein synthesis. *Nat Chem Biol* 2(3):139–143.

Berman HM, Bhat TN, Bourne PE, Feng Z, Gilliland G, Weissig H, Westbrook J (2000): The Protein Data Bank and the challenge of structural genomics. *Nature structural biology* 7(Suppl):957–959.

Beynon RJ (2004): Sequential exoproteolysis as a structural probe: a cautionary note. *J Mass Spectrom* 39(2):188–192.

Boobbyer DN, Goodford PJ, et al. (1989): New hydrogen-bond potentials for use in determining energetically favorable binding sites on molecules of known structure. *J Med Chem* 32(5): 1083–1094.

Bordo D, Argos P (1994): The role of side-chain hydrogen bonds in the formation and stabilization of secondary structure in soluble proteins. *J Mol Biol* 243(3):504–519.

Branden C, Tooze J (1991): *Introduction to Protein Structure.* New York: Garland Publishing.

Cerda-Costa N, Esteras-Chopo A, et al. (2007): Early kinetics of amyloid fibril formation reveals conformational reorganisation of initial aggregates. *J Mol Biol* 366(4):1351–1363.

Chacon MA, Barria MI, et al. (2004): Beta-sheet breaker peptide prevents Abeta-induced spatial memory impairments with partial reduction of amyloid deposits. *Mol Psychiatr* 9(10): 953–961.

Che Y, Brooks BR, et al. (2007): Protein recognition motifs: design of peptidomimetics of helix surfaces. *Biopolymers* 86(4).288–297.

Colloc'h N, Etchebest C, et al. (1993): Comparison of three algorithms for the assignment of secondary structure in proteins: the advantages of a consensus assignment. *Protein Eng* 6(4):377–382.

Cornell WC, Cieplak P, Bayly CI, Gould IR, Merz KM, Ferguson DM, Spellmeyer DC, Fox T, Caldwell JW, Kollman PA (1995): A second generation force field for the simulation of proteins, nucleic acids, and organic molecules. *J Am Chem Soc* 117:5179–5197.

Creighton T (1993): *Protein: Structures and Molecular Properties.* New York: W.H. Freeman.

Cubellis MV, Cailliez F, et al. (2005): Secondary structure assignment that accurately reflects physical and evolutionary characteristics. *BMC Bioinform* 6(Suppl. 4):S8

Cuff JA, Barton GJ (1999): Evaluation and improvement of multiple sequence methods for protein secondary structure prediction. *Proteins* 34(4):508–519.

Decatur SM (2000): IR spectroscopy of isotope-labeled helical peptides: probing the effect of *N*-acetylation on helix stability. *Biopolymers* 54(3):180–185.

Doerr A (2007): Ultrafast spectroscopy: timing is everything. *Nat Methods* 4(2):111

Dupuis F, Sadoc JF, et al. (2004): Protein secondary structure assignment through Voronoi tessellation. *Proteins* 55(3):519–528.

Fesinmeyer RM, Peterson ES, et al. (2005): Studies of helix fraying and solvation using 13C′ isotopomers. *Protein Sci* 14(9):2324–2332.

Fodje MN, Al-Karadaghi S (2002): Occurrence, conformational features and amino acid propensities for the pi-helix. *Protein Eng* 15(5):353–358.

Frishman D, Argos P (1995): Knowledge-based protein secondary structure assignment. *Proteins* 23 (4):566–579.

Gallo M, Paludi D, et al. (2005): Identification of a conserved N-capping box important for the structural autonomy of the prion alpha 3-helix: the disease associated D202N mutation destabilizes the helical conformation. *Int J Immunopathol Pharmacol* 18(1):95–112.

Holm L, Sander C (1998): Touring protein fold space with Dali/FSSP. *Nucleic Acids Res* 26(1):316–319.

Hutchinson EG, Thornton JM (1996): PROMOTIF—a program to identify and analyze structural motifs in proteins. *Protein Sci* 5(2):212–220.

Hvidt A, Westh P (1998): Different views on the stability of protein confirmations and hydrophobic effect. *J Sol Chem* 27 395–402.

Janes RW (2005): Bioinformatics analyses of circular dichroism protein reference databases. *Bioinformatics* 21(23):4230–4238.

Jeffrey GA, Saenger W (1994): *Hydrogen Bonding in Biological Structures.* Berlin: Springer-Verlag.

Jones DT (1999a): GenTHREADER: an efficient and reliable protein fold recognition method for genomic sequences. *J Mol Biol* 287(4):797–815.

Jones DT (1999b): Protein secondary structure prediction based on position-specific scoring matrices. *J Mol Biol* 292(2):195–202.

Kabsch W, Sander C (1983a): Dictionary of protein secondary structure: pattern recognition of hydrogen-bonded and geometrical features. *Biopolymers* 22(12):2577–2637.

Kabsch W, Sander C (1983b): How good are predictions of protein secondary structure? *FEBS Lett* 155(2):179–182.

Kajava AV, Squire JM, et al. (2006): Beta-structures in fibrous proteins. *Adv Protein Chem* 73:1–15.

King SM, Johnson WC (1999): Assigning secondary structure from protein coordinate data. *Proteins* 35(3):313–320.

Kolano C, Helbing J, et al. (2006): Watching hydrogen-bond dynamics in a beta-turn by transient two-dimensional infrared spectroscopy. *Nature* 4447118 469–472.

Kutzki O, Park HS, et al. (2002): Development of a potent Bcl-x(L) antagonist based on alpha-helix mimicry. *J Am Chem Soc* 124(40):11838–11839.

Labesse G, Colloc'h N, et al. (1997): P-SEA: a new efficient assignment of secondary structure from C alpha trace of proteins. *Comput Appl Biosci* 13(3):291–295.

Lees JG, Miles AJ, et al. (2006): A reference database for circular dichroism spectroscopy covering fold and secondary structure space. *Bioinformatics* 22(16):1955–1962.

Linderstrøm-Lang K (1952): *Proteins and Enzymes*. Stanford University Press.

Maiti NC, Apetri MM, et al. (2004): Raman spectroscopic characterization of secondary structure in natively unfolded proteins: alpha-synuclein. *J Am Chem Soc* 126(8):2399–2408.

Majumdar I, Krishna SS, et al. (2005): PALSSE: a program to delineate linear secondary structural elements from protein structures. *BMC Bioinform* 6:202

Manikandan K, Ramakumar S (2004): The occurrence of $C-H \cdots O$ hydrogen bonds in alpha-helices and helix termini in globular proteins. *Proteins* 56(4):768–781.

Martin J, Letellier G, et al. (2005): Protein secondary structure assignment revisited: a detailed analysis of different assignment methods. *BMC Struct Biol* 5:17

Nelson R, Sawaya MR, et al. (2005): Structure of the cross-beta spine of amyloid-like fibrils. *Nature* 435(7043):773–778.

Offmann B, Tyagi M, de Brevern AG (2007): Local protein structure. *Curr Bioinform* 2(3):165–202.

Parisien M, Major F (2005): A new catalog of protein beta-sheets. *Proteins* 61(3):545–558.

Pauling L (1939): *The Nature of the Chemical Bond*. New York: Cornell University Press.

Pauling L, Corey RB (1951): Configurations of polypeptide chains with favored orientations around single bonds: two new pleated sheets. *Proc Natl Acad Sci USA* 37(11):729–740.

Pauling L, Corey RB, et al. (1951): The structure of proteins;two hydrogen-bonded helical configurations of the polypeptide chain. *Proc Natl Acad Sci USA* 37(4):205–211.

Pelton, JT, McLean LR, (2000): Spectroscopic methods for analysis of protein secondary structure. *Anal Biochem* 277(2):167–176.

Qiu J, Elber R (2006): SSALN: an alignment algorithm using structure-dependent substitution matrices and gap penalties learned from structurally aligned protein pairs. *Proteins* 62(4):881–891.

Quiocho FA, Spurlino JC, et al. (1997): Extensive features of tight oligosaccharide binding revealed in high-resolution structures of the maltodextrin transport/chemosensory receptor. *Structure* 5 (8):997–1015.

Ramachandran GN, Sasisekharan V (1968): Conformation of polypeptides and proteins. *Adv Protein Chem* 23:283–438.

Richards FM, Kundrot CE (1988): Identification of structural motifs from protein coordinate data: secondary structure and first-level super secondary structure. *Proteins* 3(2):71–84.

Richardson JS, Richardson DC (1988): Amino acid preferences for specific locations at the ends of alpha helices. *Science* 240(4859):1648–1652.

Rose GD (2006): Lifting the lid on helix-capping. *Nat Chem Biol* 2(3):123–124.

Rost B (1995): TOPITS: threading one-dimensional predictions into three-dimensional structures. *Proc Int Conf Intell Syst Mol Biol* 3:314–321.

Rost B (1996): PHD: predicting one-dimensional protein structure by profile-based neural networks. *Methods Enzymol* 266:525–539.

Rzepecki P, Nagel-Steger L, et al. (2004): Prevention of Alzheimer's disease-associated Abeta aggregation by rationally designed nonpeptidic beta-sheet ligands. *J Biol Chem* 279(46):47497–47505.

Shi J, Blundell TL, et al. (2001): FUGUE: sequence-structure homology recognition using environment-specific substitution tables and structure-dependent gap penalties. *J Mol Biol* 310(1):243–257.

Sklenar H, Etchebest C, et al. (1989): Describing protein structure: a general algorithm yielding complete helicoidal parameters and a unique overall axis. *Proteins* 6(1):46–60.

Song M, Rajesh S, et al. (2001): Design and synthesis of new inhibitors of HIV-1 protease dimerization with conformationally constrained templates. *Bioorg Med Chem Lett* 11(18):2465–2468.

Sreerama N, Venyaminov SY, et al. (1999): Estimation of the number of alpha-helical and beta-strand segments in proteins using circular dichroism spectroscopy. *Protein Sci* 8(2):370–380.

Stromer T, Serpell LC (2005): Structure and morphology of the Alzheimer's amyloid fibril. *Microsc Res Tech* 67(3–4):210–217.

Sugiyama S, Matsuo Y, et al. (1996): The 1.8-A X-ray structure of the *Escherichia coli* PotD protein complexed with spermidine and the mechanism of polyamine binding. *Protein Sci* 5(10): 1984–1990.

Taylor WR (2001): Defining linear segments in protein structure. *J Mol Biol* 310(5):1135–1150.

Teeter MM (1984): Water structure of a hydrophobic protein at atomic resolution: Pentagon rings of water molecules in crystals of crambin. *Proc Natl Acad Sci USA* 81(19):6014–6018.

Tetin SY, Prendergast FG, et al. (2003): Accuracy of protein secondary structure determination from circular dichroism spectra based on immunoglobulin examples. *Anal Biochem* 321(2):183–187.

Villanueva J, Villegas V, et al. (2002): Protein secondary structure and stability determined by combining exoproteolysis and matrix-assisted laser desorption/ionization time-of-flight mass spectrometry. *J Mass Spectrom* 37(9):974–984.

Wade RC, Clark KJ, et al. (1993): Further development of hydrogen bond functions for use in determining energetically favorable binding sites on molecules of known structure. 1. Ligand probe groups with the ability to form two hydrogen bonds. *J Med Chem* 36(1):140–147.

Wagner G (2005): Ending the prolonged life of cancer cells. *Nat Chem Biol* 1(1):8–9.

Wallqvist A, Fukunishi Y, et al. (2000): Iterative sequence/secondary structure search for protein homologs: comparison with amino acid sequence alignments and application to fold recognition in genome databases. *Bioinformatics* 16(11):988–1002.

Watson D, Castano E, et al. (2005): Physicochemical characteristics of soluble oligomeric Abeta and their pathologic role in Alzheimer's disease. *Neurol Res* 27(8):869–881.

Wilmot CM, Thornton JM (1990): Beta-turns and their distortions: a proposed new nomenclature. *Protein Eng* 3(6):479–493.

20

IDENTIFYING STRUCTURAL DOMAINS IN PROTEINS

Stella Veretnik, Jenny Gu, and Shoshana Wodak

INTRODUCTION

Analysis of protein structures typically begins with decomposition of the structure into more basic units called structural domains. The underlying goal is to reduce a complex protein structure to a set of simpler, yet structurally meaningful units, each of which can be analyzed independently. Structural semi-independence of domains is their hallmark: domains often have compact structure that can fold (and sometimes function) independently. The total number of distinct structural domains is currently hovering around one thousand: they are represented by the unique folds in SCOP classification (Murzin et al., 1995) or unique topologies in CATH classification (Orengo et al., 1997). Interestingly, this is what Chothia predicted at a rather early stage of the Structural Genomics era (Chothia, 1992).

A significant fraction of these domains is universal to all life forms, others are kingdom-specific and yet others are confined to subgroups of species (Ponting and Russell, 2002; Yang, and Doolittle, and Bourne, 2005). The enormous variety of protein structures is then achieved through combination of various domains within a single structure. This "combining" of domains can be achieved by combining together single domain polypeptide chains within a noncovalently linked structure or by combining domains (via gene fusions/recombination) on a single polypeptide chain that folds into the final structure (Bennett, Choe, and Eisenberg, 1994). There are benefits to both strategies: the former can be seen as a more economic and modular approach in which many different structures can be put together with relatively few components within the cell. The latter, combining a specific set of modules within a single polypeptide chain, ensures that they are expressed together and localized in the same cells or cellular compartments (Tsoka and Ouzounis, 2000). This chapter is concerned with this latter case: the decomposition of the multidomain polypeptide chain into structural domains.

Structural Bioinformatics, Second Edition   Edited by Jenny Gu and Philip E. Bourne
Copyright © 2009 John Wiley & Sons, Inc.

Sequence information is often insufficient for identifying structural domains in the protein because the same structure can be reached from widely divergent sequence space (typically down to 30% sequence identity). Therefore, knowledge of protein structure is often a prerequisite to the delineation of structural domains. With the current rapid increase in the number of solved structures, a fast and consistent delineation of structural domains has great potential as well as great importance. Yet, in spite of the attention given over the last 30 years to the problem of domain delineation, it is not completely resolved as of today. While there are many methods for domain identification, the consistency of agreement among methods is 80% or less with the complex multidomain structures presenting most of the challenges (Holland et al., 2006). Some of the methods for domain identification are based exclusively, or partially, on the knowledge of human experts, while the other ones (computational methods) attempt to solve this problem in a completely automated way (Veretnik et al., 2004; Veretnik and Shindyalov, 2006). Expert methods are nearly always superior to the computational methods because experts can employ myriads of algorithms based on prior knowledge, biological and structural sensibility that ultimately will lead to a structure partitioning which might be inconsistent with some of the prior cases. Computational methods, on the contrary, are not able to reconcile the contradictions within the structural data set that might be internally inconsistent due to its biological complexity. Hence, current computational methods have difficulties in defining domain boundaries if the new examples require ingenious solutions that are contradictory to those previously acquired. The advantage of computational methods, however, is that they are fast and consistent—a feature of great importance in the era of structural genomics.

This chapter presents an overview of computational methods for parsing experimentally determined protein 3D structures into structural domains. The basic concepts underlying the domain parsing methods have been developed nearly 30 years ago, and has changed little since then. The actual algorithms, on the contrary, have undergone some new developments recently in terms of speed, generality, as well as fine-tuning. More important still, because of the increased complexity of the structures solved, we know much more today about the world of protein folds and their diversity. This makes the domain assignment problem more challenging, yet it also provides a broader data set of structures for tuning/improving algorithms. Before proceeding, we make a short note on terminology: the terms "partitioning," "delineation," "decomposition," "cutting," and "assignment" are used interchangeably here and have no special meaning associated with a particular term when referring to the partitioning of a protein chain into structural domains.

DEFINITIONS OF STRUCTURAL DOMAINS

Structural domains can be thought of as the most fundamental units of the protein structure that capture the basic features of the entire protein. Among such features are (1) stability, (2) compactness, (3) presence of the hydrophobic core, and (4) ability to fold independently. These structural/thermodynamic properties of domains suggest that atomic interactions within domains are more extensive than that between the domains (Wetlaufer, 1973; Richardson, 1981). From this, it follows that domains can be identified by looking for groups of residues with a maximum number of atomic contacts within a group, but a minimum number of contacts between the groups, as illustrated in Figure 20.1a. The spatial compactness of domains sometimes results in noncontiguous domains where stretches of residues that are distant on the polypeptide chain are found in close proximity in the folded

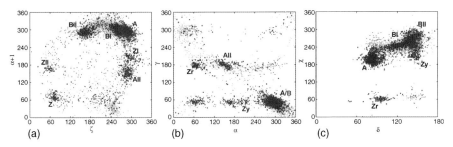

Figure 3.9. Backbone conformations in DNA and RNA. 2D scattergrams show conformations of 1861 nucleotides from 186 structures of uncomplexed DNA (dark blue), 5878 nucleotides from 261 structures of DNA in complexes with proteins or drugs (yellow), and 3752 nucleotides from 132 structures of RNA (red). A, AII, BI, BII, ZI, and ZII are respective double helical forms, Zr/Zy are purine and pyrimidine steps, respectively, in the Z-form. (a) Conformations at the phosphodiester linkage O3'-P-O5', ζ versus α; (b) torsions α versus γ; (c) relationship between the sugar pucker and the base orientation, δ versus χ. Torsion angles are defined in Figure 3.2.

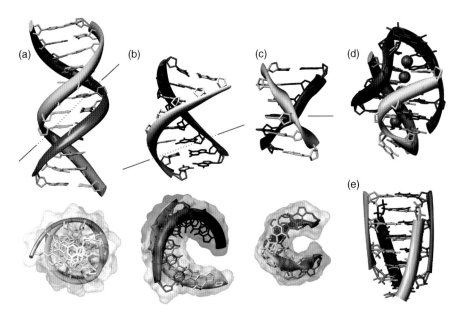

Figure 3.11. Important structural forms of DNA. Molecules are colored by strand. (a) B-DNA dodecamer (Drew et al., 1989); (b) A-DNA octamer (McCall et al., 1985); (c) Z-DNA hexamer (Gessner et al., 1989); (d) DNA guanine tetraplex. Four guanine tetramers are flanked by two T-T-T-T loops, potassium cations are shown in magenta (Haider, Parkinson, and Neidle, 2003); (e) cytosine tetraplex, or i-motif (Chen et al., 1994). In (a), (b), and (c), two perpendicular views are shown. The minor groove is shown in the red line.

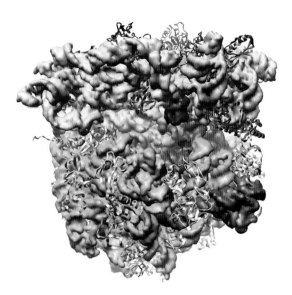

Figure 3.14. The structure of the complete 70s ribosome (Selmer et al., 2006). (*See text for full caption.*)

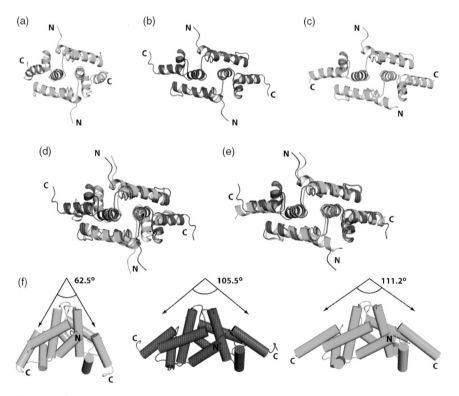

Figure 5.4. Automated structure calculation of the ZNF24 SCAN homodimer using a seed model. (*See text for full caption.*)

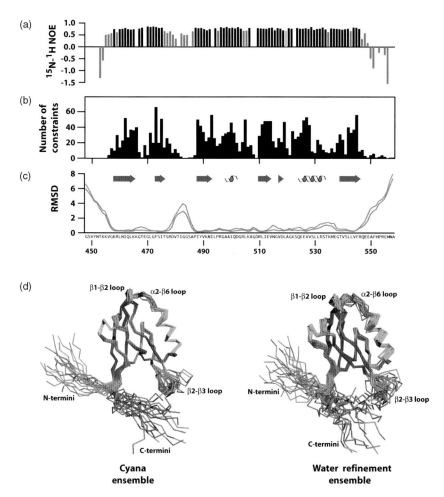

Figure 5.5. Refinement and validation of structural models. (a) Heteronuclear $^{15}N^1H$ NOE values for the second PDZ domain from Par-3. NOE values below 0.7 are shown in light gray. (b) NOE distance constraint density plotted by residue number. Distance constraints used in the final structure calculation were derived from 3000 assigned peaks observed in three 3D NOESY spectra. Elimination of redundant NOEs and meaningless constraints (distances that impose no conformational restriction or are held fixed by covalent geometry) yielded a final set of 1222 NOE distance constraints. Including dihedral angle constraints from TALOS, a total of 1339 NOE and dihedral angle constraints were used in the structure calculation corresponding to ~12 restraints per residue. However, if residues with $^{15}N^1H$ NOE values below 0.7 are excluded from consideration, the restraint density increases to ~17 restraints per residue. Structures defined by <10, 10–15, and >15 constraints per residue typically correspond to low, medium, and high-quality structures, respectively. (c) The root mean squared deviation for the structural ensembles before (green line) and after (orange line) water refinement in explicit water plotted as a function of residue number. (d) Structural ensembles produced in CYANA using NOEASSIGN before (left panel) and after water refinement (right panel) show differences in conformational sampling for flexible loop regions as reflected by increased RMSD values in panel c. The helix, strand and loop elements in each ensemble are colored cyan, magenta, and salmon, respectively.

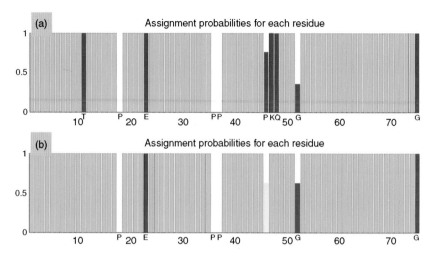

Figure 5.10. Representation of automated assignments determined for the protein ubiquitin from a set of input data generated by HIFI-NMR. The heights of the vertical bars indicate assignment probabilities. Correct assignments are indicated by green (high probability) and yellow (low probability) bars; incorrect assignments, as determined from the known structure, are indicated by red bars. (*See text for full caption.*)

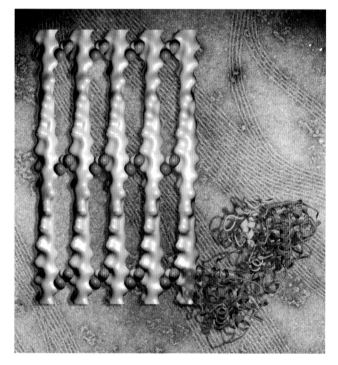

Figure 6.4. Example of a hybrid study that combines elements of electron crystallography and helical reconstruction with homology modeling and molecular docking approaches to elucidate the structure of an actin–fimbrin cross-link (Volkmann et al., 2001b).

Color files of all figures from this book are available for download from the following web address: ftp://ftp.wiley.com/public/sci_tech_med/structural_bioinformatics

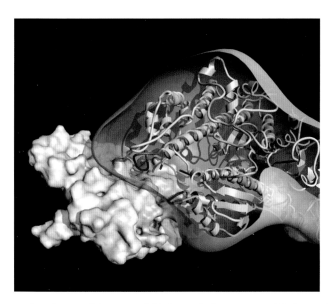

Figure 6.5. Example of a combination of high-resolution structural information from X-ray crystallography and medium-resolution information from electron cryomicroscopy (here 2.1 nm). Actin and myosin were docked into helical reconstructions of actin decorated with smooth-muscle myosin (Volkmann et al., 2000). Interaction of myosin with filamentous actin has been implicated in a variety of biological activities including muscle contraction, cytokinesis, cell movement, membrane transport, and certain signal transduction pathways. Attempts to crystallize actomyosin failed due to the tendency of actin to polymerize. Docking was performed using a global search with a density correlation measure (Volkmann and Hanein, 1999). The estimated accuracy of the fit is 0.22 nm in the myosin portion and 0.18 nm in the actin portion. One actin molecule is shown on the left as a molecular surface representation. (*See text for full caption.*)

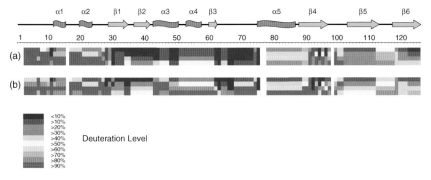

Figure 7.5. Amide H/D exchange for the PYP ground state (a) and the I_2 intermediate (b). Each block represents protein sequence segments defined by overlapping peptides and consists of four time points (from top: 30, 300, 3000, and 30,000 s). The deuteration levels of each such segment at each time point are shown by different colors from blue (<10% deuteration) through red (>90% deuteration), as indicated in the figure. Residues 1–2, 16, 76–77, and 98 are not covered by any sequence segment. The secondary structure elements are shown on top of the protein sequence; α stands for α-helix, β for β-sheet.

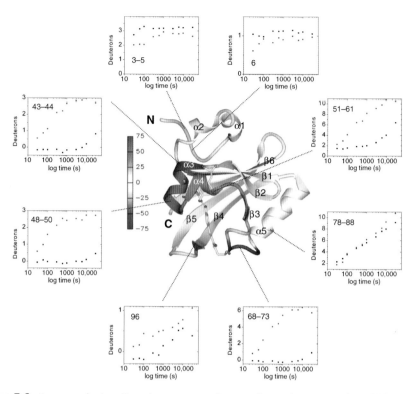

Figure 7.6. Average deuteration changes upon I$_2$ state formation mapped on PYP structure. Average deuterium incorporation changes upon I$_2$ formation from Figure 7.4 are mapped on the PYP ground-state ribbon diagram colored from −75% to +75% change according to the color bar. α-Helices 1–5, β-strands I–VI, and N- and C-termini are labeled. The chromophore and active-site residues TYR 42 and GLU 46, which form hydrogen bonds (green dotted lines) to the phenolate oxygen of the chromophore in the ground state, are shown for better orientation as well.

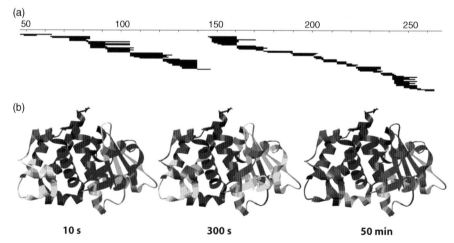

Figure 7.7. Pepsin fragmentation and H/D exchange behavior for LF$_N$. (*See text for full caption.*)

Color files of all figures from this book are available for download from the following web address: ftp://ftp.wiley.com/public/sci_tech_med/structural_bioinformatics

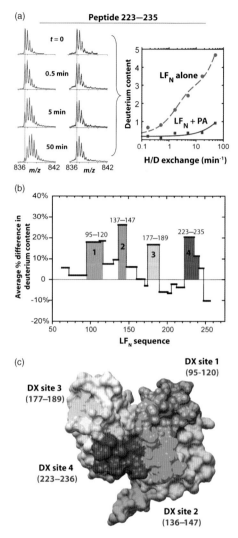

Figure 7.8. Deuterium uptake for LF$_N$ in the absence and presence of PA. (a) The mass spectra of a peptide corresponding to residues 223–235 in the absence (left) and presence (right) of PA. In the absence of PA, the centroid of the peaks increases steadily, whereas in the presence of PA, the centroid of peaks does not shift significantly. Gray lines are arbitrary reference points used to visualize the peak shift. Data from spectra were then used to calculate the deuterium content for LF$_N$ peptides in the absence (●) and presence of PA (■) over six time points. Data are fit to the sum of double exponentials and shown for LF$_N$ alone as a dashed red line, and for LF$_N$ + PA as a solid blue line. (b) The average percent difference in deuterium content over the six time points was plotted for representative peptides that spanned the LF$_A$ sequence. Positive values are taken to represent decreases in deuterium uptake in the presence of PA, relative to the absence of PA. The four sites with significant decreases in deuterium uptake in the presence of PA, denoted as DX sites 1–4, are colored blue, green, yellow, and red, respectively. (c) Surface representation of the proposed binding surface of LF$_N$ with the four DX sites (1–4) identified by H/D exchange colored in blue, green, yellow, and red, respectively.

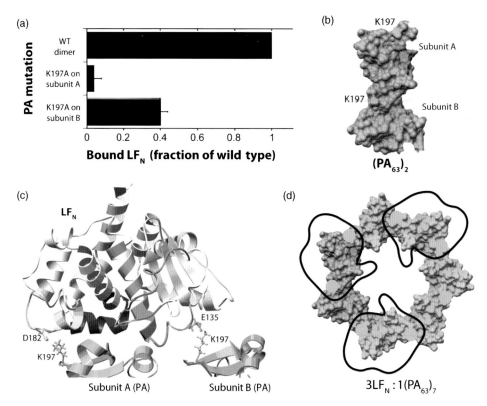

Figure 7.9. H/D exchange and mutagenesis define a model for the LF_N–PA interaction. (a) Data represent the fraction of mutant ^{35}S-LFN bound to PA and mutants of PA on cells relative to that seen for the WT ternary complex. Error bars represent SE of the mean. (b) Surface representation of dimeric PA with K197 shown in yellow on subunits A and B. (c) LF_N–PA_2 model adapted from Lacy et al. with the two subunits (A and B) of PA shown in pink and cyan. DX sites 1–4 identified using H/D exchange are colored blue, green, yellow, and red, respectively. D182 and E135 side chains are shown for LF_N and K197 side chains on both PA subunits are shown. (d) Top view of LF_N binding to domain 1 of heptameric PA. Black trace represents outline of LF_N. Yellow circles indicate K197 positions. Figures were generated using Chimera.

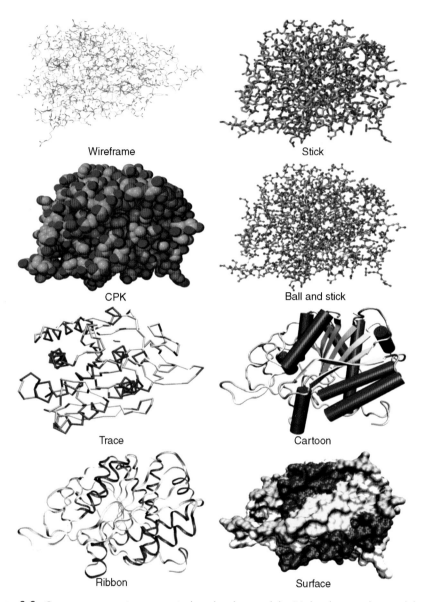

Wireframe

Stick

CPK

Ball and stick

Trace

Cartoon

Ribbon

Surface

Figure 9.6. Common computer-generated molecular models. Molecules can be modeled or represented in various ways. Each model gives a different perspective and insight into molecular structure and function. Carboxypeptidase A (3CPA.pdb) is used as an example. The CPK, stick, ball and stick were drawn by jAMVLE. The remainder were drawn by VMD. All files rendered by Pov-Ray.

Color files of all figures from this book are available for download from the following web address: ftp://ftp.wiley. com/public/sci_tech_med/structural_bioinformatics

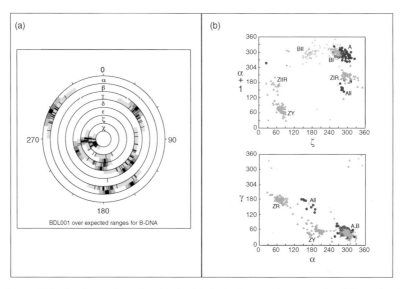

Figure 12.3. Conformations for the double helical DNA. (*See text for full caption.*)

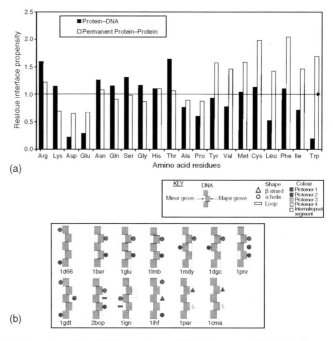

Figure 12.4. Interfaces of protein–DNA complexes. (a) Histogram of the interface residue propensities calculated for 26 protein–DNA complexes and compared to those for permanent protein–protein complex (Jones and Thornton, 1996). A propensity of >1 indicates that a residue occurs more frequently at the interface than on the protein surface. (b) Schematic drawings that model protein–DNA complexes for double-headed binding proteins. (*See text for full caption.*)

Color files of all figures from this book are available for download from the following web address: ftp://ftp.wiley.com/public/sci_tech_med/structural_bioinformatics

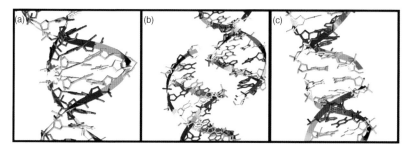

Figure 12.5. Examples of packing motifs in B-DNA duplexes. From left to right: (a) minor groove–minor groove interactions in BDL042 (Leonard and Hunter, 1993); (b) major groove–backbone interactions in BDJ060 (Goodsell, Grzeskowiak, and Dickerson, 1995); (c) stacking interactions in BDJ025 (Grzeskowiak et al., 1991). The bases are colored green for guanine, yellow for cytosine, red for adenine, and blue for thymine. Reprinted from Berman, Gelbin, and Westbrook (1996) with permission from Elsevier Science.

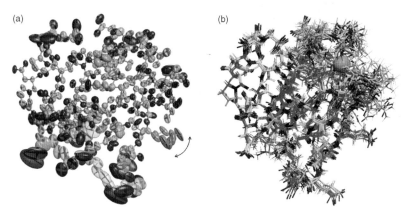

Figure 14.1. The different types of model generated by X-ray crystallography and NMR spectroscopy. Both are representations of the same protein: rubredoxin. (a) In X-ray crystallography the model of a protein structure is given in terms of atomic coordinates, occupancies and B-factors. The side chain of Glu50 has two alternative conformations, with the change from one conformation to the other identified by the double-headed arrow. The B-factors on all the atoms are illustrated by "thermal ellipsoids," which give an idea of each atom's anisotropic displacement about its mean position. The larger the ellipsoid, the more disordered the atom. Note that the main chain atoms tend to be better defined than the side chain atoms, some of which exhibit particularly large uncertainty of position. The coordinates and B-factors come from PDB entry 1IRN, which has a zinc ion bound and was solved at 1.2 Å resolution and refined with anisotropic B-factors. (b) The result of an NMR structure determination is a whole ensemble of model structures, each of which is consistent with the experimental data. The ensemble shown here corresponds to 10 of the 20 structures deposited as PDB code 1BFY. In this case, the metal ion is iron. The more disordered regions represent either regions that are more mobile, or regions with a paucity of experimental data, or a combination of both. The region around the iron-binding site appears particularly disordered, in stark contrast with the same region in the X-ray model in (a), where the B-factors are low and the model well-ordered. Both diagrams were generated with the help of the Raster3D program (Merritt and Bacon, 1999).

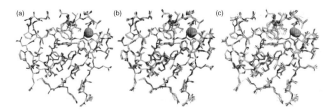

Figure 14.7. Examples of typical uncertainties in atomic positions for (a) an s.u. of 0.2 Å (b) 0.3 Å, and (c) 0.39 Å. (*See text for full caption.*)

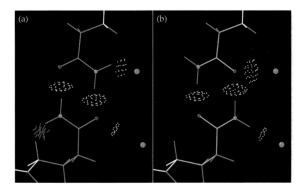

Figure 15.1. Resolving the ambiguity in a pair of doubly H-bonded side chain amides that have four equivalent H-bonds to each other and to waters, in the two best of four possible flip states. (*See text for full caption.*)

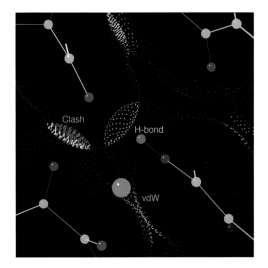

Figure 15.2. Slice through a small section of protein structure showing the relation of all-atom contact surfaces where the 0.25 Å-radius probe sphere left a contact dot only when it touched another noncovalently bonded atom as it rolled on the atomic van der Waals surfaces. (*See text for full caption.*)

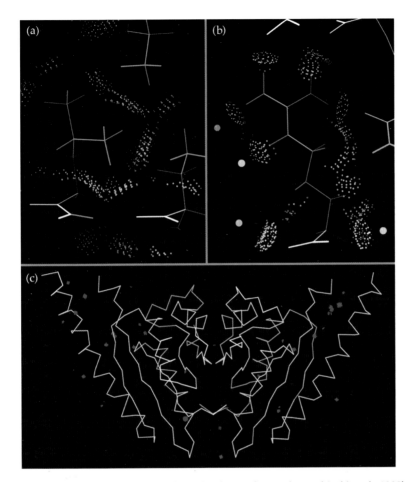

Figure 15.3. All-atom contact examples from the dimer of 1MJH (Zarembinski et al., 1998), a well-determined structural-genomics protein at 1.7 Å resolution. (a) All contacts for one of the typically well-packed and well-fit regions of aliphatic side chains, with the green of close van der Waals contacts predominant. (b) All contacts for an ARG side chain, with the five planar H-bonds (lens-shaped groups of pale green dots) of its guanidinium NHs formed either to protein O atoms or to waters (pink balls). (c) An overview of the dimer, with only the Cα backbone and the serious clashes ≥0.4 Å (red spikes) shown. When interactively displayed in KiNG or MAGE, it is easy to locate and fix the small number of isolated problems, including two flipped-over HIS rings at the putative active site and a LYS with a high B factor value squeezed into insufficient space between two hydrophobic side chains.

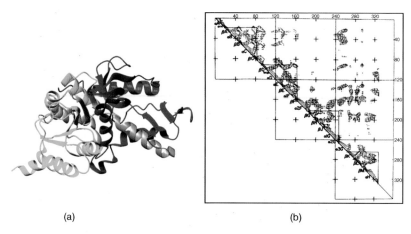

(a)   (b)

Figure 20.2. Domain structure of dogfish lactate dehydrogenase, determined using the Cα–Cα distance map. (a) Ribbon diagram of lactate dehydrogenase, showing the NAD binding (green) and catalytic domains (red). In gray is part of the helix, spanning residues 164–180, linking the two domains. (b) Distance map and structural domains in lactate dehydrogenase. Contours represent Cα–Cα distances of 4, 8, and 16 Å within the subunit of dogfish lactate dehydrogenase. Elements of secondary structure are identified along the diagonal. Triangles enclose regions where short Cα–Cα distances are abundant. The NAD binding domain comprises the first two triangles (counting from the N-terminus) that are subdomains. The catalytic domain comprises the last two triangles (the C-terminal domain). Taken from Rossman and Liljas (2006) and reproduced by permission of Academic Press (London) Ltd.

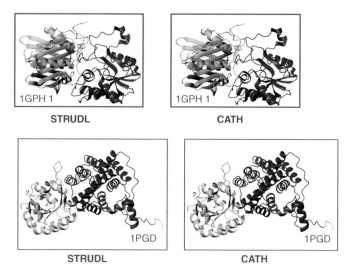

Figure 20.7. Different assignments by STRUDL (Wernisch, Hunting, and Wodak, 1999) and CATH (Jones et al., 1998), illustrating the effect of "noise," or decorations, in the protein chain trace. The STRUDL assignments are displayed on the left hand side, and the CATH assignments are displayed on the right. The short chain segments, which CATH assigns to separate domains, are shown in blue. Some of the discrepancies may be due to simple "slips" in the CATH assignments that have been, or will be, corrected.

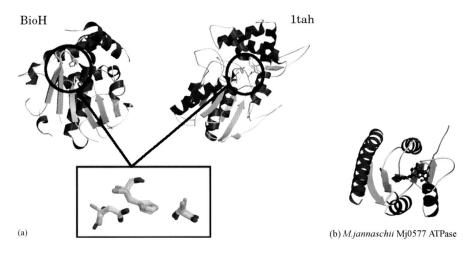

Figure 21.4. Prediction of function in earnest: three structures solved in the absence of functional information. (*See text for full caption.*)

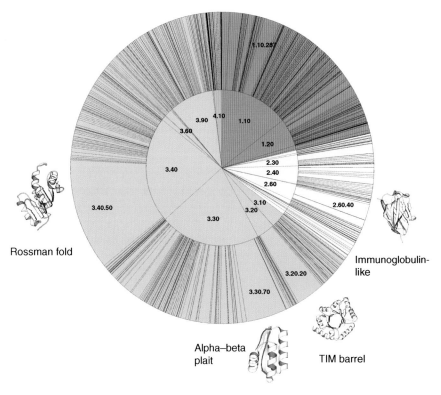

Figure 22.4. The distribution of CATH domains in 535 completed genomes in Gene3D. CATH classes are illustrated by color: purple for mainly alpha structures; yellow for mainly beta structures; green for alpha–beta structures. The inner wheel shows different architectures and the outer wheel different folds. For the most numerous folds, example domains are shown on the outside of the wheel.

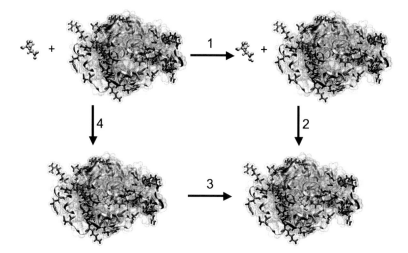

Figure 24.2. Titration calculation for a GLU 35 in the active site of hen egg white lysozyme (PDB ID 2LZT). The free energy of protonating GLU 35 (blue ball and stick moiety) in the presence of a biomolecule (cyan object) is calculated by a thermodynamic cycle. Specifically, the free energy of interest ΔG_3 is calculated in terms of the other steps in the cycle, ΔG_1 the energy of protonating the isolated GLU in solution (related to the "model pK_a"), ΔG_2 the energy of "binding" the protonated GLU to the lysozyme structure, and ΔG_4 the energy of "binding" the unprotonated GLU to the lysozyme structure, such that ΔG_3 can be calculated from a thermodynamic cycle as $\Delta G_3 = \Delta G_1 + G_2 - \Delta G_4$. Figure from *Reviews in Computational Chemistry*, N. Baker.

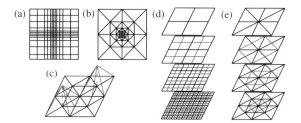

Figure 24.3. Meshes and hierarchies used in Poisson–Boltzmann solvers. (a) Cartesian mesh suitable for finite difference calculations; nonuniform mesh spacing can be used to provide a limited degree of adaptivity. (b) Finite element mesh exhibiting adaptive refinement. (c) Examples of typical piecewise linear basis functions used to construct the solution in finite element methods. (d) The multilevel hierarchy used to solve the PBE for a finite difference discretization; red lines denote the additional unknowns added at each level of the hierarchy. (e) The multilevel hierarchy used to solve the PBE for a finite element discretization; red lines denote simplex the subdivisions used to introduce additional unknowns at each level of the hierarchy.

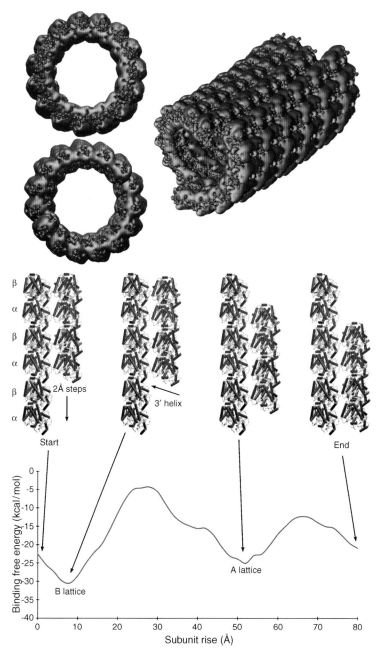

Figure 24.4. Microtubule electrostatics. (a) Electrostatic properties of a 1.2 million atom 400 × 300 × 300 Å microtubule fragment illustrating the current state of the art for continuum electrostatics calculations. The potential was calculated using APBS to solve the PBE at 150 mM ionic strength; more details are available in Baker et al. (2001b). (b) A potential of mean force, calculated using continuum electrostatics and a simple nonpolar function, describing microtubule protofilament assembly; more details are available in Sept, Baker and McCammon (2003). Part B is from Sept, Baker, and McCammon.

COLOR PLATES

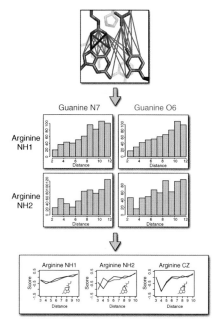

Figure 25.1. Creation of a distance-dependent statistical potential function. The process used to create a simple, distance-dependent statistical potential function is illustrated. (*See text for full caption.*)

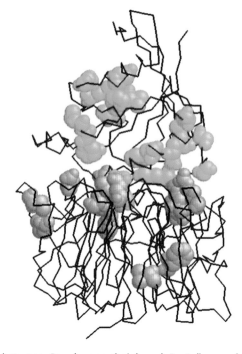

Figure 26.2. Complex between *Ran* (upper chain) and *Rcc1* (lower chain) marking the tree-determinant residues (in spacefill) for the two proteins (PDB ID: 1I2M). (*See text for full caption.*)

Color files of all figures from this book are available for download from the following web address: ftp://ftp.wiley.com/public/sci_tech_med/structural_bioinformatics

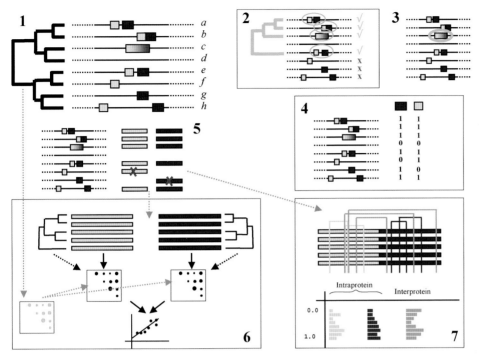

Figure 26.4. Theme of strategies implemented in computational methods for assessing the possible interaction between two proteins. (1) Sequence and genomic information about two proteins (yellow and blue) is used to assess their possible interaction. The inputs for these methods are the sequences and genomic positions of the orthologs of the two proteins, as defined by a given orthology criteria, in a number of organisms (related by a phylogeny). (2) Conservation of gene neighboring: the number of genomes where both proteins are close (according with a given distance cutoff) and their phylogeny are used to assess whether the proteins are interacting or not. (3) Gene fusion: a search for genomes where both proteins appear as part of a single polypeptide is performed. (4) Phylogenetic profiling: phylogenetic profiles for both proteins are constructed by assessing the presence [1] or absence [0] of the two proteins in the set of genomes, and the similarity between these profiles is evaluated. (5) Multiple sequence alignments for the two proteins are built using the sequences of the organisms where both are present. These MSA are used by the following two methods. (6) Similarity of phylogenetic trees: the multiple sequence alignments of the proteins (5) are used to generate distance matrices for both sets of orthologs. Alternatively, these multiple sequence alignments can be used to generate the actual phylogenetic trees and extract the distance matrices from them. The similarity of these distance matrices is used as an indicator of interaction. Eventually, the phylogenetic distances between the species involved can be incorporated in the method for correcting the background similarity expected between the trees due to the underlying speciation events. (7) Correlated mutation: the accumulation of correlated mutations between the two multiple sequence alignments is used as an indicator of interaction.

COLOR PLATES

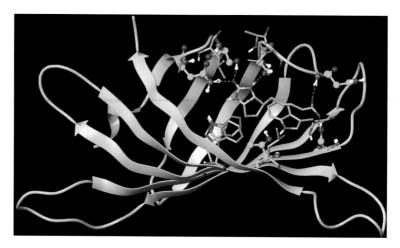

Figure 27.2. Streptavidin bound to biotin. Picture highlighting interactions between streptavidin (orange carbon atoms) and biotin. Amino acid residues that interact through hydrogen bonds with the ligand are shown with cyan-colored carbons, residues providing hydrophobic interactions have gray carbons.

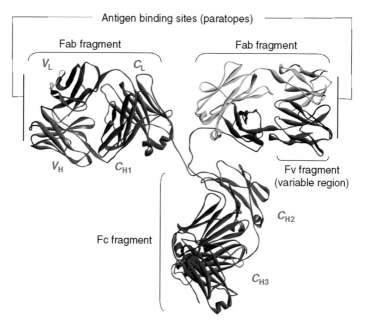

Figure 35.1. The structure of intact anticanine lymphoma monoclonal IgG2A mouse antibody [PDB : 1IGT]. The antibody heavy chains are shown in blue and red, light chains are colored in green and magenta; CDR loops are black on heavy chains and cyan on light chains. The image of the molecule was produced by J. Ponomarenko using the WebLabViewer software (Accelrys Inc.).

Color files of all figures from this book are available for download from the following web address: ftp://ftp.wiley.com/public/sci_tech_med/structural_bioinformatics

Figure 35.2. Two orthogonal views of the structure of influenza A virus (strain A/Aichi/2/68 H3N2 (X31)) hemagglutinin HA1 chain [PDB : 1EO8] with epitopes known from X-ray structures in complexes with antibodies. Chain HA1 is shown in light gray upon which are mapped residues of one linear B-cell epitope 100-YDVPDYASL-108 recognized by 17/9 Fab [PDB : 1HIM] (teal) and four structural B-cell epitopes inferred from protein structures in complexes with antibody fragments: HC45 Fab [PDB : 1 QFU] (blue and orange), BH151 Fab [PDB : 1EO8] (magenta and orange), HC63 Fab [PDB : 1KEN] (green and yellow), and HC19 Fab [PDB : 2VIR] (red and yellow). (*See text for full caption.*)

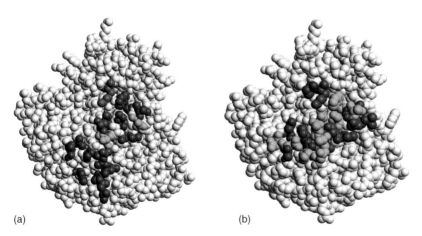

(a)   (b)

Figure 35.3. Structural and functional B-cell epitopes on influenza virus N9 neuraminidase. (a) NC10 structural epitope derived from [PDB : 1NMB]. Residue K432, which was not a part of the structural epitope but has markedly reduced antibody binding after substitution, is shown in magenta. Residues of the structural epitope N329, A369, and S370, which contributed significantly in binding of NC10 antibody, are shown in red. Residues of the structural epitope for which mutations I368R and N400K have not markedly reduced antibody binding are shown in cyan. (b) NC41 structural epitope derived from [PDB : 1NCA]. (*See text for full caption.*)

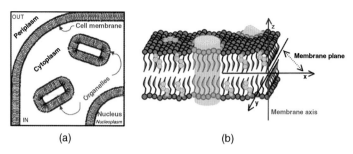

(a)   (b)

Figure 36.1. (a) Schematic representation of a section of a cell: IN, inside of the cell; OUT, outside of the cell. (b) Schematic representation of a cellular membrane. Glycerophospholipids are shown in gray, sphingolipids in green, cholesterol in orange, and proteins in light blue. The membrane plane and the membrane axis are also shown with dark gray and red lines, respectively.

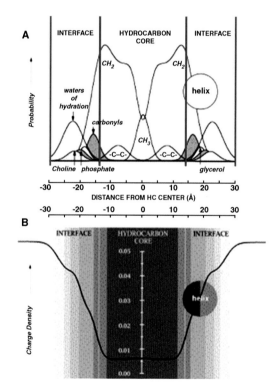

Figure 36.2. (a) The structure of a fluid dioleoylphosphocoline (DOPC) calculated as the time-averaged spatial distributions of the chemical groups methyl (−CH3), carbonyl (−CH2−), −CH=CH−, and water projected onto an axis normal to the bilayer plane. The bilayer can be easily divided into three regions on the basis of the distribution of the water of hydration of phospolipid headgroups: a 30 Å thick central hydrocarbon core (red vertical lines) including most of −CH2−, −CH3 and −CH=CH−, and two side 15 Å interfaces (pink lines) including most of the polar group with their water of hydration. (b) The structure of a fluid dioleoylphosphocoline (DOPC) represented as the variation of charge density along the membrane axis.

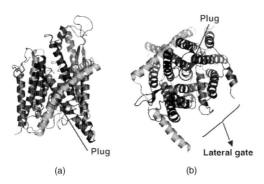

(a)   (b)

Figure 36.4. Structure of the SecY/Sec61 translocon complex from *M. jannaschii* as in the PDB (code 1RHZ). Left: side view. Right: top view. Subunits α, β, and γ are colored in blue, green, and pink, respectively. The plug (transmembrane helix 2, subunit α, aa 59–64), which allows the insertion of the helix in the membrane, is colored in light blue. The figure was generated from the original PDB file using the software PyMol.

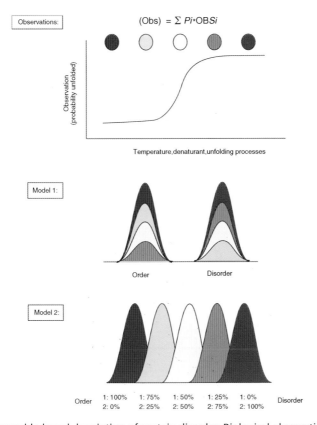

Figure 38.5. Ensemble-based description of protein disorder. Biological observations are the sum contribution of the different states in solution. The unfolding process of proteins, for example, can be described with one of the two models. (*See text for full caption.*)

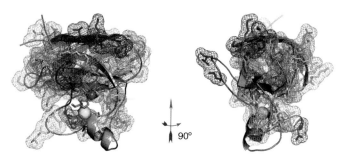

Figure 39.1. Single nucleotide polymorphisms in the human Cu, Zn superoxide dismutase (SOD1) associated with the amyotrophic lateral sclerosis. Over 70 mutations are scattered throughout the tertiary structure affecting multiple structurally distinct sites, marked by colored meshes. For simplicity only one monomer of the homodimeric SOD1 is shown in two projections.

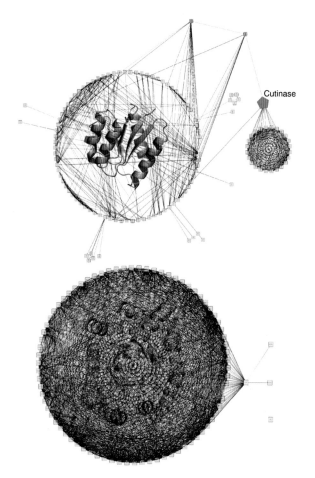

Figure 39.3. The first and the second largest PDUG subclusters. Representative structures for the first (upper graph) and the second largest (lower graph) subclusters of the PDUG are overlayed with the graphs.

Color files of all figures from this book are available for download from the following web address: ftp://ftp.wiley.com/public/sci_tech_med/structural_bioinformatics

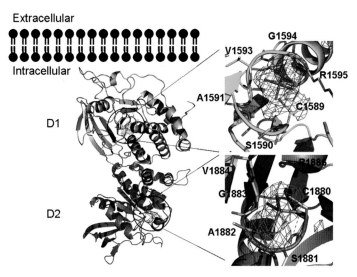

Figure 40.3. Structure of the human PTPσ tandem phosphatase domains. The structure of the PTPσ tandem phosphatase domains D1 and D2 is shown as a ribbon diagram with bound tungstate ions as stick and overlapping anomalous difference electron density in red. Domain D1 is shown in dark green and D2 in magenta. Interactions with the tungstate ion in the D1 and D2 active sites are magnified with hydrogen bonds represented as black dashes.

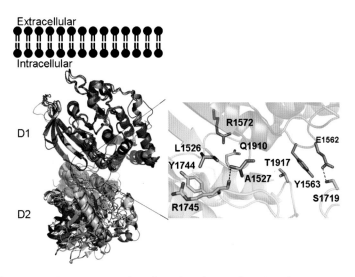

Figure 40.4. Structural comparison of tandem phosphatase domains of RPTPs. Superpositions of the structures of the tandem phosphatase domains of PTPσ (green), LAR (purple), CD45 (blue), and PTPγ (orange). Amino acids involved in interdomain interactions for PTPσ are shown to the right of the D1–D2 structures. For all structures, the D1 and D2 domains are shown in dark and light shades of color, respectively.

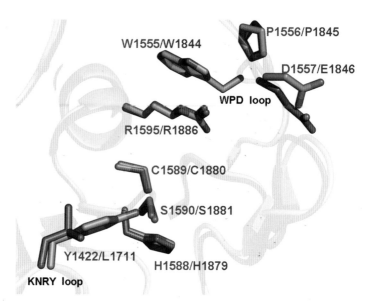

Figure 40.5. Comparison of the PTPσ D1 and D2 domain active sites. Superposition of the PTPσ D1 (green) and D2 (magenta) domain active sites. Active site residues and residues making up the WPD and KNRY loops are shown as stick figures. Root mean square deviation of D1/D2 superposition for 254 structurally equivalent Cα atoms is ~1.0 Å.

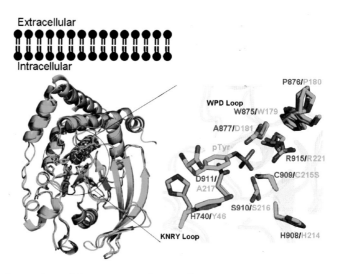

Figure 40.6. Comparison of the structures of IA-2 and PTP1B. Superposition of the structures of IA2 (green) and PTP1B (cyan), with active site residues shown as stick figures. Active site residues of IA2 and PTP1B bound to phosphotyrosine are magnified, highlighting differences responsible for the lack of catalytic activity of IA-2.

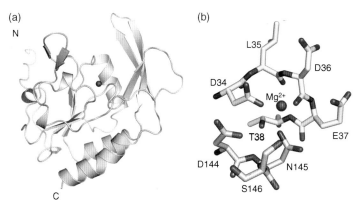

Figure 40.7. Structure of SCP3. (a) Ribbon diagram of SCP3 showing the DXDX catalytic loop (yellow) and the catalytic Mg^{2+} ion modeled from SCP1 (magenta). (b) Atomic details of the SCP3 catalytic site, again with the Mg^{2+} ion modeled from SCP1.

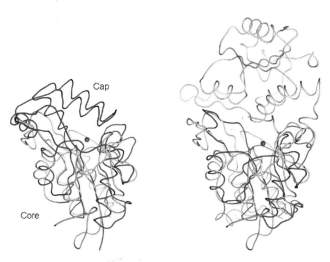

Figure 40.8. Structural comparisons of SCP3. (a) Superposition of SCP3 (green) with *Methanococcus jannaschii* phosphoserine phosphatase (red, PDB ID: 1F5S). The SCP3 catalytic site is freely accessible to solvent, whereas the alpha-helical capping domain in phosphoserine phosphatase shields its active site. (b) Superposition of SCP3 (green) with a dimer of the tetrameric *Haemophilus influenzae* deoxy-D-mannose-oculosonate 8-phosphatase (red and gray, PDB ID: 1K1E). Mg^{2+} ions are shown as pink spheres, and conserved phosphate-binding loops are shown in yellow. The capping domain of 1F5S occludes the active site entrance. In 1K1E, the second subunit of the dimer plays a similar role.

COLOR PLATES

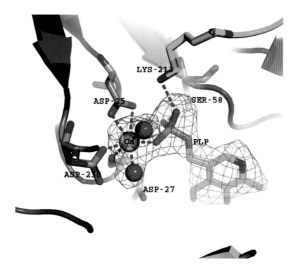

Figure 40.11. Chronophin catalytic site. The active site of chronophin with its ligand PLP and inhibitory Ca^{2+}. The Ca^{2+} (green sphere) is hepta-coordinated and participates in a bidentate interaction with the active site nucleophile Asp-25.

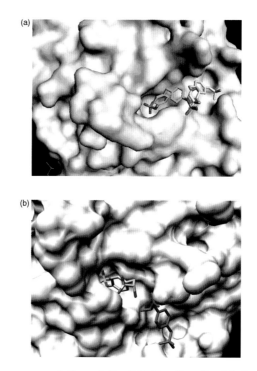

Figure 40.15. Surface representation of the PTP1B active site. (a) Compound **1** binds to a catalytically incompetent form of PTP1B via one of the two mutually incompatible binding modes, which encompass either the active site (multicolor) or a secondary peripheral site (black and white). (b) Phosphotyrosine simultaneously binds to the active site (multicolor) and a secondary peripheral site (red).

Color files of all figures from this book are available for download from the following web address: ftp://ftp.wiley.com/public/sci_tech_med/structural_bioinformatics

(a)

(b)

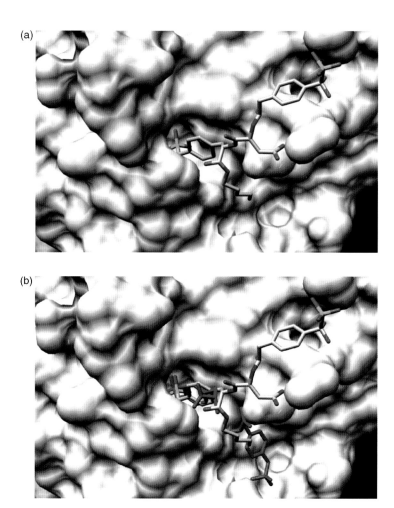

Figure 40.16. Structure of the PTP1B/compound **3** complex (a) The bidentate ligand, compound **3**, is bound to the active site and a secondary peripheral site different from that observed with compound **1** and phosphotyrosine. (b) Overlay of the double binding mode of phosphotyrosine (red) and the bidentate ligand compound **3** (multicolor).

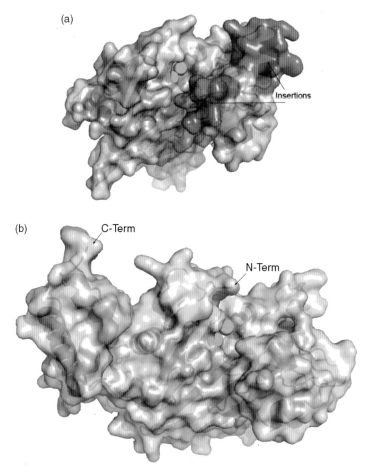

Figure 40.19. Surface representations of PP2Cs. (a) PP2Ctg: Ca^{2+} shown as green spheres; surface corresponding to amino acid insertions 207–213 and 218–232 colored orange. (b) PP2Cα in similar orientation as PP2Ctg: Mn^{2+} shown as green spheres.

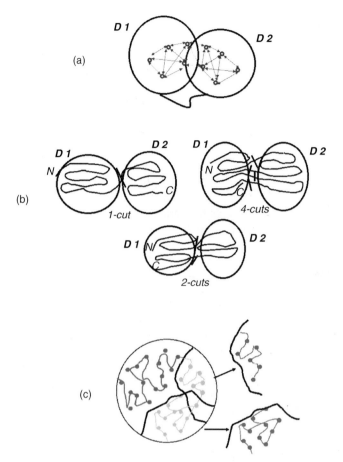

Figure 20.1. Illustration of the problem of parsing the protein 3D structure into structural domains. (a) Most commonly, structural domains are defined as groups of residues with a maximum number of contacts within each group and a minimum number of contacts between the groups. Domains may be composed of one or more chain segments. Any domain assignment procedure must therefore be able to cut the polypeptide chains as many times as necessary. (b) When both domains are composed of contiguous chain segments (continuous domains), only a one-chain cut (1-cut) is required. When one domain is continuous and the other discontinuous, a situation that may arise as a result of gene insertion, then the chain has to be cut in two places (2-cuts). When both domains are discontinuous, additional chain cuts may be required. In the example shown, the chain is cut in four places (4-cuts), and thus the domain on the left hand side contains three chain segments, whereas that on the right contains two chain segments. (c) This drawing shows two solutions to the problem of partitioning the protein 3D structure into substructures. To partition the 3D structure into domains, many such solutions need to be examined to single out the one that satisfies the criterion given in (a). Figure also appears in the Color Figure section.

structure (Figure 20.1b). This situation may arise as a result of gene insertion events or domain swapping (Bennett, Choe, and Eisenberg, 1994). Domains are the building blocks of the proteins: different combinations of domains result in variety of protein structures. Frequently, domains are associated with specific functions, such as binding a ligand, DNA or RNA or interacting with other protein domains.

ALGORITHMS FOR IDENTIFYING STRUCTURAL DOMAINS: INSIGHT INTO HISTORY AND METHODOLOGY

Using the above definitions for domains as guidelines, systematic partitioning of structures into domains has been undertaken. This task can be done manually by experts in protein structure or automatically by encoding basic rules and assumptions in an algorithm. The former has the advantage of employing human expertise with myriads of algorithms integrating prior experience, common sense, topological sensibility, and other knowledge. The latter, while far less sophisticated, has the advantage of speed and consistency that are critical in the current era of structural genomics, when the sheer number of solved structures overwhelms human experts.

The first *manual* survey of structural domains in proteins was carried out nearly 30 years ago (Wetlaufer, 1973) using visual inspection of the then-available X-ray structures. Wetlaufer defined domains as regions of the polypeptide chain that form compact globular units, sometimes loosely connected to one another. At about the same time, the first so-called Cα–Cα distance plots were computed (Phillips, 1970; Ooi and Nishikawa, 1973) and shown to be useful for identifying structural domains. Domains were identified visually in these plots by looking for series of short Cα–Cα distances in triangular regions near the diagonal, separated by regions outside the diagonal where few short distances occur, as illustrated in Figure 20.2. Manual and semimanual surveys became more extensive and systematic after 1995; currently there are three major sources of information regarding domains: SCOP

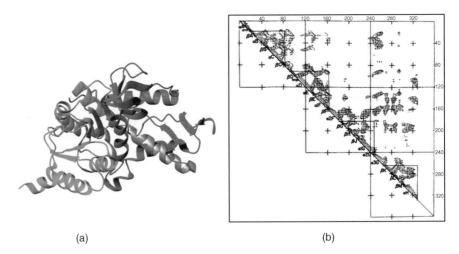

(a)   (b)

Figure 20.2. Domain structure of dogfish lactate dehydrogenase, determined using the Cα–Cα distance map. (a) Ribbon diagram of lactate dehydrogenase, showing the NAD binding (green) and catalytic domains (red). In gray is part of the helix, spanning residues 164–180, linking the two domains. (b) Distance map and structural domains in lactate dehydrogenase. Contours represent Cα–Cα distances of 4, 8, and 16 Å within the subunit of dogfish lactate dehydrogenase. Elements of secondary structure are identified along the diagonal. Triangles enclose regions where short Cα–Cα distances are abundant. The NAD binding domain comprises the first two triangles (counting from the N-terminus) that are subdomains. The catalytic domain comprises the last two triangles (the C-terminal domain). Taken from Rossman and Liljas (1974) and reproduced by permission of Academic Press (London) Ltd. Figure also appears in the Color Figure section.

(Murzin et al., 1995, Chapter 17), AUTHORS (Islam, Luo, and Sternberg, 1995), and CATH (Orengo et al., 1997, Chapter 18). We will discuss the evaluating procedure for domain assignment later in the chapter.

Rossman and Liljas performed the first systematic *algorithmic* survey of domains for a set of protein 3D-structures (Rossman and Liljas, 1974) by analyzing Cα–Cα distance maps. This was followed a few years later by three other studies conducted by Crippen (1978) and Rose (1979) and by Wodak and Janin (1981b), and Janin and Wodak (1983) (Table 20.1). The methods described in these studies involved different algorithms and produced different results, but had one major aspect in common: the protein 3D structure was partitioned in a hierarchical fashion that yields domains and smaller substructures, as illustrated in Figure 20.3. Surveys of these smaller substructures in a set of proteins revealed recurrent structural motifs comprising of two or three secondary structure elements joined by loops (Wodak and Janin, 1981a; Zehfus and Rose, 1986). Interestingly,

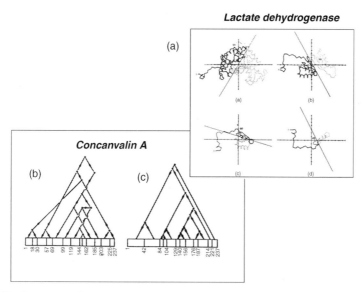

Figure 20.3. Illustration of the hierarchic partitioning of protein 3D structures into substructures performed by three of the early automatic domain analysis methods. (a) The partitioning procedure of Rose et al. (1979) is applied to dogfish lactate dehydrogenase. The chain tracing of the subunit is projected into a "disclosing plane" passing through its principal axes of inertia. A line is then found that divides the projection into two parts of about equal size. The different projections represent (from left to right and from the bottom down) the whole subunit (residues 1–331) of the N-terminal domain (1–161) and the subdomain (residues 1–90 and 1–49). Taken from Rose (1979) and reproduced with permission of Academic Press (London) Ltd. (b) Domain and subdomain hierarchies in concanavalin A adapted from Crippen (1978). It represents an ascending hierarchy of clusters starting from 13 "reasonably straight segments" at the bottom of the hierarchy and culminating with the entire concanavalin A subunit. (c) The domain and subdomain hierarchy in concanavalin A, adapted from Wodak and Janin (1981b). It describes the descending hierarchy derived from identifying the minimum interface area between substructures in successive interface area scans. Unlike the hierarchy in (b), this one generates continuous substructures. The agreement with the hierarchy in (b) is therefore poor.

TABLE 20.1. Summary of Domain Decomposition Methods 1974–2007

Method	Year Generation	Strategy	Type of Domains Generated	Approaches/Models Used	Comparison to Representative Data Set	Cross-Validation	Novel/Interesting Features
Rossman and Liljas (1974)	1974 First generation	Top-down	Contiguous	Distance plots of the structure against itself using Cα distances, search for strong interactions close to diagonal	No	No	
Crippen (1978)	1978 First generation	Bottom-up	Contiguous and noncontiguous	Clustering of small structural units	No	No	The only first generation method to produce noncontiguous domains
Rose (1979)	1979 First generation	Top-down	Contiguous	Cutting the projection of 3D structure onto 2D domain disclosing plain	No	No	
Wodak and Janin (1981a) Wodak and Janin (1981b)	1981 First generation	Top-down	Contiguous	Finding minimum in the interface between two domains	No	No	

Method	Year/Generation	Approach	Domain type	Description			Notes
PUU (Holm and Sander, 1994)	1994 Second generation	Top-down	Contiguous and noncontiguous	Rendering of the contact matrix, constructed using rigid body oscillation	Yes	No	
DETECTIVE (Swindells, 1995a)	1995 First generation	Bottom-up	Contiguous and noncontiguous	Building of the hydrophobic core	No	No	
Islam, Luo, and Sternberg (1995)	1995 Second generation	Top-down	Contiguous and noncontiguous	Finding minima in the interdomain contact density	Yes	No	
DOMAK (Siddiqui and Barton, 1995)	1995 Second generation	Top-down	Contiguous and noncontiguous	Splitting structure by maximizing intradomain/interdomain contacts	Yes	No	
Sowdhamini and Blundell (1995)	1995 Second generation	Bottom-up	Contiguous and noncontiguous	Clustering of secondary structures	Yes	No	
NCBI method (Madej et al., 1995)	1995 Second generation	Top-down	Contiguous and noncontiguous	Similar to PUU, also intradomain contact density is at least two times the interdomain	Yes	No	Approach similar to PUU, but the performance is significantly better
Taylor (1999)	1999 Second generation	Bottom-up	Contiguous and noncontiguous	Clustering of residues in spatial proximity using Ising model	Yes	No	Careful analysis of splitting β-strands between domains
STRUDL (Wernisch, Hunting, and Wodak, 1999)	1999 Second generation +	Top-down	Contiguous and noncontiguous	Finding minimum interdomain contacts using Kernighan–Lin graph heuristics	Yes	Yes	

(continued)

TABLE 20.1 (*Continued*)

Method	Year Generation	Strategy	Type of Domains Generated	Approaches/Models Used	Comparison to Representative Data Set	Cross-Validation	Novel/ Interesting Features
DomainParser (Xu, Xu, and Gabow, 2000)	2000 Second generation +	Top-down	Contiguous and noncontiguous	Finds minimum interdomain contacts using graph theoretical approach with maximum flow/ minimum cut using Ford–Fulkerson algorithm	Yes	Yes	Careful tuning of the method during postprocessing step
(Xuan, Ling, and Chen, 2000)	2000 Second generation	Bottom-up	Contiguous and noncontiguous	Assemble domains from rudimentary fragments using fuzzy clustering	Yes	No	
PDP (Alexandrov and Shindyalov, 2003)	2003 Second generation	Top-down	Contiguous and noncontiguous	Finding partitioning with minimal number of contacts between domains	Yes	No	Unconstrained splitting of secondary structures among domains, still the best method tested
HVdWD (Hierarchy of van der Waals Domains) (Berezovsky, 2003)	2003 Second generation	Bottom-up	Contiguous and noncontiguous	Clustering of short segments. Both initial segments and the clustering threshold are based primarily on van der Waals interactions among atoms	Yes	No	Considers decomposition under different thresholds. Results can contain different levels of decomposition

Method	Year / Generation	Approach	Domain type	Description			Notes
Kundu, Sorensen, and Phillips (2004)	2004 Second generation	Top down	Contiguous and noncontiguous	Decomposition of the structure using Gaussian Network Model; assumes semi-independent motion of domains	Yes	No	
Sistla, K V, and Vishveshwara 2005	2005	Top-down	Contiguous and noncontiguous	Identifies strongly interacting units by applying spectral analysis to the contact matrices [idea similar to PUU]	Yes	No	Thresholds for decomposition may vary depending on the compactness of the protein structure
DePot (Taylor and Vaisman, 2006)	2006 Second generation	Bottom-up	Contiguous and noncontiguous	Hierarchical clustering of residues starting with individual residues and using Delaunay tessellation of the protein structure [idea similar to Taylor, 1999]	Yes	No*	Different thresholds of residue proximity are considered during decomposition
DDOMAIN (Zhou et al., 2007)	2007 Second generation +	Top-down	Contiguous only	Similar to PDP, finding partition with lowest overall interaction between domains	Yes	Yes	Most extensive use or training/ testing data sets
DomainICA (Emmert-Streib and Mushegian, 2007)	2007 Second generation +	Top-down	Contiguous only	Partitioning the graph of nodes representing secondary structures, by maximizing the cycle distribution on the graph	Yes	Yes	Uses exclusively topological data of secondary structures. No postprocessing step

the systematic identification of similarly defined smaller substructures in proteins was recognized years later as a useful way for predicting regions in proteins that would form first during folding (nucleation sites) (Moult and Unger, 1991; Berezovsky, 2003). Over 20 different algorithmic approaches have been applied to this problem in the subsequent three decades (this survey covers methods up to 2007). The overarching theme of structural partitioning is based on *contact density*, that is, the simple fact that there are more residue-to-residue contacts within a structural unit than between the structural units. The implementations of this principle can be very different in different methods as can be seen in Table 20.1, but the underlying idea always comes back to finding regions with a high density of interactions.

Methodologically, algorithms can be separated into first generation: methods published during the period 1974–1995; and second generation algorithms: methods published from 1994 until now. The first generation of algorithms was based on a very limited set of available structural data at the time thus, some rather general principles were difficult to capture due to the paucity of structural examples. For the same reason, neither a systematic comparison against a representative data set nor separation of the data into training and testing subsets was possible until the middle 1990s. The second generation methods, which began around 1994, relied on a large set of structures for training and they routinely compare performance of the algorithm against a representative data set, typically assigned by human experts. Surprisingly, the methodological rigor is often absent and the algorithms are often trained and tested on the same set of data. Methods that incorporated cross-validation to test the algorithm during development by using nonoverlapping training and testing data sets are more balanced in their approach and we assign them as "second generation + " (Table 20.1). The majority of the second generation methods are able to deal with noncontiguous domains (Zehfus, 1994).

Domain identification can be divided into two fundamental approaches: *top-down* (starting from the entire structure and proceeding through iterative partitioning into smaller units) and *bottom-up* (defining very small structural units and assembling them into domains). Some methods use both the approaches within their algorithm by first decomposing the structure and then reassembling them, or vice versa. Here, we classify each method based on its chief or overall approach to the domain identification. Generally, the process of domain decomposition is performed in two steps: (1) tentative domains are constructed either by splitting the structure into domains or building domains from smaller units; (2) these tentatively defined domains are evaluated in a postprocessing step. Overall, an amazing array of approaches has been put forward over the years to solve the domain decomposition problem; the strategies vary from maximizing intradomain contacts to minimizing domain–domain interface, from semi-independent motion of domains to clustering residues in spatial proximity. Other techniques applied for domain definition come from graph theoretical approaches (i.e., Kernighan–Lin graph heuristics, Network flow, Delaunay tessellation), rigid body oscillation, spectral analysis, Ising model, Gaussian Network Model, and so on (Table 20.1). In spite of this variety of approaches and techniques, an interesting observation emerges when one carefully compares the performance of the algorithms: most of the algorithms solve correctly 70–80% of the available structures, faltering on the remaining ones because of their complex, multidomain structures. Typically, all methods correctly identify potential boundaries between the structurally compact regions. However, in some cases the number of boundaries are overpredicted leading to too many assigned domains (overcut) or underpredicted leading to fewer assigned domains (undercut). Thus, the problem still remaining is not "where does the boundary of domains fall?", but rather "is the identified boundary a true domain boundary?" Some of the difficulties come from very complex and contradictory scenarios that are presented by existing protein structures, making it nearly impossible to capture by a simple set of rules. This, in turn, is part of the complexity inherent in biology in general. In this case, complexity arises in that some domains are not globular, lack a hydrophobic core in the absence of the binding partner, have relatively low intradomain contact density, or have extensive interactions along the domain–domain interface. Identifying a correct subset of domain boundaries from the set of potential boundaries is the next (and maybe the final) frontier in solving the domain assignment problem for structures.

ALGORITHMS FOR IDENTIFYING STRUCTURAL DOMAINS: IN-DEPTH

Of the more recent second generation methods for domain assignment, several make elegant use of graph theoretical methods. In this section, we discuss two such methods that use somewhat different techniques to address the problem of partitioning protein structure into structurally meaningful domains. First, we will discuss STRUctural Domain Limits (STRUDL) (Wernisch, Hunting, and Wodak, 1999) followed by DomainParser (Guo et al., 2003).

The procedure implemented in STRUDL views the protein as a 3D graph of interacting residues, with no reference to any covalent structure. The problem of identifying domains then becomes that of partitioning this graph into sets of residues such that the interactions between the sets is minimum. Since, this problem is NP-hard, efficient heuristic procedures—procedures capable of approximating the exact solution with reasonable speed—are an attractive alternative. The algorithm used in this case was a slightly modified version of the Kernigan–Lin heuristic for graphs (Kernighan, 1970). The application of this heuristic to domain assignment is summarized in Figure 20.5.

A useful, though not essential, aspect of this application is that the interactions between residue subsets were evaluated using contact areas between the atoms. This area was defined as the area of intersection of the van der Waals sphere around each atom and the faces of its weighted Voronoi polyhedron. This contact measure is believed by the authors to be more robust than counting atomic contacts due to its lower sensitivity to distance thresholds.

To identify domains for which the limits and size are not known in advance, the partitioning procedure described in Figure 20.4 is repeated k times, with k representing all the relevant values of the domain size, ranging from 1 to $N/2$, and N being the total number of residues in the protein. The partition with minimum contact area, identified for each value of k, is recorded. This information is then used to compute a *minimum contact density profile*. In this profile, the minimum contact area found for each k is normalized by the product of the sizes of the corresponding domains, to reduce noise (Holm and Sander, 1994;Islam, Luo, and Sternberg, 1995) and plotted against k. The domain definition algorithm then searches for the global minimum in this profile. Figure 20.5a illustrates the profile obtained for a variant of the p-hydroxy-benzoate hydroxylase mutant (PDB-RCSB code 1dob), a 394-residue protein composed of two discontinuous domains. The global minimum in the *minimum contact density profile*, although quite shallow, is clearly visible at $k = 172$, and yields the correct solution. The corresponding partition cuts the chain in five distinct locations yielding two domains, comprising six chain segments.

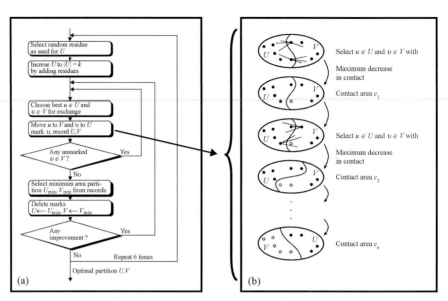

Figure 20.4. Domain identification using the graph heuristic procedure implemented in STRUDL (Wernisch, Hunting, and Wodak, 1999). (a) Overview of the major steps in the STRUDL algorithm (b) The residue exchange procedure in STRUDL. Residues $u \, \varepsilon \, U$ and $v \, \varepsilon \, V$ (filled circles) are selected so as to produce a maximal decrease or, if that is not possible, a minimal increase in the contact area between U and V upon exchange. Once moved to V, residue u is flagged (empty circle) and can hence not be moved back to U. The exchange procedure stops when V contains only flagged residues. Among all partitions with contact area C_i, with $i = 0, \ldots, n$ the one with minimum contact area $C_{\min}$ is selected.

The smaller of the two domains contains 172 residues; the largest contains 222 (394–172) residues.

Once the global minimum is identified in the *minimum contact density profile*, a decision must be taken to either accept or reject the corresponding partition, with a rejection corresponding to classifying the structure as a one-domain protein. An obvious criterion on which to base such decision is the actual value of the contact area density in profiles such as that of Figure 20.5a. If this value is below a given threshold, the partition is accepted; otherwise, it is rejected. But this simple criterion is unfortunately not reliable enough. Following other authors, additional criteria that represent expected properties of domains and of the interfaces between them were therefore used to guide the decision (Rashin, 1981; Wodak and Janin, 1981b). The choice of these criteria was carefully optimized on a training set of 192 proteins using a discriminant analysis, and tested on a different, much larger set of proteins, as summarized in Figure 20.5c. Finally, once a partition is accepted, the entire procedure is repeated recursively on each of the generated substructures until no further splits are authorized. This recursive approach was shown to successfully handle proteins composed of any number of continuous or discontinuous domains.

As always, assessing the performance of the method is a crucial requirement. Fortunately, the significant increase in the number of different proteins of known structure presently offers a more extensive testing ground. STRUDL was applied to a set of 787

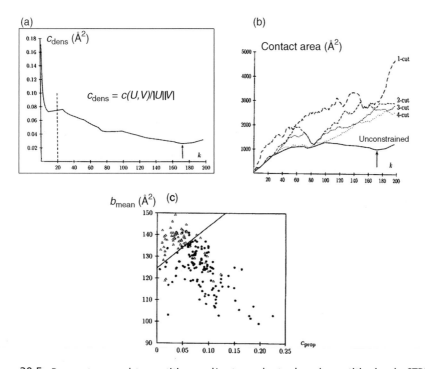

Figure 20.5. Parameters used to partition and/or to evaluate domain partitioning in STRUDL. (a) The minimum contact density profile for the *p*-hydroxy-benzoate hydroxylase mutant (1DOB). The C_{dens} value in Å, is computed using the formula given in the figure. In this formula, $c(U,V)$ is the contact area between the two residue groups U and V, and |U| and |V| are the number of residues in U and V, respectively. The plotted values represent the minimum of C_{dens} computed for a given domain size k, where $k = 1$, $N/2$, with N being the total number of residues in the protein chain. Hence |U| and |V| equal k and $N-k$, respectively. The arrow at $k = 172$ indicates the global minimum of this profile. The dashed line delimits the value of $k = 20$ below which splits are not allowed to avoid generating domains containing less than 20 residues. (b) The profiles of minimum contact area $c(U, V)$ in Å computed as a function of k, the number of residues in the smallest substructure U, for the same protein as in (a). Shown are the profile with no constraints on the number of chain cuts, as in the STRUDL procedure, and four other p-cut profiles, obtained by limiting the number of allowed chain cuts p (see Wernisch et al., 1999 for details). Those are 1-cut (- - -), 2-cuts (- - -), 3-cuts (- - -), and 4-cuts (· · ·). The global minimum of the contact area (arrow) can only be located in the unconstrained profile, illustrating the advantages of STRUDL over other procedures in which the number of allowed chain cuts is fixed. (c) Plot of the mean burial b_{mean} in Å, versus the contact area ratio c_{prop}. Both quantities are evaluated for domain partitions corresponding to the global minimum in the contact area density profiles computed by STRUDL. b_{mean} is the average interresidue contact area of a given substructure. C_{prop} is the ratio of the contact area of the putative domains to the sum of the interresidue contact areas in the entire proteins. The straight line optimally separating the single (filled circles) from the multidomain proteins (empty circles). This optimal separation entails however, 19 errors—proteins classified in the wrong category—out of the total of 192 protein considered in the set.

representative protein chains from the PDB (Bernstein et al., 1977; Berman et al., 2000), and the results were compared with the domain definitions that were used as the basis for the CATH protein structure classification (Orengo et al., 1997). This definition was based on a consensus definition produced by Jones et al. (1998) using three automatic procedures, PUU (Holm and Sander, 1994), DETECTIVE (Swindells, 1995a), and DOMAK (Siddiqui and Barton, 1995), as well as by manual assignments. Results showed that domain limits computed by STRUDL coincide closely with the CATH domain definitions in 81% of the tested proteins and, hence, that it performs as well as the best of the above-mentioned three methods. In contrast to these methods, however, it uses no information on secondary structures to prevent the splitting β-sheets, for instance. The 19% or so protein for which the domain limits did not coincide, represent interesting cases which could, for the most part, be rationalized either on the basis of the intrinsic differences between the approaches (in this case, STRUDL versus CATH) or by the variability and complexity inherent in real proteins.

Several of these cases are illustrated in Figures 20.6 and 20.7. Figure 20.6 shows cases where STRUDL splits proteins into two domains that are considered a single architecture by CATH. Among the shown examples are the DNA polymerase processivity factor PCNA (1PLQ), which clearly shows an internal duplication, and the lant seed protein narbonin (1NAR), which adopts a TIM barrel fold that many automatic domain assignment procedures tend to split into two domains. Differences of this type could be rationalized by the fact that CATH imposes criteria based on chain architecture and topology, whereas STRUDL does not. Figure 20.7 illustrates two cases, where the results of STRUDL and CATH differ by the assignment of a single relatively short protruding chain segment. Such cases illustrate the inherent noisiness in the backbone chain trace, which invariably affects domain assignments.

DomainParser uses a top-down graph theoretical approach for domain decomposition and an extensive postprocessing step. During the training stage of the algorithm, multiple

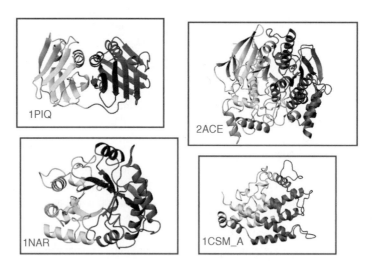

Figure 20.6. Examples of single domain, single architecture proteins in CATH (Jones et al., 1998), which STRUDL (Wernisch, Hunting, and Wodak, 1999) splits into two domains. Shown are the domain assignments produced by STRUDL. For the exact domain limits, the reader is referred to the STRUDL Web site. The displayed protein ribbons belong to *Torpedo californica* acetylcholinesterase monomer (1ACE), the lant seed protein narbonin (1NAR), the eukaryotic DNA polymerase processivity factor PCNA (1PLQ), and Chorismate mutase chain A monomer (1CSM_A).

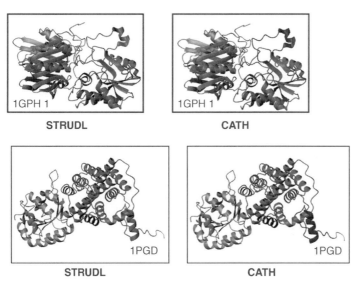

Figure 20.7. Different assignments by STRUDL (Wernisch, Hunting, and Wodak, 1999) and CATH (Jones et al., 1998), illustrating the effect of "noise," or decorations, in the protein chain trace. The STRUDL assignments are displayed on the left hand side, and the CATH assignments are displayed on the right. The short chain segments, which CATH assigns to separate domains, are shown in blue. Some of the discrepancies may be due to simple "slips" in the CATH assignments that have been, or will be, corrected. Figure also appears in the Color Figure section.

parameters are tuned; the method is then validated on a SCOP data set. Domain decomposition is addressed by modeling the protein structure as a network consisting of nodes (residues) and edges (connections between residues). A connection between any two residues is drawn when they are adjacent in the sequence or, alternatively, are in physical proximity in the structure. The strength of the interaction between two residues is expressed as the capacity of the edge to connect the two nodes. This edge capacity is a function of (a) the number of atom–atom contacts between residues; (b) the number of backbone contacts between residues; (c) the existence of backbone interactions across a beta-sheet; and (d) whether both residues belong to the same beta-strand. The values for all parameters involved in edge capacity are optimized during the training stage of the algorithm.

The partitioning of the network into two parts is then equivalent to decomposing a given structure into two domains. Ideally, partitioning should be done using the edges with least capacity, which will result in partitioning a structure along the least dense interactions among the residues. The problem of partitioning the network is solved using the *maximum flow/ minimum cut* theorem by Ford–Fulkerson as implemented by Edmond and Karp. Briefly, the approach is as follows: artificial source and sink nodes are added to the network (Figure 20.8a). A "bottleneck"—a set of critical edges in the network flow—is found by gradually increasing the flow of all edges in a network. Removing the set of critical edges from the network prevents flow from the source to the sink. At this point, nodes that are connected to the source represent one interconnected part of the network, while nodes connected to the sink are the second interconnected part of the network. Since the node capacity is increased gradually, it is expected that nodes with least capacity (least residue–residue contacts) will be the ones contributing to the bottleneck. The process of subdividing the network into two parts is repeated multiple times by connecting the source and sink to

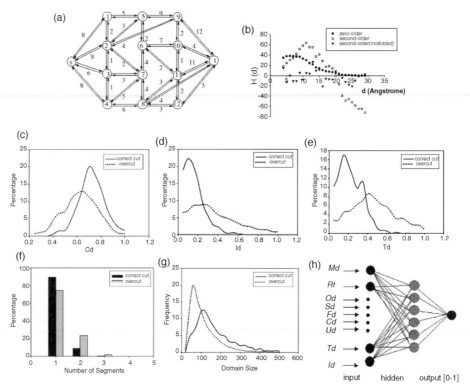

Figure 20.8. Domain decomposition using the DomainParser algorithm. (a) Schematic representation of protein structure as a directed flow network. The value on each edge represents the edge's capacity. Artificial nodes "start" and "sink" are denoted by **s** and **t**, respectively. (b–g) are examples of parameters used by the algorithms. (b) zero- and second-order spherical moment profile of structure 2ilb, (c) compactness of structure, (d) size of the domain interface relative to the domain's volume, (e) measure of relative motions between domains, (f) distribution of the number of segments per domain, and (g) distribution of domain sizes. (h) Neural network architecture for evaluation of decomposed individual domains.

different parts of the network; a set of minimal cuts is collected and evaluated during a postprocessing step. The entire procedure is then repeated in each of the resulting domains until either the domain's size drops below 80 residues or the partitioning produces domains that do not meet necessary conditions of domain definitions.

The stopping criteria are multifaceted and defined by (1) domain size (no less than 35 residues), (2) beta-sheets kept intact, (3) compactness of domain above threshold g_m, (4) size of the domain–domain interface below threshold f_m, (5) the ratio of the number of residues and the number of segments in the domain is above threshold l_s. The values for g_m, f_m, l_s and minimum domain size are determined during the training stage of the algorithm. A suite of additional parameters exists—these are involved in a postprocessing step of the algorithm in which an assessment is made about whether the substructure meets the additional criteria of a structural domain. These parameters are (1) hydrophobic moment (Figure 20.8b), (2) the number of segments in the partitioned domain (Figure 20.8f), (3) compactness (Figure 20.8c), (4) the size of domain interface relative to the domain's volume (Figure 20.8d), and (5) relative motion between compact domains (Figure 20.8e). Distribution for each of the parameters for true versus false domains is collected during the training stage using 633 correctly partitioned domains and 928 incorrectly partitioned domains. Multiple neural networks are then investigated; the best one has nine input nodes, six nodes in the hidden layer, and one output node (Figure 20.8h). Performance of the DomainParser method is then evaluated using a set of 1317 protein chains in which domains are defined by SCOP.

DOMAIN ASSIGNMENTS: EVALUATING AUTOMATIC METHODS

A chief reason for the existence of over two dozen methods for domain decomposition is the complexity of the problem itself: it is nearly impossible to capture succinctly the principles of domain decomposition and apply them successfully to the entire universe of protein structures. Thus, every new method strives to reach a bit further beyond existing methods in its ability to decompose complex structures. With so many different methods available, it is essential to be able to compare the performance of the algorithms, to determine the fraction, as well as the type, of successfully partitioned proteins. Evaluation of automatic methods is an essential part of algorithm development process and the performance of a method is often reported in the second generation of algorithms. To evaluate the performance of a new domain assignment method, authors usually apply it to a set of proteins for which experts who are either crystallographers or protein classifiers have produced domain assignments manually. However, authors of each method use different data sets for evaluation of their algorithms; thus, it is nearly impossible to cross-compare the performance of various methods. Only one consistent benchmark data set existed until recently and it is a set of 55 protein chains consistently assigned by Jones et al. (1998). This data set was assembled by finding a consensus using the Islam et al. data set among four automatic methods: PUU (Holm and Sander, 1994), DETECTIVE (Swindells, 1995b), DOMAK (Siddiqui and Barton, 1995), and the method developed by Islam, Luo, and Sternberg (1995). For the evaluation, any two assignments (one from the benchmark set, another by the algorithmic method) were considered similar if they had the same number of domains and at least 85% of the residues were assigned to the same domain. Fifty-five proteins is a very small benchmark set, thus it is likely that its resolution will be insufficient to detect differences between the methods. New benchmark data sets were compiled recently (Veretnik et al., 2004; Holland et al., 2006) specifically for the purpose of cross-comparing the performance of the algorithms as well as highlighting the strengths and weaknesses of each algorithmic method. The data set is assembled using a principle of consensus approach among the human experts: it includes proteins for which three expert methods (CATH, SCOP, and AUTHORS) produce similar domain decompositions. Such consensus approach eliminates the more complicated structures for which no agreement can be reached among human experts. This fraction of "controversial" proteins is intentionally left out, so as not to further complicate the issue. The 315 proteins in this new benchmark (Benchmark_2 in Holland et al., 2006) are realistically distributed between single-domain and multidomain proteins to avoid the typical bias toward one-domain proteins (Table 20.2). Furthermore, each type of topology combination, as determined by CATH classification on the level of Topology, occurs only once per data set to ensure that a broad range of topologies found in the protein universe are equally represented. The above benchmark data set was used to evaluate four recent publicly available automatic methods: PUU (Holm and Sander, 1996), DomainParser (Xu, Xu, and

TABLE 20.2. Benchmark Data Sets Constructed for Evaluation of Computational Methods for Domain Assignments from 3D Structure

Type of Chain	Jones et al. (1998) Benchmark	Islam, Luo, and Sternberg (1995) Benchmark	Veretnik et al. (2004) Benchmark_1	Holland et al. (2006) Benchmark_2	Holland et al. (2006) Benchmark_3	Coverage of Structural Space by Chains in Benchmark_2 (Using CATH Topologies)				
						Class 1 Arch:5 Topol: 227	Class 2 Arch:19 Topol: 139	Class 3 Arch:12 Topol: 368	Class 4 Arch:1 Topol: 86	Total Arch:37 Topol: 820
1-Domain	30 (55%)	1530 (64.7%)	318 (85%)	106 (33.7%)	106 (39.1%)	Arch: 2 Topol: 14	Arch: 9 Topol: 26	Arch: 6 Topol: 23	Arch: 1 Topol: 7	Arch: 18 Topol: 70
2-Domain	20 (36%)	720 (30.5%)	40 (10.7%)	140 (44.4%)	108 (39.9%)	Arch: 5 Topol: 30	Arch: 8 Topol: 20	Arch: 8 Topol: 27	Arch: 1 Topol: 2	Arch: 22 Topol: 79
3-Domain	3 (5.4%)	85 (3.6%)	15 (4.0%)	54 (17.1%)	45 (16.6%)	Arch: 4 Topol: 16	Arch: 9 Topol: 18	Arch: 7 Topol: 33	none	Arch: 20 Topol: 67
4-Domain	2 (3.6%)	25 (1.1%)	1 (0.3%)	8 (2.5%)	7 (2.6%)	Arch: 2 Topol: 2	Arch: 3 Topol: 4	Arch: 5 Topol: 15	none	Arch: 10 Topol: 21
5-Domain	0	2 (0.1%)	0	5 (1.6%)	5 (1.9%)	None	Arch: 5 Topol: 6	Arch: 4 Topol: 6	none	Arch: 9 Topol: 12
6-Domain	0	0	0	2 (0.6%)	0	Arch: 1 Topol: 1	none	Arch: 3 Topol: 3	none	Arch: 2 Topol: 2
Total	55	2363	374	315	271					

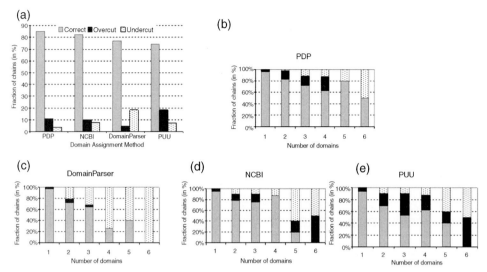

Figure 20.9. Comparison of the performance of domain assignment methods using Benchmark_2 data set (built on the principle of consensus among expert-based assignments: expert consensus). Four currently available algorithmic methods are evaluated: DomainParser, PDP, PUU, and NCBI. Structure is considered to be assigned correctly if the algorithm predicts the same number of domains as predicted by expert consensus. Correctly assigned structures are represented in grey, structures with too many assigned domains (overcut) are in black, and the structures with too few domains (undercut) are indicated by the dotted pattern. (a) Performance of each method on the entire Benchmark_2 data set. (b–e) Performance of individual methods on structures grouped by the number of domains they contain. (b) PDP method. (c) DomainParser method. (d) NCBI method. (e) PUU method.

Gabow, 2000; Guo et al., 2003), PDP (Alexandrov and Shindyalov, 2003), and NCBI (Madej, Gibrat, and Bryand, 1995). The evaluation includes information about the success rate of each algorithm and an analysis of errors in terms of predicting fewer domains (undercut) or too many domains (overcut) (Figure 20.9a). A more scrutinized examination was also conducted by inspecting (1) the performance of domain boundary prediction for each example of multidomain structures (Figure 20.9b–e), (2) tendencies for methods to fragment domains into noncontiguous stretches of polypeptide chain (Figure 20.10a), and (3) the precision of overlap of domain boundaries (Figure 20.10b).

Finally, a composite evaluation that accounts for the correct number of domains, fragments, and domain boundary assignment can be a more comprehensive approach to conducting a cross-comparison of performances of various algorithms (Figure 20.10c and d). By changing the weighing schema of individual performance components of the methods, it allows for comparisons to be conducted from different angles that stress particular features of interest. In our analysis, the PDP method is a clear winner regardless of the weight that is attributed to individual parameters. On the contrary, the performance of DomainParser improves to the point of outperforming the NCBI method when a higher weight is given to the precision of boundaries rather than prediction of the correct number of domains.

This type of systematic analysis can link performance of an algorithm in terms of its strength and weaknesses to particular assumptions found in the algorithm, as well as to the specific data sets on which algorithms had been trained. For example, the superior performance of PDP can be partially linked to its ability to cut through secondary structures,

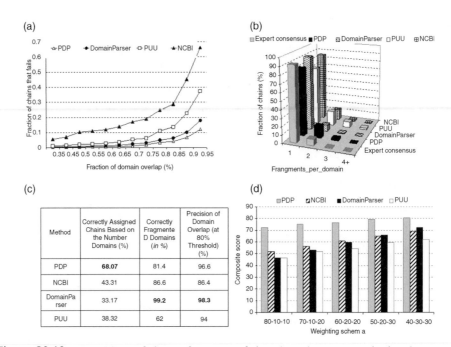

(c)

Method	Correctly Assigned Chains Based on the Number Domains (%)	Correctly Fragmente D Domains (in %)	Precision of Domain Overlap (at 80% Threshold) (%)
PDP	**68.07**	81.4	96.6
NCBI	43.31	86.6	86.4
DomainParser	33.17	**99.2**	**98.3**
PUU	38.32	62	94

Figure 20.10. Comparison of the performance of domain assignment methods using Bench-mark_2.(a) Evaluation of correct placement of domain boundaries using different levels of stringency of overlap of the predicted domain assignment and expert consensus assignment. As stringency of overlap increases, a larger fraction of structures fails the domain overlap test. (b) Evaluation of assignment of noncontiguous domains by different methods. Most of the methods predict more noncontiguous domains than experts. (c) Performance of each method using three main criteria: (1) fraction of structures with correctly assigned number of domains, (2) correctly assigned noncontiguous domains, and (3) correct placement of domain boundaries (at 80% precision threshold). (d) Composite evaluation of performance, which combines three components described in (e), using different weighting schema. The first component in the schema is the fraction of correctly assigned domains; second component: correctly fragmented domains; third component: correct placement of domain boundaries (at 80% precision threshold). Regardless of the weighting schema, PDP always appears the best method, while PUU is worse. The performance of DomainParser improves significantly when a larger weight is given to correct fragmentation and boundary placement, so that eventually it surpasses the performance of NCBI method.

particularly β-sheets, and assign them to different domains. This particular property is interpreted differently in all other tested methods. On the contrary, a complete "taboo" on cutting any secondary structure element, as imposed in the NCBI method, results in poorly matched domain boundaries in cases where the proper domain boundaries do occasionally fall within secondary structure elements rather than loops. A tendency toward very compact domains forces the PUU algorithm to build domains out of many discontinuous fragments that often do not make sense from the evolutionary perspective. Rigorous tuning of multiple parameters in the DomainParser algorithm results in excellent prediction on the level of domain boundaries and the number of domain fragments, but at the same time it produces a large fraction of undercut structures. This might partially be due to the training and testing data set that is based on SCOP, an expert database with significant level of undercut predictions compared to CATH and AUTHORS. Another issue is that the threshold for splitting β-strands and β-sheets might be set incorrectly. In summary, analysis of algorithm performance offers both a relative ranking of the methods among themselves as well as insights into particular areas of difficulties for each method; these may lead to new insights and further improvements.

DOMAIN PREDICTION BASED ON SEQUENCE INFORMATION

Identifying protein domains by using only sequence information is even a more challenging task. However, this approach is not limited to proteins with solved 3D structures. For the vast majority of protein sequences, the structure is unknown thus, sequence-based methods are of great importance and, indeed, this area of research is very active. The detriments of predicting boundaries of structural domains without knowing the structure are (1) the level of correct assignments is lower than that for structure-based methods; this difference becomes particularly sharp in the cases of multidomain proteins with complex architectures, and (2) only a few of the sequence-based methods can predict noncontiguous domains that constitute about 10% of all domains (Holland et al., 2006). There is almost no overlap or collaboration between sequence-based and structure-based methods, as we will see from terminology and evaluation methods. The very subject of prediction of structural domains is coached here in terms of predicting domain boundaries.

Sequence-based methods can be divided into several categories: (1) prediction in the absence of sequence information, (2) methods that use information from single sequence (when no homologous sequences are available), (3) methods based on multiple homologous sequences, and (4) methods that incorporate structure prediction from sequences.

Algorithms Based Purely on Domain and Protein Sizes

"Domain Guess by Size" (DGS) (Wheelan, Marchler-Bauer, and Bryant, 2000) is an example of an algorithm that does not use any sequence or structure information. Rather, it predicts protein partitioning based exclusively on the size of the protein. The algorithm predicts several probable outcomes by using likelihood functions based on empirical distributions of protein length, domain length, and segment number as observed in the PDB. This is the only method in the sequence-based category that can predict noncontiguous domains. DGS identifies domain boundaries with 57% accuracy and an error margin of ±20 residues for two-domain proteins. If proteins were simply chopped in half, this approach will only yield an 8% success rate.

Algorithms Based on Single Sequence Information

When prediction of domain boundaries is based on a single sequence, the amino acid propensity is typically calculated for linker regions associated with domain boundaries. Each residue in the sequence is scored with linker index to construct a linker profile and boundaries are identified with a threshold for the scores. DOMcut (Suyama and Ohara, 2003) identifies domain linkers with 53.5% sensitivity and 50.1% selectivity using a linker index deduced from a data set of domain/linker segments. Armadillo (Dumontier et al., 2005) uses a linker index enhanced by positional confidence scores leading to a performance of 56% and 37% sensitivity for two- and multidomain proteins, respectively. These simple strategies indicate that amino acid propensities in linker regions are distinct and differ from the rest of the protein. Such potential patterns in linker regions have further been exploited by using more advanced machine learning techniques such as neural networks. CHOPnet (Liu and Rost, 2004b), PPRODO (Sim 2005), and a newly developed neural network (Ye et al., 2008) report ∼69% accuracy for two-domain proteins.

Algorithms Based on Multiple (Homologous) Sequences

The challenges of identifying boundary regions are diminished when there is detectable sequence homology to the characterized domains. Several algorithms use linkages between the similarities in sequences such as CHOP (Liu and Rost, 2004a). These similar sequences can also be aligned to generate multiple sequence alignments (MSA) that are subsequently analyzed to identify domain boundaries. Domaination (George and Heringa, 2002a) is an example of an algorithm that analyzes features in the MSA to partition proteins into domain units. Nagarajan et al. developed an approach to identify transition positions between domains within the MSA using several probabilistic models (Nagarajan and Yona, 2004). Clustering of the sequences is also a popular approach in identifying domains and is a strategy employed by DIVCLUS (Park and Teichmann, 1998), MKDOM (Gouzy, Corpet, and Kahn, 1999), and ADDA (Heger and Holm, 2003). Finally, the use of taxonomic information has been shown to be useful to help identify domains as well (Coin, Bateman, and Durbin, 2004).

Adding Predicted Structure to Sequence Prediction

While domain identification can be achieved using a purely sequence-based strategy, the use of structural information either integrated during the training or prediction process when developing sequence-based domain identification can impart added advantage. As an example, a comparison between the sequence and structural alignments of domains illustrates the added advantage of using the information content inherent in structural data leading to an improvement over sequence alignments (Marchler-Bauer et al., 2002). We will discuss the strategies that have been employed to improve sequence-based identification of domains and highlight how the use of structural data has been helpful in the development of these tools.

Incorporating *ab initio* structure prediction, either at the secondary or tertiary structure level, has also been used to identify domain boundaries. Sequence-based secondary structure predictors have reached performances as high as 78% accuracy and can be used to help delineate domain boundaries. DomSSEA (Marsden, McGuffin, and Jones, 2002) aligns secondary structure predictions to known templates to identify continuous domains. This approach correctly predicted the number of domains with 73.3% accuracy. Domain boundary predictions were at 24% accuracy for multidomain proteins. Although DomS-SEA does not identify boundaries at a high performance rate, this strategy has not been abandoned and has recently been used in SSEP-Domain (Gewehr and Zimmer, 2006). SSEP-Domain incorporates other profile information and improves predictions to a reported 91.28% overlap between the predicted and true values for single- and two-domain proteins.

Incorporation of generated tertiary structural models has also been explored for domain boundary predictions. SnapDragon (George and Heringa, 2002b) generates several structural models based on the assumption that hydrophobic residues cluster together in space.

Boundaries are predicted based on observed consistencies between structural models of other proteins for a given multiple sequence alignment. The structures may not be atomically accurate models, but they provide an approximation of where the boundaries may be. SnapDragon reports an accuracy of 63.9% for domain boundary predictions in multidomain proteins. Ginzu and RosettaDOM have both incorporated to make automated prediction of domain boundaries (Kim et al., 2005). Ginzu has been utilized for protein sequences with detected homology to a protein with structural data. RosettaDOM generates structures using *ab initio* structural prediction to generate 400 models to identify domain boundaries, a similar concept to SnapDragon. For difficult targets requiring *de novo* structures, Rosetta-DOM reports an 82% overlap in prediction for domains and 54.6% accuracy for domain boundary prediction in multidomain proteins with an error margin of ± 20 residues. According to CASP6 evaluators, the reported performance is comparable to domain boundary predictions conducted by experts. For a more comprehensive overview of the field of adding structure predictions to sequence information, see a review written by Kong and Ranganathan (2004).

Combining Multiple Methods: MetaMethods

Methods that incorporate multiple sources of information result in improved predictions, which is shown by the results of including the predictions of secondary and tertiary structures to the pure sequence information. Several hybrid and consensual approaches have been implemented to leverage the strength of multiple independent predictions. Meta-DP is a metaserver that couples 10 domain predictors and can be extended to include more predictors (Saini and Fischer, 2005). DOMAC integrates template-based and *ab initio* methods, a concept similar to the strategy that is used by Ginzu and RosettaDOM (Cheng, 2007). The difference is that DOMAC invokes the *ab initio* domain predictor DOMpro, another neural network based predictor. Finally, the Dom-Pred server has been recently set up to provide users with results from two different methods, domains predicted from sequence (DPS) and DomSEA (Marsden et al., 2002), with the idea that domain boundary prediction is a computer-assisted task requiring user intervention for this difficult problem (Bryson, Cozzetto, and Jones, 2007).

Evaluating the Performance of Sequence-Based Methods

Similar to structure-based methods, the sequence-based methods still lack a well-established data set on which methods can be evaluated consistently and cross-compared. Various data sets are coming from SCOP (Lo Conte et al., 2000), PFAM (Bateman et al., 2002), DOMO (Gracy and Argos, 1998), SMART (Schultz et al., 1998), and 3Dee (Siddiqui, Dengler, and Barton, 2001). While for structure-based methods the assignment is attempted on a relatively small group of the structures, the sequence-based methods have the "luxury" of using the entire data sets of sequences. Thus, a more conventional analysis of performance can be adopted. The number of missed boundaries (false negatives) and the number of predicted incorrect boundaries (false positives) are calculated and reported using the terms of the *sensitivity* and *selectivity* of the methods. This evaluation approach, while statistically appealing, does not lend itself to an intuitive understanding of what is happening with respect to the individual protein structures (as we do have for structure-based methods). Does a method predict the correct number of domains with the wrong positions of the boundaries or do they tend to predict too many/too few domains?

The success of domain boundary placement is also a variable parameter: some methods assume that the boundary is placed correctly if it falls within an error window of ± 10 residues of the true boundary (which is usually a point between two residues in a sequence), while others tolerate errors as large as ± 20 residues. Some construct compromises by varying the error margin based on size, which is defined as 10% of the length of the protein sequence. This lack of uniform standards for prediction evaluation makes it even more difficult to cross-compare sequence-based methods. A different measure to evaluate success is to calculate the overlap score between the predicted and actual definition of the domain region rather than basing it on the definition of the domain boundary (Jones et al., 1998). The performances of the predictors have been reported to perform as well as up to $\sim 94\%$ for single domains. Currently, most predictors report performances at an accuracy of $\sim 70\%$ with $\sim 50\%$ sensitivity and selectivity. The performances of these algorithms decrease when the number of domains found within a protein increases. Lastly, domains formed by discontinuous segments are difficult to identify and most performance reports are for continuous segments. As the sequence-based domain identifiers improve, we challenge the field to reduce the error margin (often reported with ± 20 residues).

CONCLUSIONS AND PERSPECTIVES

In this chapter, we presented an overview of the principles underlying the detection of the structural domains in proteins and of the computational procedures that implement these principles to assign domains from the atomic coordinates of complete proteins. This overview showed that significant progress has been achieved over the years in the generality and reliability of the algorithms for domain detection. One should add that progress has also been achieved in calculation speed. The more recent second generation methods that cut the polypeptide chain in multiple places simultaneously are orders of magnitude faster than the older methods that produce these cuts sequentially, so much so that the computational demands of domain assignment methods have ceased to be an issue with the present-day computer speeds.

On the other hand, we showed that some important limitations in performance remain and there is definitely room for improvement. While the second generation algorithms elegantly solve the problem of partitioning the structures into domains composed of several chain segments and can detect any number of domains, an additional postprocessing step or additional criteria are needed to deal with the inherent variability of natural protein structures, as well as with the inherent fuzziness in domain definition. Another constructive point to consider in future algorithm development is the merging of consensus results for domain assignments performed on multiple structures of related proteins (Taylor, 1999). This strategy has improved prediction performances for secondary structures and identification of the relationships between proteins using sequence data; therefore, an improvement in domain boundary assignment should also be expected.

The use of domain–domain interfaces is another source of potential improvement. A study by Jones, Marin, and Thornton (2000) reports the properties of residues of domain–domain boundaries to resemble intermolecular interface residues rather than core residues. Ofran and Rost (2003) also reported distinct amino acid compositions and residue–residue interaction preferences between homo- and hetero-domain interacting regions. These additional biophysical and sequence selection knowledge criteria can add value to the existing computational methods.

An improved understanding of the sequence rules that define domains will have a significant impact on structural genomics efforts (Chapter 40) and provide many potential applications. An example of an application could be the priority assignment of target proteins with structures that potentially contain a compact domain and should be easier to crystallize, in principle. Alternatively, a better understanding of the sequence space and how it defines the domain boundaries will provide an insight into the evolution and diversification of protein structure and function.

Critical for improved domain boundary determination is the availability of proper benchmark data sets incorporated into the evaluation process to allow for a cross-comparison of different computational methods to highlight strengths and weaknesses that will lead to improved methodologies. However, construction of a proper benchmark data set requires agreement among human experts that is sometimes difficult to achieve— thus the most difficult and contentious cases of architectures are currently excluded (Veretnik et al., 2004). The very existence of the architectures for which multiple plausible domain decompositions exist refutes our simple-minded tendency to fit one approach for partitioning to all structures. As more protein structures are solved, the fraction of such "controversial" proteins is likely to increase. The best way to address this inherent complexity of the protein structures may be by accepting the possibility of alternative domain decompositions and implementing this feature in new algorithms.

Some of the latest algorithms have such a capacity already (Berezovsky, 2003; Taylor and Vaisman, 2006) and simply use a series of thresholds instead of a single threshold during structure decomposition. In general, the main difficulty encountered by current algorithms is not the lack of ability to find potential domain boundaries, rather it is the ability to identify a subset of "true" domain boundaries out of the pool of potential domain boundaries. Thus, performing domain decomposition using multiple thresholds could possibly overcome this challenge by allowing the exploration of a single solution for relatively simple structures and multiple solutions for proteins with more complex architectures. The future success of algorithms for domain decomposition may require a shift in our thinking about what constitutes a good solution for this complex problem and is likely to promote the consideration of alternative decomposition scenarios as an essential part of the solution.

WEB RESOURCES

CHOP: http://www.rostlab.org/services/CHOP/.
DomPred: http://bioinf.cs.ucl.ac.uk/dompred/
Meta-DP: http://meta-dp.cse.buffalo.edu/
DOMAC: http://www/bioinfotool.org/domac.html
AUTHORS (as reported in Islam et al.): http://www.bmm.icnet.uk/~domains/test/dom-rr
.html
CATH: http://www.cathdb.info/latest/index.html
Consensus benchmark: http://pdomains.sdsc.edu/dataset.php
DDomain (source code for download): http://sparks.informatics.iupui.edu/
(interactive Web site): http://sparks.informatics.iupui.edu/hzhou/ddomain.html
DomainParser (interactive Web site and source code): http://compbio.ornl.gov/structure/
domainparser/
NCBI (interactive Web site, enter PDB code in the small window on the right side of the
page): http://www.ncbi.nlm.nih.gov/Structure/MMDB/mmdb.shtml

PDP algorithm (interactive Web site, no source code): http://123d.ncifcrf.gov/pdp.html
PUU (a static list of domains): http://iubio.bio.indiana.edu/soft/iubionew/molbio/protein/
analysis/Dali/domain/3.0/DaliDomainDefinitions
Note: be aware that the file Domain Definition at http://ekhidna.biocenter.helsinki.fi/dali/
downloads does not define domains according to PUU, rather according to DomainParser.
SCOP: http://scop.mrc-lmb.cam.ac.uk/scop/

REFERENCES

Alexandrov N, Shindyalov I (2003): PDP: protein domain parser. *Bioinformatics* 19:429–430.

Bateman A, Birney E, Cerruti L, Durbin R, Etwiller L, Eddy SR, Griffiths-Jones S, Hower KL, Marshall M, Sonnhammer EL (2002): The Pfam protein families database. *Nucleic Acid Research* 30:276–280.

Bennett MJ, Choe S, Eisenberg D (1994): Domain swapping: entangling alliances between proteins. *Proc Natl Acad Sci U S A* 91:3127–3131. [The original paper describing the domain swapping principle and suggesting it as a general principle for the evolution of multimeric proteins.]

Berezovsky IN (2003): Discrete structure of van der Waals domains in globular proteins. *Protein Eng* 16:161–167. [One of the first methods to introduce multiple possible outcomes based on progressively relaxing the threshold for compactness.]

Berman HM, Westbrook J, Feng Z, Gilliland G, Bhat TN, Weissig H, Shindyalov IN, Bourne PE (2000): The Protein Data Bank. *Nucleic Acids Res* 28:235–242.

Bernstein FC, Koetzle TF, Williams GJ, Meyer EE Jr, Brice MD, Rodgers JR, Kennard O, Shimanouchi T, Tasumi M (1977): The Protein Data Bank: a computer-based archival file for macromolecular structures. *J Mol Biol* 112:535–542.

Bryson K, Cozzetto D, Jones DT (2007): Computer-assisted protein domain boundary prediction using the DomPred server. *Curr Protein Pept Sci* 8:181–188.

Cheng J (2007): DOMAC: an accurate, hybrid protein domain prediction server. *Nucleic Acids Res* 35: W354–356.

Chothia C (1992): Proteins. One thousand families for the molecular biologist. *Nature* 357:543–544. [Before any complete genome sequences were available, this short note presents arguments in support of the idea that the number of protein folds shared by all organisms is limited to about 1000.]

Coin L, Bateman A, Durbin R (2004): Enhanced protein domain discovery using taxonomy. *BMC Bioinform* 5:56.

Crippen GM (1978): The tree structural organization of proteins. *J Mol Biol* 126:315–332. [One of the very first systematic procedures for identifying structural domains in proteins. The only first-generation method that includes non-contiguous domains.]

Dumontier M, Yao R, Feldman HJ, Hogue CW (2005): Armadillo: domain boundary prediction by amino acid composition. *J Mol Biol* 350:1061–1073.

Emmert-Streib F, Mushegian A (2007): A topological algorithm for identification of structural domains in proteins. *BMC Bioinform* 8:237. *This algorithm uses only topological information of secondary structures and is more a proof of principle than an actual method.*

George RA, Heringa J (2002a) An analysis of protein domain linkers: their classification and role in protein folding. *Protein Eng* 15:871–879.

George RA, Heringa J (2002b) SnapDRAGON: a method to delineate protein structural domains from sequence data. *J Mol Biol* 316:839–851.

Gewehr JE, Zimmer R (2006): SSEP-domain: protein domain prediction by alignment of secondary structure elements and profiles. *Bioinformatics* 22:181–187.

Gouzy J, Corpet F, Kahn D (1999): Whole genome protein domain analysis using a new method for domain clustering. *Comput Chem* 23:333–340.

Gracy J, Argos P (1998): DOMO: a new database of aligned protein domains. *Trends Biochem Sci* 23:495–497.

Guo JT, Xu D, Kim D, Xu Y (2003): Improving the performance of DomainParser for structural domain partition using neural network. *Nucleic Acids Res* 31:944–952.

Heger A, Holm L (2003): Exhaustive enumeration of protein domain families. *J Mol Biol* 328: 749–767.

Holland TA, Veretnik S, Shindyalov IN, Bourne PE (2006): Partitioning protein structures into domains: why is it so difficult? *J Mol Biol* 361:562–590. [This paper introduces most comprehensive benchmark dataset available as of today. It also discusses the principles behind each computational method and how the encoded assumptions affect the performance of the method.]

Holm L, Sander C (1994): Parser for protein folding units. *Proteins* 19:256–268. [The first truly general algorithm for defining structural domains that can identify domains comprised of multiple chain segments.]

Holm L, Sander C (1996): Mapping the protein universe. *Science* 273:595–603.

Islam SA, Luo J, Sternberg MJ (1995): Identification and analysis of domains in proteins. *Protein Eng* 8:513–525.

Janin J, Wodak SJ (1983): Structural domains in proteins and their role in the dynamics of protein function. *Prog Biophys Mol Biol* 42:21–78.

Jones S, Stewart M, Michie A, Swindells MB, Orengo C, Thornton JM (1998): Domain assignment for protein structures using a consensus approach: characterization and analysis. *Protein Sci* 7:233–242. [A useful systematic comparison between different methods for assigning structural domains in proteins.]

Jones S, Marin A, Thornton JM (2000): Protein domain interfaces: characterization and comparison with oligomeric protein interfaces. *Protein Eng* 13:77–82.

Kernighan BW (1970): An efficient heuristic procedure for partitioning graphs. *Bell Sys Tech J* 49:291–307.

Kim DE, Chivian D, Malmstrom L, Baker D (2005): Automated prediction of domain boundaries in CASP6 targets using Ginzu and RosettaDOM. *Proteins* 61:(Suppl 7): 193–200.

Kong L, Ranganathan S (2004): Delineation of modular proteins: domain boundary prediction from sequence information. *Brief Bioinform* 5:179–192.

Kundu S, Sorensen DC, Phillips GN Jr (2004): Automatic domain decomposition of proteins by a Gaussian Network Model. *Proteins* 57:725–733.

Liu J, Rost B (2004a) CHOP: parsing proteins into structural domains. *Nucleic Acids Res* 32: W569–W571.

Liu J, Rost B (2004b) Sequence-based prediction of protein domains. *Nucleic Acids Res* 32:3522–3530.

Lo Conte L, Ailey B, Hubbard TJ, Brenner SE, Murzin AG, Chothia C (2000): SCOP: a structural classification of proteins database. *Nucleic Acids Res* 28:257–259.

Madej T, Gibrat J-F, Bryand SH, (1995): Threading a database of protein cores. *Proteins* 23:356–369.

Marchler-Bauer A, Panchenko AR, Ariel N, Bryant SH, (2002): Comparison of sequence and structure alignments for protein domains. *Proteins* 48:439–446.

Marsden RL, McGuffin LJ, Jones DT, (2002): Rapid protein domain assignment from amino acid sequence using predicted secondary structure. *Protein Sci* 11 2814–2824.

Moult J, Unger R (1991): An analysis of protein folding pathways. *Biochemistry* 30:3816–3824.

Murzin AG, Brenner SE, Hubbard T, Chothia C (1995): SCOP: a structural classification of proteins database for the investigation of sequences and structures. *J Mol Biol* 247:536–540.

Nagarajan N, Yona G (2004): Automatic prediction of protein domains from sequence information using a hybrid learning system. *Bioinformatics* 20:1335–1360.

Ofran Y, Rost R (2003): Predicted protein–protein interaction sites from local sequence information. *J Mol Biol* 325:377–387.

Ooi T, Nishikawa K, (1973). In: Bergmann A, Pullmann B, editors. *Conformation of Biological Molecules and Polymers*. New York: Academic Press, pp 173–187.

Orengo CA, Michie AD, Jones S, Jones DT, Swindells MB, Thornton JM (1997): CATH—a hierarchic classification of protein domain structures. *Structure* 5:1093–1108.

Park J, Teichmann SA (1998): DIVCLUS: an automatic method in the GEANFAMMER package that finds homologous domains in single- and multidomain proteins. *Bioinformatics* 14:144–150.

Phillips DC (1970): Past and present. Goodwin TW, editor. *British Biochemistry*. London: Academic Press, 11–28. [The pioneering study in which the idea of structural domains has first been described.]

Ponting CP, Russell RR (2002): The natural history of protein domains. *Annu Rev Biophys Biomol Struct* 31:45–71. [An evolutionary overview of the field of protein domains.]

Rashin AA (1981): Location of domains in globular proteins. *Nature* 291:85–87.

Richardson JS (1981): The anatomy and taxonomy of protein structure. *Adv Protein Chem* 34:167–339.

Rose GD, (1979): Hierarchic organization of domains in globular proteins. *J Mol Biol* 134:447–470. [One of the first systematic procedures for defining structural domains in proteins from the atomic coordinates.]

Rossman MG, Liljas A (1974): Letter: recognition of structural domains in globular proteins. *J Mol Biol* 85:177–181. [One of the pioneering studies in which the idea of structural domains has first been described.]

Saini HK, Fischer D (2005): Meta-DP: domain prediction meta-server. *Bioinformatics* 21:2917–2920.

Schultz J, Milpetz F, Bork P, Ponting CP (1998): SMART, a simple modular architecture research tool: identification of signaling domains. *Proc Natl Acad Sci U S A* 95:5857–5864.

Siddiqui AS, Barton GJ (1995): Continuous and discontinuous domains: an algorithm for the automatic generation of reliable protein domain definitions. *Protein Sci* 4:872–884.

Siddiqui AS, Dengler U, Barton GJ (2001): 3Dee: a database of protein structural domains. *Bioinformatics* 17:200–201.

Sim J, Kim SY, Lee J (2005): PPRODO: prediction of protein domain boundaries using neural networks. *Proteins* 59:627–632.

Sistla RK, K V B, Vishveshwara S (2005): Identification of domains and domain interface residues in multidomain proteins from graph spectral method. *Proteins* 59:616–626.

Sowdhamini R, Blundell TL (1995): An automatic method involving cluster analysis of secondary structures for the identification of domains in proteins. *Protein Sci* 4:506–520.

Suyama M, Ohara O (2003): DomCut: prediction of inter-domain linker regions in amino acid sequences. *Bioinformatics* 19:673–674.

Swindells MB (1995a): A procedure for detecting structural domains in proteins. *Protein Sci* 4:103–112.

Swindells MB (1995b): A procedure for the automatic determination of hydrophobic cores in protein structures. *Protein Sci* 4:93–102.

Taylor WR (1999): Protein structural domain identification. *Protein Eng* 12:203–216. [An elegant heuristic procedure, inspired by the Ising model of solid-state physics for assigning structural domains in proteins.]

Taylor TJ, Vaisman II (2006): Graph theoretic properties of networks formed by the Delauney tessellation of protein structures. *Phys Rev E Stat Nonlin Soft Matter Phys* 73:041925.

Tsoka S, Ouzounis CA (2000): Prediction of protein interactions: metabolic enzymes are frequently involved in gene fusion. *Nat Genet* 26:141–142.

Veretnik S, Bourne PE, Alexandrov NN, Shindyalov IN (2004): Toward consistent assignment of structural domains in proteins. *J Mol Biol* 339:647–678. [This paper describes the principle behind the construction of the consensus benchmark dataset. It also discusses the reason for disagreements among expert methods.]

Veretnik S, Shindyalov IN, (2006): Computational methods for domain partitioning of protein structures. In: Xu Y, Xu D, Liang J, *Computational Methods for Protein Structure Prediction and Modeling.* Springer 125–145. [This chapter describes the computational principles behind individual methods for domain delineation.]

Wernisch L, Hunting M, Wodak SJ (1999): Identification of structural domains in proteins by a graph heuristic. *Proteins* 35:338–352. [This paper describes a novel graph theoretical procedure for assigning structural domains in proteins. It handles any number of non-contiguous chain segments and uses no information on secondary structure. A discriminant analysis is used to derive a set of criteria that define physically meaningful domains.]

Wetlaufer DB (1973): Nucleation, rapid folding, and globular intrachain regions in proteins. *Proc Natl Acad Sci U S A* 70:697–701. [The first definition and systematic analysis of structural domains in proteins.]

Wheelan SJ, Marchler-Bauer A, Bryant SH (2000): Domain size distributions can predict domain boundaries. *Bioinformatics* 16:613–618.

Wodak SJ, Janin J (1981a): Defining compact domains in globular proteins. In: Balaban editor. *Structural Aspects of Recognition and Assembly in Biological Macromolecules.* Rehovot: International Science Services 149–167. [One of the first approaches for identifying small compact sub-structures in proteins that are likely to be stable.]

Wodak SJ, Janin J (1981b): Location of structural domains in protein. *Biochemistry* 20:6544–6552. [This paper describes the first approach for defining structural domains in proteins which relies on the evaluation of some physical property (size of the domain interface) as opposed to purely geometric criteria.]

Xu Y, Xu D, Gabow HN (2000): Protein domain decomposition using a graph-theoretic approach. *Bioinformatics* 16:1091–1104.

Xuan ZY, Ling LJ, Chen RS (2000): A new method for protein domain recognition. *Eur Biophys J* 29:7–16.

Yang S, Doolittle RF, Bourne PE (2005): Phylogeny determined by protein domain content. *PNAS* 102:373–378.

Ye L, Liu T, Wu Z, Zhou R (2008): Sequence-based protein domain boundary prediction using BP neural network with various property profiles. *Proteins* 71:300–307.

Zehfus MH, Rose GD (1986): Compact units in proteins. *Biochemistry* 25:5759–5765.

Zehfus MH (1994): Binary discontinuous compact protein domains. *Protein Eng* 7:335–340.

Zhou H, Xue B, Zhou Y (2007): DDOMAIN: Dividing structures into domains using a normalized domain-domain interaction profile. *Protein Science* 16:947–955.

21

INFERRING PROTEIN FUNCTION FROM STRUCTURE

James D. Watson, Gail J. Bartlett, and Janet M. Thornton

INTRODUCTION

The Importance of Predicting Protein Function from Structure

The various genome-sequencing projects around the globe have provided us with massive amounts of information detailing all the genes in a number of organisms required for survival. The amount of sequence data is set to rapidly expand as a result of the very large-scale metagenomics projects such as the Global Ocean Survey (Yooseph et al., 2000). By comparison, the protein structure data falls far behind. Structural genomics aims to close the gap by experimentally determining a large number of protein structures as rapidly and accurately as possible using high-throughput methods. There are several consortia working on this across the globe and each has its individual goals, but one of the key aims is to increase the coverage of protein fold space and hence the proportion of protein sequences amenable to homology modeling methods. As a result, an increased number of protein structures have been released with little or no functional annotation. This is a reversal of the usual experimental investigation of proteins, which involves taking a protein of interest, carrying out biochemical experiments to determine functional information about it, and then using the structure to rationalize this functional information (Thornton et al., 2000). For example, the tyrosine kinases were known to be signaling molecules long before the crystal structure revealed molecular mechanisms of their function (Hubbard et al., 1994). The recent ramp-up of structural genomics projects to full-scale production and investment in new technologies to address the more challenging problems of complexes and membrane proteins means that the numbers of structures deposited in macromolecular structure databases will continue to rise. One of the major goals in modern bioinformatics is the development of computational methods to accurately and automatically assign functions to these proteins.

Structural Bioinformatics, Second Edition Edited by Jenny Gu and Philip E. Bourne
Copyright © 2009 John Wiley & Sons, Inc.

Definition of Function

The function of a protein is not always well defined (Skolnick and Fetrow, 2000) and can be described at a number of levels: from highly specific enzyme reactions to biochemical pathways and up to the organism as a whole. The problems are compounded by the fact that a protein's function can vary as a consequence of changes in expression and environment (Jeffery, 1999). Oligomerization and cellular localization are examples of such changes. Phosphoglucose isomerase acts as a neuroleukin, a cytokine and a differentiation and maturation mediator in its monomeric, extracellular form, but as a dimer inside the cell, it has a role in glucose metabolism, catalyzing the interconversion of glucose-6-phosphate and fructose-6-phosphate. Different experimental techniques can elucidate these different aspects of function and therefore there is a great deal of inconsistency in the specificity of annotations.

In order to catalog the diversity of functions adopted by proteins, much work has been done to standardize functional descriptions. The most well-known being the Enzyme Commission (EC) numbering scheme (http://www.chem.qmul.ac.uk/iubmb/enzyme/). This is a four-level hierarchy that classifies different aspects of chemical reactions catalyzed by enzymes. The first digit denotes the class of the reaction, and subsequent levels classify the substrate, the type of bond involved, cofactors, and other specificities (Table 21.1).

TABLE 21.1. Description of the Different Levels in the EC Classification

Oxidoreductase
  Substrate is oxidized, regarded as the hydrogen or electron donor
    Describes substrate acted on by enzyme
      Type of acceptor
Transferase
  Transfer of a group from one substrate to another
    Describes group transferred
      Further information on the group transferred
Hydrolase
  Hydrolytic cleavage of a bond
    Describes type of bond
      Nature of substrate
Lyase
  Cleavage of bonds by elimination
    Type of bond
      Further information on the group eliminated
Isomerase
  Type of reorganization
    Type of substrate
Ligase
  Enzyme catalyzing the joining of two molecules in concert with hydrolysis of ATP
    Type of bond formed
      Type of compound formed

An enzyme reaction is assigned a four-digit EC number, where the first digit denotes the class of reaction. Note that the meaning of subsequent levels depends on the primary number; for example, the substrate acted upon by the enzyme is described at the second level for oxidoreductase, whereas it is described at the third level for hydrolase. Different enzymes clustered together at the third level are given a unique fourth number, and these enzymes may differ in substrate/product specificity or cofactor dependency, for example. Note that the EC is a classification of overall enzyme reactions and not enzymes.

TABLE 21.2. Gene Ontology Functional Classifications

Category	Description
Biological process	A biological objective to which the gene product contributes. A process (which often involves a chemical or physical transformation) is accomplished via one or more assemblies of molecular functions. A biological process can be high level (or less specific), for example, cell growth and maintenance, or low level (or more specific), for example, glycolysis.
Molecular function	The biochemical activity of a gene product, describing what it actually does without alluding to where or when. A molecular function can be broad (or less specific), for example, enzyme, or narrow (or more specific), for example, hexokinase.
Cellular component	Refers to the place in the cell where a gene product is active

The EC nomenclature is limited to enzymes and has problems when dealing with multi-functional proteins such as methylenetetrahydrofolate dehydrogenase/cyclohydrolase. This protein catalyzes the conversion of methylenetetrahydrofolate to formylfolate in two separate reactions that are thought to proceed using the same or overlapping active sites (Allaire et al., 1998). This enzyme has two EC numbers associated with it: 1.5.1.5 and 3.5.4.9.

A more recent development has been the Gene Ontology (GO) scheme (http://www.geneontology.org, Ashburner et al., 2000), which attempts to standardize the description and definition of biological terms through three structured, controlled vocabularies. The three major sections are Cellular Component, Biological Process and Molecular Function (see Table 21.2). GO has become one of the most commonly used ontologies in bioinformatics and has three key advantages: it uses a controlled vocabulary, is an open source, and is machine-readable. The last feature makes it ideally suited for use in automated comparison and function prediction servers. The one problem is that it is not a linear hierarchy and therefore the relationship between gene product and biological process, molecular function, and cellular component is often one to many, making direct comparisons more complicated. The main advantage is that, by separating and independently assigning these attributes, relationships between gene product and function can be clarified more easily.

WHAT INFORMATION CAN BE OBTAINED FROM THREE-DIMENSIONAL PROTEIN STRUCTURES?

The basic structure comes in the form of a Protein Data Bank (PDB) file (Berman et al., 2000), which is a list of 3D coordinates of all the atoms in the protein. The PDB file itself contains little, if any, functional data. There is sometimes a 'SITE' record, but this is used for various purposes such as ligand binding sites, metal binding sites, and 'active sites' and is not consistent. As discussed previously, many PDB files contain no functional information and many of the structural genomics deposits are labeled only as "hypothetical protein." However, from the structure we can derive information related to biological function; this information is summarized in Figure 21.1.

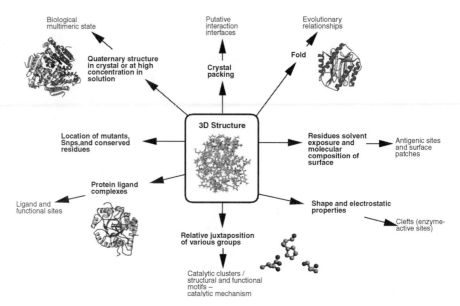

Figure 21.1. From structure to function: a summary of information that can be derived from three-dimensional structure, relating to biological function.

Starting at more coarse general observations, we can see the quaternary structure present in the crystal, which can tell us the oligomeric state of the protein, and may throw light on protein–protein interactions. We can then identify buried residues making up the core of the protein and residues on the surface exposed to the solvent. By looking more closely, we can also see the shape and molecular composition of the surface, clefts and pockets, and finally the arrangement of amino acid side chains and the juxtaposition of individual groups.

INFERRING FUNCTION FROM STRUCTURE

Computational methods to infer a function of an individual protein, such as its enzymatic activity, generally fall into two main types: those that are sequence-based and those that are structure-based. The different structure-based methods are discussed individually below and example servers (with web links) are detailed in Table 21.3. The sequence-based approaches are usually the first port of call as significant similarities in sequence can provide strong indication of function. It has been shown (Todd, Orengo, and Thornton, 2001) that above 40% sequence identity, homologous proteins tend to have the same function; but below this threshold, conservation of function falls rapidly, although there are always exceptions to this rule (Whisstock and Lesk, 2003) and if the domain composition is retained, function can be inferred at even lower sequence identities (Orengo, 2007). The most commonly used sequence-based methods involve BLAST (Altschul et al., 1997) or FASTA (Pearson, 1991) runs to perform direct sequence–sequence comparisons of the protein sequence against large sequence databases such as UniProt (Apweiler et al., 2004) or GenBank (Benson et al., 2005) in order to identify similarity with proteins of known function. More powerful and sensitive profile/pattern-based methods, such as those using

TABLE 21.3. Commonly Used Servers for Predicting Protein Function from Structure, Grouped by Type of Method

Method Class	Server	URL	Comments
Fold	Dali	http://www.ebi.ac.uk/dali/	Optimal pair-wise alignment of protein structures using α-carbon residue–residue distance matrices
	MSDfold (SSM)	http://www.ebi.ac.uk/msd-srv/ssm/	Secondary structure matching program using graph theory for comparing secondary structure elements
	CATHEDRAL	http://cathwww.biochem.ucl.ac.uk/cgi-bin/ cath/CathedralServer.pl	Provides a fast method to find the closest fold group for any given structure by comparing secondary structure elements
	VAST	http://www.ncbi.nlm.nih.gov/Structure/ VAST/vastsearch.html	Structure–structure similarity search using vector alignment
	CE	http://cl.sdsc.edu/ce.html	Combinatorial extension method
	FAST	http://biowulf.bu.edu/FAST/	A rapid algorithm for aligning three-dimensional structures of proteins. Executable can be downloaded for stand-alone use
Clefts and pockets	SURFNET	http://www.biochem.ucl.ac.uk/~roman/ surfnet/surfnet.html	Generates surfaces and void regions between surfaces from coordinate data supplied in PDB format
	SURFACE	http://cbm.bio.uniroma2.it/surface/	A database of protein surface patches annotated with sequence- and structure-derived information about function or interaction abilities
	pvSOAR	http://pvsoar.bioengr.uic.edu/	Web server to detect surface similarities in proteins. Searches against the CASTp database or nonredundant PDB sets. Can also compare pockets against one another
	CASTp	http://cast.engr.uic.edu/cast/	A database providing identification and measurements of surface-accessible pockets and interior-inaccessible cavities
	Q-SiteFinder	http://www.bioinformatics.leeds.ac.uk/ qsitefinder/	Ligand binding site prediction using hydrophobic probes to identify clusters of probes with the most favorable binding energy. Clusters are placed in rank order of the likelihood of being a binding site
	Pocket-Finder	http://www.bioinformatics.leeds.ac.uk/ pocketfinder/	A pocket detection algorithm based on the Ligsite algorithm and uses the same interface as Q-SiteFinder
	HotPatch	http://hotpatch.mbi.ucla.edu	Finds unusual patches on the surface of proteins, identifying how unusual they are and how likely each patch is to be of functional importance

(continued)

TABLE 21.3. (*Continued*)

Method Class	Server	URL	Comments
	FEATURE	http://feature.stanford.edu/http://feature.stanford.edu/webfeature/	Recognizes protein binding sites using statistical descriptions of their 3D environments. Available for download or as a web server
	S-BLEST	http://sblest.org/	Matches physicochemical features of pockets and surfaces to find structurally similar proteins in a database as well as associating environments with structural and functional roles
	Relibase	http://relibase.ebi.ac.uk/reli-cgi/rll?/reli-cgi/general_layout.pl + home	A program for searching protein–ligand databases
	SiteEngine	http://bioinfo3d.cs.tau.ac.il/SiteEngine/	Searches the structure of the complete molecule for regions similar to the binding site of interest. This is defined by the protein surface that interacts with the ligand and hence requires a bound structure
	SURF'S UP!	http://asia.genesilico.pl/surfs_up/	Identifies similar patterns of physicochemical features on compared protein surfaces. Can work with models obtained from comparative modeling
Active sites and templates	MSDsite	http://www.ebi.ac.uk/msd-srv/msdsite/index.jsp	Active site searches based on ligand or active site information. Can also search an uploaded file. Searches can be restricted by author, keywords, experiment, resolution, and so on
	MSDmotif	http://www.ebi.ac.uk/msd-srv/msdmotif/	Service for protein sequence motif detection. Small 3D-structure motifs such as beta-turns, Asx-motifs, and nests can characterize active sites and take part in enzyme mechanisms
	CSS	http://www.ebi.ac.uk/thornton-srv/databases/cgi-bin/CSS/makeEbiHtml.cgi?file=form.html	Catalytic site search: uses JESS algorithm to search a bank of structural templates in the CSA
	CSA	http://www.ebi.ac.uk/thornton-srv/databases/CSA/	A database documenting enzyme active sites and catalytic residues in enzyme three-dimensional structures
	PINTS	http://www.russell.embl.de/pints/	Allows comparison of a protein structure against a database of patterns, a structural pattern against a database of protein structures or two proteins directly. Hits are evaluated using rmsd and *E*-value

Category	Name	URL	Description
	SPASM/RIGOR	Available from http://xray.bmc.uu.se/usf/	SPASM takes a user-defined motif of the main chain and/or side chains and compares it against a database of structures. RIGOR searches a database of predefined motifs to see if any of them occur in the submitted protein structure
	SuMo	http://sumo-pbil.ibcp.fr/cgi-bin/sumo-welcome	Uses graph theory to match triangles of chemical groups rather than amino acids
	PDBSiteScan	http://wwwmgs.bionet.nsc.ru/mgs/gnw/pdbsitescan/	Searches the PDBSite database (a database of functional sites extracted from the SITE records in all PDB files)
	DRESPAT	Program available for download on request to ashish@it.iitb.ac.in	Graph theory approach to identify functional sites in proteins
	SARIG	http://bioinfo2.weizmann.ac.il/~pietro/SARIG/V3/index.html	The protein structure is represented as a residue interaction graph (RIG). Network analysis is then performed to identify active site residues
Phylogenetic relationships	ET	http://www-cryst.bioc.cam.ac.uk/~jiye/evoltrace/evoltrace.html	One example of the evolutionary trace method
	ET	http://www.cmpharm.ucsf.edu/~marcinj/JEvTrace/	A Java implementation of the evolutionary trace method available for download
	ConSurf	http://consurf.tau.ac.il/	Identifies functional regions in proteins by mapping phylogenetic information onto the surface
	FunShift	http://funshift.cgb.ki.se/	Provides functional shift (divergence) analysis among the subfamilies of a protein domain family
DNA binding	PDNA-pred server	http://www.ebi.ac.uk/thornton-srv/databases/PDNA-pred/	HTH structural motif detection. Additional data on solvent accessibility and electrostatic potentials help improve sensitivity
	Protein–DNA interaction PreDS server	http://pre-s.protein.osaka-u.ac.jp/~preds	Uses protein surface and electrostatic potential to detect double stranded DNA binding sites
	Protein–Nucleic Acid Server	http://www.biochem.ucl.ac.uk/bsm/DNA/server/	A tool to analyze the protein interface of any protein–nucleic acid complex
	eF-Site	http://ef-site.hgc.jp/eF-site/	A database of protein functional site molecular surfaces, displaying the electrostatic potentials and hydrophobic properties of each
	PSSM	http://www.netasa.org/dbs-pssm/	Neural network-based algorithm using position specific substitution matrices to predict DNA binding sites

(continued)

TABLE 21.3. (*Continued*)

Method Class	Server	URL	Comments
Combinatorial function prediction server	ProFunc	http://www.ebi.ac.uk/thornton-srv/databases/profunc/index.html	Automated function prediction server utilizing several sequence- and structure-based methods
	ProKnow	http://www.doe-mbi.ucla.edu/Services/ProKnow/	Extracts a variety of features from the submitted structure; significance is weighted using Bayes' theorem before using GO annotation to estimate the most likely overall function
	Gene3D	http://cathwww.biochem.ucl.ac.uk:8080/Gene3D/	Cluster proteins from complete genomes functionally annotated using CATH, Pfam, KEGG, GO, and others. Can upload a protein sequence for function prediction through identification of domain architectures
	PDBfun	http://pdbfun.uniroma2.it/	Residue-centric approach integrating information from a number of sources. Useful for the identification of local structural similarities between annotated residues in proteins
	AnaGram	http://jaguar.genetica.uma.es/anagram.htm	Assigns function to protein sequences by finding correlations between sequence signals and functional annotations in a database of protein

hidden Markov models, have been developed to identify very distant relationships. These models can be constructed from sequences in whole protein families where the family is defined by a similarity threshold using their 3D structure (e.g., Gene3D (Yeats et al., 2006) and SUPERFAMILY (Gough and Chothia, 2002), or in terms of sequence similarity and function (e.g., Pfam; Sonnhammer et al., 1998).

Other aspects of proteins that can be analyzed to provide functional clues include phylogenetic profiles, residue conservation patterns, genome structure, and gene location. However, when the sequence-based methods fail or provide few functional clues, examining the protein's 3D structure can provide further insight. The structure-based methods can be classified according to the level of protein structure and specificity at which they operate, ranging from analysis of the global fold of the protein down to the identification of highly specific 3D clusters of functional residues (Stark and Russell, 2003; Laskowski, Watson, and Thornton, 2005). The underlying strategies for functional annotation of proteins using structural information are highlighted in the following sections.

Fold Matching

Within defined bounds, we can compare the structure of a protein of unknown function with the structures of proteins of known function in structural databases such as CATH (Orengo et al., 1997) or SCOP (Lo Conte et al., 2000) to inherit the functional information from the closest match. This is usually the first step in assigning function from structural information as proteins with a similar structure and sequence identity are evolutionarily related and likely to share a similar function. However, caution must be exercised in transferring function from one homologous protein to another. If two proteins share local structural similarity but do not share any sequence identity, this may be the result of convergent evolution such as in the classic example of subtilisin and chymotrypsin that have no sequence similarity and have very different folds (as shown in Figure 21.2) yet are found to have an identical Ser–His–Asp catalytic triad and can also be inhibited by the same enzyme

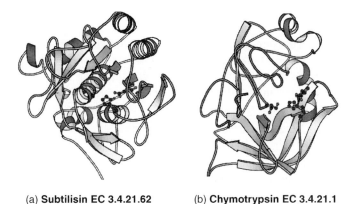

(a) **Subtilisin EC 3.4.21.62**   (b) **Chymotrypsin EC 3.4.21.1**

Figure 21.2. Subtilisin and chymotrypsin are both serine endopeptidases. They share no sequence identity, and their folds are unrelated. However, they have an identical, three-dimensionally conserved Ser–His–Asp catalytic triad, which catalyzes peptide bond hydrolysis. These two enzymes are a classic example of convergent evolution.

(a) Lysozyme EC 3.2.1.17   (b) Alpha-lactalbumin (nonenzyme)

Figure 21.3. Lysozyme and α-lactalbumin share 40% sequence identity between them, and similar structures, but they have different functions. Lysozyme is an *O*-glycosyl hydrolase, but α-lactalbumin does not have this catalytic activity. Instead, it regulates the substrate specificity of galactosyl transferase through its sugar binding site, which is common to both α-lactalbumin and lysozyme. Both the sugar binding site and the catalytic residues have been retained by lysozyme during evolution, but in α-lactalbumin, the catalytic residues have changed and it is no longer an enzyme.

inhibitors (Henschel, Kim, and Schroeder, 2006). This serves as an example where the use of global protein features at the sequence and structural level may preclude functional annotation that can only be observed using local structural comparisons. The opposite is also true when two structurally homologous and evolutionarily related proteins have different functions, the classic example being that of lysozyme and α-lactalbumin (Acharya et al., 1991). These two proteins have high sequence identity between them but vary in their function (Figure 21.3). These types of problems are very difficult to identify in an automated manner and as a consequence are usually identified by manual assessment of the predictions through experiment. Improvements in high-throughput experimental functional assays (Kuznetsova et al., 2005; Proudfoot et al., 2004) and mass ligand binding screens (Vedadi et al., 2006) may help detect these errors in a large-scale data analysis, but problems can be avoided in the first place through careful annotation of function on a case-by-case basis.

Several methods exist for fold searching, the best known being DALI (Holm and Sander, 1995). Some popular methods used by the research community include MSDfold (SSM; Krissinel and Henrick, 2004) and CATHEDRAL (Pearl et al., 2003), both of which use graph theory to match secondary structure elements. Other approaches use different algorithms to match structures: VAST uses vector alignment of secondary structures, CE (Shindyalov and Bourne, 1998) uses combinatorial extension, and FAST (Zhu and Weng, 2005) uses a directionality-based scoring scheme to compare intramolecular residue–residue relationships. All of the methods, generally speaking, agree when assessed with the same data set but major disagreements can be a problem so the best approach is to use two or three methods to identify the consensus match (for a review of fold comparison servers see Novotny, Madsen, and Kleywegt, 2004).

Surface Clefts and Binding Pockets

One of the main factors determining how a protein interacts with other molecules is the size of clefts on the protein surface. Clefts provide an increased surface area from which the solvent may be excluded and therefore an increased opportunity for the protein to form complementary hydrogen bonds and hydrophobic contacts with small ligands.

A protein ligand binding site (active site) is often found to be the largest cleft in the protein, and this cleft is often significantly larger than other clefts in the protein (Laskowski et al., 1996). The identification and analysis of surface clefts can therefore lead to a cofactor or ligand identification and hence putative functions for the protein are suggested. These surface clefts and pockets can be identified in a number of ways, but one of the most commonly methods used is the SURFNET algorithm (Laskowski, 1995), which detects gap regions within the protein by fitting spheres of a certain range of sizes between the protein atoms. Once clefts and pockets are identified, they can be compared against databases of known clefts to identify local similarities. The pvSOAR web server (Binkowski, Freeman, and Liang, 2004) allows searches to be performed against the CASTp database or against a nonredundant set of the PDB, returning statistically significant similar pockets to the user for detailed analysis.

A different approach to pocket identification is that of Q-SiteFinder (Laurie and Jackson, 2005) that uses the interaction energy between a protein and a van der Waals probe to find energetically favorable binding sites. This method is more effective than a more geometric method, Pocket-Finder, at identifying pockets that accurately overlap with bound ligands.

The analysis of the physicochemical properties of protein local environments is key to understanding protein–ligand interactions and the chemistry of active sites. As such a number of different methods have been developed to compare physicochemical features of pockets and surfaces. One such approach, FEATURE (Bagley and Altman, 1995), creates descriptions of the localized microenvironments using a variety of physical and chemical properties from detailed atomic or chemical groups, up through residue-specific properties to secondary structure level. This type of representation is used by another program, S-BLEST (Mooney et al., 2005), to rapidly search databases containing vectors of local structural properties. The matched environments can be used to find structurally similar proteins in a database as well as associating environments with structural or functional roles.

A new method developed by Najmanovich and coworkers (Najmanovich et al., 2007) uses physicochemical properties in a slightly different way as part of a two-stage graph-matching process. The first stage involves an initial superimposition by identifying the largest clique in an association graph constructed using only the Ca atoms of identical residues in the two clefts. In the second stage, all nonhydrogen atoms are used and the resultant transformation matrix allows an all atom superposition on the two clefts. One of the greatest advantages to the approach is that, by considering all nonhydrogen atoms, it can use large sets of atoms as an input that allows the analysis of large, overpredicted, and apoform binding sites.

Two other physicochemical approaches are those of SiteEngine (Shulman-Peleg, Nussinov, and Wolfson, 2004) and the recently released "SURF'S UP!" service (Sasin, Godzik, and Bujnicki, 2007). SiteEngine takes a different view of binding sites from that of FEATURE; instead of utilizing various levels of environmental properties, it represents residues as pseudocenters of any given physicochemical property. This allows a rapid comparison and site matching using a geometrical hashing technique. "SURF'S UP!" is another method that identifies similar patterns of physicochemical features on compared protein surfaces, its unique aspect being that it returns a "spherical coordinates" representation of the proteins (in PDB format) and an associated graphical image. In addition to this, the service returns a matrix of the values of similarities between the surfaces and a clustering of similar surfaces. The key advantage of the method is that it uses coarse-grained surface features that can be used on comparative models. This is especially relevant in the context of structural genomics in which one goal is to solve enough structures to allow entire genomes to be modeled by homology.

A powerful new method (Morris et al., 2005) for comparing ligand binding pockets uses spherical harmonics to essentially reduce the pockets to unique strings of integers, allowing rapid comparisons to be made. An analysis using this method (Kahraman et al., 2007) identified "buffer zones" in binding pockets filled with water. It is believed that these regions may allow the conformational flexibilities of the ligand and the protein when in the bound state or may even allow the binding of metals alongside the ligand. This suggests that much is still to be learned about ligand–protein recognition.

Care should be taken when attempting to infer function from binding site similarity as demonstrated by an analysis of human cytosolic sulfotransferases (Allali-Hassani et al., 2007). The study found that although proteins with similar small-molecule binding profiles tend to have a higher degree of binding site similarity, and small molecules with similar protein binding profiles tend to be topologically similar, the similarity is not enough to allow one to predict the small-molecule binding patterns.

Residue Templates

The three-dimensional arrangement of enzyme active-site residues is often more conserved than the overall fold, a classic example being that of the catalytic triad where a serine, histidine, and aspartate show a highly conserved geometry essential to their catalytic function. The identification of such structural motifs can be used to identify functionally important local similarities in proteins with different folds. A variety of approaches have been developed to identify these local geometric patterns of residues or atoms in protein structures.

TESS stands for template search and superposition, and is a geometric hashing algorithm for deriving three-dimensional template coordinates from structures deposited in the PDB (Wallace, Borkakoti, and Thornton, 1997) but has been superseded by the rapid JESS algorithm (Barker and Thornton, 2003). The templates used by TESS and JESS can be derived manually by mining the primary literature and assessing which residues form the active site, or they can be generated automatically. The Catalytic Site Atlas (Porter, Bartlett, and Thornton, 2004) was originally a database based on manually curated templates but has recently been automatically expanded to include homologues identified by PSI-BLAST, and a new web server (Catalytic Site Search) allows users to query the database directly using the JESS algorithm. Other template-centric approaches include fuzzy functional forms (FFFs; Fetrow and Skolnick, 1998) and SPASM/RIGOR (Kleywegt, 1999). The key difference between these two methods is that FFF uses the distances between alpha carbons (with a small variance) instead of 3D-atomic coordinates whereas SPASM/RIGOR uses C-alpha and side chain pseudoatoms as its template.

As mentioned earlier, templates can be generated automatically and there are several examples of methods that use these. PDBSiteScan (Ivanisenko et al., 2004) uses the SITE records in the PDB files and protein–protein interaction data from heterocomplexes to generate their templates. DRESPAT (Wangikar et al., 2003) is a graph theory approach to detect recurring three-dimensional patterns in the known protein structures. This approach has the advantage of being a fully automated process but can miss biologically significant motifs if they are uncommon. SuMo (Jambon et al., 2003) is quite similar in many ways with the major exception that it uses "stereochemical groups" instead of amino acids for its searches.

A different approach is shown by the PINTS (Patterns in Nonhomologous Tertiary Structures) server (Stark and Russell, 2003). Rather than operating on a preformed template, PINTS detects the largest common three-dimensional arrangement of residues between any two structures, the assumption being that a similar arrangement of residues implies similar function.

Phylogenetic Relationships

In general, functionally important areas on protein surfaces such as active sites are more highly conserved than surrounding regions. There are numerous methods to identify pockets and clefts and compare physicochemical features, but these approaches can be improved further by taking into account the residue conservation. There are a variety of methods available to calculate raw residue conservation scores from sequence alignments, but more powerful approaches take into account conservation over evolutionary time. The most popular of such methods is the evolutionary trace (ET) approach (Madabushi et al., 2002), which combines the use of a phylogenetic tree to rank residues in a protein by their evolutionary importance and structural mapping of these residues for functional inference based on clustering at the surface. It is observed that those with the highest rank tend to cluster on the protein surface in functionally important sites. A number of servers have been set up to offer the ET method and there is also a Java implementation (JevTrace) that can be downloaded (Joachimiak and Cohen, 2002).

ConSurf (Glaser et al., 2003) is another evolutionary approach that automatically generates a multiple sequence alignment (or uses one supplied by the user), creates a phylogenetic tree from this alignment, calculates the degree of conservation, and finally maps this onto the structure. This approach is useful for identifying regions of variability as well as conservation, which can be important when investigating proteins that interact with many effectors. ConSurf has recently been combined with SURFNET to enable the reduction in size of predicted clefts to only the most evolutionary conserved regions (Glaser et al., 2006). This approach shows that the largest clefts remaining after the reduction tend to be those where ligands bind.

One problem with mapping conservation onto the surface is that it relies on there being sufficient homologues in the sequence databases. Many new structures have little or no sequence relatives and therefore the ability to detect conserved surface patches or interfaces is poor. The protein interface recognition (PIER) method (Kufareva et al., 2007) aims to solve this problem by predicting a protein's interfaces from structure using local statistical properties of atoms on the protein surface.

DNA Binding Prediction

One important aspect of protein function involves nucleic acid binding. It has been known for sometime that a limited repertoire of key structural motifs exists to recognize DNA (Harrison, 1991). The most common motif is the helix-turn-helix (HTH) where the recognition helix lies in the major groove of the DNA, but there are various subfamilies and elaborations on this theme such as the winged-HTH motif (Aravind et al., 2005). The problem with the structural motifs is that they occur regularly in proteins that do not recognize DNA, so how can the actual DNA binders be correctly identified? One key feature of DNA binding proteins is that the interacting surface tends to be the most positively charged parts of the protein surface. The introduction of electrostatic parameters can greatly improve the ability to distinguish binders from nonbinders. The HTHquery server (Ferrer-Costa et al., 2005) uses electrostatic calculations to distinguish true positive from false positive matches to its database of DNA binding structural templates.

Instead of calculating the electrostatics as an addition to structural matches, the eF-Site database (Kinoshita and Nakamura, 2004) is a searchable database of electrostatic surfaces extracted from the PDB. By taking this approach, the server is not limited to DNA binding proteins and includes ligand binding, active sites, and protein–protein interaction surfaces. Another more statistical method (Tsuchiya, Kinoshita, and Nakamura, 2004) uses the shape and electrostatic potential of the surface of the protein and DNA to construct an evaluation function to predict DNA interaction sites. One of the major differences in this approach is that it uses a combination of local and global curvature to describe the protein surface at DNA binding sites.

The use of statistical approaches in the prediction of DNA binding proteins is more commonly seen in sequence-based approaches to the problem, such as the use of position-specific scoring matrices (PSSMs; Ahmad and Sarai, 2005). Information on any given residue and its sequence neighbor can be used to predict the likelihood of DNA binding and can be applied even when no sequence homology to known DNA binding proteins is observed. Through the use of a neural network-based algorithm, evolutionary information from amino acid sequences in terms of their PSSMs can be used to make better predictions of DNA binding sites (the method is available at http://www.netasa.org/dbs-pssm/).

A new method developed by Szilágyi and Skolnick (Szilagyi and Skolnick, 2006) combines sequence features and structural parameters to predict DNA binding proteins. By using information such as the relative proportions of particular amino acids in the protein sequence, the spatial distribution of other amino acids, and structural information such as the dipole moment of the molecule, a formula is constructed and DNA binding predictions are based on whether or not the composite score is above a certain threshold. The advantages of the method are its robustness when using low-resolution protein models and its ability to predict DNA binding proteins in their unbound state.

Other Methods

Two important resources provided by the MSD group at the European Bioinformatics Institute are MSDsite (Golovin et al., 2005) and MSDmotif. Both are database-centric approaches but look at different aspects of active sites. MSDsite is a search and retrieval system focused on the bound ligands and active sites of protein structures in the PDB, whereas MSDmotif concentrates on small 3D-structure motifs such as Beta-turns, Asx-motifs, and Nests (Watson and Milner-White, 2002). These motifs are seen to characterize active sites, have structural roles in protein folding and protein stability, and even play roles in enzyme mechanisms. Both databases can be used to identify local similarities that can point to functional or distant evolutionary relationships.

A novel method, THEMATICS (http://pfweb.chem.neu.edu/thematics/submit.html), uses theoretical microscopic titration curves to identify active-site residues (Ondrechen, Clifton, and Ringe, 2001). Titration curves of the most ionizable residues in a protein would be expected to go from protonated to unprotonated forms within a relatively narrow pH range. Analysis of the theoretical titrations shows a small fraction of the curves possess perturbations in the curve where the residue is partially protonated over a wide pH range. Further analysis revels that the residues showing these unusual properties tend to occur in the active site.

Combining Methods

No one method will work in all cases, and each method may suggest a number of different functions with equal probability, so a more sensitive and sensible approach is to use as many different methods as possible. When multiple methods agree on a putative function, it lends greater likelihood to the prediction. This has led to the development of various servers that integrate various sequence- and structure-based methods.

The ProFunc server (Laskowski, Watson, and Thornton, 2005a), developed in collaboration with the Midwest Center for Structural Genomics (MCSC), combines analyses from a wide variety of sources some of which are accessed through SOAP-based web services. The analyses offered range from sequence searches (such as BLAST searches against UniProt or the PDB, and motif matching against Pfam and InterPro), through genome neighborhood and gene location analysis, to wholly structure-based methods. The structure-based approaches fall into almost all of the aforementioned categories and include fold matching (SSM and DALI), surface cleft analysis (SURFNET) and mapping of conservation onto surface, structural motifs (HTH motif and "nest" identification), down to the highly specific three-dimensional template approaches. These last template approaches utilize known catalytic site templates, automatically generated ligand and DNA binding templates, and a novel "reverse template" approach (Laskowski, Watson, and Thornton, 2005b). A recent analysis (Watson et al., 2007) of the server's ability to predict protein function from structure suggested that the reverse template and fold-matching methods are particularly successful and identify local similarities that are difficult to spot using current sequence-based methods. A method based on the gene ontology schema using GO-slims is also presented that can allow the automatic assessment of hits.

The use of GO terms is also a feature of other integrative approaches. The ProKnow server (Pal and Eisenberg, 2005), like ProFunc, combines information from a combination of sequence and structural approaches including fold similarity (DALI), templates (RIGOR), and functional links taken from the DIP database (Xenarios et al., 2000) of protein interactions. The most important difference is that ProKnow uses Bayes' theorem to weight the significance of all results, presenting an overall function to the user in terms of GO annotation. Another approach, PHUNCTIONER, automatically extracts the sequence signal responsible for protein function through structural alignments and 3D profiles of conserved residues (Pazos and Sternberg, 2004). The GO database is once again used but this time to provide the functional features used to train the method. Using the method, 3D profiles were constructed for over 120 GO annotations that can be used for function prediction. The matching of a new protein to one of these profiles is a strong indicator of matching the associated GO annotation and hence from this function may be inferred.

The Query3D (Ausiello, Via, and Helmer-Citterich, 2005a) method and associated database, PDBfun (Ausiello et al., 2005), takes a residue-centric approach when integrating information from multiple sources. These range from protein cavity location and solvent accessibility, through ligand binding ability, secondary structure type to residue, sequence, and domain-level properties. The software also acts as a database management system and allows complex queries to be constructed based on residue function and can compare subsets of residues. This allows residues sharing certain properties in all proteins of known structure to be identified at the same time as finding the query protein's structural neighbors in the PDB.

STRUCTURAL GENOMICS: HIGH-THROUGHPUT FUNCTION PREDICTION

Large-scale genome-sequencing projects led to the birth of structural genomics (Blundell and Mizuguchi, 2000), a large-scale project aimed at experimentally determining a large number of protein 3D structures as rapidly and accurately as possible using high-throughput methods. The initial trial projects have completed and the full-scale production of protein structures has now started along with investment in new technologies to address the more challenging problems of complexes and membrane proteins (Service, 2005). As a consequence, the numbers of structures deposited in the PDB with little or no functional information has steadily increased.

An early review of 15 hypothetical proteins of known structure and their functional assignment (Teichmann, Murzin, andChothia, 2001) provided some glimpses of the quality of functional assignments that can be made from structure. The structure and sequences of homologues of known function were used to find surface cavities and grooves in which conserved residues indicated an active site. This, along with bound cofactors in the structure, in combination with experimental work, was assessed according to the depth of functional information that could be obtained. For the 15 proteins, detailed functional information was obtained for a quarter of them. For another half, some functional information was obtained, and for another quarter, no functional information could be obtained.

A more recent study (Watson et al., 2007) assessing function prediction from structure looked at 92 proteins of known function from the MCSC. The results were backdated to the release date of each query to see how the ProFunc server would have performed at the time. In approximately 55% of cases, the top fold match was able to provide the correct functional assignment. The standard template methods were shown to provide some success but the most accurate structure-based method was identified as the reverse template approach, providing the correct function in 60% of the cases.

Specific Examples of Functional Assignment from Structure

BioH: A New Carboxylesterase. BioH was known to be involved in biotin biosynthesis in *Escherichia coli* but had no identified biochemical function. The three-dimensional structure of BioH was determined by the MCSC to 1.7 Å (Figure 21.4a) (Sanishvili et al., 2003). Analysis of the structure using 189 manually curated residue-based templates from a variety of known enzyme-active sites identified a putative catalytic triad similar to that of hydrolases. The best matching template was derived from PDB entry 1tah (triacylglycerol lipase from *Psuedomonas glumae*) and the structural match is illustrated in Figure 21.4a. The prediction was tested experimentally using a panel of hydrolase assays and the function was confirmed as a carboxylesterase with a preference for short acyl chain substrates.

Mj0577: Putative ATP Molecular Switch. Mj0577 is an open reading frame (ORF) of previously unknown function from *Methanococcus jannaschii*. Its structure was

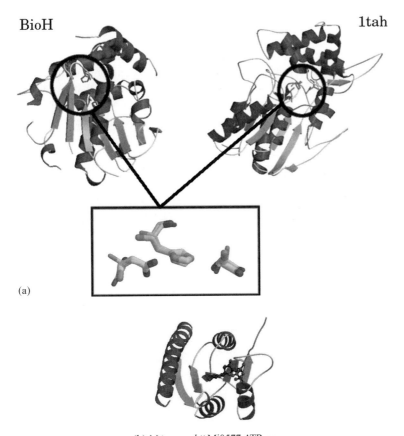

BioH   1tah

(a)

(b) *M.jannaschii* Mj0577 ATPase

Figure 21.4. Prediction of function in earnest: three structures solved in the absence of functional information. (a) BioH: a new carboxylesterase. The BioH structure showed a significant match to a known active site template from PDBentry 1tah (*Pseudomonas glumae* lipase). The structures are shown side by side with the equivalent residues shown in green sticks. The template residues from 1tah (Ser 87, His 285, and Asp 263) and the matched residues from BioH (Ser 87, His 235, and Asp 207) are shown superposed as thick and thin sticks, respectively, in the box below. (b) Putative archaeal ATPase molecular switch. The bound ATP molecule, key to the unlocking of this protein's function, is shown in a ball-and-stick form. Figure also appears in the Color Figure section.

determined at 1.7 Å (Figure 21.4b) (Zarembinski et al., 1998). The structure was found to contain a bound ATP molecule, picked up from the *E. coli* host. The presence of bound ATP led to the proposition that Mj0577 is either an ATPase or an ATP binding molecular switch. This is an example where a ligand-based approach is used to predict the function of a protein in a manner related to the surface cleft strategies described earlier. Further experimental work showed that Mj0577 cannot hydrolyze ATP by itself and can only do so in the presence of *M. jannaschii* crude cell extract. Therefore, it is more likely to act as a molecular switch, in a process analogous to ras-GTP hydrolysis in the presence of GTPase activating protein.

CONCLUSIONS

We have seen in this chapter that structure–function relationships are key to understand in molecular terms how a protein works. Structural data can complement experimental work; for example, if it is known from biochemical experiments that a particular protein of interest binds ATP, the structure of the protein complexed with an analogue of ATP will reveal exactly where ATP binds. It will also identify the residues on the protein that might stabilize the interaction between ligand and protein, and the potential structural consequences of ligand binding.

Structural data can also guide experimental work in eliciting the function of a protein. For example, if one can infer from structural homology that a particular protein is a hydrolase with a nucleotide binding domain, one can carry out experiments to confirm this and even identify possible substrates.

As structural genomics projects progress, determining protein function from structure with no prior knowledge of the function will become increasingly important. As described, a great number of methods exist to do this, and several servers have been developed to integrate the information from multiple sources in an attempt to improve predictions. However, there are still many proteins for which no functional information can be inferred from any existing method and therefore there is still a need to develop new methods to address these problematic structures. To date the methods that work best rely on the recognition of different homologues—identified either by whole fold matching or by local template matching.

When automatically assigning function to proteins, great care must be taken annotating those that are distantly related, so as to maintain accuracy in the databases. This is made more difficult by the fact that functions can differ in proteins that are almost identical and the converse is also true with examples of similar function in very different proteins. Ultimately, however, function will need to be assessed experimentally and this is a time-consuming process. In order to address this potential bottleneck, several high-throughput approaches to enzyme function assays and ligand binding assays have been developed to allow mass screening for specific functions.

REFERENCES

Acharya K, Ren J, Stuart D, Phillips D, Fenna R (1991): Crystal structure of human alpha-lactalbumin at 17 Å resolution. *J Mol Biol* 221(2):571–581.

Ahmad S, Sarai A (2005): PSSM-based prediction of DNA binding sites in proteins. *BMC Bioinformatics* 6:33.

Allaire M, Li Y, MacKenzie R, Cygler M (1998): The 3-D structure of a folate-dependent dehydrogenase/cyclohydrolase bifunctional enzyme at 15 Å resolution. *Structure* 6(2):173–182.

Allali-Hassani A, Pan PW, Dombrovski L, Najmanovich R, Tempel W, Dong A, Loppnau P, Martin F, Thornton J, Edwards AM, Bochkarev A, Plotnikov AN, Vedadi M, Arrowsmith CH (2007): Structural and chemical profiling of the human cytosolic sulfotransferases. *PLoS Biol* 5(5):e97.

Altschul SF, Madden TL, Schaffer AA, Zhang J, Zhang Z, Miller W, Lipman DJ (1997): Gapped BLAST and PSI-BLAST: a new generation of protein database search programs. *Nucleic Acids Res* 25(17):3389–3402.

Apweiler R, Bairoch A, Wu CH, Barker WC, Boeckmann B, Ferro S, Gasteiger E, Huang H, Lopez R, Magrane M, Martin MJ, Natale DA, O'Donovan C, Redaschi N, Yeh LS (2004): UniProt: the Universal Protein knowledgebase. *Nucleic Acids Res* 32:(Database issue):D115–D119.

Aravind L, Anantharaman V, Balaji S, Babu MM, Iyer LM (2005): The many faces of the helix-turn-helix domain: transcription regulation and beyond. *FEMS Microbiol Rev* 29(2):231–262. [An excellent overview of the HTH-domain from a structural and evolutionary viewpoint. A classification of the HTH-domain and all of its structural variation is presented with a discussion of the functional diversity seen in the domain.]

Ashburner M, Ball CA, Blake JA, Botstein D, Butler H, Cherry JM, Davis AP, Dolinski K, Dwight SS, Eppig JT, Harris MA, Hill DP, Issel-Tarver L, Kasarskis A, Lewis S, Matese JC, Richardson JE, Ringwald M, Rubin GM, Sherlock G (2000): Gene ontology: tool for the unification of biology. The Gene Ontology Consortium. *Nat Genet* 25(1):25–29. [This paper describes the role of GO in producing a vocabulary of functional information that can be applied to all eukaryotes, and its application to genome annotation and transfer of functional information between genomes.]

Ausiello G, Via A, Helmer-Citterich M (2005): Query3d: a new method for high-throughput analysis of functional residues in protein structures. *BMC Bioinformatics* 6(Suppl 4):S5.

Ausiello G, Zanzoni A, Peluso D, Via A, Helmer-Citterich M (2005): pdbFun: mass selection and fast comparison of annotated PDB residues. *Nucleic Acids Res* 33 (Web server issue):W133–W137.

Bagley SC, Altman RB (1995): Characterizing the microenvironment surrounding protein sites. *Prot Sci* 4(4):622–635.

Barker JA, Thornton JM (2003): An algorithm for constraint-based structural template matching: application to 3D templates with statistical analysis. *Bioinformatics* 19(13):1644–1649. [This paper describes the JESS algorithm for template matching. This is a faster and more flexible approach than the original TESS algorithm.]

Benson DA, Karsch-Mizrachi I, Lipman DJ, Ostell J, Wheeler DL (2005): GenBank. *Nucleic Acids Res* 33:(Database issue):D34–D38.

Berman HM, Westbrook J, Feng Z, Gilliland G, Bhat TN, Weissig H, Shindyalov IN, Bourne PE (2000): The Protein Data Bank. *Nucleic Acids Res* 28(1):235–242.

Binkowski TA, Freeman P, Liang J (2004): pvSOAR: detecting similar surface patterns of pocket and void surfaces of amino acid residues on proteins. *Nucleic Acids Res* 32:(Web server issue): W555–W558.

Blundell TL, Mizuguchi K (2000): Structural genomics: an overview. *Prog Biophys Mol Biol* 73 (5):289–295.

Ferrer-Costa C, Shanahan HP, Jones S, Thornton JM (2005): HTHquery: a method for detecting DNA-binding proteins with a helix-turn-helix structural motif. *Bioinformatics* 21(18):3679–3680.

Fetrow JS, Skolnick J (1998): Method for prediction of protein function from sequence using the sequence-to-structure-to-function paradigm with application to glutaredoxins/thioredoxins and T1 ribonucleases. *J Mol Biol* 281(5):949–968.

Glaser F, Morris RJ, Najmanovich RJ, Laskowski RA, Thornton JM (2006): A method for localizing ligand binding pockets in protein structures. *Proteins* 62(2):479–488.

Glaser F, Pupko T, Paz I, Bell RE, Bechor-Shental D, Martz E, Ben-Tal N (2003): ConSurf: identification of functional regions in proteins by surface-mapping of phylogenetic information. *Bioinformatics* 19(1):163–164.

Golovin A, Dimitropoulos D, Oldfield T, Rachedi A, Henrick K (2005): MSDsite: a database search and retrieval system for the analysis and viewing of bound ligands and active sites. *Proteins* 58(1):190–199.

Gough J, Chothia C (2002): SUPERFAMILY: HMMs representing all proteins of known structure SCOP sequence searches, alignments and genome assignments. *Nucleic Acids Res* 30(1):268–272.

Harrison SC (1991): A structural taxonomy of DNA-binding domains. *Nature* 353(6346):715–719.

Henschel A, Kim WK, Schroeder M (2006): Equivalent binding sites reveal convergently evolved interaction motifs. *Bioinformatics* 22(5):550–555.

Holm L, Sander C (1995): Dali: a network tool for protein structure comparison. *Trends Biochem Sci* 20(11):478–480.

Hubbard S, Wei L, Ellis L, Hendrickson W (1994): Crystal structure of the tyrosine kinase domain of the human insulin receptor. *Nature* 372(6508):746–754.

Ivanisenko VA, Pintus SS, Grigorovich DA, Kolchanov NA (2004): PDBSiteScan: a program for searching for active, binding and posttranslational modification sites in the 3D structures of proteins. *Nucleic Acids Res* 32:(Web server issue):W549–W554.

Jambon M, Imberty A, Deleage G, Geourjon C (2003): A new bioinformatic approach to detect common 3D sites in protein structures. *Proteins* 52(2):137–145.

Jeffery C (1999): Moonlighting proteins. *Trends Biochem Sci* 24(1):8–11. [This review discusses multifunctional proteins and mechanisms by which cells switch between these different functions.]

Joachimiak MP, Cohen FE (2002): JEvTrace: refinement and variations of the evolutionary trace in JAVA. *Genome Biol* 3(12):RESEARCH0077.

Kahraman A, Morris RJ, Laskowski RA, Thornton JM (2007): Shape variation in protein binding pockets and their ligands. *J Mol Biol* 368(1):283–301. [This paper investigates the structural variation of binding pockets using a spherical harmonics approach. The study finds that pockets binding the same ligand show greater variation in their shapes than can be accounted for by the conformational variability of the ligand. The study shows there is still much to know about ligand binding, and shape complementarity may not be sufficient to drive molecular recognition.]

Kinoshita K, Nakamura H (2004): eF-site and PDBjViewer: database and viewer for protein functional sites. *Bioinformatics* 20(8):1329–1330.

Kleywegt G (1999): Recognition of spatial motifs in protein structures. *J Mol Biol* 285(4):1887–1897. [This work describes the programs SPASM and RIGOR and their application to function prediction, comparative structural analysis, and design of novel functional sites.]

Krissinel E, Henrick K (2004): Secondary-structure matching (SSM), a new tool for fast protein structure alignment in three dimensions. *Acta Crystallogr D* 60:(Pt 12 Pt 1):2256–2268.

Kuznetsova E, Proudfoot M, Sanders SA, Reinking J, Savchenko A, Arrowsmith CH, Edwards AM, Yakunin AF (2005): Enzyme genomics: application of general enzymatic screens to discover new enzymes. *FEMS Microbiol Rev* 29(2):263–279. [This paper describes a high-throughput set of experimental assays designed to identify the subclass or sub-subclasses of enzymes (phosphatase, phosphodiesterase/nuclease, protease, esterase, dehydrogenase, and oxidase) to which an unknown protein belongs. This is followed up by secondary screens with natural ligands to identify specific substrates for the reaction.]

Kufareva I, Budagyan L, Raush E, Totrov M, Abagyan R (2007): PIER: protein interface recognition for structural proteomics. *Proteins* 67(2):400–417.

Laskowski RA (1995): SURFNET: a program for visualizing molecular surfaces, cavities, and intermolecular interactions. *J Mol Graph* 13(5):323–328.

Laskowski R, Luscombe N, Swindells M, Thornton J (1996): Protein clefts in molecular recognition and function. *Prot Sci* 5(12):2438–2452. [This analysis shows how cleft volume relates to molecular interaction and function. The authors found that both ligand binding and protein–protein interactions usually involved the largest cleft in the protein.]

Laskowski RA, Watson JD, Thornton JM (2005): ProFunc: a server for predicting protein function from 3D structure. *Nucleic Acids Res* 33(Web server issue):W89–W93.

Laskowski RA, Watson JD, Thornton JM (2005): Protein function prediction using local 3D templates. *J Mol Biol* 351(3):614–626. [This paper describes the template-based approaches used in the ProFunc server and discusses a novel "reverse-template" method and its application to function prediction.]

Laurie AT, Jackson RM (2005): Q-SiteFinder: an energy-based method for the prediction of protein–ligand binding sites. *Bioinformatics* 21(9):1908–1916.

Lo Conte L, Ailey B, Hubbard T, Brenner S, Murzin A, Chothia C (2000): SCOP: a structural classification of proteins database. *Nucleic Acids Res* 28(1):257–259.

Madabushi S, Yao H, Marsh M, Kristensen DM, Philippi A, Sowa ME, Lichtarge O (2002): Structural clusters of evolutionary trace residues are statistically significant and common in proteins. *J Mol Biol* 316(1):139–154.

Mooney SD, Liang MH, DeConde R, Altman RB (2005): Structural characterization of proteins using residue environments. *Proteins* 61(4):741–747.

Morris RJ, Najmanovich RJ, Kahraman A, Thornton JM (2005): Real spherical harmonic expansion coefficients as 3D shape descriptors for protein binding pocket and ligand comparisons. *Bioinformatics* 21(10):2347–2355.

Najmanovich RJ, Allali-Hassani A, Morris RJ, Dombrovski L, Pan PW, Vedadi M, Plotnikov AN, Edwards A, Arrowsmith C, Thornton JM (2007): Analysis of binding site similarity, small-molecule similarity and experimental binding profiles in the human cytosolic sulfotransferase family. *Bioinformatics* 23(2):e104–e109.

Novotny M, Madsen D, Kleywegt G J (2004): Evaluation of protein fold comparison servers. *Proteins* 54(2):260–270. [An excellent review of several protein fold comparison servers.]

Ondrechen MJ, Clifton JG, Ringe D (2001): THEMATICS: a simple computational predictor of enzyme function from structure. *Proc Natl Acad Sci USA* 98(22):12473–12478.

Orengo, CA, (15th May 2007): University College London, London Personal communication.

Orengo CA, Michie AD, Jones S, Jones DT, Swindells MB, Thornton JM (1997): CATH: a hierarchic classification of protein domain structures. *Structure* 5(8):1093–1108.

Pal D, Eisenberg D (2005): Inference of protein function from protein structure. *Structure* 13 (1):121–130. [This paper describes the ProKnow server, which annotates proteins with Gene Ontology functional terms. The method uses a variety of structure- and sequence-based approaches to extract features of the protein; these are then weighted using Bayes' theorem to determine the most likely function. The assigned function is quantified by the ontology depth from 1 (general) to 9 (specific).]

Pazos F, Sternberg MJ (2004): Automated prediction of protein function and detection of functional sites from structure. *Proc Natl Acad Sci USA*, 101(41):14754–14759.

Pearl FM, Bennett CF, Bray JE, Harrison AP, Martin N, Shepherd A, Sillitoe I, Thornton J, Orengo CA (2003): The CATH database: an extended protein family resource for structural and functional genomics. *Nucleic Acids Res* 31(1):452–455.

Pearson WR (1991): Searching protein sequence libraries: comparison of the sensitivity and selectivity of the Smith–Waterman and FASTA algorithms. *Genomics* 11(3):635–650.

Porter CT, Bartlett GJ, Thornton JM (2004): The Catalytic Site Atlas: a resource of catalytic sites and residues identified in enzymes using structural data. *Nucleic Acids Res* 32:(Database issue): D129–D133.

Proudfoot M, Kuznetsova E, Brown G, Rao NN, Kitagawa M, Mori H, Savchenko A, Yakunin AF (2004): General enzymatic screens identify three new nucleotidases in *Escherichia coli*: biochemical characterization of SurE, YfbR, and YjjG. *J Biol Chem* 279(52):54687–54694.

Sanishvili R, Yakunin AF, Laskowski RA, Skarina T, Evdokimova E, Doherty-Kirby A, Lajoie GA, Thornton JM, Arrowsmith CH, Savchenko A, Joachimiak A, Edwards AM (2003): Integrating structure, bioinformatics, and enzymology to discover function: BioH, a new carboxylesterase from *Escherichia coli*. *J Biol Chem* 278(28):26039–26045.

Sasin JM, Godzik A, Bujnicki JM (2007): SURF'S UP!: protein classification by surface comparisons. *J Biosci* 32(1):97–100.

Service R (2005): Structural biology structural genomics, round 2. *Science* 307:(5715):1554–1558.

Shindyalov IN, Bourne PE (1998): Protein structure alignment by incremental combinatorial extension (CE) of the optimal path. *Protein Eng* 11(9):739–747.

Shulman-Peleg A, Nussinov R, Wolfson HJ (2004): Recognition of functional sites in protein structures. *J Mol Biol* 339(3):607–633.

Skolnick J, Fetrow J (2000): From genes to protein structure and function: novel applications of computational approaches in the genomic era. *Trends Biotechnol* 18(1):34–39. [A good review of computational approaches to structure and function prediction, which emphasizes the need for structural descriptors of functional sites to use the sequence and structural information.]

Sonnhammer EL, Eddy SR, Birney E, Bateman A, Durbin R (1998): Pfam: multiple sequence alignments and HMM-profiles of protein domains. *Nucleic Acids Res* 26(1):320–322.

Stark A, Russell RB (2003): Annotation in three dimensions PINTS: patterns in non-homologous tertiary structures. *Nucleic Acids Res* 31(13):3341–3344.

Szilagyi A, Skolnick J (2006): Efficient prediction of nucleic acid binding function from low-resolution protein structures. *J Mol Biol* 358(3):922–933.

Teichmann S, Murzin A, Chothia C (2001): Determination of protein function, evolution and interactions by structural genomics. *Curr Opin Struct Biol* 11(3):354–363. [A review of the quality of functional assignments that can be made from structure.]

Thornton J, Todd A, Milburn D, Borkakoti N, Orengo C (2000): From structure to function: approaches and limitations. *Nat Struct Biol*, 7(Suppl):991–994. [A short review, presenting the functional information obtainable from a protein structure. It also describes some examples of prediction of function in earnest and discusses implications for rational drug design.]

Todd A, Orengo C, Thornton J (2001): Evolution of function in protein superfamilies, from a structural perspective. *J Mol Biol* 307(4):1113–1143. [This work combines sequence and structural information to identify superfamily relatives and assesses variation in enzyme function at different levels of sequence identity. In addition, with reference to the 31 diverse enzyme superfamilies, the paper presents a detailed review of how functional variation is implemented in terms of sequence and structural changes.]

Tsuchiya Y, Kinoshita K, Nakamura H (2004): Structure-based prediction of DNA-binding sites on proteins using the empirical preference of electrostatic potential and the shape of molecular surfaces. *Proteins* 55(4):885–894.

Vedadi M, Niesen FH, Allai-Hassani A, Fedorov OY, Finerty PJ Jr Wasney GA, Yeung R, Arrowsmith C, Ball LJ, Berglund H, Hui R, Marsden BD, Nordlund P, Sundstrom M, Weigelt J, Edwards AM (2006): Chemical screening methods to identify ligands that promote protein stability, protein crystallization, and structure determination. *Proc Natl Acad Sci* 103 (43):15835–15840.

Wallace A, Borkakoti N, Thornton J (1997): TESS: a geometric hashing algorithm for deriving 3D coordinate templates for searching structural databases. Application to enzyme active sites. *Prot Sci* 6(11):2308–2323.

Wangikar PP, Tendulkar AV, Ramya S, Mali DN, Sarawagi S (2003): Functional sites in protein families uncovered via an objective and automated graph theoretic approach. *J Mol Biol* 326 (3):955–978.

Watson JD, Milner-White EJ (2002): A novel main-chain anion-binding site in proteins: the nest. A particular combination of phi,psi values in successive residues gives rise to anion-binding sites that occur commonly and are found often at functionally important regions. *J Mol Biol* 315:(2) 171–182.

Watson JD, Sanderson S, Ezersky A, Savchenko A, Edwards A, Orengo C, Joachimiak A, Laskowski RA, Thornton JM (2007): Towards fully automated structure-based function prediction in structural genomics: a case study. *J Mol Biol* 367(5):1511–1522.

Whisstock JC, Lesk AM (2003): Prediction of protein function from protein sequence and structure. *Q Rev Biophys* 36:(3):307–340. [This is an excellent review of approaches for predicting protein function. Several methods are discussed and interesting examples illustrate the problems with inferring function from sequence and structural similarity.]

Xenarios I, Rice DW, Salwinski L, Baron MK, Marcotte EM, Eisenberg D (2000): DIP: the database of interacting proteins. *Nucleic Acids Res* 28(1):289–291.

Yeats C, Maibaum M, Marsden R, Dibley M, Lee D, Addou S, Orengo CA (2006): Gene3D: modelling protein structure, function and evolution. *Nucleic Acids Res* 34:(Database issue): D281–D284.

Yooseph S, Sutton G, Rusch DB, Halpern AL, Williamson SJ, Remington K, Eisen JA, Heidelberg KB, Manning, G, Li, W, Jaroszewski L, Cieplak P, Miller CS, Li H, Mashiyama ST, Joachimiak MP, Van Belle C, Chandonia JM, Soergel DA, Zhai Y, Natarajan K, Lee S, Raphael BJ, Bafna V, Friedman R, Brenner SE, Godzik A, Eisenberg D, Dixon JE, Taylor SS, Strausberg RL, Frazier M, Venter JC (2007): The Sorcerer II Global Ocean Sampling Expedition: expanding the universe of protein families. *PLoSBiol* 5(3):e16.

Zarembinski T, Hung L, Mueller-Dieckmann H, Kim K, Yokota H, Kim R, Kim S (1998): Structure-based assignment of the biochemical function of a hypothetical protein: a test case of structural genomics. *Proc Natl Acad Sci USA* 95(26):15189–15193.

Zhu J, Weng Z (2005): FAST: a novel protein structure alignment algorithm. *Proteins* 58(3):618–627.

22

STRUCTURAL ANNOTATION OF GENOMES

Adam J. Reid, Corin Yeats, Jonathan Lees, and Christine A. Orengo

INTRODUCTION

Physical techniques such as X-ray crystallography and NMR allow the determination of individual protein structures on the atomic level. Such information is invaluable for detailed studies of protein function. However, directly determining the structure of proteins by physical methods is laborious and expensive. Although there are structures for ~8000 different proteins in the Protein Data Bank (http://www.rcsb.org/pdb/) as of June 2007, there are currently ~5 million protein sequences in UniProt (http://www.ebi.uniprot.org) and the number of sequences grows at a faster rate than that of solved structures.

Therefore, time and cost constraints make it impossible to directly determine the structures of proteins in all genomes with current technology. Fortunately however, homology modeling allows reasonably accurate prediction of structure for sequences with >40% sequence identity over their whole length (Marti-Renom et al., 2000). In addition, fold recognition techniques allow structures to be predicted at much lower levels of sequence similarity and this can often give helpful insights into protein functions. Consequently, only a sampling of structures needs to be solved, although it is not simple to determine how this should be done (see Section "Can We Determine All the Structures Present in the Genomes?—Structural Annotation of Genomes and Structural Genomics"). Structural genome annotation can help determine which genome sequences we can already describe structurally and which we need to target for structure determination.

Another reason why structural annotation of genomes is so useful is that structures are much more conserved than sequence during evolution, annotation allows us to merge sequence-based families into larger superfamilies of homologues that could not have been identified using purely sequence data. Structural data can be exploited through classifications such as those of CATH and SCOP (see Chapters 17 and 18), which classify

Structural Bioinformatics, Second Edition Edited by Jenny Gu and Philip E. Bourne
Copyright © 2009 John Wiley & Sons, Inc.

domains into homologous superfamilies based on their structural similarity. Reading Chapters 17 and 18 is highly recommended to fully appreciate and understand concepts presented in this chapter.

Using sensitive homology recognition methods, it is then possible to model these large, structural superfamilies and thereby determine distant homologous relationships between genomic sequences without knowledge of their structure. These approaches have sometimes allowed insights into protein evolution and function that would not have been possible with sequence data alone (see Section "What can Structural Genome Annotation Tell Us About Evolution?").

In this chapter we discuss the following:

- The availability of completed genomes (Section "Availability of Completed Genomes").
- The most commonly used structure-based methodologies for annotating genomes (Section "Methodologies for Identifying Structural Protein Domains in Genomes").
- How much of the genomes can be accounted for with these data (Section "How Well are Genomes Covered by Structural Domain Annotation?").
- How structural genomics (SG) is seeking to increase structural coverage of the genomes (Section "Can We Determine All the Structures Present in the Genomes?— Structural Annotation of Genomes and Structural Genomics").
- What these structural annotations can tell us about evolution (Section "What can Structural Genome Annotation Tell Us About Evolution?").
- What resources are available for acquiring the relevant data (Section "Structural Genome Annotation Resources").

AVAILABILITY OF COMPLETED GENOMES

Integr8, hosted at the European Bioinformatics Institute, provides access to proteomes from all completely sequenced organisms derived from UniProt and other resources. Similar data are available from elsewhere such as NCBI. According to the NCBI genomes resource (http://www.ncbi.nlm.nih.gov/Genomes/), as of June 2007, there are currently 27 eukaryotic, 483 bacterial, and 41 archaeal complete genomes. Those in progress, largely microbial but many eukaryotic as well, total over a thousand more genomes to be completed. Currently, completed genomes contain an estimated 800,000 open reading frames.

METHODOLOGIES FOR IDENTIFYING STRUCTURAL PROTEIN DOMAINS IN GENOMES

A simple protocol for identifying protein structures in genomes might be to take the sequences of known structures from the Protein Data Bank and BLAST (Altschul et al., 1997) them against the genomes. In what ways can this be improved upon?

First, BLAST is not the most sensitive method of homology detection; there are also profile methods such as PSI-BLAST (Altschul et al., 1997) and Hidden Markov Models (HMMs) (Eddy, 1996; Karplus, Barrett, and Hughey, 1998) that additionally use evolutionary information to detect more distant homologues and allow greater coverage of the genomes than could otherwise be achieved with BLAST (Park et al., 1998). Even more sensitive than profile methods is using an approach that threads sequences through known structural folds to find the most optimal structure (Jones, Taylor, and Thornton, 1992). Rather than asking whether sequences have a common ancestor, threading attempts to determine whether a sequence is likely to adopt a particular structure and therefore is a member of a known fold group. However, as profile methods are slower than BLAST, threading is even more time demanding. Consequently, most structural annotation is performed using profile methods, which are both sensitive and reasonably fast. The implementation of profile-based methods is discussed in Sections 2.2 and 2.3.

The second major improvement gained from using structures for genome annotation results from using structural classifications that define groups of structurally similar domains. These domains may have very little sequence similarities but share structural similarities and are therefore classified together. From these classifications, it is easier to group very diverse representatives of a superfamily, as defined by CATH, and generate an optimum set of representative profiles to further detect other homologues.

Classifying Structural Domains

Structural protein domains are often described as compact, independently folding units within a protein chain (Richardson, 1981). By analyzing and comparing the structures and sequences of the different independent units, it is possible to group them into sets of homologous structures (superfamilies) and those with a similar secondary structure composition (architecture) and overall topology (fold). By using diverse examples of domains from each superfamily, a set of sequence profiles that represents each superfamily can then be generated. Since structural similarities are typically easier to recognize than amino acid sequence similarity, this enables the detection of more distant homologies than simply through sequence-based grouping, as exemplified by Pfam.

The CATH (Greene et al., 2007) and SCOP (Andreeva et al., 2004) resources both use a similar hierarchy and approach and are discussed in Chapters 18 and 17, respectively. The initial boundaries and relationships defined by these classifications are determined with a suite of sequence and structure comparison tools and then reviewed by an expert curator, before being placed within the classification. FSSP (Holm and Sander, 1994) is another useful resource that provides a similarity measure (Z-score) between known protein structures, rather than classifying them in an explicit hierarchy.

Using Homology Detection to Predict Structural Domains

Given the sequences from a particular superfamily, the essential steps in identifying structural domains in genomes, or for that matter any other kind of protein domain, are sequence profiling, detection, and resolution. First, it is necessary to determine which features of the sequences are important in defining the superfamily (profiling). Second, these sequence profiles or models must be used to find potential examples of the superfamily in genomes (detection). Finally, it is necessary to determine which are real domains, compared to false positive hits, and resolve any overlapping domains (resolution).

The most common approach is to use remote homology detection methods for the profiling and detection steps. Chief among these methods are HMMs (Krogh et al., 1994; Eddy, 1996).

The structural genome annotation resources, such as Gene3D (Yeats et al., 2006) based on CATH and SUPERFAMILY (Wilson et al., 2007) based on SCOP, use HMMs to profile families and assign domain superfamilies to genomes. Purely sequence-based domains such as Pfam domains are modeled using similar protocols to the structure-based resources (Finn et al., 2006). The 3D-GENOMICS resource (Fleming et al., 2004) uses PSI-BLAST and HMMer to assign SCOP superfamilies, with PSI-BLAST optimized for genome annotation (Muller, MacCallum, and Sternberg, 1999). The Genomic Threading Database(GTD) (McGuffin et al., 2004) uses PSI-BLAST and a threading-related method to validate matches (discussed in Section 3.3). Gene3D will be used as the main example in this chapter for the discussion of structure-based approaches to genome annotation with differences between the methods highlighted subsequently.

The Gene3D Protocol for Predicting CATH Domains in Proteins of Unknown Structure. Hidden Markov Models are perhaps the most successful approach for modeling protein domain families (Park et al., 1998). For an alignment of a family of homologous protein domain sequences, HMMs capture the distribution of residues at each conserved position as well as the likelihood of deletions and insertions at/or between the conserved positions, respectively. It has been shown that HMMs best model whole superfamilies, while multiple models should be built for subclusters within each superfamily (Gough et al., 2001; Sillitoe et al., 2005). The full diversity of a superfamily cannot be adequately captured in a single model as superfamilies frequently contain sequences with low pair-wise identity that cannot be successfully aligned. Therefore, multiple models are generally used—one for each subfamily within a superfamily. However, full profiles capture more remote homologues than within the subfamiliy, so models for each subcluster may therefore overlap, but not completely.

In CATH, sequences within homologous superfamilies are clustered at 35% identity and the resulting sequence families are ideal for building alignments and training HMMs (Reid, Yeats, and Orengo, 2007). As shown in Figure 22.1, the representative sequences from each cluster are chosen as the seeds from which to build HMMs. The representative for each cluster is the cluster member with the highest resolution structure and has the sequence length closest to average for the cluster. In Gene3D, the *target2k* program, distributed as part of the SAM HMM package (Karplus, Barrett, and Hughey, 1998), is used to generate the alignments based on these representatives. *Target2k* uses an iterative HMM procedure to build an alignment from, in this case, the GenBank nonredundant protein database. These alignments provide an enrichment of additional sequence data not present in the structural databases.

The sequence alignments produced by *target2k* are converted into HMMs using the w0.5 program from the SAM package. Gene3D uses HMMer rather than SAM to find domains; therefore, the SAM models are converted into HMMer models. Wistrand and Sonnhammer (2005) showed that SAM has superior model building performance, while HMMer scores are deemed more accurate than SAM. In addition, HMMer scores HMMs against sequences much faster than SAM.

Each CATH domain superfamily is therefore modeled by one or more HMMs. It is then necessary to use these models to determine where domains occur in the genomes. Each HMM (7794 in CATH v3.1 from 2091 superfamilies) is scanned against all the sequences from UniProt and RefSeq, including all completely sequenced genomes. In some cases, there are overlapping domains and these need to be resolved to produce coherent multidomain architectures (MDAs). The program DomainFinder performs this

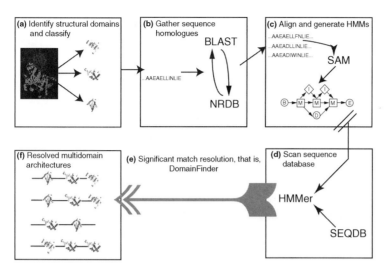

Figure 22.1. Workflow for creating CATH multidomain architectures in Gene3D. (a) Chains are chopped into domains and classified in CATH. (b) Sequence relatives are gathered using SAM T2K (incorporating BLAST and an iterative HMM procedure). (c) Sequence alignments are converted into HMM. (d) Scanned against UniProt and RefSeq. (e,f) DomanFinder is used to resolve domain architectures based on E-value and overlap.

task (Lee et al., 2005). DomainFinder works by iteratively assigning the most high scoring, nonoverlapping domain to each protein, up to an E-value of 0.01. Allowing overlapping domains increases the sensitivity and specificity, but reduces the certainty over domain boundaries.

Variations in the SUPERFAMILY Protocol. The SUPERFAMILY procedure for modeling, detecting, and resolving structural domain families predicted for SCOP superfamilies is much the same as for Gene3D (Gough et al., 2001). It differs, however, in using SAM for detection of structural domain families rather than HMMer and in allowing a 20% overlap between adjacent domains.

In the most recent version of SUPERFAMILY, a family-level classification has been introduced (Wilson et al., 2007). In addition to specifying the superfamily of protein domains identified in genomes, the more specific family is also specified if it can be determined. This has been achieved using a hybrid pair-wise-profile method (Gough, 2006). As in the original SUPERFAMILY procedure, a query sequence and other domains from each family within that superfamily are aligned to a particular HMM. From this, an alignment can be inferred between the query domain sequence and the members of each family by aligning those residues that align to the same position in the HMM. An E-value score is calculated based on whether the query sequence scores highly to one family representative relative to the others. A low E-value indicates clear membership to one particular family.

Using Fold Recognition to Predict Structural Domains

The approaches described above recognize homologues by using sequence patterns indicative of a common evolutionary ancestor. Fold recognition methods such as threading have also been used for structural annotation of genomes and have a similar aim to homology detection. Threading does not primarily exploit evolutionary information to recognize structural domains in sequences, instead it determines how well a sequence fits a particular fold (Jones, Taylor, and Thornton, 1992). Because very different sequences can form similar structures, threading allows very distant homologous domains, and even analogous domains, to be recognized. Full structural threading requires large amounts of computing power and it has been shown that using a more heuristic approach with data derived from threading, as implemented in GenTHREADER (McGuffin and Jones, 2003) and 3D-PSSM (Kelley, MacCallum, and Sternberg, 2000), produces similar results (Cherkasov and Jones, 2004).

The GTD (McGuffin et al., 2004) uses GenTHREADER (McGuffin and Jones, 2003) to annotate genomes with fold level SCOP annotations. GenTHREADER combines threading and homology-based sequence alignment scores generated by PSI-BLAST (Altschul et al., 1997) with scores based on secondary structure predictions (Marsden, McGuffin, and Jones, 2002). Results from each of these methods are combined using a neural network. The Genomic Threading Database does not however provide resolution of multidomain chains.

Predicting Other Structural Features

Various other structural and functional features of proteins can be predicted when annotating genomes, including coiled-coil regions (involved in protein oligomerization), transmembrane helices (membrane-spanning regions, which are present in $\sim$50% of drug targets), transport peptides, protein modification sites, and regions of low complexity—representing a significant source of information for functional prediction. These annotations are also useful for determining regions of proteins that may not contain globular domains or are otherwise unsuitable (e.g., containing transmembrane helices) as structural genomics targets for stucture elucidation.

Coiled-coils, protein structures involving intertwined alpha helices, can be predicted by sequence comparison to known coiled-coils and by recognizing heptad repeats of side chain chemistries necessary to form these structures. This is implemented in the program COILS (Lupas, Van Dyke, and Stock, 1991). Transmembrane helices can be predicted by looking for patterns of residue types, principally $\sim$20 residue stretches of hydrophobic residues (Cuthbertson, Doyle, and Sansom, 2005). Transmembrane helix prediction is implemented in several programs of which SPLIT4 (Juretic, Zoranic, and Zucic, 2002), TMHMM2 (Krogh et al., 2001), HMMTOP2 (Tusnady and Simon, 1998), and TMAP (Persson and Argos, 1997) are thought to perform the best. Low-complexity regions can be predicted using SEG (Wootton and Federhen, 1996).

Accuracy of Methods

There has been much discussion in the literature on the accuracy of remote homologue detection and fold recognition methods for single domains (Park et al., 1998; Rychlewski and Fischer, 2005; Reid, Yeats, and Orengo, 2007). For example, Reid, Yeats, and Orengo (2007) showed that profile methods perform very well at annotating remote homologues (<35% identical); PSI-BLAST, HMMer, and SAM all achieve at least 80% sensitivity with less than 1% errors. There has however been little work published on the accuracy of methods in resolving multidomain architectures. Resources such as SUPERFAMILY provide confidence values for individual domain assignments (E-values), but not for the multidomain architecture as a whole. One exception is work of Muller, MacCallum, and Sternberg (1999) in which genome annotation using PSI-BLAST with SCOP domains was benchmarked. In a benchmark set of 1254 sequences with 1621 domains, 652 domains ($\sim$40%) were correctly assigned with 16 false positives at E-value of 5×10^{-4}.

There is a need for further work in this area resolving structure-based genome annotation. The problem of how to resolve a multidomain architecture from multiple hits is not a simple one and reliable benchmarks would be of great use to the structural annotation community.

HOW WELL ARE GENOMES COVERED BY STRUCTURAL DOMAIN ANNOTATION?

Using the methods outlined above, how much of known genomes can be annotated? Using the current version of Gene3D (v6), it can be seen that coverage between the three superkingdoms is not even (Figure 22.2). This reflects biases in structures deposited in the PDB and also the different distributions of folds in the different superkingdoms. In archaea, $46 \pm 8\%$ of the genes in each genome have structural annotation, $52 \pm 7\%$ for bacteria and $44 \pm 10\%$ for eukaryotes. Figure 22.2 also shows that Pfam (sequence-based) domains have higher coverage than CATH (structure-based) domains due to the limited number of known structures. The combined coverage created by assigning CATH domains and nonoverlapping Pfam domains shows that CATH domains can be assigned to regions that are not covered by the sequence-based domains. This is because structural domain classifications allow more distantly related domains to be recognized. The two approaches are therefore complementary.

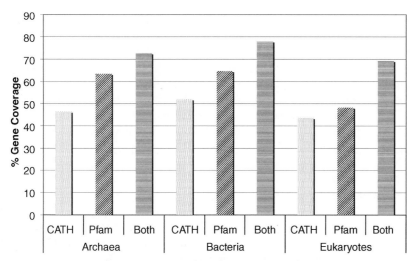

Figure 22.2. Percentage of genomes covered by CATH domains and Pfam domains, and using both together for the three superkingdoms. Created using complete genomes from Gene3D v6.

CAN WE DETERMINE ALL THE STRUCTURES PRESENT IN THE GENOMES?—STRUCTURAL ANNOTATION OF GENOMES AND STRUCTURAL GENOMICS

Looking at the structural annotation of genomes, we see that although there are a limited number of experimentally determined structures, many proteins without solved structures are likely to adopt known folds. Those remaining unannotated fall into two classes: those that have a known fold but the sequence has diverged too far to be detected and those that have novel folds. Structural genomics initiatives (e.g., http://www.nigms.nih.gov/Initiatives/PSI/) aim to solve structures for unannotated sequences as efficiently as possible. Novel structures are predicted and put forward for experimental characterization, and bioinformatic methods are used to identify these remote homologues.

The precise aims of SG initiatives vary (Todd et al., 2005; Chandonia and Brenner, 2006), but all of them are working toward solving protein structures to fill the annotation gap. Ideally, the result would be a sufficient sampling of all protein structures such that those remaining unsolved could be successfully modeled using the solved structures as templates. Currently, reliable structural models can be achieved using a template of >30% sequence identity with >60% overlap; however, resolving important functional features often requires >60% sequence identity. Thus, some SG initiatives aim to solve new folds where there are no relatives of known structure, while others aim to increase the knowledge of structural and functional diversity in existing folds/superfamilies.

Using annotation resources, such as Gene3D and SUPERFAMILY, unannotated regions can be identified and targeted by structural genomics initiatives. These may be Pfam families that are not related to CATH or SCOP superfamilies, or families created by clustering the sequences of structurally unannotated regions in the genomes (e.g., NewFams) (Marsden et al., 2006). Solving the larger families/clusters identified in the unannotated regions will result in a greater increase in structural genome coverage. Unfortunately, determining these potentially novel proteins is not as simple as finding unannotated regions.

Having accounted for known structural (e.g., CATH) domains, it is necessary to exclude domains that are less useful or unlikely to be amenable to structure determination. There are many factors that should be taken into account, but two that are most important involve singletons and proteins intractable to crystallization. Singleton domains are unannotated domain-like regions ($\geq$50 residues) that have no homologues as determined by sequence-based methods. These domains are likely to be novel but not useful for modeling other domains. These species-specific domains are thought to make up between 7% and 22% of domains in genomes (Marsden, Lewis, and Orengo, 2007).

Regions of proteins that will be intractable to high-throughput structural characterization also need to be excluded. Typically, between 13% and 18% of proteins in a genome are thought to contain problematic regions such as transmembrane sequences, coiled-coils, or low-complexity regions (Marsden, Lewis, and Orengo, 2007). These regions cannot easily be solved with current techniques; transmembrane sequences are very hydrophobic and low-complexity regions tend to be unstructured. Disordered regions of proteins (those without a unique structure) are increasingly seen as important for protein function but are intractable to traditional methods of structure determination. In some structurally solved proteins, disordered regions are thought to make up between 5% and 30% of their entire length (Linding et al., 2003) while others may be entirely disordered (Uversky, 2002). Figure 22.3 (Marsden, Lewis, and Orengo, 2007) shows that we currently achieve less than

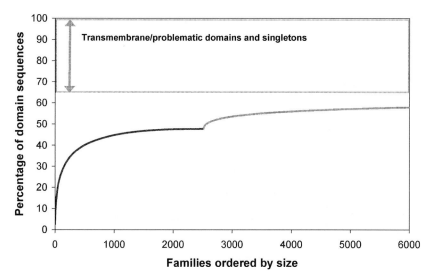

Figure 22.3. Structural coverage of domain sequences in Swiss-Prot-TrEMBL families, ranked in order of number of members in family, largest to smallest. The black line represents coverage of domain sequences by 2486 CATH and Pfam-A structural families, whereas the gray line represents additional coverage that would be achieved by solving a structure for structurally uncharacterized Pfam-A and NewFam domain families (Marsden, Lewis, and Orengo, 2007).

50% structural coverage of known domain families in UniProt. Solving structurally uncharacterized Pfam-A and completely uncharacterized NewFam (Marsden et al., 2006) families will increase this coverage by ~10% and that 35% of domain sequences are not tractable to structural genomics.

Although bioinformatic methods can identify families of proteins that are more likely to have novel structures, there are limitations in this approach. Currently, only 10% of targets solved by structural genomics initiatives have novel structures (Bourne et al., 2004) that can be explained by several factors. First, not all initiatives aim to solve novel structures but target functionally distinct members of known folds. Second, homology detection and fold recognition methods are not perfect. Third and perhaps most important, known experimental conditions for structure determination may favor known structures and their homologues. The effect of the experimental condition can be observed in the targeted structures analyzed by Bourne et al. (2004) showing half of this set of proteins are homologous to a subset of structures, presumably repeatedly selected by experimental conditions, compared to 90% of the solved structures.

WHAT CAN STRUCTURAL GENOME ANNOTATION TELL US ABOUT EVOLUTION?

The Distribution of Structural Domain Families in the Genomes Follows a Power Law

Several groups have shown that protein domain families are not evenly distributed within genomes (Huynen and van Nimwegen, 1998; Apic, Gough, and Teichmann, 2001

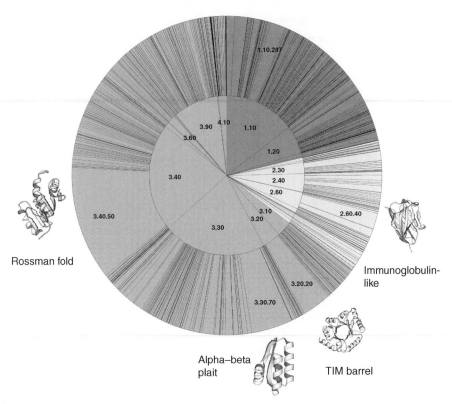

Figure 22.4. The distribution of CATH domains in 535 completed genomes in Gene3D. CATH classes are illustrated by color: purple for mainly alpha structures; yellow for mainly beta structures; green for alpha–beta structures. The inner wheel shows different architectures and the outer wheel different folds. For the most numerous folds, example domains are shown on the outside of the wheel. Figure also appears in the Color Figure section.

Qian, Luscombe, and Gerstein, 2001). As shown in Figure 22.5, the distribution of protein domain superfamilies follows a power law, indicating that relatively few families account for a large proportion of domains and most families are very small. This uneven distribution is even more apparent when the domains are grouped according to folds. The most common folds are so frequent that they have been termed "superfolds" (Orengo, Jones, and Thornton, 1994). Structures in these fold groups tend to carry out a wide range of functions and are believed to maintain favorable residue packing while accommodating extensive residue substitution (Shakhnovich et al., 2003).

The Rossman fold, for instance, makes up a large proportion of known structural domains in completed genomes. Figure 22.4 shows a CATH wheel with the distribution of known architectures and folds in completed genomes, derived from Gene3D. It shows a distinct bias toward certain architectures (e.g., 1.10—orthogonal bundle, 3.30—two-layer sandwich, and 3.40—three-layer sandwich) and certain folds (e.g., 2.60.40—immunoglobulin-like, 3.30.70—αβ plait, 3.20.20—TIM barrel, and 3.40.50—Rossman fold). Within individual genomes, there is the same bias toward the superfolds; however, the smallest families are specific to certain species or subkingdoms. This is the tail of the power law distribution, where

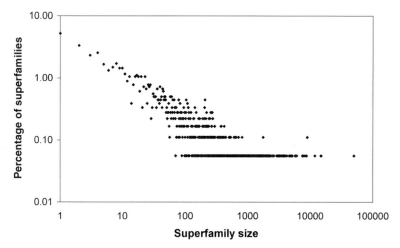

Figure 22.5. Power–law distribution of protein domain families in Gene3D (v6). Superfamily size is the number of members in the superfamily.

we find many families with very few instances (Figure 22.5). This fact has rather negative consequences for structural genomics as discussed above.

Using Protein Structural Domain Annotations to Explore the Evolution of Diverse Multidomain Protein Architectures in Nature

Most proteins contain multiple domains—perhaps greater than 90% in eukaryotes, somewhat lower in prokaryotes (Apic, Gough, and Teichmann, 2001). It has been known for some time that new functions can be achieved by mixing and matching existing domains—domains are thought to be the fundamental functional units of proteins (Edelman and Gall, 1969). Various aspects of such combinations of domains or multidomain architectures have been studied, revealing the nature of evolution at the multidomain level.

There are far fewer domain combinations/protein architectures observed in Nature than are possible. Around 5000 CATH and Pfam families can be mapped to the genomes. There are then $>10^7$ possible two-domain proteins and $>10^{11}$ possible three-domain proteins. However, we observe only 50,000 protein families (i.e., distinct multidomain protein families) using Gene3D (Marsden et al., 2006), and indeed Chothia et al. (2003) estimated that less than 0.5% of possible combinations are observed. Most domains then are seen in very few multidomain contexts, whereas more frequently occurring domains are seen in partnership with many different domain families (Apic, Gough, and Teichmann, 2001; Lee et al., 2005). These highly recurring domains are mostly performing important generic functions such as providing energy or redox equivalents for reactions or binding DNA (Apic, Gough, and Teichmann, 2001). This suggests that certain domain families are very large because they perform functions that are required in many different contexts (Vogel et al., 2004a) rather than simply because they are energetically favored folds.

Another feature of domain combinations is that some pairs of domain families are highly overrepresented in proteins. These domain pairs have been called supradomains (Vogel et al., 2004b). The 200 supradomains discovered by Vogel and coworkers using SUPERFAMILY were found in 75,000 genome sequences. These are thought to occur because a functional site is created between the domains or simply because it is functionally beneficial to have both types of domains together.

Ancient Domains and LUCA

Structural protein domain classifications bring together more distantly related domains than sequence-based ones and therefore are ideal for identifying ancient domain families (those common to all kingdoms) that were present in the Last Universal Common Ancestor (LUCA). LUCA is a hypothetical organism that gave rise to all modern-day species. By determining which domain families were present in LUCA, we can infer how complex it was and which processes went on its cells. Using only sequence data, the most common domain families to all kingdoms are involved in protein biosynthesis (Koonin, 2003). However, when using structural data to bring together more remotely related domains, a more interesting picture emerges (Ranea et al., 2006).

Using restrictive criteria, Ranea et al. (2006) found 140 ancestral domain families common to almost all completely sequenced members of the three kingdoms of life. The domain families thus discovered are known to be involved in all the essential functional systems present in extant organisms, supporting the theory that life prior to the divergence of the three kingdoms was composed of cells in much the same way as it is today (Doolittle, 2000).

Winstanley, Abeln, and Deane (2005) used occurrence of SCOP domains via SUPER-FAMILY to determine the relative age of protein folds. Using a species tree and the presence or absence of a particular fold in each organism with a completed genome, the position at which each fold first emerged was mapped onto the evolutionary tree. This approach estimated where new folds occurred relative to the appearance of different groups of species. It was noted that folds from the alpha/beta class tend to be far older than those of other architectures.

Several groups have studied the usefulness of structural domain content in constructing phylogenies and refining tree structures representing the evolutionary relationships between species (Lin and Gerstein, 2000; Yang, Doolittle, and Bourne, 2005). Yang, Doolittle, and Bourne (2005) used SUPERFAMILY annotation to determine whether the presence or absence of SCOP superfamilies in 174 complete genomes gave rise to a phylogeny consistent with more sophisticated methods. They found a good correspondence in separating the major clades and achieved greater accuracy with fewer genomes over earlier attempts suggesting that this approach to phylogeny will improve further in the future.

Organismal Complexity and the Evolution of Protein Domain Superfamilies

Several studies have probed into the concept of organismal complexity using structural genome annotation (Vogel, Teichmann, and Chothia, 2003; Ranea et al., 2004; Ranea et al., 2005). Vogel, Teichmann, and Chothia (2003) identified two types of expansions of protein family repertoires by comparing the relative sizes of SCOP domain superfamilies in worm and fly. They termed these progressive and conservative expansions. Conservative expansion increases the ability of an organism to adapt to its environment, whereas progressive expansion leads to significant changes with possible selective advantages, but not necessarily the case. Conservative expansion was seen in two chemoreceptor families in the less complex worm. Progressive expansion was observed in the fly where expanded families tend to be involved in cell–cell communication such as receptors involved in neuronal development. These findings highlight the fact that expansions in different domain families are not equal in terms of their impact on organismal complexity. Instead, some families contribute much more to the development of complex multicellular organisms.

Ranea et al. (2004, 2005) investigated relative expansions in protein structural domain families across 100 completely sequenced bacterial genomes identifying 200 domain families that were common to all species; 27 families exhibited no variation in the number of relatives between genomes; 30 mostly metabolism-related families had expanded linearly with increased genome size; and 27 mostly regulation-related families expanded non-linearly. Hence, increasing metabolic complexity was associated with an increasingly greater cost in terms of regulation, thus allowing limits to be suggested for optimal bacterial genome sizes (Ranea et al., 2004).

Structural Domain Annotation and Phylogenetic Profiling

Phylogenetic profiling, introduced by Pellegrini et al. (1999), allows the identification of coevolving proteins by determining those proteins that are always present or absent together in different species. Coevolving proteins tend to be functionally related and therefore phylogenetic profiles allow for improved functional annotation of proteins.

Pagel, Wong, and Frishman (2004) were the first to use domains rather than whole protein sequences in this context, and more recently Ranea et al. (2007) used CATH structural domains from Gene3D to construct phylogenetic profiles capturing protein coevolution. One of the main problems in using phylogenetic profiles is identifying orthologs, a key factor in the success of phylogenetic profile methods. Ranea et al. (2007) used the subclusterings of domains available in Gene3D to aid the identification of distinct functional subgroups within domain families, narrowing down the identification of orthologs.

Exploiting Structural Annotation of Genomes to Study the Evolution of Metabolic Networks

The two principal models of metabolic network evolution are the retrograde model first proposed by Horowitz (1945) and the patchwork model introduced by Ycas (1974) later expanded by Jensen (1976). The retrograde model proposes that when a metabolite A is scarce in the environment, it is advantageous for an organism to recruit an enzyme that synthesizes A from environmentally available precursors (B, C). If B (or C) becomes scarce, then another enzyme might be recruited to synthesize B (or C) from its precursors. Generally associated with this model is the proposition that each novel enzyme is recruited by duplication from the previous enzyme in the pathway, having similar chemistry but different substrate specificity. Conversely, in the patchwork evolution model, enzymes with broad substrate specificities have duplicated and specialized with paralogs appearing in different pathways. If the retrograde model was correct, we would expect to see clustering of homologues in pathways with the last metabolic enzyme being the most ancestral. The patchwork model instead predicts that homologues would be spread between different pathways. Therefore, the use of effective homologue detection methods on whole sets of enzymes from different genomes is very useful in identifying which evolutionary model is being utilized.

Teichmann et al. (2001a, 2001b) used HMMs built on SCOP superfamilies (before SUPERFAMILY) while others have used PSI-BLAST-assigned SCOP domains (Saqi and Sternberg, 2001; Alves, Chaleil, and Sternberg, 2002) or Gene3D (Rison, Teichmann, and Thornton, 2002) to determine which evolutionary mechanism hypothesis is correct. Homologues were mostly found distributed across pathways supporting the patchwork model of evolution (Teichmann et al., 2001b; Rison, Teichmann, and Thornton, 2002), although some superfamilies were shown to be specialized for particular pathways (Saqi and Sternberg, 2001).

Transcription Factors and Regulatory Networks

Structural domain assignments from the SUPERFAMILY database have also been used to study gene regulatory networks (Teichmann and Babu, 2004). Gene regulatory networks consist of interactions between genes and the transcription factors that regulate their expression. It was shown using SUPERFAMILY that most of these interactions (90%) have arisen by duplication of pre-existing transcription factors or target genes and that half of these duplications led to inheritance of the interaction while the other half gained new interactions.

Kummerfeld and Teichmann (2006) have produced a structurally annotated database of transcription factors using SUPERFAMILY HMMs. Their prediction method involves a curated subset of HMMs that pertain to only those transcription factor superfamilies that are involved in sequence-specific DNA binding. These are the transcription factors that are important for differential expression within organisms.

STRUCTURAL GENOME ANNOTATION RESOURCES

Gene3D Resource for Structural and Functional Annotation of Genomes

Gene3D (Yeats et al., 2006) is a genome annotation resource that provides access to the predicted CATH protein domain architectures and other sequence features (transmembrane helices, coiled-coils, etc.), as well as functional information (e.g., GO (Harris et al., 2004) and KEGG (Kanehisa et al., 2006)), protein–protein interaction data (e.g., Intact (Kerrien et al., 2007)), and microarray expression profiles (from ArrayExpress (Parkinson et al., 2007)). All of UniProt and RefSeq sequences are annotated with CATH structural domains, including complete genomes identified by Integr8. Pfam annotation is also included, increasing the number of proteins that can be annotated. To facilitate functional grouping, protein sequences are clustered into hierarchical families using TribeMCL clustering (Enright, Van Dongen, and Ouzounis, 2002). As discussed above, this resource has been successfully used in studying evolution, identifying sequence relatives, and determining structural genomics targets.

Gene3D can be accessed through a Web site (see Table 22.1) that provides access to all the annotation via a single search box, which accepts UniProt, PDB, RefSeq, CATH, and Pfam identifiers. BLAST searches and HMM scans can also be used to retrieve annotation for homologues of proteins or families of interest. Completed genomes can be browsed and the data can be downloaded as flat files for more intensive investigations.

A recent addition to the Gene3D resource is PhyloTuner profiles (Ranea et al., 2007). PhyloTuner makes use of the hierarchically clustered CATH domains in Gene3D to identify coevolving domain families and interfunctional relationships.

TABLE 22.1. Details of Structural Genome Annotation Resources

Resource	Location and Author	Structural Domain Type	Assignment Method	URL
Gene3D	UCL, London, UK	CATH	Profiles (SAM/HMMer)	http://cathww.biochem.ucl.ac.uk:8080/Gene3D/
SUPERFAMILY	LMB-MRC, Cambridge, UK	SCOP	Profiles (SAM)	http://supfam.mrc-lmb.cam.ac.uk/SUPERFAMILY/
3D-GENOMICS	Imperial College, London, UK	SCOP	Profiles (PSI-BLAST and HMMer)	http://www.sbg.bio.ic.ac.uk/3dgenomics/
Genomic Threading Database	UCL, London, UK	SCOP	Threading and Profiles (GenTHREADER and PSI-BLAST)	http://bioinf.cs.ucl.ac.uk/GTD/
InterPro	EBI, Hinxton, UK	CATH/SCOP/Consensus	Profiles (Gene3D HMMs and superfamily HMMs)	http://www.ebi.ac.uk/interpro/

SUPERFAMILY

SUPERFAMILY (Wilson et al., 2007) provides access to predicted SCOP domains. The data can be accessed using keyword searches for sequence identifiers (e.g., UniProt), PDB identifiers, organism names, SCOP superfamily names and identifiers, and SUPERFAMILY model numbers. Sequences can be submitted for domain assignment and, if not found in the database, are scanned *de novo* against SUPERFAMILY HMMs. If no significant hits are found, the powerful profile–profile method PRC is used to search for more remote hits. SCOP family-level assignments are presented in addition to the SUPERFAMILY-level assignment. Statistics for each genome, such as domain composition, can be browsed, and simple genome comparisons can be performed. Domain assignments, alignments, and models can be downloaded.

3D GENOMICS

This resource from the Sternberg group at Imperial College (Fleming et al., 2004) provides SCOP and Pfam domains, secondary structure predictions, and sequence features, such as low-complexity regions, coiled-coils, and transmembrane helices, for completed genomes. It also determines homologous features between genomes on the fly using BLAST. Another useful feature is the ability to perform genome comparisons based on statistics of domain features.

Genomics Threading Database (GTD)

The GTD (McGuffin et al., 2004) uses GenTHREADER (McGuffin and Jones, 2003) to obtain fold-level annotation with SCOP domains for completed genomes. GenTHREADER involves a threading-based approach to the structure prediction in combination with PSI-BLAST and secondary structure prediction. GTD allows keyword searches using PDB and SCOP codes, gene ids and description, and also BLAST searches. Useful summary statistics on fold coverage of the genomes are provided.

InterPro

InterPro (Mulder et al., 2007) brings together data from many different domain annotation resources, allowing users to compare their predictions. It includes both Gene3D (CATH) and SUPERFAMILY (SCOP) structural domain predictions as well as Pfam, Prosite, SMART, Panther, PRINTS, ProDom, and TIGR sequence domains/motifs. It produces a consensus report of these resources, where possible, as InterPro domains.

Structural Genome Annotation with DAS and Web Services

The way in which genome annotation resources are accessed is changing. The need for more integrated and flexible data sources has led to the innovations of Web services and Distributed Annotation System (DAS). Web services allow, for example, those queries that one might perform using a Web browser to be performed within code run locally. For example, large queries can be retrieved and integrated into pipelines without installing and running software on local compute farms. EBI Web services, for example, provide structural genome annotation from InterPro (http://www.ebi.ac.uk/Tools/webservices/). Web services are soon to be released for Gene3D.

DAS allows multiple sources of genome annotation, which are held on different servers around the world, to be drawn together into different applications. The Gene3D DAS server, for example, provides lists of Gene3D feature ids and protein family clusterings for all UniProt entries. SUPERFAMILY, 3D-GENOMICS, GTD, and InterPro also have DAS services. Information on accessing DAS services can be found at the DAS registration server (http://www.dasregistry.org/index.jsp).

SUMMARY

Knowledge of protein structure is useful for two primary reasons. First, it is essential for understanding protein function because the arrangement of secondary structures and the nature of binding site environments determine substrates and interaction partners. Second, structure is generally more conserved than sequence; therefore, more distant evolutionary relationships can be recognized using structure rather than sequence. Structural genome annotation has been used in diverse areas of research, including studies of protein evolution, taxonomy, organismal complexity, and in the development of new methods for predicting protein functions.

Although it would be desirable to have a catalog of all protein structures, it is not feasible to determine all of these structures directly. Structural domain annotation exploits structural protein domain classifications, such as CATH and SCOP, to determine remotely related domains in genomes. In this way, it expands structural coverage of genomes and also defines those regions that are currently not accounted for by known structures.

It is currently unclear how many of the remaining structural gaps will be filled. Novel structures will continue to be solved for some time yet, and methods of remote homology detection continue to improve, uniting seemingly disparate domain families. We also await technical breakthroughs for solving many membrane proteins and are only just beginning to appreciate the importance of disordered proteins (those with little or no regular structure!).

REFERENCES

Altschul SF, et al. (1997): Gapped BLAST and PSI-BLAST: a new generation of protein database search programs. *Nucleic Acids Res* 25:3389–3402.

Alves R, Chaleil RA, Sternberg MJ (2002): Evolution of enzymes in metabolism: a network perspective. *J Mol Biol* 320:751–770.

Andreeva A, et al. (2004): SCOP database in 2004: refinements integrate structure and sequence family data. *Nucleic Acids Res* 32:D226–D229.

Apic G, Gough J, Teichmann SA (2001): An insight into domain combinations. *Bioinformatics* 17 (Suppl 1):S83–S89.

Bourne PE, et al. (2004): The status of structural genomics defined through the analysis of current targets and structures. *Pac Symp Biocomput* 375–386.

Chandonia JM, Brenner SE (2006): The impact of structural genomics: expectations and outcomes. *Science* 311:347–351.

Cherkasov A, Jones SJ (2004): Structural characterization of genomes by large scale sequence–structure threading. *BMC Bioinformatics* 5:37.

Chothia C, et al. (2003): Evolution of the protein repertoire. *Science* 300:1701–1703.

Cuthbertson JM, Doyle DA, Sansom MS (2005): Transmembrane helix prediction: a comparative evaluation and analysis. *Protein Eng Des Sel* 18:295–308.

Doolittle WF (2000): The nature of the universal ancestor and the evolution of the proteome. *Curr Opin Struct Biol* 10:355–358.

Eddy SR (1996): Hidden Markov models. *Curr Opin Struct Biol* 6:361–365.

Edelman GM, Gall WE (1969): The antibody problem. *Annu Rev Biochem* 38:415–466.

Enright AJ, Van Dongen S, Ouzounis CA (2002): An efficient algorithm for large-scale detection of protein families. *Nucleic Acids Res* 30:1575–1584.

Finn RD, et al. (2006): Pfam: clans, web tools and services. *Nucleic Acids Res* 34:D247–D251.

Fleming K, et al. (2004): 3D-GENOMICS: a database to compare structural and functional annotations of proteins between sequenced genomes. *Nucleic Acids Res* 32:D245–D250.

Gough J, et al. (2001): Assignment of homology to genome sequences using a library of hidden Markov models that represent all proteins of known structure. *J Mol Biol* 313:903–919.

Gough J (2006): Genomic scale sub-family assignment of protein domains. *Nucleic Acids Res* 34:3625–3633.

Greene LH, et al. (2007): The CATH domain structure database: new protocols and classification levels give a more comprehensive resource for exploring evolution. *Nucleic Acids Res* 35: D291–D297.

Harris MA, et al. (2004): The Gene Ontology (GO) database and informatics resource. *Nucleic Acids Res* 32:D258–D261.

Holm L, Sander C (1994): The FSSP database of structurally aligned protein fold families. *Nucleic Acids Res* 22:3600–3609.

Horowitz NH (1945): On the evolution of biochemical syntheses. *Proc Natl Acad Sci USA* 31:153–157.

Huynen MA, van Nimwegen E (1998): The frequency distribution of gene family sizes in complete genomes. *Mol Biol Evol* 15:583–589.

Jensen RA (1976): Enzyme recruitment in evolution of new function. *Annu Rev Microbiol* 30:409–425.

Jones DT, Taylor WR, Thornton JM (1992): A new approach to protein fold recognition. *Nature* 358:86–89.

Juretic D, Zoranic L, Zucic D (2002): Basic charge clusters and predictions of membrane protein topology. *J Chem Inf Comput Sci* 42:620–632.

Kanehisa M, et al. (2006): From genomics to chemical genomics: new developments in KEGG. *Nucleic Acids Res* 34:D354–D357.

Karplus K, Barrett C, Hughey R (1998): Hidden Markov models for detecting remote protein homologies. *Bioinformatics* 14:846–856.

Kelley LA, MacCallum RM, Sternberg MJ (2000): Enhanced genome annotation using structural profiles in the program 3D-PSSM. *J Mol Biol* 299:499–520.

Kerrien S, et al. (2007): IntAct–open source resource for molecular interaction data. *Nucleic Acids Res* 35:D561–D565.

Koonin EV (2003): Comparative genomics, minimal gene-sets and the last universal common ancestor. *Nat Rev Microbiol* 1:127–136.

Krogh A, et al. (1994): Hidden Markov models in computational biology. Applications to protein modeling. *J Mol Biol* 235:1501–1531.

Krogh A, et al. (2001): Predicting transmembrane protein topology with a hidden Markov model: application to complete genomes. *J Mol Biol* 305:567–580.

Kummerfeld SK, Teichmann SA (2006): DBD: a transcription factor prediction database. *Nucleic Acids Res* 34:D74–D81.

Lee D, et al. (2005): Identification and distribution of protein families in 120 completed genomes using Gene3D. *Proteins* 59:603–615.

Lin J, Gerstein M (2000): Whole-genome trees based on the occurrence of folds and orthologs: implications for comparing genomes on different levels. *Genome Res* 10:808–818.

Linding R, et al. (2003): Protein disorder prediction: implications for structural proteomics. *Structure* 11:1453–1459.

Lupas A, Van Dyke M, Stock J (1991): Predicting coiled coils from protein sequences. *Science* 252:1162–1164.

Marsden RL, et al. (2006): Comprehensive genome analysis of 203 genomes provides structural genomics with new insights into protein family space. *Nucleic Acids Res* 34: 1066–1080.

Marsden RL, Lewis TA, Orengo CA (2007): Towards a comprehensive structural coverage of completed genomes: a structural genomics viewpoint. *BMC Bioinformatics* 8:86.

Marsden RL, McGuffin LJ, Jones DT (2002): Rapid protein domain assignment from amino acid sequence using predicted secondary structure. *Protein Sci* 11:2814–2824.

Marti-Renom MA, et al. (2000): Comparative protein structure modeling of genes and genomes. *Annu Rev Biophys Biomol Struct* 29:291–325.

McGuffin LJ, Jones DT (2003): Improvement of the GenTHREADER method for genomic fold recognition. *Bioinformatics* 19:874–881.

McGuffin LJ, et al. (2004): The Genomic Threading Database: a comprehensive resource for structural annotations of the genomes from key organisms. *Nucleic Acids Res* 32:D196–D199.

Mulder NJ, et al. (2007): New developments in the InterPro database. *Nucleic Acids Res* 35: D224–D228.

Muller A, MacCallum RM, Sternberg MJ (1999): Benchmarking PSI-BLAST in genome annotation. *J Mol Biol* 293:1257–1271.

Orengo CA, Jones DT, Thornton JM (1994): Protein superfamilies and domain superfolds. *Nature* 372:631–634.

Pagel P, Wong P, Frishman D (2004): A domain interaction map based on phylogenetic profiling. *J Mol Biol* 344:1331–1346.

Park J, et al. (1998): Sequence comparisons using multiple sequences detect three times as many remote homologues as pairwise methods. *J Mol Biol* 284:1201–1210.

Parkinson H, et al. (2007): ArrayExpress: a public database of microarray experiments and gene expression profiles. *Nucleic Acids Res* 35:D747–D750.

Pellegrini M, et al. (1999): Assigning protein functions by comparative genome analysis: protein phylogenetic profiles. *Proc Natl Acad Sci USA* 96:4285–4288.

Persson B, Argos P (1997): Prediction of membrane protein topology utilizing multiple sequence alignments. *J Protein Chem* 16:453–457.

Qian J, Luscombe NM, Gerstein M (2001): Protein family and fold occurrence in genomes: power-law behaviour and evolutionary model. *J Mol Biol* 313:673–681.

Ranea JA, et al. (2004): Evolution of protein superfamilies and bacterial genome size. *J Mol Biol* 336:871–887.

Ranea JA, et al. (2005): Microeconomic principles explain an optimal genome size in bacteria. *Trends Genet* 21:21–25.

Ranea JA, et al. (2006): Protein superfamily evolution and the last universal common ancestor (LUCA). *J Mol Evol* 63:513–525.

Ranea JA, Yeats C, Grant A, Orengo CA (2007): Predicting protein function with hierarchical phylogenetic profiles: the Gene3D Phylo-Tuner method applied to eukaryotic genomes. *PLoS Comput Biol* 3:e237.

Reid AJ, Yeats C, Orengo CA (2007): Methods of remote homology detection can be combined to increase coverage by 10% in the midnight zone. *Bioinformatics* 23 (18):2353–2360.

Richardson JS (1981): The anatomy and taxonomy of protein structure. *Adv Protein Chem* 34: 167–339.

Rison SC, Teichmann SA, Thornton JM (2002): Homology, pathway distance and chromosomal localization of the small molecule metabolism enzymes in *Escherichia coli. J Mol Biol* 318:911–932.

Rychlewski L, Fischer D (2005): LiveBench-8: the large-scale, continuous assessment of automated protein structure prediction. *Protein Sci* 14:240–245.

Sillitoe I, Dibley M, Bray J, Addou S, Orengo C (2005): Assessing strategies for improved superfamily recognition. *Protein Sci* 14(7):1800–1810.

Saqi MA, Sternberg MJ (2001): A structural census of metabolic networks for *E. coli. J Mol Biol* 313:1195–1206.

Shakhnovich BE, et al. (2003): Functional fingerprints of folds: evidence for correlated structure-function evolution. *J Mol Biol* 326:1–9.

Teichmann SA, et al. (2001a): Small-molecule metabolism: an enzyme mosaic. *Trends Biotechnol* 19:482–486.

Teichmann SA, et al. (2001b): The evolution and structural anatomy of the small molecule metabolic pathways in *Escherichia coli. J Mol Biol* 311:693–708.

Teichmann SA, Babu MM (2004): Gene regulatory network growth by duplication. *Nat Genet* 36:492–496.

Todd AE, et al. (2005): Progress of structural genomics initiatives: an analysis of solved target structures. *J Mol Biol* 348:1235–1260.

Tusnady GE, Simon I (1998): Principles governing amino acid composition of integral membrane proteins: application to topology prediction. *J Mol Biol* 283:489–506.

Uversky VN (2002): Natively unfolded proteins: a point where biology waits for physics. *Protein Sci* 11:739–756.

Vogel C, Teichmann SA, Chothia C (2003): The immunoglobulin superfamily in *Drosophila melanogaster* and *Caenorhabditis elegans* and the evolution of complexity. *Development* 130:6317–6328.

Vogel C, et al. (2004a): Structure, function and evolution of multidomain proteins. *Curr Opin Struct Biol* 14:208–216.

Vogel C, et al. (2004b): Supra-domains: evolutionary units larger than single protein domains. *J Mol Biol* 336:809–823.

Wilson D, et al. (2007): The SUPERFAMILY database in 2007: families and functions. *Nucleic Acids Res* 35:D308–D313.

Winstanley HF, Abeln S, Deane CM (2005): How old is your fold? *Bioinformatics* 21 (Suppl. 1): i449–i458.

Wistrand M, Sonnhammer EL (2005): Improved profile HMM performance by assessment of critical algorithmic features in SAM and HMMer. *BMC Bioinformatics* 6:99.

Wootton JC, Federhen S (1996): Analysis of compositionally biased regions in sequence databases. *Methods Enzymol* 266:554–571.

Yang S, Doolittle RF, Bourne PE (2005): Phylogeny determined by protein domain content. *Proc Natl Acad Sci USA* 102:373–378.

Ycas M (1974): On earlier states of the biochemical system. *J Theor Biol* 44:145–160.

Yeats C, et al. (2006): Gene3D: modelling protein structure, function and evolution. *Nucleic Acids Res* 34:D281–D284.

23

EVOLUTION STUDIED USING PROTEIN STRUCTURE

Song Yang, Ruben Valas, and Philip E. Bourne

One of the principle goals of evolutionary biology is to generate phylogeny that best represents the evolutionary histories of all organisms on earth. Aside from directly investigating the fossil records of ancestor species, all phylogenetic methods depend on the comparison of specific features (homologous characteristics) of contemporary organisms to determine the evolutionary relationships between different organisms. Among the features are morphological, physiological, genetic, and genomic which changed as the organisms evolved. The study of evolution changed dramatically with the discovery of DNA and the evolutionary fingerprint it represents. Evolutionary relationships between organisms can be studied by comparing their DNA sequences (Zuckerkandl and Pauling, 1965). Gene mutation is the primary cause of evolution, so utilizing the universal carrier of genetic information as the characteristic by which phylogenetic comparison is made makes sense. This approach has significant advantages over the classical approach in which morphological and physiological characteristics are used. This is exemplified by the discovery of a third branch of life, the archaea, which have no substantial morphological or physiological differences to other prokaryotes. Archaea were discovered to be a separate domain of life by analyzing small subunit ribosomal RNAs (SSU rRNA) (Woese and Fox, 1977).

While studying phylogeny using DNA sequence data has proven very successful, it has its limitations. Since individual genes have different evolutionary rates in different lineages, phylogenies built from individual genes do not always agree (Doolittle, 1995a; Doolittle, 1995b). As a consequence, although efforts have been made to generate a universal tree of life (Woese, 1998a; Woese, 1998b; Forterre and Philippe, 1999), many parts of the tree are still unresolved and highly debated, especially at the root of the tree where the three superkingdoms—archaea, bacteria, and eukaryotes—diverged (Mayr, 1998; Woese, 1998a; Woese, 1998b).

Advances in large-scale sequencing technology enable the acquisition of the complete genome of an organism. There are currently hundreds of complete genomes available from

Structural Bioinformatics, Second Edition   Edited by Jenny Gu and Philip E. Bourne
Copyright © 2009 John Wiley & Sons, Inc.

species across the tree of life. The growing numbers of genomes being sequenced have led to a new field of research, phylogenomics, where not only one or a few gene sequences, but the whole genomes of different organisms are compared and used as metrics for phylogenetic inference (Delsuc et al., 2005; Snel et al., 2005). The complete genomes do not only contain the primary sequences of genes; functional sites in the noncoding regions and the overall genome structure are also under evolutionary pressure and can potentially be used as comparative features. These whole genome features include gene content—the numbers and types of genes found in a genome (Snel et al., 1999; Tekaia et al., 1999; Lin and Gerstein, 2000; House and Fitz-Gibbon, 2002; Wolf et al., 2002), and gene order—the relative position of genes on the chromosomes (Dandekar et al., 1998; Wolf et al., 2001). These approaches are able to recover the three superkingdoms of life and verify the main groupings of the SSU rRNA tree, regardless of the potential prevailing horizontal gene transfer (HGT) events. However, incongruence still exists, and the search for new homologous features and tree construction algorithms continues (Philippe et al., 2005).

STRUCTURES AS EVOLUTIONARY UNITS

The three-dimensional structures of proteins have different evolutionary rates than the sequences from which they are derived. As described in detail in previous chapters, structures are more conserved than sequences. Changes in sequence can be tolerated provided they do not perturb the physical and chemical properties that define secondary structure and the organization of those secondary structures into tertiary units. It is not surprising then that there are examples of proteins with no apparent sequence similarity and sometimes with no apparent functional relationship that can have almost identical 3D structures (Pastore and Lesk, 1990; Flaherty et al., 1991; Schnuchel et al., 1993). This feature of protein structure makes it a better evolutionary marker than sequence to recognize more distant ancestral relationships, provided divergent evolution can be separated from convergent evolution. Convergent evolution implies there is no ancestral relationship, just convergence on a stable structural arrangement. Given that structure infers distant evolutionary relationships, how do we recover those relationships? The answer lies in protein domains.

As discussed in Chapter 2, the basic elements of protein structure are protein domains, which are compact and spatially distinct parts of a protein that can fold independently of neighboring sequences (Branden and Tooze, 1999). Each domain has its own unique 3D structure (also called fold) and corresponds to a series of amino acid sequences that can fold into the domain structure. Protein domains are the building blocks of proteins; combinations of different numbers and types of domains form structurally complex proteins with novel functions. Sharing of common domains by different proteins may infer common ancestry.

It is widely accepted that the number of protein folds is limited; estimates of absolute number range from 1,000 to 10,000 (Zhang, 1997; Govindarajan et al., 1999; Wolf et al., 2000), a remarkably small number given the almost infinite possible number of sequences. Intuitively then, the emergence of a new fold might constitute a significant evolutionary event. It is timely that structure-based evolutionary studies are being enabled by the structural genomics initiative (Chapter 40) which aims to solve 3D structures covering all unique folds and thus provide a complete view of protein structure space (Burley et al., 1999; Chandonia and Brenner, 2006). Although some scientists would argue that this goal will remain elusive (Xie and Bourne, 2005; Marsden et al., 2007).

Protein domains are not only structural units, but also evolutionary units. Combinations of different domains and domain duplication are the major evolutionary processes in the acquisition of novel functions (Doolittle, 1995a; Doolittle, 1995b). During evolution, novel domains can evolve by means of random mutation. Domains can also be lost or horizontally transferred. Therefore, protein structural domains can be used as structural features to study the evolutionary history of organisms.

Previous phylogenetic methods were based on the primary sequence of the DNA or protein, not the 3D structural features of proteins. Part of the reason is that 3D structural information is so limited; there are less than 50,000 structures in the PDB (Berman et al., 2000) as of 2007 and those are highly redundant with respect to sequence and structure. Unlike sequence and genomic data that can be generated by high-throughput techniques, structure determination, structure genomics notwithstanding, is still relatively slow. However, the accumulation of complete genome sequences and advances in gene finding and homology modeling algorithms provide an alternative approach to determine the structure content of an organism on a genome-wide scale. As discussed in previous chapters, using domains with existing 3D structures as templates, all the homologous domains in complete genomes are found by sequence comparison methods. Although whole genome domain recognition relies on sequence similarity, the resulting protein domain content, nonetheless, inherits 3D structure information whose evolution is conserved beyond sequence. Current techniques can reliably assign protein domains that cover over 50% of the genome for a given organism, making it possible to study evolution through structure (Buchan et al., 2002; Gough and Chothia, 2002).

PHYLOGENY BY PROTEIN DOMAIN CONTENT

Pioneering work on the study of evolution utilizing genomic structural information originated with Gerstein (1997), when only one species from each of the three super-kingdoms had been sequenced. Using a fold recognition method, it was possible to annotate only 10–20% of the genome, yet this attempt successfully showed that the approach of studying evolution using structure held much promise. Work in this area continued as more and more 3D structures became available and sequence comparison algorithms became more sophisticated (Wolf et al., 1999; Caetano-Anolles and Caetano-Anolles, 2003). Simultaneously, since the completion of the human genome project in 2001, there has been an increase in the number of the complete genomes from a wide spectrum of organisms. As of July 2004, 212 complete genomes (20 archaea, 154 bacteria, and 38 eukaryotes) were available. At the same time, the number of 3D structures of proteins increased, resulting in approximately 800 unique folds and 1300-fold superfamilies according to SCOP (Murzin et al., 1995). Structural annotation of these complete genomes was performed by automatic homology search algorithms (such as hidden Markov models), and put in public databases, such as Superfamily (Apic et al., 2001) and Gene3D (Buchan et al., 2002). These databases contain the number, type and position of protein domains along the chromosomes for every completed genome. Protein domain content data could thus be used to study phylogeny.

Similar to gene-content methods, the method based on protein domain content recon-struct phylogenetic trees from the distance between organisms, where the distance is calculated from the proportion of shared protein domains between genomes (Yang et al., 2005). A neighbor-joining (NJ) phylogenetic tree for a total of 174 taxa (19 archaea, 119 bacteria, and 36 eukaryotes) readily groups all organisms into the three superkingdoms

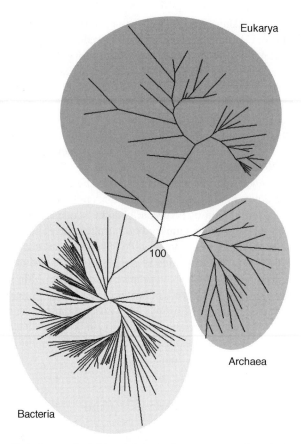

Figure 23.1. Overall phylogeny (neighbor-joining) of 174 organisms for which complete genomes have been determined. Bootstrap number was limited to the major branch points.

with high bootstrap values, in accordance with the phylogeny based on the comparison of SSU rRNA (Figure 23.1). The major phyla within each superkingdom were also recovered, although better phylogenies were generated when the taxa were restricted to a single superkingdom.

The NJ tree of 36 eukaryotes based on the presence and absence of SCOP fold superfamilies contained the major clades, animals, plant, and fungi. The groupings within each clade were mostly correct. The 19 archaea genomes were readily divided into the 4 Crenarchaeota and 15 Euryarchaeota. The small-genomed and enigmatic Nanoarchaeum equitans, reportedly the only known archaeal parasite, appeared near the root of the branch leading to the Pyrococci. This method was able recover the monophyly of most of the principle bacterial groups defined by classical taxonomy, including Actinobacteria, Cyanobacteria, Proteobacteria, Frimicutes, and Chlamydiae, and so on, although the relationship between groups was still unresolved. One major anomaly was the grouping of parasitic organisms that have extremely reduced genomes. This so called "big genome attraction" artifact (Lake and Rivera, 2004) is caused by massive gene loss in certain genomes that induce great differences in genome size and gene content among bacteria. When an empirical weighting factor aimed to compensate for the artifact was used, the small-genome taxa retreated to their correct locations along with their full-sized counterparts (Yang et al., 2005).

In this approach, the mere presence or absence of protein domains in genomes more accurately reconstructed most of the phylogenies than that constructed using the overall abundance of each domain in a genome. Domain abundance is greatly affected by gene and chromosome duplication, which is a contributor to the evolutionary distance between genomes, yet not a uniform process. As a result, excessive duplication can lead to inflated distances that mask the more crucial differences in the form of gain or loss of individual domains. The protein domain content of a given genome is changed whenever (i) a new fold evolves during a long-term divergence, (ii) a fold is lost as a result of deletion of all or part of a gene, or (iii) a new fold is acquired by horizontal transfer. Ordinarily, gene duplication on its own does not give rise abruptly to new folds.

Why is protein domain content better than gene content in constructing phylogenies? One answer is that proteins (gene products) are modular, and many of them are mosaics of different domains. Indeed, duplicated and/or shuffled domains are fundamental to establishing functional diversity. Genes may be retained even when the domain content changes and vice versa. Certainly protein domain content measures evolutionary change differently from gene content.

There are other intrinsic advantages in using the simple presence or absence of a structural attribute for phylogenetic purposes. For one, there is less concern about mistaken paralogy as so often occurs when comparing protein sequences. Moreover, the rate of sequence change and its attendant problems of site-specific variation do not play a role and arbitrary decisions about gene designation and function are not issues. Lastly, as we have seen above, three-dimensional structures are more highly conserved than primary sequences, allowing one to see further into the evolutionary past.

In summary, a simple scheme that uses only the presence or absence of protein domains in genomes was able to reconstruct the phylogenetic relationship of 174 organisms spanning the tree of life, achieving comparable results to those methods using sophisticated sequence analysis and/or combinations of gene content and gene order. This approach demonstrated that structural information can be used to study evolution.

Variations of this approach have also been tested by other groups. Instead of SCOP, the Pfam protein domain and family collection, based purely on sequence comparison has been used (Bentley and Parkhill, 2004). In addition to the presence or absence of single domains, combinations of domains or domain organizations within a protein are considered as structural attributes to reconstruct the tree of life (Wang and Caetano-Anolles, 2006; Fukami-Kobayashi et al., 2007). In general these methods achieve comparable results to those methods based on protein domain content.

THE LAST UNIVERSAL COMMON ANCESTOR (LUCA)

In 1977 Carl Woese used 16s rRNA to show cellular life can be divided into three superkingdoms: the eukaryotes, bacteria, and archaea (Woese and Fox, 1977). The archaea were initially believed to be the oldest superkingdom. Thirty years later the relationship between these three groups still remains open to debate. At the heart of this debate lies the origin of the eukaryotes, which appear to be a mix of archaea and bacteria. There are numerous theories for the origin of the eukaryotes (reviewed in (Embley and Martin, 2006)). Most of these theories agree that the eukaryotes originated from a symbiosis of an archaea and an α-proteobacteria (the mitochondrial ancestor). They disagree on what the archaea was and where the archaea descend from the bacteria. The sequencing of hundreds of genomes was supposed to generate enough sequence data to decipher the relationship between the three superkingdoms, but the debate continues.

Work based upon domain combinations has further fuelled the debate; some trees show the eukaryotes are more closely related to the bacteria than archaea (Wang and Caetano-Anolles, 2006), while others show the archaea and eukaryotes are sister clades (Fukami-Kobayashi et al., 2007). Despite this disagreement structure holds promise for ending this debate. The mitochondrial transfer has left a stronger sequence signal than the archaeal ancestor of the eukaryotes (Pisani et al., 2007), so sequence based methods will continue to be biased towards results which place the eukaryotes closer to the bacteria. Structure, on the contrary, is not biased in this way.

It has recently been argued that trees are the wrong representation for evolutionary histories given the large number of horizontal transfers within the prokaryotes (Doolittle and Bapteste, 2007). The assumption of a tree structure may be the reason the relationship between the three superkingdoms is not clear. If we use superfamilies as structure representatives, there are superfamilies that the eukaryotes acquired from mitochondria and some that they acquired from archaea. This moves the eukaryotes closer to both prokaryotic superkingdoms in a tree, but blurs which bacteria and archaea contributed superfamilies to the eukaryotes. In a network representation, the eukaryotic root would have two major branches; folds contributed from the archaea and others from the bacteria. Horizontal transfer of superfamilies will have some effect on tree reconstruction, but this effect should be lessened by using structure instead of sequence data. Sometimes the transfer of superfamilies does not matter because the receiving species already had a copy of that superfamily in another gene. This becomes more of a problem when dealing with domain combinations, as it is less likely that a particular domain combination is already in the recipient genome.

Previous studies have attempted to predict the gene content of the last universal common ancestor (LUCA) (Ouzounis et al., 2006), here we present findings of a different approach using structural information. Since structure is more conserved than sequence it should be easier to construct LUCA's domain content than gene content. Obviously any structural domains that are universal across cellular life were in LUCA. There are about 40 totally universal superfamilies (Yang et al., 2005). Many genes that were present in LUCA have been lost in many species so the universal set would greatly undercount LUCA's domain content. A recent study estimated that LUCA had at minimum 140 superfamilies (Ranea et al., 2006). They considered a superfamily to be in LUCA if it was in 90% of all extant species, and in at least 70% of the archaea and 70% of the eukaryotes.

The problem with this approach is the lack of a defined relationship between the three superkingdoms. A domain's presence in the archaea and eukaryotes is irrelevant to that domain being in LUCA if both of these superkingdoms are derived from bacteria. Any proposed root for the tree of life infers a LUCA fold set. For example, if LUCA was chloroflexus like, then LUCA's fold set can be accurately estimated using parsimony methods and a tree of the chloroflexus species. Folds that were present in LUCA are not necessarily essential. LUCA could not live as a parasite because there would be no host for it, so parasitic bacteria are free to lose structures that were essential in LUCA. Without careful assumptions about structure loss and the relationship of the three superkingdoms any LUCA fold set will be misleadingly small. LUCA probably had a wide repertoire of protein superfamilies, which infers a significant amount of protein evolution occurred before the emergence of any of the three superkingdoms.

ANCIENT GEOCHEMICAL ENVIRONMENT REFLECTED BY THE MODERN STRUCTURE REPERTOIRE

During evolution the genesis of novel protein domain was constrained by the geochemical environment of that moment, such as the temperature, pH, redox environment, element availability, and so on. It has recently been suggested that these constraints are reflected in some structural features of proteins and observed in the current structure repertoire.

Disulfide bonds are covalent bonds formed between two cysteine residues, which, in addition to other intramolecular interactions, can stabilize structural domains and contribute to the variability of structural space. However, the disulfide bond itself is volatile under reducing conditions. Since the oxygen content of the earth's atmosphere was gradually increasing during evolution, we can expect a correlation between the emergence of disulfide bond-dependent domains and the evolution of the earth's environment (Yang, 2007).

The divergence of the three superkingdoms was estimated to be at about 1.8–2.2 billion years ago (Doolittle et al., 1996), which is approximately the time when the oxygen content became significant. The emergence of disulfide bond-containing domains can be illustrated by a Venn diagram that contains the numbers and percentages of disulfide bond-containing domains in each superkingdom (Figure 23.2). Only 4.7% of the folds common to all superkingdoms contain disulfide bonds, which may have originated before the divergence of the three superkingdoms. By contrast, 31.9% of the domains unique to eukaryotes are disulfide bond-containing domains. The result largely confirms that most folds containing disulfide bonds formed after the oxygen level had increased in the atmosphere.

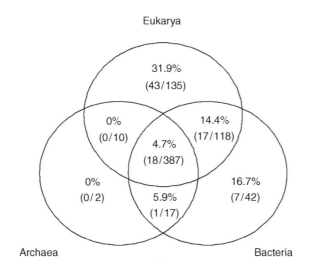

Figure 23.2. Numbers and percentages of disulfide bond-dependent domains in each super-kingdom. The two numbers in the parenthesis are the number of disulfide bond-containing fold and total fold within each region, respectively. SCOP (version 1.63) contains 765-folds, of which 708-folds are found in at least one species in this study, and the remaining 57-folds are mainly virus-related.

Another example of geochemical properties that influenced the evolution of proteins comes in the form of metal ions that are incorporated into a protein's 3D structure (Dupont et al., 2006). Many protein domains require the existence of metal ions, such as Fe, Zn, Mn, and so on, in order to fold and function properly. Sometimes, the same protein fold can accommodate different types of metal ions. Similar to the disulfide bonds, the emergence and evolution of those metal-containing domains are expected to correlate with changes in ion availability and redox state within the ancient ocean. This is seen in the distribution of modern metal-containing domains in whole taxa. In a study by Dupont et al. (2006), a correlation was observed between the proportion of metal-containing proteins in each of the three superkingdoms and the trace metal bioavailability in the ancient ocean at the time each superkingdom emerged. Overall this indicates a major evolutionary shift in biological metal ion usage, both in how a specific metal ion is used and which metal ion is used. For instance, eukaryotes have significantly more folds that incorporate zinc, which is consistent with their evolution in a more oxygen-rich and hence zinc-rich environment. Likewise, iron has moved from a predominance of iron-sulfur clusters, stable in the redox conditions of early earth, to heme-like structures more stable under oxidizing conditions. It is interesting to note that while the earth's geochemistry and life on earth are to some degree symbiotic, these subjects are rarely studied together.

THE EVOLUTIONARY HISTORY OF PROTEIN DOMAINS

As evolutionary units used in a very granular way, protein structural domains are able to decipher phylogenetic relationships among species in the tree of life, address questions concerning the origin of the three superkingdoms and explore the impact of the ancient geochemical environment on evolution. At a more detailed level, the evolution of each individual protein domain and the formation of the protein domain repertoire are also of great interest. The evolutionary origins of protein domains, identification of protein domain loss, transfer, duplication and combination with other domains to form new proteins make up the history of protein domains. The influence of the evolution of domains on the evolution of proteins and functions as well as the organisms as a whole, remain fundamental and challenging topics in evolutionary biology.

The evolution of protein domains consists of two different but related aspects: the changes in protein domains themselves, and the presence or absence of domain and domain combinations in individual genomes. The former includes important events such as the innovation of new domains and the gradual changes in the sequences, structures and functions of protein domains during evolution. Although structure is more conserved than sequence, during long term evolution, local insertions/deletions/substitutions, circular permutations, and rearrangements can gradually change the structure and give rise to protein folds with more structural variation, as indicated in Grishin's work on the gradual evolutionary path between different structures (Grishin, 2001).

Scheeff and Bourne investigated the evolutionary history of the protein kinase-like superfamily, which contains a variety of kinases that phosphorylate different substrates and play important roles in all three superkingdoms of life (Scheeff and Bourne, 2005). The comparison of the superfamily through a structural alignment revealed a "universal core" domain consisting only of regions required for ATP binding and the phosphotransfer reaction. Substantial structural and sequence revisions over long evolutionary timescales, mainly to accommodate different substrates, were identified and used to construct a phylogenetic tree of the protein kinase-like superfamily. Surprisingly, it was found that atypical protein kinases (those that phosphorylate nonproteins) emerged early as a group and were not derived through divergence from individual typical protein kinase families. Structural features such as conserved hydrogen bonding patterns found in the analyzed kinases amidst significant rearrangements of secondary structural components, provided insights not achievable from sequence alone.

Evolutionary events such as the duplication of a domain or domain combination in a genome, domain loss, domain transfers between species, will change the genomic content of domains or domain combinations, but not their identities. In the last 10 years, with the accumulation of complete genomes and the improvement of homology detection algorithms, scientists have started to investigate the distribution of protein domains in the three superkingdoms of life and other aspects of protein domain evolution (Doolittle, 1995a; Doolittle, 1995b; Copley et al., 2002; Koonin et al., 2002; Ponting and Russell, 2002).

Cyrus Chothia, Sarah Teichmann and their colleagues investigated the existence of multi-domain proteins in the three superkingdoms of life and estimated that 2/3 of prokaryote proteins have two or more domains, whereas 4/5 of proteins in eukaryotes are multi-domain (Teichmann et al., 1998). In 2001, the investigation of domain combinations in 40 genomes showed a power-law distribution of the tendency of domains to form combinations (Apic et al., 2001); some two-domain or three-domain combinations frequently recur in different protein contexts, which were called "supra-domains" (Vogel et al., 2004). A simulation of the processes of domain duplication and combination suggests that domain combinations are stochastic processes followed by duplication to varying extents (Vogel et al., 2005). During the evolution of domains, gene fusion is more common than gene fission (Kummerfeld and Teichmann, 2005), and convergent evolution is a rare event (Gough, 2005). A recent analysis suggested that the abundance of protein domains and domain combinations are correlated with the complexity of the organism, as characterized by the numbers of cell types each organism contains (Vogel and Chothia, 2006).

Christine Orengo and colleagues approach the topic of domain combination using the CATH (Orengo et al., 1997; Orengo et al., 2002) protein classification scheme and the Gene3D genomic domain assignment database (Chapter 22). They found a correlation between domain abundance and genome size (Ranea et al., 2004). The abundances of some domains within a genome are independent of genome size, others are proportional, and still others are nonlinearly distributed. Each type of abundance is roughly correlated with function, namely: protein translation and biosynthesis; metabolism; and gene regulation, respectively. For further reading on domain rearrangement the work of Arne Elfsson and his group is a good place to start: for domain recombination (Bjorklund et al., 2005; Ekman et al., 2005); and for duplication (Bjorklund et al., 2006).

FILLING IN FOLD SPACE: CURRENT LIMITATIONS

Many of the methods discussed in this chapter rely on the ability to predict the presence of fold superfamilies in a genome. The Superfamily database only predicts superfamilies for which there is a solved structure, so coverage is limited. For example, the human genome has at least one domain assignment for 63% of its proteins, but only 38% of complete protein sequence is assigned (many proteins have multiple domains). The unassigned regions are considered to be gaps. Methods that use domain combinations are going to be less accurate due to the large number of gaps. The key to improving trees built from this data is to decrease the number of domain combinations that have gaps. Recent work has argued that SCOP already covers the vast majority of fold space (Levitt, 2007). It is possible that many of these gaps represent known superfamilies whose sequences are too far from any known structure. However, the PDB and SCOP are both biased towards structures that can be crystallized and are missing many membrane folds. Therefore the Superfamily database will continue to be unable to assign large portions of the genome as it is unlikely that structures of new superfamilies to fill these gaps are going to be solved in the near future. The Pfam database (Finn et al., 2006) based on sequence has much better coverage, but lacks distant homologues as seen from structure. Trees built from a mixture of Superfamily and Pfam domains may be a good solution for furthering our understanding of evolution.

CONCLUSION

Protein domains are structural building blocks that define the function of a protein through their various combinations; they are also the units that encode the evolutionary history of proteins as well as genomes that contain them. Illustrating the entire evolutionary history of each protein domain, from the genesis of a new domain or domain combination to the loss and transfer of a domain from a specific genome, can further our understanding of some of the fundamental unsolved problems in evolution, such as events in the early evolution of life.

REFERENCES

Apic G, Gough J, et al. (2001): Domain combinations in archaeal, eubacterial and eukaryotic proteomes. *J Mol Biol* 310(2):311–325.

Bentley SD, Parkhill J (2004): Comparative genomic structure of prokaryotes. *Annu Rev Genet* 38:771–792.

Berman HM, Westbrook J, et al. (2000): The protein data bank. *Nucleic Acids Res* 28(1):235–242.

Bjorklund AK, Ekman D, et al. (2005): Domain rearrangements in protein evolution. *J Mol Biol* 353(4):911–923.

Bjorklund AK, Ekman D, et al. (2006): Expansion of protein domain repeats. *PLoS Comput Biol* 2(8): e114.

Branden C-I, Tooze J (1999): *Introduction to Protein Structure*. New York: Garland Publishing.

Buchan DW, Shepherd AJ, et al. (2002): Gene3D: structural assignment for whole genes and genomes using the CATH domain structure database. *Genome Res* 12(3):503–514.

Burley SK, Almo SC, et al. (1999): Structural genomics: beyond the human genome project. *Nat Genet* 23(2):151–157.

Caetano-Anolles G, Caetano-Anolles D (2003): An evolutionarily structured universe of protein architecture. *Genome Res* 13(7):1563–1571.

Chandonia JM, Brenner SE (2006): The impact of structural genomics: expectations and outcomes. *Science* 311(5759):347–351.

Copley RR, Doerks T, et al. (2002): Protein domain analysis in the era of complete genomes. *FEBS Lett* 513(1):129–134.

Dandekar T, Snel B, et al. (1998): Conservation of gene order: a fingerprint of proteins that physically interact. *Trends Biochem Sci* 23(9):324–328.

Delsuc F, Brinkmann H, et al. (2005): Phylogenomics and the reconstruction of the tree of life. *Nat Rev Genet* 6(5):361–375.

Doolittle RF (1995a): The multiplicity of domains in proteins. *Annu Rev Biochem* 64:287–314.

Doolittle RF (1995b): Of archae and eo: what's in a name? *Proc Natl Acad Sci U S A* 92(7):2421–2423.

Doolittle RF, Feng DF, et al. (1996): Determining divergence times of the major kingdoms of living organisms with a protein clock. *Science* 271(5248):470–477.

Doolittle WF, Bapteste E (2007): Pattern pluralism and the tree of life hypothesis. *Proc Natl Acad Sci U S A* 104(7):2043–2049.

Dupont CL, Yang S, et al. (2006): Modern proteomes contain putative imprints of ancient shifts in trace metal geochemistry. *Proc Natl Acad Sci U S A* 103(47):17822–17827.

Ekman D, Bjorklund AK, et al. (2005): Multi-domain proteins in the three kingdoms of life: orphan domains and other unassigned regions. *J Mol Biol* 348(1):231–243.

Embley TM, Martin W (2006): Eukaryotic evolution, changes and challenges. *Nature* 440 (7084): 623–630.

Finn RD, Mistry J, et al. (2006): Pfam: clans, web tools and services. *Nucleic Acids Res* 34(Database issue):D247–D251.

Flaherty KM, McKay DB, et al. (1991): Similarity of the three-dimensional structures of actin and the ATPase fragment of a 70-kDa heat shock cognate protein. *Proc Natl Acad Sci U S A* 88(11): 5041–5045.

Forterre P, Philippe H (1999): Where is the root of the universal tree of life? *Bioessays* 21(10): 871–879.

Fukami-Kobayashi K, Minezaki Y, et al. (2007): A tree of life based on protein domain organizations. *Mol Biol Evol* 24(5):1181–1189.

Gerstein M (1997): A structural census of genomes: comparing bacterial, eukaryotic, and archaeal genomes in terms of protein structure. *J Mol Biol* 274(4):562–576.

Gough J, Chothia C (2002): SUPERFAMILY: HMMs representing all proteins of known structure. SCOP sequence searches, alignments and genome assignments. *Nucleic Acids Res* 30 (1):268–272.

Gough J (2005): Convergent evolution of domain architectures (is rare). *Bioinformatics* 21(8): 1464–1471.

Govindarajan S, Recabarren R, et al. (1999): Estimating the total number of protein folds. *Proteins* 35(4): 408–414.

Grishin NV (2001): Fold change in evolution of protein structures. *J Struct Biol* 134(2–3):167–185.

House CH, Fitz-Gibbon ST (2002): Using homolog groups to create a whole-genomic tree of free-living organisms: an update. *J Mol Evol* 54(4):539–547.

Koonin EV, Wolf YI, et al. (2002): The structure of the protein universe and genome evolution. *Nature* 420(6912):218–223.

Kummerfeld SK, Teichmann SA (2005): Relative rates of gene fusion and fission in multi-domain proteins. *Trends Genet* 21(1):25–30.

Lake JA, Rivera MC (2004): Deriving the genomic tree of life in the presence of horizontal gene transfer: conditioned reconstruction. *Mol Biol Evol* 21(4):681–690.

Levitt M (2007): Growth of novel protein structural data. *Proc Natl Acad Sci U S A* 104(9): 3183–3188.

Lin J, Gerstein M (2000): Whole-genome trees based on the occurrence of folds and orthologs: implications for comparing genomes on different levels. *Genome Res* 10(6):808–818.

Marsden RL, Lewis TA, et al. (2007): Towards a comprehensive structural coverage of completed genomes: a structural genomics viewpoint. *BMC Bioinform* 8:86.

Mayr E (1998): Two empires or three? *Proc Natl Acad Sci U S A* 95(17):9720–9723.

Murzin AG, Brenner SE, et al. (1995): SCOP: a structural classification of proteins database for the investigation of sequences and structures. *J Mol Biol* 247(4):536–540.

Orengo CA, Michie AD, et al. (1997): CATH–a hierarchic classification of protein domain structures. *Structure* 5(8):1093–1108.

Orengo CA, Bray JE, et al. (2002): The CATH protein family database: a resource for structural and functional annotation of genomes. *Proteomics* 2(1):11–21.

Ouzounis CA, Kunin V, et al. (2006): A minimal estimate for the gene content of the last universal common ancestor–exobiology from a terrestrial perspective. *Res Microbiol* 157(1): 57–68.

Pastore A, Lesk AM (1990): Comparison of the structures of globins and phycocyanins: evidence for evolutionary relationship. *Proteins* 8(2):133–155.

Philippe H, Delsuc F, et al. (2005): Phylogenomics. *Annu Rev Ecol Evol Syst* 36:541–562.

Pisani D, Cotton JA, et al. (2007): Supertrees disentangle the chimeric origin of eukaryotic genomes. *Mol Biol Evol.* 24(8):1752–1760.

Ponting CP, Russell RR (2002): The natural history of protein domains. *Annu Rev Biophys Biomol Struct* 31:45–71.

Ranea JA, Buchan DW, et al. (2004): Evolution of protein superfamilies and bacterial genome size. *J Mol Biol* 336(4):871–887.

Ranea JA, Sillero A, et al. (2006): Protein superfamily evolution and the last universal common ancestor (LUCA). *J Mol Evol* 63(4):513–525.

Scheeff ED, Bourne PE (2005): Structural evolution of the protein kinase-like superfamily. *PLoS Comput Biol* 1(5):e49.

Schnuchel A, Wiltscheck R, et al. (1993): Structure in solution of the major cold-shock protein from *Bacillus subtilis*. *Nature* 364(6433):169–171.

Snel B, Bork P, et al. (1999): Genome phylogeny based on gene content. *Nat Genet* 21(1):108–110.

Snel B, Huynen MA, et al. (2005): Genome trees and the nature of genome evolution. *Annu Rev Microbiol* 59:191–209.

Teichmann SA, Park J, et al. (1998): Structural assignments to the *Mycoplasma genitalium* proteins show extensive gene duplications and domain rearrangements. *Proc Natl Acad Sci U S A* 95(25): 14658–14663.

Tekaia F, Lazcano A, et al. (1999): The genomic tree as revealed from whole proteome comparisons. *Genome Res* 9(6):550–557.

Vogel C, Berzuini C, et al. (2004): Supra-domains: evolutionary units larger than single protein domains. *J Mol Biol* 336(3):809–823.

Vogel C, Teichmann SA, et al. (2005): The relationship between domain duplication and recombination. *J Mol Biol* 346(1):355–365.

Vogel C, Chothia C (2006): Protein family expansions and biological complexity. *PLoS Comput Biol* 2(5):e48.

Wang M, Caetano-Anolles G (2006): Global phylogeny determined by the combination of protein domains in proteomes. *Mol Biol Evol* 23(12):2444–2454.

Woese CR, Fox GE (1977): Phylogenetic structure of the prokaryotic domain: the primary kingdoms. *Proc Natl Acad Sci U S A* 74(11):5088–5090.

Woese CR (1998a): The universal ancestor. *Proc Natl Acad Sci U S A* 95 (12):6854–6859.

Woese CR (1998b): Default taxonomy: Ernst Mayr's view of the microbial world. *Proc Natl Acad Sci U S A* 95(19):11043–11046.

Wolf YI, Brenner SE, et al. (1999): Distribution of protein folds in the three superkingdoms of life. *Genome Res* 9(1):17–26.

Wolf YI, Grishin NV, et al. (2000): Estimating the number of protein folds and families from complete genome data. *J Mol Biol* 299(4):897–905.

Wolf YI, Rogozin IB, et al. (2001): Genome trees constructed using five different approaches suggest new major bacterial clades. *BMC Evol Biol* 1:8.

Wolf YI, Rogozin IB, et al. (2002): Genome trees and the tree of life. *Trends Genet* 18(9):472–479.

Xie L, Bourne PE (2005): Functional coverage of the human genome by existing structures, structural genomics targets, and homology models. *PLoS Comput Biol* 1(3):e31.

Yang S, Doolittle RF, et al. (2005): Phylogeny determined by protein domain content. *Proc Natl Acad Sci U S A* 102(2):373–378.

Yang S (2007): *Evolution Studied Through Protein Structural Domains.* San Diego: Department of Chemistry and Biochemistry, La Jolla, University of California, p 174.

Zhang CT (1997): Relations of the numbers of protein sequences, families and folds. *Protein Eng* 10(7):757–761.

Zuckerkandl E, Pauling L (1965): Molecules as documents of evolutionary history. *J Theor Biol* 8(2): 357–366.

Section V

MACROMOLECULAR INTERACTIONS

24

ELECTROSTATIC INTERACTIONS

Nathan A. Baker and J. Andrew McCammon

INTRODUCTION

An understanding of electrostatic interactions is essential for the full development of structural bioinformatics. The structures of proteins and other biopolymers are being determined at an increasing rate through structural genomics and other efforts. Specific linkages of these biopolymers in cellular pathways or supramolecular assemblages are being detected by genetic and other experimental efforts. To integrate this information in physical models for drug discovery or other applications requires the ability to evaluate the energetic interactions within and among biopolymers. Among the various components of molecular energetics, the electrostatic interactions are of special importance due to the long range of these interactions and the substantial charges of typical components of biopolymers. Indeed, electrostatics can be used to help assign biopolymers, such as proteins to functional families, since particular kinds of ligand binding sites may be indicated by the spatial distribution of the charges in the proteins.

In what follows, we provide a brief overview of the roles of electrostatics in biopolymers and supramolecular assemblages, and then outline some of the methods that have been developed for analyzing electrostatic interactions.

OVERVIEW OF FUNCTIONAL ROLES OF ELECTROSTATICS

Electrostatic interactions help to determine the structure and flexibility of biopolymers, and the strength and kinetics of their associations with small molecules, other biopolymers, and biological membranes. Such interactions are of obvious importance for highly charged biomolecules such as nucleic acids, sugars, and anionic lipids. However, proteins are also rich in charged groups, and the cumulative contributions to the electrostatic potential of a protein from its dipolar groups (such as the peptide linkages) can be substantial.

Structural Bioinformatics, Second Edition   Edited by Jenny Gu and Philip E. Bourne
Copyright © 2009 John Wiley & Sons, Inc.

In typical *in vitro* settings, biopolymers are immersed in a solution comprising water and small, diffusible ions. The high dielectric coefficient of water, together with the tendency of diffusible ions to move toward biopolymer charges of opposite sign, reduces the effective interactions among the biopolymer charges. Nevertheless, these "solvent-screened" interactions strongly influence biopolymer behavior, especially within the physiological "Debye length" of about 1 nm. For biopolymers such as DNA that have high charge densities, counterions "condense" near the surface of the biopolymer. The resulting effective charge of DNA, for example, is about 25% of what it would be in the absence of condensation.

The general tendencies of charges to prefer a high dielectric environment (due to favorable free energy of solvation), and of opposite charges to attract, are reflected in the structures of most globular proteins: charged side chains are typically at the surface of the protein, and the relatively few buried charges often are "salt bridged" with opposite charges. Similar principles influence the structure and thermodynamics of protein–protein complex formation. Although the advantage of ion pairing in the formation of protein folds or complexes is substantially offset by the disadvantage of loss of aqueous solvation, the thermodynamic penalty of charge desolvation dictates that ion pairing and other favorable electrostatic interactions within or between proteins are common features of protein structure.

For kinetics, it has been firmly established that the rates of association of many biopolymers with one another or with small ligands are greatly increased by "electrostatic steering" of the diffusional encounters. This is commonly observed in situations where an evolutionary advantage has likely been conferred by great speed. Even with the combined dielectric and ionic screening expected in a typical physiologic (150 mM ionic strength) solution, electrostatic steering effects can lead to increases in the rate constant of association by two orders of magnitude.

BRIEF HISTORY

The importance of electrostatic interactions in protein behavior was recognized early in the twentieth century by Linderstrom-Lang, who introduced a simple spherical model for protein titration in 1924 (Linderstrom-Lang, 1924). In this model, the protein was regarded as impenetrable, and the charges of the acidic and basic groups were treated as being uniformly distributed on the surface of the protein sphere. Thus, substantial cancellation of charge occurred, and the work of charging the protein sphere was approximated as the self-interaction energy of the net charge on the spherical surface. During subsequent decades, more detailed models that retained the approximation of spherical symmetry were developed. The first model that included discrete locations for the interacting charges, still located within a spherical body but now including such features as dielectric heterogeneity and a nonzero ionic strength, was presented by Tanford and Kirkwood in 1957 (Tanford and Kirkwood, 1957). Such models were used to account for the titration properties of proteins, the effects of pH and ionic strength on the activity of enzymes, and—as late as 1981, in work by Flanagan et al. (1981)—the electrostatic contributions to the energetics of dimer–tetramer assembly in hemoglobin.

A new era of electrostatic models was ushered in by a 1982 study by Warwicker and Watson (1982). Drawing upon the increased knowledge of the three-dimensional structure of proteins, and especially on increased computer power, Warwicker and Watson introduced a grid-based finite difference approach for calculating the electrostatic potential of a nonspherical protein. The interior of the protein had a dielectric coefficient of 2, and the surrounding solvent had a dielectric coefficient of 80. This work provided the first hints concerning the possible functional importance of the shaping of the electrostatic potential and its gradient by the topography of the protein. Zauhar and Morgan introduced a boundary element approach for the analysis of this model in 1985 (Zauhar and Morgan, 1985). An important advance was described by Klapper et al. in 1986 (Klapper et al., 1986), who allowed for the inclusion of ionic strength effects by finite difference solution of the linearized Poisson–Boltzmann equation (PBE).

A much simpler model for describing electrostatic contributions to solvation energies and forces was introduced by Still et al. in 1990 (Still et al., 1990). This method is based on the Born ion, a canonical electrostatics model problem describing the electrostatic potential and solvation energy of a spherical ion (Born, 1920). The generalized Born method of Still et al. uses an analytical expression based on the Born ion model to approximate the electrostatic potential and solvation energy of small molecules. Although it fails to capture all the details of molecular structure and ion distributions provided by more rigorous models, such as the PBE, it has gained popularity as a very rapid method for evaluating approximate forces and energies for solvated molecules and continues to be vigorously developed.

The kinetic effects of electrostatics in steering biomolecular encounters are usually studied in the context of the diffusion equation, since the motions of the solutes are overdamped. The most detailed studies make use of the Brownian dynamics simulation method of Ermak and McCammon (1978), which allows for structure and flexibility of the biomolecules, and hydrodynamic, as well as potential-derived interactions. Rate constants for diffusion-controlled encounters of a protein with other small or large molecules can be determined by simulating their Brownian motion and analyzing their trajectories using a procedure introduced by Northrup, Allison and McCammon (1984).

THE NEED FOR MORE EFFICIENT AND SCALABLE ELECTROSTATICS METHODS

The era of structural bioinformatics has created an urgent need for faster methods to solve problems in biomolecular electrostatics. As the structures of more proteins and other biopolymers become available through the structural genomics and other initiatives, there will be a corresponding need to calculate the physical properties of these molecules to help assign them to families and functions. The need is even greater when one considers that any given biopolymer typically acts in concert with many others. Thus, there are combinatorial factors that increase the number of calculations that must be done, either to assess the thermodynamics of association of biopolymers, or—even more dramatically—to model the dynamics of association of such molecules, for example, with frequent updates of the electrostatic forces in the course of molecular or Brownian dynamics simulations.

Specific applications of biomolecular electrostatics are described in further detail in Section "Applications." All highlighted applications can be extended to high-throughput (e.g., analyzing large numbers of biomolecular binding partners) or large-scale supramolecular systems. However, such extension requires fast and scalable methods for solving the electrostatic equations. The remainder of this chapter outlines the corresponding theory and methods providing such scalable methods and illustrates recent progress in this area.

POISSON–BOLTZMANN THEORY

Although methods such as generalized Born have found uses in several aspects of structural bioinformatics, we will confine the remainder of this discussion to Poisson–Boltzmann types of methods because of their relatively rigorous framework for inclusion of biomolecular topology and ionic strength effects.

Introduction to the Equation

The canonical expression for the electrostatic potential in a continuum setting is the Poisson equation

$$-\nabla \cdot \varepsilon(\mathbf{x})\nabla\phi(\mathbf{x}) = \frac{1}{\varepsilon_0}\rho(\mathbf{x}), \qquad (24.1)$$

where $\varepsilon(\mathbf{x})$ is a spatially varying dielectric coefficient, $\phi(\mathbf{x})$ is the electrostatic potential (in units of V), $\varepsilon_0 = 8.854 \times 10^{-12}$ F/m is the permittivity of a vacuum, and $\rho(\mathbf{x})$ is the charge density (in units of C/m^3) that generates $\phi(\mathbf{x})$. The dielectric coefficient $\varepsilon(\mathbf{x})$ typically assumes different values inside the solute and in the bulk solvent to reflect the relative polarizabilities of the two media. For biomolecules in an aqueous environment, ε is generally given a value of 2–20 inside the solute and a value of 80 in the solvent. Figure 24.1a shows the traditional definition of $\varepsilon(\mathbf{x})$, which includes a jump discontinuity across the molecular surface while $\varepsilon(\mathbf{x})$ changes between the protein and solvent dielectric values. However, more recent work (Im, Beglov, and Roux, 1998; Grant, Pickup, and Nicholls, 2001; Schnieders et al., 2007) has proposed smoother definitions for $\varepsilon(\mathbf{x})$ to reduce artifacts arising from the rapidly changing coefficient.

Likewise, the charge distribution $\rho(\mathbf{x})$ has typically been given a very discontinuous definition, which can pose numerical difficulties for solution of the Poisson equation. In the

(a)   (b)

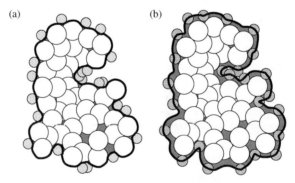

Figure 24.1. Popular definitions of the Poisson–Boltzmann equation coefficients. (a) The molecular surface (solid black line) often used to define the dielectric coefficient $\varepsilon(\mathbf{x})$. This surface can be constructed by rolling solvent probes (small hatched circles) over the macromolecule (large white spheres). Gray regions show areas outside the atomic volume that are treated as "inside" the molecular surface. (b) The ion-accessible volume is the region outside the solid black line. This volume is defined as the region of space accessible to ion probe spheres (hashed circles).

absence of mobile counterions, $\rho(\mathbf{x})$ is often treated as a collection of Dirac delta functions that model the N_f atomic partial charges of the solute:

$$\rho(\mathbf{x}) = \rho_f(\mathbf{x}) = \sum_i^{N_f} q_i \delta(\mathbf{x}-\mathbf{x}_i), \tag{24.2}$$

where q_i are the magnitudes of the atomic partial charges (in units of C), and $\mathbf{x}_i$ are the partial charge positions. The Dirac delta function is a point distribution function with the property $\int f(\mathbf{x})\delta(y-\mathbf{x})d\mathbf{x} = f(\mathbf{y})$ and hence has units of m^{-3}.

The PBE is a variant of the Poisson equation where mobile counterion charges are introduced to the charge distribution function such that $\rho(\mathbf{x}) = \rho_f(\mathbf{x}) + \rho_m(\mathbf{x})$, where $\rho_m(\mathbf{x})$ denotes the mobile charge distribution. Mean field, or Debye-Hückel, electrolyte theory describes the distribution of each counterion species i as $\rho_i(\mathbf{x}) = \bar{\rho}_i e^{-\beta(q_i\phi(\mathbf{x})-V_i(\mathbf{x}))}$, where $\bar{\rho}_i$ is the bulk concentration (in m^{-3}) of species i, $\beta = (k_BT)^{-1}$ is the inverse thermal energy, k_B is Boltzmann's constant, T is the temperature, $q_i\phi(\mathbf{x})$ is the electrostatic energy of placing a counterion with charge q_i at position $\mathbf{x}$ in the potential $\phi(\mathbf{x})$, and $V_i(\mathbf{x})$ is a steric energy function that prevents mobile charges from entering the interior of the solute. This representation allows the mobile charge distribution function for N_m counterion species to be written as

$$\rho_m(\mathbf{x}) = \beta \sum_i^{N_m} q_i \rho_i(\mathbf{x}) = \sum_i^{N_m} q_i \bar{\rho}_i e^{-\beta(q_i\phi(\mathbf{x})-V_i(\mathbf{x}))}. \tag{24.3}$$

In the case of a 1:1 monovalent ion distribution where the steric functions V_i are equal (e.g., $V_i = V$), Eq. 24.3 can be simplified to $\rho_m(\mathbf{x}) = -\bar{\kappa}^2(\mathbf{x})\sinh\phi(\mathbf{x})$, where $\bar{\kappa}^2(\mathbf{x}) = \kappa^2\varepsilon_s e^{-V(\mathbf{x})}$, ε_s is the dielectric constant of the bulk solvent, and κ is the Debye-Hückel parameter, defined for a general N_m-component electrolyte solution as $\kappa^2 = \frac{2}{\varepsilon_s\varepsilon_0}\sum_i^{N_m} \bar{\rho}_i q_i^2$. As illustrated in Figure 24.1b, the function $e^{-\beta V(\mathbf{x})}$ is usually treated as a discontinuous characteristic function equal to unity within an ion-accessible volume (typically slightly larger than the protein volume) and zero otherwise. The PBE for a 1:1 monovalent electrolyte is therefore

$$-\nabla \cdot \varepsilon(\mathbf{x})\nabla\phi(\mathbf{x}) + \bar{\kappa}^2(\mathbf{x})\sinh\phi(\mathbf{x}) = \frac{1}{\varepsilon_0}\sum_i^{N_f} q_i \delta(\mathbf{x}-\mathbf{x}_i). \tag{24.4}$$

For sufficiently small values of $\phi(\mathbf{x})$, the approximation $\sinh(\phi(\mathbf{x})) \sim \phi(\mathbf{x})$ is often applied to this equation to give the linearized PBE:

$$-\nabla \cdot \varepsilon(\mathbf{x})\nabla\phi(\mathbf{x}) + \bar{\kappa}^2(\mathbf{x})\phi(\mathbf{x}) = \frac{1}{\varepsilon_0}\sum_i^{N_f} q_i \delta(\mathbf{x}-\mathbf{x}_i). \tag{24.5}$$

All these equations are solved in conjunction with a Dirichlet boundary condition, which specifies the value of potential at the boundary of some domain. For a sufficiently large domain, this condition is typically zero or some asymptotic form of the solution, such as the Debye–Hückel potential.

The PBE can also be derived from statistical mechanics using a continuum representation of the solvent dielectric properties (Netz and Orland, 2000; Holm, Kekicheff and Podgornik, 2001). While such treatments are too complicated to present here, one important aspect of these derivations is the development of a field theoretic framework for electrostatic interactions with the construction of the PBE as the "saddle point" equation for the potential that minimizes a relevant electrostatic "action" or energy.

Energies

As discussed above, the PBE defines an electrostatic energy that can be derived from physical chemistry arguments (Sharp and Honig, 1990) or field theory saddle point approximations (Netz and Orland, 2000; Holm, Kekicheff and Podgornik, 2001). The free energy is a functional of the electrostatic potential as well as the atomic positions, charges, and radii. For a $1:1$ monovalent electrolyte, this functional has the form

$$G = \int \left[\rho_f \phi - \frac{\varepsilon}{2} (\nabla \phi)^2 - \bar{\kappa}^2 (\cosh \phi - 1) \right] d\mathbf{x}. \qquad (24.6)$$

The first term, $\int \rho_f \phi \, d\mathbf{x}$, is the energy of inserting the protein charges into the electrostatic potential and can be interpreted as the energy of interaction for the fixed charges. The second term, $-\frac{1}{2} \int \varepsilon (\nabla \phi)^2 d\mathbf{x}$, represents electrostatic stresses in the dielectric medium. Finally, the third term includes the effects of the mobile charge configuration and can be interpreted in terms of the excess osmotic pressure of the system. The subtraction of unity from the exponential in this term makes this an excess osmotic pressure and is necessary to cause the energy to vanish in the absence of a potential. Like the PBE, this energy expression can be linearized (see Eq. 24.5) for sufficiently small ϕ by assuming cosh $(\phi(\mathbf{x})) \sim 1 + \phi(\mathbf{x})^2/2$. This linearized form of the energy leads to an additional simplification; Gauss' Law allows the second term to be rewritten as $-\frac{1}{2} \int \varepsilon (\nabla \phi)^2 d\mathbf{x} = \frac{1}{2} \int \phi \nabla \cdot \varepsilon \nabla \phi d\mathbf{x}$ and gives two equivalent free energy expressions

$$G = \frac{1}{2} \int \rho_f \phi \, d\mathbf{x} = \frac{1}{2} \int \left[\varepsilon (\nabla \phi)^2 + \bar{\kappa}^2 \phi^2 \right] d\mathbf{x}. \qquad (24.7)$$

These free energy expressions can be used for a variety of static calculations on biomolecules, including the determination of binding constants, pK_as, and solvation energies. These calculations are typically performed from a series of Poisson–Boltzmann energy evaluations that are then analyzed by free energy cycles. Figure 24.2 shows the specific case of pK_a calculations, where the energy of protonating a functional group in a biomolecule is calculated in a stepwise fashion by determining the energies of the isolated biomolecule without the functional group, the isolated functional group in its protonated and unprotonated state, the biomolecule with the protonated functional group, and the biomolecules with the unprotonated functional group. These energies are then combined (as shown in Figure 24.2) to give the free energy of protonating the functional group in the biomolecular environment, which can be converted to a pK_a value. Similar cycles are used to calculate binding and solvation energies.

Forces

Poisson–Boltzmann calculations have also found an increasingly important role in force evaluation for implicit solvent dynamics simulations. In such simulations, the dynamical

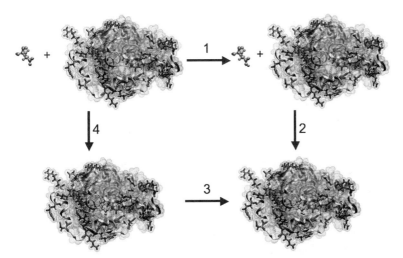

Figure 24.2. Titration calculation for a GLU 35 in the active site of hen egg white lysozyme (PDB ID 2LZT). The free energy of protonating GLU 35 (blue ball and stick moiety) in the presence of a biomolecule (cyan object) is calculated by a thermodynamic cycle. Specifically, the free energy of interest ΔG_3 is calculated in terms of the other steps in the cycle, ΔG_1 the energy of protonating the isolated GLU in solution (related to the "model pK_a"), ΔG_2 the energy of "binding" the protonated GLU to the lysozyme structure, and ΔG_4 the energy of "binding" the unprotonated GLU to the lysozyme structure, such that ΔG_3 can be calculated from a thermodynamic cycle as $\Delta G_3 = \Delta G_1 + G_2 - \Delta G_4$. Figure from *Reviews in Computational Chemistry*, N. Baker. Figure also appears in the Color figure section.

trajectory of a solute is calculated without the inclusion of the numerous explicit solvent molecules required for traditional molecular dynamics simulations. Instead, the solvent effects are modeled by stochastic forces applied to the biomolecule to mimic solvent buffeting, damping forces to provide the effect of solvent viscosity, continuum approximations of apolar interactions, and continuum electrostatics calculations (such as PBE) to include the effects of implicit solvent and salt on electrostatic forces in the solute.

To derive forces from the PBE, we simply differentiate the free energy with As mentioned previously, a potential ϕ, which solves the PBE, is a saddle point of G, that is, $\partial G/\partial \phi = 0$. Therefore, the force $\mathbf{F}_i$ on atom i can be written solely in terms of variations in the coefficients with respect to atomic displacements $\partial \mathbf{y}_i$

$$\mathbf{F}_i = -\int \left[\phi \left(\frac{\partial \rho_f}{\partial \mathbf{y}_i} \right) - \frac{1}{2} (\nabla \phi)^2 \left(\frac{\partial \varepsilon}{\partial \mathbf{y}_i} \right) - (\cosh \phi - 1) \left(\frac{\partial \bar{\kappa}^2}{\partial \mathbf{y}_i} \right) \right] d\mathbf{x}. \qquad (24.8)$$

The terms of the integrand in this force expression have the same interpretation as for the free energy. The first term is the force density for atomic displacements in the potential ϕ, the second is the dielectric boundary pressure on atom i, and the third is osmotic pressure on atom i.

The mechanics of evaluating atomic forces from Eq. 24.8 have been discussed in detail in the literature (Gilson et al., 1993; Im, Beglov and Roux, 1998). These authors present excellent reviews of this topic, including the effects of discontinuities in the PBE coefficients ε and $\bar{\kappa}^2$ on the methods for force evaluation and the accuracy of the numerical results.

NUMERICAL SOLUTION OF THE POISSON–BOLTZMANN EQUATION

Very few analytical solutions of the Poisson–Boltzmann equation exist for realistic biomolecular geometries and charge distributions. Therefore, this equation is usually solved numerically by a variety of computational methods. These methods typically rely on a discretization to project the continuous solution down onto a finite-dimensional set of basis functions. In the case of the linearized PBE (Eq. 24.5), the resulting equations are the usual linear matrix–vector equation, which can be solved directly. However, the nonlinear equations obtained from the full PBE require more specialized techniques, such as Newton methods, to determine the solution to the discretized algebraic Specifically, Newton methods start with an initial solution guess and iteratively improve this guess by solving related linear equations for corrections to the current solution. Newton methods, as well as other popular methods for solution of nonlinear equations, have been reviewed by Holst and Saied (1995).

Multilevel Finite Difference and Finite Element Methods

Some of the most popular discretization techniques employ Cartesian meshes to subdivide the domain in which the Poisson–Boltzmann equation is to be solved. Of these, the finite difference method has been at the forefront of PBE solvers. In its most general form, the finite difference method solves the PBE on a nonuniform Cartesian mesh, as shown in Figure 24.3a for a two-dimensional domain. While Cartesian meshes offer relatively simple problem setup, they provide little control over how unknowns are placed in the solution domain. Specifically, as shown by Figure 24.3a, the Cartesian nature of the mesh makes it impossible to locally increase the accuracy of the solution in a specific region without increasing the number of unknowns across the entire grid.

Differential operators for problems discretized by finite difference methods are typically approximated using Taylor expansions. For example, a discretized one-

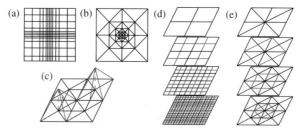

Figure 24.3. Meshes and hierarchies used in Poisson–Boltzmann solvers. (a) Cartesian mesh suitable for finite difference calculations; nonuniform mesh spacing can be used to provide a limited degree of adaptivity. (b) Finite element mesh exhibiting adaptive refinement. (c) Examples of typical piecewise linear basis functions used to construct the solution in finite element methods. (d) The multilevel hierarchy used to solve the PBE for a finite difference discretization; red lines denote the additional unknowns added at each level of the hierarchy. (e) The multilevel hierarchy used to solve the PBE for a finite element discretization; red lines denote simplex the subdivisions used to introduce additional unknowns at each level of the hierarchy. Figure also appears in the Color figure section.

dimensional Laplacian operator has the form

$$-\nabla^2 u(x_i) \approx \frac{-u(x_{i+1}) + 2u(x_i) - u(x_{i-1})}{h^2}, \tag{24.9}$$

where x_j denotes the grid point coordinates and h is the mesh spacing. Given vectors $\mathbf{u}$ and $\mathbf{f}$ representing the values of the solution and source terms at the grid points, it is straightforward to develop a matrix form of the problem $\mathbf{Au} = \mathbf{f}$, where $\mathbf{A}$ is a sparse symmetric matrix with 2 on the main diagonal, -1 on the first off-diagonal elements, and 0 elsewhere in the matrix. Discretization of the differential operator for the PBE yields a matrix with a similar sparse symmetric structure, but more nonzeros per row.

Unlike finite difference methods, adaptive finite element discretizations offer the ability to place computational effort on specific regions of the problem domain. Finite element meshes (see Figure 24.3b) are composed of simplices that are joined at edges and vertices. The solution is constructed from piecewise polynomial basis functions (Figure 24.3c) that are associated with mesh vertices and typically are nonzero only over a small set of neighboring simplices. Solution accuracy can be increased in specific areas by locally increasing the number of vertices through simplex refinement (subdivision). As shown in Figure 24.3b, the number of unknowns (vertices) is generally increased only in the immediate vicinity of the simplex refinement and not throughout the entire problem domain as in finite difference methods. This ability to locally increase the solution resolution is called "adaptivity" and is the major strength of finite element methods applied to the PBE (Holst et al., 2000; Baker, Holst, and Wang, 2000; Baker, et al., 2001a).

Typically, the algebraic system is assembled using a Galerkin discretization, where a "weak" form of the PBE is imposed for each basis function in the system. Specifically, the original differential form of the PBE is transformed by integration with a basis function v to give an integral equation

$$\int (\varepsilon \nabla \phi \cdot \nabla v + \bar{\kappa}^2 v \sinh\phi - \rho_f v) \mathrm{d}x = 0. \tag{24.10}$$

The algebraic system is implicitly assembled by representing the solution as a linear combination of finite element basis functions $\phi = \sum_i u_i \psi_i$ and imposing the weak PBE (Eq. 24.10) using every basis function ψ_j as test functions. As with the finite difference method, this discretization scheme leads to sparse symmetric matrices with a small number of nonzero entries in each row.

Multilevel solvers, in conjunction with the Newton methods described above, have been shown to provide most efficient solution of the algebraic equations obtained by discretization of the PBE with either finite difference or finite element techniques. Most sizable algebraic equations are solved by iterative methods that repeatedly apply a set of operations to improve an initial guess until a solution of the desired accuracy is reached. However, the speed of traditional iterative methods has been limited by their inability to quickly reduce low-frequency (long-range) error in the solution. Multilevel methods overcome this problem by projecting the discretized system onto meshes (or grids) at multiple resolutions (see Figure 24.3). The advantage of this multiscale representation is that the slowly converging low-frequency components of the solution on the finest mesh are quickly resolved on coarser levels of the system. This gives rise to a multilevel solver algorithm, where the algebraic system is solved directly on the coarsest level and then used to accelerate solutions on finer levels of the mesh.

As shown in Figure 24.3, the assembly of the multiscale representation, or "multilevel hierarchy," depends on the method used to discretize the PBE. For finite difference types of methods, the nature of the grid lends itself to the assembly of a hierarchy with little additional work. In the case of adaptive finite element discretizations, the most natural multiscale representation is constructed by refinement of an initial mesh that typically constitutes the coarsest level of the hierarchy.

Boundary Element Solvers

The finite difference and finite element methods described above have substantial computational requirements for large systems, due in part to the storage and computation of data for points that are distributed throughout the volume of the system. In principle, boundary element methods have the advantage of limiting the analysis to the molecular surfaces, so that substantially fewer grid points need be considered; the number of unknowns is therefore greatly reduced in the corresponding algebraic system. Boundary element methods have their own limitations, however. First, because of their physical formulation, they are limited to solving the linearized PBE. Second, the potential advantages in numerical efficiency were lost in early applications due to the use of Gaussian elimination in the treatment of the dense matrix in the algebraic system; storage and computational operations were scaled with the square and the cube of the number of boundary nodes, respectively. Fast multipole methods (FMM) offer a way to offset these inefficiencies by introducing reduced, collective descriptions of contiguous elements on distant boundaries. A recent implementation of FMM has achieved linear scaling in the number of boundary nodes, and additional efficiencies have been achieved by careful placement of nodes and boundary elements (see Lu et al., 2006; Lu and McCammon, 2007). Continuing efforts along these lines should be helpful in reducing the cost of electrostatic calculations in the context of high-throughput proteomics studies and simulations of interacting protein molecules with frequent updating of the electrostatic forces.

APPLICATIONS

Applications of biomolecular electrostatics methods can be found in many different areas of computational biophysics and structural bioinformatics. Note that, in all of these applications, electrostatics is but one component of the overall system that also includes nonpolar effects, conformational energetics, and entropic changes. The scope of our discussion will necessarily be limited to electrostatics; however, interested readers should refer to work by others for a more detailed description of binding phenomena (Kollman et al., 2000; Swanson, Henchman and McCammon, 2004; Woo and Roux, 2005). The following sections highlight some of the major applications of these methods to biomolecular thermodynamics, kinetics, and bioinformatics.

Thermodynamics

The application of electrostatics methods to thermodynamic processes in biophysics necessarily involves the calculation of free energies, as outlined in Section "Energies." In general, such applications are interested in *changes* in energy associated with a particular biomolecular transformation such as protonation, ligand binding, folding, or other conformational changes. There are far too many examples of these applications to discuss in detail, however, a few particular examples illustrate some of the basic principles.

As mentioned above (Section "Energies") and illustrated in Figure 24.2, electrostatics methods are commonly used to evaluate pK_a values for amino acid and nucleic acid residues in the context of specific biomolecular structures (Bashford and Karplus, 1990; Yang et al., 1993; Antosiewicz, McCammon, and Gilson, 1996). Such applications not only provide important insight into biomolecular function but also serve as a critical test for the accuracy of continuum electrostatics methods. As a result of continual comparison between calculated and experimental pK_a values, new methods for pK_a calculations have evolved to include important contributions from local hydrogen bonding (Nielsen and Vriend, 2001) and environment (Li, Robertson and Jensen, 2005), conformational flexibility (Georgescu et al., 2002), and large-scale rearrangement in biomolecular structures (Whitten and García-Moreno, 2005).

Biomolecular electrostatics can also provide insight into macromolecular assembly and function through analysis of protein–protein interactions as well as larger-scale assemblages. In particular, electrostatics calculations on macromolecular complexes demonstrate some of the demands for scalability and efficiency in electrostatics software. Some examples of such applications include analysis of microtubule electrostatics (Baker et al., 2001b) and proto-filament assembly (Sept, Baker and McCammon, 2003) (see Figure 24.4), viral capsid electrostatics (Zhang et al., 2004; Konecny et al., 2006), and ribosome electrostatic properties

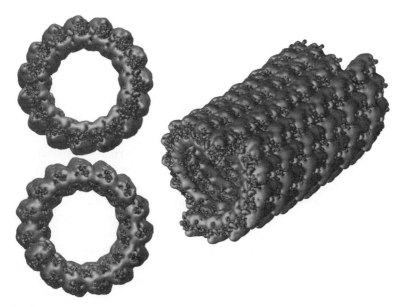

Figure 24.4. Microtubule electrostatics. (a) Electrostatic properties of a 1.2 million atom $400 \times 300 \times 300\,\text{Å}$ microtubule fragment illustrating the current state of the art for continuum electrostatics calculations. The potential was calculated using APBS to solve the PBE at 150 mM ionic strength; more details are available in Baker et al. (2001b). (b) A potential of mean force, calculated using continuum electrostatics and a simple nonpolar function, describing microtubule protofila-ment assembly; more details are available in Sept, Baker and McCammon (2003). Part B is from Sept, Baker, and McCammon. Figure also appears in the Color figure section.

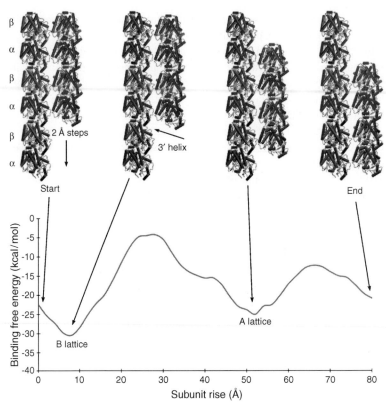

Figure 24.4. (*Continued*).

(Baker et al., 2001b; Trylska et al., 2004; Trylska et al., 2005). Other work on molecular assemblages investigates electrostatic properties of lipid bilayers. An excellent example of such membrane studies is provided in work by Murray and Honig (2002). This work focused on C2 domains, a large group of sequentially varied but structurally conserved modules that targets the binding of proteins involved in signal transduction, membrane trafficking and fusion, and other cellular activities. Murray and Honig demonstrated that the targeting of C2 modules is determined in large part by electrostatics. For example, the membrane binding faces of C2 modules from protein kinase C become positively charged when the modules coordinate calcium ions, causing a calcium-triggered binding to negatively charged patches of membrane. By contrast, the corresponding face of C2 modules from cytosolic phospholipase A2 switch from a negative to neutral character upon coordination of calcium ions; the binding of these ions triggers binding to neutral membranes by reducing the unfavorable free energy of dehydration of the charged face upon contact with the membrane. Murray and Honig show how these principles can be used to rationalize or predict the binding properties of other C2 modules, including ones whose structures are based on homology modeling.

Kinetics

A very wide variety of kinetic processes involving proteins are governed in large part by electrostatic interactions. These include the electrostatically steered binding of substrates or macromolecules to proteins and the association of proteins with membranes. Faster and more accurate methods for calculating electrostatic forces are therefore critical to the computational analysis of such processes. The actual dynamics can be simulated in various ways. The Brownian dynamics method, mentioned earlier, treats the diffusing solute molecules explicitly as structured objects with one or more centers of interaction. Several reviews of work in this area are available (Elcock, Sept, and McCammon, 2001; Gabdoulline and Wade, 2002). It has been possible with such calculations to replicate the experimentally observed rates of association of such protein pairs as barnase–barstar and fasciculin 2-acetylcholinesterase, including the variations in the rates as functions of ionic strength and protein mutagenesis. Similar calculations, by Sept and McCammon (2001), have provided a basis for understanding the nucleation and growth of polar actin filaments.

Brownian dynamics has been employed recently in a ground breaking simulation of the diffusion of 1000 interacting protein molecules (McGuffee and Elcock, 2006). Here, the electrostatic interactions among the proteins were treated by use of effective charges, whose magnitudes were calibrated by Poisson–Boltzmann calculations. When the diffusing solutes are small enough to be treated as structureless, they can be represented implicitly as a concentration or density distribution. Their dynamics can then be described by solution of the diffusion or Smoluchowski equation, instead of stepping the solutes through time individually as in Brownian dynamics. Recent progress in this approach allows simulation of the time-dependent diffusion and reaction of solutes in this continuum representation, including the effects of electrostatics (Cheng et al., 2007). As discussed in a recent review by Stein, Gabdoulline and Wade (2007), algorithmic advances in simulation methodology are enabling investigation of biomolecular encounter kinetics at the scales of biochemical networks and systems.

Electrostatic Comparisons

Effort has also been made to pursue more "informatics"-based approaches to the interpretation of electrostatic properties. Much of this work includes identification of functionally relevant residues in biomolecules by looking at electrostatic destabilization of conserved residues (Elcock, 2001), characteristics of titration curves (Ondrechen, Clifton and Ringe, 2001), or localization of charged groups (Zhu and Karlin, 1996). Other research has focused on comparisons of electrostatic potentials including global analyses of potential both in three-dimensional space over the entire biomolecular structure and at localized regions such as active sites (Blomberg et al., 1999; Livesay et al., 2003; Livesay and La, 2005; Saito, Go, and Shirai, 2006; Zhang et al., 2006, in the reference section). Characterization of electrostatic properties of biomolecules has provided insight into a variety of biomolecular properties, including functional similarities, ligand binding specificities, as well as protein–protein and protein–DNA binding sites. Potential uses for these methods are likely to grow with increases in both biomolecular structural data and information about biomolecular interaction networks. With such growth, tools to facilitate the analysis of electrostatic properties across thousands of biomolecular structures will continue to be very important.

FUTURE DIRECTIONS

In the future, it should be possible to apply continuum electrostatics methods to help determine and investigate macromolecular interactions within increasingly larger

TABLE 24.1. Some of the Software that Implements the Concepts Described in this Chapter

Program	Description	URL
DelPhi	Solves the PBE using highly optimized finite difference methods.	http://trantor.bioc.colum bia.edu/delphi/
APBS	Solves the PBE using parallel multigrid and parallel adaptive finite element methods.	http://apbs.sf.net/
MEAD	Solves the PBE using finite difference methods and determines pK_a values while incorporating conformational flexibility of the macromolecule.	http://www.scripps.edu/bash ford/
UHBD	Solves the PBE using finite difference methods, calculates binding and solvation energies, determines pK_as, and performs Brownian dynamics simulations.	http://mccammon.ucsd.edu/uhbd.html
AMBER	In addition to providing explicit solvent simulation tools, this package implements generalized Born and PBE-based implicit solvent methods in both dynamics and free energy evaluation simulations.	http://amber.scripps.edu
CHARMM	In addition to providing explicit solvent simulation tools, this package implements generalized Born and PBE-based implicit solvent methods in both dynamics and free energy evaluation simulations.	http://charmm.org/
Qnifft	Poisson–Boltzmann finite difference solver developed by the Sharp lab.	http://crystal.med.upenn.edu/software.html

assemblages and, eventually, at the cellular scale. Some future research is likely to focus on the computational evaluation of the thermodynamics and kinetics of biomolecular interactions, moving beyond pairs of interacting proteins to biochemical networks of proteins, nucleic acids, metabolites, and many other biochemical species. New computational methods suitable for much larger systems will allow for studies the structure and dynamics of such macromolecular assemblages as ribosomes and elongation factors; the machinery of DNA replication, repair, and transcription; the nuclear pore complex; and a host of other cellular components. In addition, research continues to improve the ability to use electrostatics to screen and compare biomolecular structures for specific function or similarity to other molecules. Finally, the computational biophysics community will continue to improve implicit solvent models to provide increasing accuracy and reliability in describing biomolecular systems with diverse geometries and charge densities. For all of these tasks, high-throughput continuum electrostatics methods will facilitate the development of computational proteomics to determine the network of biomolecular reactions in living organisms.

ACKNOWLEDGMENTS

Support for this work was provided in part by HHMI and the National Biomedical Computation Resource. Additional support has been provided by grants to N.A.B. from NIH and the Sloan Foundation and by grants to J.A.M. from NIH, NSF, the NSF Supercomputer Centers, the Center for Theoretical Biological Physics, and Accelrys.

REFERENCES

Antosiewicz, J, McCammon, JA, Gilson, MK (1996): The determinants of pK_as in proteins. *Biochemistry* 35(24):7819–7833. [Classic papers describing the use of Poisson–Boltzmann calculations to predict pK_a values in proteins.]

Baker, N, Holst, M, Wang, F (2000): Adaptive multilevel finite element solution of the Poisson–Boltzmann equation II. Refinement at solvent-accessible surfaces in biomolecular systems. *J Comput Chem* 21(15):1343–1352.

Baker, NA, Sept, D, Holst, MJ, McCammon, JA (2001a): The adaptive multilevel finite element solution of the Poisson–Boltzmann equation on massively parallel computers. *IBM J Res Dev* 45(3–4):427–438. [Papers describing the use of adaptive finite element methods to solve the Poisson–Boltzmann equation.]

Baker, NA, Sept, D, Joseph, S, Holst, MJ, McCammon, JA (2001b): Electrostatics of nanosystems: application to microtubules and the ribosome. *Proc Natl Acad Sci USA* 98(18):10037–10041. [Application of parallel multigrid methods to examine the electrostatics of ribosomal and microtubule systems.]

Bashford, D, Karplus, M (1990): pK_a's of ionizable groups in proteins: atomic detail from a continuum electrostatic model. *Biochemistry* 29(44):10219–10225.

Blomberg, N, Gabdoulline, RR, Nilges, M, Wade, RC (1999): Classification of protein sequences by homology modeling and quantitative analysis of electrostatic similarity. *Proteins* 37 (3):379–387.

Born, M (1920): Volumen und hydratationswärme der ionen. *Z Phys* 1:45–48. [The original continuum model of ion solvation.]

Cheng, Y, Suen, JK, Zhang, D, Bond, SD, Zhang, Y, Song, Y, Baker, NA, Bajaj, CL, Holst, MJ, McCammon, JA (2007): Finite element analysis of the time-dependent Smoluchowski equation for acetylcholinesterase reaction rate calculations. *Biophys J* 92(10):3397–3406. [Application of Poisson–Boltzmann electrostatics to continuum diffusion encounter rate calculations.]

Elcock, AH (2001): Prediction of functionally important residues based solely on the computed energetics of protein structure. *J Mol Biol* 312(4):885–896. [Screening method for functionally-important residues using Poisson–Boltzmann electrostatic energy calculations.]

Elcock, AH, Sept, D, McCammon, JA (2001): Computer simulation of protein-protein interactions. *J Phy Chem B* 105(8):1504–1518.

Ermak, DL, McCammon, JA (1978): Brownian dynamics with hydrodynamic interactions. *J Chem Phys* 69:1352–1360. [Classic paper on use of Brownian dynamics to study molecular diffusion.]

Flanagan, MA, Ackers, GK, Matthew, JB, Hanania, GIH, Gurd, FRN (1981): Electrostatic contributions to the energetics of dimer-tetramer assembly in human hemoglobin: pH dependence and effect of specifically bound chloride ions. *Biochemistry* 20:7439–7449. [Continuum electrostatics approach to understanding assembly energetics in hemoglobin.]

Gabdoulline, RR, Wade, RC (2002): Biomolecular diffusional association. *Curr Opin Struct Biol* 12(2):204–213. [Reviews of simulation methods for studying biomolecular encounter rates.]

Georgescu, RE, Alexov, EG, Gunner, MR (2002): Combining conformational flexibility and continuum electrostatics for calculating pK_as in proteins. *Biophy J* 83(4):1731–1748. [Methods demonstrating the importance of biomolecular conformational changes in pK_a calculations.]

Gilson, MK, Davis, ME, Luty, BA, McCammon, JA (1993): Computation of electrostatic forces on solvated molecules using the Poisson–Boltzmann equation. *J Phys Chem* 97(14):3591–3600.

Grant, JA, Pickup, BT, Nicholls, A (2001): A smooth permittivity function for Poisson–Boltzmann solvation methods. *J Comput Chem* 22(6):608–640.

Holm, C, Kekicheff, P, Podgornik, R (2001): *Electrostatic Effects in Soft Matter and Biophysics. NATO Science Series.* Boston: Kluwer Academic Publishers (Detailed descriptions of a variety of electrostatics models including several extensions of and corrections to Poisson–Boltzmann theory).

Holst, MJ, Saied, F (1995): Numerical solution of the nonlinear Poisson–Boltzmann equation: developing more robust and efficient methods. *J Comput Chem* 16(3):337–364. [Paper describing the use of multigrid methods for solving the Poisson–Boltzmann equation.]

Holst, M, Baker, N, Wang, F (2000): Adaptive multilevel finite element solution of the Poisson–Boltzmann equation I. Algorithms and examples. *J Comput Chem* 21(15):1319–1342.

Im, W, Beglov, D, Roux, B (1998): Continuum solvation model: electrostatic forces from numerical solutions to the Poisson–Boltzmann equation. *Comput Phys Commun* 111(1–3):59–75. [Two papers presenting very different derivations of forces from Poisson–Boltzmann calculations.]

Klapper, I, Hagstrom, R, Fine, R, Sharp, K, Honig, B (1986): Focusing of electric fields in the active site of Cu-Zn superoxide dismutase: effects of ionic strength and amino-acid modification. *Proteins* 1(1):47–59. [Classic paper on solution methods for the linearized Poisson–Boltzmann equation.]

Kollman, PA, Massova, I, Reyes, C, Kuhn, B, Huo, S, Chong, L, Lee, M, Lee, T, Duan, Y, Wang, W, Donini, O, Cieplak, P, Srinivasan, J, Case, DA, Cheatham, TE III 2000 Calculating structures and free energies of complex molecules: combining molecular mechanics and continuum models. *Acc Chem Res* 33(12):889–897.

Konecny, R, Trylska, J, Tama, F, Zhang, D, Baker, NA, Brooks, CL, III McCammon, JA (2006): Electrostatic properties of cowpea chlorotic mottle virus and cucumber mosaic virus capsids. *Biopolymers* 82(2):106–120. [Papers describing the application of large-scale Poisson–Boltzmann electrostatics calculations for understanding viral capsid properties.]

Li, H, Robertson, AD, Jensen, JH (2005): Very fast empirical prediction and rationalization of protein pK_a values. *Proteins* 61(4):704–721. [A very fast heuristic method for pK_a prediction with accuracy nearly equal to Poisson–Boltzmann methods.]

Linderstrom-Lang, K (1924): *C R Travl Lab Carlsberg* 15(7), Classic paper using a simple spherical model for protein titration.

Livesay, DR, Jambeck, P, Rojnuckarin, A, Subramaniam, S (2003): Conservation of electrostatic properties within enzyme families and superfamilies. *Biochemistry* 42(12):3464–3473.

Livesay, DR, La, D (2005): The evolutionary origins and catalytic importance of conserved electrostatic networks within TIM-barrel proteins. *Protein Sci* 14(5):1158–1170.

Lu, B, Cheng, X, Huang, J, McCammon, JA (2006): Order N algorithm for computation of electrostatic interactions in biomolecular systems. *Proc Natl Acad Sci USA* 103(51):19314–19319.

Lu, B, McCammon, JA (2007): Improved boundary element methods for Poisson–Boltzmann electrostatic potential and force calculations. *J Chem Theory Comput* 3:1134–1142. [Several papers describing new highly efficient boundary element methods for Poisson and Poisson–Boltzmann calculations.]

McGuffee, SR, Elcock, AH (2006): Atomically detailed simulations of concentrated protein solutions: The effects of salt, pH, point mutations, and protein concentration in simulations of 1000-molecule systems. *J Am Chem Soc* 128(37):12098–12110. [State-of-the-art simulations of thousands of proteins molecules in a concentrated solution.]

Murray, D, Honig, B (2002): Electrostatic control of the membrane targeting of C2 domains. *Mol Cell* 9(1):145–154. [Electrostatics calculations used to describe the mechanism of protein-membrane association.]

Netz, RR, Orland, H (2000): Beyond Poisson–Boltzmann: fluctuation effects and correlation functions. *Eur Phys J E* 1(2–3):203–214.

Nielsen, JE, Vriend, G (2001): Optimizing the hydrogen-bond network in Poisson–Boltzmann equation-based pK_a calculations. *Proteins* 43(4):403–412. [Paper describing the importance of hydrogen placement and rearrangement in pK_a prediction.]

Northrup, SH, Allison, SA, McCammon, JA (1984): Brownian dynamics simulation of diffusion-influenced biomolecular reactions. *J Chem Phys* 80:1517–1524. [Classic paper describing the use of Brownian dynamics simulations, together with long-range electrostatic interactions, to calculate bimolecular encounter rates.]

Ondrechen, MJ, Clifton, JG, Ringe, D (2001): THEMATICS: a simple computational predictor of enzyme function from structure. *Proc Natl Acad Sci USA* 98(22):12473–12478. [Prediction of important protein residues from titration curves.]

Saito, M, Go, M, Shirai, T (2006): An empirical approach for detecting nucleotide-binding sites on proteins. *Protein Eng Des Sel* 19(2):67–75.

Schnieders, MJ, Baker, NA, Ren, P, Ponder, JW (2007): Polarizable atomic multipole solutes in a Poisson–Boltzmann continuum. *J Chem Phys* 126:124114. [Several papers describing smoother Poisson–Boltzmann coefficient definitions for various applications.]

Sept, D, McCammon, JA (2001): Thermodynamics and kinetics of actin filament nucleation. *Biophys J* 81(2):667–674. [Poisson–Boltzmann and Brownian dynamics methods used to simulation actin filament assembly.]

Sept, D, Baker, NA, McCammon, JA (2003): The physical basis of microtubule structure and stability. *Protein Sci* 12:2257–2261. [Poisson–Boltzmann methods used to describe microtubule protofilament arrangements.]

Sharp, KA, Honig, B (1990): Calculating total electrostatic energies with the nonlinear Poisson–Boltzmann equation. *J Phys Chem* 94(19):7684–7692. [Description of how meaningful energies can be calculated from Poisson–Boltzmann electrostatics.]

Stein, M, Gabdoulline, RR, Wade, RC (2007): Bridging from molecular simulation to biochemical networks. *Curr Opin in Struct Biol* 17(2):166–172.

Still, WC, Tempczyk, A, Hawley, RC, Hendrickson, T (1990): Semianalytical treatment of solvation for molecular mechanics and dynamics. *J Am Chem Soc* 112(16):6127–6129. [Development of the generalized Born model of continuum electrostatics.]

Swanson, JMJ, Henchman, RH, McCammon, JA (2004): Revisiting free energy calculations: A theoretical connection to MM/PBSA and direct calculation of the association free energy. *Biophy J* 86(1):67–74.

Tanford, C, Kirkwood, JG (1957): Theory of protein titration curves. I. General equations for impenetrable spheres. *J Am Chem Soc* 79:5333–5339. [Early continuum electrostatics model of protein titration.]

Trylska, J, Konecny, R, Tama, F, Brooks, CL, III McCammon, JA (2004): Ribosome motions modulate electrostatic properties. *Biopolymers* 74(6):423–431.

Trylska, J, McCammon, JA, Brooks, CL III* (2005): Exploring assembly energetics of the 30S ribosomal subunit using an implicit solvent approach. *J Am Chem Soc* 127(31):11125–11133. [Applications of continuum electrostatics methods to ribosome function.]

Warwicker, J, Watson, HC (1982): Calculation of the electric potential in the active site cleft due to alpha-helix dipoles. *J Mol Biol* 157(4):671–679. [The first application of continuum electrostatics methods to atomically-detailed biomolecular structures.]

Whitten, ST, García-Moreno, E B (2005): Local conformational fluctuations can modulate the coupling between proton binding and global structural transitions in proteins. *Proc Natl Acad*

Sci USA 102(12):4282–4287. [Model for conformational effects on biomolecular titration states.]

Woo, HJ, Roux, B (2005): Calculation of absolute protein-ligand binding free energy from computer simulations. *P Natl Acad Sci USA* 102(19):6825–6830. [Several papers describing modern methods, applicable to continuum electrostatics calculations, for evaluating binding free energies.]

Yang, AS, Gunner, MR, Sampogna, R, Sharp, K, Honig, B (1993): On the calculation of pK_as in proteins. *Proteins* 15(3):252–265.

Zauhar, RJ, Morgan, RS (1985): A new method for computing the macromolecular electric potential. *J Mol Biol* 186(4):815–820. [An early boundary element method for continuum electrostatics calculations.]

Zhang, D, Konecny, R, Baker, NA, McCammon, JA (2004): Electrostatic interaction between RNA and protein capsid in cowpea chlorotic mottle virus simulated by a coarse-grain RNA model and a Monte Carlo approach. *Biopolymers* 75(4), *325–*: 37.

Zhang, X, Bajaj, CL, Kwon, B, Dolinsky, TJ, Nielsen, JE, Baker, NA (2006): Application of new multiresolution methods for the comparison of biomolecular electrostatic properties in the absence of global structural similarity. *Multiscale Model Simul* 5(4):1196–1213. [Comparisons of electrostatic potentials around and on the surfaces of biomolecules.]

Zhu, ZY, Karlin, S (1996): Clusters of charged residues in protein three-dimensional structures. *Proc Natl Acad Sci USA* 93(16):8350–8355. [Analysis of charge group localization and relationship to functional and structural characteristics of biomolecules.]

25

PREDICTION OF PROTEIN–NUCLEIC ACID INTERACTIONS

Timothy Robertson and Gabriele Varani

INTRODUCTION

The specific recognition of nucleic acid sequences by nucleic acid binding proteins is of critical importance to the biological function of every living species. As a result, the phenomena responsible for this recognition process have long been of considerable interest to scientists. Beginning with the pioneering work of Seeman, research into sequence-specific DNA recognition focused on the search for a "recognition code"—a collection of simple rules that would pair particular amino acids to specific bases (Seeman, Rosenberg, and Rich, 1976). However, it was soon realized that any recognition code would degenerate (Pabo and Sauer, 1984) and a general "code" may not exist at all (Matthews, 1988; Pabo and Nekludova, 2000) (Chapter 12).

Given that a sequence-based recognition code appears increasingly unlikely, the key to understanding the process of protein–DNA and protein–RNA recognition lies in the structural, physical, and chemical details of the molecular interactions themselves; these structural mechanisms have received renewed attention with the increased availability of high-resolution structures for protein/nucleic acid complexes. Computational studies of these structures have classified their interactions (Luscombe et al., 2000), describing features of their binding sites (Jones et al., 1999; Jones et al., 2001; Luscombe, Laskowski, and Thornton, 2001) and the evolutionary conservation of their interface residues (Luscombe and Thornton, 2002; Mirny and Gelfand, 2002a).

In other areas of research, structural information has been used extensively to create potentials for prediction of protein structures (Sippl, 1990; Samudrala and Moult, 1998; Xu, Xu, and Uberbacher, 1998; Lu and Skolnick, 2001; Zhou and Zhou, 2002; Zhang et al., 2004a; Skolnick, 2006), as well as protein–ligand (DeWitte and Shakhnovich, 1996; Ischchenko and Shakhnovich, 2002; Velec, Gohlke, and Klebe, 2005; Zhang, et al., 2005),

Structural Bioinformatics, Second Edition   Edited by Jenny Gu and Philip E. Bourne
Copyright © 2009 John Wiley & Sons, Inc.

and protein–protein interactions (Jiang et al., 2002; Lu, Lu, and Skolnick, 2003; Zhang et al., 2004a; Zhang, et al., 2005). However, until recently, relatively little effort was devoted to the application of this structural knowledge to the development of computational models for the prediction of protein–nucleic acid interactions. Here, we describe progress toward the application of computational models to the prediction and simulation of protein/nucleic acid interactions.

MOTIVATION

It is self-evident that the ability to accurately and reliably predict the sequence specificity of nucleic acid binding proteins from structure would represent a great intellectual achievement, but the practical applications would also be substantial. Of course, the value of any structure-based method depends upon its generality—if a model can only predict the preferred DNA or RNA binding sequences of a known structure, it is interesting but not especially useful (most structurally characterized protein/nucleic acid complexes have well-known binding sites). If a model can also predict the binding sequences of homologous (though structurally uncharacterized) proteins, it has more value; if it can be used to design entirely novel nucleic acid binding proteins (or to otherwise predict the consequences of mutations to a complex), it is certain to find many applications. Nevertheless, even the most obvious applications—the structure-based identification of DNA or RNA binding sites from known structures—can be practically useful, because many of the "known" recognition sequences of nucleic acid binding proteins are neither absolutely defined nor perfectly characterized (Schneider and Stephens, 1990; Matys et al., 2003). Furthermore, through the use of protein homology modeling techniques (Chapter 30), structure-based methods may one day provide crucial insights into the range and flexibility of recognition sequences for proteins with structures and/or nucleic acid binding properties that are not currently well known.

These applications are analogous to the roles of structure-based models in other areas of biology. The prediction of preferred DNA/RNA binding sequences for a protein has parallels to the problem of structure-based drug design or to the prediction of binding sites for protein–protein interactions (Chapter 26). As noted above, the prediction of nucleic acid recognition sequences for homologous protein sequences is an obvious application of protein homology modeling. Finally, the development of novel nucleic acid binding proteins or the redesign of their specificity is a special case of the rational protein design problem (Chapter 39).

POTENTIAL FUNCTIONS FOR PROTEIN–NUCLEIC ACID INTERACTIONS

Physical Potential Functions

Extending from techniques used to computationally simulate the dynamics of proteins and other molecules (Chapter 37), the same theoretical concepts have been applied to the prediction of energies and specificity for protein/nucleic acid interactions. In particular, the molecular dynamics (MD) simulation of nucleic acid molecules and protein–nucleic acid complexes was routinely performed, long before the first efforts to computationally predict the DNA/RNA recognition sequences of proteins from structure (Weiner et al., 1984;

Weiner et al., 1986; Cornell et al., 1995; Langley, 1998; Cheatham, Cieplak, and Kollman, 1999; Foloppe and MacKerell, 2000; MacKerell and Banavali, 2000; Wang, Cieplak, and Kollman, 2000; MacKerell, Banavali, and Foloppe, 2001). Thus, MD force fields were among the first potential functions to be applied to the problem of predicting the energetic properties of sequence-specific protein/nucleic acid complexes (Lebrun and Lavery, 1999; Pastor, MacKerell, and Weinstein, 1999; Lebrun, Lavery, and Weinstein, 2001; Marco, Garcia-Nieto, and Gago, 2003; Gorfe, Caflisch, and Jelesarov, 2004; Gutmanas and Billeter, 2004).

Of course, the prediction of nucleic acid binding sequence specificity is a larger problem than simulating the dynamics (or estimating the binding energy) of a given protein/nucleic acid complex. Prediction of DNA or RNA binding sequences requires the ability to accurately estimate the energetic consequence of mutations to the binding interface, which itself requires the ability to simulate (i.e., predict) the structural changes that result from these mutations. Thus, the challenge is closer to the structure-based protein design problem (Chapter 39).

In one of the earliest attempts to apply a molecular dynamics force field to this area of research, Pichierri et al. used the AMBER package to calculate free energy, enthalpy, and entropy "maps" of base–amino acid interactions (Pichierri et al., 1999), a study followed by similar efforts from other groups (Sayano et al., 2000; Yoshida et al., 2002). By sampling protein side-chain conformations at grid points surrounding canonical nucleotide structures, these efforts were able to identify energetically favorable conformations for particular amino acid/base pair interactions. Somewhat later, Thayer and Beveridge applied their group's ongoing MD simulation research into sequence-dependent DNA deformation to the prediction of binding sites for the catabolite activator protein (CAP) (Thayer and Beveridge, 2002). They demonstrated a hybrid approach, wherein a hidden Markov model (HMM) was trained using both binding sequence data and nucleotide roll/tilt data obtained from MD simulation of DNA molecules, and found that the use of structural information improved the quality of binding site predictions. Thus, both direct and indirect recognition mechanisms have been explored using MD potentials.

Ultimately, however, molecular dynamics simulations are not well suited to the prediction of cognate binding sequences for nucleic acid binding proteins on a large scale. There is no formal MD analogue to the process of mutation, and the effect of DNA/RNA sequence changes must be approximated through time-consuming simulations of the bound and unbound forms of the different nucleic acid sequences (i.e., simulation of the thermodynamic cycle for every possible combination of nucleotide mutation), and this process is computationally prohibitive for all but the smallest protein–nucleic acid complexes.

For this reason, many applications of physics-based potentials to the problem of binding site prediction have used molecular mechanics techniques, wherein the intent is not to follow conformational changes over time but to directly estimate the free energies of interaction between protein and nucleic acid molecules. For example, Paillard and Lavery (2004) used an energy minimization strategy, based on the AMBER force field, to predict the binding free energies and optimal binding sites for a set of 18 DNA binding proteins. In the process, the authors provided insight into how the recognition of DNA sequences by proteins depends variably (i.e., in a complex-dependent manner) on both protein–DNA interactions both the sequence-specific energy of DNA deformation ("indirect" recognition). The CHARMM package has also been successfully used, particularly for MD simulations of the sequence-dependent flexibility of protein-bound DNA (Pastor, MacKerell, and Weinstein, 1999;

Foloppe and MacKerell, 2000; MacKerell and Banavali, 2000; MacKerell, Banavali, and Foloppe, 2001; Huang and MacKerell, 2005).

A somewhat different example is the hybrid physical/statistical potential function used by the ROSETTA protein design software package. Havranek et al. have extended the ROSETTA potential function to protein–DNA systems (Havranek, Duarte, and Baker, 2004), while Chen et al. applied the method to the prediction of protein–RNA interactions (Chen et al., 2004). The ROSETTA potential function not only incorporates many terms common to physics-based potentials (e.g., the Lennard–Jones model of the van der Waals force), but also makes use of a unique statistical model of hydrogen bonding geometry that appears to confer greater sensitivity to the method (Kortemme, Morozov, and Baker, 2003). In the application to protein–RNA interactions, the statistical hydrogen bonding potential function was able to significantly outperform a Coulomb electrostatics model in a number of different decoy discrimination experiments (Chen et al., 2004).

Statistical Potential Functions

In contrast to the complexity of physics-inspired potential functions, some of the earliest efforts to predict sequence-specific protein/nucleic acid interactions from structure involved the use of simple, knowledge-based ("statistical") potentials (Table 25.1). This class of methods makes use of the database of known structures to derive probability-based scores that can be used to predict protein/nucleic acid interaction energies. While there is a great deal of variation between these methods, all share a common theme: the database of *known* protein–nucleic acid structures is assumed to adequately represent the distributions of particular "features" that can be observed in *any* real biological structure. Examples of such "features" include interatomic distances, torsion angles, bond lengths, and (in the case of protein/nucleic acid interactions) the spatial distributions of protein atoms and residues around the nucleotide bases.

T A B L E 2 5 . 1 . Classification of Methods for Modeling Protein/Nucleic Acid Interactions

Class	Description	Examples
Physical	*Potential functions derived from physical (or physics-inspired) models of atomic interaction (e.g., Lennard–Jones, Coulomb Electrostatics, Born Solvation, etc.)*	Cornell et al. (1995), Pichierri et al. (1999), MacKerell, Banavali, and Foloppe (2001), Endres et al. (2004), Paillard and Lavery (2004)
Statistical	*Potential functions derived from statistical formalisms (e.g., Bayes' Networks, Hidden Markov Models, Support Vector Machines, etc.), and parameterized using data from known protein/nucleic acid structures.*	Kono and Sarai (1999), Liu et al. (2005), Zhang et al. (2005), Robertson and Varani (2007), Donald et al. (2007)
Hybrid	*Methods that incorporate both physical and statistical components.*	Thayer and Beveridge (2002), Chen et al. (2004), Havranek et al. (2004), Morozov et al. (2005)

From a biophysical perspective, knowledge-based potential functions are rooted in the assumption that individual atomic structures represent low-energy molecular conformations, reflecting the optimal contributions of many different microscopic forces. If this assumption holds true, then a sufficiently large database of randomly sampled structures would capture the physically realistic range of any particular structural feature. Moreover, these features would be expected to occur in proportion to their energies, with high-energy features observed far less frequently than low-energy features. A full discussion of the theory behind knowledge-based potentials is beyond the scope of this chapter, but the basic concept is straightforward: the more a given structure "resembles" the database of known structures (which are presumably correct), the "better" that structure is likely to be (Sippl, 1995; Godzik, 1996; Rojnuckarin and Subramaniam, 1999).

Broadly speaking, most statistical potential functions follow a simple formula: the structural feature of interest is quantified, this measurement is divided into bins (creating a histogram), and the "training set" (i.e., structures chosen from the Protein Data Bank specifically to represent the class of molecules being scored) is examined to see how the feature of interest is distributed. For example, if intermolecular atomic distances are being measured across a protein–DNA interface, it makes sense to divide the continuous range of realistic values (e.g., up to 10 Å) into 10 bins with 1 Å widths. If a diverse training set of protein–DNA structures were then examined to count the number of intermolecular contacts that fell within these bins, a histogram would be the result (Figure 25.1), and a simple score for an atom–atom pair is generated by taking the logarithm of the likelihood of each distance bin:

$$s(i,j,d_{ij}) = -\log\left(\frac{f_{\text{observed}}(i,j,d_{ij})}{f_{\text{expected}}(i,j,d_{ij})}\right) \tag{25.1}$$

Here, i and j represent atom types on opposite sides of the protein–DNA interface, d_{ij} is the distance between them, and the numerator is the observed frequency of pairs between atom types i and j (separated by distance d_{ij}) in the training set. The denominator, in contrast, represents the ideal (or, in statistical terminology, the "expected") value for this parameter and is commonly referred to as the "reference state" for the potential function.

The reference state greatly impacts score performance (Sippl, 1995), but its choice is somewhat arbitrary, and it is often impossible to know what this function should be *a priori*. The choice is usually justified empirically and as a result, a large research literature has focused on the reference state, exploring its impact on score performance (Sippl, 1990; DeWitte and Shakhnovich, 1996; Samudrala and Moult, 1998; Zhou and Zhou, 2002; Zhang et al., 2004a; Zhang et al., 2004b; Velec, Gohlke, and Klebe, 2005; Zhang, et al., 2005; Donald, Chen, and Shakhnovich, 2007; Robertson and Varani, 2007). For example, a naïve implementation might assume that all distances are equally likely for any given atom pair (i.e., the reference state is a uniform distribution), whereas a more sophisticated approach may use the distribution of distances observed for all atom pairs as the reference distribution for any particular atom pair.

In practice, the existing statistical potentials for protein–nucleic acid interactions can be broadly grouped into two categories based on the structural features that they consider: orientation-dependent potentials (which exploit three-dimensional spatial and angular distributions) and distance-dependent potentials (which instead use one-dimensional data, such as interatomic and residue–residue distances). There are a small number of methods that do not fit cleanly into either category; these methods have thus far been targeted to

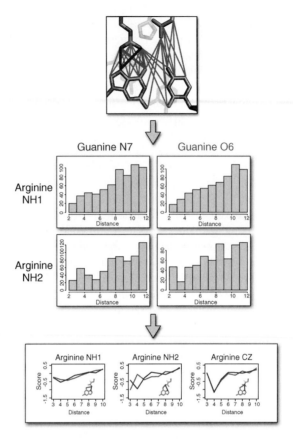

Figure 25.1. Creation of a distance-dependent statistical potential function. The process used to create a simple, distance-dependent statistical potential function is illustrated. Top: pair-wise atomic distances are measured between protein (blue) and nucleic acid (gray) atoms across the binding interfaces of protein/DNA or protein/RNA complexes. Middle: atomic distances are tabulated for a large number of structures, and histograms are created for each unique atom-type pair. Bottom: the distance histograms are converted to scores using method-specific mathematical formalisms (e.g., a log-likelihood model, as in Eq. 25.1). Figure also appears in the Color figure section.

specific classes of problems, such as the prediction of sequence targets for particular families of nucleic acid binding domains.

Orientation-Dependent Statistical Potential Functions. The earliest example of the application of a knowledge-based potential to protein/nucleic acid interactions can perhaps be found in the work of Kono and Sarai (1999), who used the geometric regularity of the DNA double helix to create local "reference frames" for the nucleotides, which were then used to count the number of amino acid alpha carbons in three-dimensional spatial bins surrounding the nucleotide bases. By converting these counts to frequencies and using these frequencies to estimate the likelihood of new protein–nucleic acid complex structures (generated by computationally "threading" new base sequences onto existing complexes), they were able to successfully predict the cognate DNA recognition sequences for a number of transcription factors of known structure. The success of this approach is largely due to the nonuniform distribution of amino acids about the DNA bases (e.g., lysine contacts to guanine tend to be clustered in the major groove of B-form DNA helices) (Luscombe, Laskowski, and Thornton, 2001), though the power of the method is striking, considering its simplicity.

More recently, Chen et al. developed an orientation-dependent hydrogen bonding function for protein–RNA interactions, and demonstrated its utility for predicting the amino acid sequences of RNA binding proteins (Chen et al., 2004). Unlike Kono and Sarai, this work used a complex, multiterm potential function that involved a mixture of statistical and physical terms. However, it further demonstrated that a database of protein–nucleic acid structures could be used to derive a statistical potential that measures the quality of hydrogen bonds across the protein–RNA interface. This objective was achieved by observing four structural features of hydrogen bonds (the donor–acceptor bond length, the bond angles at the donor and the acceptor atoms, and the planarity of the bond) and using the structural database to find the angular and distance distributions of these features (Kortemme, Morozov, and Baker, 2003). These distributions were converted to frequencies and used to infer the likelihood of new protein–RNA structures (in this case, generated by substituting new amino acids onto the backbones of known RNA binding proteins).

Distance-Dependent Statistical Potential Functions. The previous examples each used three-dimensional features of the protein–nucleic acid interface to develop statistical potential (scoring) functions. More recently, however, several groups have demonstrated that even a one-dimensional representation of the structural data is sufficient to discriminate native-like protein–DNA and protein–RNA complexes from nonnative structures. In particular, it appears that the interresidue and interatomic distances observed in the database of protein–nucleic acid structures contain sufficient information to identify cognate DNA and RNA recognition sequences, to discriminate native complex structures from sets of near-native "decoys," and even to estimate the experimentally determined energetics of protein–nucleic acid binding. Despite their extreme simplicity, these "distance-dependent" methods are proving as powerful as the far more complicated, physically inspired potentials discussed previously.

A simple example of this distance-dependent approach was provided by Liu et al., who developed a potential function based on the observed separations between the beta carbons of amino acid residues and the geometric centers of nucleotide triplets, doublets and single nucleotides in protein–DNA structures (Liu et al., 2005). Liu et al. demonstrated that their approach could successfully predict the binding energies and cognate recognition sequences of a large number of DNA binding proteins; they further hypothesized that the use of nucleotide triplets allowed for the capture of higher order interactions, despite the use of one-dimensional distance information.

In a more complex approach, Zhang et al. (2005) developed a statistical potential function based on the observed intermolecular atomic distances in protein–DNA structures. This method mapped all the protein and DNA atoms in protein–DNA complex structures to 19 chemically derived atom types and determined a unique intermolecular distance potential for each of the 361 ($=19^2$) possible atom-type pairs. Despite training this method on a set of protein–ligand complexes (instead of protein–nucleic acid complexes), Zhang et al. demonstrated that their potential produced scores that correlate well with the experimentally determined dissociation (K_D) constants for a collection of 45 protein–DNA complexes.

Most recently, two groups have independently demonstrated that all-atom, distance-dependent statistical potentials for protein–DNA interactions can achieve great predictive power by using only intermolecular distances at the protein/nucleic acid interface. Donald et al. showed that a quasichemical potential (wherein the reference state is determined by the relative frequencies of the different atom types) can accurately predict cognate DNA recognition sequences and relative binding energies ($\Delta\Delta G$ values) for a large number of experimentally characterized protein–DNA complexes (Donald, Chen, and Shakhnovich, 2007); Robertson and Varani (2007) demonstrated that an all-atom, distance-dependent potential (wherein a unique distance-dependent score was defined for each of nearly 14,000 intermolecular atom-type pairs) can achieve cognate binding sequence discrimination performance on par with physical potential functions. Both of these scoring functions rely on the large number of pair-wise atomic interactions made across a protein/nucleic acid interface (a number that grows proportionally to the square of the number of atoms in the interface) to derive a highly redundant description of the structural phenomena responsible for nucleic acid recognition. Thus, despite the simple one-dimensional data used to create the scores (i.e., interatomic distances), these methods are able to model chemical and structural phenomena that are actually three dimensional in nature. In other words, the massive redundancy of these scores allows them to capture higher order structural motifs with surprising fidelity.

Other Statistical Models. In addition to the orientation- and distance-dependent potentials discussed above, a few groups have explored the use of methods that do not fit cleanly into either category. In particular, Ge et al. developed a knowledge-based method for predicting the sequence-specific binding energies of polyamide molecules to DNA double helices, based on the observed positions of water molecules and amino acid atoms in known protein–DNA structures (Ge, Schneider, and Olson, 2005). This method clustered the atoms (water or amino acid) observed to hydrogen bond to DNA bases in their structure training set and used these structural clusters to determine 3D ellipsoid regions of likely drug–DNA hydrogen bonding interactions.

Finally, in an example of a prediction method targeted to a specific family of DNA binding proteins, Kaplan et al. developed a knowledge-based potential for Cys_2His_2 zinc finger proteins and used it to scan for transcription factor binding sites in the *Drosophila* genome (Kaplan, Friedman, and Margalit, 2005). The method is not general (it applies only to a single class of zinc finger proteins), nor is it truly a structure-based potential (it does not make direct use of structure data), but it exploits the well-understood regularity of the Cys_2His_2 family of zinc finger proteins to construct a probability model of DNA recognition. Thus, it represents an interesting application of structural knowledge to the prediction of transcription factor binding sites.

FLEXIBILITY IN PROTEIN/NUCLEIC ACID COMPLEXES

An important, largely unsolved problem inherent to the structure-based prediction of protein–nucleic acid interactions lies in the treatment of molecular flexibility, which is frequently observed in protein–RNA complexes, but is also important to protein–DNA interactions. Although some simulation methods (such as molecular dynamics or gradient-based minimization techniques) explicitly allow for the representation of small motions at the protein–nucleic acid interface, there is good reason to believe that the molecular movements involved in nucleic acid sequence recognition are larger than can be accurately modeled using these methods alone (Dickerson, 1998). For this reason, several groups have introduced methods to incorporate protein flexibility into the problem of protein–DNA and protein–RNA interface prediction. For example, Endres et al. (2004) used the AMBER force field, in combination with a dead-end elimination (DEE) algorithm for protein side-chain packing, and demonstrated that this method can predict the consensus binding sequence of the Zif268 zinc finger transcription factor. However, their results also suggested that the use of rotamer packing may have negatively impacted the performance of their method, leaving open the question of the effectiveness of the flexibility model.

In a second example, when Havranek et al. applied the ROSETTA protein design algorithm (Kuhlman and Baker, 2000) to protein–DNA complexes, they demonstrated that their Monte Carlo-based approach to side-chain rotamer packing can predict native-like protein sequences from the structures of protein–DNA complexes (Havranek, Duarte, and Baker, 2004), and they later successfully used the method to guide a localized redesign of the I-MsoI homing endonuclease protein (Ashworth et al., 2006). However, when Morozov et al. used the same approach to predict $\Delta\Delta G$ values for a large set of experimentally characterized mutations to protein–DNA complexes, the performance of the method was negatively impacted by the introduction of protein flexibility. When side-chain packing was enabled, the ROSETTA potential did no better than a control method, which counted the number of intermolecular contacts observed at the protein–DNA interface. Also, an attempt to incorporate a limited model of DNA flexibility into their software was found to further degrade the performance of the approach (Morozov et al., 2005).

Therefore, to date, no group has successfully demonstrated a method that improves the prediction of sequence-specific protein–nucleic acid interactions through the incorporation of molecular flexibility. Even relatively conservative methods (such as algorithms for the prediction of protein side-chain conformations) can achieve only limited accuracy (Dunbrack, 2002). Thus, the modeling of molecular flexibility represents not only a fertile area of future research, but also a significant obstacle. This issue must be addressed if current scoring functions are to be reliably applied to computationally intensive problems, such as the structure-based annotation of genome sequence, or the rational design of nucleic acid binding proteins with new sequence specificities.

For protein flexibility, the path toward progress in this area is at least somewhat clear—in addition to the large research literature surrounding the molecular dynamics simulation of proteins (Chapter 37), there have also been a number of efforts to investigate the use of protein flexibility models for computational protein design (Desjarlais and Handel, 1999; Larson et al., 2002; Kuhlman et al., 2003; Kuhlman and Baker, 2004; Saunders and Baker, 2005; Ambroggio and Kuhlman, 2006a; Ambroggio and Kuhlman, 2006b; Bui et al., 2006; Georgiev and Donald, 2007), as well as for small-molecule docking and structure-based drug design (Claussen et al., 2001; Cavasotto and Abagyan, 2004; Polgar and Keseru, 2006). However, research into models of nucleic acid flexibility is less well established, and aside from MD simulations of protein/nucleic acid complexes (for example, see Lebrun and Lavery, 1999; Lebrun, Lavery, and Weinstein, 2001; Thayer and Beveridge, 2002; Marco, Garcia-Nieto, and Gago, 2003; Gorfe, Caflisch, and Jelesarov, 2004; Gutmanas and Billeter, 2004; Arauzo-Bravo et al., 2005; Zhao, Huang, and Sun, 2006; Mendieta et al., 2007), there have been relatively few efforts to model the flexibility of nucleic acid molecules or their complexes with proteins.

Encouragingly, several groups have made careful studies of the observed flexibility of DNA molecules in protein/DNA complex structures (Olson et al., 1995; Dickerson, 1998; Olson et al., 1998; Lafontaine and Lavery, 2001; Steffen et al., 2002a; Gromiha et al., 2005). Furthermore, Murray et al. (2003) recently observed that the RNA backbone appears to adopt a limited number of "rotameric" conformations that might facilitate the automated prediction of three-dimensional structures for RNA molecules. These studies may represent fruitful starting points for future research into nucleic acid flexibility, but there is a great deal of work left to be done before flexibility models can be incorporated successfully into predictive models of protein/nucleic acid interactions.

APPLICATIONS

The structure-based prediction of protein–nucleic acid interactions is a new field, and there are still few examples of practical applications—particularly those with large-scale experimentally validated results. Nevertheless, research in the field is growing rapidly, and existing publications already offer promise that the structure-based analysis and prediction of protein–nucleic acid interactions will become a valuable tool for biologists seeking new sources of insight and guidance for experimental design.

Molecular Docking

An application of structural analysis that has been successfully demonstrated is the prediction of protein–DNA and protein–RNA interactions through the computational docking of protein and nucleic acid structures. Computational docking is a well-established technique in the study of protein/protein and protein/small-molecule interactions (Chapters 26 and 27), and because it is relatively easy to conduct rigid-body binding simulations of molecular structures, this was one of the earliest tests applied to protein–nucleic acid systems. Nearly a decade ago, Knegtel et al. applied their Monte Carlo-based docking method (MONTY) to the prediction of protein–DNA interactions. Their research incorporated both protein side-chain (Knegtel et al., 1994a) and DNA flexibility (Knegtel, Boelens, and Kaptein, 1994b) but demonstrated their methods successfully on only a few protein–DNA complexes. Later, Aloy et al. (1998) used the innovative rigid-body search algorithm of FTDock to predict protein–DNA interactions for a larger number of complexes and subsequently extended the method to incorporate protein side-chain flexibility at the protein–DNA interface (Sternberg et al., 1998). Most recently, Robertson and Varani (2007) used the FTDock rigid-body method to validate a knowledge-based potential for protein– DNA interactions and showed that the score was able to significantly improve upon the FTDock results. Chen et al. also achieved good docking decoy discrimination performance for protein–RNA complexes using the ROSETTA docking method (Monte Carlo-based rigid-body docking, coupled with protein side-chain flexibility (Gray et al., 2003)). Finally, in one of the most advanced examples of protein/nucleic acid docking to date, van Dijk et al. (2006) used their HADDOCK method to conduct fully flexible docking simulations of three protein–DNA complexes. By alternating rounds of rigid-body docking with computationally intensive MD simulations of the best-scoring docking decoys, they were able to predict reasonably accurate models of protein–DNA complexes starting from structures of their unbound components.

Despite these initial efforts, there are considerable practical limitations to the prediction of protein–nucleic acid interactions using docking simulations. In particular, only a few attempts have been made to dock unbound structures of protein and DNA molecules with mixed results. Because of the importance of conformational change in the process of protein/nucleic acid recognition, our understanding of molecular motion limits our ability to accurately model protein–nucleic acid interactions. The flexible nature of DNA and RNA molecules, the prevalence of induced fit (wherein the protein and/or the nucleic acid molecule change shape to accommodate binding), and the importance of indirect recognition mean that any generally useful docking method for protein–DNA or protein–RNA inter-actions will have to incorporate models of molecular flexibility that are more advanced and robust than currently available.

Analysis of Direct and Indirect Recognition

The computational analysis of protein–DNA and protein–RNA structures can be used to develop and explore hypotheses about the mechanisms of protein–nucleic acid recognition that cannot be easily addressed via experiment. For example, a number of studies have used computational models to quantitatively investigate the importance of direct and indirect recognition in sequence-specific protein/nucleic acid interactions. In this case, the ability to computationally model protein/nucleic acid complexes allows for investigation of a problem that is very difficult to directly assess using experimental approaches. Direct recognition (recognition due to direct atomic interactions across a protein/nucleic acid interface) can, in theory, be distinguished from indirect readout (recognition of sequence-dependent distortions in nucleic acid structure) by quantifying the relative contributions of inter- and intramolecular interactions to protein–DNA binding free energy. Paillard and Lavery (2004) and Gromiha et al. (2005) have both performed large-scale analyses of this important problem, using different potential functions, and found that the contributions of direct and indirect recognition vary significantly from structure to structure. Previously, Steffen et al. showed that they could partially predict the preferred binding sites of the integration host factor (IHF) protein by calculating the energies of bending for different DNA sequences (Steffen et al., 2002a; Steffen et al., 2002b), using a statistical potential function based on the DNA geometric parameters described by Olson et al. (1995, 1998). A number of other structure-based analyses have investigated this question for different protein–DNA com-plexes (Gorfe, Caflisch, and Jelesarov, 2004; Arauzo-Bravo et al., 2005; Dixit, Andrews, Beveridge, 2005; Aeling et al., 2006; Napoli et al., 2006; Aeling et al., 2007).

Structure Refinement

Another interesting yet largely unexplored application—particularly for statistical poten-tials—lies in the refinement of protein–nucleic acid structures. Traditionally, the software used to create crystallographic and NMR structures has employed physics-based potentials during refinement. However, Kuszewski et al. recently developed a statistical potential function (DELPHIC) that describes the relative orientations of dinucleotide steps. They used this potential to refine several NMR structures of DNA and RNA molecules (Kuszewski, Schwieters, and Clore, 2001; Clore and Kuszewski, 2003), and demonstrated that the potential was able to improve the quality of the refined structures. They also demonstrated that the use of a database-derived potential did not reduce their ability to obtain correct refinement for noncanonical nucleic acid structures. These results suggest that statistical potentials may find application in the refinement of molecular structures, particularly when the experimental data are of limited quality (structures with low crystallographic resolution, poorly defined NMR structures, and so on).

Structure-Based Genome Annotation

One of the most compelling applications of structure-based methods for the prediction of protein–nucleic acid interactions is also one of the most obvious: if it were possible to accurately predict the sequence recognition preferences of a protein from its structure, it should be possible to use that structure to predict the binding sites of the protein within a genome. In fact, the idea of using structures of protein–nucleic acid complexes to annotate genomes for transcription factor (and other) binding sites is so compelling that nearly every publication in the field has incorporated some sort of binding sequence scanning experiment as a methodological test. For example, Kono and Sarai (1999) used their statistical potential to predict the recognition sequences of 25 different DNA binding proteins (to varying degrees of success), and also demonstrated the method's ability to identify five of the six known binding motifs of the MAT a1/α2 protein in the upstream region of a known target gene. Liu et al. used their distance-dependent statistical potential to find the known binding sites of the cyclic AMP regulatory protein (CRP) in the *Escherichia coli* genome (Liu et al., 2005). Paillard and Lavery (2004) used their physics-based potential to reproduce the consensus binding sites of multiple protein–DNA complexes, while Morozov et al. (2005) used their method to predict position weight matrices for a number of DNA binding proteins. Finally, Robertson and Varani used their statistical potential function to recapitulate the cognate recognition sequences of 52 different DNA binding proteins (Robertson and Varani, 2007) (Figure 25.2). While many important aspects of protein/nucleic acid interfaces were ignored in these early tests (for example, most of these efforts did not consider interfacial water molecules in their analyses, and few incorporated any type of interface flexibility), it is nonetheless clear that structure-based genome annotation is rapidly becoming a reality.

More recent research is beginning to tackle the even more difficult—but considerably more important—problem of predicting nucleic acid binding sequences for proteins for which only homology models are available. Morozov and Siggia (2007) recently employed a simple, contact counting method to predict the binding preferences of 57 *Saccharomyces cerevisiae* transcription factors of previously unknown structure. This approach represents a leap forward in the application of structural models to biological research by enormously expanding the number of interesting targets well beyond those proteins for which high-resolution structures are available.

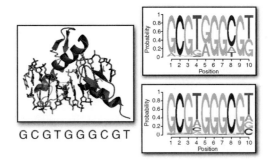

GCGTGGGCGT

Figure 25.2. A successful structure-based prediction of a DNA recognition sequence. An example of a successful prediction of a DNA recognition sequence from structure is shown. Left: the crystal structure of the Zif268 DNA binding protein, along with its consensus DNA recognition sequence. Right, top: the sequence logo (Schneider and Stephens, 1990) derived from experimentally determined DNA binding preferences for the protein (Morozov et al., 2005; Swirnoff and Milbrandt, 1995). Right, bottom: a sequence logo calculated directly from the protein/DNA complex structure using a statistical potential function (Robertson and Varani, 2007).

Rational Protein Design

If a structure-based model of a DNA or RNA binding protein is accurate enough to be used to predict the cognate nucleic acid recognition sequences for that protein, then it may be possible to use the model to solve the converse problem—the prediction of a protein sequence that will optimally bind a given nucleic acid sequence. This is simply a restatement of the protein design problem, as applied to nucleic acid binding proteins. While there has been an enormous amount of research dedicated to the structure-based design (or directed evolution) of certain classes of nucleic acid binding proteins (Pomerantz, Sharp, and Pabo, 1995b), particularly for zinc finger proteins (Pomerantz, Pabo, and Sharp, 1995a; Isalan, Choo, and Klug, 1997; Kim et al., 1997; Liu et al., 1997; Kim and Pabo, 1998; Pomerantz, Wolfe, and Pabo, 1998; Wolfe, Ramm, and Pabo, 2000; Peisach and Pabo, 2003; Wolfe, Grant, and Pabo, 2003; Mandell and Barbas, 2006; Papworth, Kolasinska, and Minczuk, 2006; Nomura and Sugiura, 2007), there are still no general-purpose methods that can successfully design protein sequences that will fold into conformations with a desired DNA or RNA binding ability. Given that designed zinc finger proteins have considerable potential as therapeutic compounds for human diseases (Klug, 2005; Papworth, Kolasinska, and Minczuk, 2006), there are clearly many potential applications for any method that can solve this more general problem.

Recently, a few groups have been making progress toward this goal using structure-based protein design algorithms. For example, Havranek et al. used the ROSETTA protein design software to recapitulate the native amino acid sequences of DNA binding proteins given their protein backbone structures, as well as the structures of their bound DNA molecules (Havranek, Duarte, and Baker, 2004), while Chen et al. (2004) successfully applied the same approach to RNA binding proteins. These results demonstrate that the ROSETTA potential can largely (though by no means perfectly) identify the native amino acid sequences of protein–nucleic acid complexes as optimal for their bound structures. This is a necessary condition for any structure-based protein design algorithm.

To date, the ROSETTA software has been used to redesign two different nucleic acid binding proteins: Dobson et al. (2006) used the method to design a variant of the U1A RNA binding protein, while Ashworth et al. (2006) redesigned the I-MsoI homing endonuclease (a DNA binding protein). Dobson et al. demonstrated that the completely redesigned U1A molecule (with approximately 30% sequence identity to the wild-type protein) was able to fold into a native-like structure, although no attempt was made to retain the RNA binding function of the redesigned protein. Ashworth et al. used a more conservative design strategy for the I-MsoI protein–DNA complex, introducing two amino acid mutations at the protein–nucleic acid interface; they demonstrated that the redesigned protein bound specifically to a DNA sequence with a single guanine/cytosine base pair transversion (relative to the cognate recognition sequence). Thus, although these early tests of protein design methods have been promising, there is still a great deal of work to be done before it is possible to reliably design nucleic acid binding proteins of all classes with arbitrary sequence specificity.

Both of these attempts at structure-based protein design relied exclusively upon structural information (coupled with the energetic analysis provided by a physical potential function) to make their predictions. However, a few groups have demonstrated a phenomenon that may be useful for a more directed type of protein design: significant correlations can be observed between patterns of protein sequence evolution within families of nucleic acid binding proteins and the evolutionary patterns of the nucleic acid recognition sequences of those same proteins (Lichtarge, Yamamoto, and Cohen, 1997; Mirny and Gelfand, 2002b; Raviscioni et al., 2005). In particular, Raviscioni et al. demonstrated that for 12 different families of transcription factors, the evolutionarily most important protein residues of the families tended to interact with the most conserved base pairs of their DNA recognition sequences. They subsequently used this information to experimentally alter the sequence specificity of a zinc finger transcription factor (Raviscioni et al., 2005). Thus, it appears that this so-called "evolutionary trace" method (Lichtarge, Bourne, and Cohen, 1996) (and other closely related approaches) can guide the rational design of nucleic acid binding proteins. The method can highlight the most relevant protein residues for nucleic acid recognition and also suggest protein mutations that will lead to desired changes in sequence specificity. Coupled with physics-based or statistical models of protein–DNA interactions, evolutionary trace methods may prove to be a powerful tool for the rational, structure-based design of DNA and RNA binding proteins.

FUTURE WORK AND CRITICAL CHALLENGES

Steady increases in computer power, the size of the structural database, and our improved understanding of protein, DNA, and RNA biochemistry are rapidly allowing the structure-based modeling and prediction of protein/nucleic acid interactions to become realistic prospects. This field has experienced impressive advancements within the last decade; 10 years ago, research into protein–nucleic acid recognition was dominated by the misguided search for a "recognition code," but today it is possible to choose between multiple successful methods to predict, design, and model protein–DNA and protein–RNA interactions at the atomic level. However, all of these methods have limitations, and successful applications and experimentally validated tests are still rare. A great deal of work remains to be done if the promise of this research is to be fulfilled. In particular, at least three major areas of research must be advanced before the modeling of protein–nucleic acid interactions can become a widely used predictive tool: (1) the accuracy of existing potential functions must be enhanced; (2) the ability to simulate molecular flexibility of protein, DNA and RNA molecules must be improved; and (3) additional details of the molecular interface (e.g., water, metals/ligands, base stacking, pi–pi and cation–pi interactions) must be incorporated into the current models.

As our understanding of the dynamics and details of protein–nucleic acid recognition advances, deficiencies in our understanding of the forces and interactions involved in these processes will undoubtedly emerge. At the moment, it appears that even the simplest statistical potentials perform as well as considerably more complex molecular dynamics force fields in many situations: Will this continue to be true as simulations increase in complexity and detail, and the sampling problem of large-scale simulations become less acute? Perhaps more complex models of electrostatics, solvent effects, and hydrogen bonding will be necessary to accurately model detailed interactions at molecular interfaces. However, it is also possible that database-derived potentials may be able to implicitly capture these effects as well as the best physics-based methods.

As noted above, even the current best methods for simulating protein–nucleic acid interactions can capture only small amounts of molecular flexibility, such as protein side-chain rearrangements or (in the case of molecular dynamics methods) small-scale molecular motions. However, even these limited techniques take prohibitive amounts of computer time and thus far have been unable to significantly improve simulation results in most circumstances. When one considers that protein–DNA and (particularly) protein–RNA interactions often involve large-scale conformational rearrangements of both protein and nucleic acid molecules (Leulliot and Varani, 2001), the need for improved conformational simulation techniques becomes acute. Clearly, new methods need to be developed to accurately sample or simulate the motions of protein and nucleic acid molecules as they interact. Moreover, these new methods must be computationally tractable for large-scale simulations if they are to be usefully applied to problems such as structure-based genome annotation or protein/nucleic acid interface redesign.

In addition to limitations in the ability to simulate molecular motions, it is clear that even the current best models miss important details of the protein/nucleic acid interface. For example, few of the methods discussed in this chapter explicitly consider the role of interface water molecules or metal ions at the molecular interface, although nearly every known protein–DNA and protein–RNA complex depends on the precise positioning of these molecules to mediate sequence-specific recognition. The associated questions are numerous: How do water molecules position themselves in a protein–nucleic acid interface? What are the enthalpic, entropic, and free energy consequences of water- or ion-mediated interactions with nucleic acid molecules? Can interactions with highly conserved interfacial water molecules be displaced through the careful redesign of hydrogen bonding networks? Water molecules represent just one area for which our understanding of the details of protein–nucleic acid recognition is poor. From the importance of intra- and intermolecular stacking interactions involving carbon ring systems (i.e., pi–pi and cation–pi interactions) to the role of conformational entropy in binding free energy, there are other important characteristics of these interfaces that remain underinvestigated.

The structure-based modeling of protein–DNA and protein–RNA interactions not only is an exciting field, with many interesting and relevant scientific implications, but also has considerable challenges. Within the last decade, our understanding of sequence-specific protein/nucleic acid recognition has leapt forward, and a number of practical applications have been demonstrated in principle, despite the relatively simple modeling techniques used so far. It is very likely that significant advances will be made in our ability to predict, design, and simulate critically important molecular interactions. As these open questions are addressed, we anticipate that the prediction of protein–nucleic acid interactions will emerge as an important, widely used application of structural bioinformatics.

REFERENCES

Aeling KA, Opel ML, Steffen NR, Tretyachenko-Ladokhina V, Hatfield GW, Lathrop RH, Senear DF (2006): Indirect recognition in sequence-specific DNA binding by Escherichia coli integration host factor: the role of DNA deformation energy. *J Biol Chem* 51:39236–39248.

Aeling KA, Steffen NR, Johnson M, Hatfield GW, Lathrop RH, Senear DF (2007): DNA deformation energy as an indirect recognition mechanism in protein–DNA interactions. *IEEE/ACM Trans Comput Biol Bioinform* 1:117–125.

Aloy P, Moont G, Gabb H, Querol E, Aviles F, Sternberg M (1998): Modelling repressor proteins docking to DNA. *Proteins* 4:535–549.

Ambroggio XI, Kuhlman B (2006a): Computational design of a single amino acid sequence that can switch between two distinct protein folds. *J Am Chem Soc* 4:1154–1161.

Ambroggio XI, Kuhlman B (2006b): Design of protein conformational switches. *Curr Opin Struct Biol* 4:525–530.

Arauzo-Bravo MJ, Fujii S, Kono H, Ahmad S, Sarai A (2005): Sequence-dependent conformational energy of DNA derived from molecular dynamics simulations: toward understanding the indirect readout mechanism in protein–DNA recognition. *J Am Chem Soc* 4:16074–16089.

Ashworth J, Havranek JJ, Duarte CM, Sussman D, Monnat RJ, Stoddard BL, Baker D (2006): Computational redesign of endonuclease DNA binding and cleavage specificity. *Nature* 7093: 656–659.

Bui H, Schiewe AJ, von Grafenstein H, Haworth IS (2006): Structural prediction of peptides binding to MHC class I molecules. *Proteins* 1:43–52.

Cavasotto CN, Abagyan RA (2004): Protein flexibility in ligand docking and virtual screening to protein kinases. *J Mol Biol* 1:209–225.

Cheatham T III, Cieplak P, Kollman P (1999): A modified version of the Cornell et al. force field with improved sugar pucker phases and helical repeat. *J Biomol Struct Dyn* 845–862.

Chen Y, Kortemme T, Robertson T, Baker D, Varani G (2004): A new hydrogen-bonding potential for the design of protein–RNA interactions predicts specific contacts and discriminates decoys. *Nucleic Acids Res* 17:5147–5162.

Claussen H, Buning C, Rarey M, Lengauer T (2001): FlexE: efficient molecular docking considering protein structure variations. *J Mol Biol* 2:377–395.

Clore GM, Kuszewski J (2003): Improving the accuracy of NMR structures of RNA by means of conformational database potentials of mean force as assessed by complete dipolar coupling cross-validation. *J Am Chem Soc* 6:1518–1525.

Cornell W, Cieplak P, Bayly C, Gould I, Merz K Jr, Ferguson D, Spellmeyer D, Fox T, Caldwell J, Kollman P (1995): A second generation force field for the simulation of proteins, nucleic acids, and organic molecules. *J Am Chem Soc* 5179–5197.

Desjarlais JR, Handel TM (1999): Side-chain and backbone flexibility in protein core design. *J Mol Biol* 1:305–318.

DeWitte R, Shakhnovich E (1996): SMoG: de novo design method based on simple, fast and accurate free energy estimates. *J Am Chem Soc* 11733–11744.

Dickerson RE (1998): DNA bending: the prevalence of kinkiness and the virtues of normality. *Nucleic Acids Res* 8:1906–1926.

Dixit SB, Andrews DQ, Beveridge DL (2005): Induced fit and the entropy of structural adaptation in the complexation of CAP and lambda-repressor with cognate DNA sequences. *Biophys J* 3147–3157.

Dobson N, Dantas G, Baker D, Varani G (2006): High-resolution structural validation of the computational redesign of human U1A protein. *Structure* 5:847–856.

Donald J, Chen W, Shakhnovich E (2007): Energetics of protein–DNA interactions. *Nucleic Acids Res* 1039–1047.

Dunbrack RL (2002): Rotamer libraries in the 21st century. *Curr Opin Struct Biol* 4:431–440.

Endres RG, Schulthess TC, Wingreen NS (2004): Toward and atomistic model for predicting transcription-factor binding sites. *Proteins* 2:262–268.

Foloppe N, MacKerell A (2000): All-atom empirical force field for nucleic acids: I. Parameter optimization based on small molecule and condensed phase macromolecular target data. *J Comput Chem* 2:86–104.

Ge W, Schneider B, Olson W (2005): Knowledge-based elastic potentials for docking drugs or proteins with nucleic acids. *Biophys J* 1166–1190.

Georgiev I, Donald BR (2007): Dead-end elimination with backbone flexibility. *Bioinformatics* 13:185–194.

Godzik A (1996): Knowledge-based potentials for protein folding: what can we learn from known protein structures? *Structure* 4:363–366.

Gorfe A, Caflisch A, Jelesarov I (2004): The role of flexibility and hydration on the sequence-specific DNA recognition by the Tn916 integrase protein: a molecular dynamics analysis. *J Mol Recognit* 120–131.

Gray JJ, Moughon S, Wang C, Schueler-Furman O, Kuhlman B, Rohl CA, Baker D (2003): Protein–protein docking with simultaneous optimization of rigid-body displacement and side-chain conformations. *J Mol Biol* 1:281–299.

Gromiha MM, Siebers JG, Selvaraj S, Kono H, Sarai A (2005): Role of inter and intramolecular interactions in protein–DNA recognition. *Gene* 108–113.

Gutmanas A, Billeter M (2004): Specific DNA recognition by the Antp homeodomain: MD simulations of specific and nonspecific complexes. *Proteins* 772–782.

Havranek JJ, Duarte CM, Baker D (2004): A simple physical model for the prediction and design of protein–DNA interactions. *J Mol Biol* 59–70.

Huang N, MacKerell A (2005): Specificity in protein–DNA interactions: energetic recognition by the (cytosine-C5)-methyltransferase from HhaI. *J Mol Biol* 2:265–274.

Isalan M, Choo Y, Klug A (1997): Synergy between adjacent zinc fingers in sequence-specific DNA recognition. *Proc Natl Acad Sci USA* 11:5617–5621.

Ischchenko A, Shakhnovich E (2002): Small Molecule Growth 2001 (SMoG2001): an improved knowledge-based scoring function for protein–ligand interactions. *J Med Chem* 2770–2780.

Jiang L, Gao Y, Mao F, Liu Z, Lai L (2002): Potential of mean force for protein–protein interactions studies. *Proteins* 190–196.

Jones S, van Heyningen P, Berman H, Thornton J (1999): Protein–DNA interactions: a structural analysis. *J Mol Biol* 5:877–896.

Jones S, Daley D, Luscombe N, Berman H, Thornton J (2001): Protein–RNA interactions: a structural analysis. *Nucleic Acids Res* 4:943–954.

Kaplan T, Friedman N, Margalit H (2005): Ab initio prediction of transcription factor targets using structural knowledge. *PLoS Comput Biol* 1.

Kim JS, Kim J, Cepek KL, Sharp PA, Pabo CO (1997): Design of TATA box-binding protein/zinc finger fusions for targeted regulation of gene expression. *Proc Natl Acad Sci USA* 8: 3616–3620.

Kim JS, Pabo CO (1998): Getting a handhold on DNA: design of poly-zinc finger proteins with femtomolar dissociation constants. *Proc Natl Acad Sci USA* 6:2812–2817.

Klug A (2005): Towards therapeutic applications of engineered zinc finger proteins. *FEBS Lett* 4:892–894.

Knegtel RM, Antoon J, Rullmann C, Boelens R, Kaptein R (1994a): MONTY: a Monte Carlo approach to protein–DNA recognition. *J Mol Biol* 1:318–324.

Knegtel RM, Boelens R, Kaptein R (1994b): Monte Carlo docking of protein–DNA complexes: incorporation of DNA flexibility and experimental data. *Protein Eng* 6:761–767.

Kono H, Sarai A (1999): Structure-based prediction of DNA target sites by regulatory proteins. *Proteins* 1:114–131.

Kortemme T, Morozov A, Baker D (2003): An orientation-dependent hydrogen bonding potential improves prediction of specificity and structure for proteins and protein–protein complexes. *J Mol Biol* 4:1239–1259.

Kuhlman B, Baker D (2000): Native protein sequences are close to optimal for their structures. *Proc Natl Acad Sci USA* 19:10383–10388.

Kuhlman B, Dantas G, Ireton GC, Varani G, Stoddard BL, Baker D (2003): Design of a novel globular protein fold with atomic-level accuracy. *Science* 5649:1364–1368.

Kuhlman B, Baker D (2004): Exploring folding free energy landscapes using computational protein design. *Curr Opin Struct Biol* 1:89–95.

Kuszewski J, Schwieters C, Clore GM (2001): Improving the accuracy of NMR structures of DNA by means of a database potential of mean force describing base–base positional interactions. *J Am Chem Soc* 17:3903–3918.

Lafontaine I, Lavery R (2001): High-speed molecular mechanics searches for optimal DNA interaction sites. *Comb Chem High Throughput Screen* 8:707–717.

Langley D (1998): Molecular dynamic simulations of environment and sequence dependent DNA conformations: the development of the BMS nucleic acid force field and comparison with experimental results. *J Biomol Struct Dyn* 487–509.

Larson SM, England JL, Desjarlais JR, Pande VS (2002): Thoroughly sampling sequence space: large-scale protein design of structural ensembles. *Protein Sci* 12:2804–2813.

Lebrun A, Lavery R (1999): Modeling DNA deformations induced by minor groove binding proteins. *Biopolymers* 341–353.

Lebrun A, Lavery R, Weinstein H (2001): Modeling multi-component protein–DNA complexes: the role of bending and dimerization in the complex of p53 dimers with DNA. *Protein Eng* 233–243.

Leulliot N, Varani G (2001): Current Topics in RNA-Protein Recognition: Control of Specificity and Biological Function through Induced Fit and Conformational Capture. *Biochemistry* 27: 7947–7956.

Lichtarge O, Bourne HR, Cohen FE (1996): An evolutionary trace method defines binding surfaces common to protein families. *J Mol Biol* 2:342–358.

Lichtarge O, Yamamoto KR, Cohen FE (1997): Identification of functional surfaces of the zinc binding domains of intracellular receptors. *J Mol Biol* 3:325–337.

Liu Q, Segal DJ, Ghiara JB, Barbas CF (1997): Design of polydactyl zinc-finger proteins for unique addressing within complex genomes. *Proc Natl Acad Sci USA* 11:5525–5530.

Liu Z, Mao F, Guo J, Yan B, Wang P, Qu Y, Xu Y (2005): Quantitative evaluation of protein–DNA interactions using and optimized knowledge-based potential. *Nucleic Acids Res* 2:546–558.

Lu H, Skolnick J (2001): A distance-dependent atomic knowledge-based potential for improved protein structure selection. *Proteins* 223–232.

Lu H, Lu L, Skolnick J (2003): Development of unified statistical potentials describing protein–protein interactions. *Biophys J* 1895–1901.

Luscombe NM, Austin SE, Berman HM, Thornton JM (2000): An overview of the structures of protein–DNA complexes. *Genome Biol* 1.

Luscombe NM, Laskowski RA, Thornton JM (2001): Amino acid-base interactions: a three-dimensional analysis of protein–DNA interactions at an atomic level. *Nucleic Acids Res* 13: 2860–2874.

Luscombe NM, Thornton JM (2002): Protein–DNA interactions: amino acid conservation and the effects of mutations on binding specificity. *J Mol Biol* 5:991–1009.

MacKerell A, Banavali N (2000): All-atom empirical force field for nucleic acids: II. Application to molecular dynamics simulations of DNA and RNA in solution. *J Comput Chem* 2:105–120.

MacKerell A, Banavali N, Foloppe N (2001): Development and current status of the CHARMM force field for nucleic acids. *Biopolymers* 4:257–265.

Mandell JG, Barbas CF (2006): Zinc Finger Tools: custom DNA-binding domains for transcription factors and nucleases. *Nucleic Acids Res* 516–523.

Marco E, Garcia-Nieto R, Gago F (2003): Assessment by molecular dynamics simulations of the structural determinants of DNA-binding specificity for transcription factor Sp1. *J Mol Biol* 9–32.

Matthews B (1988): Protein–DNA interaction. No code for recognition. *Nature* 6188:294–295.

Matys V, Fricke E, Geffers R, Gossling E, Haubrock M, Hehl R, Homishcer K, Karas D, Kel A, Kel-Margoulis O (2003): TRANSFAC: transcriptional regulation, from patterns to profiles. *Nucleic Acids Res* 374–378.

Mendieta J, Perez-Lago L, Salas M, Camacho A (2007): DNA sequence-specific recognition by a transcriptional regulator requires indirect readout of A-tracts. *Nucleic Acids Res* 35(10): 3252–3261.

Mirny L, Gelfand M (2002a): Structural analysis of conserved base pairs in protein–DNA complexes. *Nucleic Acids Res* 7:1704–1711.

Mirny LA, Gelfand MS (2002b): Using orthologous and paralogous proteins to identify specificity-determining residues in bacterial transcription factors. *J Mol Biol* 1:7–20.

Morozov AV, Havranek JJ, Baker D, Siggia ED (2005): Protein-DNA binding specificity predictions with structural models. *Nucleic Acids Res* 1:5781–5798.

Morozov AV, Siggia ED (2007): Connecting protein structure with predictions of regulatory sites. *PNAS* 17:7068–7073.

Murray LJW, Arendall WB, Richardson DC, Richardson JS (2003): RNA backbone is rotameric. *Proc Natl Acad Sci USA* 24:13904–13909.

Napoli AA, Lawson CL, Ebright RH, Berman HM (2006): Indirect readout of DNA sequence at the primary-kink site in the CAP–DNA complex: recognition of pyrimidine–purine and purine–purine steps. *J Mol Biol* 1:173–183.

Nomura W, Sugiura Y (2007): Design and synthesis of artificial zinc finger proteins. *Methods Mol Biol* 83–93.

Olson WK, Babcock MS, Gorin A, Liu G, Marky NL, Martino JA, Pedersen SC, Srinivasan AR, Tobias I, Westcott TP (1995): Flexing and folding double helical DNA. *Biophys Chem* 1:7–29.

Olson WK, Gorin AA, Lu XJ, Hock LM, Zhurkin VB (1998): DNA sequence-dependent deformability deduced from protein–DNA crystal complexes. *Proc Natl Acad Sci USA* 19:11163–11168.

Pabo C, Sauer R (1984): Protein-DNA recognition. *Annu Rev Biochem* 293–321.

Pabo C, Nekludova L (2000): Geometric analysis and comparison of protein–DNA interfaces: why is there no simple code for recognition. *J Mol Biol* 3:597–624.

Paillard G, Lavery R (2004): Analyzing protein-DNA recognition mechanisms. *Structure* 1:113–122.

Papworth M, Kolasinska P, Minczuk M (2006): Designer zinc-finger proteins and their applications. *Gene* 1:27–38.

Pastor N, MacKerell AD, Weinstein H (1999): TIT for TAT: the properties of inosine and adenosine in TATA box DNA. *J Biomol Struct Dyn* 787–810.

Peisach E, Pabo CO (2003): Constraints for zinc finger linker design as inferred from X-ray crystal structure of tandem Zif268–DNA complexes. *J Mol Biol* 1:1–7.

Pichierri F, Aida M, Gromiha MM, Sarai A (1999): Free-energy maps of base-amino acid interactions for DNA-protein recognition. *J Am Chem Soc* 26:6152–6157.

Polgar T, Keseru GM (2006): Ensemble docking into flexible active sites. Critical evaluation of FlexE against JNK-3 and beta-secretase. *J Chem Inf Model* 4:1795–1805.

Pomerantz JL, Pabo CO, Sharp PA (1995a): Analysis of homeodomain function by structure-based design of a transcription factor. *Proc Natl Acad Sci USA* 21:9752–9756.

Pomerantz JL, Sharp PA, Pabo CO (1995b): Structure-based design of transcription factors. *Science* 5194:93–96.

Pomerantz JL, Wolfe SA, Pabo CO (1998): Structure-based design of a dimeric zinc finger protein. *Biochemistry* 4:965–970.

Raviscioni M, Gu P, Sattar M, Cooney AJ, Lichtarge O (2005): Correlated evolutionary pressure at interacting transcription factors and DNA response elements can guide the rational engineering of DNA binding specificity. *J Mol Biol* 3:402–415.

Robertson TA, Varani G (2007): An all-atom, distance-dependent scoring function for the prediction of protein–DNA interactions from structure. *Proteins* 359–374.

Rojnuckarin A, Subramaniam S (1999): Knowledge-based interaction potentials for proteins. *Proteins* 1:54–67.

Samudrala R, Moult J (1998): An all-atom distance-dependent conditional probability discriminatory function for protein structure prediction. *J Mol Biol* 895–916.

Saunders CT, Baker D (2005): Recapitulation of protein family divergence using flexible backbone protein design. *J Mol Biol* 2:631–644.

Sayano K, Kono H, Gromiha M, Sarai A (2000): Multicanonical Monte Carlo calculation of the free-energy map of the base-amino acid interaction. *J Comput Chem* 11:954–962.

Schneider T, Stephens R (1990): Sequence logos: a new way to display consensus sequences. *Nucleic Acids Res* 6097–6100.

Seeman N, Rosenberg J, Rich A (1976): Sequence-specific recognition of double helical nucleic acids by proteins. *Proc Natl Acad Sci USA* 804–808.

Sippl M (1990): Calculation of conformational ensembles from potentials of mean force. *J Mol Biol* 859–883.

Sippl MJ (1995): Knowledge-based potentials for proteins. *Curr Opin Struct Biol* 2:229–235.

Skolnick J (2006): In quest of an empirical potential for protein structure prediction. *Curr Opin Struct Biol* 2:166–171.

Steffen NR, Murphy SD, Lathrop RH, Opel ML, Tolleri L, Hatfield GW (2002a): The role of DNA deformation energy at individual base steps for the identification of DNA-protein binding sites. *Genome Inform* 153–162.

Steffen N, Murphy S, Tolleri L, Hatfield G, Lathrop R (2002b): DNA sequence and structure: direct and indirect recognition in protein-DNA binding. *Bioinformatics* 22–22.

Sternberg M, Aloy P, Gabb H, Jackson R, Moont G, Querol E, Aviles F (1998): A computational system for modelling flexible protein–protein and protein–DNA docking. *Proc Int Conf Intell Syst Mol Biol* 183–192.

Swirnoff AH, Milbrandt J (1995): DNA-binding specificity of NGFI-A and related zinc finger transcription factors. *Mol Cell Biol* 4:2275–2287.

Thayer KM, Beveridge DL (2002): Hidden Markov models from molecular dynamics simulations on DNA. *Proc Natl Acad Sci USA* 1:8642–8647.

van Dijk M, van Dijk ADJ, Hsu V, Boelens R, Bonvin AMJJ (2006): Information-driven protein-DNA docking using HADDOCK: it is a matter of flexibility. *Nucleic Acids Res* 11:3317–3325.

Velec H, Gohlke H, Klebe G (2005): DrugScoreCSD—knowledge-based scoring function derived from small molecule crystal data with superior recognition rate of near-native ligand poses and better affinity prediction. *J Med Chem* 6296–6303.

Wang J, Cieplak P, Kollman P (2000): How well does a restrained electrostatic potential (RESP) model perform in calculating conformational energies of organic and biological molecules? *J Comput Chem* 1049–1074.

Weiner S, Kollman P, Case D, Singh U, Ghio C, Alagona G, Profeta S Jr, Weiner P (1984): A new force field for molecular mechanical simulation of nucleic acids and proteins. *J Am Chem Soc* 765–784.

Weiner S, Kollman P, Nguyen D, Case D (1986): An all-atom force field for simulations of proteins and nucleic acids. *J Comput Chem* 230–252.

Wolfe SA, Ramm EI, Pabo CO (2000): Combining structure-based design with phage display to create new Cys(2)His(2) zinc finger dimers. *Structure* 7:739–750.

Wolfe SA, Grant RA, Pabo CO (2003): Structure of a designed dimeric zinc finger protein bound to DNA. *Biochemistry* 46:13401–13409.

Xu Y, Xu D, Uberbacher E (1998): An efficient computational method for globally optimal threading. *J Comput Biol* 597–614.

Yoshida T, Nishimura T, Aida M, Pichierri F, Gromiha MM, Sarai A (2002): Evaluation of free energy landscape for base–amino acid interactions using ab initio force field and extensive sampling. *Biopolymers* 84–95.

Zhang C, Liu S, Zhou H, Zhou Y (2004a): An accurate, residue-level, pair potential of mean force for folding and binding based on the distance-scaled, ideal-gas reference state. *Protein Sci* 400–411.

Zhang C, Liu S, Zhou H, Zhou Y (2004b): The dependence of all-atom statistical potentials on structural training database. *Biophys J* 3349–3358.

Zhang C, Liu S, Zhu Q, Zhou Y (2005): A knowledge-based energy function for protein–ligand, protein–protein, and protein–DNA complexes. *J Med Chem* 7:2325–2335.

Zhao X, Huang X, Sun C (2006): A molecular dynamics analysis of the GCC-box binding domain in ethylene-responsive element binding factors. *J Struct Biol* 537–545.

Zhou H, Zhou Y (2002): Distance-scaled, finite ideal-gas reference state improves structure-derived potentials of mean force for structure selection and stability prediction. *Protein Sci* 2714–2726.

26

PREDICTION OF PROTEIN–PROTEIN INTERACTIONS FROM EVOLUTIONARY INFORMATION

Alfonso Valencia and Florencio Pazos

INTRODUCTION

The more we know about the molecular biology of the cell, the more we see genes and proteins as part of networks or pathways instead of as isolated entities and understand their function as a variable of the cellular context. To fulfill this new paradigm, genomic sequences offer a catalog of building blocks, and protein interactions provide the first information for the challenging task of understanding the functions of these genes and proteins within their situation in networks and pathways. To further refine the information on the context and timing in which these observed biological functions take place, additional insight from genetic regulatory mechanisms and genetic specificity (cell-type specificity, individual differences, etc.) will become increasingly important.

The study of protein interactions can be divided into two complementary aspects, the determination of the residues or regions implicated in the interactions and the deciphering of the identity of the interaction partners (which proteins interact with which ones). These two problems have been typically addressed by biophysical and biochemical techniques, such as binding studies (chromatographic isolation of complexes, coimmunoprecipitation, protection, cross-linking studies, etc.) and indirect genetic methods (gene suppression studies, systematic mutagenesis, and interspecies exchanges). The development of genomic and postgenomic technologies has changed the panorama considerably, with the possibility of obtaining systematically massive amounts of data about protein interactions. Progress has been done in the automation of experimental approaches such as yeast-two-hybrid based methods and mass spectrometry determination of components of macromolecular complexes. At the same time, a number of new bioinformatics techniques have been developed

Structural Bioinformatics, Second Edition   Edited by Jenny Gu and Philip E. Bourne
Copyright © 2009 John Wiley & Sons, Inc.

based on the considerable amount of sequence and genomic information that is being accumulated in databases. In the following, we review the status of these bioinformatics approaches to the study of protein interactions, including the prediction of interaction partners and protein–protein binding regions.

Evolutionary Features Related with Structure and Function

Multiple sequence alignments (MSA) represent the main source of evolutionary information available. The relation between equivalent positions of homologous proteins compiles the result of the changes allowed by the evolution of the corresponding proteins at different positions. The interplay of structural and functional requirements contributing to the stabilization of natural variants creates complex patterns of mutational behavior of the positions that have been exploited by a number of computational methods in the search for structural and/or functional constraints (Figure 26.1).

Conserved Positions. The information most widely extracted from multiple sequence alignments is related with conserved positions (Zuckerkandl and Pauling, 1965). These invariable positions are interpreted as important residues for the structure or function of the protein since apparently changes were disallowed during evolution (Figure 26.1). Conserved positions usually are located in structural cores (i.e., structurally important positions) and active/binding sites (i.e., positions directly involved in

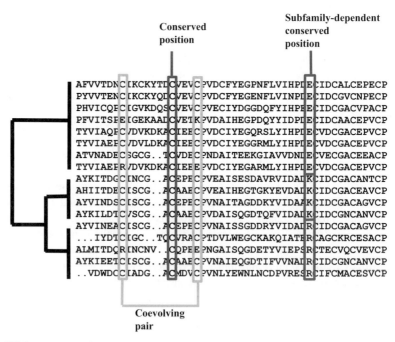

Figure 26.1. Sequence features related with structure and function extracted from multiple sequence alignments. MSA and its implicit phylogeny are shown. Conserved positions are invariant throughout all the sequences. In the family-dependent conserved positions, a different amino acid type is conserved within each one of the subfamilies. For coevolving pairs of positions, changes in one of the positions are correlated with the corresponding changes in the other.

protein function). Although the problem of locating conserved positions could look trivial at first sight, there are many different approaches that produce different results depending, for example, on the treatment of conservative changes (Valdar, 2002). Modern methods for detecting conserved sites incorporate complex evolutionary models taking into account the phylogeny of the sequences in the MSA (Pupko et al., 2002), so as to avoid artifacts due to the peculiarities or uneven distribution of sequences in the alignment (i.e., highly similar sequences resulting in most of the positions being conserved). Despite a number of attempts to analyze the relationship between binding sites, conserved positions, and type of amino acid in those positions (Villar and Kauvar, 1994; Ouzounis et al., 1998), the complex nature of the relations indicates that much remains to be done in this area.

Family-Dependent Conserved Positions. There is a more subtle kind of conservation, the family-dependent conservation. These positions are typically conserved in the subfamilies that form well-defined branches of the phylogenetic tree and differ in the chemical type of the amino acids that characterize each subfamily (Figure 26.1). We have previously used the term *tree-determinants* to describe these positions given their drastic in fluence in the structure of the corresponding phylogenetic trees. Family-dependent conserved positions would be most likely related with the specific features of each subfamily that in most cases turn out to be the differential binding to other proteins and substrates.

One of the first approaches to the detection of this kind of functional residues, implemented in the *sequencespace* program, was developed by Casari, Sander, and Valencia (1995) and later followed by a plethora of other methods (Lichtarge, Bourne, and Cohen, 1996; Andrade et al., 1997; Mesa, Pazos, and Valencia, 2003; Kalinina et al., 2004; Mihalek, Res, and Lichtarge, 2004). The underlying principle in all these methods is the detection of positions in multiple sequence alignments with a differential conservation in the different groups of sequences that form part of a larger protein family. All these methods use the subfamily classification implicit in the phylogeny represented by the MSA (Figure 26.1). Recently, some methods were developed that do not rely on that implicit classification but are able to incorporate an external classification of the sequences according to some known functional property (Hannenhalli and Russell, 2000; Pazos, Rausell, and Valencia, 2006).

Coevolving Positions. Another sequence-based approach for the prediction of protein structure and molecular complexes is based on the detection of correlated mutations in multiple sequence alignments and their use as distance constraints between residues belonging to the same or different proteins. Correlated mutations correspond to pairs of positions with a clear pattern of covariation (Figure 26.1). The underlying evolutionary model for explaining their relation with space neighboring assumes that part of the detected correlated pairs correspond to compensatory mutations, where in particular sequences of the multiple sequence alignments the mutation of one residue was compensated along the evolution by a mutation of a neighbor residue, most likely to keep proteins (or protein complexes) in permissible limits of protein stability. Coevolving positions have been demonstrated to have an important functional role together with conserved positions (Socolich et al., 2005).

The method proposed in 1994 (Göbel et al., 1994) for detecting these types of positions was a weak predictor of proximity between residues in protein structures. Despite their low accuracy, these predicted contacts have been demonstrated to be useful, for example, in filtering structural models (Olmea, Rost, and Valencia, 1999) or driving *ab initio* simulations (Ortiz et al., 1999).

PREDICTION OF INTERACTING REGIONS

Structure and Sequence Characteristics of the Interaction Sites

In general, the interaction surface (interface) is difficult to distinguish from the rest of the surface, from either a sequence or structure point of view. The interaction sites are not clearly enriched in any particular type of amino acid (Chakrabarti and Janin, 2002). Nevertheless, some particular types of complexes do have preferences for some amino acids at the interfaces, that is, hydrophobic residues in homodimers (Jones and Thornton, 1997; Ofran and Rost, 2003a). Similarly, structural characteristics are only discriminative for some type of complexes, that is, nonplanar interfaces for enzyme inhibitor and large planar interfaces for homodimers (Jones and Thornton, 1997; Lo Conte, Chothia, and Janin, 1999). For these reasons, complex machine learning systems are required for dealing with the problem of predicting interaction sites from sequence or structure information.

Structure-Based Methods

The problem of determining the physical structure of protein complexes when the structure of the members is known is part of the problem of docking molecules, such as proteins with small molecules (Chapter 27). Despite the considerable efforts that have gone in solving this problem, directed at the design of new drugs, the solutions in the case of protein–protein interaction are still far from optimal.

Despite these difficulties, various groups have achieved considerable progress in the development of physical docking programs. Most of the current approaches consider proteins as rigid bodies and have the physical matching of the surfaces as their main guide. Only a few approaches integrate conformational flexibility, allowing the interacting surfaces to adapt one to the other, normally at the expenses of reducing the search space for possible interacting surfaces. Some programs take into account other features for predicting regions of interaction, like hydrophobicity or electrostatics (Chapter 24). For recent reviews on docking, see Schneidman-Duhovny, Nussinov, and Wolfson (2004) and Gray (2006). An interesting effort to compare the various protein docking approaches in a blind test is organized by Janin et al. (CAPRI: http://capri.ebi.ac.uk/). The results of this experiment, if enough structures of protein complexes become available for the comparison, will be important for updating our view of the capability of current docking methods. A pressing question now is that the structural genomics efforts are on the way to solve a substantially larger number of isolated proteins and protein domains (Chapter 40).

There are other structure-based techniques for predicting interacting regions apart from physical docking. Some approaches are based on the fact that in many cases binding or active sites are energetically unstable. This is due to many factors, including the need to put together energetically unfavorable combinations of residues due to functional requirements (e.g., two residues with the same charge in a catalytic site) or to maintain unstable interacting regions that are stabilized upon binding, favoring in this way the bound form in the binding equilibrium. The uneven distribution of the stabilization energy in proteins has been related with binding sites and cooperativity effects (Luque and Freire, 2000). Relationships between functionally important residues and those that negatively contribute to the electrostatic stabilization energy have also been found (Elcock, 2001).

New concepts taken from graph theory have been applied to the detection of interaction sites in 3D structures. Within these methods, a protein structure is represented as an undirected graph where the nodes are the residues and the edges represent contacts between them. Within this graph, residues (nodes) with special connectivity characteristics such as high "centrality" have been associated with functional and binding sites (Del Sol and O'Meara, 2005).

The combined data on massive protein interactions (see the section "Experimental Techniques and High-Throughput Applications") and structural information can be mined in the search for short motifs or patterns frequently involved in interactions (Neduva and Russell, 2006). These motifs can be used for predicting new interacting regions.

With the increasing availability of protein 3D structures, methods that use structural alignments for predicting binding and functional sites are emerging. These methods are conceptually similar to the ones based on sequence alignments (see the section "Sequence-Based Methods") but use structural alignments to relate more distant proteins (Pazos and Sternberg, 2004; Chung, Wang, and Bourne, 2006).

Sequence-Based Methods

Ab Initio Methods. We term *ab initio* methods in this context as methods that only use the sequence of a protein to predict residues involved at the binding interface for intermolecular recognition.

This category of methods includes the detection of intrinsically unstructured segments in protein sequences, also termed "disordered regions"; see Bracken et al. (2004) and Chapter 38. To some extent opposing the paradigm that structure determines function, what determines the function of these proteins is actually their lack of structure. The functional importance of these regions arises from their frequent involvement in protein–protein interactions. In many cases, these unstructured segments become structured on binding, favoring the bound form of the interaction equilibrium. This type of entropy-driven interaction has the unique specificity/affinity characteristics required for some biological complexes (Vucetic et al., 2003). These intrinsically unstructured regions are to some extent related to the low complexity segments in the primary sequence (Wootton and Federhen, 1996; Romero et al., 2001). Nevertheless, specific predictors have been developed to locate these regions (Iakoucheva and Dunker, 2003; Ward et al., 2004).

Another method that uses a single sequence as the primary input to identify interacting regions is the neural network (NN) developed by Ofran and Rost (2003b). This NN is trained with sequence segments of interaction sites, and it correctly predicts at least one interaction site for 20% of the complexes in its test set.

Methods Based on Multiple Sequence Alignments. Most sequence-based methods for predicting protein interaction regions use evolutionary information in the form of multiple sequence alignment. For example, Ofran and Rost's method commented in the previous section obtains better results when MSA-derived information is incorporated (Ofran and Rost, 2007).

The use of evolutionary conservation, a strategy often employed to identify residues of functional importance, has also been used for detecting interaction sites. Although the best results are obtained when combining conservation with structural information (see the section "Hybrid Methods"), we include in this section of sequence-based methods predictors that combine residue conservation with *predicted* structural features that are often based on MSA. For example, *ConSeq* uses a sophisticated calculation of conservation together with predicted solvent accessibility for predicting functional residues and binding sites (Berezin et al., 2004).

The relation between tree-determinant residues (see the section "Family-Dependent Conserved Positions") and interacting surfaces has been analyzed in detail in some well-characterized systems (Casari, Sander, and Valencia, 1995; Lichtarge, Bourne, and Cohen, 1996; Atrian et al., 1997; Pazos et al., 1997b; Pereira-Leal and Seabra, 2001). More importantly, it has also been demonstrated by direct experiments how exchanging tree-determinant residues between protein subfamilies lead to a switch in the observed specificity of interaction of the corresponding proteins. This demonstration has been performed on two different proteins in the *ras* family of small GTPases (Stenmark et al., 1994; Azuma et al., 1999). In Figure 26.2, the tree-determinant residues extracted from the multiple sequence alignments of *Ran* and *Rcc1* are marked in the structure of the complex formed by these two proteins. It can be seen that many of these residues, predicted from sequence

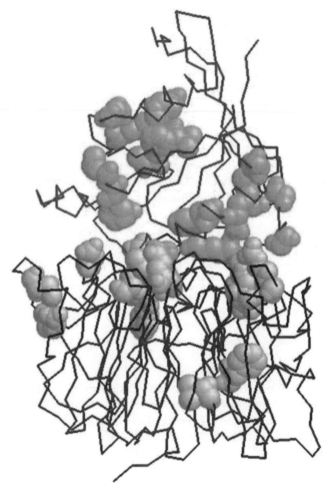

Figure 26.2. Complex between *Ran* (upper chain) and *Rcc1* (lower chain) marking the tree-determinant residues (in spacefill) for the two proteins (PDB ID: 1I2M). Tree-determinant residues were calculated with the *sequencespace* algorithm (Casari, Sander, and Valencia, 1995). Figure courtesy of Juan A.G. Ranea. Figure also appears in the Color Figure section.

information alone and before the structure of the complex was solved, map in the interaction surface, especially in *Rcc1*.

Coevolving positions (see the section "Coevolving Positions") have also been used for detecting interfaces (Pazos et al., 1997a; Yeang and Haussler, 2007; Madaoui and Guerois, 2008). In spite of the weak power of correlated mutations in predicting neighboring residues, it was demonstrated that, in many cases, it is enough to discriminate the right arrangement of two protein chains from many alternatives (Pazos et al., 1997a). Recently, it has been argued that the observed correlation appears only in permanent (not transient) complexes (Mintseris and Weng, 2005).

Hybrid Methods

Methods that incorporate both structural and sequence information to predict interaction surfaces demonstrated significant improvement from the equivalent ones that use only sequence information. The simplest way of combining sequence and structural information for predicting binding regions is by simply mapping the sequence-based predicted sites (conserved, family-dependent conserved, correlated, etc.) into the 3D structure of the protein to assess whether the predicted positions have the structural characteristics expected for a functional or binding site (clustered or accessible to the solvent) (Figure 26.3). For example,

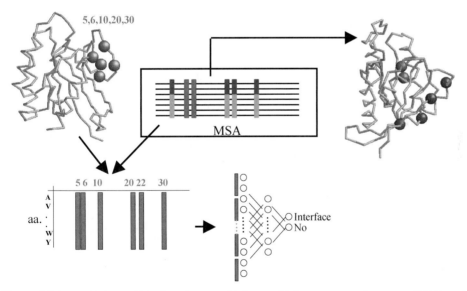

Figure 26.3. Prediction of binding sites combining multiple sequence alignments (MSA) and structural information. Conserved positions in the MSA (black) and family-dependent conserved positions (gray) have been extensively used as predictors of interaction and functional sites. If available, structural information can be used *a priori* or *a posteriori*. If used *a priori* (left), the structure is used to define surface patches (gray). The sequence features of members in this patch are encoded (i.e., amino acid frequency vectors in this case) and used as input features to a machine learning system such as a neural network (bottom). The machine learning system is trained to distinguish patches involved in interaction from non-interacting ones. *A posteriori* (right), the structure can be used to assess whether a set of positions predicted from sequence has the characteristic 3D features of a binding site such as being clustered in proximity and exposed to the solvent.

the *ConSurf* program looks for conserved residues that map in the surface of the protein, taking into account the distribution of sequences in the MSA for calculating conservation (Armon, Graur, and Ben-Tal, 2001). Panchenko et al.'s method also looks for clusters of conserved residues (Panchenko, Kondrashov, and Bryant, 2004). The method developed by Aloy et al. searches for conserved apolar residues that cluster in the surface of the protein (Aloy et al., 2001). Distant sequences in the MSA are successively removed until the set of conserved residues forms a cluster. Landgraf et al.'s method looks for surface regions of the protein following the same phylogeny as the whole family (Landgraf, Xenarios, and Eisenberg, 2001). Structural information is used to define those regions (as sets of residues close in the surface of the protein). 3D information can also be incorporated in other methods for detecting family-dependent conserved positions simply by imposing clustering (Madabushi et al., 2002; Yao et al., 2003). Some methods map sequence conservation on 3D structures but using a simplified alphabet that reflects the functional groups of the amino acids instead of considering all the 20 (Innis, Anand, and Sowdhamini, 2004).

A much direct way of combining sequence and structural information for predicting binding sites is by using machine learning systems, such as neural networks or support vector machines (SVM). This general strategy was introduced by Zhou and Shan (2001) and Fariselli et al. (2002), and later followed by many other authors (Koike and Takagi, 2004; Bordner and Abagyan, 2005; Bradford and Westhead, 2005). These methods use the structure of a protein to define surface patches of neighbor residues and the multiple sequence alignment to obtain the sequence profiles (or other sequence-based features) for the members of the patch (Figure 26.3). The machine learning system is then trained with that information for a training set of proteins with known interaction surfaces. After the training process, surface patches (plus their corresponding sequence features) of proteins with known structure but unknown interaction surface are presented to the system to identify which surface patches are involved in intermolecular recognition. The reported accuracy of these methods is higher than 70%. Nevertheless, it is difficult to test these methods and compare their performance due to the absence of a good test set of protein interfaces (nonredundant in sequence and structure, covering all types of complexes, etc).

PREDICTION OF INTERACTION PARTNERS

Experimental Techniques and High-Throughput Applications

The experimental approaches for the determination of interaction partners have undergone dramatic improvements during the last few years, particularly through the systematic application of different strategies based on the yeast-two-hybrid protocol (Fields and Song, 1989) and the tandem affinity purification (TAP) of complexes followed by mass spectrometry (MS) analysis (Gavin et al., 2002). These approaches were used to determine large proportions of the interactomes of a number of model organisms, ranging from bacteria such as *H. pylori* (Rain et al., 2001) to human (Stelzl et al., 2005), covering unicelular eukaryotes such as yeast (Gavin et al., 2002; Ito et al., 2000; Uetz et al., 2000) or multicelular organisms such as *C. elegans* (Li et al., 2004) and *D. melanogaster* (Giot et al., 2003). These interactomes, in spite of having provided a lot of information on living systems (Uetz and Finley, 2005), still have a high degree of error (von Mering et al., 2002; Aloy and Russell, 2002b) and their determination is very expensive in terms of time and money (see the section "Future Trends").

Computational Methods Based on Genomic Information

In parallel to these experimental approaches for the detection of interacting pairs of proteins, a number of bioinformatics techniques have also been developed. These methods, based on genomic and sequence features intuitively related with protein interaction, usually detect pairs of proteins that may interact physically or functionally (i.e., proteins that form part of the same signaling or metabolic pathway).

The results of many of these methods can be accessed online at repositories such as STRING (von Mering et al., 2003) (http://string.embl.de). These data can be used for inferring the functional role of a protein by identifying its potential interactors. This is called "context-based" function prediction and is orthogonal and complementary to the classic sequence-based functional transfer.

The main methods based on sequence and genomic features for predicting interacting pairs of proteins are depicted in Figure 26.4.

Phylogenetic Profiling. This method is based on the detection of pairs of genes that have a similar species distribution. That is, they are present/absent in the same species (Pellegrini et al., 1999). The idea behind this approach is that proteins that need each other to perform a given function will either be simultaneously present or absent. In the second case, loss of interacting partners can be observed as a consequence of "reductive evolution" where the organism (especially bacteria) would eliminate genes when the corresponding interacting partner is not present. As a result, interacting or functionally related genes would tend to have similar phylogenetic profiles, that is, similar patters of presence/absence (Figure 26.4, panel 4).

In their first versions, phylogenetic profiles were coded as binary vectors with "1" coding for the presence of a given gene and "0" coding for its absence (Figure 26.4, panel 4). Later, quantitative information was incorporated by encoding the similarity of a protein sequence in a given organism with respect to an organism of reference into the positions of the vector, representing in this way not only the presences/absences of the genes but their relative divergences as well (Date and Marcotte, 2003). Another improvement came from the incorporation of information on the phylogeny of the species involved, together with an evolutionary model of gene gain and loss. This allows to naturally exclude profile similarities not due to functional reasons but rather to the underlying evolutionary (Zhou et al., 2006; Barker, Meade, and Pagel, 2007).

Not only are similar profiles informative but also the "anticorrelated" ones (one protein is present whenever the other is absent and the other way around). These anticorrelated profiles have been related with enzyme "displacement" in metabolic pathways (Morett et al., 2003). Furthermore, this versatile technique has recently been extended to triplets of proteins, allowing the search for more complicated patterns of presence/absence (e.g., "protein C is present if A is absent and B is also absent"), which allows the detection of interesting cases representing biological phenomena beyond binary functional interactions, like complementation (Bowers et al., 2004).

This successful approach has two main limitations. First, it cannot be applied to many essential proteins, since they are present in all organisms and hence they create noninformative profiles. Second, this methodology requires fully sequenced genomes in order to accurately determine whether a given gene is present in an organism or not.

Conservation of Gene Neighboring. The conservation of the proximity of genes along the genome between distantly related species to predict interaction is also being

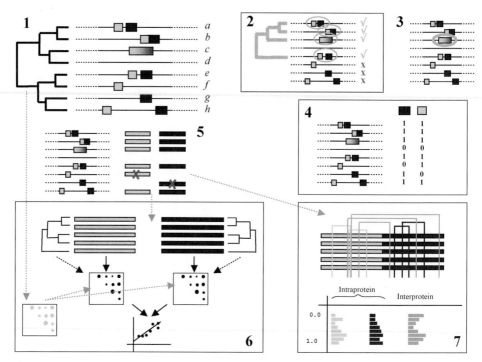

Figure 26.4. Theme of strategies implemented in computational methods for assessing the possible interaction between two proteins. (1) Sequence and genomic information about two proteins (yellow and blue) is used to assess their possible interaction. The inputs for these methods are the sequences and genomic positions of the orthologs of the two proteins, as defined by a given orthology criteria, in a number of organisms (related by a phylogeny). (2) Conservation of gene neighboring: the number of genomes where both proteins are close (according with a given distance cutoff) and their phylogeny are used to assess whether the proteins are interacting or not. (3) Gene fusion: a search for genomes where both proteins appear as part of a single polypeptide is performed. (4) Phylogenetic profiling: phylogenetic profiles for both proteins are constructed by assessing the presence [1] or absence [0] of the two proteins in the set of genomes, and the similarity between these profiles is evaluated. (5) Multiple sequence alignments for the two proteins are built using the sequences of the organisms where both are present. These MSA are used by the following two methods. (6) Similarity of phylogenetic trees: the multiple sequence alignments of the proteins (5) are used to generate distance matrices for both sets of orthologs. Alternatively, these multiple sequence alignments can be used to generate the actual phylogenetic trees and extract the distance matrices from them. The similarity of these distance matrices is used as an indicator of interaction. Eventually, the phylogenetic distances between the species involved can be incorporated in the method for correcting the background similarity expected between the trees due to the underlying speciation events. (7) Correlated mutation: the accumulation of correlated mutations between the two multiple sequence alignments is used as an indicator of interaction. Figure also appears in the Color Figure section.

utilized by several methods (Dandekar et al., 1998; Overbeek et al., 1999). A typical example would be proteins encoded by genes within a conserved operon where, if present together in another bacteria, would contain the sufficient, necessary components to execute the functional mechanism related with the activity of that operon.

The limitation of this technique is that this approach is reserved for bacterial genomes where there is a clear tendency for the clustering of functionally related genes in operons to allow cotranscription. Inferences can only be made for eukaryotic proteins only if homologues are clearly established in a prokaryotic organism.

Gene Fusion. A third group of methods is related to the presence of fused genes in various genomes (Enright et al., 1999; Marcotte et al., 1999). This strategy uses the observation that some interacting protein pairs are encoded by two independent genes in some organisms whereas they are encoded by a single gene in other organisms. In such cases, it will be logical to conclude that the two proteins, as independent entities or as domains of the same protein, would be functionally related.

Marcotte et al. (1999) proposed an evolutionary hypothesis for explaining such fusion events: if two proteins have to interact in order to perform a given function, the concentration of the active complex would be much higher if the two proteins are fused together than if the two proteins are separated and hence having to rely on random diffusion to find each other and form the active complex.

The fact that two proteins are fused is a clear indication of their functional relationship (except for "promiscuous domains"). Hence, this approach produces almost no false positives. The disadvantage, however, is the range of applicability because these fusion events, in spite of being very informative, are not very frequent.

Similarity of Phylogenetic Trees. The observation that interacting proteins tend to have topologically similar phylogenetic trees has been used by some methods to predict interaction partners. This similarity was qualitatively observed in sporadic examples of interacting families such as insulin and insulin receptors (Fryxell, 1996), and first quantified for two proteins by Goh et al. (2000). Pazos and Valencia (2001) statistically demonstrated the relation between the similarity of phylogenetic trees and interaction in large sets of interacting proteins. Figure 26.5 shows an example of two interacting proteins (the nuoE and nuoF subunits of the NADH dehydrogenase complex) that have similar phylogenetic trees.

The hypothesis behind this approach is that interacting proteins would be subject to a process of coevolution that would be translated into a stronger than expected similarity between their phylogenetic trees (Juan et al., 2008a; Pazos and Valencia, 2008). The first generation algorithms of this approach measure similarity between the phylogenetic trees by indirectly evaluating it as the similarity between the distance matrices of the two families, using a correlation formulation (Pazos and Valencia, 2001).

Since it was proposed, this simple and intuitive method (*mirrortree*) has been applied to many proteins, and different implementations and variations of it were developed (Juan et al., 2008a; Pazos and Valencia, 2008). For example, this concept of similarity of trees was used to look for the correct mapping between two families of interacting proteins (e.g., to select which ligand within a family interacts with which receptor within another family). The idea is that the correct mapping (set of relationships between the leaves of both trees) will be the one maximizing the similarity between both trees (Ramani and Marcotte, 2003; Izarzugaza et al., 2006). Another obvious extension of the method has been to incorporate information on the phylogeny of the species involved in the trees in order to correct the "background" similarity expected between any pair of trees due to the underlying speciation process (Pazos et al., 2005; Sato et al., 2005). Recently, it has also been shown that incorporating information on the whole network of interprotein co-evolutions drastically improves the accuracy of this methodology (Juan et al., 2008b).

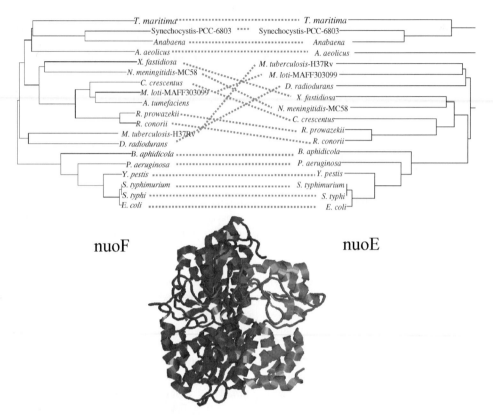

Figure 26.5. Example of two interacting proteins that have similar phylogenetic trees. The *E. coli* proteins are the *nuoE* (gray) and *nuoF* (black) subunits of the N module of the NADH reductase. The structure corresponds to the *T. thermophilus* orthologs (chains B and A in PDB 2FUG). The phylogenetic trees of these two proteins are shown at the top. The species correspondence is indicated with dotted lines.

One important limitation of this method is that it can only be applied to pairs of proteins where the orthologs are found in many common species. Only the leaves of the trees corresponding to species where both proteins are present can be used (Figure 26.4, panel 6).

Other Sequence-Based Methods. There are a number of other methods for inferring protein interactions from genomic information.

Coevolving positions (see the section "Coevolving Positions") have been used for predicting interacting pairs of proteins (Pazos and Valencia, 2002), extending their previous usage for detecting interacting surfaces (see the section "Methods based on Multiple Sequence Alignments"). The idea is that pairs of multiple sequence alignments corresponding to interacting proteins would be more enriched in interprotein correlated mutations than pairs that do not interact.

The methods described so far do not involve training, that is, they do not "learn" from examples of known interactions and noninteractions. On the contrary, there are other methods that are trained with examples (Sprinzak and Margalit, 2001; Jansen et al., 2003; Yamanishi, Vert, and Kanehisa, 2004; Ben-Hur and Noble, 2005; Chen and Liu, 2005). In general, the input is a set of characteristics (descriptors) of the proteins or protein pairs (including experimental data in some cases). Based on a set of known interactions, a machine learning classifier "learns" to distinguish interacting from noninteracting pairs using the values of these descriptors. For example, Sprinzak and Margalit (2001) use pairs of sequence signatures extracted from known interactions to predict new ones. One of the descriptors used by some of these methods is the domain composition of the proteins (Bornberg-Bauer et al., 2005; Chen and Liu, 2005). The idea is that some combinations of domains are more prone to interact.

Computational Methods Based on Structural Information

There are also computational methods for the prediction of interaction partners that use structural information. Most of these methods are intended to predict whether the homologues of two proteins known to interact will interact too or not. Aloy et al. derived statistical potentials from known interactions and then used them to score the possible interactions between the homologues of the members of a given complex (Aloy and Russell, 2002a; Aloy and Russell, 2003). In a similar way, the energetic feasibility of different complexes between members of the Ras family and different families of Ras effectors was evaluated using the FOLD-X program (Kiel et al., 2007).

FUTURE TRENDS

Deciphering the complex network of protein interactions behind cellular processes is crucial for understanding many aspects of living systems and is being tackled with ongoing experimental and computational techniques in need of further developments. Currently, the experimental techniques for the massive characterization of these networks still have several important drawbacks. First, there is a surprisingly low overlap between the results of similar experiments (Uetz and Finley, 2005). Second, experimental techniques such as yeast two hybrid often yield a large number of false positives with an accuracy estimated to be as low as ∼10% in some cases (von Mering et al., 2002). Third, experimental approaches are still far from reaching a "high-throughput" state since the intrinsic drawbacks of the methodologies allow them to test a fraction of all possible pairs of proteins (Uetz and Finley, 2005). Finally, other limitations of these techniques paradoxically arise from their experimental nature. These include the tendency to preferentially detect interactions between highly expressed proteins or between proteins belonging to a particular cellular compartment in detriment of others (von Mering et al., 2002).

Computational methods, while having their own set of limitations, are free from many of these drawbacks faced by experimental approaches. Thus, these two classes of methods complement each other. A common limitation of the computational methods for detecting interaction partners is the dependency on the quantity and quality of the genomic data. For this reason, we expect the accuracy of these methods to improve as more sequences and genomes become available. The use of structural information can help improving the performance of current methodologies. For example, HMM derived from structural alignments (Chapters 16 and 22) are widely used for detecting remote homologues. The localization of homologues is common to many of these computational techniques.

We are already seeing a combination of results from the new experimental powerful techniques, the new bioinformatics approaches for the prediction of interacting partners, the relations established between genes with similar expression patterns in DNA array experiments, and the systematic mining of available information about protein interactions in databases and literature repositories (Hoffmann and Valencia, 2004). Combining and finding consensus within these accumulated information obtained from different sources, we will quickly approach a situation in which it will be possible to glimpse the structure of the complex network of protein interactions that govern cell life in relatively simple model systems such as yeast and bacteria.

The methods for predicting interaction partners and detecting interacting regions are helping in the interpretation of the massive amounts of sequence and genomic information in functional terms. The promise is that the combination of improved experimental techniques and computational methods along with an effective consensus interpretation of accumulated data from widely disparate sources would lead not only to a better understanding of cell function but also to better ways of manipulating it with new and more specialized drugs designed specifically for that purpose.

REFERENCES

Aloy P, Querol E, Aviles FX, Sternberg MJE (2001): Automated structure-based prediction of functional sites in proteins: applications to assessing the validity of inheriting protein function from homology in genome annotation and to protein docking. *J Mol Biol* 311:395–408.

Aloy P, Russell RB (2002a): Interrogating protein interaction networks through structural biology. *Proc Natl Acad Sci USA* 99:5896–5901.

Aloy P, Russell RB (2002b): Potential artefacts in protein-interaction networks. *FEBS Lett* 530:253–254.

Aloy P, Russell RB (2003): InterPreTS: protein interaction prediction through tertiary structure. *Bioinformatics* 19:161–162.

Andrade MA, Casari G, Sander C, Valencia A (1997): Classification of protein families and detection of the determinant residues with an improved self-organizing map. *Biol Cybern* 76:441–450.

Armon A, Graur D, Ben-Tal N (2001): ConSurf: an algorithmic tool for the identification of functional regions in proteins by surface mapping of phylogenetic information. *J Mol Biol* 307:447–463.

Atrian S, Sanchez-Pulido L, Gonzàlez-Duarte R, Valencia A (1997): Shaping of *Drosophila* alcohol dehydrogenase through evolution. relationship with enzyme functionality. *J Mol Evol* 47:211–221.

Azuma Y, Renault L, Garcia-Ranea JA, Valencia A, Nishimoto T, Wittinghofer A (1999): Model of the Ran–RCC1 interaction using biochemical and docking experiments. *J Mol Biol* 289: 1119–1130.

Barker D, Meade A, Pagel M (2007): Constrained models of evolution lead to improved prediction of functional linkage from correlated gain and loss of genes. *Bioinformatics* 23:14–20.

Ben-Hur A, Noble WS (2005): Kernel methods for predicting protein–protein interactions. *Bioinformatics* 21:i38–46.

Berezin C, Glaser F, Rosenberg J, Paz I, Pupko T, Fariselli P, Casadio R, Ben-Tal N (2004): ConSeq: the identification of functionally and structurally important residues in protein sequences. *Bioinformatics* 20:1322–1324.

Bordner AJ, Abagyan R (2005): Statistical analysis and prediction of protein–protein interfaces. *Proteins* 60:353–366.

Bornberg-Bauer E, Beaussart F, Kummerfeld SK, Teichmann SA, Weiner J 3rd (2005): The evolution of domain arrangements in proteins and interaction networks. *Cell Mol Life Sci* 62:435–445.

Bowers PM, Cokus SJ, Eisenberg D, Yeates TO (2004): Use of logic relationships to decipher protein network organization. *Science* 306:2246–2249.

Bracken C, Iakoucheva LM, Romero PR, Dunker AK (2004): Combining prediction, computation and experiment for the characterization of protein disorder. *Curr Opin Struct Biol* 14:570–576.

Bradford JR, Westhead DR (2005): Improved prediction of protein–protein binding sites using a support vector machines approach. *Bioinformatics* 21:1487–1494.

Casari G, Sander C, Valencia A (1995): A method to predict functional residues in proteins. *Nat Struct Biol* 2:171–178.

Chakrabarti P, Janin J (2002): Dissecting protein–protein recognition sites. *Proteins* 47:334–343.

Chen XW, Liu M (2005): Prediction of protein–protein interactions using random decision forest framework. *Bioinformatics* 21:4394–4400.

Chung JL, Wang W, Bourne PE (2006): Exploiting sequence and structure homologs to identify protein–protein binding sites. *Proteins* 62:630–640.

Dandekar T, Snel B, Huynen M, Bork P (1998): Conservation of gene order: a fingerprint of proteins that physically interact. *Trends Biochem Sci* 23:324–328.

Date SV, Marcotte EM (2003): Discovery of uncharacterized cellular systems by genome-wide analysis of functional linkages. *Nat Biotechnol* 21:1055–1062.

Del Sol A, O'Meara P (2005): Small-world network approach to identify key residues in protein–protein interaction. *Proteins* 58:672–682.

del Sol Mesa A, Pazos F, Valencia A (2003): Automatic methods for predicting functionally important residues. *J Mol Biol* 326:1289–1302.

Elcock A (2001): Prediction of functionally important residues based solely on the computed energetics of protein structure. *J Mol Biol* 312:885–896.

Enright AJ, Iliopoulos I, Kyrpides NC, Ouzounis CA (1999): Protein interaction maps for complete genomes based on gene fusion events. *Nature* 402:86–90.

Fariselli P, Pazos F, Valencia A, Casadio R (2002): Prediction of protein–protein interaction sites in heterocomplexes with neural networks. *Eur J Biochem* 269:1356–1361.

Fields S, Song O (1989): A novel genetic system to detect protein–protein interactions. *Nature* 340:245–246.

Fryxell KJ (1996): The coevolution of gene family trees. *Trends Genet* 12:364–369.

Gavin AC, Bösche M, Krause R, Grandi P, Marcioch M, Bauer A, Schultz J, Rick JM, Michon AM, Cruciat CM, et al. (2002): Functional organisation of the yeast proteome by systematic analysis of protein complexes. *Nature* 415:141–147.

Giot L, Bader JS, Brouwer C, Chaudhuri A, Kuang B, Li Y, Hao YL, Ooi CE, Godwin B, Vitols E, et al. (2003): A protein interaction map of *Drosophila melanogaster*. *Science* 302:1727–1736.

Göbel U, Sander C, Schneider R, Valencia A (1994): Correlated mutations and residue contacts in proteins. *Proteins* 18:309–317.

Goh C-S, Bogan AA, Joachimiak M, Walther D, Cohen FE (2000): Co-evolution of proteins with their interaction partners. *J Mol Biol* 299:283–293.

Gray JJ (2006): High-resolution protein–protein docking. *Curr Opin Struct Biol* 16:183–193.

Hannenhalli SS, Russell RB (2000): Analysis and prediction of functional sub-types from protein sequence alignments. *J Mol Biol* 303:61–76.

Hoffmann R, Valencia A (2004): A gene network for navigating the literature. *Nat Genet* 36:664.

Iakoucheva LM, Dunker AK (2003): Order, disorder, and flexibility: prediction from protein sequence. *Structure* 11:1316–1317.

Innis CA, Anand AP, Sowdhamini R (2004): Prediction of functional sites in proteins using conserved functional group analysis. *J Mol Biol* 337:1053–1068.

Ito T, Tashiro K, Muta S, Ozawa R, Chiba T, Nishizawa M, Yamamoto K, Kuhara S, Sakaki Y (2000): Toward a protein–protein interaction map of the budding yeast: a comprehensive system to examine two-hybrid interactions in all possible combinations between the yeast proteins. *Proc Natl Acad Sci USA* 97:1143–1147.

Izarzugaza JM, Juan D, Pons C, Ranea JA, Valencia A, Pazos F (2006): TSEMA: interactive prediction of protein pairings between interacting families. *Nucleic Acids Res* 34:W315–319.

Jansen R, Yu H, Greenbaum D, Kluger Y, Krogan NJ, Chung S, Emili A, Snyder M, Greenblatt JF, Gerstein M (2003): A Bayesian networks approach for predicting protein–protein interactions from genomic data. *Science* 302:449–453.

Jones S, Thornton JM (1997): Analysis of protein–protein interaction sites using surface patches. *J Mol Biol* 272:121–132.

Juan D, Pazos F, Valencia A (2008a): Co-evolution and co-adaptation in protein networks. *FEBS Lett* 582:1225–1230.

Kalinina OV, Mironov AA, Gelfand MS, Rakhmaninova AB (2004): Automated selection of positions determining functional specificity of proteins by comparative analysis of orthologous groups in protein families. *Protein Sci* 13:443–456.

Kiel C, Foglierini M, Kuemmerer N, Beltrao P, Serrano L (2007): A genome-wide Ras-effector interaction network. *J Mol Biol* 370:1020–1032.

Koike A, Takagi T (2004): Prediction of protein–protein interaction sites using support vector machines. *Protein Eng Des Sel* 17:165–173.

Landgraf R, Xenarios I, Eisenberg D (2001): Three-dimensional cluster analysis identifies interfaces and functional residue clusters in proteins. *J Mol Biol* 307:1487–1502.

Li S, Armstrong CM, Bertin N, Ge H, Milstein S, Boxem M, Vidalain PO, Han JD, Chesneau A, Hao T, et al. (2004): A map of the interactome network of the metazoan *C. elegans. Science* 303:540–543.

Lichtarge O, Bourne HR, Cohen FE (1996): An evolutionary trace method defines binding surfaces common to protein families. *J Mol Biol* 257:342–358.

Lo Conte L, Chothia C, Janin J (1999): The atomic structure of protein–protein recognition sites. *J Mol Biol* 285:2177–2198.

Luque I, Freire E (2000): Structural stability of binding sites: consequences for binding affinity and allosteric effects. *Proteins* S4:63–71.

Madabushi S, Yao H, Marsh M, Kristensen DM, Philippi A, Sowa ME, Lichtarge O (2002): Structural clusters of evolutionary trace residues are statistically significant and common in proteins. *J Mol Biol* 316:139–154.

Madaoui H, Guerois R (2008): Coevolution at protein complex interfaces can be detected by the complementarity trace with important impact for predictive docking. *Proc Natl Acad Sci USA* 105:7708–7713.

Marcotte EM, Pellegrini M, Thompson MJ, Yeates TO, Eisenberg D (1999): A combined algorithm for genome-wide prediction of protein function. *Nature* 402:83–86.

Mihalek I, Res I, Lichtarge O (2004): A family of evolution–entropy hybrid methods for ranking protein residues by importance. *J Mol Biol* 336:1265–1282.

Mintseris J, Weng Z (2005): Structure, function, and evolution of transient and obligate protein–protein interactions. *Proc Natl Acad Sci USA* 102:10930–10935.

Morett E, Korbel JO, Rajan E, Saab-Rincon G, Olvera L, Olvera M, Schmidt S, Snel B, Bork P (2003): Systematic discovery of analogous enzymes in thiamin biosynthesis. *Nat Biotechnol* 21:790–795.

Neduva V, Russell RB (2006): Peptides mediating interaction networks: new leads at last. *Curr Opin Biotechnol* 17:465–471.

Ofran Y, Rost B (2003a): Analysing six types of protein–protein interfaces. *J Mol Biol* 325:377–387.

Ofran Y, Rost B (2003b): Predicted protein–protein interaction sites from local sequence information. *FEBS Lett* 544:236–239.

Ofran Y, Rost B (2007): ISIS: interaction sites identified from sequence. *Bioinformatics.* 23:e13–e16.

Olmea O, Rost B, Valencia A (1999): Effective use of sequence correlation and conservation in fold recognition. *J Mol Biol* 293:1221–1239.

Ortiz A, Kolinski A, Rotkiewicz P, Ilkowski B, Skolnick J (1999): *Ab initio* folding of proteins using restraints derived from evolutionary information. *Proteins* S3:177–185.

Ouzounis C, Perez-Irratxeta C, Sander C, Valencia A (1998): Are binding residues conserved? *Pac Symp Biocomput* 401–412.

Overbeek R, Fonstein M, D'Souza M, Pusch GD, Maltsev N (1999): Use of contiguity on the chromosome to predict functional coupling. *In Silico Biol* 1:93–108.

Panchenko AR, Kondrashov F, Bryant S (2004): Prediction of functional sites by analysis of sequence and structure conservation. *Protein Sci* 13:884–892.

Pazos F, Helmer-Citterich M, Ausiello G, Valencia A (1997a): Correlated mutations contain information about protein–protein interaction. *J Mol Biol* 271:511–523.

Pazos F, Sanchez-Pulido L, García-Ranea JA, Andrade MA, Atrian S, Valencia A (1997b): Comparative analysis of different methods for the detection of specificity regions in protein families. In: Lundh D, Olsson B, Narayanan A, editors. *Biocomputing and Emergent Computation.* Singapore: World Scientific, pp 132–145.

Pazos F, Valencia A (2001): Similarity of phylogenetic trees as indicator of protein–protein interaction. *Protein Eng* 14:609–614.

Pazos F, Valencia A (2002): *In silico* two-hybrid system for the selection of physically interacting protein pairs. *Proteins* 47:219–227.

Pazos F, Valencia A (2008): Protein co-evolution, co-adaptation and interactions. *Embo J* 27:2648–2655.

Pazos F, Sternberg MJE (2004): Automated prediction of protein function and detection of functional sites from structure. *Proc Natl Acad Sci USA* 101:14754–14759.

Pazos F, Ranea JAG, Juan D, Sternberg MJE (2005): Assessing protein co-evolution in the context of the tree of life assists in the prediction of the interactome. *J Mol Biol* 352:1002–1015.

Pazos F, Rausell A, Valencia A (2006): Phylogeny-independent detection of functional residues. *Bioinformatics* 22:1440–1448.

Pellegrini M, Marcotte EM, Thompson MJ, Eisenberg D, Yeates TO (1999): Assigning protein functions by comparative genome analysis: protein pylogenetic profiles. *Proc Natl Acad Sci USA* 96:4285–4288.

Pereira-Leal JB, Seabra MC (2001): Evolution of the Rab family of small GTP-binding proteins. *J Mol Biol* 313:889–901.

Pupko T, Bell RE, Mayrose I, Glaser F, Ben-Tal N (2002): Rate4Site: an algorithmic tool for the identification of functional regions in proteins by surface mapping of evolutionary determinants within their homologues. *Bioinformatics* 18:S71–S77.

Rain JC, Selig L, De Reuse H, Battaglia V, Reverdy C, Simon S, Lenzen G, Petel F, Wojcik J, Schächter V, et al. (2001): The protein–protein interaction map of *Helicobacter pylori. Nature* 409:211–215.

Ramani AK, Marcotte EM (2003): Exploiting the co-evolution of interacting proteins to discover interaction specificity. *J Mol Biol* 327:273–284.

Romero P, Obradovic Z, Li X, Garner EC, Brown CJ, Dunker AK (2001): Sequence complexity of disordered protein. *Proteins* 42:38–48.

Sato T, Yamanishi Y, Kanehisa M, Toh H (2005): The inference of protein–protein interactions by co-evolutionary analysis is improved by excluding the information about the phylogenetic relationships. *Bioinformatics* 21:3482–3489.

Schneidman-Duhovny D, Nussinov R, Wolfson HJ (2004): Predicting molecular interactions *in silico*: II. Protein–protein and protein–drug docking. *Curr Med Chem* 11:91–107.

Socolich M, Lockless SW, Russ WP, Lee H, Gardner KH, Ranganathan R (2005): Evolutionary information for specifying a protein fold. *Nature* 437:512–518.

Sprinzak E, Margalit H (2001): Correlated sequence-signatures as markers of protein–protein interactions. *J Mol Biol* 311:681–692.

Stelzl U, Worm U, Lalowski M, Haenig C, Brembeck FH, Goehler H, Stroedicke M, Zenkner M, Schoenherr A, Koeppen S, et al. (2005): A human protein–protein interaction network: a resource for annotating the proteome. *Cell* 122:957–968.

Stenmark H, Valencia A, Martinez O, Ulrich O, Goud B, Zerial M (1994): Distinct structural elements of rab5 define its functional specificity. *EMBO J* 13:575–583.

Uetz P, Giot L, Cagney G, Mansfield TA, Judson RS, Knight JR, Lockshon D, Narayan V, Srinivasan M, Pochart P, et al. (2000): A comprehensive analysis of protein–protein interactions in *Saccharomyces cerevisiae*. *Nature* 403:623–631.

Uetz P, Finley RL Jr (2005): From protein networks to biological systems. *FEBS Lett* 579:1821–1827.

Valdar WS (2002): Scoring residue conservation. *Proteins* 48:227–241.

Villar HO, Kauvar LM (1994): Amino-acid preferences at protein binding sites. *FEBS Lett* 349:125–130.

von Mering C, Krause R, Snel B, Cornell M, Oliver SG, Fields S, Bork P (2002): Comparative assessment of large scale data sets of protein–protein interactions. *Nature* 417:399–403.

von Mering C, Huynen M, Jaeggi D, Schmidt S, Bork P, Snel B (2003): STRING: a database of predicted functional associations between proteins. *Nucleic Acids Res* 31:258–261.

Vucetic S, Brown CJ, Dunker AK, Obradovic Z (2003): Flavors of protein disorder. *Proteins* 52: 573–584.

Ward JJ, Sodhi JS, McGuffin LJ, Buxton BF, Jones DT (2004): Prediction and functional analysis of native disorder in proteins from the three kingdoms of life. *J Mol Biol* 337:635–645.

Wootton JC, Federhen S (1996): Analysis of compositionally biased regions in sequence databases. *Methods Enzymol* 266:554–571.

Yamanishi Y, Vert JP, Kanehisa M (2004): Protein network inference from multiple genomic data: a supervised approach. *Bioinformatics* 20:I363–I370.

Yao H, Kristensen DM, Mihalek I, Sowa ME, Shaw C, Kimmel M, Kavraki L, Lichtarge O (2003): An accurate, sensitive, and scalable method to identify functional sites in protein structures. *J Mol Biol* 326:255–261.

Yeang CH, Haussler D (2007): Detecting coevolution in and among protein domains. *PLoS Comput Biol* 3:e211.

Zhou H-X, Shan Y (2001): Prediction of protein interaction sites from sequence profile and residue neighbor list. *Proteins* 44:336–343.

Zhou Y, Wang R, Li L, Xia X, Sun Z (2006): Inferring functional linkages between proteins from evolutionary scenarios. *J Mol Biol* 359:1150–1159.

Zuckerkandl E, Pauling L (1965): Evolutionary divergence and convergence in proteins. In: Bryson V, Vogel HJ, editors. *Evolving Genes And Proteins*. New York: Academic Press, pp 97–166.

27

DOCKING METHODS, LIGAND DESIGN, AND VALIDATING DATA SETS IN THE STRUCTURAL GENOMICS ERA

Natasja Brooijmans

INTRODUCTION

Since the beginning of civilization, humans have searched for substances that can cure or alleviate the symptoms of disease. In the early stages, extracts from plants and animal parts were used to treat disease, and the discovery of such remedies was driven empirically. Starting in the early 1900s, drug discovery has increasingly focused on discovering and developing chemical entities that on their own have a desired pharmaceutical effect. Initially this was fueled by attempts to extract and identify the active component in extracts from natural products that were used. A number of developments, however, have resulted in the multidisciplinary science that drug discovery is today, including the traditional fields of chemistry and pharmacology, along with contributions from biochemistry, molecular biology, and biophysics. Increasingly, computational tools are used in the drug discovery process from target identification and validation to the designing of new molecules.

The discoveries of several different concepts played a pivotal role in the development of drug discovery as an industry. In the nineteenth century, Paul Ehrlich developed the idea that different parasites, microorganisms, and cancer cells have dissimilar chemoreceptors resulting in different susceptibility to dyes. He hypothesized that these differences can be exploited therapeutically (Ehrlich, 1957). In 1905, Langley expanded the concept of receptors to being binding sites for different substances. Binding of the substrate to the binding site can result in either stimulation or blocking of the receptor depending on the substance used (Langley, 1905).

Emil Fisher developed the "lock-and-key" concept for enzymes, which stipulates that the substrate has to fit exactly into the binding site for the enzyme to perform its function on

Structural Bioinformatics, Second Edition   Edited by Jenny Gu and Philip E. Bourne
Copyright © 2009 John Wiley & Sons, Inc.

the substrate (Fischer, 1894). Koshland put a slightly more flexible point of view forward in the "induced-fit" model that states that both ligand and enzyme undergo conformational changes to fit optimally (Koshland, 1958). More recently, Foote and Milstein put forward the idea that optimally fitting conformers of both receptor and ligand exist in solution, and that in the presence of each other, there is a shift in the equilibrium toward the best-fitting conformers for both, resulting in binding (Foote and Milstein, 1994).

The field of structure-based drug design (Chapter 34) most explicitly exploits the concept of three-dimensional binding sites that interact with ligands; computational methods and models play herein a critical role (Figure 27.1). In structure-based design, the known or predicted shape of the binding site is used to optimize the ligand to best fit the receptor. The driving forces of these specific interactions in biological systems are driven by complementarities in both shape and electrostatics of the binding site surfaces and the ligand or substrate. Van der Waals interactions thus play a role, in addition to Coulombic interactions and the formation of hydrogen bonds (Figure 27.2). Docking is one of the tools used to understand and explore the steric and electrostatic complementarity between the receptor and the ligand.

The interactions between the receptor and ligand are quantum mechanical in nature, but due to the complexity of biological systems, quantum theory cannot be applied directly. Consequently, most methods used in docking and computational drug discovery are more empirical in nature and usually lack generality. Quantum mechanical phenomena, such as the formation of a covalent bond between the protein and the ligand upon binding during the transition state of the reaction, cannot be predicted and/or evaluated using these empirical methods.

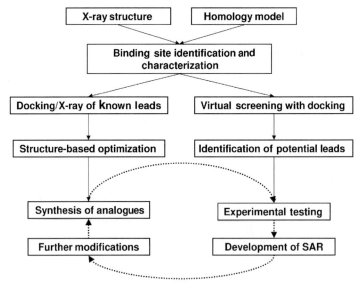

Figure 27.1. Structure-based drug discovery cycle. Either an X-ray structure or a homology model can be used for binding mode predictions of known leads and virtual screening to find new leads. The known or predicted binding mode of leads can be used to design analogues with better and more interactions with the protein. Prioritized analogues have to be synthesized, experimentally tested, and the structure–activity relationships obtained can be used to further optimize the ligands.

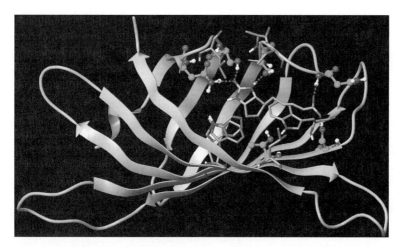

Figure 27.2. Streptavidin bound to biotin. Picture highlighting interactions between streptavidin (orange carbon atoms) and biotin. Amino acid residues that interact through hydrogen bonds with the ligand are shown with cyan-colored carbons, residues providing hydrophobic interactions have gray carbons. Figure also appears in the Color Figure section.

The strength of the interaction between the ligand and the receptor can be measured experimentally and is often reported as the dissociation constant, K_d, or by the concentration of ligand that inhibits activity by 50%, the IC_{50}. The binding free energy is the thermodynamic quantity that is of interest in computational structure-based design and is defined by Eq. 27.1:

$$\Delta G_{bind} = \Delta G_{complex} - (\Delta G_{ligand} - \Delta G_{receptor}). \qquad (27.1)$$

The relationship between the binding free energy ΔG and the experimentally determined K_d or IC_{50} is shown in Eq. 27.2:

$$\Delta G_{bind} = -RT \ln K_{eq} = -RT \ln K_d = -RT \ln 1/IC_{50}. \qquad (27.2)$$

With Eq. 27.2, it is possible to calculate the binding free energy where the K_d is defined with respect to the standard state (Atkins, 1997), and measurements done in different laboratories can only be compared when performed under the same standard state, that is, experimental conditions. The pressure, 1 atm, and the activity of the solutions, namely 1 M, define standard state conditions.

Modern drug discovery cycle usually starts with the identification of a biological target (Figure 27.3), most often a protein, known to play a critical role in a particular disease. Biological assays are developed that can measure inhibition (or activation) of the target of interest by small molecules *in vitro* or *in vivo*. If amenable to high-throughput screening, pharmaceutical companies usually screen their entire corporate collection of molecules in the assay against the target to identify possible leads. Often several hundred thousand molecules are tested in the high-throughput screening (HTS). During HTS each compound is generally tested once at a single dose yielding a percent inhibition value. Numerous hits are identified, based on statistical significance of the measured percent inhibition, and confirmed in subsequent assays. Follow-up tests include dose–response assays and confirmation that

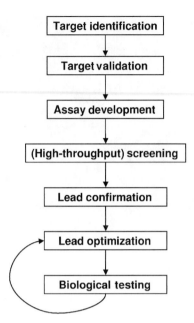

Figure 27.3. Modern drug discovery cycle. Modern drug discovery starts with identifying potential targets that play a role in disease. The target has to be validated, for example through gene-chip analyses, siRNA experiments, gene knockout experiments, and so on. Once the target has been validated, a biological assay has to be developed that can be used to identify potential leads. After lead identification through screening, hits are confirmed by reassaying, IC_{50} and K_d determinations, NMR experiments that measure protein binding of the hits, analogue testing, and so on. Once hits have been confirmed, analogues are synthesized to further optimize the lead series for potency and ADME/Tox properties.

the compounds inhibit the target by a specific mechanism rather than through nonspecific binding (McGovern et al., 2003). The HTS liquid samples of the hits are also tested on purity and integrity to ensure that the measured activity is due to the structure assumed to be in the sample. As the number of hits gets smaller, similarity searches will usually be performed and similar structures will be tested for activity. Analogue testing is a way to develop structure–activity relationships (SAR) early on and further validate the hits. The hits are narrowed to a limited number of different chemical lead series (usually <10) that will be explored by making chemical modifications. As the project continues, the number of lead series that is investigated becomes smaller and more analogues are made for potential leads to further SAR investigations. In addition to optimization of potency, physical–chemical properties that influence absorption, distribution, metabolism, excretion (ADME) properties, and toxicity are also modulated. Eventually, of the hundreds of thousands of molecules tested and thousands of molecules synthesized, one molecule *might* make it into clinical trials and perhaps become an FDA-approved drug.

The receptor-based drug design cycle starts with the availability of a crystal structure or homology model of the target of interest (Figure 27.1). Similarly to HTS, a virtual screen (VS) is usually performed to identify molecules that are compatible with the active site of the target. Just as in HTS, hundreds of thousands, and even over a million, compounds can be screened with current hardware and software. Top-scoring hits are tested in a biological assay to determine activity and confirmed hits from both VS and HTS can be optimized based on the predicted binding modes and ideal protein–ligand complex structures. Numerous cycles of optimization are applied to these series of leads.

In this chapter, we discuss the basis of docking algorithms and focus on their use, along with other approaches, in structure-based drug design that has served as one of the main driving forces to improve docking models. We also cover strategies to identify binding sites and protein targets that are important for rational small molecule and fragment-based drug design. With the increasing availability of crystal structures of receptors that are pharmacologically important, with or without ligands bound, numerous analyses have been performed to extract reoccurring patterns. Consequently, new methods and tools taking on a more heuristic, bioinformatics approach have been developed to take advantage of these data in an aggregate fashion which are complementary to docking algorithms.

DOCKING AND SCORING

Docking can be used to predict how a ligand interacts with the binding site of a receptor (Figure 27.2). The ligand is usually a small molecule, but can be a protein as well. The receptor is most often a protein of pharmacological interest, but can be a particular sequence of DNA or RNA as well. The prediction of the protein–ligand complex is usually done by searching the translational and rotational degrees of freedom of the ligand within the receptor binding site, and by searching the conformational degrees of freedom of the ligand itself. Binding conformations generated by docking programs are thus defined by both a position of the ligand on the receptor surface and a particular ligand conformer. The first docking algorithm for small molecules was developed by the Kuntz group at UCSF in 1982 (Kuntz et al., 1982). Subsequent application of this program to find new leads against HIV-1 protease highlighted the potential of using computer-based screening in drug discovery (DesJarlais et al., 1990). Since then, many different docking algorithms have been developed, and structure-based drug design has become a standard tool in drug discovery.

Before ligands can be docked against a receptor, generally the binding site has to be identified first. This is done to limit the search space on the receptor surface and thus minimize the degrees of freedom that have to be searched. The active site is often known from crystal structures of ligand-bound receptors, but it can also be predicted. The largest cavity on a protein surface is frequently the active site, but this is not always the case and different active site prediction and analysis methods have been developed.

The basis for the DOCK search algorithm can be found in distance geometry and consists of sets of ligand atoms that are being matched to sets of receptor spheres. The receptor spheres describe the binding site and can be viewed as a "negative image" of the active site and thus match what the ideal ligand shape would look like. The matching of ligand atoms to receptor spheres is used to search the translational and rotational degrees of freedom of a rigid ligand in the active site (Kuntz et al., 1982).

The second part of a docking program consists of a scoring function, which will distinguish among the generated binding modes the best solution that closely matches the actual mode of binding. The most-favorable predicted protein–ligand complex is considered to be the biologically relevant one. A large number of different search algorithms and scoring functions have been developed to model protein–ligand docking and are discussed in this section. The various applications of docking, namely binding mode prediction and virtual screening, are also discussed. While the receptor is often considered to be rigid in docking experiments, this is not realistic for the binding sites are flexible. Different approaches that incorporate limited receptor flexibility will also be highlighted.

Search Algorithms

Docking algorithms generally contain two search engines, one is responsible for the placement and exploration of the ligand in the binding site, while the other searches the internal degrees of freedom of the ligand, that is, angles and dihedral angles. Some docking programs only dock rigid ligands and use pregenerated ligand conformers, this approach serves to conserve computational resources but may not necessarily provide the best solution. Pregenerated ligand conformers can be generated with some of the following methods: OMEGA (OpenEye Scientific Software; Verdonk et al., 2003a), Corina (Gasteiger, Rudolph, and Sadowski, 1990), MacroModel (Schrödinger Inc.), and Concord and Conform (Tripos Discovery Informatics).

For ligands without any degrees of freedom, it is feasible to search the six translational and rotational degrees of freedom in the receptor binding site (or even the full receptor surface). However, as the number of rotatable bonds of the ligand increases, the number of degrees of freedom increases exponentially when finding solutions to fit into the binding site. Search algorithms try to find a balance between searching the energy landscape as thoroughly as possible, while still finding a solution within a reasonable amount of time.

Several different orientation algorithms are available and can be divided into three major categories. The first method is a grid search that exhaustively searches the binding site by moving the rigid ligand through the available 6 degrees of freedom in a systematic fashion in the binding site. Search time and accuracy are determined by the size of the three-dimensional grid and the size of the steps that are taken. Fast Fourier transform (FFT) methods have been applied to speed up the search. FFT performs the search in Fourier rather than Cartesian space, which leads to significant enhancements of speed. The original algorithm only looked for surface complementarity (Katchalski-Katzir et al., 1992), but later implementations also use electrostatic complementarity, as in FTDOCK (Gabb, Jackson, and Sternberg, 1997) for example. Unfortunately, because most ligands are flexible, the number of degrees of freedom to be searched with a grid search is too large, and this approach is not used in small-molecule docking algorithms. However, it is applied in protein–protein docking algorithms, which generally treat both proteins as rigid molecules.

The second and significantly more efficient algorithm places ligands in the binding site by descriptor matching (Ewing and Kuntz, 1997). Descriptors to find a solution are generally points, sometimes with properties associated to them, placed in the receptor site. Ligand atoms are matched with the receptor descriptors with some tolerance. The descriptor-matching algorithm orients the ligand in the binding pocket, which subsequently has to be optimized and scored. Usually a large number of orientations are possible, and they all have to be explored and optimized. Careful pruning of the search tree is necessary to ensure efficiency and avoid an exponential growth of the systematic search (Ewing et al., 2001; Moustakas et al., 2006b). When binding site descriptors are associated with physical–chemical properties, such as hydrophobicity and hydrogen-bonding capabilities, the ligand atoms have to match geometrically as well as chemically. DOCK (Shoichet and Kuntz, 1993; Moustakas et al., 2006b), FlexX (Rarey, Wefing, and Lengauer, 1996) and PhDOCK (Joseph-McCarthy et al., 2003) use descriptor matching to orient the ligand in the binding site. Usually sets of three or four receptor spheres, minimally, are matched to the same number of atoms in the ligand. If the intersphere distances are similar to the interatom distances a match has occurred and the ligand can be placed in the binding site.

In the third class of algorithms are energy-based search methods using molecular mechanics force fields (Table 27.1) to explore the energy surface of the ligand with molecular dynamics (MD) or energy minimization algorithms. The minima of the ligand on the surface have to be located and assessed in terms of their complementarity to the receptor. Energy minimization is a local search, and is only used to optimize binding conformations generated by other search engines. MD searches are global searches in principle, but also tend to get stuck in local minima. Consequently, simulation search times have to be long to ensure full coverage of the binding site and, consequently, limits their applicability as a high-throughput solution to computationally identify binding targets (David, Luo, and Gilson, 2001). Some alternatives to help speed up the search time include genetic algorithms, as implemented in AutoDock (Morris et al., 1998) or GOLD (Jones, Willett, and Glen, 1995), and Monte Carlo searches (QXP; McMartin and Bohacek, 1997) which are stochastic and help prevent simulations from being stuck in a local minima when searching for solutions. This search strategy generally explores all degrees of freedom simultaneously, and is therefore not strictly considered as orientation algorithms.

Molecules that contain rotatable bonds can occupy different conformers in solution, but usually occupy a single conformer when bound to the protein. Different conformers of a ligand have different internal energies. The internal energies of the ligand and the protein are generally ignored during the search for the most optimal ligand pose in the binding site, but docking algorithms do eliminate physically unrealistic conformers for the ligand, for example, conformers with overlapping atoms.

In summary, the presence of rotatable bonds in the ligand increases the degrees of freedom that have to be explored as well. Ligand search algorithms can be divided into three types of methods, namely systematic, stochastic, and deterministic. Molecular dynamics and energy minimization are deterministic, and genetic algorithms and Monte Carlo searches (see above) are stochastic search algorithms. As described for the grid search, a systematic search for all degrees of freedom and different ligand orientations in the binding site would be too time consuming. One can limit the degrees of freedom to be searched by docking a rigid part of the ligand first, the so-called anchor, and incrementally adding the bonds one at a time ("grow"). The "anchor-and-grow" algorithm has been implemented in a number of docking programs, namely DOCK (Ewing and Kuntz, 1997) and FlexX (Kramer, Rarey, and Lengauer, 1999). The anchor-and-grow algorithms are not guaranteed to be exhaustive.

Ligand flexibility can also be dealt with by rigidly docking pregenerated conformer libraries. Especially for large compound libraries used in virtual screening, this is a very efficient method for handling ligand degrees of freedom. The cost of generating the conformers has to be only incurred once, and the library can subsequently be docked quickly against any target of interest. The Shoichet laboratory implemented this strategy in the DOCK algorithm (Lorber and Shoichet, 1998), and pregenerated conformer libraries are also used in PhDOCK (Joseph-McCarthy et al., 2003) and EUDOC (Pang et al., 2001a). Key to successful convergence on a solution is that the generated conformer ensemble for each ligand contains a conformer that is close, probably <1 Å (1 Å $= 1 \times 10^{-10}$ m) RMSD (root mean square deviation) to the actual solution. Usually the bioactive conformer is not the conformation found at the global energy minimum in the energetic landscape. Therefore it is necessary to generate an ensemble of low-energy conformers that will hopefully contain a conformer that is close enough to the bioactive conformer and will dock in the correct binding mode. As mentioned earlier, different conformer generation algorithms are

TABLE 27.1. Different Scoring Functions Used for Docking

$$E_{total} = \sum_{bonds} K_r(r-r_{eq})^2 + \sum_{angles} K_\vartheta(\vartheta-\vartheta_{eq})^2$$
$$+ \sum_{dihedrals} \frac{V_n}{2}[1+\cos(n\phi-\gamma)] + \sum_{i<j}\left[\frac{A_{ij}}{R_{ij}^{12}} - \frac{B_{ij}}{R_{ij}^6} + \frac{q_iq_j}{DR_{ij}}\right]$$

The molecular mechanics force field includes contributions from the bonds, angles, and dihedral angles of the molecule, essentially representing the internal energy of the molecules. Intermolecular forces are calculated using the "6–12" potential, representing the van der Waals interactions, and interactions between the atomic charges, calculated using Coulomb's equation.

$$\Delta G_{bind} = \Delta G_o + \Delta G_{hb} \sum_{h\text{-}bonds} f(\Delta R, \Delta\alpha) + \Delta G_{ionic} \sum_{ionicint.} f(\Delta R, \Delta\alpha)$$
$$+ \Delta G_{lipo}|A_{lipo}| + \Delta G_{rot}N_{rot}$$

The Ludi empirical scoring function estimates the binding free energy based on hydrogen-bonding interactions, charge-charge interactions and the loss of entropy of the ligand based on the number of rotatable bonds. ΔG_o is a measure of the loss in translational and rotational degrees of freedom.

$$Fitness = \sigma_o + \sum_w o(w)(\sigma_p + \sigma_i(w))$$

The GOLD fitness function is a measure of the ChemScore or GOLD score, represented by σ_o, which is similar to the empirical scoring function shown above, and a measure of the water contribution. The water contribution takes both the loss in entropy of the water molecule into account (σ_l) and the intrinsic binding affinity of the water (σ_w).

available to efficiently generate low-energy conformers for ligands. A number of studies have investigated the ability of these algorithms to identify bioactive conformers. As expected, the chances of retrieving a biologically active conformer decrease as the number of rotatable bonds increases. However, for lead-like ligands (see the section "Ligand-Based Perspectives") with less than eight rotatable bonds, a conformer within 1 Å of the bioactive conformer can be found in >80% of the cases. Macromodel was shown to have the best success, but at a much more significant computational cost than rule-based OMEGA (Bostroem, 2001; Good and Cheney, 2003).

Some weaknesses with docking approaches need to be considered when applying these algorithms to find optimal docking solutions. A recent study by the Vertex group highlighted that the bioactive conformer of a ligand is rarely found in the low-energy conformer. A large-scale study was performed using 150 protein–ligand complexes of pharmaceutical interest, with available IC_{50} and/or K_d measurements. Several different methods were used to generate the global minimum conformation of each ligand in solution. The strain energy for each ligand was calculated using the partially minimized crystallographic conformer and the global minimum energy of the ligand. As expected, the majority of the ligands do not bind to the receptor in the global-minimum conformation. Somewhat surprisingly, however, was the observation that 60% of the ligands do not bind in a local energy conformation either. As a consequence, ligand strain energies are quite high and can even exceed 9 kcal/mol (Perola and Charifson, 2004).

Furthermore, observations by Perola and Charifson showed that interaction energies estimated, without taking into account the internal ligand energy, are off with an error of 0–10 kcal/mol. Tirado-Rives and Jorgensen have calculated the cost of "conformer focusing" upon protein–ligand binding using both force field and quantum mechanical calculations. With the accuracy of current force fields, even inclusion of the ligand strain energy results in an uncertainty in the binding free energy calculation of between 5 and 10 kcal/mol. Thus, to accurately obtain free energies of binding, more accurate force fields are required (Tirado-Rives and Jorgensen, 2006). Their study also highlights that using a constant penalty per rotatable bond is not correct.

Scoring Functions

As mentioned earlier, we are interested in the binding free energy ΔG_{bind} of the ligand with the protein. ΔG_{bind} is a function of both enthalpic and entropic energies as in Eq. 27.3:

$$\Delta G_{bind} = \Delta H - T\Delta S, \tag{27.3}$$

where ΔH, enthalpy, is a result of the changes in van der Waals and Coulombic interactions as water is removed from the binding site and the ligand surface and replaced with van der Waals and Coulombic interactions between the receptor and the ligand. Changes in internal energy of the receptor and the ligand have to be considered also. Entropic changes ΔS upon binding are a result of restricting the internal degrees of freedom of the receptor, especially in the binding site, and the ligand, as well as changes in translational and rotational degrees of freedom. The loss of flexibility and other entropic changes are generally unfavorable to complex formation. Another large contributor to the entropy is the solvent, which generally gains entropy as it is released from the protein binding site, and the ligand upon complex formation. This so-called "hydrophobic effect" (Kauzmann, 1959) is often the driving force for protein–ligand interactions. The binding free energy is thus a complex term, which can be calculated directly, but at a considerable computational cost (Head, Given, and Gilson, 1997; Woo and Roux, 2005). Because of the complex nature of binding free energies, the quest for generally applicable and highly accurate scoring functions that can be evaluated quickly is still on going.

Relative binding free energies can be calculated more easily, and is applied more frequently in drug discovery. The methods used are either thermodynamic integration (TI) or free energy perturbation (FEP) (Kollman, 1993), and the relative binding free energy is obtained by slowly "mutating" one ligand into another one. This is generally only feasible for small changes to the ligand. One thus only obtains the energy difference between a ligand and a close analogue. In lead optimization, this can be used to prioritize what molecule to synthesize next from a pool of possible ligands to make. However, even calculation of relative binding free energies still comes at a considerable cost. Part of the reason FEP/TI calculations are time consuming is that (unphysical) intermediate states have to be sampled extensively in going from one ligand to its analogue.

More practical approaches to calculating binding free energies can be divided into four major approaches: first-principles scoring functions, semiempirical methods, empirical scoring functions, and knowledge-based potentials. First-principles methods generally use a molecular mechanics force field (Table 27.1), although often the intramolecular forces between the atoms are not calculated to save time. Since the molecular mechanics force field only accounts for enthalpic forces between the protein and the ligand, the estimated binding free energy is only an approximation to the true binding free energy. DOCK (Meng, Shoichet, and Kuntz, 1992) and the original AutoDock (Goodsell and Olson, 1990) use molecular mechanics scoring functions. With the rigid receptor assumption, the energies contributed by the receptor can be precalculated and stored on a grid. This significantly speeds up the calculations and makes use of a molecular mechanics scoring function in docking feasible. Interactions with solvent are generally not accounted for.

A commonly observed phenomenon is the high ranking of charged ligands due to an overestimation of charge-charge interactions to the binding free energy in energy-based scoring functions. A newer strategy uses a tiered approach to score and post-process the generated binding conformations to estimate the contribution of the solvent with a continuum solvent model such as the Poisson–Boltzmann (Warwicker and Watson, 1982; Gilson and Honig, 1988) or Generalized Born (Bashford and Case, 2000) methods. These enhanced scoring functions generally show better agreement with true binding free energies (as measured by the K_d/IC$_{50}$) and improved rank ordering of compounds (Shoichet, Leach and Kuntz, 1999; Zou, Sun, and Kuntz, 1999; Lyne et al., 2004;Huang et al., 2006). At this point, it is not feasible to include the continuum solvent models directly into the docking calculation.

The linear interaction energy (LIE) method (Aqvist, Medina, and Samuelsson, 1994) is a semiempirical method that was developed to calculate absolute binding free energies while avoiding the need for the extensive sampling of nonphysical transition states as in FEP or TI. Due to its semiempirical nature, the parameters are not generally applicable and need to be derived for every new target and/or new chemical series. This requires the availability of reliable binding data. Although only the initial and final states are simulated with either molecular dynamics or Monte Carlo simulations, the method is still time-consuming and is only used in lead optimization and not in lead identification (Ljungberg et al., 2001; Rizzo, Tirado-Rives, and Jorgensen, 2001).

Numerous empirical scoring functions have been developed (Table 27.1), and a significant advantage is that they can be evaluated rapidly and thus can be used in high-throughput applications such as docking. An empirical scoring function consists of a number of terms known to play a role in protein–ligand complex formation such as hydrogen bonding and hydrophobic contacts between the protein and the ligand. The weights for each of the terms are derived by fitting the equation using regression methods to a training set of protein–ligand complexes for which both structural information and experimentally determined interaction energies are known. While the terms used represent the physical principles underlying complex formation, there are several weaknesses to consider. First, due to the fitting, much of the atomic detail is lost. Second, the scarce sampling of the chemical space limited by available protein–ligand complexes with reliable measured IC_{50} or K_d values results in empirical scoring functions that are general rather than specific (Tame, 1999). Third, most empirical scoring functions consider interactions by simply looking at the distances and angles (in the case of hydrogen bonds) of the atoms that are involved. However, it is quite likely that the environment in which the interaction occurs plays a role also. To address these limitations, the program Glide now has a new scoring function that accounts for the environment of hydrogen bonding interactions and therefore differences in contributions from different hydrogen bonds are more properly modeled (discussed later; Friesner et al., 2006).

Based on work done in the protein folding community (Sippl, 1990), knowledge-based potentials have been developed to score protein–ligand interactions (Mooij and Verdonk, 2005; Muegge, 2006). Knowledge-based functions are derived using protein–ligand complexes alone, based on the statistical assumption that interactions that are more frequently observed are more important for complex stability. This is the basic assumption in statistical mechanics and the Boltzmann principle. The advantage of this approach over the derivation of empirical scoring functions is that more protein–ligand complexes can be used for deriving potentials since the prerequisite for experimentally determined IC_{50}/K_d for each complex has been eliminated. This approach is more computationally efficient, less time demanding, and therefore can be used in docking programs.

Water molecules play a critical role in the formation of protein–ligand interactions and are a significant determinant in the binding free energy. One approach to account for solvent effects is by using a continuum solvent model. While this is improvement over accounting only for direct protein–ligand interactions, in many cases it is not enough. The continuum solvent models do not account for two explicit water molecule contributions. First, some water molecules serve to bridge interactions through hydrogen bonding between the ligand and protein and therefore affect the binding affinity. Second, some well-defined water molecules can be displaced by the ligand and significantly contributes to the binding affinity due to the increase in translational and rotational entropy upon release. With the improved computational power, more realistic scoring functions have been implemented in recent years to account for the role of water molecules in more detail.

FlexX is an example of an algorithm that accounts for contributions from water molecules using a particle approach. In the first step, phantom particles are generated using the protein structure of the active site; Particles with more than two interactions with the protein are retained. The second step, ligand docking, involves turning "on" and "off" the phantom particles to determine if interactions are formed with the ligand. Particles interacting with the ligand are then used in the scoring function to determine ligand binding capacity (Rarey, Kramer, and Lengauer, 1999). The energy change associated with the release of waters from the active site is not accounted for explicitly with this approach.

GOLD is another algorithm with scoring functions that accounts for the contribution of both bridging and displaced waters. The user specifies which water molecules are to be retained during the docking search and GOLD will further explore whether the water molecule should be included in the ligand binding process. For each water molecule, GOLD will consider the bridging potential of the water molecule, or alternatively, the displacement by the ligand. Both the Chemscore and Goldscore scoring functions (Table 27.1) have been parameterized to take both functions of water molecules into account (Verdonk et al., 2005).

Extra Precision Glide (GlideXP) has the Chemscore scoring function at its basis, but with additional terms that better incorporates the underlying physics of protein–ligand interactions. Group-based lipophilic interactions, rather than atom-based, and the protein environment in which these interactions take place is included as a term called the hydrophobic-enclosure term. Additionally, a number of other hydrogen-bonding terms have been incorporated along with pi–pi and cation–pi interactions. Accounting for desolvation, explicit waters are docked for top-scoring XP conformations and polar ligand or protein groups that are inadequately solvated will be penalized (Friesner et al., 2006).

Binding Site Identification and Mapping

Docking methods generally rely on a demarcation of the search space on the receptor surface to address the computationally intensive nature of the search problem in finding the correct ligand binding conformation. The space searched spans only the active site and requires some knowledge about the position of the binding site on the receptor. Often the active site of a receptor is known from an X-ray crystallography structure, or perhaps NMR, of a protein–ligand complex. However, sometimes the active site, or additional binding sites, has to be predicted. It is often assumed that the largest cleft on a protein surface corresponds to the active site, but this is not always the case. A number of methods are available for protein binding site predictions. The simplest methods rely on assessing structural features of the receptor alone by measuring the volumes of different cavities using alpha shapes (Liang, Edelsbrunner, and Woodward, 1998) or by placing probes on the protein surface.

DOCK adopts this strategy to reduce the sampling search space with the SPHGEN algorithm that creates docking spheres to identify potential active sites. After placement of spheres on the protein surface, these spheres can be clustered to identify the binding site, which is often defined by the largest cluster of spheres (Kuntz et al., 1982). An alternate strategy uses the physical-chemical properties of surface patches and cavities to identify active sites. For example, Jain et al. identify "sticky spots" on protein surfaces by identifying regions of the protein where different probes are predicted to have favorable interaction energies. The probes that have high interaction energies with the region define the sticky spots. The sticky spots will subsequently be extended to the largest continuous cavity possible (Ruppert, Welch, and Jain, 1997). A third strategy combines hydrophobic and hydrophilic probes with a number of additional physical parameters to identify favorable interaction sites on the protein surface as implemented in SiteMap and SiteScore. The SiteScore is defined relative to a set of 155 known tight binding sites and also serves as a measure of "drugability." Over 95% of known binding sites in 230 proteins has been predicted correctly (Halgren, 2007).

With the ever-increasing amount of available structural information on protein–ligand complexes, the accumulating information and derived analysis can also be used to predict protein active sites. These heuristic based methods will be discussed in the section "Drug Design in the Structural Proteomics Era".

Assessing Performance of Docking Algorithms

After drug candidates have been identified with the search algorithms that have been discussed, optimization of these leads begins. For structure-based drug design to have an impact on lead optimization, the binding mode of the lead with the receptor has to be known. Docking tools play a critical role in predicting binding modes, and the ability of a docking program to correctly identify the biologically relevant binding mode from the other possibilities is the most basic test of docking algorithms.

The success rate in retrieving binding modes of known protein–ligand complexes is an important validation for docking programs. The measure that is usually used to determine whether a binding mode prediction is a success is the RMSD. The RMSD is defined by Eq. 27.4 as

$$\text{RMSD} = \sqrt{\frac{\sum_{i=1}^{N_{\text{atoms}}} d_i^2}{N_{\text{atoms}}}}, \tag{27.4}$$

where d_i is the distance between the coordinates of atoms i in the predicted structure and the known crystal structure when overlaid. One generally considers an RMSD ≤ 2 Å as a successful prediction. With larger deviations, many of the observed interactions between the protein and the ligand will not have been predicted correctly.

The current availability of protein–ligand complex structures provides opportunities to construct data sets needed for large validation studies to evaluate and improve docking methods. For flexible ligand docking, success rates of $\sim 70\%$ have been reported (Kramer, Rarey, and Lengauer, 1999; Pang et al., 2001a; Verdonk et al., 2003b; Friesner et al., 2004a; Moustakas et al., 2006a), although it is not always predictable which program will give good predictions for a particular target and/or chemical series (Kontoyianni, McClellan, and Sokol, 2004; Warren et al., 2006b). In recent years, a number of different test sets have also appeared in the literature. The first well-curated test set was assembled by the GOLD developers which focused on crystallographic reliability (Jones et al., 1997). Recently, test sets with complexes of pharmacological interest (Perola, Walters, and Charifson, 2004), reasonable number of rotatable bonds (Moustakas et al., 2006b), and a wider diversity of representative complexes (Hartshorn et al., 2007) have been published.

The most recent CCDC/Astex set was compiled using six criteria and a final manual selection step. The six criteria are (1) date of deposition; (2) resolution; (3) availability of structure factors; (4) presence of interesting ligands; (5) drug-likeness of ligand; and (6) non-existence of overlap between protein and ligand. The manual selection process involves an assessment of the pharmacological or agrochemical interest of the ligand, presence of covalent bonds between the protein and the ligand, and the presence of incomplete side chains or missing loops near the binding site. The final set contains 85 high-quality, diverse, protein–ligand crystal structures (Hartshorn et al., 2007). The current CCDC/Astex set has no overlap with the previous one (Nissink et al., 2002).

Virtual Screening

The goal of virtual screening is to identify ligands within a large collection that can bind to the target of interest and induce the desired biological effect. The collection of compounds that is screened can be compounds in a corporate collection of a (bio)pharmaceutical company, compounds available for purchase from commercial vendors, or virtual compounds that have yet to be synthesized. In large, diverse collections of compounds, very few compounds will generally exhibit interactions with the target of interest (high-throughput screening hit rates are usually <0.5%). Screening with docking algorithms can be used to identify compounds that are most likely to show interactions with the target, thus reducing the number of compounds to be tested experimentally for activity.

In validation studies, a library of decoys is usually enriched with known inhibitors of the target from the literature. One measure of success is the hit rate, which is defined as the percentage of true hits found among the top-scoring compounds. Another measure of success is the enrichment factor (EF). The EF is defined as

$$EF = \frac{a/n}{A/N},\qquad(27.5)$$

where a is the number of active compounds among the top n compounds, and A is the total number of active compounds in the library of N compounds. Many virtual screening validation studies use decoys from commercial collections that have not been tested against the target of binding. Therefore, the EF scoring function is not reflective of the true scenario since some decoys may exhibit interactions with the target. A downside of the EF function as a measure of success is that the maximally achievable value depends on the number of active and inactive compound in the library. Similar to binding mode predictions, the EF scoring function can differ greatly from target to target for particular algorithms (Bissantz, Folkers, Rognan, 2000; Kontoyianni, McClellan, and Sokol, 2004; Warren et al., 2006a). Identifying the best program to use for the target of interest is not straightforward.

Receiver operating characteristic (ROC) curves have also been used to evaluate virtual screening experiments (Klon et al., 2004; Triballeau et al., 2005). The benefit of this approach it the independence from the ratio of actives and decoys. Additionally, it lends itself for easy interpretation with the range of values between 0 and 1, a value of 0.5 equates to random selection from the library.

Truchon and Bayly criticizes the use of ROC curves, mainly because it is not sensitive to "early recognition." With library sizes of 500,000 compounds and more, this is an important consideration because active compounds have to rank highly to be considered for experimental screening. Usually only a few thousand compounds are screened from virtual screening, which means that active compounds have to be within the top 0.1%. The authors propose a new metric, BEDROC, to better recognize these active compounds among a large set of decoys earlier on (Truchon and Bayly, 2007).

The compilation of compound libraries can also have a significant impact on virtual screening hit rates. When the true hits have physical–chemical properties that are significantly different from the properties of the decoys, artificially high EF scores can be observed (Verdonk et al., 2004). To this end, Shoichet's group has developed the directory of useful decoys (DUD) database, which contains 2950 actives for 40 different targets. For each active compound 36 "decoys" are chosen from the ZINC database. Decoys are similar to active compounds in terms of physical–chemical properties, but have dissimilar chemical structures. A comparison of EF scores with DUD and more biased libraries has been made, showing again that EF scores are artificially high when decoys are chosen to be dissimilar from actives (Huang, Shoichet, and Irwin, 2006).

Protein Flexibility

Although limited flexibility in the active sites are accounted for in some docking algorithms, for example FLO and GOLD (McMartin and Bohacek, 1997; Verdonk et al., 2003b), most treat proteins as rigid structures. This assumption is a seriously limitation because induced fit effects are often critical in complex formation and the crystal structure or homology model used for docking is only a single representation of the many accessible protein conformations found in solution. Minor adjustments to flexible side chains in the active site are often necessary to accommodate the ligand, which is underscored by results from "cross-docking" experiments. In cross-docking experiments, the flexible ligand is docked into a receptor crystal structure that is bound to a different ligand. Success rates in reproducing the known bound conformation of the ligand is significantly lower in cross-docking experiments than in regular docking experiments (Bissantz et al., 2000; Jain, 2003; Friesner et al., 2004b; Kontoyianni, McClellan, and Sokol, 2004; Warren et al., 2006a).

Although the need for full receptor flexibility in the active site is recognized, this is unfortunately restricted with the current computational power of available technologies. Different methods have been developed that allow for receptor flexibility while limiting the increase in computational demand. A number of groups implemented "ensemble" grids that use averaging approaches to represent a number of receptor structures (Knegtel, Kuntz, and Oshiro, 1997; Claussen et al., 2001; Oesterberg et al., 2002). The ensemble grid does not increase the computational time needed for docking the molecule, and yields significantly better cross-docking results as well as better enrichment rates (Knegtel, Kuntz, and Oshiro, 1997; Claussen et al., 2001; Oesterberg et al., 2002). Different methods have been used, including energy-weighted and geometry-weighted average grids (Knegtel, Kuntz, and Oshiro, 1997). Energy-weighted grids use a Boltzmann weighting scheme (Oesterberg et al., 2002). Another mixed representation approach merged variations in structural conformations into a single representation with dissimilar areas of the protein treated as alternatives (Claussen et al., 2001).

Several groups have explored the use of structural ensembles, usually generated from MD simulations, and docking compounds against each individual conformer (Pang et al., 2001b; Lin et al., 2002; Wong et al., 2005). The predicted binding mode is chosen based on the best-scoring protein–ligand complex. Due to inherent inaccuracies in scoring, further optimization of the generated binding modes and reevaluation of scores is sometimes needed (Wong et al., 2005). Docking time will increase linearly with the number of structures used. The use of structural ensembles in the development of pharmacophores has also been explored (Meagher, Lerner, and Carlson, 2006).

The induced fit docking (IFD) protocol combines Glide docking with the protein optimization protocols implemented in Prime. The van der Waals radii of the protein and ligand are scaled during the docking process and significant overlap is allowed in the generated protein–ligand binding conformations. These binding conformations are then optimized with a fixed ligand and the minimum energy structure of the protein is retained. Finally, the ligand is re-docked into the optimized receptor with the regular scoring function and resulting conformations are scored and ranked (Sherman et al., 2006).

DRUG DESIGN IN THE STRUCTURAL PROTEOMICS ERA

The contributions of structural genomics efforts to drug discovery based research are many. First, the increasing number of crystal structures in the PDB (Chapter 11) forms a data rich resource with protein–ligand complexes capturing the interactions formed between targets and ligands. This wealth of information has influenced the development of binding site prediction and annotation methods. Multiple structures provide a selection of choice to identify the most appropriate structure for virtual screening of compounds. Additionally, the accumulating binding data has also resulted in pilot studies to predict the selectivity of ligands for their target compared to off-targets. Feasible high-throughput crystallization experiments are another product of structural genomics making new avenues for research by stimulating fragment-based drug discovery. Finally, a number of ligand-based assessments that rely on the availability of protein–ligand complexes will be highlighted.

Binding Site Prediction and Annotation

Binding site prediction is not only critical in docking to limit the search space, but also plays a critical role for the annotation of protein function. The structural genomics consortia (Chapter 40) aims to solve crystal structures of novel folds and identifying active sites can play a critical role in understanding protein function. Binding site annotation can be used to investigate differences and similarities within a protein family, to either identify potential off-targets or guide optimization of leads. Drug toxicity is usually a result of a drug binding to other protein targets (off-targets) than the intended target the compound was designed for. Being able to identify potential off-targets early on in the development of a drug can be critical in the development of the drug.

While functional assignment is relatively straightforward for proteins with sequence or structural homology to proteins of known function, this is not the case for proteins without a close homologue. Binding sites can be experimentally identified through systematic alanine scanning (DeLano, 2002). Unfortunately, this is a time-consuming and costly process which is why the development of computational methods that can identify active sites and critical residues for protein function is of interest.

A number of groups have used large-scale structural analyses of known protein–ligand complexes to distinguish feature properties of binding sites from the rest of the protein surface. Herzberg and Moult discovered that functional sites often contain residues that are constrained in uncommon conformations. The restricted conformation imposed on the residue is a constraint demanded by the function, whether for catalytic or structural purposes within the binding site (Herzberg and Moult, 1991). Similarly, Zhu and Karlin found that functionally important sites often contain clusters of charged residues. Again, these charged residues could either be required functionally or structurally (Karlin and Zhu, 1996). As such, charged residue such as these can be identified using continuum electrostatics methods based on the assumption that charged residues of functional importance are destabilizing to the protein as shown experimentally through site-directed mutagenesis (Elcock, 2001).

Other distinguishing features include amino acid composition and size of surface cavities. Amino acid composition in binding sites is another feature that can also be used to distinguish the true binding site with α-spheres. Analysis of 15,232 binding sites shows that binding sites are enriched in aromatic residues and methionine. In 611 of 756 proteins the true binding site had the highest protein–ligand binding (PLB) index, a success rate of 79% (Soga et al., 2007). Probes for protein surface cavities have been developed by Nayal and Honig using two molecular surfaces created by different probe sizes. Drug binding cavities are detected based on a random forest classifier trained on 408 properties of cavities. They also performed a large-scale analysis of the properties of cavities (Nayal and Honig, 2006).

Finally, the CPASS database was developed to store the active site residues of ~42,000 known protein–ligand complexes which can subsequently be used to identify similar binding sites in proteins of unknown function when the structure becomes available (Powers et al., 2006). This compilation can also serve to further identify other structural features and properties of binding sites that can be leveraged to improve current binding site prediction methods.

Rather than rely on the availability of structural information to identify binding sites, sequence information can provide additional insight. Amino acid patterns have been developed for different types of functional sites using known structures, and these patterns can subsequently be used to identify similar binding sites in other proteins (Atwood, 2000). However, due to sequence variation, these methods are not that reliable. Lichtarge developed the evolutionary trace (ET) method, which defines conserved residues within clusters of proteins within protein families. The conserved clusters are related to function and generally occur in active sites. When mutations are observed, this often concur with novel functionality (Lichtarge, Bourne, and Cohen, 1996). The ET method was extended to improve the identification of binding sites in proteins and eliminate the requirement of a manually curated sequence alignment. The method strictly relies on sequence information, derived from multiple sequence alignments of protein families. Residues are identified through multiple sequence alignments to be highly correlated with functional divergences when mutated. With the addition of the Rank Information measure, functional sites can be identified in a fully automated fashion, making it available for large-scale identification of binding sites (Yao, Mihalek, and Lichtarge, 2006).

The annotation of binding sites is also an active area of research, because it allows for the identification of similar binding sites and distinguishes differences between related targets that are important for lead optimization. The GRID program, developed by Goodford (Goodford, 1985), identifies favorable interaction points for different types of small probes. The low-energy positions of these probes can be used to optimize small-molecule ligands. The molecular interaction fields (MIF) generated by GRID have been used to distinguish different kinase families using PCA analysis (Naumann and Matter, 2002). The differences between related binding sites identified using MIF and PCA analysis has been used to design selective ligands (Vulpetti et al., 2005). GRID is also used by Mason and Cheney to generate site points for a number of receptors and generated all four-point pharmacophores for the binding sites. By calculating the overlap between all four-point pharmacophores of a ligand to all four-point pharmacophores of an active site, specificity of a ligand to its target versus off-targets could be ascertained (Mason and Cheney, 1999).

Cavbase derives pseudo-centers based on the underlying amino acid properties for binding sites directly to annotate binding sites. The pseudo-centers represent potential hydrogen bonding and aromatic interaction points in the binding site. The pseudo-centers are mapped onto the cavity surface and a clique-detection algorithm is applied to compare different binding sites. Using Cavbase binding sites with similar function, but stemming from proteins with different folds can be retrieved. Different protein families and subfamilies within a protein family can be clustered using the Cavbase description also (Kuhn et al., 2006).

Singh and coworkers developed the structural interaction fingerprint (SIFt) methodology (Deng, Chuaqui, and Singh, 2004). SIFt is simply a one-dimensional binary string derived from protein–ligand crystal structures and represents how the ligand interacts with the protein. An extension, profile-SIFt (p-SIFT), can be used to highlight differences in the active site within a protein family based on protein–ligand interactions observed complex structures (Chuaqui et al., 2005).

Earlier we mentioned the term "drugability" of protein binding sites. Drugability is defined as the potential of the target to respond to a small molecule designed to elicit a biological response. A critical parameter for a potential drug is the affinity, which can be measured by the K_d or IC_{50} for example, for the target. The size and depth of the binding site play a critical role in determining whether a high-affinity molecule can be developed, but other factors play a role also. Cheng et al. (2007) developed a method to calculate the maximal achievable binding energy for a binding site. The method relies on estimating the desolvation of the binding site, which is related to the size of the binding site and the curvature of the surface. This simple model was able to distinguish known drugable targets from known undrugable targets.

The predetermination of whether a target is drugable or not is becoming more important in this genomic and proteomic era where novel potential targets are identified on a large scale. This is underscored by the fact that most drugs currently on the market only target a limited number of receptors. It is estimated that ~1200 small molecule drugs only target 324 molecular targets for both human targets and pathogenic organisms (Overington, Al-Lazikani, and Hopkins, 2006).

Structure Selection for Virtual Screening

Many targets have several crystal structures that can be used as a template for drug screening. Some structures might be holo (ligand bound), others might be apo (no ligand bound) structures. Choosing the structure that is most appropriate for virtual screening might not be straightforward. Shoichet showed for 10 systems that the holo form of the protein is generally superior to the apo-form in virtual screening enrichments. In addition, usually both the holo- and the apo-form of the protein have higher enrichments than homology models built for the same receptors. Encouragingly, homology models do show enrichments better than random in most cases (McGovern and Shoichet, 2003). A kinase-specific study of the use of homology models in virtual screening showed good enrichment for five models as well, but one model gave negative enrichment (Diller and Li, 2003).

There appears to be a relationship between the success of virtual screening with the sequence identity or homology of the target sequence with the template used to build the homology model. A retrospective study showed that similar enrichment factors can be achieved for homology models versus crystal structures as long as similarity is greater than 60% (Oshiro et al., 2004). In contrast, Gilson's laboratory showed by docking against multiple homology models built for the same target, enrichment correlated poorly with sequence identity between the target and the template (Fernandes, Kairys, and Gilson, 2004). The shape of the binding site in the template possibly plays a more important role for a more accurate screen.

Rockey and Elcock assessed whether the cocrystallized ligand plays in role in predicting the correct binding modes against the homology model. A number of homology models were built using kinase structures bound to staurosporine. The results of re-docking staurosporine (a pan-kinase inhibitor) into homology models built from a staurosporine-bound template structure was compared to docking results in models built from templates without a cocrystallized ligand or bound to a different ligand. The authors demonstrated that the binding mode prediction using models generated from templates bound to the same ligand is often better than docking the apo structures or templates bound to a chemically different ligand. Virtual screening experiments have not been performed to assess the potential improvements based on these findings, but the observations are in line with studies highlighted above that showed ligand-bound structures are better for virtual screening than the apo structures (Rockey and Elcock, 2006).

While these retrospective studies show the potential utility of virtual screening, a number of prospective, or truly predictive, studies have also appeared. Again, the use of virtual screening to identify leads is highlighted by these studies (Alvarez, 2004).

Receptor-Based Ligand Specificity Predictions

Binding site annotation can be used to identify similar binding sites, and thus off-targets. More recently, a number of methods have been developed to identify additional targets for ligands by interaction energy calculations. Elcock's laboratory developed the SCR scoring function to identify off-targets of kinase ligands. Models for potential off-target kinases were built using the backbone coordinates of the target structure. Amino acid side chains in the binding site for the off-targets are optimized around the ligand and the interaction energy for each protein–ligand complex is calculated using a simple empirical scoring function. High-affinity targets for a number of kinase ligands were successfully identified, although the calculations were not successful for each ligand (Rockey and Elcock, 2005).

Page and Bates applied the MM-PBSA method (Kollman et al., 2000) to the identification of off-targets of kinase inhibitors. The MM-PBSA method involves the generation of molecular dynamics trajectories of the complexes and calculates average free energies over the trajectory. The aim of this strategy is to account for conformational changes that may occur in a protein structure leading to the formation of an appropriate binding pocket creating the potential for ligand interaction. Unfortunately, the results were not very encouraging, possibly due to the inherent approximations made by the MM-PBSA method (Page and Bates, 2006).

INVDOCK was designed to screen ligands against multiple protein cavities and limited receptor flexibility is allowed. Upon flexible ligand docking against a rigid receptor, both the ligand and the receptor are allowed to move during the subsequent energy minimization. To identify potential off-targets, the score of a ligand against other receptors is compared to the score of the ligand against its known target. Initial results are encouraging in that a number of confirmed targets for tamoxifen and vitamin E were identified based on docking scores against these targets, although a number of known targets of these molecules were missed. In addition, a number of unconfirmed targets were identified using the INVDOCK procedure (Chen and Zhi, 2001).

The use of structurally conservative and nonconservative amino acid mutations is also explored to compare binding site similarities between kinases and how these impact the specificity of kinase inhibitors. It was shown that targets of the same inhibitor have similar residues at all positions that are important for binding of the ligand (Sheinerman, Giraud, and Laoui, 2005).

Fragment-Based Drug Discovery

Drug discovery has traditionally focused on identifying molecules that exhibit activity in a biochemical assay for a particular target. The library sizes that are screened in the biochemical assay have increased significantly over the years with the development of high-throughput screening and combinatorial chemistry. However, these techniques have not led to a significant change in the productivity of drug discovery (Hird, 2000). Compound libraries of pharmaceutical companies generally consist of historical compounds synthesized for previous projects attenuated by compounds synthesized and purchased to enhance the diversity of the collection. While the number of compounds in these collections has increased significantly (up to 2 million compounds) this still covers only a small fraction of accessible chemical space.

Lipinski assessed the physical–chemical properties of clinically tested drug molecules and was able to define the rule-of-five, which consists of physical–chemical properties shared by these molecules (see Chapter 34). After elimination of potential candidates based on these rules, the size of the chemical drug-like space consists of $\sim 10^{60}$ molecules (Bohacek, McMartin, and Guida, 1996) and screening ~ 1 million compounds leaves 10^{55} drug-like compounds unscreened. In addition, HTS is used to identify starting points for further optimization, not for the identification of the final drugs. Based on retrospective analyses, it has been shown that ligand properties such as molecular weight, log P, and the number of hydrogen-bond donating and accepting groups increases as the leads are optimized for potency against their target (Teague et al., 1999; Hann, Leach, and Harper, 2001). Log P is a measure of the hydrophobicity of a molecule and is determined by measuring the concentration of the compound in octanol and water. The increase in these different properties during lead optimization led to a modified definition of lead-like properties that are often used in the analysis of HTS data, to design virtual screening libraries and in the design of libraries for synthesis in library enhancement efforts.

While chemical space for lead-like compounds (molecular weight ~ 350 Da) is significantly smaller than the drug-like space, it still consists of $\sim 10^{25}$ compounds. Fragment-space however, becomes significantly more accessible, as there are only ~ 14 million compounds at <160 Da (Fink, Bruggesser, and Reymond, 2005). However, the affinity of these fragments for the target is generally between $100\,\mu$M and 100 mM and identifying these fragments in biochemical assays is not straightforward. The limitations in designing a suitable biochemical assay are partially effected by detection limits, but also because the screening concentration has to be very high to detect inhibition. Most compounds are not soluble enough to be screened at such high concentrations. A number of non-biochemical approaches have been used to pursue fragment-based drug discovery in recent years, among these are NMR, X-ray crystallography, and the tethering approach, to be further discussed later.

Fragment-based design can be traced back to the SAR-by-NMR technique developed by Abbott (Shuker et al., 1996). It was shown that binding of low-affinity compounds (mM range) could be detected by identifying chemical shifts in the target NMR spectrum. By either observing different chemical shifts induced by different fragments, or by screening fragments in the presence of an excess amount of a previously identified fragment, compounds that bind in different binding pockets can be identified. Fragments interacting with different parts of the protein can be subsequently linked. Proper linking of the weak fragments can result in non-additive enhancements in potency, leading to sub-micromolar compounds (Shuker et al., 1996).

In the SAR-by-NMR technique the binding modes of the active fragments are generally not identified. A number of groups have instead adopted high-throughput crystallography to identify fragments that bind to the target of interest. Generally mixtures of fragments are soaked into existing crystals of the target of interest. Both hardware and software developments have improved the practicality of using X-ray crystallography to identify fragments for further optimization (Hartshorn et al., 2005; Blaney, Nienaber, Burley, 2006). Upon identification of the exact binding mode of the fragment(s), structure-based lead optimization approaches can be used to increase the potency of the fragment for the target of interest and to develop a drug-like molecule.

The tethering strategy to identify binding partners developed by Sunesis involves identifying fragments that have formed a covalent bond to a cysteine in the protein. The formation of the covalent bond enhances the affinity of the fragment for the protein surface, enabling detection. Detection is done using mass spectrometry (MS). If a cysteine is not present near the area of interest, mutagenesis can be used to produce protein with a cysteine in the right position (Erlanson, Wells, and Braisted, 2004).

Computationally, most *de novo* design tools use fragments to design ligands from scratch that are complementary to the site of interest. A number of tools have been developed that explore protein surfaces with small probes (Goodford, 1985; Miranker and Karplus, 1991) which can also be used as starting points for drug discovery. Docking can of course also be used to screen libraries that contain only fragments. However, most fragment-based drug discovery programs use a tiered approach, starting with initial identification of active fragments through experimental techniques and subsequently optimizing fragments using input from structure-based design.

A number of different approaches can be used to find leads from fragments. The first approach, fragment evolution, uses the identified fragment as a platform to add interactions with nearby pockets on the protein surface (Figure 27.4). The second approach, fragment linking, uses two or more fragments that are known to bind to different pockets on the protein surface and construct linkers that can optimally connect the fragments (Figure 27.4). At a later stage, it might be necessary to replace the originally found fragment with some other fragment to optimize other properties than affinity, for example specificity or solubility. This is called fragment-based optimization.

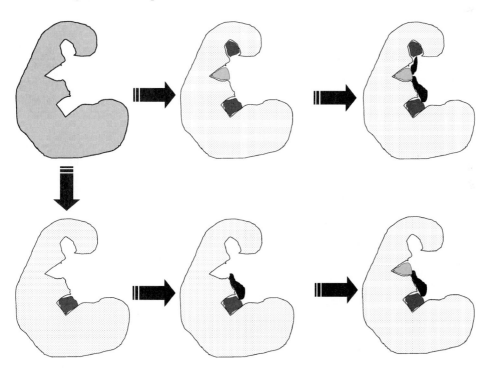

Figure 27.4. Fragment-based drug discovery. Illustrated are the two main approaches to fragment-based drug discovery. The top path shows how different fragments are identified that form interactions with different pockets in the binding site. In the next step, linkers are designed that can link the fragments together without disturbing the interactions made by the fragments. The bottom path illustrates the growth approach to fragment-based discovery in which one starts from a single fragment and grows the fragment to form interactions with nearby pockets in the binding site.

Ligand-Based Perspectives

In this section, a number of different studies are highlighted that have used large-scale ligand-based analyses that can aid in the further optimization of new leads. Just as similar proteins, either based on structure or sequence, often exhibit similar biological functions, one can investigate this for ligands as well. Usually, similar ligands are assumed to bind to similar proteins and thus to have similar biological activities. This is the driving force behind the use of ligand-based similarity measures to identify additional ligands for a particular target. Shoichet's group developed the similarity ensemble approach (SEA), which relates receptors to each other based on the similarity of their ligands. Sixty-five thousand annotated ligands were used and 246 receptors. Using this method, a network of pharmacological space could be constructed and showed clustering of for example the nuclear hormone receptors and the distinct pharmacology of otherwise related proteins. This method can be used to form hypotheses of potential off-targets of ligands, which can then be confirmed experimentally by measuring the affinity of that ligand for the identified potential target (Keiser et al., 2007).

Jain and coworkers also used ligand-based methods to show that they can be used to identify ligands that have the same biological target. Similarity between ligands targeting different receptors can potentially highlight off-target effects, and could thus possibly be used to predict side effects before entering the clinic (Cleves and Jain, 2006).

Three molecules that are found in a large number of cocrystal structures, namely ATP, NAD, and FAD, were investigate to assess the conformational diversity of ligands bounds to proteins. These ligands bind to evolutionarily unrelated proteins, which allows an assessment of conservation of the binding mode of these ligands. The authors observed large conformational variations in the ligands bound to different proteins, and even within homologous superfamilies, each of the studied molecules can adopt a wide range of conformations (Stockwell and Thornton, 2006).

Although a number of studies have shown that chemically similar ligands can occasionally adopt different binding modes depending on either the experimental conditions (Davis, Teague, Kleywegt, 2003) or changes to the ligand, no systematic study had been done to solidify this generalization until recently. To address this issue, Boström and coworkers conducted a systematic study that exhaustively searched all available protein–ligand structures in the PDB. The search retrieved 1484 binding sites of interest with 206 binding site pairs that bind similar ligands. A comparison was made between the binding sites, to analyze changes in ligand conformation based on conserved water molecules, side chain positions, and other potential factors. Water molecules can have a profound influence of ligand binding since they may mediate hydrogen bonds between the ligand and the protein. In $\sim$70% of the cases, the positioning of waters molecules in the binding site pairs differs. Amino acid side chains play an important role in forming interactions with the ligand and in determining the shape of the binding site. In $\sim$50% of the binding site pairs, different rotamers for amino acid side chains were observed, but backbone movements were rare (Bostroem, Hogner, and Schmitt, 2006). Both these observations underscore the necessity of developing methods for docking that take into account the role of explicit water molecules and binding site flexibility. Although structural changes in the binding site are observed for similar ligand pairs, it is important to note from this study that the orientation of the ligands in the binding site was conserved (Bostroem, Hogner, and Schmitt, 2006).

Rather than select for leads based on initial potency, other parameters such as molecular weight of the compound and ligand efficiency are being considered during the selection process. High-molecular-weight compounds often have poor solubility and permeability, which can lead to poor *in vivo* activity. During lead optimization, the molecular weight of the compounds generally increases therefore it is important to start with leads of smaller, reasonable molecular weight. Ligand efficiency has also become a parameter for lead consideration defined as the binding free energy per non-hydrogen atom $\Delta G_{efficiency}$ (Hopkins, Groom, and Alex, 2004; Abad-Zapatero and Metz, 2005):

$$\Delta G_{efficiency} = \Delta G / N_{non\text{-}hydrogen\ atom}. \qquad (27.6)$$

In essence this metric normalizes the potency by the size of the ligand, and can be substituted with other metrics of size, such as molecular weight or the polar surface area (PSA; Abad-Zapatero and Metz, 2005). The use of this metric results in highlighting low molecular weight compounds that are less potent than larger molecules, but have a larger contribution of each atom to the overall potency.

Ligand-based analyses have also critically contributed to improving the design and augmentation of chemical libraries when purchasing commercial compounds to expand the corporate collection. On the one hand compounds should be purchased that are not represented in the current collection in order to increase the amount of chemical space that is being sampled by the general purpose HTS collection. Diversity calculations can be used for this. On the contrary, each chemical series needs to be represented by more than a single molecule in the collection to improve the likelihood of identifying that series in a biological screen against the relevant target. Using a retrospective analysis of HTS data based on a number of assumptions, it was estimated that between 50 and 100 analogues are needed for each chemical series to identify each active series (Nilakantan, Immermann, and Haraki, 2002).

SUMMARY

To conclude, while docking is important for many other facets of research, the importance of docking is placed in the context of drug screening and development. Mainly because the advantages of coupling computational screens and modeling of biologically active compounds help facilitate an otherwise time consuming and expensive process. The hope is that these *in silico* approach will help reduce and better manage resources. The need for such virtual screening also drives improved technology to identify suitable matches between the compounds and targets that collectively have pharmacological impact. As demonstrated in this chapter, there are strengths and limitations associated with both physics or heuristic based algorithms. However, with improvements in computational resources and our growing knowledge of the binding process and drug efficacy, docking and virtual screening will play an increasingly important role in expediting drug discovery.

REFERENCES

Abad-Zapatero C, Metz J (2005): Ligand efficiency indices as guideposts for drug discovery. *Drug Discov Today* 10:464–469.

Alvarez JC (2004): High-throughput docking as a source of novel drug leads. *Curr Opin Chem Biol* 8:365–370.

Aqvist J, Medina C, Samuelsson J-E (1994): A new method for predicting binding affinity in computer-aided drug design. *Protein Eng* 7:385–391.

Atkins PW (1997): *Physical Chemistry*, 6th edn. New York: W. H. Freeman & Co.

Atwood TK (2000): The quest to deduce protein function from sequence: the role of pattern databases. *Int J Biochem Cell Biol* 32:139–155.

Bashford D, Case DA (2000): Generalized Born models of macromolecular solvation effects. *Annu Rev Phys Chem* 51:129–152.

Bissantz C, Folkers G, Rognan D (2000): Protein-based virtual screening of chemical databases. 1. Evaluation of different docking/scoring combinations. *J Med Chem* 43:4759–4767.

Blaney JM, Nienaber V, Burley SK (2006): Fragment-based lead discovery and optimization using X-ray crystallography, computational chemistry, and high-throughput organic synthesis. In: Jahnke W, Erlanson DA, editors. *Fragment-Based Approaches in Drug Discovery*. Weinheim: Wiley-VCH Verlag, pp 215–248.

Bohacek RS, McMartin C, Guida WC (1996): The art and practice of structure-based drug design: a molecular modeling perspective. *Med Res Rev* 16:3–50.

Bostroem J (2001): Reproducing the conformations of protein-bound ligands: a critical evaluation of several popular conformational searching tools. *J Comput Aid Mol Des* 15:1137–1152.

Bostroem J, Hogner A, Schmitt S (2006): Do structurally similar ligands bind in a similar fashion?. *J Med Chem* 49:6716–6725.

Chen YZ, Zhi DG (2001): Ligand–protein inverse docking and its potential use in the computer search of protein targets of a small molecule. *Proteins* 43:217–226.

Cheng AC, Coleman RG, Smyth KT, Cao Q, Soulard P, Caffrey DR, Salzberg AC, Huang ES (2007): Structure-based maximal affinity model predicts small-molecule druggability. *Nat Biotechnol* 25:71–75.

Chuaqui C, Deng Z, Singh J (2005): Interaction profiles of protein-kinase-inhibitor complexes and their application to virtual screening. *J Med Chem* 48:121–133.

Claussen H, Buning C, Rarey M, Lengauer T (2001): FlexE: efficient molecular docking considering protein structure variations. *J Mol Biol* 308:377–395.

Cleves AE, Jain AN (2006): Robust ligand-based modeling of the biological targets of known drugs. *J Med Chem* 49:2921–2938.

David L, Luo R, Gilson MK (2001): Ligand–receptor docking with the mining minima optimizer. *J Comput Aid Mol Des* 15:157–171.

Davis AM, Teague SJ, Kleywegt GJ (2003): Application and limitations of X-ray crystallographic data in structure-based ligand and drug design. *Angew Chem Int Ed Engl* 42: 2718–2736.

DeLano WL (2002): Unraveling hot spots in binding interfaces: progress and challenges. *Curr Opin Struct Biol* 12:14–20.

Deng Z, Chuaqui C, Singh J (2004): Structural interaction fingerprint (SIFt): a novel method for analyzing three-dimensional protein–ligand binding interactions. *J Med Chem* 47:337–344.

DesJarlais RL, Seibel GL, Kuntz ID, Furth PS, Alvarez JC, Ortiz de Montellano PR, DeCamp DL, Babe LM, Craik CS (1990): Structure-based design of nonpeptide inhibitors specific for human immunodeficiency virus 1 protease. *Proc Natl Acad Sci USA* 87:6644–6648.

Diller DJ, Li R (2003): Kinases, homology models, and high throughput docking. *J Med Chem* 46:4638–4647.

Ehrlich P (1957): Himmelweit F, editor. *Gesammelte Arbeiten*. Berlin: Springer-Verlag.

Elcock AH (2001): Prediction of functionally important residues based solely on the computed energetics of protein structure. *J Mol Biol* 312:885–896.

Erlanson DA, Wells JA, Braisted AC (2004): Tethering: fragment-based drug discovery. *Annu Rev Biophys Biomol Struct* 33:199–223.

Ewing TJA, Kuntz ID (1997): Critical evaluation of search algorithms for automated molecular docking and database screening. *J Comp Chem* 18:1175–1189.

Ewing TJA, Makino S, Skillman AGJ, Kuntz ID (2001): DOCK 4.0: search strategies for automated molecular docking of flexible molecule databases. *J Comput Aid Mol Des* 15:411–428.

Fernandes MX, Kairys V, Gilson MK (2004): Comparing ligand interactions with multiple receptors via serial docking. *J Chem Inf Comput Sci* 44:1961–1970.

Fink T, Bruggesser H, Reymond J-L (2005): Virtual exploration of the small-chemical universe below 160 Dalton. *Angew Chem Int Ed Engl* 44:1504–1508.

Fischer E (1894): Einfluss der configuration auf die wirkung der enzyme. *Ber Dt Chem Ges* 27:2985–2993.

Foote J, Milstein C (1994): Conformational isomerism and the diversity of antibodies. *Proc Natl Acad Sci USA* 91:10370–10374.

Friesner RA, Banks JL, Murphy RB, Halgren TA, Klicic JJ, Mainz DT, Repasky MP, Knoll EH, Shelley M, Perry JK, et al. (2004a): Glide: a new approach for rapid, accurate docking and scoring. 1. Method and assessment of docking accuracy. *J Med Chem* 47:1739–1749.

Friesner RA, Banks JL, Murphy RB, Halgren TA, Klicic JJ, Mainz DT, Repasky MP, Knoll EH, Shelley M, Perry JK, et al. (2004b): Glide: a new approach for rapid, accurate docking and scoring. 1. Method and assessment of docking accuracy. *J Med Chem* 47:1739–1749.

Friesner RA, Murphy RB, Repasky MP, Frye LL, Greenwood JR, Halgren TA, Sanschagrin PC, Mainz DT (2006): Extra precision glide: docking and scoring incorporating a model of hydrophobic enclosure for protein–ligand complexes. *J Med Chem* 49:6177–6196.

Gabb HA, Jackson RM, Sternberg MJE (1997): Modelling protein docking using shape complementarity, electrostatics, and biochemical information. *J Mol Biol* 272:106–120.

Gasteiger J, Rudolph C, Sadowski J (1990): Automatic generation of 3D-atomic coordinates for organic molecules. *Tetrahedron Comput Methodol* 3:537–547.

Gilson MK, Honig B (1988): Calculation of the total electrostatic energy of a macromolecular system: solvation energies, binding energies, and conformational analysis. *Proteins* 7:7–18.

Good AC, Cheney DL (2003): Analysis and optimization of structure-based virtual screening protocols (1): exploration of ligand conformational sampling techniques. *J Mol Graph Mod* 22:23–30.

Goodford PJ (1985): A computational procedure for determining energetically favorable binding sites on biologically important macromolecules. *J Med Chem* 28:849–857.

Goodsell DS, Olson AJ (1990): Automated docking of substrates to proteins by simulated annealing. *Proteins* 8:195–202.

Halgren TA (2007): New method for fast and accurate binding-site identification and analysis. *Chem Biol Drug Des* 69:146–148.

Hann MM, Leach AR, Harper G (2001): Molecular complexity and its impact on the probability of finding leads for drug discovery. *J Chem Inf Comput Sci* 41:856–864.

Hartshorn MJ, Murray CW, Cleasby A, Frederikson M, Tickle IJ, Jhoti H (2005): Fragment-based lead discovery using X-ray crystallography. *J Med Chem* 48:403–413.

Hartshorn M, Verdonk ML, Chessari G, Brewerton SC, Mooij WTM, Mortenson PN, Murray CW (2007): Diverse, high-quality test set for the validation of protein–ligand docking performance. *J Med Chem* 50:726–741.

Head MS, Given JA, Gilson MK (1997): "Mining minima": direct computation of conformational free energies. *J Phys Chem A* 101:1609–1618.

Herzberg O, Moult J (1991): Analysis of the steric strain in the polypeptide backbone of protein molecules. *Proteins* 11:223–229.

Hird N (2000): Isn't combinatorial chemistry just chemistry? *Drug Discov Today* 5:307–308.

Hopkins AL, Groom CR, Alex A (2004): Ligand efficiency: a useful metric for lead selection. *Drug Discov Today* 9:430–431.

Huang N, Kalyanaraman C, Irwin JJ, Jacobson MP (2006): Physics-based scoring of protein–ligand complexes: enrichment of known inhibitors in large-scale virtual screening. *J Chem Inf Model* 46:243–253.

Huang N, Shoichet BK, Irwin JJ (2006): Benchmarking sets for molecular docking. *J Med Chem* 49:6789–6801.

Jain AN (2003): Surflex: fully automated flexible molecular docking using a molecular similarity-based search engine. *J Med Chem* 46:499–511.

Jones G, Willett P, Glen RC (1995): Molecular recognition of receptor sites using a genetic algorithm with a description of desolvation. *J Mol Biol* 245:43–53.

Jones G, Willett P, Glen RC, Leach AR, Taylor R (1997): Development and validation of a genetic algorithm for flexible docking. *J Mol Biol* 267:727–748.

Joseph-McCarthy D, Thomas BE, IV, Berlmarsh M, Moustakas D, Alvarez JC (2003): Pharmacophore-based molecular docking to account for ligand flexibility. *Proteins* 51:172–188.

Karlin S, Zhu Z-Y (1996): Characterizations of diverse residue clusters in protein three-dimensional structures. *Proc Natl Acad Sci USA* 93:8344–8349.

Katchalski-Katzir E, Shariv I, Eisenstein M, Friesem AA, Aflalo C, Vakser IA (1992): Molecular surface recognition: determination of geometric fit between proteins and their ligands by correlation techniques. *Proc Natl Acad Sci USA* 89:2195–2199.

Kauzmann W (1959): Some factors in the interpretation of protein denaturation. *Adv Protein Chem* 14:1–63.

Keiser MJ, Roth BL, Armbruster BN, Ernsberger P, Irwin JJ, Shoichet BK (2007): Relating protein pharmacology by ligand chemistry. *Nat Biotechnol* 25:197–206.

Klon AE, Glick M, Thoma M, Acklin P, Davies JW (2004): Finding more needles in the haystack: a simple and efficient method for improving high-throughput docking results. *J Med Chem* 47:2643–2749.

Knegtel RMA, Kuntz ID, Oshiro CM (1997): Molecular docking to ensembles of protein structures. *J Mol Biol* 266:424–440.

Kollman PA (1993): Free energy calculations: applications to chemical and biochemical phenomena. *Chem Rev* 93:2395–2417.

Kollman PA, Massova I, Reyes C, Kuhn B, Huo S, Chong L, Lee M, Lee T, Duan Y, Wang W, et al. (2000): Calculating structures and free energies of complex molecules: combining molecular mechanics and continuum models. *Account Chem Res* 33:889–897.

Kontoyianni M, McClellan LM, Sokol GS (2004): Evaluation of docking performance: comparative data on docking algorithms. *J Med Chem* 47:558–565.

Koshland DE Jr (1958): Application of a theory of enzyme specificity to protein synthesis. *Proc Natl Acad Sci USA* 44:98–104.

Kramer B, Rarey M, Lengauer T (1999): Evaluation of the FlexX incremental construction algorithm for protein–ligand docking. *Proteins* 37:228–241.

Kuhn D, Weskamp N, Schmitt S, Huellermeier E, Klebe G (2006): From the similarity analysis of protein cavities to the functional classification of protein families using Cavbase. *J Mol Biol* 359:1023–1044.

Kuntz ID, Blaney JM, Oatley SJ, Langridge R, Ferrin TE (1982): A geometric approach to macromolecular-ligand interactions. *J Mol Biol* 161:269–288.

Langley JN (1905): On the reaction of cells and of nerve endings to certain poisons, chiefly as regards the reaction of striated muscle to nicotine and to curari. *J Physiol* 33:374–413.

Liang J, Edelsbrunner H, Woodward C (1998): Anatomy of protein pockets and cavities: measurement of binding site geometry and implications for ligand design. *Protein Sci* 7:1884–1897.

Lichtarge O, Bourne HR, Cohen FE (1996): An evolutionary trace method defines binding surfaces common to protein families. *J Mol Biol* 257:342–358.

Lin J-H, Perryman AL, Schames JR, McCammon JA (2002): Computational drug design accommodating receptor flexibility: the relaxed complex scheme. *J Am Chem Soc* 124:5632–5633.

Ljungberg KB, Marelius J, Musil D, Svensson P, Norden B, Aqvist J (2001): Computational modelling of inhibitor binding to human thrombin. *Eur J Pharm Sci* 12:441–446.

Lorber DM, Shoichet BK (1998): Flexible ligand docking using conformational ensembles. *Protein Sci* 7:938–950.

Lyne PD, Kenny PW, Cosgrove DA, Deng C, Zabludoff S, Wendoloski JJ, Ashwell S (2004): Identification of compounds with nanomolar binding affinity for checkpoint kinase-1 using knowledge-based virtual screening. *J Med Chem* 47:1962–1968.

Mason JS, Cheney DL (1999): Ligand–receptor 3D similarity using multiple 4-point pharmacophores. *Proc Pac Symp Biocomput* 4:456–467.

McGovern SL, Helfand BT, Feng B, Shoichet BK (2003): A specific mechanism of nonspecific inhibition. *J Med Chem* 46:4265–4272.

McGovern SL, Shoichet BK (2003): Information decay in molecular docking screen against holo, apo, and modeled conformations of enzymes. *J Med Chem* 46:2895–2907.

McMartin C, Bohacek RS (1997): QXP: Powerful, rapid computer algorithms for structure-based drug design. *J Comput Aid Mol Des* 11:333–344.

Meagher KL, Lerner MG, Carlson HA (2006): Refining the multiple protein structure pharmacophore method: consistency across three independent HIV-1 protease models. *J Med Chem* 49:3478–3484.

Meng EC, Shoichet BK, Kuntz ID (1992): Automated docking with grid-based energy evaluation. *J Comp Chem* 13:505–524.

Miranker A, Karplus M (1991): Functionality maps of binding sites: a multiple copy simultaneous search method. *Proteins* 11:29–34.

Mooij WTM, Verdonk ML (2005): General and targeted statistical potentials for protein–ligand interactions. *Proteins* 61:272–287.

Morris GM, Goodsell DS, Halliday RS, Huey R, Hart WE, Belew RK, Olson AJ (1998): Automated docking using a Lamarckian genetic algorithm and an empirical binding free energy function. *J Comp Chem* 19:1639–1662.

Moustakas D, Lang PT, Pegg S, Pettersen E, Kuntz ID, Brooijmans N, Rizzo RC (2006a): Development and validation of a modular, extensible docking program: DOCK 5. *J Comput Aid Mol Des* 20:601–619.

Moustakas DT, Lang PT, Pegg S, Pettersen E, Kuntz ID, Brooijmans N, Rizzo RC (2006b): Development and validation of a modular, extensible docking program: DOCK5. *J Comput Aid Mol Des* 20:601–619.

Muegge I (2006): PMF scoring revisited. *J Med Chem* 49:5895–5902.

Naumann T, Matter H (2002): Structural classification of protein kinases using 3D molecular interaction field analysis of their ligand binding sites: target family landscapes. *J Med Chem* 45:2366–2378.

Nayal M, Honig B (2006): On the nature of cavities on protein surfaces: application to the identification of drug-binding sites. *Proteins* 63:892–906.

Nilakantan R, Immermann F, Haraki K (2002): A novel approach to combinatorial library design. *Combin Chem High Throughput Screen* 5:105–110.

Nissink JW, Murray CW, Hartshorn M, Verdonk ML, Cole JC, Taylor R (2002): A new test set for validating predictions of protein–ligand interactions. *Proteins* 49:457–471.

Oesterberg F, Morris GM, Sanner MF, Olson AJ, Goodsell DS (2002): Automated docking to multiple target structures: incorporation of protein mobility and structural water heterogeneity in AutoDock. *Proteins* 46:34–40.

Oshiro CM, Bradley EK, Eksterowicz J, Evensen E, Lamb ML, Lanctot JK, Putta S, Stanton R, Grootenhuis PDJ, (2004): Performance of 3D-database molecular docking studies into homology models. *J Med Chem* 47:764–767.

Overington J, Al-Lazikani B, Hopkins AL (2006): How many drug targets are there? *Nat Rev Drug Discov* 5:993–996.

Page CS, Bates PA (2006): Can MM-PBSA calculations predict the specificities of protein kinase inhibitors? *J Comp Chem* 27:1990–2007.

Pang Y-P, Perola E, Xu K, Prendergast FG (2001a): EUDOC: a computer program for identification of drug interaction sites in macromolecules and drug leads from chemical databases. *J Comp Chem* 22:1750–1771.

Pang YP, Perola E, Xu K, Prendergast FG (2001b): EUDOC: a computer program for identification of drug interaction sites in macromolecules and drug leads from chemical databases. *J Comp Chem* 22:1750–1771.

Perola E, Charifson PS (2004): Conformational analysis of drug-like molecules bound to proteins: an extensive study of ligand reorganization upon binding. *J Med Chem* 47:2499–2510.

Perola E, Walters WP, Charifson PS (2004): A detailed comparison of current docking and scoring methods on systems of pharmaceutical relevance. *Proteins* 56:235–249.

Powers R, Copeland JC, Germer K, Mercier KA, Ramanathan V, Revesz P (2006): Comparison of protein active site structures for functional annotations of proteins and drug design. *Proteins* 65:124–135.

Rarey M, Wefing S, Lengauer T (1996): Placement of medium-sized molecular fragments into active sites of proteins. *J Comput Aid Mol Des* 10:41–54.

Rarey M, Kramer B, Lengauer T (1999): The particle concept: placing discrete water molecules during protein–ligand docking predictions. *Proteins* 34:17–28.

Rizzo RC, Tirado-Rives J, Jorgensen WL (2001): Estimation of binding affinities for HEPT and nevirapine analogues with HIV-1 reverse transcriptase via Monte Carlo simulations. *J Med Chem* 44:145–154.

Rockey WM, Elcock AH (2005): Rapid computational identification of the targets of protein kinase inhibitors. *J Med Chem* 48:4138–4152.

Rockey WM, Elcock AH (2006): Structure selection for protein kinase docking and virtual screening: homology models or crystal structures? *Curr Protein Pept Sci* 7:437–457.

Ruppert J, Welch W, Jain AN (1997): Automatic identification and representation of protein binding sites for molecules docking. *Protein Sci* 6:524–533.

Sheinerman FB, Giraud E, Laoui A (2005): High affinity targets of protein kinase inhibitors have similar residues at the positions energetically important for binding. *J Mol Biol* 352:1134–1156.

Sherman W, Day T, Jacobson MP, Friesner RA, Farid R (2006): Novel procedure for modeling ligand/receptor induced fit effects. *J Med Chem* 49:534–553.

Shoichet BK, Kuntz ID (1993): Matching chemistry and shape in molecular docking. *Protein Eng* 6:723–732.

Shoichet BK, Leach AR, Kuntz ID (1999): Ligand solvation in molecular docking. *Proteins* 34: 4–16.

Shuker SB, Hajduk PJ, Meadows RP, Fesik SW (1996): Discovery high-affinity ligands for proteins: SAR by NMR. *Science* 274:1531–1535.

Sippl MJ (1990): Calculation of conformational ensembles from potentials of mean force. *J Mol Biol* 213:859–883.

Soga S, Shirai H, Kobori M, Hirayama N (2007): Use of amino acid composition to predict ligand-binding sites. *J Chem Inf Model* 47:400–406.

Stockwell GR, Thornton JM (2006): Conformational diversity of ligands bound to proteins. *J Mol Biol* 356:928–994.

Tame JRH (1999): Scoring functions: a view from the bench. *J Comput Aid Mol Des* 13:99–108.

Teague SJ, Davis AM, Leeson PD, Oprea TI (1999): The design of leadlike combinatorial libraries. *Angew Chem Int Ed Engl* 38:3743–3748.

Tirado-Rives J, Jorgensen WL (2006): Contribution of conformer focusing to the uncertainty in predicting free energies for protein–ligand binding. *J Med Chem* 49:5880–5884.

Triballeau N, Acher F, Brabet I, Pin J-P, Bertrand H-O (2005): Virtual screening workflow development guided by the "receiver operating characteristic" curve approach. Application to high-throughput docking on metabotropic glutamate receptor subtype 4. *J Med Chem* 48:2534–2547.

Truchon J-F, Bayly CI (2007): Evaluating virtual screening methods: good and bad metrics for the "early recognition" problem. *J Chem Inf Model* 47:488–508.

Verdonk ML, Cole JC, Hartshorn MJ, Murray CW, Taylor R (2003a): Improved protein–ligand docking using GOLD. *Proteins* 52:609–623.

Verdonk ML, Cole JC, Hartshorn MJ, Murray CW, Taylor RD (2003b): Improved protein–ligand docking using GOLD. *Proteins* 52:609–623.

Verdonk ML, Berdini V, Hartshorn M, Mooij WTM, Murray CW, Taylor RD, Watson P (2004): Virtual screening using protein–ligand docking: avoiding artificial enrichment. *J Chem Inf Comput Sci* 44:793–806.

Verdonk ML, Chessari G, Cole JC, Hartshorn MJ, Murray CW, Nissink JW, Taylor RD, Taylor R (2005): Modeling water molecules in protein–ligand docking using GOLD. *J Med Chem* 48:6504–6515.

Vulpetti A, Crivori P, Cameron A, Bertrand J, Brasca MG, D'Alessio R, Pevarello P (2005): Structure-based approaches to improve selectivity: CDK2-GSK3b binding site analysis. *J Chem Inf Model* 45:1282–1290.

Warren GL, Andrews CW, Capelli A-M, Clarke B, LaLonde J, Lambert MH, Lindvall M, Nevins N, Semus SF, Senger S, et al. (2006a): A critical assessment of docking programs and scoring functions. *J Med Chem* 49:5912–5931.

Warren GL, Webster Andrews C, Capelli A-M, Clarke B, LaLonde J, Lambert MH, Lindvall M, Nevins N, Semus SF, Senger S, et al. (2006b): A critical assessment of docking programs and scoring functions. *J Med Chem* 49:5912–5931.

Warwicker J, Watson HC (1982): Calculation of the electric potential in the active site cleft due to α-helix dipoles. *J Mol Biol* 157:671–679.

Wong CF, Kua J, Zhang Y, Straatsma TP, McCammon JA (2005): Molecular docking of balanol to dynamics snapshots of protein kinase A. *Proteins* 61:850–858.

Woo H-J, Roux B (2005): Calculation of absolute protein–ligand binding free energies from computer simulations. *Proc Natl Acad Sci USA* 102:6825–6830.

Yao H, Mihalek I, Lichtarge O (2006): Rank information: a structure-independent measure of evolutionary trace quality that improves identification of protein functional sites. *Proteins* 65:111–123.

Zou X, Sun Y, Kuntz ID (1999): Inclusion of solvation in ligand binding free energy calculations using the generalized-Born model. *J Am Chem Soc* 121:8033–8043.

Section VI

STRUCTURE PREDICTION

28

CASP AND OTHER COMMUNITY-WIDE ASSESSMENTS TO ADVANCE THE FIELD OF STRUCTURE PREDICTION

Jenny Gu and Philip E. Bourne

A MEASURE FOR SUCCESS

In the early 1990s, the community recognized that methods for structure determination from sequence information have been proliferating, creating the need to benchmark these developments to gauge the utility of the algorithms to the biological community and to measure the progress in this developing field of structure prediction. In part, this need also came from users of these tools concerned by over optimistic claims of prediction performance. In 1994, John Moult pioneered the idea that the only way to objectively assess the utility of these tools was to conduct a blind experiment in which predictions were made on protein structures not yet publicly available, but with available sequences (Moult et al., 1995). critical assessment of structure prediction (CASP) was born and, through this assessment, confidence from the user community can be said to have been reclaimed. Even a prediction of limited accuracy can be useful if the user knows what to expect and the method has been cross-validated. The CASP experiment consists of three parts: defining the types of predictions to be performed (this has changed over time), the collection of prediction targets, and the evaluation of the performance of each predictor. For the first competition, the field was evaluated in three categories: comparative modeling, threading, and *ab initio* prediction.

To clarify for individuals interested in reading the original background papers, there have been changes in nomenclature over the years of the CASP experiment. The term "comparative modeling" has become interchangeable with "homology modeling" and "threading" has been replaced by the term "fold recognition." "*Ab initio* structure prediction" is sometimes referred to as "new fold recognition" to reflect the underlying methodologies. Part of the reason for this change, as we shall see, has come about by the role that the ever-increasing number of experimental structures plays in just about every case.

Stated simply, the categories are based on the level of sequence identity between the protein sequence to be modeled and the potential structural homologue (template). If the protein has high sequence identity, homology modeling is used. Conversely, *ab initio* structure prediction is reserved for cases where no known structural homologues exist. The *ab initio* category has also come under recent scrutiny because successful methods use knowledge-based approaches, where that knowledge is derived from existing structures. Thus, the *ab initio* category has recently been redefined as "new fold recognition" (Moult et al., 2005). True *ab initio* approaches are now limited to numerical simulation techniques using traditional empirical potentials.

Nomenclature aside, CASP has served as a metric by which advances in structure prediction is measured and has undoubtedly accelerated developments in the field. So much so that similar assessments (some would say competitions) now take place in the field of docking (CAPRI; Janin et al., 2003).

The primary goal of CASP is to evaluate the performance of bona fide blind predictions of structures. Participating teams are given a period of several weeks to complete their model while automatic servers are given 48 h. Targets are obtained from different experimental groups with PDB structures not yet released to the public. Several different evaluation measures are employed to measure success and the performance of structure prediction methods. Evaluation is conducted by expert assessors in the field and conclusions are then released to the public and published in special editions of the journal *Proteins: Structure Function and Bioinformatics* summarizing results for the year that include strengths, weaknesses and where improvements can be made. Submissions for assessment are handled by the Protein Structure Prediction Center (Zemla et al., 2001) dedicated to this community-wide effort.

While this may appear to be a competition between structural predictors to identify the best performing methods, it should be considered instead to be a collaborative effort that provides the community with a set of principles for improving the standards in the field of structure prediction (Moult, 2006). Significant advances are still needed to predict protein structures that are comparable to those obtained experimentally. The four major challenges that the field is confronted with at this time are (i) when high sequence homology exists, to produce models close to those obtained experimentally; (ii) to improve alignments between the unknown and the template; (iii) to develop a better refinement process to construct models of remotely related proteins; and (iv) to construct a reliable scheme to discriminate possible model solutions generated from a template-free algorithm.

How can the best approach to structure prediction and the best models be judged? Devising appropriate statistical measures is important to firmly establish progress in the field and to establish that the proposed models do indeed represent the true protein *in vitro* and *in vivo*. Through the cumulative lessons gained from each CASP, several criteria are deemed necessary to conduct successful benchmarking experiments. First, a large test set that is agreed to by the community should be used. This test set must be independent of the training set that has been used to train the methods in question. Second, error estimation must be reliable and continuously questioned and improved to develop proper measures. Third, all measures have inherent bias and therefore independent tests of accuracy must be performed. Finally, the results must be open and freely available to the community as well as the source code for participating methods. Even though there may be intellectual

TABLE 28.1. Available Community-Wide Benchmarking Services

Benchmarking Service	Brief Description	Reference
CASP	Critical assessment for structural prediction. Several categories include homology modeling, fold recognition, *de novo* structure prediction. Disorder, domain, and functional predictions also included	Moult et al. (1995)
CAFASP	Critical assessment of fully automated structure prediction. Evaluation of methods that do now utilize expert interpretations	Fischer et al. (1999)
EVA	Continuously and automatically analyses protein structure prediction servers in 'real time'	Eyrich et al. (2001)
EVAcon	Continuous evaluation of inter-residue contacts in structure prediction	Grana et al. (2005a), Grana et al. (2005)
LiveBench	Similar to EVA but differs in methods of evaluation	Bujnicki et al. (2001)
CAPRI	Comparative evaluation of docking interactions in protein structure prediction	Janin et al. (2003)
FORCASP	Discussion forum for CASP enthusiasts	

property issues, access to the software is necessary to rigorously check the performance of the methods to be sure that they perform as stated.

As the value of CASP has come to be appreciated by the community over the years, the project expanded benchmarking efforts to evaluate docking, domain boundary identification, protein disorder prediction, and functional prediction. An important addition that was introduced along the way was that of fully automated servers as opposed to predictions requiring significant human intervention. Table 28.1 summarizes the major classes of benchmarks employed by CASP.

The result of this ongoing community experiment has yielded interesting snapshots of performance and progress in the field over the years. Here, we review the history of CASP since its inception and highlight some important events that have helped to advance and fuel the field. Details of the technical strategies themselves, such as homology modeling, fold recognition, *ab initio* structure prediction, disorder prediction, domain boundary identification, and functional annotation are found in Chapters 30–32, 38, 20, and 21, respectively.

COMMUNITY BENCHMARK HISTORY AND FINDINGS

Participation from the community is absolutely necessary for the success of CASP (Moult et al., 1995). In CASP1, 34 groups participated from March through October 1994 with targets expiring at different dates during that period. A total of 34 predictions were submitted for 7 test candidates in comparative modeling; 20 targets were used for threading and *ab initio* structure predictions with 66 and 29 predictions submitted, respectively.

The meeting to evaluate the performance of the first CASP1 experiment was held at the Asilomar conference center, California, in December 1994. It was determined during this first evaluation that there was a need to develop better measures to gauge the success of predictions. Simple statistical tests did not evaluate all the features needed for successful model prediction. Moreover, it was established that results were biased by the time each group spent on a single prediction, which is not a true measure of the method itself. To circumvent this issue, in CASP2 onwards, participants were permitted to submit several predictions for each target.

Comparative modeling approaches were found to be challenged by obtaining the proper sequence alignment or target sequence to template and loop regions, insertions and deletions were not well modeled (Mosimann, Meleshko, and James, 1995). Inclusion of models containing fundamental errors suggested that minimal testing of model coordinates should be required before papers referencing these models are to be accepted for journal publication. The most disturbing finding was the presence of D-chirality for some of the $C\alpha$ carbons and incorrect chirality for some beta-branched threonine and isoleucine residues. Out of the 43 predicted structures, 14 contained an example of this incorrect enantiomorph most likely arising during the energy minimization step used to refine these models. A significant number of predictions had the planar peptide dihedral angle deviating by more than $\pm15°$ from the plane (well-refined structures did better with $\pm6°$ from the plane. The distribution of ϕ, ψ dihedral angles often deviated significantly from the groupings typically seen in a Ramachandran plot that is made using experimental protein structures. As expected, the root-mean-squared deviation (RMSD) between the models and the template was lowest for models with high sequence identity to the target.

Structure prediction of proteins with little or no sequence homology to proteins with known structure required the use of threading methods. Assessment of threading methods indicated that, in many cases, they were capable of identifying the correct fold but alignment was an issue (Lemer, Rooman, and Wodak, 1995). A major contributor to the success of fold recognition was attributed to the importance of hydrophobic interactions in defining the protein core. Fold identification was considered correct if the predicted structure was aligned to the template with an RMSD ≤ 3 Å based on $C\alpha$ atoms. Different methods were better depending on the type of fold. This raised the idea of using consensus methods that we will come back to shortly.

Ab initio structure prediction to determine novel protein folds faced significant challenges during CASP1 (Defay and Cohen, 1995). Accurate tertiary structure prediction was not possible with these methods. However, limited success was attained when generating protein folds and motifs that were recognizably similar to another known structure, presence. A useful development was the identification of sequence similarity within the sequence that often translates to structural symmetry and is likely to be the result of ancient duplication and fusion events. Approximate *ab initio* predictions could be made for very small proteins through exhaustive conformational searches. multiple sequence alignments (MSA) also helped improve distance matrix approaches that use sequence variability information to make inferences about specific contact potentials.

Finally, for CASP1, although the focus was on tertiary structure prediction, the performance of secondary structure predictors could be inferred since they were a part of many methods (Rost and Sander, 1995). The secondary structure of helices was predicted well, although sometimes too long, with beta and coil underperforming. The secondary structure of helices was overdetermined and nonsoluble proteins were also poorly determined.

CASP1 was "a cathartic experience, and we have emerged from it with a new and sharper sense of direction" (Moult et al., 1995). With newly identified limitations in the various prediction methods, better evaluation strategies were once again re-evaluated 2 years later during CASP2 (Moult et al., 1997). Organizers collected 42 structures and participants submitted solutions for 34 of these. Finding suitable targets was something of a problem as the era of high throughput structure determination was yet to begin. More protein targets were available for evaluating *ab initio* structure prediction, therefore permitting proper statistical evaluation. A new docking category was introduced. For its debut, seven small ligands that bind to four different proteins (protein–ligand docking) and one protein–protein complex (protein–protein docking) were obtained. While this data set was too small to properly evaluate the state of docking methods at the time, 13 groups did participate and a total of 56 predictions were submitted. Overall participation doubled from 34 to 70 groups with the total number of submitted predictions approaching 1000. Evaluation measures were improved and were more sensitive in judging poor quality models and singling out particular features of the model that reflected the strengths and weaknesses of specific steps in the algorithms, such as loop modeling, alignment accuracy, or correct topology.

The development of new evaluation criteria aimed not to identify the best performing predictor, but to identify the strengths and weaknesses of each method (Venclovas et al., 1997). The three considerations were (1) gauging the performance at each different stages of modeling; (2) distinguishing results from easily modeled versus difficult regions; and (3) eliminating effects of possible experimental uncertainties. It was shown that there had been some improvement since CASP1, particularly in side chain accuracy. Large loop errors were still found in comparative modeling that often reflected poor alignments in these regions. The performance of threading methods was the hardest to estimate as it was being plagued by alignment issues. Finally, the geometry of the models was significantly improved with no D-amino acids (Martin, MacArthur, and Thornton, 1997). No method came out as clearly superior, but assessors were deemed to provide good justification of their decisions (Levitt, 1997).

Threading methods were difficult to evaluate because they involved two separate criteria: fold recognition and model accuracy. Attempts to gauge performance used weighted averages of fold recognition and alignment accuracy with and without normalization for target difficulty (Levitt, 1997). Using these measures, it was clear that threading predictions had improved significantly based on the number of successful predictions from a large number of groups using easier targets. Generally, fold recognition methods outperformed simple sequence alignment against the database of known folds. The average RMSD of one predictor between model and target structures was 5.1 Å. The best models for two easy targets had RMSDs of 2.9 and 4.2 Å, respectively, with a structural alignment containing more than 80% of the same residues.

Submission of *ab initio* predictions included secondary structure, three-dimensional coordinate sets, modes of oligomerizaton, and residue and secondary structure segment contact patterns (Lesk, 1997). Secondary structure predictions continued to perform well with the new blind test set; however, tertiary structure predictions were limited to fragmentary success. Predictions of contacts between residues and elements of secondary structure were not consistent.

Many of the small molecule–ligand complexes used to assess docking methods involved serine proteases (Dixon, 1997). Overall, the results for the small molecule targets performed well within a RMSD of 3 Å for at least one of the submitted predictions made for each target.

The predictors, however, did not perform consistently and correct predictions close to the target solution were not necessarily the top ranking solution. The protein–protein target proved to be more difficult in that there were significantly larger conformational orientations to consider and the identification of surface binding sites were more difficult to identify compared to binding clefts.

By CASP2, the community of predictors had grown so large that an automated system that could process and verify predictions according to several different evaluation criteria was required. This gave rise to the Prediction Center where predictions could be uploaded and evaluated with the results presented in tabular and graphical form (Zemla et al., 1999). By the time of CASP3 (Moult et al., 1999), other benchmarking services had appeared. These included CAFASP for fully automated prediction (Fischer et al., 1999), EVA (Eyrich et al., 2001), and LiveBench (Bujnicki et al., 2001) (Table 28.1).

EVA (Koh et al., 2003) is a Web server providing automated evaluation of the accuracy of automated protein structure prediction methods, but differs in procedural details to CAFASP. Rather than conducting evaluations every 2 years, evaluations are conducted and updated automatically each week. Secondary structure prediction, contact prediction, comparative protein structure modeling, and folding recognition methods are evaluated. Target proteins to be predicted are collected daily from newly submitted proteins to the Protein Data Bank (PDB) and compared once a week to experimental structures. The results are published on the Web server where a measure of sustained performance as well as the ranking of methods is published. It is argued that this approach is preferable because larger data sets are used and therefore better statistics and ranking schemes can be obtained (Eyrich et al., 2003).

LiveBench (Bujnicki et al., 2001; Rychlewski, Fischer, and Elofsson, 2003) also provides continuous benchmarking for automated, publicly available protein fold recognition servers to measure progress in the field. This evaluation approach differs from CAFASP in that it assesses structure prediction performance using newly released targets that do not show significant sequence similarity to proteins already in the PDB. In the strict sense, this is not a blind test, but the fold recognition algorithms tested presumably have not seen these new structures before in the training set. The advantage of LiveBench over CAFASP is that it provides a larger test set from which to make comparisons between the different algorithms.

The introduction of other benchmarking experiments indicated the growing interest in sharing research findings in the pursuit of improved structural predictions. The conclusion from CAFASP1 was that no automated method at the time proved to be markedly superior (Fischer et al., 1999). More important, this conclusion highlights the wide gap in quality model produced by automated servers compared to those derived with expert interpretations at the time of CAFASP1. This gap has narrowed since then, and like the beginnings of CASP, CAFASP have also identified requirements for improved blind tests to go forward.

CASP3 saw further improvements, particularly in the alignment of target and template sequences. Encouragingly, *ab initio* prediction showed improvements in the accuracy of prediction of fragments up to 60 residues. Fold Recognition, on the contrary, seemed to show a greater improvement between CASP1 and CASP2, but less so between CASP2 and CASP3 (Sippl et al., 1999). As the previous CASPs, numerical evaluation methods continued to be refined to provide better measures of progress and success.

Heralded as the "2000 Olympic Games of protein structure prediction" (Fischer, Elofsson, and Rychlewski, 2000) CASP4 participants were faced with more stringent evaluation criteria. In summary, alignment of sequences by comparative modeling continued to be a problem (Venclovas et al., 2001). Secondary structure prediction, according to the three state accuracy measure (Q3), appeared to have reached a limit. Contact predictions were approaching a useful level of accuracy and overall success rate for remote homologue detection had improved, successfully identifying a large fraction of the folds in the blind set. Significant improvements in *de novo* predictions were with CASP4 showing particularly good predictions for small proteins with the correct topology (Schonbrun, Wedemeyer, and Baker, 2002). The success of the algorithms could be attributed to a combination of knowledge and physics-based methods.

The evaluation methods used by CASP4 and related services continued to be challenged (Marti-Renom et al., 2002; Moult et al., 2002). The reliability of the rankings of protein structure modeling methods were assessed using the parameteric Student's t test and the nonparameteric Wilcox signed rank test of statistical significance of the difference between the paired samples. It was determined that with these tests, the top eight methods of prediction could not be distinguished. The target sequences used for CASP4 were analyzed and shown not to distinguish between the top eight methods given the standard deviation of the difference in model quality. The results suggested that CASP needed to be supplemented by an assessment made by other evaluation services that are automated, continuous in time, and based on several criteria applied to a large number of methods.

By CASP4, while the best human-assisted methods continued to outperform automated servers (Sippl et al., 2001), automated consensus metapredictors were very successful in fold recognition. CAFASP2 revealed that the most significant progress in fold recognition came with the development of metaservers incorporating prediction results from several independent methods to generate consensus models (Kinch et al., 2003; Venclovas et al., 2003). The performance gap between automated and manual methods narrowed with about one fourth of the top 30 performing groups using fully automated servers in the fold recognition category. Moreover, the consensus servers that incorporated predictions from multiple fold recognition servers outperformed individual servers alone (Fischer et al., 2001; Schonbrun, Wedemeyer, and Baker, 2002). These metapredictors performed approximately 30% better than the best of 60 independent servers participating in CAFASP3 (Fischer et al., 2003). Top performing metapredictors were comparable to the best 5–10 human CASP predictors. Prediction of multidomain proteins, however, remained a challenge (Kinch et al., 2003). Nevertheless, the potential of consensus methods recognized earlier, clearly had the potential to advance the field further (Schonbrun, Wedemeyer, and Baker, 2002).

CASP5 saw the recognition of the importance of protein disorder predictors (Melamud and Moult, 2003). Structural disorder observed in structures is reported as missing atoms. Several groups had postulated that this property was encoded in the protein sequence and were able to develop tools to recognize such signals using disorder in existing structures as a training set. While the exact definition of disorder was in question, six participating groups successfully identified disordered regions within the blind test set without too much over prediction.

CASP6 saw a record number of groups making predictions. Starting with 34 groups during CASP1, the number of groups increased to 70, 163, 98, 216, and then to 266 for CASP6 (Moult et al., 2005). At CASP6, a total of 41,283 models were deposited of which 32,703 could be assessed. 23,119 had coordinate sets with 4484 alignments converted to coordinates for assessment. The total number of submitted predictions included 1397 with residue contacts, 1293 with domain assignments, and 990 with function predictions—all new classes of prediction. Finally, 1769 disorder predictions were made.

Assessment of the comparative modeling category (Tress et al., 2005) indicated that identification of the best structural template to use for modeling remained a big challenge. Predictors still produce incorrect models that contain tangles in the backbone and beta whorls that are not observed in Nature. A statistically significant difference in performance between the best performing methods compared to the rest of the participants was observed. Once again, the differences stemmed from the methods used to select and align templates as well as the use of expert knowledge. The technique shared by many successful groups was the detection of templates with 3D-Jury (Ginalski et al., 2003) followed by alignment improvement before direct modeling.

The most important new evaluation introduced with CASP6 was an assessment of side chain orientations and identification of biologically important sites. Conclusions from this analysis showed that side chain packing improvements could come only at the expense of rotamer accuracy, therefore indicating the need for improved refinement techniques. Predictions for functionally important sites showed, surprisingly, an overall better performance in determining structural orientations. Closer inspection of these sites suggested that local structural factors dominantly contribute to the observed orientation in the final target structure and matches well with the model template that were used. Steady but modest progress was observed for comparative modeling and homologous fold recognition for difficult targets. Sequence relationships for the superposition between the model and target still affects the alignment accuracy.

New measures to evaluate the sequence-dependent and sequence-independent alignment methods were used to gauge the success of fold recognition algorithms (Wang, Jin, and Dunbrack, 2005). Alignment, once again, remained the bottleneck in successful prediction; however, it was noted that more time to include biological and functional information would in all likelihood improve predictions.

Disordered regions were once again included, with twenty participating groups in this category (Jin and Dunbrack, 2005). Assessment of protein disorder segments in otherwise ordered structures in the blind set are limited to segments that are often short. One group clearly performed better than other methods identifying 75% of disordered residues with high over prediction. Overall, about 17% of the residues were identified as being disordered. Other groups predicted disordered residues at specificities higher than 90%, but only correctly identifying half the disordered residues in the blind set.

Results of 3D contact predictions were reported with measures that are accepted as standards in the field: accuracy, coverage, and a score representing the average distances between strict contacts (Xd) (Grana et al., 2005a). Top performing methods used genetic programming and neural networks trained with different input information. The blind test set was too small to make conclusions about progress and limits of performance. The reader is referred to other services such as EVAcon (Grana et al., 2005) for a better interpretation. A potentially interesting conclusion was that the contact prediction methods perform better on average than 3D prediction methods when applied to difficult targets.

Two new components of CASP, domain boundaries and functional prediction, were added to address the needs of structural genomics. Domain boundary prediction is crucial for modeling larger protein structures and proved to be a difficult task and can only be achieved when a related structural template is available to use as a reference (Chapter 20). Functional predictions were evaluated to address the growing need of structural genomics in understanding the functional role of uncharacterized protein structures (Soro and Tramontano, 2005). The task of functional annotation is challenged by findings that show that common evolutionary origin does not necessarily confer shared function (Devos and Valencia, 2000;

Rost, 2002). The discovery of proteins that perform multiple functions further complicates the situation (Jeffery, 2003a; Jeffery, 2003b). Last, this aspect of CASP differs from structural prediction in that the function of the target protein may likely still not be known and thus functional predictions remain speculative.

An objective of functional prediction is to provide experimentalists working on the target proteins with some useful information. Each predictor is required to provide the following information for each target: (1) GO category of molecular function, biological process, and cellular components; (2) binding information; (3) location of binding site; (4) role of residues; and finally, (5) posttranslational modifications. Free text comments are allowed at the end of the submitted prediction file. Twenty-six groups participated with each group allowed to submit up to 5 ranked functional predictions for each target. A total of 1235 predictions were made. Conclusions from CASP6 were (1) the experiment should be limited to enzymes or proteins predicted to be enzymes; (2) function should be described in terms of EC numbers that are less ambiguous than GO annotations; (3) a general description of the method should be made available to the biological community to facilitate evaluation and further methods development.

OVERALL PROGRESS

One metric to summarize the history of CASP is known as GDT_TS (Zemla et al., 1999) that analyzes the superposition of structures (Kryshtafovych et al., 2005). This metric has approximately doubled since the start of CASP. The most difficult targets with sequence identities of less than 20% identity remain difficult to model with only 20% of sequences correctly aligned for the best models. Accuracy was limited by three factors: differences in main-chain conformation, the number of targets in the mid-range of difficulty (30–50% sequence identity) for structure prediction, and remote evolutionary relationships. Progress for predictions using targets with high sequence similarities to know templates was difficult to quantify, but automated server performance has improved. Perhaps most important is the improvement over time in fold recognition. CASP6 included the first report of a successful model for a small protein refined from a backbone RMSD of 2.2 Å down to 1.6 Å with many core side chains correctly oriented (Moult, 2005). This significant improvement is dominated by methods from Baker's Group, using the software Rosetta (Bradley et al., 2005) and Robetta (Chivian et al., 2005).

CASP7

At the time of writing, CASP7 had concluded but the outcomes were yet to be published. CASP7 introduced four new challenges devised during CASP6 (Moult et al., 2005). These challenges are (1) to model structures of single-residue mutants; (2) to model structure changes associated with specificity changes within protein families; (3) to directly focus on improving refinement methods and thus produce a 0.5 Å RMSD improvement in the $C\alpha$ accuracy of models based on sequences with greater than 30% identity; and (4) to devise scoring functions that will reliably pick the most accurate models from a set of candidate structures produced by the current new fold methods.

The latter point is of particular interest since it introduces a new category that is the ability for groups to successfully assess their own models as opposed to using techniques defined by the CASP organizers (Cozzetto et al., 2007). Participants were asked to provide an index of quality for individual models as well as an index for the expected correctness of each residue. The method to predict model quality is useful for two purposes: first, to select the best model among the plausible choice, and second, to assign an absolute quality value to each individual model. Results suggest that it is possible to create methods to distinguish the best solution within plausible models.

Concerning the previously used categories, CASP7 showed that intramolecular resi-due–residue contacts inferred from 3D structure predictions are similar in accuracy to those predicted by contact prediction methods. The latter approach does not construct a protein structure model; nevertheless, performs better for some targets in identifying interacting residues (Izarzugaza et al., 2007). Domain boundary predictions were more consistent when the target has a suitable structural template to use as a reference (Tress et al., 2007). Disorder predictors continue to be an interest to the community even though developing a proper evaluation criterion remains elusive (Bordoli, Kiefer, and Schwede, 2007). Overall, partici-pating methods have generally improved their accuracy in identifying disorder predictors, but the improvements are not significantly better than the best method in the last round, CASP6.

In the functional prediction category, submissions were made for GO molecular function terms, Enzyme Commission numbers, and ligand binding sites (Lopez et al., 2007). The results were disappointing in that there were few participants in this category and the test was not a purely blind study. As a relatively new category, some improvements in organization need to be made before a true assessment of the value of functional predictors can be made.

WHERE DO WE GO FROM HERE?

As with any scientific endeavor, CASP is an evolving process and will continue to serve the community by creating new standards to be met by those in the field of structure prediction. We have tried to convey that CASP has undoubtedly accelerated the field through a focused effort that has continued to challenge participants. The adoption of this type of critical assessment effort in other fields is the best testament to the success of CASP.

ACKNOWLEDGMENT

A process such as that undertaken during the seven CASP experiments is possible only through the participation of the community, both as predictors and service providers in assessing the work of others. CASP has been a great success and a testament to everyone in the field.

WEB SITES

PredictProtein: http://www.predictprotein.org/
    CASP—Protein Structure Prediction Center: http://predictioncenter.gc.ucdavis.edu/
    CAFASP: http://www.cs.bgu.ac.il/~dfischer/CAFASP5/
    CAPRI: http://capri.ebi.ac.uk/

FORCASP: http://www.forcasp.org/

EVA: http://cubic.bioc.columbia.edu/eva/

EVAcon—Continuous evaluation service for protein contact prediction (Grana et al., 2005).: http://cubic.bioc.columbia.edu/eva/con/index.html

LiveBench: http://meta.bioinfo.pl/livebench.pl

REFERENCES

Bordoli L, Kiefer F, Schwede T (2007): Assessment of disorder predictions in CASP7. *Proteins* 69 (Suppl. 8):129–136.

Bradley P, Malmstrom L, Qian B, Schonbrun J, Chivian D, Kim DE, Meiler K, Misura KMS, Baker D (2005): Free modeling with Rosetta in CASP6. *Proteins* 61:128–134

Bujnicki JM, Elofsson A, Fischer D, Rychlewski L (2001): LiveBench-1: continuous benchmarking of protein structure prediction servers. *Protein Sci* 10:352–361.

Chivian D, Kim DE, Malmstrom L, Schonbrun J, Rohl CA, Baker D (2005): Prediction of CASP6 structures using automated Robetta protocols. *Proteins* 61(Suppl. 7):157–166.

Cozzetto D, Kryshtafovych A, Ceriani M, Tramontano A (2007): Assessment of predictions in the model quality assessment category. *Proteins* 69(Suppl. 8):175–183.

Defay T, Cohen FE (1995): Evaluation of current techniques for *ab initio* protein–structure prediction. *Proteins* 23:431–445.

Devos D, Valencia A (2000): Practical limits of function prediction. *Proteins* 41:98–107.

Dixon JS (1997): Evaluation of the CASP2 docking section. *Proteins* 29:198–204.

Eyrich VA, Marti-Renom MA, Przybylski D, Madhusudhan MS, Fiser A, Pazos F, Valencia A, Sali A, Rost B (2001): EVA: continuous automatic evaluation of protein structure prediction servers. *Bioinformatics* 17:1242–1243.

Eyrich VA, Przybylski D, Koh IY, Grana O, Pazos F, Valencia A, Rost B (2003): CAFASP3 in the spotlight of EVA. *Proteins* 53(Suppl. 6):548–560.

Fischer D, Barret C, Bryson K, Elofsson A, Godzik A, Jones D, Karplus KJ, Kelley LA, MacCallum RM, Pawowski K (1999): CAFASP-1: critical assessment of fully automated structure prediction methods. *Proteins* (Suppl. 3):209–217.

Fischer D, Elofsson A, Rychlewski L (2000): The 2000 Olympic Games of protein structure prediction;fully automated programs are being evaluated vis-a-vis human teams in the protein structure prediction experiment CAFASP2. *Protein Eng* 13:667–670.

Fischer D, Elofsson A, Rychlewski L, Pazos F, Valencia A, Rost B, Ortiz AR, Dunbrack RL Jr (2001): CAFASP2: the second critical assessment of fully automated structure prediction methods. *Proteins* (Suppl. 5):171–183.

Fischer D, Rychlewski L, Dunbrack RL Jr, Ortiz AR, Elofsson A (2003): CAFASP3: the third critical assessment of fully automated structure prediction methods. *Proteins* 53(Suppl. 6): 503–516.

Ginalski K, Elofsson A, Fischer D, Rychlewski L (2003): 3D-Jury: a simple approach to improve protein structure predictions. *Bioinformatics* 19:1015–1018.

Grana O, Baker D, MacCallum RM, Meiler J, Punta M, Rost B, Tress ML, Valencia A (2005a): CASP6 assessment of contact prediction. *Proteins* 61(Suppl. 7):214–224.

Grana O, Eyrich VA, Pazos F, Rost B, Valencia A (2005): EVAcon: a protein contact prediction evaluation service. *Nucleic Acids Res* 33:W347–W351.

Izarzugaza JMG, Grana O, Tress ML, Valencia A, Clarke ND (2007): Assessment of intramolecular contact predictions for CASP7. *Proteins* 69(Suppl. 8):152–158.

Janin J, Henrick K, Moult J, Eyck LT, Sternberg MJ, Vajda S, Vakser I, Wodak SJ (2003): CAPRI: a critical assessment of predicted interactions. *Proteins* 52:2–9.

Jeffery CJ (2003a): Moonlighting proteins: old proteins learning new tricks. *Trends Genet* 19: 415–417.

Jeffery CJ (2003b): Multifunctional proteins: examples of gene sharing. *Ann Med* 35:28–35.

Jin Y, Dunbrack RL Jr (2005): Assessment of disorder predictions in CASP6. *Proteins* 61(Suppl. 7): 167–175.

Kinch LN, Wrabl JO, Krishna SS, Majumdar I, Sadreyev RI, Qi Y, Pei J, Cheng H, Grishin NV (2003): CASP5 assessment of fold recognition target predictions. *Proteins* 53(Suppl. 6):395–409.

Koh IY, Eyrich VA, Marti-Renom MA, Przybylski D, Madhusudhan MS, Eswar N, Grana O, Pazos F, Valencia A, Sali A, Rost B (2003): EVA: evaluation of protein structure prediction servers. *Nucleic Acids Res* 31:3311–3315.

Kryshtafovych A, Venclovas C, Fidelis K, Moult J (2005): Progress over the first decade of CASP experiments. *Proteins* 61:225–236.

Lemer CMR, Rooman MJ, Wodak SJ (1995): Protein–structure prediction by threading methods— evaluation of current techniques. *Proteins* 23:337–355.

Lesk AM (1997): CASP2: report on *ab initio* predictions. *Proteins* 29:151–166.

Levitt M (1997): Competitive assessment of protein fold recognition and alignment accuracy. *Proteins* (Suppl. 1):92–104.

Lopez G, Rojas A, Tress M, Valencia A (2007): Assessment of predictions submitted for the CASP7 function prediction category. *Proteins* 69(Suppl. 8):165–174.

Marti-Renom MA, Madhusudhan MS, Fiser A, Rost B, Sali A (2002): Reliability of assessment of protein structure prediction methods. *Structure* 10:435–440.

Martin AC, MacArthur MW, Thornton JM (1997): Assessment of comparative modeling in CASP2. *Proteins* (Suppl. 1):14–28.

Melamud E, Moult J (2003): Evaluation of disorder predictions in CASP5. *Proteins* 53(Suppl. 6): 561–565.

Mosimann S, Meleshko R, James MNG (1995): A critical assessment of comparative molecular modeling of tertiary structures of proteins. *Proteins* 23:301–317.

Moult J, Pedersen JT, Judson R, Fidelis K (1995): A large-scale experiment to assess protein structure prediction methods. *Proteins* 23:ii–v.

Moult J, Hubbard T, Bryant SH, Fidelis K, Pedersen JT (1997): Critical assessment of methods of protein structure prediction (CASP): round II. *Proteins* (Suppl. 1):2–6.

Moult J, Hubbard T, Fidelis K, Pedersen JT (1999): Critical assessment of methods of protein structure prediction (CASP): round III. *Proteins* (Suppl. 3):2–6.

Moult J, Fidelis K, Zemla A, Hubbard T, Tramontano A (2002): The significance of performance ranking in CASP-response to Marti-Renom et al. *Structure* 10:291–292.

Moult J (2005): A decade of CASP: progress, bottlenecks and prognosis in protein structure prediction. *Curr Opin Struct Biol* 15:285–289.

Moult J, Fidelis K, Rost B, Hubbard T, Tramontano A (2005): Critical assessment of methods of protein structure prediction (CASP): round 6. *Proteins* 61(Suppl. 7):3–7.

Moult J (2006): Rigorous performance evaluation in protein structure modeling and implications for computational biology. *Phil Trans* 361:453–458.

Rost B (2002): Enzyme function less conserved than anticipated. *J Mol Biol* 318:595–608.

Rost B, Sander C (1995): Progress of 1d protein structure prediction at last. *Proteins* 23:295–300.

Rychlewski L, Fischer D, Elofsson A (2003): LiveBench-6: large-scale automated evaluation of protein structure prediction servers. *Proteins* 53(Suppl. 6):542–547.

Schonbrun J, Wedemeyer WJ, Baker D (2002): Protein structure prediction in 2002. *Curr Opin Struct Biol* 12:348–354.

Sippl MJ, Lackner P, Domingues FS, Koppensteiner WA (1999): An attempt to analyse progress in fold recognition from CASP1 to CASP3. *Proteins* (Suppl. 3):226–230.

Sippl MJ, Lackner P, Domingues FS, Prlic A, Malik R, Andreeva A, Wiederstein M (2001): Assessment of the CASP4 fold recognition category. *Proteins* (Suppl. 5):55–67.

Soro S, Tramontano A (2005): The prediction of protein function at CASP6. *Proteins* 61:201–213.

Tress M, Ezkurdia L, Grana O, Lopez G, Valencia A (2005): Assessment of predictions submitted for the CASP6 comparative modeling category. *Proteins* 61:27–45.

Tress M, Cheng J, Baldi P, Joo K, Lee J, Seo J, Lee J, Baker D, Chivian D, Kim D, Ezkurdia L (2007): Assessment of predictions submitted for the CASP7 domain prediction category. *Proteins* 69(Suppl. 8):137–151.

Venclovas C, Zemla A, Fidelis K, Moult J (1997): Criteria for evaluating protein structures derived from comparative modeling. *Proteins* (Suppl. 1):7–13.

Venclovas C, Zemla A, Fidelis K, Moult J (2001): Comparison of performance in successive CASP experiments. *Proteins* (Suppl. 5):163–170.

Venclovas C, Zemla A, Fidelis K, Moult J (2003): Assessment of progress over the CASP experiments. *Proteins* 53(Suppl. 6):585–595.

Wang G, Jin Y, Dunbrack RL Jr (2005): Assessment of fold recognition predictions in CASP6. *Proteins* 61(Suppl. 7):46–66.

Zemla A, Venclovas C, Moult J, Fidelis K (1999): Processing and analysis of CASP3 protein structure predictions. *Proteins* (Suppl. 3):22–29.

Zemla A, Venclovas C Moult J, Fidelis K (2001): Processing and evaluation of predictions in CASP4. *Proteins* (Suppl. 5):13–21.

29

PREDICTION OF PROTEIN STRUCTURE IN 1D: SECONDARY STRUCTURE, MEMBRANE REGIONS, AND SOLVENT ACCESSIBILITY

Burkhard Rost

INTRODUCTION

No General Prediction of 3D Structure from Sequence, Yet

The hypothesis that the 3D structure[1] of a protein (the fold) is uniquely determined by the specificity of the sequence has been verified for many proteins (Anfinsen, 1973). While particular proteins (chaperones) often play an important role in folding (Ellis, Dobson, and Hartl, 1998), it is still generally assumed that the final structure is at the free-energy minimum (Dobson and Karplus, 1999). Thus, all information about the native structure of a protein is coded in the amino acid sequence and its native solution environment. Can we decipher the code; that is, can we predict 3D structure from sequence? In principle, the code could by deciphered from physicochemical principles (Levitt and Warshel, 1975; Hagler and Honig, 1978). In practice, the inaccuracy in experimentally determining the basic parameters and the limited computing resources prevent prediction of protein structure from first principles (van Gunsteren, 1993). Hence, the only successful structure prediction tools are knowledge-based, using a combination of statistical theory and empirical rules. The field of protein structure prediction has advanced significantly over the past 15 years

---

[1] Abbreviations used: 1D structure, one-dimensional, for example, sequence or strings of secondary structure or solvent accessibility; 2D structure, two-dimensional (e.g., inter-residue distances); 3D structure, three-dimensional coordinates of protein structure. Symbols used: secondary structure: H = helix, E = strand, L = other; transmembrane helices: T = transmembrane, N = globular; solvent accessibility: e = exposed ($\geq$16% relative accessible surface), b = buried (<16%).

Structural Bioinformatics, Second Edition Edited by Jenny Gu and Philip E. Bourne
Copyright © 2009 John Wiley & Sons, Inc.

(see Chapter 28) with improved methods (Kryshtafovych et al., 2005; Moult et al., 2005). However, possibly the most important change was the growth of the PDB (Berman et al., 2002) that increases the odds of finding local 3D matches to build comparative models (Chance et al., 2004). We can, however, still not predict structure from sequence reliably without finding related sequences of known structure in the databases (Moult et al., 2005).

Structure Prediction in 1D has become Increasingly Accurate and Important

An extreme simplification of the prediction problem is to project 3D structure onto strings of structural assignments. For example, we can assign a secondary structure state, marked by one symbol, for each residue, or we can assign a number for the accessibility of that residue. Such strings of per-residue assignments are essentially one-dimensional (1D). In fact, arguably the most surprising improvements in bioinformatics over the last decade may have been achieved by methods predicting protein structure in 1D. The key to this breakthrough was the wealth of evolutionary information contained in ever-growing databases. Moreover, prediction accuracy continues to rise (Rost, 2001). This success is crucial for target selection in structural genomics (Liu et al., 2004), for using structure prediction to get clues about function (Rost et al., 2003; Ofran et al., 2005), and for using simplified predictions for more sensitive database searches and predictions of higher dimensional aspects of protein structure (see below).

Apologies to Developers!

This brief synopsis of methods predicting protein structure in 1D has no chance of being fair to all developing methods for 1D protein structure prediction. Even a MEDLINE search restricted to "secondary structure prediction" revealed more than 100 publications over the last year. Consequently, the review will be somehow unfair to the majority of developers. Instead, the focus lies on the small subset of most accurate or most widely used methods.

METHODS

Secondary Structure Prediction Methods

Basic Concept. The principal idea underlying most secondary structure prediction methods is the fact that segments of consecutive residues have preferences for certain secondary structure states (Bränden and Tooze, 1991; Rost, 1996). Thus, the prediction problem becomes a pattern classification problem tractable by pattern recognition algorithms. The goal is to predict whether a residue is in a helix, in a strand, or in none of the two (no regular secondary structure, often referred to as the "coil" or "loop" state). The first-generation prediction methods in the 1960s and '70s were all based on single amino acid propensities (Chou and Fasman, 1974; Robson, 1976; Garnier, Osguthorpe, and Robson, 1978; Schulz and Schirmer, 1979; Fasman, 1989). Basically, these methods compiled the probability of a particular amino acid for a particular secondary structure state. The second-generation methods dominated the scene until the early 1990s; they extended the principal concept to compiling propensities for segments of adjacent residues; that is, taking the local environment of the residues into consideration. Typically methods used segments of 3–51 adjacent residues (Nishikawa and Ooi, 1982;

Nishikawa and Ooi, 1986; Deleage and Roux, 1987; Biou et al., 1988; Bohr et al., 1988; Gascuel and Golmard, 1988; Levin and Garnier, 1988; Qian and Sejnowski, 1988; Garnier and Robson, 1989). Basically, any imaginable theoretical algorithm had been applied to the problem of predicting secondary structure from sequence: physicochemical principles, rule-based devices, expert systems, graph theory, linear and multilinear statistics, nearest-neighbor algorithms, molecular dynamics, and neural networks (Schulz and Schirmer, 1979; Fasman, 1989; Rost and Sander, 1996; Rost and Sander, 2000). However, it seemed that prediction accuracy stalled at levels around 60% of all residues correctly predicted in either of the three states helix, strand, or other. It was argued that the limited accuracy resulted from the fact that all methods used only information local in sequence (input: about 3–51 consecutive residues). Local information was estimated to account for roughly 65% of the secondary structure formation. Two additional problems were common to most methods developed from 1957 to 1993. First, predicted secondary structure segments were, on average, only half as long as observed segments. Historically, this problem was solved for the first time through a particular combination of neural networks (Rost and Sander, 1992; Rost and Sander, 1993). Second, strands were predicted at levels of accuracy only slightly superior to random predictions. Again, the argument for this deficiency was that the hydrogen bonds determining the formation of sheets (note: paired strands form a sheet) are less local in sequence than the bonds responsible for helices (Chapter 19). Again, this problem was first solved through neural networks (Rost and Sander, 1992; Rost et al., 1993). The solution was rather simple: we realized that about 20% of the correctly predicted residues were in strands, about 30% in helices, and about 50% in nonregular secondary structure. These values are similar to the percentage of the respective classes in proteins. This observation prompted us to simply bias the database used for training neural networks by presenting each class equally often. The result was a prediction well balanced between the three classes, that is, about 60% of the strand residues were correctly predicted. In practice, this was an important advance. However, it also cast an important spotlight onto the explanation that secondary structure formation is partially determined by nonlocal interactions. Clearly, sheets are nonlocal structures. Nevertheless, the preferences for a segment to form a strand or a helix appear similarly strong because both can be predicted at similar levels of accuracy designing the appropriate prediction method (Rost and Sander, 1993; Rost and Sander, 1994a; Rost, 1996).

Evolutionary Information Key to Significantly Improved Predictions. On the one hand, about 67 out of 100 residues can be exchanged in a protein without changing structure (Rost, 1999). On the other hand, exchanges of very few residues often destabilize a protein structure. The explanation for this ostensible contradiction is simple: evolution has realized the unlikely by exploring all neutral mutations that do not prevent structure formation. Thus, the residue exchange patterns extracted from an aligned protein family are highly indicative of specific structural details. This also implies that a profile of a few consecutive residues taken from alignments implicitly contains nonlocal information. The source of this information is that evolutionary selection acts on a 3D object (the protein) rather than on an abstracted 1D construct (the sequence). Early on, it was realized that this information could improve predictions (Dickerson, Timkovich, and Almassy, 1976; Maxfield and Scheraga, 1979; Zvelebil et al., 1987). However, the breakthrough of the third-generation methods to levels above 70% accuracy required a combination of larger databases with more advanced algorithms (Rost et al., 1993; Rost and Sander, 2000). It was also recognized very early on that information from the position-specific evolutionary exchange profile of a particular protein family facilitates discovering more distant members of that family (Dickerson, Timkovich, and Almassy, 1976). Automatic database search methods successfully used position-specific profiles for searching (Barton, 1996). However, the breakthrough to large-scale routine searches has been achieved by the development of PSI-BLAST (Altschul et al., 1997) and hidden Markov models (HMMs) (Eddy, 1998; Karplus, Barrett, and Hughey, 1998). Since the improvement of secondary structure prediction relies significantly on the information content of the family profile used, today's larger databases and better search techniques resulted in pushing prediction accuracy even higher. The current top-of-the-line secondary structure prediction methods are all based on extended profiles (Rost, 2001; Przybylski and Rost, 2002).

Key Players. PHD was the program that surpassed the level of 70% accuracy first (Rost and Sander, 1993; Rost and Sander, 1994). It uses a system of neural networks to achieve a performance well balanced between all secondary structure classes (Figure 29.1). Although still widely used, PHD is no longer the most accurate method (Rost and Eyrich, 2001; Przybylski and Rost, 2002). David Jones pioneered using automated, iterative PSI-BLAST searches (Jones, 1999b). The most important step climbed by the resulting method PSIPRED has been the detailed strategy to avoid polluting the profile through unrelated proteins. To avoid this trap, the database searched has to be filtered first (Jones, 1999). Other than the advanced use of PSI-BLAST, PSIPRED achieves its success through a neural network system similar to that implemented in PHD. At the CASP meeting at which David Jones introduced PSIPRED, Kevin Karplus and colleagues presented their prediction method (SAM-T99sec/SAM-T04) finding more diverged profiles through HMMs (Karplus et al., 1999). The most important prediction method used by SAM-T99sec is a simple neural network with two layers of hidden units. However, the major strength of the method appears to be the quality of the alignment used. Some recent methods improve accuracy significantly not through more divergent profiles but through the particular algorithm; examples are the SSpro series of programs, the latest is SSpro4 (Pollastri et al., 2002a; Pollastri et al., 2002b), and Porter (Pollastri and McLysaght, 2005). The principal idea is to overcome the limitations of feed-forward neural networks with an input window of relative small and fixed size with bidirectional recurrent neural networks (BRNN) capable of taking the entire protein chain as input (Pollastri et al., 2006). The latest implementation SSpro4 uses this concept in combination with advanced PSI-BLAST profiles to become one of the best existing methods. Porter (Pollastri and McLysaght, 2005) goes beyond SSpro4 by also using details in the alignment more cleverly and by combining a variety of different networks. Porter may be the most accurate existing method for the prediction of secondary structure. Another method that reaches high accuracy is SABLE2 (Adamczak, Porollo, and Meller, 2004; Adamczak, Porollo, and Meller, 2005) that combines the prediction of secondary structure and solvent accessibility; PROFsec and PROFacc realize the same idea in a similar way (Rost, 2005). One recent method reached very high levels of prediction accuracy without paying for this achievement by complexity: YASPIN combines HMMs with neural networks in a very simple and efficient way (Lin et al., 2005). Quite a different route toward secondary structure prediction is taken by the HMMSTR/I-sites programs (Bystroff et al., 2000).

Specialized Methods: Coiled Coil Predictions. A coiled coil is a bundle of several helices assuming a side-chain packing geometry often referred to as "knobs-into-holes" (Crick, 1953). The "knobs" are the side chains of one helix that pack into the hole created

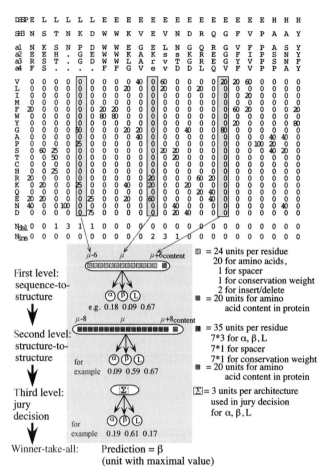

Figure 29.1. Neural network system for secondary structure prediction (PHDsec). From the multiple alignment (here guide sequence SH3 plus four other proteins a1–a4; note that lower case letters indicate deletions in the aligned sequence), a profile of amino acid occurrences is compiled. To the resulting 20 values at one particular position μ in the protein (one column), three values are added: the number of deletions and insertions, as well as the conservation weight (CW). Thirteen adjacent columns are used as input. The whole network system for secondary structure prediction consists of three layers: two network layers and one layer averaging over independently trained networks.

by four side chains surrounding the facing helix. This supercoil slightly alters the helix periodicity from 3.6 to 3.5 and results in the coiled-coil-specific symmetry in which every seventh residue occupies a similar position on the helix surface. The first and fourth of the seven residues are typically hydrophobic, the other four hydrophilic, frequently exposing the helix to solvent. These specific sequence features are at the base of accurate predictions for coiled coil helices (Lupas, 1996, 1997). The most widely used program is COILS that is based on amino acid preferences compiled for the few coiled coil proteins that were known at high resolution a decade ago (Lupas et al., 1991). The program detects coiled coil preferences in windows of 14, 21, and 28 residues. The longer the window the better the distinction between proteins that have coiled coil regions and those that do not (Lupas, 1996).

If we know the precise location of the coiled coil regions and the multimeric state, we can predict 3D structure for coiled coil regions at levels of accuracy that resemble experimentally determined structures (below 2.5 Å; Nilges and Brünger, 1993; O'Donoghue and Nilges, 1997). O'Donoghue and Nilges used the experimentally known boundaries of the coiled coil regions and their known multimeric state for prediction. It remains to be tested how sensitive that 3D prediction is with respect to errors in predicting the coiled coil regions. Recently, Wolf and coworkers developed a method to predict the multimeric state of a coiled coil region (Wolf, Kim, and Berger, 1997; Newman, Wolf, and Kim, 2000). When labeling all likely coiled coil proteins in entire proteomes, we found that about 8–10% of all eukaryotic proteins and 2–10% of all proteins in archaebacteria and prokaryotes contain at least one coiled coil region (Liu and Rost, 2001).

Other Alphabets. Most secondary structure prediction methods focus on the standard DSSP alphabet of eight states. Many groups have tried to directly predict all eight states and have found that this approach did not bear fruit, possibly because this detailed separation is not supported by data, or more importantly is not conserved in evolution. In fact, we possibly should define the goals of simplifying structure prediction exactly by what is evolutionarily conserved (Rost, Sander, and Schneider, 1994; Zemla et al., 1999). This is exactly what Kevin Karplus and colleagues realized when sampling the space of local alphabets to identify those that are most conserved in evolution and to develop methods that predict exactly these alphabets (Karchin, Cline, and Karplus, 2004). The concept is successful. So far it has remained an exceptional case possibly only because the approach has remained so unique that very few groups have worked on ways of how to use the information predicted by such methods directly.

Solvent Accessibility Prediction Methods

Basic Concept. It has long been argued that if the segments of secondary structure could be accurately predicted, the 3D structure could be predicted by simply trying different arrangements of the segments in space (Mumenthaler and Braun, 1995; Cohen and Presnell, 1996). One criterion for assessing each arrangement could be to use predictions of residue solvent accessibility (Lee and Richards, 1971; Chothia, 1976; Connolly, 1983). The principal goal is to predict the extent to which a residue embedded in a protein structure is accessible to solvent. Solvent accessibility can be described in several ways (Lee and Richards, 1971; Chothia, 1976; Connolly, 1983). The most detailed fast method compiles solvent accessibility by estimating the volume of a residue embedded in a structure that is exposed to solvent (Figure 29.2; this method was developed by Connolly (1983) and later implemented in DSSP (Kabsch and Sander, 1983)). Different residues have a different possible accessible area. The most extreme simplification for accessibility accounts for this by normalizing (dividing observed value by maximally possible value) to a two-state description distinguishing between residues that are buried (relative solvent accessibility less than 16%) and exposed (relative solvent accessibility ≥16%). The precise choice of the threshold is not well defined (Hubbard and Blundell, 1987; Rost and Sander, 1994b). The classical method to predict accessibility is to assign either of the two states, buried or exposed, according to residue hydrophobicity, that is, very hydrophobic stretches are predicted to be buried (Kyte and Doolittle, 1982; Sweet and Eisenberg, 1983). However, more advanced methods have been shown to be superior to simple hydrophobicity analyses (Holbrook, Muskal, and Kim, 1990; Mucchielli-Giorgi, Hazout, and Tuffery, 1999;

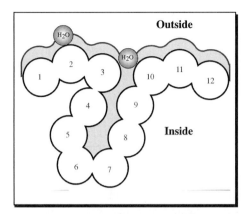

Figure 29.2. Measure accessibility. Residue solvent accessibility is usually measured by rolling a spherical water molecule over a protein surface and summing the area that can be accessed by this molecule on each residue (typical values range from 0 to 300 $Å^2$). To allow comparisons between the accessibility of long extended and spherical amino acids, typically relative values are compiled (actual area as percentage of maximally accessible area). A more simplified description distinguishes two states: buried (here residues numbered 1–3 and 10–12) and exposed (here residues 4–9) residues. Since the packing density of native proteins resembles that of crystals, values for solvent accessibility provide upper and lower limits to the number of possible inter-residue contacts.

Carugo, 2000; Li and Pan, 2001; Naderi-Manesh et al., 2001). Typically, these methods use similar ways of compiling propensities of single residues or segments of residues to be solvent accessible, as secondary structure prediction methods. For particular applications, such as using predicted solvent accessibility to predict glycosylation sites, it seems beneficial to train neural networks on different definitions of accessibility (Hansen et al., 1998; Gupta et al., 1999). In particular, Hansen et al. (1998) realized alternative compilations by changing the size of the water molecule used in DSSP (Figure 29.2). In contrast to the situation for secondary structure, most of the information needed to predict accessibility is contained in the preference of single residues (Rost and Sander, 1994). Nevertheless, using windows of adjacent residues also improves solvent accessibility prediction significantly (Rost and Sander, 1994; Thompson and Goldstein, 1996).

Evolutionary Information Improves Accessibility Prediction. Solvent accessibility at each position of the protein structure is evolutionarily conserved within sequence families. This fact has been used to develop methods for predicting accessibility using multiple alignment information (Rost, 1994; Thompson and Goldstein, 1996; Cuff and Barton, 2000). The two-state (buried, exposed) prediction accuracy is above 75%, that is, more than four percentage points higher than for methods not using alignment information. Predictions of solvent accessibility have also been used successfully for prediction-based threading, as a second criterion toward 3D prediction, by packing secondary structure segments according to upper and lower bounds provided by accessibility predictions, and as a basis for predicting functional sites (Rost and O'Donoghue, 1997).

Available Key Players. Possibly, the exclusion of methods predicting solvent accessibility from the CASP meetings (see Section "Practical Aspects") slowed down the progress of the field. In particular, few of the methods developed are readily available through public servers. Prominent exceptions are the solvent accessibility predictions by PHD (Rost and Sander, 1994b) and PROFphd (Rost, 2005) available through the PredictProtein server (Rost, Yachdav, and Liu, 2004; Rost, 1996). Both use systems of neural networks with alignment information. The improvement of PROFphd over PHD was achieved by (1) training the neural networks only on high-resolution structures and by (2) using predicted secondary structure as additional input (Rost, 2001; Rost, 2005). Technically, both PROFphd and PHD are the only available methods predicting real values for relative solvent accessibility on a grid of 0, 1, 4, 9, 16, 25, 36, 49, 64, 81 (percentage relative accessibility). Another method that improved prediction accuracy considerably over older programs is embedded in the JPred2 server (Cuff and Barton, 2000). It uses PSI-BLAST profiles as input to neural networks predicting accessibility in two states (buried/exposed). FKNNacc is based on the identification of related local stretches (k-fuzzy nearest-neighbor methods) (Sim, Kim, and Lee, 2005). More recently several methods have combined the prediction of secondary structure and solvent accessibility; for examples, PROFacc, SABLE2 (Adamczak, Porollo, and Meller, 2004), and ACCpro4 (analogue of SSpro4) (Pollastri et al., 2002a; 2002b).

Related Task: Prediction of Ooi/Coordination Number. The prediction of the residue coordination or Ooi number (Nishikawa and Ooi, 1986b) is directly related to that of the solvent accessibility. This number estimates how many residues surround a particular residue in a sphere. Thereby, the Ooi number provides two boundaries: one for the number of internal bounds and the other for the degree of accessibility to solvent. Several of the recent methods that target the prediction of the Ooi number are based on neural networks (Fariselli et al., 2001; Pollastri et al., 2002a; Pollastri et al., 2002b). A related method simply estimated the contact density based on inter-residue contact predictions succeeded in predicting folding rates (Punta and Rost, 2005). Thus suggesting that the direct relationship between residue coordination and solvent accessibility can also be used to make inferences about the protein folding rates.

Transmembrane Region Prediction Methods

The Task. Even in the optimistic scenario that in the near future most protein structures will be experimentally determined, one class of proteins will still represent a challenge for experimental determination of 3D structure: transmembrane proteins. The major obstacle with these proteins is that they do not crystallize and are hardly tractable by NMR spectroscopy. Consequently, for this class of proteins structure prediction methods are needed even more than for globular water-soluble proteins. Fortunately, the prediction task is simplified by strong environmental constraints on transmembrane proteins: the lipid bilayer of the membrane reduces the degrees of freedom making the prediction almost a 2D problem (Taylor, Jones, and Green, 1994; Bowie, 2005). Two major classes of integral membrane proteins are known: the first insert transmembrane helices (TMH) into the lipid bilayer (Figure 29.3) and the second form pores by transmembrane beta-strand barrels (TMB) (von Heijne, 1996; Seshadri et al., 1998; Buchanan, 1999). While predicting transmembrane helices is simpler than predicting globular helices, there is ample evidence that prediction accuracy has been significantly overestimated (Möller, Croning, and Apweiler, 2001; Chen, Kernytsky, and Rost, 2002).

Basic Concept. We can use a number of observations that constrain the problem of predicting membrane helices. (1) TM helices are predominantly apolar and between 12 and

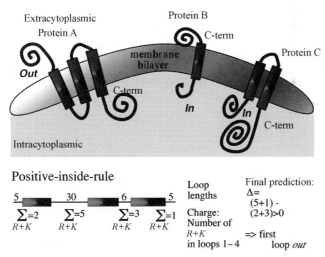

Figure 29.3. Topology of helical membrane proteins. In one class of membrane proteins, typically apolar helical segments are embedded in the lipid bilayer oriented perpendicular to the surface of the membrane. The helices can be regarded as more or less rigid cylinders. The orientation of the helical axes, that is, the topology of the transmembrane protein, can be defined by the orientation of the first N-terminal residues with respect to the cell. Topology is defined as *out* when the protein N-term (first residue) starts on the extracytoplasmic region (protein A) and as *in* if the N-term starts on the intracytoplasmic side (proteins B and C). The lower part explains the "inside-out rule." The differences between the positive charges are compiled for all even and odd nonmembrane regions. If the even loops have more positive charges, the N-term of the protein is predicted outside. This rule holds for most proteins of known topology.

35 residues long (Chen and Rost, 2002). (2) Globular regions between membrane helices are typically shorter than 60 residues (Wallin and von Heijne, 1998; Liu and Rost, 2001). (3) Most TMH proteins have a specific distribution of the positively charged amino acids arginine and lysine, coined the *positive-inside rule* by Gunnar von Heijne (von Heijne, 1986; von Heijne, 1989). Connecting *loop* regions at the inside of the membrane have more positive charges than *loop* regions at the outside (Figure 29.3). (4) Long globular regions (more than 60 residues) differ in their composition from those globular regions subject to the *positive-inside* rule. Most methods simply compile the hydrophobicity along the sequence and predict a segment to be a transmembrane helix if the respective hydrophobicity exceeds some given threshold (Tanford, 1980; Eisenberg et al., 1984; Klein, Kanehisa, and De Lisi, 1985; Engelman, Steitz, and Goldman, 1986; Jones, Taylor, and Thornton, 1994; von Heijne, 1994; Hirokawa, Boon-Chieng, and Mitaku, 1998; Phoenix, Stanworth, and Harris, 1998; Tusnady and Simon, 1998; Harris, Wallace, and Phoenix, 2000; Lio and Vannucci, 2000). In addition, some methods also explore the hydrophobic moment (Eisenberg et al., 1984; von Heijne, 1996; Liu and Deber, 1999) or other membrane-specific amino acid preferences (Ben-Tal et al., 1997; Monne, Hermansson, and von Heijne, 1999; Pasquier et al., 1999; Pilpel, Ben-Tal, and Lancet, 1999). The most important step is to adequately average hydrophobicity values over windows of adjacent residues (von Heijne, 1992; von Heijne, 1994). One of the major problems of hydrophobicity-based methods appears the poor distinction between membrane and globular proteins (Rost et al., 1995; Möller, Croning, and Apweiler, 2001). A number of methods use the

positive-inside rule to also predict the orientation of membrane helices (Sipos and von Heijne, 1993; Jones, Taylor, Thornton, 1994; Persson and Argos, 1997; Sonnhammer, von Heijne, and Krogh, 1998; Harris, Wallace, and Phoenix, 2000; Tusnady and Simon, 2001).

Evolutionary Information Improves Prediction Accuracy. Using evolutionary information also improves TMH predictions, significantly (Neuwald, Liu, and Lawrence, 1995; Rost et al., 1995; Rost, Casadio, and Fariselli, 1996; Persson and Argos, 1997). However, the growth of the sequence databases seems to have reversed the advantage of using evolutionary information (Chen, Kernytsky, and Rost, 2002). Until around 1997, most membrane helices were conserved in the following sense. Assume protein A has a TMH at positions N1–N2. Since the number of membrane helices is important for the function of the protein, we expect that all proteins A' that are found to be similar to A in a database search will also have a membrane helix at the corresponding positions N1–N2. However, this assumption is no longer correct (Chen and Rost, 2002). The practical result is that alignment-based predictions are much less accurate when based on the large merger of SWISS-PROT and TrEMBL (Bairoch and Apweiler, 2000) than when based on the smaller SWISS-PROT only (Chen and Rost, 2002). Interestingly, we can explore the power of using evolutionary information by carefully filtering the results from PSI-BLAST searches (Chen, Kernytsky, and Rost, 2002).

Available Key Players. TopPred2 is one of the classics in the field. It averages the GES-scale of hydrophobicity (Engelman, Steitz, and Goldman, 1986) using a trapezoid window (von Heijne, 1992; Sipos and von Heijne, 1993). MEMSAT (Jones, Taylor, Thornton, 1994) introduced a dynamic programming optimization to find the most likely prediction based on statistical preferences. TMAP (Persson and Argos, 1996) uses statistical preferences averaged over aligned profiles. PHD combines a neural network using evolutionary information with a dynamic programming optimization of the final prediction (Rost et al., 1995; Rost, Casadio, and Fariselli, 1996). DAS optimizes the use of hydrophobicity plots (Cserzö et al., 1997). SOSUI (Hirokawa, Boon-Chieng, and Mitaku, 1998) uses a combination of hydrophobicity and amphiphilicity preferences to predict membrane helices. TMHMM is the most advanced, and seemingly most accurate, current method to predict membrane helices (Sonnhammer, von Heijne, and Krogh, 1998). It embeds a number of statistical preferences and rules into a hidden Markov model to optimize the prediction of the localization of membrane helices and their orientation (note: similar concepts are used for HMMTOP; Tusnady and Simon, 1998). MINNOU (Cao et al., 2006) is based on a combination of the predictions of secondary structure and solvent accessibility, such as SABLE, with the objective of predicting transmembrane helices. Phobius (Kall, Krogh, and Sonnhammer, 2004; Kall, Krogh, and Sonnhammer, 2005) combines the prediction of signal peptides through HMMs and thereby improves in both tasks over many other methods. THUMPUP is a toolkit that, amongst other tasks, also predicts membrane helices (Zhou et al., 2005).

The group of Rita Casadio pushed the application of machine-learning devices to the task of predicting beta-barrel membrane proteins (TMB) (Jacoboni et al., 2001; Martelli et al., 2002). They developed two methods: B2TMR, based on neural networks (Jacoboni et al., 2001), and HMM-B2TMR, based on hidden Markov models (Martelli et al., 2002). PROFtmb applied this concept and improved by a detailed, accurate distinction between upward and downward strands, that is, by effectively predicting the topology of TMBs (Bigelow et al., 2004; Bigelow and Rost, 2006b). transFold (Waldispuhl et al., 2006b) is also based on HMMs; however, it is restricted to short TMB proteins as it performs additional energy calculations. This limitation comes with an important added benefit: tranFold also predicts strand pairings. TMB-HUNT applied a variant of the k-nearest neighbor method to the task, and BOMP (Berven et al., 2004) utilized simple statistics; another statistics/Bayesian network-based approach appears rather accurate (Gromiha and Suwa, 2006).

Integral Membrane or not?. The first question when scanning large data sets often is which proteins are likely associated with the membrane. The general methods are motif and domain based, and potentially identify the protein as one of a subtype of TMH or TMB proteins. TMH- or TMB-specific methods are designed to identify features common to all TMH (or all TMB) proteins and do not identify subtypes. InterProScan (Zdobnov et al., 2002) is a portal that allows querying the general methods at once. UniProt (Bairoch et al., 2005) provides a comprehensive view of previously analyzed results on many proteins and accompanying experimental information on structure or function.

TMB-Specific Methods. BOMP (beta-barrel outer membrane protein predictor) (Berven et al., 2004, 2006), TMB-HUNT (Garrow, Agnew, and Westhead, 2005), and PROFtmb (Bigelow and Rost, 2006b) are specially designed to identify TMB proteins in a pool. They have all been evaluated for accuracy in discriminating TMBs from background. Unfortunately, a definitive comparison is complicated by the fact that the evaluations are all done on different data sets. It is recommended that you submit your query to all three and scrutinize the results. Taking a consensus of predictors has been found consistently to yield better accuracy than relying on one individual predictor.

TMH-Specific Methods. Of the six TMH-specific methods, only TMHMM has been rigorously evaluated for accuracy in discriminating TMH proteins from others. While all methods implicitly predict whether a protein is a TMH by the presence of one or more predicted TM-helices, because the others are not evaluated for accuracy, it is not recommended to use these methods to screen a pool for potential TMH proteins.

PROGRAMS AND PUBLIC SERVERS

All methods described are available through public servers. A list of URLs and the contact addresses is summarized in Table 29.1. Most programs, except HMMSTR and PSIPRED, listed in Table 29.1 are also available by single click: META-PP allows you to fill out a form with the sequence and your e-mail address once and to simultaneously submit your protein to a number of high-quality servers (Eyrich and Rost, 2003). This concept of accessing many servers through one has been pioneered by the BCM-Launcher (Smith et al., 1996) supposedly accessing the largest number of different methods. Other combinations are given by NPSA (Combet et al., 2000), META-Poland (Rychlewski, 2000), and ProSAL (Kleywegt, 2001) In contrast to all others, META-PP attempts to (i) return as few results as possible by filtering out technical messages and to (ii) combine only high-quality methods. Note that both the BCM launcher and the current GCG package (Devereux, Haeberli, and Smithies, 1984) return predictions of secondary structure from methods that are neither *state of the art* nor competitive with the best method from a decade ago without indicating this to the user.

TABLE 29.1. Availability of Prediction Methods[a]

Method	Type	Server	Program
ACCpro	acc	www.ics.uci.edu/~baldig/scratch/index.html	Gianluca Pollastri
FKNNacc	acc	hinja.ssu.ac.kr/FKNNacc	Julian Lee
PHDacc	acc	www.predictprotein.org/	Guy Yachdav
PROFacc	acc	www.predictprotein.org/	Guy Yachdav
SABLE	acc	sable.cchmc.org	Jarek Meller
ASP	sec +	www.predictprotein.org/	Malin Young
COILS	sec	www.predictprotein.org/	Andrei Lupas
HMMSTR	sec +	www.bioinfo.rpi.edu/~bystrc/hmmstr/ server.php	Chris Bystroff
Porter	sec	distill.ucd.ie/porter	Gianluca Pollastri
PROFphd	sec	www.predictprotein.org/	Guy Yachdav
PSIPRED	sec	insulin.brunel.ac.uk/psiform.html	David Jones
SABLE	sec	sable.cchmc.org	Jarek Meller
SAM-T99sec	sec	www.cse.ucsc.edu/research/compbio/ HMM-apps/T99-query.html	Kevin Karplus
SSpro2	sec	promoter.ics.uci.edu/BRNN-PRED	Pierre Baldi, Gianluca Pollastri
YASPIN	sec	zeus.cs.vu.nl/programs/yaspinwww	Jaap Heringa
B2TMR	tmb	gpcr.biocomp.unibo.it/biodec	Rita Casadio
BOMP	tmb	www.bioinfo.no/tools/bomp	Ingvar Eidhammer
PROFtmb	tmb	www.predictprotein.org/	Henry Bigelow
TMB-HUNT	tmb	www.bioinformatics.leeds.ac.uk/~andy/ betaBarrel/AACompPred/aaTMB_Hunt.cgi	Andrew Garrow, DR Westhead
HMMTOP	tmh	www.enzim.hu/hmmtop	Gábor E. Tusnády
MEMSAT	tmh	insulin.brunel.ac.uk/psipred	David Jones
MINNOU	tmh	minnou.cchmc.org	Jarek Meller
PHDhtm	tmh	www.predictprotein.org/	Guy Yachdav
Phobius	tmh	phobius.cgb.ki.se	Erik Sonnhammer
SOSUI	tmh	sosui.proteome.bio.tuat.ac.jp/sosuiframe0.html	Takatsugu Hirokawa
SPLIT	tmh	www.mbb.ki.se/tmap/index.html	Davor Juretic
TMAP	tmh	www.mbb.ki.se/tmap/index.html	Bengt Persson
TMHMM	tmh	www.cbs.dtu.dk/services/TMHMM-1.0	Anders Krogh
TMpred	tmh	www.ch.embnet.org/software/TMPRED_ form.html	Philipp Bucher
TopPred2	tmh	www.sbc.su.se/~erikw/TopPred22	Gunnar von Heijne

[a]Acronyms: prediction of: acc—solvent accessibility; sec—secondary structure; tmb—transmembrane beta strands; tmh—transmembrane helices.

PRACTICAL ASPECTS

Evaluation of Prediction Methods Correctly Evaluating Protein Structure Prediction is Difficult

Developers of prediction methods in bioinformatics may significantly overestimate their performance because of the following reasons. First, it is difficult and time consuming to correctly separate data sets used for developing and testing. Second, estimates of performance of the different methods are often based on different data sets. This problem frequently originates from the rapid growth of the sequence and structure databases. Third, single scores are usually not sufficient to describe the performance of a method. The lack of clarity is particularly unfortunate at a time when an increasing number of tools are made easily available through the Internet and many of the users are not experts in the field of protein structure prediction. Two prominent examples illustrate this problem. (1) Transmembrane helix predictions have been estimated to yield levels above 95% per-residue accuracy that is more than 18 percentage points above actual performance on subsequent test cases (Chen and Rost, 2002). It is not unlikely that the recent methods for the prediction of beta-barrel membrane proteins are also overestimated. (2) Many publications on predicting the secondary structural class from amino acid composition allowed correlations between "training" and testing sets. Consequently, levels of prediction accuracy published, close to 100%, exceeded by far the theoretical possible margins—around 60% (Wang and Yuan, 2000).

Although overestimates of performance can slow down experimental and theoretical progress substantially, sometimes methods perform more than anticipated by the authors. For instance, PROFtmb was published as a method that fails to identify TMB proteins in eukaryotes when used with default thresholds. Nevertheless, experimental groups have used this tool to identify eukaryotic TMB proteins. Similarly, PHDsec, PSIPRED, and PROFsec were published with the warning that they work exclusively for globular, nonmembrane proteins. There is evidence that these tools actually also work for integral membrane proteins (Forrest, Tang, and Honig, 2006).

CASP: How Well do Experts Predict Protein Structure?.

Altogether seven CASP experiments have attempted to address the problem of overestimated performance (Moult et al., 1995, 1997, and 1999; Zemla, Venclovas, and Fidelis, 2001; Moult et al., 2003; Moult et al., 2005). The procedure used by CASP is the following: (1) Experimentalists who are about to determine the structure of a protein send the sequence to the CASP organizers (Zemla, Venclovas, and Fidelis, 2001). (2) Sequences are distributed to the predictors. The deadline for returning results is given by the date that the structure will be published. (3) All predictions are evaluated in a meeting at Asilomar. CASP resolves the bias resulting from using known protein structures as targets. However, it often cannot provide statistically significant evaluations since the number of proteins tested is too small (Rost and Eyrich, 2001; Eyrich et al., 2001; Marti-Renom et al., 2002). Nevertheless, CASP provides valuable insights into the performance of prediction methods and has become the major source of development in the field of protein structure prediction. Due to the fact that "failing at CASP is bad for the CV," most predictions are submitted only after experts have studied the data in detail. Thus, CASP intrinsically evaluates how well the best experts in the field can predict structure.

EVA and LiveBench: Automatic, Large-Scale Evaluation of Performance.

The limitations of CASP prompted two efforts at creating large-scale and continuously running tools that automatically assess protein structure prediction servers: EVA (Eyrich et al., 2001, 2003; Koh et al., 2003; Grana et al., 2005) and LiveBench (Rychlewski, Fischer, and Elofsson, 2003; Rychlewski and Fischer, 2005). LiveBench specializes in the evaluation of fold recognition, while EVA analyzes comparative modeling, contact prediction, fold recognition, and secondary structure prediction. The EVA results for secondary structure prediction methods were essential to conclude that these methods have improved significantly and to isolate the particular reasons for the improvements (mostly due to growing databases) (Rost, 2001; Rost and Eyrich, 2001; Przybylski and Rost, 2002).

Secondary Structure Prediction in Practice

77% Right Means 33% Wrong!. The best current methods (Porter, SABLE, SAM-T04, PSIPRED, PROFphd, SSpro) reach levels around 77–80% accuracy (percentage of residues predicted correctly in one of the three states: helix, strand, or other) (Rost, 2001; Rost and Baldi, 2001; Eyrich et al., 2003; Rost, 2005). Five observations are important for using prediction methods. (1) Levels of accuracy are averages over many proteins (Figure 29.4a). Hence, the accuracy for the prediction of your protein may be much lower, or much higher, than 77%. (2) Stronger predictions are usually more accurate (Figure 29.4b). This allows, to some extent, to find out whether or not the prediction for your protein is more likely to be above or below average. (3) Often predictions go badly wrong, that is, helices are incorrectly predicted as strands and vice versa. In fact, the best current methods confuse helices and strands for, on average, about 3% of all residues (Eyrich et al., 2003). Encouragingly, some of these "bad errors" are in fact not so severe, after all, since some of these are due to regions that can switch structural conformations in response to environmental changes (see below). (4) Prediction accuracy is rather sensitive to the

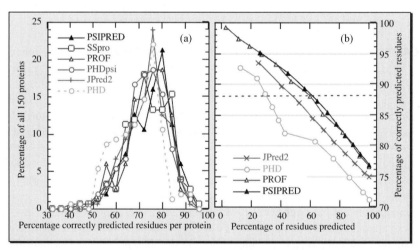

Figure 29.4. Prediction accuracy varies but stronger predictions are better! All results are based on 150 novel protein structures not used to develop any of the method shown (Eyrich et al., 2001, 2003). For all methods shown, the three-state per-residue accuracy varies significantly between these proteins, with one standard deviation in the order of 10% (a). This implies that it is difficult for users to estimate the "actual" accuracy for their protein. However, most methods now provide an index measuring the reliability of the prediction for each residue. Shown is the accuracy versus the cumulative percentages of residues predicted at a given level of reliability (coverage versus accuracy). For example, PSIPRED and PROFphd reach a level above 88% for about 60% of all residues (dashed line). This particular line is chosen since secondary structure assignments by DSSP agree to about 88% for proteins of similar structure. Although JPred2 is only marginally less accurate than PSIPRED and PROFphd, it reaches this level of accuracy for less than half of all residues.

information contained in the alignment used for the prediction: differences between single-sequence-based predictions and optimal alignment-based predictions can exceed more than 25 percentage points (Przybylski and Rost, 2002). (5) If on average 77% of the residues are correctly predicted, this trivially implies that 33% are wrong. Often, it is extremely instructive to form an expert opinion about where these wrong predictions are (Hubbard et al., 1996).

Latest Improvement: 4 Parts Database Growth, 3 Better Search, 4 Other. Jones solicited two causes for the improved accuracy of PSIPRED: (1) training and (2) testing the method on PSI-BLAST profiles. Cuff and Barton (2000) examined in detail how different alignment methods improve. However, which fraction of the improvement results from the mere growth of the database, which from using more diverged profiles, and which from training on larger profiles? Using the PHD version from 1994 to separate the effects (Przybylski and Rost, 2002), we first compared a noniterative standard BLAST (Altschul and Gish, 1996) search against SWISS-PROT (Bairoch and Apweiler, 2000) with one against SWISS-PROT + TrEMBL (Bairoch and Apweiler, 2000) + PDB (Berman et al., 2002). The larger database improves performance by about two percentage points (Przybylski and Rost, 2002). Secondly, we compared the standard BLAST against the big database with an iterative PSI-BLAST search. This yielded less than two percentage points additional improvement (Przybylski and Rost, 2002). Thus, overall, the more divergent profile search against today's databases supposedly improves any method using alignment information by almost four percentage points. The improvement through using PSI-BLAST profiles to develop the method is relatively small: PHDpsi was trained on a small database of not very divergent profiles in 1994; for example, PROFphd was trained on PSI-BLAST profiles of a 20 times larger database in 2000. The two differ by only one percentage point, and part of this difference resulted from implementing new concepts into PROF (Rost, 2005).

Averaging Over Many Methods May Help. All methods predict some proteins at lower levels of accuracy than others (Rost et al., 1993; Lesk et al., 2001; Rost and Eyrich, 2001). Nevertheless, for most proteins there is a method that predicts secondary structure at a level higher than average (Rost and Eyrich, 2001). The latter is applied when averaging over prediction methods. In fact, such averages are helpful as long as compiled over *good* methods (Rost et al., 2001). Thus, using ALL available programs is a rather bad idea!

Solvent Accessibility Prediction in Practice

Very few of the seemingly more accurate methods predicting solvent accessibility are publicly available. Furthermore, there is no EVA-like evaluation of methods based on large sets and identical conditions. The first case scenario in which methods are compared based on different data sets, and different ways to define accessibility, is the rule rather than the exception. Thus, it is important to view values published with caution. Most methods predict accessibility in two states (exposed and buried). Levels of prediction accuracy vary significantly according to choice of the thresholds to distinguish between the two states (Cuff and Barton, 2000). If we define all residues that are less than 16% solvent accessible as "exposed," the best current methods reach levels around $75 \pm 10\%$ accuracy (Rost and Sander, 1994a; Lesk, 1997; Cuff and Barton, 2000; Przybylski and Rost, 2002).

Using alignment information significantly improves prediction accuracy. However, accessibility predictions are more sensitive to alignment errors than secondary structure predictions (Rost and Sander, 1994a; Przybylski and Rost, 2002). A reason for this may be that accessibility is evolutionarily less well conserved than is secondary structure (Przybylski and Rost, 2002).

Transmembrane Region Prediction in Practice

Caution: No Appropriate Estimate of Performance Available!. The appropriate evaluation of methods predicting membrane helices is even more difficult than the evaluation of other categories of structure prediction. Three major problems prevent adequate analyses. (1) We do not have enough high-resolution structures to allow a statistically significant analysis (Chen and Rost, 2002). (2) Low-resolution experiments (gene fusion) differ from high-resolution experiments (crystallography) almost as much as prediction methods do (Chen and Rost, 2002). Thus, low-resolution experiments do not suffice to evaluate prediction accuracy. (3) All methods optimize some parameters. Since there are so few high-resolution structures, all methods use as many of the known ones, as possible. However, those methods perform much better on proteins for which they were developed than on new proteins was impressively demonstrated—and overlooked in a recent analysis of prediction methods (Möller, Croning, and Apweiler, 2001).

Crude Estimates for Where We Are at in the Field. The best current methods (HMMTOP2, PHDhtm, and TMHMM2) predict all helices correct for about 70% of all proteins (Chen and Rost, 2002). MINNOU (Cao et al., 2006) and Phobius (Kall, Krogh, and Sonnhammer, 2004, 2005) are likely to reach similar or even higher performance. For more than 60% of the proteins, the topology is also predicted correctly (Möller, Croning, and Apweiler, 2001; Chen and Rost, 2002). PHDhtm getting about 70% of the observed TMH residues right (Chen and Rost, 2002) achieves the most accurate per-residue prediction. All methods based on advanced algorithms tend to underestimate transmembrane helices; thus, about 86% of the TMH residues predicted by the best methods in this category PHD and DAS are correct (Chen and Rost, 2002). Most methods tend to confuse signal peptides with membrane helices; the best separation is achieved by a system predicting subcellular localization ALOM2 (Nakai and Kanehisa, 1992). Almost as accurate are PHD and TopPred2 (followed by TMHMM) (Möller, Croning, and Apweiler, 2001). Surprisingly, most methods have also been overestimated in their ability to distinguish between globular and helical membrane proteins; particularly, most methods based only on hydrophobicity scales incorrectly predict membrane helices in over 90% of a representative set of globular proteins (Chen, Kernytsky, and Rost, 2002). TMHMM, SOSUI, and PHDhtm appear to yield the most accurate distinction between membrane and globular proteins (less than 2% false positives) (Chen and Rost, 2002). The most accurate hydrophobicity index appears to be the one recently developed in the Ben-Tal group (Kessel and Ben-Tal, 2002). All methods fail to distinguish membrane helices from signal peptides to the extent that the best methods still falsely predict membrane helices for 25% (PHDhtm) to 34% (TMHMM2) of all signal peptides tested. The good news for the practical application are that we have an accurate method detecting signal peptides (Nielsen et al., 1997) and that most incorrectly predicted membrane helices start closer than 10 residues to N-terminal methionine residues, that is, could be corrected by experts.

Genome Analysis: Many Proteins Contain Membrane Helices. Despite the overestimated performance, predictions of transmembrane helices are valuable tools to quickly scan entire genomes for membrane proteins. A few groups base their results only on hydrophobicity scales, known to have extremely high error rates in distinguishing globular and membrane proteins. Nevertheless, the averages published for entire genomes are surprisingly similar between different authors (Goffeau et al., 1993; Rost, Casadio, and Fariselli, 1996; Arkin, Brünger, and Engelman, 1997; Frishman and Mewes, 1997; Jones, 1998; Wallin and von Heijne, 1998; Liu and Rost, 2001). About 10–30% of all proteins appear to contain membrane helices. One crucial difference, however, is that more cautious estimates do not perceive a statistically significant difference in the percentage of TMH proteins across the three kingdoms: eukaryotes, prokaryotes, and archaea (Liu, Tan, and Rost, 2002). However, preferences between particular types of membrane proteins differ; in particular, eukaryotes have more 7TM proteins (receptors), whereas prokaryotes have more 6- and 12TM proteins (ABC transporters) (Wallin and von Heijne, 1998; Liu and Rost, 2001).

EMERGING AND FUTURE DEVELOPMENTS

Regions Likely to Undergo Structural Change Predicted Successfully

Young et al. (1999) have unraveled an impressive correlation between local secondary structure predictions and global conditions. The authors monitor regions for which second-ary structure prediction methods give equally strong preferences for two different states. Such regions are processed combining simple statistics and expert rules. The final method is tested on 16 proteins known to undergo structural rearrangements and on a number of other proteins. The authors report no false positives, and they identify most known structural switches. Subsequently, the group applied the method to the myosin family identifying putative switching regions that were not known before but appeared reasonable candidates (Kirshenbaum, Young, and Highsmith, 1999). I find this method most remarkable in two ways: (1) it is the most general method using predictions of protein structure to predict some aspects of function and (2) it illustrates that predictions may be useful even when structures are known (as in the case of the myosin family). Recently, several group have begun to predict local flexibility directly;for example, Wiggle (Gu, Gribskov, and Bourne, 2006) identifies flexible regions through a rather unique, innovative combination of molecular dynamics and machine learning. PROFbval (Schlessinger, Yachdav, and Rost, 2006) predicts normalized B-values directly from sequence alignments through neural networks; like many recent state-of-the-art methods, PROFbval uses other 1D predictions as input. GlobPlot (Linding et al., 2003a) uses the correlation between low-complexity and flexibility.

Aspects of Protein Function Predicted by Expert Analysis of Secondary Structure

The typical scenario in which secondary structure predictions help us to learn more about function comes from experts combining predictions and their intuition, most often to find similarities to proteins of known function but insignificant sequence similarity (Brautigam et al., 1999; Davies et al., 1999; de Fays et al., 1999; Di Stasio et al., 1999; Gerloff et al., 1999;

Juan et al., 1999; Laval et al., 1999; Seto et al., 1999; Xu et al., 1999; Jackson and Russell, 2000; Paquet et al., 2000; Shah et al., 2000; Stawiski et al., 2000). Usually, such applications are based on very specific details about predicted secondary structure (some examples in Table 29.1). Thus, successful correlations of secondary structure and function appear difficult to incorporate into automatic methods.

Exploring Secondary Structure Predictions to Improve Database Searches

Initially, three groups independently applied secondary structure predictions to fold recognition, that is, the detection of structural similarities between proteins of unrelated sequences (Rost, 1995; Fischer and Eisenberg, 1996; Russell, Copley, and Barton, 1996). A few years later, almost every other fold recognition/threading method has adopted this concept (Ayers et al., 1999; de la Cruz and Thornton, 1999; Di Francesco, Munson, and Garnier, 1999; Hargbo and Elofsson, 1999; Jones, 1999a; Jones et al., 1999; Koretke et al., 1999; Ota et al., 1999; Panchenko, Marchler-Bauer, Bryant, 1999; Kelley, MacCallum, and Sternberg, 2000). Two recent methods extended the concept by not only refining the database search but also by actually refining the quality of the alignment through an iterative procedure (Heringa, 1999; Jennings, Edge, and Sternberg, 2001; Simossis and Heringa, 2005; Simossis, Kleinjung, and Heringa, 2005; Sammeth and Heringa, 2006). A related strategy has been explored by Ng and Henikoffs to improve predictions and alignments for membrane proteins (Ng, Henikoff, and Henikoff, 2000); the concept has been adopted by others (Muller et al., 2001; Hedman et al., 2002; Sutormin, Rakhmaninova, and Gelfand, 2003; Yu and Altschul, 2005; Sadovskaya, Sutormin, and Gelfand, 2006).

From 1D Predictions to 2D, and 3D Structure

Are secondary structure predictions accurate enough to help predict higher order aspects of protein structure automatically? 2D (inter-residue contacts) predictions: Baldi et al. (2000) have recently improved the level of accuracy in predicting beta-strand pairings over earlier work (Hubbard and Park, 1995) by using another elaborate neural network system. 3D predictions: the following list of five groups exemplifies that secondary structure predictions have now become a popular first step toward predicting 3D structure. (1) Ortiz et al. (1999) successfully use secondary structure predictions as one component of their 3D structure prediction method. (2) Eyrich and coworkers (Eyrich et al., 1999a; Eyrich, Standley, Friesner, 1999) minimize the energy of arranging predicted rigid secondary structure segments. (3) Lomize et al. (1999) also start from secondary structure segments. (4) Chen, Singh, and Altman, 1999 suggest using secondary structure predictions to reduce the complexity of molecular dynamics simulations. (5) Levitt and coworkers (Samudrala et al., 1999, 2000) combine secondary structure-based simplified presentations with a particular lattice simulation attempting to enumerate all possible folds.

Using Accessibility to Predict Aspects of Function

Features of multiple alignments can reveal aspects of protein function (Casari, Sander, and Valencia, 1995; Lichtarge, Bourne, and Cohen, 1996; Lichtarge, Yamamoto, and Cohen 1997; Pazos et al. 1997; Marcotte et al. 1999; Irving et al., 2000; Wodak and Mendez, 2004). To simplify the story, residues can be conserved because of structural and functional reasons. If we could distinguish between the two, we could predict the functional residues. Obviously, residues that are exposed *and* conserved are likely to reveal functional constraints. This suggests using predicted accessibility and combination with alignments to predict functional residues. One particular application is the identification of residues involved in transient protein–protein interactions (Yao et al. 2003; Wodak and Mendez 2004; Ofran and Rost 2007a; Porollo and Meller 2007). Once again, methods combining evolutionary information and 1D predictions as input perform best (Ofran and Rost 2007a); in fact, they are accurate enough to even predict interaction hot spots (Ofran and Rost 2007b). Another possible application of predicted accessibility is the prediction of subcellular localization: the surface compositions differ significantly between extracellular, cytoplasm, and nuclear proteins (Andrade, O'Donoghue, and Rost, 1998). Many methods have recently used this correlation to better the prediction of localization (Nair and Rost, 2003a; Nair and Rost, 2003b; Nair and Rost, 2004; Yu et al., 2004; Nair and Rost, 2005; Lee et al., 2006; Pierleoni et al. 2006; Rossi, Marti-Renom, and Sali, 2006). The combination of 1D predictions and predictions of localization can be used to annotate experimental protein–protein interactions as, for example, exemplified in the PiNAT system (Ofran et al., 2006) that annotates protein networks.

Using 1D Predictions for Target Selection in Structural Genomics

Structural genomics proposes to experimentally determine one high-resolution structure for every known protein (Gaasterland, 1998; Rost, 1998; Sali, 1998; Burley et al., 1999; Shapiro and Harris, 2000; Thornton, 2001). Obviously, this goal could be reached faster if we could avoid all proteins of known structure. This is relatively straightforward (Sali, 1998; Liu and Rost, 2002; Liu, Tan, and Rost, 2002). More difficult is the task of avoiding proteins that do not express, do not purify, or are longer than 200 residues, and do not crystallize (or do not diffract well enough). One way toward this goal is to exclude all proteins with membrane helices (Liu and Rost, 2001; Liu and Rost, 2002). Can bioinformatics do more than that? A whole battery of approaches is being applied to steer the target selection and structure exploitation in the operating structural genomics centers (Yao et al., 2003; Bray et al., 2004; Chelliah et al., 2004; DeWeese-Scott and Moult, 2004; Grant, Lee, and Orengo, 2004; Jones and Thornton, 2004; Liu and Rost, 2004a; Liu and Rost, 2004b; Liu et al., 2004; Nair and Rost, 2004; Chandonia and Brenner, 2005; Gutteridge and Thornton, 2005; Kifer, Sasson, and Linial, 2005; Nimrod et al., 2005; Oldfield et al., 2005b; Pal and Eisenberg, 2005; Sasson and Linial, 2005; Bhattacharya, Tejero, and Montelione, 2006; Gu, Gribskov, and Bourne, 2006; Kouranov et al., 2006; Rossi, Marti-Renom, and Sali, 2006; Levitt, 2007; Liu, Montelione, and Rost, 2007; Marti-Renom et al., 2007). Many of these approaches are based on 1D predictions.

Eukaryotes Full of Disorder?

At the point of the first edition of this book, it had been shown that regions of low-complexity, as predicted by the program SEG (Wootton and Federhen, 1996), are the rule rather than the exception in the protein universe (Saqi, 1995; Garner et al., 1998; Romero et al., 1998; Wright and Dyson, 1999; Dunker et al., 2001; Dunker and Obradovic, 2001; Romero et al., 2001). Using predictions of secondary structure, we had found that there are many proteins that do not have low-complexity regions but nevertheless appear to have long (more than 70 residues) regions without regular secondary structure (helix/strand, dubbed NORS). Such NORS proteins appear to be significantly more abundant in eukaryotes than in all other kingdoms, reaching levels around 25% of the entire genome (Liu, Tan, and Rost, 2002). We found many of the NORS to be evolutionarily conserved, suggesting that these may in fact be proteins with induced structure rather than without structure. These findings had confirmed earlier analyses (Dunker et al., 2001) and have been confirmed by subsequent work (Ward et al., 2004).

Since the publication of the first edition of the book, the field has exploded. Over the last years, many studies have shown that often the lack of a unique, native 3D structure in physiological conditions can be crucial for function (Dyson and Wright, 2004; Fuxreiter et al., 2004; Dunker et al., 2005; Dyson and Wright, 2005; Tompa, 2005; Vucetic et al., 2005; Liu et al., 2006; Radivojac et al., 2006). Such proteins are often referred to as *disordered, unfolded, natively unstructured,* or *intrinsically unstructured* proteins. A typical example is a protein that adopts a unique 3D structure only upon binding to an interaction partner and thereby performs its biochemical function (Dyson and Wright, 2004; Fuxreiter et al., 2004; Dyson and Wright, 2005). The better our experimental and computational means of identifying such proteins, the more we realize that they come in a great variety: some adopt regular secondary structure (helix or strand) upon binding, some remain loopy; some proteins are almost entirely unstructured, others have only short unstructured regions. The more our ability to recognize short unstructured regions increases, the more we realize that the label "unstructured protein" would be rather misleading, as most "unstructured proteins" have relatively short unstructured regions. There is no single one way to define "unstructured regions." Many methods have been developed that predict such natively unstructured or disordered regions based on a great diversity of aspects (Tompa, 2002; Jones and Ward, 2003; Linding et al., 2003a; Vucetic et al., 2003; Radivojac et al., 2004; Beltrao and Serrano, 2005; Coeytaux and Poupon, 2005; Dosztanyi et al., 2005; Lise and Jones, 2005; Oldfield et al., 2005a; Gu, Gribskov, and Bourne, 2006; Vullo et al., 2006; Schlessinger, Liu, and Rost, 2007a; Schlessinger, Punta, and Rost, 2007b). Increasingly, it becomes apparent that unstructured regions are an important means of increasing complexity of molecular interactions. It is rather remarkable that methods developed for the prediction of regular 1D structure become once again crucial for the exploration of regions that appear to pose an altogether different challenge to the advance of structural biology over the next decade.

FURTHER READING

Review articles:

- Secondary structure, old (Fasman 1989): Goldmine for finding citations for very early prediction methods.
- Secondary structure, new (Rost, 2001): Brief walk through the recent highlights in the field of protein secondary structure prediction.
- Secondary structure for alignments (Simossis and Heringa 2004): Reviews using secondary structure predictions to improve alignment methods.
- Coiled coil helices (Lupas, 1997): Analysis of the performance of various programs predicting coiled coil helices based on new structures.
- Transmembrane predictions (Bowie, 2005; Punta et al., 2006; von Heijne, 1996): Reviews of methods predicting transmembrane regions from different perspectives.

- Using 1D predictions for function prediction (Rost et al., 2003; Ofran et al., 2005): Reviews methods that extend beyond database comparisons to predict aspects of protein function.
- Prediction of natively unstructured/disordered regions (Dyson and Wright, 2005; Fink, 2005; Oldfield et al., 2005; Tompa, 2005)

Original articles (sorted by subject/year):

(1) Secondary structure, PHDsec (Rost and Sander, 1993; Rost and Sander, 1994a): Original papers describing the first method that surpassed the threshold of 70% prediction accuracy through combining neural networks and evolutionary information.

(2) Secondary structure, PSIPRED (Jones, 1999b): The alignments used by PHD are replaced by PSI-BLAST alignments. This improves prediction accuracy significantly. However, possibly the most important aspect is the description of ways to run PSI-BLAST automatically without finding too many wrong hits.

(3) Secondary structure, SSpro/Porter (Baldi et al., 1999; Pollastri et al., 2002a; Pollastri et al., 2002b; Pollastri and McLysaght, 2005): The most complicated and seemingly most successful architecture for using neural networks to predict secondary structure is presented here.

(4) Secondary structure, switches (Young et al., 1999): A data set of 16 protein sequences having functions that involve substantial backbone rearrangements are analyzed with respect to the ambivalence of predicted secondary structure. They find all segments involved in conformational switches to have ambivalent predictions, measured by the similarity in prediction probabilities for helix, sheet, and loop, as reported by PHD.

(5) Secondary structure applied to classify genomes (Przytycka, Aurora, and Rose, 1999): Does a protein's secondary structure determine its three-dimensional fold? This question is tested directly by analyzing proteins of known structure and constructing a taxonomy based solely on secondary structure. The taxonomy is generated automatically, and it takes the form of a tree in which proteins with similar secondary structure occupy neighboring leaves.

(6) Solvent accessibility, PHDacc (Rost and Sander, 1994b): Analysis of the evolutionary conservation of solvent accessibility and description of an alignment-based neural network prediction method.

(7) Transmembrane helices, TMHMM (Sonnhammer, von Heijne, and Krogh, 1998): Most advanced and seemingly most accurate method predicting transmembrane helices through cyclic hidden Markov models.

(8) Transmembrane helices, genome analysis (Wallin and von Heijne, 1998): The authors analyze all membrane helix predictions for a number of entirely sequenced genomes. They conclude that more complex organisms use more helical membrane proteins than simpler organisms.

(9) Using 1D predictions for genome analysis (Liu and Rost, 2001): 28 entirely sequenced genomes are compared based on predictions of coiled coil proteins (COILS; Lupas, 1996), membrane helices (PHD; Rost, 1996), and functional

classes (EUCLID; Tamames et al., 1998). In contrast to many other publications, the correlation between the complexity of an organism and its use of helical membrane proteins is not confirmed.

ACKNOWLEDGMENTS

Thanks to the EVA team that enabled to quote some of the numbers given: Volker Eyrich (Schroedinger, USA), Marc Marti-Renom (Valencia, Spain), Andrej Sali (UCSF, USA), and Osvaldo Grana and Alfonso Valencia (Madrid, Spain). The work of BR was supported by the grants 2-R01-LM07329, U54-GM074958, GM75026, and U54-GM072980 from the National Institutes of Health (NIH). His work was made possible by the atmosphere created by a dream team at Columbia, to thank just a few of those: Jinfeng Liu (now Genentech, USA), Yanay Ofran (now Tel Aviv University, Israel), Dariusz Przybylski, Rajesh Nair, Marco Punta, Guy Yachdav, Yana Bromberg, Henry Bigelow, Sven Mika, Nancy Sosa, Eyal Mozes, Avner Schlessinger, Andrew Kernytsky, Kazimierz Wrzeszczynski, Ta-Tsen Soong, Phil Carter (now London, England), Claus Andersen (now Siena Biotech, Italy), and Volker Eyrich (now Schroedinger, USA). Last, not the least, thanks to all those who deposit their experimental data in public databases and to those who maintain these databases, in particular thanks to the teams around Phil Bourne, Helen Berman, Rolf Apweiler, and Amos Bairoch who started to clean it up!

REFERENCES

Adamczak R, Porollo A, Meller J (2004): Accurate prediction of solvent accessibility using neural networks-based regression. *Proteins* 56(4):753–767.

Adamczak R, Porollo A, Meller J (2005): Combining prediction of secondary structure and solvent accessibility in proteins. *Proteins* 59(3):467–475.

Altschul SF, Gish W (1996): Local alignment statistics. *Methods Enzymol* 266:460–480.

Altschul SF, Madden TL, Schaeffer AA, Zhang J, Zhang Z, Miller W, Lipman DJ (1997): Gapped Blast and PSI-Blast: a new generation of protein database search programs. *Nucleic Acids Res* 25:3389–3402.

Andrade MA, O'Donoghue SI, Rost B (1998): Adaptation of protein surfaces to subcellular location. *J Mol Biol* 276:517–525.

Anfinsen CB (1973): Principles that govern the folding of protein chains. *Science* 181:223–230.

Arkin IT, Brünger AT, Engelman DM (1997): Are there dominant membrane protein families with a given number of helices? *Prot Struct Funct Genet* 28:465–466.

Ayers DJ, Gooley PR, Widmer-Cooper A, Torda AE (1999): Enhanced protein fold recognition using secondary structure information from NMR. *Protein Sci* 8(5):1127–1133.

Bairoch A, Apweiler R (2000): The SWISS-PROT protein sequence database and its supplement TrEMBL in 2000. *Nucleic Acids Res* 28(1):45–48.

Bairoch A, Apweiler R, Wu CH, Barker WC, Boeckmann B, Ferro S, Gasteiger E, Huang H, Lopez R, Magrane M, Martin MJ, Natale DA, O'Donovan C, Redaschi N, Yeh LS (2005): The Universal Protein Resource (UniProt). *Nucleic Acids Res* 33(Database issue):D154–D159.

Baldi P, Brunak S, Frasconi P, Soda G, Pollastri G (1999): Exploiting the past and the future in protein secondary structure prediction. *Bioinformatics* 15(11):937–946.

Baldi P, Pollastri G, Andersen CA, Brunak S (2000): Matching protein beta-sheet partners by feedforward and recurrent neural networks. *Ismb* 8:25–36.

Barton GJ (1996): Protein sequence alignment and database scanning. In: Sternberg MJE, editor. *Protein structure prediction*, 1st edn. , Vol. 170. Oxford: Oxford University Press, pp. 31–64.

Beltrao P, Serrano L (2005): Comparative genomics and disorder prediction identify biologically relevant SH3 protein interactions. *PLoS Comput Biol* 1(3):e26.

Ben-Tal N, Honig B, Miller C, McLaughlin S (1997): Electrostatic binding of proteins to membranes. Theoretical predictions and experimental results with charybdotoxin and phospholipid vesicles. *Biophys J* 73(4):1717–1727.

Berman HM, Battistuz T, Bhat TN, Bluhm WF, Bourne PE, Burkhardt K, Feng Z, Gilliland GL, Iype L, Jain S, Fagan P, Marvin J, Padilla D, Ravichandran V, Schneider B, Thanki N, Weissig H, Westbrook JD, Zardecki C (2002): The Protein Data Bank. *Acta Crystallogr D* 58 (Pt 6 No 1): 899–907.

Berven FS, Flikka K, Jensen HB, Eidhammer I (2004): BOMP: a program to predict integral beta-barrel outer membrane proteins encoded within genomes of Gram-negative bacteria. *Nucleic Acids Res* 32(Web server issue):W394–W399.

Berven FS, Karlsen OA, Straume AH, Flikka K, Murrell JC, Fjellbirkeland A, Lillehaug JR, Eidhammer I, Jensen HB (2006): Analysing the outer membrane subproteome of *Methylococcus capsulatus* (Bath) using proteomics and novel biocomputing tools. *Arch Microbiol* 184(6): 362–377.

Bhattacharya A, Tejero R, Montelione GT (2006): Evaluating protein structures determined by structural genomics consortia. *Proteins* 66(4):778–795.

Bigelow H, Petrey D, Liu J, Przybylski D, Rost B (2004): Prediction of transmembrane beta-barrels for entire proteomes. *Nucleic Acids Res* 32:2566–2577.

Bigelow H, Rost B (2006b): PROFtmb: a web server for predicting bacterial transmembrane beta barrel proteins. *Nucleic Acids Res* 34(Web server issue):W186–W188.

Biou V, Gibrat JF, Levin JM, Robson B, Garnier J (1988): Secondary structure prediction: combination of three different methods. *Prot Engin* 2:185–191.

Bohr H, Bohr J, Brunak S, Cotterill RMJ, Lautrup B, Nørskov L, Olsen OH, Petersen SB (1988): Protein secondary structure and homology by neural networks. *FEBS Letters* 241: 223–228.

Bowie JU (2005): Solving the membrane protein folding problem. *Nature* 438(7068):581–589.

Brändén C, Tooze J (1991): *Introduction to Protein Structure*, New York: Garland Publ.

Brautigam C, Steenbergen-Spanjers GC, Hoffmann GF, Dionisi-Vici C, van den Heuvel LP, Smeitink JA, Wevers RA (1999): Biochemical and molecular genetic characteristics of the severe form of tyrosine hydroxylase deficiency. *Clin Chem* 45(12):2073–2078.

Bray JE, Marsden RL, Rison SC, Savchenko A, Edwards AM, Thornton JM, Orengo CA (2004): A practical and robust sequence search strategy for structural genomics target selection. *Bioinformatics*.

Buchanan SK (1999): β-Barrel proteins from bacterial outer membranes: structure, function and refolding. *Curr Opin Struct Biol* 9:455–461.

Burley SK, Almo SC, Bonanno JB, Capel M, Chance MR, Gaasterland T, Lin D, Sali A, Studier FW, Swaminathan S (1999): Structural genomics: beyond the human genome project. *Nat Genet* 23(2): 151–157.

Bystroff C, Thorsson V, and Baker D (2000). HMMSTR: a hidden Markov model for local sequence-structure correlations in proteins. *J Mol Bio* 301:173–190.

Cao B, Porollo A, Adamczak R, Jarrell M, Meller J (2006): Enhanced recognition of protein transmembrane domains with prediction-based structural profiles. *Bioinformatics* 22(3):303–309.

Carugo O (2000): Predicting residue solvent accessibility from protein sequence by considering the sequence environment. *Protein Engin* 13(9):607–609.

Casari G, Sander C, Valencia A (1995): A method to predict functional residues in proteins. *Nat Struct Biol* 2:171–178.

Chance MR, Fiser A, Sali A, Pieper U, Eswar N, Xu G, Fajardo JE, Radhakannan T, Marinkovic N (2004): High-throughput computational and experimental techniques in structural genomics. *Genome Res* 14(10B):2145–2154.

Chandonia J-M, Brenner SE (2005): Implications of structural genomics target selection strategies: Pfam5000, whole genome, and random approaches. *Proteins* 58(1):166–179.

Chelliah V, Chen L, Blundell TL, Lovell SC (2004): Distinguishing structural and functional restraints in evolution in order to identify interaction sites. *J Mol Biol* 342(5):1487–1504.

Chen CC, Singh JP, Altman RB (1999): Using imperfect secondary structure predictions to improve molecular structure computations. *Bioinformatics* 15(1):53–65.

Chen CP, Kernytsky A, Rost B (2002): Transmembrane helix predictions revisited. *Protein Sci* 11:2774–2791.

Chen CP, Rost B (2002): State-of-the-art in membrane prediction. *Appl Bioinf* 1:21–35.

Chothia C (1976): The nature of the accessible and buried surfaces in proteins. *J Mol Biol* 105:1–12.

Chou PY, Fasman GD (1974): Prediction of protein conformation. *Biochemistry* 13:211–215.

Coeytaux K, Poupon A (2005): Prediction of unfolded segments in a protein sequence based on amino acid composition. *Bioinformatics* 21(9):1891–1900.

Cohen FE, Presnell SR (1996): The combinatorial approach. In: Sternberg MJE, editor. *Protein Structure Prediction*, 1st edn., Vol. 170: Oxford: Oxford University Press, pp. 207–228.

Combet C, Blanchet C, Geourjon C, Deléage G (2000): NPS@: network protein sequence analysis. *Trends Bio Sci* 25(3):147–150.

Connolly ML (1983): Solvent-accessible surfaces of proteins and nucleic acids. *Science* 221: 709–713.

Crick FHC (1953): The packing of α-helices: simple coiled-coils. *Acta Crystallogr A* 6:689–697.

Cserzö M, Wallin E, Simon I, von Heijne G, Elofsson A (1997): Prediction of transmembrane α-helices in prokaryotic membrane proteins: the dense alignment surface method. *Protein Engin* 10:673–676.

Cuff JA, Barton GJ (2000): Application of multiple sequence alignment profiles to improve protein secondary structure prediction. *Prot Struct Funct Genet* 40(3):502–511.

Davies GP, Martin I, Sturrock SS, Cronshaw A, Murray NE, Dryden DT (1999): On the structure and operation of type I DNA restriction enzymes. *J Mol Biol* 290(2):565–579.

de Fays K, Tibor A, Lambert C, Vinals C, Denoel P, De Bolle X, Wouters J, Letesson JJ, Depiereux E (1999): Structure and function prediction of the *Brucella abortus* P39 protein by comparative modeling with marginal sequence similarities. *Protein Engin* 12(3):217–223.

de la Cruz X, Thornton JM (1999): Factors limiting the performance of prediction-based fold recognition methods. *Protein Sci* 8(4):750–759.

Deleage G, Roux B (1987): An algorithm for protein secnodary structure prediction based on class prediction. *Protein Engin* 1:289–294.

Devereux J, Haeberli P, Smithies O (1984): GCG package. *Nucleic Acids Res* 12:387–395.

DeWeese-Scott C, Moult J (2004): Molecular modeling of protein function regions. *Proteins* 55(4): 942–961.

Di Francesco V, Munson PJ, Garnier J (1999): FORESST: fold recognition from secondary structure predictions of proteins. *Bioinformatics* 15(2):131–140.

Di Stasio E, Sciandra F, Maras B, Di Tommaso F, Petrucci TC, Giardina B, Brancaccio A (1999): Structural and functional analysis of the N-terminal extracellular region of beta-dystroglycan. *Biochem Biophys Res Commun* 266(1):274–278.

Dickerson RE, Timkovich R, Almassy RJ (1976): The cytochrome fold and the evolution of bacterial energy metabolism. *J Mol Biol* 100:473–491.

Dobson CM, Karplus M (1999): The fundamentals of protein folding: bringing together theory and experiment. *Curr Opin Struct Biol* 9(1):92–101.

Dosztanyi Z, Csizmok V, Tompa P, Simon I (2005): The pairwise energy content estimated from amino acid composition discriminates between folded and intrinsically unstructured proteins. *J Mol Biol* 347(4):827–839.

Dunker AK, Cortese MS, Romero P, Iakoucheva LM, Uversky VN (2005): Flexible nets. The roles of intrinsic disorder in protein interaction networks. *Febs J* 272(20):5129–5148.

Dunker AK, Lawson JD, Brown CJ, Williams RM, Romero P, Oh JS, Oldfield CJ, Campen AM, Ratliff CM, Hipps KW, Ausio J, Nissen MS, Reeves R, Kang C, Kissinger CR, Bailey RW, Griswold MD, Chiu W, Garner EC, Obradovic Z (2001): Intrinsically disordered protein. *J Mol Graph Model* 19(1):26–59.

Dunker AK, Obradovic Z (2001): The protein trinity-linking function and disorder. *Nat Biotechol* 19 (9):805–806.

Dyson HJ, Wright PE (2004): Unfolded proteins and protein folding studied by NMR. *Chem Rev* 104 (8):3607–3622.

Dyson HJ, Wright PE (2005): Intrinsically unstructured proteins and their functions. *Nat Rev Mol Cell Biol* 6(3):197–208.

Eddy SR (1998): Profile hidden Markov models. *Bioinformatics* 14(9):755–763.

Eisenberg D, Schwartz E, Komaromy M, Wall R (1984): Analysis of membrane and surface protein sequences with the hydrophobic moment plot. *J Mol Biol* 179:125–142.

Ellis RJ, Dobson C, Hartl U (1998): Sequence does specify protein conformation. *Trends Biochem Sci* 23(12):468.

Engelman DM, Steitz TA, Goldman A (1986): Identifying nonpolar transbilayer helices in amino acid sequences of membrane proteins. *Ann Rev Biophys Biophys Chem* 15:321–353.

Eyrich V, Martí-Renom MA, Przybylski D, Fiser A, Pazos F, Valencia A, Sali A, Rost B (2001): EVA: continuous automatic evaluation of protein structure prediction servers. *Bioinformatics* 17:1242–1243.

Eyrich VA, Koh IYY, Przybylski D, Graña O, Pazos F, Valencia A, Rost B (2003): CAFASP3 in the spotlight of EVA. *Prot Struct Funct Bio* 53 (Suppl 6):548–560.

Eyrich VA, Rost B (2003): META-PP: single interface to crucial prediction servers. *Nucleic Acids Res* 31(13):3308–3310.

Eyrich VA, Standley DM, Felts AK, Friesner RA (1999a): Protein tertiary structure prediction using a branch and bound algorithm. *Prot Struct Funct Genet* 35(1):41–57.

Eyrich VA, Standley DM, Friesner RA (1999b): Prediction of protein tertiary structure to low resolution: performance for a large and structurally diverse test set. *J Mol Biol* 288(4): 725–742.

Fariselli P, Olmea O, Valencia A, Casadio R (2001): Prediction of contact maps with neural networks and correlated mutations. *Protein Engin* 14(11):835–843.

Fasman GD (1989): The development of the prediction of protein structure. In: Fasman GD, editor. *Prediction of Protein Structure and the Principles of Protein Conformation*, New York, London: Plenum Press, pp. 193–303.

Fink AL (2005): Natively unfolded proteins. *Curr Opin Struct Biol* 15(1):35–41.

Fischer D, Eisenberg D (1996): Fold recognition using sequence-derived properties. *Protein Sci* 5:947–955.

Forrest LR, Tang CL, Honig B (2006): On the accuracy of homology modeling and sequence alignment methods applied to membrane proteins. *Biophys J* 91(2):508–517.

Frishman D, Mewes HW (1997): Protein structural classes in five complete genomes. *Nat Struct Bio* 4:626–628.

Fuxreiter M, Simon I, Friedrich P, Tompa P (2004): Preformed structural elements feature in partner recognition by intrinsically unstructured proteins. *J Mol Biol* 338(5):1015–1026.

Gaasterland T (1998): Structural genomics taking shape. *Trends Genet* 14(4):135.

Garner E, Cannon P, Romero P, Obradovic Z, Dunker AK (1998): Predicting disordered regions from amino acid sequence: common themes despite differing structural characterization. *Genome Inform* 9:201–214.

Garnier J, Osguthorpe DJ, Robson B (1978): Analysis of the accuracy and implications of simple methods for predicting the secondary structure of globular proteins. *J Mol Biol* 120: 97–120.

Garnier J, Robson B (1989): The GOR method for predicting secondary structure in proteins. In: Fasman, GD editor. *Prediction of Protein Structure and the Principles of Protein Conformation*, New York: Plenum Press, pp. 417–465.

Garrow AG, Agnew A, Westhead DR (2005): TMB-Hunt: a web server to screen sequence sets for transmembrane beta-barrel proteins. *Nucleic Acids Res* 33(Web server issue): W188–W192.

Gascuel O, Golmard JL (1988): A simple method for predicting the secondary structure of globular proteins: implications and accuracy. *CABIOS* 4:357–365.

Gerloff DL, Cannarozzi GM, Joachimiak M, Cohen FE, Schreiber D, Benner SA (1999): Evolutionary, mechanistic, and predictive analyses of the hydroxymethyldihydropterin pyrophosphokinase family of proteins. *Biochem Biophys Res Commun* 254(1):70–76.

Goffeau A, Slonimski P, Nakai K, Risler JL (1993): How many yeast genes code for membrane-spanning proteins? *Yeast* 9:691–702.

Grana O, Eyrich VA, Pazos F, Rost B, Valencia A (2005): EVAcon: a protein contact prediction evaluation service. *Nucleic Acids Res* 33(Web server issue):W347–W351.

Grant A, Lee D, Orengo C (2004): Progress towards mapping the universe of protein folds. *Genome Biol* 5(5):107.

Gromiha MM, Suwa M (2006): Discrimination of outer membrane proteins using machine learning algorithms. *Proteins* 63(4):1031–1037.

Gu J, Gribskov M, Bourne PE (2006): Wiggle-predicting functionally flexible regions from primary sequence. *PLoS Comput Biol* 2(7):e90.

Gupta R, Jung E, Gooley AA, Williams KL, Brunak S, Hansen J (1999): Scanning the available *Dictyostelium discoideum* proteome for O-linked GlcNAc glycosylation sites using neural networks. *Glycobiology* 9(10):1009–1022.

Gutteridge A, Thornton JM (2005): Understanding nature's catalytic toolkit. *Trends Biochem Sci* 30 (11):622–629.

Hagler AT, Honig B (1978): On the formation of protein tertiary structure on a computer. *Proc Natl Acad Sci USA* 75:554–558.

Hansen J, Lund O, Tolstrup N, Gooley AA, Williams KL, Brunak S (1998): NetOglyc: prediction of mucin type O-glycosylation sites based on sequence context and surface accessibility. *Glycoconj J* 15:115–130.

Hargbo J, Elofsson A (1999): Hidden Markov models that use predicted secondary structures for fold recognition. *Prot Struct Funct Genet* 36(1):68–76.

Harris F, Wallace J, Phoenix DA (2000): Use of hydrophobic moment plot methodology to aid the identification of oblique orientated alpha-helices. *Mol Memb Biol* 17(4):201–207.

Hedman M, Deloof H, Von Heijne G, Elofsson A (2002): Improved detection of homologous membrane proteins by inclusion of information from topology predictions. *Protein Sci* 11(3): 652–658.

Heringa J (1999): Two strategies for sequence comparison: profile-preprocessed and secondary structure-induced multiple alignment. *Comput Chem* 23(3–4):341–364.

Hirokawa T, Boon-Chieng S, Mitaku S (1998): SOSUI: classification and secondary structure prediction system for membrane proteins. *Bioinformatics* 14(4):378–379.

Holbrook SR, Muskal SM, Kim S-H (1990): Predicting surface exposure of amino acids from protein sequence. *Protein Engin* 3:659–665.

Hubbard T, Tramontano A, Barton G, Jones D, Sippl M, Valencia A, Lesk A, Moult J, Rost B, Sander C, Schneider R, Lahm A, Leplae R, Buta C, Eisenstein M, Fjellström O, Floeckner H, Grossmann JG, Hansen J, Helmer-Citterich M, Joergensen FS, Marchler-Bauer A, Osuna J, Park J, Reinhardt A, Ribas de Pouplana L, Rojo-Dominguez A, Saudek V, Sinclair J, Sturrock S, Venclovas C, Vinals C (1996): Update on protein structure prediction: results of the 1995 IRBM workshop. *Folding & Design* 1:R55–R63.

Hubbard TJP, Blundell TL (1987): Comparison of solvent-inaccessible cores of homologous proteins: definitions useful for protein modelling. *Protein Engin* 1:159–171.

Hubbard TJP, Park J (1995): Fold recognition and *ab initio* structure predictions using hidden Markov models and β-strand pair potentials. *Prot Struct Funct Genet* 23:398–402.

Irving JA, Pike RN, Lesk AM, Whisstock JC (2000): Phylogeny of the serpin superfamily: implications of patterns of amino acid conservation for structure and function. *Genome Res* 10(12): 1845–1864.

Jackson RM, Russell RB (2000): The serine protease inhibitor canonical loop conformation: examples found in extracellular hydrolases, toxins, cytokines and viral proteins. *J Mol Biol* 296(2):325–334.

Jacoboni I, Martelli PL, Fariselli P, De Pinto V, Casadio R (2001): Prediction of the transmembrane regions of beta-barrel membrane proteins with a neural network-based predictor. *Protein Sci* 10:779–787.

Jennings AJ, Edge CM, Sternberg MJ (2001): An approach to improving multiple alignments of protein sequences using predicted secondary structure. *Protein Engin* 14(4):227–231.

Jones DT (1998): Do transmembrane protein superfolds exist? *FEBS Letters* 423(3):281–285.

Jones DT (1999a): GenTHREADER: an efficient and reliable protein fold recognition method for genomic sequences. *J Mol Biol* 287(4):797–815.

Jones DT (1999b): Protein secondary structure prediction based on position-specific scoring matrices. *J Mol Biol* 292(2):195–202.

Jones DT, Taylor WR, Thornton JM (1994): A model recognition approach to the prediction of all-helical membrane protein structure and topology. *Biochemistry* 33:3038–3049.

Jones DT, Tress M, Bryson K, Hadley C (1999): Successful recognition of protein folds using threading methods biased by sequence similarity and predicted secondary structure. *Prot Struct Funct Genet* 37(S3):104–111.

Jones DT, Ward JJ (2003): Prediction of disordered regions in proteins from position specific score matrices. *Prot Struct Funct Genet* 53(Suppl 6):573–578.

Jones S, Thornton JM (2004): Searching for functional sites in protein structures. *Curr Opin Chem Biol* 8(1):3–7.

Juan HF, Hung CC, Wang KT, Chiou SH (1999): Comparison of three classes of snake neurotoxins by homology modeling and computer simulation graphics. *Biochem Biophys Res Commun* 257(2): 500–510.

Kabsch W, Sander C (1983): Dictionary of protein secondary structure: pattern recognition of hydrogen bonded and geometrical features. *Biopolymers* 22:2577–2637.

Kall L, Krogh A, Sonnhammer EL (2004): A combined transmembrane topology and signal peptide prediction method. *J Mol Biol* 338(5):1027–1036.

Kall L, Krogh A, Sonnhammer EL (2005): An HMM posterior decoder for sequence feature prediction that includes homology information. *Bioinformatics* 21 (Suppl 1):i251–i257.

Karchin R, Cline M, Karplus K (2004): Evaluation of local structure alphabets based on residue burial. *Proteins* 55(3):508–518.

Karplus K, Barrett C, Hughey R (1998): Hidden Markov models for detecting remote protein homologies. *Bioinformatics* 14:846–856.

Karplus K, Barrett C, Cline M, Diekhans M, Grate L, Hughey R (1999): Predicting protein structure using only sequence information. *Prot Struct Funct Genet* S3:121–125.

Kelley LA, MacCallum RM, Sternberg MJ (2000): Enhanced genome annotation using structural profiles in the program 3D-PSSM. *J Mol Biol* 299(2):499–520.

Kessel A, Ben-Tal N (2002): Free energy determinants of peptide association with lipid bilayers. In: Simon S, McIntosh T, editors. *Peptide-Lipid Interactions*, Vol. 52. San Diego, NY: Academic Press,

Kifer I, Sasson O, Linial M (2005): Predicting fold novelty based on ProtoNet hierarchical classification. *Bioinformatics* 21(7):1020–1027.

Kirshenbaum K, Young M, Highsmith S (1999): Predicting allosteric switches in myosins. *Protein Sci* 8:1806–1815.

Klein P, Kanehisa M, De Lisi C (1985): The detection and classification of membrane-spanning proteins. *Biochim Biophys Acta* 815:468–476.

Kleywegt G, (2001): *ProSAL: Protein sequence analysis launcher. Swedish Structural Biology Network.*<http//alpha2bmcuuse/~gerard/srf/prosalhtml>.

Koh IYY, Eyrich VA, Marti-Renom MA, Przybylski D, Madhusudhan MS, Narayanan E, Graña O, Valencia A, Sali A, Rost B (2003): EVA: evaluation of protein structure prediction servers. *Nucleic Acids Res* 31(13):3311–3315.

Koretke KK, Russell RB, Copley RR, Lupas AN (1999): Fold recognition using sequence and secondary structure information. *Prot Struct Funct Genet* 37(S3):141–148.

Kouranov A, Xie L, de la Cruz J, Chen L, Westbrook J, Bourne PE, Berman HM (2006): The RCSB PDB information portal for structural genomics. *Nucleic Acids Res* 34(Database issue): D302–D305.

Kryshtafovych A, Venclovas C, Fidelis K, Moult J (2005): Progress over the first decade of CASP experiments. *Proteins* 61 (Suppl 7):225–236.

Kyte J, Doolittle RF (1982): A simple method for displaying the hydrophathic character of a protein. *J Mol Biol* 157:105–132.

Laval V, Chabannes M, Carriere M, Canut H, Barre A, Rouge P, Pont-Lezica R, Galaud J (1999): A family of Arabidopsis plasma membrane receptors presenting animal beta-integrin domains. *Biochim Biophys Acta* 1435(1–2):61–70.

Lee BK, Richards FM (1971): The interpretation of protein structures: estimation of static accessibility. *J Mol Biol* 55:379–400.

Lee S, Lee B, Jang I, Kim S, Bhak J (2006): Localizome: a server for identifying transmembrane topologies and TM helices of eukaryotic proteins utilizing domain information. *Nucleic Acids Res* 34(Web server issue):W99–103.

Lesk AM (1997): CASP-2: report on *ab initio* predictions. *Prot Struct Funct Genet* 151–166.

Lesk AM, Lo Conte L, Hubbard TJP (2001): Assessment of novel folds targets in CASP4: predictions of three-dimensional structures, secondary structures, and interresidue contacts. *Prot Struct Funct Genet* 45(Suppl 5):98–118.

Levin JM, Garnier J (1988): Improvements in a secondary structure prediction method based on a search for local sequence homologies and its use as a model building tool. *Biochim Biophys Acta* 955:283–295.

Levitt M, Warshel A (1975): Computer simulation of protein folding. *Nature* 253:694–698.

Levitt M (2007): Growth of novel protein structural data. *Proc Natl Acad Sci USA* 104(9):3183–3188.

Li X, Pan XM (2001): New method for accurate prediction of solvent accessibility from protein sequence. *Prot Struct Funct Genet* 42(1):1–5.

Lichtarge O, Bourne HR, Cohen FE (1996): Evolutionarily conserved Galphabetagamma binding surfaces support a model of the G protein-receptor complex. *Proc Natl Acad Sci USA* 93(15): 7507–7511.

Lichtarge O, Yamamoto KR, Cohen FE (1997): Identification of functional surfaces of the zinc binding domains of intracellular receptors. *J Mol Biol* 274(3):325–337.

Lin K, Simossis VA, Taylor WR, Heringa J (2005): A simple and fast secondary structure prediction method using hidden neural networks. *Bioinformatics* 21(2):152–159.

Linding R, Jensen LJ, Diella F, Bork P, Gibson TJ, Russell RB (2003a): Protein disorder prediction: implications for structural proteomics. *Structure* 11(11):1453–1459.

Linding R, Russell RB, Neduva V, Gibson TJ (2003b): GlobPlot: exploring protein sequences for globularity and disorder. *Nucleic Acids Res* 31(13):3701–3708.

Lio P, Vannucci M (2000): Wavelet change-point prediction of transmembrane proteins. *Bioinformatics* 16(4):376–382.

Lise S, Jones DT (2005): Sequence patterns associated with disordered regions in proteins. *Proteins* 58(1):144–150.

Liu LP, Deber CM (1999): Combining hydrophobicity and helicity: a novel approach to membrane protein structure prediction. *Bioorg Med Chem* 7(1):1–7.

Liu J, Rost B (2001): Comparing function and structure between entire proteomes. *Protein Sci* 10(10): 1970–1979.

Liu J, Rost B (2002): Target space for structural genomics revisited. *Bioinformatics* 18:922–933.

Liu J, Tan H, Rost B (2002): Loopy proteins appear conserved in evolution. *J Mol Biol* 322:53–64.

Liu J, Rost B (2003): NORSp: predictions of long regions without regular secondary structure. *Nucleic Acids Res* 31(13):3833–3835.

Liu J, Rost B (2004a): CHOP proteins into structural domains. *Prot Struct Funct Bio* 55(3):678–688.

Liu J, Rost B (2004b): Sequence-based prediction of protein domains. *Nucleic Acids Res* 32(12): 3522–3530.

Liu J, Hegyi H, Acton TB, Montelione GT, Rost B (2004): Automatic target selection for structural genomics on eukaryotes. *Prot Struct Funct Bio* 56:188–200.

Liu J, Perumal NB, Oldfield CJ, Su EW, Uversky VN, Dunker AK (2006): Intrinsic disorder in transcription factors. *Biochemistry* 45(22):6873–6888.

Liu J, Montelione GT, Rost B (2007): Novel leverage of structural genomics. *Nat Biotechnol* 25:849–851.

Lomize AL, Pogozheva ID, Mosberg HI (1999): Prediction of protein structure: the problem of fold multiplicity. *Prot Struct Funct Genet* Suppl 3: (3) 199–203.

Lupas A (1996): Prediction and analyis of coiled-coil structures. *Method Enzymol* 266:513–525.

Lupas A, Van Dyke M, Stock J (1991): Predicting coiled coils from protein sequences. *Science* 252:1162–1164.

Lupas A (1997): Predicting coiled-coil regions in proteins. *Curr Opin Struct Biol* 7:388–393.

Marcotte EM, Pellegrini M, Ng HL, Rice DW, Yeates TO, Eisenberg D (1999): Detecting protein function and protein–protein interactions from genome sequences. *Science* 285(5428): 751–753.

Martelli PL, Fariselli P, Krogh A, Casadio R (2002): A sequence-profile-based HMM for predicting and discriminating beta barrel membrane proteins. *Bioinformatics* 18 (Suppl 1):S46–S53.

Marti-Renom MA, Madhusudhan MS, Fiser A, Rost B, Sali A (2002): Reliability of assessment of protein structure prediction methods. *Structure* 10:435–440.

Marti-Renom MA, Pieper U, Madhusudhan MS, Rossi A, Eswar N, Davis FP, Al-Shahrour F, Dopazo J, Sali A (2007): DBAli tools: mining the protein structure space. *Nucleic Acids Res*

Maxfield FR, Scheraga HA (1979): Improvements in the prediction of protein topography by reduction of statistical errors. *Biochemistry* 18:697–704.

Möller S, Croning DR, Apweiler R (2001): Evaluation of methods for the prediction of membrane spanning regions. *Bioinformatics* 17(7):646–653.

Monne M, Hermansson M, von Heijne G (1999): A turn propensity scale for transmembrane helices. *J Mol Biol* 288(1):141–145.

Moult J, Pedersen JT, Judson R, Fidelis K (1995): A large-scale experiment to assess protein structure prediction methods. *Prot Struct Funct Genet* 23(3):ii–v.

Moult J, Hubbard T, Bryant SH, Fidelis K, Pedersen JT (1997): Critical assessment of methods of protein structure prediction (CASP): round II. *Prot Struct Funct Genet* (Suppl 1): 2–6.

Moult J, Hubbard T, Bryant SH, Fidelis K, Pedersen JT (1999): Critical assessment of methods of protein structure prediction (CASP): round III. *Prot Struct Funct Genet* (Suppl 3): 2–6.

Moult J, Fidelis K, Zemla A, Hubbard T (2003): Critical assessment of methods of protein structure prediction (CASP): round V. *Prot Struct Funct Genet* 53 (Suppl 6):334–339.

Moult J, Fidelis K, Rost B, Hubbard T, Tramontano A (2005): Critical assessment of methods of protein structure prediction (CASP): round 6. *Proteins* 61(S7):3–7.

Mucchielli-Giorgi MH, Hazout S, Tuffery P (1999): PredAcc: prediction of solvent accessibility. *Bioinformatics* 15(2):176–177.

Muller T, Rahmann S, Rehmsmeier M (2001): Non-symmetric score matrices and the detection of homologous transmembrane proteins. *Bioinformatics* 17(Suppl 1):S182–S189.

Mumenthaler C, Braun W (1995): Predicting the helix packing of globular proteins by self-correcting distance geometry. *Protein Sci* 4:863–871.

Naderi-Manesh H, Sadeghi M, Arab S, Moosavi Movahedi AA (2001): Prediction of protein surface accessibility with information theory. *Prot Struct Funct Genet* 42(4):452–459.

Nair R, Rost B (2003): Better prediction of sub-cellular localization by combining evolutionary and structural information. *Prot Struct Funct Bio* 53(4):917–930.

Nair R, Rost B (2003): LOC3D: annotate sub-cellular localization for protein structures. *Nucleic Acids Res* 31(13):3337–3340.

Nair R, Rost B (2004): LOCnet and LOCtarget: sub-cellular localization for structural genomics targets. *Nucleic Acids Res* 32 (Supplement 2):W517–W521.

Nair R, Rost B (2005): Mimicking cellular sorting improves prediction of subcellular localization. *J Mol Biol* 348(1):85–100.

akai K, Kanehisa M (1992): A knowledge base for predicting protein localization sites in eukaryotic cells. *Genomics* 14:897–911.

Neuwald AF, Liu JS, Lawrence CE (1995): Gibbs motif sampling: detection of bacterial outer membrane protein repeats. *Protein Sci* 4:1618–1631.

Newman JR, Wolf E, Kim PS (2000): A computationally directed screen identifying interacting coiled coils from *Saccharomyces cerevisiae*. *Proc Natl Acad Sci USA* 97(24):13203–13208.

Ng P, Henikoff J, Henikoff S (2000): PHAT: a transmembrane-specific substitution matrix. *Bioinformatics* 16(8):760–766.

Nielsen H, Engelbrecht J, Brunak S, von Heijne G (1997): Identification of prokaryotic and eukaryotic signal peptides and prediction of their cleavage sites. *Protein Engin* 10:1–6.

Nilges M, Brünger AT (1993): Successful prediction of coiled coil geometry of the GCN4 leucine zipper domain by simulated annealing: comparison to the X-ray. *Prot Struct Funct Genet* 15:133–146.

Nimrod G, Glaser F, Steinberg D, Ben-Tal N, Pupko T (2005): *In silico* identification of functional regions in proteins. *Bioinformatics* 21(Suppl 1):i328–i337.

Nishikawa K, Ooi T (1982): Correlation of the amino acid composition of a protein to its structural and biological characteristics. *J Biochem* 91:1821–1824.

Nishikawa K, Ooi T (1986a): Amino acid sequence homology applied to the prediction of protein secondary structure, and joint prediction with existing methods. *Biochim Biophys Acta* 871:45–54.

Nishikawa K, Ooi T (1986b): Radial location of amino acid residues in a globular protein: correlation with the sequence. *J Biochem* 100:1043–1047.

O'Donoghue SI, Nilges M (1997): Tertiary structure prediction using mean-force potentials and internal energy functions: successful prediction for coiled-coil geometries. *Folding & Design* 2(4): S47–S52.

Ofran Y, Punta M, Schneider R, Rost B (2005): Beyond annotation transfer by homology: novel protein function prediction methods that can assist drug discovery. *Drug Disc Today* 10(21): 1475–1482.

Ofran Y, Yachdav G, Mozes E, Soong T-t, Nair R, Rost B (2006): Create and assess protein networks through molecular characteristics of individual proteins. *Bioinformatics* 22(12):e402–e407.

Ofran Y, Rost B (2007a): ISIS: Interaction sites identified from sequence. *Bioinformatics* 23(ECCB): e5–e12.

Ofran Y, Rost B (2007b): Protein–protein interaction hot spots carved into sequences. *PLoS Computat Biol* 3(7):e119.

Oldfield CJ, Cheng Y, Cortese MS, Brown CJ, Uversky VN, Dunker AK (2005a): Comparing and combining predictors of mostly disordered proteins. *Biochemistry* 44(6):1989–2000.

Oldfield CJ, Ulrich EL, Cheng Y, Dunker AK, Markley JL (2005b): Addressing the intrinsic disorder bottleneck in structural proteomics. *Prot Struct Funct Bio* 59(3):444–453.

Ortiz AR, Kolinski A, Rotkiewicz P, Ilkowski B, Skolnick J (1999): *Ab initio* folding of proteins using restraints derived from evolutionary information. *Prot Struct Funct Genet* Suppl 3:177–185.

Ota M, Kawabata T, Kinjo AR, Nishikawa K (1999): Cooperative approach for the protein fold recognition. *Proteins* 37(S3):126–132.

Pal D, Eisenberg D (2005): Inference of protein function from protein structure. *Structure (Camb)* 13 (1):121–130.

Panchenko A, Marchler-Bauer A, Bryant SH (1999): Threading with explicit models for evolutionary conservation of structure and sequence. *Prot Struct Funct Genet* (Suppl 3): 133–140.

Paquet JY, Vinals C, Wouters J, Letesson JJ, Depiereux E (2000): Topology prediction of *Brucella abortus* Omp2b and Omp2a porins after critical assessment of transmembrane beta strands prediction by several secondary structure prediction methods. *J Biomol Struct Dyn* 17(4):747–757.

Pasquier C, Promponas VJ, Palaios GA, Hamodrakas JS, Hamodrakas SJ (1999): A novel method for predicting transmembrane segments in proteins based on a statistical analysis of the SwissProt database: the PRED-TMR algorithm. *Protein Engin* 12(5):381–385.

Pazos F, Sanchez-Pulido L, Garcia-Ranea JA, Andrade MA, Atrian S, Valencia A (1997): Comparative analysis of different methods for the detection of specificity regions in protein families. Paper presented at BCEC97: Bio-Computing and Emergent Computation, 1–2 September 1997, Skövde, Sweden.

Persson B, Argos P (1996): Topology prediction of membrane proteins. *Protein Sci* 5:363–371.

Persson B, Argos P (1997): Prediction of membrane protein topology utilizing multiple sequence alignments. *J Protein Chem* 16(5):453–457.

Phoenix DA, Stanworth A, Harris F (1998): The hydrophobic moment plot and its efficacy in the prediction and classification of membrane interactive proteins and peptides. *Membr Cell Biol* 12 (1):101–110.

Pierleoni A, Martelli PL, Fariselli P, Casadio R (2006): BaCelLo: a balanced subcellular localization predictor. *Bioinformatics* 22(14):e408–e416.

Pilpel Y, Ben-Tal N, Lancet D (1999): kPROT: a knowledge-based scale for the propensity of residue orientation in transmembrane segments. Application to membrane protein structure prediction. *J Mol Biol* 294(4):921–935.

Pollastri G, Baldi P, Fariselli P, Casadio R (2002a): Prediction of coordination number and relative solvent accessibility in proteins. *Proteins* 47(2):142–153.

Pollastri G, Przybylski D, Rost B, Baldi P (2002b): Improving the prediction of protein secondary structure in three and eight classes using recurrent neural networks and profiles. *Prot Struct Funct Bio* 47:228–235.

Pollastri G, McLysaght A (2005): Porter: a new, accurate server for protein secondary structure prediction. *Bioinformatics* 21(8):1719–1720.

Pollastri G, Vullo A, Frasconi P, Baldi P (2006): Modular DAG-RNN architectures for assembling coarse protein structures. *J Comput Biol* 13(3):631–650.

Porollo A, Meller J (2007): Prediction-based fingerprints of protein-protein interactions. *Prot Struct Funct Genet* 66(3):630–645.

Przytycka T, Aurora R, Rose GD (1999): A protein taxonomy based on secondary structure. *Nat Struct Biol* 6(7):672–682.

Przybylski D, Rost B (2002): Alignments grow, secondary structure prediction improves. *Prot Struct Funct Bio* 46:195–205.

Punta M, Rost B (2005): Protein folding rates estimated from contact predictions. *J Mol Biol* 348 (3):507–512.

Punta M, Forrest LR, Bigelow H, Kernytsky A, Liu J, Rost B (2006): Membrane protein prediction methods. *Methods* 41(4):460–474.

Qian N, Sejnowski TJ (1988): Predicting the secondary structure of globular proteins using neural network models. *J Mol Biol* 202:865–884.

Radivojac P, Obradovic Z, Smith DK, Zhu G, Vucetic S, Brown CJ, Lawson JD, Dunker AK (2004): Protein flexibility and intrinsic disorder. *Protein Sci* 13:71–80.

Radivojac P, Vucetic S, O'Connor TR, Uversky VN, Obradovic Z, Dunker AK (2006): Calmodulin signaling: analysis and prediction of a disorder-dependent molecular recognition. *Proteins* 63(2): 398–410.

Robson B (1976): Conformational properties of amino acid residues in globular proteins. *J Mol Biol* 107:327–356.

Romero P, Obradovic Z, Kissinger C, Villafranca JE, Garner E, Guilliot S, Dunker AK (1998): Thousands of proteins likely to have long disordered regions. *Pac Symp Biocomput* 3:437–448.

Romero P, Obradovic Z, Li X, Garner EC, Brown CJ, Dunker AK (2001): Sequence complexity of disordered protein. *Prot Struct Funct Genet* 42(1):38–48.

Rossi A, Marti-Renom MA, Sali A (2006): Localization of binding sites in protein structures by optimization of a composite scoring function. *Protein Sci* 15(10):2366–2380.

Rost B, Sander C (1992): *Exercising Multi-Layered Networks on Protein Secondary Structure. Paper presented at the Neural Networks: From Biology to High Energy Physics, Elba, Italy.*

Rost B, Sander C (1993): Prediction of protein secondary structure at better than 70% accuracy. *J Mol Biol* 232:584–599.

Rost B, Sander C, Schneider R (1993): Progress in protein structure prediction? *TIBS* 18:120–123.

Rost B (1994): Conservation and prediction of solvent accessibility in protein families. *Prot Struct Funct Genet* 20:216–226.

Rost B, Sander C (1994a): Combining evolutionary information and neural networks to predict protein secondary structure. *Prot Struct Funct Genet* 19:55–72.

Rost B, Sander C (1994b): Conservation and prediction of solvent accessibility in protein families. *Prot Struct Funct Genet* 20(3):216–226.

Rost B, Sander C, Schneider R (1994): Redefining the goals of protein secondary structure prediction. *J Mol Biol* 235:13–26.

Rost B, Yachdav G, Liu J (2004): The PredictProtein server. *Nucleic Acids Res* 32 (Supplement 2): W321–W326.

Rost B (1995): *TOPITS: Threading One-dimensional Predictions into Three-dimensional Structures. Paper presented at the Third International Conference on Intelligent Systems for Molecular Biology, Cambridge, England.*

Rost B, Casadio R, Fariselli P, Sander C (1995): Prediction of helical transmembrane segments at 95% accuracy. *Protein Sci* 4:521–533.

Rost B (1996): PHD: predicting one-dimensional protein structure by profile based neural networks. *Method Enzymol* 266:525–539.

Rost B, Sander C (1996): Bridging the protein sequence-structure gap by structure predictions. *Annual Review of Biophysics and Biomolecular Structure* 25:113–136.

Rost B, Casadio R, Fariselli P (1996): Topology prediction for helical transmembrane proteins at 86% accuracy. *Protein Sci* 5:1704–1718.

Rost B, O'Donoghue SI (1997): Sisyphus and prediction of protein structure. *Comput Appl Biol Sci* 13:345–356.

Rost B (1998): Marrying structure and genomics. *Structure* 6:259–263.

Rost B (1999): Twilight zone of protein sequence alignments. *Protein Engin* 12(2):85–94.

Rost B, Sander C (2000): Third generation prediction of secondary structure. *Methods Mol Biol* 143:71–95.

Rost B (2001): Protein secondary structure prediction continues to rise. *J Struct Biol* 134:204–218.

Rost B, Baldi P (2001): *New Improvements In Protein Secondary Structure Prediction. CUBIC_2001_11, Columbia University.*

Rost B, Eyrich V (2001): EVA: large-scale analysis of secondary structure prediction. *Prot Struct Funct Genet* 45 (Suppl 5):S192–S199.

Rost B, Baldi P, Barton G, Cuff J, Eyrich V, Jones D, Karplus K, King R, Ouali M, Pollastri G, Przybylski D (2001): *Simple Jury Predicts Protein Secondary Structure Best. CUBIC_2001_10, Columbia University.*

Rost B, Liu J, Nair R, Wrzeszczynski KO, Ofran Y (2003): Automatic prediction of protein function. *Cell Mol Life Sci* 60(12):2637–2650.

Rost B (2005): How to use protein 1D structure predicted by PROFphd. In: Walker JE, editor. *The Proteomics Protocols Handbook*, Totowa, NJ: Humana, pp. 875–901.

Russell RB, Copley RR, Barton GJ (1996): Protein fold recognition by mapping predicted secondary structures. *J Mol Biol* 259:349–365.

Rychlewski L (2000): *META server.* Warsaw: IIMCB.

Rychlewski L, Fischer D, Elofsson A (2003): LiveBench-6: large-scale automated evaluation of protein structure prediction servers. *Prot Struct Funct Genet* 53 (Suppl 6):542–547.

Rychlewski L, Fischer D (2005): LiveBench-8: the large-scale, continuous assessment of automated protein structure prediction. *Protein Sci* 14(1):240–245.

Sadovskaya NS, Sutormin RA, Gelfand MS (2006): Recognition of transmembrane segments in proteins: review and consistency-based benchmarking of Internet servers. *J Bioinform Comput Biol* 4(5):1033–1056.

Sali A (1998): 100 000 protein structures for the biologist. *Nat Struct Biol* 5(12):1029–1032.

Sammeth M, Heringa J (2006): Global multiple-sequence alignment with repeats. *Proteins* 64(1): 263–274.

Samudrala R, Xia Y, Huang E, Levitt M (1999): *Ab initio* protein structure prediction using a combined hierarchical approach. *Prot Struct Funct Genet*Suppl 3 (3):194–198.

Samudrala R, Huang ES, Koehl P, Levitt M (2000): Constructing side chains on near-native main chains for *ab initio* protein structure prediction. *Protein Engin* 13(7):453–457.

Saqi M (1995): An analysis of structural instances of low complexity sequence segments. *Protein Engin* 8:1069–1073.

Sasson O, Linial M (2005): ProTarget: automatic prediction of protein structure novelty. *Nucleic Acids Res* 33(Web server issue):W81–W84.

Schlessinger A, Yachdav G, Rost B (2006): PROFbval: predict flexible and rigid residues in proteins. *Bioinformatics* 22(7):891–893.

Schlessinger A, Liu J, Rost B (2007a): Natively unstructured loops differ from other loops. *PLoS Comput Biol* 3(7):e140.

Schlessinger A, Punta M, Rost B (2007b): Natively unstructured regions in proteins identified from contact predictions. *Bioinformatics* 23(18):2376–2384.

Schulz GE, Schirmer RH, (1979): Prediction of secondary structure from the amino acid sequence. In *Principles of Protein Structure*, Berlin: Springer-Verlag, pp. 108–130.

Seshadri K, Garemyr R, Wallin E, von Heijne G, Elofsson A (1998): Architecture of beta-barrel membrane proteins: analysis of trimeric porins. *Protein Sci* 7(9):2026–2032.

Seto MH, Liu HL, Zajchowski DA, Whitlow M (1999): Protein fold analysis of the B30.2-like domain. *Prot Struct Funct Genet* 35(2):235–249.

Shah PS, Bizik F, Dukor RK, Qasba PK (2000): Active site studies of bovine alpha1 → 3-galactosyltransferase and its secondary structure prediction. *Biochim Biophys Acta* 1480 (1–2):222–234.

Shapiro L, Harris T (2000): Finding function through structural genomics. *Curr Opin Biotechol* 11 (1):31–35.

Sim J, Kim SY, Lee J (2005): Prediction of protein solvent accessibility using fuzzy k-nearest neighbor method. *Bioinformatics* 21(12):2844–2849.

Simossis VA, Heringa J (2004): Integrating protein secondary structure prediction and multiple sequence alignment. *Curr Protein Pept Sci* 5(4):249–266.

Simossis VA, Heringa J (2005): PRALINE: a multiple sequence alignment toolbox that integrates homology-extended and secondary structure information. *Nucleic Acids Res* 33(Web server issue): W289–W294.

Simossis VA, Kleinjung J, Heringa J (2005): Homology-extended sequence alignment. *Nucleic Acids Res* 33(3):816–824.

Sipos L, von Heijne G (1993): Predicting the topology of eukaryotic membrane proteins. *Eur J Biochem* 213:1333–1340.

Smith RF, Wiese BA, Wojzynski MK, Davison DB, Worley KC (1996): BCM Search Launcher: an integrated interface to molecular biology data base search and analysis services available on the World Wide Web. *Genome Res* 6(5):454–462.

Sonnhammer ELL, von Heijne G, Krogh A (1998): A hidden Markov model for predicting transmembrane helices in protein sequences', paper presented to Sixth International Conference on Intelligent Systems for Molecular Biology (ISMB98), Montreal, Canada.

Stawiski EW, Baucom AE, Lohr SC, Gregoret LM (2000): Predicting protein function from structure: unique structural features of proteases. *Proc Natl Acad Sci USA* 97(8):3954–3958.

Sutormin RA, Rakhmaninova AB, Gelfand MS (2003): BATMAS30: amino acid substitution matrix for alignment of bacterial transporters. *Proteins* 51(1):85–95.

Sweet RM, Eisenberg D (1983): Correlation of sequence hydrophobicities measures similarity in three-dimensional protein structure. *J Mol Biol* 171:479–488.

Tamames J, Ouzounis C, Casari G, Sander C, Valencia A (1998): EUCLID: automatic classification of proteins in functional classes by their database annotations. *Bioinformatics* 14(6):542–543.

Tanford C (1980): *The Hydrophobic Effect: Formation of Micelles and Biological Membranes*. New York: John Wiley & Sons.

Taylor WR, Jones DT, Green NM (1994): A method for α-helical integral membrane protein fold prediction. *Prot Struct Funct Genet* 18:281–294.

Thompson MJ, Goldstein RA (1996): Predicting solvent accessibility: higher accuracy using Bayesian statistics and optimized residue substitution classes. *Prot Struct Funct Genet* 25:38–47.

Thornton J (2001): Structural genomics takes off. *Trends Biochem Sci* 26(2):88–89.

Tompa P (2002): Intrinsically unstructured proteins. *Trends Biochem Sci* 27(10):527–533.

Tompa P (2005): The interplay between structure and function in intrinsically unstructured proteins. *FEBS Letters* 579(15):3346–3354.

Tusnady GE, Simon I (1998): Principles governing amino acid composition of integral membrane proteins: application to topology prediction. *J Mol Biol* 283(2):489–506.

Tusnady GE, Simon I (2001): Topology of membrane proteins. *J Chem Inf Comput Sci* 41(2):364–368.

van Gunsteren WF (1993): Molecular dynamics studies of proteins. *Curr Opin Struct Biol* 3:167–174.

von Heijne G (1986): The distribution of positively charged residues in bacterial inner membrane proteins correlates with the trans-membrane topology. *EMBO J* 5:3021–3027.

von Heijne G (1989): Control of topology and mode of assembly of a polytopic membrane protein by positively charged residues. *Nature* 341:456–458.

von Heijne G (1992): Membrane protein structure prediction. *J Mol Biol* 225:487–494.

von Heijne G (1994): Membrane proteins: from sequence to structure. *Ann Rev Biophys Biomol Struct* 23:167–192.

von Heijne G (1996): Prediction of transmembrane protein topology. In: Sternberg MJE, editor., *Protein Structure Prediction*, 1st edn., Vol. 170: Oxford: Oxford University Press, pp. 101–110.

Vucetic S, Brown CJ, Dunker AK, Obradovic Z (2003): Flavors of protein disorder. *Proteins* 52 (4):573–584.

Vucetic S, Obradovic Z, Vacic V, Radivojac P, Peng K, Iakoucheva LM, Cortese MS, Lawson JD, Brown CJ, Sikes JG, Newton CD, Dunker AK (2005): DisProt: a database of protein disorder. *Bioinformatics* 21(1):137–140.

Vullo A, Bortolami O, Pollastri G, Tosatto SC (2006): Spritz: a server for the prediction of intrinsically disordered regions in protein sequences using kernel machines. *Nucleic Acids Res* 34(Web server issue):W164–W168.

Waldispuhl J, Berger B, Clote P, Steyaert JM (2006): Predicting transmembrane beta-barrels and interstrand residue interactions from sequence. *Proteins* 65(1):61–74.

Wallin E, von Heijne G (1998): Genome-wide analysis of integral membrane proteins from eubacterial, archaean, and eukaryotic organisms. *Protein Sci* 7:1029–1038.

Wang Z-X, Yuan Z (2000): How good is prediction of protein structural class by the component-coupled method? *Prot Struct Funct Genet* 38:165–175.

Ward JJ, Sodhi JS, McGuffin LJ, Buxton BF, Jones DT (2004): Prediction and functional analysis of native disorder in proteins from the three kingdoms of life. *J Mol Biol* 337(3):635–645.

Wodak SJ, Mendez R (2004): Prediction of protein-protein interactions: the CAPRI experiment, its evaluation and implications. *Curr Opin Struct Biol* 14(2):242–249.

Wolf E, Kim PS, Berger B (1997): MultiCoil: a program for predicting two- and three-stranded coiled coils. *Protein Sci* 6(6):1179–1189.

Wootton JC, Federhen S (1996): Analysis of compositionally biased regions in sequence databases. *Method Enzymol* 266:554–571.

Wright PE, Dyson HJ (1999): Intrinsically unstructured proteins: re-assessing the protein structure–function paradigm. *J Mol Biol* 293(2):321–331.

Xu H, Aurora R, Rose GD, White RH (1999): Identifying two ancient enzymes in Archaea using predicted secondary structure alignment. *Nat Struct Biol* 6(8):750–754.

Yao H, Kristensen DM, Mihalek I, Sowa ME, Shaw C, Kimmel M, Kavraki L, Lichtarge O (2003): An accurate, sensitive, and scalable method to identify functional sites in protein structures. *J Mol Biol* 326(1):255–261.

Young M, Kirshenbaum K, Dill KA, Highsmith S (1999): Predicting conformational switches in proteins. *Protein Sci* 8:1752–1764.

Yu CS, Lin CJ, Hwang JK (2004): Predicting subcellular localization of proteins for Gram-negative bacteria by support vector machines based on *n*-peptide compositions. *Protein Sci* 13(5): 1402–1406.

Yu YK, Altschul SF (2005): The construction of amino acid substitution matrices for the comparison of proteins with non-standard compositions. *Bioinformatics* 21(7):902–911.

Zdobnov EM, Lopez R, Apweiler R, Etzold T (2002): The EBI SRS server: recent developments. *Bioinformatics* 18(2):368–373.

Zemla A, Venclovas C, Fidelis K, Rost B (1999): A modified definition of SOV, a segment-based measure for protein secondary structure prediction assessment. *Prot Struct Funct Genet* 34:220–223.

Zemla A, Venclovas C, Fidelis K (2001): *Protein structure prediction center. Lawrence Livermore National Laboratory,<http//PredictionCenterllnlgov/>*.

Zhou H, Zhang C, Liu S, Zhou Y (2005): Web-based toolkits for topology prediction of transmembrane helical proteins, fold recognition, structure and binding scoring, folding-kinetics analysis and comparative analysis of domain combinations. *Nucleic Acids Res* 33(Web server issue): W193–W197.

Zvelebil MJ, Barton GJ, Taylor WR, Sternberg MJE (1987): Prediction of protein secondary structure and active sites using alignment of homologous sequences. *J Mol Biol* 195:957–961.

30

HOMOLOGY MODELING

Hanka Venselaar, Elmar Krieger, and Gert Vriend

INTRODUCTION

The goal of protein modeling is to predict a structure from its sequence with an accuracy that is comparable to the best results achieved experimentally. This would allow users to safely use *in silico* generated protein models in scientific fields where today only experimental structures provide a solid basis: structure-based drug design, analysis of protein function, interactions, antigenic behavior, or rational design of proteins with increased stability or novel functions. Protein modeling is the only way to obtain structural information when experimental techniques fail. Many proteins are simply too large for NMR analysis and cannot be crystallized for X-ray diffraction.

Among the three major approaches to 3D structure prediction described in this and the following two chapters, homology modeling is the "easiest" approach based on two major observations:

- The structure of a protein is uniquely determined by its amino acid sequence (Epstain, Goldberger, and Anfinsen, 1963), and therefore the sequence should, in theory, contain suffice information to obtain the structure.

- During evolution, structural changes are observed to be modified at a much slower rate than sequences. Similar sequences have been found to adopt practically identical structures while distantly related sequences can still fold into similar structures. This relationship was first identified by Chothia and Lesk (1986) and later quantified by Sander and Schneider (1991), as summarized in Figure 30.1. Since the initial establishment of this relationship, Rost et al. were able to derive a more precise limit for this rule with accumulated data in the PDB (Rost, 1999). Two protein sequences are highly likely to adopt a similar structure provided that the percentage identity between these proteins for a given length is above the threshold shown in Figure 30.1.

Structural Bioinformatics, Second Edition   Edited by Jenny Gu and Philip E. Bourne
Copyright © 2009 John Wiley & Sons, Inc.

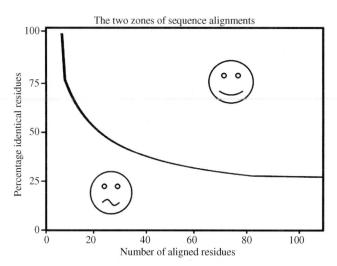

Figure 30.1. The two zones of sequence alignments defining the likelihood of adopting similar structures. Two sequences are highly likely to fold into the same structure if their length and percentage sequence identity fall into the region above the threshold, indicated with the smiling icon (the "safe" zone). The region below the threshold indicates the zone where inference of structural similarity cannot be made, thus making it difficult to determine if model building will be possible. Figure based on Sander and Schneider (1991).

The identity between the sequences can be obtained by doing a simple BLAST run with the sequence of interest, the model sequence, against the PDB, see Figure 30.2. The sequence that aligns with the model sequence is called the "template." When the percentage identity in the aligned region of the structure template and the sequences to be modeled falls in the safe modeling zone of Figure 30.1, a model can be built. It is shown in Figure 30.1 that the threshold for safe homology modeling can be as low as 25%, especially for longer sequences. However, the quality of the model is dependent on the sequence identity and should be considered since sequences with 25% sequence identity tend to yield poor structural models with high uncertainty in side chain positioning.

The impact of identities between the model sequence and the template on the process of protein homology modeling can be broken down as follows. At sequence identities greater than 75%, protein homology modeling can easily be achieved requiring minor manual intervention. Consequently, the time needed to finalize such a model is limited only by the interpretation of the model to answer the biological question at hand, see Figure 30.3. The level of accuracy achieved with these models is comparable to structures that were solved by

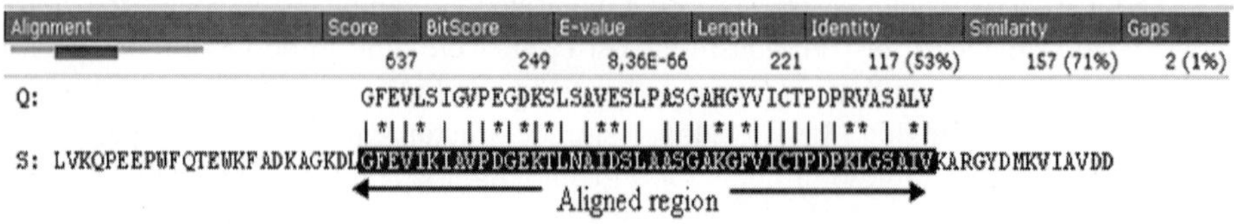

Alignment	Score	BitScore	E-value	Length	Identity	Similarity	Gaps
	637	249	8,36E-66	221	117 (53%)	157 (71%)	2 (1%)

Q: GFEVLSIGVPEGDKSLSAVESLPASGAHGYVICTPDPRVASALV
                        |*||* | ||*|*|*| |**|| ||||*|*|||||||** | *|
S: LVKQPEEPWFQTEWKFADKAGKDLGFEVIKIAVPDGEKTLNAIDSLAASGAKGFVICTPDPKLGSAIVKARGYDMKVIAVDD
                        ◄————————— Aligned region —————————►

Figure 30.2. Typical blast output of a model sequence run against the PDB sequences, where Q represents the model sequence with unknown structure and S is the sequence for the PDB structure template.

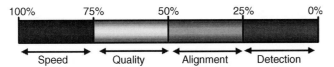

Figure 30.3. The limiting steps in homology modeling as function of percentage sequence identity between the model sequence and the sequence of the template. Figure based on Rodriguez and Vriend (1997).

NMR. For sequence identities between 50% and 75%, more time is needed to fine-tune the details of the model and correct the alignment if needed. Between 25% and 50% identity obtaining the best possible alignment becomes the concern in the process of homology modeling. Finally, sequence identities lower than 25% often mean that no template structure can be detected with a simple BLAST search, thus requiring the use of more sensitive alignment techniques (Chapter 31), to find a potential template structure.

In practice, homology modeling is a multistep process that can be summarized as follows:

1. Template recognition and initial alignment
2. Alignment correction
3. Backbone generation
4. Loop modeling
5. Side chain modeling
6. Model optimization
7. Model validation
8. Iteration

These steps are all illustrated in Figure 30.4 and will be discussed in detail in the rest of this chapter.

Decisions need to be made by the modeler at almost every step of this model building process. The choices are not always obvious, thus subjecting model building to a serious thought about how to gamble between multiple seemingly similar choices. To reduce possible errors introduced by a subjective decision making process, algorithms have been developed to automate the model building process (Table 30.1).

Current techniques allow modelers to construct models for about 25–65% of the amino acids coded in the genome, thereby supplementing the efforts of structural genomics projects (Xiang, 2006). This value differs significantly between individual genomes and increases steadily with the continuous growth of the PDB. The remaining 75–35% of these genomes have no identified template that can be used for homology modeling and therefore modelers will need to resort to fold recognition (Chapter 31), *ab initio* structure prediction (Chapter 32), or simply the traditional NMR or X-ray experiment to obtain structural data (Chapters 4 and 5). While automated model building provides a high-throughput solution, the evaluation of these automated methods during CASP (Chapter 28) indicated that human expertise is still helpful, especially if the sequence identity of the alignment is close to the zone of uncertainty regarding the feasibility of building a proper model (25%, see Figure 30.1) (Fischer et al., 1999). The eight steps of homology modeling will be discussed below in more detail.

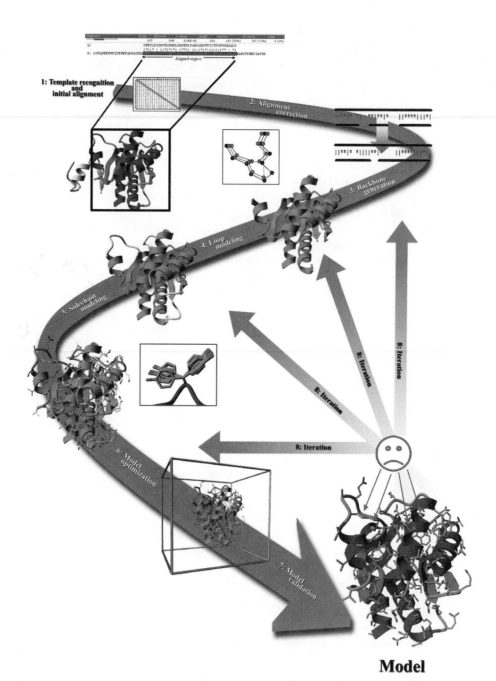

Model

Figure 30.4. The process of building a "model" by homology to a "template." The numbers in the plot correspond to the step numbers in the subsequent section. (Colored version of this Figure is available for viewing at http://swift.cmbi.ru.nl/material/)

T A B L E   3 0 . 1 A Few Examples of the Online Available Homology Modeling Servers

Server Name	URL
Automatic Homology Modeling Servers	
3D-Jigsaw	http://www.bmm.icnet.uk/servers/3djigsaw/
CPHModels	http://www.cbs.dtu.dk/services/CPHmodels/
EsyPred3D	http://www.fundp.ac.be/urbm/bioinfo/esypred/
Robetta	http://robetta.bakerlab.org/
SwissModel	http://swissmodel.expasy.org/
TASSER-lite	http://cssb.biology.gatech.edu/skolnick/webservice/tasserlite/index.html
Semiautomatically Homology Modeling Servers	
HOMER	http://protein.cribi.unipd.it/homer/help.html
WHAT If	http://swift.cmbi.kun.nl/WIWWWI/

STEP 1—TEMPLATE RECOGNITION AND INITIAL ALIGNMENT

Sequences in the safe homology modeling zone (Figure 30.1) share high percentage identity to a possible template and therefore can easily be paired with simple sequence alignment programs such as BLAST (Altschul et al., 1990) or FASTA (Pearson, 1990).

To identify the template, the program compares the query sequence to all the sequences of known structures in the PDB mainly using two matrices:

- *A Residue Exchange Matrix* (Figure 30.5). This matrix defines the likelihood that any 2 of the 20 amino acids ought to be aligned. Exchanges between different residues with similar physicochemical properties (e.g., F → Y) get a better score than exchanges between residues that widely differ in their properties. Conserved residues generally obtain the highest score.

- *An Alignment Matrix* (Figure 30.6). The axes of this matrix correspond to the two sequences to be aligned, and the matrix elements are simply the values from the residue exchange matrix (Figure 30.5) for a given pair of residues. During the

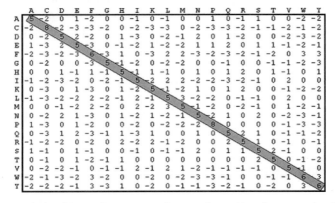

Figure 30.5. A typical residue exchange or scoring matrix used by alignment algorithms. Because the score for aligning residues A and B is normally the same as for B and A, this matrix is symmetric.

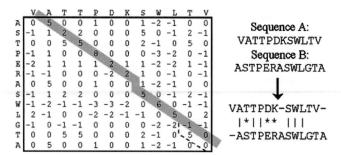

Figure 30.6. The alignment matrix for the sequences VATTPDKSWLTV and ASTPERASWLGTA, using the scores from Figure 30.3. The optimum path corresponding to the alignment on the right side is shown in gray. Residues with similar properties are marked with a star "*". The dashed line marks an alternative alignment that scores more points but requires to open a second gap.

alignment process, one tries to find the best path through this matrix, starting from a point near the top left and going down to the bottom right. To make sure that no residue is used twice, one must always take at least one step to the right and one step down. A typical alignment path is shown in Figure 30.6. At first sight, the dashed path in the bottom right corner would have led to a higher score. However, it requires the opening of an additional gap in sequence A (Gly of sequence B is skipped). By comparing thousands of sequences and sequence families, it became clear that the opening of gaps is about as unlikely as at least a couple of nonidentical residues in a row. The jump roughly in the middle of the matrix on the other hand is justified, because after the jump we earn lots of points (5, 6, 5) that otherwise would only have been (1, 0, 0). The alignment algorithm therefore subtracts an "opening penalty" for every new gap and a much smaller "gap extension penalty" for every residue that is additionally skipped once the gap has already been made. The gap extension penalty is much smaller than the gap open penalty because one gap of three residues is much more likely than three gaps of one residue each.

In practice, template structures can be easily retrieved by submitting the query sequence to one of the countless BLAST servers on the web, using the PDB as the database to search. Usually, the template structure with the highest sequence identity will be the first option, see Figure 30.2, but other considerations should also be made. For example, the conformational state (i.e., active or inactive), present cofactors, other molecules, or multimeric complexes will have an impact on model building. Nowadays, the increasing amount of CPU power makes it possible to choose multiple templates and build multiple models giving the investigator the opportunity to select the best model for further study. It has also become possible to combine multiple templates into one structure that is used for modeling. The online Swiss-Model and the Robetta servers, for example, use this approach (Peitsch, Schwede, and Guex, 2000; Kim, Chivian, and Baker, 2004).

STEP 2—ALIGNMENT CORRECTION

Having identified one or more possible modeling templates using the initial screen described above, more sophisticated methods are needed to arrive at a better alignment.

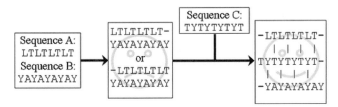

Figure 30.7. Addressing a challenging alignment problem with homologous sequences. Sequences A and B are impossible to align, unless one considers a third sequence C from a homologous protein.

Sometimes it may be difficult to align two sequences in a region where the percentage sequence identity is very low. A potential strategy is to use other sequences from other homologous proteins to find a solution. An example of using this strategy to address a challenging alignment is shown in Figure 30.7. Suppose the sequence LTLTLTLT needs to be aligned with YAYAYAYAY. There are two equally poor possibilities, and the use of third sequence, TYTYTYTYT, that aligns to the two sequences can resolve the issue.

The example mentioned above introduced a very powerful concept called "multiple sequence alignment." Many programs are available to align a number of related sequences, for example, ClustalW (Thompson, Higgins, and Gibson, 1994), and the resulting alignment contains a lot of additional information. Think about an Ala $\rightarrow$ Glu mutation. Relying on the matrix in Figure 30.5, this exchange always gets a score of 1. In the three-dimensional structure of the protein, it is however very unlikely to see such an Ala $\rightarrow$ Glu exchange in the hydrophobic core, but on the surface this mutation is perfectly normal. The multiple sequence alignment implicitly contains information about this structural context. If at a certain position only exchanges between hydrophobic residues are observed, it is highly likely that this residue is buried. To consider this knowledge during the alignment, one uses the multiple sequence alignment to derive position specific scoring matrices, also called "profiles" (Taylor, 1986; Dodge, Schneider, and Sander, 1998). During the past years, new programs such as MUSCLE and T-Coffee have been developed that use these profiles to generate and refine the multiple sequence alignments. (Notredame, Higgins, and Heringa, 2000; Edgar, 2004). Structure-based alignments programs, such as 3DM, also include structural information in combination with profiles to generate multiple sequence alignments (Folkertsma et al., 2004). The use of 3DM on a specific class of proteins can result in entropy versus variability plots. The location of a residue in this plot is directly related to function in the protein. This information can in turn be added to the profile and used to correct the alignment or to optimize position specific gap penalties.

When building a homology model, we are in the fortunate situation of having an almost perfect profile—the known structure of the template. We simply know that a certain alanine sits in the protein core and must therefore not be aligned with a glutamate. Multiple sequence alignments are nevertheless useful in homology modeling, for example, to place deletions or insertions only in areas where the sequences are strongly divergent. A typical example for correcting an alignment with the help of the template is shown in Figures 30.8 and 30.9. Although a sequence alignment gives the highest score for alignment 1 in Figure 30.8, a simple look at the structure of the template reveals that alignment 2 is actually a better alignment, because it leads to a small gap compared to a huge hole associated with alignment 1.

Template	I	C	R	L	P	G	S	A	E	A
1: Model (bad)	V	C	R	M	P	-	-	-	E	A
2: Model (good)	V	C	R	-	-	-	M	P	E	A

Figure 30.8. Example of a sequence alignment where a three-residue deletion must be modeled. While alignment 1, dark gray, appears better when considering just the sequences (a matching proline at position 5), a look at the structure of the template leads to a different conclusion (Figure 30.9).

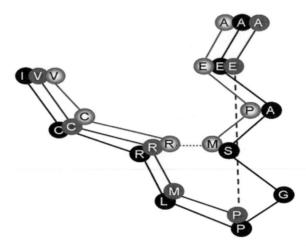

Figure 30.9. Correcting an alignment based on the structure of the modeling template (Cα trace shown in black). While the alignment with the highest score (dark gray, also in Figure 30.8) leads to a big gap in the structure, the second option (light gray) creates only a tiny hole. This can easily be accommodated by small backbone shifts.

STEP 3—BACKBONE GENERATION

When the alignment is ready, the actual model building can start. Creating the backbone is trivial for most of the models: one simply transfers the coordinates of those template residues that show up in the alignment with the model sequence (see Figure 30.2). If two aligned residues differ, the backbone coordinates for N, Cα, C, and O, and often also Cβ can be copied. Conserved residues can be copied completely to provide an initial guess.

Experimentally determined protein structures are not perfect (but still better than models in most cases). There are countless sources of errors, ranging from poor electron density in the X-ray diffraction map to simple human errors when preparing the PDB file for submission. A lot of work has been spent on writing software to detect these errors (correcting them is even harder), and the current count is at more than 25,000,000 errors in the approximately 50,000 structures deposited in the PDB by the end of 2007. The current PDB-redo and RECOORD projects have, respectively, shown that re-refinement of X-ray and NMR structures normally improves the quality of the structure, suggesting that re-refinement before modeling seems to be a wise option (Nederveen, 2005; Joosten, 2007).

A straightforward way to build a good model is to choose the template with the fewest errors (the PDBREPORT database (Hooft et al., 1996) at www.cmbi.ru.nl/gv/pdbreport can be very helpful). But what if two templates are available, and each has a poorly determined region, but these regions are not the same? One should clearly combine the good parts of both templates in one model—an approach known as multiple template modeling. (The same applies if the alignments between the model sequence and possible templates show good matches in different regions.) Although this is simple in principle (and used by automated modeling servers like Swiss-Model (Peitsch et al., 2000)), it is hard in practice to achieve results that are closer to the true structure than all the templates. Nevertheless, the feasibility of this strategy has already been demonstrated by Andrej Šali's group in CASP4 (Chapter 28).

One extreme example of a program that combines multiple templates is the Robetta server. This server uses several different algorithms to predict domains in the sequence. The regions of the model sequence that contain a homologous domain in the PDB are modeled while those parts without are predicted *de novo* by using the Rosetta method. This method compares small fragments of the sequence with the PDB and inserts them with the same local conformation into the model. The Robetta server can generate models of complete sequences even without a known template (Kim et al., 2004). Skolnick's TASSER follows a same strategy but combines larger fragments, found by threading, and folds them into a complete structure (Zhang and Skolnick, 2004).

STEP 4—LOOP MODELING

For the majority of homology model building cases, the alignment between model and template sequence contains gaps, either in the model sequence (deletions as shown in Figures 30.8 and 30.9) or in the template sequence (insertions). Gaps in the model sequence are addressed by simply omitting residues from the template, thus creating a hole in the model that must be closed. Gaps in the template sequences are treated by inserting the missing residues into the continuous backbone. Both cases imply a conformational change of the backbone. The good news is that conformational changes often do not occur within regular secondary structure elements. Therefore, it is often safe to shift all insertions or deletions in the alignment out of helices and strands and place them in structural elements that can accommodate such changes in the alignment such as loops and turns. The bad news is that changes in loop conformation are notoriously hard to predict (a big unsolved problem in homology modeling). To make things worse, even without insertions or deletions a number of different loop conformations can be observed between the template and the model. Three main reasons for this difficulty can be attributed to the following reasons (Rodriguez and Vriend, 1997):

- Surface loops tend to be involved in crystal contacts and therefore a significant conformational change is expected between the template in crystal form and the final structure modeled in the absence of crystallization constraints.
- The exchange of small to bulky side chains underneath the loop induces a structural change by pushing it away from the protein.
- The mutation of a loop residue to proline or from glycine to any other residue will result in a decrease in conformational flexibility. In both cases, the new residue must

fit into a more restricted area in the Ramachandran plot, which normally requires a conformational change of the loop.

There are three main approaches to model loops:

- *Knowledge Based*: A search is made through the PDB for known loops containing end points that match the residues between which the loop is to be inserted. The identified coordinates of the loop are then transferred. All major molecular modeling programs and servers support this approach (e.g., 3D-Jigsaw (Bates and Sternberg, 1999), Insight (Dayringer, Tramontano, and Fletterick, 1986), Modeller (Sali and Blundell, 1993), Swiss-Model (Peitsch et al., 2000), or WHAT IF (Vriend, 1990)).
- *In Between or Hybrid*: The loop is divided in small fragments that are all separately compared to the PDB. This strategy has been described earlier as implemented for the Rosetta method (Kim et al., 2004). The local conformation of all small fragments results in an *ab initio* modeled loop but is still based on known protein structures. This method is reminiscent of the very old ECEPP software by the Scheraga group (Zimmerman et al., 1977).
- *Energy Based*: As in true *ab initio* fold prediction, an energy function is used to judge the quality of a loop. This is followed by a minimalization of the structure, using Monte Carlo (Simons et al., 1999) or molecular dynamics techniques (Fiser, Do, and Sali, 2000) to arrive at the best loop conformation. Often the energy function is modified (e.g., smoothed) to facilitate the search (Tappura, 2001).

For short loops (up to 5–8 residues), the various methods have a reasonable chance of predicting a loop conformation close to the true structure. As mentioned above, surface loops tend to change their conformation due to crystal contacts. So if the prediction is made for an isolated protein and then found to differ from the crystal structure, it might still be correct.

STEP 5—SIDE CHAIN MODELING

When we compare the side chain conformations (rotamers) of residues that are conserved in structurally similar proteins, we find that they often have similar χ_1-angles (i.e., the torsion angle about the C_α–C_β bond). It has been shown by Summers et al. that in homologous proteins (over 40% identity) at least 75% of the $C\gamma$ occupy the same orientation (Summers, Carlson, and Karplus, 1987). It is therefore possible to simply copy conserved residues entirely from the template to the model (see also Step 3) and achieve a very good starting point for structure optimization. In practice, this rule of thumb holds only at high levels of sequence identity, when the conserved residues form networks of contacts. When they get isolated (<35% sequence identity), the rotamers of conserved residues may differ in up to 45% of the cases (Sanchez and Sali, 1997).

In practice, all successful approaches to side chain placement are at least partly knowledge based. Libraries of common rotamers extracted from high-resolution X-ray structures are often used to position side chains. The various rotamers are successively explored and scored with a variety of energy functions. Intuitively, one might expect rotamer prediction to be computationally demanding due to the combinatorial explosion of the search space—the choice of a certain rotamer automatically affects the rotamers of all neighboring residues, which in turn affect their neighbors and the effect propagates continuously. For a sequence of 100 residues and an average $\sim$5 rotamers per residue, the rotamer space would yield 5^{100} different possible conformations to score. Significant research efforts have been invested to develop algorithms to address this issue and make the search through the rotamer conformation space more tractable (e.g., Desmet et al., 1992; Canutescu, Shelenkov, and Dunbrack, 2003).

Aside from directly extracting conserved rotamers from the template, the key to handling the combinatorial explosion of conformational possibilities lies in the protein backbone. Instead of using a "fixed" library that stores all possible rotamers for all residue types, an alternative "position specific" library can be used. These libraries utilize information contained in the backbone to select the correct rotamer. A simple form of a position-specific library classifies the backbone based on secondary structure since residues found in helices often favor a rotamer conformation that is not observed in strands or turns. More sophisticated position-specific libraries can be built by classifying the backbone according to its Phi/Psi angles or by analyzing high-resolution structures and collecting all stretches of 5–9 residues (depending on the method) with reference amino acid at the center of the stretch. These collected examples in the template are superposed on the corresponding backbone to be modeled. The possible side chain conformations are selected from the best backbone matches (Chinea et al., 1995). Since certain backbone conformations will be strongly favored for certain rotamers to be adopted (allowing, for example, a hydrogen bond between side chain and backbone), this strategy greatly reduces the search space. For a given backbone conformation, there may be a residue that strongly populates a specific rotamer conformation and therefore can be modeled immediately. This residue would then serve as an anchor point to model surrounding side chains that may be more flexible and adopt a number of other conformations. An example for a backbone conformation that favors two different tyrosine rotamers is shown in Figure 30.10. Position-specific rotamer libraries are widely used today for drug docking purposes to visualize all possible shapes of the active site (de Filippis, Sander, and Vriend, 1994; Dunbrack and Karplus, 1994; Stites, Meeker, and Shortle, 1994). The study by Chinea et al. shows that the search space is even considerably smaller than assumed by Desmet et al.

Figure 30.10. Example of a backbone-dependent rotamer library. The current backbone conformation favors two different rotamers for tyrosine (shown as sticks), which appear about equally often in the database.

Although search space for rotamer prediction initially presented a combinatorial problem to be explored exhaustively, it is far smaller than originally believed as evidenced in a study conducted in 2001. Xiang and Honig first removed a single side chain from known structures and repredicted it. In a second step, they removed all the side chains and added them again using the same method. Surprisingly, it turned out that the accuracy was only marginally higher in the much easier first case (Xiang and Honig, 2001).

Rotamer prediction accuracy is usually quite high for residues in the hydrophobic core where more than 90% of all χ_1-angles fall within $\pm 20°$ from the experimental values, but much lower for residues on the surface where the percentage is often even below 50%. There are three reasons for this:

- *Experimental Reasons*: Flexible side chains on the surface tend to adopt multiple conformations, which are additionally influenced by crystal contacts. So even experiments cannot provide one single correct answer.
- *Theoretical Reasons*: The energy functions used to score rotamers can easily handle the hydrophobic packing in the core (mainly Van der Waals interactions). The calculation of electrostatic interactions on the surface, including hydrogen bonds with water molecules and associated entropic effects, is more complicated. Nowadays, these calculations are being included in more force fields that are used to optimize the models (Vizcarra and Mayo, 2005).
- *Biological Reasons*: Loops on the surface often adopt different conformations as part of the biological function, for example, to let the substrate enter the protein.

It is important to note that the rotamer prediction accuracies given in most publications cannot be reached in real-life applications. This is simply due to the fact that the methods are evaluated by taking a known structure, removing the side chains, and repredicting them. The algorithms thus rely on the correct backbone, which is not available in homology modeling since the backbone of the template often differs significantly from the model structure. The rotamers must thus be predicted based on a "wrong" backbone, and reported prediction accuracies tend to be higher than what would be expected for the modeled backbone.

STEP 6—MODEL OPTIMIZATION

The problem of rotamer prediction mentioned above leads to a classical "chicken and egg" situation. To predict the side chain rotamers with high accuracy, we need the correct backbone, which in turn depends on the rotamers and their packing. The common approach to address this problem is to iteratively model the rotamers and backbone structure. First, we predict the rotamers, remodel the backbone to accommodate rotamers, followed by a round of refitting the rotamers to the new backbone. This process is repeated until the solution converges. This boils down to a series of rotamer prediction and energy minimization steps. Energy minimization procedures used for loop modeling are applied to the entire protein structure, not just an isolated loop. This requires an enormous accuracy in the energy functions used, because there are many more paths leading away from the answer (the model structure) than toward it. This is why energy minimization must be used carefully. At every minimization step, a few big errors (like bumps, i.e., too short atomic distances) are removed while at the same time many small errors are introduced. When the big errors are gone, the

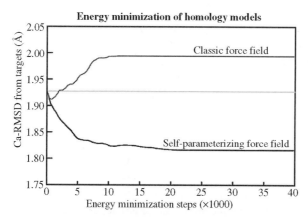

Figure 30.11. The average RMSD between models and the real structures during an extensive energy minimization of 14 homology models with two different force fields. Both force fields improve the models during the first ~500 energy minimization steps, but then the small errors sum up in the classic force field and guide the minimization in the wrong direction, away from the real structure, while the self-parameterizing force field goes in the right direction. To reach experimental accuracy, the minimization would have to proceed all the way down to ~0.5 Å, which is the uncertainty in experimentally determined coordinates.

small ones start accumulating and the model moves away from its correct structure, see Figure 30.11.

As a rule of thumb, today's modeling programs therefore either restrain the atom positions and/or apply only a few hundred steps of energy minimization. In short, model optimization does not work until energy functions (force fields) become more accurate. Two ways to achieve better energetic models for energy minimization are currently being pursued:

- *Quantum Force Fields*: Protein force fields must be fast to handle these large molecules efficiently; energies are therefore normally expressed as a function of the positions of the atomic nuclei only. The continuous increase of computer power has now finally made it possible to apply methods of quantum chemistry to entire proteins, arriving at more accurate descriptions of the charge distribution (Liu et al., 2001). It is however still difficult to overcome the inherent approximations of today's quantum chemical calculations. Attractive Van der Waals forces are, for example, so hard to treat, that they must often be completely omitted. While providing more accurate electrostatics, the overall accuracy achieved is still about the same as in the classical force fields.

- *Self-Parameterizing Force Fields*: The accuracy of a force field depends to a large extent on its parameters (e.g., Van der Waals radii, atomic charges). These parameters are usually obtained from quantum chemical calculations on small molecules and fitting to experimental data, following elaborate rules (Wang, Cieplak, and Kollman, 2000). By applying the force field to proteins, one implicitly assumes that a peptide chain is just the sum of its individual small molecule building blocks—the amino acids. Alternatively, one can just state a goal—for example, improve the models during an energy minimization—and then let the force field parameterize itself while trying to optimally fulfill this goal (Krieger, Koraimann, and Vriend, 2002; Krieger et al., 2004). This leads to a computationally rather expensive procedure. Take initial parameters (e.g., from an existing force field), change a parameter randomly, energy minimize models, see if the result improved, keep the new force field if yes, otherwise go back to the previous force field. With this procedure, the force field accuracy increases enough to go in the right direction during an energy minimization (Figure 30.11), but experimental accuracy is still far out of reach.

The most straightforward approach to model optimization is to simply run a molecular dynamics simulation of the model. Such a simulation follows the motions of the protein on a femtosecond (10^{-15} s) time scale and mimics the true folding process. One thus hopes that the model will complete its folding and converges to the true structure during the simulation. The advantage is that a molecular dynamics simulation implicitly contains entropic effects that are otherwise hard to treat; the disadvantage is that the force fields are again not accurate enough to make it work.

Different distributed computing projects, including folding@home (http://folding. stanford.edu), Rosetta@home (http://boinc.bakerlab.org/rosetta/), and Models@home (http://www.cmbi.kun.nl/models), have been developed to use many personal computers in a network to run molecular dynamics simulations and to mimic protein folding. Model optimization becomes more and more important. Even the focus of the CASP competition (Chapter 28) is changing to a comparison of the optimization of initial models provided by these online servers instead of building the initial models from scratch.

STEP 7—MODEL VALIDATION

Every protein structure contains errors, and homology models are no exception. The number of errors (for a given method) mainly depends on two values:

- *The Percentage Sequence Identity Between Template and Model Sequence.* If the identity is greater than 90%, the accuracy of the model can be compared to crystallographically determined structures, except for a few individual side chains (Chothia and Lesk, 1986; Sippl, 1993). From 50% to 90% identity, the root mean square error in the modeled coordinates can be as large as 1.5 Å, with considerably larger local errors. If the sequence identity drops to 25%, the alignment turns out to be the main bottleneck for homology modeling, often leading to very large errors rendering it a meaningless effort to model these regions, see also Figure 30.3.
- *The Number of Errors in the Template.* Errors in the template structure can be reduced by an additional re-refinement of the template structure as mentioned earlier. Errors in a model become less of a problem if they can be localized. For example, a loop located distantly from a functionally important region such as the active site of an enzyme can tolerate some inaccuracies. Nevertheless, it is in the best interest to model all regions as accurately as possible since seemingly unimportant residues may be important for protein–protein interactions or a yet unassigned function.

There are two principally different ways to estimate errors in a structure:

(a) *Calculating the Model's Energy Based on a Force Field.* This checks if the bond lengths and bond angles are within normal ranges, and if there are lots of clashing side chains in the model (corresponding to a high Van der Waals energy). Essential questions like "Is the model folded correctly?" cannot yet be answered this way, because completely misfolded but well-minimized models often reach the same force field energy as the correct structure (Novotny, Rashin, and Bruccoleri, 1988). This is mainly due to the fact that molecular dynamics force fields are lacking several contributing terms to the energy function, most notably those related to solvation effects and entropy.

(b) *Determination of Normality Indices* that describe how well a given characteristic of the model resembles the same characteristic in real structures. Many features of protein structures are well suited for normality analysis. Most of them are directly or indirectly based on the analysis of interatomic distances and contacts. Some published examples are the following:

- General checks for the normality of bond lengths, bond, and torsion angles (Morris et al., 1992; Czaplewski et al., 2000) are good checks for the quality of experimentally determined structures, but are less suitable for the evaluation of models because the better model building programs simply do not make this kind of errors.

- Inside/outside distributions of polar and apolar residues can be used to detect completely misfolded models (Baumann, Frommel, and Sander, 1989).

- The radial distribution function for a given type of atom (i.e., the probability to find certain other atoms at a given distance) can be extracted from the library of known structures and converted into an energy-like quantity called a "potential of mean force" (Sippl, 1990). Such a potential can easily distinguish good contacts (e.g., between a $C\gamma$ of valine and a $C\delta$ of isoleucine) from bad ones (e.g., between the same $C\gamma$ of valine and the positively charged amino group of lysine).

- The direction of atomic contacts can also be accounted for in addition to interatomic distances. The result is a three-dimensional distribution of functions that can also easily identify misfolded proteins and are good indicators of local model building problems (Vriend and Sander, 1993).

Most methods used for the verification of models can also be applied to experimental structures (and hence to the templates used for model building). A detailed verification is essential when trying to derive new information from the model, either to interpret or predict experimental results or to plan new experiments.

STEP 8—ITERATION

When errors in the model are recognized and located, they can be corrected by iterating portions of the homology modeling process. Small errors that are introduced during the optimization step can be removed by running a shorter molecular dynamics simulation. A error in a loop can be corrected by choosing another loop conformation in the loop modeling step. Large mistakes in the backbone conformation sometimes require the complete process to be repeated with another alignment or even with a different template.

In summary, it is safe to say that homology modeling is unfortunately not as easy as stated in the beginning. Ideally, homology modeling also uses threading (Chapter 31) to improve the alignment, *ab initio* folding (Chapter 32) to predict the loops, and molecular dynamics simulations with a perfect energy function to converge into the true structure. Doing all this correctly will keep researchers busy for a long time, leaving lots of fascinating discoveries.

ACKNOWLEDGMENTS

We thank Rolando Rodriguez, Chris Spronk, Sander Nabuurs, Robbie Joosten, Maarten Hekkelman, David Jones, and Rob Hooft for stimulating discussions, practical help, and critically reading the document. We apologize to the numerous crystallographers who made all this work possible by depositing structures in the PDB for not referring to each of the 50,000 very important articles describing these structures.

REFERENCES

Altschul SF, Gish W, Miller W, Myers EW, Lipman DJ (1990): Basic local alignment search tool. *J Mol Biol* 215:403–410.

Bates PA, Sternberg MJE (1999): Model building by comparison at CASP3: using expert knowledge and computer automation. *Proteins* (Suppl. 3):47–54.

Baumann G, Frommel C, Sander C (1989): Polarity as a criterion in protein design. *Protein Eng* 2:329–334.

Canutescu AA, Shelenkov AA, Dunbrack RL (2003): A graph-theory algorithm for rapid protein side-chain prediction. *Protein Sci* 12(9):2001–2014.

Chinea G, Padron G, Hooft RWW, Sander C, Vriend G (1995): The use of position specific rotamers in model building by homology. *Proteins* 23:415–421.

Chothia C, Lesk AM (1986): The relation between the divergence of sequence and structure in proteins. *EMBO J* 5:823–836.

Czaplewski C, Rodziewicz-Motowidlo S, Liwo A, Ripoll DR, Wawak RJ, Scheraga HA (2000): Molecular simulation study of cooperativity in hydrophobic association. *Protein Sci* 9:1235–1245.

Dayringer HE, Tramontano A, Fletterick RJ (1986): Interactive program for visualization and modelling of proteins, nucleic acids and small molecules. *J Mol Graph* 4:82–87.

de Filippis V, Sander C, Vriend G (1994): Predicting local structural changes that result from point mutations. *Protein Eng* 7:1203–1208.

Desmet J, De Maeyer M, Hazes B, Lasters I (1992): The dead-end elimination theorem and its use in protein side-chain positioning. *Nature* 356:539–542.

Dodge C, Schneider R, Sander C (1998): The HSSP database of protein structure–sequence alignments and family profiles. *Nucleic Acids Res* 26:313–315.

Dunbrack RLJR, Karplus M (1994): Conformational analysis of the backbone dependent rotamer preferences of protein side chains. *Nat Struct Biol* 5:334–340.

Edgar RC (2004): MUSCLE: multiple sequence alignment with high accuracy and high throughput. *Nucleic Acids Res* 5(32):1792–1797.

Epstain CJ, Goldberger RF, Anfinsen CB (1963): Cold Spring Harbor Symposium on Quantum Biology, Vol. 28, p 439.

Fischer D, Barret C, Bryson K, Elofsson A, Godzik A, Jones D, Karplus KJ, Kelley LA, MacCallum RM, Pawowski K, Rost B, Rychlewski L, Sternberg MJE (1999): CAFASP-1: critical assessment of fully automated structure prediction methods. *Proteins* (Suppl. 3): 209–217.

Fiser A, Do RK, Sali A (2000): Modeling of loops in protein structures. *Protein Sci* 9:1753–1773.

Folkertsma S, van Noort P, van Durme J, Joosten H, Bettler E, Fleuren W, Oliveira L, Horn F, de Vlieg J, Vriend G (2004): A family based approach reveals the function of residues in the nuclear receptor ligand-binding domain. *J Mol Biol* 341:321–336.

Hooft RWW, Vriend G, Sander C, Abola EE (1996): Errors in protein structures. *Nature* 381:272–272.

Joosten RP (2007): PDB improvement starts with data deposition. *Science* 317(5835):195–196.

Kim DE, Chivian D, Baker D (2004): Protein structure prediction and analysis using the Robetta server. *Nucleic Acids Res* 32:W526–W531.

Krieger E, Koraimann G, Vriend G (2002): Increasing the precision of comparative models with YASARA NOVA—a self-parameterizing force field. *Proteins* 47(3):393–402.

Krieger E, Darden T, Nabuurs SB, Finkelstein A, Vriend G (2004): Making optimal use of empirical energy functions: force-field parameterization in crystal space. *Proteins* 57(4):678–683.

Liu H, Elstner M, Kaxiras E, Frauenheim T, Hermans J, Yang W (2001): Quantum mechanics simulation of protein dynamics on long timescale. *Proteins* 44:484–489.

Morris AL, MacArthur MW, Hutchinson EG, Thorton JM (1992): Stereochemical quality of protein structure coordinates. *Proteins* 12:345–364.

Nederveen AJ, Doreleijers JF, Vranken W, Miller Z, Spronk CA, Nabuurs SB, Guntert P, Livny M, Markley JL, Nilges M, et al. (2005): RECOORD: a recalculated coordinate database of 500+ proteins from the PDB using restraints from the BioMagResBank. *Proteins* 59, 662–672.

Notredame C, Higgins DG, Heringa J (2000): T-Coffee: a novel method for fast and accurate multiple sequence alignment. *J Mol Biol* 302:205–217.

Novotny J, Rashin AA, Bruccoleri RE (1988): Criteria that discriminate between native proteins and incorrectly folded models. *Proteins* 4:19–30.

Pearson WR (1990): Rapid and sensitive sequence comparison with FASTP and FASTA. *Methods Enzymol* 183:63–98.

Peitsch MC, Schwede T, Guex N (2000): Automated protein modelling—the proteome in 3D. *Pharmacogenomics* 1:257–266.

Rodriguez R, Vriend G (1997): Professional gambling. Proceedings of the NATO Advanced Study Institute on Biomolecular Structure and Dynamics: Recent Experimental and Theoretical Advances.

Rost B (1999): Twilight zone of protein sequence alignments. *Protein Eng* 12:85–94.

Sali A, Blundell TL (1993): Comparative protein modelling by satisfaction of spatial restraints. *J Mol Biol* 234:779–815.

Sanchez R, Sali A (1997): Evaluation of comparative protein structure modeling by MODELLER-3. *Proteins* (Suppl. 1):50–58.

Sanchez R, Sali A (1999): ModBase: a database of comparative protein structure models. *Bioinformatics* 15:1060–1061.

Sander C, Schneider R (1991): Database of homology-derived protein structures and the structural meaning of sequence alignment. *Proteins* 9:56–68.

Simons KT, Bonneau R, Ruczinski I, Baker D (1999): *Ab initio* structure prediction of CASP III targets using ROSETTA. *Proteins* (Suppl. 3):171–176.

Sippl MJ (1990): Calculation of conformational ensembles from potentials of mean force. *J Mol Biol* 213:859–883.

Sippl MJ (1993): Recognition of errors in three dimensional structures of proteins. *Proteins* 17:355–362.

Stites WE, Meeker AK, Shortle D (1994): Evidence for strained interactions between side-chains and the polypeptide backbone. *J Mol Biol* 235:27–32.

Summers NL, Carlson WD, Karplus M (1987): Analysis of side-chain orientations in homologous proteins. *J Mol Biol* 196:175–198.

Tappura K (2001): Influence of rotational energy barriers to the conformational search of protein loops in molecular dynamics and ranking the conformations. *Proteins* 44:167–179.

Taylor WR (1986): Identification of protein sequence homology by consensus template alignment. *J Mol Biol* 188:233–258.

Thompson JD, Higgins DG, Gibson TJ (1994): ClustalW: improving the sensitivity of progressive multiple sequence alignments through sequence weighting, position-specific gap penalties and weight matrix choice. *Nucleic Acids Res* 22:4673–4680.

Vizcarra CL, Mayo SL (2005): Electrostatics in computational protein design. *Curr Opin Chem Biol* 9(6):622–626.

Vriend G (1990): WHAT IF—a molecular modeling and drug design program. *J Mol Graph* 8:52–56.

Vriend G, Sander C (1993): Quality control of protein models: directional atomic contact analysis. *J Appl Crystallogr* 26:47–60.

Wang J, Cieplak P, Kollman PA (2000): How well does a restrained electrostatic potential (RESP) model perform in calculating conformational energies of organic and biological molecules? *J Comp Chem* 21:1049–1074.

Xiang Z, Honig B (2001): Extending the accuracy limits of prediction for side-chain conformations. *J Mol Biol* 311:421–430.

Xiang Z (2006): Advances in homology protein structure modeling. *Curr Protein Pept Sci* 7(3): 217–227.

Zhang Y, Skolnick J (2004): Tertiary structure predictions on a comprehensive benchmark of medium to large size proteins. *Biophys J* 87(4):2647–2655.

Zimmerman SS, Pottle MS, Némethy G, Scheraga HA (1977): Conformational analysis of the 20 naturally occurring amino acid residues using ECEPP. *Macromolecules* 10(1):1–9.

31

FOLD RECOGNITION METHODS

Adam Godzik

INTRODUCTION

Despite a good qualitative understanding of the forces that shape the folding process, present knowledge is not enough to be used for direct prediction of protein structure from first principles, such as the fundamental equations of physics. A related but easier problem is to recognize which of the known protein folds is likely to be similar to the (unknown) fold of a new protein when only its amino acid sequence is known. This has been variously called an inverse folding problem (find a sequence fitting a structure), threading (since a sequence is being threaded through a known structure), and finally a fold recognition problem. Solution of the fold recognition problem is a necessary prerequisite to the solution of the general folding problem. If we are unable to recognize a structure similar to the correct one, how could we possibly arrive at the correct structure starting for a random one? At the same time, even the partial solution to the fold recognition problem offers the immediate advantage of an efficient and fast structure prediction tool.

Such considerations lead to the development of methods, which attempt to recognize possible structural similarities even in the absence of recognizable sequence similarity. Of course, advances in sequence analysis are constantly changing the threshold between fold recognition and sequence-based homology recognition. Therefore, in this contribution, both sequence and structure/energy-based fold recognition methods will be discussed together.

The practical importance of the fold recognition approach to protein structure prediction stems from the fact that very often apparently unrelated proteins adopt similar folds. As discussed elsewhere in this book, more than half of the newly solved proteins thought to be unrelated to any of the known proteins turn out to have a well-known fold. Only a few years ago, such cases were viewed as a curiosity, but now they become almost a rule. Some folds such as a beta barrel TIM fold, were discovered in over 20 protein superfamilies thought to be unrelated. Other popular folds often found in apparently unrelated protein families are Greek-key beta barrels, which are found in immunoglobulins, copper binding proteins,

Structural Bioinformatics, Second Edition   Edited by Jenny Gu and Philip E. Bourne
Copyright © 2009 John Wiley & Sons, Inc.

several families of receptors, adhesion molecules, and so on. Similarly, the ferrodoxin fold is present in 36 superfamilies (SCOP, 1995). Overall, only 100 folds account for about half of all protein superfamilies in one of the more popular protein structure classifications, the SCOP database (discussed in detail in Chapter 12). There are several possible explanations for this phenomenon and it is almost certain that some examples may be found for each:

Divergent Evolution: Proteins with similar folds are actually related, but our current sequence analysis tools are not sensitive enough to recognize very distant homologies. Several new algorithms, such as PSI-BLAST (Altschul et al., 1997), hidden Markov models (HMM) (Bateman et al., 2000), and profile–profile alignment tools (Rychlewski et al., 2000) redefined sequence similarity. Many protein families that were thought to be unrelated a few years ago are now firmly in the distant homology class. With the continuous improvement of such algorithms, we can expect that a majority of cases of "unexplained structural similarity" will eventually be included in this class.

Convergent Evolution: Common functional requirements, such as binding to the same classes of substrates, lead to similar structural solutions. There are very few undisputed examples of convergent evolution and most involve similarities of small subfragments, such as active site residues (serine proteases) or binding patterns (DNA binding proteins).

Limited Number of Folds: Unrelated proteins end up having similar folds because the space of possible folds is small and Nature is simply running out of solutions. Despite strong theoretical arguments (Ptitsyn and Finkelstein, 1980), there are not many examples for such "accidentally" similar structures except for very small proteins, such as three and four helical bundles. In fact, many theoretically predicted arrangements of secondary structure elements (Chothia and Finkelstein, 1990) are still not seen in Nature despite rapid growth of known protein structures.

Misguided Analysis: Apparent structural similarity may result from deficiencies in our analysis tools and not from any actual similarity between protein structures. For instance, SCOP, CATH, and FSSP structure classification agree in only about 60% of cases (Hadley and Jones, 1999). Detection of structural similarity is usually based on empirical criteria, most often on fitting similarity score to an empirical distribution of random similarity scores. Therefore, a false positive is possible when two structures are very different from all other structures, but not really similar. Of course, predicting such similarities is not really feasible, and also not particularly useful.

A correct choice between these possibilities is fundamental because it not only influences our thinking about the protein sequence/structure/function relationship, but also indicates the most efficient structure prediction strategies. Even more important, the first two possibilities suggest that there could be functional similarities between proteins with similar structures, either due to the evolutionary relationship between proteins or due to convergent evolution. This in turn increases the practical importance of fold recognition that can be now viewed not only as a structure prediction, but also a function prediction tool.

Closer analysis suggests that the first mechanism is responsible for most of the known examples of "unexpected" structural similarity. Often the homology was postulated only after a structural similarity was discovered (Taylor, 1986; Russell and Barton, 1992), but later accepted based on other similarities, including similarities in function. The case of the "enolase superfamily" (Babbitt et al., 1995) is particularly interesting because it illustrates the practical importance of establishing a homology relationship between protein families.

In this case, distant homology between enolase and mandelate racemase was postulated based on extensive structural similarities between both enzymes, and despite the lack of (then) recognizable sequence similarity and significant differences between their biochemical function. This hypothesis led to the reevaluation of the enzymatic mechanisms of both proteins and the discovery that a crucial step in two different reactions catalyzed by these two proteins is highly similar, involving the abstraction of the α-proton of a carboxylic acid to form an enolic intermediate. This firmly established the homology of both enzymes and added to our understanding of two different enzymatic reactions, opening a new venue in designing inhibitors for both enzymes. At the same time, it made possible the structure and function prediction for several newly discovered enzymes. At the time of this discovery, fold recognition was in its infancy and none of the then available algorithms was able to recognize a homology that distant, now this would be considered a medium difficulty problem.

Our current understanding of the evolution at the molecular level is good enough to describe the process of small changes and adaptations in proteins. This understanding allows us to reliably assess relationships between proteins when their sequences are similar. It also means that in proteins for which reliable evolutionary relationships can be established, functions and structures had no time to diverge dramatically. At the same time, there is no general consensus about how new protein folds and completely new functions have emerged. Consequently, it is not clear how to study relations between proteins when there is no clear similarity between their sequences and only other arguments suggest that they might be related. The confusion extends to the nomenclature used to describe specific cases. Terms such as superfamily, fold family, and so on are used by different groups in different contexts (Doolittle, 1994). Authors variously claim distant evolutionary relationships (Farber and Petsko, 1990), convergent structural evolution (Lesk, 1995), or random similarities (Ptitsyn and Finkelstein, 1980) in seemingly similar cases.

Fold recognition can be successful in each of the first three scenarios discussed above, but it is the first one that makes it particularly useful. Protein structure prediction is rarely a goal in itself and the real questions usually concern possible functions of new proteins. If the predicted fold comes from a homologous (even distant) protein, there is a good chance that some aspects of function could also be conserved. More recent analyses of fold families suggest that usually the arrangement of active site residues, identity of cofactors, and general features of the reaction being catalyzed are often conserved for enzymes sharing the same fold (Todd, Orengo, and Thornton, 2001). The possibility of predicting even some aspects of function for new proteins adds practical importance to fold recognition, despite its humble origins as a poor man version of the protein folding problem.

THEORETICAL BACKGROUND FOR FOLD RECOGNITION

Two different perspectives dominate theoretical studies of proteins and as a result, there are two different classes of fold recognition algorithms. Roughly, they could be called biological and physical because to some extent they embody the research philosophies of these two disciplines. A biologist strives to explain the natural world in terms of patterns of evolution. In a tradition that reaches to Aristotle and Carolus Linnaeus, biologists identify, describe, and classify the diversity of life to reveal the patterns of evolution. Following in this spirit, molecular biologists study proteins very much like eighteenth and nineteenth century naturalists studied plant and animal species. Proteins are described and classified into families and analyzed for patterns of mutations at various positions along the sequence.

The sequence similarity between proteins from different species forms the basis of molecular phylogenetics, which now rivals traditional morphological phylogenetics in terms of analysis of the relationships between species. It is often extended to study relations between entire processes, such as regulatory networks and metabolic pathways or between sub-populations within species. In contrast, a physicist seeks to explain Nature in terms of fundamental laws, with similarities between systems being just manifestations of the same laws at work. In this spirit, protein structure is seen as a complex shape defined by specific interactions between amino acids along the chain. Different sequences adopting similar folds are viewed as multiple solutions to the same minimization problem and fold recognition problem can be formulated as a constrained minimization where only some points in conformational space are being considered.

There are many problems that are easily understood and studied within one approach. For closely related proteins, it is possible to build reliable evolutionary trees and analyze the relationships between the organisms from which they came without any reference to the fact that these proteins fold to the same structure adopting a free energy minimum in solution. For studying the short-time dynamics of side chain movements in proteins or for predicting the results of a point mutation, we may not care about how this enzyme evolved. But many problems require insights from both perspectives. Why do some proteins have very similar structures and yet perform very different functions? Why do others have similar functions but their structures are different? How to design a good drug that would bind to the target even if the target would undergo a mutation? How can we predict when and how function would diverge between distantly related proteins?

In this contribution, we show how these two views of the protein lead to the development of two distinct classes of fold recognition algorithms. We discuss how the further progress of this and other areas of protein structure prediction relies on successfully merging these two approaches. The first part discusses the molecular evolution of proteins and how understanding of this process lead to the development of more sensitive algorithms for comparison of protein sequences. The second part presents a "physicist" view of the protein world, concentrating on ideas of energy, potentials of mean force, and free energy and describes prediction methods that use this language to recognize a possible fold of a new sequence.

In most of this chapter, we would study questions of similarities and differences between proteins and the question of relations between them. Therefore, the term "homology" will be used only to denote an implied evolutionary relationship between proteins. In contrast, the term "analogy" or "similarity" will be used to describe the similarity between two sequences or structures without implying any relationship between them.

PROTEINS AS SEEN BY A BIOLOGIST

Molecular Evolution, Sequence Similarity, and Protein Homology

Since protein sequences first became available, researchers realized that sequences of homologous proteins from related organisms are very similar. The difference grows with an increasing evolutionary distance between species, which corresponds very well to the known mechanisms of DNA replication and repair. This observation led to the development of sequence alignment methods (Doolittle et al., 1996), which attempt to find an optimal alignment between the sequences of two (or more) proteins being compared. Since homologous proteins by definition evolved from a common ancestor, we expect that a series of mutations, deletions, and insertions lead from the common ancestor to both modern sequences. A list of such elementary steps is equivalent to a unique alignment. When possible homology between two proteins is considered, the null hypothesis is that the two sequences do not come from the common ancestor. The comparison of their sequences should be similar to that of a comparison between two random strings of letters. Therefore, if a larger than random similarity is found, it is generally assumed that such proteins are homologous. This connection is so strong that sequence similarity is *de facto* used as a synonym of homology, despite the fact that homology is a much stronger concept and the two are not equivalent. At the same time, two proteins may be homologous despite lack of an easily recognizable sequence similarity.

Once the homology is established, we can predict the structure and function of new proteins reasoning by analogy assuming that in evolution at least some aspects of function are conserved. The rule that strong sequence similarity is equivalent to strong structure and function similarity is the only reliable prediction rule discovered so far. The homology modeling discussed in Chapter 25 turned this general rule into a powerful prediction method. Over the past 40 years, the efforts of X-ray crystallographers and, more recently, NMR-spectroscopists yielded thousands of protein structures, and biochemists have characterized tens of thousands of proteins. These proteins, with their sequences and/or structures available in public databases, form a rich source of knowledge that can be used to identify newly sequenced proteins. To apply analogy reasoning, two problems must be solved. First, the similarity must be recognized, which for distant homologues may not be trivial. Second, the detailed alignment, that is, residue-by-residue equivalence table between the two proteins must be constructed. Interestingly, the former problem turns out to be much harder that the first (Jaroszewski, Rychlewski, and Godzik, 2000).

Protein Sequence Analysis

Sequence comparison is a well-developed scientific field (Doolittle et al., 1990; Gribskov and Dereveux, 1991; Waterman, 1995; Doolittle et al., 1996). With rigorous mathematical techniques similar to those in telecommunication signal analysis, two protein sequences are treated as two strings of characters and the similarity between them is compared to that expected by chance between random strings. If it is larger than that expected by chance, common ancestry is assumed and both proteins are identified as homologous, with subsequent assertions concerning their structure and function.

The similarity between two sequence strings is usually defined as the sum of similarities between residues in both proteins at equivalent positions. In the example illustrated in Figure 31.1, identical residues are denoted by vertical bars. A scoring matrix, giving a numerical score for aligning any two amino acids, defines a similarity. The similarity score ranges from large and positive (for the same residue in both positions) to smaller and positive (for residues with similar features, such as valine and leucine) to large and negative (for very different residues). Scoring based on minimizing the difference between two protein sequences is also possible.

$$S = \sum_{i}^{n} \mathrm{MM}(A_i, B_{AB(i)}),$$

(31.1)

```
1 : PLGEAALKGPMMKKEQAYSLTFTEAGTYDYHITPHP--EFMRGKVVV
            |      | || | || ||||              |
2 : GAEK--FKSKINE---NYVLTVTQPGAYLVKITPHYAMGMIALIAVG
```

Figure 31.1. An example of a sequence alignment between two proteins.

where MM is the mutation matrix and $AB(i)$ is the residue equivalent to i in sequence B. Many different similarity matrices are used in literature with several of the best quite similar to each other despite different assumptions used in their derivation (Frishman and Argos, 1996; Tomii and Kanehisa, 1996). An important feature of the scoring function such as in (Eq. 31.1) is that it is additive or local; in other words, score for one position does not depend on the alignment (or on residues) in another position.

Gaps and insertions in both sequences, such as seen in Figure 31.1, are necessary for the optimal alignment. This is in full agreement with our knowledge of evolution at the molecular level, where mutations as well as deletions and insertions in DNA sequences are possible. The optimal alignment with gaps can be found by dynamic programming (Needelman and Wunsch, 1970; Smith and Waterman, 1981). Alternatively, similar sequences can be identified by searching for high-scoring fragments (HSF) (Altschul et al., 1990), the uninterrupted alignment fragments, presented as highlighted boxes in Figure 31.2. Software tools, such as BLAST (Altschul et al., 1990) (now updated to PSI-BLAST (Altschul et al., 1997)) or FASTA (Pearson and Miller, 1992) became standards in searching for similar sequences in protein databases and are easily available as software packages (Group, 1991) or Web servers. Other sets of tools, such as CLUSTALW (Higgins and Gibson, 1995) or PileUp (Group, 1991), address questions of organizing a family of homologous proteins into a family tree.

Unfortunately, despite their solid theoretical foundations, all methods and algorithms used in sequence analysis face the same problem. With increasing evolutionary distance, sequence similarity between homologous proteins fades. Using simple alignment tools and mutation matrix scoring, it is increasingly difficult to distinguish the homology from the null hypothesis of random similarity. This is referred to as the "twilight zone" of sequence similarities and corresponds to about 25% of identical amino acids in an optimal alignment between protein pairs. In other words, at the level of about 25% sequence identity, it is equally likely to find a spurious as it is to find a true homology. This value strongly depends on the length of the alignment, as illustrated by well-known examples of identical pentapeptides with different structures (Kabsch and Sander, 1985; Argos, 1987) and analyzed in detail for pairs of similar structures (Sander and Schneider, 1991). Even for whole proteins, there are spurious sequence similarities at such levels. For instance, hypoxantine guanine phosphosibosyltransferase (PDB code 1hmp) and the coat protein of a poliovirus (1piv)

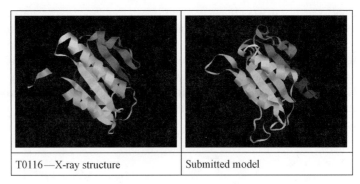

| T0116—X-ray structure | Submitted model |

Figure 31.2. A successful example of a fold prediction using a profile–profile alignment program FFAS. A comparison of the predicted and experimental structure of CASP4 target 116 (see the text for the discussion of the CASP experiment). The score of the alignment was statistically significant with the e-value of e-2, despite the very low sequence similarity between the target and the template of 10% identical residues.

share an 80-amino acid fragment with over 40% sequence identity, despite the lack of any structure or function similarity. This and many other examples of spurious similarities around 25% sequence identity illustrate how dangerous it is to assume homology using sequence similarity as the only argument.

In general, homologous proteins that can be reliably identified using simple sequence similarity searches are usually closely related, with little or no variation in function and generally very similar structures.

Protein Families and Multiple Alignments

The diversity of sequences in a family of homologous proteins captures successful biological experiments in mutating a protein coding sequence without destroying its function and, what follows, its structure. We can assume that with very few exceptions of pseudogenes or dramatic changes of function between paralogous proteins, the mutations destroying a structure of a protein would not be represented among proteins existing in Nature. Therefore, the analysis of a pattern of mutations in homologous families can provide us with information about the importance of various positions along the sequence and, indirectly, of types of restrictions placed on a given position by the protein function and structure. For instance, we can expect that positions that are easily mutated are not important neither for function nor for structure and are most likely located in exposed loops or turns. In contrast, a position in the hydrophobic core of a protein would easily accommodate only some types of mutations (hydrophob–hydrophob) but not others (hydrophob–hydrophil). In the same way, residues in active sites, on the surface, on the interface between protein domains, all have their own rules, stemming from the fact that similar mutations at different positions would lead to different effects for the entire protein and some would be easily accepted, while others will not. For this reason, a uniform mutation matrix, the same at every position along the sequence does not provide a good description of the evolutionary process under strong pressure of preserving the structure and function of a protein. A set of position-specific mutation rules can be derived from the analysis of a multiple alignment of a homologous family and subsequently used to align new sequences in this family. This idea, in various forms and under different names introduced by several groups (profile, position-specific mutation matrix, or hidden Markov models of protein families), allowed sequence analysis methods to break through the twilight zone and reliably recognize distant homologues even when their sequence identity was much below the 25% identity and comparison of single sequences appeared random.

An entire class of distant homology recognition methods evolved from the analysis of mutation patterns in homologous families. A pattern of sequence variation along the sequence can be used to identify positions where some specific structural and/or functional requirements restrict variation, even without a full understanding of these restrictions. It is important to note that techniques used to recognize protein folds by comparing sequences (or sequence profiles), while often treated as part of fold recognition field, can be also used for more general distant homology recognition problem whether or not the distant homologues have known structure. When applied to fold recognition, these methods explicitly search for proteins from the first of the groups discussed above, the diverging homologous proteins. Distant homology recognition methods closely compete with threading, that is, energy-based fold recognition and in recent years seems to be gaining the upper hand (see later in the chapter) (Figure 31.3).

From the time this idea was introduced in 1987 (Gribskov, McLachlan, and Eisenberg, 1987), it remained on the forefront of the sequence analysis field. For instance, several

```
LRRLLPDDTHIMAVVKANAYGHGDVQVARTALEAGASRLAVAFLDEALALREKGIEAP
    pdb|1SFT|A
FRQYVGPKTNLMAVVKADAYGHGAVRVAQTALQAGADWLAIATLGEGIELREAGITAP
    ALR_SYNY3
MKKHIGEHVHLMAVEKANAYGHGDAETAKAALDAGASCLAMAILDEAISLRKKGLKAP
    ALR_BACSU
LRE-LAPASKLVAVVKANAYGHGLLETART-LPD-ADAFGVARLEEALRLRAGGITQP
    ALR1_SALTY
LRE-LAPASKMVAVVKANAYGHGLLETART-LPD-ADAFGVARLEEALRLRAGGITKP
    ALR1_ECOLI
```

Figure 31.3. An example of a multiple alignment: the (small part of the) family of alanine racemase.

top algorithms in the last CASP4 meeting belong to this category. In recent years, it gained even more popularity as it was implemented in PSI-BLAST, the newest variant of the most popular sequence alignment program BLAST (Altschul et al., 1997). There are many variants and specific implementations of this basic idea (see Table 31.1) with most differences occurring in the following areas:

Multiple Alignment Construction   Simultaneous alignment of several sequences is a NP-hard computational problem (Just, 2001), so most algorithms use heuristic approaches, ranging from hierarchical build-up procedures (PSI-BLAST, PileUp) through constructing an approximate phylogenetic tree and using it as a guide in alignment calculation (CLUS-TALW (Higgins and Gibson, 1995; Jeanmougin et al., 1998)) to stochastic minimization techniques, such as simulated annealing (Godzik and Skolnick, 1994) or hidden Markov models (Karplus et al., 1997).

How to Analyze the Multiple Alignment?   How to extract the most relevant information from the multiple alignment? For instance, there are large groups of closely related proteins that do not add much information. Some algorithms simply average the composition at the aligned positions (GCG-Profile) or try to maximize the information content at each position (PSI-BLAST) while others calculate sequence weights from the matrix of interfamily similarities (FFAS).

How is the Similarity Between a Representation of a Family and a Sequence (or a Second Family) Calculated?   Some methods compare a representation of a family (profile, position-specific mutation matrix, and hidden Markov model) to a sequence (GCG-Profile or PSI-BLAST), while others compare two families (BLOCKS, FFAS). Also specifics of the scoring methods vary between methods.

Table 31.2 below summarizes differences between several leading profile alignment algorithms. It is interesting to note that despite very different mathematical formulation (profile methods, position-specific mutation matrix methods, or hidden Markov model-based methods), methods are essentially equivalent and use very similar concepts despite very different mathematical notation.

PROTEINS AS SEEN BY A PHYSICIST

All the fold recognition methods discussed so far are based on homology recognition, that is, they assume that structural similarity results from the distant relation between the two

TABLE 31.1. A Short Overview of Major Sequence-Only Fold Recognition/Distant Homology Recognition Algorithms

	Profiles (Gribskov, McLachlan, and Eisenberg, 1987)	PSI-BLAST (Altschul et al., 1997)	Hidden Markov Models (Karplus et al., 1997)	Intermediate Sequence Search (Park et al., 1997)	(Yona and Levitt, 2002)	FFAS (Rychlewski et al., 2000)
Multiple alignment	Hierarchical, user-controlled iterations	Hierarchical, user-controlled iterations and e-value threshold	CLUSTALW, edited by hand	No multiple alignment built	PSI-BLAST on specially prepared database	PSI-BLAST five iterations with 10^{-3} e-value threshold
Profile	Simple averaging	Preclustering with 98% identity cutoff	Stochastic search for an optimal model describing position-specific a.a. distributions	Iterative search for homologues of already identified homologues of the prediction target	Profiles treated as distributions	Preclustering with 97% identity cutoff
		Pseudocount-based variability estimation		Run until convergence		Amino acid composition filter
		Background amino acid frequencies				Sequence diversity-based weight
Template database	Database of nonredundant sequences (nr)	Database of nonredundant sequences (nr)	Database of HMMs for PDB proteins	Database of nonredundant sequences	Profiles of PDB proteins	Profiles of proteins from PDB, PFAM, COG, and several genomes

TABLE 31.2. A Short Overview of Several Threading Algorithms

	Model	Alignment	Energy	S	PSS	Other Features	Use of Homologues
Bryant	Cα	MC	Distance dependent	Y	N	Library of conserved cores of families	In "position-specific scoring matrices"
Sippl	Interaction centers	Dynamic programming	Distance dependent	N	N		Yes— independent runs
Jones (Threader)	Cα	2D dynamic programming	Distance dependent	Y	Y		N
Jones (GenThreader)	Cα, N, C, O	Dynamic programming	None in alignment	Y	N	Neural network measuring alignment and model quality	Yes— as a profile
Eisenberg	Residue surface	Dynamic programming	Environment classes	Y	Y		N
Godzik (I)	Contacts	Dynamic programming + thawing	Contact based	N	N		N
Honig	Cα	Mean-field dynamic programming	Distance dependent	N	N	Hydrogen bonding term	N
Torda	N, Cα, C, O, Cβ	Dynamic programming	Neighbor nonspecific score	N	N	Two scoring functions, for alignment and ranking	N

[a] Columns S and PSS denote use of sequence and predicted secondary structure, respectively.

proteins. Thus, the hypothesis being tested was whether or not a new protein sequence belongs to a given family of proteins with a specific set of mutation rules. The structure was not used directly and entered the picture only by restricting accepted mutations in different ways at different positions. At the same time, most proteins fold on their own (sometimes with the help of chaperones acting as catalysts of folding), without checking what the structure of their homologues is in databases but following physical laws governing their behavior.

According to the widely accepted "thermodynamic hypothesis," the native conformation of a protein corresponds to a global free energy minimum of the protein/solvent system (Anfinsen, 1973; Privalov and Gill, 1988). Therefore, having a correct energy function, one could use the tools of computational physics to search for the native structure in conformational space. Despite many important advances (Bonneau et al., 2001), this approach is still unable to reliably predict a previously unknown structure of a protein for which only a sequence is known. Two principal problems facing the *ab initio* prediction of protein structure are the lack of adequate molecular potentials and the enormous size of the conformational space of even the smallest protein. Comparing the energy of the same system in two (or more) conformations, as done in fold recognition methods, avoids the latter problem, but unfortunately, as will be discussed later, introduces many new complications.

Energy-based fold recognition methods can be compared to minimization by a grid search, where the grid points where the energy is being calculated are based on known protein structures. Because of the visual analogy of energy calculations using a sequence of one protein forced (threaded through) to adopt a structure of another, energy-based fold recognition was called threading (Godzik and Skolnick, 1992; Bryant and Lawrence, 1993). Since large structural databases must be scanned, energy calculations in threading algorithms by necessity must be optimized for speed. Many different threading algorithms have been developed (Finkelstein and Reva, 1990; Bowie, Luethy, and Eisenberg, 1991; Godzik, Skolnick, and Kolinski, 1992; Jones, Taylor, and Thornton, 1992; Maiorov and Crippen, 1992; Sippl and Weitckus, 1992; Bryant and Lawrence, 1993; Ouzounis et al., 1993; Matsuo and Nishikawa, 1994; Yi and Lander, 1994; Selbig, 1995; Thiele, Zimmer, and Lengauer, 1995; Wilmanns and Eisenberg, 1995; Alexandrov, Nussinov, and Zimmer, 1996; Koretke, Luthey-Shulten, and Wolynes, 1996; Lathrop and Smith, 1996; Russell, Copley, and Barton, 1996; Tropsha et al., 1996; Jaroszewski et al., 1998b). In all cases, threading algorithms followed the paradigm of sequence alignment with its basic steps of identifying the possible template and building the alignment. As a result, the threading approach to structure prediction has limitations similar to sequence-based fold recognition. First and foremost, an example of the correct structure must exist in the structural database that is being screened. If not, the method will fail. Then, the quality of the model is limited by the extent of actual structural similarity between the template and the probe structure.

Force Fields for Simulations and Threading

To speed up energy calculations, the full three-dimensional structure of a protein is usually simplified. Each level of simplification affects the way the energy of the system is calculated. There are less possible interaction centers and more degrees of freedom are averaged. The interaction energy between generalized centers becomes a potential of mean force (Hill, 1960). By averaging over fast changing degrees of freedom, such as bond vibrations and positions of solvent molecules, potentials of mean force are more adequate to describe long-time processes such as folding, despite some loss of accuracy because of the loss of many details. For instance, it can be shown that potentials of mean force can easily distinguish grossly misfolded proteins from their correctly folded counterparts, something that atom–atom molecular potentials are unable to do (Novotny, Brucolleri, and Karplus, 1984).

In principle, it is possible to derive potentials of mean force from simulation by explicitly averaging fast degrees of freedom. This is done routinely in simulations for simple molecular liquids where accurate potentials of mean force can be obtained by averaging vibrational degrees of freedom. However, for complicated systems such as proteins, this is not possible and parameters are usually obtained from the analysis of regularities in experimentally determined protein structures. There have been many derivations of empirical interaction parameter sets, starting from 1976 (Tanaka and Scheraga, 1976) and continuing until today. Several detailed reviews were published recently (Godzik, Kolinski, and Skolnick, 1995; Rooman and Wodak, 1995; Tobi et al., 2000) and a compilation of existing parameter sets is available through the authors Webpage at bioinformatics.burnham. org. There are still many unanswered questions lingering over the theoretical foundations of derivations of knowledge-based interaction parameters and we can expect significant progress in this area.

Despite the lack of complete success as measured by the ability to predict protein structures from their amino acid sequence alone, existing energy parameters adequately capture many features of interactions within proteins. Potentials of mean force derived from the statistical analysis of interaction regularities in proteins can reliably recognize grossly misfolded structures or wrong crystallographic models (Luethy, Bowie, and Eisenberg, 1992), and access the quality of models prepared in homology modeling (Jaroszewski et al., 1998a). And of course, the same potentials can be used in fold recognition.

Threading Approximations

Using energy to recognize similarity between distant homologues leads to several unique challenges. One of the most important ones is that the energy stabilizing a protein structure comes from interactions between side chains distant in sequence. Scoring of alignments in a sequence-based comparison is based on Eq. 31.1 or its variants, where all contributions to the total score come from comparing residues (or PSMMs or single steps in HMM) at single positions and do not depend on gaps or deletions introduced elsewhere in the alignment. In other words, the score is local and this allows the fast and powerful dynamic programming algorithm to be used for alignments. Energy-based scores are not local and alignment with nonlocal functions is an NP-complete problem, that is, it has the same level of computational complexity as the traveling salesman problem and other famous minimization problems (Lathrop, 1994). From the early days of threading, this forced the use of many approximations.

The most obvious approach is to use alignment technique that could work with nonlocal scoring functions. This was a solution used by a few groups (Bryant and Lawrence, 1993) because of the enormous computational cost and slow convergence. Even then it was necessary to limit the space of possible alignments by eliminating deletions and insertions inside secondary structure elements and restricting lengths. By making these approximations a little stronger, it was possible to use combinatorial brand-and-bound minimization algorithms to find a global alignment minimum (Lathrop and Smith, 1996).

Another solution was to use two-level dynamic programming to optimize interaction partners for each possible pairs of aligned residues (Jones, Taylor, and Thornton, 1992). By explicitly considering only the most important interactions between strongly interacting residues, the computational overhead was manageable and the Threader algorithm that used this approach was one of the most successful early threading algorithms.

Most other groups used approximations to energy calculations that allowed them to use it in dynamic programming. The most common approximation was a "frozen approximation" (Godzik, Skolnick, and Kolinski, 1992) where interaction partners for energy calculations were "frozen" to be the same as in the template and were updated only after the alignment was made. Several other groups adopted this approach, which could be iterated (interaction partners updated after alignment is calculated to calculate the alignment and so on (Godzik, Skolnick, and Kolinski, 1992; Wilmanns and Eisenberg, 1995) or relaxed for some interactions (Thiele, Zimmer, and Lengauer, 1995)). This allowed fast alignment calculations but for a price of introducing yet another simplification to the energy calculations. A detailed analysis of various approximations and errors made in a specific threading algorithm is discussed in the following section.

Differences between various threading algorithms are usually found in three areas:

Protein Model and Interaction Description: To speed up energy calculations, the full three-dimensional structure of a protein is usually simplified, which profoundly affects the way the energy of the system is calculated. Side chains are described by interaction points, which could be located at $C\alpha$ or $C\beta$ positions, special interaction points, or can encompass the entire side chain. The interaction energy can be distant dependent or not, and only some parts of the protein molecule can be included in the energy calculations.

Energy Parameterization: There are many variants of the empirical energy parameter derivation that mostly differ in the assumptions about the reference state (Godzik, 1996).

Alignment Algorithms: Threading energy is a nonlocal function of the alignment between the prediction target sequence and the template structure. Dynamic programming with frozen approximation (Godzik, Skolnick, and Kolinski, 1992), 2D dynamic programming (Jones, Taylor, and Thornton, 1992), Monte Carlo minimization (Bryant and Lawrence, 1993), branch-and-bound algorithm (Lathrop and Smith, 1996), and various hybrid approaches can be used for the alignment.

Table 31.2 brings together a short summary and comparison of various threading algorithms, with emphasis on "pure threading" algorithms. However, in practice many of these algorithms still rely heavily on sequence information mixing elements of classical threading and homology recognition algorithms. This is especially true for most of the recently developed algorithms or most recent updates of old algorithms, such as 3D PSSM (Kelley, MacCallum, and Sternberg, 2000), GenThreader (Jones, 1998), Bioinbgu (Fischer, 2000), and others. Also many other technical choices influence relative performance of different algorithms, that are compared as "package deals" and it is difficult to establish relative importance of various specific choices. Therefore, despite significant success of many of these algorithms in fold prediction competitions (CASP meetings) and in providing structural insights in many specific biological problems, they have not

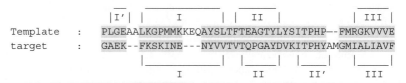

Figure 31.4. An example of a structural alignment between two proteins. Some specific interactions, discussed in the text, are identified in each structure.

contributed significantly to our understanding of folding and forces that influence protein structures.

What are the Major Sources of Errors in Threading (Zhang et al., 1997)?

In an attempt to study the limits of the topology fingerprint threading, we studied in detail the effects of various approximations on the threading results for the small benchmark of 68 pairs. Structural alignments of all pairs were prepared using the CE algorithm (Shindyalov and Bourne, 1998)—an example of a structural alignment is given in Figure 31.4, with some interactions identified by lines above and below the sequence.

The correct energy of the target protein within its own structure is the reference point used to compare all other values. Figure 31.4 is used to illustrate the discussion with the target protein residues identified by bold:

(1) The correct self-threading energy, but calculated only for structure fragments that do have corresponding fragments in the template protein with their entire interaction environment. Only interactions that are entirely within nonaligned fragments will be omitted.

(2) The same as (1), but only interactions with structure fragments that were aligned to the template protein were used. Therefore, the interaction **IA** from Figure 31.4 is now omitted. To contrast it with the following approximations, we call it the "correct partners–correct interactions" (CC) approximation.

(3) Interactions from the target protein and the interaction partners from the template protein according to the structural alignment are used to calculate the energy. Therefore, pair interactions contributing to the energy would be (**F**L + **V**L, **T**L + **D**T, and **I**E + **F**R). This is the "wrong partners–correct interactions" (WC) approximation.

(4) Interactions from the template protein and the interaction partners from the target protein according to the structural alignment are used to calculate the energy. In this calculation, interactions contributing to the energy would be (**FV**, **VD**, and **IF**). This is the "correct partners–wrong interactions" (CW) approximation.

(5) Finally, both interactions and interaction partners from the template protein were used, that is, interactions (**F**L + **V**L, **V**L + **D**F, and **I**E + **F**R).). This is the "wrong partners–wrong interactions" (WW) approximation.

The correct energy as well as approximations 1–3 could only be calculated if the experimental structure of the target is known. Approximations 4 and 5 require the correct structural alignment, so indirectly they also rely on knowing the target structure. Therefore, all these energies are unknown for genuine predictions and again can only be estimated using models of the target structure obtained by comparative modeling. In practice, both the comparative modeling and the alignment procedure based only on the target sequence are likely to introduce errors of their own.

Approximations 1 and 2 are introduced to analyze the extend of target–template structural similarity. Approximation 3 is not particularly interesting (once we know the correct structure, what is the point in using wrong partners). All three approximations are included here only for the sake of completeness, and would not be used much in the subsequent analysis.

At first the energy of one of the proteins in the pair was calculated using its own structure and, in several steps, interaction information as supplemented by information derived from the structure of the other protein with the same fold. The last step was equivalent to the energy as calculated in threading. The goal of this experiment was to identify the source of errors made in threading energy calculations. The important point is that approximations 1–4 can be calculated only in the context of a benchmark. Specific residue names are used in examples that follow to identify and differentiate different approximations:

All interactions and amino acid partners from the target protein sequence and structure (FV, TD, IA, and IF) were used. This is the correct self-threading energy.

Interactions and interaction partners from the prediction target protein were used. However, now only interactions with structural fragments aligned to the template protein are allowed. Therefore, the interaction *IA* from Figure 31.2 is now omitted. This is the "correct amino acid partners–correct interactions" approximation.

Interactions from the target protein are used, but the amino acid partners are taken from the template protein according to the structural alignment. Therefore, interactions contributing to the energy would be (F*L* + V*L*, T*L* + D*T*, and I*E* + F*R*). This is the "wrong amino acid partners–correct interactions" approximation.

Interactions present in the target protein are used, but the amino acid partners from the probe protein were used according to the structural alignment. In this calculation, interactions contributing to the energy would be (FV, VD, and IF). This is the "correct amino acid partners–wrong interactions" approximation. Note that the "wrong" interaction II from the template is used.

Finally, interactions and interaction partners from the target protein were used, that is, interactions (F*L* + V*L*, V*L* + D*F*, and I*E* + F*R*). Note that the "wrong" interaction II from the template is used. This is the "wrong amino acid partners–wrong interactions" approximation.

Approximation 5 is equivalent to the "frozen approximation" introduced to eliminate the nonlocal character of the scoring function in threading (Godzik, Skolnick, and Kolinski, 1992). "Thawing" the interactions (updating the interaction environment) can bring the energy calculation resulting from 5 into 4, but only if the alignment is correct. Approximation 2 could be used if the correct alignment was known and the structure correctly repacked to allow for changes in the interaction patterns along with changes in sequence. Finally, the self-threading energy 1 corresponds to the stability test of a complete, correct structure of the probe sequence. Our current generation of the topology fingerprint threading algorithm calculates the energy according to approximation 5. It is possible to iteratively converge to 4.

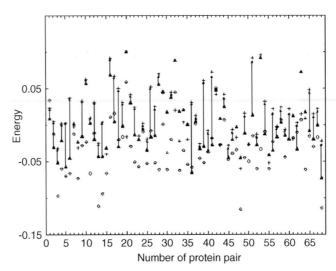

Figure 31.5. Differences between various approximations for energy calculations. Open circles correspond to the real energy, triangles to approximation 4, and crosses to approximation 5 (see text for details).

The results for the 68 pairs are presented in Figure 31.5. Of crucial importance is the observation that the use of the correct partners–wrong interactions (approximation 4), gives a very good approximation of the correct energy. It differs, on the average, by only 1.2 energy units per alignment fragment and 0.25 energy units per residue. Clearly, the interaction patterns in the conserved structural fragments are close enough to be used for energy calculations. Finally, the approximation used as a first step in the topology fingerprint threading algorithm (approximation 5) is clearly the worst, resulting in errors that are about 6 energy units per aligned fragment on average. In other words, the frozen approximation introduced for computational speed in the context of the current interaction definition is terrible. Basically, the pair contribution to the interaction energy using the frozen approximation is wrong, and the error is of the same magnitude as the pair interaction itself.

These results suggest at least two possible ways of improving the sensitivity of threading. One is to move beyond the "frozen approximation" to at least the "correct partners–wrong interactions" approximation. Unfortunately, the scoring function becomes nonlocal, prohibiting the use of dynamic programming alignment (Lathrop, 1994). Another possibility is to change the interaction definition so that the energy differences resulting from approximations 2–5 are smaller. The set of aligned proteins can be used to select the protein representation and associated interaction scheme that minimizes this difference.

Comparing and Assessing Various Fold Recognition Algorithms

The ultimate test of fold recognition methods is the prediction of the folds of new proteins, when only their sequences are known and before any structural information is available. Dedicated meetings, such as the Critical Assessment of Techniques for Protein Structure Prediction (CASP) meeting in Asilomar, California bring together almost all groups actively developing fold recognition algorithms. In these meetings, structure prediction groups are provided with sequences of proteins and the structures that are about to be solved but are not yet publicly available. Therefore, all structure predictions are done blind, without any knowledge of the actual structure. This provides a perfect opportunity to compare the performance of various structure prediction algorithms. The last CASP meeting took place in December 2000. The most interesting result from the last CASP meetings is that methods based only on the sequence information, such as HMM methods (Karplus et al., 1997) or profile–profile alignment methods (Rychlewski et al., 2000) compete with energy only threading methods (Sippl and Weitckus, 1992; Bryant and Lawrence, 1993) as well as with hybrid methods combining contributions both from sequence and structure (Jones et al., 1999; Koretke et al., 1999). In regular CASP meeting, the predictions are submitted by research groups that are free to combine results from fold prediction algorithms with other approaches, their own intuition, biochemical knowledge, and so on. To focus on direct comparison of algorithms, an automated comparison of fold prediction servers (CAFASP experiment) was initiated at CASP3 (Fischer et al., 2001). To avoid comparison based on small number of examples, an ongoing test and comparison of fold prediction algorithms LiveBench (Bujnicki et al., 2001) was initiated and is now in its fourth year.

However, for development of new methods, another choice is to use benchmarks or sets of proteins, whose structure is predicted and is known. We call them prediction targets. For each target, its sequence is matched against a large number of proteins with known structures (templates). The goal is to identify the most appropriate template protein. In a benchmark, the quality of a given prediction method can be measured by the number of targets for which the template chosen by the algorithm was indeed similar to its real structure. One of the early benchmarks was based on 68 proteins identified by Fischer et al. (1995) and was used to evaluate several variants of 3D profile methods developed at UCLA (1996), the RFSRV

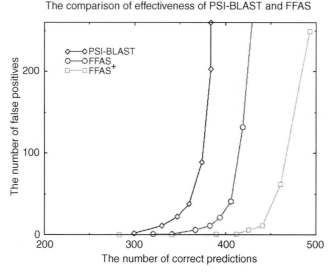

Figure 31.6. Sensitivity plot for several sequence only fold recognition algorithms on a benchmark of 929 proteins identified by SCOP and DALI as being structurally similar. Number of correct predictions for every method is shown as a function false predictions with the same level of significance. PDB-BLAST is a specific strategy of using PSI-BLAST for fold recognition, FFAS is a profile–profile alignment described previously (Rychlewski et al., 2000).

method (Fischer, 2000) and the GeneFold algorithm (Jaroszewski et al., 1998a). Progress in automated structure comparison and easy availability of fold classification databases make it possible to develop larger benchmarks—most popular benchmarks are based on existing classifications of protein structures, such as SCOP.

An example of such comprehensive benchmark of over 900 proteins pairs was built using structure clustering from DALI (1995) and SCOP (1995) databases. The DALI database was used for selection of protein pairs of significant structural similarity but low sequence similarity and SCOP was used to verify the structural similarity of the pair and to assess the level of similarity (fold, superfamily, and family). The full benchmark list (as well as a full list of results for all methods discussed here) is available from our Web server at bioinformatics.burnham-inst.org/benchmarks the FFAS page (Figure 31.6).

SUMMARY

We are still missing a basic understanding of sequence/structure/function relationships in proteins. Analogy-based prediction algorithms remain the only reliable fold prediction tools. New methods, such as threading and hybrid threading/sequence fold recognition, can often recognize even the most distant homologues and, in some cases, even unrelated proteins with similar overall structures. This pushed the envelope of analogy-based function analysis to the point that the majority of newly sequenced genomes can be tentatively assigned to already characterized protein superfamilies. However, at this evolutionary distance, fold prediction is no longer equivalent to function prediction. Instead of having the same exact function, distantly related proteins might share some functional analogy that is not obvious to the casual observer. The main challenge facing fold recognition field is to develop tools to follow the structure prediction with function prediction and analysis.

REFERENCES

Alexandrov NN, Nussinov R, Zimmer RM (1996): Fast fold recognition via sequence to structure alignment and contact capacity potentials. Pacific Symposium on Biocomputing, 96:Hawaii: World Scientific.

Altschul SF, Gish W, Miller W, Myers EW, Lipman DJ (1990): Basic local alignment search tool. *J Mol Biol* 215:403–410.

Altschul SF, Madden TL, Schaeffer AA, Zhang J, Zhang Z, Miller W, Lipman DJ (1997): Gapped BLAST and PSI-BLAST: a new generation of protein database search programs. *Nucleic Acid Res* 25:3389–3402.

Anfinsen CB (1973): Principles that govern the folding of protein chains. *Science* 181:223–230.

Argos P (1987): Analysis of sequence-similar pentapeptides in unrelated protein tertiary structures. *J Mol Biol* 197:331–348.

Babbitt PC, Mrachko GT, Hasson MS, Hiusman GW, Kolter R, Ringe D, Petsko GA, Kenyon GL, Gerlt JA (1995): Functionally diverse enzyme superfamily that abstracts the α proton of carboxylic acids. *Science* 267:1159–1161.

Bateman A, Birney E, Durbin R, Eddy SR, Howe KL, Sonnhammer EL (2000): The Pfam protein families database. *Nucleic Acids Res* 28(1):263–266.

Bonneau R, Tsai J, Ruczinski I, Chivian D, Rohl C, Strauss CE, Baker D (2001): Rosetta in CASP4: progress in *ab initio* protein structure prediction. *Proteins* 45(Suppl 5):119–126.

Bowie JU, Luethy R, Eisenberg D (1991): A method to identify protein sequences that fold into a known three-dimensional structure. *Science* 253:164–170.

Bryant SH, Lawrence CE (1993): An empirical energy function for threading protein sequence through folding motif. *Proteins* 16:92–112.

Bujnicki JM, Elofsson A, Fischer D, Rychlewski L (2001): LiveBench-1: continuous benchmarking of protein structure prediction servers. *Protein Sci* 10(2):352–361.

Chothia C, Finkelstein A (1990): The classification and origins of protein folding patterns. *Annu Rev Biochem* 59:1007–1039.

DALI (1995): *Protein Structure Comparison by Alignment of Distance Matrices.* Heidleberg: EMBL.

Doolittle RF (1990): Molecular evolution: computer analysis of protein and nucleic acids sequences. In: Abelson JN, Simon MI, editors. *Methods in Enzymology.* Vol. 183. San Diego: Academic Press.

Doolittle RF (1994): Convergent evolution: the need to be explicit. *Trends Biochem Sci* 19: 15–18.

Farber GK, Petsko GA (1990): The evolution of α/β enzymes. *Trends Biochem Sci* 15:228–234.

Finkelstein AV, Reva BA (1990): Determination of globular protein chain fold by the method of self-consistent field. *Biofizika* 35:402–406.

Fischer D, Tsai C-J, Nussinov R, Wolfson H (1995): A 3D sequence independent representation of the Protein Data Bank. *Protein Eng* 8:981–997.

Fischer D (2000): Hybrid fold recognition: combining sequence derived properties with evolutionary information. *Pacific Symposium Biocomputing*, 119–130.

Fischer D, Elofsson A, Rychlewski L, Pazos F, Valencia A, Rost B, Ortiz AR, Dunbrack RL Jr (2001): CAFASP2: the second critical assessment of fully automated structure prediction methods. *Proteins* 45(Suppl 5):171–183.

Frishman D, Argos P (1996): Incorporation of non-local interactions in protein secondary structure prediction from the amino acid sequence. *Protien Eng* 9:133–142.

Godzik A, Skolnick J (1992): Sequence structure matching in globular proteins: application to supersecondary and tertiary structure prediction. *Proc Natl Acad Sci USA* 89:12098–12102.

Godzik A, Skolnick J, Kolinski A (1992): A topology fingerprint approach to the inverse folding problem. *J Mol Biol* 227:227–238.

Godzik A, Skolnick J (1994): Flexible algorithm for direct multiple alignment of protein structures and sequences. *Comput Appl Biosci* 10:587–596.

Godzik A, Kolinski A, Skolnick J (1995): Are proteins ideal mixtures of amino acids? Analysis of energy parameter sets. *Protein Sci* 4:2107–2117.

Godzik A (1996): Knowledge-based potentials for protein folding: What can we learn from protein structures? *Structure* 4:363–366.

Gribskov M, McLachlan M, Eisenberg D (1987): Profile analysis: detection of distantly related proteins. *Proc Natl Acad Sci USA* 84:4355–4358.

Gribskov M, Dereveux J (1991): *Sequence Analysis Primer. UWBC Biotechnical Resource Series.* New York: Stocton Press.

Group GC (1991): *Program Manual for the GCG Package.*

Hadley C, Jones DT (1999): A systematic comparison of protein structure classifications: SCOP, CATH and FSSP. *Structure* 7(9):1099–1112.

Higgins JD, Gibson TJ (1995): CLUSTAL W: improving the sensitivity of progressive multiple sequence alignment through sequence weighting, position-specific gap penalties and weight matrix choice. *Nucleic Acid Res* 2:4673–4680.

Hill TL (1960): *An Introduction to Statistical Thermodynamics*. New York: Dover Publications, Inc.

Jaroszewski L, Rychlewski L, Zhang B, Godzik A (1998a): Fold prediction by a hierarchy of sequence and threading methods. *Protein Sci* 7:1431–1440.

Jaroszewski L, Pawlowski K, Godzik A (1998b): Multiple model approach: exploring the limits of comparative modeling. *J Mol Model* 4:294–309.

Jaroszewski L, Rychlewski L, Godzik A (2000): Improving the quality of twilight-zone alignments. *Protein Sci* 9 (8):1487–1496.

Jeanmougin F, Thompson JD, Gouy M, Higgins DG, Gibson TJ (1998): Multiple sequence alignment with Clustal X. *Trends Biochem Sci* 23:403–405.

Jones DT, Taylor WR, Thornton JM (1992): A new approach to protein fold recognition. *Nature* 358:86–89.

Jones D, (1998) GenTHREADER. http://globin.bio.warwick.ac.uk/genome: Warwick.

Jones DT, Tress M, Bryson K, Hadley C (1999): Successful recognition of protein folds using threading methods biased by sequence similarity and predicted secondary structure. *Proteins* (Suppl 3): 104–111.

Just W (2001): Computational complexity of multiple sequence alignment with SP-score. *J Comput Biol* 8(6):615–623.

Kabsch W, Sander C (1985): Identical pentapeptides with different backbones. *Nature* 317:207

Karplus K, Sjolander K, Barret C, Cline M, Haussler D, Hughey R, Holm L, Sander C (1997): Predicting protein structure using hidden Markov models. *Proteins* (Suppl 1): 134–139.

Kelley LA, MacCallum RM, Sternberg MJ (2000): Enhanced genome annotation using structural profiles in the program 3D- PSSM. *J Mol Biol* 299(2):499–520.

Koretke KK, Luthey-Shulten Z, Wolynes PG (1996): Self-consistently optimized statistical mechanical energy functions for sequence structure alignment. *Protein Sci* 5:1043–1059.

Koretke KK, Russell RB, Copley RR, Lupas AN (1999): Fold recognition using sequence and secondary structure information. *Proteins* (Suppl 3): 141–148.

Lathrop RH (1994): The protein threading problem with sequence amino acid interaction preferences is NP-complete. *Protein Eng* 7:1059–1068.

Lathrop R, Smith TF (1996): Global optimum protein threading with gapped alignment and empirical pair scoring function. *J Mol Biol* 255:641–665.

Lesk AM (1995): Systematic representation of protein folding patterns. *J Mol Graph* 13:159–164.

Luethy R, Bowie JU, Eisenberg D (1992): Assessment of protein models with three-dimensional profiles. *Nature* 356:83–85.

Maiorov VN, Crippen GM (1992): Contact potential that recognizes the correct folding of globular proteins. *J Mol Biol* 277:876–888.

Matsuo Y, Nishikawa K (1994): Protein structural similarities predicted by a sequence–structure compatibility method. *Protein Sci* 3:2055–2063.

Needelman SB, Wunsch CD (1970): A general method applicable to the search for similarities in the amino acid sequence of two proteins. *J Mol Biol* 48:443–453.

Novotny J, Brucolleri R, Karplus M (1984): An analysis of incorrectly folded protein models. Implications for structure prediction. *J Mol Biol* 177:787–818.

Ouzounis C, Sander C, Scharf M, Schneider R (1993): Prediction of protein structure by evaluation of sequence–structure fitness: aligning sequences to contact profiles derived from 3D structures. *J Mol Biol* 232:805–825.

Park J, Teichmann SA, Hubbard T, Chothia C (1997): Intermediate sequences increase the detection of homology between sequences. *J Mol Biol* 273(1):349–354.

Pearson WR, Miller W (1992): Dynamic programming algorithms for biological sequence comparison. *Methods Enzymol* 210:575–601.

Privalov PL, Gill SJ (1988): Stability of protein structure and hydrophobic interaction. *Adv Protein Chem* 39:191–235.

Ptitsyn OB, Finkelstein AV (1980): Similarities of protein topologies: evolutionary divergence, functional convergence or principles of folding? *Quart Rev Biophys* 13:339–386.

Rooman MJ, Wodak SJ (1995): Are database derived potentials valid for scoring both forward and inverted protein folding? *Protein Eng* 3:849–858.

Russell R, Barton G (1992): Multiple protein sequence alignment from tertiary structure comparison: assignment of global and residue confidence levels. *Proteins* 14:309–323.

Russell RB, Copley RR, Barton GJ (1996): Protein fold recognition by mapping predicted secondary structures. *J Mol Biol* 259:349–365.

Rychlewski L, Jaroszewski L, Li W, Godzik A (2000): Comparison of sequence profiles. Strategies for structural predictions using sequence information. *Protein Sci* 9:232–241.

Sander C, Schneider R (1991): Database of homology-derived protein structures and the structural meaning of sequence alignment. *Proteins* 9:56–68.

SCOP (1995): *Structural Classification of Proteins.* MRC: Cambridge.

Selbig J (1995): Contact pattern induced pair potentials for protein fold recognition. *Protein Eng* 8:339–351.

Shindyalov IN, Bourne PE (1998): Protein structure alignment by incremental combinatorial extension (CE) of the optimal path. *Protein Eng* 1:739–747.

Sippl MJ, Weitckus S (1992): Detection of native-like models for amino acid sequences of unknown three-dimensional structure in a database of known protein conformations. *Proteins* 13:258–271.

Smith TF, Waterman MS (1981): Comparison of biosequences. *Adv Appl Math* 46:473–500.

Tanaka S, Scheraga HA (1976): Medium and long range interaction parameters between amino acids for predicting three-dimensional structures of proteins. *Macromolecules* 9:945–950.

Taylor WR (1986): Identification of protein sequence topology by consensus template alignment. *J Mol Biol* 188:233–258.

Thiele R, Zimmer R, Lengauer T (1995): Recursive dynamic programming for adaptive sequence and structure alignment. *Int Conf Intell Syst Mol Biol* 3:384–392.

Tobi D, Shafran G, Linial N, Elber R (2000): On the design and analysis of protein folding potentials. *Proteins* 40(1):71–85.

Todd AE, Orengo CA, Thornton JM (2001): Evolution of function in protein superfamilies, from a structural perspective. *J Mol Biol* 307(4):1113–1143.

Tomii K, Kanehisa M (1996): Analysis of amino acid indices and mutation matrices for sequence comparison and structure prediction of proteins. *Protein Eng* 9:27–36.

Tropsha A, Singh RK, Vaisman II, Zheng W (1996): An algorithm for prediction of structural elements in small proteins. Pacific Symposium on Biocomputing, 96:Hawaii: World Scientific.

UCLA (1996): *The UCLA-DOE Benchmark to Assess the Performance of Fold Recognition Methods.* Los Angeles.

Waterman MS (1995): *Introduction to Computational Biology: Maps, Sequences and Genomes (Interdisciplinary Statistics).* Chapman & Hall.

Wilmanns M, Eisenberg D (1995): Inverse protein folding by the residue pair preference profile method. *Protein Eng* 8:626–639.

Yi TM, Lander ES (1994): Recognition of related proteins by iterative template refinements. *Protein Sci* 3:1315–1328.

Yona G, Levitt M (2002): Within the twilight zone: a sensitive profile–profile comparison tool based on information theory. *J Mol Biol* 315(5):1257–1275.

Zhang B, Jaroszewski L, Rychlewski L, Godzik A (1997): Similarities and differences between non-homologous proteins with similar folds. Evaluation of Threading Strategies. *Folding & Design* 12:307–317.

32

DE NOVO PROTEIN STRUCTURE PREDICTION: METHODS AND APPLICATION

Kevin Drew, Dylan Chivian, and Richard Bonneau

INTRODUCTION

The goal of *de novo* structure prediction is to predict the native structure of a protein given only its amino acid sequence. It is generally assumed that a protein sequence folds to a well-defined ensemble average commonly referred to as the native conformation or native ensemble of conformations. It is also commonly assumed that the native state is at or near the global free energy minimum and that this minimum is accessible, leading to folding rates spanning not longer than minutes (albeit spanning >6 orders of magnitude). Exciting exceptions to these assumptions exist, but these assumptions remain quite safe for a large majority of proteins. The problem of finding this native state can be broken into two smaller problems: first, developing an accurate potential and, second, developing an efficient method for searching the energy landscape arising from the potential.

Many methods that are today referred to as *de novo* have previously and/or alternately been referred to "new folds" or *ab initio* methods. For the purpose of this review, we will classify a method as *de novo* structure prediction if that method does not rely on homology between the query sequence and a sequence in the Protein Data Bank (PDB) to create a template for structure prediction. *De novo* methods, by this definition, are forced to consider much larger conformational landscapes than fold recognition and comparative modeling techniques that limit the conformational space by exploring regions near an initial structural template. Sampling this large conformational space is computationally intense and has inhibited *de novo* structure prediction from being useful in certain applications. Recent advances are just now, for example, proving *de novo* structure prediction useful in full genome annotation.

Structural Bioinformatics, Second Edition   Edited by Jenny Gu and Philip E. Bourne
Copyright © 2009 John Wiley & Sons, Inc.

To date the most successful methods for structure prediction have been homology-based comparative modeling and fold recognition (Moult et al., 1999). These methods rely on homologous or weakly homologous sequences of known structure that are frequently not available. When homologous sequences are not available, successful methods for local structure have included predicting secondary structure and local structure patterns (Rost and Sander, 1993; King et al., 1997; Bystroff and Baker, 1998; Jones, 1999; Karplus et al., 1999). Local prediction methods are valuable but are limited and cannot accurately model a whole protein. Our review will focus on current methods for predicting tertiary structure in the absence of homology to a known structure and will discuss these local prediction methods only in the context of tertiary structure prediction.

Many of the methods used today to predict protein structure use information from the PDB. Information extracted from the PDB can be found in the parameters of knowledge-based scoring functions, the training sets of machine learning approaches, and the coordinate libraries of methods that use fragments or templates from the PDB. To test the performance of any one of these approaches, one must carefully remove sequences that are homologous to proteins in the test set from all databases used by the method in question. Any errors or oversights made at this stage could lead to overestimates of success. These concerns are addressed by the critical assessment of structure prediction (CASP) and will be discussed later on.

In spite of recent progress, many issues must still be resolved if a consistently reliable *de novo* prediction scheme is to be developed. There is still no one method that performs consistently across all classes of proteins (most methods perform worse on all-beta proteins), and all methods examined seem to fail totally on sequences longer than 150 residues in length (although in some rare cases larger alpha and alpha–beta proteins have been successfully predicted by *de novo* methods, the methods are in no way reliable at these lengths).

A common pedagogical distinction between prediction methods has been the distinction between methods based on statistical principles on the one hand and physical or first principles on the other hand. Most current successful *de novo* structure prediction methods fall into the statistics camp. We will not discuss this distinction here at great length except for noting that one of the shortcomings of this artificial division is that most effective structure prediction methodologies are in fact a combination of these two camps. For example, several methods that are described as based on physical or first principles employ energy functions and parameters that are statistical approximations of data (e.g., the Lennard–Jones representation of van der Waals forces is often thought of as a physical potential but is a heuristic fit to data). A more useful distinction may be the distinction between reduced complexity models (RCMs) and models that use atomic detail. In this review, we will discuss low-resolution (models containing drastic reductions in complexity such as unified atoms and centroid representations of side-chain atoms) and high-resolution methods (methods that represent protein and sometimes solvent in full atomic detail) focusing on this practical classification/division of methods in favor of distinctions based on a given method's derivation or parameterization.

Overview of this Chapter

Some of the topics covered in this review will be discussed in the context of applications of *de novo* structure prediction, specifically the applications of genome annotation and function prediction. Also, many examples in the text will center on the *de novo* structure prediction method Rosetta; a section describing the Rosetta method is thus presented. Domain prediction is a critical preprocessing step that expands the reach of *de novo* methods to larger proteins. This step, domain prediction, is necessary for annotating full genomes allowing *de novo* methods increased coverage; although these methods are generally sequenced based, they are discussed briefly due to this high relevance. As important as increasing the coverage of structure prediction methods is reducing the complexity of folding models, which lessens the computational time needed to obtain a low-resolution structure while still allowing the process to compute some functional information. The scoring functions of common methods are introduced followed by a discussion of high-resolution structure prediction methods. Using high-resolution methods is difficult with genome annotation because of the computational complexity, but it adds tremendous power to function prediction due to its modeling of atomic detail. Next, the topic of postprocessing *de novo* structures into clusters is covered. Finally, some history and milestones of CASP are discussed and then a more in depth look into applications of *de novo* structure predictions.

Domain Prediction is a Critical Prerequisite to Structure Prediction

De novo methods have run times that increase dramatically with sequence length. As the size of a protein increases, so does the size of the conformational space associated with that protein. Current *de novo* methods are limited to proteins and protein domains less than 150 amino acids in length (with Rosetta the limit is ~150 residues for alpha/beta proteins, 80 for beta folds, and ~150 residues for alpha only folds). This limit means that roughly half of the protein domains seen so far in the PDB are within the size limit of *de novo* structure prediction. Two approaches to circumventing this size limitation are (1) increasing the size range of *de novo* structure prediction and (2) dividing proteins into domains prior to attempting to predict structure. A domain is generally defined as a portion of a protein that folds independently of the rest of the protein. Dividing query sequences into their smallest component domains prior to folding is one straightforward way to dramatically increase the reach of *de novo* structure prediction. For many proteins, domain divisions can be easily found (as would be the case for a protein where one domain was unknown and one domain was a member of a well-known protein family), while several domains remain beyond our ability to correctly detect them. The determination of domain family membership and domain boundaries for multidomain proteins is a vital first step in annotating proteins on the basis of primary sequence and has ramifications for several aspects of protein sequence annotation; multiple works describe methods for detecting such boundaries. In short, most protein domain parsing methods rely on hierarchically searching for domains in a query sequence with a collection of primary sequence methods, domain library searches, and matches to structural domains in the PDB (Chivian et al., 2003; Kim, Chivian, and Baker, 2004; Liu and Rost, 2004).

Recent works use coarse-grained structural simulations and predictions coupled with methods for assigning structural domain boundaries to three-dimensional structures to detect protein domains from sequence. The guiding principle behind this approach is that very low-resolution predictions will pick up overall patterns of the polypeptide packing into distinct structural domains. An interesting recent work attempted to use local sequence signals to detect structure domain boundaries under the assumption that there would be detectable differences in local sequence propensities at domain boundaries (Galzitskaya and Melnik, 2003). As of yet these methods have unacceptably high error rates and are far too computationally demanding for use in genome-wide predictions (David Kim, personal communication) (George and Heringa, 2002). In spite of the limitations mentioned above,

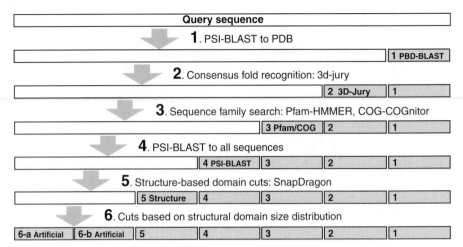

Figure 32.1. Schematic outline of the Ginzu program for domain parsing. Methods with higher reliability are used at the top of the hierarchy, with sequence matches to the PDB being the highest quality information. As higher reliability methods are exhausted, noisier methods are used (such as parsing multiple sequence alignments, step 4, and guessing domain boundaries based on the distribution of domain sizes in the PDB). Sequence regions hit by higher confidence methods (represented as gray rectangles) are masked and the remaining sequence (represented by white rectangles) are forwarded onto the remaining methods. Steps 1–4 and 6 are currently implemented in the Ginzu program, and step 5 (adding sequence homologue independent methods such as structure-based domain parsing from sequence to the procedure) represents possible future work.

these methods (that depend on detecting sequence homologues for a given query sequence) are attractive for proteins that have no detectable homologues or matches to protein domain families, and future work on this front could increase the number of proteins within reach of *de novo* methods considerably. It is likely that a method that successfully combines these coarse structure-based methods with existing sequence-based methods into a hierarchically organized domain detection program (e.g., Ginzu) will eventually outperform any existing method at domain parsing and greatly increase the accuracy of downstream structure prediction. Figure 32.1 shows a schematic domain detection program (this schematic is implemented as the program Ginzu that is developed in conjunction with Robetta, a publicly available server that does domain prediction, homology modeling, and *de novo* modeling: http://robetta.org (Bradley et al., 2005)).

REDUCED COMPLEXITY MODELS

To overcome the sampling problems mentioned in the introduction, most methods for fold prediction to date have involved some significant reduction in the detail that is used to model the polypeptide chain. Methods for reducing protein structure to discrete low-complexity models can be divided into two major classes: lattice and off-lattice models.

Lattice models entail confining possible special coordinates to a predefined three-dimensional grid, and off-lattice models reduce degrees of freedom along the backbone to a set of discrete angles. For a more exhaustive description of these methods than we provide below and their reduced-complexity movesets, we refer the reader to earlier reviews of *de novo* structure prediction methods (Bonneau and Baker, 2001).

Lattice Models

Lattice models have a long history in modeling the behavior of polymers due to their analytical and computational simplicity (Dill et al., 1995). The evaluation of energies on a lattice can be achieved quite efficiently (integer math can be made quite fast) and methods involving exhaustive searches of the available conformational space become feasible (Skolnick and Kolinski, 1991; Hinds and Levitt, 1994; Ishikawa, Yue, and Dill, 1999). However, lattice methods have a somewhat restricted ability to represent subtle geometric considerations (strand twist, secondary structure propensities, and packings) and can reproduce the backbone with accuracies no greater than approximately half the lattice spacing (Reva et al., 1996). The most common systematic error observed for a variety of lattice models is their inability to reproduce helices, and most lattice models exhibit various degrees of secondary structure bias (Park and Levitt, 1995). Given the importance of regular secondary structure in proteins, this is clearly a problem. Recent successful tests, however, have shown that the computational advantages of lattice models may outweigh the problems associated with their systematic biases (Hinds and Levitt, 1994; Kolinski and Skolnick, 1994a; Kolinski and Skolnick, 1994b; Samudrala et al., 1999a).

Discrete State Off-Lattice Models

Most off-lattice reduced complexity models fix all side-chain degrees of freedom and all bond lengths. The most common practice is to limit the side chain to a single rotamer or further to the C_b or to one or more centroids plus backbone atoms (Sternberg, Cohen, and Taylor, 1982; Simons et al., 1997). Discrete state models of the protein backbone usually fix all side-chain degrees of freedom and limit the backbone to specific phi/psi pairs: models containing from 4 to 32 phi/psi states representing various strand, helix, and loop conformations have been described in the literature (Park and Levitt, 1995). It has been shown that properly optimized six state models (i.e., models that account for local features observed in proteins like strands, helices, and canonical loops) can reproduce native contacts, preserve secondary structure, and fit the overall coordinates of the native state as well as the 18-state lattice models that do not account for such protein-specific information (Park and Levitt, 1995).

Narrowing the Search with Discrete State Local Structure Prediction

Local sequences excised from protein structures often have stable structures in the absence of their global contacts, demonstrating that local sequences can have a strong, sequence-dependent structural bias toward one or more well-defined structures (Marqusee, Robbins, and Baldwin, 1989; Blanco, Rivas, and Serrano, 1994; Munoz and Serrano, 1996; Yi et al., 1998; Callihan and Logan, 1999). Several examples also exist of excised fragments having little observable structure, indicating that the strength and the multiplicity of these local biases are highly sequence dependant. Still other studies have focused on sequences that are observed to fold to different conformations depending on their global sequence context, again demonstrating the possible multiplicity of local structure biases (Kabsch and Sander, 1984; Cohen, Presnell, and Cohen, 1993).

Despite the unavoidable ambiguities in local sequence–structure relationships, secondary structure prediction methods have been steadily improving (Orengo et al., 1999a). The prediction methods that accurately predict the type, strength, and possible multiplicity of these local structure biases for any given query sequence segment drastically reduce the size of the available conformational landscape.

Bystroff et al. developed a method that recognizes sequence motifs (ISITES) with strong tendencies to adopt a single local conformation that was used to make good local structure predictions in CASP2 (Bystroff et al., 1996; Han, Bystroff, and Baker, 1997; Bystroff and Baker, 1998; Bystroff, Thorsson, and Baker, 2000; Bystroff and Shao, 2002). I-sites is an HMM (hidden Markov Model) method designed to detect strong relationships between sequence and structure as defined by a library of local structure–sequence relationships. One potential advantage of this method is that the I-sites method is not constrained to fragments of a fixed length (Rosetta is constrained to 3- and 9-length fragments) (Bystroff, Thorsson, and Baker, 2000). Thus, larger patterns of local structure bias will be detected more often by this method. Karplus et al. (2003) also used a similar approach to detecting fragments of local structure (a two-stage HMM) as part of their *de novo* method. These methods have the primary advantage of better performance when local sequence–structure bias is high (e.g., when local structure is strongly and/or uniquely determined by sequence). The TASSER method smoothly combines fragments of aligned protein structure (from threading runs) with regions of unaligned proteins (represented on a lattice for computational efficiency) to effectively scale between the fold recognition and *de novo* regime (Zhang and Skolnick, 2004).

The above-mentioned experiments and observations suggest that any method attempting to use local sequence–structure biases to guide complexity reductions will have to be adaptive to the strength and the multiplicity of different local sequence–structure patterns. The majority of methods proving successful at CASP3 through CASP6 used secondary structure predictions in one way or another, most often three-state secondary structure predictions were used. In one case, predicted secondary structure elements were fit to the results of initial lattice-based exhaustive enumeration, thus erasing any possible secondary structure bias in initial lattice model prior to all-atom refinement (Samudrala et al., 1999b). The Rosetta method uses secondary structure to bias the selection of fragments of known structure from the PDB. Yet another paradigm is to, given a secondary structure string, reduce the problem of predicting the tertiary fold to the problem of how to assemble rigid secondary structure elements (Eyrich, Standley, and Friesner, 1999). Methods have also been described of determining local structure biases independent of secondary structure prediction algorithms (Srinivasan and Rose, 1995).

Using either fragment substitution (assembling fragments of local structure) as a moveset or local structure constraints derived from predicted local structure also has the advantage that the subsequent global search is limited to protein-like regions of the conformational landscape (helices, correct chirality of secondary strand packing, strands and sheets with correct twist, etc).

There are two main ways to use local structure prediction as an overriding/hard constraint on the global search: (1) using fragments to build up global structures (local structure defining the moveset) and (2) using local structure as a hard constraint (local structure heavily modifying the objective function).

There is likely to be an upper limit on the accuracy of secondary structure prediction methods, due to their failure to account for nonlocal interactions. The best secondary structure prediction algorithms have three-state accuracies of 76–78%, and any *de novo* method must account for this error rate to make consistently successful predictions (Rost and Sander, 1993; Jones, 1999). A milestone for *de novo* structure prediction, which takes nonlocal interactions into account, will be the production of models with secondary structure predicted more accurately than possible with traditional secondary structure methods.

SCORING FUNCTIONS FOR REDUCED COMPLEXITY MODELS

Once a model for representing the protein is chosen that sufficiently reduces the complexity of the conformational search, a scoring or energy function that works in the chosen low-complexity space must be developed. The energy function must adequately represent the forces responsible for protein structure: solvation, strand hydrogen bonding, and so on. Given that most low-complexity models do not explicitly represent all atoms and can reproduce even the native state backbone with only limited accuracy, any energy function designed to work in the low-complexity regime must represent these forces in a manner robust to such systematic error (the systematic limitations of the model). Finally, these functions must be computationally efficient, for the initial stages of any conformational search require huge numbers of energy evaluations. Because of the shortcomings of molecular mechanics-based potentials and the considerations above, many methods developed in the past 10 years utilize scoring functions derived from the PDB, which favor arrangements of residues that are commonly found in known protein structures and minimize the contribution of rarely seen arrangements. The effect scoring functions have on our applications of genome annotation and function prediction is not as obvious and direct as others, but efficient and accurate scoring functions nevertheless have a large impact on how scalable our methods are to a genome level.

Solvation-Based Scores

It has been long thought that the hydrophobic effect is the principal driving force behind protein folding (Baldwin, 1999). There are many diverse methods for determining the fitness of backbone conformations based on solvation or hydrophobic packing, and the debate over the proper functional form for representing solvation effects represents an open question of considerable importance and interest (Park, Huang, and Levitt, 1997). A common approach is to classify sites in the protein according to the degree of solvent exposure. This is done by classifying sites either by the exposed surface area or by the number of nearby residues and then determining the frequencies of occurrences of the amino acids in each type of sites (Bowie and Eisenberg, 1994). The score of an amino acid at a site is then taken to be the logarithm of the amino acid's frequency of occurrence at that type of site (see Table 32.1 for equation). This type of residue environment term favors placement of hydrophobic amino acids at buried positions and hydrophilic amino acids at exposed positions. The final score is a weighted sum of the scores.

Another commonly used class of functional forms for solvation consists of global measures of hydrophobic arrangement. One simple global quantity is a residue's distance from the entire conformation's center of mass, which can be used to calculate quantities analogous to the hydrophobic radius of gyration. Bowie and Eisenberg (1994) used this type of function, coined hydrophobic contrast, in combination with other terms, including a surface area-based term, to fold small alpha helical proteins using an evolutionary algorithm.

TABLE 32.1. Components of the Rosetta Energy Function[a]

Name	Description (Putative Physical Origin)	Functional Form	Parameters (Values)
env[b]	Residue environment (solvation)	$\sum_i -\ln[P(\mathrm{aa}_i\|\mathrm{nb}_i)]$	i = residue index aa = amino acid type nb = number of neighboring residues[c] (0, 1, 2 … 30, > 30)
pair[b]	Residue pair interactions (electrostatics, disulfides)	$\sum_i \sum_{j>i} -\ln\left[\dfrac{P(\mathrm{aa}_i,\mathrm{aa}_j\|s_{ij}d_{ij})}{P(\mathrm{aa}_i\|s_{ij}d_{ij})P(\mathrm{aa}_j\|s_{ij}d_{ij})}\right]$	I, j = residue indices aa = amino acid type d = centroid–centroid distance (10–12 Å, 7.5–10 Å, 5 Å–7.5 Å, <5 Å) s = sequence separation (>8 residues) i, j = residue (or centroid) indices
vdw[d]	Steric repulsion	$\sum_i \sum_{j>i} \dfrac{(r_{ij}^2 - d_{ij}^2)^2}{r_{ij}}; d_{ij} < r_{ij}$	d = interatomic distance r = summed van der Waal radii[e] i, j = residue indices
rg	Radius of gyration (vdw attraction; solvation)	$\sqrt{\langle d_{ij}^2 \rangle}$	d = distance between residue centroids I = residue index
cbeta	Cβ density (solvation; correction for excluded volume effect introduced by simulation)	$\sum_i \sum_{\mathrm{sh}} -\ln\left[\dfrac{P_{\mathrm{compact}}(\mathrm{nb}_i,\mathrm{sh})}{P_{\mathrm{random}}(\mathrm{nb}_i,\mathrm{sh})}\right]$	sh = shell radius (6 Å, 12 Å) nb = number of neighboring residues within shell[f] P_{compact} = probability in compact structures assembled from fragments P_{random} = probability in structures assembled randomly from fragments

SS[g]	Strand pairing	Scheme A: $SS_{\phi, \theta} + SS_{hb} + SS_d$	m, n = strand dimer indices; dimer is two consecutive strand residues
	(hydrogen bonding)	Scheme B: $SS\text{-}\phi, \theta + SS_{hb} + SS_{d\sigma}$	$\hat{V}$ = vector between first N and last C atom of dimmer
		where	$\hat{m}$ = unit vector between $\hat{V}_m$ and $\hat{V}_n$ midpoints
			$\hat{x}$ = unit vector along carbon-oxygen bond of first dimer residue
		$SS\phi, \theta = \sum_m \sum_{n>m}$ $-\ln[P(\phi_{mn}, \theta_{mn} \vert d_{mn}, sp_{mn}, s_{mn})]$	$\hat{y}$ = unit vector along oxygen-carbon bond of second dimer residue
		$SS_{hb} = \sum_m \sum_{n>m} -\ln[P(hb_{mn} \vert d_{mn}, s_{mn})]$	ϕ, θ = polar angles between $\hat{V}_m$ and $\hat{V}_n$ (36° bins)
		$SS_d = \sum_m \sum_{n>m} -\ln[P(d_{mn} \vert s_{mn})]$	hb = dimer twist, $\sum_{k=m,n} 0.5(\vert \hat{m} \cdot \hat{x}_k \vert + \vert \hat{m} \cdot \hat{y}_k \vert)$
		$SS_{d\sigma} = \sum_m \sum_{n>m} -\ln[P(d_{mn}\sigma_{mn} \vert \rho_m, \rho_n)]$	(<0.33, 0.33–0.66, 0.66–1.0, 1.0–1.33, 1.33–1.6, 1.6–1.8, 1.8–2.0)
			d = distance between $\hat{V}_m$ and $\hat{V}_n$ midpoints (<6.5 Å)
			σ = angle between $\hat{V}_m$ and $\hat{M}$ (18° bins)
			sp = sequence separation between dimer containing strands (<2, 2–10, >10 residues)
			s = sequence separation between dimers (>5 or >10)
			ρ = mean angle between vectors $\hat{m}$, $\hat{x}$ and $\hat{m}$, $\hat{y}$ (180° bins)
			n_{sheets} = number of sheets
sheet[h]	Strand arrangement into sheets	$-\ln[P(n_{sheet} n_{lone_{strands}} \vert n_{strands})]$	$n_{lone\_strands}$ = number of unpaired strands
			$n_{strands}$ = total number of strands
HS	Helix–strand packing	$\sum_m \sum_{n>m} -\ln[P(\phi_{mn}, \psi_{mn} \vert sp_{mn}, d_{mn})]$	m = strand dimer index; dimer is two consecutive strand residues
			n = helix dimer index; dimmer is central two residues of four consecutive helical residues

(*continued*)

TABLE 32.1. (Continued)

Name	Description (Putative Physical Origin)	Functional Form	Parameters (Values)
			$\hat{V}$ = vector between first N and last C atom of dimer N and last C atom of dimer
			ϕ, θ = polar angles between $\hat{V}_m$ and $\hat{V}_n$ (36° bins)
			sp = sequence separation between dimer containing helix and strand (binned <2, 2–10, >10 residues)
			d = distance between $\hat{V}_m$ and $\hat{V}_n$ midpoints (<12 Å)

[a] All terms originally described in references Simons et al., 1997, 1999.

[b] Binned function values are linearly interpolated, yielding analytic derivatives.

[c] Neighbors within a 10 Å radius. Residue position defined by Cβ coordinates (Cα for glycine).

[d] Not evaluated for atom (centroid) pairs whose interatomic distance depends on the torsion angles of a single residue.

[e] Radii determined from (1) 25th closest distance seen for atom pair in pdbselect25 structures, (2) the fifth closest distance observed in X-ray structures with better than 1.3 Å resolution and <40% sequence identity, or (3) X-ray structures of <2 Å resolution, excluding i, $i + 1$ contacts (centroid radii only).

[f] Residue position defined by Cβ coordinates (Cα for glycine).

[g] Interactions between dimers within the same strand are neglected. Favorable interactions are limited to preserve pair-wise strand interactions, that is, dimer m can interact favorably with dimers from at most one strand on each side, with the most favorable dimer interaction ($SS_{\phi, \sigma} + SS_{hb} + SS_d$) determining the identity of the interacting strand. $SS_{d\sigma}$ is exempt from the requirement of pair-wise strand interactions. SS_{hb} is evaluated only for m, n pairs for which $SS_{\phi, \theta}$ is favorable. $SS_{d\sigma}$ is evaluated only for m, n pairs for which $SS_{\phi, \theta}$ and SS_{hb} are favorable. A bonus is awarded for each favorable dimer interaction for which $|m - n| > 11$ and strand separation is >8 residues.

[h] A sheet is constituted of all strands with dimer pairs <5.5 Å apart, allowing each strand having at most one neighboring strand on each side. Discrimination between alternate strand pairings is determined according the most favorable dimer interaction. Probability distributions fitted to $c(n_{\text{strands}}) - 0.9n_{\text{sheets}} - 2.7n_{\text{lone\_strands}}$, where $c(n_{\text{strands}}) = (0.07, 0.41, 0.43, 0.60, 0.61, 0.85, 0.86, 1.12)$.

764

Huang and Levitt used this type of function to recognize native structures (Huang, Subbiah, and Levitt, 1995; Samudrala et al., 1999b). One problem with the above global functions is that they assume that proteins are ideally spherical in shape when in actuality native proteins exhibit a much larger range of shapes. A more flexible approach uses an ellipsoidal approximation of the shape of the hydrophobic core that does not require a significant increase in computation and was shown to aid in the selection of near-native conformations from decoy sets containing a high number of protein-like yet incorrect compact conformations. The problem associated with these functions is that they will inevitably exclude a small percentage of protein structures that deviate from their assumptions concerning shape and thus fail when a protein is divided into small subdomains or contains large invaginations (1HQI is a toroid). In spite of this potential downfall, they have demonstrated their usefulness in several methods due to their ability to recognize the majority of small hydrophobic cores, their simplicity, and the speed with which they can be computed.

Pair Interactions

Many low-resolution scoring functions utilize an empirically derived pair potential in place of or in addition to the residue environment term described above. The most common of these potentials are functions of the position of a single center per residue (C_a, C_b, or centroid/united atom center) and are thus quite computationally efficient; all-atom functions have also been used (Samudrala et al., 1999c). Many variations of pair terms have been developed, with the two main branches of methods being distance-dependant and contact-based methods (Sippl, 1995; Miyazawa and Jernigan, 1999). Like the sequence–structure bias mentioned above, these scoring functions are sometimes justified by positing that the arrangements of residues in proteins follow a Boltzman distribution $E(x) = kT \ln P(x)$, where x is a feature such as the occurrence of two residues separated by a distance less than r. Alternatively, these scoring functions may be seen purely as probability distribution functions (Domingues et al., 1999; Simons et al., 1999c). In the former case, the optimization may be viewed as a search for the lowest energy configurations and, in the latter, a search for the highest probability configurations. For most applications, there is little practical difference between the two viewpoints. The issue becomes more substantive, however, when such database-derived scoring functions are combined with physics-based potentials, as will likely become increasingly useful.

There are several problems associated with statistically derived pair potentials. The assumption that summing over component interactions can represent free energies is not generally valid across all interactions present in proteins, and thus the basic functional form may not be adequate to represent the free energy of a decoy conformation (Mark and van Gunsteren, 1994; Dill, 1997). The most significant problem with pair potentials is that they are dominated by hydrophobic/polar partitioning that gives rise to anomalous effects, such as a long-range repulsion between hydrophobic residues (Thomas and Dill, 1996). This can be corrected by conditioning the pair distributions on the environments of the two residues, which largely eliminates these undesired effects (Simons et al., 1997). With the elimination of the otherwise overwhelming influence of hydrophobic partitioning, specific interactions such as electrostatic attraction between oppositely charged residues dominate the pair scoring/energy functions, and hydrophobic interactions make relatively modest contributions. The pair term, in this case, is perhaps best viewed as the second term in a series expansion for the residue–residue distributions in the database in which the residue environment distributions are the first term.

Some of the earliest comprehensive tests of the discriminatory power of these pair potentials were done in the context of threading self-recognition (Novotny, Bruccoleri, and Karplus, 1984). Later work demonstrated that the self-recognition problem was not a sufficiently challenging test of scoring functions and focused on the performance of multiple pair-wise energy functions on larger more diverse sets of conformations (Jernigan and Bahar, 1996; Jones and Thornton, 1996; Park and Levitt, 1996; Park, Huang, and Levitt, 1997; Eyrich, Standley, and Friesner, 1999). The performance of the various energy functions at recognizing native-like structures in large ensembles of incorrect "decoys" highly depends on the methods used to create the decoy sets, highlighting the fact that an energy function that works well in the context of one method will not necessarily work well given a decoy set created using an orthogonal method.

Sequence-Independent Terms/Secondary Structure Arrangement

Many features of proteins, such as the association of beta strands into sheets, can be described by sequence-independent scoring functions. Several early approaches to folding all-beta proteins were protocols dominated by initial low-resolution combinatorial searches of possible strand arrangements and were concerned only with the probability of different strand arrangements (Cohen, Sternberg, and Taylor, 1980; Cohen, Sternberg, and Taylor, 1982; Chothia, 1984; Reva and Finkelstein, 1996). These early methods narrow the conformational space by only considering sequence-specific effects in the context of highly probable strand arrangements. Several of the relatively successful methods at CASP3 incorporated secondary structure packing terms (Lomize, Pogozheva, and Mosberg, 1999; Simons et al., 1999b). Ortiz et al. used an explicit hydrogen bond term in combination with a "bab" and a "bba" chirality term to ensure protein-like secondary structure formation. It cannot be expected that a low-complexity lattice model produce the correct chirality or subtle higher order effects like strand twist and these rules sensibly correct for these expected shortcomings (Ortiz et al., 1999). The Rosetta method used three terms that monitored strand–strand pairing, sheet formation, and helix strand interactions to ensure protein-like secondary structure arrangements.

Structures from Limited Constraint Sets

The obvious drawback of current low-resolution *de novo* structure prediction methods is their relatively low accuracy and reliability. Even limited amounts of experimental data on the structure of a protein can remedy this considerably. For example, Rosetta produced accurate structures in conjunction with NMR chemical shift data (to enhance fragment selection) and sparse NOE constraints (Bowers, Strauss, and Baker, 2000). Distance constraints from cross-linking followed by mass spectrometry could also be readily incorporated into such an approach and potentially could be obtained on a high-throughput scale (see Chapter 7).

ROSETTA *DE NOVO* STRUCTURE PREDICTION

Throughout this work, we have used and will use examples drawn from or centered on the Rosetta *de novo* structure prediction protocol and will thus provide a brief overview of Rosetta before continuing to discuss key elements of the procedure in greater detail

(Simons et al., 1997; Simons et al., 1999a; Simons et al., 1999b; Bonneau, Strauss, Baker, 2001a; Rohl et al., 2004b) (see Figures 32.2 and 32.3). Results from the fourth and fifth critical assessments of structure prediction (CASP4, CASP5, CASP6) have shown that Rosetta is currently one of the best methods for *de novo* protein structure prediction and distant fold recognition (Bonneau et al., 2001c; Lesk, Lo Conte, and Hubbard, 2001; Bradley et al., 2003; Chivian et al., 2003). Rosetta was initially developed as a computer program for *de novo* fold prediction but has been expanded to include design, docking, experimental determination of structure from partial data sets, protein–protein interaction, and protein–DNA interaction prediction (Kuhlman and Baker, 2000; Chevalier et al., 2002; Gray et al., 2003a; Gray et al., 2003b; Kuhlman et al., 2003; Kortemme et al., 2004; Rohl et al., 2004a; Rohl, 2005). When we refer to Rosetta in this work, we will be primarily referring to the *de novo* or *ab initio* mode of the Rosetta code base. Early progress in high-resolution structure prediction has been achieved via combinations of low-resolution approaches (for initially searching the conformational landscape) and higher resolution potentials (where atomic detail and physically derived energy functions are employed). Thus, Rosetta structure prediction is carried out in two phases: (1) a low-resolution phase where overall topology is searched using a statistical scoring function and fragment assembly and (2) an atomic detail refinement phase using rotamers and small backbone angle moves and a more physically relevant (detailed) scoring function.

Rosetta *de novo* (Rosetta) uses information from the PDB to estimate the possible conformations for local sequence segments. The procedure first generates libraries of local sequence fragments excised from a nonredundant version of the PDB on the basis of local sequence similarity (3- and 9-residue matches between the query sequence and a given structure in the PDB) (Simons et al., 1997).

Using the PDB for local sequence information was inspired by careful studies of the relationship between local sequence and local structure (Han, Bystroff, and Baker, 1997), which demonstrated that this relationship was highly variable on a sequence-specific basis and that there is a great deal of sequence-specific local structure that could be recognized even in the absence of global homology. The selection of fragments of local structure on the basis of local sequence matches dramatically reduces the size of the accessible conformational landscape. In practice we see that, as desired, for some local sequence segments there is a strong bias toward a single local structure in the computed local structure fragments, while other local sequences exhibit a wide range of local conformations in the fragment library.

Using fragment substitution as a set of allowable moves to optimize Rosetta's objective function does have drawbacks. As the structure collapses late in the simulation and forms contacts favorable according to the energy function, the acceptance rate of fragment moves becomes unworkably small. This is due to the fact that the substitution of 6 or 18 backbone dihedral angles creates large perturbations to the Cartesian coordinates of parts of the protein distant along the polypeptide chain. The likelihood that such perturbations cause steric clashes and break energetically favorable contacts late in a given simulation is exceedingly large. To recover effective minimization of the Rosetta score after initial collapse, several additional move types have been added to the Rosetta moveset. The simplest move type consists of small-angle moves within populated regions of the Ramachandran map. Additional moves, descriptively named "chuck," "wobble," and "gunn," aim to perform fragment insertions that have small effects far from the insertion limiting the possibility of breaking favorable contacts. These additional move types are also critical to the modeling of loops in homology modeling and are described in detail elsewhere (Rohl et al., 2004).

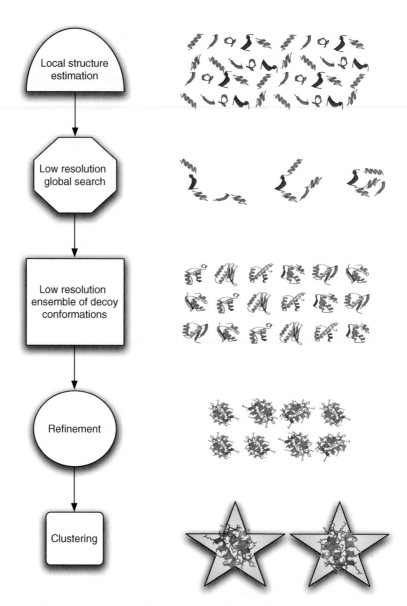

Figure 32.2. Rosetta structure prediction protocol: Rosetta begins by determining local structure conformations (fragments) for 3- and 9-mer stretches of the input sequence. Then multiple fragment substitution simulated annealing searches are done to find the best arrangement of the fragments according to Rosetta's low-resolution scoring function (Table 32.1). The resulting structures then undergo a high-resolution refinement step based on a more physically based scoring function (Table 32.2). Finally, the structures are clustered by RMSD (to each other, as the native is unknown) and the centers of the largest clusters are chosen as representative folds (the centers of the largest clusters are likely to be correct fold predictions).

Rosetta fragment generation works well even for sequences that have no homologues in the known sequence databases owing to the PDB structure coverage of possible local sequence at the 3- and 9-residue length. Tertiary structures are generated using a Monte Carlo search of the possible combinations of likely local structures minimizing a scoring function

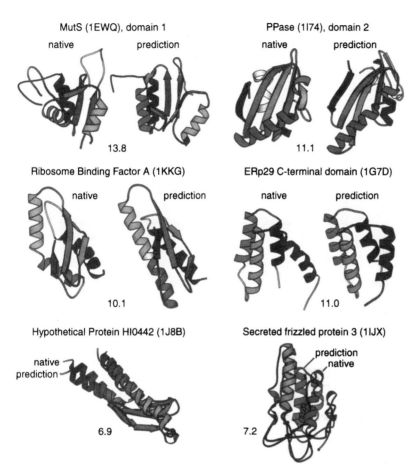

Figure 32.3. Examples of *de novo* structure predictions generated using Rosetta. A, B and C are examples from our genome-wide prediction of domains of unknown function in *Halobacterium* NRC-1 (Bonneau et al., 2004). In each case, the predicted structure is shown next to the closest structure–structure match to the PDB. For A, B, and C, only the backbone ribbons are shown, as these predictions were not refined using the all-atom potential and are examples of the utility of low-resolution prediction in determining function. Panel D shows a recent prediction where high-resolution refinement subsequent to the low-resolution search produced the lowest energy conformation, a prediction of unprecedented accuracy (provided by Phil Bradley).

that accounts for nonlocal interactions such as compactness, hydrophobic burial, specific pair interactions (disulfides and electrostatics), and strand pairing.

The Rosetta score for this initial low-resolution stage is described in its entirety in Table 32.1. For the second stage, refinement, the centroid representations of amino acid side chains used in the low-resolution phase are replaced with atomic detail rotamer representations. Rotamers are generally defined as the different acceptable conformations of atoms in the amino acid side chains. The scoring function used during this refinement phase includes solvation terms, hydrogen bond terms, and other terms with direct physical interpretation. See Table 32.2 for a full description of the all-atom Rosetta score.

TABLE 32.2. Components of the Rosetta All-Atom Energy Function[a]

Name	Description (Physical Origin)	Functional Form	Parameters	References
rama	Ramachandran torsion preferences	$\sum_i -\ln[P(\phi_i, \psi_i \mid aa_i, ss_i)]$	i = residue index; ϕ, ψ = backbone torsion angles (10°, 36° bins); aa = amino acid type; ss = secondary structure type[b]	Bowers, Strauss, and Baker (2000)
LJ[c]	Lennard–Jones interactions	$\sum_i \sum_{j>i} \left\{ \begin{array}{ll} \left[\left(\dfrac{r_{ij}}{d_{ij}}\right)^{12} - 2\left(\dfrac{r_{ij}}{d_{ij}}\right)^6\right]e_{ij}, & \text{if } \dfrac{d_{ij}}{r_{ij}} > 0.6 \\ \left[-8759.2\left(\dfrac{r_{ij}}{d_{ij}}\right) + 5672.0\right]e_{ij}, & \text{else} \end{array}\right.$	i,j = residue indices; d = interatomic distance; e = geometric mean of atom well depths[d]	Kuhlman and Baker (2000)
hb[f]	Hydrogen bonding	$\sum_i \sum_j (-\ln[P(d_{ij}\mid h_j ss_{ij})]$ $-\ln[P(\cos\theta_{ij}\mid d_{ij}h_j ss_{ij})] - \ln[P(\cos\psi_{ij}\mid d_{ij}h_j ss_{ij})]$	r = summed van der Waal radii[e]; i = donor residue index; j = acceptor residue index; d = acceptor–proton interatomic distance; h = hybridization (sp^2, sp^3); ss = secondary structure type[g]; θ = proton–acceptor–acceptor base bond angle; ψ = donor–proton–acceptor bond angle	Kortemme et al. (2004)
solv	Solvation	$\sum_i \left[\Delta G_i^{\mathrm{ref}} - \sum_j \left(\dfrac{2\Delta G_i^{\mathrm{free}}}{4\pi^{3/2}\lambda_i r_{ij}^2}\, e^{-d_{ij}^2} V_j + \dfrac{2\Delta G_j^{\mathrm{free}}}{4\pi^{3/2}\lambda_j r_{ij}^2}\, e^{-d_{ij}^2} V_i \right) \right]$	i,j = atom indices; d = distance between atoms; r = summed van der Waal radii[e]; λ = correlation length[h]; V = atomic volume[h]; ΔG^{ref}, ΔG^{free} = energy of a fully solvated atom[h]	Lazaridis and Karplus (1999)

pair	Residue pair interactions (electrostatics, disulfides)	$\sum_i \sum_{j>i} -\ln\left[\dfrac{P(\mathrm{aa}_i,\mathrm{aa}_j \mid d_{ij})}{P(\mathrm{aa}_i \mid d_{ij})P(\mathrm{aa}_j \mid d_{ij})}\right]$	i, j = residue indices aa = amino acid type d = distance between residues[i]	Kuhlman and Baker (2000)
dun	Rotamer self-energy	$\sum_i -\ln\left[\dfrac{P(\mathrm{rot}_i \mid \phi_i, \psi_i)P(\mathrm{aa}_i \mid \phi_i, \psi_i)}{P(\mathrm{aa}_i)}\right]$	i, j = residue indices rot = Dunbrack backbone-dependent rotamer aa = amino acid type ϕ, ψ = backbone torsion angles aa = amino acid type	Dundbrack (1997)
ref	Unfolded state reference energy	$\sum_i n_{aa}$	n = number of residues	Kuhlman and Baker (2000)

Also described in Rohl (2005).

[a] All binned function values are linearly interpolated, yielding analytic derivatives, except as noted.

[b] Three-state secondary structure type as assigned by DSSP Kabsch and Sander (1983).

[c] Not evaluated for atom pairs whose interatomic distance depends on the torsion angles of a single residue.

[d] Well depths taken from CHARMm19 parameter set, Neria et al. (1991).

[e] Radii determined from fitting atom distances in protein X-ray structures to the 6–12 Lennard Jones potential using CHARMm19 well depths.

[f] Evaluated only for donor acceptor pairs for which $1.4 \leq d \leq 3.0$ and $90° \leq \psi, \theta \leq 180°$. Side-chain hydrogen bonds in involving atoms forming main-chain hydrogen bonds are not evaluated. Individual probability distributions are fitted to eighth-order probability distributions and analytically differentiated.

[g] Secondary structure types for hydrogen bonds are assigned as helical ($j - i = 4$, main chain), strand: ($|j - i| > 4$, main chain) or other.

[h] Values taken from Lazaridis and Karplus (1999).

[i] Residue position defined by Cβ coordinates (Cα of glycine).

Using Rosetta-generated structure predictions, we were able to recapitulate many functional insights not evident from sequence-based methods alone (Bonneau et al., 2001b; Bonneau et al., 2002). We have reported success in annotating proteins and protein families without links to known structure with Rosetta (Bateman et al., 2000; Bonneau et al., 2002). Various aspects of this overall protocol will be reviewed in greater detail below. We also encourage the reader to refer to several prior works where the Rosetta method is described in its entirety.

HIGH-RESOLUTION STRUCTURE PREDICTION

We now move to the subject of creating high-resolution protein structures. We have already seen the precursor that reduces the complexity of the protein model and its corresponding scoring functions. High-resolution protein folding builds on these topics but no doubt requires a more accurate all-atom model and a stronger root in physically based scoring functions. With these added requirements, as we shall see, the computational complexity increases.

The most natural starting point for simulating high-resolution protein folding is standard molecular dynamics (MD) simulation (numerically integrating Newton's equations of motion for the polypeptide chain) using a physically reasonable potential function. There are several problems that have limited the success of such approaches thus far. First, MD is computationally expensive—with explicit representation of sufficient water molecules to minimally solvate the folding chain, a nanosecond MD simulation of a 100-residue protein takes $\sim$400 h on a current single processor. Advances in simulation strategy and increases in available computer power have considerably extended simulation times; for example, Kollman and coworkers carried out a microsecond simulation of a 36-residue peptide using a considerable amount of supercomputer time (Lee, Baker, and Kollman, 2001a). However, simulating the folding of a 100-residue protein for the typical $\sim$1 s required for a single folding transition will require more than six orders of magnitude more computing time. The second, and perhaps more serious, class of problems associated with MD are the inadequacies in current potential functions for macromolecules in water. While important progress has been made, there is still a lack of consensus as to the best computationally tractable yet physically realistic model for water (a number of quite different models for water, both polarizable and nonpolarizable, have been used in current simulation methods) and some uncertainty in the values of the parameters used in molecular mechanics potentials (partial atomic charges, Lennard–Jones well depths, and radii). Accurate representation of electrostatics is also a considerable challenge given the high degree of polarizability of water; there is a large difference in the dielectric properties of the solvent and the protein, and the uncertainties in the magnitude and location of atomic charges density contribute to the error of calculations. Because the free energy of a protein represents a delicate balance of large and opposing contributions, these problems significantly reduce the likelihood that the native state will be found at the global free energy minimum using current potentials (Badretdinov and Finkelstein, 1998). The best current use of MD methods may be in refining and discriminating among models produced by lower resolution methods (Vejda, Lazaridis, Kollman and Lee, personal communication) (Lee, Duan, and Kollman, 2000).

The Balance Between Resolution and Sampling, Prospects for Improved Accuracy and Atomic Detail

As evident by the description of MD above, every *de novo* structure prediction procedure must strike a delicate balance between the computational efficiency of the procedure and the level of physical detail used to model protein structure within the procedure. Low-resolution models can be used to predict protein folds and sometimes suggest function (Bonneau et al., 2001b). Low-resolution models have also been remarkably successful at predicting features of the folding process, such as folding rates and phi values (Alm and Baker, 1999a; Alm and Baker, 1999b). It is clear, however, that modeling proteins (and possibly bound water and other cofactors) at atomic detail and scoring these higher resolution models with physically derived detailed potentials is a needed development if higher resolution structure prediction is to be achieved.

Initial focus was on the use of low-resolution approaches for searching the conformational landscape followed by a refinement step where atomic detail and physical scoring functions are used to select and/or generate higher resolution structures. This is a practical methodological mix when one considers the vastness of the conformational landscape and has resulted in good predictions in many cases. For example, several studies have illustrated the usefulness of using *de novo* structure prediction methods as part of a two-stage process in which low-resolution methods are used for fragment assembly and the resulting models are refined using a more physical potential and atomic detail (e.g., rotamers) (Dunbrack, 1997) to represent side chains (Bradley et al., 2003; Tsai et al., 2003; Misura and Baker, 2005). In the first step, Rosetta is used to search the space of possible backbone conformations with all side chains represented as centroids. This process is well described and has well-characterized error rates and behavior. High-confidence or low-scoring models are then refined using potentials that account for atomic detail such as hydrogen bonding, van der Waals forces, and electrostatics.

One major challenge that faces methods attempting to refine *de novo* methods is that the addition of side-chain degrees of freedom combined with the reduced length scale (reduced radius of convergence; one must get much closer to the correct answer before the scoring function recognizes the conformation as correct) of the potentials employed requires the sampling of a much larger space of possible conformations. Thus, one has to correctly determine roughly twice the number of bond angles to a higher tolerance if one hopes to succeed. An illustrative example of the difference in length scale (radius of convergence) between low-resolution methods and high-resolution methods is the scoring of hydrogen bonds. In the low-resolution Rosetta procedure, backbone hydrogen bonding is scored indirectly by a term designed to pack strands into sheets under the assumption that correct alignment of strands satisfies hydrogen bonds between backbone atom along the strand and that intrahelix backbone hydrogen bonds are already well accounted for by the local structure fragments. This low-resolution method first reduces strands to vectors and then scores strand arrangement (and the correct hydrogen bonding implicit in the relative positions/arrangement of all strand vector pairs) via functions dependent on the angular and distance relationships between the two vectors. Thus, the scoring function is robust to a rather large amount of error in the coordinates of individual electron donors and acceptors participating in backbone hydrogen bonds (as large numbers of residues are reduced to the angle and the distance between the two vectors representing a given pair of strands). In the high-resolution, refinement, mode of Rosetta, an empirical hydrogen bond term with angle and distance dependence between individual electron donors and acceptors is used (Rohl, 2005). This, more detailed hydrogen bond term has a higher fidelity and a more straightforward connection to the calculation of physically realistic energies (meaningful units) but requires more sampling, as smaller changes in the orientation of the backbone can cause large fluctuations in computed energy.

Another major challenge with high-resolution methods is the difficulty of computing accurate potentials for atomic detail protein modeling in solvent. Currently, electrostatic and solvation terms are among the most difficult terms to accurately model. Full treatment of the free energy of a protein conformation (with correct treatment of dielectric screening) is complicated by the fact that some waters are detectably bound to the surface of proteins and mediate interactions between residues (Finney, 1977). Another challenge is the computational cost of full treatment of electrostatic free energy by solving the Poisson–Boltzmann or linearized Poisson–Boltzmann equations for large numbers of conformations. In spite of these difficulties, several studies have shown that refinement of *de novo* structures with atomic detail potentials can increase our ability to select and/or generate near-native structures (Neves-Petersen and Petersen, 2003). These methods can correctly select near-native conformations from these ensembles and improve near-native structures but still rely heavily on the initial low-resolution search to produce an ensemble containing good starting structures (Lee et al., 2001; Tsai et al., 2003; Misura and Baker, 2005). Some recent examples of high-resolution predictions are quite encouraging, and an emerging consensus in the field is that higher resolution *de novo* structure prediction (structure predictions with atomic detail representations of side chains) will begin to work if sampling is dramatically increased.

Progress in high-resolution structure prediction will invariably be carried out in parallel with methods including, but not limited to, predicting protein–protein interactions, designing proteins, and distilling structures from partially assigned experimental data sets. Indeed many of the scoring and search strategies that high-resolution *de novo* structure refinement methods employ were initially developed in the context of homology modeling and protein design (Kuhlman et al., 2002; Rohl et al., 2004a).

Clustering: A Heuristic Approach to Approximating Entropic Determinants of Protein Folding

Several protein structure prediction methods are effectively two-step procedures involving the generation of large ensembles of conformations (each being the result of a minimization or simulation) followed by the clustering of the generated ensemble to produce one or more cluster centers that are taken to be the predicted models. Individual conformations are considered in the same cluster if their RMSD (root-mean-square distance) is within a given cutoff. Regardless of how one justifies the use of clustering as a means of selecting small numbers of predictions or models from ensembles of decoys conformations, the justification is indirectly supported by the efficacy of the procedure and the resultant observation that clustering has become a central, seemingly required, feature of successful *de novo* prediction methods. Starting with CASP3, the field has witnessed a proliferation of clustering methods as postsimulation processing steps in protein structure prediction methods (Simons et al., 1999a; Bonneau et al., 2002; Hung and Samudrala, 2003; Zhang and Skolnick, 2004).

Prediction of protein structure *de novo* using Rosetta relies heavily on a final clustering stage. In the first step, a large ensemble of potential protein structures is generated, and each conformation being the result of an extensive Monte Carlo search designed to minimize the

Rosetta scoring function (see Table 32.1). Clustering is then applied, and the clusters are ranked according to size. The tightness of clustering in the ensemble is also used as a measure of method success. Larger tight clusters indicate a higher probability that the method produced correct fold predictions for a given protein. Representatives from each cluster are chosen that are called cluster centers.

Each Rosetta simulation involving Monte Carlo runs can be thought of as a fast quench starting from a random point on the conformational landscape. Many of these individual simulations results in incorrect conformations that score nearly as well as any correct conformations generated in the full ensemble of decoy conformations, as judged by the Rosetta score. A number of other potentials tested also lack discriminative power at this stage. This lack of discrimination by *de novo* scoring functions is partially the result of inaccuracies in the scoring function, limitations in our ability to search the landscape, and the fact that entropic terms are a major contributor to the free energy of folding. In any case, this lack of discrimination is mitigated by the final clustering step, and it has been shown that the centers of the largest clusters in a clustered Rosetta decoy ensemble are in most cases the conformations closest to native.

The ubiquitous use of clustering can be justified in several ways: (1) Clustering can be thought of as a heuristic way to approximate the entropy of a given conformation given the full ensemble of decoy conformations generated for a given protein. (2) Clustering can be thought of as a signal averaging procedure, averaging out errors and noise in the low-resolution scoring function (3) Clustering can be thought of as taking advantage of foldable protein specific energy landscape features such as broad energy wells that are the result of proteins evolving to be robust to sequence and conformational changes from the native sequence or structure (a mix of sequence and configurational entropy) (Shortle, Simons, and Baker, 1998).

An interesting alternative to the strategy of clustering ensembles of results from independent minimizations is the use of replica exchange methods. These methods employ large numbers of simulations spanning a range of temperatures (defined physically if one uses a physical potential or simply as a constant in the exponent of the Boltzmann equation for probabilistic scoring functions). These independent simulations are carried out in parallel and are allowed to exchange temperatures throughout the run. This simulation strategy ideally allows for a random walk in energy space (and thus better sampling) and can be used to calculate entropic term *ex post facto*. Replica exchange Monte Carlo has been used successfully in the simulation and prediction of protein structure and is interesting due to its explicit connection to a physical description of the system and its ability to search low-energy states without getting trapped (Okamoto, 2004).

CASP: EVALUATION OF STRUCTURE PREDICTIONS

In general, the most effective methods for predicting structure *de novo* depend on parameters ultimately derived from the PDB. Several methods use the PDB directly to estimate local sequence and even explicitly use fragments of local sequence from the PDB to build global conformations. These uses of the PDB require that methods be tested using structures not present in the sets of protein structures used to train these methods (or present in the sets of structures used to predict local structure fragments). For these reasons, the CASP, a biannual community-wide blind test of prediction methods, was conceived and implemented (Benner, Cohen, and Gerloff, 1992; Barton and Russell, 1993; Fischer et al., 1999). The first such evaluation of structure predictions showed that published estimates of prediction error were smaller than prediction error measured on a set of novel proteins outside the training set. This is not surprising given the difficulty of avoiding overfitting is as complex a data space as protein structure (Lesk, 1997). Indeed, early experiments showed that no methods for *de novo* structure prediction were effective outside of carefully chosen benchmarks containing only the smallest proteins. Spurred on by these early evaluations, the field returned to the drawing board and produced multiple methods with much higher accuracies in the new folds or *de novo* category (Moult et al., 1999; Murzin, 1999; Orengo et al., 1999a). Thus, the CASP experiments proved to be invaluable to the field at that point in the development of the field, provoking a renewed interest in the *de novo* structure prediction and properly realigned interest in techniques according to effectiveness.

Arguably, CASP has the flaw that experts are allowed to intervene and manually curate their predictions prior to submission to the CASP evaluators. Thus, the results of CASP are a convolution of (1) the art of prediction (each group's intuition and skill using their tools) and (2) the relative performance of the core methods (the performance of each method in an automatic setting). Although this convolution reflects the reality when researchers aim to predict proteins of high interest, such as proteins involved in a specific function or proteins critical to a given disease or process being experimentally studied, it does not reflect the demands placed on a method when trying to predict whole genomes, where the shear number of predictions does not allow for much manual intervention. Several additional tests similar to CASP (in that they are blind tests of structure prediction) have been organized in response to the concerns of many that it is important to remove the human aspects of CASP. The critical assessment of fully automatic structure prediction (CAFASP) is an experiment running parallel with CASP that aims to test fully automated methods' performance on CASP targets, mainly testing servers instead of groups (Fischer et al., 1999; Fischer et al., 2003).

Several groups have also raised concerns that there are problems associated with the small numbers of proteins tested in each CASP experiment, and thus EVA and LiveBench were organized to test methods using larger numbers of proteins (Bujnicki et al., 2001; Rost and Eyrich, 2001; Rychlewski and Fischer, 2005). Both use proteins that have structures that are unknown to the participating prediction groups but that have been recently submitted to the PDB and are not open to the public at the time their sequences are released to those participating in LiveBench or EVA. The participating groups then have the time it takes for the new PDB entries to be validated to predict the structures.

All four of these tests of prediction methods, as well as benchmarks carried out by authors of any methods in question, are valuable ways of judging the performance of *de novo* methods. The methods, and elements of methods, we described above are generally accepted to be the best performers by the five above measures (four blind tests and author benchmarks).

BIOLOGICAL APPLICATIONS OF STRUCTURE PREDICTION

The Role of Structure Prediction in Biology

The main application of protein structure prediction is an open question that will take many years to develop. The answer depends on the relative rate of progress in several fields. We, however, consider that the most fruitful application of structure prediction in biology lies in understanding protein function and the ability to understand protein function on a genome scale. Structure predictions can offer meaningful biological insights at several functional levels depending on the method used for structure prediction, the resolution of the prediction, and the comprehensiveness or scale on which predictions are available for a given system.

At the highest levels of detail/accuracy (comparative modeling), there are several similarities between the uses of experimental and predicted protein structure and the types of functional information that can be extracted from models generated by both methods (Baker and Sali, 2001). For example, experimentally determined structures and structures resulting from comparative modeling can be used to help understand the details of protein function at an atomic scale. They can also be used to map conservation and mutagenesis data onto a structural framework and explore detailed functional relationships between proteins with similar folds or active sites.

At the other end of the prediction resolution spectrum, *de novo* structure prediction and fold recognition methods produce models of lower resolution than comparative models. These models can be used to assign putative functions to proteins for which little is known (Bonneau et al., 2001b). At the most basic level, we can use structural similarities between a predicted structure and known structures to explore possible distant evolutionary relationships between query proteins of unknown function and other well-studied proteins for which structures have been experimentally determined. A query protein is likely to share some functional aspects with proteins in the PDB that show strong structure–structure matches to a predicted structure with high confidence for that protein. This is based on the assumption that detectable structure relationships are conserved across a greater evolutionary distance than are detectable through sequence similarities. This assumption is well supported by multiple surveys of the distributions of folds and their related functions in the PDB (Murzin et al., 1995; Holm and Sander, 1997; Lo Conte et al., 2002; Orengo, Pearl, and Thornton, 2003). Promising preliminary results have been obtained with the approach of matching known structures to models produced by Rosetta: Dali, a server that compares structure through multiple structure alignments, frequently matches Rosetta models to protein structures related to the native structure for the sequence (Holm and Sander, 1993; Holm and Sander, 1995; Holm and Sander, 1997). The relationship between fold and function, however, is by no means a simple subject, and we refer the reader to several works that discuss this relationship in greater detail (Martin et al., 1998; Orengo, Todd, and Thornton, 1999b; Zhang et al., 1999; Kinch and Grishin, 2002).

A second way of exploring the functional significance of high-confidence predicted structures is to use libraries of three-dimensional functional motifs to search for conserved active site or functional motifs on the predicted structures (Moodie, Mitchell, and Thornton, 1996; Wallace et al., 1996; Fetrow and Skolnick, 1998).

Third, structures may be used to increase the reliability of matches to sequence motif libraries such as PROSITE—Taylor and Thornton's groups have shown that structural consistency can be used quite effectively to filter through weak sequence matches to PROSITE patterns (Martin et al., 1998; Kasuya and Thornton, 1999).

These basic methods, fold–fold matching, the use of small three-dimensional functional motif searches, and increasing confidence of motif searches, can in principle be combined to form the basis for deriving functional hypothesis from predicted structure, and thereby extending the completeness of genome annotations based only on primary sequence.

Structure Prediction as a Road to Function

The relationship between protein structure and protein function is discussed in detail in other reviews but will be reviewed briefly here in the context of *de novo* structure prediction. One paradigm for predicting the function of proteins of unknown function in the absence of homologues, sometimes referred to as the "sequence-to-structure-to-structure-to-function" paradigm, is based on the assumption that three-dimensional structure patterns are conserved across a much greater evolutionary distance than recognizable primary sequence patterns (Fetrow and Skolnick, 1998). This assumption is based on the results of several structure–function surveys that show that structure similarities (fold matches between different proteins in the PDB) in the absence of sequence similarities imply some shared function in the majority of cases (Holm and Sander, 1997; Martin et al., 1998; Orengo, Todd, and Thornton, 1999b; Lo Conte et al., 2000; Todd, Orengo, and Thornton, 2001). One protocol, alluded to above, for predicting protein function based on this observation is to predict the structure of a query sequence of interest and then use the predicted structure to search for fold or structural similarities between the predicted protein structure and experimentally determined protein structures in the PDB or a nonredundant subset of the PDB (Holm and Sander, 1993; Murzin et al., 1995; Ortiz, Strauss, and Olmea, 2002; Orengo, Pearl, and Thornton, 2003). There are several problems associated with deriving functional annotation from fold similarity; for example, fold similarities can occur through convergent evolution and thus have no functional implications. Also, aspects of function can change throughout evolution leaving only general function intact across a given fold superfamily (Rost, 1997; Grishin, 2001; Kinch and Grishin, 2002). Fold matches between the predicted structures and the PDB are thus treated as sources of putative general functional information and are functionally interpreted primarily in combination with other methods, such as global expression analysis and the predicted protein association network. To circumvent these ambiguities one can (1) use *de novo* structure prediction and/or fold recognition to generate a confidence-ranked list of possible structures for proteins or protein domains of unknown function, (2) search each of the ranked structure predictions against the PDB for fold similarities and possible three-dimensional motifs (3) calculate confidences for the fold predictions and three-dimensional motif matches and finally (4) evaluate possible functional roles in the context of the other systems biology data, such as expression analysis, protein interactions, metabolic networks, and comparative genomics.

Genome Annotation

To date, the annotation of protein function in newly sequenced genomes relies on a large array of tools based ultimately on primary sequence analysis (Altschul et al., 1997; Brenner, Chothia, and Hubbard, 1998; Tatusov et al., 2003; Bateman et al., 2004). These tools have afforded great progress in genome annotation including large improvements in gene detection, sequence alignment, and detection of homologous sequences across genomes as well as the creation of databases of common protein families and primary sequence functional motifs. Comparative modeling methods have been highly successful on many fronts, creating large databases of highly accurate structure predictions for many organisms, but are based on primary sequence matches between PDB and query sequences (Pieper et al., 2002). Primary sequence methods also exist for the prediction of basic local structure qualities (some of these patterns being lower complexity patterns) of sequences such as the location of coiled coil, transmembrane, and disordered regions (Sonnhammer, von Heijne, and Krogh, 1998; Jones, 1999; Nielsen, Brunak, and von Heijne, 1999; Ward et al., 2004).

However, there are many factors that reduce the ability of sequence homology searches to identify distant homologues (Russell and Ponting, 1998). Domain insertions and extensions, circular permutations, and the exchange of secondary structure elements have all been observed in cases where structural and functional relationships were not clear based on sequence homology. To reliably interpret the flood of sequence information currently entering databases, we must have at our disposal methods that can deal with these difficult cases as well as the clearer evolutionary relationships detectable at the sequence level. One recently solved genome (*Mycoplasma genitalium*) showed sequence homology to proteins of known function and/or structure for 38% of its proteins (Rychlewski, Zhang, and Godzik, 1998). For *S. cerevisiae*, approximately one-thirds of the open reading frames (ORFs) in the genome show homology to proteins of known structure. Annotation of ORFs lacking sequence homology to proteins of known function represents one of the most promising potential uses for *de novo* prediction. Current methods can make reasonable predictions for small alpha and alpha–beta proteins, and the Rosetta method in particular has been successful in blind tests and extensively used in house tests on this class of proteins. Of the ~6000 ORFs in bakers yeast, ~300 have at least 15% of their residues predicted to be helical with a total length less than 110 residues and no link to proteins of known structure (220 of these 300 also lack functional annotation) (Mewes et al., 2000). Note that these figures exclude transmembrane proteins, which are still out of reach for most structure prediction methods. Models can also be produced for modular domains of up to ~150 residues that occur in sufficiently diverse sequence contexts for their boundaries to be readily evident from multiple sequence alignments.

Efforts to use *de novo* structure prediction (and/or fold recognition) for genome annotations efforts must still employ sequence-based methods, as they provide a solid foundation on which all *de novo* methods discussed herein are reliant (see Figure 32.4). Any organization of these methods into an annotation pipeline must properly account for the fact that the accuracy/reliability is quite different between sequence and structure-based methods. One approach is to use structure prediction as part of a hierarchy where methods yielding high-confidence results are exhausted prior to computationally expensive and less accurate *de novo* structure prediction and fold recognition (Bonneau et al., 2004).

The need for methods that predict transmembrane protein structures and understanding membrane–protein interactions is not discussed in this chapter. The focus here is instead on soluble domains (including soluble domains excised from proteins containing transmembrane regions). Please refer to Chapter 36 for a more detailed discussion about research advances in this area. Part of the difficulty in predicting transmembrane protein structure lies in the paucity of membrane protein structures deposited on the PDB (Sonnhammer, von Heijne, and Krogh, 1998; Deshpande et al., 2005). Again, it is only with access to the PDB, an ideal and comprehensive gold standard by many criteria, that we can approach the problem of predicting soluble protein structure on a genome-wide scale.

Data Integration as a Means of Improving Structure Prediction Coverage and Error Rate

Genome-wide measurements of mRNA transcripts, protein concentrations, protein–protein interactions, and protein–DNA interactions generate rich sources of data on proteins, both those with known and those with unknown functions (Baliga et al., 2002; Baliga et al., 2004).

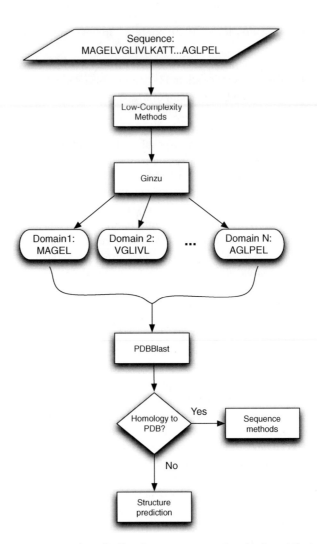

Figure 32.4. Structure annotation pipeline: Sequences enter the pipeline at the top of this figure, where they go through the masking of low-complexity regions such as transmembrane helices, signal peptides, and disordered regions. The resulting sequence (that not masked by step 1) is passed to Ginzu, a domain prediction pipeline (starting with PSIBLAST to PDB).

These systems-level measurements seldom suggest a unique function for a given protein of interest, but often suggest their association with or perhaps their direct participation in a previously known cellular process. Investigators using genome-wide experimental techniques are thus routinely generating data for proteins of hitherto unknown function that appear to play pivotal roles in their studies.

The first full-genome applications of *de novo* structure prediction was to the genome of *Halobacterium* NRC-1 (Bonneau et al., 2004). This archaeon is an extreme halophile that thrives in saturated brine environments such as the Dead Sea and solar salterns. It offers a versatile and easily assayed system with several well-coordinated physiologies that are necessary for survival in its harsh environment. The completely sequenced genome of

Halobacterium NRC-1 (containing ~2600 genes) has provided insights into many of its physiological capabilities; however, nearly half of all genes encoded in the halobacterial genome had no known function prior to reannotation (Rost and Valencia, 1996; Devos and Valencia, 2000; Ng et al., 2000; Devos and Valencia, 2001). A multiinstitutional effort is currently underway to study the genome-wide response of *Halobacterium NRC-1* to its environment, elevating the need for applying improved methods for annotating proteins of unknown function found in the *Halobacterium NRC-1* genome. Rosetta *de novo* structure prediction was used to predict three-dimensional structures for 1185 proteins and protein domains (<150 residues in length) found in *Halobacterium NRC-1*. Predicted structures were searched against the PDB to identify fold matches (Ortiz, Strauss, and Olmea, 2002) and were analyzed in the context of a predicted association network composed of several sources of functional associations such as predicted protein interactions, predicted operons, phylogenetic profile similarity, and domain fusion. This annotation pipeline was also applied to the recently sequenced genome of *Haloarcula marismortui* with similar rates of correct fold identification.

An application of *de novo* structure prediction to yeast has also been described (Malmstrom et al., 2007). This study focused on the application and integration of several methods (ranging from experimental methods to *de novo* structure prediction) to 100 essential open reading frames in yeast (Hazbun et al., 2003). For these 100 proteins, the group applied affinity purification followed by mass spectrometry (to detect protein binding partners), two-hybrid analysis, fluorescence microscopy (to localize proteins), with *de novo* structure prediction—Ginzu to separate domains (Chivian et al., 2003; Kim, Chivian, and Baker, 2004) and Rosetta to build structures for domains of unknown function. Due to the cost of experiments and the computational cost of Rosetta *de novo* structure prediction, the group was initially able to prototype the method on just these 100 proteins. Function was assigned to 48 of the proteins (as defined by assignment to Gene Ontology categories). In total, 77 of the 100 proteins were annotated (had confident hits) by the methods employed. Given that the starting set represented a difficult set of ORFs of no known function, this represents a significant milestone (Malmstrom et al., 2007).

Scaling this sort of approach up to whole genomes (including large eukaryotic genomes) is still a significant challenge. A grid computing solution (described below) has been employed to complete this study and fold the remaining ORFs in the yeast genome. Of 14934 domains parsed from yeast proteins, 3338 domains were folded with Rosetta and 581 were annotated with SCOP superfamilies. SCOP superfamilies were assigned using a Bayesian approach that integrated Gene Ontology annotations of process, component, and function. Also, 7094 domains were assigned structural annotation using homology modeling and fold recognition. Due to the wide use of yeast as a model organism, we can expect this complete resource to be a major step in crossing the social and technical barrier that has so far prevented the wide application of *de novo* structure prediction to biology. (Malmstrom et al., 2007)

A similar approach has also been applied to the Y chromosome of *Homo sapiens* (Ginalski et al., 2004). By integrating fold recognition with *de novo* structure, prediction folds were assigned to ~42 of the 60 recognized domains examined (these 60 domains originated from the 27 proteins thought to be encoded on this chromosome at the time of the study). In both these application, yeast and human, careful thought was put into reducing the set of proteins examined, and scaling up *de novo* structure prediction remains a critical bottleneck (the introduction of all-atom or high-resolution refinement of these predicted structures will only exacerbate this critical need for computing).

The Human Proteome Folding Project: Scaling up *De Novo* Structure Prediction, Rosetta on the World Community Grid

There are several strategies one can use to limit the number of protein domains for which computationally expensive *de novo* structure prediction needs to be carried out, allowing for the calculation of useful *de novo* structure predictions for only the most relevant subsets of larger genomes, as discussed above. In spite of these strategies, finding the required compute resources has been a constant challenge for the application of *de novo* structure prediction to functional annotation and has limited the application of the method. To circumvent this problem, we are currently applying a grid, distributed computing, solution to folding over 100,000 domains with the full Rosetta *de novo* structure prediction protocol (worldcommunitygrid.org). These domains were chosen by applying Ginzu (Chivian et al., 2003; Kim, Chivian, and Baker, 2004) to over 60 complete genomes as well as several other appropriate sequences in public sequence databases. The results will be integrated with data types that are appropriate/available for a given organism in collaboration with several other groups (Hazbun et al., 2003; Bonneau et al., 2004). This work is ongoing in collaboration with David Baker, Lars Malmstroem (University of Washington), Rick Alther, Bill Boverman and Viktors Berstis (IBM), and United Devices (Austin, TX). Currently, there are over a million volunteers (people who have downloaded the client to run grid Rosetta) comprising a virtual grid of over 3 million devices. Interested parties wishing to participate (donate idle CPU time on your desktop computer to this project) can download the grid-enabled Rosetta client at worldcommunitygrid.org. This amount of computational power will enable us to remove the barrier represented by the computational cost of *de novo* methods.

FUTURE DIRECTIONS

Structure Prediction and Systems Biology, Data Integration

Even with dramatically improved accuracy, we still face challenges due to the ambiguities of the relationship between fold and function seen for many fold families (indeed even close sequence homology is not always trivial to interpret as functional similarity). Thus, the full potential of *de novo* structure prediction in a systems biology context can only be realized if structure predictions are integrated into larger analysis and subsequently made accessible to biologists through better data integration, analysis, and visualization tools. One clear example of this is provided by the bacterial transcription factors for which even strong sequence similarity can imply several possible functions, and system-wide information is required to determine a meaningful function (the target of a given transcription factor).

Need for Improved Accuracy and Extending the Reach of *De Novo* Methods

Although we have argued that data integration is as critical a bottleneck as any other and that there are current applications of *de novo* structure prediction, it is clear that improved accuracy is also essential for progress in the field and for the acceptance of *de novo* structure methods by the end users of whole genome annotations. There is still a significant amount of error in predictions generated using current structure prediction and domain parsing methods. Extending the size limit of protein folding methods is a promising area of active research as is the development of higher resolution refinement methods. *De novo* structure prediction requires large amounts of CPU time compared to sequence-based and fold recognition methods (although the use of distributed computing and Moore's law continue to make this less of a bottleneck). Integrating *de novo* predictions with orthogonal sources of general and putative functional information, both experimental and computational, will likely facilitate the annotation of significant portions of the protein sequences resulting from ongoing sequencing efforts, as well as proteins in currently sequenced genomes.

REFERENCES

Alm E, Baker D (1999a): Matching theory and experiment in protein folding. *Curr Opin Struct Biol* 9(2):189–196.

Alm E, Baker D (1999b): Prediction of protein-folding mechanisms from free-energy landscapes derived from native structures. *Proc Natl Acad Sci USA* 96(20):11305–11310.

Altschul SF, Madden TL, Schaffer AA, Zhang J, Zhang Z, Miller W, Lipman DJ (1997): Gapped BLAST and PSI-BLAST: a new generation of protein database search programs. *Nucleic Acids Res* 25(17):3389–3402.

Badretdinov A, Finkelstein AV (1998): How homologs can help to predict protein folds even though they cannot be predicted for individual sequences. *J Comput Biol* 5(3):369–376.

Baker D, Sali A (2001): Protein structure prediction and structural genomics. *Science* 294 (5540):93–96.

Baldwin RL (1999): Protein folding from 1961 to 1982. *Nat Struct Biol* 6(9):814–817.

Baliga NS, Pan M, Goo YA, Yi EC, Goodlett DR, Dimitrov K, Shannon P, Aebersold R, Ng WV, Hood L (2002): Coordinate regulation of energy transduction modules in Halobacterium sp. analyzed by a global systems approach. *Proc Natl Acad Sci USA* 99(23):14913–14918.

Baliga NS, Bjork SJ, Bonneau R, Pan M, Iloanusi C, Kottemann MC, Hood L, DiRuggiero J (2004): Systems level insights into the stress response to UV radiation in the halophilic archaeon *Halobacterium* NRC-1. *Genome Res* 14(6):1025–1035.

Barton GJ, Russell RB (1993): Protein structure prediction. *Nature* 361(6412):505–506. [letter; comment]

Bateman A, Birney E, Durbin R, Eddy SR, Howe KL, Sonnhammer EL (2000): The Pfam protein families database. *Nucleic Acids Res* 28(1):263–266.

Bateman A, Coin L, Durbin R, Finn RD, Hollich V, Griffiths-Jones S, Khanna A, Marshall M, Moxon S, Sonnhammer EL, Studholme DJ, Yeats C, Eddy SR (2004): The Pfam protein families database. *Nucleic Acids Res* 32(Database issue):D138–D141

Benner SA, Cohen MA, Gerloff D (1992): Correct structure prediction? [letter]. *Nature* 359 (6398):781

Blanco FJ, Rivas G, Serrano L (1994): A short linear peptide that folds into a native stable beta-hairpin in aqueous solution. *Nat Struct Biol* 1(9):584–590.

Bonneau R, Baker D (2001): Ab initio protein structure prediction: progress and prospects. *Annu Rev Biophys Biomol Struct* 30:173–189.

Bonneau R, Strauss CE, Baker D (2001a): Improving the performance of Rosetta using multiple sequence alignment information and global measures of hydrophobic core formation. *Proteins* 43(1):1–11.

Bonneau R, Tsai J, Ruczinski I, Baker D (2001b): Functional inferences from blind ab initio protein structure predictions. *J Struct Biol* 134(2–3):186–190.

Bonneau R, Tsai J, Ruczinski I, Chivian D, Rohl C, Strauss CE, Baker D (2001c): Rosetta in CASP4: progress in ab initio protein structure prediction. *Proteins* (Suppl. 5):119–126.

Bonneau R, Strauss CE, Rohl CA, Chivian D, Bradley P, Malmstrom L, Robertson T, Baker D (2002): De novo prediction of three-dimensional structures for major protein families. *J Mol Biol* 322(1):65–78.

Bonneau R, Baliga NS, Deutsch EW, Shannon P, Hood L (2004): Comprehensive de novo structure prediction in a systems-biology context for the archaea *Halobacterium* sp. NRC-1. *Genome Biol* 5(8):R52

Bowers P, Strauss CE, Baker D (2000): De Novo Protein Structure Determination Using Sparse NMR Data. *J Biomol NMR* 18(4):311–318.

Bowie JU, Eisenberg D (1994): An evolutionary approach to folding small alpha-helical proteins that uses sequence information and an empirical guiding fitness function. *Proc Natl Acad Sci USA* 91 (10):4436–4440.

Bradley P, Chivian D, Meiler J, Misura KM, Rohl CA, Schief WR, Wedemeyer WJ, Schueler-Furman O, Murphy P, Schonbrun J, Strauss CE, Baker D (2003): Rosetta predictions in CASP5: successes, failures, and prospects for complete automation. *Proteins* 53(Suppl. 6):457–468.

Bradley P, Malmstrom L, Qian B, Schonbrun J, Chivian D, Kim DE, Meiler J, Misura KM, Baker D (2005): Free modeling with Rosetta in CASP6. *Proteins* 61(Suppl. 7):128–134.

Brenner SE, Chothia C, Hubbard TJ (1998): Assessing sequence comparison methods with reliable structurally identified distant evolutionary relationships. *Proc Natl Acad Sci USA* 95 (11):6073–6078.

Bujnicki JM, Elofsson A, Fischer D, Rychlewski L (2001): LiveBench-1: continuous benchmarking of protein structure prediction servers. *Protein Sci* 10(2):352–361.

Bystroff C, Simons KT, Han KF, Baker D (1996): Local sequence–structure correlations in proteins. *Curr Opin Biotechnol* 7(4):417–421.

Bystroff C, Baker D (1998): Prediction of local structure in proteins using a library of sequence–structure motifs. *J Mol Biol* 281(3):565–577.

Bystroff C, Thorsson V, Baker D (2000): HMMSTR: a hidden Markov model for local sequence-structure correlations in proteins. *J Mol Biol* 301(1):173–190.

Bystroff C, Shao Y (2002): Fully automated ab initio protein structure prediction using I-SITES, HMMSTR and ROSETTA. *Bioinformatics* 18(Suppl. 1):S54–61.

Callihan DE, Logan TM (1999): Conformations of peptide fragments from the FK506 binding protein: comparison with the native and urea-unfolded states. *J Mol Biol* 285(5):2161–2175.

Chevalier BS, Kortemme T, Chadsey MS, Baker D, Monnat RJ, Stoddard BL (2002): Design, activity, and structure of a highly specific artificial endonuclease. *Mol Cell* 10(4):895–905.

Chivian D, Kim DE, Malmstrom L, Bradley P, Robertson T, Murphy P, Strauss CE, Bonneau R, Rohl CA, Baker D (2003): Automated prediction of CASP-5 structures using the Robetta server. *Proteins* 53(Suppl. 6):524–533.

Chothia C (1984): Principles that determine the structure of proteins. *Annu Rev Biochem* 53:537–572.

Cohen FE, Sternberg MJ, Taylor WR (1980): Analysis and prediction of protein beta-sheet structures by a combinatorial approach. *Nature* 285(5764):378–382.

Cohen FE, Sternberg MJ, Taylor WR (1982): Analysis and prediction of the packing of alpha-helices against a beta-sheet in the tertiary structure of globular proteins. *J Mol Biol* 156(4):821–862.

Cohen BI, Presnell SR, Cohen FE (1993): Origins of structural diversity within sequentially identical hexapeptides. *Protein Sci* 2(12):2134–2145.

Deshpande N, Addess KJ, Bluhm WF, Merino-Ott JC, Townsend-Merino W, Zhang Q, Knezevich C, Xie L, Chen L, Feng Z, Green RK, Flippen-Anderson JL, Westbrook J, Berman HM, Bourne PE (2005): The RCSB Protein Data Bank: a redesigned query system and relational database based on the mmCIF schema. *Nucleic Acids Res* 33(Database Issue):D233–D237

Devos D, Valencia A (2000): Practical limits of function prediction. *Proteins* 41(1):98–107.

Devos D, Valencia A (2001): Intrinsic errors in genome annotation. *Trends Genet* 17(8):429–431.

Dill KA, Bromberg S, Yue K, Fiebig KM, Yee DP, Thomas PD, Chan HS (1995): Principles of protein folding—a perspective from simple exact models. *Protein Sci* 4(4):561–602.

Dill KA (1997): Additivity principles in biochemistry. *J Biol Chem* 272(2):701–704.

Domingues FS, Koppensteiner WA, Jaritz M, Prlic A, Weichenberger C, Wiederstein M, Floeckner H, Lackner P, Sippl MJ (1999): Sustained performance of knowledge-based potentials in fold recognition. *Proteins* 37(S3):112–120.

Dunbrack RL Jr, Cohen FE (1997): Bayesian statistical analysis of protein side-chain rotamer preference. *Proteins Sci* 6:1661–1681.

Eyrich VA, Standley DM, Friesner RA (1999): Prediction of protein tertiary structure to low resolution: performance for a large and structurally diverse test set. *J Mol Biol* 288(4):725–742.

Fetrow JS, Skolnick J (1998): Method for prediction of protein function from sequence using the sequence-to-structure-to-function paradigm with application to glutaredoxins/thioredoxins and T1 ribonucleases. *J Mol Biol* 281(5):949–968.

Finney JL (1977): The organization and function of water in protein crystals. *Philos Trans R Soc Lond B Biol Sci* 278(959):3–32.

Fischer D, Barret C, Bryson K, Elofsson A, Godzik A, Jones D, Karplus KJ, Kelley LA, MacCallum RM, Pawowski K, Rost B, Rychlewski L, Sternberg M (1999): CAFASP-1: critical assessment of fully automated structure prediction methods [In Process Citation]. *Proteins* (Suppl. 3): 209–217.

Fischer D, Rychlewski L, Dunbrack RL Jr, Ortiz AR, Elofsson A (2003): CAFASP3: the third critical assessment of fully automated structure prediction methods. *Proteins* 53(Suppl. 6):503–516.

Galzitskaya OV, Melnik BS (2003): Prediction of protein domain boundaries from sequence alone. *Protein Sci* 12(4):696–701.

George RA, Heringa J (2002): SnapDRAGON: a method to delineate protein structural domains from sequence data. *J Mol Biol* 316(3):839–851.

Ginalski K, Rychlewski L, Baker D, Grishin NV (2004): Protein structure prediction for the male-specific region of the human Y chromosome. *Proc Natl Acad Sci USA* 101(8):2305–2310.

Gray JJ, Moughon S, Wang C, Schueler-Furman O, Kuhlman B, Rohl CA, Baker D (2003a): Protein-protein docking with simultaneous optimization of rigid-body displacement and side-chain conformations. *J Mol Biol* 331(1):281–299.

Gray JJ, Moughon SE, Kortemme T, Schueler-Furman O, Misura KM, Morozov AV, Baker D (2003b): Protein–protein docking predictions for the CAPRI experiment. *Proteins* 52(1):118–122.

Grishin NV (2001): Fold change in evolution of protein structures. *J Struct Biol* 134(2–3):167–185.

Han KF, Bystroff C, Baker D (1997): Three-dimensional structures and contexts associated with recurrent amino acid sequence patterns. *Protein Sci* 6(7):1587–1590.

Hazbun TR, Malmstrom L, Anderson S, Graczyk BJ, Fox B, Riffle M, Sundin BA, Aranda JD, McDonald WH, Chiu CH, Snydsman BE, Bradley P, Muller EG, Fields S, Baker D, Yates JR, III Davis TN (2003): Assigning function to yeast proteins by integration of technologies. *Mol Cell* 12(6):1353–1365.

Hinds DA, Levitt M (1994): Exploring conformational space with a simple lattice model for protein structure. *J Mol Biol* 243(4):668–682.

Holm L, Sander C (1993): Protein structure comparison by alignment of distance matrices. *J Mol Biol* 233(1):123–138.

Holm L, Sander C (1995): Dali: a network tool for protein structure comparison. *Trends Biochem Sci* 20(11):478–480.

Holm L, Sander C (1997): Dali/FSSP classification of three-dimensional protein folds. *Nucleic Acids Res* 25(1):231–234.

Huang ES, Subbiah S, Levitt M (1995): Recognizing native folds by the arrangement of hydrophobic and polar residues. *J Mol Biol* 252(5):709–720.

Hung LH, Samudrala R (2003): PROTINFO: Secondary and tertiary protein structure prediction. *Nucleic Acids Res* 31(13):3296–3299.

Ishikawa K, Yue K, Dill KA (1999): Predicting the structures of 18 peptides using Geocore. *Protein Sci* 8(4):716–721.

Jernigan RL, Bahar I (1996): Structure-derived potentials and protein simulations. *Curr Opin Struct Biol* 6(2):195–209.

Jones DT, Thornton JM (1996): Potential energy functions for threading. *Curr Opin Struct Biol* 6(2):210–216.

Jones DT (1999): Protein secondary structure prediction based on position-specific scoring matrices. *J Mol Biol* 292(2):195–202.

Kabsch W, Sander C (1983): Dictionary of protein secondary structure: pattern recognition of hydrogen-bonded and geometrical features. *Biopolymers* 22:2577–2637.

Kabsch W, Sander C (1984): On the use of sequence homologies to predict protein structure: identical pentapeptides can have completely different conformations. *Proc Natl Acad Sci USA* 81 (4):1075–1078.

Karplus K, Barrett C, Cline M, Diekhans M, Grate L, Hughey R (1999): Predicting protein structure using only sequence information. *Proteins* 37(S3):121–125.

Karplus K, Karchin R, Draper J, Casper J, Mandel-Gutfreund Y, Diekhans M, Hughey R (2003): Combining local-structure, fold-recognition, and new fold methods for protein structure prediction. *Proteins* 53(Suppl. 6):491–496.

Kasuya A, Thornton JM (1999): Three-dimensional structure analysis of PROSITE patterns. *J Mol Biol* 286(5):1673–1691.

Kim DE, Chivian D, Baker D (2004): Protein structure prediction and analysis using the Robetta server. *Nucleic Acids Res* 32(Web Server issue):W526–W531

Kinch LN, Grishin NV (2002): Evolution of protein structures and functions. *Curr Opin Struct Biol* 12 (3):400–408.

King RD, Saqi M, Sayle R, Sternberg MJ (1997): DSC: public domain protein secondary structure predication. *Comput Appl Biosci* 13(4):473–474.

Kolinski A, Skolnick J (1994a): Monte Carlo simulations of protein folding. I, Lattice model and interaction scheme. *Proteins* 18(4):338–352.

Kolinski A, Skolnick J (1994b): Monte Carlo simulations of protein folding. II. Application to protein A, ROP, and crambin. *Proteins* 18(4):353–366.

Kortemme T, Joachimiak LA, Bullock AN, Schuler AD, Stoddard BL, Baker D (2004): Computational redesign of protein–protein interaction specificity. *Nat Struct Mol Biol* 11(4):371–379.

Kuhlman B, Baker D (2000): Native protein sequences are close to optimal for their structures. *Proc Natl Acad Sci USA* 97(19):10383–10388.

Kuhlman B, ONeill JW, Kim DE, Zhang KY, Baker D (2002): Accurate computer-based design of a new backbone conformation in the second turn of protein L. *J Mol Biol* 315(3):471–477.

Kuhlman B, Dantas G, Ireton GC, Varani G, Stoddard BL, Baker D (2003): Design of a novel globular protein fold with atomic-level accuracy. *Science* 302(5649):1364–1368.

Lazaridis T, Karplus M (1999): Effective energy function for proteins in solution. *Proteins* 35: 133–152.

Lee MR, Duan Y, Kollman PA (2000): Use of MM-PB/SA in estimating the free energies of proteins: application to native, intermediates, and unfolded villin headpiece. *Proteins* 39(4):309–316.

Lee MR, Baker D, Kollman PA (2001a): 2.1 and 1.8 Å average C(alpha) RMSD structure predictions on two small proteins, HP-36 and s15. *J Am Chem Soc* 123:1040–1046.

Lee MR, Tsai J, Baker D, Kollman PA (2001b): Molecular dynamics in the endgame of protein structure prediction. *J Mol Biol* 313(2):417–430.

Lesk AM (1997): CASP2: report on ab initio predictions. *Proteins* (Suppl. 1):151–166.

Lesk AM, Lo Conte L, Hubbard TJ (2001): Assessment of novel fold targets in CASP4: predictions of three-dimensional structures, secondary structures, and interresidue contacts. *Proteins* (Suppl. 5): 98–118.

Liu J, Rost B (2004): CHOP: parsing proteins into structural domains. *Nucleic Acids Res* 32(Web Server issue):W569–W571

Lo Conte L, Ailey B, Hubbard TJ, Brenner SE, Murzin AG, Chothia C (2000): SCOP: a structural classification of proteins database. *Nucleic Acids Res* 28(1):257–259.

Lo Conte L, Brenner SE, Hubbard TJ, Chothia C, Murzin AG (2002): SCOP database in 2002: refinements accommodate structural genomics. *Nucleic Acids Res* 30(1):264–267.

Lomize AL, Pogozheva ID, Mosberg HI (1999): Prediction of protein structure: the problem of fold multiplicity. *Proteins* 37(S3):199–203.

Malmstrom L, Riffle M, Strauss CE, Chivian D, Davis TN, Bonneau R, Baker D (2007): Superfamily assignments for the yeast proteome through integration of structure prediction with the gene ontology. *PLoS Biol* 5(4):e76

Mark AE, van Gunsteren WF (1994): Decomposition of the free energy of a system in terms of specific interactions. *J Mol Biol* 240(1):167–176.

Marqusee S, Robbins VH, Baldwin RL (1989): Unusually stable helix formation in short alanine-based peptides. *Proc Natl Acad Sci USA* 86(14):5286–5290.

Martin AC, Orengo CA, Hutchinson EG, Jones S, Karmirantzou M, Laskowski RA, Mitchell JB, Taroni C, Thornton JM (1998): Protein folds and functions. *Structure* 6(7):875–884.

Mewes HW, Frishman D, Gruber C, Geier B, Haase D, Kaps A, Lemcke K, Mannhaupt G, Pfeiffer F, Schuller C, Stocker S, Weil B (2000): MIPS: a database for genomes and protein sequences. *Nucleic Acids Res* 28(1):37–40.

Misura KM, Baker D (2005): Progress and challenges in high-resolution refinement of protein structure models. *Proteins* 59(1):15–29.

Miyazawa S, Jernigan RL (1999): An empirical energy potential with a reference state for protein fold and sequence recognition. *Proteins* 36(3):357–369.

Moodie SL, Mitchell JB, Thornton JM (1996): Protein recognition of adenylate: an example of a fuzzy recognition template. *J Mol Biol* 263(3):486–500.

Moult J, Hubbard T, Fidelis K, Pedersen JT (1999): Critical assessment of methods of protein structure prediction (CASP): round III [In Process Citation]. *Proteins* (Suppl. 3):2–6.

Munoz V, Serrano L (1996): Local versus nonlocal interactions in protein folding and stability–an experimentalists point of view. *Fold Des* 1(4):R71–R77

Murzin AG, Brenner SE, Hubbard T, Chothia C (1995): SCOP: a structural classification of proteins database for the investigation of sequences and structures. *J Mol Biol* 247(4):536–540.

Murzin AG (1999): Structure classification-based assessment of CASP3 predictions for the fold recognition targets [In Process Citation]. *Proteins* (Suppl. 3):88–103.

Neria E, Nitzan A, Barnett RN, Landman U (1991): Quantum dynamical simulations of nonadiabatic processes: Solvation dynamics of the hydrated electron. *Phys Rev Lett* 67:1011–1014.

Neves-Petersen MT, Petersen SB (2003): Protein electrostatics: a review of the equations and methods used to model electrostatic equations in biomolecules—applications in biotechnology. *Biotechnol Annu Rev* 9:315–395.

Ng WV, Kennedy SP, Mahairas GG, Berquist B, Pan M, Shukla HD, Lasky SR, Baliga NS, Thorsson V, Sbrogna J, Swartzell S, Weir D, Hall J, Dahl TA, Welti R, Goo YA, Leithauser B, Keller K, Cruz R, Danson MJ, Hough DW, Maddocks DG, Jablonski PE, Krebs MP, Angevine CM, Dale H, Isenbarger TA, Peck RF, Pohlschroder M, Spudich JL, Jung KW, Alam M, Freitas T, Hou S,

Daniels CJ, Dennis PP, Omer AD, Ebhardt H, Lowe TM, Liang P, Riley M, Hood L, DasSarma S (2000): Genome sequence of *Halobacterium* species NRC-1. *Proc Natl Acad Sci USA* 97(22): 12176–12181.

Nielsen H, Brunak S, von Heijne G (1999): Machine learning approaches for the prediction of signal peptides and other protein sorting signals. *Protein Eng* 12(1):3–9.

Novotny J, Bruccoleri R, Karplus M (1984): An analysis of incorrectly folded protein models. Implications for structure predictions. *J Mol Biol* 177(4):787–818.

Okamoto Y (2004): Generalized-ensemble algorithms: enhanced sampling techniques for Monte Carlo and molecular dynamics simulations. *J Mol Graph Model* 22(5):425–439.

Orengo CA, Bray JE, Hubbard T, LoConte L, Sillitoe I (1999a): Analysis and assessment of ab initio three-dimensional prediction, secondary structure, and contacts prediction [In Process Citation]. *Proteins* (Suppl. 3):149–170.

Orengo CA, Todd AE, Thornton JM (1999b): From protein structure to function. *Curr Opin Struct Biol* 9(3):374–382.

Orengo CA, Pearl FM, Thornton JM (2003): The CATH domain structure database. *Methods Biochem Anal* 44:249–271.

Ortiz AR, Kolinski A, Rotkiewicz P, Ilkowski B, Skolnick J (1999): Ab initio folding of proteins using restraints derived from evolutionary information [In Process Citation]. *Proteins* (Suppl. 3): 177–185.

Ortiz AR, Strauss CE, Olmea O (2002): MAMMOTH (matching molecular models obtained from theory): an automated method for model comparison. *Protein Sci* 11(11):2606–2621.

Park BH, Levitt M (1995): The complexity and accuracy of discrete state models of protein structure. *J Mol Biol* 249(2):493–507.

Park B, Levitt M (1996): Energy functions that discriminate X-ray and near native folds from well-constructed decoys. *J Mol Biol* 258(2):367–392.

Park BH, Huang ES, Levitt M (1997): Factors affecting the ability of energy functions to discriminate correct from incorrect folds. *J Mol Biol* 266(4):831–846.

Pieper U, Eswar N, Stuart AC, Ilyin VA, Sali A (2002): MODBASE, a database of annotated comparative protein structure models. *Nucleic Acids Res* 30(1):255–259.

Reva BA, Finkelstein AV (1996): Search for the most stable folds of protein chains: II. Computation of stable architectures of beta-proteins using a self-consistent molecular field theory. *Protein Eng* 9(5):399–411.

Reva BA, Finkelstein AV, Sanner MF, Olson AJ (1996): Adjusting potential energy functions for lattice models of chain molecules. *Proteins* 25(3):379–388.

Rohl CA, Strauss CE, Misura KM, Baker D (2004): Protein structure prediction using Rosetta. *Methods Enzymol* 383:66–93.

Rohl CA, Strauss CE, Chivian D, Baker D (2004a): Modeling structurally variable regions in homologous proteins with rosetta. *Proteins* 55(3):656–677.

Rohl CA, Strauss CE, Misura KM, Baker D (2004b): Protein structure prediction using Rosetta. *Methods Enzymol* 383:66–93.

Rohl CA (2005): Protein structure estimation from minimal restraints using rosetta. *Methods Enzymol* 394:244–260.

Rost B, Sander C (1993): Prediction of protein secondary structure at better than 70% accuracy. *J Mol Biol* 232(2):584–599.

Rost B, Valencia A (1996): Pitfalls of protein sequence analysis. *Curr Opin Biotechnol* 7(4):457–461.

Rost B (1997): Protein structures sustain evolutionary drift. *Fold Des* 2(3):S19–24.

Rost B, Eyrich VA (2001): EVA: large-scale analysis of secondary structure prediction. *Proteins* (Suppl. 5): 192–199.

Russell RB, Ponting CP (1998): Protein fold irregularities that hinder sequence analysis. *Curr Opin Struct Biol* 8(3):364–371.

Rychlewski L, Zhang B, Godzik A (1998): Fold and function predictions for *Mycoplasma genitalium* proteins. *Fold Des* 3(4):229–238.

Rychlewski L, Fischer D (2005): LiveBench-8: the large-scale, continuous assessment of automated protein structure prediction. *Protein Sci* 14(1):240–245.

Samudrala R, Xia Y, Huang E, Levitt M (1999a): Ab initio protein structure prediction using a combined hierarchical approach. *Proteins* (Suppl. 3):194–198.

Samudrala R, Xia Y, Huang E, Levitt M (1999b): Ab initio protein structure prediction using a combined hierarchical approach [In Process Citation]. *Proteins* (Suppl. 3):194–198.

Samudrala R, Xia Y, Levitt M, Huang ES (1999c): A combined approach for ab initio construction of low resolution protein tertiary structures from sequence. *Pac Symp Biocomput* 505–516.

Shortle D, Simons KT, Baker D (1998): Clustering of low-energy conformations near the native structures of small proteins. *Proc Natl Acad Sci USA* 95(19):11158–11162.

Simons KT, Kooperberg C, Huang E, Baker D (1997): Assembly of protein tertiary structures from fragments with similar local sequences using simulated annealing and Bayesian scoring functions. *J Mol Biol* 268(1):209–225.

Simons KT, Bonneau R, Ruczinski I, Baker D (1999a): Ab initio protein structure prediction of CASP III targets using ROSETTA. *Proteins* (Suppl. 3):171–176.

Simons KT, Bonneau R, Ruczinski I, Baker D (1999b): Ab initio protein structure prediction of CASP III targets using ROSETTA. *Proteins* 37(S3):171–176.

Simons KT, Ruczinski I, Kooperberg C, Fox BA, Bystroff C, Baker D (1999c): Improved recognition of native-like protein structures using a combination of sequence-dependent and sequence-independent features of proteins. *Proteins* 34(1):82–95.

Sippl MJ (1995): Knowledge-based potentials for proteins. *Curr Opin Struct Biol* 5(2):229–235.

Skolnick J, Kolinski A (1991): Dynamic Monte Carlo simulations of a new lattice model of globular protein folding, structure and dynamics [published erratum appears in J Mol Biol, 1992 Jan 20; 223(2): 583]. *J Mol Biol* 221(2):499–531.

Sonnhammer EL, von Heijne G, Krogh A (1998): A hidden Markov model for predicting transmembrane helices in protein sequences. *Proc Int Conf Intell Syst Mol Biol* 6:175–182.

Srinivasan R, Rose GD (1995): LINUS: a hierarchic procedure to predict the fold of a protein. *Proteins* 22(2):81–99.

Sternberg MJ, Cohen FE, Taylor WR (1982): A combinational approach to the prediction of the tertiary fold of globular proteins. *Biochem Soc Trans* 10(5):299–301.

Tatusov RL, Fedorova ND, Jackson JJ, Jacobs AR, Kiryutin B, Koonin EV, Krylov DM, Mazumder R, Mekhedov SL, Nikolskaya AN, Rao BS, Smirnov S, Sverdlov AV, Vasudevan S, Wolf YI, Yin JJ, Natale DA (2003): The COG database: an updated version includes eukaryotes. *BMC Bioinformatics* 4(1):41

Thomas PD, Dill KA (1996): Statistical Potentials Extracted from Protein Structures: how accurate are they? *J Mol Biol* 257(1):457–469.

Todd AE, Orengo CA, Thornton JM (2001): Evolution of function in protein superfamilies, from a structural perspective. *J Mol Biol* 307(4):1113–1143.

Tsai J, Bonneau R, Morozov AV, Kuhlman B, Rohl CA, Baker D (2003): An improved protein decoy set for testing energy functions for protein structure prediction. *Proteins* 53(1):76–87.

Wallace AC, Laskowski RA, Thornton JM (1996): Derivation of 3D coordinate templates for searching structural databases: application to Ser-His-Asp catalytic triads in the serine proteinases and lipases. *Protein Sci* 5(6):1001–1013.

Ward JJ, McGuffin LJ, Bryson K, Buxton BF, Jones DT (2004): The DISOPRED server for the prediction of protein disorder. *Bioinformatics* 20(13):2138–2139.

Yi Q, Bystroff C, Rajagopal P, Klevit RE, Baker D (1998): Prediction and structural characterization of an independently folding substructure in the src SH3 domain. *J Mol Biol* 283(1):293–300.

Zhang B, Rychlewski L, Pawlowski K, Fetrow JS, Skolnick J, Godzik A (1999): From fold predictions to function predictions: automation of functional site conservation analysis for functional genome predictions. *Protein Sci* 8(5):1104–1115.

Zhang Y, Skolnick J (2004): Tertiary structure predictions on a comprehensive benchmark of medium to large size proteins. *Biophys J* 87(4):2647–2655.

33

RNA STRUCTURAL BIOINFORMATICS

Magdalena A. Jonikas, Alain Laederach, and Russ B. Altman

INTRODUCTION

RNA molecules play a range of roles in cells, including key roles in both transcription and translation. Functional RNAs can have complex three-dimensional structures or simply rely on the properties of their primary sequence to perform their function. Among the highly structured RNAs are transfer RNAs (tRNA), ribosomal RNAs (rRNA), ribozymes and riboswitches. These classes of RNAs play catalytic or structural roles in the cell, facilitated by their complex three-dimensional structures. Other RNAs, such as micro-RNAs and sno-RNAs, perform their regulatory function by relying on simple base pairing to target RNAs.

Three levels of structure can exist in RNA, the primary sequence, secondary structure and tertiary structure, with each structural level serving an essential role in the function of RNA. The structure of RNA is covered extensively in Chapter 3; however, we will discuss briefly these different structural levels using tRNA to illustrate these principles. At the primary sequence level, four bases serve as the main components: adenine, guanine, cytosine, and uracil (a pyrimidine similar to the thymine found in DNA). In addition to this basic set, many modified bases such as pseudouridine and even the DNA base thymine are frequently present in RNA molecules. Figure 33.1a shows several primary sequences of molecules in the tRNA family in a multiple sequence alignment.

The secondary structure units of RNA are duplexes and loops. RNA duplexes, also called helices or stems, are formed when complementary regions within the molecule bind to form a right-handed helix. Hydrogen bonds between two complementary bases, as well as stacking interactions, stabilize the helical structure. Loops are regions of the RNA molecule that do not form a duplex. Figure 33.1b is a dot plot showing interacting residues and Figure 33.1c shows a secondary structure map for a tRNA molecule where groups of base pairs form four helices.

The tertiary structure of an RNA molecule is the interaction of duplexes and loops to form a compact three-dimensional structure. Intramolecular base pairing between

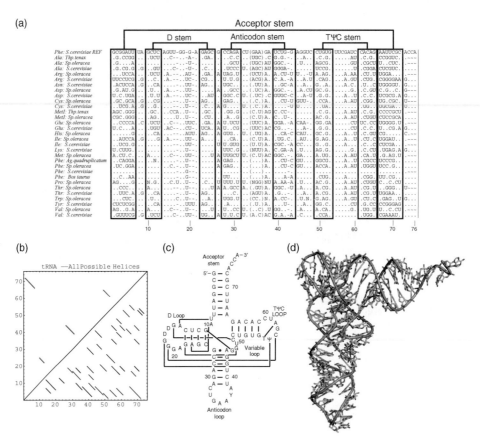

Figure 33.1. (a) A multiple sequence alignment of the tRNA family (Cannone et al., 2002). The four conserved helices are labeled in the alignment. (b) A dot plot for tRNA. The lower right triangle shows all the possible helices in tRNA. The upper left triangle shows the actual helices present in tRNA. (Cannone et al., 2002) (c) The secondary structure representation of yeast phenylalanine tRNA. Long ranged tertiary interactions found by covariation analysis are shown in addition to the four conserved helices (Cannone et al., 2002). (d) Tertiary structure of the yeast phenylalanine tRNA molecule (6TNA.pdb).

nucleotides can result in structures more complex than helices. These structures include pseudoknots that occur when a loop at the end of a hairpin helix participates in another helical region. These hydrogen bonded pairing interactions are not always part of a helix and often form long-range tertiary interactions instead. Occasionally, three different nucleotides will be observed to form hydrogen bonding interactions with each other resulting in a triple pair. Figure 33.1d shows the tertiary structure of a tRNA molecule, solved by X-ray crystallography.

The history of RNA structure models begins with Michael Levitt's remarkable 1969 prediction of the tRNA molecule's three-dimensional structure, which was largely based on phylogenetic analysis (Levitt, 1969). Since then, a number of structured RNAs, including tRNA, have been solved by X-ray crystallography and Nuclear Magnetic Resonance (NMR) (Kim et al., 1972; Kim et al., 1973; Laederach, 2007). Today, several hundred X-ray and NMR elucidated RNA structures, mostly of small RNA fragments, can be found in both the

Protein Data Bank (PDB) and the Nucleic Acid Data Bank (NDB) (Berman et al., 2002). However, because they are large and flexible, determining the structures of larger RNAs by X-ray crystallography is challenging, leaving computational modeling of RNA structure as an attractive alternative.

Prior to discussing RNA structure models, we must first address the key similarities and differences between RNA, DNA, and protein structure. Although RNA is similar to DNA in composition, one small difference in the chemistry of the sugar results in significant structural differences. The RNA backbone contains ribose, instead of the deoxyribose found in DNA, which contains one more hydroxyl group than deoxyribose, attached in the $2'$ position, allowing more flexibility in the RNA backbone. This flexibility contributes to the large variety of structures that RNA can adopt whereas DNA is generally found only in double-stranded helical structures. In addition, RNA is typically a single-stranded molecule with duplex structures forming between regions within the same molecule. These duplexes can be found throughout an RNA molecule, packing together to form a three-dimensional structure. As a result, RNA molecules can adopt highly complex structures that enable them to participate in cell regulation functions.

Functional RNA molecules are similar to proteins in their ability to adopt complex three-dimensional structures in order to catalyze chemical reactions. However, there are significant differences between RNA and protein structures. Proteins are made up of 20 amino acids that vary in size, shape, polarity, and charge whereas RNA is made up of four very similar bases. This is a significant difference, because we can use our understanding of the interactions between the diverse protein building blocks, such as hydrophobic collapse, to gain insight into protein folding and structure. However, in RNA the negatively charged backbone interacts with cations in solution to form a compact structure (Heilman-Miller, et al., 2001 Heilman-Miller, Thirumalai, and Woodson, 2001; Russell et al., 2002; Williamson, 2005).

METHODS FOR PREDICTING SECONDARY STRUCTURES

As new noncoding DNA segments of the genome are discovered, scientists typically want to know the secondary structure of the corresponding RNA molecule in an attempt to elucidate their functional roles. Furthermore, there is growing evidence that most genomic sequences are transcribed resulting in a large amount of functional RNA in the cell (Eddy, 2002). Although experimental assays such as DMS footprinting are a gold standard for secondary structure determination, they are also difficult, noisy, and expensive, rendering them prohibitive for such high-throughput applications (Galas and Schmitz, 1978; Inoue and Cech, 1985; Tullius, 1988). Computational prediction offers an alternative to experimental methods and opens up the opportunity to scan new RNA sequences for secondary structures (Table 33.1).

Phylogenetic Analysis of Base Pair Covariation

Phylogenetic analysis reveals the selective pressure to maintain the base pair relationships and therefore analysis of base pair covariation can be used to define intramolecular topology. Phylogenetic analysis of RNA secondary structure uses a multiple sequence alignment of an RNA family to determine base pairs through their covariance (Fox and Woese, 1975; Woese et al., 1980; Woese et al., 1983). Figure 33.1a shows an example of a multiple sequence

TABLE 33.1. Summary of Computational Methods for RNA Secondary Structure Prediction

Approach	Description	Able to Predict Nonnested Base Pairs Such as Pseudoknots?
Phylogenetic analysis of base pair covariation	Uses multiple sequence alignment to determine both secondary structure and long-range tertiary interactions. When enough sequences are available, this method is the gold standard for secondary structure prediction	Yes
Energy-based dynamic programming	Determines secondary structure based on free energy calculations	No
Grammar-based	Builds statistical models of RNA secondary structure using probabilistic production rules	No
Genetic algorithm	Uses concepts from evolution to generate possible solutions to a molecule's secondary structure	Yes
Other methods	Many other approaches combine the strengths of different approaches mentioned above	

alignment of molecules in the tRNA family. This method is based on the assumption that within a multiple sequence alignment, the consistent covariation of two positions in a Watson–Crick manner (e.g., if one position is a U, the other is always an A; when the former position mutated to a C, the latter position is always a G) suggests that there may be selective pressures to maintain the pairing between these two positions. This strategy assumes a Watson–Crick base-pairing interaction to make a structural prediction. This assumption is reasonable, because the covariation of base pairs preserves the structure of a molecule even if the sequence changes, thus maintaining the stability and function of the molecule. This method is useful for families of molecules that are similar enough to be aligned reliably, but a baseline diversity of sequences must be met to detect covariations. Generally, sequence similarities of 60–80% are optimal (James, Olsen, and Pacem, 1989).

To detect covariation between nucleotides in the RNA primary sequences, the mutual information between two columns of an aligned set of sequences if first calculated. This measure serves as an indicator of how much information about one column is contained in another. The mutual information M_{ij} between two columns i and j is calculated using the following equation:

$$M_{ij} = \sum_{xi,xj} f_{xi,xj} \log \frac{f_{xi,xj}}{f_{xi} f_{xj}}, \tag{33.1}$$

where f_{xi}, f_{xj} is the frequency of each four bases in column i and j respectively, and $f_{xi,xj}$ is the frequency of the 16 possible pairs observed in i and j.

Two conditions must be satisfied for a helix to be predicted using phylogenetic analysis. First, the complementary sequences of a helix must occur in homologous parts of the multiple sequences. Second, two or more independent covariations must occur in the complementary sequences in such a way that the base pairing is preserved. Although only canonical (A/U, G/C) pairs are considered initially, noncanonical pairs (e.g., G/U) can be considered after a helical region has been determined.

When enough sequences are available, covariation analysis is considered the gold standard for secondary structure prediction methods with successful application to many families of RNA, including 5S RNA (Fox and Woese, 1975) and rRNAs (Woese et al., 1983). The multiple sequence alignments and covariance models used in these predictions can be found in several databases such as Rfam built by Griffiths-Jones and coworkers that currently contains 574 families (Griffiths-Jones et al., 2003), and the Comparative RNA Web (CRW) created by Gutell and coworkers that contains detailed secondary structure models for several RNA families (Cannone et al., 2002). We have included links to these sites at the end of this chapter.

The success of this powerful technique stems from the inherent unrestrictive model-free approach to the prediction of RNA secondary structure, thus allowing predictions of pseudoknots, triple pairs, and even long-range tertiary single pairs. In a 2002 paper, Gutell and coworkers measured the accuracy of secondary structure models predicted by covariation analysis for the 16S and 23S rRNA molecules by comparing them with information from the solved crystal structures. They found that covariation analysis was able to predict all of the standard secondary structure base pairings and helices, as well as some of the tertiary interactions (Gutell, Lee, and Cannone, 2002). However, this method is limited by its need for a family of molecules with 60–80% identity, as well as manual intervention for editing the multiple sequence alignment.

Energy-Based Dynamic Programming Methods

These methods for predicting RNA secondary structure are based on the observation that the stability of a helical region is the ensemble of individual energy contributions from local interactions. To predict secondary structure within a molecule, all the possible helical regions are given an energy score based on their components. Dot plots are a simple approach to finding possible helices by marking all possible base pairs on a two-dimensional grid of the primary sequence versus itself (Jacobson and Zuker, 1993) with diagonal regions revealing potential helix formation. Figure 33.1b shows such a plot for the yeast phenylalanine tRNA with the lower right triangle showing all the possible helices, while the top left triangle shows the helices that are present in the molecule.

Putative helices are given an energy score based on interactions within the helix. The three main interactions that contribute favorably to the stability of a helix are (1) hydrogen bonds of base pairs, (2) stacking interactions between bases, and finally, (3) irregular structures, such as tetraloops at the ends of helices. In addition to these favorable contributions, unfavorable contributions from both asymmetric and symmetric bulges within a helix, large loops at the ends of helices, and multiple branches stemming from a single loop are accounted for in the energy function. Each of these phenomena is assigned an energy contribution, which is added up for each helix to determine its likelihood. Lower energies correlate with more stable helices, whereas a high energy means that the helix is unlikely to exist in the structure (Zuker and Stiegler, 1981).

An example of a simple energy model is to assign the following energy values to base pairs in a potential helix:

G–C = −3 energy units (this is the most favorable Watson–Crick pairing.)
A–U = −2 energy units
G–U = −1 energy unit (this is a noncanonical pairing.)

Turner and coworkers enabled more complicated thermodynamic parameters based on experimental measurements to calculate the contribution of each base pair (Xia et al., 1998) and to incorporate position-specific parameters (Mathews et al., 1999). Zuker and coworkers incorporated these parameters into a secondary structure prediction tool called Mfold (Zuker and Stiegler, 1981), which is now available online as a web server (Mathews et al., 1999; Zuker, 2003). Instead of generating all the possible helices using the dot plot method described earlier, Mfold uses a dynamic programming approach that calculates the lowest free energy for each possible sequence fragment, starting with the shortest fragment. This method uses a recursion on the determined energies of smaller fragments to calculate free energies of longer fragments. Other popular energy-based secondary structure prediction programs include RNAfold and RNAstructure, which are also based on dynamic programming algorithms.

Energy-based methods are limited by the quality of their parameters. Furthermore, pseudoknot structures are difficult to predict. For example, the effects on stability, such as bulge loops and single noncanonical pairs, are nonnearest-neighbors and are not included in these models (Longfellow, Kierzek, and Turner, 1990; Burkard et al., 1999; Kierzek, Burkard, and Turner, 1999). Energy-based methods assume that RNA sequences are at equilibrium and do not consider the role of folding kinetics in the formation of secondary structure. Moreover, some RNA molecules have more than one secondary structure conformation. For example, ribos-witches can take on different secondary structures in the presence and absence of a metabolite.

To address some of these limitations, energy-based prediction methods now report suboptimal structures since the lowest energy structure may not be the correct (or only) one. For example, Mfold and similar programs calculate both the energetically optimal structure and the diverse set of suboptimal structures that have free energy values similar to the optimal structure (Mathews, 2006).

Unlike the phylogenetic analysis strategy, energy-based methods do not take a model-free approach and assume that helices are well nested and therefore are not able to predict structures such as pseudoknots and triple bases. This is a serious limitation because structured functional RNAs often contain these nonnested structural units. Rivas and Eddy addressed this limitation in 1999 with their dynamic programming algorithm for predicting RNA secondary structure, including pseudoknots, with thermodynamic parameters from energy-based methods (Rivas and Eddy, 1999). However, because it is prohibitively computationally expensive, it can only be used for small RNA molecules.

Grammar-Based Methods

Grammar-based methods examine the palindrome-like primary sequence as a consequence of conserving these base-pairing interactions. This strategy utilizes tools created by research advances in computational linguistics to examine palindrome in languages, in particular the context-free grammars (CFGs) in Chomsky's hierarchy of transformational grammars (Chomsky, 1956; Chomsky, 1959). These CFGs are developed and used for the specific purpose of predicting RNA secondary structures.

A transformational grammar is made up of symbols and rewriting rules $\alpha \to \beta$ (also called productions) where α and β are both strings of symbols. Symbols may be either nonterminal or terminal, and the left-hand side α must contain at least one nonterminal, which is rewritten into a new string of symbols. Terminals are generally represented with lower case letters, while nonterminals are represented with upper case letters.

To illustrate transformational grammar, we will give an example of a very simple regular grammar that can be used to generate strings of a and b. Regular grammars are the simplest and most restrictive type of grammars, allowing only production rules of the form W → aW or W → a. Using the terminal letters a and b, a single nonterminal letter S, and a blank terminal symbol ε, a simple example of a transformational grammar is

$$S \rightarrow aS, \ S \rightarrow bS, \ S \rightarrow \varepsilon.$$

This can also be represented as

$$S \rightarrow aS|S \rightarrow bS|S \rightarrow \varepsilon.$$

This grammar can be used to generate strings of a and b. For example,

$$S \Rightarrow bS \Rightarrow bbS \Rightarrow bbaS \Rightarrow bbabS \Rightarrow bbab.$$

Context-free grammars are the second of the four types of grammars in the Chomsky hierarchy and are slightly less restrictive than regular grammars, which are the first type. Context-free grammars allow any production rule of the form W → β, where the left-hand side must consist of just one nonterminal, but the right-hand side can be any string. This allows the grammar to create nested, long-distance pair-wise correlations between terminal symbols. This type of grammar is appropriate for RNA secondary structure prediction because it can generate RNA primary sequences with nested base pairs. For example, the sequence *accggaaacggu*, which is an RNA stem with four base pairs and a loop, can be built using the following simple grammar:

$$S \rightarrow aW_1u|cW_1g|uW_1a|gW_1c$$
$$W_1 \rightarrow aW_2u|cW_2g|uW_2a|gW_2c$$
$$W_2 \rightarrow aW_3u|cW_3g|uW_3a|gW_3c$$
$$W_3 \rightarrow aW_4u|cW_4g|uW_4a|gW_4c$$
$$W_4 \rightarrow gaaa|gcau.$$

Here is the derivation that results in our example sequence:

$$S \Rightarrow aW_1u \Rightarrow acW_2gu \Rightarrow accW_3ggu \Rightarrow accgW_4cggu \Rightarrow accggaaacggu.$$

Grammar-based methods of RNA secondary structure prediction use stochastic context-free grammars, in which each production rule is assigned a probability. These methods have elaborate production rules and are used to build statistical models of RNA families, perform multiple sequence alignments, and predict secondary structures (Eddy and Durbin, 1994; Sakakibara et al., 1994; Grate, 1995; Lefebvre, 1995; Lefebvre, 1996). Implementations of stochastic context-free grammars, including covariance models, are well covered in *Biological Sequence Analysis* by Durbin et al. (1998).

Like the energy-based methods described earlier, SCFGs are also unable to handle nonnesting occurrences of base pairs, preventing them from predicting pseudoknots, triple pairs, and long-range interactions.

Genetic Algorithm Approaches

Generally, the genetic algorithm is a stochastic method that uses concepts from evolution and the survival-of-the-fittest individual to find a solution. This algorithm proceeds stepwise and yields a set of probable solutions that it attempts to improve at each step. Each step consists of three procedures:

1. mutation, which introduces random changes into the solutions,
2. crossover, which exchanges parts of solutions with each other, and
3. selection, which uses a fitness criterion to select the best solutions.

Several programs implement the genetic algorithm to solve RNA secondary structures. These include a method by van Batenburg, Gultyaev, and Pleij (1995) and MPGAfold (Shapiro and Navetta, 1994; Shapiro and Wu, 1997; Shapiro et al., 2001). These methods generate initial possible solutions from the set of all possible stems, and allow them to evolve, using free energy as a measure of fitness in each selection step. These methods permit the formation of pseudoknots and are also able to capture significant intermediate RNA secondary structure states, giving insight into the RNA folding process (Shapiro et al., 2001; Kasprzak, Bindewald, and Shapiro, 2005).

Other Approaches

CONTRAfold, a secondary structure prediction tool developed by Do and coworkers, uses a class of probabilistic machine-learning methods called conditional log-linear methods (CLLM). CLLMs are more flexible than SCFGs and allow the incorporation of thermodynamic parameters determined in energy-based methods. By using the strengths of both SCFG and energy-based methods, CONTRAfold is able to outperform both, although it inherits their limitation of not being able to predict nonnested structures such as pseudoknots (Do, Woods, and Batzoglou, 2006).

Several methods incorporate both thermodynamics and covariation data from multiple sequence alignments into RNA secondary structure predictions. These methods include iterated loop matching (Ruan, Stormo, and Zhang, 2004a; Ruan, Stormo, and Zhang, 2004b), KNetFold (Bindewald and Shapiro, 2006;Bindewald, Schneider, and Shapiro, 2006) and HXMATCH (Hofacker, Fekete, and Stadler, 2002; Witwer, Hofacker, and Stadler, 2004). These methods generally start with pseudoknot-free structures, but are able to make predictions that include pseudoknots by combining thermodynamics with covariation data. These methods are reviewed and described in greater detail in Shapiro et al. (2007).

Several other successful methods include Kinefold, a web server that implements stochastic folding simulations using a Monte Carlo methodology (Xayaphoummine et al., 2003; Xayaphoummine, Bucher, and Isambert, 2005) and HotKnots, a heuristic approach that iteratively forms stable stems (Ren et al., 2005). Both of these methods allow alternative secondary structures to form and are able to predict pseudoknots.

THREE-DIMENSIONAL MODELING METHODOLOGY

RNA Motifs

By studying RNA crystal structures, researchers have identified and characterized numerous RNA secondary and tertiary structure motifs: recurring structural building blocks from which RNA molecules are built. When the crystal structures of the large and small subunits of

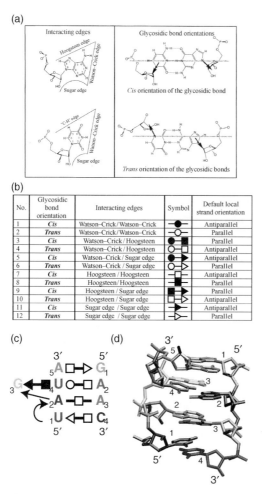

Figure 33.2. (a) Left: The interacting edges of purine and pyminidine bases. Right: The relative bond orientations of base pairs: *cis* and *trans* (Leontis et al., 2002b). (b) The 12 geometric families of edge-to-edge base pairs (Leontis et al., 2002b). (c) The generic sarcin motif. (d) The sarcin motif in the 23S rRNA of *H. marismortui* (RR0033).

the ribosome were solved, researchers found them to be mostly made up of known RNA structural elements. Motifs are classically defined and described as a pattern of interactions between elements of RNA structure. Motifs consistently fold into the same 3D structure, and the formation of these motifs affects both the folding and the final structure of RNA molecules by facilitating the formation of additional contacts such as hydrogen bonds between backbones, bases, and backbones and bases (Lescoute et al., 2005; Leontis, Lescoute, and Westhof, 2006).

RNA motifs are often the result of non-Watson–Crick base pairing. In addition to the classical Watson–Crick base pairing described in Chapter 3, there are additional families of base–base interactions frequently found in RNA structures. Figure 33.2a shows the three interacting edges of a base, Hoogsteen, Watson–Crick, and sugar, as well as the two relative bond orientations, *cis* and *trans*. Figure 33.2 lists the 12 geometric families of edge-to-edge base pairs. Figure 33.2c shows an example of a symbolic representation of the sarcin motif.

Figure 33.2d shows the three-dimsensional structure of the sarcin motif found in the 23S rRNA of *H. marismortui* (Leontis, Stombaugh, and Westhof, 2002b; Leontis and Westhof 2003).

A number of tools search through RNA sequences for known motifs. One of these, the RNAMotif tool, translates a motif into a search tree based on nesting within the motif, and then performs a depth first search to find motifs within a sequence (Macke et al., 2001). Some motif search tools use information from high-quality secondary structure predictions that include non-Watson–Crick base pairs. These methods use the predicted secondary structure to identify potential regions for motifs and check them against known consensus motifs (Leontis, Stombaugh, and Westhof, 2002a). A review by Holbrook from 2005 lists occurrences of known RNA motifs in solved RNA structures, as well as novel motifs found in each structure (Holbrook, 2005).

Constraint Satisfaction

The program MC-SYM (Macromolecular Conformation by SYMbolic generation) written by Major and coworkers in 1991 can produce atomic resolution three-dimensional models of RNA molecules (Major et al., 1991). The inputs to this program are a sequence of nucleotides and their relationships, such as base pairing, and a set of constraints, such as distances between specified residues.

MC-SYM uses a constraint satisfaction problem (CSP) algorithm to model RNA structures at the atomic resolution. A CSP algorithm looks for values of $X = \{x_1, x_2 \ldots x_n\}$ that satisfy a set of constraints $C = \{c_{p,q}, \ldots | p \in \{1 \ldots n\}, q \in \{1 \ldots p-1\}\}$, where the values of X are taken from a set of allowed values $D = \{d_1, d_2 \ldots d_n\}$. In the application to RNA structure, the values of X are the atomic coordinates of nucleotides 1 through n. The constraints in C are relationships between nucleotides such as base pairing or stacking. The set of allowed values D for the geometry of each nucleotide was chosen from a limited set of internucleotide and internal nucleotide torsion angles found in the crystal structures of tRNA molecules. Relationships between pairs of nucleotides are indicated by the user and determine the set of allowed geometries. These relationships include free connections, stacked connections, Watson–Crick and reverse Hoogsteen base pairs, and A-helix form. For a free connection to the 3′ end of another nucleotide, there are 10 combinations of the four internal torsion angles α, β, δ, and χ, combined with three torsions of the internucleotide bond ζ, resulting in 30 possible geometries of a nucleotide (refer to Figure 3.2 in Chapter 3 for locations of torsion angles).

The CSP algorithm generates a search tree, where each node corresponds to the assignment of a geometry to a nucleotide. The program evaluates the consistency of each assignment with the constraints, and continues with the next assignment if the constraints are satisfied. If the constraints are not satisfied, the program backtracks by removing the current node and its branches, returning to the previous node. The result of this method is a set of possible geometries for the given sequence of nucleotides and their interactions and constraints. This method does not directly prevent the overlapping of atoms, so a round of energy minimizations, which can remove these collisions, usually follows it.

This method does very well with small RNA structures such as the motifs discussed in this chapter. It has successfully predicted loops and pseudoknots to within 2.0–3.0 Å RMSd to the known or consensus structures. In addition, MC-SYM generated full atomic structures of RNA substructures such as hairpin loops that can be used as building blocks for modeling larger structures. In 1993, Gautheret and coworkers used MC-SYM to generate full atomic models of the tRNA T-loop, UUCG tetraloop, tRNA anticodon, and hairpin loops (Gautheret, Major, and Cedergren, 1993). Major and coworkers then used these substructure models to build a full atomic model of tRNA (Major, Gautheret, and Cedergren, 1993).

Altman and coworkers have also used a constraint satisfaction method to produce a model of the 16S ribosomal RNA subunit (Altman, Weiser, and Noller, 1994). Constraints were based mostly on chemical probing results. Furthermore, an analysis of the structural information content of the different experimental modalities used in these models was carried out after the ribosomal structure was solved by X-ray crystallography (Whirl-Carrillo et al., 2002).

Fragment Assembly of 3D RNA Structures

Tools that model large RNA structures generally treat RNA molecules as assemblies of smaller fragments. These tools use known structures of small RNA fragments and require user interaction to assemble them into larger structures.

The MANIP tool allows a user to assemble an RNA structure from a set of smaller RNA fragments (Massire and Westhof, 1998). MANIP uses a database of known fragment structures, assembled from RNA structures solved by NMR or X-ray crystallography. MANIP begins its modeling process by refining the molecule's secondary structure using an alignment of available RNA sequences. MANIP then recognizes structure motifs within the secondary structure and constructs these fragments automatically based on the database of known fragment structures. MANIP then assembles these fragments and integrates available experimental data. The program NUCLIN/NUCLSQ further refines the resulting structure. This tool has successfully modeled the RNAse P molecule (Massire and Westhof, 1998; Tsai et al., 2003).

Another interactive tool for modeling large RNA structures is ERNA-3D, which produces 3D representations of molecules from known secondary structures, using knowledge of motif structures from other RNAs (Zwieb and Muller, 1997). This tool allows a user to manipulate the 3D-motif structures and position them manually. This tool was used to build a high-resolution 3D model of several transfer-messenger RNAs (Burks et al., 2005).

Other 3D Modeling Tools

The Nucleic Acid Builder (NAB) is a computer language developed to construct models of nucleic acids up to a few hundred residues in size (Macke and Case, 1998). This tool uses a combination of rigid body transformations and distance geometry to generate structures that are then refined by molecular dynamics. The RNA2D3D program can generate, view, and compare 3D RNA structures as well as generate a first-order approximation of a small RNA molecule or fragment of a molecule. This tool has been used to predict the 3D structure of a pseudoknot (Yingling and Shapiro, 2006). Finally, S2S is a tool that links several levels of RNA data, including multiple sequence alignments, secondary structure, and tertiary structure (Jossinet and Westhof, 2005).

CONCLUSION

The RNA world is rich with computational opportunities both at the 2D and 3D levels. The unique chemical nature of the RNA sugar–phosphate, combined with the base-pairing promiscuity yields a highly complex biopolymer that plays a critical role in the cell. As the multifaceted roles of RNA in the cell become more apparent, computational approaches that assist the biological community in deciphering these roles will be critical.

WEB RESOURCES

RNA World Website	http://www.imb-jena.de/RNA.html
Functional RNA database	http://www.ncrna.org/
Rfam database of RNA alignments and covariance models	http://rfam.janelia.org
Comparative RNA Web	http://www.rna.ccbb.utexas.edu
Mfold web server	http://www.bioinfo.rpi.edu/applications/mfold/

SUGGESTED READINGS

1. Secondary structure prediction methods:
   - Chapter 8 of *Bioinformatics: Sequence and Genome Analysis* by David W. Mount (Mount, 2004).
   - Chapters 9 and 10 of *Biological Sequence Analysis* by R. Durbin, S. Eddy, A. Krogh, and G. Mitchison (Durbin et al., 1998).
2. An excellent review on RNA structure prediction:
   - A 2007 review by Shapiro and coworkers on RNA structure prediction: bridging the gap in RNA structure prediction (Shapiro et al., 2007).

REFERENCES

Altman RB, Weiser B, Noller HF (1994): Constraint satisfaction techniques for modeling large complexes: application to the central domain of 16S ribosomal RNA. *Proc Int Conf Intell Syst Mol Biol* 2:10–18.

Berman HM, Battistuz T, Bhat TN, Bluhm WF, Bourne PE, Burkhardt K, Feng Z, Gilliland GL, Iype L, Jain S, Fagan P, Marvin J, Padilla D, Ravichandran V, Schneider B, Thanki N, Weissig H, Westbrook JD, Zardecki C (2002): The Protein Data Bank. *Acta Crystallogr D* 58(Pt 6 No 1):899–907.

Bindewald E, Shapiro BA (2006): RNA secondary structure prediction from sequence alignments using a network of k-nearest neighbor classifiers. *RNA* 12(3):342–352.

Bindewald E, Schneider TD, Shapiro BA (2006): CorreLogo: an online server for 3D sequence logos of RNA and DNA alignments. *Nucleic Acids Res* 34(Web server issue):W405–W411.

Burkard ME, Kierzek R, Turner DH (1999): Thermodynamics of unpaired terminal nucleotides on short RNA helixes correlates with stacking at helix termini in larger RNAs. *J Mol Biol* 290 (5):967–982.

Burks J, Zwieb C, Muller F, Wower I, Wower J (2005): Comparative 3-D modeling of tmRNA. *BMC Mol Biol* 6(1):14.

Cannone JJ, Subramanian S, Schnare MN, Collett JR, D'Souza LM, Du Y, Feng B, Lin N, Madabusi LV, Muller KM, Pande N, Shang Z, Yu N, Gutell RR (2002): The comparative RNA web (CRW) site: an online database of comparative sequence and structure information for ribosomal, intron, and other RNAs. *BMC Bioinformatics* 3:2.

Chomsky N (1956): Three models for the description of language. *IRE Transact Inf Theory* 2:113–124.

Chomsky N (1959): On certain formal properties of grammars. *Inf Control* 2:137–167.

Do CB, Woods DA, Batzoglou S (2006): CONTRAfold: RNA secondary structure prediction without physics-based models. *Bioinformatics* 22(14):e90–e98.

Durbin R, Eddy S, Krogh A, Mitchison G (1998): *Biological Sequence Analysis*. Cambridge University Press.

Eddy SR (2002): Computational genomics of noncoding RNA genes. *Cell* 109(2):137–140.

Eddy SR, Durbin R (1994): RNA sequence analysis using covariance models. *Nucleic Acids Res* 22(11):2079–2088.

Fox GW, Woese CR (1975): 5S RNA secondary structure. *Nature* 256(5517):505–507.

Galas DJ, Schmitz A (1978): DNAse footprinting: a simple method for the detection of protein–DNA binding specificity. *Nucleic Acids Res* 5(9):3157–3170.

Gautheret D, Major F, Cedergren R (1993): Modeling the three-dimensional structure of RNA using discrete nucleotide conformational sets. *J Mol Biol* 229(4):1049–1064.

Grate L (1995): Automatic RNA secondary structure determination with stochastic context-free grammars. *Proc Int Conf Intell Syst Mol Biol* 3:136–144.

Griffiths-Jones S, Bateman A, Marshall M, Khanna A, Eddy SR (2003): Rfam: an RNA family database. *Nucleic Acids Res* 31(1):439–441.

Gutell RR, Lee JC, Cannone JJ (2002): The accuracy of ribosomal RNA comparative structure models. *Curr Opin Struct Biol* 12(3):301–310.

Heilman-Miller SL, Thirumalai D, Woodson SA (2001): Role of counterion condensation in folding of the tetrahymena ribozyme. I. Equilibrium stabilization by cations. *J Mol Biol* 306(5):1157–1166.

Hofacker IL, Fekete M, Stadler PF (2002): Secondary structure prediction for aligned RNA sequences. *J Mol Biol* 319(5):1059–1066.

Holbrook SR (2005): RNA structure: the long and the short of it. *Curr Opin Struct Biol* 15(3):302–308.

Inoue T, Cech TR (1985): Secondary structure of the circular form of the Tetrahymena rRNA intervening sequence: a technique for RNA structure analysis using chemical probes and reverse transcriptase. *Proc Natl Acad Sci USA* 82(3):648–652.

Jacobson AB, Zuker M (1993): Structural analysis by energy dot plot of a large mRNA. *J Mol Biol* 233(2): 261–269.

James BD, Olsen GJ, Pace NR (1989): Phylogenetic comparative analysis of RNA secondary structure. *Methods Enzymol* 180:227–239.

Jossinet F, Westhof E (2005): Sequence to structure (S2S): display, manipulate and interconnect RNA data from sequence to structure. *Bioinformatics* 21(15):3320–3321.

Kasprzak W, Bindewald E, Shapiro BA (2005): Structural polymorphism of the HIV-1 leader region explored by computational methods. *Nucleic Acids Res* 33(22):7151–7163.

Kierzek R, Burkard ME, Turner DH (1999): Thermodynamics of single mismatches in RNA duplexes. *Biochemistry* 38(43):14214–14223.

Kim SH, Quigley G, Suddath FL, McPherson A, Sneden D, Kim JJ, Weinzierl J, Blattmann P, Rich A (1972): The three-dimensional structure of yeast phenylalanine transfer RNA: shape of the molecule at 5.5-A resolution. *Proc Natl Acad Sci USA* 69(12):3746–3750.

Kim SH, Quigley GJ, Suddath FL, McPherson A, Sneden D, Kim JJ, Weinzierl J, Rich A (1973): Three-dimensional structure of yeast phenylalanine transfer RNA: folding of the polynucleotide chain. *Science* 179(70):285–288.

Laederach A (2007): Informatics challenges in structured RNA. *Brief Bioinform*.

Lefebvre F (1995): An optimized parsing algorithm well suited to RNA folding. *Proc Int Conf Intell Syst Mol Biol* 3:222–230.

Lefebvre F (1996): A grammar-based unification of several alignment and folding algorithms. *Proc Int Conf Intell Syst Mol Biol* 4:143–154.

Leontis NB, Westhof E (2001): Geometric nomenclature and classification of RNA base pairs. *RNA* 7(4):499–512.

Leontis NB, Stombaugh J, Westhof E (2002a): Motif prediction in ribosomal RNAs: lessons and prospects for automated motif prediction in homologous RNA molecules. *Biochimie* 84(9):961–973.

Leontis NB, Stombaugh J, Westhof E (2002b): The non-Watson–Crick base pairs and their associated isostericity matrices. *Nucleic Acids Res* 30(16):3497–3531.

Leontis NB, Westhof E (2003): Analysis of RNA motifs. *Curr Opin Struct Biol* 13(3):300–308.

Leontis NB, Lescoute A, Westhof E (2006): The building blocks and motifs of RNA architecture. *Curr Opin Struct Biol* 16(3):279–287.

Lescoute A, Leontis NB, Massire C, Westhof E (2005): Recurrent structural RNA motifs, isostericity matrices and sequence alignments. *Nucleic Acids Res* 33(8):2395–2409.

Levitt M (1969): Detailed molecular model for transfer ribonucleic acid. *Nature* 224(5221):759–763.

Longfellow CE, Kierzek R, Turner DH (1990): Thermodynamic and spectroscopic study of bulge loops in oligoribonucleotides. *Biochemistry*, 29(1):278–285.

Macke TJ, Case D (1998): Modeling unusual nucleic acid structures In: Leontis N, SantaLucia J, Jr editors. *Molecular Modeling of Nucleic Acids*. Washington, DC: American Chemical Society, pp. 379–393.

Macke TJ, Ecker DJ, Gutell RR, Gautheret D, Case DA, Sampath R (2001): RNAMotif, an RNA secondary structure definition and search algorithm. *Nucleic Acids Res* 29(22):4724–4735.

Major F, Turcotte M, Gautheret D, Lapalme G, Fillion E, Cedergren R (1991): The combination of symbolic and numerical computation for three-dimensional modeling of RNA. *Science* 253(5025):1255–1260.

Major F, Gautheret D, Cedergren R (1993): Reproducing the three-dimensional structure of a tRNA molecule from structural constraints. *Proc Natl Acad Sci USA* 90(20):9408–9412.

Massire C, Westhof E (1998): MANIP: an interactive tool for modelling RNA. *J Mol Graph Model* 16(4–6):197–205, 55-7.

Mathews DH, Sabina J, Zuker M, Turner DH (1999): Expanded sequence dependence of thermodynamic parameters improves prediction of RNA secondary structure. *J Mol Biol* 288(5):911–940.

Mathews DH (2006): Revolutions in RNA secondary structure prediction. *J Mol Biol* 359(3):526–532.

Mount DW (2004): *Bioinformatics: Sequence and Genome Analysis*. 2nd edn, Cold Spring Harbor, NY: Cold Spring Harbor Laboratory Press.

Ren J, Rastegari B, Condon A, Hoos HH (2005): HotKnots: heuristic prediction of RNA secondary structures including pseudoknots. *RNA* 11(10):1494–1504.

Rivas E, Eddy SR (1999): A dynamic programming algorithm for RNA structure prediction including pseudoknots. *J Mol Biol* 285(5):2053–2068.

Ruan J, Stormo GD, Zhang W (2004a): An iterated loop matching approach to the prediction of RNA secondary structures with pseudoknots. *Bioinformatics* 20(1):58–66.

Ruan J, Stormo GD, Zhang W (2004b): ILM: a web server for predicting RNA secondary structures with pseudoknots. *Nucleic Acids Res* 32(Web server issue):W146–W149.

Russell R, Millett IS, Tate MW, Kwok LW, Nakatani B, Gruner SM, Mochrie SG, Pande V, Doniach S, Herschlag D, Pollack L (2002): Rapid compaction during RNA folding. *Proc Natl Acad Sci USA* 99(7):4266–4271.

Sakakibara Y, Brown M, Hughey R, Mian IS, Sjolander K, Underwood RC, Haussler D (1994): Stochastic context-free grammars for tRNA modeling. *Nucleic Acids Res* 22(23):5112–5120.

Shapiro BA, Bengali D, Kasprzak W, Wu JC (2001): RNA folding pathway functional intermediates: their prediction and analysis. *J Mol Biol* 312(1):27–44.

Shapiro BA, Navetta J (1994): A massively parallel genetic algorithm for RNA secondary structure prediction. *J Supercomput,* 8(3):195–207.

Shapiro BA, Wu JC (1997): Predicting RNA H-type pseudoknots with the massively parallel genetic algorithm, *Comput Appl Biosci* 13(4):459–471.

Shapiro BA, Wu JC, Bengali D, Potts MJ (2001): The massively parallel genetic algorithm for RNA folding: MIMD implementation and population variation. *Bioinformatics* 17(2):137–148.

Shapiro BA, Yingling YG, Kasprzak W, Bindewald E (2007): Bridging the gap in RNA structure prediction. *Curr Opin Struct Biol* 17(2):157–165.

Tsai HY, Masquida B, Biswas R, Westhof E, Gopalan V (2003): Molecular modeling of the three-dimensional structure of the bacterial RNase P holoenzyme. *J Mol Biol* 325(4):661–675.

Tullius TD (1988): DNA footprinting with hydroxyl radical. *Nature* 332(6165):663–664.

van Batenburg FH, Gultyaev AP, Pleij CW (1995): An APL-programmed genetic algorithm for the prediction of RNA secondary structure. *J Theor Biol* 174(3):269–280.

Whirl-Carrillo M, Gabashvili IS, Bada M, Banatao DR, Altman RB (2002): Mining biochemical information: lessons taught by the ribosome. *RNA* 8(3):279–289.

Williamson JR (2005): Assembly of the 30S ribosomal subunit. *Q Rev Biophys* 38(4):397–403.

Witwer C, Hofacker IL, Stadler PF (2004): Prediction of consensus RNA secondary structures including pseudoknots. *IEEE/ACM Trans Comput Biol Bioinform* 1(2):66–77.

Woese CR, Gutell R, Gupta R, Noller HF (1983): Detailed analysis of the higher-order structure of 16S-like ribosomal ribonucleic acids. *Microbiol Rev* 47(4):621–669.

Woese CR, Magrum LJ, Gupta R, Siegel RB, Stahl DA, Kop J, Crawford N, Brosius J, Gutell R, Hogan JJ, Noller HF (1980): Secondary structure model for bacterial 16S ribosomal RNA: phylogenetic, enzymatic and chemical evidence. *Nucleic Acids Res* 8(10):2275–2293.

Xayaphoummine A, Bucher T, Thalmann F, Isambert H (2003): Prediction and statistics of pseudoknots in RNA structures using exactly clustered stochastic simulations. *Proc Natl Acad Sci USA* 100(26):15310–15315.

Xayaphoummine A, Bucher T, Isambert H (2005): Kinefold web server for RNA/DNA folding path and structure prediction including pseudoknots and knots. *Nucleic Acids Res* 33(Web server issue): W605–W610.

Xia T, SantaLucia J, Jr Burkard ME, Kierzek R, Schroeder SJ, Jiao X, Cox C, Turner DH (1998): Thermodynamic parameters for an expanded nearest-neighbor model for formation of RNA duplexes with Watson–Crick base pairs. *Biochemistry* 37(42):14719–14735.

Yingling YG, Shapiro BA (2006): The prediction of the wild-type telomerase RNA pseudoknot structure and the pivotal role of the bulge in its formation. *J Mol Graph Model* 25(2):261–274.

Zuker M, Stiegler P (1981): Optimal computer folding of large RNA sequences using thermodynamics and auxiliary information. *Nucleic Acids Res* 9(1):133–148.

Zuker M (2003): Mfold web server for nucleic acid folding and hybridization prediction. *Nucleic Acids Res* 31(13):3406–3415.

Zwieb C, Muller F (1997): Three-dimensional comparative modeling of RNA. *Nucleic Acids Symp Ser* (36):69–71.

Section VII

THERAPEUTIC DISCOVERY

34

STRUCTURAL BIOINFORMATICS IN DRUG DISCOVERY

William R. Pitt, Alícia Pérez Higueruelo, and Colin R. Groom

The pharmaceutical industry had its origins in the beginning of the twentieth century. Scientific advancement has since seen the discovery of DNA and sequencing of the human genome, the understanding of proteins as specific molecular entities, the harnessing of X-rays to understand proteins at the atomic level, and the introduction of computational technology. All these advancements have allowed structural bioinformatics to begin to do its part to further aid drug discovery. This chapter will take a brief look at the development of the drug discovery process before discussing the current and future role of structural bioinformatics in this context.

HISTORIC DEVELOPMENT OF DRUG DISCOVERY

The current dominant paradigm the pharmaceutical drug discovery follows is to search for a particular small-molecule modulator of a specific macromolecular target. A protein with a known or a suspected role in a disease process is selected and small-molecule modulators of the protein are identified and optimized. Our ability to pursue this paradigm rests on the scientific and technological achievements of the twentieth century, particularly with regard to our ability to manipulate organic small molecules on the one hand and to study the biological targets on the other (Drews, 2000). Humanity has, of course, been looking for remedies for its ailments long before there was a drug discovery industry (Sneader, 1995; Sneader, 2005). The use of willow bark as a treatment for pain relief, for example, can be traced back to Hippocrates and earlier. Such use is entirely empiric—a certain recipe gave relief to certain symptoms. Many such folk remedies were known, the progeny of some of which, such as willow bark, have found their place in our modern medicine chests.

Structural Bioinformatics, Second Edition   Edited by Jenny Gu and Philip E. Bourne
Copyright © 2009 John Wiley & Sons, Inc.

The first step toward our modern approach to drug discovery was the suggestion that these remedies generally contained an active ingredient that could be isolated and purified. This idea can be traced back to 1530 to Paracelsus, a Swiss physician; however, the active ingredient salicin in willow bark could not be purified until 1829, another 300 years later. The synthesis of urea by Fredrich Wohler the year before ushered in organic synthesis and gave chemists the ability to manipulate small organic compounds of this type. Salicylic acid itself was first synthesized in 1852. Along with the power to create a specific molecule came the ability to create many closely related compounds. These techniques were first put to profitable use in the dye industry in the mid-1800s, creating for the first time numerous low-cost dye compounds.

Using such dyes for histological staining, Paul Ehrlich recognized that related molecules often exhibit related biological effects, a concept referred to today as structure–activity relationships (SAR). This concept was applied to derivatives of salicylic acid to try and discover forms of the drug that were less unpleasant for the patient. This research eventually led to the development of acetylsalicylic acid in 1897 by Felix Hoffmann at Bayer. Bayer named this compound Aspirin: "a" for acetyl, "spir" from *Spiraea ulmaria*, the meadowsweet plant, and "in" a common suffix for medicines at the time. In noticing how some compounds more readily stained bacterial cells than human cells, Paul Ehrlich eventually developed another major cornerstone of modern drug discovery, the concept of a *therapeutic index*. All drugs have a minimal dose at which they demonstrate beneficial effects and a minimal dose at which they demonstrate harmful effects. The therapeutic index is simply the ratio of these two doses.

The interplay between trying to make compounds more effective and keeping drugs safe for use continues to be at the center of pharmaceutical discovery till date. The recognition of activity being associated with specific molecular entities was paralleled (much later) by John Langley's suggestion in 1878 that there must be specific "receptors" for such compounds in the host, which bind to these entities. Knowing that there is a host receptor however is not the same as knowing what that receptor is. It would be another 100 years, for example, before John Vane and his colleagues discovered the link between aspirin and prostaglandin synthesis, establishing cyclooxygenase (COX) as aspirin's site of action (Vane, 1971). This work earned John Vane the 1982 Nobel Prize in Physiology and Medicine. Cyclooxygenase was first given a structural face by Michael Garavito and colleagues in 1994 (Picot, Loll, and Garavito, 1994). This structure, and that of the inducible COX-2 (Kurumbail et al., 1996; Luong et al., 1996), have made possible the first forays into structure-based design against this venerated target (Marnett and Kalgutkar, 1999). Recognition of different isoforms of cyclooxygenase (Vane et al., 1994) and their roles has led us to specific COX-2 inhibitors, such as Celecoxib (Penning et al., 1997), with reduced side effects of peptic ulcers compared to early COX inhibitors. The story does not have its happy ending however, with evidence pointing to an increase in frequency of undesirable cardiovascular effects with COX-2 inhibitors (Kearney et al., 2006).

MODERN DRUG DISCOVERY

At present, most pharmaceutical drug discovery programs begin with a known macromolecular target, and seek to identify a suitable small-molecule modulator (Rang, 2005). The process is depicted in Figure 34.1. Typically, the target (usually a protein) has already been identified, through biological or genetic investigations, to be important in the disease of

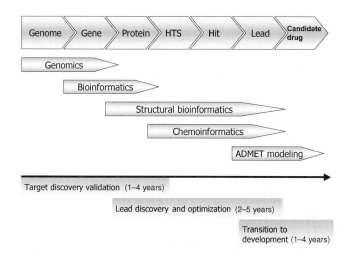

Figure 34.1. The role of informatics in the target-based drug discovery process.

interest. The approach of modern drug discovery is rational and reductionistic with a defined hypothesis of how the chosen mechanism of action could be beneficial against disease. Following the identification of the target of interest, confidence in the approach is built with a variety of genetic and chemical target validation experiments.

The process to discover a lead molecule begins with the development of an assay to look for modulators (inhibitors, antagonists, or agonists) of the target's activity, followed by a high-throughput screen (HTS) of a large number of small molecules, in some cases several millions (Fox et al., 2006; Macarron, 2006). This may seem like a very large number of molecules but there are an estimated 10^{60}–10^{200} possible organic small molecules (Bohacek, Martin, and Guida, 1996; Weininger, 1998; Gorse, 2006; Fink and Reymond, 2007). Therefore, even with a HTS collection of 10^6, one cannot hope to "cover chemical space" in any meaningful way. However, HTS campaigns are occasionally successful pointing perhaps to the promiscuity of protein binding sites. In the best cases, this method identifies one or more small-molecule "hits" in the micromolar range, that is, for a competitive inhibitor, having binding constants, often expressed as Ki's or IC50's from 10 micromolar to the low nanomolar range. Medicinal chemistry is then applied to elaborate the initial small molecule to improve the potency, ideally lowering the Ki to the low nanomolar range to produce a potent lead molecule (Foye, 1989).

The process of optimizing the lead molecule into a "candidate" drug is often the longest and most expensive preclinical stage in the drug discovery process (although this is still a fraction of the costs of drug development). The candidate is usually an analogue of the original lead, but it is still considered an art to successfully synthesize and select the exact compound that fulfills all the required properties of potency, absorption, bioavailability, metabolism, safety, and, of course, efficacy. The lead-to-candidate stage of drug discovery is a multidimensional optimization problem, a search within the relatively limited chemical space of analogues of the lead compound (Chapter 27).

Following the selection of the candidate molecule, the drug development scientists develop large-scale production methods and conduct preclinical animal safety studies. investigational new drugs (INDs) must pass through a set of three clinical trials: Phase I, a small study on healthy subjects to confirm safety; Phase II, a slightly larger study on a patient population to confirm efficacy; and Phase III, a large study of patients to gather additional information about safety and efficacy (CDER Handbook, 1998). However, even after the medicinal chemist has carefully crafted and balanced the properties of potency, bioavailability, and metabolism, only 11% of the compounds entering clinical trials make it to market, most often failure is due to poor biopharmaceutical properties, toxicity, or lack of efficacy (Kola and Landis, 2004). Due to the attrition of so many potential pharmaceuticals and the rising costs of drug discovery, the average cost to bring each new chemical entity (NCE) to market in 2001 was estimated to be $804 million (DiMasi, Hansen, and Grabowski, 2003). This has caused many to reassess the approach taken by the pharmaceutical industry. Possible alternative approaches are briefly overviewed in the following sections.

Thus far, the discussion has focused on small-molecule drugs, that is, organic molecules of molecular weight of the order of hundreds of Daltons; however, the role of biological therapeutics, primarily enzymes and antibodies, should not be underestimated (Brekke and Sandlie, 2003; Adair and Lawson, 2005). Structural bioinformatics have a major impact in both areas; however, the focus of this chapter will primarily be on its role in developing small-molecule drugs that are NCEs with a known mode of action. However, this approach requires the availability of experimental macromolecular structures and we will therefore first address the generation of protein structures in a pharmaceutical context.

GENERATING PROTEIN STRUCTURES

The optimal structure-based drug design (SBDD) program involves repeated cycles of determining the structure of the target in complex with a number of lead compounds and their analogues (Congreave, Murray, and Blundell, 2005). One hurdle in establishing a rapid cycle of crystallography and structure-based design is, of course, obtaining crystals of the target protein of sufficient quality. A genetic construct encoding the exact full-length sequence of a protein is not necessarily the ideal strategy to use to obtain material for screening or structural studies. The reason being is that the full-length protein may be large and contain protein domains that are not relevant to the studies performed. The protein may also be poorly expressed, unfolded, or insoluble, and thus create challenges that preclude protein crystallization.

To avoid such problems, structural bioinformatics can be used to design suitable constructs. As described in Chapter 20 (domain assignments), proteins are frequently composed of discrete domains. The computational techniques of domain detection can be combined with experimental techniques of domain assignment (e.g., limited proteolysis followed by mass spectroscopy analysis). The designed constructs can be evaluated on the basis of their expression levels, solubility, activity, and crystallizability (Kim, Chivian, and Baker, 2004).

The selection of an appropriate domain for structural studies often begins by aligning the sequence of the target protein with that of a protein of known structure. One can then determine where it may be possible to truncate the protein at the amino- and carboxy-termini regions. Where structures are not available, secondary structure prediction can be used (see Chapter 19). Combining secondary structure prediction, based on multiple sequence alignments, with analysis of sequence conservation can be particularly successful. One can often express a protein from a construct designed to start and finish where sequence conservation is high and at the end of predicted elements of secondary structure. It should be noted that the same process could be used to guide construct design for the production of the target protein for HTS assays. In fact, in the best cases, the same construct can be designed for both screening and structural biology.

Rational construct design is now usually complemented with high-throughput, triage-type approaches to generate material for structural biology studies. Improvements in automation and analysis techniques allow a library of constructs to be designed and evaluated. Subsequently, the most promising constructs are pursued, while others are disregarded (Goh et al., 2004). Where no crystals or poor crystals are obtained, these can be modified to improve properties, applying the so-called crystal engineering (Derewenda, 2004).

As described in Chapter 4 (X-ray crystallography), a second hurdle in structure determination by crystallography is the so-called phase problem. One common solution to this problem is molecular replacement. In molecular replacement, a model representing some or all of the new protein is rotated and translated into the new unit cell in an attempt to find a solution. Originally, molecular replacement was only used for cases of a specific protein in different space groups or in cases of high sequence identity. Greater computational power, as well as the greater wealth of known folds, has resulted in molecular replacement being successfully applied even in cases where the starting model exhibited only 20% sequence identity or less (Storici et al., 1999; Hong et al., 2000).

The extent to which the core of a protein is distorted with greater sequence divergence (Chothia and Lesk, 1986) will presumably place a limit on when molecular replacement can be used. However, even when phases are determined experimentally (e.g., through heavy atoms or selenomethionine techniques), identification of a suitable structural homologue at even 10% sequence identity can greatly accelerate the interpretation of the electron density maps and of the protein structure (Bugl et al., 2000). Multiple homology models can be used instead of crystal structures in this context and they have provided structure solutions where the use of homologue structures has failed (Claude et al., 2004).

Most protein crystals are actually quite soft and contain a high proportion of water. Not only does this give us some confidence that many structures are representative of the protein in solution, but it also allows crystal structures of protein with ligands to be obtained, simply by soaking a pregrown crystal in a solution of the ligand of interest. This can be of great help in an iterative drug design cycle, where speed is very important; however, care must be taken to ensure that the soaking procedure does allow a ligand and protein to adopt relevant conformations.

When a particular target is inaccessible to structural biology, a project may rely on the use of a related protein for structure determinations. At the simplest, this situation may involve using the orthologous protein from another species or selecting a similar member of the same gene family. More complex uses are those where there is similarity at the structural and functional level that does not extend to sequence, for example, in the case of thermolysin and NEP (Holland et al., 1994).

DRUG TARGETS

Biological systems contain only four types of macromolecules with which we can interfere using small-molecule therapeutic agents: proteins, polysaccharides, lipids, and nucleic acids. Toxicity, specificity, and the difficulty in obtaining potent compounds against the latter three types mean that the majority of successful drugs achieve their activity by modifying the activity of a protein, usually by competing for a binding site on that protein with an endogenous small molecule. There have been a number of attempts to catalog the protein targets of known drugs (Drews, 1996; Drews and Ryser, 1997; Hopkins and Groom, 2002; Golden, 2003; Russ and Lampel, 2005; Imming, Sinning, and Meyer, 2006; Wishart, 2006; Zheng et al., 2006a; Zheng et al., 2006b), with several of these resulting in target databases.

A consensus survey of current drug targets for all therapeutic drug classes in the industry has now been arrived at and is reported by Overington (Overington, Al-Lazikani, and Hopkins, 2006). The 21,000 drugs captured in the U.S. FDA's Orange Book and Center for Biologics Evaluation and Research (CBER) were reduced to 1357 unique drugs, of which 1204 are small-molecule drugs and 166 are biological agents. Targets were then assigned to 1065 of the compiled drugs (a far from trivial exercise), giving a number of just 324 drug targets for all classes of approved therapeutic drugs. This captures the molecular targets of all approved drugs to treat human disease, whether the target is of human origin or present in an infecting organism. When one considers proteins encoded by the human genome alone, targeted by small-molecule drugs, the figure is further reduced to 207.

Almost 70% of drug targets fall into just 10 protein families, with over 50% of drugs exerting their activity via 4 families. These families are class 1 GPCRs, nuclear receptors, ligand, and voltage-gated ion channels. Categorizing drug targets by their structural domains, using the CATH (Chapter 18), SCOP (Chapter 17), and PFAM databases, reveals that there are 130 "privileged druggable domains." Of particular interest to the structural bioinformatician is the degree of structural characterization of drug targets. The PDB (see Chapter 11) contains structures of 105 drug targets, with over 92% (300/324) of targets having some structure in the PDB.

Drug Target Discovery

Structural bioinformatics plays an important role in "target-based" drug discovery. However, "traditional" approaches to drug discovery are still very much with us; indeed, many argue that such approaches remain the most cost-effective way of generating new medicines (Brown and Superti-Furga, 2003). Such approaches can be described as "letting the compounds choose the target" as opposed to identifying compounds to interact with preselected targets. The former approach has recently been rebranded as "chemical genetics" (Stockwell, 2000; Stockwell, 2004). Assays of direct functional relevance are chosen for such approaches and while knowledge of the molecular target can be helpful, it is certainly not essential. This is exemplified by the development of the antiepileptic drug levetiracetam, the molecular target of which was discovered sometime after the drug had demonstrated clinical efficacy (Lynch et al., 2004). Indeed, the precise molecular function of this target protein remains unknown.

The ability to test compounds in a wide variety of biological assays has also provided data that have led many to question the effectiveness of single target-led drug discovery programs (Hopkins, Mason, and Overington, 2006). As more "cross-screening" takes place, additional activities continue to be observed in what were once described as selective drugs (Paolini et al., 2006). The activity of compounds against multiple targets can sometimes be correlated with the sequence and structural similarity of those targets (Keiser et al., 2007).

Informatics and knowledge-based methods play an important role in the framework of the postgenomic, target-based drug discovery paradigm, in support of screening and medicinal chemistry. The ability of chemoinformatics to process the properties of millions of virtual compounds for selection for synthesis and screening is an enabling technology for combinatorial chemistry and HTS. Biological structural information can be usefully

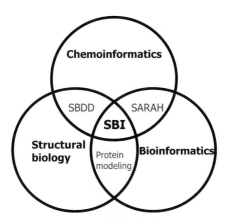

Figure 34.2. The relationship between structural bioinformatics (SBI) and other disciplines in drug discovery. Achievements from both structural biology and chemoinformatics yield methodology for structure-based drug design (SBDD). The interaction between bioinformatics and chemoinformatics can be exemplified with structure activity relationship and homology (SARAH) (Frye, 1999; Schäfer and Egner, 2007). Protein modeling is at the interface between bioinformatics and structural biology.

exploited from the identification of the target protein to all the way to the design of a bioavailable drug. As depicted in Figure 34.2, structural bioinformaticians in a pharmaceutical company also serve to link resources and results stemming from bioinformatics, structural biology, chemoinformatics, and the structure-based drug design groups. The following sections will describe how structural bioinformatics can be employed to help make informed choices on which proteins to target to identify a molecule of medicinal benefit.

Target Identification and Assessment

As structural inferences become available at the very earliest stages of a drug discovery program, structural bioinformaticians can provide *a priori* assessment of the ability of a target to be inhibited by a drug-like molecule. Designing compounds with appropriate biopharmaceutical properties that are still able to bind to their targets with an appropriate affinity is the challenge for the medicinal chemist. A current challenge for structural bioinformaticians is to determine the magnitude and difficulty of the medicinal chemists' task to synthesize the molecule designed *in silico*.

The availability of the sequences from complete genomes has revealed many more potential targets than could possibly be pursued using current experimental technologies. As such, it is sometimes necessary to prioritize targets from a large potential subset that has been identified. Such a subset may arise, for example, from a gene expression study (Sallinen et al., 2000) by analyzing the genomes of disease-causing organisms (McDevitt and Rosenberg, 2001) or by assessing a particular metabolic or signaling pathway (Sivachenko and Yuryev, 2007). Properties of interest when considering large numbers of targets go beyond the assessment of the active site and include those summarized in Table 34.1. The pharmaceutical industry has had different degrees of success on various target families; this may be due to a number of these factors and has been discussed particularly for protein kinases (Groom and Hopkins, 2002).

TABLE 34.1. Considerations When Selecting Molecular Targets

Property	Consideration
Confidence in rational	How strong is the evidence to indicate that modulating the activity of the target will produce the desired response?
Sense of modulation	Is inhibition or agonism necessary to correctly modulate disease process? Note that although GPCRs can effectively be turned on by agonists, this is difficult with most enzymes.
Ability to bind drug-like molecules	Does the target have the potential to recognize compounds with appropriate properties to a reasonable affinity?
Ease of screening	Can the target protein be obtained in quantities sufficient to facilitate high-throughput screening?
Availability of protein structure	Is an X-ray or NMR structure of the protein available to allow structure-assisted drug design?
Pathway	Is the target protein in a redundant signaling or metabolic pathway where it can be bypassed?
Potential for resistance	Do pathogens contain mutant isoforms of the target protein?
Availability of chemical leads	Are there chemical leads available with suitable properties?
Selectivity	Are there related proteins (in the host) that might be affected by inhibitors against the target?

A number of these factors can be analyzed computationally. For example, researchers at Bristol–Myers Squibb identified genes conserved across 10 fungal genomes, which may encode protein targets for antifungal therapies (Liu et al., 2006). Of the 1049 essential yeast genes, 240 had a homologue in all 10 species with a sequence identity of $\geq 40\%$. This level of similarity was chosen as the authors identified that the few known, broad-spectrum antifungal drugs targeted proteins with around this level of cross-species similarity. Ideal targets may be those that also had no human homologue, as this may minimize side effects, 41% of the target proteins fitted this criteria. This "concordance analysis" was used as the starting point for assays to both confirm the essentiality of these genes across the organisms of interest and to identify chemical hits.

The following section describes the characteristics of a binding site that may make a particular protein target amenable to modulation by a small-molecule drug; however, one is not always aware as to the location of the binding site or if indeed such a site exists. Fortunately, a number of structure-based binding site (or pocket) prediction algorithms are available (Chapter 27) (An, Totrov, and Abagyan, 2005; Glaser et al., 2006). Not only is it possible to identify drug binding pockets, but also it is possible to compare them (Kuhn et al., 2006) and identify channels between pockets and to the surface of a protein for which the small molecule can pass through (Petřek et al., 2006).

Principles of Druggability

Not all small molecules can be drugs; similarly, not all proteins can be drug targets. A small molecule must have certain properties, and a protein must contain a binding site that is complementary or compatible with these properties. Binding sites on proteins usually exist out of functional necessity. Due to hydrophobic forces, the energetically optimal protein would be spherical, with all its hydrophobic residues pointing inward (see Chapter 2). As such, drug binding pockets seem to be quite rare, most of those identified to date exist out of functional necessity, that is, to bind an endogenous small molecule. Therefore, the majority of drugs achieve their activity by competing for a binding site on a protein with an endogenous small molecule. Drugs exploiting allosteric binding sites, with no known natural endogenous ligand, are relatively rare (e.g., the nonnucleoside binding site on HIV-1 reverse transcriptase and some protein kinases) (Noble, Endicott, and Johnson, 2004), and these binding sites are usually not observed in the absence of the ligand, since they are generally transitory in nature.

Example: Kinases and Other ATPases. Examination of the natural ligands of a protein can be valuable in assessing the capacity of a binding site to bind a drug-like molecule. The numerous types of ATPases present an interesting example. By definition, ATP is a common cofactor for many enzymes. It is recognized by a number of protein folds in a variety of ways. The adenosine portion of ATP (the adenine and ribose rings) has properties one would expect to be able to mimic in a drug. In contrast, it would be difficult to mimic the three charged phosphate groups in a drug-like molecule, because charged compounds typically cannot penetrate cell membranes. Thus, when one is considering an ATPase as a potential target, it is helpful to determine the way in which the ATP is recognized.

In protein kinases, the adenine ring of ATP fits into a well-defined, relatively hydrophobic pocket, forming a number of important hydrogen bonds. The phosphate groups play a relatively minor role in this recognition (Johnson et al., 1998). It has proven to be relatively straightforward to generate potent, drug-like inhibitors of protein kinases that are competitive for ATP, making use of this attractive binding pocket (Bridges, 2001; Dumas, 2001). In ATPases that rely on coordination of the phosphate group and contains the so-called Walker A or B motif (Walker et al., 1982), for example, inhibition by drug-like molecules has proven difficult by and large; however, exceptional cases was noted with inhibitors bound to two such motifs (Wigley et al., 1991). In many cases, the exact structure of the ATPase may not be known. Even when the structural motif used to recognize ATP is not known, one can gain clues as to the attractiveness of the ATP site by inference from the binding affinities of ATP, ADP, AMP, adenosine, and adenine using biochemical information from the literature. Differences in dissociation constants in this series may allow one to predict which regions of ATP are important recognition features.

Example: Proteases. The proteases present additional subtleties involved in assessing the tractability of individual molecular targets. The substrates of proteases, that is, peptides, represent reasonable chemical leads for a drug discovery process. Crucial to success, however, is the ability to "depeptidize" these leads for peptidomimetic design. That is, to remove the peptide bonds to avoid absorption and metabolic stability issues. For some proteases, this is relatively straightforward. One example is the serine protease thrombin, where many drug-like inhibitors, barely resembling peptides, have been developed (Steinmetzer, Hauptmann, and Sturzebecher, 2001; Kwaan and Samama, 2004). In the case of the aspartyl protease renin, it has proven much more challenging to develop nonpeptidic inhibitors (Stanton, 2003), although the drug aliskiren has recently been developed and is probably the least peptidic and most promising renin drug (Wood et al., 2003).

When one analyzes the way in which substrates are recognized by these two types of proteases, the reasons become clear. Serine proteases typically recognize both main-chain and side-chain features in their substrates, forming hydrogen bonds to relatively few main-chain groups on only one side of the scissile bond. Aspartyl proteases are typically tolerant of side-chain substitutions in their substrates and rely on main-chain hydrogen bonds to bind their substrates. In this case, it has proven difficult to retain potency in an inhibitor while reducing the peptidic character of leads. This has also been the case for many viral proteases, where substrate peptides are often weakly bound in shallow pockets on the enzyme surface (Chen et al., 1996).

Protein families such as kinases and proteases are well-precedented drug targets. Choosing whether to target more novel targets or unusual binding sites is a crucial process, particularly for a small research group with limited resources. For nonprecedented targets, one can predict their druggability (Chapter 27) via a visual inspection of their binding sites. The following section describes how this important process can be carried out in a more automated and objective manner.

Quantitative Assessment of Druggability

The foregoing discussion assumes the opportunity to perform an in-depth, individual analysis of the relevant protein structures. For a quick assessment of a large number of potential targets, a more quantitative approach is more appropriate. Such a quantitative approach is already well established for assessing the drug-like properties of a small molecule. The rule-of-five (Table 34.2) is a set of properties applied to suggest which compounds are likely to show good oral bioavailability (Lipinski et al., 1997). Although not intended as such, the preference of the pharmaceutical industry for such compounds makes this a useful "rule of thumb" to describe drug-likeness. Only 20% of orally dosed small-molecule drugs fail the rule-of-five test (Overington, Al-Lazikani, and Hopkins, 2006).

Further work has refined what distinguishes drugs from other compounds. The reader is referred to relevant work (Ajay, Walters, and Murcko, 1998; Sadowski and Kubinyi, 1998; Gillet et al., 1999; Blake, 2000; Clark and Pickett, 2000; Lipinski, 2000; Walters and Murcko, 2002; Lajiness, Vieth, and Erickson, 2004; Vieth and Sutherland, 2006; Oprea et al., 2007).

Given that the properties of a drug are complementary to those of the binding site, analysis of the calculated physicochemical properties of the putative drug binding pocket on the target protein can provide an important guide to the medicinal chemist in predicting the likelihood of discovering a drug against the particular target site (Chapter 27). Equivalent rules have been developed to describe physicochemical properties of binding sites with the potential to bind rule-of-five compliant inhibitors with a potent binding constant (e.g., $K_i < 100\,nM$). A number of properties complementary to the rule-of-five can be calculated; for example, the surface area and volume of the pocket, hydrophobic and hydrophilic character, and the curvature and shape of the pocket.

Programs such as SURFACE (Lee and Richards, 1971; CCP4, 1994), CAST (Liang, Edelsbrunner, and Woodward, 1998), MS (Connolly, 1993), and GRASP (Nicholls, Sharp, and Honig, 1991) can be used to calculate these and other parameters. Indeed, it has been shown that parameters such as these are quite predictive of the likelihood of a

TABLE 34.2. The Rule-of-Five

A compound is likely to show poor absorption or permeation if
   It has more than five hydrogen bond donors
   The molecular weight is over 500
   The Clog P (calculated octanol/water partition coefficient) is over five
   The sum of nitrogens and oxygens is over 10

high-throughput screen to identify suitable chemical hits (Cheng et al., 2007). Experimental determination of druggability can be performed using NMR-based fragment binding assays (Hajduk, Huth, and Fesik, 2005a; Hajduk, Huth, and Tse, 2005b). Underlying these experiments and predictions is the concept of ligand efficiency (Hopkins et al., 2004), discussed in further detail in Section 6.1. This captures the concept that druggability relates to the amount of potential binding energy available in a binding site, as a function of the size of the site.

The computational predictions of druggability of the binding sites, as described above, are normally carried out on X-ray crystal structure treated as static coordinates. Of course proteins are not static entities and in the search for nonobvious druggable binding sites, their flexible nature is increasingly being taken into account.

Protein Flexibility and Drug Design

The plastic nature of biological molecules is well documented with corresponding flexibility and conformational changes (Chapter 38) shown to be important for biological function. Moreover, molecular recognition and complementarity is in itself a dynamic event. Certainly, the idea of a rigid receptor is now out of date (Teague, 2003). The degree in which one can explore this flexibility depends on the technology and information available. Recent advances in biological and structural techniques coupled with the increasing power of computation are enabling scientists to understand and predict flexibility in biological systems.

Binding site mobility can be thought of in two ways: conformational selection of protein targets or induced fit (Teague, 2003). In the former, the ligand enriches the equilibrium of protein conformers toward the one that fits the ligand best. In the induced fit model, the ligand binds and then the protein accommodates it. In the case of many synthetic drugs, which are often fairly hydrophobic and rigid molecules, binding is an entropically favorable process, often showing unfavorable enthalpy. The dogma is that this is disadvantageous because a drug is prone to lose affinity when the target protein mutates, due to normal drug resistance mechanisms or genetic polymorphisms (Teague, 2003). One can take advantage of protein mobility in ligand design and attempt to exploit allosteric sites. Allosteric sites are able to bind ligands, which shift the conformational equilibrium of the protein partners toward conformations that cannot function (Christopoulos, 2002). Exploring possible conformations of the protein can open new sites or expose new shapes for small molecules to bind, and there are a number of cases exemplifying this (Erlanson et al., 2000; Papadopoulos, 2006; Thanos, DeLano, and Wells, 2006; Toth, Mukhyala, and Wells, 2007).

Targeting Protein–Protein Interactions

We have seen in the previous sections how drug discovery has focused on a handful of targets that meet the classical druggable criteria: being linked to disease and having a "beautiful" pocket to which a small, drug-like molecule binds (Hopkins and Groom, 2002). In a recent review, Whitty and Kumaravel (2006) classified targets in terms of chemical and biological risk. The probability of finding a small molecule to modulate the target under evaluation is its chemical risk. The biological risk covers the therapeutic effect of its modulation. Classical drug targets, such as enzymes and GPCRs, present both low chemical and low biological risk. Although ubiquitous and pivotal to many biological processes, multiprotein complexes have not been considered to be in this class. However, due to their fundamental role in living organisms, their modulation is likely to be therapeutically relevant, thus they should have low biological risk. This has been confirmed by the successful development of antibody drugs modulating extracellular targets (Adair and Lawson, 2005).

In contrast, multiprotein complexes present high chemical risk. Classical hit identification campaigns have struggled to deliver small chemical starting points for protein–protein interaction (PPI) targets. Many have assumed that a small molecule has little chance to satisfy and complement the huge and flat surface of a protein–protein interaction, which have surface areas of around 1600 Å^2 and an average of 22 residues per binding partner (Gonzalez-Ruiz and Gohlke, 2006).

The experimental evidence of the existence of "hot spots" in the complementary surface of protein–protein complexes challenges these beliefs and opens the door for small molecules to modulate these interactions. It has been found that the stabilization energy of the multiprotein complex comes mainly from small areas of the interface between the proteins (Clackson and Wells, 1995). Therefore, small molecules may be able to recreate similar stabilization energy if they are able to interact with these "hot spot" regions in one of the partners of the complex. Again this relates to inhibitors having fragments of high ligand efficiency.

Small molecules modulating protein–protein interactions are now emerging not only from pharmaceutical companies but also from academic research (Arkin and Wells, 2004; Pagliaro et al., 2004; Whitty and Kumaravel, 2006). Examples include the inhibition of p53-MDM2 interaction (Murray and Gellman, 2007), antagonists of the Bcl-2 antiapoptotic family proteins (Papadopoulos, 2006), and the blockade of the binding of IL2 to its receptor IL-2Ra (Thanos, DeLano, and Wells, 2006).

Example: IL-2/IL-2Ra: In 1997, Hoffmann-la Roche scientists presented the first nonpeptide small-molecule inhibitor of a cytokine/cytokine receptor interaction (Tilley et al., 1997). IL-2/IL-2Ra is a system with low biological risk as a drug target, as antibodies for the alpha subunit of the IL-2 receptor were clinically effective as immunosuppressive agents. The design of small molecules substitutes for these antibodies, based on structural information of IL-2, enhanced by mutagenesis experiments to map the binding site of the IL-2Ra. Compounds synthesized to mimic IL-2 were indeed inhibitors of the cytokine/cytokine receptor interaction. Surprisingly, NMR studies showed that the small molecules bind to IL-2 itself. Scientists at Sunesis then cocrystallized the IL-2/Roche inhibitor and applied their Tethering® technology (Erlanson et al., 2000) to characterize the IL-2 binding site and identify chemical fragment binders (Arkin et al., 2003). They found that the IL-2 interacting surface is highly adaptive, as there is a substantial difference between the apo and holo crystal complexes. This example highlights the importance of exploring flexibility in the interacting regions to identify possible adaptable hot spots that could create grooves for a small molecule to bind and disrupt protein–protein interactions (Thanos, DeLano, and Wells, 2006).

Protein–Protein Interaction Hot Spots. Systematic alanine mutation (alanine scanning) at protein–protein interfaces can be used to measure the contribution to the interaction energy of the residues side chains at the surface (Janin, 2000) to identify hot spots. But one should exercise caution when interpreting alanine scanning data. The change in energy can be due to factors other than the atomic interactions of the residue side chain, for example, changes in the conformation of the unbound and bound mutant (DeLano, 2002).

Starting with the structure of the protein–protein complex, three main computational approaches can be used to predict hot spots in PPI (Gonzalez-Ruiz and Gohlke, 2006). Computational alanine scanning performs energy calculations to estimate the difference in free energy between the wild-type complex and the complex with a single mutation to alanine in the interface. Mutations that lead to a significant change in the free energy of the complex are predicted to be at a hot spot. Energies are calculated using molecular mechanics coupled with solvation models and molecular dynamics (MD) to sample the conformational space of the complex (Massova and Kollman, 1999).

Alternatively, the difference in free energy between the wild-type complex and the alanine mutant is estimated with regression-based scoring functions, which are less computationally demanding. These functions are similar to those used for ligand–protein interactions (see scoring functions for docking in Chapter 27), where linear combination of terms including van der Waals, electrostatics and hydrogen bond interactions, water bridges, solvation, and molecular flexibility are optimized against experimental values of $\Delta\Delta G$ (Kortemme and Baker, 2002).

A third approach applies graph theory to protein–protein complex residues. It has been found that small-world networks describe the topology of the interfaces (Del Sol and O'Meara, 2005). These small-world networks are graphs where each node can be reached with small number of edges due to the existence of highly central node or hub (by analogy, on average everyone in the world has $6°$ of separation to everyone else via highly popular people). Furthermore, these central nodes correlate with experimentally determined hot spots.

Protein–Protein Interaction Flexibility. From the viewpoint of developing PPIs as drugs, predicting protein flexibility has two main applications: site adaptability and allosterism. Appropriate ligand binding pockets may be "preformed" at the interaction surface (e.g., MDM2), however they may form in response to protein modification, or be induced to form by a ligand (e.g., in the case of IL-2). Prediction of site adaptability would be invaluable for ligand design. In fact, the assessment of target druggability can change dramatically if one takes into account protein flexibility. Brown and Hajduk have shown recently how their druggability index (Hajduk, Huth, and Fesik, 2005a; Hajduk, Huth, and Tse, 2005b) varies when molecular dynamics simulations (Chapter 37) are included in the prediction (Brown and Hajduk, 2006). Of course, one way to bypass the inherent chemical risk in modulating protein–protein interactions is by finding allosteric inhibitors that indirectly affect the association.

Although binding site flexibility has a very important role in molecular recognition, it is still very challenging to predict. However, there is already an arsenal of structural bioinformatics tools, which can sample molecular conformational space. They can be divided into three main techniques: molecular dynamics, normal mode analysis, and constrained geometric simulation (Gonzalez-Ruiz and Gohlke, 2006). Molecular dynamics generates conformational pathways between states. This approach uses an all-atom model and therefore is computationally expensive; Gromacs (van der Spoel et al., 2005) is a commonly used software for these simulations.

The second technique to sample flexibility is through normal mode analysis (NMA—Chapter 37). Structural perturbations around a local minimum are explored by calculating vibrational modes of the system. In recent years, elastic network models are being widely used. In these models atoms are considered as network nodes connected by harmonic springs (Liu and Karimi, 2007). The third computational technique uses graph theory flexibility concepts for an efficient sampling of the protein mobility. constrained geometric simulation (CGS) considers atoms as nodes and edges as distance constraints defined by the covalent and hydrogen bonds between atoms. Rigid and flexible regions are determined by the geometric constraints that these edges impose on the nodes (Jacobs et al., 2001).

Chemical Genetics

The term "chemical genetics" describes the analysis of phenotypic function by the use of chemical probes. To dissect the function of a specific protein, a tool compound or probe should be a selective inhibitor (or activator) of this protein above all others within the cell. Compounds with this characteristic are rare or possibly even nonexistent. To overcome this difficulty, Shokat and coworkers (Knight and Shokat, 2007) have used a technique, sometimes termed orthogonal chemical genetics (Belshaw et al., 1995; Smukste and Stockwell, 2005), whereby a protein is altered in such a way as to create a unique binding site. At the same time a compound that will fit that pocket and only that pocket is designed and synthesized.

Cells are then engineered to express the mutant protein in place of the naturally occurring version. The cognate-designed compound can then be used to selectively inhibit that protein. Cell-based assays can be carried out in the presence of the inhibitor to understand the function of the protein. Several protein and lipid kinases have been examined in this way, including cdc28, HER3, and PI3Kα (Knight and Shokat, 2007 and references therein).

In each case, the "gatekeeper" residue was mutated to a small residue such as glycine to open up the "gatekeeper" pocket. This pocket, in conjunction with other specific interactions elsewhere in the ATP binding site, is used to provide the unique pocket required for this technique to be successful. In the example of p90 ribosomal protein S6 kinase (RSK), Cohen et al. (2005) targeted the naturally open gatekeeper pocket (gatekeeper = threonine) and a second selectivity providing interaction of a covalent attachment to a proximal nonconserved cysteine. This combination of two selectivity features is only found in RSK1, RSK2, and RSK4, which in turn allowed the authors to selectively target this family. Then they engineered in this cysteine other protein kinases containing the open gatekeeper pocket and used their RSK inhibitor to selectively knockdown their activity.

This approach has the advantage over gene knockout or catalytically dead knock-in approaches to functional characterization in that the protein is disabled reversibly and on a timescale that can be physiologically and medicinally relevant. Gene knockouts also have the added disadvantages of often causing developmental problems, compensatory proteins to be upregulated, and a loss of normal protein–protein interactions.

HIT IDENTIFICATION

Once one has identified a target binding site and structural information is available, structure-based virtual screening (SBVS—Chapter 27) can be employed to identify chemical starting points for a drug lead identification program. Such techniques are now used routinely for the identification of focused screening sets from either commercial supplier's catalogs or internal compound collections (Oprea and Matter, 2004; Davis et al., 2005). These sets are tested experimentally to identify hit molecules that can form the starting points for a medicinal chemistry lead identification program.

Various VS techniques can be employed to this end and it is sensible to use all the information available about a target, drawing on the full range of structural bioinformatics techniques. Structure-based methodologies have the advantage over some ligand-based methods in that they do not explicitly bias the compound selection toward any particular chemotype, and therefore they can be a good way of identifying novel chemical matter (Kitchen et al., 2004).

The target information gathering process can be extended to include proteins in the same family or those that have a similar binding site to the target. The selection of libraries that are focused on whole gene families, for example, protein kinases or ATPases in general, can be aided by the use of structural information (Bohm and Stahl, 2000). As with ligand-based virtual screening, SBVS need not be limited to available compounds or even to compounds that have ever been made. Virtual compounds can be fragments of real compounds or whole imaginary compounds. The search for hits can therefore extend further into unexplored areas of "chemical space" (Weininger, 1998; Fink and Reymond, 2007).

While the use of protein structures has the potential to greatly aid the hit finding process, in order for the full potential to be realized, a number of factors have to be considered. First, the nature of the protein structure itself. For instance, is it a crystal structure or an NMR structure? What is the quality and resolution of the structure? What is known about the binding site location and protein flexibility? Is it an apostructure or does it have an inhibitor or natural ligand in the binding site? Are the locations of hydration sites known?

Software applications need to assess the electronic and shape complementarities of flexible ligands for flexible proteins and to do this quickly enough to allow searching of large collections of compounds in a timely manner.

Measures of Virtual Screening Success

It is important to try and understand the ability of the available programs to make useful predictions concerning molecular recognition as they can be very expensive and time consuming. The extent of the predictive power is never known with any precision because truly objective tests of such software are rare and performance varies not only between operators but also between protein families. Prospective users perform in-house validation of software using proprietary data sets to gain an understanding of performance and relevance in their working environments.

The success of VS is usually measured by the number of real hits identified, compared to the number expected to be achieved by random selection of compounds for testing (the "enrichment"). Some researchers quote the area under the curve of the "receiver-operating characteristic" (ROC) plot, while others use the enrichment rate (the hit rate/the random hit rate) for the initial selections (Chapter 27). The random hit rate is simply estimated by multiplying the number of compounds selected by the proportion of true hits with the entire pool of compounds. Assessment of virtual screening methodologies is usually done in hindsight. Of course, this is far from ideal and truly blind predictions like those used in the CASP (Moult et al., 2005) and CAPRI (Méndez et al., 2005) (Chapter 28) competitions would be preferable. McMaster University did hold a blind high-throughput virtual screening competition using *Escherichia coli* dihydrofolate reductase (DHFR) as a test case. Unfortunately, no competitive inhibitors were identified in the real HTS and therefore the study was not a good assessment of the techniques used (Lang et al., 2005). Interestingly, 50% of the 32 competitors used only ligand-based methods, 10% only used protein structure-based methods, and the remaining 40% used a combination. A comparison of the enrichment achieved by a variety of methods applied to the same data sets have been shown to be useful (e.g., Nettles et al., 2006; Hawkins, Skillman, and Nicholls, 2007). In this way, the timings and ease of use can also be compared, but again, bias is difficult to avoid.

Protein Structure in Hit Identification

Nearly all SBVS is done using X-ray crystal structures rather than NMR structures. The former are more readily available in the public domain (the ratio of X-ray to NMR structure in the pdb is approximately 6 : 1) and more commonly produced within pharmaceutical companies, but the latter may have advantages when considering an ensemble of structures to take protein flexibility into account. See Chapters 4 and 5 for more information on the complementarity of NMR and X-ray structure determination.

The resolution of a crystal structure is more critical for some SBVS techniques than others. For instance, docking can be highly dependent on the atomic coordinates and very small differences can result in a completely different binding mode. For structures with a resolution above 3.5 Å, it may be advisable to use a virtual screening method other than high-throughput docking. High-resolution structures have the advantage of revealing the sites of ordered water molecules. Water molecules that form three or more hydrogen bonds to the protein are considered to be integral parts of the protein structure (the so-called structural waters).

The inclusion of such water molecules can be very important to achieve the correct binding mode with docking (e.g., Pickett et al., 2003). However, even if absent from a crystal structure, these structural waters can be predicted with good accuracy (Pitt, Murray-Rust, and Goodfellow, 1993; Verdonk et al., 2001; Verdonk et al., 2005). Water molecules with two or less hydrogen bonds to the protein can sometimes be displaced by a ligand with the advantage of increasing water molecule disorder. If the hydrogen bonds are replaced with those to the ligand, compounds with increased potency and selectivity can result. Displacing water molecules from hydrophobic surfaces can often result in an affinity gain for a small molecule due to the hydrophobic effect, but these waters are usually too disordered to be observed in an X-ray crystal structure, except perhaps when they are confined to a small cavity (Garcia-Sosa, Mancera, and Dean, 2003).

Having a ligand in the crystal structure of the targeted binding site is extremely useful because it provides information on a successful set of binding interactions and a ligand conformation. In many cases, the structures of apoproteins are more open and disordered than respective ligand-bound structures. In other cases, allosteric pockets can open up only in the presence of a small molecule. Information on such binding sites is highly valued because they can provide a competitive advantage to a pharmaceutical company.

Binding of a ligand causes a conformational change in the protein to maximize complementarity. Crystal structures of protein–ligand complexes can therefore provide a negative image of the ligand that results in SBVS hits that are biased toward similar molecules. It can therefore be an advantage to use ligand–protein crystal structures generated with a diverse set of compounds.

Protein Modeling

It is often the case that a crystal structure of the exact protein one targeting is yet to be solved, but a structure of a homologous protein has been. If the homology is high or there are many related structures, useful models can be produced (Chapter 30) and even used for docking

(Hillisch, Pineda, and Hilgenfeld, 2004). It is highly advisable, if docking is to be attempted, to create a model with a tightly binding and preferably rigid compound in the binding site. This is possible in the comparative modeling package Modeller (Marti-Renom et al., 2000 and Chapter 30). Protein–ligand complex model construction has the same effect as cocrystallization in producing a negative image of the compound, and thus can help to find compounds with similar shape and hydrogen bonding functionality as the modeled compound. Another case where docking into homology models is productive is if there is a metal binding site or similar conserved interaction for which constraints can be used to anchor a compound in the binding site and thus remove three degrees of freedom. On the whole, when working with more speculative protein models and very low resolution or partially disordered crystal structures, a ligand-based virtual screen (LBVS) technique such as pharmacophore searching can be more appropriate than docking.

Structural Bioinformatics in Target Families

Almost any potential target will have related host proteins whose function must not be affected by a successful therapeutic. This will be true, for example, if one wishes to inhibit a protein from one of the larger gene families. In considering selectivity issues, certainly one must first look at related sequences in the same protein family. However, there is not necessarily a direct correlation between the similarities one would infer from homology and those one would infer from compound action. This is because only a small fraction of the residues in a protein interact directly with any one ligand. Although long-range interactions and conformational changes can play a role in inhibitor binding, in general it is those residues lining a ligand binding site that are of most importance.

The ATP cleft of the protein kinases serves as a good example. Although all protein kinases bind ATP in the same region and conformation, only a few of the residues facing the cleft are highly conserved. The remaining residues are subject to random drift, and thus distantly related kinases can actually end up having clefts more similar than more closely related kinases.

This use of noncontiguous regions of sequences to establish relationships was first introduced by Sandberg et al. (1998) and has more recently been described as structure–activity relationships and homology (SARAH) (Frye, 1999; Schäfer and Egner, 2007). In SARAH, protein targets are grouped both by their sequence similarities and by their ability to bind an array of compounds, such as may be determined from pharmaceutical HTS. SARAH builds on earlier ideas around affinity fingerprinting (Kauvar et al., 1995). As Frye points out, protein targets were often grouped according to their response to ligands before we had sequence information at our disposal (Lefkowitz, Hoffman, and Taylor, 1990).

Although conceptually very attractive, it has proven difficult to correlate the similarity of binding sites of protein kinases with the ligand they have been shown to bind, demonstrating the importance of long-range interactions and subtle complex conformational changes (Scapin, 2006a; Scapin, 2006b; Fedorov et al., 2007; Verkhivker, 2007).

Docking and Scoring

Programs such as GOLD (Jones et al., 1997; Verdonk et al., 2005) and FlexX (Kramer, Rarey, and Lengauer, 1999) are very good at predicting the binding mode of compounds removed from a crystal structure and after alternate conformations have been generated. However, because the changes to the shape of binding sites are often induced by compounds (as mentioned above), docking diverse compounds into protein structures is relatively unreliable. Some docking programs, such as Flo + (McMartin and Bohacek, 1997), FlexE (Claussen et al., 2001), and Glide (Friesner et al., 2004; Halgren et al., 2004) can take protein binding site flexibility into account and can often successfully predict a binding mode, even when a conformational change in the protein is required. When used for SBVS, docking is performed in a high-throughput mode (HTD). This can mean that factors such as protein flexibility have to be ignored to speed up the calculation.

Once docking is completed, ranking of potential binders is performed. A scoring function can be used to rank docked compounds by how well they fit into the binding sites. The development of these scoring functions is still a challenging area of research. The best that high-throughput compound ranking functions can currently hope to achieve is to filter out the nonsense results allowing a knowledgeable human to select from the remaining compounds visually. See Chapter 27 for more information on docking and scoring.

It has been argued that achieving the correct binding mode is not essential for VS, and that the most important measure of success is the number of hits identified. It is highly possible that HTD can achieve good enrichment rates purely by filtering out compounds that are too big to fit into the active site or too small to be detected by a biochemical assay. However, this type of enrichment could be achieved by simply filtering by molecular weight. Correctly identifying hits using HTD could be achieved by using the negative image effect mentioned above even if these hits do not bind in the same way as the query molecule removed from the protein–ligand complex crystal structure or homology model. However, it is possible that enrichment rates achieved this way might be attained much more efficiently than using a ligand-based similarity search program. In order for the HTD to make more use of the binding site structure, predicting the correct binding mode is essential.

Fragment-based approaches are proving to be effective in producing hit molecules (Hajduk and Greer, 2007). Fragments are typically molecules below a molecular weight of 250 Da that bind with micromolar to millimolar affinity and need to be extended in size to be turned into a "hit." Relatively simple structures are more likely to bind to a given protein binding site and therefore hit rates are typically higher (Hopkins, Mason, and Overington, 2006) However, the relatively few interactions they make with the protein can make measuring binding affinity more difficult (Hann, Leach, and Harper, 2001) as well as docking and scoring more problematic. Although fragment screening has recently received much attention, it should be pointed out that drug discoverers have for some time used small molecules, often endogenous ligands, as starting points for medicinal chemistry.

Docking can also be used simply as an alignment method for conformation and alignment 3D QSAR methods, such as CoMFA (Cramer, Patterson, and Bunce, 1988), which can then be used to find new hits. This has the advantage of placing the QSAR model in the context of the receptor, thus facilitating interpretation. However, small variations in binding mode of conserved substructures, although likely to occur in reality, generate noise for CoMFA models and ligand-based overlays can achieve more predictive results (Cramer and Wendt, 2007).

Example: Protein Kinases. A realistic example of the virtual screening of a commercial compound collection is quoted in Kubinyi (2006): Forino et al. (2005) chose to search a commercial vendor's catalog of 50,000 compounds using FlexX with a crystal structure of protein kinase B alpha (PKBα/Akt1). Compounds were selected for testing using a consensus scoring scheme followed by visual inspection. They identified 3 hits with low micromolar affinity after testing 100 compounds. This represents a hit rate of 3% and an enrichment over random selection of 15 (maximum enrichment possible 500). Further examples can be seen in Dubinina et al. (2007).

Pharmacophores

Structure-based pharmacophore (SBP) VS is an extension of methods originally developed for ligand-based compound selection. In its simplest form, it uses the active site volume as a means to limit the size of the resulting virtual hits. An added benefit is that the conformational space of flexible compounds is restricted, thus resulting in a more realistic fit to a pharmacophore. When using a homology model or low-resolution experimental structure, this method works well, as the size of the pocket can be expanded in areas of uncertain protein structure giving a larger margin for error. This approach is implemented very efficiently by the use of the so-called receptor exclusion spheres in the program Catalyst (Greene et al., 1994; Greenidge et al., 1998). Good results can be achieved by generating a 3D query from pharmacophoric features and receptor exclusion spheres, derived from a crystal structure of a protein–ligand complex. Receptor exclusion spheres and similar approaches in the programs Unity (Tripos, United States) and MOE (CCG Inc., Canada) do not serve to positively select compounds that fit the shape of the active site but complementarity can be used to help rank the resulting virtual hits. The Rocs program (Rush et al., 2005) does allow for a search of compounds that fit the shape of a binding site, but does not match predefined pharmacophoric features.

Neither SBP nor HTD approaches need small molecule starting points before they can be employed. Pharmacophores are generated based purely on the character of the binding site. This is an advantage, but it also presents a new problem as to how to choose the features that are essential for binding. The classic program GRID (Goodford, 1985) and knowledge-based system SuperStar (Verdonk et al., 2001) calculate "site points," which are the positions of optimum interactions for chemical groups of different types. These points can be converted into pharmacophoric features, but typically there are too many points and choices must be made on which to be included in a pharmacophore. At least one method has been developed to automate this process (Chen and Lai, 2006).

An alternative approach is to use an exhaustive ensemble of pharmacophores. The program FLAP (Baroni et al., 2007) generates a fingerprint from all possible combinations of four-point pharmacophores. Once the generation of pharmacophores is automated, one can produce pharmacophores for whole families of proteins with the aim of producing an activity profile (Steindl et al., 2007) or selecting gene family focused library.

Using pharmacophore points to aid docking (e.g., FlexPharm from BioSolveIT, Germany) or to filtering docking results (e.g., Silver from CCDC, United Kingdom) is a good way of combining the two techniques. One can also use a pharmacophore search as a filter from which to go on and do a more CPU-intensive docking study.

Example: Protein Kinases. Again taken from Kubinyi (2006), Singh et al. (2003) of Biogen Inc. constructed a pharmacophore based on a crystal structure of SB 203580 (a well-known p38 inhibitor) bound to type I TGFbeta receptor kinase, using the program Catalyst. They did not use receptor exclusion spheres but the query compound to define the shape of the active site. Out of 200,000 commercial compounds, 87 fitted their pharmacophore. They do not quote a hit rate, but simply focus on one novel 5 nM hit. It subsequently turned out that Eli Lilly and Co. had independently discovered the same compound using a high-throughput cellular assay (Sawyer et al., 2003).

De Novo Design

In addition to the virtually screening preconstructed virtual compound libraries, construct ligands can be designed to fit a binding site from scratch (Schneider and Fechner, 2005). In theory, this is a far more efficient way to search chemical space, especially if a fragment growth strategy is employed. However, it is difficult for the developers of these methods to get the balance right between producing molecules that are so simple that they are uninteresting and those that are so complicated that synthesizing them would be unrealistic. Also, the algorithms are in competition with the inventiveness of the human mind, which can make leaps across huge distances in chemical space with confidence. Perhaps the greatest difficultly (as with docking) is that *de novo* design scoring functions cannot predict activity with any accuracy. In addition, a chemist has to undertake a costly custom synthesis based on an unreliable prediction. This can be a far higher hurdle to leap than simply purchasing an available compound. Even the availability of a low potency, but small fragment hit, can give the necessary confidence to pursue difficult synthetic chemistry approaches.

Combining Virtual Screening Methods

It has already been mentioned that docking and structure-based pharmacophores are increasingly mixed to produce hybrid methods. If small-molecule binders and the structure of the target are known, then SBVS techniques can be combined with ligand-based techniques (e.g., Taylor et al., 2007). If different approaches are used, then a consensus approach gives confidence in the results. In addition, different techniques will undoubtedly find different hits and therefore provide a greater range of starting points for a medicinal chemistry program. Once hit molecules have been identified, the next task is to structurally alter them with the aim of converting them into a drug leads.

LEAD IDENTIFICATION

A lead molecule for a drug discovery project can be defined as one having low nanomolar potency against its target, showing *in vivo* efficacy and having some selectivity over undesirable proteins. For more than 30 years, structure-based drug design has been used to speed up the process of optimizing small molecules to this end (Congreave, Murray, and Blundell, 2005). The fundamental impact of having protein structural information is in allowing medicinal chemists to select their next molecule to synthesize based on a clear visualization of the molecular recognition event. By bridging and building on resources in bioinformatics, structural biology, and structure-based drug design, structural bioinformatics can accelerate the quest for a potential drug by helping to provide guide design and medicinal chemistry. The next sections provide an introduction of each of the main components of structure-based design to convert a hit molecule into a lead.

Improving Ligand Affinity and Selectivity

The approach of optimizing potency and selectivity against a protein target using iterative steps of crystal structure analysis, compound design, synthesis, and testing is commonplace. Of the utmost importance is the introduction and retention of drug-like properties, while improving the activity profile. Another concept that has been taken up by medicinal chemistry teams is that of ligand efficiency (Hopkins, Groom, and Alex, 2004 Abad-Zapatero and Metza, 2005). Here, the affinity of the compound is compared to its molecular weight or other properties the medicinal chemists are seeking to optimize, such as lipophilicity, with the aim of generating potent molecules of minimum size.

It has been recognized that larger, more lipophilic molecules are relatively rare among drugs (Wenlock et al., 2003) and harder to optimize (Davis et al., 2005). However, the minimum size and lipophilicity of a lead will not only depend on the quality of the initial hit molecule from which it was derived, but also on the properties of the binding site targeted. For example, it has been found that synthetic ligands for some targets, such as peptide GPCRs, tend to be much larger and lipophilic than others, for example, ion channels (Paolini et al., 2006). In contrast, the adenine binding site of protein kinase ATP binding pockets, at least in the active ligand-bound conformation, is narrow and hydrophobic and consequent inhibitors tend to contain a key flat aromatic heterocyclic ring that results in poor aqueous solubility.

In addition to drug-like physical properties, the ever-present concerns of synthetic accessibility and novelty mean that SBDD is about more than simply improving the fit of a small-molecule hit for its target active site. That said, the use of structural information can be invaluable in revealing areas where physical properties can be improved without affecting potency or, much better still, while improving potency. Solubilizing groups are often added to the aromatic core of protein kinase inhibitors so that they extend out of the pocket and into the surrounding solvent or occupy the ribose or phosphate binding sites, which are more hydrophilic in character than the adenine binding site.

Many of the techniques used for hit identification (described above), such as docking, SBP, and *de novo* design, can also be applied to optimize the potency of these hits. For instance, small, focused virtual libraries can be designed around a single molecule to expand SAR. Docking or pharmacophore analysis can then be used to help prioritize synthesis. Similarly, *de novo* design software can be used to replace or extend a particular substituent of a hit molecule. In addition, because of the smaller number of molecules being considered at this stage, more computationally intensive techniques can be employed, such as flexible protein docking and molecular mechanics techniques such as energy minimization, molecular dynamics, and Monte Carlo (MC).

As already stated, the computationally efficient scoring functions used for virtual screening are unable to predict activities in a reliable way. A number of methods that are more computationally intensive but also more theoretically rigorous have been developed to try and predict binding affinity or relative activity from a structure of a ligand–protein complex. Free energy perturbation (FEP) is perhaps the most well known of this class of techniques and, in theory, can be used to calculate the difference in binding affinity of two closely related compounds (Gohlke and Klebe, 2002). The ability to do this would obviously allow medicinal chemists to have an accurate estimate of the activity of the next compound they planned to make based on the measured activity of its parent molecule.

Unfortunately, this scenario is far from becoming a reality, with practical considerations of the calculation preventing it thus far. These include difficulties in setting up the system, the length of time the calculations take to run, and the assumptions inherent in the system setup. Progress is being made toward ameliorating the two former problems, but the latter issue may well be system and compound dependent, and therefore will always lead to uncertainty. Examples of this type of consideration are the inclusion of explicit solvent in the calculation, changes in protonation state, and the large-scale or high-energy molecular motions.

Example: Protein Kinases and Free Energy Perturbation. Pearlman and
Charifson (2001) applied the related technique, thermodynamic integration (TI) to a set
of p38 MAP kinase inhibitors and obtained a very good correlation between calculated
binding energies and measured IC50s for 12 of their 16 molecule set, although the range of
activities was only slightly over 1 order of magnitude. These authors created the predictive
index (PI) to aid comparison of binding energy calculations (+ 1 for perfect, 0 for random,
and −1 for a perfectly incorrect prediction). Their p38 test set achieved a PI of 0.843. Michel,
Verdonk, and Essex (2006) describe the application of FEP to CDK2 and other drug targets.
Although they did not obtain useful predictions for their CDK inhibitors (PI $< = 0.36$), they
did achieve high PIs for sets of neuraminidase (PI $< =0.95$) and cyclooxygenase 2 inhibitors
(PI $= 0.96$). The authors also offer the opinion that it is now feasible to study 24 compounds a
day using FEP on a 150 CPU cluster.

***Example: Protein Kinases and Molecular Mechanics Poisson–Boltzmann
Surface Area Method.*** Another technique that has recently been explored is the
molecular mechanics Poisson–Boltzmann surface area (MM-PBSA) method (Kollman
et al., 2000; Foloppe and Hubbard, 2006). Unlike FEP, this method is designed to cope with
more diverse chemical structures. When Pearlman (2005) applied MM-PBSA to the p38 set
used for the TI experiment (see above), the results were inferior ($-0.04 \geq PI \leq 0.45$). Kuhn
et al. (2005) also applied MM-PBSA to the Pearlman and Charifson (2001) p38 data set and
also to another diverse set of 12 p38 inhibitors. Results for the first set were on a par with those
produced by Pearlman ($-0.22 \geq PI \leq 0.31$) and the second set produced even less predictive
results.

The authors then went on to test whether MM-PBSA scoring could produce good
enrichment in docking-based virtual screening. The results they obtained for their p38 test set
were only slightly better than random but better than the conventional scoring method that
they used (ChemScore). In contrast, all methods applied to their estrogen receptor test set
produced good enrichments. In the case of COX-2, MM-PBSA did produce a reasonable
enrichment over random selection while ChemScore did not. The authors state that MM-
PBSA can produce results quickly (∼15 min/compound on a single processor) and imply
that the technique may have utility in the hit identification stage of a drug discovery project
but not in lead optimization.

As can be surmised from the discussion above, predicting ligand affinity in the hit to lead
phase of a project is at best difficult, even when good structural information is available.

Once at least one series of molecules has been created, the process of fine-tuning them in
such a way that at least one example is fit enough to be tested in humans begins.

LEAD OPTIMIZATION

Potency and simple property filters, such as the rule-of-five, are often the main criteria in the
lead discovery stage of a drug. To design a candidate medicine from the initial lead, however,
one needs to consider a host of additional parameters that can affect the biopharmaceutical
and safety properties of the drug such as the *in vivo* absorption, distribution, metabolism,
excretion, and toxicology (ADMET). Tools of structural bioinformatics, namely, sequence–structure relationships and protein homology modeling, can be usefully employed in the
field of ADMET modeling.

The most developed work in this field, relevant to structural bioinformatics, is in the area of cytochrome P450 modeling for the prediction of drug metabolism (de Groot, 2006). More than 90% of the drugs currently used are metabolized by only 7 of the 57 known human P450 isoforms. Iterative docking and refinement of structures and models has led to a structural rationalization of the metabolic sites of known drugs (Marechal and Sutcliffe, 2006) and these insights are used to generate pharmacophores and other *in silico* models predictive of P450 metabolism or inhibition. More recently, structures of some of the cytochrome P450 enzymes have been determined, allowing this to be integrated with previous predictive models for binding and metabolism (Williams et al., 2004; Yano et al., 2004).

In addition to metabolism, the effect of drug molecules on endogenous transporters and channels is also of paramount importance. Of particular interest to the pharmaceutical industry is unwanted inhibition of the human *ether-à-go-go*-related-gene (hERG) K($+$) channels. Blockade of hERG can lead to cardiac arrhythmia, with the most common problem being an acquired long QT syndrome, that is, a delay of cardiac repolarization that increases the risk of torsades de pointes (TdP) arrhythmia. Although an uncommon side effect, a number of drugs in very common use have been shown to carry this liability. Although the structure of the hERG channel is not yet known, structural bioinformatics studies (Mitcheson et al., 2000; Farid et al., 2006), site-directed mutagenesis (Chen, Seebohm, and Sanguinetti, 2002; Kamiya et al., 2006; Perry et al., 2006), and analysis of known inhibitors have allowed some understanding of the characteristics of compounds binding to hERG (Firouzi and Sanquinetti, 2003; Sanguinetti and Mitcheson, 2005; Jamieson et al., 2006) and the particular features of the hERG channel responsible for binding compounds and channel activity (Fernandez et al., 2004; Stansfeld et al., 2007).

Protein structure analysis is also beginning to have an impact on the study of drug binding by plasma proteins (Hajduk et al., 2003). Computational approaches cannot yet substitute for screening technologies; however, they can help better direct experimental resources and allow one to assess compounds not yet synthesized (Eddershaw, Beresford, and Bayliss, 2000).

STRUCTURAL BIOINFORMATICS DATABASES

From the drug design point of view, it would be very helpful to have structural bioinformatics data integrated with synthetic ligand SAR data. As described in this chapter, data on the protein sequence, protein structure and function, small-molecule structure and function, and protein–ligand complex structure are all employed in the discovery of new drugs. The collation, annotation, storage, and visualization of this ever-increasing mass of data from a variety of different disciplines present a huge problem. An added complication is that some of the data are proprietary, but must be put somehow into the wider context of the publicly available data. Currently, drug discovery computational chemists and informaticians collect the data they need from various sources, store them in a variety of ways (usually flat files), and visualize them in their favorite molecular graphics, sequence analysis, and SAR data mining packages.

Many very useful sources of data facilitate this process. Some data sources do provide access to information on interactions between drug size molecules and proteins; these include Relibase (Hendlich, 1998), MSDChem (Dimitropoulos, Ionides, and Henrick, 2006), and PDBsum (Laskowski et al., 2001), which are described in Chapter 13. The commercial package Chematica™ (Inpharmatica, United Kingdom) and the publicly available ProFunc (Laskowski, Watson, and Thornton, 2005) are likely to be very useful for target identification and validation. KiBank (Zhang et al., 2004) contains binding affinity, structures of chemicals and target proteins, and many useful links and information. SMID (Snyder et al., 2006) is a database of small molecule–protein interactions with visualizations of structures from the PDB. BindingDB (Chen, Liu, and Gilson, 2001) contains small molecule and protein affinity data measured using ITC and enzyme activity assays and also provides online virtual screening tools.

TOWARD PERSONALIZED MEDICINE

Genetic variation is a major cause of the differential effect that drugs can have on individuals. The term pharmacogenetics, coined by Vogel in 1959 and more recently extended to pharmacogenomics, frames the study of the relationship between drug response and the patient's genetic profile. This field encompasses the study of environmental, age-related, as well as inherited variability in patients. An example where this area is bearing fruit is in the prediction of the outcome for breast cancer patients based upon the expression profiles of 70 risk genes (Lacal, 2007).

The diversity in a patient's drug response is usually characterized by the observation of phenotypic changes. Pharmacogenomicists, facilitated by the human genome project and using the modern genomics tools for DNA analysis, are now linking these phenotypic observations to underlying genetic causes. One approach they have taken is to study single nucleotide polymorphisms (SNPs). SNPs are single nucleotide mutations in the DNA sequence that occur in at least 1% of the population (Pfost, Boyce-Jacino, and Grant, 2000).

Structural bioinformatics helps to understand the relationship between genetic variability and drug response by modeling protein–drug complex structure in the presence and absence of SNPs. The aim is to predict variation in the putative binding site of the drug or other more complex effects such as variation in the expression levels, solubility, and interaction with other macromolecules or stability. Nonsynonymous SNPs (nsSNPs) generate an amino acid point mutation and are the focus of bioinformatics analysis. Often this analysis is based on the fact that an nsSNP can modify the function of the protein by altering its stability (Yue, Li, and Moult, 2005). For a list of tools available to facilitate this analysis, see Table 34.3. Moult and coworkers take observations of this phenomenon that are thought to cause monogenetic disease and make predictions about the effect of snSNPs, based upon changes in structural descriptors (Wang and Moult, 2001). When experimentally determined structures are not available, comparative models can be used (Karchin et al., 2005). Perhaps the most highly studied areas are the polymorphisms of the CYP P450 enzymes and drug transporters (Eichelbaum, Ingelman-Sundberg, and Evans, 2006; Daly et al., 2007). For further reading on stratified medicine, please see Trusheim, Berndt, and Douglas (2007).

CONCLUSION AND FUTURE DIRECTIONS

With the advent of the genome era and the increase in the availability of protein structures due to structural genomics, most drug discovery programs directed against a soluble protein have a structure or model of the target protein available. With higher throughput cloning and expression strategies and the careful design of constructs with the aid of structural

TABLE 34.3. Online Resources to Study SNPs

Name	Description	URL	References
dbSNP	The NCBI database of genetic variation	http://www.ncbi.nlm.nih.gov/projects/SNP/	Sherry et al., 2001
SNPs3D	Genes linked with disease, gene–gene network and nsSNPs related with protein function	http://www.snps3d.org/	Yue, Melamud, and Moult, 2006
MutDB	Human variation data with protein structural and functional information	http://www.mutdb.org/	Dantzer et al., 2005
SIFT	Prediction of how a mutation affects protein function based on sequence homology and the physical properties of amino acids	http://blocks.fhcrc.org/sift/SIFT.html	Ng and Henikoff, 2002
LS-SNP	Genomic scale software pipeline to annotate snSNPs	http://alto.compbio.ucsf.edu/LS-SNP/	Karchin et al., 2005
SNPeffect	Computational tools to predict the effect caused by the mutations on the physicochemical and biological properties of the affected proteins	http://snpeffect.vib.be/	Reumers et al., 2005
StSNP	A tool for modeling nsSNPs locations on proteins and presenting their metabolic pathways	http://glinka.bio.neu.edu/StSNP/	Uzun et al., 2005
PolyPhen	Prediction of functional effectof human nsSNPs	http://genetics.bwh.harvard.edu/pph/	n/a
SNP@Domain	SNPs annotated with protein domain structures and sequences	http://snpnavigator.net/	Han et al., 2006
Pmut	Annotation and prediction of pathological mutations	http://mmb2.pcb.ub.es:8080/PMut/	n/a
PolyDoms	Mapping of human coding SNPs onto protein domains	http://polydoms.cchmc.org/polydoms/	Jegga et al., 2007
Panther	Estimates the likelihood of a particular nsSNP to cause a functional impact on the protein	http://www.pantherdb.org/tools/csnpScoreForm.jsp	Thomas and Kejariwal, 2004

833

bioinformatics, this will continue to increase. The early use of structural information in a drug discovery project can provide a great deal of insight for the lead discovery program. Quantitative target assessment plays a role in the decision of whether to invest in a project or not. Many medicinal chemistry projects have floundered for the simple reason that the binding pocket of the target did not have the physicochemical properties complementary to binding a small molecule. Thus, initial analysis of the binding site can provide a significant guide to the ultimate success of the target.

Automatic target assessment of a large number of protein structures is now a possibility.

In addition to assessing targets, comparative structural bioinformatics can identify new drug targets from a combined study of pathogen and human genomes. In comparing active sites, computational target triaging methods can simultaneously assess target druggability and species selectivity. Structural bioinformatics also allows one to probe the complementarity between a compound and its target not only by medicinal chemistry approaches, but also by site-directed mutagenesis. Identifying interspecies differences in the binding sites of drug targets can also be exploited in *in vivo* models or by specifically engineering animal receptors by site-directed mutagenesis to mimic the human receptor binding pocket. The combination of site-directed mutagenesis and medicinal chemistry to produce exquisitely selective chemical probes is a superb example of the value of structural bioinformatics. Protein structures are now exploited in a myriad of ways for drug discovery, going much further than serving as rigid models of an inhibitor binding site allowing one to design molecules of a higher affinity. Combinations of structure-based and ligand-based virtual screening can now select enriched sets of compounds to test for biological activity.

Knowledge-based approaches, combined with the ubiquitous availability of sequence and structure data, have moved us to a new *prospective* paradigm, in which researchers often discover drug-like modulators of targets long before any clinical utility is demonstrated. Indeed, this has been so seductive that drug researchers are reminding themselves of the need to be driven by assays that are genuinely relevant for a disease of interest.

This chapter highlights how our frustration at the complexities of molecular recognition is beginning to be compensated for by the broader use of protein structures and sequence data. As we move toward a situation where drug discovery projects are bathed in structural and sequence information, it is the role of the structural bioinformatician to integrate this into the drug discovery process.

ACKNOWLEDGMENTS

The contribution of Eric Fauman and Andrew Hopkins to the basis of this document is acknowledged (Fauman, Hopkins, and Groom, 2003).

REFERENCES

Abad-Zapatero C, Metza JT (2005): Ligand efficiency indices as guideposts for drug discovery. *Drug Discov Today* 10(7):464–469.

Adair JR, Lawson ADG (2005): Therapeutic antibodies. *Drug Design Reviews* 2(3):209–217.

Ajay A, Walters WP, Murcko MA (1998): Can we learn to distinguish between "drug-like" and "non-drug-like" molecules? *J Med Chem* 41:3314–3324.

An J, Totrov M, Abagyan R (2005): Pocketome via comprehensive identification and classification of ligand binding envelopes. *Mol Cell Proteomics* 4(6):752–761.

Arkin MR, Randal M, DeLano WL, Hyde J, Luong TN, Oslob JD, Raphael DR, Taylor L, Wang J, McDowell RS, Wells JA, Braisted AC (2003): Binding of small molecules to an adaptive protein–protein interface. *Proc Natl Acad Sci USA* 100(4):1603–1608.

Arkin MR, Wells JA (2004): Small-molecule inhibitors of protein–protein interactions: progressing towards the dream. *Nat Rev Drug Discov* 3(4):301–317.

Baroni M, Cruciani G, Sciabola S, Perruccio F, Mason JS (2007): A common reference framework for analyzing/comparing proteins and ligands. Fingerprints for ligands and proteins (FLAP): theory and application. *J Chem Inf Model* 47(2):279–294.

Belshaw PJ, Schoepfer JG, Liu KQ, Morrison KL, Schreiber SL (1995): Rational design of orthogonal receptor–ligand combinations. *Angew Chem Int Ed Engl* 34:2129–2132.

Blake JF (2000): Chemoinformatics—predicting the physicochemical properties of "drug-like" molecules. *Curr Opin Biotechnol* 11(1):104–107.

Bohacek RS, Martin C, Guida WC (1996): The art and practice of structure-based drug design: a molecular modelling approach. *Med Res Rev* 16(1):3–50.

Bohm HJ, Stahl M (2000): Structure-based library design: molecular modelling merges with combinatorial chemistry. *Curr Opin Chem Biol* 4(3):283–286.

Brekke OH, Sandlie I (2003): Therapeutic antibodies for human diseases at the dawn of the twenty-first century. *Nat Rev Drug Discov* 2(1):52–62.

Bridges AJ (2001): Chemical inhibitors of protein kinases. *Chem Rev* 101(8):2541–2571.

Brown D, Superti-Furga G (2003): Rediscovering the sweet spot in drug discovery. *Drug Discov Today* 8(23):1067–1077.

Brown SP, Hajduk PJ (2006): Effects of conformational dynamics on predicted protein druggability. *Chem Med Chem* 1(1):70–72.

Bugl H, Fauman EB, Staker BL, Zheng F, Kushner SR, Saper MA, Bardwell JC, Jakob U (2000): RNA methylation under heat shock control. *Mol Cell* 6(2):349–360.

CDER Handbook (1998): Available at http://www.fda.gov/cder/handbook/index.htm. Accessed 2007 May 2.

Chen P, Tsuge H, Almassy RJ, Gribskov CL, Katoh S, Vanderpool DL, Margosiak SA, Pinko C, Matthews DA, Kan C-C (1996): Structure of the human cytomegalovirus protease catalytic domain reveals a novel serine protease fold and catalytic triad. *Cell* 86(5):835–843.

Chen X, Liu M, Gilson MK (2001): Binding DB: a Web-accessible molecular recognition database. *J Comb Chem High-Throughput Screen* 4:719–725.

Chen J, Seebohm G, Sanguinetti MC (2002): Position of aromatic residues in the S6 domain, not inactivation, dictates cisapride sensitivity of hERG and eag potassium channels. *Proc Natl Acad Sci USA* 99(19):12461–12466.

Chen J, Lai L (2006): Pocket v.2: further developments on receptor-based pharmacophore modeling. *J Chem Inf Model* 46(6):2684–2691.

Cheng AC, Coleman RG, Smyth KT, Cao Q, Soulard P, Caffret DR, Salzberg AC, Huang ES (2007): Structure-based maximal affinity model predicts small-molecule druggability. *Nat Biotechnol* 25 (1):71–75.

Chothia C, Lesk AM (1986): The relation between the divergence of sequence and structure in proteins. *EMBO J* 5(4):823–826.

Christopoulos A (2002): Allosteric binding sites on cell-surface receptors: novel targets for drug discovery. *Nat Rev Drug Discov* 1(3):198–210.

Clackson T, Wells JA (1995): A hot spot of binding energy in a hormone–receptor interface. *Science* 267(5196):383–386.

Clark DE, Pickett SD (2000): Computational methods for the prediction of "drug-likeness." *Drug Discov Today* 5(2):49–58.

Claude JB, Suhre K, Notredame C, Claverie JM, Abergel C (2004): CaspR: a Web server for automated molecular replacement using homology modelling. *Nucleic Acids Res* 32:W606–W609.

Claussen H, Buning C, Rarey M, Lengauer T (2001): FlexE: efficient molecular docking considering protein structure variations. *J Mol Biol* 308(2):377–395.

Cohen MS, Zhang C, Shokat KM, Taunton J (2005): Structural bioinformatics-based design of selective, irreversible kinase inhibitors. *Science* 308(5726):1318–1321.

Collaborative Computational Project (1994): Number 4. The CCP4 suite: programs for protein crystallography. *Acta Crystallogr D Biol Crystallogr* 50(Pt 5):760–763.

Congreave M, Murray CW, Blundell TL (2005): Structural biology and drug discovery. *Drug Discov Today* 10(13):895–907.

Connolly ML (1993): The molecular surface package. *J Mol Graph* 11(2):139–141.

Cramer RD, Patterson D, Bunce JD (1988): Comparative molecular field analysis (CoMFA). *J Am Chem Soc* 110(18):5959–5967.

Cramer RD, Wendt B (2007): Pushing the boundaries of 3D-QSAR. *J Comput Aided Mol Des* 21(1–3):23–32.

Daly TM, Dumaual CM, Miao X, Farmen MW, Njau RK, Fu DJ, Bauer NL, Close S, Watanabe N, Bruckner C, Hardenbol P, Hockett RD (2007): Multiplex assay for comprehensive genotyping of genes involved in drug metabolism, excretion, and transport. *Clin Chem* 53(7):1222–1230.

Dantzer J, Moad C, Heiland R, Mooney S (2005): MutDB services: interactive structural analysis of mutation data. *Nucleic Acids Res* 33 (Web Server Issue):W311–W314.

Davis AM, Keeling DJ, Steele J, Tomkinson NP, Tinker AC (2005): Components of successful lead generation. *Curr Top Med Chem* 5(4):421–439.

de Groot MJ (2006): Designing better drugs: predicting cytochrome P450 metabolism. *Drug Discov Today* 11(13–14):601–606.

Del Sol A, O'Meara P (2005): Small-world network approach to identify key residues in protein–protein interaction. *Proteins* 58(3):672–682.

DeLano WL (2002): Unraveling hot spots in binding interfaces: progress and challenges. *Curr Opin Struct Biol* 12(1):14–20.

Derewenda Z (2004): Rational protein crystallization by mutational surface engineering. *Structure* 12 (4):529–535.

DiMasi JA, Hansen RW, Grabowski HG (2003): The price of innovation: new estimates of drug development costs. *J Health Econ* 22:151–185.

Dimitropoulos D, Ionides J, Henrick K (2006): UNIT 14.3: using MSDchem to search the PDB Ligand Dictionary. In: Baxevanis AD, Page RDM, Petsko GA, Stein LD, Stormo GD, editors. *Current Protocols in Bioinformatics*. Hoboken, NJ: John Wiley & Sons, pp 14.3.1–14.3.3.

Drews J (1996): Genomic sciences and the medicine of tomorrow. *Nat Biotechnol* 14:1516–1518.

Drews J, Ryser ST (1997): Classic drug targets. *Nat Biotechnol* 15.

Drews J (2000): Drug discovery: a historical perspective. *Science* 287:1960–1964.

Dubinina GG, Chupryna OO, Platonov MO, Borisko PO, Ostrovska GV, Tolmachov AO, Shtil AA (2007): *In silico* design of protein kinase inhibitors: successes and failures. *Anticancer Agents Med Chem* 7(2):171–188.

Dumas J (2001): Protein kinase inhibitors: emerging pharmacophores. *Expert Opin Ther Pat* 11(3):405–429.

Eddershaw PJ, Beresford AP, Bayliss MK (2000): ADME/PK as part of a rational approach to drug discovery. *Drug Discov Today* 5(9):409–414.

Eichelbaum M, Ingelman-Sundberg M, Evans WE (2006): Pharmacogenomics and individualized drug therapy. *Annu Rev Med* 57:119–137.

Erlanson DA, Braisted AC, Raphael DR, Randal M, Stroud RM, Gordon EM, Wells JA (2000): Site-directed ligand discovery. *Proc Natl Acad Sci USA* 97(17):9367–9372.

Farid R, Day T, Friesner RA, Pearlstein RA (2006): New insights about hERG blockade obtained from protein modeling, potential energy mapping, and docking studies 1. *Bioorg Med Chem* 14(9):3160–3173.

Fauman EB, Hopkins AL, Groom CR (2003): Structural bioinformatics in drug discovery. In: Bourne PE, Weissig H, editors. *Structural Bioinformatics* 1st ed. New Jersey: Wiley-Liss, Inc., pp 476–497.

Fedorov O, Sundström M, Marsden B, Knapp S (2007): Insights for the development of specific kinase inhibitors by targeted structural genomics. *Drug Discov Today* 12(9/10):365–372.

Fernandez D, Ghanta A, Kauffman GW, Sanguinetti MC, (2004): Physicochemical features of the hERG channel drug binding site. *J Biol Chem* 279(11):10120–10127.

Fink T, Reymond J (2007): Virtual exploration of the chemical universe up to 11 atoms of C, N, O, F: assembly of 25.4 million structures (110.9 million stereoisomers) and analysis for new ring systems, stereochemistry, physicochemical properties, compound classes and drug discovery. *J Chem Inf Model* 47(2):342–353.

Firouzi TM, Sanquinetti MC (2003): Structural determinants and biophysical properties of hERG and KCNQ1 channel gating. *J Mol Cell Cardiol* 35(1):27–35.

Foloppe N, Hubbard R (2006): Towards predictive ligand design with free-energy based computational methods? *Curr Med Chem* 13(29):3583–3608.

Forino M, Jung D, Easton JB, Houghton PJ, Pellecchia M (2005): Virtual docking approaches to protein kinase B inhibition. *J Med Chem* 48(7):2278–2281.

Fox S, Farr-Jones S, Sopchak L, Boggs A, Wang Nicely H, Khoury R, Biros M (2006): High-throughput screening: update on practices and success. *J Biomol Screen* 11(7):864–869.

Foye WO (1989): *Principles of Medicinal Chemistry*, 3rd ed. Philadelphia: Lea & Febiger.

Friesner RA, Banks JL, Murphy RB, Halgren TA, Klicic JJ, Mainz DT, Repasky MP, Knoll EH, Shelley M, Perry JK, Shaw DE, Francis P, Shenkin PS (2004): Glide: a new approach for rapid, accurate docking and scoring. 1. Method and assessment of docking accuracy. *J Med Chem* 47(7):1739–1749.

Frye SV (1999): Structure–activity relationship homology (SARAH) a conceptual framework for drug discovery in the genomic era. *Chem Biol* 6:R3–R7.

Garcia-Sosa AT, Mancera RL, Dean PM (2003): WaterScore: a novel method for distinguishing between bound and displaceable water molecules in the crystal structure of the binding site of protein–ligand complexes. *J Mol Model* 9(3):172–182.

Gillet VJ, Willet P, Bradshaw J, Green DVS (1999): Selecting combinatorial libraries to optimize diversity and physical properties. *J Chem Inf Comput Sci* 39(1):169–177.

Glaser F, Morris RJ, Najmanovich RJ, Laskowski RA, Thornton JM (2006): A method for localizing ligand binding pockets in protein structures. *Proteins* 62:479–488.

Goh CS, Lan N, Douglas SM, Wu B, Echols N, Smith A, Milburn D, Montelione GT, Zhao H, Gerstein M (2004): Mining the structural genomics pipeline: identification of protein properties that affect high-throughput experimental analysis. *J Mol Biol* 336(1):115–130.

Gohlke H, Klebe G (2002): Approaches to the description and prediction of the binding affinity of small-molecule ligands to macromolecular receptors. *Angew Chem Int Ed Engl* 41(15):2644–2676.

Golden JB (2003): Prioritizing the human genome: knowledge management for drug discovery. *Curr Opin Drug Discov Devel* 6(3):310–316.

Gonzalez-Ruiz D, Gohlke H (2006): Targeting protein–protein interactions with small molecules: challenges and perspectives for computational binding epitope detection and ligand finding. *Curr Med Chem* 13(22):2607–2625.

Goodford PJ (1985): A computational procedure for determining energetically favorable binding sites on biologically important molecules. *J Med Chem* 28(7):849–857.

Gorse AD (2006): Diversity in medicinal chemistry space. *Curr Top Med Chem* 6(1):3–18.

Greene J, Kahn S, Savoj H, Sprague P, Teig S (1994): Chemical function queries for 3D database search. *J Chem Inf Comput Sci* 34(6):1297–1308.

Greenidge PA, Carlsson B, Bladh LG, Gillner M (1998): Pharmacophores incorporating numerous excluded volumes defined by X-ray crystallographic structure in three-dimensional database searching: application to the thyroid hormone receptor. *J Med Chem* 41(14):2503–2512.

Groom CR, Hopkins AL (2002): Protein kinase drugs: optimism doesn't wait on facts. *Drug Discov Today* 7(11):601–611.

Hajduk PJ, Mendoza R, Petros AM, Huth JR, Bures M, Fesik SW, Martin YC (2003): Ligand binding to domain-3 of human serum albumin: a chemometric analysis. *J Comput Aided Mol Des* 17(2–4):93–102 [Erratum in *J Comput Aided Mol Des*, 17(10):711.]

Hajduk PJ, Huth JR, Fesik SW (2005a): Druggability indices for protein targets derived from NMR-based screening data. *J Med Chem* 48(7):2518–2525.

Hajduk PJ, Huth JR, Tse C (2005b): Predicting protein druggability. *Drug Discov Today* 10(23–24):1675–1682.

Hajduk PJ, Greer J (2007): A decade of fragment-based drug design: strategic advances and lessons learned. *Nat Rev Drug Discov* 6(3):211–219.

Halgren TA, Murphy RB, Friesner RA, Beard HS, Frye LL, Pollard WT, Banks JL (2004): Glide: a new approach for rapid, accurate docking and scoring. 2. Enrichment factors in database screening. *J Med Chem* 47(7):1750–1759.

Han A, Kang HJ, Cho Y, Lee S, Kim YJ, Gong S (2006): SNP@Domain: a Web resource of single nucleotide polymorphisms (SNPs) within protein domain structures and sequences. *Nucleic Acids Res* 34 (Web Server Issue):W642–W644.

Hann MM, Leach AR, Harper G (2001): Molecular complexity and its impact on the probability of finding leads for drug discovery. *J Chem Inf Comput Sci* 41:856–864.

Hawkins PC, Skillman AG, Nicholls A (2007): Comparison of shape-matching and docking as virtual screening tools. *J Med Chem* 50(1):74–82.

Hendlich M (1998): Databases for protein–ligand complexes. *Acta Crystallogr D Biol Crystallogr* 54(Pt 6 Pt 1):1178–1182.

Hillisch A, Pineda LF, Hilgenfeld R (2004): Utility of homology models in the drug discovery process. *Drug Discov Today* 9(15):659–669.

Holland DR, Barclay PL, Danilewicz JC, Matthews BW, James K (1994): Inhibition of thermolysin and neutral endopeptidase 24.11 by a novel glutaride derivative. *Biochemistry* 33:51–56.

Hong L, Koelsch G, Lin X, Wu S, Terzyan S, Ghosh AK, Zhang XC, Tanj J (2000): Structure of the protease domain of memapsin 2 (beta-secretase) complexed with inhibitor. *Science* 290(5489):150–153.

Hopkins AL, Groom CR (2002): The druggable genome. *Nat Rev Drug Discov* 1(9):727–730.

Hopkins AL, Groom CR, Alex A (2004): Ligand efficiency: a useful metric for lead selection. *Drug Discov Today* 9(10):430–431.

Hopkins AL, Mason JS, Overington JP (2006): Can we rationally design promiscuous drugs? *Curr Opin Struct Biol* 16(1):127–136.

Imming P, Sinning C, Meyer A (2006): Drugs, their targets and the nature and number of drug targets. *Nat Rev Drug Discov* 5(10):821–834.

Jacobs DJ, Rader AJ, Kuhn LA, Thorpe MF (2001): Protein flexibility predictions using graph theory. *Proteins* 44(2):150–165.

Jamieson C, Moir EM, Rankovic Z, Wishart G (2006): Medicinal chemistry of hERG optimizations: highlights and hang-ups. *J Med Chem* 49(17):5029–5046.

Janin J (2000): Kinetics and thermodynamics of protein–protein interactions. In: Kleanthous C, editor. *Protein–Protein Recognition.* Oxford University Press, pp 1–32.

Jegga AG, Gowrisankar S, Chen J, Aronow BJ (2007): PolyDoms: a whole genome database for the identification of non-synonymous coding SNPs with the potential to impact disease. *Nucleic Acids Res* 35 (Database Issue):D700–D706.

Johnson LN, Lowe ED, Noble MEM, Owen DJ (1998): The Eleventh Datta Lecture. The structural basis for substrate recognition and control by protein kinases. *FEBS Lett* 430(1–2):1–11.

Jones G, Willett P, Glen RC, Leach AR, Taylor R (1997): Development and validation of a genetic algorithm for flexible docking. *J Mol Biol* 267(3):727–748.

Kamiya K, Niwa R, Mitcheson JS, Sanguinetti MC (2006): Molecular determinants of hERG channel block. *Mol Pharmacol* 69(5):1709–1716.

Karchin R, Diekhans M, Kelly L, Thomas DJ, Pieper U, Eswar N, Haussler D, Sali A (2005): LS-SNP: large-scale annotation of coding non-synonymous SNPs based on multiple information sources. *Bioinformatics* 21(12):2814–2820.

Kauvar LM, Higgins DL, Villar HO, Sportsman JR, Engqvist-Goldstein A, Bukar R, Bauer KE, Dilley H, Rocke DM (1995): Predicting ligand binding to proteins by affinity fingerprinting. *Chem Biol* 2(2):107–118.

Kearney PM, Baigent C, Godwin J, Halls H, Emberson JR, Patrono C (2006): Do selective cyclo-oxygenase-2 inhibitors and traditional non-steroidal anti-inflammatory drugs increase the risk of atherothrombosis? Meta-analysis of randomised trials. *BMJ* 332(7553):1302–1308.

Keiser MJ, Roth BL, Armbruster BN, Ernsberger P, Irwin JJ, Shoichet BK (2007): Relating protein pharmacology by ligand chemistry. *Nat Biotechnol* 25(2):197–206.

Kim DE, Chivian D, Baker D (2004): Protein structure prediction and analysis using the Robetta server. *Nucleic Acids Res* 32:526–531.

Kitchen DB, Decornez H, Furr JR, Bajorath J (2004): Docking and scoring in virtual screening for drug discovery: methods and applications. *Nat Rev Drug Discov* 3(11):935–949.

Knight ZA, Shokat KM (2007): Chemical genetics: where genetics and pharmacology meet. *Cell* 128 (3):425–430.

Kola I, Landis J (2004): Can the pharmaceutical industry reduce attrition rates? *Nat Rev Drug Discov* 3 (8):711–715.

Kollman PA, Massova I, Reyes C, Kuhn B, Huo S, Chong L, Lee M, Lee T, Duan Y, Wang W, Donini O, Cieplak P, Srinivasan J, Case DA, Cheatham TEJ 3rd (2000): Calculating structures and free energies of complex molecules: combining molecular mechanics and continuum models. *Acc Chem Res* 33(12):889–897.

Kortemme T, Baker D (2002): A simple physical model for binding energy hot spots in protein–protein complexes. *Proc Natl Acad Sci USA* 99(22):14116–14121.

Kramer B, Rarey M, Lengauer T (1999): Evaluation of the FLEXX incremental construction algorithm for protein–ligand docking. *Proteins* 37(2):228–241.

Kubinyi H (2006): Success stories of computer-aided design. In: Ekins S, editor. *Computer Applications in Pharmaceutical Research and Development [Wiley Series in Drug Discovery and Development (Wang, B. Ed.)].* Wiley-Interscience, pp 377–424.

Kuhn B, Gerber P, Schulz-Gasch T, Stahl M (2005): Validation and use of the MM-PBSA approach for drug discovery. *J Med Chem* 48(12):4040–4048.

Kuhn D, Weskamp N, Schmitt S, Hullermeier E, Klebe G (2006): From the similarity analysis of protein cavities to the functional classification of protein families using cavbase. *J Mol Biol* 359 (4):1023–1044.

Kurumbail RG, Stevens AM, Gierse JK, McDonald JJ, Stegeman RA, Pak JY, Gildehaus D, Miyashiro JM, Penning TD, Seibert K, Isakson PC, Stallings WC (1996): Structural basis for selective inhibition of cyclooxygenase-2 by anti-inflammatory agents. *Nature* 384(6610): 644–648.

Kwaan HC, Samama MM (2004): Anticoagulant drugs: an update. *Expert Rev Cardiovasc Ther* 2(4):511–522.

Lacal JC (2007): How molecular biology can improve clinical management: the MammaPrint experience. *Clin Transl Oncol* 9(4):203.

Lajiness MS, Vieth M, Erickson J (2004): Molecular properties that influence oral drug-like behavior. *Curr Opp Drug Disc and Development* 7(4):470–477.

Lang PT, Kuntz ID, Maggiora GM, Bajorath J (2005): Evaluating the high-throughput screening computations. *J Biomol Screen* 10(7):649–652.

Laskowski RA, Hutchinson EG, Michie AD, Wallace AC, Jones ML, Thornton JM (2001): PDBsum: a Web-based database of summaries and analyses of all PDB structures. *Trends Biochem Sci* 22(12):488–490.

Laskowski RA, Watson JD, Thornton JM (2005): ProFunc: a server for predicting protein function from 3D structure. *Nucleic Acids Res* 33:W89–W93.

Lee B, Richards FM (1971): The interpretation of protein structures: estimation of static accessibility. *J Mol Biol* 55(3):379–400.

Lefkowitz RJ, Hoffman BB, Taylor P (1990): Neurotransmission: the autonomic and somatic motor nervous systems. In: Gilman AG, Rall TW, Nies AS, Taylor P, editors. *Goodman and Gilman's The Pharmacological Basis of Therapeutics*. New York: Pergamon.

Liang J, Edelsbrunner H, Woodward C (1998): Anatomy of protein pockets and cavities: measurement of binding site geometry and implications for ligand design. *Protein Sci* 7(9):1884–1897.

Lipinski CA, Lombardo F, Dominy BW, Feeney PJ (1997): Experimental and computational approaches to estimate solubility and permeability in drug discovery and development settings. *Adv Drug Deliv Rev* 46(1–3):3–26.

Lipinski CA (2000): Drug-like properties and the causes of poor solubility and poor permeability. *J Pharmacol Toxicol Methods* 44(1):235–249.

Liu M, Healy MD, Dougherty BA, Esposito KM, Maurice TC, Mazzucco CE, Bruccoleri RE, Davison DB, Frosco M, Barrett JF, Wang YK (2006): Conserved fungal genes as potential targets for broad-spectrum antifungal drug discovery. *Eukaryot Cell* 5(4):638–649.

Liu X, Karimi HA (2007): A high-throughput modeling and analysis of protein structural dynamics. *Brief Bioinform* 8:432–445.

Luong C, Miller A, Barnett J, Chow J, Ramesha C, Browner MF (1996): Flexibility of the NSAID binding site in the structure of human cyclcooxygenase-2. *Nat Struct Biol* 3(11):927–933.

Lynch BA, Lambeng N, Nocka K, Kensel-Hammes P, Bajjalieh SM, Matagne A, Fuks B (2004): The synaptic vesicle protein SV2A is the binding site for the antiepileptic drug levetiracetam. *Proc Natl Acad Sci USA* 101(26):9861–9866.

Macarron R, Critical review of the role of HTS in drug discovery. *Drug Discov Today* 11(7–8): (2006):277–279.

Marechal JD, Sutcliffe MJ (2006): Insights into drug metabolism from modelling studies of cytochrome P450-drug interactions. *Curr Top Med Chem* 6(15):1619–1626.

Marnett LJ, Kalgutkar AS (1999): Cyclooxygenase 2 inhibitors: discovery, selectivity and the future. *Trends Pharmacol Sci* 20(11):465–469.

Marti-Renom MA, Stuart A, Fiser A, Sánchez R, Melo F, Sali A (2000): Comparative protein structure modeling of genes and genomes. *Annu Rev Biophys Biomol Struct* 29:291–325.

Massova I, Kollman PA (1999): Computational alanine scanning to probe protein–protein interactions: a novel approach to evaluate binding free energies. *J Am Chem Soc* 121(36):8133–8143.

McDevitt D, Rosenberg M (2001): Exploiting genomics to discover new antibiotics. *Trends Microbiol* 9(12):611–617.

McMartin C, Bohacek RS (1997): QXP: powerful, rapid computer algorithms for structure-based drug design. *J Comput Aided Mol Des* 11(4):333–344.

Méndez R, Leplae R, Lensink MF, Wodak S (2005): Assessment of CAPRI predictions in rounds 3–5 shows progress in docking procedures. *Proteins* 60(2):150–169.

Michel J, Verdonk ML, Essex JW (2006): Protein–ligand binding affinity predictions by implicit solvent simulations: a tool for lead optimization? *J Med Chem* 49(25):7427–7439.

Mitcheson JS, Chen J, Lin M, Culberson C, Sanguinetti MC (2000): A structural basis for drug-induced long QT syndrome. *Proc Natl Acad Sci USA* 97(22):12329–12333.

Moult J, Fidelis K, Rost B, Hubbard T, Tramontano A (2005): Critical assessment of methods of protein structure prediction (CASP): round 6. *Proteins* (Suppl 7):3–7.

Murray JK, Gellman SH (2007): Targeting protein–protein interactions: lessons from p53/MDM2. *Biopolymers* 88(5):657.

Nettles JH, Jenkins JL, Bender A, Deng Z, Davies JW, Glick M (2006): Bridging chemical and biological space: "target fishing" using 2D and 3D molecular descriptors. *J Med Chem* 49(23):6802–6810.

Ng PC, Henikoff S (2002): Accounting for human polymorphisms predicted to affect protein function. *Genome Res* 12(3):436–346.

Nicholls A, Sharp KA, Honig B (1991): Protein folding and association: insights from the interfacial and thermodynamic properties of hydrocarbons. *Proteins* 11(4):281–296.

Noble ME, Endicott JA, Johnson LN (2004): Protein kinase inhibitors: insights into drug design from structure. *Science* 303(5665):1800–1805.

Oprea TI, Matter H (2004): Integrating virtual screening in lead discovery. *Curr Opin Chem Biol* 8(4):349–358.

Oprea TI, Allu TK, Fara DC, Rad RF, Ostopovici L, Bologa CG (2007): Lead-like, drug-like or "Pub-like": how different are they? *J Comput Aided Mol Des* 21(1–3):113–119.

Overington JP, Al-Lazikani B, Hopkins AL (2006): How many drug targets are there? *Nat Rev Drug Discov* 5(12):993–996.

Pagliaro L, Felding J, Audouze K, Nielsen SJ, Terry RB, Krog-Jensen C, Butcher S (2004): Emerging classes of protein–protein interaction inhibitors and new tools for their development. *Curr Opin Chem Biol* 8(4):442–449.

Paolini GV, Shapland RH, van Hoorn WP, Mason JS, Hopkins AL (2006): Global mapping of pharmacological space. *Nat Biotechnol* 24(7):805–815.

Papadopoulos K (2006): Targeting the Bcl-2 family in cancer therapy. *Semin Oncol* 33(4):449–456.

Pearlman DA, Charifson PS (2001): Are free energy calculations useful in practice? a comparison with rapid scoring functions for the p38 MAP kinase protein system. *J Med Chem* 44(21):3417–3423.

Pearlman DA (2005): Evaluating the molecular mechanics Poisson–Boltzmann surface area free energy method using a congeneric series of ligands to p38 MAP kinase. *J Med Chem* 48(24):7796–7807.

Penning TD, Talley JJ, Bertenshaw SR, Carter JS, Collins PW, Docter S, Graneto MJ, Lee LF, Malecha JW, Miyashiro JM, Rogers RS, Rogier DJ, Yu SS, Anderson DG, Burton EG, Cogburn JN,

Gregory SA, Koboldt CM, Perkins WE, Seibert K, Veenhuizen AW, Zhang YY, Isakson PC (1997): Synthesis and biological evaluation of the 1,5-diarylpyrazole class of cyclooxygenase-2 inhibitors: identification of 4-[5-(4-methylphenyl)-3-(trifluoromethyl)-1H-pyrazol-1-yl]benzenesulfonamide (SC-58635, celecoxib). *J Med Chem* 40(9):1347–1365.

Perry M, Stansfeld PJ, Leaney J, Wood C, de Groot MJ, Leishman D, Sutcliffe MJ, Mitcheson JS (2006): Drug binding interactions in the inner cavity of hERG channels: molecular insights from structure–activity relationships of clofilium and ibutilide analogs. *Mol Pharmacol* 69(2):509–519.

Petřek M, Otyepka M, Banáš P, Košinová P, Koča J, Damborský J (2006): CAVER: a new tool to explore routes from protein clefts, pockets and cavities. *BMC Bioinformatics* 7:316.

Pfost DR, Boyce-Jacino MT, Grant DM (2000): A SNPshot: pharmacogenetics and the future of drug therapy. *Trends Biotechnol* 18(8):334–338.

Pickett SD, Sherborne BS, Wilkinson T, Bennett J, Borkakoti N, Broadhurst M, Hurst D, Kilford I, McKinnell M, Jones PS (2003): Discovery of novel low molecular weight inhibitors of IMPDH via virtual needle screening. *Bioorg Med Chem Lett* 3(10):1691–1694.

Picot D, Loll PJ, Garavito RM (1994): The X-ray crystal structure of the membrane protein prostaglandin H2 synthase-1. *Nature* 367(6460):243–249.

Pitt WR, Murray-Rust J, Goodfellow JM (1993): AQUARIUS2: knowledge-based modeling of solvent sites around proteins. *J Comput Chem* 14(9):1007–1018.

Rang HR (2005): *Drug Discovery and Development: Technology in Transition.* Churchill Livingstone: Elsevier.

Reumers J, Schymkowitz J, Ferkinghoff-Borg J, Stricher F, Serrano L, Rousseau F (2005): SNPeffect: a database mapping molecular phenotypic effects of human non-synonymous coding SNPs. *Nucleic Acids Res* 33 (Database Issue):D527–D532.

Rush TS, Grant JA, Mosyak L, Nicholls A (2005): A shape-based 3-D scaffold hopping method and its application to a bacterial protein–protein interaction. *J Med Chem* 48(5):1489–1495.

Russ AP, Lampel S (2005): The druggable genome: an update. *Drug Discov Today* 10(23–24): 1607–1610.

Sadowski J, Kubinyi H (1998): A scoring scheme for discriminating between drugs and nondrugs. *J Med Chem* 41(18):3325–3329.

Sallinen SL, Sallinen PK, Haapasalo HK, Helin HJ, Helen PT, Schraml P, Kallioniemi OP, Kononen J (2000): Identification of differentially expressed genes in human gliomas by DNA microarray and tissue chip techniques. *Cancer Res* 60(23):6617–6622.

Sandberg M, Eriksson L, Jonsson J, Sjostrom M, Wold S (1998): New chemical descriptors relevant for the design of biologically active peptides. A multivariate characterization of 87 amino acids. *J Med Chem* 41(14):2481–2491.

Sanguinetti MC, Mitcheson JS (2005): Predicting drug-hERG channel interactions that cause acquired long QT syndrome. *Trends Pharmacol Sci* 26(3):119–124.

Sawyer JS, Anderson BD, Beight DW, Campbell RM, Jones ML, Herron DK, Lampe JW, McCowan JR, McMillen WT, Mort N, Parsons S, Smith EC, Vieth M, Weir LC, Yan L, Zhang F, Yingling JM (2003): Synthesis and activity of new aryl- and heteroaryl-substituted pyrazole inhibitors of the transforming growth factor-beta type I receptor kinase domain. *J Med Chem* 46 (19):3953–3956.

Scapin G (2006a): Structural biology and drug discovery. *Curr Pharm Des* 12(17):2087–2097.

Scapin G (2006b): Protein kinase inhibition: different approaches to selective inhibitor design. *Curr Drug Targets* 7(11):1443–1454.

Schäfer M, Egner U (2007): Structural aspects of druggability and selectivity of protein kinases in inflammation. *Anti-Inflamm Anti-Allergy Agents Med Chem* 6(1):5–17.

Schneider G, Fechner U (2005): Computer-based *de novo* design of drug-like molecules. *Nat Rev Drug Discov* 4(8):649–663.

Sherry ST, Ward MH, Kholodov M, Baker J, Phan L, Smigielski EM, Sirotkin K (2001): dbSNP: the NCBI database of genetic variation. *Nucleic Acids Res* 29(1):308–311.

Singh J, Chuaqui CE, Boriack-Sjodin PA, Lee WC, Pontz T, Corbley MJ, Cheung HK, Arduini RM, Mead JN, Newman MN, Papadatos JL, Bowes S, Josiah S, Ling LE (2003): Successful shape-based virtual screening: the discovery of a potent inhibitor of the type I TGFbeta receptor kinase (TbetaRI). *Bioorg Med Chem Lett* 13(24):4355–4359 [Erratum in: *Bioorg Med Chem Lett,* 14(11):2991.]

Sivachenko AY, Yuryev A (2007): Pathway analysis software as a tool for drug target selection, prioritization and validation of drug mechanism. *Expert Opin Ther Targets* 11(3):411–421.

Smukste I, Stockwell BR (2005): Advances in chemical genetics. *Annu Rev Genomics Hum Genet* 6:261–286.

Sneader W (1995): *Drug Prototypes and Their Exploitation.* John Wiley & Sons.

Sneader W (2005): *Drug Discovery: A History.* John Wiley & Sons.

Snyder KA, Feldman HJ, Dumontier M, Salama JJ, Hogue CW (2006): Domain-based small molecule binding site annotation. *BMC Bioinformatics* 7:152.

Stansfeld PJ, Gedeck P, Gosling M, Cox B, Mitcheson JS, Sutcliffe MJ (2007): Drug block of the hERG potassium channel: insight from modeling. *Proteins* 68(2):568–580.

Stanton A (2003): Therapeutic potential of renin inhibitors in the management of cardiovascular disorders. *Am J Cardiovasc Drugs* 3(6):389–394.

Steindl TM, Schuster D, Laggner C, Chuang K, Hoffmann RD, Langer T (2007): Parallel screening and activity profiling with HIV protease inhibitor pharmacophore models. *J Chem Inf Model* 47(2):563–571.

Steinmetzer T, Hauptmann J, Sturzebecher J (2001): Advances in the development of thrombin inhibitors. *Expert Opin Invest Drugs* 10(5):845–864.

Stockwell BR (2000): Chemical genetics: ligand-based discovery of gene function. *Nat Rev Genet* 1(2):117–125.

Stockwell BR (2004): Exploring biology with small organic molecules. *Nature* 432(7019):846–854.

Storici P, Capitani G, De Biase D, Moser M, John RA, Jansonius JN, Schirmer T (1999): Crystal structure of GABA-aminotransferase, a target for antiepileptic drug therapy. *Biochemistry* 38(27):8628–8634.

Taylor JD, Gilbert PJ, Williams MA, Pitt WR, Ladbury JE (2007): Identification of novel fragment compounds targeted against the pY pocket of v-Src SH2 by computational and NMR screening and thermodynamic evaluation. *Proteins* 67(4):981–990.

Teague SJ (2003): Implications of protein flexibility for drug discovery. *Nat Rev Drug Discov* 2(7):527–541.

Thanos CD, DeLano WL, Wells JA (2006): Hot-spot mimicry of a cytokine receptor by a small molecule. *Proc Natl Acad Sci USA* 103(42):15422–15427.

Thomas PD, Kejariwal A (2004): Coding single-nucleotide polymorphisms associated with complex vs. Mendelian disease: evolutionary evidence for differences in molecular effects. *Proc Natl Acad Sci USA* 101(43):15398–15403.

Tilley JW, Chen L, Fry DC, Emerson SD, Powers GD, Biondi D, Varnell T, Trilles R, Guthrie R, Mennona F, Kaplan G, LeMahieu RA, Carson M, Han RJ, Liu CM, Palermo R, Ju G (1997): Identification of a small molecule inhibitor of the IL-2/IL-2Rα receptor interaction which binds to IL-2. *J Am Chem Soc* 119(32):7589–7590.

Toth G, Mukhyala K, Wells JA (2007): Computational approach to site-directed ligand discovery. *Proteins* 68(2):551.

Trusheim MR, Berndt ER, Douglas FL (2007): Stratified medicine: strategic and economic implications of combining drugs and clinical biomarkers. *Nat Rev Drug Discov* 6(4):287–293.

Uzun A, Leslin C, Abyzov A, Ilyin V (2005): *Structure SNP (StSNP) database: a tool for modeling nsSNPs locations on proteins and presenting their metabolic pathways*. Available at http://glinka.bio.neu.edu/StSNP/. Accessed 2007, May 2.

van der Spoel D, Lindahl E, Hess B, Groenhof G, Mark AE, Berendsen HJ (2005): GROMACS: fast, flexible, and free. *J Comput Chem* 26(16):1701–1718.

Vane JR (1971): Inhibition of prostaglandin synthesis as a mechanism of action for aspirin-like drugs. *Nature* 231(25):232–235.

Vane JR, Mitchell JA, Appleton I, Tomlinson A, Bishop-Bailey D, Croxtall J, Willoughby DA (1994): Inducible isoforms of cyclooxygenase and nitric-oxide synthase in inflammation. *Proc Natl Acad Sci USA* 91(6):2046–2050.

Verdonk ML, Cole JC, Watson P, Gillet V, Willett P (2001): SuperStar: improved knowledge-based interaction fields for protein binding sites. *J Mol Biol* 307(3):841–859.

Verdonk ML, Chessari G, Cole JC, Hartshorn MJ, Murray CW, Nissink JWM, Taylor RD, Taylor R (2005): Modeling water molecules in protein–ligand docking using GOLD. *J Med Chem* 48(20):6504–6515.

Verkhivker GM (2007): Computational proteomics of biomolecular interactions in the sequence and structure space of the tyrosine kinome: deciphering the molecular basis of the kinase inhibitors selectivity. *Proteins* 66(4):912–929.

Vieth M, Sutherland JJ (2006): Dependence of molecular properties on proteomic family for marketed oral drugs. *J Med Chem* 49(12):3451–3453.

Vogel F (1959): Moderne problem der humangenetik. *Ergeb Inn Med U Kinderheilk* 12:52–125.

Walker JE, Saraste M, Runswick MJ, Gay NJ (1982): Distantly related sequences in the α- and β-subunits of ATP synthase, myosin, kinases and other ATP-requiring enzymes and a common nucleotide binding fold. *EMBO J* 1(8):945–951.

Walters WP, Murcko MA (2002): Prediction of "drug-likeness". *Adv Drug Deliv Rev* 54(3):255–271.

Wang Z, Moult J (2001): SNPs, protein structure and disease. *Hum Mutat* 17:263–270.

Weininger D (1998): *Enc Comp Chem* 1:425–430.

Wenlock MC, Austin RP, Barton P, Davis AM, Leeson PD (2003): A comparison of physiochemical property profiles of development and marketed oral drugs. *J Med Chem* 46(7):1250–1256.

Whitty A, Kumaravel G (2006): Between a rock and a hard place? *Nat Chem Biol* 2(3):112–118.

Wigley DB, Davies GJ, Dodson EJ, Maxwell A, Dodson G (1991): Crystal structure of an Nterminal fragment of the DNA gyrase B protein. *Nature* 351(6328):624–629.

Williams PA, Cosme J, Vinkovic DM, Ward A, Angove HC, Day PJ, Vonrhein C, Tickle IJ, Jhoti H (2004): Crystal structures of human cytochrome P450 3A4 bound to metyrapone and progesterone. *Science* 305:683–686.

Wishart DS (2006): Drug-target discovery *in silico*: using the Web to identify novel molecular targets for drug action. *SEB Exp Biol Ser* 58:145–176.

Wood JM, Maibaum J, Rahuel J, Grutter MG, Cohen NC, Rasetti V, Ruger H, Goschke R, Stutz S, Fuhrer W, Schilling W, Rigollier P, Yamaguchi Y, Cumin F, Baum HP, Schnell CR, Herold P, Mah R, Jensen C, O'Brien E, Stanton A, Bedigian MP (2003): Structure-based design of aliskiren, a novel orally effective renin inhibitor. *Biochem Biophys Res Commun* 308(4):698–705.

Yano JK, Wester MR, Schoch GA, Griffin KL, Stout CD, Johnson EF (2004): The structure of human microsomal cytochrome P450 3A4 determined by X-ray crystallography to 2.05-Å resolution. *J Biol Chem* 279(37):38091–38094.

Yue P, Li Z, Moult J (2005): Loss of protein structure stability as a major causative factor in monogenic disease. *J Mol Biol* 353(2):459–473.

Yue P, Melamud E, Moult J (2006): SNPs3D: candidate gene and SNP selection for association studies. *BMC Bioinformatics* 7:166.

Zhang J-W, Aizawa M, Amari S, Iwasawa Y, Nakano T, Nakata K (2004): Development of KiBank, a database supporting structure-based drug design. *Comput Biol Chem* 28(5–6):401–407.

Zheng C, Han L, Yap CW, Xie B, Chen Y (2006a): Progress and problems in the exploration of therapeutic targets. *Drug Discov Today* 11:412–420.

Zheng CJ, Han LY, Yap CW, Ji ZL, Cao ZW, Chen YZ (2006b): Therapeutic targets: progress of their exploration and investigation of their characteristics. *Pharmacol Rev* 58(2):259–279.

FURTHER READING

Hubbard RE, Campbell SF, Clore M, Lilley DM, editors. (2006): *Structure-Based Drug Discovery: An Overview (Biomolecular Sciences)*. Royal Society of Chemistry.

Jhoti H, Leach A (2007): *Structure-Based Drug Discovery*. 1st ed. Kluwer Academic Publishers.

35

B-CELL EPITOPE PREDICTION

Julia V. Ponomarenko and Marc H.V. van Regenmortel

INTRODUCTION

When a living organism encounters a pathogenic virus or microbe, the B cells of the immune system recognize the pathogen's antigens by their membrane-bound immunoglobulin receptors and, in response, produce antibodies specific to these antigens. The term *antigen* refers to any entity—a cell, a macromolecular assembly, or a molecule—that may be bound by either a B-cell receptor or an antibody molecule. The binding portion of an antigen is called a *B-cell epitope* or an antigenic determinant. If an antigen is a protein, an epitope may be either a short peptide from the protein sequence or a patch of atoms on the protein surface in the three-dimensional space. Since other types of epitopes, such as T-cell epitopes, are not discussed in this chapter, B-cell epitopes will be referred to as *epitopes*.

The property of an antigen to bind specifically complementary antibodies is known as the antigen's *antigenicity*; likewise, the ability of an antigen to induce an immune response is called its *immunogenicity*. Neither antigenicity nor immunogenicity is an intrinsic feature of the antigen. Antigenicity is defined with respect to a specific antibody, and an epitope thus acquires an identity only because an antibody is able to bind to it. The entire accessible surface of an antigen is likely to be recognized by a large panel of antibodies that is large enough (Berzofsky, 1985). Although an antigen's antigenicity is defined by antibody–antigen interactions, its immunogenicity depends on characteristics of the immune system of the organism encountering the antigen. For example, rabbit albumin is immunogenic in the mouse but not in the rabbit.

The main topic of this chapter is the problem of epitope prediction in protein antigens. The story began more than 40 years ago when Anderer in Germany showed that short C-terminal peptides of the coat protein of tobacco mosaic virus could elicit antibodies that recognized the virus and neutralized its infectivity. At that time, there were no data on three-dimensional structures of viral proteins; therefore, attempts to discover peptides that could mimic protein epitopes and possess the same immunogenicity as the whole protein remained entirely empirical. Subsequently, theoretical methods for epitope prediction have been developed leading to synthesis of such peptides that are important for development of immunodiagnostic tests and vaccines.

THE PROBLEM OF B-CELL EPITOPE PREDICTION

The main objective of epitope prediction is to design a molecule that can replace an antigen in the process of either antibody production or antibody detection. Such a molecule can be synthesized or, in case of a protein, its gene can be cloned into an expression vector. Designed molecules are preferable to use because they are inexpensive and noninfectious in contrast to viruses or bacteria, which may be harmful to a researcher or experimental animal. Since scientists have mainly focused on peptides as molecules mimicking antigen epitopes, the applications of epitope prediction and identification methods will be discussed here for peptides only.

A synthetic peptide may correspond to a short continuous stretch from a protein sequence and bind an antibody raised against a protein; such a peptide is called a *continuous epitope* of the protein. However, as crystallographic studies of antibody–protein complexes have shown, most epitopes are *discontinuous*, since they consist of atoms from distant residues joined on the protein surface in the three-dimensional space. A synthetic peptide can also be designed to mimic the structure of a discontinuous epitope of the protein. A synthetic peptide need not necessarily mimic either continuous or discontinuous protein epitopes since it can also mimic nonprotein epitopes—polysaccharides, DNA, glycoproteins, and other molecules—thereby expanding the range of potential pathogens that can be targeted by vaccines and diagnostics. Antibodies and synthetic peptides representing epitopes have many applications in the diagnosis of infectious and autoimmune diseases. In addition, many attempts have been made to develop peptide-based synthetic vaccines.

Diagnostic immunoassays utilize either peptides or antipeptide antibodies for detection, isolation, and characterization of molecules associated with various disease states. Synthetic peptides can be used for detection of antibodies produced as a result of infections, allergies, autoimmune diseases, or cancers. To be used as a diagnostic tool, a peptide should be antigenic, that is, able to bind a specific antibody. The prediction of such peptides is one of the goals of the epitope prediction methods. Another goal is the prediction of immunogenic peptides that can be used for production of antipeptide antibodies. Antibodies can detect proteins and various disease marker molecules, including viral proteins and bacterial lyposaccharides, present at the early stages of infections. Antipeptide antibodies used in diagnostic immunoassays can be obtained either *in vivo* by immunizing animals or *in vitro* by developing hybridoma cell lines, that is, engineered cells designed to produce high volumes of antibodies, or using combinatorial libraries. For example, a panel of recombinant antibodies can be obtained using a phage display selection of antibody fragments that bind a certain protein or peptide. Practically, any 10–15 residue-long peptide can stimulate the immune system of an animal to produce antibodies that bind the peptide. The problem is to find a peptide possessing *cross-reactive immunogenicity*, that is, a peptide that is able to elicit cross-reactive antibodies binding the specific molecule from which the peptide was derived. The prediction of such peptides is vital not only for design of diagnostics but also for vaccines.

Design of potential synthetic vaccines is another major application of epitope prediction and identification methods. There are concerns about the safety of many of today's effective vaccines as they are live attenuated or killed bacteria or viruses (examples include polio, measles, cholera, pertussis); if they are incompletely attenuated or killed, they may revert their pathogenicity or cause undesirable immune reactions. In contrast, synthetic peptides are considered as candidates for safe and inexpensive vaccines. Methods predicting immunogenic peptides, which can elicit antibodies neutralizing a pathogen, could lead to rational vaccine design. However, scientists do not have sufficient knowledge of how the immune system responds to a particular pathogen; therefore, it is difficult to predict which peptides are likely to possess cross-neutralizing immunogenicity and provide protection against the pathogen (Van Regenmortel, 2006). Synthetic vaccine candidates must still be tested experimentally to demonstrate their ability to generate neutralizing antibodies.

Although a theoretical prediction of epitopes is a highly challenging task, significant progress has been made in finding synthetic peptides representing epitopes that are useful diagnostic and research tools. Research efforts are also underway to develop synthetic peptide vaccines against HIV, human T-cell leukemia virus type 1 (HTLV-1), and *Streptococcus pyogenes* infections, malaria, and severe acute respiratory syndrome (SARS). Also, synthetic peptides are considered as therapeutic vaccines to cure cancer, Alzheimer, and autoimmune diseases. Bioinformatics input into further development of epitope prediction methods is highly desired and anticipated, and it may play a major role in helping the development of synthetic vaccines.

ANTIBODY STRUCTURE AND FUNCTION

Among distinct classes of antibodies, IgG immunoglobulin is the most common class in higher mammals. IgG contains two identical heavy chains (length of about 500 amino acids), each comprised of one variable (V_H) and three constant (C_{H1}, C_{H2}, and C_{H3}) immunoglobulin domains, and two identical light chains (length of about 250 amino acids), each comprised of one variable (V_L) and one constant (C_L) immunoglobulin domain (Figure 35.1). Each IgG molecule possesses two identical antigen binding sites (*paratopes*) situated in variable regions of the molecule and comprised of *complementarity determining regions* (*CDRs*). The CDRs comprise a total of about 50–70 amino acid residues that form six loops (three in both heavy and light chains) and vary greatly in sequence and length among antibodies that bind different antigens (Collis, Brouwer, and Martin, 2003). Hypervariability of CDRs gives the more than 10^9 different antibodies circulating in an organism the capacity to bind virtually any antigen likely to be encountered.

Antibodies demonstrate high specificity for antigens, which means that an antibody raised against an antigen may sometimes not recognize it even after a single residue substitution. At the same time, antibodies are commonly cross-reactive, that is, bind different antigens. For example, cross-reactivity with structurally similar and structurally distinct antigens has been demonstrated for the monoclonal antibodies raised against hen egg-white lysozyme (HEL) (Bentley, Boulot, and Chitarra, 1994). In the case of eight different avian lysozymes that are structurally similar antigens, the variable fragment of D11.15 antibody bound all lysozymes with high affinity, while the variable fragment of D1.3 antibody was highly specific to only two lysozymes, HEL and bobwhite quail

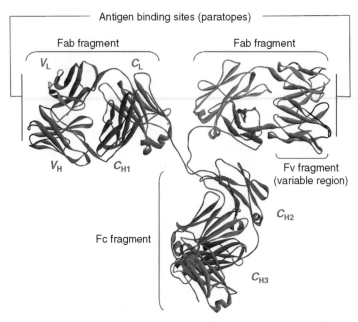

Figure 35.1. The structure of intact anticanine lymphoma monoclonal IgG2A mouse antibody [PDB : 1IGT]. The antibody heavy chains are shown in blue and red, light chains are colored in green and magenta; CDR loops are black on heavy chains and cyan on light chains. The image of the molecule was produced by J. Ponomarenko using the WebLabViewer software (Accelrys Inc.). This figure also appears in Color Figure section.

egg-white lysozyme. Antibody cross-reactivity with structurally distinct antigens has been demonstrated in complexes of D1.3 antibody with HEL and E225 antibody. Two D1.3 paratopes shared half of their residues, thus allowing other residues to make exclusive contacts with either E225 or HEL (Bentley, Boulot, and Chitarra, 1994). Usually, only about 10–25 of the 50–70 CDR residues of a certain antibody participate in the interaction with any given epitope, meaning that an antibody may harbor a large number of different paratopes.

Another additional factor mediating antibody cross-reactivity is its conformational diversity, as demonstrated, for example, by analyses of X-ray structures and binding kinetics of SPE7 antibody raised against DNP hapten (James, Roversi, and Tawfik, 2003). CDRs of SPE7 antibody adopted different conformations in its free states while binding different antigens. "Conformational diversity, whereby one sequence adopts multiple structures and multiple functions, increases the effective size of the antibody repertoire" (James, Roversi, and Tawfik, 2003).

Unique structural and functional properties of immunoglobulins allow an antibody to bind many different antigens. Also, an individual organism produces billions of different unique antibodies. Hence, the immune system in higher mammals can combat virtually any infection. For more detailed reviews on antibody structure, kinetics, and energetics of antibody–protein interactions, and structural basis of antibody affinity maturation see (Sundberg and Mariuzza, 2004) and the reviews by the authors of the first X-ray structures of an intact antibody Fab fragments in free state and in complex with lysozyme (Braden et al., 1998; Braden and Poljak, 2000).

EXPERIMENTAL METHODS USED FOR B-CELL EPITOPE IDENTIFICATION

Experimental approaches for B-cell epitope identification can be divided into two categories—structural and functional. They lead to two different perceptions of what constitutes an epitope (Van Regenmortel, 1989). Structural methods include X-ray crystallography, nucleic magnetic resonance (NMR), and electron microscopy (EM) of antibody–antigen complexes. Functional methods for detecting and characterizing antibody–antigen binding utilize various techniques, such as surface plasmon resonance (Pattnaik, 2005), mass spectrometry (Hager-Braun and Tomer, 2005), and NMR spectroscopy (Johnson and Pinto, 2004), as well as immunoassays, including ELISA, ELISPOT, Western blot, and so on. The full list of assays used for epitope identification is available at the Immune Epitope Database web site (Peters et al., 2005). For a review of experimental methods used for B-cell epitope identification see (Morris, 1996).

Structural studies, particularly X-ray crystallography of antibody–antigen complexes, identify the so-called *structural epitopes* defined by the set of antigen atoms considered to be in contact with atoms of the antibody. If a structural epitope is defined as consisting of antigen residues in which any atom is separated from any antibody atom by a distance ≤ 4 Å, epitopes encompass 10–22 residues (average of 16 residues, calculated on a data set of 59 epitopes from structures of one-chain proteins in complex with two-chain antibody fragments (Ponomarenko and Bourne, 2007)). Examples of structural epitopes known in influenza A virus hemagglutinin HA1 chain are shown in Figure 35.2.

Functional epitopes are usually delineated by functional assays that measure the change in antibody–antigen binding affinity resulting from antigen modifications, such as site-directed mutagenesis (Benjamin and Perdue, 1996; Sundberg and Mariuzza, 2004). Functional epitopes in proteins are usually smaller than structural epitopes: only three–five residues of the structural epitope contribute significantly to the antibody–antigen binding energy (Cunningham and Wells, 1993).

Functional epitopes derived from mutagenesis studies and structural epitopes obtained from structures of antibody–protein complexes are shown in Figure 35.3 for NC10 and NC41 epitopes of influenza virus N9 neuraminidase (Tulip et al., 1994). Seldom do all residues included in the structural epitope contribute to the energy of antibody–protein interaction. Thus, NC10 and NC41 structural epitopes of N9 neuraminidase include 16 and 21 residues, respectively. However, only three epitope residues contributed significantly in binding either NC10 or NC41 (Figure 35.3, residues in red), while mutations of many other residues of the structural epitopes had no effect on antibody binding (Figure 35.3, residues in cyan). At the same time, NC10 functional epitope included K432 residue that was not involved in contacts with the antibody (Figure 35.3a, residue in magenta) (Tulip et al., 1994).

Conformational changes in antibody and antigen molecules are often observed in the structures of antibody–antigen complexes in comparison to uncomplexed structures of antibody and antigen (Berger et al., 1999; Bosshard, 2001). Conformational changes take place when antibodies bind both short peptides and whole proteins. For example, Fab 17/9 undergoes major conformational rearrangements in CDR H3 upon binding a peptide from influenza virus hemagglutinin HA1(75–110) (Rini, Schulze-Gahmen, and Wilson, 1992). Another example is the structure of HEL lysozyme in complex with D44.1 Fab [PDB : 1MLC] in which both molecules show conformational changes when compared with uncomplexed structures; moreover, two complexes in the crystal unit

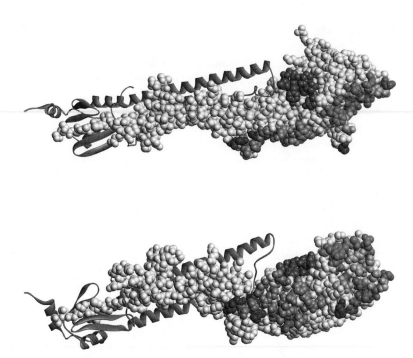

Figure 35.2. Two orthogonal views of the structure of influenza A virus (strain A/Aichi/2/68 H3N2 (X31)) hemagglutinin HA1 chain [PDB : 1EO8] with epitopes known from X-ray structures in complexes with antibodies. Chain HA1 is shown in light gray upon which are mapped residues of one linear B-cell epitope 100-YDVPDYASL-108 recognized by 17/9 Fab [PDB : 1HIM] (teal) and four structural B-cell epitopes inferred from protein structures in complexes with antibody fragments: HC45 Fab [PDB : 1QFU] (blue and orange), BH151 Fab [PDB : 1EO8] (magenta and orange), HC63 Fab [PDB : 1KEN] (green and yellow), and HC19 Fab [PDB : 2VIR] (red and yellow). The hemagglutinin HA2 chain is shown in cyan. Residues common to HC45 and BH151 epitopes are shown in orange; residues common to HC63 and HC19 epitopes are shown in yellow. Structural epitope consists of protein antigen residues in which any atom of the epitope residue is separated from any antibody atom by a distance $\leq 4\,\text{Å}$. The images were produced by J. Ponomarenko using the WebLabViewer software (Accelrys Inc.). This figure also appears in Color Figure section.

have slightly different antibody–antigen contacts and also slightly different paratope conformations (Braden et al., 1994). Investigating differences in complexes present in the same crystal asymmetric unit, Decanniere and colleagues (Decanniere et al., 2001) found that two independent complexes of the variable fragment of cAb-Lys3 antibody and HEL show a significant difference in relative orientation between antibody and antigen, and this difference reflects the flexibility of the antibody–antigen interactions rather that an artifact of crystallization.

Antigen and antibody residues that are not part of a paratope or an epitope may also undergo conformational changes during antibody–antigen binding. For example, the main chain of the residue Thr87, which is a part of neither CDRs nor the paratope of the antibody TP7, moved 3.2 Å when TP7 bound Taq DNA polymerase I [PDB : 1BGX] (Murali et al., 1998). The structure of human angiogenin in complex with 26-2F Fab [PDB : 1H0D] demonstrates an example of conformational changes of antigen residues

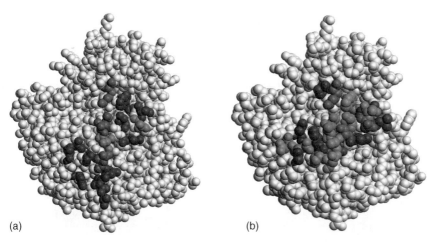

(a)   (b)

Figure 35.3. Structural and functional B-cell epitopes on influenza virus N9 neuraminidase. (a) NC10 structural epitope derived from [PDB : 1NMB]. Residue K432, which was not a part of the structural epitope but has markedly reduced antibody binding after substitution, is shown in magenta. Residues of the structural epitope N329, A369, and S370, which contributed significantly in binding of NC10 antibody, are shown in red. Residues of the structural epitope for which mutations I368R and N400K have not markedly reduced antibody binding are shown in cyan. (b) NC41 structural epitope derived from [PDB : 1NCA]. Residues of the structural epitope S367, N400, and K432, which contributed significantly in binding of NC41 antibody, are shown in red. Residues of the structural epitope, for which mutations P328K, N329D, N329K, N344I, N345G, I368R, I368Y, T401L, and W403G have not markedly reduced antibody binding, are shown in cyan. Residues A369, S370, and S372, for which mutations to smaller residues (Gly, Ala) have not markedly reduced antibody binding, are shown in green. All colored residues are part of the structural epitopes defined as antigen residues in which any atom is separated from any antibody atom by a distance $\leq 4\,\text{Å}$. Residues, for which mutations have markedly reduced antibody binding, are shown in red; the others are shown in cyan. Residues, for which mutagenesis data are not available, are shown in blue. Residues are colored according to data provided in (Tulip et al., 1994). The images were produced by J. Ponomarenko using the WebLabViewer software (Accelrys Inc.). This figure also appears in Color Figure section.

that are not a part of the epitope (Chavali et al., 2003). The 26-2F epitope of angiogenin is located at two loops (residues 34–41 and 85–91) and their conformations do not change when the antibody binds angiogenin. Meanwhile, conformation of the cell binding region located between these two loops (residues 59–68) changes dramatically: rmsd for 10 Cα atoms is 13.2 Å (Chavali et al., 2003). In free angiogenin [PDB : 1B1I], residues 62–68 are part of a β-hairpin formed by the 2nd and 3rd β-strands and the intervening loop 66–68; residues 59–61 lie on a loop that connects the hairpin structure with the 3rd α-helix. In the complex with antibody, the 2nd β-strand is lost, and the entire 58–67 segment forms an extended loop. Hence, the antibody binding changes the native angiogenin conformation that is critical for the functioning of the protein (Chavali et al., 2003).

In case the three-dimensional structure of the protein or its homologue is known, a functional epitope can be mapped on the protein structure for further study of antigenic properties of the epitope and antigen. Thus, using the homology modeling approach, Kolaskar and colleagues modeled the structure of the envelope glycoprotein of the

Japanese encephalitis virus and mapped discontinuous epitopes known from functional assays on the modeled structure (Kolaskar and Kulkarni-Kale, 1999). Analyzing 49 functional epitopes identified by site-directed mutagenesis, Blythe (Blythe, 2006) mapped them to known 3D structures of 22 proteins and demonstrated that buried residues prevailed in functional epitopes. It was also observed that most residues in mapped epitopes were separated too far from each other to form a single patch on the surface of the protein (Blythe, 2006). This observation highlights the difference between functional and structural epitopes.

A frequently applied method of epitope identification is the screening of peptide libraries for antibody binding (Geysen, Meloen, and Barteling 1984; Geysen et al., 1987; Folgori et al., 1994; Tribbick, 2002; Rowley, O'Connor, and Wijeyewickrema, 2004). A variety of inexpensive methods of peptide synthesis allow production of large libraries of peptides corresponding to overlapping or randomized regions of a certain protein. When antigenic peptides are selected, they can be mapped on the protein three-dimensional structure obtained from either X-ray analysis or homology modeling. For example, 17B and CG10 discontinuous epitopes of HIV-1 envelope protein gp120 and 13b5 epitope of HIV-1 capsid protein p24 have been reconstituted using 12mer peptides selected by screening the combinatorial phage display peptide libraries (Enshell-Seijffers et al., 2003). Thus reconstituted 17B and 13b5 epitopes corresponded to the structural epitopes. The structure of CG10 antibody in complex with gp120 was not available; however, using the ELISA assay, the authors demonstrated that the reconstituted CG10 epitope corresponded to the functional epitope (Enshell-Seijffers et al., 2003). Examples of epitopes mapped using libraries of overlapping peptides include epitopes of the M. bovis secretory protein MPB70 (Radford et al., 1990) and the Sta58 major outer membrane protein of R. tsutsugamushi (Lachumanan et al., 1993). The algorithms used for reconstituting discontinuous epitopes from the selected peptides are discussed later in this chapter.

Experimental methods allow the identification of either structural or functional epitopes that may not correspond exactly to the antigenic or immunogenic epitopes of a protein. The reason for this is that antibodies may recognize not only the epitopes against which they were elicited but also bind to a variety of related epitopes that possess some structural similarity with the immunogen. Antibodies raised against a native protein may thus cross-react with a linear peptide that only partially mimics the epitope present in the protein. Furthermore, it is possible that the protein preparation used to raise antibodies contained at least some denatured molecules and that the observed cross-reactivity with peptides was due to antibodies specific to the denatured protein. Many investigators believe that the majority of continuous epitopes described in literature actually correspond to unfolded regions of the denatured proteins and are not genuine epitopes present in native proteins (Laver et al., 1990; Van Regenmortel, 1996). In the case of functional epitopes, residues attributed to them on the basis that their replacement by other residues abolishes the antigenic activity may in fact not be a part of the epitope since the residue substitution may have caused conformational changes in the protein altering the binding activity. It was demonstrated, for example, that substitutions of charged and buried residues in E. coli toxin CcdB, can lead to the protein misfolding (Bajaj, Chakrabarti, and Varadarajan, 2005). Experimental evidence for individual residues to be a part of an epitope should therefore be considered with caution, taking into account the types of immunological assays used as well as possible conformational changes that may occur in the antigen.

HISTORY OF ATTEMPTS AT B-CELL EPITOPE PREDICTION

Reliable prediction of potentially immunogenic epitopes in a given protein or a whole pathogen genome may significantly reduce experimental efforts in discovering epitopes needed for the design of vaccines and immunodiagnostics. Prediction of discontinuous epitopes and design of synthetic mimics of discontinuous epitopes require knowledge of protein structures. In 1980, there were only 57 structures of proteins in the Protein Data Bank (PDB) (Berman et al., 2000). Consequently, experimental techniques and prediction methods have focused primarily on continuous epitopes. Currently, when more than 40,000 protein structures are available, including more than 250 structures of antibody–protein complexes, further development of structure-based methods for epitope prediction is anticipated. The historical landmarks in the study of B-cell epitopes are listed in Table 35.1.

Prediction of Continuous Epitopes

The rationale behind attempts at continuous epitope prediction lies in the following assumption that synthetic peptides corresponding to linear segments in proteins may elicit antipeptide antibodies that, in turn, may cross-react with the native protein antigen (so-called cross-reactive antibodies) and, in addition, neutralize the infectivity of the pathogen

TABLE 35.1. Selected Landmarks in the Study of B-Cell Epitopes

Date	Landmark	Reference
1958	The first 3D structure of protein myoglobin	(Kendrew et al., 1958)
1963	First antipeptide antibody raised against the C-terminus of tobacco mosaic virus coat protein and able to neutralize the infectivity of the virus	(Anderer and Schlumberger, 1965)
1973	The first high-resolution (2.8 Å) structure of an intact immunoglobulin Fab fragment	(Poljak et al., 1973)
1980	Antipeptide cross-reactive antibodies were produced without *a priori* knowledge of discontinuous epitope in proteins and proteins 3D structures; antipeptide antibodies have become a standard tool in biology	(Walter, 1986)
1980	3D structures are known for 57 proteins	(Berman et al., 2000)
1981	The first method for linear epitope prediction based on amino acid properties	(Hopp and Woods, 1981)
1984	First attempts at epitope prediction based on 3D protein structure	(Tainer et al., 1984; Westhof et al., 1984)
1986	The first 3D structure of antibody-protein complex (Fab-lysozyme, 2.8 Å)	(Amit et al., 1986)
1990	The first 3D structure of peptide–antibody complex (2.8 Å, the antigen is a synthetic 19-amino acid peptide homologue of the C-helix of myohemerythrin (Mhr), antipeptide antibody Fab' fragment)	(Stanfield et al., 1990)
2000	The first database collecting B-cell epitopes: FIMM	(Schonbach et al., 2000)
2005	AntiJen database	(Toseland et al., 2005)
2005	Bcipepe database	(Saha, Bhasin, and Raghava, 2005)
2005	Immune Epitope Database and Analysis Resource	(Peters et al., 2005)

harboring the antigen (so-called neutralizing cross-reactive antibodies (Van Regenmortel, 2006)). In 1965, the first antipeptide neutralizing antibody was produced, but widespread use of the technique of antipeptide antibodies production occurred only 15 years later (Lerner, 1984; Walter, 1986). Antipeptide cross-reactive and neutralizing antibodies can be used as immunochemical reagents for protein identification in research, medical diagnostics, and therapies. A peptide possessing cross-neutralizing immunogenicity can be used as a vaccine. In either case, the presence of cross-reactive antigenicity or immunogenicity does not mean that the peptide exactly reproduces the structure of the epitope in the protein—the only requirement is a sufficient degree of epitope resemblance to allow cross-reactivity to be observed. More details and examples of the design of synthetic peptide-based vaccines are discussed in the section "Applications."

The history of attempts at linear epitope prediction began in 1981, when Hopp and Woods (Hopp and Woods, 1981) proposed a method for predicting epitopes based on calculation of residue hydrophilicity and on the assumption that hydrophilic regions in the protein are predominantly located on the surface and are therefore potentially antigenic. In 1983, Hopp and Woods published the first computer program for epitope prediction implementing the method (Hopp and Woods, 1983). The predictive power of the method was demonstrated by the authors on 12 proteins for which immunochemical data existed. In the case of the hepatitis B surface protein antigen (HBsAg), the synthesized peptide corresponding to residues 138–149 of the protein, which was predicted to be an epitope, was able to bind antibodies directed against HBaAg and to elicit an anti-HBsAg immune response in animals (Hopp and Woods, 1983). Among eight epitopes predicted in influenza hemagglutinin HA1 subunit, one epitope was later shown to be recognized by the anti-HA monoclonal antibody 17/9 (Rini, Schulze-Gahmen, and Wilson, 1992) (Figure 35.3).

Since 1981, various attempts at linear epitope prediction were published that implemented different amino acid properties, such as hydrophilicity (Hopp and Woods, 1981; Parker, Guo, and Hodges, 1986), hydrophobicity (Kyte and Doolittle, 1982; Eisenberg, Weiss, and Terwilliger, 1984), solvent accessibility (Emini et al., 1985), secondary structure (Chou and Fasman, 1974; Garnier, Osguthorpe, and Robson, 1978; Levitt, 1978; Pellequer, Westhof, and Van Regenmortel, 1993), antigenicity (Welling et al., 1985; Jameson and Wolf, 1988), flexibility (Karplus and Schulz, 1985), and many others. Subsequently, attempts were made to compare the success rate of these scale-based methods (Van Regenmortel and Daney de Marcillac, 1988; Pellequer, Westhof, and Van Regenmortel, 1991; Hopp, 1993). A comparison of 12 scales applied to 85 identified linear epitopes in 14 proteins showed that most scales gave 50–62% correct predictions (Pellequer, Westhof, and Van Regenmortel, 1991), while a β-turn scale was slightly better (70%) (Pellequer, Westhof, and Van Regenmortel, 1993). Attempts to combine amino acid properties did not significantly improve the prediction success rate (Pellequer and Westhof, 1993; Alix, 1999; Odorico and Pellequer, 2003).

Recently, several methods using machine learning approaches have been published. For example, BepiPred method is based on a combination of Parker hydrophilicity scale (Parker, Guo, and Hodges, 1986), Levitt secondary structure scale (Levitt, 1978), and hidden Markov models (HMM) (Larsen, Lund, and Nielsen, 2006). The machine learning classifier developed by Sollner (Sollner, 2006; Sollner and Mayer, 2006) uses the common single amino acid propensity scales together with parameters reflecting the probability of close neighborhood for a given stretch of amino acids. ABCpred algorithm (Saha and Raghava, 2006) that applied artificial neural network (ANN) achieved a maximum accuracy of 66%.

Methods using single amino acid propensity scales give area under the receiver operating characteristic curve (A_{ROC}) values of 0.60 (Greenbaum et al., 2007). A recent assessment of 484 amino acid scales (in the range of window sizes and cutoff values) on 50 proteins with linear epitopes from AntiJen database (Toseland et al., 2005) demonstrated that "single-scale amino acid propensity profiles cannot be used to predict epitope location reliably" (Blythe and Flower, 2005). "The combination of scales and experimentation with several machine learning algorithms showed little improvement over single scale-based methods, which were considered to perform inadequately" (Greenbaum et al., 2007). The conclusions demonstrate the need for evaluation of methods based on more reliable data sets of immunogenic epitopes and for methods applying more sophisticated propensity scales.

Prediction of Discontinuous Epitopes

Prediction of discontinuous epitopes in a protein is based on the knowledge of the protein three-dimensional structure. First attempts at epitope prediction based on 3D protein structure began in 1984 when a correlation was established between crystallographic temperature factors and several known continuous epitopes of tobacco mosaic virus protein, myoglobin, and lysozyme (Westhof et al., 1984). For myohemerythrin, highly mobile regions were determined based on X-ray crystallographic temperature factors, and it was shown that antibodies raised against peptides, which corresponded to mobile regions, reacted strongly with myohemerythrin, while antibodies raised against peptides from less mobile regions did not (Tainer et al., 1984). Thornton and colleagues (Thornton et al., 1986) proposed a method to locate potential discontinuous epitopes based on a protrusion of protein regions from the protein's globular surface. The protruded parts of proteins corresponded to the experimentally determined continuous epitopes in myoglobin, lysozyme and myohemerythrin (Thornton et al., 1986). A correlation between antigenicity, solvent accessibility, and flexibility of antigenic regions in proteins was also found (Novotny et al., 1986). However, at the time, protein structural data were mostly used for prediction of continuous epitopes rather than discontinuous epitopes.

When the first X-ray structure of an antibody–protein complex was solved in 1986 (Amit et al., 1986), a completely different approach to epitope prediction became possible. The first attempts to predict epitopes were based on structural properties of a whole discontinuous epitope, which was considered to be an area on the protein surface favorable for antibody binding. Thus, Jones and Thornton, using their method for predicting protein–protein binding sites (Jones and Thornton, 1997), were able to successfully predict epitopes on the surface of the β-subunit of human chorionic gonadotropin (βhCG) (Jones and Thornton, 2000).

At the same time, scientists began exploiting protein–protein docking algorithms for epitope prediction. Most docking algorithms are based on the "lock-and-key" model, which assumes the existence of predefined complementarity of three-dimensional shapes of epitope and paratope. However, conformational changes observed in the structures of antibody–antigen complexes may be the main reason why unbound docking (when antibody and antigen structures are taken from different structures) poorly performs on antibody–protein complexes (Halperin et al., 2002), and bound docking (when antibody and antigen structures are taken from the same structure of their complex) substantially outperforms unbound docking (Duhovny, Nussinov, and Wolfson, 2002). Simulated docking of paratopes based on their X-ray structures did not reveal the specific atomic interactions responsible for antibody–antigen recognition (Carneiro and Stewart, 1994). For the detailed review of docking algorithms and their performance on antibody–protein complexes see (Halperin et al., 2002).

Discontinuous epitopes in a protein with a known 3D structure can be reconstituted from the antibody binding peptides selected from randomized peptide libraries. Interpreting such peptides may be problematic when there is no sequence similarity between peptides and the protein. Several bioinformatics tools address the problem: 3D-Epitope-Explorer (3DEX) (Schreiber et al., 2005), MIMOX (Huang et al., 2006), Epitope Mapping Tool (EMT) (Batori et al., 2006), EPIMAP (Mumey et al., 2003), MIMOP (Moreau et al., 2006), PepSurf (Mayrose et al., 2007), and Mapitope (Bublil et al., 2007). Also, to facilitate a structure-based design of peptides representing the whole surface or a particular region of a protein, the SUPERFICIAL method (Goede, Jaeger, and Preissner, 2005) and the MEPS server (Castrignano et al., 2007) have been developed. These peptides can be further selected for binding a specific antibody and mapped on the protein structure to reconstitute the epitope.

Although many attempts to predict epitopes have been made, scientists are still unable to reliably predict epitopes, both continuous and discontinuous. Despite some successful predictions achieved for particular epitopes in certain proteins, existing methods perform poorly overall even if the three-dimensional structure of a protein is known.

BIOINFORMATICS METHODS FOR B-CELL EPITOPE PREDICTION

To discuss bioinformatics methods for epitope prediction, 9 web servers available on the Internet and 14 different epitope prediction methods have been chosen. The methods have been tested on six epitopes of influenza A H3N2 hemagglutinin HA1 (strains A/Aichi/2/68 and X31) including four structural discontinuous epitopes (Ponomarenko and Bourne, 2007) and the only two known protective continuous epitopes (Bui et al., 2007). The epitopes are shown in Figure 35.2 and listed in Tables 35.2 and 35.3. Other web servers for B-cell epitope prediction, which were not evaluated in this chapter, include ePitope (http://epitope-informatics.com), Antigenic (http://liv.bmc.uu.se/cgi-bin/emboss/antigenic), and VaxiJen (http://www.jenner.ac.uk/VaxiJen/).

Sequence-Based Methods

Sequence-based methods are limited to the prediction of continuous epitopes. The majority of the sequence-based methods assume that epitopes have to be accessible for antibody binding and, hence, are based on using epitope properties related to surface exposure. In the current review, sequence-based methods have been tested on prediction of two protective epitopes known in influenza A virus hemagglutinin HA1 (Bui et al., 2007). The first continuous epitope is the 91–108 epitope (SKAFSNCYPYDVPDYASL), which is a protective epitope in rabbit; that is, an epitope able to elicit in a rabbit antibodies neutralizing infectivity of influenza viruses of the strains A/Achi/2/68(H3N2) and A/Texas/1/77(H3N2) (Muller, Shapira, and Arnon, 1982). The 100–108 epitope known from the X-ray structure in complex with 17/9 antibody, which is a part of the protective epitope, is shown in Figure 35.2 in teal. The second continuous epitope is the 127–133 epitope (WTGVTQN) protective against the influenza strain A/Achi/2/68 (H3N2) in mouse (Naruse et al., 1994).

TABLE 35.2. Prediction of Protective Continuous B-Cell Epitopes in HA1

Method	Method Parameters	Number of Predicted Epitopes in HA1	Positions: Score (Rank) (If Applied) of the Best Prediction			
			91–108 Epitope	100–108 Epitope	127–133 Epitope	
Chou and Fasman's β-turn scale	Window size 7	313 (maximum score 1.39, minimum score 0.72)	95–101: 1.35 (7)	101–107: 1.17 (57)	127–133: 1.07 (128)	
Chou and Fasman's β-turn scale	Window size 9	311 (maximum score 1.28, minimum score 0.86)	96–104: 1.28 (6)	100–108: 1.1 (86)	n/a	
Chou and Fasman's β-turn scale	Window size 18	303 (maximum score 1.28, minimum score 0.86)	91–108: 1.14 (41)	n/a	n/a	
Emini's surface accessibility scale	Window size 6	314 (minimum score 0.08, maximum score 5.53)	100–105: 1.95 (37)	100–105: 1.95 (37)	128–133: 1.06 (129)	
Karplus and Schulz's scale	Window size 7	312 (minimum score 0.9, max score 1.13)	100–106: 1.01 (138)	100–106: 1.01 (138)	127–133: 1.04 (74)	
Kolaskar's antigenicity scale	Window size 7	313 (minimum score 1.02, maximum score 1.17)	97–103: 1.16 (2)	102–108: 1.11 (25)	127–133: 0.97 (260)	
Parker's hydrophobicity scale	Window size 7	313 (min score 4.19, maximum score 6.24)	101–107: 3.59 (49)	101–107: 3.59 (49)	127–133: 2.2 (130)	
Bepipred	Score threshold 0.35	16 (maximum score 1.94, minimum score 1.28)	99–105	99–105	127–146	
Bepipred	Score threshold 0.90	5 (maximum score 1.94, minimum score 1.28)	100–103	100–103	130–144	
Bepipred	Score threshold 1.30	4 (maximum score 1.94, minimum score 1.28)	Not predicted	Not predicted	Not predicted	
ABCPred	Window size 16	31 (max score 0.93, threshold 0.51)	93–109: 0.70 (24)	93–109: 0.70 (24)	123–139: 0.90 (4)	
ABCPred	Window size 10	28 (maximum score 0.80, threshold 0.51)	95–105: 0.70 (14)	99–109: 0.63 (20)	125–135: 0.61 (23)	
CEP	default	17	91–96, 101–107	101–107	121–137	

TABLE 35.3. Prediction of Structural Discontinuous B-Cell Epitopes in HA1

Epitope	BH151 Epitope [PDB : 1EO8] 15 Residues			HC63 Epitope [PDB : 1KEN] 16 Residues			HC45 Epitope [PDB : 1QFU] 19 Residues			HC19 Epitope [PDB : 2VIT] 18 Residues		
Epitope Size Method	Sensitivity	ppv	Specificity	Sensitivity	ppv	Specificity	Sensitivity	ppv	Specificity	Sensitivity	ppv	Specificity
CEP	**0.27**	**0.22**	**0.96**	**0.56**	**0.17**	**0.86**	**0.47**	**0.24**	**0.91**	**0.72**	**0.3**	**0.9**
DiscoTope (−3.1)	0	0	0.99	0	0	0.99	0	0	0.99	0	0	0.99
DiscoTope (−7.7)	0	0	0.78	**0.56**	**0.13**	**0.8**	0.11	0.03	0.78	0.17	0.04	0.78
DiscoTope (−10.5)	0.07	0.01	0.54	**0.81**	**0.09**	**0.58**	0.21	0.03	0.55	0.5	0.06	0.56
PatchDock 1st model	0.06	0.04	0.91	0.05	0.04	0.87	0	0	0.89	0	0	0.89
PatchDock best model of 10	0	0	0.89	0	0	0.88	**0.21**	**0.17**	**0.94**	**0.22**	**0.02**	**0.36**
DOT 1st model	0	0	0.95	0	0	0.97	0	0	0.95	0	0	0.95
DOT best model of 10	0	0	0.95	0	0	0.97	**0.21**	**0.27**	**0.96**	**0.22**	**0.33**	**0.97**
ProMate	0	0	0.99	0	0	0.99	0	0	0.99	0	0	0.99
PPI-PRED (1st patch)	0	0	0.85	**0.69**	**0.23**	**0.88**	0	0	0.85	**0.33**	**0.13**	**0.87**
PPI-PRED (best patch)	**0.87**	**0.23**	**0.86**	**0.69**	**0.23**	**0.88**	**0.84**	**0.29**	**0.87**	**0.33**	**0.13**	**0.87**

Sensitivity (recall) $= TP/(TP + FN)$—a proportion of correctly predicted epitope residues (TP) with respect to the total number of epitope residues (TP + FN).

Specificity $= 1 - FP/(TN + FP)$—a proportion of correctly predicted nonepitope residues (TN) with respect to the total number of nonepitope residues (TN + FP).

Positive predictive value (ppv) (precision) $= TP/(TP + FP)$—a proportion of correctly predicted epitope residues (TP) with respect to the total number of predicted epitope residues (TP + FN).

The statistical significance of a prediction, that is, the difference between observed and expected frequencies of an actual epitope/nonepitope residue in the predicted epitope/nonepitope was determined by Fisher's exact test (right tailed). The prediction was considered significant if the significance level was ≥95%, that is, the *P-value* was ≤0.05. Significant predictions are shown in bold font.

860

Amino acid scale-based methods apply amino acid scales to compute the scores of a residue i in a given protein sequence. The $i - (n - 1)/2$ neighboring residues on each side of residue i are used to compute the score for residue i in a window of size n. The final score for residue i is the average of the scale values for n amino acids in the window.

Chou and Fasman's method (Chou and Fasman, 1978) is based on calculating the probability of a stretch of residues to be a part of secondary structure β-turn. The first of the two tested protective epitopes was predicted with ranks seven and six when window size was seven and nine, respectively (Table 35.2).

Emini's solvent accessibility scores (Emini et al., 1985) were calculated based on surface accessibility scale, which has been determined by Janin and Wodak (Janin and Wodak, 1978) and reflected surface exposure probabilities for amino acids as calculated on X-ray structures of 28 proteins. The method defines a surface probability (S_n) at a sequence position n as a product of fractional surface probabilities for amino acids at positions from $n - 2$ to $n + 3$. Thus, S_n for a random hexapeptide is equal to 1.0 with probabilities greater than 1.0 indicating an increased probability for being found on the surface. In HA1, epitope 100–105 had score 1.951 (rank 37), while 128–133 epitope had score 1.058 (rank 129) (Table 35.2).

Karplus and Schulz's flexibility scale (Karplus and Schulz, 1985) was constructed on the basis of the mobility of protein segments derived from known temperature B factors of the C$\langle$ atoms of 31 proteins of known structure. The method did not predict the protective epitopes in HA1 (Table 35.2).

Kolaskar and Tongaonkar (Kolaskar and Tongaonkar, 1973) calculated the frequency of occurrence of each type of amino acid (f_{Ag}) in 156 experimentally determined epitopes from 34 different proteins. Then, using Parker's (Parker, Guo, and Hodges, 1986) scales, averaged values for hydrophilicity, accessibility, and flexibility were calculated for every overlapping heptapeptide in the 156 epitopes, and the frequency of occurrence of amino acids on the surface (f_s) was calculated. The antigenic propensy (A_p) value for each amino acid was calculated as $A_p = f_{Ag}/f_s$. About 75% of epitopes have been correctly predicted applying the antigenic propensity scale to the epitopes on which the scale has been built (Kolaskar and Tongaonkar, 1973). In HA1, the method predicted the 91–108 epitope, while the second protective epitope was not predicted (Table 35.2).

Parker's hydrophobicity scale (Parker, Guo, and Hodges, 1986) was determined experimentally using high-performance liquid chromatography (HPLC) on a set of 20 synthetic peptides accounting for each of the 20 amino acids. Hydrophobicity value for each amino acid was assigned according to the retention time for each peptide. In the method of calculating a score of a residue i, a window of seven residues was used. The corresponding value of the scale was introduced for each of the seven residues and the arithmetical mean of the seven residues values was assigned to the fourth, ($i + 3$), residue in the segment. In HA1, the method predicted the 101–107 epitope with the rank 49 (of 313 predictions) and the 127–133 epitope with the rank 130 (Table 35.2).

The aforementioned five scale-based methods have been evaluated on the data sets from Pellequer (Pellequer, Westhof, and Van Regenmortel, 1993), AntiJen (Toseland et al., 2005), and HIV (Los Alamos NL) databases. The values of the area under the receiver operating characteristic curve (A_{ROC}) for these methods did not exceed 0.60 (Blythe and Flower, 2005; Greenbaum et al., 2007). In addition, there was no significant correlation found between the location of known continuous epitopes and the sequence profiles generated using either single-scale-based methods or their combinations (Blythe and Flower, 2005).

The web server that implements the considered five scale-based methods is available at the IEDB web site (Peters et al., 2005).

ABCpred (Saha and Raghava, 2006) method applies artificial neural network for predicting continuous epitopes in protein sequences. The method has been trained on a nonredundant set of 700 epitopes from Bcipep database (Saha, Bhasin, and Raghava, 2005) and 700 randomly derived peptides from Swiss-Prot database. Tested on the trained data sets using fivefold cross-validation, the method achieved an accuracy of ~66% (Saha and Raghava, 2006). The same accuracy has been observed for the independent data set of 187 epitopes from the structural database of allergenic proteins (SDAP) (Ivanciuc, Schein, and Braun, 2003) and epitopes excluded from training. In comparison to other methods, ABCpred performed significantly better than the methods based on Karplus and Schulz's flexibility scale (Karplus and Schulz, 1985) and Parker's hydrophobicity scale (Parker, Guo, and Hodges, 1986); the last two methods demonstrated the best accuracies of 59.4% and 61.5%, respectively, on the independent data set (Saha and Raghava, 2006). ABCpred predicted both HA1 protective epitopes among 31 and 28 predictions at the window sizes 16 and 10, respectively (Table 35.2).

BepiPred (Larsen, Lund, and Nielsen, 2006) method is based on a combination of hidden Markov model and two amino acid scales, Parker's hydrophilicity scale (Parker, Guo, and Hodges, 1986) and Levitt's secondary structure scale (Levitt, 1978). These scales have been chosen as the best performers on the Pellequer's (Pellequer, Westhof, and Van Regenmortel, 1993) data set among five scales: Parker's, Levitt's, Emini's accessibility (Emini et al., 1985), hydrophibicity (Kyte and Doolittle, 1982), and antigenicity (Welling et al., 1985). The estimated *P-values* for the hypothesis that Parker's and Levitt's scale-based methods performed as random were below 0.1%. The HMM method built on the epitopes from AntiJen (Toseland et al., 2005) database gave $A_{roc} = 0.66 \pm 0.01$ on the Pellequer's data set. On the HIV data set, BepiPred performed the best among amino acid scale-based methods, HMM, and BepiPred. Thus, Levitt's and Parker's scales alone and their combination gave areas under ROC curve of 0.57, 0.59, and 0.58, respectively; HMM gave 0.59; and BepiPred gave 0.6 that was significantly better than random prediction (Larsen, Lund, and Nielsen, 2006).

BepiPred assigns a score value to each protein residue; for example, for HA1 the maximum score was 1.94, and the minimum score was -1.28 (the average score was 0.25). The method's specificity and sensitivity depend on the score threshold. Thus, on the benchmark data set of 85 epitopes and 85 random sequences (nonepitopes), the method gave at thresholds of $-0.20, 0.35$, and 1.30 sensitivities and specificities, 75% and 50%, 49% and 75%, and 13% and 96%, respectively. Sensitivity was calculated as a proportion of correctly predicted epitopes with respect to the number of all epitopes; specificity was calculated as a proportion of correctly predicted nonepitopes with respect to the number of all nonepitopes. At the thresholds of 0.35 and 0.90, both protective epitopes in HA1 were predicted among 16 and 5 predictions, respectively, while no epitope was predicted at the threshold of 1.30 (Table 35.2).

Currently, the achieved accuracy of continuous epitope prediction is ~60–66%, applying either combinations of amino acid scales or sophisticated machine-learning techniques. The higher accuracy could possibly be achieved by improving the quality of existing B-cell epitope data sets, which contain many erroneously delineated epitopes (Greenbaum et al., 2007). Thus, the location of continuous epitopes, which were taken from the AntiJen database and mapped to 42 proteins with known 3D structures, correlated poorly with solvent accessibilities, secondary structure elements, predicted locations of disordered regions, and post-translational modifications present on the protein surfaces (Blythe, 2006). Such a finding confirms earlier suggestions (Laver et al., 1990) that experimental methods of epitope delineation often lead to incorrect epitope identification.

Structure-Based Methods

Six different prediction methods based on three-dimensional structures of antigens and implemented as web servers are discussed here, including the only two existing methods developed specifically for epitope prediction, CEP (Kulkarni-Kale, Bhosle, and Kolaskar, 2005) and DiscoTope (Haste Andersen, Nielsen, and Lund, 2006); two protein–protein docking methods, DOT (Mandell et al., 2001) and PatchDock (Schneidman-Duhovny et al., 2003); and two structure-based methods for protein–protein binding site prediction, PPI-PRED (Bradford, David and Westhead, 2005) and ProMate (Neuvirth, Raz, and Schreiber, 2004). The results of prediction are demonstrated in the example of four structural epitopes in HA1 (Figure 35.2, Table 35.3).

CEP web server (Kulkarni-Kale, Bhosle, and Kolaskar, 2005) predicts both continuous and discontinuous epitopes in a given three-dimensional structure. The algorithm performs calculation of solvent accessibility of residues using the Voronoi polyhedron and delineates continuous epitopes based on the presence of at least three contiguous accessible residues. Discontinuous epitopes are predicted by collapsing predicted continuous epitopes for which Cα atoms are within a distance of 6 Å. The authors claimed (Kulkarni-Kale, Bhosle, and Kolaskar, 2005) that CEP correctly predicted structural epitopes in 60% structures of 21 analyzed antibody–protein complexes. In HA1, CEP predicted (with no scores assigned) 17 nonoverlapping continuous epitopes of different lengths varying 4–28 residues and covering 55% of HA1 protein sequence. Both protective HA1 epitopes were predicted (Table 35.2).

The number of discontinuous epitopes predicted by CEP in a protein varied and was related to the protein size: for more than half of 48 proteins analyzed in the work (Ponomarenko and Bourne, 2007), CEP gave more than 5 predictions and for a quarter of proteins, more than 10 predictions. Since CEP predictions were not ranked, the best (by *P-value*) prediction was reported for each epitope in Table 35.3. In HA1, CEP predicted 14 epitopes, along with the tested structural epitopes (Table 35.3).

DiscoTope (Haste Andersen, Nielsen, and Lund, 2006) is a method focusing explicitly on discontinuous epitope prediction. The method has been tested and trained on a data set of 76 X-ray structures of antibody–protein complexes divided into 25 groups by protein sequence homology, which were further split into 5 data sets used for cross-validated training and evaluation of the method. DiscoTope method is based on a linear combination (with equal weights) of the normalized values (obtained by subtracting the mean and dividing with the standard deviation) of the following measures for each residue: Parker's hydrophilicity value (Parker, Guo, and Hodges, 1986), amino acid statistics, number of contacts, and area of relative solvent accessibility. To measure amino acid statistics, the authors used log-odds ratios of each of 20 amino acids calculated based on their representation in all 9mer peptides in the analyzed antigens. A_{ROC} values averaged over all analyzed antigens measured the methods performance. DiscoTope gave A_{ROC} value of 0.71 and performed significantly better than the methods based on either single measure or any combination of two measures.

DiscoTope calculates a score value for each protein residue and assigns a residue being a part of a predicted epitope depending on the score threshold; one epitope is predicted per protein. The specificity and sensitivity of the method were measured for each protein counting epitope and nonepitope residues using formulas provided in Table 35.3 footnote and then averaged over all proteins. Thus, at the thresholds of -3.1 the sensitivity and specificity were 15.5% and 95%, at the threshold of -7.7, 47.3%, and 75%, respectively (Haste Andersen, Nielsen, and Lund, 2006). At the threshold of -7.7, DiscoTope predicted in HA1 the only epitope, HC63, which was not used for the method development while, surprisingly, it did not predict the BH151 and HC45 epitopes, which were in the training data set (Table 35.3).

Protein–protein docking algorithms generate models of antibody–protein complexes for given structures of an antibody and a protein and assign a score for each model; the best model that closely represents the original complex will have the highest score. When antibody and antigen structures are used from the same original complex, the docking is called "bound", otherwise it is "unbound." Unbound docking usually performs poorly on antibody–protein complexes (Duhovny, Nussinov, and Wolfson, 2002).

The DOT algorithm (Mandell et al., 2001) performs rigid-body docking based on the electrostatic and van der Waals energy calculation of interacting molecules and fast Fourier Transform correlation approach, which is used to produce a convolution of calculated energies. The method is implemented at the ClusPro web server (Comeau et al., 2004). PatchDock (Schneidman-Duhovny et al., 2003) is also a rigid-body protein–protein docking algorithm based on local shape feature matching. The algorithm filters out the models in which the paratope does not intersect the CDR regions, which are detected by aligning the sequence of a given antibody to a consensus sequence from a library of antibodies (Duhovny, Nussinov, and Wolfson, 2002). In HA1, using "bound" docking, DOT and PatchDock performed similarly: epitopes are often not predicted by the first models; by taking into account the top 10 models, both methods predicted only two epitopes, HC45 and HC19 with, however, low sensitivity and positive-predictive values (ppvs) (Table 35.3).

Protein–Protein Binding Site Prediction Methods. The motivation to review and test protein–protein binding site prediction methods on structural epitopes lies in the observation that antibody–protein interactions are similar to other transient protein–protein interactions by their physical–chemical nature: they can be categorized as intermediate by affinity, transient nonobligate interactions (Janin and Wodak, 1978; Jones and Thornton, 1996; Lo Conte, Chothia, and Janin, 1999). In the past, general methods for the prediction of intermediate transient nonobligate protein–protein interactions have been applied to structural epitopes (Jones and Thornton, 1996; Lo Conte, Chothia, and Janin, 1999). Also, epitopes and other protein–protein interfaces share many properties. Thus, Blythe (Blythe, 2006) compared 57 protein–protein binding interfaces of 44 proteins (Neuvirth, Raz, and Schreiber, 2004) with epitopes and paratopes inferred from X-ray structures of 37 complexes and calculated the following interface properties: amino acid composition, hydrophobicity by the Eisenberg's scale (Eisenberg, Weiss, and Terwilliger, 1984), amino acid contribution to form intermolecular hydrogen bonds, residue evolutionary conservancy, and geometrical parameters, such as planarity and complementarity of interfaces. Epitopes and nonobligate heterodimer interfaces were similar considering all aforementioned properties except residue conservancy; epitope residues were more variable than heterodimer interfaces (Blythe, 2006). Independent analysis of 55 representative structural epitopes showed that by residue evolutionary conservation scores calculated by ConSurf (Landau et al., 2005), epitope residues were significantly less evolutionarily conserved than average protein surfaces (Ponomarenko and Bourne, 2007).

ProMate (Neuvirth, Raz, and Schreiber, 2004) is a method based on structural and sequence properties of protein interfaces from a manually curated data set of 67 proteins involved in heterodimeric transient interactions. ProMate is built on the following properties of interfaces: (i) preference of hydrophobic and polar residues; (ii) preference of β-structure, long loops and coil regions and disfavoring α-helices; (iii) higher degree of evolutionary conservancy of residues at the interface compared to other surface residues; (iv) lower temperature factor than for protein surfaces in general; and (v) preference of interfaces to be solvated by more bound water molecules than protein surfaces in general. The authors reported a success rate of 70% (Neuvirth, Raz, and Schreiber, 2004). In HA1, no epitope was predicted by ProMate (Table 35.3).

PPI-PRED (Bradford, David and Westhead, 2005) is a protein–protein binding site prediction method that is also based on characteristic properties of protein interfaces from manually curated data set of 180 proteins from 149 obligate and transient complexes. Antibody–protein complexes were excluded from both ProMate and PPI-PRED data sets, which, surprisingly, shared only 15 proteins. PPI-PRED is based on a support vector machine (SVM) training on seven properties of protein interfaces: residue interface propensity, amino acid hydrophobicity scores, residue evolutionary conservancy, solvent accessibility, electrostatic potential, interface surface shape index, and curvedness. The reported success rate of the method was 76% (Bradford, David and Westhead, 2005). In HA1, PPI-PRED successfully predicted all structural epitopes (Table 35.3).

The considered six structure-based methods have been evaluated on the benchmark data set of 59 epitopes inferred from X-ray structures of 48 one-chain (monomer) proteins in complexes with variable fragments of two-chain antibodies (Ponomarenko and Bourne, 2007). The overall performance of the methods did not exceed 40% ppv (precision) at 46% sensitivity (recall) and A_{ROC} value of 0.70 (Ponomarenko and Bourne, 2007). When the top 10 models were considered, the docking methods performed the best. Despite the fact that structural epitopes and protein–protein nonobligate transient heterodimer interfaces share many properties (Blythe, 2006), protein–protein binding site prediction methods, ProMate and PPI-PRED, poorly predicted structural epitopes (Ponomarenko and Bourne, 2007).

Another evaluation of PPI-PRED and four docking algorithms (DOT, PatchDock, ZDOCK, and webGRAMM) have been done on 37 structural epitopes from the Blythe's data set (Blythe, 2006). Using the Matthew's Correlation Coefficient (MCC), Blythe measured the correlation between predicted and structural epitopes and paratopes. For the first models, all evaluated methods demonstrated near-random correlations. Likewise, when the top 10 models for each complex were considered, low and negative MCC values prevailed over positive values for all algorithms except PatchDock, which gave average MCC values of 0.540 for epitopes and 0.694 for paratopes. The last method succeeded because of application of the CDR filter, which other algorithms do not use; if they did, the results would be comparable. Thus, using predefined CDRs for antibodies, DOT method showed MCC values of 0.526 for epitopes and 0.783 for paratopes compared with 0.293 and 0.488, respectively, when no information on CDRs was used (Blythe, 2006). Overall, docking algorithms gave a ~50% likelihood of resembling the native antibody–protein complex structure when the top 10 models were considered (Blythe, 2006).

There are several potential strategies that can be explored to improve current epitope prediction methods. First, prediction methods would be through training of the methods on data sets containing only well-defined epitopes inferred from X-ray structures of antibody–protein complexes. Second, the performance of docking algorithms could be improved by developing them specifically for antibody–antigen complexes. Third, features that discriminate epitopes from nonepitopes, for example, the evolutionary conservation score can be used as additional features for discrimination. Finally, the observation that the average frequencies of amino acids in epitopes and randomly selected protein surface patches are not similar can be used as leverage for discrimination. Although, Blythe analyzed single epitopes in 37 protein antigens (Blythe, 2006) and found no differences in the frequency of amino acid types in epitopes, the study did not account for randomly selected patches that may contain other epitopes. Instead, analysis of multiple epitopes was performed for 52 representative antigens with consideration of whether the surface residue is a part of any structural epitope inferred from all known X-ray structures of antibodies complexed with the same antigen (187 X-ray structures were analyzed; the data set is provided in Ponomarenko and Bourne, 2007). This study shows that hydrophobic residues were infrequent among epitope residues in comparison to surface residues in general, while charged residues were abundant in epitopes (95% certainty); also, epitope residues are frequently located in β-turns and bends, but not in coil regions (68% certainty; comparison with antigen surface residues). Thus, as more X-ray structures on antibody–protein complexes become available, more discriminatory features will be revealed through a more precise analysis of epitopes and, hence, better epitope predictors may become available.

APPLICATIONS

Prediction of epitopes has two major applications: immunodiagnostics and vaccines. Diagnostic peptides representing antigenic epitopes of a molecule are utilized for the detection of antibodies that cross-react with this molecule. These peptides can also be used for production of antipeptide cross-reactive antibodies that, in turn, can serve as diagnostic tools. Peptides representing cross-reactive immunogenic epitopes are attractive candidates for prophylactic and therapeutic vaccines.

Diagnostic Peptides Representing Antigenic Epitopes

Peptide-based tests have many advantages over immunoassays based on whole molecules because such tests are safe, standardized, specific, reproducible, and suitable for large-scale analysis *in vitro*. Synthetic peptides that mimic epitopes have been used in diagnostic systems for various human diseases, including cancers, allergies, autoimmune diseases, viral, bacterial, and parasitic infections (Leinikki et al., 1993; Meloen et al., 2003; Gomara and Haro, 2007).

The safety of peptides compared to pathogens' molecules is the main advantage of using peptides in the diagnostics of infectious agents. Thus, a peptide-based enzyme-linked immunosorbent assay (ELISA) has been applied for early diagnostics of severe acute respiratory syndrome (SARS) (Hsueh et al., 2004). Demonstrating 100% specificity, this assay was based on synthetic peptides representing continuous epitopes from the spike, membrane, and nucleocapsid proteins of SARS-associated coronavirus. Synthetic peptides have also opened new perspectives for diagnosis of parasitic infections, such as malaria, Chagas' disease, leishmaniasis, and many others (Noya et al., 2003). Diagnostic peptides can also be isolated from phage-displayed libraries by affinity selection with serum antibodies from patients, for example, with Lyme disease (Hamby et al., 2005). The sequence similarity search between such isolated peptides and the pathogen proteins uncovered a new epitope in the VlsE antigen of *Borrelia burgdorferi*. The identification of this epitope demonstrated the potential of peptide library screening, even though its diagnostic sensitivity was relatively low compared to the experimentally identified epitopes (Hamby et al., 2005).

Synthetic peptides are also attractive diagnostic tools for allergies and autoimmune diseases. Epitopes of autoantigens may detect autoantibodies against intracellular autoantigens present during autoimmune diseases; autoantibodies help establish disease prognosis (Fournel and Muller, 2003; Routsias, Vlachoyiannopoulos, and Tzioufasm, 2006). One of the proposed therapeutics for autoimmune diseases is based on the design of molecules that are complementary to autoantigen epitopes (Routsias, Vlachoyiannopoulos, and Tzioufasm, 2006). Synthetic peptides are also considered useful *in vitro* diagnostic tools for allergies since they allow a diagnosis without the risks associated with tests based on *in vivo* allergen challenges (Eigenmann, 2004).

Synthetic Vaccines

A major reason for analyzing and predicting epitopes is because they may lead to the development of peptide-based synthetic vaccines. Many studies have been carried out to develop synthetic prophylactic vaccines against numerous viral, bacterial, and parasitic infections as well therapeutic vaccines for chronic infections and noninfectious diseases, including autoimmune diseases, various neurological disorders, allergies, and cancers (Arnon, 1987; Nicholson, 1994; Arnon and Ben-Yedidia, 2003; Hans et al., 2006). Thousands of peptides have been preclinically examined; over 100 of them have progressed to phase I clinical trials and about 30 to phase II. However, not a single peptide vaccine has passed phase III and became available to the public (Hans et al., 2006). Some notable successes have been obtained in the development of vaccines based on T-cell epitopes (De Groot and Martin, 2003; Flower et al., 2003; Larche and Wraith, 2005; Jiang et al., 2006), which are not discussed in the current review. The software and database developed to facilitate vaccine design are reviewed in (Davies and Flower, 2007; Taylor and Flower, 2007).

In order to be considered a vaccine candidate, a peptide must first demonstrate antigenicity. Furthermore, its possible immunogenicity must be experimentally tested. However, adequate immunogenicity is a substantially rarer phenomenon than antigenicity for short peptides. To induce an immune response, peptides shorter than 10–15 amino acids usually require to be covalently bound to a carrier protein or a liposome. Also, to achieve sufficient immunogenicity, the synthetic vaccine should include T-cell epitopes. Furthermore, short peptides tend to rapidly degrade. This problem can be solved through stabilization of peptide conformations by cyclization or other chemical procedures (Sundaram et al., 2004; Dakappagari et al., 2005; Hans et al., 2006). Peptide stability can also be improved by replacing L-peptides by "retroinverso" (RI) peptides composed of D-amino acids assembled in reverse order (Muller et al., 1998). For example, the RI peptide analogue of gonadotropin-releasing hormone (GnRH) was shown to be a potential contraceptive vaccine (Fromme et al., 2003).

The majority of peptides examined as vaccine candidates correspond to continuous epitopes. However, most epitopes in proteins are discontinuous, and methods to predict such epitopes identify only a certain number of residues that are part of the discontinuous structure without indicating how these residues could be assembled by synthesis to reconstitute an active epitope. Every attempt at reconstituting a discontinuous epitope by synthesis is therefore a unique endeavor, with little success so far (Enshell-Seijffers et al., 2003; Oomen et al., 2003; Villen et al., 2004; Dakappagari et al., 2005; Timmerman et al., 2005), and it will remain to be seen whether the prediction of discontinuous epitopes will be useful for developing synthetic vaccines. Several attempts to develop synthetic vaccines against viral infections are briefly described in this chapter. For the peptide-based approaches to the development of cancer vaccines see a review (Sundaram, Dakappagari, and Kaumaya, 2002).

Foot-and-mouth virus (FDMV) infection of cattle and pigs occurs throughout the world and causes huge economic losses. Most studies to replace conventional FDMV vaccines by peptide-based vaccines have concentrated on two continuous epitopes of the viral VP1 protein located on a loop (residues 140–160; site A) and the C-terminal region (residues 201–213; site C). Conjugated to carriers, peptides corresponding to these epitopes elicited neutralizing antibodies in guinea pigs (Bittle et al., 1982; DiMarchi et al., 1986), while immunization of cattle with these peptide epitopes gave viral protection lower than 50% (Taboga et al., 1997). The discontinuous epitope corresponding to the site D formed by residues from the VP1, VP2, and VP3 proteins has also been actively studied; however, the synthetic construction that mimicked this epitope did not show the protection level required for an effective vaccine (Villen et al., 2004). The high rate of mutations present in FMDV constitutes a major problem for vaccine development.

A major worldwide health threat is influenza, which in the United States alone kills over 40,000 people yearly. The virus escapes the immune system by undergoing continuous antigenic variation, especially in the hemagglutinin (HA) surface glycoprotein. Two continuous epitopes of HA1 located in residues 91–108 and 139–147 have been studied as vaccine candidates. Conjugated to the tetanus toxoid, the 91–108 epitope elicited an immune response in rabbits and mice that partially protected those animals against influenza (Muller, Shapira, and Arnon, 1982). The addition of two conserved T-cell epitopes of the influenza nucleoprotein to the experimental vaccine increased the protection level (Levi et al., 1995; Levi and Arnon, 1996). Studies with human/mouse radiation chimeras showed that the recombinant vaccine based on the mixture of these peptides had considerable potential for use in humans (Ben-Yedidia and Arnon, 2005).

Canine parvovirus causing enteritis and myocarditis in dogs and minks is the first veterinary virus for which an effective synthetic peptide vaccine has been developed. Several peptides from the N-terminal region of the viral VP2 protein (residues 1–15, 7–1, and 3–19) coupled to a carrier induced an immune response in dogs and minks and protected them against infection (Langeveld et al., 1994; Casal et al., 1995; Langeveld et al., 1995). The success achieved with these peptides is probably due to the high flexibility of the N-terminal region of VP2 that facilitates the cross-reactivity between peptides and virus particles.

In spite of global research efforts, no vaccine against HIV has yet been developed (Koff, Kahn, and Gust, 2007). Many attempts have been made to design HIV peptide vaccines using structural data on HIV proteins in complexes with neutralizing monoclonal antibodies (Burton et al., 2004). One of them involves the continuous epitope ELDKWAS corresponding to the membrane-proximal external region of the gp41 protein. This epitope is widely regarded as a promising vaccine candidate because it is recognized by the anti-HIV broadly cross-reactive neutralizing monoclonal antibody 2F5 and is located in the conserved region of the gp41, which is necessary for the fusion of the virus with the host cell membrane. When incorporated into various synthetic constructs, the ELDKWAS peptide improved its binding to the 2F5 antibody, although it did not induce neutralizing antibodies (Ho et al., 2005). In efforts to assess which structural elements close to the epitope influenced the neutralizing properties of the antibody, the crystal structures of 2F5 complexed with 7mer, 11mer, and 17mer peptides of the gp41 incorporating the ELDKWAS sequence have been solved (Ofek et al., 2004). In all complexes, the conformations of the peptides were significantly different from the conformation of the epitopic region in the gp41. Moreover, the epitopic region in the gp41 demonstrates a high degree of conformational flexibility, and its conformation depends on the fusogenic state of the protein. Therefore, it is not clear which conformation should be stabilized in the synthetic constructs intended for vaccination (Van Regenmortel, 2007). Such transient epitopes with variable conformations represent a major challenge in the development of synthetic vaccines (Zolla-Pazner, 2004).

This lack of success in developing synthetic peptide-based vaccines suggests that some of the assumptions underlying this type of research are probably unsound and that novel approaches should be investigated if synthetic vaccines are to become a reality.

CONCLUSION

This chapter summarized current attempts at prediction of epitopes using bioinformatics tools. Although combining sequence-based prediction methods with 3D structural information was found to lead to slightly better predictions, the improvement was still rather modest.

There are three main reasons for the low success rate of epitope predictions. First, it is difficult to determine to what extent a protein reacting with an antibody has retained its native conformation, meaning that even structural information is an unreliable guide for epitope prediction. Furthermore, proteins are dynamic structures able to undergo significant conformational rearrangements when binding to antibodies.

Second, peptides, called continuous epitopes, are usually poor mimics of discontinuous epitopes—the actual atomic configurations that antibodies recognize at the surface of proteins. Furthermore, the existing databases of continuous epitopes are known to be of dubious value for assessing the reliability of epitope prediction.

Third, discontinuous epitopes, which are the most common ones found in proteins, can be defined only in terms of residues in contact with the antibody, and they cannot be isolated as independent entities that possess binding activity outside of the protein context in which they are embedded. The so-called prediction of discontinuous epitopes usually consists in predicting that certain surface residues are likely to be a part of an epitope, but it does not entail predicting the minimum set of combined residues in a defined configuration that is likely to possess immunological activity. Furthermore, attempts to reconstitute active discontinuous epitopes by chemical assembly of a limited number of their constituent residues have so far met with little success.

As a result, predicting discontinuous epitopes has few applications at present. It can only be hoped that the development of new bioinformatics tools will improve our ability to predict which synthetic peptides are likely to be useful constructs for replacing intact proteins as antigens and immunogens.

ABBREVIATIONS

CDR	Complementarity determining region of the antibody
Fab	antigen-binding fragment of antibody that includes one complete light chain paired with one heavy chain fragment containing the variable domain and the first constant domain
A_{ROC}	area under the receiver operating characteristic curve
TP, FN, TN, FP	true positives, false negatives, true negatives, and false positives, respectively.

ACKNOWLEDGMENT

JP work was supported by the National Institutes of Health Contract HHSN26620040006C.

REFERENCES

Alix AJ (1999): Predictive estimation of protein linear epitopes by using the program PEOPLE. *Vaccine* 18(3–4):311–314.

Anderer FA, Schlumberger HD (1965): Properties of different artificial antigens immunologically related to tobacco mosaic virus. *Biochim Biophys Acta* 97:503–509.

Amit AG, Mariuzza RA, Phillips SE, Poljak RJ (1986): Three-dimensional structure of an antigen–antibody complex at 2.8 Å resolution. *Science* 233(4765):747–753.

Arnon R (1987): *Synthetic Vaccines*. Boca Raton, FL: CRC Press.

Arnon R, Ben-Yedidia T (2003): Old and new vaccine approaches. *Int Immunopharmacol* 3(8): 1195–1204.

Bajaj K, Chakrabarti P, Varadarajan R (2005): Mutagenesis-based definitions and probes of residue burial in proteins. *Proc Natl Acad Sci USA* 102(45):16221–16226.

Batori V, Friis EP, Nielsen H, Roggen EL (2006): An *in silico* method using an epitope motif database for predicting the location of antigenic determinants on proteins in a structural context. *J Mol Recognit* 19(1):21–29.

Ben-Yedidia T, Arnon R (2005): Towards an epitope-based human vaccine for influenza. *Hum Vaccin* 1(3):95–101.

Benjamin DC, Perdue SS (1996): Site-directed mutagenesis in epitope mapping. *Methods* 9(3): 508–515.

Bentley GA, Boulot G, Chitarra V (1994): Cross-reactivity in antibody–antigen interactions. *Res Immunol* 145(1):45–48.

Berger C, Weber-Bornhauser S, Eggenberger J, Hanes J, Pluckthun A, Bosshard HR (1999): Antigen recognition by conformational selection. *FEBS Lett* 450(1–2):149–153.

Berman HM, Westbrook J, Feng Z, Gilliland G, Bhat TN, Weissig H, Shindyalov IN, Bourne PE (2000): The Protein Data Bank. *Nucleic Acids Res* 28(1):235–242.

Berzofsky JA (1985): Intrinsic and extrinsic factors in protein antigenic structure. *Science* 229(4717): 932–940.

Bittle JL, Houghten RA, Alexander H, Shinnick TM, Sutcliffe JG, Lerner RA, Rowlands DJ, Brown F (1982): Protection against foot-and-mouth disease by immunization with a chemically synthesized peptide predicted from the viral nucleotide sequence. *Nature* 298(5869):30–33.

Blythe MJ, Flower DR (2005): Benchmarking B cell epitope prediction: underperformance of existing methods. *Protein Sci* 14(1):246–248.

Blythe MJ (2006): Computational Characterisation of B Cell Epitopes. PhD thesis. School of Animal and Microbial Sciences, The Edward Jenner Institute for Vaccine Research, The University of Reading. p. 243.

Bosshard HR (2001): Molecular recognition by induced fit: how fit is the concept?*News Physiol Sci* 16:171–173.

Braden BC, Souchon H, Eisele JL, Bentley GA, Bhat TN, Navaza J, Poljak RJ (1994): Three-dimensional structures of the free and the antigen-complexed Fab from monoclonal anti-lysozyme antibody D44.1. *J Mol Biol* 243(4):767–781.

Braden BC, Goldman ER, Mariuzza RA, Poljak RJ (1998): Anatomy of an antibody molecule: structure, kinetics, thermodynamics and mutational studies of the antilysozyme antibody D1.3. *Immunol Rev* 163:45–57.

Braden BC, Poljak RJ (2000): Structure and energetics of anti-lysozyme antibodies, *Protein–Protein Recognition*. In: Kleanthous C, editor. New York: Oxford University Press, pp. 126–161.

Bradford JR, David WD, Westhead R (2005): Improved prediction of protein–protein binding sites using a support vector machines approach. *Bioinformatics* 21(8):1487–1494.

Bublil EM, Freund NT, Mayrose I, Penn O, Roitburd-Berman A, Rubinstein ND, Pupko T, Gershoni JM (2007): Stepwise prediction of conformational discontinuous B-cell epitopes using the mapitope algorithm. *Proteins* 68(1):294–304.

Bui HH, Peters B, Assarsson E, Mbawuike I, Sette A (2007): Ab and T cell epitopes of influenza A virus, knowledge and opportunities. *Proc Natl Acad Sci USA* 104(1):246–251.

Burton DR, Desrosiers RC, Doms RW, Koff WC, Kwong PD, Moore JP, Nabel GJ, Sodroski J, Wilson IA, Wyatt RT (2004): HIV vaccine design and the neutralizing antibody problem. *Nat Immunol* 5(3): 233–236.

Carneiro J, Stewart J (1994): Rethinking "shape space": evidence from simulated docking suggests that steric shape complementarity is not limiting for antibody–antigen recognition and idiotypic interactions. *J Theor Biol* 169(4):391–402.

Casal JI, Langeveld JP, Cortes E, Schaaper WW, van Dijk E, Vela C, Kamstrup S, Meloen RH (1995): Peptide vaccine against canine parvovirus: identification of two neutralization subsites in the N terminus of VP2 and optimization of the amino acid sequence. *J Virol* 69 (11):7274–7277.

Castrignano T, De Meo PD, Carrabino D, Orsini M, Floris M, Tramontano A (2007): The MEPS server for identifying protein conformational epitopes. *BMC Bioinformatics* 8:(Suppl 1): S6.

Chavali GB, Papageorgiou AC, Olson KA, Fett JW, Hu G, Shapiro R, Acharya KR (2003): The crystal structure of human angiogenin in complex with an antitumor neutralizing antibody. *Structure* 11(7):875–885.

Chou PY, Fasman GD (1974): Conformational parameters for amino acids in helical, beta-sheet, and random coil regions calculated from proteins. *Biochemistry* 13(2):211–222.

Chou PY, Fasman GD (1978): Prediction of the secondary structure of proteins from their amino acid sequence. *Adv Enzymol Relat Areas Mol Biol* 47:45–148.

Collis AV, Brouwer AP, Martin AC (2003): Analysis of the antigen combining site: correlations between length and sequence composition of the hypervariable loops and the nature of the antigen. *J Mol Biol* 325(2):337–354.

Comeau SR, Gatchell DW, Vajda S, Camacho CJ (2004): ClusPro: an automated docking and discrimination method for the prediction of protein complexes. *Bioinformatics* 20(1):45–50.

Cunningham BC, Wells JA (1993): Comparison of a structural and a functional epitope. *J Mol Biol* 234(3):554–563.

Dakappagari NK, Lute KD, Rawale S, Steele JT, Allen SD, Phillips G, Reilly RT, Kaumaya PT (2005): Conformational HER-2/neu B-cell epitope peptide vaccine designed to incorporate two native disulfide bonds enhances tumor cell binding and antitumor activities. *J Biol Chem* 280(1):54–63.

Davies MN, Flower DR (2007): Harnessing bioinformatics to discover new vaccines. *Drug Discov Today* 12(9–10):389–395.

De Groot AS, Martin W (2003): From immunome to vaccine: epitope mapping and vaccine design tools. *Novartis Found Symp* 254:57–72; discussion 72–76, 98–101, 250–252.

Decanniere K, Transue TR, Desmyter A, Maes D, Muyldermans S, Wyns L (2001): Degenerate interfaces in antigen–antibody complexes. *J Mol Biol* 313(3):473–478.

DiMarchi R, Brooke G, Gale C, Cracknell V, Doel T, Mowat N (1986): Protection of cattle against foot-and-mouth disease by a synthetic peptide. *Science* 232(4750):639–641.

Duhovny D, Nussinov R, Wolfson HJ (2002): Efficient unbound docking of rigid molecules. 2nd Workshop on Algorithms in Bioinformatics(WABI) Rome, Italy, Lecture Notes in Computer Science 2452: pp. 185–200.

Eigenmann PA (2004): Do we have suitable *in-vitro* diagnostic tests for the diagnosis of food allergy? *Curr Opin Allergy Clin Immunol* 4(3):211–213.

Eisenberg D, Weiss RM, Terwilliger TC (1984): The hydrophobic moment detects periodicity in protein hydrophobicity. *Proc Natl Acad Sci USA* 81(1):140–144.

Emini EA, Hughes JV, Perlow DS, Boger J (1985): Induction of hepatitis A virus: neutralizing antibody by a virus-specific synthetic peptide. *J Virol* 55(3):836–839.

Enshell-Seijffers D, Denisov D, Groisman B, Smelyanski L, Meyuhas R, Gross G, Denisova G, Gershoni JM (2003): The mapping and reconstitution of a conformational discontinuous B-cell epitope of HIV-1. *J Mol Biol* 334(1):87–101.

Flower DR, McSparron H, Blythe MJ, Zygouri C, Taylor D, Guan P, Wan S, Coveney PV, Walshe V, Borrow P, Doytchinova IA (2003): Computational vaccinology: quantitative approaches. *Novartis Found Symp* 254:102–120; discussion 120–225, 216–222, 250–252.

Folgori A, Tafi R, Meola A, Felici F, Galfre G, Cortese R, Monaci P, Nicosia A (1994): A general strategy to identify mimotopes of pathological antigens using only random peptide libraries and human sera. *Embo J* 13(9):2236–2243.

Fournel S, Muller S (2003): Synthetic peptides in the diagnosis of systemic autoimmune diseases. *Curr Protein Pept Sci* 4(4):261–274.

Fromme B, Eftekhari P, Van Regenmortel M, Hoebeke J, Katz A, Millar R (2003): A novel retro-inverso gonadotropin-releasing hormone (GnRH) immunogen elicits antibodies that neutralize the activity of native GnRH. *Endocrinology* 144(7):3262–3269.

Garnier J, Osguthorpe DJ, Robson B (1978): Analysis of the accuracy and implications of simple methods for predicting the secondary structure of globular proteins. *J Mol Biol* 120(1):97–120.

Geysen HM, Meloen RH, Barteling SJ (1984): Use of peptide synthesis to probe viral antigens for epitopes to a resolution of a single amino acid. *Proc Natl Acad Sci USA* 81(13): 3998–4002.

Geysen HM, Rodda SJ, Mason TJ, Tribbick G, Schoofs PG (1987): Strategies for epitope analysis using peptide synthesis. *J Immunol Methods* 102(2):259–274.

Goede A, Jaeger IS, Preissner R (2005): SUPERFICIAL: surface mapping of proteins via structure-based peptide library design. *BMC Bioinformatics* 6:223.

Gomara MJ, Haro I (2007): Synthetic peptides for the immunodiagnosis of human diseases. *Curr Med Chem* 14(5):531–546.

Greenbaum JA, Andersen PH, Blythe M, Bui HH, Cachau RE, Crowe J, Davies M, Kolaskar AS, Lund O, Morrison S, Mumey B, Ofran Y, Pellequer JL, Pinilla C, Ponomarenko JV, Raghava GP, van Regenmortel MH, Roggen EL, Sette A, Schlessinger A, Sollner J, Zand M, Peters B (2007):

Towards a consensus on datasets and evaluation metrics for developing B-cell epitope prediction tools. *J Mol Recognit* 20(2):75–82.

Hager-Braun C, Tomer KB (2005): Determination of protein-derived epitopes by mass spectrometry. *Expert Rev Proteomics* 2(5):745–756.

Halperin I, Ma B, Wolfson H, Nussinov R (2002): Principles of docking: an overview of search algorithms and a guide to scoring functions. *Proteins* 47(4):409–443.

Hamby CV, Llibre M, Utpat S, Wormser GP (2005): Use of peptide library screening to detect a previously unknown linear diagnostic epitope: proof of principle by use of lyme disease sera. *Clin Diagn Lab Immunol* 12(7):801–807.

Hans D, Young PR, Fairlie DP (2006): Current status of short synthetic peptides as vaccines. *Med Chem* 2(6):627–646.

Haste Andersen P, Nielsen M, Lund O (2006): Prediction of residues in discontinuous B-cell epitopes using protein 3D structures. *Protein Sci* 15(11):2558–2567.

Ho J, Uger RA, Zwick MB, Luscher MA, Barber BH, MacDonald KS (2005): Conformational constraints imposed on a pan-neutralizing HIV-1 antibody epitope result in increased antigenicity but not neutralizing response. *Vaccine* 23(13):1559–1573.

Hopp TP, Woods KR (1981): Prediction of Protein Antigenic Determinants from Amino Acid Sequences. *Proc Natl Acad Sci USA* 78:3824–3828.

Hopp TP, Woods KR (1983): A computer program for predicting protein antigenic determinants. *Mol Immunol* 20(4):483–489.

Hopp TP (1993): Retrospective: 12 years of antigenic determinant predictions, and more. *Pept Res* 6(4):183–190.

Hsueh PR, Kao CL, Lee CN, Chen LK, Ho MS, Sia C, Fang XD, Lynn S, Chang TY, Liu SK, Walfield AM, Wang CY (2004): SARS antibody test for serosurveillance. *Emerg Infect Dis* 10(9):1558–1562.

Huang J, Gutteridge A, Honda W, Kanehisa M (2006): MIMOX: a web tool for phage display based epitope mapping. *BMC Bioinformatics* 7:451.

Ivanciuc O, Schein CH, Braun W (2003): SDAP: database and computational tools for allergenic proteins. *Nucleic Acids Res* 31(1):359–362.

James LC, Roversi P, Tawfik DS (2003): Antibody multispecificity mediated by conformational diversity. *Science* 299(5611):1362–1367.

Jameson BA, Wolf H (1988): The antigenic index: a novel algorithm for predicting antigenic determinants. *Comput Appl Biosci* 4(1):181–186.

Janin J, Wodak S (1978): Conformation of amino acid side-chains in proteins. *J Mol Biol* 125(3):357–386.

Jiang S, Song R, Popov S, Mirshahidi S, Ruprecht RM (2006): Overlapping synthetic peptides as vaccines. *Vaccine* 24(37–39):6356–6365.

Johnson MA, Pinto BM (2004): NMR spectroscopic and molecular modeling studies of protein–carbohydrate and protein–peptide interactions. *Carbohydr Res* 339(5):907–928.

Jones S, Thornton JM (1996): Principles of protein–protein interactions. *Proc Natl Acad Sci USA* 93(1):13–20.

Jones S, Thornton JM (1997): Prediction of protein–protein interaction sites using patch analysis. *J Mol Biol* 272(1):133–143.

Jones S, Thornton J (2000): Analysis and classification of protein–protein interactions from a structural perspective. In: Kleanthous C, editor. *Protein–Protein Recognition*. New York: Oxford University Press, pp. 33–59.

Karplus PA, Schulz GE (1985): Prediction of chain flexibility in proteins. A tool for the selection of peptide antigens. *Naturwissenschaften* 72:S.212.

Kendrew JC, Bodo J, Dintzis HM, Parrish RG, Wyckoff H, Phillips DC (1958): 3-Dimensional model of the myoglobulin molecule obtained by X-ray analysis. *Nature* 181:662–666.

Koff WC, Kahn P, Gust ID (2007): *AIDS Vaccine Development: Challenges and Opportunities.* Caister Academic Press.

Kolaskar AS, Tongaonkar PC (1990): A semi-empirical method for prediction of antigenic determinants on protein antigens. *FEBS Lett* 276(1–2):172–174.

Kolaskar AS, Kulkarni-Kale U (1999): Prediction of three-dimensional structure and mapping of conformational epitopes of envelope glycoprotein of Japanese encephalitis virus. *Virology* 261(1):31–42.

Kulkarni-Kale U, Bhosle S, Kolaskar AS (2005): CEP: a conformational epitope prediction server. *Nucleic Acids Res* 33 (Web server issue):W168–W171.

Kyte J, Doolittle RF (1982): A simple method for displaying the hydropathic character of a protein. *J Mol Biol* 157(1):105–132.

Lachumanan R, Devi S, Cheong YM, Rodda SJ, Pang T (1993): Epitope mapping of the Sta58 major outer membrane protein of *Rickettsia tsutsugamushi. Infect Immun* 61(10): 4527–4531.

Landau M, Mayrose I, Rosenberg Y, Glaser F, Martz E, Pupko T, Ben-Tal N (2005): ConSurf 2005: the projection of evolutionary conservation scores of residues on protein structures. *Nucleic Acids Res* 33 (Web server issue):W299–302.

Langeveld JP, Casal JI, Osterhaus AD, Cortes E, de Swart R, Vela C, Dalsgaard K, Puijk WC, Schaaper WM, Meloen RH (1994): First peptide vaccine providing protection against viral infection in the target animal: studies of canine parvovirus in dogs. *J Virol* 68(7):4506–4513.

Langeveld JP, Kamstrup S, Uttenthal A, Strandbygaard B, Vela C, Dalsgaard K, Beekman NJ, Meloen RH, Casal JI (1995): Full protection in mink against mink enteritis virus with new generation canine parvovirus vaccines based on synthetic peptide or recombinant protein. *Vaccine* 13(11): 1033–1037.

Larche M, Wraith DC (2005): Peptide-based therapeutic vaccines for allergic and autoimmune diseases. *Nat Med* 11 (Suppl 4):S69–S76.

Larsen JE, Lund O, Nielsen M (2006): Improved method for predicting linear B-cell epitopes. *Immunome Res* 2:2.

Laver WG, Air GM, Webster RG, Smith-Gill SJ (1990): Epitopes on protein antigens: misconceptions and realities. *Cell* 61(4):553–556.

Leinikki P, Lehtinen M, Hyoty H, Parkkonen P, Kantanen ML, Hakulinen J (1993): Synthetic peptides as diagnostic tools in virology. *Adv Virus Res* 42:149–186.

Lerner RA (1984): Antibodies of predetermined specificity in biology and medicine. *Adv Immunol* 36:1–44.

Levi R, Aboud-Pirak E, Leclerc C, Lowell GH, Arnon R (1995): Intranasal immunization of mice against influenza with synthetic peptides anchored to proteosomes. *Vaccine* 13(14):1353–1359.

Levi R, Arnon R (1996): Synthetic recombinant influenza vaccine induces efficient long-term immunity and cross-strain protection. *Vaccine* 14(1):85–92.

Levitt M (1978): Conformational preferences of amino acids in globular proteins. *Biochemistry* 17(20): 4277–4285.

Lo Conte L, Chothia C, Janin J (1999): The atomic structure of protein–protein recognition sites. *J Mol Biol* 285(5):2177–2198.

Mandell JG, Roberts VA, Pique ME, Kotlovyi V, Mitchell JC, Nelson E, Tsigelny I, Ten Eyck LF (2001): Protein docking using continuum electrostatics and geometric fit. *Protein Eng* 14(2):105–113.

Mayrose I, Shlomi T, Rubinstein ND, Gershoni JM, Ruppin E, Sharan R, Pupko T (2007): Epitope mapping using combinatorial phage-display libraries: a graph-based algorithm. *Nucleic Acids Res* 35(1):69–78.

Meloen RH, Puijk WC, Langeveld JP, Langedijk JP, Timmerman P (2003): Design of synthetic peptides for diagnostics. *Curr Protein Pept Sci* 4(4):253–260.

Moreau V, Granier C, Villard S, Laune D, Molina F (2006): Discontinuous epitope prediction based on mimotope analysis. *Bioinformatics* 22(9):1088–1095.

Morris GE (1996): *Epitope Mapping Protocols.* Humana Press.

Muller GM, Shapira M, Arnon R (1982): Anti-influenza response achieved by immunization with a synthetic conjugate. *Proc Natl Acad Sci USA* 79(2):569–573.

Muller S, Benkirane N, Guichard G, Van Regenmortel MH, Brown F (1998): The potential of retro-inverso peptides as synthetic vaccines. *Expert Opin Investig Drugs* 7(9): 1429–1438.

Mumey BM, Bailey BW, Kirkpatrick B, Jesaitis AJ, Angel T, Dratz EA (2003): A new method for mapping discontinuous antibody epitopes to reveal structural features of proteins. *J Comput Biol* 10(3–4):555–567.

Murali R, Sharkey DJ, Daiss JL, Murthy HM (1998): Crystal structure of Taq DNA polymerase in complex with an inhibitory Fab: the Fab is directed against an intermediate in the helix-coil dynamics of the enzyme. *Proc Natl Acad Sci USA* 95(21):12562–12567.

Naruse H, Ogasawara K, Kaneda R, Hatakeyama S, Itoh T, Kida H, Miyazaki T, Good RA, Onoe K (1994): A potential peptide vaccine against two different strains of influenza virus isolated at intervals of about 10 years. *Proc Natl Acad Sci USA* 91(20):9588–9592.

Neuvirth H, Raz R, Schreiber G (2004): ProMate: a structure based prediction program to identify the location of protein–protein binding sites. *J Mol Biol* 338(1):181–199.

Nicholson BH (1994): *Synthetic Vaccines.* Oxford: Blackwell Scientific Publ,.

Novotny J, Handschumacher M, Haber E, Bruccoleri RE, Carlson WB, Fanning DW, Smith JA, Rose GD (1986): Antigenic determinants in proteins coincide with surface regions accessible to large probes (antibody domains). *Proc Natl Acad Sci USA* 83(2):226–230.

Noya O, Patarroyo ME, Guzman F, Alarcon de Noya B (2003): Immunodiagnosis of parasitic diseases with synthetic peptides. *Curr Protein Pept Sci* 4(4):299–308.

Odorico M, Pellequer JL (2003): BEPITOPE: predicting the location of continuous epitopes and patterns in proteins. *J Mol Recognit* 16(1):20–22.

Ofek G, Tang M, Sambor A, Katinger H, Mascola JR, Wyatt R, Kwong PD (2004): Structure and mechanistic analysis of the anti-human immunodeficiency virus type 1 antibody 2F5 in complex with its gp41 epitope. *J Virol* 78(19):10724–10737.

Oomen CJ, Hoogerhout P, Bonvin AM, Kuipers B, Brugghe H, Timmermans H, Haseley SR, van Alphen L, Gros P (2003): Immunogenicity of peptide-vaccine candidates predicted by molecular dynamics simulations. *J Mol Biol* 328(5):1083–1089.

Parker JM, Guo D, Hodges RS (1986): New hydrophilicity scale derived from high-performance liquid chromatography peptide retention data: correlation of predicted surface residues with antigenicity and X-ray-derived accessible sites. *Biochemistry* 25(19): 5425–5432.

Pattnaik P (2005): Surface plasmon resonance: applications in understanding receptor–ligand interaction. *Appl Biochem Biotechnol* 126(2):79–92.

Pellequer JL, Westhof E, Van Regenmortel MH (1991): Predicting location of continuous epitopes in proteins from their primary structures. *Methods Enzymol* 203:176–201.

Pellequer JL, Westhof E (1993): PREDITOP: a program for antigenicity prediction. *J Mol Graph* 11(3): 204–210 191-2.

Pellequer JL, Westhof E, Van Regenmortel MH (1993): Correlation between the location of antigenic sites and the prediction of turns in proteins. *Immunol Lett* 36(1):83–99.

Peters B, Sidney J, Bourne P, Bui HH, Buus S, Doh G, Fleri W, Kronenberg M, Kubo R, Lund O, Nemazee D, Ponomarenko JV, Sathiamurthy M, Schoenberger S, Stewart S, Surko P, Way S, Wilson S, Sette A (2005): The immune epitope database and analysis resource: from vision to blueprint. *PLoS Biol* 3(3):e91.

Poljak RJ, Amzel LM, Avey HP, Chen BL, Phizackerley RP, Saul F (1973): Three-dimensional structure of the Fab' fragment of a human immunoglobulin at 2.8-Å resolution. *Proc Natl Acad Sci USA* 70:3305–3310.

Ponomarenko JV, Bourne PE (2007): Antibody–protein interactions: benchmark datasets and prediction tools evaluation. *BMC Struct Biol* 7. (64).

Radford AJ, Wood PR, Billman-Jacobe H, Geysen HM, Mason TJ, Tribbick G (1990): Epitope mapping of the *Mycobacterium bovis* secretory protein MPB70 using overlapping peptide analysis. *J Gen Microbiol* 136(2):265–272.

Rini JM, Schulze-Gahmen U, Wilson IA (1992): Structural evidence for induced fit as a mechanism for antibody–antigen recognition. *Science* 255(5047):959–965.

Routsias JG, Vlachoyiannopoulos PG, Tzioufas AG (2006): Autoantibodies to intracellular auto-antigens and their B-cell epitopes: molecular probes to study the autoimmune response. *Crit Rev Clin Lab Sci* 43(3):203–248.

Rowley MJ, O';Connor K, Wijeyewickrema L (2004): Phage display for epitope determination: a paradigm for identifying receptor–ligand interactions. *Biotechnol Annu Rev* 10:151–188.

Saha S, Bhasin M, Raghava GP (2005): Bcipep: a database of B-cell epitopes. *BMC Genomics* 6(1):79.

Saha S, Raghava GP (2006): Prediction of continuous B-cell epitopes in an antigen using recurrent neural network. *Proteins* 65(1):40–48.

Schneidman-Duhovny D, Inbar Y, Polak V, Shatsky M, Halperin I, Benyamini H, Barzilai A, Dror O, Haspel N, Nussinov R, Wolfson HJ (2003): Taking geometry to its edge: fast unbound rigid (and hinge-bent) docking. *Proteins* 52(1):107–112.

Schönbach C, Koh JLY, Sheng X, Wong L, Brusic V (2000): FIMM, a database of functional molecular immunology. *Nucl Acids Res* 28:222–224.

Schreiber A, Humbert M, Benz A, Dietrich U (2005): 3D-Epitope-Explorer (3DEX): localization of conformational epitopes within three-dimensional structures of proteins. *J Comput Chem* 26(9): 879–887.

Sollner J (2006): Selection and combination of machine learning classifiers for prediction of linear B-cell epitopes on proteins. *J Mol Recognit* 19(3):209–214.

Sollner J, Mayer B (2006): Machine learning approaches for prediction of linear B-cell epitopes on proteins. *J Mol Recognit* 19(3):200–208.

Stanfield RL, Fieser TM, Lerner RA, Wilson IA (1990): Crystal structures of an antibody to a peptide and its complex with peptide antigen at 2.8 Å. *Science* 248:712–719.

Sundaram R, Dakappagari NK, Kaumaya PT (2002): Synthetic peptides as cancer vaccines. *Biopolymers* 66(3):200–216.

Sundaram R, Lynch MP, Rawale SV, Sun Y, Kazanji M, Kaumaya PT (2004): *De novo* design of peptide immunogens that mimic the coiled coil region of human T-cell leukemia virus type-1 glycoprotein 21 transmembrane subunit for induction of native protein reactive neutralizing antibodies. *J Biol Chem* 279(23):24141–24151.

Sundberg EJ, Mariuzza RA (2004): Antibody structure and recognition of antigen. In: Honjo AFT, Neuberger M, editor. *Molecular Biology of B Cells*. Elsevier Science (USA). pp. 491–509.

Taboga O, Tami C, Carrillo E, Nunez JI, Rodriguez A, Saiz JC, Blanco E, Valero ML, Roig X, Camarero JA, Andreu D, Mateu MG, Giralt E, Domingo E, Sobrino F, Palma EL (1997): A large-scale evaluation of peptide vaccines against foot-and-mouth disease: lack of solid protection in cattle and isolation of escape mutants. *J Virol* 71(4):2606–2614.

Tainer JA, Getzoff ED, Alexander H, Houghten RA, Olson AJ, Lerner RA, Hendrickson WA (1984): The reactivity of anti-peptide antibodies is a function of the atomic mobility of sites in a protein. *Nature* 312(5990):127–134.

Taylor PD, Flower DR (2007): Immunoinformatics and computational vaccinology: a brief introduction, In: Flower DR, Timmis J, editors. *Silico Immunology*. Springer, pp. 23–46.

Thornton JM, Edwards MS, Taylor WR, Barlow DJ (1986): Location of 'continuous' antigenic determinants in the protruding regions of proteins. *Embo J* 5(2):409–413.

Timmerman P, Beld J, Puijk WC, Meloen RH (2005): Rapid and quantitative cyclization of multiple peptide loops onto synthetic scaffolds for structural mimicry of protein surfaces. *Chembiochem* 6(5):821–824.

Toseland CP, Clayton DJ, McSparron H, Hemsley SL, Blythe MJ, Paine K, Doytchinova IA, Guan P, Hattotuwagama CK, Flower DR (2005): AntiJen: a quantitative immunology database integrating functional, thermodynamic, kinetic, biophysical, and cellular data. *Immunome Res* 1(1):4.

Tribbick G (2002): Multipin peptide libraries for antibody and receptor epitope screening and characterization. *J Immunol Methods* 267(1):27–35.

Tulip WR, Harley VR, Webster RG, Novotny J (1994): N9 neuraminidase complexes with antibodies NC41 and NC10: empirical free energy calculations capture specificity trends observed with mutant binding data. *Biochemistry* 33(26):7986–7997.

Van Regenmortel MHV, Daney de Marcillac G (1988): An assessment of prediction methods for locating continuous epitopes in proteins. *Immunol Lett* 17(2):95–107.

Van Regenmortel MHV (1989): Structural and functional approaches to the study of protein antigenicity. *Immunol Today* 10(8):266–272.

Van Regenmortel MHV (1996): Mapping epitope structure and activity: from one-dimensional prediction to four-dimensional description of antigenic specificity. *Methods* 9(3):465–472.

Van Regenmortel MHV (2006): Immunoinformatics may lead to a reappraisal of the nature of B cell epitopes and of the feasibility of synthetic peptide vaccines. *J Mol Recognit* 19(3):183–187.

Van Regenmortel MHV (2007): The rational design of biological complexity: a deceptive metaphor. *Proteomics* 7(6):965–975.

Villen J, de Oliveira E, Nunez JI, Molina N, Sobrino F F Andreu D (2004): Towards a multi-site synthetic vaccine to foot-and-mouth disease: addition of discontinuous site peptide mimic increases the neutralization response in immunized animals. *Vaccine* 22(27–28):3523–3529.

Walter G (1986): Production and use of antibodies against synthetic peptides. *J Immunol Methods* 88(2):149–161.

Welling GW, Weijer WJ, van der Zee R, Welling-Wester S (1985): Prediction of sequential antigenic regions in proteins. *FEBS Lett* 188(2):215–218.

Westhof E, Altschuh D, Moras D, Bloomer AC, Mondragon A, Klug A, Van Regenmortel MH (1984): Correlation between segmental mobility and the location of antigenic determinants in proteins. *Nature* 311(5982):123–126.

Zolla-Pazner S (2004): Identifying epitopes of HIV-1 that induce protective antibodies. *Nat Rev Immunol* 4(3):199–210.

Section VIII

FUTURE CHALLENGES

36

METHODS TO CLASSIFY AND PREDICT THE STRUCTURE OF MEMBRANE PROTEINS

Marialuisa Pellegrini-Calace and Janet M. Thornton

GLOSSARY

- Cell membrane: lipid bilayer common to all living cells that separates the cell from the external environment
- Cell wall: rigid layer external to the cell membrane that provides the cell with structural support, protection, and a filtering mechanism
- Outer membrane: the outside membrane (external to the cell membrane) of Gram-negative bacteria, choroplasts, and mytochondria
- Membrane core: very hydrophobic slab in the middle of the membrane made of the alkyl chain of phospholipids
- Membrane interfaces: polar areas that separate the membrane core from aqueous solutions and are made of phospholipid polar heads
- Membrane plane: any plane running parallel to the bilayer slab
- Membrane axis: any direction perpendicular to the membrane plane
- Membrane protein topology: number of transmembrane secondary structure elements (helices or strands) + location of the protein N-terminus at the cytoplasmic side of the membrane ("IN") or at the opposite side ("OUT")
- Positive-inside rule: basic residues Arg and Lys are about four times more frequent in cytosolic than in periplasmic loops
- Lily-pad effect: the stretch of hydrophobic amino acids interacting with the membrane core is flanked by aromatic belts made of Tyr and Trp
- Re-entrant loops: loops that penetrate the membrane but do not cross it

---

Membrane proteins represent about 20–30% of the expressed genes as predicted from the genome sequencing of bacterial, archaean, and eukaryotic organisms (Jones, 1998; White and Wimley, 1998; Popot and Engelman, 1990). They are responsible for maintaining the homeostasis and responsiveness of cells by mediating a wide range of fundamental biological processes, such as cell signaling, transport of molecules impermeable to the membrane, intercellular communication, cell recognition, and cell adhesion. Understanding the structure and function of membrane proteins and studying their properties and biochemical mechanisms are therefore among the most important targets in biological and pharmaceutical research. The main difference between membrane and globular proteins is that they function in environments that have very different physicochemical properties. Processes involving membrane proteins happen within biological membranes, composed of phospholipid bilayers impermeable to polar molecules, whereas globular proteins function in aqueous media. Therefore, membrane proteins are highly insoluble and unstable in aqueous solutions leading to severe experimental problems associated with their study and the resulting amount of available structural data is scarce. Consequently, the membrane protein world is still less well known than the globular protein world despite its biological and physiological importance. (Jones, 1998).

THE BIOLOGICAL MEMBRANE

Biological membranes serve as permeability barriers between cells, organelles, and their "external" environment (Figure 36.1a). These membranes are made of fluid phospholipidic bilayers that create a peculiar chemicophysical environment distinct from the surrounding aqueous solution and interact with many of the aforementioned membrane proteins (Figure 36.1). In three dimensions, a membrane bilayer can be described by a membrane plane and a membrane axis. The membrane plane is defined as the plane parallel to the bilayer slab (xy plane, according to the notation in Figure 36.1); the membrane axis is the perpendicular vector to the membrane plane (z-axis, according to the notation in Figure 36.1).

From a chemical point of view, biological membranes are mainly composed of glycerol-based phospholipids, such as phosphatidylcholine (PC), phosphatidylserine (PS), and phosphatidylethanolamine (PE). However, other lipids are pivotal and specific for different types of membranes. Sphingolipids, which are ceramide-based phospholipids

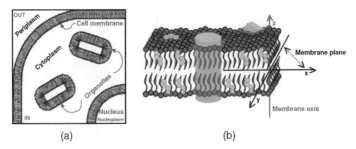

(a)   (b)

Figure 36.1. (a) Schematic representation of a section of a cell: IN, inside of the cell; OUT, outside of the cell. (b) Schematic representation of a cellular membrane. Glycerophospholipids are shown in gray, sphingolipids in green, cholesterol in orange, and proteins in light blue. The membrane plane and the membrane axis are also shown with dark gray and red lines, respectively. Figure also appears in the Color Figure section.

and are slightly less apolar than glycerol-based lipids, and sterols, which are chemically unsaturated polycycles, are found in all eukaryotic membranes. In contrast, glycolipids with bulky sugar portions are found in the outer membrane of Gram-negative bacteria and mitochondria. Glycerophospholipids and sphingolipids are made of a long and apolar tail, capable of hydrophobic interactions, and a polar phosphate group head, capable of hydrophilic interactions. These molecules are therefore flexible and amphipathic. Sterols are rather rigid compounds and are responsible for the increased rigidity of eukaryotic membranes—the higher fraction of sterols produces more rigid membranes and vice versa.

Fluid lipid bilayers have a very high degree of intrinsic thermal disorder essential for accomplishing their biological function and therefore it is not possible to derive three-dimensional crystallographic images of their structure at atomic level. However, a high periodicity is present along the membrane axis and this allows the distribution of the principal chemical groups of lipids (such as methylenes, carbonyls, phosphates, and so on) along the axis to be calculated by diffraction studies (White and Wimley, 1998). The structure of a fluid dioleoylphosphocholine (DOPC) in Figure 36.2a shows that a

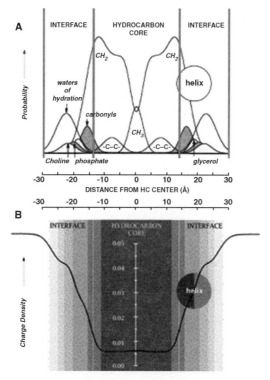

Figure 36.2. (a) The structure of a fluid dioleoylphosphocoline (DOPC) calculated as the time-averaged spatial distributions of the chemical groups methyl (−CH3), carbonyl (−CH2−), −CH=CH−, and water projected onto an axis normal to the bilayer plane. The bilayer can be easily divided into three regions on the basis of the distribution of the water of hydration of phospolipid headgroups: a 30Å thick central hydrocarbon core (red vertical lines) including most of −CH2−, −CH3 and −CH=CH−, and two side 15Å interfaces (pink lines) including most of the polar group with their water of hydration. (b) The structure of a fluid dioleoylphosphocoline (DOPC) represented as the variation of charge density along the membrane axis. Figure also appears in the Color Figure section.

phospholipid bilayer is a complex environment composed of nonpolar phases with different features. In fact, the bilayer can be intuitively divided into three parts: a central hydrocarbon core (red lines, 30 Å thick) and two interfaces (pink lines, 15 Å thick) that include most of the polar groups and play a pivotal role in membranes. These interfaces represent half of the membrane total thickness, undergo significant changes in polarity over very small distances, and are responsible for all polar and specific noncovalent interactions between the membrane and protein side chains.

The biological membrane is therefore a very complex and dynamic environment where the protein function is driven by mechanisms specific and different from those observed in globular proteins in aqueous solutions such as in the cytosol and periplasm.

THE FOLDING PROCESS OF MEMBRANE PROTEINS

The folding of helical membrane proteins has been extensively studied and can be seen as a two-stage biological process, as proposed by Popot and Engelman in 1990 (Popot and Engelman, 1990). In a first step, transmembrane helices insert into the membrane and only afterward do they assemble to form the final tertiary or quaternary structure.

More recent studies have shown that helical membrane proteins follow a folding route parallel to secretory proteins (Figure 36.3). The very first hydrophobic segment translated by the ribosome, the transmembrane signal peptide, (Figure 36.3a), is recognized by the signal recognition particle (SRP), which docks to the ribosome and blocks the chain elongation process (Figure 36.3b). The new complex is then recognized by the SRP receptor (SR, Figure 36.3c), which targets it to the plasma membrane and makes it bind to the translocation machinery (Figure 36.3d). Subsequently, the SRP and SR separate from the complex, the elongation restarts, and the translocon releases the membrane protein inserting it directly in the membrane where it assumes its final fold (Figure 36.3e).

Translocons are known as Sec61complexes in eukaryotes and SecY in eubacteria and archaea complexes. A structure of the SecY/Sec61 translocon complex from *Methanococcus jannaschii* is available at the Protein Data Bank (PDB) with the code 1RHZ (Figure 36.4; Van Den Berg et al., 2004) and provides deeper insights on how the folding process occurs. During synthesis and integration of a nascent transmembrane protein, the translocon allows the nascent chain to pass through its pore, which is hourglass shaped and hydrophilic. Thus, it decides whether to move the sequence to the other side of the membrane (secretory proteins) or to release it into the hydrophobic core of the phospholipid bilayer by a lateral gate (integral membrane proteins). Finally, it deciphers the protein topology and decides if the protein needs to be flipped relatively to the membrane bilayer (Wallin et al., 1997; Seshadri et al., 1998). The release of the chain into the bilayer core is driven by a simple partitioning between the translocon and the membrane itself. Basically, the translocon and the membrane work cooperatively toward protein insertion, so that hydrophobic helices sufficiently prefer the bilayer while more polar helices are moved toward the aqueous compartment at the other end of the channel. However, the exact biological mechanism underlying the partitioning is still unknown (Von Heijne, 1986; Ulmschneider, Sansom, and Di Nola, 2005).

β-barrel membrane proteins have been shown to follow a very similar path. They are synthesized in the cytoplasm and targeted to the translocon, which releases them into the periplasm separating the outer membrane from the inner membrane. Subsequently, they

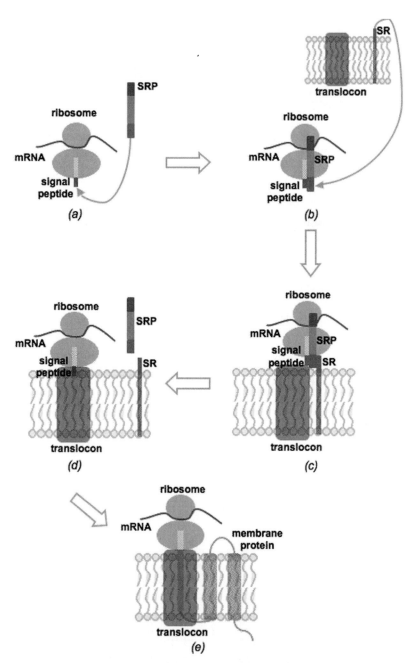

Figure 36.3. Schematic diagram of the membrane protein synthesis and folding. (a) The ribosome synthesizes the first transmembrane segment (signal); (b) the signal is recognized by the signal recognition particle (SRP), which binds the ribosome blocking the chain elongation; (c) the SRP receptor (SR) recognizes the SRP complex and targets it to the membrane making it bind the translocation machinery; (d) SRP and SR are displaced and the chain elongation restarts; and (e) the membrane protein nascent chain is released and folds within the bilayer.

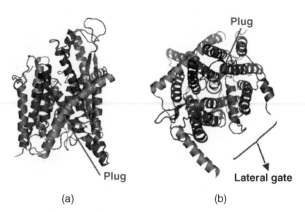

(a)   (b)

Figure 36.4. Structure of the SecY/Sec61 translocon complex from *M. jannaschii* as in the PDB (code 1RHZ). Left: side view. Right: top view. Subunits α, β, and γ are colored in blue, green, and pink, respectively. The plug (transmembrane helix 2, subunit α, aa 59–64), which allows the insertion of the helix in the membrane, is colored in light blue. The figure was generated from the original PDB file using the software PyMol. Figure also appears in the Color Figure section.

are transported to the outer membrane by an unknown mechanism involving periplasmic chaperones (Bernstein, 2000; Ruiz, Kahne, and Silhavy, 2006).

WHY IS IT DIFFICULT TO SOLVE MEMBRANE PROTEIN 3D STRUCTURES?

As mentioned in the introduction, membrane proteins constitute some of the most important targets in biological research and represent some of the most important drug targets for new pharmaceutical development. However, attempts to determine their 3D structures have been much less effective than the structure determination of globular proteins. Until recently, any structure-based drug design on membrane proteins had to rely on either homology modeling and/or site-directed mutagenesis studies. Although the big picture is improving, as shown in Figure 36.5, the PDB still includes tens of thousands of globular structures versus only

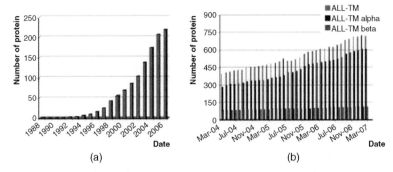

(a)   (b)

Figure 36.5. Histogram of the variation of available 3D structures of membrane proteins over time. (a) Increase of membrane protein structures as in Stephen White's Web resource from 1988 until now. (b) Increase of the total number of membrane protein structures in the past 3 years as in the PDB_TM database.

hundreds of membrane protein structures. In July 2007, there were 775 membrane protein structures in the PDB, representing 125 different proteins. A similar gap can also be observed in the understanding of protein folding and assembly mechanisms—*in vitro*, *in vivo*, and computational studies have revealed much more information for globular rather than membrane proteins. The discrepancy can be easily explained considering that from an experimental point of view, membrane proteins are generally much more difficult to handle than their soluble counterparts. In fact, many studies fail at the first main task that is to obtain a sufficient amount of the required membrane protein.

Structure determination of a protein involves three main experimental steps: expression, purification, and structure determination itself. Generally speaking, the isolation of membrane proteins is very difficult because of protein unfolding, instability, and aggregation, which occur very often (Booth and Curnow, 2006). There are several reasons for the poor success of such experiments. First of all, with only few exceptions, membrane proteins are poorly abundant in native tissues and therefore need to be expressed in heterologous systems. Moreover, they are more difficult to express as recombinant proteins because they have a complex folding mechanism, can cause toxic effects when inserted in host membranes, can be surrounded by a nonphysiological lipid environment due to the organism-specific membrane lipid composition, and experience nonnative posttranslational modifications typical of the host. For instance, a comparative analysis of overexpression yields for several G-protein-coupled receptors (GPCRs) in four main expression systems, *Escherichia coli*, yeast, insect, and mammalian cells, showed that in general GPCRs are only slightly overexpressed. Moreover, different GPCRs gave different expression levels in the same expression system and similarly the expression level of a given GPCR varied in different expression systems. Thus, predicting whether a membrane protein will be efficiently overexpressed in a given system is not a trivial task. On the other hand, and fairly obviously, the closer the host is to the native environment of the protein, the more likely it is to gain reasonable overexpression (Sarramegna et al., 2003). A further hurdle comes from the requirement of detergents during protein purification to improve protein solubility and prevent damage to the intrinsic transmembrane topology. In fact, detergents have a negative effect on the yield and stability of purified proteins and also reduce crystal contacts, hampering crystallization and structure determination.

Finally, although X-ray crystallography and NMR methods have improved significantly in the last years, available technologies are still more effective for water-soluble proteins. A key indicator of the extent of these problems is the lack of structure information available for the family of GPCRs, which represent over 50% of current drug targets (Lundstrom, 2006)—high-resolution structures have been determined for only one GPCR, bovine rhodopsin. Although many specific proteins have been studied individually in specific expression systems, leading to some remarkable successes (i.e., the rat neurotensin receptor in *E. coli*), overcoming all these experimental challenges for single protein often requires the engineering of multiple gene constructs with various deletions, fusion partners, and purifications, that is, it requires a considerable amount of time, effort, and funding (Grisshammer et al., 2005) and even then success is not always realized.

STRUCTURAL GENOMICS AND MEMBRANE PROTEINS

Structural genomics (see Chapter 29) pipelines have recently been established to improve the success rate of membrane protein structure determination. They require a broad area of expertise and a large handling capacity to allow the study of a large number of targets in parallel, thus providing a few advantages. First, as part of the systematic approach to expression, bioinformatics tools may reveal at an early stage which proteins represent better targets than others. Moreover, studies on a large number of targets in parallel can produce an adequate supply of protein to carry on purification and crystallization studies. Furthermore, large consortia of laboratories often have the capacity to handle alternative expression systems in parallel, improving the success of production of material suitable for further structural biology studies (Lundstrom, 2006). In the last few years, the efforts of 11 structural genomics networks focused on nonmembrane proteins have made a significant contribution to the increased coverage of globular protein superfamilies and folds (Todd et al., 2005). These networks target "novel" proteins, with the result that 67% of the protein domains they have determined are unique and lack a characterized close homologue in the PDB, while only 21% of domains from nonstructural genomics laboratories are unique. Consequently, this success has generated high expectations for the extension of such approaches to membrane proteins.

Current structural genomics initiatives that focus on membrane proteins focus either on whole genomes, on selected water-soluble and membrane protein targets, or exclusively on membrane proteins. The Centre for Eukaryotic Structural Genomics (CESG, http://www.uwstructuralgenomics.org/) deals with the whole *Arabidopsis thaliana* genome. The Joint Center for Structural Genomics (JCSG, http://www.jcsg.org/) focuses on *Thermotogota maritima* proteome, mouse genome, and selected human GCPRs as specific membrane protein targets. The European Membrane Protein Consortium (E-MeP, http://www.e-mep.org/), funded by European Union, the privately funded Membrane Protein Network (MePNet, http://www.mepnet.org/), and the Membrane Protein Structure Initiative (MPS*i*, http://www.mpsi.ac.uk/), funded by the UK Biotechnology and Biological Sciences Research Council (BBSRC), are fully dedicated to membrane proteins and involve world leaders in membrane protein structural biology. Basically, all these projects aim at developing high-throughput methods for the expression, purification, crystallization, and structure determination of integral membrane proteins. In addition, MePNet specializes on GPCRs and MPS*i* on ion channels and transporters. The major challenge facing these membrane protein structural genomics initiatives is the production of sufficient protein in the form necessary for successful crystallization. Although it is now relatively straightforward to synthesize many different constructs, the number of possible constructs is very large and improved methods to define the lengths of loops and the topology are needed to help improve the design of successful constructs. Moreover, as these projects are beginning to determine novel membrane protein structures, such as the structures of divalent magnesium ion transporter CorA (PDB code 2BBJ) and of the open and closed forms of the spinach aquaporin SOP1P2;1 (PDB codes 2B5F and 1Z98, respectively), structural bioinformatics tools are needed to analyze and exploit this knowledge at the genome scale.

COMPUTATIONAL METHODS FOR THE IDENTIFICATION OF MEMBRANE PROTEINS AND THE PREDICTION OF THEIR STRUCTURES

In this section, computational methods for the identification of membrane proteins and prediction of their structures will be presented. Many methods have been developed over the years for the prediction of membrane protein topology and structure. However, more effort has been invested in studying helical membrane proteins than their beta counterparts, because identifying long and very hydrophobic sequence stretches, such as transmembrane helices, represents an easier task than detecting the shorter and less hydrophobic strands.

Generally speaking, the effectiveness of prediction methods strongly depends on available information and knowledge. Therefore, an overview of the current knowledge about membrane protein structures will first be provided; then, the most well-known computational methods for the study of membrane proteins will be described. Finally, the main online resources fully dedicated to membrane proteins will be summarized.

What is Known About Membrane Protein Sequences and Structures?

Biological membranes provide a discontinuous environment, and therefore the amino acid composition of membrane proteins varies along the membrane axis. For both helical and β-barrel proteins, the stretch of hydrophobic amino acids interacting with the membrane core is flanked by aromatic belts made of Tyr and Trp that generate the so-called "lily-pad" effect (Wallin et al., 1997; Seshadri et al., 1998; Ulmschneider, Sansom, and Di Nola, 2005). In addition, for helical membrane proteins, the basic residues Arg and Lys are about four times more frequent in cytosolic than in periplasmic loops, whereas there is no comparable effect for negatively charged residues (the "positive-inside" rule) (Heijne, 1986). The GxxxG sequence motif, which is thought to facilitate close helix–helix interactions, and other periodic patterns have also been observed in transmembrane sequences with a frequency higher than random (Elofsson and Von Heijne, 2007).

The membrane asymmetry and discontinuity is also responsible for a key feature of membrane proteins, their topology. The topology of a membrane protein describes both the combination of the number of secondary structure elements, for example, the number of helices or strands that cross the membrane, and the location of the N-terminus of the first transmembrane secondary structure element. The cytoplasm side is designated as "in," whereas the opposite side (periplasm, organelle media, etc.) is designated as "out." It is worth noting here that in β-barrels the N-terminus has only been observed on the cytoplasmic side of the membrane ("in").

In terms of 3D structure, due to constraints imposed by the membrane, integral membrane proteins show only two main architectures, β-barrels in the outer membrane and α-helix bundles in all cellular membranes. This is because structure and stability of membrane proteins depend on the thermodynamic cost of transferring proteins, which are very polar or even charged compounds, into the hydrophobic hydrocarbon core of the membrane. To lower this cost, polar backbone groups ($-CO-$ and $-NH-$) must be shielded and are usually involved in hydrogen bonds, forming stable secondary structures, that is, either α-helices or β-strands.

Helical membrane protein structures can be made of long hydrophobic transmembrane helices (15–30 amino acids in length) lying at various tilt angles from the membrane axis; interfacial helices lying parallel to the membrane plane, re-entrant loops, which are loops that penetrate the membrane but do not cross it; and cytoplasmic or periplasmic globular domains. Moreover, helix kinks occur when prolines are found in the sequence.

β-barrel structures are made of an even number of shorter β-strands that lie in an antiparallel orientation and are usually connected by alternating long and short loops. Moreover, they show a typical dyad sequence repeat motif, that is, hydrophobic and hydrophilic residues alternate so that hydrophobic side chains point out toward the membrane and hydrophilic side chains point toward the inside of the protein (Wimley, 2003).

Identification of Membrane Proteins from Sequence

The identification of membrane protein from sequence first involves the detection of sequence fragments that are hydrophobic enough to be transmembrane, leading to a simple discrimination between membrane proteins and water-soluble proteins. However, membrane protein sequences contain more information than just the hydrophobicity and can be used to predict protein secondary structure and topology.

The earliest but still most widely used method to identify helical membrane proteins is the sliding-window analysis of the hydrophobicity (Hϕ) of protein sequences (Rose, 1978; Kyte and Doolittle, 1982), which is based on the assumption that on average transmembrane helices are more hydrophobic than water-soluble regions. The method consists of plotting the average hydrophobicity of a 19 residue segment of a protein sequence against the "central" residue number, so that the most hydrophobic segments of a chain can be detected. The segment length value was optimized by comparing the average hydropathy of a set of transmembrane helices with a set of hydrophobic helices belonging to globular proteins (Kyte and Doolittle, 1982). Obviously, the results of this Hϕ method strongly depend on the hydropathy scales used (Jayasinghe, Hristova, and White, 2001). The most commonly used scales include the Kyte and Doolittle (KD) scale (Kyte and Doolittle, 1982), based on water vapor transfer free energies and interior–exterior distributions of amino acids; the Eisenberg Consensus (EC) scale (Eisenberg et al., 1982), which is based on protein hydrophobic dipole; the semitheoretical Goldman–Engelman–Steitz (GES) scale (Engelman, Steitz, and Goldman, 1986), which combines separate experimental values for the polar and nonpolar characteristics of groups in the amino acid side chains; and the White and Wimley (WW) scale (White and Wimley, 1999), composed of experimentally derived transfer free energies between water and a phosphatidylcholine bilayer interface. The Hϕ plot of bovine rhodopsin calculated using the WW scale is shown in Figure 36.6.

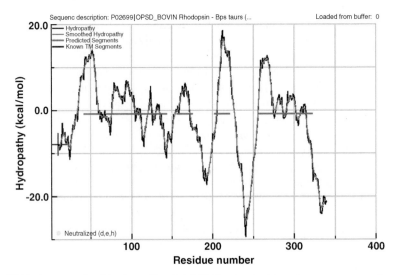

Figure 36.6. An example of hydropathy plot (Hf). The bovin rhosopin sequences analyzed using the WW scale within the Membrane Protein Explorer (MPEx) program (http://blanco.biomol.uci.edu/mpex/). Red lines represent the hydrophobicity threshold that defines protein segments as transmembrane or globular.

A significant amount of new information has become available since the development of the Hφ method. Experimental studies indicate the presence of topogenic signals in helical integral membrane proteins, such as sequence patterns that correlate with the topology of the membrane-spanning segments (see positive-inside rule and lily-pad effect, described in Section "What is Known About Membrane Protein Sequences and Structures"). First and early attempts to integrate topogenic rules into the heart of the secondary prediction process allowed the development of automated methods for the prediction of transmembrane protein secondary structure and topology (Claros and Von Heijne, 1994; Jones, Taylor, and Thornton, 1994). MEMSAT (Jones, Taylor, and Thornton, 1994) was the first method to integrate the prediction of topology with the prediction of transmembrane segments, but since then many other methods have been published, implementing more sophisticated machine learning techniques such as neural networks and hidden-Markov-Models, such as TMHMM (Sonnhammer, Von Heijne, and Krogh, 1998) and HMMTOP (Tusnady and Simon, 2001). A summary of the most well-known methods available online is provided in Table 36.1. A typical machine learning method for the prediction of membrane protein topology and transmembrane segments classifies each amino acid according to four states, cytoplasmic (inside) loop, transmembrane, noncytoplasmic (outside) loop, and globular, and is therefore able to assign transmembrane segments and the location of the N-terminus of the first transmembrane helix. Benchmarking studies published in early 2000s showed that the most effective methods for the prediction of transmembrane helices had accuracies in the 55–60% range for eukaryotic proteins (Chen, Kernytsky, and Rost, 2002; Cuthbertson, Doyle, and Sansom, 2005). These results proved that despite the number of new methods, there was remarkably little progress in terms of the prediction accuracy and suggested that the limitation in prediction accuracy was not due to the algorithm being employed but due to the lack of new biological information. In fact, prediction methods analyzed in the studies rely only on two main basic features, the hydrophobicity of transmembrane helices and the positive-inside rule.

With the advent of the genomic era, large-scale computational analyses have been performed for membrane proteins (Lehnert et al., 2004) and a recent benchmarking study showed that methods that include evolutionary information, such as the newest versions of Phobius (Kall, Krogh, and Sonnhammer, 2007) and MEMSAT, are able to predict the correct topology for more than 70% of the membrane proteins in the benchmark set (Jones, 2007). Moreover, global experimental data of the topology of the *E. coli* and *Saccharomyces cerevisiae* inner membrane proteomes have recently been published (Daley et al., 2005; Kim et al., 2006). The inclusion of experimental information significantly improves the prediction of membrane protein topology, leading to correct assignments for about 80% of the proteins (Kim, Melen, and Von Heijne, 2003a; Bernsel and Von Heijne, 2005; Daley et al., 2005; Kim et al., 2006).

The topology prediction for β-barrel proteins is more complicated. Although many structural "rules" are known, the discrimination of β-barrel membrane proteins from water-soluble proteins is more difficult because hydrophobic amino acids are hidden in the dyad repeat and there are fewer apolar amino acids in a β-strand than in an α-helix due to their intrinsic geometry (the average interresidue distance for α-helices and β-strands is 1.5 Å and 2.7 Å, respectively) (Wimley, 2003). The main methods for predicting the topology of β-barrel membrane proteins are listed in Table 36.1. A recent benchmarking evaluation (Bagos, Liakopoulos, and Hamodrakas, 2005) showed HMM-based methods, such as ProfTMB, to be the best available predictors, claiming accuracy higher than 80%.

TABLE 36.1. Available Resources for the Prediction of Membrane Protein Topology

α–Helical Membrane Proteins

Resource	Web Site	Information Used for Prediction
TopPred2 (Claros and Von Heijne, 1994)	http://www.genomics.purdue.edu/Pise/5.a/toppred-simple.html	Hydrophobicity
MEMSAT3 (Jones, 2007)	http://www.psipred.net	aa statistical preferences and evolutionary information
TMAP (Cuthbertson, Doyle, and Sansom, 2005)	http://bioweb.pasteur.fr/seqanal/interfaces/tmap.html	aa statistical preferences and evolutionary information
PHDhtm (Rost, Yachdav, and Liu, 2004)	http://www.predictprotein.org	ANN and evolutionary information
TMpred (Hofmann and Stoffel, 1993)	http://www.ch.embnet.org/software/TMPRED_form.html	aa statistical preferences
TMHMM/PHOBIUS (Krogh et al., 2001; Kall, Krogh, and Sonnhammer, 2007)	http://www.cbs.dtu.dk/services/TMHMM/	Various, HMM
HMMTOP (Tusnady and Simon, 2001)	http://www.enzim.hu/hmmtop/	Various, HMM
OrientTM (Liakopoulos, Pasquier, and Hamodrakas, 2001)	http://athina.biol.uoa.gr/orienTM/	Position-specific statistical information
SPLIT4 (Juretic, Zoranic, and Zucic, 2002)	http://split.pmfst.hr/split/4/	aa statistical preferences
THUMBUP (Zhou and Zhou, 2003)	http://sparks.informatics.iupui.edu/Softwares-Services_files/thumbup.htm	aa mean burial propensities
PONGO (Amico et al., 2006)	http://pongo.biocomp.unibo.it/pongo/	Metaserver
β-Barrel Membrane Proteins		
PROFtmb (Bigelow and Rost, 2006)	http://cubic.bioc.columbia.edu/services/proftmb/query.php	HMM
ConBBPRED (Bagos, Liakopoulos, and Hamodrakas, 2005)	http://biophysics.biol.uoa.gr/ConBBPRED	Metaserver
PRED-TMBB (Bagos et al., 2004b)	http://biophysics.biol.uoa.gr/PRED-TMBB/	HMM
Both α-Helical and β-Barrel Membrane Proteins		
MINNOU (Cao et al., 2006)	http://minnou.cchmc.org/	Prediction-based structural profiles

Finally, a few methods have been published aiming at the identification of all β-barrel membrane proteins in whole genomes from the analysis of sequence composition, secondary structure prediction, and architecture of known structures (Wimley, 2002; Liu et al., 2003; Bagos et al., 2004a; Bagos et al., 2004b; Berven et al., 2004; Gromiha, 2005; Park et al., 2005). Some of these methods are available as online resources (see Section "Web-Available Data Resources for Membrane Proteins" and Table 36.1).

Identification of Membrane Proteins Using 3D Structural Data

Despite the structural constraints imposed by the membrane, distinguishing membrane proteins from water-soluble proteins using their 3D structures can still be difficult, since they share more similarities than differences. The cores of both membrane and globular proteins show a comparable hydrophobicity and the surface of water-soluble proteins often includes hydrophobic atoms and can also exhibit hydrophobic patches if protein–protein interactions occur. Likewise, transmembrane protein segments that mutually interact or are involved in enzymatic activity or transport can contain polar residues. Moreover, in oligomers, the interfaces between multichain water-soluble complexes are mostly hydrophobic, while they are likely to be polar in transmembrane oligomers and this jeopardizes the analysis of single monomers extracted from oligomeric structures. Therefore, methods for the identification of membrane protein using structures, rather than relying only on 3D structural differences with water-soluble protein structures, exploit the detection of the optimal position of the protein relative to the membrane bilayer.

The position of a protein in a lipid bilayer can be defined by four simple parameters (see Figure 36.7): (1) The location of the center of mass of the protein along the membrane axis (d), that is, the distance of its projection on the bilayer axis from the center of the bilayer; (2) the angles of rotation (ϕ) and (3) tilt (τ) of the main protein axis relative to the orientation of the membrane axis; and (4) the thickness of the portion of the protein that spans and interact with the bilayer (D). To predict the optimal location of a protein in the membrane, the energy for each position needs to be evaluated by a specific function. Two automated algorithms are currently available to locate the position of the protein structure

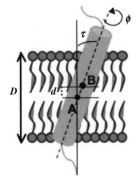

Figure 36.7. Diagram of the parameters used to describe the position of a protein with respect to the membrane. d distance of the protein center of mass (B) from the center of mass of the bilayer (A); ϕ: angle of rotation of the protein axis relative to the orientation of the membrane axis; τ: angle of tilt of the protein axis relative to the orientation of the membrane axis; and D: bilayer thickness.

relative to the membrane bilayer and to detect putative membrane protein structures. They are implemented in the PDB_TM and OPM membrane protein databases (Tusnady, Dosztanyi, and Simon, 2004;Tusnady, Dosztanyi, and Simon, 2005; Lomize et al., 2006a; Lomize et al., 2006b). PDB_TM uses the TMDET algorithm, which scans protein 3D coordinates, builds the possible biological complex, analyzes the internal symmetry of the structure, calculates the membrane-exposed area, and searches for the best membrane plane. Different positions of proteins in the membrane are evaluated by applying a specific function, which depends on a hydrophobic factor, that is, the relative hydrophobic membrane-exposed surface, and a structure factor mainly determined by the position of the protein axis relative to the membrane axis. The protein is cut in 1 Å thick slices along the membrane axis, and the function is calculated for each slice as the product of these two factors. If a 15 slice portion (corresponding to a putative 15 Å membrane core) along a given membrane axis is found with a function value above a predefined threshold, the axis is accepted as the best membrane axis for that protein and the structure is classified as membrane protein. Otherwise, it is classified as globular protein. If there are many orientations that score above the threshold, the highest scoring orientation is adopted.

The second approach (OPM) calculates the optimal spatial arrangement by minimizing the protein transfer energy from water to the membrane. The membrane is represented as a model in which the bilayer core is a planar hydrophobic slab of adjustable thickness and interfaces are defined by a polarity profile that takes into account the effective water concentration that varies gradually along the bilayer normally between the polar headgroups and the apolar core. The energy function is based on solvation parameters calculated from water-decadiene partition of uncharged amino acids and neglects electrostatic interactions and contributions of amino acids that are part of the protein surface but line hydrophilic pores and channels.

Identification of Structural Features from Sequence

As mentioned in Section "What is Known About Membrane Protein Sequences and Structures" several recently determined structures of helical membrane proteins showed that helix bundles, rather than being made simply of transmembrane helices perpendicularly penetrating the bilayer, also have irregular structural features, such as interfacial helices, re-entrant loops, and proline-induced kinks. In addition, it is clear that the degree of amino acid exposure to lipid and residue distances from the middle of the membrane are very important for the interaction of proteins with the membrane and their function. Consequently, a new and more subtle definition of topology as a two-and-a-half dimension (2.5D) structure has been proposed by von Heijne and Eloffson (Elofsson and Von Heijne, 2007).

Methods to predict 2.5D structural features from sequence have been recently developed. TOP-MOD is a novel HMM-based predictor of re-entrant loops (Viklund, Granseth, and Elofsson, 2006). ZPRED is a method to predict the Z-coordinates of amino acids along the membrane axis, that is, their distance from the middle of the membrane, based on HMM and neural network techniques that also include evolutionary information (Granseth, Viklund, and Elofsson, 2006). The degree of lipid exposure for a residue can be predicted by LIPS, which is available as a Web server. It is based on a canonical model of coiled coil helices where the surface of each helix is partitioned into seven surface patches (faces) that can interact with lipids or other helices (Adamian and Liang, 2006). Finally, proline-induced helix kinks can be confidently predicted from the degree of conservation of proline in multiple alignments (Yohannan et al., 2004a; Yohannan et al., 2004b).

2.5D predictions obviously represent a useful step forward toward more reliable predictions of membrane protein 3D structures and will also be helpful in the process of structural classification to discriminate between protein families with the same overall topology.

Prediction of Membrane Protein 3D Structure

Compared to the prediction of the tertiary structure of water-soluble proteins, the prediction of the structure of membrane proteins is, in principle, an easier problem, since the constraint provided by the lipid bilayer to align transmembrane structure elements reduces the complexity of the problem. Despite this relative simplicity, little attention has been focused on the prediction of membrane protein structure. This neglect may reflect the limited variety of anticipated forms or more probably a lack of experimental data from which to derive rules to assess reliability.

As for the prediction of 3D structures of globular proteins, homology modeling, fold recognition, or *de novo* modeling methods should be used for membrane proteins as dictated by the available experimental information. These three techniques have been extensively described in Chapters 25–27. However, their application to membrane proteins needs particular care because of the physicochemical properties and constraints of the membrane.

None of the commonly used homology modeling programs has been explicitly modified for membrane proteins. However, the reliability of the application of standard homology modeling techniques, developed for globular proteins, to membrane proteins has recently been evaluated on a benchmark data set composed of homologous membrane protein structures (Forrest, Tang, and Honig, 2006). In particular, three issues have been addressed: the accuracy of amino acid substitution matrices used for the sequence alignment, the effectiveness of secondary structure prediction, and the relationship between sequence identity and structure similarity.

Results show that the construction of homology models of membrane proteins and globular proteins follows similar general rules and comparable trends are observed with respect to sequence–structure relationship (Chothia and Lesk, 1986; Flores et al., 1993) and alignment accuracy (Tang et al., 2003). Moreover, among the many modeling programs available, SegMOD/ENCAD (Levitt, 1992) and Modeller 7v7 (Sali and Blundell, 1993) have been showed to produce the highest quality models in terms of stereochemistry, whereas NEST (Petrey et al., 2003) provides models that are more similar to the target native structure. Among the existing loop prediction programs, PLOP (Xiang, Soto, and Honig, 2002; Jacobson et al., 2004) is reported as the most accurate, whereas Loopy (Xiang, Soto, and Honig, 2002) as the fastest (Punta et al., 2007).

The *ab initio/de novo* modeling of membrane proteins is unfortunately far more complicated. The *ab initio* modeling of β-barrel proteins has not been reported, whereas for helical proteins some progress has been achieved. In the *ab initio* prediction, a very important role is played by the prediction of the interactions between pairs of transmembrane helices, which lead to the final protein fold. Interhelical interactions are particularly difficult to identify since they occur within protein chains but also between different oligomers in multimers. Several interhelical interaction studies have been published (Senes, Gerstein, and Engelman, 2000; Fleishman and Ben-Tal, 2002; Degrado, Gratkowski, and Lear, 2003; Liang et al., 2003), mostly aiming at the detection of key motifs that recur in transmembrane helix interfaces, although no Web servers for the prediction of these interactions are currently available.

Several methods for the *de novo* modeling of integral membrane proteins have been developed, although most of them have been tested on very few structures due to the paucity of the available experimental data. Known methods implement different approaches, such as potential smoothing and search algorithms (Pappu, Marshall, and Ponder, 1999), fragment–assembly approaches combined with knowledge-based potentials (Pellegrini-Calace, Carotti, and Jones, 2003; Hurwitz, Pellegrini-Calace, and Jones, 2006; Yarov-Yarovoy, Schonbrun, and Baker, 2006), and oligomer symmetry combined with Monte Carlo searches (Kim, Chamberlain, and Bowie, 2003b). However, they only work for proteins with less than 7 transmembrane helices and, to our knowledge, are not available as online tools. The only Web-based resource available is SWISS-MODEL 7TM (http://swissmodel.ncifcrf.gov/cgi-bin/sm-gpcr.cgi) (Arnold et al., 2006), which lies on the boundary between homology modeling and *de novo* modeling. It was designed to extend the homology modeling of 7 transmembrane helical proteins by allowing the choice of "canonical" templates, that is, templates with clear sequence similarity with the protein target, and "computational" templates, which are group-specific templates (GSTs) generated from low-resolution structural data for a number of selected 7TM receptors for which sufficient relevant mutagenesis data are available.

WEB-AVAILABLE DATA RESOURCES FOR MEMBRANE PROTEINS

Currently there are several Web resources, as summarized in Table 36.2, which are dedicated to membrane proteins and include a significant amount of information, from sequence to 3D structures, crystallization details, and relationships with diseases.

Web Collections of Membrane Protein Structures

Five collections of membrane protein structures are available online. The Membrane Protein Data Bank (MPDB, http://www.mpdb.ul.ie/) (Raman, Cherezov, and Caffrey, 2006) is a relational database of structural and functional information on integral, anchored, and peripheral membrane proteins based on the PDB. Each MPDB entry includes protein name, size, function, and a number of features, such as the family designation according to the Transporter Classification (TC) system (see Section "Classification of Membrane Proteins") (Saier, Tran, and Barabote, 2006), the family annotation as in PFAM, the secondary structure of the transmembrane domain, as observed in the crystal structure, and some experimental details.

The best known data bank "Membrane Proteins of Known 3D Structure" from Stephen White (http://blanco.biomol.uci.edu/Membrane_Proteins_xtal.html) is the most comprehensive and provides useful information about integral membrane proteins, whose 3D structures have been determined to a resolution sufficient to identify transmembrane helices of helix-bundle membrane proteins (4–4.5 Å). In its July 2007 writing, it includes 259 entries of which 125 are "unique" structures that are taken as the representative of the set of proteins of the same "type" but from different species. The "Membrane Proteins of Known Structure" (http://www.mpibp-frankfurt.mpg.de/michel/public/memprotstruct.html) from Hartmut Michel is a list of membrane proteins with known structure and also includes crystallization conditions and key references for structure determination. However, at the time of writing, the resource has not been updated for more than one year. The recent PDB_TM (http://pdbtm.enzim.hu/) (Tusnady, Dosztanyi, and Simon, 2005) contains structures of all known transmembrane proteins and fragments along with their most likely position relative to the membrane planes as calculated by the TMDET algorithm, which uses a combination of hydrophobicity scores and

TABLE 36.2. Available Resources for Membrane Protein Identification and Classification

Resource	Sequence, Structure, and Function Information Sources	
	Web Site	Information Type Included
MPDB (Raman, Cherezov, and Caffrey, 2006)	http://www.mpdb.ul.ie/	3D structure and function, all membrane proteins
Membrane protein of known 3D structures	http://blanco.biomol.uci.edu/Membrane_Proteins_xtal.html	Structure and function, all membrane proteins
Membrane protein of known structures	http://www.mpibp-frankfurt.mpg.de/michel/public/memprotstruct.html	Structure and experimental conditions, all membrane proteins
PDB_TM (Tusnady, Dosztanyi, and Simon, 2005)	http://pdbtm.enzim.hu/	Structure and membrane location, all membrane proteins
OPM (Lomize et al., 2006b)	http://opm.phar.umich.edu	Structure and membrane location, all membrane proteins
GPCRDB (Horn et al., 2003)	http://www.gpcr.org/7tm/index.html	Fully comprehensive, GPCR family only
LGIC (Donizelli, Djite, and Le Novere, 2006)	http://www.ebi.ac.uk/compneur-srv/LGICdb/LGICdb.php	Sequence and structure, ligand-gated ion channels only
TCDB (Saier, Tran, and Barabote, 2006)	http://www.tcdb.org/	Sequence and structure transporters only
PRNDS (Katta et al., 2004)	http://bicmku.in:8081/PRNDS/	Sequence-based information, porins only
Automated Server for Identification and Phylogenetic Classification of Membrane Proteins		
GPCR subfamily (Karchin, Karplus, and Haussler, 2002) classifier	http://www.soe.ucsc.edu/research/compbio/gpcr-subclass/	SVM
GPCRpred (Bhasin and Raghava, 2004)	http://www.imtech.res.in/raghava/gpcrpred/	SVM
PRED-GPCR (Papasaikas et al., 2004)	http://athina.biol.uoa.gr/bioinformatics/PRED-GPCR/	HMM
SOSUI (Hirokawa, Boon-Chieng, and Mitaku, 1998)	http://bp.nuap.nagoya-u.ac.jp/sosui/	Physicochemical parameters (hydropathy, amphiphilicity, charges, and sequence length)
PRED-TMBB (Bagos et al., 2004b)	http://biophysics.biol.uoa.gr/PRED-TMBB/	HMM
TMBETA-GENOME (Gromiha et al., 2007)	http://tmbeta-genome.cbrc.jp/annotation/home.html	SVM
BOMP (Berven et al., 2004)	http://www.bioinfo.no/tools/bomp	Pattern matching and aa preferences

structural features (see Section "Identification of Membrane Proteins Using 3D Structural Data"). The PDB_TM database is organized similarly to the PDB and the properties described in each entry include the TMDET score on which the selection of the position in the membrane was made, the type of protein according to the secondary structure of its membrane-spanning residues determined by DSSP (α-helical, β-barrel, or unstructured), further information derived from keyword searches in the corresponding UniProt and PDB files, the definition of the membrane planes, and the localization of each transmembrane sequence portion.

The Orientations of Proteins in Membranes database (OPM, http://opm.phar.umich.edu/) (Lomize et al., 2006b) provides a collection of membrane proteins from the PDB whose spatial arrangements in the lipid bilayer are calculated theoretically and compared with experimental data. Like PDB_TM, OPM provides a list of transmembrane proteins with their hydrophobic boundaries. Entries with the most complete quaternary structure or determined with the highest resolution are selected as representative models, whereas the other structures, such as mutants or different conformational states of the same protein, are included as "related PDB entries." Complexes with many unassigned residues, such as low-resolution electron microscopy models, incomplete or nonfunctional assemblies (i.e., fragments), monomeric units of transmembrane functional multimers, NMR models derived from orientational constraints, such as ionophores, and theoretical models are excluded from the database.

Other Web Resources for Membrane Proteins

A few other Web-based resources are available that describe and collect membrane protein features, which are more comprehensive but more family specific (see Table 36.2). The GPCR database (GPCRDB (Horn et al., 2003)) is a comprehensive information system for G-protein-coupled receptors that provides much information, from cDNA and amino acid sequences to ligand binding constants and three-dimensional experimental and theoretical models. The Ligand-Gated Ion Channels database (LGIC) (Donizelli, Djite, and Le Novere, 2006) lists the nucleic acid and protein sequences of the extracellularly activated ligand-gated ion channels and provides multiple sequence alignments, some superfamily phylogenetic studies and available atomic coordinates of subunits. The Transporter Classification database (TCDB) (Saier, Tran, and Barabote, 2006) details a comprehensive classification system for membrane transport proteins and provides descriptions, TC numbers, and examples of over 360 families of transport proteins. The PoRiN Database Server (PRNDS) (Katta et al., 2004) collects data obtained by a statistically validated sequence-based search from UniProt and TrEMBL.

Finally, the MPtopo database collects membrane protein topologies that have been experimentally verified. It is therefore a very relevant and powerful resource because the reliability of predicted transmembrane sequence assignments for membrane proteins in the standard sequence databases has a significant intrinsic uncertainty.

CLASSIFICATION OF MEMBRANE PROTEINS

Classification of Membrane Protein Structures

The classification of membrane protein structures is difficult and still not completely resolved. In fact, to our knowledge, no current database classifies the available membrane protein structures into "homologous" structural families. Recent attempts to classify membrane protein structures systematically into families and superfamilies are available in the CATH (Pearl et al., 2005) and SCOP (Murzin et al., 1995) databases, both going beyond the simple division into α-helical and β-barrel proteins. However, the classification process is difficult, with many structures falling into singleton fold classes, especially channels and transporters, which are oligomeric transmembrane α-helical proteins with a variable number of transmembrane helices and/or oligomerization states, which defy simple clustering into structural families. As pointed out in a very recent review, a major obstacle to the development of effective methods to identify and classify distantly related membrane proteins is the lack of a structure-based "gold standard," provided by SCOP and CATH for globular proteins (Elofsson and Von Heijne, 2007).

Most of the Web resources dedicated to membrane protein structure described in Section "Why is It Difficult to Solve Membrane Protein 3D Structures?" provide a partial or full classification of their entries, although none is based on 3D structural features. Proteins are usually classified according to a variety of features, including the position in the membrane (peripheral, if only the membrane surface is penetrated; monotopic, if proteins sink into the core but do not cross the membrane; bitopic, if proteins span the membrane only once; polytopic, if proteins span the membrane with more than one transmembrane segment), the number of transmembrane segments, the biological functions, the species, or the type of membrane.

Alternatively, membrane protein structures are classified according to membrane protein sequence subfamilies in various other resources. The GPCRDB classifies receptors in subfamilies using a chemical definition, that is, according to which ligand they bind. The LGIC database defines three superfamilies of channels according to their subunit structure: the receptors of the cys-loop families, such as nicotinic receptor and serotonin-activated anionic channels, made of five homologous subunits each with four transmembrane segments; the ATP-gated channels, made of three homologous subunits, each with two transmembrane segments; and lastly the glutamate-activated cationic channels, such as NMDA or AMPA receptors, that are made of four homologous subunits with three nonhomologous transmembrane segments. In the MPtopo database, entries are divided into three data sets, according to the source of the information used (i.e., 3D_Helix set—helix-bundle membrane proteins whose 3D structures have been solved and transmembrane segments are known precisely; 1D_Helix set—helix-bundle membrane proteins whose transmembrane helices have been identified by experimental techniques such as gene fusion, proteolytic degradation, and amino acid deletion; 3D_Other set—contains β-barrel and monotopic membrane proteins whose three-dimensional structures have been determined crystallographically). Finally, the TCDB provides an IUBMB classification scheme called the Transporter Classification system. The TC system is analogous to the Enzyme Commission (EC) system for classification of enzymes but also incorporates phylogenetic information, structural information, and even relationships with diseases. Transport systems are classified on the basis of five criteria, and each criterion corresponds to one of the five fields constituting the TC number for a particular type of transporter (e.g., A,B,C,D, and E). "A" is usually a number and corresponds to the transporter class (i.e., channel, carrier, etc); "B" is a letter and corresponds to the transporter subclass; "C" is a number and corresponds to the transporter family or superfamily; "D" is a number, corresponding to the subfamily in which a transporter is found, and finally the "E" corresponds to the substrate or range of substrates transported. Any two transport proteins in the same subfamily of a transporter family that transport the same substrate are given the same TC number.

Other Types of Classification of Integral Membrane Proteins

GPCRs represent the most explored class of membrane proteins and several studies involving their analysis, identification, and classification from sequence have been published, with corresponding programs often freely available for academic use (Inoue, Ikeda, and Shimizu, 2004; Filizola and Weinstein, 2005; Gao and Wang, 2006). Moreover, three Web-based resources for GPCR detection and family assignments are currently available. They employ pattern recognition machine learning approaches, that is, hidden-Markov-Models and support vector machines (SVMs) and are listed in Table 36.2. The GPCR subfamily classifier (Karchin, Karplus, and Haussler, 2002) is an SVM method developed to determine GPCR function from sequence and classify entries according to the subfamily to which they are predicted to belong. GPCRpred (Bhasin and Raghava, 2004) is SVM based as well and predicts protein families and subfamilies from the sequence dipeptide composition. Data sets used for the training of both GPCR Subfamily Classifier and GPCRpred were derived from the GPCRDB information system, which organizes GPCRs into a hierarchy of classes, class subfamilies, and types, according to the ligand they bind. PRED-GPCR (Papasaikas et al., 2004) is a program for the recognition and classification of GPCR members at the family level. It is HMM based and includes profiles for 67 well-characterized GPCR families. The employed classification system is based on the TiPs pharmacological receptor classification (Alexander et al., 1999) and on GPCRDB. In addition, PRINTS (Mulder et al., 2007), which is a compendium of protein fingerprints (conserved sequence motifs) used to characterize protein families, has proven to be an effective classifier for GPCRs.

Finally, one Web-based resource, called SOSUI, is available for the classification and secondary structure prediction of helical membrane proteins (Hirokawa, Boon-Chieng, and Mitaku, 1998). However, in this case the classification only provides a discrimination of membrane proteins from water-soluble proteins.

A few studies have been published to identify all β-barrel membrane proteins in a genome, based on the analysis of sequence composition, secondary structure prediction, and architecture of known structures (Liu et al., 2003; Bagos et al., 2004a; Bagos et al., 2004b; Berven et al., 2004; Gromiha, 2005; Park et al., 2005). In particular, BOMP, PRED-TMBB, and TMBETA-GENOME are available as online resources (Table 36.2). BOMP uses a simple pattern matching associated to an amino acid likelihood score. PRED-TMBB and TMBETA-GENOME implement machine learning techniques, HMM and SVM, respectively, and are based on sequence and secondary structure information. Moreover, TMBETA-GENOME includes a collection of the annotated β-barrel membrane proteins for all the completed genomes.

CONCLUSIONS

Membrane proteins represent a very important class of proteins and are responsible for maintaining the homeostasis and responsiveness of cells by mediating a wide range of fundamental biological processes. They are found in biological membranes, which are composed of phospholipid bilayers impermeable to polar molecules and inhabit a physico-chemical environment very different from polar media like the cytoplasm and periplasm. Membrane proteins are highly insoluble and unstable in aqueous solutions, so that severe experimental problems are associated with their study. Consequently, the amount of available structural data is still scarce, although the picture recently being improving, with

775 membrane protein structures corresponding to 125 different proteins currently available in the PDB.

The recent determination of several new structures has allowed some progress to be achieved in the development of bioinformatics methods for the prediction of membrane protein 2D and 3D structures. Current bioinformatics tools mainly focus on the identification of transmembrane proteins and the prediction of their 2D structure and topology (see Table 36.1) and are based on machine learning techniques. In particular, HMMs have proven to be the best performing methods for the prediction of both helical and β-barrel protein topology with accuracy higher than 70%. Moreover, new experimental topology data have been shown to increase the prediction accuracy by about 20% but its use is still limited due the paucity of available information. Very few methods, mainly based on *ab initio/de novo* approaches, have been developed specifically for the prediction of membrane protein 3D structures. However, methods for the prediction of membrane protein 3D structural features, such as re-entrant loops and proline-induced helix kinks, have been recently developed (2.5D structure prediction), allowing a step forward toward more reliable predictions of membrane protein 3D structures.

Currently available Web-based resources for membrane proteins, as summarized in Table 36.2, include lists and collections of data, from cDNA sequences and protein sequences to 3D structures, and sequence- and/or function-based classification schemes. They are often family-centric, and to our knowledge, a fully comprehensive resource including the most important features for all membrane protein families is not yet available. However, many new structures are expected in the next few years, in part due to the efforts of the Structural Genomics projects dedicated to membrane proteins. It is our belief that the new experimental information will improve existing methods and will encourage the development of novel prediction methods and automated and complete resources for a full computational analysis and classification of membrane protein structures.

REFERENCES

Adamian L, Liang J (2006): Prediction of transmembrane helix orientation in polytopic membrane proteins. *BMC Struct Biol* 6:13.

Alexander S, Peters J, Mead A, Lewis S (1999): TiPS receptor and ion channel nomenclature supplement 1999. *Trends Pharmacol Sci* 19.

Amico M, Finelli M, Rossi I, Zauli A, Elofsson A, Viklund H, Von Heijne G, Jones D, Krogh A, Fariselli P, Luigi Martelli P, Casadio R (2006): PONGO: a Web server for multiple predictions of all-alpha transmembrane proteins. *Nucleic Acids Res* 34:W169–W172.

Arnold K, Bordoli L, Kopp J, Schwede T (2006): The SWISS-MODEL workspace: a Web-based environment for protein structure homology modelling. *Bioinformatics* 22:195–201.

Bagos PG, Liakopoulos TD, Spyropoulos IC, Hamodrakas SJ (2004a): A hidden Markov model method, capable of predicting and discriminating beta-barrel outer membrane proteins. *BMC Bioinformatics* 5:29.

Bagos PG, Liakopoulos TD, Spyropoulos IC, Hamodrakas SJ (2004b): PRED-TMBB: a Web server for predicting the topology of beta-barrel outer membrane proteins. *Nucleic Acids Res* 32: W400–W404.

Bagos PG, Liakopoulos TD, Hamodrakas SJ (2005): Evaluation of methods for predicting the topology of beta-barrel outer membrane proteins and a consensus prediction method. *BMC Bioinformatics* 6:7.

Bernsel A, Von Heijne G (2005): Improved membrane protein topology prediction by domain assignments. *Protein Sci* 14:1723–1728.

Bernstein HD, (2000): The biogenesis and assembly of bacterial membrane proteins. *Curr Opin Microbiol* 3:203–209.

Berven FS, Flikka K, Jensen HB, Eidhammer I (2004): BOMP: a program to predict integral beta-barrel outer membrane proteins encoded within genomes of Gram-negative bacteria. *Nucleic Acids Res* 32:W394–W399.

Bhasin M, Raghava GP (2004): GPCRpred: an SVM-based method for prediction of families and subfamilies of G-protein coupled receptors. *Nucleic Acids Res* 32:W383–W389.

Bigelow H, Rost B (2006): PROFtmb: a Web server for predicting bacterial transmembrane beta barrel proteins. *Nucleic Acids Res* 34:W186–W188.

Booth PJ, Curnow P (2006): Membrane proteins shape up: understanding *in vitro* folding. *Curr Opin Struct Biol* 16:480–488.

Cao B, Porollo A, Adamczak R, Jarrell M, Meller J (2006): Enhanced recognition of protein transmembrane domains with prediction-based structural profiles. *Bioinformatics* 22:303–309.

Chen CP, Kernytsky A, Rost B (2002): Transmembrane helix predictions revisited. *Protein Sci* 11:2774–2791.

Chothia C, Lesk AM (1986): The relation between the divergence of sequence and structure in proteins. *EMBO J* 5:823–826.

Claros MG, Von Heijne G (1994): TopPred II: an improved software for membrane protein structure predictions. *Comput Appl Biosci* 10:685–686.

Cuthbertson JM, Doyle DA, Sansom MS (2005): Transmembrane helix prediction: a comparative evaluation and analysis. *Protein Eng Des Sel* 18:295–308.

Daley DO, Rapp M, Granseth E, Melen K, Drew D, Von Heijne G (2005): Global topology analysis of the *Escherichia coli* inner membrane proteome. *Science* 308:1321–1323.

Degrado WF, Gratkowski H, Lear JD (2003): How do helix–helix interactions help determine the folds of membrane proteins? Perspectives from the study of homo-oligomeric helical bundles. *Protein Sci* 12:647–665.

Donizelli M, Djite MA, Le Novere N (2006): LGICdb: a manually curated sequence database after the genomes. *Nucleic Acids Res* 34:D267–D269.

Eisenberg D, Weiss RM, Terwilliger TC, Wilcox W (1982): Hydrophobic moments and protein structure. *Faraday Symp Chem Soc* 17:109–120.

Elofsson A, Von Heijne G (2007): Membrane protein structure: prediction vs reality. *Annu Rev Biochem* 76:125–140.

Engelman DM, Steitz TA, Goldman A (1986): Identifying nonpolar transbilayer helices in amino acid sequences of membrane proteins. *Annu Rev Biophys Biophys Chem* 15:321–353.

Filizola M, Weinstein H (2005): The study of G-protein coupled receptor oligomerization with computational modeling and bioinformatics. *FEBS J* 272:2926–2938.

Fleishman SJ, Ben-Tal N (2002): A novel scoring function for predicting the conformations of tightly packed pairs of transmembrane alpha-helices. *J Mol Biol* 321:363–378.

Flores TP, Orengo CA, Moss DS, Thornton JM (1993): Comparison of conformational characteristics in structurally similar protein pairs. *Protein Sci* 2:1811–1826.

Forrest LR, Tang CL, Honig B (2006): On the accuracy of homology modeling and sequence alignment methods applied to membrane proteins. *Biophys J* 91:508–517.

Gao QB, Wang ZZ (2006): Classification of G-protein coupled receptors at four levels. *Protein Eng Des Sel* 19:511–516.

Granseth E, Viklund H, Elofsson A (2006): ZPRED: predicting the distance to the membrane center for residues in alpha-helical membrane proteins. *Bioinformatics* 22:e191–e196.

Grisshammer R, White JF, Trinh LB, Shiloach J (2005): Large-scale expression and purification of a G-protein-coupled receptor for structure determination: an overview. *J Struct Funct Genomics* 6:159–163.

Gromiha MM, (2005): Motifs in outer membrane protein sequences: applications for discrimination. *Biophys Chem* 117:65–71.

Gromiha MM, Yabuki Y, Kundu S, Suharnan S, Suwa M (2007): TMBETA-GENOME: database for annotated beta-barrel membrane proteins in genomic sequences. *Nucleic Acids Res* 35: D314–D316.

Heijne GV, (1986): The distribution of positively charged residues in bacterial inner membrane proteins correlates with the transmembrane topology. *EMBO J* 5:3021–3027.

Hirokawa T, Boon-Chieng S, Mitaku S (1998): SOSUI: classification and secondary structure prediction system for membrane proteins. *Bioinformatics* 14:378–379.

Hofmann K, Stoffel W (1993): TMbase: a database of membrane spanning proteins segments. *Biol Chem Hoppe-Seyler* 374:166.

Horn F, Bettler E, Oliveira L, Campagne F, Cohen FE, Vriend G (2003): GPCRDB information system for G protein-coupled receptors. *Nucleic Acids Res* 31:294–297.

Hurwitz N, Pellegrini-Calace M, Jones DT (2006): Towards genome-scale structure prediction for transmembrane proteins. *Philos Trans R Soc Lond B Biol Sci* 361:465–475.

Inoue Y, Ikeda M, Shimizu T (2004): Proteome-wide classification and identification of mammalian-type GPCRs by binary topology pattern. *Comput Biol Chem* 28:39–49.

Jacobson MP, Pincus DL, Rapp CS, Day TJ, Honig B, Shaw DE, Friesner RA (2004): A hierarchical approach to all-atom protein loop prediction. *Proteins* 55:351–367.

Jayasinghe S, Hristova K, White SH (2001): Energetics, stability, and prediction of transmembrane helices. *J Mol Biol* 312:927–934.

Jones DT, Taylor WR, Thornton JM (1994): A model recognition approach to the prediction of all-helical membrane protein structure and topology. *Biochemistry* 33:3038–3049.

Jones DT, (1998): Do transmembrane protein superfolds exist? *FEBS Lett* 423:281–285.

Jones DT, (2007): Improving the accuracy of transmembrane protein topology prediction using evolutionary information. *Bioinformatics* 23:538–544.

Juretic D, Zoranic L, Zucic D (2002): Basic charge clusters and predictions of membrane protein topology. *J Chem Inf Comput Sci* 42:620–632.

Kall L, Krogh A, Sonnhammer EL (2007): Advantages of combined transmembrane topology and signal peptide prediction: the Phobius Web server. *Nucleic Acids Res* 35:W429–W432.

Karchin R, Karplus K, Haussler D (2002): Classifying G-protein coupled receptors with support vector machines. *Bioinformatics* 18:147–159.

Katta AV, Marikkannu R, Basaiawmoit RV, Krishnaswamy S (2004): Consensus based validation of membrane porins. *In Silico Biol* 4:549–561.

Kim H, Melen K, Von Heijne G (2003a): Topology models for 37 *Saccharomyces cerevisiae* membrane proteins based on C-terminal reporter fusions and predictions. *J Biol Chem* 278:10208–10213.

Kim S, Chamberlain AK, Bowie JU (2003b): A simple method for modeling transmembrane helix oligomers. *J Mol Biol* 329:831–840.

Kim H, Melen K, Osterberg M, Von Heijne G (2006): A global topology map of the *Saccharomyces cerevisiae* membrane proteome. *Proc Natl Acad Sci USA* 103:11142–11147.

Krogh A, Larsson B, von Heijne G, Sonhammer EL (2001): Predicting transmembrane protein topology with a hidden Markov model: application to complete genomes. *J Mol Biol* 305:567–580

Kyte J, Doolittle RF (1982): A simple method for displaying the hydropathic character of a protein. *J Mol Biol* 157:105–132.

Lehnert U, Xia Y, Royce TE, Goh CS, Liu Y, Senes A, Yu H, Zhang ZL, Engelman DM, Gerstein M (2004): Computational analysis of membrane proteins: genomic occurrence, structure prediction and helix interactions. *Q Rev Biophys* 37:121–146.

Levitt M (1992): Accurate modeling of protein conformation by automatic segment matching. *J Mol Biol* 226:507–533.

Liakopoulos TD, Pasquier C, Hamodrakas SJ (2001): A novel tool for the prediction of transmembrane protein topology based on a statistical analysis of the SwissProt database: the OrienTM algorithm. *Protein Eng* 14:387–390.

Liang Y, Fotiadis D, Filipek S, Saperstein DA, Palczewski K, Engel A (2003): Organization of the G-protein-coupled receptors rhodopsin and opsin in native membranes. *J Biol Chem* 278:21655–21662.

Liu Q, Zhu Y, Wang B, Li Y (2003): Identification of beta-barrel membrane proteins based on amino acid composition properties and predicted secondary structure. *Comput Biol Chem* 27:355–361.

Lomize AL, Pogozheva ID, Lomize MA, Mosberg HI (2006a): Positioning of proteins in membranes: a computational approach. *Protein Sci* 15:1318–1333.

Lomize MA, Lomize AL, Pogozheva ID, Mosberg HI (2006b): OPM: orientations of proteins in membranes database. *Bioinformatics* 22:623–625.

Lundstrom K, (2006): Structural genomics for membrane proteins. *Cell Mol Life Sci* 63:2597–2607.

Mulder NJ, Apweiler R, Attwood TK, Bairoch A, Bateman A, Binns D, Bork P, Buillard V, Cerutti L, Copley R, Courcelle E, Das U, Daugherty L, Dibley M, Finn R, Fleischmann W, Gough J, Haft D, Hulo N, Hunter S, Kahn D, Kanapin A, Kejariwal A, Labarga A, Langendijk-Genevaux PS, Lonsdale D, Lopez R, Letunic I, Madera M, Maslen J, Mcanulla C, Mcdowall J, Mistry J, Mitchell A, Nikolskaya AN, Orchard S, Orengo C, Petryszak R, Selengut JD, Sigrist CJ, Thomas PD, Valentin F, Wilson D, Wu CH, Yeats C (2007): New developments in the InterPro database. *Nucleic Acids Res* 35:D224–D228.

Murzin AG, Brenner SE, Hubbard T, Chothia C (1995): SCOP: a structural classification of proteins database for the investigation of sequences and structures. *J Mol Biol* 247:536–540.

Papasaikas PK, Bagos PG, Litou ZI, Promponas VJ, Hamodrakas SJ (2004): PRED-GPCR: GPCR recognition and family classification server. *Nucleic Acids Res* 32:W380–W382.

Pappu RV, Marshall GR, Ponder JW (1999): A potential smoothing algorithm accurately predicts transmembrane helix packing. *Nat Struct Biol* 6:50–55.

Park KJ, Gromiha MM, Horton P, Suwa M (2005): Discrimination of outer membrane proteins using support vector machines. *Bioinformatics* 21:4223–4229.

Pearl F, Todd A, Sillitoe I, Dibley M, Redfern O, Lewis T, Bennett C, Marsden R, Grant A, Lee D, Akpor A, Maibaum M, Harrison A, Dallman T, Reeves G, Diboun I, Addou S, Lise S, Johnston C, Sillero A, Thornton J, Orengo C (2005): The CATH Domain Structure Database and related resources Gene3D and DHS provide comprehensive domain family information for genome analysis. *Nucleic Acids Res* 33:D247–D251.

Pellegrini-Calace M, Carotti A, Jones DT (2003): Folding in lipid membranes (FILM): a novel method for the prediction of small membrane protein 3D structures. *Proteins* 50:537–545.

Petrey D, Xiang Z, Tang CL, Xie L, Gimpelev M, Mitros T, Soto CS, Goldsmith-Fischman S, Kernytsky A, Schlessinger A, Koh IY, Alexov E, Honig B (2003): Using multiple structure alignments, fast model building, and energetic analysis in fold recognition and homology modeling. *Proteins* 53(Suppl 6):430–435.

Popot JL, Engelman DM (1990): Membrane protein folding and oligomerization: the two-stage model. *Biochemistry* 29:4031–4037.

Punta M, Forrest LR, Bigelow H, Kernytsky A, Liu J, Rost B (2007): Membrane protein prediction methods. *Methods* 41:460–474.

Raman P, Cherezov V, Caffrey M (2006): The Membrane Protein Data Bank. *Cell Mol Life Sci* 63:36–51.

Rose GD, (1978): Prediction of chain turns in globular proteins on a hydrophobic basis. *Nature* 272:586–590.

Rost B, Yachdav G, Liu J (2004): The PredictProtein server. *Nucleic Acids Res* 32:W321–W326.

Ruiz N, Kahne D, Silhavy TJ (2006): Advances in understanding bacterial outer-membrane biogenesis. *Nat Rev Microbiol* 4:57–66.

Saier MH, Tran CV Jr, Barabote RD (2006): TCDB: the Transporter Classification Database for membrane transport protein analyses and information. *Nucleic Acids Res* 34:D181–D186.

Sali A, Blundell TL (1993): Comparative protein modelling by satisfaction of spatial restraints. *J Mol Biol* 234:779–815.

Sarramegna V, Talmont F, Demange P, Milon A (2003): Heterologous expression of G-protein-coupled receptors: comparison of expression systems from the standpoint of large-scale production and purification. *Cell Mol Life Sci* 60:1529–1546.

Senes A, Gerstein M, Engelman DM (2000): Statistical analysis of amino acid patterns in transmembrane helices: the GxxxG motif occurs frequently and in association with beta-branched residues at neighboring positions. *J Mol Biol* 296:921–936.

Seshadri K, Garemyr R, Wallin E, Von Heijne G, Elofsson A (1998): Architecture of beta-barrel membrane proteins: analysis of trimeric porins. *Protein Sci* 7:2026–2032.

Sonnhammer EL, Von Heijne G, Krogh A (1998): A hidden Markov model for predicting transmembrane helices in protein sequences. *Proc Int Conf Intell Syst Mol Biol* 6:175–182.

Tang CL, Xie L, Koh IY, Posy S, Alexov E, Honig B (2003): On the role of structural information in remote homology detection and sequence alignment: new methods using hybrid sequence profiles. *J Mol Biol* 334:1043–1062.

Todd AE, Marsden RL, Thornton JM, Orengo CA (2005): Progress of structural genomics initiatives: an analysis of solved target structures. *J Mol Biol* 348:1235–1260.

Tusnady GE, Simon I (2001): The HMMTOP transmembrane topology prediction server. *Bioinformatics* 17:849–850.

Tusnady GE, Dosztanyi Z, Simon I (2004): Transmembrane proteins in the Protein Data Bank: identification and classification. *Bioinformatics* 20:2964–2972.

Tusnady GE, Dosztanyi Z, Simon I (2005): PDB_TM: selection and membrane localization of transmembrane proteins in the Protein Data Bank. *Nucleic Acids Res* 33:D275–D278.

Ulmschneider MB, Sansom MS, Di Nola A (2005): Properties of integral membrane protein structures: derivation of an implicit membrane potential. *Proteins* 59:252–265.

Van Den Berg B, Clemons WM Jr, Collinson I, Modis Y, Hartmann E, Harrison SC, Rapoport TA (2004): X-ray structure of a protein-conducting channel. *Nature* 427:36–44.

Viklund H, Granseth E, Elofsson A (2006): Structural classification and prediction of reentrant regions in alpha-helical transmembrane proteins: application to complete genomes. *J Mol Biol* 361:591–603.

Von Heijne G, (1986): Mitochondrial targeting sequences may form amphiphilic helices. *EMBO J* 5:1335–1342.

Wallin E, Tsukihara T, Yoshikawa S, Von Heijne G, Elofsson A (1997): Architecture of helix bundle membrane proteins: an analysis of cytochrome c oxidase from bovine mitochondria. *Protein Sci* 6:808–815.

White SH, Wimley WC (1998): Hydrophobic interactions of peptides with membrane interfaces. *Biochim Biophys Acta* 1376:339–352.

White SH, Wimley WC (1999): Membrane protein folding and stability: physical principles. *Annu Rev Biophys Biomol Struct* 28:319–365.

Wimley WC, (2002): Toward genomic identification of beta-barrel membrane proteins: composition and architecture of known structures. *Protein Sci* 11:301–312.

Wimley WC, (2003): The versatile beta-barrel membrane protein. *Curr Opin Struct Biol* 13:404–411.

Xiang Z, Soto CS, Honig B (2002): Evaluating conformational free energies: the colony energy and its application to the problem of loop prediction. *Proc Natl Acad Sci USA* 99:7432–7437.

Yarov-Yarovoy V, Schonbrun J, Baker D (2006): Multipass membrane protein structure prediction using Rosetta. *Proteins* 62:1010–1025.

Yohannan S, Faham S, Yang D, Whitelegge JP, Bowie JU (2004a): The evolution of transmembrane helix kinks and the structural diversity of G protein-coupled receptors. *Proc Natl Acad Sci USA* 101:959–963.

Yohannan S, Yang D, Faham S, Boulting G, Whitelegge J, Bowie JU (2004b): Proline substitutions are not easily accommodated in a membrane protein. *J Mol Biol* 341:1–6.

Zhou H, Zhou Y (2003): Predicting the topology of transmembrane helical proteins using mean burial propensity and a hidden-Markov-model-based method. *Protein Sci* 12:1547–1555.

PROTEIN MOTION: SIMULATION

Ilan Samish, Jenny Gu, and Michael L. Klein

"If we walk far enough," said Dorothy, "we shall some time come to some place, I am sure."
L. Frank Baum, in *The Wonderful Wizard of Oz*, 1900.

INTRODUCTION

Molecular motions of biological macromolecules underlie folding, stability, and function. Crystallographic structural snapshots (Chapter 4) as well as nuclear magnetic resonance (NMR) structural (Chapter 5) ensembles enable, with the help of bioinformatics methods, the identification of functional sites, hinge and dynamic regions, interacting partners, remote homologues, and much more valuable biophysical insights. Yet, this experimental structural view is often constrained by the lack of understanding of the underlying time-dependent mechanism bridging structure and the execution of biological function. Understanding protein motion is pivotal for deciphering functional mechanisms including folding (Daggett and Fersht, 2003; Vendruscolo and Paci, 2003; Munoz, 2007) along with the complimentary misfolding (Dobson, 2003; Chiti and Dobson, 2006; Daggett, 2006), allosteric regulation (Popovych et al., 2006; Swain and Gierasch, 2006; Gilchrist, 2007; Gu and Bourne, 2007), catalysis (Hammes-Schiffer and Benkovic, 2006), functional plasticity (Kobilka and Deupi, 2007), protein–DNA interactions (Sarai and Kono, 2005), and protein–protein interactions (Bonvin, 2006; Fernandez-Ballester and Serrano, 2006; Reichmann et al., 2007). While information from different static structures can be morphed yielding a dynamic path between the structures (Flores et al., 2006), it is often required to simulate a static structure or substructure in a time-dependent manner in order to provide a dynamic understanding.

Simulating protein motions began in 1969 with the first energy minimization of proteins using a general force field (FF; Levitt and Lifson, 1969). In this study, Levitt and Lifson applied 50 steepest descent iterations (described later) to lysozyme and myoglobin. For the small 964-atom lysozyme it took 3 h of computing time on a room-size computer whose

Structural Bioinformatics, Second Edition Edited by Jenny Gu and Philip E. Bourne
Copyright © 2009 John Wiley & Sons, Inc.

name was "Golem." Interestingly, the semiempirical FF applied in this early work is not much different than today's classical mechanics common FF. The nontrivial shift from minimization to molecular dynamics (MD) simulation that follow the trajectories of each atom of a protein utilizing classical Newtonian molecular mechanics (MM) was conducted in 1977 by McCammon, Gelin, and Karplus (1977). Their *in vacuo* MD simulation of the 58-residue bovine pancreatic trypsin inhibitor spanned 9.2 ps (10^{-12} s). It took 11 more years for a full-atom (namely, including hydrogens) simulation of this protein to be carried out in the presence of solvent water (Levitt and Sharon, 1988) molecules to be realized, setting the stage for modern MD protein simulations. This classical approach to describe systems of biological interest provides insight regarding the molecular conformational changes that give rise to the observed functional consequences; and all this in the presence of solvent water and counter ions as appropriate. Today, the most powerful available computers, for example, Blue-Gene, are being utilized for MD simulations, with size, resolution, and time length of the simulations still growing. Actually, the very name "Blue-Gene" was given in 1999 due to the specific aim of this computer to solve the protein folding problem via the simulation of protein motion (IBM, 1999). Indeed, the importance of this theoretical field is reflected by a growing number of publications contributing $\sim$1200 publications on protein MD in 2007 alone (ISI Web of Science query "TS = (protein) and TS = (molecular dynamics)").

In this chapter we will focus solely on computational methods aimed at understanding protein motion. While most such tools are applicable for other biomacromolecules as well, for example, nucleic acids, we will not dwell into methods tailored for them. MD simulation yields the history of the motion (so-called trajectory) of each atom comprising the protein of interest and can therefore provide us with a microscopic characterization of the system. The conversion of the microscopic information to macroscopic thermodynamic observables, such as pressure, energy, heat capacities, and so on, requires statistical mechanics. The ergodic hypothesis (Chapter 8) enables the derivation of general conclusions on the system from a sufficiently long MD simulation of a protein molecule. Briefly, the hypothesis claims that a large number of observations made on a single system at many arbitrary instances in time have the same statistical properties as observing many systems, arbitrarily chosen from the phase space at one time point. One of the key challenges of a proper MD simulation of protein motion is the design of a time-dependent simulation that will obey ergodicity. One of the methods to assess whether the system indeed obeys ergodicity is to conduct multiple MD runs and thereby assess convergence. As demonstrated by the Blue-Gene team by analyzing 26 independent 100 ns MD runs of rhodopsin, much care should be applied when assuming ergodicity of a single trajectory (Grossfield, Feller, and Pitman, 2007). The challenge includes two aspects–what is a "sufficiently long" time and how can we ensure that the "relevant" phase space was indeed fully investigated (sampled). Notably, the latter aspect requires the nontrivial definition of a relevant phase space. Most often, not all phase space is of interest but only the low-energy wells within the complex (rugged) energy landscape of the system that are involved in executing the biological function. The movement between energy wells is specifically challenging, as on the one hand small time steps are needed to follow high-frequency motions of light atoms like hydrogen whereas on the other hand, long time (many time steps) is required in order to grasp transitions that underlie the slow timescale motions of protein domains undergoing conformational changes (Figure 37.1). In order to solve these challenges, different resolutions of physical properties and time steps are used to address macromolecular motion occurring at different timescales.

The study of molecular motion via computational methods has advanced to a state where there are many resources available to a wider scientific community. Understanding protein

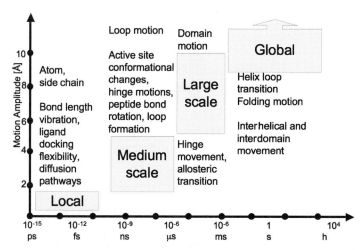

Figure 37.1. Characteristic division of protein motion to time and amplitude scales (division after Becker (2001)). This schematic division is convenient for motion analysis and simulation. In reality, the timescale and amplitude of many of the described phenomena are far more disperse. For example, some allosteric transition may take up to a second or alternatively, helix-coil transitions may initiate in the submicrosecond time regime. Designed, self-folding mini-proteins fold on the fast microsecond timescale (Bunagan et al., 2006). Other movements highly depend on local environment; for example, side-chain rotation is typically quick when solvent exposed (10^{-11}–10^{-10} s) and slow when buried (10^{-4}–1 s). The timescale division is also relevant from the aspect of disorder (Chapter 38).

motion has been an important research field in itself and will become increasingly recognized in the field of structural bioinformatics since protein flexibility and dynamics is inextricably linked to its function. The importance is exemplified by new databases such as ProMode (Wako, Kato, and Endo, 2004) containing a collection of NMA simulations and MolMovDB (Flores et al., 2006) containing extrapolated motion as well as simulations for single structures. These resources allow researchers to appreciate the full range of protein motion diversity and will continue to become of growing utility to the community (Gerstein and Echols, 2004). More importantly, these databases set a foundation in which to develop an improved systematic approach to categorize and expand our understanding of protein motion in a large-scale fashion as more structures become available to study the function and dynamics of biomolecules.

Many of the off-the-shelf packages can be operated without requiring deep fundamental understanding of the underlying theory. Yet, it is important to have a basic appreciation for the theory behind the different techniques in order to receive correct results and in order to correctly interpret the results. Rather than providing an exhaustive compilation of resources available to the community and rather than providing thorough mathematical foundations, this chapter aims to provide a brief introduction to the fundamental terms and theoretical rationale behind computational approaches aimed at understanding protein motion. The differences between various simulation approaches, along with their applicative strengths and limitations for interpretation, will be highlighted. For further understanding of this established yet evolving field, readers are referred to recent textbooks (Becker, 2001; Leach, 2001; Schlick, 2002; Cui and Bahar, 2006; Holtje, 2008), meetings lecture notes (Leimkuhler, 2006), as well as reviews on biomolecular simulation in general (Norberg and Nilsson, 2003; Schleif, 2004; Jorgensen and Tirado-Rives, 2005; Karplus and Kuriyan, 2005;

Adcock and McCammon, 2006;Scheraga, Khalili, and Liwo, 2007). Several reviews cover specific aspects of the methodology such as long time-scale simulations (Elber, 2005), coarse-grained models (Tozzini, 2005; Levy and Onuchic, 2006), and so-called QM–MM methods (Friesner and Guallar, 2005; Dal Peraro et al., 2007a; Senn and Thiel, 2007). Other reviews focus on MD simulation utilized for specific applications such as membrane and membrane-protein simulations (Saiz and Klein, 2002; Roux and Schulten, 2004; Gumbart et al., 2005;Brannigan, Lin, and Brown, 2006; Feller, 2007;Kandt, Ash, and Tielman, 2007; Lindahl and Sansom, 2008), steered MD applications (Sotomayor and Schulten, 2007), drug design (Alonso, Bliznyuk, and Gready, 2006), and adaptation to extremophile conditions (Tehei and Zaccai, 2005).

PROTEIN MOTION TIMESCALES, SIZE, AND SIMULATION

The structural changes underlying protein motions occur on timescales ranging over several orders of magnitude (Figure 37.1). Each timescale range contributes to different biophysical aspects of the protein. The complimentary biophysical forces involved in different types of motion scales can be batched according to their resolution and extent of affect they have on the motion. The very existence of a wide range of timescales is a major component of the multidimensional complexity that has to be addressed when studying protein motion. Intrinsic to this complexity is the manner in which the different types of motions are coupled. For example, the allosteric R to T transition of hemoglobin takes tens of micro-seconds to complete. This major change requires the relaxation of the heme cofactor (picosecond range), tertiary structure relaxation (nanosecond range), and so forth.

The relationship between theoretical and experimental study of protein motion is not limited to cross-validation but expands to the very parameterization of the many knowledge-based theoretical methods. On the experimental side, a wide plethora of biophysical spectroscopic methods are tailored to explore motions of different scales. Hydrogen deuterium exchange explored via NMR (Krishna et al., 2004) and via mass spectroscopy (Tsutsui and Wintrode, 2007; and Chapter 7), time-resolved Förster (commonly replaced by the word "fluorescence") resonance energy transfer (FRET) (Förster, 1948), laser-induced temperature jumps (Hofrichter, 2001), site-directed spin labeling (Fanucci and Cafiso, 2006), and time-resolved crystallography (Moffat, 2001) are just a few of the common methods used to study the kinetic aspect of protein motion. Other methods were developed to study thermodynamic aspects such as barrier heights and the corresponding free energy associated with specific types of motions. Similarly, a large number of distinct computational methods were developed to study different timescales of computational motions. The gaps between the complimentary experimental and theoretical methods are closing with an increasing degree of supporting cross-referencing, for example, as reviewed for the study of protein folding (Vendruscolo and Paci, 2003).

Ideally, all motion studies should account for all the physical forces of the system and the surrounding solvent when simulating the fastest identified discrete movement. However, solving the full Schrödinger equation is too complex and approximations are embedded, even in quantum mechanical (QM) calculations. In addition, as the time evolution of each computed variable is affected by many other variables, the complexity of the computation scales exponentially with the size of the studied system. Nontrivial challenges also arise with the definition of the basic time unit to be simulated and the distance after which the effect of the surrounding forces can be neglected. Lastly, the basic time-step unit that can be stably

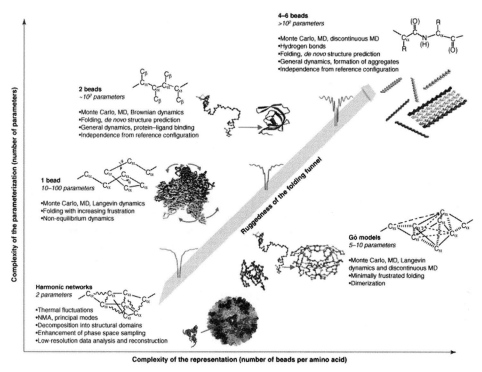

Figure 37.2. The features of bead models. Schematic representation of the model, indicative number of parameters, methods of solution, main characteristics, and applications are shown for each model. Sample applications are also illustrated with representative pictures (PDB codes 1HHP, 1CWP, 1MWR, 486D) showing the size of a studied system and the kind of study that can be done. The location of the models in the x–y plane qualitatively illustrates their complexity in system size (x-axis) and parameterization (y-axis). Reprinted with permission from Tozzini, 2005.

integrated should be considered. As shown in Figure 37.1, such a time step should be around 1 fs giving rise to intractable computation time for slow processes of the large biomacro-molecules. Consequently, simulations of macromolecules represented at different resolutions are used for the study of motions of different timescales. As such, alternative approaches involving a reduced or more detailed representation of protein structures have been developed (Figure 37.2).

Protein motion is studied utilizing methods ranging from detailed QM to coarse-grained models. The main driving force to the development of these models is to circumvent issues arising from computational demands resulting in the intractability of simulating very large biomolecules coupled to the varying demands of resolution requested from the simulation. Resolution includes the size of the system involved in the motion, the duration of the motion assessed, and the details of the factors affecting the motion. Different aspects of the resolution challenge are addressed by different approaches. For example, often the area of prime interest is a very small segment within the full system. Consequently, several simulation methods enable a differential resolution of different regions of the system. These include modeling the atoms of the macromolecule explicitly while modeling the solvent implicitly. In this context, the very term "explicitly" has different meanings in different motion simulation subfields: The classical FF (Figure 37.3) is considered explicit when

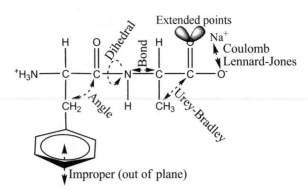

Figure 37.3. Common force field (FF) energy function components exemplified on a Phe-Ala dipeptide. Cross-terms such as Urey-Bradley and improper out-of-plane terms are part of higher order force fields. The external energy Coulombic electrostatic term and van der Waals Lennard-Jones terms are shown here for a sodium ion but, like all other depicted terms, are computed for all relevant atoms. Likewise, the out of plane improper term can be described for any group of three or more atoms and not only for aromatic rings. Elements such as lone pairs can be described via "extended point" pseudoatoms or by adding polarizability to the FF.

compared to implicit solvent. However, while using QM–MM methods to investigate enzyme catalysis, the area of prime interest, for example, the active site, is modeled explicitly via QM methods and the rest of the system (protein and solvent) is described with classical MM.

As MD simulations yield a trajectory of the motion, it is appealing to data mine this "movie" with every relevant bioinformatics approach. Yet, it is pivotal to fit the resolution of the analysis and conclusions to the resolution of the data. In this chapter the computational study of motion is presented in a top-down approach. The presentation of the coarse-grained methods will be followed by MM approaches, which in turn will be followed by QM–MM methods. Thus, each section will give insight on the resolution-related limitations of the previous section.

COARSE-GRAINED METHODS

As suggested in Figure 37.2, slow and large movements are often sufficiently hard-wired into the protein to be well represented by the use of a reduced coarse-grained representation of the structure (Tozzini, 2005; Levy and Onuchic, 2006). Alternatively, it can be stated that groups of adjacent atoms, for example, residues, move sufficiently in concert to be parameterized as a single point in space. Complementary, the expansion of motion simulation into the mesoscale (microsecond) time regime as well as into large interacting protein–protein complexes can be facilitated by coarse graining. In MD simulations, coarse-grained models are gaining increased attention. As not all atoms contribute similarly to the assessed question, the coarse graining can focus on the less significant atoms, for example, solvent and nonpolar hydrogens. The exclusion of nonpolar hydrogens, referred to as united atom FFs, "unifies" the hydrogen's additional mass and movement constraint into the atom to which it is attached forming a larger pseudoatom. The united atom approach is especially meaningful for membrane proteins where a large portion of the simulation computes the nonpolar bilayer.

The reduced representation of the structural space, also termed a bead- or minimalist-model, can be applied in several levels. Intuitively, each residue can be represented by a single bead, typically modeled at the coordinates of the Cα central atom. To get a better appreciation to the protein structure and without a large computational increase, a representation of the Cβ atom (Micheletti, Carloni, and Maritan, 2004) or centroid pseudoatoms representing a group of atoms with similar properties, for example, side-chains (Becker et al., 2004) or atom groups within a lipid (Saiz and Klein, 2002; Marrink et al., 2007; Saiz et al., 2002), may be embedded. Another important coarse-graining approach includes a differential resolution of different parts of the simulated system. Most broadly, this includes applying a hierarchy of up to four different resolutions of simulations: (a) simulating the active site of an enzyme with a high-resolution (in this case electrons) QM scheme, (b) embedding the active site within a segment of the protein framework treated via a classical Newtonian dynamics and an empirical interatomic FF, (c) linking the protein segment to a coarse-grained treatment of the protein framework, and (d) representing the solvent implicitly. These applications are discussed in the relevant sections below.

Coarse graining of the protein model can vary also in the type and character of the connectors modeled between the beads. Thirty years ago, Gō and coworkers studied folding via placing repulsive and attractive connectors representing nonbonded interactions between the one-residue beads (Ueda, Taketomi, and Gō, 1978). The resulting Gō models reproduce several aspects of the thermodynamics and kinetics of folding, largely due to the fact that the folding process has generally evolved to satisfy the principle of minimal frustration (Baker, 2000). The *minimal frustration* principle states that naturally occurring proteins have been evolutionarily designed to have sequences that achieve efficient folding to a structurally organized ensemble of structures with few traps arising from discordant energetic signals. Frustration, for example, via solvation terms, can be added to the model in order to roughen the energy surface, enabling the appearance of stable intermediate states. As reviewed by Levy and Onuchic, the complexity and character of minimalist models can be tailored to the question addressed, for example the detailed assembly mechanism of a protein complex (Levy and Onuchic, 2006). Notably, enhanced connectivity may be applied between interacting regions or electrostatics may be represented by providing different parameterization such as beads representing charged amino acids (Levy and Onuchic, 2006).

As functional motions of proteins are usually cooperative and include fluctuations around native states, *normal mode analysis* (NMA) methods are well fit to gain insight into these important dynamics. This topic is often viewed as part of the coarse-grained dynamic description via an *elastic network models* (ENMs), which is often sufficient to achieve valuable results. Although normal modes do not provide a fully detailed motion trajectory, they do yield the direction and relative magnitude of movement of regions within the protein. While here we focus on ENM-related methods, it is important to remember that normal modes are a more general framework frequently utilized as an analysis tool to understand dynamical trajectories of proteins produced by other types of simulation methods.

By definition, NMA is the study of harmonic potential wells by analytical means. NMA is the most direct quick way to obtain large amplitude, low-frequency motions (Gō, 1978; Brooks and Karplus, 1983). This rapidly growing field is maturing into a standard toolbox for the analysis of macromolecular motion (Cui and Bahar, 2006). It is assumed that the energy surface near the minimum can be approximated using a multidimensional parabola (assumption of harmonic dynamics) that is characterized by the second-derivative matrix (the Hessian). Solutions to such a system are vectors of periodic functions (normal modes) vibrating in unison at the characteristic frequency of the mode. Consequently, it is important that the protein structure is minimized to a single energetic minimum before a simulation is initiated. The most important distinction from MD simulations is that all variants of NMA approaches assume that the protein fluctuation is sufficiently described by harmonic dynamics. In other words, while MD simulations use (a relatively) exact description of the forces between atoms and solve these equations in an approximate manner, NMA uses approximate equations for motion that can be solved in a precise manner.

The use of a "single parameter model" for NMA by Tirion (1996) has opened the door for the expansion of more drastically simplified energy potentials used by the ENM approach. The ENM maintains the structures as springs between adjacent atoms, or atom-group centers, consistent with their solid-like nature. Tirion applied a full atom NMA with a uniform harmonic potential. Bahar and coworkers introduced a simplified ENM method termed the *Gaussian network model* (GNM) to study the contribution of topological constraints on the collective protein dynamics (Bahar, Atilgan, and Erman, 1997). GNM uses the Cα carbon acting as nodes in this polymer network theory undergoing a Gaussian distributed fluctuation. The fluctuation is influenced by neighboring residues typically within a 7 Å distance (r_c). The connectivity (or Kirchoff) matrix of interresidue contacts is defined by

$$\Gamma = \begin{cases} -1 & \text{if} & i \neq j & \text{and} & R_{ij} \leq r_c \\ -1 & \text{if} & i \neq j & \text{and} & R_{ij} \leq r_c \\ -\sum_{j, j \neq i} \Gamma_{ij} & \text{if} & i = j \end{cases}. \tag{37.1}$$

Notably, the diagonal elements of the matrix are equal to the residue connectivity number, namely the degree of elastic nodes in graph theory. This measure is in essence the local packing density around each residue. The contacts are assumed to have a spring-like behavior with a uniform force constant that has been established by fitting to experimental data. Thus, the potential of GNM is given by

$$V_{\text{GNM}} = \frac{\gamma}{2} \left[\sum_{i,j}^{N} (R_{ij} - R_{ij}^0)^2 H(r_c - R_{ij}) \right], \tag{37.2}$$

where γ is the uniform force constant taken for all springs and R_{ij}^0 and R_{ij} are the distance vectors between residues i and j in equilibrium and after a fluctuation, respectively. This dot product is calculated only for the distances defined by the Kirchoff matrix as defined by H $(r_c - R_{ij})$, a step function that equals 1 if the argument is positive and 0 otherwise.

In GNM, the probability distribution of all fluctuations is isotropic and Gaussian. Alternatively, GNM was extended to the anisotropic network model (ANM) (exemplified in Figure 37.4 insert B; Atilgan et al., 2001). The potential of ANM is equivalent to that of GNM (Eq. 37.2) except that in ANM R_{ij}^0 and R_{ij} are scalar distances rather than vectors. For ANM, the scalar product results in anisotropic fluctuations upon taking the second derivatives of the potential with respect to the displacements along the X-, Y-, and Z-axes. Physically, in the GNM potential, any change in the direction of the interresidue vector R_{ij}^0 is resisted, that is, penalized. On the contrary, the ANM potential is dependent upon the magnitude of the interresidue distances without penalizing for orientation changes. Due to the role of orientation deformations, ANM gives rise to excessively high fluctuations when compared to the GNM or experimental results. Due to this lower

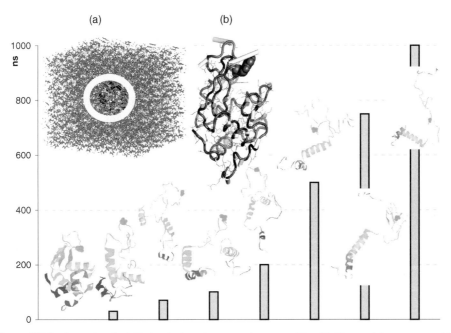

Figure 37.4. Example of a MD simulation. A mutant lysozyme (Trp-62-Gly—residue presented by spacefill) exhibits misfolding along a 1 µs (y-axis, time in nanoseconds) MD simulation trajectory. Snapshots of the protein structures are displayed as the simulation progresses (bars denoting time of each snapshot). The room temperature simulation mimics experimental conditions (Klein-Seetharaman, 2002), for example, 8 M urea (insert A: water box with urea presented via sticks and protein circled) and provide atomic resolution dynamic insight into the role of the mutated residue as a key nucleation site. As the second mode of an ANM standard simulation shows (insert B), different regions around this site move toward different directions (vectors represented by sticks with stick length corresponding to the motion amplitude). In the misfolding simulations some tertiary structure contacts disassemble rapidly while part of the secondary structure remains stable throughout the simulation. Simulation box and trajectory snapshots modified with permission after (Zhou 2007). ANM simulation of the wild-type structure conducted using (Eyal, Yang, and Bahar, 2006).

accuracy of ANM, a higher cutoff distance for interactions is required and the prediction of local relative degrees of flexibility may be less accurate. Yet, ANM is superior for assessing the directional mechanism of motion. Since the Kirchhoff matrix is of size of $3N \times 3N$ for ANM compared to $N \times N$ for GNM, GNM is more applicable for supramolecular structure analysis.

The reproduction of the intrinsic thermal fluctuations within crystal structures (Debye–Waller B-factors) as well as NMR relaxation measurements and H/D exchange free energies provide substantial validation to the ENM approach. Indeed, in recent years, the field of NMA has established itself as a key method for studying protein motion with validation extending comparison to MD studies and numerous other known functional movements, for example, allosteric transitions (Bahar and Rader, 2005; Elber, 2005; Tozzini, 2005; Cui and Bahar, 2006). Interestingly, Gerstein and coworkers have shown that over half of the 3800 known protein motions (inferred from different conformations of the same protein as analyzed from the Database of Macromolecular Movements http://molmovdb.org/) can be approximated by perturbing the original structures along the direction of their two lowest frequency normal modes (Krebs et al., 2002). In addition, the concerted motion of atom groups can be utilized for domain identification or to refine low-resolution models (Ma, 2006). Alternatively, applying NMA on the same domain in different conformations, for example, closed versus open ion channels, enables one to quickly infer a dynamic view of possible pathways between the conformations. Using such an approach as well as benefiting from the ease of applying NMA to nonprotein moieties, Vemparala, Domene, and Klein (2008) recently assessed the interactions of anesthetics with the open and closed conformations of a potassium channel. As a generality, the simplicity of the reduced representation is directly correlated to the smoothing of the local energy surface roughness. The success of the method exemplifies that motion under near native-state conditions is often simple and robust, dominated by the interresidue contact topology of the folded state.

Finally, in addition to traditional molecular simulation techniques strongly rooted in biophysics, the growing availability of structures have provided an alternative and less computationally intensive approach to globally address motion used across different proteins. Proteins solved in multiple conformations has allowed for the inference of biomolecular motion by restrained extrapolation between different structures serving as end points (Krebs and Gerstein, 2000) and is collected in the Database of Molecular Motions (MolMovDB) (Gerstein and Krebs, 1998; Flores et al., 2006). The database framework allows for the construction of a hierarchical categorization for biomolecular movements and the development of a systematic nomenclature for protein motion. First, protein motions are categorized based on molecular size of the biomolecule and the basis of packing that leads to subsequent categorization into the types of motion such as shear and hinge. Categorization based on protein size distinguishes between movements of subunits, domains, and fragments. This resource also contains references to supporting experimental evidence of the extrapolated motions when available. More importantly, it presents a survey of putative protein motions that are available in the "mechanistic toolbox" used by Nature rather than an isolated case interpretation of motion that is modeled in high resolution for a specific system. Hence, the survey allows for bioinformatics-derived interpretations to gain insight and possibly identify new generalized or fundamental rules governing protein dynamics.

CLASSICAL MOLECULAR MECHANICS

System Representation and Energy Function

A prerequisite to the actual time-dependent classical macromolecular motion simulation is the construction of (a) a system for simulation, and (b) a FF that will include the parameters of the system components along with an energy function to compute the system's energy. In doing so, the contradictory demands for accurate description along with computational efficiency should be addressed.

One of the most critical factors in a time-dependent simulation is the choice of a proper energy function. In classical MM dynamics, hereafter termed MD (except for the QM–MM section) a pseudoempirical potential energy function is utilized. Rather than first principles, additive knowledge-based terms are utilized in the context of the Newtonian laws of mechanics. These constrain the structure to the native form while properly representing the energy required for distorting the structure (Figure 37.3) within the natural range of fluctuations. The potential energy includes additive terms for the "internal" intramolecular energy between covalently bonded atoms and the "external" intermolecular energy found among nonbonded interactions that is

$$V(R)\text{total} = V(R)\text{internal} + V(R)\text{external}. \tag{37.3}$$

Ideally, each component and its additive constituents would be sufficiently accurate to be analyzed separately. Yet, in the current state of simulation research, the parameterization and resulting fit of the full function to experimental data is the most accurate. Consequently, cross-FF comparison of discrete components of the energy function is not advised. Still, the additive nature of the energy function implies separability, enabling to simulate each part of the function in different time steps (termed "multiple time step dynamics" see section below).

Focusing on a computationally efficient mathematical representation, the internal energy is modeled via a set of springs capturing the bond, angle and dihedral angles:

$$V(R)_{\text{internal}} = \sum_{\text{bonds}} K_b(b-b_0)^2 + \sum_{\text{angles}} K_\theta(\theta-\theta_0)^2 + \sum_{\text{dihedral}} K_\chi[1+\cos(n\chi-\sigma)]. \tag{37.4}$$

The b_0, θ_0, and χ represent the ideal reference parameters for the bond distance, angle, and dihedral angle respectively. For each such term the respective "K" represents the spring constant. The reference parameters of each atom depend on the hybridization, for example, sp^2 versus sp^3, as well as on the immediate environment such as, aromaticity and electronegativity of adjacent groups. Consequently, FFs include many atom types for each atom depending on the immediate surrounding and chemical nature. As an example, there are over 20 types of carbons.

Equation 37.5 contains the most basic terms but other equations are also in use, for example, bond vibrations are better represented by a double-exponential Morse function. Harmonic bond stretching terms and anharmonic cubic and quartic terms can be used to then reproduce the Morse function potential via a Taylor expansion. The accuracy of bond extension is important for small molecules resulting in the use of such methods in FFs attempting to focus on small molecules, for example, MM4 (Allinger, Chen, and Lii, 1996) and MMFF (Halgren, 1996).

Two additional constraining forces are often added: first, the Urey-Bradley (UB) crossterm represents the distance between two atoms connected through a third atom—the $1-3$ distance corresponding to the angle, namely, the stretch–bend cross-term. Other cross-terms between stretching, bending, and dihedrals are used, especially in FFs focusing on small molecules. Second, the so-called "improper" term aims at fitting computational results with observable vibrational spectra. Unlike the other terms, the improper term is used solely for some of the atoms, for example, improving out-of-plane distortions associated with planar groups—in-plane scissoring and rocking and out-of-plane wagging and twisting. As with the bond and angle terms, these components are modeled by the classical Hooke law for describing a spring, namely,

$$\sum_{\text{impropers}} K_\varphi(\varphi-\varphi_0)^2 + \sum_{\text{UB}} K_{\text{UB}}(S-S_0)^2. \tag{37.5}$$

The external energy includes the van der Waals and the electrostatic Coulomb contributions:

$$V(R)_{\text{external}} = \sum_{\substack{\text{nonbonded} \\ \text{atom pairs}}} \left\{ \varepsilon_{ij} \left[\left(\frac{R_{ij}^{\text{min}}}{r_{ij}} \right)^{12} - \left(\frac{R_{ij}^{\text{min}}}{r_{ij}} \right)^{6} \right] + \frac{q_i q_j}{\varepsilon_d r_{ij}} \right\}. \qquad (37.6)$$

The short-range interatomic repulsion and van der Waals dispersion contributions are expressed here via the commonly used Lennard-Jones 12–6 potential (Lennard-Jones, 1931). The R_{ij}^{min} represents the r_{ij} distance where the Lennard-Jones potential is zero and ε_{ij} corresponds to the depth of the well found at the minimum in the energy landscape. Notably, while the attractive exponent is experimentally and theoretically justifiable, the r^{-12} dependence is too steep for hydrocarbons but is often used due to the computational simplicity (square of r^{-6}). Consequently, some FFs dampen the repulsive exponent, for example, down to r^{-9}. Due to the short range of this force, van der Waals forces are computed only for atoms within a threshold distance.

Electrostatics (see also Chapter 24) are represented by Eq. 37.6 using partial charges q_i and q_j of the corresponding atoms. The equation is normalized to ε_d, the effective dielectric constant of the milieu, typically ranging from 2 in membranous environments to 80 in the solvent-exposed regions. The large change between these two numbers is usually conducted using a simple step function. Accurate modeling of the local effective charge density as it gradually changes between different microenvironments is one of the yet unresolved challenges of the field. Assigning partial charges to the many atom groups is a key challenge in constructing the FF. Some FFs were parameterized using a single class of molecules while others, termed "general" FFs, were parameterized using atom groups that should be representative of all the molecules. Thus, one of the main distinctions between FFs is the data set on which it was parameterized and the "transferability" of the parameterization to molecules that are not part of the original data set for which the FF was parameterized. The process of partial charge parameterization, often done via QM methods, is beyond the scope of this introductory chapter. While in the past knowledge-based FFs for specific chemical groups (proteins, nucleic acids, different organometallo groups) were common, QM-derived parameters of general FFs are becoming more and more accurate. Using a set of molecules that is representative of chemical space, general FFs can be applied to a wide variety of molecules—a feature that is becoming increasingly relevant for macromolecules subjected to drug design (Chapter 34). In the parameterization process different aspects have to be addressed including the generality and transferability of the parameterization as well as the fit of the partial charge to the condensed phase (rather than vacuum) in which the atom group is expected to be simulated.

The most computationally intensive component within the FF is the electrostatic one because it has the slowest decay with distance and long-range interactions must be included. Beyond the complexity of the function, while the bonded interactions scale linearly, the number of pair interactions scales by N^2, with N denoting the number of atoms. Moreover, the electrostatic solute–solvent interactions are important, resulting in much more interactions to compute. Thus, one of the ways to reduce the number of computed atoms is to reduce the number of solvent molecules to a minimum by choosing the correct periodic boundary condition. This condition implies an "image" of the system at the boundaries so a molecule "exiting" the system from one side will "enter" the system from the other side thus avoiding boundary effects. Cubic periodic boundary conditions are most intuitive to apply and include immersing the solute in a box of solvent (Figure 37.4 A). However, to add efficiency by decreasing the number of computed atoms without loss of accuracy, other conditions are often applied, for example, rhombic dodecahedron or truncated octahedron. For instance, a hexagonal prism may be more appropriate for a long macromolecule such as DNA. Different programs, such as, PBCAID (Qian, Strahs, and Schlick, 2001), can compute the optimal structural periodic domain.

Historically, the Coulomb law was solved using a spherical truncation of the electrostatic interactions conducted in distances in the range of 12 Å. Truncation-based approximation schemes (Seibel, Singh, and Kollman, 1985; van Gunsteren et al., 1986; Miaskiewicz, Osman, and Weinstein, 1993; Steinbach and Brooks, 1994; MacKerell, 1997) are gaining less popularity due to not only the availability of enhanced computing power, but rather the availability of efficient numerical methods. The numerically exact methods include *fast multipole methods*, which distribute the charge into subcells and then compute the interactions between the subcells by multipole expansion in a hierarchical fashion (Schmidt and Lee, 1991). The *Ewald method*, originally suggested in 1921 (Ewald, 1921), has evolved into a group of increasingly popular methods (Sagui and Darden, 1999). The Ewald summation divides the electrostatic computation to atom pairs within the main cell and those within the reciprocal, periodic boundary cells. The latter are computed using fast Fourier transform. The summation trick includes representing the point charges by Gaussian charge densities, producing an exponentially decaying function that can be summed in the reciprocal space. A correction term is required to ensure the effect of the overall net charges. A major breakthrough, termed particle-mesh Ewald (PME) includes using a grid for a smooth interpolation of the potential for the values in the Fourier series representing the reciprocal state (Darden, York, and Perdersen, 1993). The smoothing interpolation can be applied via Lagrangian (Darden, York, and Pedersen, 1993) or B-spline (Essmann et al., 1995) equations with the latter enabling smoother and more accurate interpolation. Further simplification can be applied by splitting the summation into short- and long-range interactions (Hockney and Eastwood, 1981). The short-range interactions can be computed via direct particle–particle summation while the latter can use the particle–mesh interpolation approach. The resulting particle–particle particle–mesh (Hockney and Eastwood, 1981; P^{3}M) approach is widely used. P^{3}M is easily applicable for parallel computing with specific optimization to parallel systems still evolving (Shaw, 2005).

FFs are available in different resolutions and foci. For example, Scheraga's ECEPP FF acts in dihedral-angle space (Zimmerman et al., 1977). For Cartesian coordinate FFs the very definition of the basic unit described can vary. The commonly used macromolecular FFs, for example, CHARMM (Brooks et al., 1983), AMBER (Weiner et al., 1984), GROMOS (Vangunsteren and Berendsen, 1977), and OPLS (Jorgensen and Tirado Rives, 1988) include differences in the functional form and in the parameterization. The differences between FFs and the resulting choice of the best FF is a controversial topic that depends on the fit to the specific macromolecule studied, the accuracy requested, as well as individual preferences. In a consensus study aimed to examine what can be inferred about protein MD through the interpretation of these FFs, it was found that some general features are FF-independent while others are more prone to FF choice (Rueda et al., 2007). In a large array of proteins studied for 10 ns, average differences in solvent accessible surface were ∼5%, average backbone root-mean-square deviations (RMSDs) between simulated and experimental structures were in the order of ∼2.0 Å with an effective distance of 1.4 Å —close to the thermal noise of MD simulations (Rueda et al., 2007). The analysis also suggests 70% overlap in structural space indicating that the different FFs sample a similar region of conformational space. The deviations in sampling appear to be dependent on the origin of the experimental structure with the largest observed deviations appearing in flexible regions of structures that were solved using NMR. The findings conclude that classic forces do indeed provide a useful approximation to interpret the real potential energy function.

The original FFs have developed into FF families with the above citations referring to the first publication for each FF. Namely, each FF had different "versions" with new version not being necessarily an improvement of an older version but often a different type of parameterization that is better fit to a specific problem/molecule. "United atom" FFs, for example, GROMOS96 (Gunsteren et al., 1996), can embed the nonpolar hydrogens into their adjacent heavy atoms and be considered as "pseudoatoms" that are sufficiently parameterized to represent the entire group. Complimentary to the increasing demand for accuracy, additional additive components are possible such as a directional hydrogen bond component. While hydrogen bonds are classically treated by the external energy terms, these prove to be not sufficiently accurate. Despite several attempts to include a hydrogen bonding term in a FF, the task seems more complex and currently there is no FF with an explicit hydrogen bonding term that is commonly used. The complexity of this term arises from the exact parameterization of the distance and angle dependency, the large differences according to the chemical nature of the atom that the hydrogen is attached to and the difficulty to synchronize the term with the other external energy terms. Hence, despite the importance of the hydrogen bond, redefined as a hydrogen bridge (Desiraju, 2002) due to its wide variation in strength and character, the full accuracy of this bond is still not within the common macromolecular FF today. Yet, proper parameterization of the orientation-dependent hydrogen bonding potential as part of protein design and structure prediction efforts provides a promising milestone towards leveraging FFs to include this component (Morozov and Kortemme, 2005).

Polarizable FFs (Halgren and Damm, 2001; Warshel, Kato, and Pisliakov, 2007; Friesner, 2005) with their wide spectrum of implementation level have been attempted since the very early simulations (Warshel and Levitt, 1976) yet are still far from mature and not widely used. In essence, such FFs try to integrate into the simulation a better view of the electron density distribution fluctuations around atoms. As classical Newtonian mechanics act on point masses they do not directly account for features such as aromaticity, directionality of lone electron pairs, for example, in solvent, and the nature of directional interactions such as hydrogen bonds, let alone bifurcated hydrogen bonds. Thus, classical FFs lack the explicit representation of induced dipoles, including dipoles of single atoms as well as groups of atoms ranging from a single aromatic group to the cumulative effect of a protein helix dipole. Unfortunately, as MacKerrel states,

> The practical issues involved along with uncertainties in the best way to approximate the physics have resulted in a distinct lack of usable polarizable FF for macromolecules. (Adcock and McCammon, 2006)

Though, efforts put into developing such FFs, the increasing computation power and the many applications that may benefit from such developments may change this situation. Calculations employing polarizable models indicate that the polarization energy is typically ~15% of the total molecular mechanics energy (Friesner, 2005). Applications of polarizable FFs are best fit for the processes that are most affected by polarizability: solvation energies, pK_as of ionizable residues, redox potentials, electron or proton transfer and ion conduction processes, as well as electrostatic effects in ligand binding and other enzyme catalysis processes.

One option to model such affects is to use a QM–MM hybrid approach (see later). Yet, in many cases an accurate view is requested for a region that is too large for other than crude semiempirical treatment. Several models have been developed for the treatment of polarizability, including the use of inducible dipoles, fluctuating charges and Drude oscillators (Rick and Stuart, 2002). Current polarizable FFs include (a) atomic multiple method-based FFs, for example, AMOEBA that is distributed with the TINKER package (Grossfield, Ren, and Ponder, 2003), fluctuating charges-based FFs, for example, as developed for OPLS (Banks et al., 1999; Liu et al., 1998), and Drude-oscillator based FFs, for example, as developed for CHARMM (Anisimov et al., 2004) and AMBER (Wang et al., 2006). Hence, the field of polarizable FFs is likely to evolve in the near future enabling better insight into key biological phenomena.

The above section described the energy function and the solute representation. In order to have a full system ready for simulation the representation of the solvent must be addressed. Historically, the system was isolated completely from the solvent, namely, was simulated *in vacuum*. In order to include solvent effects an explicit or implicit solvent is added. To ensure that solvent effects are fully represented the simulation of the solvent should mimic a solvent without a solute. Consequently, a major part of the computation involves the solvent molecules. Different explicit water molecules were parameterized to represent this condense phase (Jorgensen and Tirado-Rives, 2005). Popular models, for example TIP3P (Jorgensen et al., 1983) and the earlier SPC (Berendsen et al., 1981) follow the transferable intermolecular potential (TIP) form—three interaction sites centered on the nuclei with a partial charge for computing the Coulombic interaction and an additional Lennard-Jones interaction between the oxygens. The TIP4P (Jorgensen et al., 1983) four-point model adds a pseudoatom 0.15 Å below the oxygen along the HOH angle thus improving the quadrupole moment and liquid structure. The fixed geometry preventing bond stretching and angle bending adds to the efficiency of these models. The TIP5P (Mahoney and Jorgensen, 2000) adds partial charges on the hydrogens as well as two lone-pair electrons. At the expense of additional computing time, the TIP5P is the only model to fully recapitulate the temperature dependence of water density. However, the 1983 efficient and sufficiently correct TIP3P and TIP4P models introduced by Jorgensen, Klein and coworkers remain the method of choice for most macromolecular simulations. Several modified versions of these water models were developed, for example, TIP4P-Ew, a reparameterization fit for the Ewald summation (Horn et al., 2004).

Due to the high computational cost of the full representation of solvent, implicit solvation models have evolved. These aim to capture the mean influence of the solvent via the direct estimation of the solvation free energy, defined as the reversible work required for transferring the solute in a fixed configuration from vacuum to solution (Chen et al., 2008; see also Chapter 24). Implicit solvation utilized in MD simulations ranges from applying the Poisson–Boltzmann equation (Baker, 2005) to the far more efficient yet approximate generalized Born (GB) model (Still et al., 1990). In both the cases, the nonpolar solvation free energy is often modeled via the solvent-accessible surface area (SA; Lee and Richards, 1971). Implicit solvation can also be applied in a partial manner, for example, a thin layer of water around the macromolecule surrounded by an implicit solvent (Im et al., 2001). The Poisson equation for macroscopic continuum media is defined as

$$\nabla[\varepsilon(r)\nabla\Phi(r)] = -4\pi\rho_u(r), \tag{37.7}$$

where, $\varepsilon(r)$, $\Phi(r)$, and $\rho_u(r)$ denote the position-dependent dielectric constant at the point r, the electrostatic potential, and the charge density of the solute, respectively. The equation can be solved numerically by mapping the system on a grid and using the finite-difference relaxation algorithm (Klapper et al., 1986). The accurate description is exemplified by the development of a polarizable multiple PB model that compares well with explicit AMOEBA water (Schnieders et al., 2007). Alternatively, the MM/PB-SA model provides an end-point free-energy method (Gilson, Given, and Head, 1997; Srinivasan et al., 1998). As recently reviewed (Gilson and Zhou, 2007), MM/PB-SA (and the companion MM/GB-SA) use MD simulations of the free ligand, free protein, and their complex as a basis for calculating the average potential and solvation energies. The MD runs typically use an empirical FF and an explicit solvent model. The resulting snapshots are postprocessed by stripping them from their explicit solvent molecules and computing their potential energies with the empirical FF and their solvation energies with an implicit model. Convergence of the energy averages and the role of specific solvent molecules are regarded as challenges of this approach. Yet, it is far more efficient compared to the free energy perturbation approach that requires equilibrium sampling of the full transformation.

The GB model is an efficient approximation to the exact Poisson–Boltzmann equation and is the main method of choice for biomolecular simulation with recent GB/SA FFs available for the major FF classes (Chen et al., 2008). GB is based on modeling the protein as a sphere whose internal dielectric constant differs from the external solvent. In essence, the effective Born radius of an atom characterizes its degree of burial inside the solute. For an atom with (internal) radius a and charge q that is located in a solvent with a dielectric constant ε, the Born radius is defined by

$$G_{\text{Born}} = -\frac{q^2}{2a}\left(1 - \frac{1}{\varepsilon}\right). \tag{37.8}$$

Qualitatively it can be thought of as the distance from the atom to the molecular surface. Accurate estimation of the effective Born radii is critical for the GB model and has undergone much advancement in recent years. Thus, recent methodological advances have pushed the level of accuracy and complexity of continuum electrostatics-based solvation theories, providing the means to efficiently and accurately expand motion simulation to longer timescales.

Minimization Schemes

Energy minimization—moving the system to a local minimum on the energy landscape is an essential step for analyzing the macromolecular structure as well as for preparing the coordinates for a time-dependent simulation. Any given structure of a macromolecule is likely to contain errors often referred to as energetic "hot spots." Even in high-resolution structures, steric clashes as well as side-chain flips of residues such as Asn, Gln, and His are not uncommon (Chapters 14–15). Other common inaccuracies include salt-bridges that are not optimized, poor packing between atoms, and missing solvent molecules within internal cavities. Moreover, macromolecular simulation is conducted using FFs that are intrinsically inaccurate. Consequently, even a structure of the highest possible resolution will have local hot spots that can be relaxed by a simple minimization. Complimentary, a structure minimized by one FF will have to be minimized again if moving to a different FF.

The definition of a "local" minimum depends on the energy minimization scheme. While there may be deep energy wells in the energy landscape that are masked by energy barriers, the aim of most energy minimization schemes is to overcome (or "polish") the energy surface roughness for the identification of a more significant local energy minima. FFs are built from components that are deliberately easily derivable. A one-dimensional minimization to the X_0 minimum can be written as a Taylor expansion:

$$F(X) = F(X_0) + (X - X_0)F'(X_0) + \tfrac{1}{2}(X - X_0)2F''(X_0) + \cdots, \qquad (37.9)$$

with F' and F'' being the first and second derivatives of the energy function described by the FF. Moving to a multidimensional function requires the replacement of the derivatives by the gradient and the minimum point by a vector. The accuracy of the minimization depends on the number of computed components within the Taylor expansion. Minimization algorithms that do not include derivation, for example, grid search, are termed zero-order methods. A first derivative of the function, termed first-order methods, utilizes the gradient of the potential energy surface. The most popular of these are the *steepest descent* minimization (SDM) and the *conjugated gradient* minimization (CGM) methods. Second derivative methods (order two algorithms), for example, the *Adapted Basis Newton–Raphson* (ABNR), include the use of information about the curvature.

SDM moves the coordinate system in a direction that corresponds to the change in the gradient of forces exerted on the system by the FF energy function. The first step of this iterative process is of a predetermined arbitrary size. If the SDM step provides a lower energy, it is assumed that the structure is moving to a lower place on the energy surface so the next step size is increased, for example, by 20%. On the contrary, if the resulting energy is higher it is assumed that a local energy barrier was encountered so in order to confirm that the system is not moving out of the local energy well, the next step size is decreased, for example, by half. SDM is efficient in relaxing local barriers and in finding the local minima basins, yet it is also likely to fluctuate out of "shallow" basins.

CGM is another first-order minimizer that is computationally slightly less efficient than SDM. The advantage of CGM is that the minimization solution for the system is closer to the local minima compared to SDM with the use of "memory" for previously calculated gradients. After the first step, the direction of the minimization is determined as a weighted average of squares of the current gradient and the direction taken in the previous step. CGM converges much better than SDM and partially overcomes the problem of fluctuating around the minima. As CGM accumulates numerical errors it is usually restarted every predetermined number of steps.

Order two minimization algorithms make use of the curvature (Hessian) and not only the slope (gradient) of the FF function. The Newton–Raphson algorithm assumes that, near the minimum, the energy can be approximated to be a quadratic function. For the one-dimensional function the minimum can be approximated as $X - F'(X_0)/F''(X_0)$. For the multidimensional minimization, the first derivative divided by the second derivative is the gradient divided by the Hessian. Beyond the nonquadratic form of the energy function, calculation of the Hessian repeatedly is inefficient. Consequently, the ABNR (Brooks et al., 1983) is used. ABNR utilizes the curvature information only for the few atoms that contributed the most to the previous move, thus reducing the computation to a small subspace and achieving efficiency that is not far from the CGM algorithm.

An optimized minimization protocol usually takes advantage of these three minimization schemes by applying them sequentially. The crude SDM brings the system quickly to a lower region on the energy landscape where it is replaced by the more accurate CGM and finally refined by ABNR. The termination criterion of each step is often defined in terms of reaching to a gradient root-mean-square-deviation (GRMSD) threshold. If the threshold is not reached, the minimization scheme is replaced after a predetermined number of steps.

Molecular Dynamics—Moving the System

In classical MD, the forces are formulated according to the classical Newtonian equations of motion. A standard Taylor expansion provides the formalism for the move of position $x(t)$ to a position after a short and finite time interval Δt:

$$x(t+\Delta t) = x(t) + \frac{dx(t)}{dt}\Delta t + \frac{d^2x(t)}{dt^2}\frac{\Delta t^2}{2} + \cdots. \qquad (37.10)$$

The Taylor series is composed of the position, the velocity and the acceleration of the system. In order to comply with Newton's third law—conservation of total energy and momentum, computation of these terms are not sufficient. Finite-difference integration algorithms are utilized in which the integration is divided into small steps of future and past time. The Verlet algorithm (Verlet, 1967) uses the time of $x(t-\Delta t)$:

$$x(t+\Delta t) = 2x(t) - x(t-\Delta t) + \frac{d^2x(t)}{dt^2}\Delta t^2. \qquad (37.11)$$

The integration requires only a single force evaluation per iteration. Further, this formalism guarantees time reversibility, as required for ergodic behavior. Yet, it is not "self-starting," requiring a lower order Taylor expansion to initiate propagation. Moreover, it must be modified to include velocity-dependent forces or temperature scaling. The computationally slightly more expensive leapfrog algorithm (Hockney, 1970) reduces numerical errors and improves velocity evaluation as it uses "half"-time steps and gradient information:

$$x(t+\Delta t) = x(t) + \frac{dx(t)}{dt}\left(t+\frac{\Delta t}{2}\right)\Delta t. \qquad (37.12)$$

$$\frac{dx(t)}{dt}\left(t+\frac{\Delta t}{2}\right) = \frac{dx(t)}{dt}\left(t-\frac{\Delta t}{2}\right) + \frac{d^2x(t)}{dt^2}\Delta t. \qquad (37.13)$$

To further increase accuracy and numerical stability, the slightly more computationally intensive Verlet velocity algorithm is utilized. This Verlet-type half-time step (like leapfrog) algorithm synchronizes the calculation of positions, velocities and accelerations without sacrificing precision.

The use of dynamic integrators may bias the system considerably. To reproduce an experimental condition, all computed coordinates should belong to a single macroscopic or thermodynamic state. Namely, the characterized ensemble should be constrained to fixed thermodynamic properties. Properties that can be fixed include the number of atoms (N), volume (V), temperature (T), pressure (P), enthalpy (H), chemical potential (μ), and energy (E). Correspondingly, the ensembles include the popular NVT canonical ensemble, the NPH isobaric–isoenthalpic ensemble, the NPT isobaric-isothermal ensemble, the μVT grand canonical ensemble, and the NVE microcanonical ensemble in which the energy conservation implies a closed system. While the Newtonian equations of motion imply energy conservation, NVE does not represent common experimental conditions and may result in an energy drift. To keep these constant conditions, the system must be monitored and corrected accordingly. For example, velocities can be rescaled to maintain constant temperature or volume can be adjusted to maintain a constant pressure. Alternatively, the Andersen (Andersen, 1980) or the Berendsen (Berendsen et al., 1984) thermostat couples an external bath of constant temperature to provide an *NVT* ensemble. These thermostats are often applied via a velocity Verlet algorithm. The Langevin thermostat (Grest and Kremer, 1986) applies stochastic forces and a dampening friction coefficient. The so-called "hot solvent—cold solute" is a known challenge that arises from the higher sensitivity of the solvent to simulation noise. This may also cause a differential effect on different parts of the macromolecule, for example, flexible tails. Consequently, care should be given to the level of coupling to the external bath and the Langevin thermostat may be preferred for inhomogeneous systems (Mor, Ziv, and Levy, 2008). Similarly, the constant pressure method of Andersen (1980) was developed by Nosé and Klein for noncubic cells. It includes an additional degree of freedom—a piston corresponding to the volume, which adjusts itself to equalize the internal and applied pressures maintaining a *NPH* ensemble (Nose and Klein, 1983). Constant pressure MD algorithms were developed by maintaining volume and temperature thus providing an *NPT* ensemble (Martyna, Tobias, and Klein, 1994). Later, Feller et al. replaced the deterministic equations of motion for the piston degree of freedom by a Langevin equation thus introducing the popular Langevin piston for maintaining an *NPT* ensemble (Feller et al., 1995).

The selection of the minimal time step (Δt) for which the dynamics are calculated affect the computational cost directly. Moving to a longer time step without affecting accuracy can be applied by freezing the fastest modes of vibration, namely, constraining the bonds involving hydrogen atoms to fixed lengths. The SHAKE (Vangunsteren and Berendsen, 1977), RATTLE (Andersen, 1983) and LINCS (Hess et al., 1997) algorithms all aim at this higher efficiency. For example, SHAKE applies a constrained Verlet method to the velocities. LINCS resets the constraints directly (rather than their velocity derivative) thus avoiding the accumulating drifts. The constraints can be solved via matrix inversion (MSHAKE) to receive higher accuracy with an additional computational cost. Other variations include the introduction of quaternion dynamics for rigid fragments (QSHAKE). Notably, the use of such efficiency adding algorithms should be used with care. For example, these algorithms will alter the evaluation of temperature that is directly computed from the atomic velocities of all atoms in the system.

The use of multiple time-step dynamics can be expanded beyond the fast fluctuations of the hydrogen atoms and easily integrated into the dynamic integrator. As certain interactions evolve more rapidly than others, forces within the system can be classified according to the variation of the force with time. Taking advantage of the additivity of the FF, each force constituent is integrated in a different time step. The *reversible reference system propagator* (r-RESPA) allows such extended time steps to be utilized (Tuckerman, Berne, and Martyna, 1992). Similarly, the r-RESPA scheme may be utilized to couple the system to the fast multiple method (Zhou and Berne, 1995). Typically, the bonded interactions, the Lennard-Jones interactions and the real-space part of the electrostatic interactions are computed every 2 fs, while the Fourier-space electrostatics is computed every 6 fs.

Another approach to overcome the limitations of a slowly evolving single trajectory is to enhance the dynamics of part of the system or the entire system. Approaches include applying an elevated temperature, using multiple copies of regions of the system or adding a biasing force to direct the dynamic propagation toward the investigated goal. The first group of methods includes an unbiased enhanced sampling of the system. In the locally enhanced sampling, introduced by Elber and coworkers (Czerminski and Elber, 1991), a fragment of the system is duplicated into noninteracting copies. The strength of each copy to the remainder of the system is downscaled according to the overall number of copies. While a single to multicopy perturbation is required prior to the simulation and then the reversible at the end of the simulation, this method demonstrates quick convergence in cases involving an unknown phase space that is confined to a small region of the system. More generally, the entire system can be represented by multiple copies—an approach well fit to distributed computing such as the folding@home project (Snow et al., 2002). Applying multiple 5–20 ns MD simulations, this approach provided a motion simulation of a large ensemble of a mini-protein providing dynamic description that was well validated via experimental tempera-ture-jump measurements. A multiple copy approach can also be applied as a mean to rapidly overcome local energy barriers. The parallel tempering method and the replica exchange MD (denoted REMD; Sugita and Okamoto, 1999) utilizes noninteracting replicas at different temperatures. Using a Monte Carlo transition probability the replicas swap by reassigning velocities. The method is well fit for small molecules but is considered to be rather slow for full conformational sampling of large macromolecules. REMD was recently applied as a tool of increased sampling of conformational space, for example, for refining homology models (Zhu et al., 2008) and to explore the zipping and assembly of a protein folding model (Ozkan et al., 2007). In essence, these *steered MD* methods are the theoretical analogy to experimental single molecule experiments such as force microscopy or optical tweezers (Sotomayor and Schulten, 2007). However, unlike the experimental procedure, the theoretical one provides a full movie of the macromolecular dynamics. Steered MD is well fit to explore significant structural changes such as ligand binding and unfolding providing insight into elasticity and local stabilizing features within the macromolecule.

Taken together, classical MD methods provide a valuable tool to explore macromolec-ular motion in timescales approaching several microseconds. In order to achieve such long timescales without compromising on simulation accuracy, software that can be run in parallel is required. Examples of such software packages include the NAMD (Nelson et al., 1996) freeware as well as IBM's Blue Matter (Germain, 2005) (customized for the Blue Gene computer) and Desmond (Bowers et al., 2006) of DE Shaw Research. Interestingly, NAMD also runs highly efficiently on the Blue Gene computer (Kumar et al., 2008). Examples of power computing conducted with these software packages include a 15 ns NAMD-based simulation of the large aquaporin tetramer embedded in a solvated membrane (Tornroth-Horsefield et al., 2006), a 2.6 μs Blue Matter-based simulation of the role of direct interactions in the modulation of rhodopsin by ω-3 polyunsaturated lipids (Grossfield, Feller, and Pitman, 2006), and a 3 μs Desmond-based simulation of the NhaA Na^+/H^+ antiporter (Arkin et al., 2007). These membrane protein simulations exemplify the new biological insight now available with classical MD simulations coupled and optimized to run on supercomputers. Hence, coarse-grained and full atomistic membrane protein MD forms a major target for the growing and increasingly important field of protein motion simulation (Saiz and Klein, 2002; Roux and Schulten, 2004; Gumbart et al., 2005; Brannigan, Lin, and Brown, 2006; Feller, 2007; Kandt, Ash, and Tielman, 2007; Lindahl and Sansom, 2008).

Langevin and Brownian Dynamics

Newton's equation of motion (second law) states that the force is the mass multiplied by the acceleration ($F = ma$). In some cases, the Newtonian equation may be replaced by a stochastic equation of dynamics for the solute–solvent interactions, thus increasing the conformational sampling. The Langevin equation adds a drag force representing friction due to the solvent and a random force that represents stochastic collisions between the solvent and the solute. Schematically, for the one-dimensional space, the Langevin equation is

$$F'(x) = F''(x) - \zeta F'(x) + R(x). \tag{37.14}$$

where ζ is the friction coefficient and $R(x)$ the random force. A correction term, termed "mean force" may be added to the equation to represent an average effect of degrees of freedom not explicitly treated by the simulation.

In cases of substantial displacement of the solute, the internal forces may be sufficiently low, relatively, so the dynamics represent a random walk within the viscous environment. Brownian dynamics, the diffusional analog of the MD method necessitates that the friction coefficient within the Langevin equation is sufficiently large resulting in a random walk diffusional trajectory. Thus, Brownian dynamics are fit for diffusion-controlled reactions. Protein–protein interactions as well as ligand biding and channeling are just some of these applications (Gabdoulline and Wade, 2002). Improving sampling for slow association events may be applied via steered Brownian dynamics (Zou and Skeel, 2003).

QM–MM METHODS

In 1976 Warshel and Levitt (1976) studied the lysozyme active site via QM–MM, establishing a field that has become widespread in the study of biomacromolecules. The simplified model for protein folding introduced by them in 1975 (Levitt and Warshel, 1975) is now emerging as one of the methods of choice for studying protein folding. While QM calculations are computationally far more intensive compared to the reduced MM representation, the large increase in accuracy often requires their utilization (Senn and Thiel, 2007; Dal Peraro et al., 2007a; Friesner and Guallar, 2005). This is especially the case for the most interesting part of the protein—the active site or the ligand/cofactor binding region. Understanding time-dependent changes within such regions in the best possible accuracy is important for understanding the key function of the macromolecule. Furthermore, some of the intrinsic characteristics of such regions are quantum effects that form intrinsic limitations for classical dynamics applications. These include (a) parameterization of highly conjugated systems or tunneling of electrons/protons, (b) bond rearrangement chemistry and other chemical processes where the change in the electronic structure, for example, through excitation is pivotal, and (c) parameterization of atom groups other than the common amino acids and nucleic acids for which most FFs have been tailored for. As such, the reactive region subjected to QM computation should be of sufficient size to encompass any important electronic structure effects.

In QM–MM simulations, a small area of interest is treated by QM electronic structure theory, the majority of the protein and aqueous solvent is represented by a FF, and the boundary between the two regions is uniquely treated to ensure that large errors are not made in coupling the two regions. The exact form of the coupling terms and the details of the boundary region define the specific QM–MM scheme.

The QM–MM scheme can be applied with most types of QM methods. In the scope of this introductory chapter the QM aspect can be only briefly introduced. In the Schrödinger equation, an electron is described as a unique wavefunction describing the quantum state of the electron. The computation of many-electron wavefunctions is complex for large molecules and so approximate schemes have been developed that have been parameterized to overcome many of their intrinsic shortcomings. One of the common approaches nowadays is to work with the electron density as the basic quantity, which is referred to as *density functional theory* (DFT). It turns out that DFT is computationally a simpler quantity to deal with. Yet, the main limitation of DFT is the inherent inability to describe the London dispersion forces underlying van der Waals interactions. However, empirical approaches are currently being developed to overcome this deficiency.

First principles QM alone can only yield the molecular structure at the local minima (or saddle point important for reaction pathway). Consequently, entropic temperature effects cannot be readily incorporated. To enable the calculation of free energy effects, the QM has to be incorporated within an MD scheme that includes temperature effects and thus can benefit from established statistical mechanics methods such as steered dynamics or umbrella sampling. DFT was first applied within MD framework by Car and Parrinello (1985), forming the commonly utilized CPMD simulation. In CPMD, the electronic degrees of freedom are treated as dynamical variables and the resulting fictitious electron dynamics are coupled to the classical dynamics of the nuclei. At present, up to hundreds of atoms may be simulated to times in the order of hundreds of picoseconds.

CPMD can be coupled to classical MM MD yielding a dynamical QM–MM scheme. The key in combining the combined electronic structure and dynamics scheme within a classical protein FF-based dynamics simulation scheme is the specification of the coupling interface between the quantum QM and classical MM regions. If the two regions are not covalently linked, the challenge is reduced to the parameterization of nonbonded interactions between the MM and QM atoms. However, commonly the regions are covalently bonded requiring more special treatment. The simplest way to treat the interface is to cap the QM-computed moiety by hydrogens or pseudoatoms and treat the two entities distinctly, possibly adding correction terms. An alternative method is the use of a frozen orbital on the bond(s) connecting the regions. Here, the electronic structure equations are solved for the QM region incorporating the electron density of the frozen orbital as well as the charge distribution originating from the MM region.

From the first study on the stabilization of the carbonium ion with the lysozyme enzyme, current QM–MM methods aid in understanding enzymes. The approach is especially applicable for nonprotein moieties integral to protein complexes such as metals. For example, Dal Peraro et al. applied a QM–MM approach to capture the mean polarization and charge transfer contributions to the interatomic forces within protein metal sites, resulting in better MD simulations enabling relaxation of local frustrations in NMR structures (Dal Peraro et al., 2007b). The free energy perturbation method to calculate the free-energy differences along the path is facilitated by the approximation of replacing the QM density by atomic charges fitted to the electrostatic potential. Enzyme reactivity is a prime target of QM–MM simulations. Specifically, it is usually difficult to model bond rearrangements and transition state stabilization via conventional MM dynamics. Similarly, QM–MM dynamics can expand the usage of MM dynamics to the simulation of electronically excited states, for example, chromophore photoactivated isomerization in the photoactive yellow protein (PYP; Groenhof et al., 2004) or in rhodopsin (Gascon, Sproviero, and Batista, 2006). Hence, with the rapid growth in computer power, the usage of QM–MM dynamics is expected to grow and provide high-resolution understanding of structure–function relationships within biomacromolecules.

CONCLUSION

The field of motion simulation has matured into a central component of understanding macromolecule dynamics. Recently, the field has benefited from evolving methodological advances along with continually increasing computational power and the concomitant increase in experimental structural and biophysical data. The combination of atomistic simulations based on detailed intermolecular FFs with longer timescale coarse-grained simulations bridges the gap in our understanding of macromolecular motion. No less important, development in experimental methods and the increasing synergy between experiments and simulation assist in validating models of macromolecules and self-assembled structures. Long MD simulations imitating experimental conditions provide a detailed dynamic understanding of experimental results, for example, simulating the folding of a mutated lyzosyme (Zhou et al., 2007; Figure 37.4). Consequently, the role of MD simulations in understanding macromolecular dynamics is expected to increase; expanding in accuracy of results, length of time studied and size of systems analyzed.

Although addressed throughout the chapter, we wish to emphasize once again that the appropriate choice for a simulation study of motion depends on the desired analysis, required resolution of dynamic information, and consideration of the system to be simulated. The main tradeoff involves the desired resolution of motion to be modeled and the computational demands. In addition, the dynamical simulation toolbox should be viewed as another component in the understanding of motion. The emerging integration of simulation techniques with established and emerging experimental and bioinformatic methods provides a way to understand the dynamics in a validated manner. As the data for protein motion accumulates from various experimental resources, this will fuel research focus aimed to understand how functional motion is encoded at the sequence level (Gu et al., 2006). Hence, the field of protein motion simulation is expected to expand in depth, breadth and importance to the biological community.

RECOMMENDED READING

See notes on relevant books at the references reading section of Chapter 8.

ACKNOWLEDGMENTS

Thanks to Christopher M. MacDermaid for critical reading of the manuscript. IS thanks the European Molecular Biology Organization (EMBO) and the Human Frontier Science Program Organization (HFSPO) for support. MLK thanks the NIH for support.

REFERENCES

Adcock SA, McCammon JA (2006): Molecular dynamics: survey of methods for simulating the activity of proteins. *Chem Rev* 106(5):1589–1615.

Allinger NL, Chen KS, Lii JH (1996): An improved force field (MM4) for saturated hydrocarbons. *J Comput Chem* 17(5–6):642–668.

Alonso H, Bliznyuk AA, Gready JE (2006): Combining docking and molecular dynamic simulations in drug design. *Med Res Rev* 26(5):531–568.

Andersen HC (1980): Molecular-dynamics simulations at constant pressure and-or temperature. *J Chem Phys* 72(4):2384–2393.

Andersen HC (1983): Rattle—a velocity version of the shake algorithm for molecular-dynamics calculations. *J Comput Phys* 52(1):24–34.

Anisimov VM, et al. (2004): CHARMM all-atom polarizable force field parameter development for nucleic acids. *Biophys J* 86(1):415a.

Arkin IT, et al. (2007): Mechanism of Na+/H+ antiporting. *Science* 317(5839):799–803.

Atilgan AR, et al. (2001): Anisotropy of fluctuation dynamics of proteins with an elastic network model. *Biophys J* 80(1):505–515.

Bahar I, Rader AJ (2005): Coarse-grained normal mode analysis in structural biology. *Curr Opin Struct Biol* 15(5):586–592.

Bahar I, Atilgan AR, Erman B (1997): Direct evaluation of thermal fluctuations in proteins using a single-parameter harmonic potential. *Fold Des* 2(3):173–181.

Baker D (2000): A surprising simplicity to protein folding. *Nature* 405(6782):39–42.

Baker NA (2005): Improving implicit solvent simulations: a Poisson-centric view. *Curr Opin Struct Biol* 15(2):137–143.

Banks JL, et al. (1999): Parametrizing a polarizable force field from ab initio data. I. The fluctuating point charge model. *J Chem Phys* 110(2):741–754.

Becker OM, editor. (2001): *Computational Biochemistry and Biophysics*. New York: Marcel Dekker, 512.

Becker OM, et al. (2004): G protein-coupled receptors: in silico drug discovery in 3D. *Proc Natl Acad Sci USA* 101(31):11304–11309.

Berendsen HJC, et al. (1981): *Intermolecular Forces*. Pullman B, editor. Dordrecht: Reidel, pp. 331–342.

Berendsen HJC, et al. (1984): Molecular-dynamics with coupling to an external bath. *J Chem Phys* 81(8):3684–3690.

Bonvin AM (2006): Flexible protein–protein docking. *Curr Opin Struct Biol* 16(2):194–200.

Bowers KJ, et al. (2006): Scalable algorithms for molecular dynamics simulations on commodity clusters. Proceedings of the (2006) ACM/IEEE conference on Supercomputing, Tampa, Flordia.

Brannigan G, Lin LC, Brown FL (2006): Implicit solvent simulation models for biomembranes. *Eur Biophys J* 35(2):104–124.

Brooks BR, et al. (1983): Charmm–a Program for macromolecular energy, minimization, and dynamics calculations. *J Comput Chem* 4(2):187–217.

Brooks B, Karplus M (1983): Harmonic dynamics of proteins: normal modes and fluctuations in bovine pancreatic trypsin inhibitor. *Proc Natl Acad Sci USA* 80(21):6571–6575.

Bunagan MR, et al. (2006): Ultrafast folding of a computationally designed Trp-cage mutant: Trp2-cage. *J Phys Chem B* 110(8):3759–3763.

Car R, Parrinello M (1985): Unified approach for molecular-dynamics and density-functional theory. *Phys Rev Lett* 55(22):2471–2474.

Chen J, Brooks CL 3rd, Khandogin J (2008): Recent advances in implicit solvent-based methods for biomolecular simulations. *Curr Opin Struct Biol* 18(2):140–148.

Chiti F, Dobson CM (2006): Protein misfolding, functional amyloid, and human disease. *Annu Rev Biochem* 75:333–366.

Cui Q, Bahar I, editors. (2006): Normal Mode Analysis. Theory and applications to biological and chemical systems. *Mathematical and Computational Biology Series*. Boca Raton, FL: Chapman & Hall/CRC.

Czerminski R, Elber R (1991): Computational studies of ligand diffusion in globins: I. Leghemo-globin. *Proteins* 10(1):70–80.

Daggett V (2006): Protein folding-simulation. *Chem Rev* 106(5):1898–1916.

Daggett V, Fersht A (2003): The present view of the mechanism of protein folding. *Nat Rev Mol Cell Biol* 4(6):497–502.

Dal Peraro M, et al. (2007a): Investigating biological systems using first principles Car-Parrinello molecular dynamics simulations. *Curr Opin Struct Biol* 17(2):149–156.

Dal Peraro M, et al. (2007b): Modeling the charge distribution at metal sites in proteins for molecular dynamics simulations. *J Struct Biol* 157(3):444–453.

Darden T, York D, Pedersen L (1993): Particle mesh ewald—an N.log(N) method for Ewald sums in large systems. *J Chem Phys* 98(12):10089–10092.

Desiraju GR (2002): Hydrogen bridges in crystal engineering: interactions without borders. *Acc Chem Res* 35:565–573.

Dobson CM (2003): Protein folding and misfolding. *Nature* 426(6968):884–890.

Elber R (2005): Long-timescale simulation methods. *Curr Opin Struct Biol* 15(2):151–156.

Essmann U, et al. (1995): A smooth particle mesh Ewald method. *J Chem Phys* 103(19):8577–8593.

Ewald P (1921): Die Berechnung optischer and elektostatisher getterpotentiale. *Ann Phys* 64: 253–287.

Eyal E, Yang LW, Bahar I (2006): Anisotropic network model: systematic evaluation and a new web interface. *Bioinformatics* 22(21):2619–2627.

Fanucci GE, Cafiso DS (2006): Recent advances and applications of site-directed spin labeling. *Curr Opin Struct Biol* 16(5):644–653.

Feller SE, et al. (1995): Constant-pressure molecular-dynamics simulation—the Langevin Piston Method. *J Chem Phys* 103(11):4613–4621.

Feller SE (2007): Molecular dynamics simulations as a complement to nuclear magnetic resonance and X-ray diffraction measurements. *Methods Mol Biol* 400:89–102.

Fernandez-Ballester G, Serrano L (2006): Prediction of protein-protein interaction based on structure. *Methods Mol Biol* 340:207–234.

Flores S, et al. (2006): The database of macromolecular motions: new features added at the decade mark. *Nucleic Acids Res* 34 (database issue):D296–D301.

Förster T (1948): Zwischenmolekulare Energiewanderung Und Fluoreszenz. *Annalen Der Physik* 2(1–2):55–75.

Friesner RA (2005): Modeling polarization in proteins and protein-ligand complexes: methods and preliminary results. *Adv Protein Chem* 72:79–104.

Friesner RA, Guallar V (2005): Ab initio quantum chemical and mixed quantum mechanics/molecular mechanics (QM/MM) methods for studying enzymatic catalysis. *Annu Rev Phys Chem* 56:389–427.

Gabdoulline RR, Wade RC (2002): Biomolecular diffusional association. *Curr Opin Struct Biol* 12 (2):204–213.

Gascon JA, Sproviero EM, Batista VS (2006): Computational studies of the primary phototransduc-tion event in visual rhodopsin. *Acc Chem Res* 39(3):184–193.

Germain RS, et al. (2005): Early performance data on the Blue Matter molecular simulation framework. *IBM J Res Dev* 49(2–3):447–455.

Gerstein M, Echols N (2004): Exploring the range of protein flexibility, from a structural proteomics perspective. *Curr Opin Chem Biol* 8(1):14–19.

Gerstein M, Krebs W (1998): A database of macromolecular motions. *Nucleic Acids Res* 26(18): 4280–4290.

Gilchrist A (2007): Modulating G-protein-coupled receptors: from traditional pharmacology to allosterics. *Trends Pharmacol Sci* 28(8):431–437.

Gilson MK, Zhou HX (2007): Calculation of protein-ligand binding affinities. *Annu Rev Biophys Biomol Struct* 36:21–42.

Gilson MK, Given JA, Head MS (1997): A new class of models for computing receptor-ligand binding affinities. *Chem Biol* 4(2):87–92.

Gō N (1978): Shape of conformational energy surface near global minimum and low-frequency vibrations in native conformation of globular proteins. *Biopolymers* 17(5):1373–1379.

Grest GS, Kremer K (1986): Molecular-dynamics simulation for polymers in the presence of a heat bath. *Phys Rev A* 33(5):3628–3631.

Groenhof et al. G, et al. (2004): Photoactivation of the photoactive yellow protein: Why photon absorption triggers a *trans*-to-*cis* Isomerization of the chromophore in the protein. *J Am Chem Soc* 126(13):4228–4233.

Grossfield A, Ren PY, Ponder JW (2003): Ion solvation thermodynamics from simulation with a polarizable force field. *J Am Chem Soc* 125(50):15671–15682.

Grossfield A, Feller SE, Pitman MC (2006): A role for direct interactions in the modulation of rhodopsin by omega-3 polyunsaturated lipids. *Proc Natl Acad Sci USA* 103(13):4888–4893.

Grossfield A, Feller SE, Pitman MC (2007): Convergence of molecular dynamics simulations of membrane proteins. *Proteins* 67(1):31–40.

Gu J, Bourne PE (2007): Identifying allosteric fluctuation transitions between different protein conformational states as applied to cyclin dependent kinase 2. *BMC Bioinformatics* 8:45.

Gu J, Gribskov M, Bourne PE (2006): Wiggle-predicting functionally flexible regions from primary sequence. *PLoS Comput Biol* 2(7):e90.

Gumbart J, et al. (2005): Molecular dynamics simulations of proteins in lipid bilayers. *Curr Opin Struct Biol* 15(4):423–431.

Gunsteren W, et al. (1996): Biomolocular Simulations: The GROMOS96 Manual and User Guide. Zurich, Switzerland: Vdf Hochschul verlag AG an der ETH Zurich, p. 1042.

Halgren TA (1996): AT Merck molecular force field. 1. Basis, form, scope, parameterization, and performance of MMFF94. *J Comput Chem* 17(5–6):490–519.

Halgren TA, Damm W (2001): Polarizable force fields. *Curr Opin Struct Biol* 11(2):236–242.

Hammes-Schiffer S, Benkovic SJ (2006): Relating protein motion to catalysis. *Annu Rev Biochem* 75:519–541.

Hess B, et al. (1997): LINCS: a linear constraint solver for molecular simulations. *J Comput Chem* 18 (12):1463–1472.

Hockney RW (1970): The potential calculation and some applications. Methods in Computational Physics. New York: Academic Press, pp. 135–211.

Hockney RW, Eastwood JW (1981): Computer Simulation Using Particles. Bristol: Institute of Physics.

Hofrichter J (2001): Laser temperature-jump methods for studying folding dynamics. *Methods Mol Biol* 168:159–191.

Holtje HD, et al. (2008): Molecular Modeling Basic Principles and Applications, 3rd ed. New York: Wiley-VCH.

Horn HW, et al. (2004): Development of an improved four-site water model for biomolecular simulations: TIP4P-Ew. *J Chem Phys* 120(20):9665–9678.

IBM (1999): http://www.research.ibm.com/bluegene/press_release.html.

Im W, Berneche S, Roux B (2001): Generalized solvent boundary potential for computer simulation. *J Chem Phys* 114:2924–2937.

Jorgensen WL, et al. (1983): Comparison of simple potential functions for simulating liquid water. *J Chem Phys* 79(2):926–935.

Jorgensen WL, Tirado Rives J (1988): The OPLS potential functions for proteins—energy minimizations for crystals of cyclic-peptides and crambin. *J Am Chem Soc* 110(6):1657–1666.

Jorgensen WL, Tirado-Rives J (2005): Potential energy functions for atomic-level simulations of water and organic and biomolecular systems. *Proc Natl Acad Sci USA* 102(19):6665–6670.

Kandt C, Ash WL, Tieleman DP (2007): Setting up and running molecular dynamics simulations of membrane proteins. *Methods* 41(4):475–488.

Karplus M, Kuriyan J (2005): Molecular dynamics and protein function. *Proc Natl Acad Sci USA* 102 (19):6679–6685.

Klapper I, et al. (1986): Focusing of electric fields in the active site of Cu-Zn superoxide dismutase: effects of ionic strength and amino-acid modification. *Proteins* 1(1):47–59.

Klein-Seetharaman J, et al. (2002): Long-range interactions within a nonnative protein. *Science* 295 (5560):1719–1722.

Kobilka BK, Deupi X (2007): Conformational complexity of G-protein-coupled receptors. *Trends Pharmacol Sci* 28(8):397–406.

Krebs WG, et al. (2002): Normal mode analysis of macromolecular motions in a database framework: developing mode concentration as a useful classifying statistic. *Proteins* 48(4):682–695.

Krebs WG, Gerstein M (2000): The morph server: a standardized system for analyzing and visualizing macromolecular motions in a database framework. *Nucleic Acids Res* 28(8):1665–1675.

Krishna MM, et al. (2004): Hydrogen exchange methods to study protein folding. *Methods* 34 (1):51–64.

Kumar S, et al. (2008): Scalable molecular dynamics with NAMD on the IBM Blue Gene/L system. *IBM J Res Dev* 52(1–2):177–188.

Leach AR (2001): *Molecular Modeling: Principles and Applications*, 2nd ed. New York: Prentice Hall.

Lee B, Richards FM (1971): The interpretation of protein structures: estimation of static accessibility. *J Mol Biol* 55(3):379–400.

Leimkuhler B,(ed.) (2006): New algorithms form macromolecular simulation. *Lecture notes in computational science and engineering*, Vol. 49. New York: Springer.

Lennard-Jones JE (1931): Cohesion. *Proc Phys Soc* 43(5):461–482.

Levitt M, Lifson S (1969): Refinement of protein conformations using a macromolecular energy minimization procedure. *J Mol Biol* 46(2):269–279.

Levitt M, Sharon R (1988): Accurate simulation of protein dynamics in solution. *Proc Natl Acad Sci USA* 85(20):7557–7561.

Levitt M, Warshel A (1975): Computer simulation of protein folding. *Nature* 253(5494):694–698.

Levy Y, Onuchic JN (2006): Mechanisms of protein assembly: lessons from minimalist models. *Acc Chem Res* 39(2):135–142.

Lindahl E, Sansom MS (2008): Membrane proteins: molecular dynamics simulations. *Curr Opin Struct Biol* 18(4):425–431.

Liu YP, et al. (1998): Constructing ab initio force fields for molecular dynamics simulations. *J Chem Phys* 108(12):4739–4755.

Ma J (2006): Applications of normal mode analysis in structural refinement of supramolecular complexes, In: Cui Q, Bahar I, editors. Normal mode analysis theory and applications to biological and chemical systems. Boca Raton, FL: Chapman & Hall/CRC, pp. 137–154.

MacKerell AD (1997): Influence of magnesium ions on duplex DNA structural, dynamic, and solvation properties. *J Phys Chem B* 101(4):646–650.

Mahoney MW, Jorgensen WL (2000): A five-site model for liquid water and the reproduction of the density anomaly by rigid, nonpolarizable potential functions. *J Chem Phys* 112(20):8910–8922.

Marrink SJ, et al. (2007): The MARTINI force field: coarse grained model for biomolecular simulations. *J Phys Chem B* 111(27):7812–7824.

Martyna GJ, Tobias DJ, Klein ML (1994): Constant-pressure molecular-dynamics algorithms. *J Chem Phys* 101(5):4177–4189.

McCammon JA, Gelin BR, Karplus M (1977): Dynamics of folded proteins. *Nature* 267(5612): 585–590.

Miaskiewicz K, Osman R, Weinstein H (1993): Molecular-dynamics simulation of the hydrated D(Cgcgaattcgcg)2 dodecamer. *J Am Chem Soc* 115(4):1526–1537.

Micheletti C, Carloni P, Maritan A (2004): Accurate and efficient description of protein vibrational dynamics: comparing molecular dynamics and Gaussian models. *Proteins* 55(3):635–645.

Moffat K (2001): Time-resolved biochemical crystallography: a mechanistic perspective. *Chem Rev* 101(6):1569–1581.

Mor A, Ziv G, Levy Y (2008): Simulations of proteins with inhomogeneous degrees of freedom: the effect of thermostats. *J Comput Chem* 29(12) 1992–1998.

Morozov AV and Kortemme T (2005): Potential functions for hydrogen bonds in protein structure prediction and design. *Adv Protein Chem* 72:1–38.

Munoz V (2007): Conformational dynamics and ensembles in protein folding. *Annu Rev Biophys Biomol Struct* 36, 395–412.

Nelson MT, et al. (1996): NAMD: a parallel, object oriented molecular dynamics program. *Int J Supercomp Appl High Perf Comput* 10(4):251–268.

Norberg J, Nilsson L (2003): Advances in biomolecular simulations: methodology and recent applications. *Q Rev Biophys* 36(3):257–306.

Nose S, Klein ML (1983): Constant pressure molecular-dynamics for molecular-systems. *Mol Phys* 50(5):1055–1076.

Ozkan SB, et al. (2007): Protein folding by zipping and assembly. *Proc Natl Acad Sci USA* 104 (29):11987–11992.

Popovych N, et al. (2006): Dynamically driven protein allostery. *Nat Struct Mol Biol* 13(9):831–838.

Qian X, Strahs D, Schlick T (2001): A new program for optimizing periodic boundary models of solvated biomolecules (PBCAID). *J Comput Chem* 22(15):1843–1850.

Reichmann D, et al. (2007): The molecular architecture of protein-protein binding sites. *Curr Opin Struct Biol* 17(1):67–76.

Rick SW, Stuart SJ (2002): Potentials and algorithms for incorporating polarizability in computer simulations. *Rev Comput Chem* 18:89–146.

Roux B, Schulten K (2004): Computational studies of membrane channels. *Structure* 12(8): 1343–1351.

Rueda M, et al. (2007): A consensus view of protein dynamics. *Proc Natl Acad Sci USA* 104 (3):796–801.

Sagui C, Darden TA (1999): Molecular dynamics simulations of biomolecules: long-range electrostatic effects. *Annu Rev Biophys Biomol Struct* 28:155–179.

Saiz L, Klein ML (2002): Computer simulation studies of model biological membranes. *Acc Chem Res* 35(6):482–489.

Saiz L, Bandyopadhyay S, Klein ML (2002): Towards an understanding of complex biological membranes from atomistic molecular dynamics simulations. *Biosci Rep* 22(2):151–173.

Sarai A, Kono H (2005): Protein–DNA recognition patterns and predictions. *Annu Rev Biophys Biomol Struct* 34:379–398.

Scheraga HA, Khalili M, Liwo A (2007): Protein-folding dynamics: overview of molecular simulation techniques. *Annu Rev Phys Chem* 58:57–83.

Schleif R (2004): Modeling and studying proteins with molecular dynamics. *Methods Enzymol* 383:28–47.

Schlick T (2002): Marsden JE, et al. editor. Molecular Modeling and Simulation: An Interdisciplinary Guide. Interdisciplinary Applied Mathematics, Vol.21. New York: Springer, p. 634.

Schmidt KE, Lee MA (1991): Implementing the fast multipole method in 3 dimensions. *J Stat Phys* 63(5–6):1223–1235.

Schnieders MJ, et al. (2007): Polarizable atomic multipole solutes in a Poisson-Boltzmann continuum. *J Chem Phys* 126(12):124114.

Seibel GL, Singh UC, Kollman PA (1985): A molecular dynamics simulation of double-helical B-DNA including counterions and water. *Proc Natl Acad Sci USA* 82(19):6537–6540.

Senn HM, Thiel W (2007): QM/MM studies of enzymes. *Curr Opin Chem Biol* 11(2):182– 187.

Shaw DE (2005): A fast, scalable method for the parallel evaluation of distance-limited pairwise particle interactions. *J Comput Chem* 26(13):1318–1328.

Snow CD, et al. (2002): Absolute comparison of simulated and experimental protein-folding dynamics. *Nature* 420(6911):102–106.

Sotomayor M, Schulten K (2007): Single-molecule experiments in vitro and in silico. *Science* 316 (5828):1144–1148.

Srinivasan J, et al. (1998): Continuum solvent studies of the stability of DNA, RNA, and phosphoramidate–DNA helices. *J Am Chem Soc* 120(37):9401–9409.

Steinbach PJ, Brooks BR (1994): New spherical-cutoff methods for long-range forces in macromolecular simulation. *J Comput Chem* 15(7):667–683.

Still WC, et al. (1990): Semianalytical treatment of solvation for molecular mechanics and dynamics. *J. Am. Chem. Soc.* 112(16):6127–6129.

Sugita Y, Okamoto Y (1999): Replica-exchange molecular dynamics method for protein folding. *Chem Phys Lett* 314(1–2):141–151.

Swain JF, Gierasch LM (2006): The changing landscape of protein allostery. *Curr Opin Struct Biol* 16 (1):102–108.

Tai K, Baaden M, Murdock S, Wu B, Ng MH, Johnston S, Boardman R, Fangohr H, Cox K, Essex JW, Sansom MS (2004): BioSimGrid: towards a worldwide repository for biomolecular simulations. *Org Biomol Chem* 2(22):3219–3221.

Tehei M, Zaccai G (2005): Adaptation to extreme environments: macromolecular dynamics in complex systems. *Biochim Biophys Acta* 3(1724):404–410.

Tirion MM (1996): Large amplitude elastic motions in proteins from a single-parameter, atomic analysis. *Phys Rev Lett* 77(9):1904–1908.

Tornroth-Horsefield S, et al. (2006): Structural mechanism of plant aquaporin gating. *Nature* 439 (7077):688–694.

Tozzini V (2005): Coarse-grained models for proteins. *Curr Opin Struct Biol* 15(2):144–150.

Tsutsui Y, Wintrode PL (2007): Hydrogen/deuterium exchange-mass spectrometry: a powerful tool for probing protein structure, dynamics and interactions. *Curr Med Chem* 14(22): 2344–2358.

Tuckerman M, Berne BJ, Martyna GJ (1992): Reversible multiple time scale molecular-dynamics. *J Chem Phys* 97(3):1990–2001.

Ueda Y, Taketomi H, Gō N (1978): Studies on protein folding, unfolding, and fluctuations by computer-simulation. 2. 3-Dimensional lattice model of lysozyme. *Biopolymers* 17(6):1531–1548.

van Gunsteren WF, et al. (1986): A molecular dynamics computer simulation of an eight-base-pair DNA fragment in aqueous solution: comparison with experimental two-dimensional NMR data. *Ann NY Acad Sci* 482:287–303.

Vangunsteren WF, Berendsen HJC (1977): Algorithms for macromolecular dynamics and constraint dynamics. *Mol Phys* 34(5):1311–1327.

Vemparala S, Domene C, Klein ML (2008): Interaction of anesthetics with open and closed conformations of a potassium channel studied via molecular dynamics and normal mode analysis. *Biophys J*

Vendruscolo M, Paci E (2003): Protein folding: bringing theory and experiment closer together. *Curr Opin Struct Biol* 13(1):82–87.

Verlet L (1967): Computer experiments on classical fluids. I. Thermodynamical properties of Lennard-Jones molecules. *Phys Rev* 159(1):98–103.

Wako H, Kato M, Endo S (2004): ProMode: a database of normal mode analyses on protein molecules with a full-atom model. *Bioinformatics* 20(13):2035–2043.

Wang ZX, et al. (2006): Strike a balance: optimization of backbone torsion parameters of AMBER polarizable force field for simulations of proteins and peptides. *J Comput Chem* 27(6):781–790.

Warshel A, Levitt M (1976): Theoretical studies of enzymic reactions: dielectric, electrostatic and steric stabilization of the carbonium ion in the reaction of lysozyme. *J Mol Biol* 103(2):227–249.

Warshel A, Kato M, Pisliakov AV (2007): Polarizable force fields: history, test cases, and prospects. *J Chem Theory Comput* 3(6):2034–2045.

Weiner SJ, et al. (1984): A new force-field for molecular mechanical simulation of nucleic-acids and proteins. *J Am Chem Soc* 106(3):765–784.

Zhou R, et al. (2007): Destruction of long-range interactions by a single mutation in lysozyme. *Proc Natl Acad Sci USA* 104(14):5824–5829.

Zhou RH, Berne BJ (1995): A new molecular-dynamics method combining the reference system propagator algorithm with a fast multipole method for simulating proteins and other complex-systems. *J Chem Phys* 103(21):9444–9459.

Zhu J, et al. (2008): Refining homology models by combining replica-exchange molecular dynamics and statistical potentials. *Proteins* 72(4) 1171–1188.

Zimmerman SS, et al. (1977): Conformational-analysis of 20 naturally occurring amino-acid residues using ECEPP. *Macromolecules* 10(1):1–9.

Zou G, Skeel RD (2003): Robust biased Brownian dynamics for rate constant calculation. *Biophys J* 85(4):2147–2157.

38

THE SIGNIFICANCE AND IMPACTS OF PROTEIN DISORDER AND CONFORMATIONAL VARIANTS

Jenny Gu and Vincent J. Hilser

INTRODUCTION

Protein disorder is a topic worth attention from the structural bioinformatics community largely for the technical challenges it presents to the field, but also for its biological and functional implications. The success of structural genomic efforts using X-ray crystallography depends on overcoming several potential bottlenecks (Chapter 40), one of which is the formation of protein crystals that can be obstructed by the presence of highly flexible and disordered regions. Despite precluding the number of structures that can be obtained, thus impacting the coverage of protein space, our current generalized understanding of disordered regions is a result of structural bioinformatics efforts that were able to extract and analyze patterns associated with these regions. These disorder predictors have been proven to be useful in advancing our understanding of disordered regions with potential impact to improve the success rate of structural genomics efforts, particularly those focused on eukaryotic proteins (Oldfield et al., 2005b).

The importance of resolving differences observed in conformational variants within protein families and understanding their impacts is also a rising issue. Most structural genomics efforts aim to solve a representative structure for each protein family to maximize the coverage of protein space with particular focus on identifying new protein folds. However, it is equally important to understand structural changes that result from sequence differences introduced by a few single point mutations, insertions, and/or deletions since it can have a large functional impact. Furthermore, the structural information recorded in the Protein Data Bank (PDB) is often overlooked as a macroscopic view of a collection of microscopic ensembles that give rise to the observed protein structure. In other words, the

Structural Bioinformatics, Second Edition   Edited by Jenny Gu and Philip E. Bourne
Copyright © 2009 John Wiley & Sons, Inc.

observed protein structure is not the only conformation adopted by the protein. In fact, most observed biological phenomena are a macroscopic consequence of the collective microscopic states. Understanding the differences in the microscopic states and how the changes impact the macroscopic event is currently addressed in several ways that will be discussed.

By exploiting the technical weakness in structural data, researchers have been able to gain insight into the potential biological significance of these otherwise poorly characterized disordered regions (Ringe and Petsko, 1986). Recognition for the importance of protein disorder in biological function came around the late 1970s when disordered regions seem to reoccur within particular features of enzymes such as the zymogens of pancreatic serine proteases and tyrosyl-tRNA synthetases (Blow, 1977). In light of these investigations, the hypothesis presented at the time was that the reactivity and specificity are associated with more rigid structures while disordered regions may be involved with control of the function. Since then, many functional roles of disordered regions including regulatory control have been implicated through experimental investigation of these regions, statistical mechanics, and structural bioinformatics approaches.

While the topics of protein disorder and conformational variations are intrinsically related to protein flexibility, these topics warranted a separate chapter from "Protein Motion: Simulation" (Chapter 37) largely because it deals with a time frame and complexity beyond what is captured by protein dynamic modeling approaches (Figure 38.1). Molecular dynamics simulations have been used to study conformational disorder and variants of proteins with limitations (Torda and Scheek, 1990; Kuriyan et al., 1991; Fuentes et al., 2005). Longer molecular dynamic simulations are reserved for smaller proteins or are otherwise restricted to a short time frame within limits of nanoseconds for larger proteins. As such, the observed conformational changes with these simulations will also be limited. The topics of disorder and conformational variations discussed here extend beyond what can be offered by molecular dynamic simulations, although various strategies such as the use of Monte Carlo sampling (Lindorff-Larsen et al., 2004) and averaging over a few samples of generated conformers while using experimental constraints (Kemmink et al., 1993; Bonvin and Brunger, 1995) have been used to address this issue. Coarse-grained dynamic modeling addresses molecular motion beyond the time frame limitations of classical molecular dynamics. However, a systematic analysis between disordered regions and the modeled large-amplitude fluctuating regions using these rigid-body based approaches needs to be conducted.

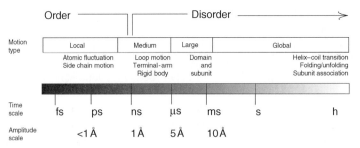

Figure 38.1. Range of protein dynamics and structural observation. Protein flexibility lies on a spectrum where the fluctuations occur at a range of different time scales. Ordered structures can be visualized with simulated motion limited to the nanosecond range. Protein fluctuations beyond this range are viewed as being disordered and lacking stable structures.

In this chapter, we briefly discuss the experimental methods used to study disordered regions and highlight the computational resources that have largely fueled the advancement of this field by providing many of the current generalized observations. The biological importance of protein disorder and conformational variations as they exist in microscopic ensembles will also be examined in more detail. We attempt to create an introductory chapter to the subject and apologize if not all research efforts are represented in this otherwise rapidly growing field.

PROTEIN DISORDER: UNDERSTANDING THE REALM OF "INVISIBLE"

Defining Protein Disorder

Before proceeding, we must first make clear that the field currently lacks a unifying definition when discussing protein flexibility, disorder, and intrinsically unstructured proteins. These terms are often used interchangeably due to the qualitative nature of the definition and can leave readers with some confusion if the slight distinctions are not clarified. Other disorder-related terms that have been coined in the field are intrinsic coils, random coils, unfolded proteins, molten globules, and premolten globules as examples to define protein states that are not natively folded. These terms are often refer to the global state of the protein rather than specific regions within the protein structures that are disordered. Without setting the standard nomenclature for the field, we will clarify by defining the usage of "disorder" in this chapter as regions in the protein structure where the equilibrium position of the backbone, along with the dihedral angles, that have no specific values and vary significantly over time.

When evaluating and using disorder predictors, it is also important to have a clarified view of how these regions were defined in the training of disorder predictors and other efforts to understand these regions. Some sequence-based disorder predictors, such as PONDR (Romero et al., 1997) and DISOPRED (Jones and Ward, 2003), were trained on disorder defined as missing regions in the X-ray crystallographic structures. This definition is also used to benchmark the performance of disorder predictors by evaluators in CASP experiments (Chapter 28). However, other predictors such as GlobPlot (Linding et al., 2003b) and DisEMBL (Linding et al., 2003a) are trained on definition based on a temperature factor (B-value) threshold to define disorder in X-ray crystal structures. Finally, other subtle differences in disorder predictors should be considered such as RONN (Yang et al., 2005) and Wiggle (Gu, Gribskov, and Bourne, 2006). RONN incorporates additional use of curated information from homologous proteins to make predictions regarding disordered regions, and Wiggle was trained on a data set where flexible regions are defined using dynamic modeling techniques. These subtle distinctions should be noted when considering which predictor would best serve the scientific question at hand.

Prevalence of Disordered Protein Regions

Flexible and disordered regions present two challenges to our understanding of protein structures. Aside from being unable to resolve atomic coordinates for these regions to understand the structure, the regions also interfere with the formation of protein crystals needed to collect X-ray diffraction data. Disordered regions are often addressed by removing them from proteins targeted for structure determination. These disordered regions can also be detected using nuclear magnetic resonance (NMR—Chapter 5), but the structure of these regions cannot be easily determined due to the increased conformational space sampled by the disordered regions. An analysis of a nonredundant subset of the PDB shows that ~7% of the complete sequences, as deposited in the Swiss-Prot Database, contained no disordered regions (Le Gall et al., 2007). A number of sequences where >95% of the protein is resolved structurally comprise about ~25% of the data set, a surprisingly small count that illustrates the prevalence of disordered regions within protein structures.

The presence of disordered regions is not a technical artifact and several different techniques have been employed to study this phenomenon. Early studies used spectroscopic techniques such as infrared circular dichroism (CD), Fourier transform infrared (FTIR), electron paramagnetic resonance (EPR), and optical rotary dispersion (ORD) to detect native and nonnative structures that may form within the disordered regions. More recently, NMR and small-angle X-ray scattering (SAXS) have been used to provide quantitative data about disordered and denatured proteins (Kern, Eisenmesser, and Wolf-Watz, 2005; Mittag and Forman-Kay, 2007; Sasakawa et al., 2007; Tsutakawa et al., 2007). These experimental approaches can provide quantitative data that can be incorporated into the calculation of the observed conformational ensembles in solution to determine the structural information about denatured, unfolded, and intrinsically disordered proteins. Hydrogen–deuterium (H/D) exchange mass spectrometry (Chapter 7) has also been used to study dynamic processes such as the role of transient structural disorder as a facilitator of protein–ligand binding (Xiao and Kaltashov, 2005). These experiments have detected structural formations within these disordered regions, and these structures have been associated with functional implications.

With the development of sequence-based predictors, the prevalence of disordered regions in organisms have been investigated across the three kingdoms of life (Oldfield et al., 2005a; Ward et al., 2004). The frequency of native disorder was calculated for several representative genomes and found to have increased content in eukaryotic proteins (33.0%) compared to 2.0% and 4.2% of archaean and eubacterial proteins, respectively (Ward et al., 2004). The analysis showed that proteins containing disorder are often located in the cell nucleus with functional association to regulations of transcription and cell signaling. In a separate study, an increase in intrinsic disorder content has been observed in regulatory cell signaling, cytoskeletal, and human cancer-associated proteins (Iakoucheva et al., 2002). Disordered regions are currently being curated into a database, DisProt (Sickmeier et al., 2007), which contains 472 proteins and 1121 disordered regions as reported for release 3.6 (June 29, 2007).

Computational Approaches to Understanding Protein Disorder

The computational tools that have been developed to predict regions of protein flexibility and disorder range from the use of simple sequence complexity profiles to complex machine learning infrastructure schemes such as the neural network and support vector machines (SVMs) (Figure 38.2). The successful development of these tools is attributed to the fact that sequence signatures of protein disorder are present. The popular choice of training set to construct these predictors often use reported missing residues in X-ray crystallographic structures, but reported temperature factors (B-factors) and NMR characterized disordered regions have also been used. First we will discuss algorithms that do not use structural information to identify and understand disordered regions. This is achieved by either examination of the sequence space only or focusing on residues in which the structure cannot be resolved. Then we will follow with alternative strategies that use temperature

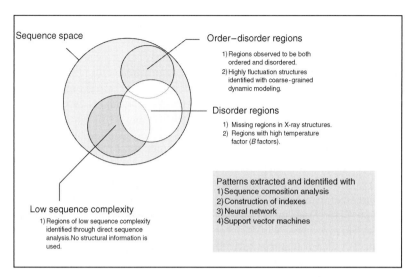

Figure 38.2. General strategies to predict the disorder sequence space. Schema of various strategies used to identify and understand the sequence space of disordered regions. The differences stem largely from how the disordered regions were defined and the underlying infrastructure for analysis and prediction tool development. Within all of the sequence space, a subset of sequence space will be associated with regions with low complexity, detected disordered, or those transitioning between an ordered and a disordered state. Overlaps can occur between the subsets.

factors in X-ray structures, the incorporation of homology information, and coarse-grained dynamic modeling to guide the training and disorder definition process. This short overview of disorder predictors will reflect the ongoing research efforts and common strategies employed to develop sequence-based identification of disordered regions.

SEG is a successful algorithm that identified unstructured regions by examination of the changing variation in sequence complexities within the sequence database (Wootton and Federhen, 1996). For a window length L and an N-residue alphabet, the compositional complexity for a given residue is

$$K_1 = \frac{1}{L}\log_N\Omega,$$

where Ω is the multinomial coefficient $(L!/\Pi_{i=1}^{N}n_i!)$. Alternative formulations that resemble Shannon's entropy to measure sequence have also been used. After identifying low sequence complexity regions, the second stage of this algorithm constructs an optimal subsequence to evaluate the probability of occurrence of the observed pattern that is calculated as

$$P_0 = \frac{1}{N^L}\Omega F,$$

where F is the combinatorial expression $N!/\Pi_{k=0}^{L}r_k!$ that yields the frequency of observing the sequence composition with this complexity and r_k is the count of the number of times the complexity state is observed in the window. This probability of occurrence has been precomputed into a table that serves efficiently as an index to identify these regions. Thus, SEQ not only identifies low complexity sequence regions but also determines whether these identified regions are significant rather than a random occurrence. The success of this approach hinges on the assumption that disordered regions have low sequence complexities. However, many disordered regions are not detected by SEG and therefore suggest that features other than sequence complexity are involved.

To detect other disordered regions, PONDR is the first disorder predictor that uses a design of two feedforward neural networks to make predictions using several attributes such as the fractional composition and hydropathy for the 20 amino acids (Romero, Obradovic, and Dunker, 1997; Romero et al., 1997). Unlike early flexible predictors, this predictor was trained on a data set of eight- and seven-residue-long disordered regions defined by X-ray and NMR experiments, respectively. The regions were defined as having either (1) no resolvable atomic coordinates and therefore declared as missing in the PDB files of X-ray structures or (2) extensively characterized as disorder with the use of NMR techniques. Predictions were made using raw input features extracted for the sequence and smoothed out with a second predictor. Since the initial development, PONDR is now available as a series of eight different predictors that identify different "flavors" of disorder (Vucetic et al., 2003) indicating that disordered features have distinctive sequence characteristics within each subclass. The success of this initial development of disorder predictors on such a small training data set is surprising and may be illuminative of sequence properties to be further discussed in the subsequent section.

DISOPRED (Jones and Ward, 2003) is another predictor that is also based on the use of a neural network but uses sequence profiles generated by PSI-BLAST (Altschul et al., 1997) as input features to make predictions with a postfilter that takes into account the confidence of secondary structure predictions. Often predictions are made based on physicochemical properties of the amino acid, but DISOPRED uses instead the amino acid identity, composition, and evolutionary conservation. Thus, the physicochemical properties are not explicitly captured in the input features although it may be implicitly represented. The inclusion of input features that represent evolutionary conservation was inspired when secondary structure predictors (Chapter 29) were improved using this information. The use of evolutionary information helps capture conserved features, or lack there of, between protein homologues. DISOPRED reports an accuracy of 90%, but the use of accuracy to measure success can sometimes be misleading, especially when the data set is unbalanced in class frequencies. In this case, the data set contains a much greater number of ordered residues than disordered examples, an important consideration when evaluating any predictor. Another measure to evaluate predictor performance is the use of Matthews' correlation coefficient (MCC) that DISOPRED reports to be 0.34, suggesting an overprediction of disordered regions.

RONN is another neural network algorithm that incorporates evolutionary information to improve disorder prediction. Instead of using multiple sequence alignments and sequence similarity directly, this algorithm compares the sequence to homologous proteins with characterized and annotated disordered features (Yang et al., 2005). The alignment scores of the sequence to a database of known order and disorder segments are used as the input features for prediction. This interesting strategy resulted in improvements that reduced the number of incorrect classification of residues in either the ordered or the disordered structural class.

More recently, POODLE-L (Hirose et al., 2007) implemented a two-layered support vector machine that uses physicochemical properties as input features and reports an improved performance with an MCC value of 0.658. The source of improvement is difficult to ascertain, although it may be safe to speculate that either the use of the SVM for extraction or a more focused training set is the underlying source. While successful discrimination does depend on the correct selection of input features that properly represent critical properties of disorder regions, the physicochemical properties used in this predictor have also been used by other predictors. Thus, it is unlikely that this would be the source of significant improvements.

Strategies that use relatively simpler algorithms compared to machine learning approach have also been used to efficiently identify these regions. GlobPlot2 (Linding et al., 2003b) uses propensities for amino acid to be in either an ordered or a disordered structure, thus creating a disorder propensity index. Different propensity indexes and scales were calculated to accommodate the different definition of disorder in the field and therefore will make predictions accordingly. IUPred (Dosztanyi et al., 2005a; Dosztanyi et al., 2005b) uses a low-resolution energetic force field based on the pair-wise interacting residues observed in structures. The total pair-wise interaction is estimated based on a quadratic form in the amino acid composition of the protein.

Developments of specialized disorder predictors to identify particular features of disordered regions have also been made. The PONDR series of predictors have effectively achieved this by constructing predictors that identify pattern subsets on features such as the length of disordered regions. Wiggle is another specialized predictor that identifies flexible regions having functional importance. Functional flexibility was defined as regions in the protein where (1) the fluctuating motions exceed the mean fluctuation by more than one standard deviation and (2) these fluctuations are involved in correlated motion. The use of this definition successfully identified regions such as recognition loops, catalytic loops, and hinges in a training data set where protein motion was obtained using a coarse dynamic modeling technique.

Finally, various consensus and integrated strategies that incorporate different predictors to improve disorder prediction have been developed. Consensus strategies have improved structure prediction methods in the recent years and therefore will likely be the case for disorder prediction. This approach often requires interpretation of results by the user to decide between prediction results from different methods since the integration of methods does not include an automated decision-making feature. Two such examples of integrated servers that are available to the community are PrDOS (Ishida and Kinoshita, 2007) and iPDA (Su, Chen, and Hsu, 2007).

Sequence Basis for the Biophysical Property of Ordered and Disordered Regions

Decoding the sequence space is at the heart of many fields, and understanding how the biophysical properties of proteins are encoded in the sequence is imperative to making inferences about the protein structural fold and function. For example, the amino acid hydrophobicities determined from several biophysical and theoretical experiments have been used to make predictions for higher order protein features such as secondary structure (Palliser and Parry, 2001). Unfortunately, advances in the field are still needed before an accurate sequence-based biophysical description of proteins is available to the community. Reoccurring amino acid bias and sequence patterns found within particular features of proteins are often examined in hopes to glean some insight into a biophysical explanation. In regard to protein disorder and flexibility, variances in protein sequences have been examined and regions of low sequence complexity have been identified to preclude structural formation (Wootton and Federhen, 1993; Wootton and Federhen, 1996). Glutamine-rich, glycine-rich, and arginine-rich sequences are often a part of this class of low sequence complexity regions with an occasional periodicity nature of repeating units. Regions enriched in proline, glutamic acid, serine, and threonine (PEST) are also associated with protein disorder.

The arrival of disordered predictors has allowed researchers to gain more insight into sequence bias and patterns associated with ordered and disordered structures through both the training process and subsequent analysis of the sequence space identified with these algorithms. Initially, there were some concerns that the disorder predictors were making predictions based on low complexity features similar to those identified by SEG. Instead, it has been demonstrated that amino acid composition differed between low complexity, disordered, and ordered regions (Romero et al., 2001). Disordered regions have been found to contain higher levels of R, K, E, P, and S amino acids with lower levels of C, W, Y, I, and V compared to ordered regions. Based on this analysis of change in amino acid frequency between ordered and disordered structures, the residues can be ranked from disorder promoting to order promoting as follows: K, E, D, P, N, S, Q, G, R, T, A, M, H, L, V, Y, I, F, C, W. Such ranking suggests that the amount of flexibility, and hence disorder, can be tuned depending on which residues are used and in what order.

Sequences that adopt both an ordered and a disordered structure depending on the observed conformational state have been investigated to understand how such a balance could be achieved (Zhang et al., 2007). These regions are coined as having "dual personality" and have been collected from proteins with multiple X-ray structures in different conformations. Regions that are invisible in one conformer but resolved in another were defined as having ambivalence for either ordered or disordered regions. Residues were clustered into three major groups based on their relative abundance for disordered, ordered, or ambivalent regions. The first group contains hydrophilic and small amino acids (K, E, S, G, and A) that are largely associated with disordered regions. The second group consists mostly of hydrophobic resides (M, H, Y, I, F, C, and W) that are found in abundance within ordered regions. Finally, the third group consists of mostly hydrophilic amino acids (D, T, Q, N, P, and R) that are found in fairly equal propensities for ordered and disordered regions. These clusters are in some agreement with what have been identified by Romero et al. although there are differences. Wiggle (Gu, Gribskov, and Bourne, 2006) shows a different scenario of amino acid preferences for these regions ranked in the following decreasing order: E, K, Q, R, D, P, N, S, G, A, L, T, W, H, M, Y, F, C, V, I. This ranking shows correlation with the consensus hydrophobicity index (Palliser and Parry, 2001). Although some general trends can be observed between the different analyses, the lack of agreement suggests that more work is needed to understand how the biophysical property of protein disorder is encoded in the sequence.

Investigation of higher order sequence associations with disordered regions has identified at least two nonrandom reoccurrence of patterns based on amino acid identity or physicochemical properties (Lise and Jones, 2005). The analysis was conducted on segments up to eight residues and repeated observations of proline-rich or charged segments in disordered regions were identified. Although rather restrictive parameters were used in this analysis, this examination shows that patterns associated with disordered regions are much simpler compared to those found in ordered regions and rarely contained two different amino acids. These patterns are not inclusive of all the possible patterns that may be found in disordered regions.

Despite the low complexity of these local sequence patterns, the nonrandom occurrence of these patterns reveals that biophysical properties of disordered structures are dependent on both the sequence composition and order. Furthermore, different types of disordered regions have been identified with a dependency on segment length (Vucetic et al., 2003). Subclasses of disordered regions have been identified with different functional associations. More recently, these thermodynamic features of unfolded regions were recently surveyed using a structure-based thermodynamic model (Wang et al., 2008). The results show that, unlike natively folded proteins, the thermodynamics of unfolded regions is dominated by local sequence contribution and is sensitive to the composition and order of the sequence. The local dependence of the disorder thermodynamics also provided some insight regarding why certain biophysical properties of the natively folded state may be retained.

In the process of understanding the sequence basis for protein disorder, it is also important to understand the contributions of evolutionary pressures that select for the disordered state. Although evolutionary conservation has been included as one of the input features to help disordered predictors discriminate between the different sequence types, strict conservation may not necessarily be a critical feature of disordered regions. Disordered regions have been cited to evolve rapidly (Brown et al., 2002) and observed to have increased alternative splice sites that generate more functional diversity in multi-cellular organisms (Romero et al., 2006). Evidence that disordered regions can be identified with a reduced set of the amino acid alphabet further supports the notion of weak evolutionary selection (Weathers et al., 2004). Consequently, the robustness to substitutions simplifies the combinatorial possibilities of amino acid patterns; thus, it may be possible to decode a direct linkage between sequence and the biophysical properties of disordered regions.

Biological Consequences

In addition to paving way to a better understanding of the sequence space, many hypotheses that generalize the biological and functional significance of disordered regions in proteins stem from computational analysis with disordered predictors. Detection of the prevalence of disordered regions across the different genomes mentioned earlier is one example of how the disorder predictors have been applied. At least 28 functional roles of disorder proteins have been identified and can be grouped into four main functional classes: (1) molecular recognition, (2) molecular assembly, (3) protein modification, and (4) entropic chain activities (Dunker et al., 2002; Radivojac et al., 2007). These functional classes largely appear to reflect intermolecular function linked to regulatory processes, but the biophysical properties of disordered regions are also important for protein folding, allostery, and catalytic processes that are not necessarily captured by this functional classification system presented by Dunker and colleagues. Patterns detected within disordered regions have also been used to functionally classify proteins, particularly in cases where there are no structural homologues (Lobley et al., 2007).

First, disordered structures provide an advantage in molecular recognition by promoting promiscuous, transient binding for several targets and therefore help to increase the complexities of protein interaction networks (Tompa, Szasz, and Buday, 2005). The resulting promiscuous bindings allow these proteins to have multiple cellular roles (Sandhu and Dash, 2006). Binding mechanisms used by disordered regions rely more on hydro-phobic–hydrophobic interactions with more intermolecular contacts that cover a much larger surface area of the target protein compared to ordered binding sites (Meszaros et al., 2007; Vacic et al., 2007). This is often achieved through a continuous segment that sometimes contains preformed structural elements where the native structure is fully induced after binding to the substrate (Fuxreiter et al., 2004). Investigation of underlying linear motifs important for recognition suggests that order favoring sequences are grafted between disordered regions that serve as a carrier (Fuxreiter, Tompa, and Simon, 2007).

Disordered regions have been identified to be important for complex formation for larger complexes such as the viral capsid, bacterial flagellar system, cytoskeleton, ribosome, and clathrin coat (Namba, 2001; Dafforn and Smith, 2004; Ward et al., 2004). A significant correlation was observed between predicted structural disorder and the number of proteins assembled into complexes conducted on *E. coli* and *S. cerevisiae* proteins (Hegyi, Schad, and Tompa, 2007). The larger complexes show a higher average of disorder content with longer predicted segments. These results are in agreement with the idea that disordered regions are involved with protein binding and molecular recognition that could be one contributing mechanism to complex formation. Alternatively, disordered regions in complex formation may simply serve as linkers between well-formed domains. The hypotheses presented here have been based on bioinformatics analysis and need to be further investigated experimentally.

Other disordered regions have been cited to act as entropic chains comprising a functional class that includes linkers, bristles, springs, and clocks (Dunker et al., 2002). This functional class is relatively less studied and appears to play a role in introducing a level of organization in time and space. For example, linkers serve to link domains while bristles help keep molecules apart through molecular exclusion. Springs are segments that have restoring forces to favor a randomized fold and become restricted when stretched as observed in titan molecules found in muscle fibers (Labeit and Kolmerer, 1995; Kellermayer et al., 2000) and elastin (Pometun, Chekmenev, and Wittebort, 2004). The property of increased flexibility for disorder regions has been hypothesized to be exploited and used as a "random generator" that effects the timing of biological process such as determining the closure of a voltage-gated channel (Wissmann et al., 1999; Wissmann et al., 2003).

The aforementioned four categories are not exclusive of each other nor are they comprehensive of all possible functions associated with disordered regions that are still yet to be fully understood. For example, a highly disordered loop in cochaperonin GroES is important for binding to GroEL and has also been suggested to facilitate the cycles observed for chaperonin-mediated protein folding through modulation of binding affinity without affecting specificity (Landry et al., 1996). This example serves to suggest that a certain level of flexibility necessary for function is being evolutionarily selected and conserved.

Retaining this balance of flexibility is also evident in the presence of disordered regions that are important for catalysis and allostery. Changes in local segmental flexibility were studied in the catalytic subunit of cAMP-dependent protein kinase using site-directed labeling and fluorescence spectroscopy (Li et al., 2002). The backbone located around the B-helix was found to have reduced flexibility only when the substrate and pseudosubstrate are bound to the catalytic domain. This stage of the catalytic cycle coincided with the phosphoryl transfer transition suggesting that internal disorder is important for this catalytic step. In another example using single-molecule enzymatic assays, DNA was hydrolyzed by lambda exonuclease with contributions from sequence-dependent factors and disorder arising from conformational changes (van Oijen et al., 2003).

Although popular competing allosteric models are based on changes observed in rigid structure bodies, alternative views propose that proteins can be regulated through changes in protein dynamics. The Cooper–Dryden model is a mathematical formulation that shows protein allostery can be achieved in the absence of structural change (Cooper and Dryden, 1984). The dimeric CAP that binds to cAMP is an example of the Cooper–Dryden model where changes were observed in the dynamics of the system but not the structure (Popovych et al., 2006). In another example, dynamics is an integral part of the allosteric response initiated by a ligand-induced disorder to order transition in the adenylate binding loop of the biotin repressor, a transcription regulatory protein (Naganathan and Beckett, 2007). Changes in internal fluctuation between different stages of the activation cycle for cyclin-dependent kinase 2 have been identified to be associated with functionally important regions for regulation and catalytic activity with possible detection of entropy compensation mechanisms being utilized (Gu and Bourne, 2007). The advantages of coupled disordered regions for allosteric control have been demonstrated through statistical mechanics (Hilser and Thompson, 2007).

Disease Impacts

Disordered regions in proteins have been implicated in several diseases such as neurodegenerative diseases, cardiovascular diseases, and cancer. These pathogenic culprits contain disordered regions making it difficult to conduct structural studies with X-ray crystallography and NMR. NACP, for example, is a natively unfolded protein that seeds the polymerization of amyloid proteins leading to Alzheimer's disease and impacts learning (Weinreb et al., 1996). It is suggested that the disorder regions allow for promiscuous binding and help potentiate protein–protein interaction that leads to the formation of these insoluble fibrils. Likewise, the tau protein found in Alzheimer's tangles is characterized to have intrinsically disordered regions and leads to the formation of amyloid fibrils connected with disease progression (Skrabana, Sevcik, and Novak, 2006; Skrabana et al., 2006).

A subset of eukaryotic proteins related to cardiovascular disease (CVD) was examined and concluded to be enriched in disorder content (Cheng et al., 2006a). The analysis was conducted with PONDR disorder predictions, cumulative distribution function analysis, and charge–hydropathy plot analysis. Predictions for α-helical molecular recognition features suggest high abundance within these proteins. The percentage of CVD containing >30 residues predicted to be disordered was $57 \pm 4\%$ compared to $47 \pm 4\%$ of eukaryotic proteins in Swiss-Prot. The role of disorder in cardiovascular diseases needs to be further validated experimentally, but the finding does not come as a surprise since disordered regions are found to be associated with $66 \pm 6\%$ of signaling molecules, proteins that often have regulatory roles. Diseases are often a result of regulated processes gone awry.

Similarly, $79 \pm 5\%$ of human cancer associated proteins have been found to contain regions of disorder that are at least 30 consecutive residues in length (Iakoucheva et al., 2002). One example of an oncoprotein is the HPV16 E7 that is an extended dimer with a stable and cooperative fold but displays properties of natively unfolded proteins (Garcia-Alai, Alonso, and de Prat-Gay, 2007). The region of disorder is located at the N-terminal region of the E7 domain that contains two important sites for regulation: (1) the retinoblastoma tumor suppressor binding site for molecular recognition and (2) casein kinase II phosphorylation site that induces stabilization with phosphorylation. The structural plasticity of this region has allowed for adaptation to binding of a variety of protein targets and regulation of protein turnover. HPV16 is one of the human papillomavirus strains associated with high frequency to cervical cancer.

Case Study: Disorder in the Glucocorticoid Receptor

We present the structural anatomy of a transcription factor in more detail as an example to show how disordered regions may play a functional role (Figure 38.3). Glucocorticoid receptor (GR) is a steroid binding nuclear receptor with well-defined domain boundaries (McEwan et al., 2007). The receptor is composed of three domains with structures available

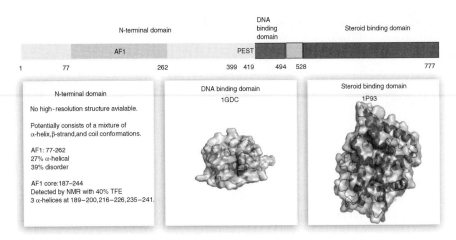

Figure 38.3. Anatomy of the glucocorticoid receptor. The structure of nearly half of the glucocorticoid receptor cannot be resolved due to intrinsic disorder in the N-terminal domain that contains the transactivating motif AF1. Low-resolution structural information shows a composition of α-helices in the AF1 core region (187–244). Residues 399–419 are found to contain the PEST motif of proline, glutamic acid, serine, and threonine that is associated with highly disordered regions. High-resolution structures are available for the DNA and steroid binding domains connected by a hinge.

for the DNA and ligand binding domains located at the C-terminal end in the bound form. The N-terminal domain (NTD), on the other hand, is highly disordered and no high-resolution structural data are available to study this region that contains the transactivating AF1 domain (residues 77–262) involved in protein–protein interactions and regulation of transcriptional activity (Lavery and McEwan, 2005). However, significant structural data have been obtained using alternative methods such as biochemical analysis, circular dichroism, NMR, fluorescence, and Fourier transform infrared spectroscopy. Predictions for the structural content of this region have also been made using secondary structure prediction algorithms. These data collectively show that GR-NTD potentially consists of a mixture of α-helix, β-strand, and coil conformations. The disordered state of this region is hypothesized to provide a mechanism for allosteric control that allows for the adoption of different conformers that subsequently create different binding interfaces to interact with a multitude of targets. This feature may be particularly important for the AF1 region that is found to be 27% α-helical and 39% disordered in GR.

The AF1 region may be an example of molecular recognition elements important for protein–protein interactions that use disordered regions as shuttles mentioned earlier. Through mutagenic studies, the induced formation of α-helical structures in this region has been correlated with the transactivation potential of GR (Dahlman-Wright et al., 1995; Dahlman-Wright and McEwan, 1996). This example demonstrates how regulation of transcriptional activity is achieved through modulating the order–disorder transition state that can be induced through a variety of factors such as DNA binding events (Lefstin and Yamamoto, 1998; Kumar et al., 1999) and even the presence of structure inducing osmolytes (Baskakov et al., 1999; Kumar et al., 2007). This strategy may be commonly used by all transcription factors as suggested by an analysis with PONDR that shows a relatively increased disorder content in transcription factors compared to other subsets of the eukaryotic proteins. Furthermore, the transcription activation regions are identified to have higher disorder content compared to the DNA binding region for the majority of the transcription factors (Liu et al., 2006).

PROTEIN CONFORMATIONAL VARIANTS AND ENSEMBLES

A discussion about protein disorder is really a discussion of protein conformational variants and the resulting ensembles that are the underlying basis for all biological phenomena and observations measured experimentally. A concept of ensemble highlights multiple possibilities that can be explored by proteins in alternative conformations rather than a single static structure, an important concept we wish to emphasize in this section. Consideration of alternative protein conformations expands not only the structural space, but also the functional space that can be regulated simply through partial unfolding that is observed as local protein disorder.

Structural variations are often appreciated when differences are observed between homologous proteins, but structural variations can also be observed for a protein at a single equilibrium state or between two states such as a ligand-bound and an unbound conformation (Figure 38.4). In the field of structural biology and structural bioinformatics, it is convenient

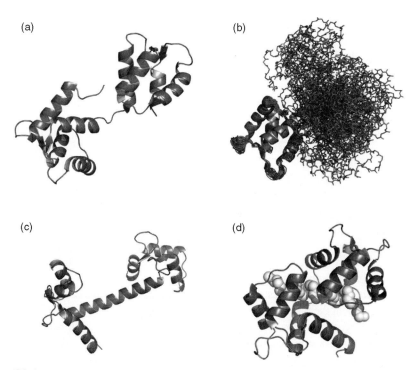

Figure 38.4. Conformational variations in calmodulin. Comparison of (a) one conformational state of yeast calmodulin and (b) 31 states aligned at the N-terminal domain in the absence of calcium ions (PDBID: 1LKJ). (c) The bovine calmodulin adopts a dumbbell-like shape with calcium binding, which is different from the more globular structure found in yeast. (d) Calmodulin bound to a substrate (white spheres).

to view high-resolution structural data as a single molecule, but we must remind ourselves that this interpretation is not the complete view. X-ray crystallographic studies are a collective contribution of all the protein molecules at equilibrium state found in the crystal lattice. Thus, the X-ray structure would represent the dominant conformation in the ensemble with regions of high temperature factors indicating a higher conformational variability. NMR, on the other hand, provides multiple solutions for conformations found in solution, thus instilling a greater appreciation in the structure interpreter for conformational variations. Protein dynamics and disorder leading to conformational variation is observed in NMR experiments as resonance overlap and peak broadening from conformational averaging and contributions from intermediate time scale dynamics.

As an example of a protein that exists in many different conformational variations, we use calmodulin to illustrate the point (Figure 38.4). This regulator responds to calcium ions and exists in three main conformational states: (1) the apo-structure, (2) bound to calcium ions, (3) and bound to the target substrate. The apo-form of calmodulin has a structure of two globular domains connected by a hinge as observed in an NMR structure of calmodulin from *Saccharomyces cerevisiae* (Figure 38.4a). Variations within a single state can be immediately observed in the apo-structure where alternative conformational states are aligned based on the N-terminal domain (Figure 38.4b). The C-terminal globular domain can exist in a different conformation relative to the N-terminal domain. Variations between homologues are observed in the calcium-bound bovine calmodulin with a helical linker region between the two domains whereas the yeast calmodulin adopts a more globular structure (Figure 38.4c). Finally, significant structural rearrangement is observed with binding to substrate. A multitude of structural conformations donned on by calmodulin represent some challenges that face the structural bioinformatics field.

A biophysical explanation for protein disorder can be described by the underlying presence of conformational variations (Figure 38.5). An important concept that must be delivered here is that most experimental measurements of proteins are not single-molecule studies and therefore are collective contributions of all protein molecules in the solution. Thus, the observed measurement can be written as the summed contribution of each conformational state in the solution:

$$\langle \text{Obs} \rangle = \sum P_i \times \text{Obs}_i.$$

With this in mind, disorder in X-ray structure, for example, arises when there are many conformational variations that do not give rise to a single converged structure that is viewed as "ordered." Sometimes highly ordered regions can be mistaken to be a disordered structure, particularly if large domain motions are involved such as those observed in calmodulin (Figure 38.4b).

The contribution of different states to the observation can be explained by one of the two models that represent the ratio of states differently (Figure 38.5). The first model assumes a discrete two-state conformation while the second allows for additional conformational states to be present. We illustrate the impact of the difference between the two models by applying it to the unfolding process of proteins, for example. In the first model, only the native (order) and denatured (disorder) states of the protein are allowed to exist in solution. The observed destabilization of proteins with increasing denaturant is then a result of the changing ratio between these two states in the solution. The probability of observing an ordered structure will decrease as the probability of observing a disordered structure will increase. In the second model, intermediate states containing partially unfolded conformers are allowed.

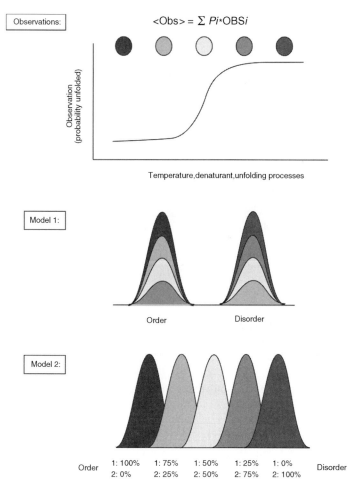

$$<Obs> = \Sigma\ P_i * OBS_i$$

Figure 38.5. Ensemble-based description of protein disorder. (Top) Biological observations are the sum contribution of the different states in solution. The unfolding process of proteins, for example, can be described with one of the two models. Model 1 assumes two discrete states in solution in the native (order) conformation or denatured (disorder) conformation. The denaturation process is the changing ratio of these two states from one spectrum to the other. Model 2 allows from other intermediate conformations to contribute to the observation. Figure also appears in the Color Figure section.

Thus, the probability of observing each intermediate state as well as the native and denatured states contributes to the observation.

The importance of an ensemble-based interpretation of the native state can be demonstrated through the use of COREX, a statistical thermodynamic model that uses free energy values that have been structurally parameterized (Hilser and Freire, 1996; Hilser et al., 2006). This experimentally validated model allows us to calculate the heat capacity (ΔC_p), enthalpy (ΔH), and entropy (ΔS) differences between the partially unfolded states and the native state (reported in kcal/K/mol). More importantly, the derivation allows for the interpretation of residue stability and contribution to the energetics of the ensemble that have provided insight into possible mechanisms for cooperative (Hilser et al., 1998) and allosteric processes (Hilser and Thompson, 2007).

Briefly, the relative Gibbs free energy of each possible conformational state adopted by the protein (ΔG_i) is expressed in terms of the standard thermodynamic equation:

$$\Delta G_i = \Delta H_i - T\Delta S_i.$$

COREX obtains the relative Gibbs free energy using (1) a high-resolution structure and (2) a statistical thermodynamic model where the variables have been parameterized based on changes in the accessible surface area (ΔASA) between the native and the partially unfolded state. The enthalpic contribution to the state can be written as the sum of enthalpic contributions from apolar (ΔH_{ap}) and polar residues (ΔH_{pol}):

$$\Delta H = \Delta H_{ap} + \Delta H_{pol}.$$

The enthalpy change is related to ΔASA (Å^2) in the following way and is parameterized at a reference temperature of $60\,°C$, which is the median unfolding temperature for the data set of model proteins used:

$$
\begin{aligned}
\Delta H(60) &= a_H(60)\Delta\text{ASA}_{ap} + b_H(60)\Delta\text{ASA}_{pol}, \\
a_H(60) &= -8.44, \\
b_H(60) &= 31.4.
\end{aligned}
$$

The entropy of the system is the sum of contributions from solvent ΔS_{solv} and conformational ΔS_{conf} entropies:

$$\Delta S = \Delta S_{solv} + \Delta S_{conf}.$$

The solvent entropy can be calculated with the knowledge of the heat capacity of the protein as derived:

$$
\begin{aligned}
\Delta S_{solv} &= \Delta S_{solv,ap} + \Delta S_{solv,pol}, \\
\Delta S_{solv} &= \Delta C_{p,ap}\ln(T/T^*_{S,ap}) + \Delta C_{p,pol}\ln(T/T^*_{S,pol}),
\end{aligned}
$$

where $T^*_{S,ap} = 385.15$ and $T^*_{S,pol} = 335.15$ are the reference temperatures at which the hydration entropy is equal to zero (Baldwin, 1986; Murphy and Freire, 1992; D'Aquino et al., 1996). The heat capacity is found to scale to ΔASA for temperatures up to $80\,°C$ as follows:

$$
\begin{aligned}
\Delta C_p &= \Delta C_{p,ap} + \Delta C_{p,pol}, \\
\Delta C_p &= a_c(T)\Delta\text{ASA}_{ap} + b_c(T)*\Delta\text{ASA}_{pol}, \\
a_c(T) &= 0.45 + 2.63 \times 10^{-4}(T-25) - 4.2 \times 10^{-5}(T-25)^2, \\
b_c(T) &= -0.26 + 2.85 \times 10^{-4}(T-25) + 4.31 \times 10^{-5}(T-25)^2.
\end{aligned}
$$

Finally, to complete the calculation of ΔS, conformational entropy is performed as follows:

$$\Delta S_{conf} = \sum \Delta S_{bu-ex} + \sum \Delta S_{ex-un} + \sum \Delta S_{bb}.$$

The three contributions to conformational entropy are (1) $\sum \Delta S_{bu-ex}$: buried residues that become exposed with partial unfolding; (2) $\sum \Delta S_{ex-un}$: exposed residues in the unfolded state; and (3) $\sum \Delta S_{bb}$: backbone entropy changes for residues that become unfolded. The entropy contributions of each amino acid have been determined and these values are used in the calculation (Lee et al., 1994; D'Aquino et al., 1996).

The relative Gibbs free energy of each state is calculated with these parameterized thermodynamic variables and will be important in determining the probability of observing such a conformational state in the ensemble. Under equilibrium conditions, statistical mechanics states that the probability of any given conformational state i (P_i) is given by the equation

$$P_i = \frac{\exp(-\Delta G_i/\mathrm{RT})}{Q},$$

where the statistical weights, also known as the Boltzmann exponents ($\exp(-\Delta G/RT)$, are defined by ΔG_i relative to the gas constant R and temperature T. Q is the conformational partition function defined as the sum of the statistical weights of all the states accessible to the protein:

$$Q = \sum_{i=0}^{N} \exp(-\Delta G_i/RT).$$

These probabilities reflect preferences for the protein to adopt a partially unfolded conformational state and can be extended to calculate the free energy contributions of each residue to the ensemble. Using the probability-weighted conformations in the generated ensemble, residue stability in the protein can be calculated as the ratio of residues in the folded and unfolded states:

$$k_{\mathrm{f},j} = \frac{\sum P_{\mathrm{f},j}}{\sum P_{\mathrm{nf},j}},$$

where $\sum P_{\mathrm{f},j}$ and $\sum P_{\mathrm{nf},j}$ are the summed probabilities of all the states in which the residue is either folded or unfolded, respectively.

The free energy contribution of each residue to the ensemble can then be calculated:

$$\Delta G_{f,j} = -RT \ln k_{\mathrm{f},j}.$$

The importance of this derived formalism is that it can be extended to study changes in energetic contributions at the residue level and provide insights into functional processes such as cooperativity (Liu, Whitten, and Hilser, 2006; Liu, Whitten, and Hilser, 2007; Pan, Lee, and Hilser, 2000) and allostery (Hilser and Thompson, 2007) by interpreting proteins as an ensemble of multiple conformations. Recent systems in which cooperativity has been identified and studied with COREX are dihydrofolate reductase and elgin C. The studies defined structural–thermodynamic linkages based on correlations in stability changes between residues in the ensemble as captured by $k_{\mathrm{f},j}$. By examining these correlations, the model helps to define a mechanism for site–site communication, particularly between ligand binding sites and distantly located regions. The analyses suggest an alternative view to energetic coupling between residues when a clear, connected pathway of intramolecular interactions between them cannot be identified. The results also further emphasize the importance of entropic contributions that is often neglected.

While it is important to produce the correct high-resolution structure using fold recognition, homology modeling, and *ab initio* structure prediction approaches (Chapter 29–32), it is also equally important to construct other physically and chemically valid conformational states that can be sampled by the protein. Generating these structural variants that collectively produce a protein ensemble can be achieved with a variety of models, some more restrictive than others. Restrictive models are those that assume disordered or partially unfolded regions of the protein to adopt only coil structures (Bernado et al., 2005; Jha et al., 2005). A less restrictive model such as TraDES (Feldman and Hogue, 2000) is an unbiased conformational sampling method that generates plausible random structures allowing for both native and nonnative contacts. Other conformer generating methods including Rosetta (Simons et al., 1997) and CNS (Brunger et al., 1998) can also be used to predict structures in these disordered and highly flexible regions. The relative probabilities of the generated conformational variants that potentially populate the ensemble can then be calculated with experimental constraints using ENSEMBLE (Choy and Forman-Kay, 2001; Marsh et al., 2007). The population weight assignment is achieved with a pseudoenergy minimization process and a Monte Carlo algorithm. With these strategies we can now possibly begin to make interpretation of the functional consequences arising from these variety of conformational states.

FUTURE DIRECTIONS

Aside from the growing amount of literature on this topic, community recognition of the importance of understanding disordered region is signified by the inclusion of disorder predictor evaluation in CASP (CH 28). The first evaluation of disorder predictors appeared in CASP5 (Melamud and Moult, 2003) in 2002 with results showing successful detection for over half of the disordered residues in the blind set with a low rate of overprediction. However, proper evaluation of these predictors remains a challenge that still needs to be refined. This is to be expected due to both the varied definition and the existence of these different types of disordered regions. Furthermore, as noted by the assessors, the data set used for evaluation is skewed toward short disordered regions identified by missing residues in X-ray crystallographic structures. As such, caution should be taken when interpreting the performance of these results. In spite of the mentioned weaknesses, the evaluation process is a necessity because the predictors serve many useful purposes. The most recent benchmarking effort conducted at CASP7 in 2006 showed that in spite of the many new generations of disorder predictors, significant improvements in the performances have not be observed and variations are seen in their sensitivity and specificity for detecting these regions (Bordoli, Kiefer, and Schwede, 2007). Improvements in disorder predictors cannot be made without a systematic study of these regions and several experimental strategies using techniques that combine heat and acid treatment with mass spectrometry and/or 2D electrophoresis have been proposed to tackle this issue (Csizmok et al., 2007).

The applications of disordered predictors are not limited to target protein identification and elimination for structural genomic efforts. New applications include improved functional categorization of newly identified proteins (Lobley et al., 2007) and a potential role in improved drug design (Cheng et al., 2006b). The power of leveraging what we know about disordered regions will prove itself to be immensely valuable for the majority of the proteins that do not adopt a native fold. Currently, the function of about 35% of proteins cannot be categorized using homology-based assignment, leaving researchers with a large set of "hypothetical protein" drug targets with unknown function (Ofran et al., 2005). A systematic characterization of protein disorder can be achieved by combining the developments in improved computational and experimental analysis (Bracken et al., 2004).

Finally, the importance of understanding subtle differences in conformational variations, due to effects such as mutational events, has always been recognized by the structural bioinformatics field. New measures to better understand these variations are indicated by the goals presented to the structure prediction community proposed at the conclusion of CASP6 (Moult et al., 2005). The four challenges to overcome are to (1) model the structure of single-residue mutants, (2) model the structural changes associated with specificity changes within protein families, (3) improve refinement methods to produce a 0.5 Å root-mean-square-deviation (RMSD) improvement in the Cα accuracy of models, and (4) devise a scoring function that will reliably pick the most accurate model of the possible candidate structures for new fold predictions. As the community addresses these challenges, an ensemble view of conformational variations should be kept in mind to understand functional consequences as well.

WEB RESOURCES

Resource	References	URL
DisProt: the database of disordered proteins	Sickmeier et al. (2007)	http://www.disprot.org/
DisProt: list of disorder predictors	Not published, a part of DisProt	http://www.ist.temple.edu/disprot/predictors.php
DISOPRED	Jones and Ward (2003)	http://bioinf.cs.ucl.ac.uk/disopred/
VLXT (PONDR)	Romero et al. (1997)	http://www.pondr.com
GlobPlot	Linding et al. (2003b)	http://globplot.embl.de
DisEMBL	Linding et al. (2003a)	http://dis.embl.de
PrDOS	Ishida and Kinoshita (2007)	http://prdos.hgc.jp/cgi-bin/top.cgi
RONN	Yang et al. (2005)	http://www.strubi.ox.ac.uk/RONN
Wiggle	Gu et al. (2006)	http://wiggle.sdsc.edu
PROFbval	Schlessinger et al. (2006)	http://cubic.bioc.columbia.edu/services/profbval/
iPDA: integrated protein disorder analyzer	Su et al. (2007)	http://biominer.bime.ntu.edu.tw/ipda/
ENSEMBLE	Choy and Forman-Kay (2001) and Marsh et al. (2007)	http://pound.med.utoronto.ca/~forman/ensemble/ensemble.html
CNS	Brunger et al. (1998)	http://helix.nih.gov/apps/structbio/cns.html
ROSETTA	Simons et al. (1997)	http://www.rosettacommons.org/
COREX/BEST server	Vertrees et al. (2005)	http://www.best.utmb.edu/BEST/

REFERENCES

Altschul SF, Madden TL, Schaffer AA, Zhang J, Zhang Z, Miller W, Lipman DJ (1997): Gapped BLAST and PSI-BLAST: a new generation of protein database search programs. *Nucleic Acids Res* 25:3389–3402.

Baldwin RL (1986): Temperature dependence of the hydrophobic interaction in protein folding. *Proc Natl Acad Sci USA* 83:8069–8072.

Baskakov IV, Kumar R, Srinivasan G, Ji YS, Bolen DW, Thompson EB (1999): Trimethylamine *N*-oxide-induced cooperative folding of an intrinsically unfolded transcription-activating fragment of human glucocorticoid receptor. *J Biol Chem* 274:10693–10696.

Bernado P, Blanchard L, Timmins P, Marion D, Ruigrok RW, Blackledge M (2005): A structural model for unfolded proteins from residual dipolar couplings and small-angle X-ray scattering. *Proc Natl Acad Sci USA* 102:17002–17007.

Blow DM (1977): Flexibility and rigidity in protein crystals. *Ciba Found Symp* 55–61.

Bonvin AM, Brunger AT (1995): Conformational variability of solution nuclear magnetic resonance structures. *J Mol Biol* 250:80–93.

Bordoli L, Kiefer F, Schwede T (2007): Assessment of disorder predictions in CASP7. *Proteins* 69:129–136.

Bracken C, Iakoucheva LM, Rorner PR, Dunker AK (2004): Combining prediction, computation and experiment for the characterization of protein disorder. *Curr Opin Struct Biol* 14:570–576.

Brown CJ, Takayama S, Campen AM, Vise P, Marshall TW, Oldfield CJ, Williams CJ, Dunker AK (2002): Evolutionary rate heterogeneity in proteins with long disordered regions. *J Mol Evol* 55:104–110.

Brunger AT, Adams PD, Clore GM, DeLano WL, Gros P, Grosse-Kunstleve RW, Jiang JS, Kuszewski J, Nilges M, Pannu NS, et al. (1998): Crystallography & NMR system: a new software suite for macromolecular structure determination. *Acta Crystallogr D* 54:905–921.

Cheng Y, LeGall T, Oldfield CJ, Dunker AK, Uversky VN (2006a): Abundance of intrinsic disorder in protein associated with cardiovascular disease. *Biochemistry* 45:10448–10460.

Cheng Y, LeGall T, Oldfield CJ, Mueller JP, Van YY, Romero P, Cortese MS, Uversky VN, Dunker AK (2006b): Rational drug design via intrinsically disordered protein. *Trends Biotechnol* 24:435–442.

Choy WY, Forman-Kay JD (2001): Calculation of ensembles of structures representing the unfolded state of an SH3 domain. *J Mol Biol* 308:1011–1032.

Cooper A, Dryden DT (1984): Allostery without conformational change. A plausible model. *Eur Biophys J* 11:103–109.

Csizmok V, Dosztanyi Z, Simon I, Tompa P (2007): Towards proteomic approaches for the identification of structural disorder. *Curr Protein Pept Sci* 8:173–179.

Dafforn TR, Smith CJ (2004): Natively unfolded domains in endocytosis: hooks, lines and linkers. *EMBO Rep* 5:1046–1052.

Dahlman-Wright K, Baumann H, McEwan IJ, Almlof T, Wright AP, Gustafsson JA, Hard T (1995): Structural characterization of a minimal functional transactivation domain from the human glucocorticoid receptor. *Proc Natl Acad Sci USA* 92:1699–1703.

Dahlman-Wright K, McEwan IJ (1996): Structural studies of mutant glucocorticoid receptor transactivation domains establish a link between transactivation activity *in vivo* and alpha-helix-forming potential *in vitro*. *Biochemistry* 35:1323–1327.

D'Aquino JA, Gomez J, Hilser VJ, Lee KH, Amzel LM, Freire E (1996): The magnitude of the backbone conformational entropy change in protein folding. *Proteins* 25:143–156.

Dosztanyi Z, Csizmok V, Tompa P, Simon I (2005a): IUPred: web server for the prediction of intrinsically unstructured regions of proteins based on estimated energy content. *Bioinformatics* 21:3433–3434.

Dosztanyi Z, Csizmok V, Tompa P, Simon I (2005b): The pairwise energy content estimated from amino acid composition discriminates between folded and intrinsically unstructured proteins. *J Mol Biol* 347:827–839.

Dunker AK, Brown CJ, Lawson JD, Iakoucheva LM, Obradovic Z (2002): Intrinsic disorder and protein function. *Biochemistry* 41:6573–6582.

Feldman HJ, Hogue CW (2000): A fast method to sample real protein conformational space. *Proteins* 39:112–131.

Fuentes G, Nederveen AJ, Kaptein R, Boelens R, Bonvin AM (2005): Describing partially unfolded states of proteins from sparse NMR data. *J Biomol NMR* 33:175–186.

Fuxreiter M, Simon I, Friedrich P, Tompa P (2004): Preformed structural elements feature in partner recognition by intrinsically unstructured proteins. *J Mol Biol* 338:1015–1026.

Fuxreiter M, Tompa P, Simon I (2007): Local structural disorder imparts plasticity on linear motifs. *Bioinformatics* 23:950–956.

Garcia-Alai MM, Alonso LG, de Prat-Gay G (2007): The N-terminal module of HPV16 E7 is an intrinsically disordered domain that confers conformational and recognition plasticity to the oncoprotein. *Biochemistry* 46:10405–10412.

Gu J, Gribskov M, Bourne PE (2006): Wiggle-predicting functionally flexible regions from primary sequence. *PLoS Comput Biol* 2:e90.

Gu J, Bourne PE (2007): Identifying allosteric fluctuation transitions between different protein conformational states as applied to cyclin dependent kinase 2. *BMC Bioinform* 8:45.

Hegyi H, Schad E, Tompa P (2007): Structural disorder promotes assembly of protein complexes. *BMC Struct Biol* 7:65.

Hilser VJ, Freire E (1996): Structure-based calculation of the equilibrium folding pathway of proteins. Correlation with hydrogen exchange protection factors. *J Mol Biol* 262:756–772.

Hilser VJ, Dowdy D, Oas TG, Freire E (1998): The structural distribution of cooperative interactions in proteins: analysis of the native state ensemble. *Proc Natl Acad Sci USA* 95:9903–9908.

Hilser VJ, Garcia-Moreno EB, Oas TG, Kapp G, Whitten ST (2006): A statistical thermodynamic model of the protein ensemble. *Chem Rev* 106:1545–1558.

Hilser VJ, Thompson EB (2007): Intrinsic disorder as a mechanism to optimize allosteric coupling in proteins. *Proc Natl Acad Sci USA* 104:8311–8315.

Hirose S, Shimizu K, Kanai S, Kuroda Y, Noguchi T (2007): POODLE-L: a two-level SVM prediction system for reliably predicting long disordered regions. *Bioinformatics* 23:2046–2053.

Iakoucheva LM, Brown CJ, Lawson JD, Obradovic Z, Dunker AK (2002): Intrinsic disorder in cell-signaling and cancer-associated proteins. *J Mol Biol* 323:573–584.

Ishida T, Kinoshita K (2007): PrDOS: prediction of disordered protein regions from amino acid sequence. *Nucleic Acids Res* 35:W460–464.

Jha AK, Colubri A, Freed KF, Sosnick TR (2005): Statistical coil model of the unfolded state: resolving the reconciliation problem. *Proc Natl Acad Sci USA* 102:13099–13104.

Jones DT, Ward JJ (2003): Prediction of disordered regions in proteins from position specific score matrices. *Proteins* 53:(Suppl. 6): 573–578.

Kellermayer MS, Smith SB, Granzier HL, Bustamante C (1997): Folding–unfolding transitions in single titin molecules characterized with laser tweezers. *Science* 276:1112–1116.

Kemmink J, van Mierlo CP, Scheek RM, Creighton TE (1993): Local structure due to an aromatic-amide interaction observed by ^{1}H-nuclear magnetic resonance spectroscopy in peptides related to the N terminus of bovine pancreatic trypsin inhibitor. *J Mol Biol* 230:312–322.

Kern D, Eisenmesser EZ, Wolf-Watz M (2005): Enzyme dynamics during catalysis measured by NMR spectroscopy. *Methods Enzymol* 394:507–524.

Kumar R, Baskakov IV, Srinivasan G, Bolen DW, Lee JC, Thompson EB (1999): Interdomain signaling in a two-domain fragment of the human glucocorticoid receptor. *J Biol Chem* 274:24737–24741.

Kumar R, Serrette JM, Khan SH, Miller AL, Thompson EB (2007): Effects of different osmolytes on the induced folding of the N-terminal activation domain (AF1) of the glucocorticoid receptor. *Arch Biochem Biophys* 465:452–460.

Kuriyan J, Osapay K, Burley SK, Brunger AT, Hendrickson WA, Karplus M (1991): Exploration of disorder in protein structures by X-ray restrained molecular dynamics. *Proteins* 10:340–358.

Labeit S, Kolmerer B (1995): Titins: giant proteins in charge of muscle ultrastructure and elasticity. *Science* 270:293–296.

Landry SJ, Taher A, Georgopoulos C, van der Vies SM (1996): Interplay of structure and disorder in cochaperonin mobile loops. *Proc Natl Acad Sci USA* 93:11622–11627.

Lavery DN, McEwan IJ (2005): Structure and function of steroid receptor AF1 transactivation domains: induction of active conformations. *Biochem J* 391:449–464.

Lee KH, Xie D, Freire E, Amzel LM (1994): Estimation of changes in side chain configurational entropy in binding and folding: general methods and application to helix formation. *Proteins* 20:68–84.

Lefstin JA, Yamamoto KR (1998): Allosteric effects of DNA on transcriptional regulators. *Nature* 392:885–888.

Le Gall T, Romero PR, Cortese MS, Uversky VN, Dunker AK (2007): Intrinsic disorder in the Protein Data Bank. *J Biomol Struct Dyn* 24:325–342.

Li F, Gangal M, Juliano C, Gorfain E, Taylor SS, Johnson DA (2002): Evidence for an internal entropy contribution to phosphoryl transfer: a study of domain closure, backbone flexibility, and the catalytic cycle of cAMP-dependent protein kinase. *J Mol Biol* 315:459–469.

Linding R, Jensen LJ, Diella F, Bork P, Gibson TJ, Russell RB (2003a): Protein disorder prediction: implications for structural proteomics. *Structure* 11:1453–1459.

Linding R, Russell RB, Neduva V, Gibson TJ (2003b): GlobPlot: exploring protein sequences for globularity and disorder. *Nucleic Acids Res* 31:3701–3708.

Lindorff-Larsen K, Kristjansdottir S, Teilum K, Fieber W, Dobson CM, Poulsen FM, Vendruscolo M (2004): Determination of an ensemble of structures representing the denatured state of the bovine acyl-coenzyme a binding protein. *J Am Chem Soc* 126:3291–3299.

Lise S, Jones DT (2005): Sequence patterns associated with disordered regions in proteins. *Proteins* 58:144–150.

Liu J, Perumal NB, Oldfield CJ, Su EW, Uversky VN, Dunker AK (2006): Intrinsic disorder in transcription factors. *Biochemistry* 45:6873–6888.

Liu T, Whitten ST, Hilser VJ (2006): Ensemble-based signatures of energy propagation in proteins: a new view of an old phenomenon. *Proteins* 62:728–738.

Liu T, Whitten ST, Hilser VJ (2007): Functional residues serve a dominant role in mediating the cooperativity of the protein ensemble. *Proc Natl Acad Sci USA* 104:4347–4352.

Lobley A, Swindells MB, Orengo CA, Jones DT (2007): Inferring function using patterns of native disorder in proteins. *PLoS Comput Biol* 3:e162.

Marsh JA, Neale C, Jack FE, Choy WY, Lee AY, Crowhurst KA, Forman-Kay JD (2007): Improved structural characterizations of the drkN SH3 domain unfolded state suggest a compact ensemble with native-like and non-native structure. *J Mol Biol* 367:1494–1510.

McEwan IJ, Lavery D, Fischer K, Watt K (2007): Natural disordered sequences in the amino terminal domain of nuclear receptors: lessons from the androgen and glucocorticoid receptors. *Nucl Receptor Signal* 5:e001.

Melamud E, Moult J (2003): Evaluation of disorder predictions in CASP5. *Proteins* 53:561–565.

Meszaros B, Tompa P, Simon I, Dosztanyi Z (2007): Molecular principles of the interactions of disordered proteins. *J Mol Biol* 372:549–561.

Mittag T, Forman-Kay JD (2007): Atomic-level characterization of disordered protein ensembles. *Curr Opin Struct Biol* 17:3–14.

Moult J, Fidelis K, Rost B, Hubbard T, Tramontano A (2005): Critical assessment of methods of protein structure prediction (CASP)—round 6. *Proteins* 61:(Suppl. 7): 3–7.

Murphy KP, Freire E (1992): Thermodynamics of structural stability and cooperative folding behavior in proteins. *Adv Protein Chem* 43:313–361.

Naganathan S, Beckett D (2007): Nucleation of an allosteric response via ligand-induced loop folding. *J Mol Biol* 373:96–111.

Namba K (2001): Roles of partly unfolded conformations in macromolecular self-assembly. *Genes Cells* 6:1–12.

Ofran Y, Punta M, Schneider R, Rost B (2005): Beyond annotation transfer by homology: novel protein-function prediction methods to assist drug discovery. *Drug Discov Today* 10:1475–1482.

Oldfield CJ, Cheng Y, Cortese MS, Brown CJ, Uversky VN, Dunker AK (2005a): Comparing and combining predictors of mostly disordered proteins. *Biochemistry* 44:1989–2000.

Oldfield CJ, Ulrich EL, Cheng Y, Dunker AK, Markley JL (2005b): Addressing the intrinsic disorder bottleneck in structural proteomics. *Proteins* 59:444–453.

Palliser CC, Parry DA (2001): Quantitative comparison of the ability of hydropathy scales to recognize surface beta-strands in proteins. *Proteins* 42:243–255.

Pan H, Lee JC, Hilser VJ (2000): Binding sites in *Escherichia coli* dihydrofolate reductase communicate by modulating the conformational ensemble. *Proc Natl Acad Sci USA* 97:12020–12025.

Pometun MS, Chekmenev EY, Wittebort RJ (2004): Quantitative observation of backbone disorder in native elastin. *J Biol Chem* 279:7982–7987.

Popovych N, Sun S, Ebright RH, Kalodimos CG (2006): Dynamically driven protein allostery. *Nat Struct Mol Biol* 13:831–838.

Radivojac P, Iakoucheva LM, Oldfield CJ, Obradovic Z, Uversky VN, Dunker AK (2007): Intrinsic disorder and functional proteomics. *Biophys J* 92:1439–1456.

Ringe D, Petsko GA (1986): Study of protein dynamics by X-ray diffraction. *Methods Enzymol* 131:389–433.

Romero P, Obradovic Z, Dunker K (1997): Sequence data analysis for long disordered regions prediction in the calcineurin family. Genome Inform Ser Workshop Genome Inform, Vol. 8, pp 110–124.

Romero P, Obradovic Z, Kissinger C, Villafranca JE, Dunker AK (1997): Identifying disordered regions in proteins from amino acid sequences. Proceedings of the IEEE. International Conference on Neural Networks, Vol. 1, pp 90–95.

Romero P, Obradovic Z, Li XH, Garner EC, Brown CJ, Dunker AK (2001): Sequence complexity of disordered protein. *Proteins* 42:38–48.

Romero PR, Zaidi S, Fang YY, Uversky VN, Radivojac P, Oldfield CJ, Cortese MS, Sickmeier M, LeGall T, Obradovic Z, Dunker AK (2006): Alternative splicing in concert with protein intrinsic disorder enables increased functional diversity in multicellular organisms. *Proc Nat Acad Sci USA* 103:8390–8395.

Sandhu KS, Dash D (2006): Conformational flexibility may explain multiple cellular roles of PEST motifs. *Proteins* 63:727–732.

Sasakawa H, Sakata E, Yamaguchi Y, Masuda M, Mori T, Kurimoto E, Iguchi T, Hisanaga SI, Iwatsubo T, Hasegawa M, Kato K (2007): Ultra-high field NMR studies of antibody binding and site-specific phosphorylation of alpha-synuclein. *Biochem Biophys Res Commun* 363:795–799.

Schlessinger A, Yachdav G, Rost B (2006): PROFbval: predict flexible and rigid residues in proteins. *Bioinformatics* 22:891–893.

Sickmeier M, Hamilton JA, LeGall T, Vacic V, Cortese MS, Tantos A, Szabo B, Tompa P, Chen J, Uversky VN, et al. (2007): DisProt: the database of disordered proteins. *Nucleic Acids Res* 35: D786–793.

Simons KT, Kooperberg C, Huang E, Baker D (1997): Assembly of protein tertiary structures from fragments with similar local sequences using simulated annealing and Bayesian scoring functions. *J Mol Biol* 268:209–225.

Skrabana R, Sevcik J, Novak M (2006): Intrinsically disordered proteins in the neurodegenerative processes: formation of tau protein paired helical filaments and their analysis. *Cell Mol Neurobiol* 26:1085–1097.

Skrabana R, Skrabanova-Khuebachova M, Kontsek P, Novak M (2006): Alzheimer's-disease-associated conformation of intrinsically disordered tau protein studied by intrinsically disordered protein liquid-phase competitive enzyme-linked immunosorbent assay. *Anal Biochem* 359:230–237.

Su CT, Chen CY, Hsu CM (2007): iPDA: integrated protein disorder analyzer. *Nucleic Acids Res* 35: W465–472.

Tompa P, Szasz C, Buday L (2005): Structural disorder throws new light on moonlighting. *Trends Biochem Sci* 30:484–489.

Torda AE, Scheek RM, Gunsteren WF, (1990): Time-averaged nuclear Overhauser effect distance restraints applied to tendamistat. *J Mol Biol* 214:223–235.

Tsutakawa SE, Hura GL, Frankel KA, Cooper PK, Tainer JA (2007): Structural analysis of flexible proteins in solution by small angle X-ray scattering combined with crystallography. *J Struct Biol* 158:214–223.

Vacic V, Oldfield CJ, Mohan A, Radivojac P, Cortese MS, Uversky VN, Dunker AK (2007): Characterization of molecular recognition features, MoRFs, and their binding partners. *J Proteome Res* 6:2351–2366.

van Oijen AM, Blainey PC, Crampton DJ, Richardson CC, Ellenberger T, Xie XS (2003): Single-molecule kinetics of lambda exonuclease reveal base dependence and dynamic disorder. *Science* 301:1235–1238.

Vertrees J, Barritt P, Whitten S, Hilser VJ (2005): COREX/BEST server: a web browser-based program that calculates regional stability variations within protein structures. *Bioinformatics* 21:3318–3319.

Vucetic S, Brown CJ, Dunker AK, Obradovic Z (2003): Flavors of protein disorder. *Proteins* 52:573–584.

Wang S, Gu J, Larson SA, Whitten ST, Hilser VJ (2008): Denatured-state energy landscapes of a protein structural database reveal the energetic determinants of a framework model for folding. *J Mol Biol* 381:1184–1201.

Ward JJ, Sodhi JS, McGuffin LJ, Buxton BF, Jones DT (2004): Prediction and functional analysis of native disorder in proteins from the three kingdoms of life. *J Mol Biol* 337:635–645.

Weathers EA, Paulaitis ME, Woolf TB, Hoh JH (2004): Reduced amino acid alphabet is sufficient to accurately recognize intrinsically disordered protein. *FEBS Lett* 576:348–352.

Weinreb PH, Zhen W, Poon AW, Conway KA, Lansbury PT Jr (1996): NACP, a protein implicated in Alzheimer's disease and learning, is natively unfolded. *Biochemistry* 35:13709–13715.

Wissmann R, Baukrowitz T, Kalbacher H, Kalbitzer HR, Ruppersberg JP, Pongs O, Antz C, Fakler B (1999): NMR structure and functional characteristics of the hydrophilic N terminus of the potassium channel beta-subunit Kvbeta1.1. *J Biol Chem* 274:35521–35525.

Wissmann R, Bildl W, Oliver D, Beyermann M, Kalbitzer HR, Bentrop D., Fakler B (2003): Solution structure and function of the "tandem inactivation domain" of the neuronal A-type potassium channel Kv1.4. *J Biol Chem* 278:16142–16150.

Wootton JC, Federhen S (1993): Statistics of local complexity in amino-acid-sequences and sequence databases. *Comp Chem* 17:149–163.

Wootton JC, Federhen S (1996): Analysis of compositionally biased regions in sequence databases. *Comp Methods Macromol Sequence Anal* 266:554–571.

Xiao H, Kaltashov IA (2005): Transient structural disorder as a facilitator of protein–ligand binding: native H/D exchange-mass spectrometry study of cellular retinoic acid binding protein I. *J Am Soc Mass Spectrom* 16:869–879.

Yang ZR, Thomson R, McNeil P, Esnouf RM (2005): RONN: the bio-basis function neural network technique applied to the detection of natively disordered regions in proteins. *Bioinformatics*.

Zhang Y, Stec B, Godzik A (2007): Between order and disorder in protein structures: analysis of "dual personality" fragments in proteins. *Structure* 15:1141–1147.

39

PROTEIN DESIGNABILITY AND ENGINEERING

Nikolay V. Dokholyan

INTRODUCTION

Rapid scientific and technological advances in biological sciences that started in the second half of the twentieth century have led to exploration and engineering at the scale of single biological molecules. DNA manipulation techniques, such as polymerase chain reaction (PCR), allowed almost arbitrary control of proteins that are expressed in living cells. These developments prompted scientists to ask increasingly ambitious questions pertaining to cell life, evolution, and the molecular origins of diseases. Rational manipulation of protein function through alteration in sequence and structure became an important aspect in the quest to answer these questions.

A protein's sequence uniquely defines its three-dimensional structure. This relationship is perhaps the most central paradigm of protein biophysics. Interestingly, however, a protein structure does not uniquely define a single sequence: multiple sequences often correspond to similar protein structures. This aspect turned out to be central for the field of structural bioinformatics. Similarly, a protein's structure defines its function, although a given structure can have several biological functions. Hence, one approach to manipulate protein function is through its sequence. Nature diversifies protein function through sequence over the course of evolution. This process is often referred to as *protein evolution*. Understanding protein evolution has become an important subject because of the temptation to manipulate protein function through rational alternation in protein structure via its sequence. This process is referred to as *protein design* or *protein engineering*, although the latter term has a grander-scale connotation, best applicable to more complex systems such as protein–protein or protein–DNA complexes.

How can the manipulation of protein function help in understanding cellular life and the molecular origins of diseases? The most obvious applications are direct examinations of how

Structural Bioinformatics, Second Edition   Edited by Jenny Gu and Philip E. Bourne
Copyright © 2009 John Wiley & Sons, Inc.

protein function alterations are coupled to phenotypic or symptomatic outcomes. More complex applications include various biosensors engineered to "sense" various target proteins' states, protein–protein interactions, protein cellular localization, and expression levels. Potential therapeutic applications of protein engineering also include design of novel antibodies and biomarkers to proteins that are elevated during a disease state with respect to the healthy state.

The appeal of protein structure-function manipulation led to a number of studies pertaining to questions about the possibilities and limitations of protein engineering. Protein design is a search procedure, either computational or experimental, that aims to determine amino acid sequences that correspond to a specific structure. Perhaps, the most central question of protein design is how "designable" is a specific structure, that is, how many sequences correspond to a specific protein structure. This question, first addressed from a purely theoretical perspective, has become one of the most intriguing questions in protein evolution and design, and is the subject of this chapter.

Before turning to the discussion of designability, we first review the intrinsic properties of naturally occurring proteins, organization of the protein structural universe, key determinants of protein structure, and protein evolution. We then discuss essential requirements for and other questions pertaining to protein design. Finally, we talk about challenges in protein design.

PROTEIN STRUCTURAL UNIVERSE

An understanding of the range of protein structures that can be adopted by Nature must first be appreciated before undertaking protein engineering efforts. Protein structures are organized hierarchically as discussed in more detail in Chapter 2. Several secondary structure elements—α-helices, β-strands, hairpins, and loops—form domains, quanta of protein structure. The domains can be defined unambiguously from the thermodynamic and folding kinetics points of view. Protein domains must be able to exist on their own, that is, isolated protein domain sequences must be able to reach their native states and be thermodynamically stable. Proteins can be further built by linking multiple domains. One must appreciate the wisdom of Nature in creating such modularity in the protein universe, because modularity permits the combinatorial multiplicity of possible proteins.

The consequence of structural modularity is that protein functions are modular as well. Indeed, let us imagine a need for a protein that would respond to a specific stimulus, cross the nucleus, associate with DNA, and initiate transcriptional response to this stimulus. Such a protein can be built by incorporating the stimulus-sensing domain (e.g., binding a specific small molecule), a nuclear transport domain, a DNA binding domain (e.g., containing a Zn-finger motif), and potentially, a catalytic domain to instigate a response. The beauty of this modular design, a strategy used by Nature, is that the need to extensively explore protein sequence and structural space is reduced. Instead, the selection and combination of a few domain architectures often produce a synergistic solution to constructing a protein with a desired biological function.

Further protein assembly into functional complexes with these modular units adds new dimensions to protein function. Thus, protein function is also hierarchical (Anantharaman, Aravind, and Koonin, 2003; Shakhnovich et al., 2003) and can be defined at multiple levels of protein structure: from the domain level to the level of molecular complexes. Perhaps the functional quantum of protein structure is a domain. Hence, it is plausible that evolution of proteins is dictated by the evolution of functional domains.

The question "how many protein folds exist?" has been an extremely hot topic of discussion. A number of answers have appeared, but all of them report the number of folds (N_F) to range between 1000 and 10,000 (Chothia, 1992; Orengo, Jones, and Thornton, 1994; Wolf, Grishin, and Koonin, 2000). How many should we expect if Nature were to exhaustively explore all possibilities? Let us count. A typical $N = 100$ residue protein has $N_{ss} = 15$ secondary structure elements, including strands, helices, loops, and turns. Then theoretically, $N_F = N_{ss}! \approx 10^{12}$ possibilities to arrange these structural elements with respect to each other (Table 39.1). This number is significantly larger than what is observed and even projected. Hence, it is clear that either evolution has not fully explored protein fold space or some folds are not designable, or both.

Importantly, the number of structures N_S that a 100-residue protein can explore is much larger than the number of folds ($N_S \gg N_F$). To estimate N_S, let us position a simplified protein model on a lattice, where each residue is represented as vertices, with the number of adjacent lattice vertices (the coordination number) z (for cubic lattice $z = 6$). The coordination number defines the number of possible interactions between residues. With this model neighboring residues can be positioned in ($z - 1$) vertices, adjacent to the current residue. Then, there are $(z - 1)^{(N-1)} = 10^{99 \log 5} \approx 10^{69}$ possibilities to build a $N = 100$ residue protein structure. We ignore self-avoidance of polymer chains (Grosberg and Khokhlov, 1997), since it is potentially compensated by our lattice considerations. However, only a fraction of 10^{69} possible protein structures will be compact. To estimate the fraction of compact proteins, we refer to a recent work by Chen, Ding, and Dokholyan (2007) who demonstrated that on average $\alpha \approx 1.7$ proximity constrains per residue define a protein structure within 3.5 Å root-mean-square distance from its native structure. Then, the total number of possible structures that a 100-residue proteins can adopt is $N_S \approx 10^{69}/\alpha^{99} \approx 10^{47}$. This number corresponds to the total number of compact 100-residue protein structures, which is also the number of sequences N_Σ that correspond to compact protein structures, that is, the number of protein sequences. Importantly, the total number of possible sequences for a 100-residue

TABLE 39.1. Estimations and Observed Numbers for the Protein Universe for $N = 100$ Residue Long Proteins

	$\log_{10} (N_X)$	
	Estimated	Naturally Observed or Projected
Estimated number of folds, N_F	12	2–3 (Holm and Sander, 1997; Martin et al., 1998; Wolf, Grishin, and Koonin, 2000; Dietmann et al., 2001)
The number of possible structures, N_S	47	5 (Berman et al., 2000)
The number of sequences corresponding to compact structures, N_Σ	47	NA
The total number of possible sequences, N_{Seq}	130	130
The average number of structures per fold, N_Σ/N_F	35	0–1 (Qian, Luscombe, and Gerstein, 2001; Dokholyan, Shakhnovich, and Shakhnovich, 2002)

protein is $20^{100} \approx 10^{130}$, much larger than the number of sequences that correspond to compact proteins.

These estimations suggest that if all folds are equivalent, that is, there are equal numbers of structures that correspond to each fold, then, there are $N_S/N_F = N_\Sigma/N_F \approx 10^{35}$ protein structures per fold. This assumption does not likely hold: Dokholyan, Shakhnovich, and Shaknovich (2002) reported that the number of naturally occurring protein domains in fold families are distributed unevenly with respect to the structures that contain them and follow a power law distribution $P(n_S) \propto n_S^{-2.5}$. While such an uneven distribution may suggest that some domains are more designable than others, it may also be the result of diverging evolutionary processes. We return to the issue of designability later in this chapter.

DETERMINANTS OF PROTEIN DOMAIN EVOLUTION

There are several different evolutionary factors that impact the designability of proteins. At the level of organisms, evolution optimizes adaptation of organisms to new or changing environments. This adaptation is accommodated through modifications of protein function that are achieved through changes of protein sequences and structures, or through changes in the expression levels of the proteins. Protein evolution is determined by the former modifications of protein sequence and consequent structural rearrangements. What are the factors that drive protein evolution? By examining mutational strategies and the effects of mutations in Nature, the examples will help guide our understanding of the underlying design rules that needs to be considered when addressing protein designability and engineering experiments.

Protein evolution proceeds through amino acid substitutions and fragmental rearrangements (deletions, insertions, truncations). These mutations can affect three aspects of proteins: (1) structure, (2) function, and (3) folding kinetics. If a mutation disrupts the folding kinetics of a protein, it may lead to misfolding and aggregation. There is a large number of misfolding-related disorders. Perhaps the most illustrative example is cystic fibrosis (Boucher, 2004) since 90% of all cases are attributed to a single amino acid deletion ΔPhe508 in the cystic fibrosis transmembrane conductance regulator (CFTR; Riordan et al., 1989; Rommens et al., 1989). The mutant protein's thermodynamic stability is not perturbed and it is not functional because it never matures as a transmembrane protein, suggesting that deletion of phenylalanine affects the folding kinetics of this protein. Clearly, this example demonstrates that preservation of protein folding kinetics may be a key factor in protein evolution.

Mutations leading to a loss of protein function can often be lethal. There are numerous examples of mutations that perturb functional sites, thereby affecting organism viability (see, for example, Akiyama et al., 2007; He et al., 2007). A number of loss-of-function mutations are embryonically lethal, especially those that involve proteins that play central roles during development. These facts suggest that the preservation of protein functional sites may also be a key factor in protein evolution.

Mutations that affect the protein stability are also associated with numerous diseases. Examples include familial amyotrophic lateral sclerosis (FALS) caused by over 100 mutations in Cu, Zn superoxide dismutase 1 (SOD1; Cleveland, 1999; Khare, Caplow, and Dokholyan, 2006; Figure 39.1) and familial amyloid polyneuropathy caused by mutations in transthyretin (Benson and Kincaid, 2007). Interestingly, in the case of FALS, many

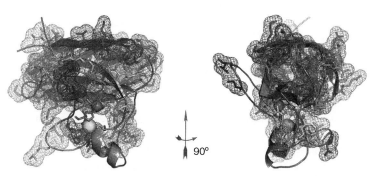

Figure 39.1. Single nucleotide polymorphisms in the human Cu, Zn superoxide dismutase (SOD1) associated with the amyotrophic lateral sclerosis. Over 70 mutations are scattered throughout the tertiary structure affecting multiple structurally distinct sites, marked by colored meshes. For simplicity only one monomer of the homodimeric SOD1 is shown in two projections. Figure also appears in the Color Figure section.

SOD1 mutants are as active as wild type, but they potentially gain toxic function due to association into pathogenic oligomeric states due to destabilization. This toxic gain-of-function may be under cellular repair machinery control until this control deteriorates due to aging. If the organisms are at or past the reproductive age when disease manifests itself, they can produce offspring, therefore organism evolution is not affected significantly. Hence, it may seem that mutations that affect protein stability are not as important as those that affect protein folding kinetics or function.

Surprisingly, this is not the case. Dokholyan and Shakhnovich (2001) showed that if one attempts to mimic protein evolution through sequence alterations that preserve protein thermodynamic stability, then amino acid positions that contribute the most to stability appear the most conserved during such alterations. This observation is further supported by examining the conservation pattern obtained in synthetic evolution, which is found to closely match that is observed naturally (Dokholyan and Shakhnovich, 2001). The correlation between the naïve synthetic and observed amino acid conservations is as high as 70%, suggesting that proteins are under a dominating evolutionary pressure to preserve their structural stability. Of course, the correlation is not perfect, suggesting that the preservation of folding kinetics and function are under evolutionary pressure as well.

Why would evolution preserve residues important for structural stability in proteins? There could be several reasons. First, the number of residues that constitute either the folding nucleus or the active site of the protein is usually much smaller than those that constitute the protein core. Hence the probability that a random substitution will affect either the folding nucleus or the active site is much smaller than the probability that it will affect the protein core. Second, function itself is not conserved in the course of evolution. Although proteins belonging to a specific fold tend to share a set of functions (Shakhnovich et al., 2003), these functions can be diverse enough that the active sites are fully altered. The folding of homologous proteins can be distinct as well, as for example in the case of two homologues Im7 and Im9 (Friel, Capaldi, and Radford, 2003), suggesting that the folding nucleus may not be under strong evolutionary pressure to be preserved. Thus, among three factors—protein structure, folding kinetics, and function—the protein structure is the most invariant. Hence, residues that maintain structural stability are under evolutionary pressure.

Mutations that impact protein stability also impact protein function. Proteins that are severely destabilized upon mutation are either eliminated due to a loss of function or retained through a compensating mutation that reverses the effect by recovering their thermodynamic stabilities. Recently Bloom et al. (2006) demonstrated that improving thermodynamic stability of marginally stable cytochrome P450 increases the chances of the enzyme to exhibit new or improved function. Mildly destabilizing mutations are perhaps the most dangerous as the proteins may still be functional, but not to the extent necessary to maintain the health of the organism. A number of recent studies support the hypothesis that mildly destabilizing mutations in the human genome are the underlying genetic origin of complex diseases (Sunyaev et al., 2001; Yampolsky et al., 2005; Eyre-Walker, Woofit, and Phelps, 2006; Kryukov, Pennacchio, and Sunyaev, 2007).

Evolutionary pressure to preserve protein structure has important implications on protein design. It is plausible that Nature has found "working" folds and exploits them by grafting necessary functions on to them. If stability is preserved and the scaffold can accommodate the new active site or a binding surface, a new protein with protein function can appear. This idea has been utilized in proof-of-principle studies: Hellinga and coworkers successfully transplanted triose phosphate isomerase activity (Dwyer et al., 2004) from one ribose-binding protein to another homologous enzyme (Allert, Dwyer, and Hellinga, 2007). However, while evolution appear to select for structural preservation of folds, Nature needs new folds to explore new functional space. Consequently, evolution is a balance of forces that conserve and diversify the protein domain universe. The mechanisms of protein domain evolution are discussed next.

MECHANISMS OF PROTEIN DOMAIN EVOLUTION

Clues to protein evolution can be observed by understanding conservation patterns imprinted in the protein sequence-structure-function space (Dokholyan and Shakhnovich, 2005; Dokholyan and Shakhnovich, 2007). Protein sequences and structures (Levitt and Chothia, 1976), as well as functions, are not distributed uniformly with respect to each other when corresponding measures of similarity between sequences, structures, or functions are used. Instead, proteins tend to fall into clusters of families with sequences sharing more than 20% sequence similarity often adopting similar structure and often function. These families of proteins are typically assumed to have originated from a common ancestor. However, it is also possible for proteins with less than 10% sequence similarity to share similar structures and functions. For these instances, convergent evolution is suspected to be involved because the sequence similarity is equal to what would be expected of two randomly selected proteins (Holm and Sander, 1993; Rost, 1997).

Convergent evolution is the result of structural exploration from two distinct origins reaching an equilibrium where the structural solution for a given function is similar to each other. The argument for a convergent evolution scenario can also be presented without the assumption of equilibrium in structural exploration. In which case, a specific force X of unknown origin that selects for specific folds to perform a given function needs to be specified, rather than exploring a multiplicity of other folds for function optimization. Moreover, this force X should be universal since it helps guide the selection of many distinct families of analogues.

There are several important arguments that are at odds with the convergent evolution theory. First, the assumption of equilibrium in structural exploration is inconsistent with the estimated number of protein folds since the number of theoretically possible folds still exceeds what is observed (see discussion in Protein Structural Universe section). Second, the origin of the mysterious force X is unknown and it is not clear what would make two presumably unrelated protein structures adopt the same structure and function. An alternative model suggests that protein designability is the guiding force of evolution and can be used to explain the uneven distribution of the number of members of protein fold families. An extension of this model, and simpler explanation for this distribution, was proposed by Dokholyan, Shakhnovich, and Shakhnovich (2002) using graph theoretical analysis that does not rely on any assumption (described in more detail later). Therefore, these models argue against the missing force X in the convergent evolution theory based on the *lex parsimoniae* principle. Third, it is arguable that protein domains presumed to be unrelated may in fact be related, but the connection between them is lost (Pei et al., 2003). The premise for establishing evolutionary relationship is often based in the amount of sequence conservation that may not necessarily hold. Indeed, sequences have been observed to diverge rapidly below the 15% similarity threshold during the course of simulated evolution while maintaining the original protein fold.

In the *divergent* protein evolution theory (Chothia and Gerstein, 1997; Grishin, 1997; Murzin, 1998; Zeldovich et al., 2007; Figure 39.2a) all proteins originate from a few protoproteins. The collective mutations accumulated across time create a repertoire of different proteins while maintaining their structural integrity. The current protein universe is a snapshot of the exploding protein universe. To reconstruct relationships between proteins, a comparison of sequences may not be sufficient. Structures are more conserved than sequence across a long evolutionary time and is useful to establish connections between seemingly unrelated protein domains. For some families, it may be possible to uncover connections between distinct fold families that are related by identifying proteins that can convert from one fold to another with few mutations (Cordes et al., 1999, 2000). Perhaps the most prominent example of such chameleon proteins is the dimeric $\alpha + \beta$ Cro protein from bacteriophage lambda, which evolved from an ancestral all-α monomeric protein (Figure 39.2b; LeFevre and Cordes, 2003). By retroevolving lambda Cro, Le Fevre et al. suggested possible mutations that are responsible for the evolution of the ancestral monomeric protein to the modern dimeric homologue.

The protein domain universe graph (PDUG) is constructed to map the relationships between proteins (Dokholyan, Shakhnovich, and Shakhnovich, 2002) and helps to highlight the mutational strategies used by Nature to evolve different folds. In PDUG, each node represents a protein domain and is connected by a bond that carries a weight corresponding to the degree of their structural similarity. PDUG is organized by a large number of subclusters (Figure 39.3) that serves to reflect the imprint of the evolutionary mechanisms. Examining the connectivities within the PDUG subclusters offers important insights into the appearance of new folds. Although chameleon proteins may be exceptions rather than the rule for the appearance of new folds, it is evident that there are certain proteins that play a role of a gatekeeper for specific folds. Elimination of these proteins during evolution separates connected families of protein folds, thereby giving birth to a new fold family (Dokholyan, 2005). One such example is a hydrolytic enzyme cutinase, which degrades cutin (Dokholyan, 2005). Cutin is a polyester composed of hydroxy and epoxy fatty acids, which serves as a protective shield of aerial plants against pathogen entry (Purdy and Kolattukudy, 1975). Cutin degradation is the first step for plant infection and is exploited by fungi expressing cutinase to invade plants. This unique mechanism of plant protection and infection may trace back to the very origin of plants. Coevolution of plants and pathogens may have spurred an observed spread of domains in the largest cluster of the PDUG (Figure 39.3).

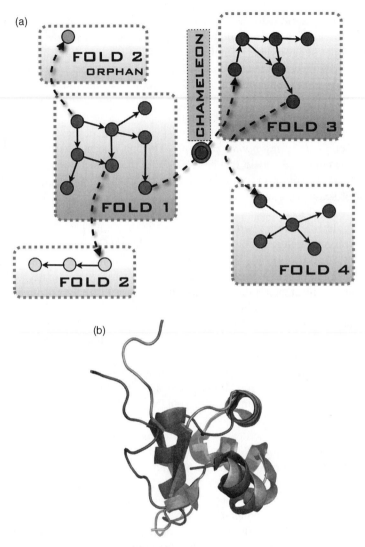

Figure 39.2. Mechanism for protein structure evolution. (a) A schematic representation of the proposed mechanisms for protein evolution (Dokholyan and Shakhnovich, 2005; Dokholyan and Shakhnovich, 2007). Minor structural changes associated with point mutations in proteins diversify the fold families. Major structural changes due to accumulated mutations result in the formation of new fold families. These fold families are populated unequally, some families are represented by just a single protein (orphan family; Dokholyan, Shakhnovich, and Shakhnovich, 2002). In some cases, the proteins may adopt structures that would "bridge" two fold families. These "chameleon" proteins are sensitive to a small number of mutations that make these proteins part of one family or another. (b) An example of the chameleon proteins λ Cro (Protein DataBank access number: 5CRO) and P22 Cro (Protein DataBank access number: 1RZS) (Newlove et al., 2006): structural alignment of these two proteins demonstrates high structural similarity despite dramatic differences in their secondary structures.

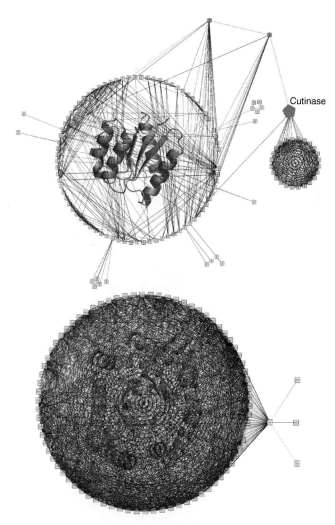

Cutinase

Figure 39.3. The first and the second largest PDUG subclusters. Representative structures for the first (upper graph) and the second largest (lower graph) subclusters of the PDUG are overlayed with the graphs. Figure also appears in the Color Figure section.

Examining the PDUG can give some support regarding whether convergent or divergent evolution is being utilized to create the current set of observed protein domains. The population of each of PDUG subcluster described above is observed to follow the power law distribution. However, the significance of this distribution is nullified when compared to a random graph (null hypothesis) that has been constructed to contain the identical number of nodes and edges found in PDUG. The control model showed the same distribution of members in fold families and therefore suggests that this distribution is a property of a random graph and do not have any further deep underlying meaning. On the contrary, the distribution of the connections between protein domains is very particular to PDUG when compare to a random graph. Using a simple model that accounts for divergent protein evolution with only two natural processes—gene duplications and point mutations—the observed distributions of domain connectivity can be reproduced (Dokholyan, Shakhnovich, and Shakhnovich, 2002) thus providing strong support for the divergent evolution theory (Figure 39.2).

In the protein domain evolution model presented by Dokholyan, Shakhnovich, and Shakhnovich (2002), proteins undergo gene duplications and new domains (paralogs and orthologs) accumulate point mutations over time (Figure 39.4). If these mutations do not significantly alter protein folding stability, that is, new proteins are stable enough to perform

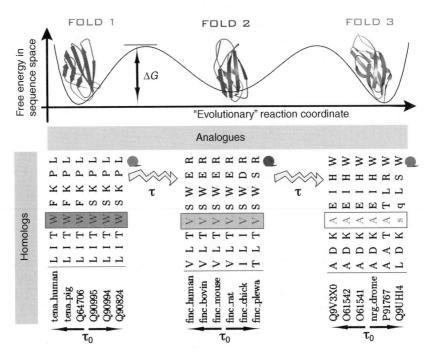

Figure 39.4. Schematic representation of the evolutionary processes that result in conservation patterns of amino acids. For a given family of folds, for example, immunoglobulin (Ig) fold in this schema, there are several minima (3) in the hypothetical free energy landscape in the sequence space as a function of the "evolutionary" reaction coordinate (such as time) (Dokholyan and Shakhnovich, 2005). Each of these minima are formed by mutations in protein sequences at typical time scales (τ_0), that do not alter the protein's thermodynamically and kinetically important sites, forming families of homologous proteins. Transitions from one minimum to another occur at longer time scales $\tau = \tau_0 \exp(\Delta G/T)$, where ΔG is the free energy barrier separating one family of homologous proteins from another. At time scale τ mutations occur that alter several amino acids at the important sites of the proteins so that the protein properties are not compromised, but the fold is changed. At time scale τ the family of analogues is formed. Three minima in this schema represent three families of homologues (1TEN, 1FNF, and 1CFB) each comprised of six homologous proteins. Only eight positions in the aligned proteins are shown: from 18 to 28. It can be observed that at position 4 (marked by blocks) in each of the families presented in the diagram amino acids are conserved within each family of homologues, but vary between these families. This position corresponds to position 21 in Ig fold alignment (to 1TEN) and is highly conserved (Dokholyan and Shakhnovich, 2001). Modified from Dokholyan NV, Shakhnovich EI. Scale-Free Evolution: From Proteins to Organisms. In: Koonin EV, Wolf YI, Karev GP, eds. Power Laws, Scale-Free Networks and Genome Biology. Austin: Landes Bioscience, 2005.

their functions, then they constitute the same fold. When a critical number of mutations accumulate, the core is potentially at risk to be disrupted or rearranged forcing significant structural changes, thereby giving birth to a new fold family. Within a fold family, sequences also diverge unequally: sequences defining homologues share 20% sequence similarity, while a fold family unifies numerous families of homologues with distinct numbers of homologue family members. Since the domain core preservation is an essential constraint in protein evolution, sequences with ~20% similarity are "guaranteed" to have structural similarity if the fraction of the shared residues constitute the critical component of the core. However, the conservation of the core may not be the only contributing factor effecting the distribution of fold families. Another contributing factor may be the physicochemical properties of the amino acids—the protein alphabet that defines protein sequence and structure. Next, we discuss the properties of the protein alphabet.

PROTEIN ALPHABET

The protein structure–sequence relationship is determined by the amino acid composition and amino acid order (see Chapter 2 for full details). The 20 amino acids have overlapping physicochemical properties resulting in a redundancy of this set of building blocks. Hypothetically the redundancy implies that proteins can be encoded with a smaller subset of the current amino acids. How would be the minimum set of amino acid residues be? To make a crude approximation of the ideal alphabet size, M_I, we compare the number of available sequences with this alphabet M_I^N for a protein of size N with the number of available protein conformations $(z-1)^{N-1}$ (estimated earlier in Protein Structural Universe section), where z is the coordination number. Equating these two numbers leads to $M_I \approx z-1 = 5$. This result is very surprising because it is fourfold smaller than the natural alphabet.

Although this estimation is extremely naïve, a number of studies have supported similar estimates. First, using information theory and bioinformatics, Strait and Dewey showed that a typical protein sequence features Shannon entropy values of approximately 2.5 bits. This number means that one needs $2^{2.5} \approx 5.7$ letters to encode protein structure. Of course, these letters are not amino acids, but rather a juxtaposition of various amino acids. Second, a study using lattice protein models and pairwise amino acid interaction potentials, demonstrated that one could design stable proteins using 20 but not 2 amino acid alphabet (Shakhnovich, 1994). The folding properties (foldability; Klimov and Thirumalai, 1996) of lattice protein models are also not significantly altered using a five-letter alphabet (Wang and Wang, 2000) but are altered using a two-letter alphabet. Third, by performing simulations of protein evolution, the amino acid conservation in protein families can be recapitulated using a six-letter amino acid alphabet (Dokholyan and Shakhnovich, 2001). Fourth, the SH3 domain can be experimentally redesigned to a sequence that consisted of only five amino acid types (Riddle et al., 1997); the expected probability of five-letter alphabet usage for a 56-residue protein is $p \sim 10^{-29}$ (Dokholyan, 2004). Last but not least, the usage of amino acid types in natural proteins is less than expected if amino acids were chosen randomly with naturally occurring frequencies (Dokholyan, 2004). Taken together, it seems that Nature has utilized a larger alphabet than it could have for the protein universe.

Why does Nature exploit a larger alphabet than necessary for creating diversity in protein structure and function? Naively, it seems that a larger alphabet size would be inefficient because of the additional metabolic machinery that is needed by each additional amino acid. However, there are other important properties of natural proteins that are only possible with the current, and larger, set of amino acids. For examples, it is important to retain robustness in the protein structure. By creating diversity within the repertoire of amino acids, to the impact of potentially damaging mutations is reduced. If Nature utilized the most efficient "ideal" set of amino acids, then proteins are not safe guarded by mutations that would be damaging to protein folds.

Another benefit for having a larger alphabet size is that the evolutionary barriers (free energy barriers with respect to some "evolutionary" reaction coordinate—Figure 39.4) that separate fold families are lowered. This property allows Nature to optimally explore fold families in the course of protein evolution. Extremely large barriers, which would be a consequence if a reduced set of amino acids were used, would "freeze" the protein universe in one state and evolution would not be able to proceed. Hence, it is expected that as evolution proceeds, the alphabet size increases (Jordan et al., 2005), although not all amino acids may be utilized at any given point during evolution.

It is plausible that Nature chooses a golden mean between robustness (tolerance to mutations) and metabolic efficiency in selecting the protein alphabet size (Figure 39.5). The redundancy within the amino acid set also has important implications on protein design: it offers a margin for error associated with the inaccuracies inherent in the current force fields used by computer-based protein design. It also allows for the diversification of sequences without altering protein fold. Next, we describe the effects of alphabet size and other physical factors on the population of a given fold.

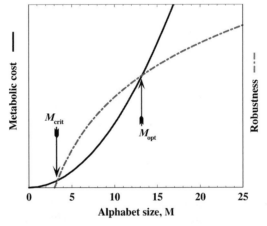

Figure 39.5. Hypothetical diagram of the benefits and drawbacks of protein alphabet size. The robustness associated with protein folds increases as the alphabet size M increases (dot–dashed line). At the same time, metabolic cost for producing M number of amino acids is also an increasing function of M (solid line). We posit that the latter function is a more rapidly growing one than that for the former. Two intercepts of these lines correspond to the critical (M_{crit}) and the optimal (M_{opt}) alphabet sizes. The critical alphabet size signifies the point at which it becomes beneficial to produce proteins tolerant to mutations. The optimal alphabet size signifies the point at which the metabolic cost and robustness are at an optimal state, above which the metabolic cost outweighs the benefits of robustness of folds to mutations.

LESSONS FOR ENGINEERING STABLE PROTEINS

Protein stability is perhaps the most important determinant of protein evolution. In which case, what lessons for rational protein design could this observation teach us? The fold does provide constraints for the number of sequences that can adopt a given structure, but how would the number of available amino acids for use affect this number? What other physical factors contribute to protein designability? Next we attempt to shed light on these questions from a simplified physical consideration.

For a protein to be stable in its native state, the potential energy $H(\Gamma)$ of its native state Γ must be smaller than that of any other alternative (*decoy*) state Γ_D, for all decoys D:

$$H(\Gamma) < H(\Gamma_D), \quad \forall D. \tag{39.1}$$

Using the random energy model (Derrida, 1980), the thermodynamic stability of proteins and the potential energy difference (*energy gap*) ΔH between the native state and the lowest potential energy decoy Γ_C can be related (Bryngelson and Wolynes, 1989; Gutin and Shakhnovich, 1993), which is then normalized by the root-mean-square deviation of the distribution of potential energies $\sigma(H)$ of decoy conformations (Dokholyan, 2004):

$$Z \equiv \frac{\Delta H}{\sigma(H)} = -1. \tag{39.2}$$

Equation 39.2 provides a measure that helps to find a sequence S that minimizes the Z-score during the process of rational design.

This Z-score has been minimized to find an *ideal* force field for a M-letter alphabet (Goldstein, Luthey-Schulten, and Wolneys, 1992), and the actual value of the minimal Z-score value has recently been derived, $Z_{\text{min}}^{(M)}$ (Dokholyan, 2004):

$$Z_{\text{min}}^{(M)} = -\left[\sum_{\sigma_a,\sigma_b}^{M} K^2(\sigma_a,\sigma_b)\right]^{1/2} + \sqrt{2N \log \gamma}, \tag{39.3}$$

where N is a protein length, γ is an effective number of degrees of freedom per atom—a parameter that reflects protein flexibility (Dokholyan, 2004). The kernel function $K(\sigma_a,\sigma_b)$ is

$$K(\sigma_a,\sigma_b) = \frac{\sum_{i\neq j=1}^{M} s_i(\sigma_a)s_j(\sigma_b)(\Delta_{ij}-f_{ij})}{\left[\sum_{i\neq j=1}^{M} s_i(\sigma_a)s_j(\sigma_b)f_{ij}(1-f_{ij})\right]^{1/2}}, \tag{39.4}$$

where f_{ij} are the frequencies of contacts between atoms i and j in decoy conformations, Δ_{ij} is the contact matrix element, which is equal to 1 or 0 depending on the presence of the contact between atoms i and j in the native state. $s_i(\sigma_a)$ is the amino acid at the position i of type σ_a.

Furthermore, Eq. 39.3 takes on a simpler form when using the Gō, model (Gō, 1983), in which atomic interactions are determined by the native protein structure (Dokholyan, 2004)

$$Z_{\text{min}}^{(Go)} = -\left[\sum_{i\neq j=1}^{N} \frac{(\Delta_{ij}-f_{ij})^2}{f_{ij}(1-f_{ij})}\right]^{1/2} + \sqrt{2N \log \gamma}. \tag{39.5}$$

Comparing Eqs 39.3 and 39.5, it is straightforward to prove that for all M

$$|Z_{\min}^{(M)}| \le |Z_{\min}^{(M+1)}| \le \cdots \le |Z_{\min}^{(\mathrm{Go})}|, \quad \forall M. \tag{39.6}$$

Equation 39.6 has profound implications on protein design: it suggests that with an increase of the size of available protein alphabet, more stable proteins can be designed. The limiting case is the Gō model, in which the alphabet size is equal to the number of possible contacts in a protein.

Another important implication of Eqs 39.3 and 39.4 is that they can answer the question regarding the smallest alphabet size M_c required to encode a protein. Interestingly, this value

$$M_c \approx 2\sqrt{\ln \gamma / \varepsilon} \tag{39.7}$$

is independent of the protein length N (Dokholyan, 2004; ε is a factor independent of N and M). Thus, with an alphabet of size $M \ge M_c$ one can design a protein of arbitrary length.

The values of the $Z_{\min}$ suggest three principal factors that govern protein designability: (i) the alphabet size M, (ii) protein flexibility γ, and (iii) properties of the unfolded or misfolded states f_{ij}. The alphabet size determines the maximum possible stability of designed proteins and, therefore, how many sequences a given fold can adopt. The protein flexibility and unfolded states are important factors to account for during rational protein design as they differentiate the designed proteins from the reference states (decoys). Hence, proper sampling of protein conformations during computational protein design is essential.

An important consequence of Eqs 39.3–39.5, is that they suggest the concept of designability (Li et al., 1996), that is, that some folds may have more sequences that make them stable than others. Indeed, since the properties of unfolded or *molten globule* (unfolded with some compact units; Ptitsyn, 1995; Finkelstein and Ptitsyn, 2002) states are determined by the native state (Pappu, Srinivasan, and Rose, 2000; Fitzkee and Rose, 2004; Kohn et al., 2004; Ding, Jha, and Dokholyan, 2005), the frequencies of contacts f_{ij} impact $Z_{\min}$. Hence, the lower the $Z_{\min}$ for a given fold, the more variation in stability due to mutations a given fold can tolerate, and the larger fold family we expect. England and Shakhnovich (2003) further uncovered role of the contact matrices on protein designability by deriving an explicit dependence of the number of designable sequence on the largest eigenvalue of the contact map.

The thermodynamic stability requirement has no implication for how stable natural proteins are (Taverna and Goldstein, 2002; Xia and Levitt, 2002). In fact, it has been established that natural proteins can be "redesigned" to be more stable than wild type (Chen et al., 2000; Dantas et al., 2003). Extremely stable proteins are not necessarily beneficial to Nature. As we show below, it is hard to evolve proteins to be extremely stable (whose Z-values are proximal to $Z_{\min}$). Additionally, extremely stable proteins are maximally sensitive to random mutations. Hence, a majority of natural proteins are "marginally" stable (Taverna and Goldstein, 2002). As long as a protein optimally performs its function, there is no evolutionary pressure to make it more stable.

Theoretically, one can always improve the stability of a protein up to $Z_{\min}$ value for a corresponding fold. However, the thermodynamic stability of the original fold and the fold limiting value $Z_{\min}$ significantly affect the ability to find such sequences. To demonstrate this, let us note that the distribution of the Z-values of amino acid sequences that stabilize a given fold is normal. Indeed, it is an assumption of the Random Energy Model that the potential energy is a sum of "random" pairwise energy terms between atoms and, therefore, the values of potential energies of protein conformations follow a Gaussian distribution, which after normalization (Eq. 39.2) becomes normal. Hence, the number of sequences Ξ that are more stable than a given one with the Z-value equals to Z_0:

$$\Xi(Z < Z_0) \propto \int_{Z_0}^{Z_{min}} \exp\left(-\frac{Z^2}{2}\right) dZ \propto \mathrm{erf}\left(-\frac{Z_{min}}{\sqrt{2}}\right) - \mathrm{erf}\left(-\frac{Z_0}{\sqrt{2}}\right). \tag{39.8}$$

If the thermodynamic stability of a protein that we wish to redesign is proximal to a fold's minimal value Z_{min}, the number of sequences becomes extremely hard to find as their number significantly decreases

$$\Xi(Z < Z_0) \propto |Z_{min} - Z_0| e^{-\frac{1}{2}Z_{min}^2} \left\{ 1 - \frac{1}{2} Z_{min} |Z_{min} - Z_0| \right\} + O(|Z_{min} - Z_0|^3), \quad Z_0 \to Z_{min}. \tag{39.9}$$

Thus, the number of sequences more stable than a given one (with the Z-value equals to Z_0) $\Xi(Z < Z_0)$ decreases dramatically as Z_0 approaches its limiting value. This observation makes computational or experimental searches for sequences that make a protein even more stable an extremely difficult task.

An important implication of Eq. 39.9 is that most random mutations are destabilizing. This consequence is especially important for many human diseases, suggesting that single nucleotide polymorphisms (SNPs) associated with diseases may decrease the stability of proteins produced from the corresponding genes. From this discussion, we can appreciate the limitation state structural stability has on shaping the allowable sequence space for a given fold.

PROTEIN ENGINEERING: US VERSUS NATURE

Learning how Nature operates has obvious advantages, but it is always important to remember the underlying motives for evolving or designing a specific protein. We often pursue goals that are distinct from Nature's goals when designing proteins. For Nature it is important to design a functional protein. Proteins are not under evolutionary pressure to be extremely stable, more active than that is required for existing cell environments, or extremely fast folders. Most existing computational algorithms for protein design, on the other hand, search specifically for the sequence that minimizes the potential energy of a given scaffold. Although the ultimate goal of the protein design field is to rationally manipulate protein function and cell life, Nature evolves proteins as a part of the complex cellular system, thereby taking into consideration a whole spectrum of factors and functions. In this perspective, the future of the protein evolution field is projected into the realm of *systems evolution*.

There are many other constraints that Nature honors in the course of evolution. For example, allostery is an important factor to account for protein function. It is currently extremely difficult to account for allosteric effects in proteins during rational design. Some success has been achieved with designing molecular switches (Ambroggio and Kuhlman, 2006). However, nobody has so far succeeded with the design of rational changes in protein structure upon ligand binding. This problem is one of the most complex in protein design since it requires design of networks of amino acid interactions throughout protein structure. While the communication of amino acids within protein structures has been established using graph theoretical analysis in a number of studies (Lockless and Ranganathan, 1999; Dokholyan et al., 2002; Vendruscolo et al., 2002), it is not clear whether it is purposely evolved or just a generic property of protein structure. For example, while there is a clear relation between topologically and kinetically important residues in proteins (Dokholyan et al., 2002; Vendruscolo et al., 2002), this relation may be a consequence of a similar one between protein structure and its folding kinetics. This observation is supported by several studies that identified significantly contributing residues to protein folding kinetics (Clementi, Nymeyer, and Onuchic, 2000; Ding et al., 2002a) using a structure-based model of protein energetics the Gō model (Gō, 1983). Thus, the success in designing allostery in proteins strongly depends on the significant progress in our understanding of amino acid interaction networks in protein structure and their evolution.

The function of some proteins is structurally achieved by disordered regions that lack secondary or any well-defined unique structure. In fact, there are a number of fully disordered proteins (Le Gall et al., 2007) and their potential impact on protein function, biology, and other functional roles are discussed in Chapter 38. Computational design of such proteins is a formidable challenge. Perhaps what makes rational design of disordered proteins even more complex is that despite the lack of well-defined structure of such proteins, they are distinct from random polymers. These proteins coexist in multiple states, which actually have unique structural properties at the ensemble level, that is, they share specific *structural signatures*. One remarkable example is α-synuclein, which lacks a unique native state, but is more compact than would be expected for a random polypeptide (Dedmon et al., 2005). Although the function of α-synuclein is unknown, disordered regions are often employed by other proteins, such as antibodies and cell signaling proteins, which have intrinsic plasticity to recognize and bind other molecules. In such proteins, disorder is only visible in free structure. Local or global structural reorganization occurs upon the binding of partner molecules. Such proteins are a significant complication for computational protein design because the stability of the complex with a partner protein, rather than the thermodynamic stability of free proteins, is the priority for Nature. Since many such proteins have a large number of partners, a further challenge for protein design is to engineer protein interactions with several of the partners, which may force a designed protein to adopt a number of alternative bound structures.

Another important focus of natural design is the active site of enzymes. Enzyme activity is extremely sensitive to the geometry and dynamic properties of the active site and its conformation in the transition states. This sensitivity is controlled at the sub-Angstrom level, making Nature the most elegant and scrupulous designer. The level of accuracy required for enzyme design makes it one of the most challenging tasks for protein designers. Although there has been significant progress in computational enzyme design, pioneered by Hellinga (Dwyer et al., 2004), we are still not in the realm of natural enzyme design.

Besides the differences in design philosophies between Nature and us, there are a number of other issues that we need to overcome to become efficient in designing functional proteins. In the next section, we discuss some of these issues.

CHALLENGES IN PROTEIN DESIGN

There are a number of experimental and computational approaches to protein design. While experimental approaches that attempt to mimic evolution *in vitro* offer a powerful search methodology for desired proteins, computational design has particular appeal since it provides rational explanation, and correspondingly a higher perspective for control, for the effect of changes on protein structure and function. Next, we discuss the main difficulties facing computational protein design.

Energetics

One of the most important components in protein design, as well as protein folding and structure prediction, is the force field. The force field is a set of parameters that contributes to the calculation of the potential energy of the protein. There are a number of strategies to derive a force field for a given molecular model: physical, empirical, knowledge based, constraint based, and their combinations. Physical force fields (e.g., CHARMM and Amber) require physical entities, that is, atoms, and therefore are most appropriate for detailed all-atom models. Empirical force fields rely on specific observations pertinent to a given molecule, and therefore are molecule type-specific. Among empirical force fields is the structure-based Gō model, which biases proteins to their native states. Knowledge-based force fields are constructed on current knowledge of molecular structure, physical/chemical properties, and evolution, and therefore are limited to molecules that resemble known molecules. Constraint-based force fields rely on experimental data, molecular structure, and dynamics.

The more details of interactions accounted by the force field (i.e., the more accurate it is), the slower the calculation of the potential energy of a given protein state. Slow computation of the potential energy, in turn, results in diminished sampling or search abilities, whether it is a search for a protein sequence during protein design, or conformational sampling during protein folding or structure prediction. Hence, there is a fine balance between the accuracy of the force field and the ability to search or sample. Finding such a balance between the potential energy calculations and the ability to search for a solution is one of the principal challenges in protein design.

Conformational sampling is the other side of the coin from estimating the free energy of a given protein state and its corresponding thermodynamic stability. The free energy estimation of a given state requires understanding the properties of the ensemble of all conformations in the free energy basin of a state of interest. During protein design, one needs to find a large number of conformations that a given sequence can adopt within such a basin. An amino acid substitution may result in atomic clashes between the newly substituted amino acid and other residues (Ding and Dokholyan, 2006; Yin et al., 2007). Such clashes can be resolved by "massaging" the protein structure via conformational sampling of the protein backbone. Surprisingly the perturbation of the backbone required for resolving a clash is often very minimal, less than 1 Å (Yin et al., 2007), although it is important to note that not all clashes can be resolved.

Estimation of protein thermodynamic stabilities is even more challenging as it requires the knowledge of the free energies of the reference (unfolded) states. The free energy of the unfolded states is predominantly determined by the accurate modeling of the ensembles of unfolded state structures. Due to difficulties in determining the ensemble of unfolded states, many design algorithms rely on approximations of the reference states or derive these parameters by training using various strategies. Rapid sampling techniques may potentially improve thermodynamic stability evaluations through explicit sampling (Ding, Jha, and Dokholyan, 2005).

Alternative Binding Sites

An important potential complication during design of protein–protein interactions is the presence of alternative binding sites. Even though interactions at the alternative binding sites are most likely to occur at lower affinity, these sites compete for the ligand and, therefore, decrease the effective binding to the target region. These complications could be alleviated by (1) improving sampling during docking of a target protein on a scaffold to identify the potential alternative binding sites, and (2) performing negative design to decrease binding affinities at the alternative binding sites.

Aggregation

Protein aggregation due to association of identical proteins is a common phenomenon and can result from multiple factors. Protein aggregation occurs when protein thermodynamic stability and concentration are such that identical species can form intermolecular hydrogen bonds between their backbones. Thermodynamic stability plays an essential role in protein aggregation, as proteins often need to unfold to expose their backbone to other proteins for hydrogen bond formation (Ding et al., 2002b).

There are two dominant reasons for protein aggregation in the course of protein design. First, the designed protein may not be stable enough. In such cases, further improvement of the protein's thermodynamic stability may alleviate the problem. Second, if the protein is rich in β-sheets, the amino acids constituting these β-strands are already in geometrical configurations consistent with those of protein aggregates. This is a serious complication during protein design as there is no clear understanding, despite a number of interesting considerations showing how Nature protects β-rich proteins from aggregation (Patki, Hausrath, and Cordes, 2006; Richardson and Richardson, 2002). It has been suggested that capping β-sheets with charged residues and creating bulges on the surface of β-sheets may protect them from associating with identical proteins. However, no successful design has been carried out to build a novel β-rich protein. Further understanding of Nature's defense against protein aggregation may offer new strategies for designing β-rich proteins.

CONCLUSIONS

Protein engineering enters a new era as rational manipulation of protein structure becomes more successful due to acquired knowledge about protein structure, energetics, and evolution. With the increase of the number of determined protein structures, we now better understand the organization of the protein structure universe. We are able to link structurally diverse protein families and understand evolutionary determinants of families of protein structures. By developing an understanding of the basic mechanisms of protein evolution, we learn to manipulate protein structure by mimicking the natural processes. From theoretical considerations, we have learned important lessons for engineering stable proteins, such as the protein alphabet size, protein flexibility, and the properties of the unfolded states.

Despite tremendous progress in protein engineering, we still face challenges. These challenges include design in the framework of cellular environments, for example when protein design must include potentially negative design against interactions with other proteins. Modeling unfolding proteins is essential to protein design and improvements in our models of unfolded states may allow for rational manipulations of naturally disordered proteins or proteins that feature functional unstructured regions. Further challenges are associated with sub-angstrom design that is necessary for building novel enzymes. Understanding protein aggregation will help us defy such natural phenomena as protein aggregation and nonspecific associations with other proteins.

Improvements may come from further development of the force fields and our ability to more exhaustively sample protein conformations. An understanding of the protein alphabet is important to computational protein design (Shakhnovich, 1998) because the underlying protein alphabet determines the parameter set for the energy function that is used for protein design. Therefore, further studies are needed to uncover the fundamental protein alphabet responsible for encoding protein three-dimensional structure. These studies may also shed light on the redundancy of the genetic code and identify the principal rules of molecular evolution.

Protein engineering is already influencing modern biology and offering help in understanding cellular life. Many of these considerations are applicable to other polymers, such as RNA and DNA molecules. Although these molecules have specific challenges associated with their design, most of the discussed aspects of protein design are also applicable to these molecules. Future engineering efforts offer remarkable promise for impacting medicine: from the engineering of antibodies and enzymes to the development of biomarkers for early diagnosis of disease. Protein engineering will also play an important role in elucidating the molecular etiologies of human diseases and the development of personalized medicine.

REFERENCES

Akiyama M, Sakai K, Ogawa M, McMillan JR, Sawamura D, Shimizu H (2007): Novel duplication mutation in the patatin domain of adipose triglyceride lipase (PNPLA2) in neutral lipid storage disease with severe myopathy. *Muscle Nerve* 36 (6):856–859.

Allert M, Dwyer MA, Hellinga HW (2007): Local encoding of computationally designed enzyme activity. *J Mol Biol* 366:945–953.

Ambroggio XI, Kuhlman B (2006): Design of protein conformational switches. *Curr Opin Struct Biol* 16:525–530.

Anantharaman V, Aravind L, Koonin EV (2003): Emergence of diverse biochemical activities in evolutionarily conserved structural scaffolds of proteins. *Curr Opin Chem Biol* 7:12–20.

Benson MD, Kincaid JC (2007): The molecular biology and clinical features of amyloid neuropathy. *Muscle Nerve* 36 (4):411–423.

Berman HM, Westbrook J, Feng Z, Gilliland G, Bhat TN, Weissig H, Shindyalov IN, Bourne PE (2000): The protein data bank. *Nucleic Acids Res* 28:235.

Bloom JD, Labthavikul ST, Otey CR, Arnold FH (2006): Protein stability promotes evolvability. *Proc Natl Acad Sci USA* 103:5869–5874.

Boucher RC (2004): New concepts of the pathogenesis of cystic fibrosis lung disease. *Eur Respir J* 23:146–158.

Bryngelson JD, Wolynes PG (1989): Intermediates and barrier crossing in a random energy-model (with applications to protein folding). *J Phys Chem* 93:6902.

Chen JM, Lu ZQ, Sakon J, Stites WE (2000): Increasing the thermostability of staphylococcal nuclease: Implications for the origin of protein thermostability. *J Mol Biol* 303:125.

Chen Y, Ding F, Dokholyan NV (2007): Fidelity of the protein structure reconstruction from inter-residue proximity constraints. *J Phys Chem B* 111:7432–7438.

Chothia C (1992): Proteins—1000 Families for the molecular biologist. *Nature* 357:543.

Chothia C, Gerstein M (1997): Protein evolution. How far can sequences diverge? *Nature* 385 (579):581.

Clementi C, Nymeyer H, Onuchic JN (2000): Topological and energetic factors: What determines the structural details of the transition state ensemble and "en-route" intermediates for protein folding? An investigation for small globular proteins. *J Mol Biol* 298:937.

Cleveland DW (1999): From Charcot to SOD1: Mechanisms of selective motor neuron death in ALS. *Neuron* 24:515.

Cordes MH, Burton RE, Walsh NP, McKnight CJ, Sauer RT (2000): An evolutionary bridge to a new protein fold. *Nat Struct Biol* 7:1129–1132.

Cordes MH, Walsh NP, McKnight CJ, Sauer RT (1999): Evolution of a protein fold in vitro. *Science* 284:325–328.

Dantas G, Kuhlman B, Callender D, Wong M, Baker D (2003): A large scale test of computational protein design: folding and stability of nine completely redesigned globular proteins. *J Mol Biol* 332:449–460.

Dedmon MM, Lindorff-Larsen K, Christodoulou J, Vendruscolo M, Dobson CM (2005): Mapping long-range interactions in alpha-synuclein using spin-label NMR and ensemble molecular dynamics simulations. *J Am Chem Soc* 127:476–477.

Derrida B (1980): The random energy-model. *Phys Rep* 67:29.

Dietmann S, Park J, Notredame C, Heger A, Lappe M, Holm L (2001): A fully automatic evolutionary classification of protein folds: Dali Domain Dictionary version 3. *Nucleic Acids Res* 29:55.

Ding F, Dokholyan NV (2006): Emergence of protein fold families through rational design. *PLoS Comput Biol* 2:e85.

Ding F, Dokholyan NV, Buldyrev SV, Stanley HE, Shakhnovich EI (2002a): Direct molecular dynamics observation of protein folding transition state ensemble. *Biophys J* 83:3525–3532.

Ding F, Dokholyan NV, Buldyrev SV, Stanley HE, Shakhnovich EI (2002b): Molecular dynamics simulation of the SH3 domain aggregation suggests a generic amyloidogenesis mechanism. *J Mol Biol* 324:851–857.

Ding F, Jha RK, Dokholyan NV (2005): Scaling behavior and structure of denatured proteins. *Structure* 13:1047–1054.

Dokholyan NV (2004): What is the protein design alphabet? *Proteins* 54:622–628.

Dokholyan NV (2005): The architecture of the protein domain universe. *Gene* 347:199–206.

Dokholyan NV, Li L, Ding F, Shakhnovich EI (2002): Topological determinants of protein folding. *Proc Natl Acad Sci USA* 99:8637–8641.

Dokholyan NV, Shakhnovich B, Shakhnovich EI (2002): Expanding protein universe and its origin from the biological Big Bang. *Proc Natl Acad Sci USA* 99:14132–14136.

Dokholyan NV, Shakhnovich EI (2001): Understanding hierarchical protein evolution from first principles. *J Mol Biol* 312:289–307.

Dokholyan NV, Shakhnovich EI (2005): Scale-free evolution: from proteins to organisms. In: Koonin EV, Wolf YI, Karev GP, editors. *Power Laws, Scale-free Networks and Genome Biology*. Austin, TX: Landes Bioscience, Eurekah.com and Springer, pp. 86–105.

Dokholyan NV, Shakhnovich EI (2007): Towards unifying protein evolution theory. In: Bastolla U, Porto M, Roman HE, Vendruscolo M, editors. *Structural Approaches to Sequence Evolution: Molecules, Networks, Populations*. Berlin: Springer, pp. 113–126.

Dwyer MA, Looger LL, Hellinga HW (2004): Computational design of a biologically active enzyme. *Science* 304:1967–1971.

England JL, Shakhnovich EI (2003): Structural determinant of protein designability. *Phys Rev Lett* 90:218101.

Eyre-Walker A, Woolfit M, Phelps T (2006): The distribution of fitness effects of new deleterious amino acid mutations in humans. *Genetics* 173:891–900.

Finkelstein AV, Ptitsyn O (2002): *Protein Physics: A Course of Lectures (Soft Condensed Matter, Complex Fluids and Biomaterials)*. Boston: Academic Press.

Fitzkee NC, Rose GD (2004): Reassessing random-coil statistics in unfolded proteins. *Proc Natl Acad Sci USA* 101:12497–12502.

Friel CT, Capaldi AP, Radford SE (2003): Structural analysis of the rate-limiting transition states in the folding of Im7 and Im9: similarities and differences in the folding of homologous proteins. *J Mol Biol* 326:293–305.

Gō N (1983): Theoretical studies of protein folding. *Annu Rev Biophys Bioeng* 12:183.

Goldstein RA, Luthey-Schulten ZA, Wolynes PG (1992): Protein tertiary structure recognition using optimized Hamiltonians with local interactions. *Proc Natl Acad Sci USA* 89:9029–9033.

Grishin NV (1997): Estimation of evolutionary distances from protein spatial structures. *J Mol Evol* 45:359–369.

Grosberg AY, Khokhlov AR (1997): *Giant Molecules.* Boston: Academic Press.

Gutin AM, Shakhnovich EI (1993): Ground-state of random copolymers and the discrete random energy-model. *J Chem Phys* 98:8174.

He X, van Waardenburg RC, Babaoglu K, Price AC, Nitiss KC, Nitiss JL, Bjornsti MA, White SW (2007): Mutation of a conserved active site residue converts tyrosyl-DNA phosphodiesterase I into a DNA topoisomerase I-dependent Poison. *J Mol Biol* 372 (4):1070–1081.

Holm L, Sander C (1993): Protein structure comparison by alignment of distance matrices. *J Mol Biol* 233:123–138.

Holm L, Sander C (1997): Dali/FSSP classification of three-dimensional protein folds. *Nucleic Acids Res* 25:231.

Jordan IK, Kondrashov FA, Adzhubei IA, Wolf YI, Koonin EV, Kondrashov AS, Sunyaev S (2005): A universal trend of amino acid gain and loss in protein evolution. *Nature* 433:633–638.

Khare SD, Caplow M, Dokholyan NV (2006): FALS mutations in Cu, Zn superoxide dismutase destabilize the dimer and increase dimer dissociation propensity: a large-scale thermodynamic analysis. *Amyloid* 13:226–235.

Klimov DK, Thirumalai D (1996): Criterion that determines the foldability of proteins. *Phys Rev Lett* 76:4070–4073.

Kohn JE, Millett IS, Jacob J, Zagrovic B, Dillon TM, Cingel N, Dothager RS, Seifert S, Thiyagarajan P, Sosnick TR, et al. (2004): Random-coil behavior and the dimensions of chemically unfolded proteins. *Proc Natl Acad Sci USA* 101:12491–12496.

Kryukov GV, Pennacchio LA, Sunyaev SR (2007): Most rare missense alleles are deleterious in humans: implications for complex disease and association studies. *Am J Hum Genet* 80:727–739.

Le Gall T, Romero PR, Cortese MS, Uversky VN, Dunker AK (2007): Intrinsic disorder in the protein data bank. *J Biomol Struct Dyn* 24:325–342.

LeFevre KR, Cordes MH (2003): Retroevolution of lambda Cro toward a stable monomer. *Proc Natl Acad Sci USA* 100:2345–2350.

Levitt M, Chothia C (1976): Structural patterns in globular proteins. *Nature* 261:552–558.

Li H, Helling R, Tang C, Wingreen N (1996): Emergence of preferred structures in a simple model of protein folding. *Science* 273:666–669.

Lockless SW, Ranganathan R (1999): Evolutionarily conserved pathways of energetic connectivity in protein families. *Science* 286:295.

Martin AC, Orengo CA, Hutchinson EG, Jones S, Karmirantzou M, Laskowski RA, Mitchell JB, Taroni C, Thornton JM (1998): Protein folds and functions. *Structure* 6:875.

Murzin AG (1998): How far divergent evolution goes in proteins. *Curr Opin Struct Biol* 8:380–387.

Newlove T, Atkinson KR, Van Dorn LO, Cordes MH (2006): A trade between similar but nonequivalent intrasubunit and intersubunit contacts in Cro dimer evolution. *Biochemistry* 45:6379–6391.

Orengo CA, Jones DT, Thornton JM (1994): Protein superfamilies and domain superfolds. *Nature* 372:631.

Pappu RV, Srinivasan R, Rose GD (2000): The Flory isolated-pair hypothesis is not valid for poly-peptide chains: implications for protein folding. *Proc Natl Acad Sci USA* 97:12565–12570.

Patki AU, Hausrath AC, Cordes MH (2006): High polar content of long buried blocks of sequence in protein domains suggests selection against amyloidogenic non-polar sequences. *J Mol Biol* 362:800–809.

Pei J, Dokholyan NV, Shakhnovich EI, Grishin NV (2003): Using protein design for homology detection and active site searches. *Proc Natl Acad Sci USA* 100:11361–11366.

Ptitsyn OB (1995): Molten globule and protein folding. *Adv Protein Chem* 47:83.

Purdy RE, Kolattukudy PE (1975): Hydrolysis of plant cutin by plant pathogens. Purification, amino acids composition, and molecular weight of two isoenzymes of cutinase and a nonspecific esterase from *Fusarium solani* f. pisi. *Biochemistry* 14 (13):2824.

Qian J, Luscombe NM, Gerstein M (2001): Protein family and fold occurrence in genomes: power-law behaviour and evolutionary model. *J Mol Biol* 313:673.

Richardson JS, Richardson DC (2002): Natural beta-sheet proteins use negative design to avoid edge-to-edge aggregation. *Proc Natl Acad Sci USA* 99:2754.

Riddle DS, Santiago JV, Bray-Hall ST, Doshi N, Grantcharova VP, Yi Q, Baker D (1997): Functional rapidly folding proteins from simplified amino acid sequences. *Nat Struct Biol* 4:805–809.

Riordan JR, Rommens JM, Kerem B, Alon N, Rozmahel R, Grzelczak Z, Zielenski J, Lok S, Plavsic N, Chou JL, et al. (1989): Identification of the cystic fibrosis gene: cloning and characterization of complementary DNA. *Science* 245:1066–1073.

Rommens JM, Iannuzzi MC, Kerem B, Drumm ML, Melmer G, Dean M, Rozmahel R, Cole JL, Kennedy D, Hidaka N, et al. (1989): Identification of the cystic fibrosis gene: chromosome walking and jumping. *Science* 245:1059–1065.

Rost B (1997): Protein structures sustain evolutionary drift. *Fold Des* 2:S19–S24.

Shakhnovich BE, Dokholyan NV, DeLisi C, Shakhnovich EI (2003): Functional fingerprints of folds: evidence for correlated structure-function evolution. *J Mol Biol* 326:1–9.

Shakhnovich EI (1994): Proteins with selected sequences fold into unique native conformation. *Phys Rev Lett* 72:3907–3910.

Shakhnovich EI (1998): Protein design: a perspective from simple tractable models. *Fold Des* 3: R45.

Sunyaev S, Ramensky V, Koch I, Lathe W 3rd, Kondrashov AS, Bork P (2001): Prediction of deleterious human alleles. *Hum Mol Genet* 10:591–597.

Taverna DM, Goldstein RA (2002): Why are proteins marginally stable? *Proteins* 46:105.

Vendruscolo M, Dokholyan NV, Paci E, Karplus M (2002): Small-world view of the amino acids that play a key role in protein folding. *Phys Rev E Stat Nonlin Soft Matter Phys* 65:061910.

Wang J, Wang W (2000): Modeling study on the validity of a possibly simplified representation of proteins. *Phys Rev E Stat Phys Plasmas Fluids Relat Interdiscip Topic* 61:6981–6986.

Wolf YI, Grishin NV, Koonin EV (2000): Estimating the number of protein folds and families from complete genome data. *J Mol Biol* 299:897.

Xia Y, Levitt M (2002): Roles of mutation and recombination in the evolution of protein thermo-dynamics. *Proc Natl Acad Sci USA* 99:10382.

Yampolsky LY, Kondrashov FA, Kondrashov AS (2005): Distribution of the strength of selection against amino acid replacements in human proteins. *Hum Mol Genet* 14:3191–3201.

Yin S, Ding F, Dokholyan NV (2007): Eris: an automated estimator of protein stability. *Nat Methods* 4:466–467.

Zeldovich KB, Chen P, Shakhnovich BE, Shakhnovich EI (2007): A first-principles model of early evolution: emergence of gene families, species, and preferred protein folds. *PLoS Comput Biol* 3: e139.

40

STRUCTURAL GENOMICS OF PROTEIN SUPERFAMILIES

Stephen K. Burley, Steven C. Almo, Jeffrey B. Bonanno, Mark R. Chance, Spencer Emtage, Andras Fiser, Andrej Sali, J. Michael Sauder, and Subramanyam Swaminathan

Under the auspices of the National Institutes of Health (NIH)—National Institute of General Medical Sciences (NIGMS) Protein Structure Initiative (PSI), the New York SGX Research Center for Structural Genomics (NYSGXRC) has applied its high-throughput X-ray crystallographic structure determination platform to systematic studies of two large protein superfamilies. Approximately, 15% of consortium resources are devoted to structural studies of protein phosphatases, which are classified as Biomedical Theme targets. A further ~15% of effort is devoted to structural studies of enolases (ENs) and amidohydrolases (AHs) as community-nominated targets. NYSGXRC efforts with the protein phosphatases have, to date, yielded structures of 21 distinct protein phosphatases: 14 from human, 2 from mouse, 2 from the pathogen *Toxoplasma gondii*, 1 from *Trypanosoma brucei*, the parasite responsible for African sleeping sickness, and 2 from the principal mosquito vector of malaria in Africa, *Anopheles gambiae*. These structures provide insights into both normal and pathophysiologic processes, including transcription regulation, signal transduction, neural development, and type 1 diabetes. In conjunction with the contributions of other international structural genomics consortia, these efforts promise to provide an unprecedented database and materials repository for structure-guided experimental and computational discovery of inhibitors for all classes of protein phosphatases. NYSGXRC efforts with members of the enolase/amidohydrolase superfmaily have yielded 58 structures of 34 distinct enolases and 18 amidohydrolases from a wide array of organisms. Our comprehensive survey of these $\alpha_8\beta_8$ barrel structures aims to provide a database and materials repository for evolutionary studies of enzyme substrate specificity.

INTRODUCTION

In 2000, the National Institutes of Health—National Institute of General Medical Sciences established the Protein Structure Initiative with the goal to "make the three-dimensional atomic-level structures of most proteins easily obtainable from knowledge of their corresponding DNA sequences" to support biological and biomedical research (http://www.nigms.nih.gov/Initiatives/PSI.htm). This pilot phase demonstrated the feasibility of the program and Phase II of the program, PSI-II, was launched in 2005, supporting four large-scale production centers to continue high-throughput structure determination efforts and six specialized centers to focus on specific bottlenecks such as membrane proteins and multicomponent assemblies (Table 40.1). More recently, these experimental efforts were supplemented by addition of two centers focused on enhancing comparative protein structure modeling, a PSI Materials Repository for centralized archiving and distribution of reagents and a PSI Knowledgebase for data sharing (Table 40.1).

Target selection represents a critical first step in the structural genomics pipeline, as it dictates the value of the ensuing structures. PSI-II employs a balanced target selection strategy that emphasizes the importance of large-scale structure determination and homology model generation, while exploiting the underlying infrastructure to address significant problems of biomedical relevance and to respond to the needs of the larger research community. Seventy percent of PSI-II efforts focus on the determination of structures with less than 30% amino acid sequence identity to an existing structure (http://www.nigms.nih.gov/Initiatives/PSI.htm). This constraint is central to the overall goals of the PSI, as it is at approximately this level of sequence identity that homology modeling begins to fail due to difficulties in obtaining accurate primary sequence alignments. Fifteen percent of PSI-II

TABLE 40.1. NIGMS Protein Structure Initiative Centers

Large-Scale Production Centers

Joint Center for Structural Genomics http://www.jcsg.org
Midwest Center for Structural Genomics http://www.mcsg.anl.gov
New York SGX Research Center for Structural Genomics http://www.nysgxrc.org/
Northeast Structural Genomics Consortium http://www.nesg.org

Specialized Technology Development Centers

Accelerated Technologies Center for Gene to 3D Structure http://www.atcg3d.org
Center for Eukaryotic Structural Genomics http://www.uwstructuralgenomics.org
Center for High-Throughput Structural Biology http://www.chtsb.org
Center for Structures of Membrane Proteins http://csmp.ucsf.edu
Integrated Center for Structure and Function Innovation http://techcenter.mbi.ucla.edu/
New York Consortium on Membrane Protein Structure http://www.nycomps.org

Homology Modeling Centers

Joint Center for Molecular Modeling
New Methods for High-Resolution Comparative Modeling

Resource Centers

PSI Materials Repository http://www.hip.harvard.edu/
PSI Knowledgebase http://www.knowledgebase.rutgers.edu/

activities are committed to projects nominated by the larger scientific community, with the remaining 15% devoted to a biomedically relevant theme developed by each of the four large-scale centers.

NYSGXRC

The New York SGX Research Center for Structural Genomics (www.nysgxrc.org) has established a cost-effective, high-throughput X-ray crystallography platform for *de novo* determination of protein structures. NYSGXRC member organizations include SGX Pharmaceuticals, Inc. (www.sgxpharma.com), the Albert Einstein College of Medicine (www.aecom.yu.edu), Brookhaven National Laboratory (www.bnl.gov), Case Western Reserve University (www.cwru.edu), and the University of California at San Francisco (www.ucsf.edu). Together, scientists from these industrial and academic organizations support all aspects of PSI-II, including (1) family classification and target selection, (2) generation of protein for biophysical analyses, (3) sample preparation for structural studies, (4) structure determination, and (5) analyses and dissemination of results. NYSGXRC production metrics during the last full grant year (July 1, 2006–June 30, 2007) are as follows: generation of ∼2060 target protein expression clones, ∼1400 successful target protein purifications (all characterized by matrix-assisted laser desorption ionization and electrospray ionization (ESI)–mass spectrometry (MS), and analytical gel filtration), >360,000 initial crystallization experiments, >106,000 crystallization optimization experiments, ∼3100 crystals harvested, >600 X-ray diffraction data sets recorded, and 158 structures deposited in the Protein Data Bank (PDB) (www.pdb.org) and released. NYSGXRC averaged ∼110 successful protein purifications per month and one structure deposition every 2–3 days. As mandated by the NIGMS, ∼15% of NYSGXRC resources are devoted to structure determinations of Biomedical Theme targets, protein phosphatases from human, mouse, and various pathogens, and another ∼15% of consortium resources are devoted to structure determinations of community-nominated targets drawn from the enolase/amidohydrolase superfamily. (The terms of the PSI-II award to the NYSGXRC explicitly forbid allocation of substantial resources to functional characterization of PSI targets.)

BIOMEDICAL THEME TARGETS: BACKGROUND AND MOTIVATION

Protein kinases and phosphatases act in counterpoint to control the phosphorylation states of proteins that regulate virtually every aspect of eukaryotic cell and molecular biology. Protein phosphorylation is a dynamic posttranslational modification, which allows for processing and integration of extra- and intracellular signals. *In vivo*, protein kinases and phosphatases play antagonistic roles, controlling phosphorylation of specific protein substrates on tyrosine, serine, and threonine side chains. These reversible phosphorylation events modulate protein function in various ways, including generation of "docking sites" that direct formation of multicomponent protein assemblies, alteration of protein localization, modulation of protein stability, and regulation of enzymatic activity. Such molecular events modulate signal transduction pathways responsible for controlling cell cycle progression, differentiation, cell–cell and cell–substrate interactions, cell motility, the immune response, ion channel and solute transporter activities, gene transcription, mRNA translation, and basic metabolism.

Aberrant regulation of protein phosphorylation results in significant perturbations of associated signaling pathways and is directly linked to a wide range of human diseases (reviewed in Tonks, 2006). PTEN, the first protein phosphatase family member identified as a tumor suppressor, is inactivated by mutations in several neoplasias, including brain, breast, and prostate cancers. CDC25A and CDC25B are potential oncogenes. Overexpression of PRL-1 and PRL-2 results in cellular transformation and PRL-3 is implicated as a metastasis factor in colorectal cancer. PTP1B is a primary target for therapeutic intervention in diabetes and obesity. CD45 is a target for graft rejection and autoimmunity. Mutations in EPMA2 are responsible for a form of epilepsy, characterized by neurological degeneration and seizures.

The importance of protein phosphatases in mammalian physiology is underscored by strategies employed by several pathogens, including *Yersinia*, *Salmonella*, and vaccinia viruses, with which pathogen-encoded protein phosphatases disrupt host-signaling pathways and are essential for virulence. Systematic structural analysis of protein phosphatases provides an opportunity to make significant progress toward (1) understanding and treating the underlying mechanisms of human diseases, (2) treating a wide range of opportunistic and infectious microorganisms, and (3) generating reagents that permit experimentation to uncover new principles in cellular and molecular biology.

The protein phosphatases encompass a range of structural families, displaying various mechanisms of action and substrate specificities. The protein tyrosine phosphatases (PTPs) represent one of the largest families in the human genome with four distinct subfamilies, including (1) the classic PTPs that recognize phosphotyrosine residues (112 human proteins), which are further divided into several subclasses of receptor-like and cytosolic PTPs, (2) the promiscuous dual-specificity phosphatases (DSPs), which recognize both phosphotyrosine and phosphoserine/phosphothreonine (33 human proteins) and include subfamiles of the phosphoinositide phosphatases (PTEN and myotubularin) and the mRNA 5′-triphosphatases (BVP and Mce1), (3) the low molecular weight phosphatases that recognize phosphotyrosine residues, and (4) the dual-specificity CDC25 phosphatases.

All members of the PTP family catalyze metal-independent dephosphorylation of phosphoamino acids, using a covalent phosphocysteine intermediate to facilitate hydrolysis. The amino acid sequence hallmark of the PTP family is the **HCXXGXXR**(S/T) motif, which contains the cysteine nucleophile. A sequence alignment showing the family conservation in the 14 amino acids surrounding this motif is shown in Figure 40.1a. It is remarkable that this active site feature represents the only amino acid sequence motif common to all PTP subfamily members.

The serine/threonine protein phosphatases are represented by two families that are distinguished by sequence homology and catalytic metal ion dependence. The PPP family members are Zn/Fe-dependent enzymes, including PP1, PP2A, and PP2B (calcineurin) (~15 human proteins). The PPM or PP2C-like family members are Mn/Mg-dependent enzymes (~16 human proteins). Despite sharing essentially no sequence similarity, members of both families utilize catalytic mechanisms involving a water nucleophile activated by a binuclear metal center (McCluskey, Sim, and Sakoff, 2002). The haloacid dehalogenase (HAD) superfamily contains a large number of magnesium-dependent phosphohydrolases, which operate through a covalent phosphoaspartic acid intermediate. Recently, a small number of HAD family members have been demonstrated to be protein phosphatases and have been implicated in a range of biological processes (Peisach et al., 2004; Meinhart et al., 2005; Wiggan, Bernstein, and Bamburg, 2005; Jemc and Rebay, 2007).

Dendrograms encompassing most of the known human phosphatases are shown in Figure 40.1 (experimental 3D public domain structures are denoted therein with

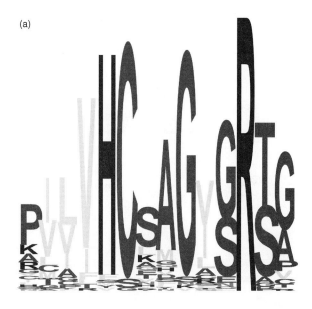

(a)

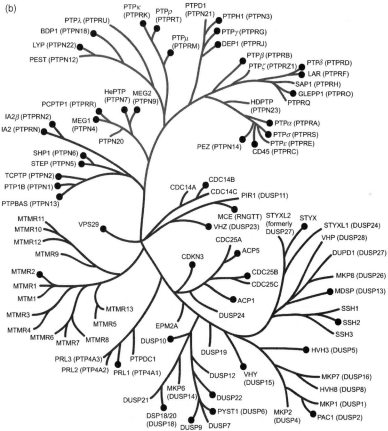

Figure 40.1. Human phosphatome phylogenetic tree. (a) Sequence logo (Schneider and Stephens, 1990) depicting the conservation of active site residues in the protein tyrosine and dual-specificity phosphatases. (b) Dendrogram of protein tyrosine and dual-specificity phosphatases based on variation in the active site motif. (c) Dendrogram of all other human protein phosphatases based on alignment of the entire catalytic domain. Phosphatases with structures in the PDB are indicated by black circles.

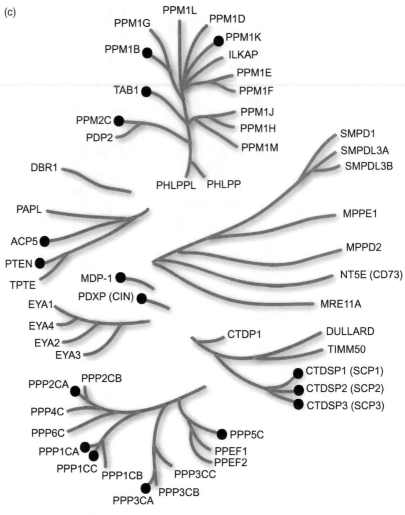

Figure 40.1. *(Continued)*.

black circles). The active site motif (Figure 40.1a) was used to construct the PTP/DSP tree (Figure 40.1b), whereas a multiple sequence alignment of the catalytic domain sequences was used to characterize homology between the remaining phosphatases (e.g., PPM and PPP families; Figure 40.1c). There are over 225 mammalian phosphatase structures in the PDB, providing coverage with either an experimental structure or a high-quality homology model for at least 64 human phosphatases (or ∼45% of the human phosphatome). Some human phosphatases have many structural representatives, such as PTP1B with over 90 structures, whereas many others have at most a single structure in the public domain.

BIOMEDICAL THEME TARGETS: SELECTION AND PROGRESS

After the start of PSI-II, the NYSGXRC established a target list of human protein phosphatases for which there was no representative in the PDB. In addition, we selected structurally uncharacterized protein phosphatases from a number of pathogens for which sequence information was available. The coding sequences of most human phosphatases were cloned from cDNA libraries, some were purchased and ~15 were synthesized. All pathogen phosphatases were codon-optimized and synthesized (Codon Devices, Inc., Cambridge, MA).

Our overall progress in this endeavor is shown in Figure 40.2 (effective October 1, 2007), wherein the number of distinct phosphatases that have progressed to each experimental stage is shown. We have observed greater attrition for the pathogen phosphatases, due in part, we believe, to the fact that many of the sequences are gene predictions that have not been experimentally verified. To compare our work on the human versus pathogen proteins, of the 58 human/mouse proteins that we successfully purified, 15 yielded structures, whereas of the 49 purified pathogen phosphatases only 5 have yielded structures.

Work on the first group of 93 pathogen phosphatases began at the end of 2005 (*A. gambiae, T. gondii,* and *Plasmodium falciparum*). In early 2007, work on an additional ~170 pathogen phosphatases was initiated, with targets selected from *Candida albicans, Encephalitozoon cuniculi, Filobasidiella neoformans, Gibberella zeae, Cryptosporidium parvum, Fusarium graminearum, Trichomonas vaginalis, T. brucei, Aspergillus nidulans, Cryptococcus neoformans, Entamoeba histolytica,* and *Giardia lamblia.*

Two years after the start of PSI-II, we have produced viable expression vectors for 304 phosphatases, and purified crystallization-grade protein for 107 of these NYSGXRC

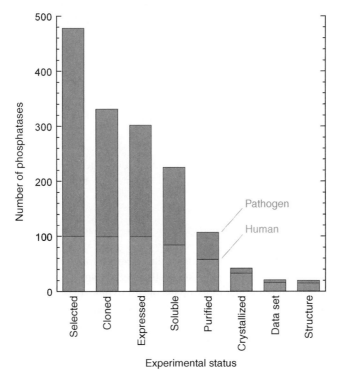

Figure 40.2. Progress on protein phosphatase structural studies. Number of protein phosphatase targets at each experimental stage of the NYSGXRC structural genomics pipeline. Human phosphatases (or mammalian orthologs) are shown in the lower part of the bars and pathogen phosphatases are shown in the upper part.

Biomedical Theme targets. We have deposited 24 X-ray crystal structures of 21 distinct protein phosphatases into the PDB (Table 40.2). Other structural biology groups and structural genomics centers have also determined significant numbers of protein phosphatase structures. Among the most productive are the SGC (Structural Genomics Consortium; http://www.sgc.utoronto.ca) and KRIBB (Korea Research Institute of Bioscience and Biotechnology; https://www.kribb.re.kr/eng/index.asp), which have deposited structures of at least 19 and 7 distinct human phosphatases, respectively, within the past 2 years. There is relatively little overlap among newly deposited structures from competing consortia.

To help minimize structure overlap among research groups, the NYSGXRC publishes its target list in the TargetDB database (http://targetdb.pdb.org/) on a weekly basis, which also provides an experimental status for each target. We compare all of our targets against the contents of the PDB on a weekly basis and typically stop work on those that have been deposited by other groups. We publish experimental protocols for every attempt on every target in the PepcDB database (http://pepcdb.pdb.org/); this information includes not only detailed general protocols but also information about each clone (DNA sequence and predicted protein sequence, mutations, whether it has been codon-optimized, and small scale expression/solubility results), fermentation (media, volume, induction time and temperature, and resulting pellet weight), protein purification (yield, concentration), purified protein quality as judged by mass spectrometry (pass/fail, exact molecular weight), and crystallization conditions. We encourage this level of transparency for all structural genomics centers, particularly for target selection. Moreover, we make all of our reagents, such as expression clones, freely available. Starting in 2008, we anticipate that all NYSGXRC expression clones will be distributed by the centralized PSI Materials Repository, located within the Harvard Institute of Proteomics (HIP; http://www.hip.harvard.edu/).

BIOMEDICAL THEME TARGETS: SELECTED EXAMPLES

The NYSGXRC Biomedical Theme project has yielded a number of important structures, which have already provided unique insights into a wide range of biological processes with direct relevance to human disease. Illustrative examples are highlighted below.

PROTEIN TYROSINE PHOSPHATASE (PTPσ)

The 21 human receptor protein tyrosine phosphatases share a common organization with extracellular ligand binding domains, a single transmembrane segment, and intracellular phosphatase catalytic domains that function in concert to regulate signaling through ligand-mediated tyrosine dephosphorylation. Twelve of these receptor PTPs possess two tandem phosphatase domains with a catalytically active membrane proximal domain (D1) and a membrane distal domain (D2) that is thought to be inactive in most family members. PTPσ belongs to the type 2A subfamily, which possess extracellular ligand binding domains composed of 3 immunoglobulin-like (Ig) domains and 4–9 fibronectin type III-like (FNIII) domains (Pulido et al., 1995; Chagnon, Uetani, and Tremblay, 2004; Tonks, 2006). Additional members of this subfamily include the human leukocyte common antigen-related PTP (LAR) and PTP-delta (PTPδ) and the invertebrate orthologs Dlar and DPTP69D in *Drosophila*, PTP-3 in *Caenorhabditis elegans*, and HmLAR1 and HmLAR2 in *Hirudo*

TABLE 40.2. NYSGXRC Protein Phosphatase Structures

Gene	Synonym	NYSGX	Species	NCBI	UniProt	Pfam Domain	PDB ID
Acp1	ACP1	8663b	Mouse	AAH39744	Q561M1	LMWPc	2p4u
CTDSP2	SCP2	8717a	Human	NP_005721	Q53ZR2	NIF	2q5e
CTDSPL	SCP3	8718a	Human	NP_001008393	Q3ZTU0	NIF	2hhl
DUSP23	LDP3	8673a	Human	NP_060293	Q9BVJ7	DSPc	2img
DUSP28	Dusp28	8736b	Mouse	NP_780327	Q8BTR5	DSPc	2hcm
DUSP9	MKP4	8638a	Human	NP_001386	Q99956	DSPc	2hxp (C290S)
PDXP	CIN	8744a	Human	NP_064711	Q96GD0	Hydrolase	2oyc, 2p27, 2p69
PPM1B	PP2CB	8702a	Human	NP_808907	Q461Q2	PP2C	2p8e
PPM1K	PPM1K	8700a	Human	NP_689755	Q8N3J5	PP2C	2iq1
PTP4A1	PRL1	8648a	Human	NP_003454	Q93096	Y_phosphatase	1rxd
PTPRD	PTPdelta	8613c	Human	NP_002830	P23468	Y_phosphatase	2nv5 rat = human
PTPRG	PTPgamma	8615a	Human	NP_002832	P23470	Y_phosphatase	2pbn, 2hy3
PTPRN	IA2	8620a	Human	NP_002837	Q16849	Y_phosphatase	2i1y
PTPRO	GLEPP1	8635a	Human	NP_109592	A0AV39	Y_phosphatase	2g59
PTPRS	PTPsigma	8623a	Human	NP_002841	Q13332	Y_phosphatase (tandem)	2fh7 (D1–D2)
STYX	STYX	8698a	Human	NP_660294	Q8WUJ0	DSPc	2r0b
PPM1	PPM1	8828z	T. gondii	N/A	N/A	PP2C	2isn
PP2C	N/A	8817z	T. gondii	CAC86553	Q8WPN9	PP2C	2i44
apaH	N/A	9095b	T. brucei	AAX70877	Q57U41	MetalHophos	2qjc
Tab1	TAB1	8880z	Mosquito	EAA07598	Q7QD46	PP2C	2irm
PPM1G	Ppm1g	8886z	Mosquito	EAA11252	Q7PP01	PP2C	2i0o (Δ125–398)

medicinalis. Expression of human receptor PTPs has been detected in all tissues examined, with the majority of PTPσ and PTPδ expression found in the brain (Pulido et al., 1995).

PTPσ and other members of the type 2A subfamily play roles in regulating the central and peripheral nervous systems by providing and responding to cues for axon growth and guidance, synaptic function, and nerve repair. These complex functions appear to utilize cell-autonomous and noncell-autonomous mechanisms, involving signals originating from both the cytoplasmic phosphatase domains and the ligand binding properties of the ectodomain (Siu, Fladd, and Rotin, 2007). Using brain lysates from PTPσ-deficient mice, in combination with substrate trapping experiments, N-cadherin and β-catenin were identified as substrates of PTPσ (Siu, Fladd, and Rotin, 2007). These findings led to a model of PTPσ-regulated axon growth involving a cadherin/catenin-dependent pathway. In this model, PTPσ directs the dephosphorylation of N-cadherin, which allows the recruitment of β-catenin. In addition, PTPσ-mediated dephosphorylatin of β-catenin permits subsequent linkage to the actin cytoskeleton, resulting in increased adhesion and reduced axon growth. In PTPσ–deficient mice, the resulting hyperphosphorylation of N-cadherin and β-catenin prevents the linkage between the cytoskeleton and the plasma membrane, resulting in reduced adhesion and enhanced axon growth. Further support for this model is provided by observations that dorsal root ganglion axon growth is accelerated in PTPσ-deficient mice. Of particular note is the enhanced rate of nerve regeneration after trauma (e.g., crush or transection) in PTPσ-deficient mice (Sapieha et al., 2005). In addition to enhanced rates of regeneration, PTPσ-deficient mice show an increased rate of errors in directional nerve growth, suggesting a role in both growth rates and the directional persistence or guidance of advancing neurons. The PTPσ ectodomain has been implicated in noncell-autonomous functions related to both optimal growth and guidance in regenerating neurons. These contributions to axon growth and regeneration make PTPσ an interesting potential target for therapeutic intervention.

We have determined the structure of the tandem phosphatase domains of human PTPσ (PTPσ-D1–D2; PDB ID: 2FH7) (Figure 40.3). As observed in the structures of the LAR and CD45 tandem phosphatases, the D1 and D2 domains of PTPσ are very similar (root mean square deviation or RMSD $\sim$1.0 Å) (Nam et al., 1999; Nam et al., 2005). The overall organization of the PTPσ tandem phosphatase domains is very similar to that observed in CD45, LAR, and PTPγ. This similarity is best appreciated by superimposing the D1 domain and examining the distribution of D2 positions (Figure 40.4). The relative domain organization is dictated by residues contributing to the D1–D2 interface, which are highly conserved among PTPσ, LAR, and PTPγ.

Defining the oligomeric state of the receptor PTPs is central to understanding ligand binding, activation, and underlying regulatory mechanisms. PTPσ-D1–D2 is a monomer in the crystalline state and also behaves as a monomer in solution as judged by analytical gel filtration chromatography (unpublished data). It is remarkable that PTPα and CD45 have been reported to be negatively regulated by homodimerization (Jiang et al., 1999; Tertoolen et al., 2001; Xu and Weiss, 2002), although this observation remains an area of intense scrutiny and some controversy (Nam et al., 1999; Nam et al., 2005). Recently, it has been suggested that PTPσ forms homodimers in the cell and that dimerization is required for ligand binding (Lee et al., 2007). The apparent discrepancy between these cell-based results and our biophysical studies may be resolved by demonstrations that dimerization depends, at least in part, on interactions involving the transmembrane segment (Lee et al., 2007), which is absent from the D1–D2 construct used for our crystallographic and biophysical studies.

Both phosphatase domains in PTPσ possess the characteristic CX_5R catalytic site motif and are capable of binding the phosphate analogue tungstate (Figure 40.3). However, as

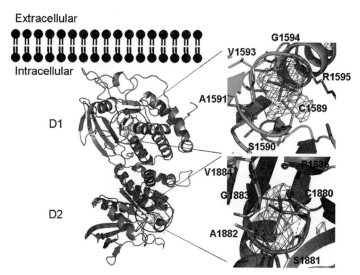

Figure 40.3. Structure of the human PTPσ tandem phosphatase domains. The structure of the PTPσ tandem phosphatase domains D1 and D2 is shown as a ribbon diagram with bound tungstate ions as stick and overlapping anomalous difference electron density in red. Domain D1 is shown in dark green and D2 in magenta. Interactions with the tungstate ion in the D1 and D2 active sites are magnified with hydrogen bonds represented as black dashes. Figure also appears in the Color Figure section.

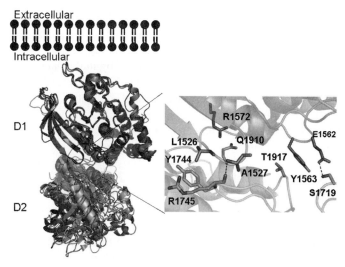

Figure 40.4. Structural comparison of tandem phosphatase domains of RPTPs. Superpositions of the structures of the tandem phosphatase domains of PTPσ (green), LAR (purple), CD45 (blue), and PTPγ (orange). Amino acids involved in interdomain interactions for PTPσ are shown to the right of the D1–D2 structures. For all structures, the D1 and D2 domains are shown in dark and light shades of color, respectively. Figure also appears in the Color Figure section.

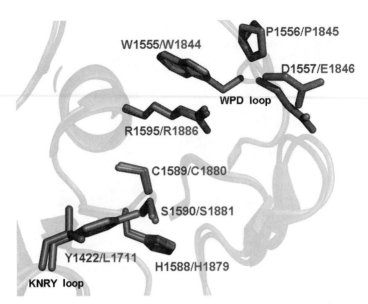

Figure 40.5. Comparison of the PTPσ D1 and D2 domain active sites. Superposition of the PTPσ D1 (green) and D2 (magenta) domain active sites. Active site residues and residues making up the WPD and KNRY loops are shown as stick figures. Root mean square deviation of D1/D2 superposition for 254 structurally equivalent Cα atoms is ~1.0 Å. Figure also appears in the Color Figure section.

observed in most other receptor tyrosine phosphatases, the D2 domain of PTPσ appears to be catalytically inactive (Wallace et al., 1998). All catalytically active D1 domains possess WPD and KNRY loops, which contain the catalytic acid (Asp) and participate in phosphotyrosine recognition, respectively. The lack of activity in the PTPσ D2 domain is almost certainly due, at least in part, to the replacement of the WPD Asp with Glu and the KNRY Tyr with Leu (Figure 40.5). Analogous differences are present in the catalytically defective D2 domains of LAR and PTPα (Buist et al., 1999; Nam et al., 1999), and restoration of the WPD and KNRY sequences in these domains results in a substantial enhancement of catalytic activity. The D2 domain is thought to play an important regulatory role as intermolecular and intramolecular binding interactions between D1 and D2 domains from the same and heterologous PTPs have been shown to modulate the catalytic activity of D1 (Wallace et al., 1998; Jiang et al., 1999; Blanchetot and den Hertog, 2000; Blanchetot et al., 2002).

INSULINOMA-ASSOCIATED PROTEIN 2 (IA-2)

Insulinoma-associated protein 2 (IA-2) is a member of the receptor-type protein tyrosine phosphatase family (Lan et al., 1994). It is enriched in the secretory granules of neuroendocrine cells, including pancreatic islet β cells, peptidergic neurons, pituitary cells, and adrenal chromaffin cells (Solimena et al., 1996). The IA-2 protein is predicted to have a lumenal domain, a single transmembrane domain, and a cytoplasmic tail containing a protein tyrosine phosphatase-like domain. However, enzymatic activity has not been demonstrated for IA-2, and several substitutions within the phosphatase-like domain appear to be responsible for its apparent inactivity against various substrates (Magistrelli, Toma, and Isacchi, 1996). Despite its apparent lack of phosphatase activity, the localization of IA-2 to the membrane of insulin secretory granules suggests that it may be involved in granule trafficking and/or maturation. Indeed, IA-2-deficient mice exhibit defects in glucose-stimulated insulin secretion (Saeki et al., 2002).

IA-2 represents a major autoantigen in type 1 diabetes, with greater than 50% of patients demonstrating circulating antibodies to the protein (Lampasona et al., 1996; Lan et al., 1996). Processing of the ~100 kD IA-2 protein involves proteolytic cleavage within the lumenal domain, resulting in a ~64 kD mature form that is immunoprecipitated by insulin-dependent diabetes mellitus in patient sera (Xie et al., 1998). Autoantibodies to IA-2 can also be detected during the prediabetic period. Measurement of autoantibodies to IA-2, insulin, and glutamic acid decarboxylase (GAD65) enables prediction of type 1 diabetes in at-risk individuals, with the presence of two or more reactivities being highly predictive of future disease (Verge et al., 1996). The humoral immune response in diabetes is primarily directed against conformational epitopes located within the cytoplasmic portion of the protein (Lampasona et al., 1996; Xie et al., 1997; Zhang, Lan, and Notkins, 1997). Distinct B cell epitopes are contained within polypeptide chain segments 605–620, 605–682, 687–979, and 777–937 (Lampasona et al., 1996). As recently reviewed (DiLorenzo, Peakman, and Roep, 2007), type 1 diabetes patients and at-risk individuals also exhibit CD4[+] T cell responses to IA-2 peptides derived from these regions of the protein. CD8[+] T cells specific for IA-2 have also recently been reported in type 1 diabetes patients (Ouyang et al., 2006).

We have determined the structure of the IA-2 phosphatase-like domain (PDB ID: 2I1Y), which reveals a classic protein tyrosine phosphatase fold that is most similar to PTP1B (RMSD ~ 1.4 Å; Figure 40.6). This protein also possesses residues that explain lack of enzymatic activity. Although the CX_5R sequence is present, several other critical residues are absent, including the catalytic acid (D) in the WPD loop and a major determinant (Y) in the phosphotyrosine-recognition loop. Lack of enzyme activity is thought to be essential for the biological function of IA-2, as it can heterodimerize with both PTPα and PTPγ and downregulate PTPα (Gross et al., 2002). Our structure is particularly noteworthy for defining

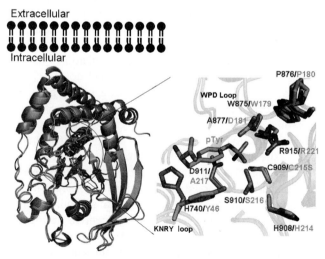

Figure 40.6. Comparison of the structures of IA-2 and PTP1B. Superposition of the structures of IA2 (green) and PTP1B (cyan), with active site residues shown as stick figures. Active site residues of IA2 and PTP1B bound to phosphotyrosine are magnified, highlighting differences responsible for the lack of catalytic activity of IA-2. Figure also appears in the Color Figure section.

the relationship between the B cell and T cell epitopes, and thereby providing insights into the pathogenesis of type 1 diabetes. We suggest that T cell response development is facilitated by antibodies to IA-2. The presence of surface antibody would allow B cells to capture fragments of the protein and present IA-2-derived peptides on MHC molecules for recognition by T cells (Ke and Kapp, 1996). Antibody-bound protein fragments could also be directed to antigen-presenting cells via Fc receptors, a targeting that allows antigens to be efficiently processed and presented to T cells (Villinger et al., 2003). A bound antibody can also modulate processing of T cell determinants, suppressing the presentation of some epitopes and enhancing the generation of others, thus influencing the process of epitope spreading (Davidson and Watts, 1989; Watts and Lanzavecchia, 1993; Simitsek et al., 1995).

SMALL C-TERMINAL DOMAIN PHOSPHATASE 3

The small C-terminal domain phosphatases (SCP) comprise a family of Ser/Thr-specific phosphatases that play a central role in mRNA biogenesis via regulation of RNA polymerase II (RNAP II). RNAP II is a large, multisubunit enzyme containing 12 components (Sayre, Tschochner, and Kornberg, 1992), the largest of which bears a unique C-terminal domain (CTD) that is flexibly linked to a region of the macromolecular machine near the RNA exit pore. The CTD consists of multiple repeats of the consensus sequence Tyr1-Ser2-Pro3-Thr4-Ser5-Pro6-Ser7 (Corden et al., 1985), with the number of repeats varying among different organisms (e.g., 26 are found in yeast versus 52 in human). Reversible phosphorylation of the CTD plays a crucial role in RNAP II progression through the transcription cycle, controlling both transcriptional initiation and elongation (Goodrich and Tjian, 1994; Dahmus, 1996). The CTD phosphorylation status also affects RNA processing events, such as 5′-capping and 3′-processing (Ho and Shuman, 1999; Rodriguez et al., 2000; Shatkin and Manley, 2000; Ahn, Kim, and Buratowski, 2004). The CTD of RNAP II is predominantly phosphorylated at Ser2 and Ser5 within the heptapeptide repeats by members of the cyclin-dependent kinase (CDK) family (e.g., CDK7, CDK8, and CDK9) (Hengartner et al., 1998; Zhou et al., 2000). Members of the SCP family work in opposition to restore the unphosphorylated state. It is remarkable that SCP family members also modulate the function of SMAD transcriptional regulators by dephosphorylating residues within the C-terminus and the linker region between the two conserved domains of SMAD (Knockaert et al., 2006; Sapkota et al., 2006).

Given their roles in transcriptional regulation, SCP family phosphatases have been implicated in a wide range of physiologic and pathologic processes. SCP3 has been identified as a tumor suppressor. The *SCP3* gene is frequently (>90%) deleted or its expression is drastically reduced in lung and other major human carcinomas (Kashuba et al., 2004). In contrast, a related family member, SCP2, was initially identified in a genomic region frequently amplified in sarcomas and brain tumors (Su et al., 1997). Members of the SCP family act as evolutionarily conserved transcriptional regulators that globally silence neuronal genes (Yeo et al., 2005). SCP2 interacts with the androgen receptor (AR) and appears to control promoter activity via RNAP II clearance during steroid-responsive transcriptional events (Thompson et al., 2006). FCP1, the first SCP related protein to be identified, interacts directly with HIV-1 TAT through its noncatalytic domain and is essential for TAT-mediated transcriptional transactivation (Abbott et al., 2005).

To date, the structures of three SCP family members have been determined, including the NYSGXRC structures of SCP2 (PDB ID: 2Q5E) and SCP3 (PDB ID: 2HHL). SCP3 is monomeric both in solution and the crystalline state, and shares high sequence identity

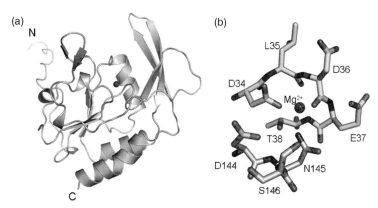

Figure 40.7. Structure of SCP3. (a) Ribbon diagram of SCP3 showing the DXDX catalytic loop (yellow) and the catalytic Mg^{2+} ion modeled from SCP1 (magenta). (b) Atomic details of the SCP3 catalytic site, again with the Mg^{2+} ion modeled from SCP1. Figure also appears in the Color Figure section.

($\sim$83%) and significant structural similarity (RMSD $\sim$0.6 Å) with both SCP1 (PDB ID: 2GHQ) (Kamenski et al., 2004; Zhang et al., 2006) and SCP2. These proteins belong to the HAD superfamily, which encompasses a large number of magnesium-dependent phospho-hydrolases characterized by the presence of a conserved DXDX motif (Figure 40.7). This signature sequence contributes to the catalytic site and is responsible for coordination of the catalytically essential magnesium cation, with the first aspartic acid serving as the nucleo-phile and phoshoryl acceptor (Kamenski et al., 2004). Most HAD family members catalyze phosphoryl transfer reactions involving small-molecule metabolites (e.g., phosphoserine). Structures of two HAD protein serine phosphatases from human (PDB ID: 1L8L) (Kim et al., 2002) and *Methanococcus jannaschii* (PDB ID: 1F5S) (Wang et al., 2001) have also been determined by others. Despite rather low sequence identity between the SCPs and small-molecule phosphatases (14% between 2HHL and 1F5S; RMSD $\sim$2.7 Å), they share significant similarities in overall topology and active site architecture (Figure 40.8).

SCP1, SCP2, and SCP3 are composed of a central five-stranded parallel β-sheet flanked by three α-helices on one side and a substantial loop-containing segment on the opposite face. An additional antiparallel three-stranded β-sheet is formed within the extended loop connecting β-strands 2 and 3 of the central β-sheet. In contrast, the *M. jannaschii* phosphoserine phosphatase (PDB ID: 1F5S) contains both the core domain and a capping domain that impinges on the active site. In the case of the tetrameric *Haemophilus influenzae* deoxy-D-mannose-octulosonate 8-phosphatase (PDB IDs: 1K1E and 1J8D), an adjacent monomer packs on top of the core domain and serves as the "cap" (Figure 40.8).

As previously discussed by Allen, Dunaway-Mariano, and colleagues (Peisach et al., 2004), the overall architecture of these active sites appear to be related to the type of substrate recognized. For example, the phosphoserine phosphatases, which recognize small molecules, typically possess small catalytic sites that are relatively sequestered from solvent. Such sequestration is generally provided by an additional "capping" domain present in the structure or by the formation of a higher order oligomeric species that occludes the catalytic site. In contrast, the SCP catalytic sites, which recognize CTD heptad repeats are larger and more accessible to solvent. This architectural variation may be of wider import. For example, in MDP-1, another putative HAD superfamily protein phosphatase that

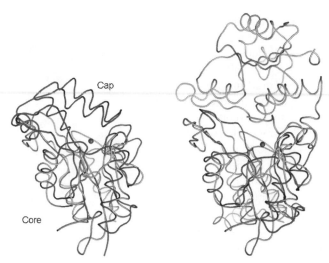

Figure 40.8. Structural comparisons of SCP3. (a) Superposition of SCP3 (green) with *Methano-coccus jannaschii* phosphoserine phosphatase (red, PDB ID: 1F5S). The SCP3 catalytic site is freely accessible to solvent, whereas the alpha-helical capping domain in phosphoserine phosphatase shields its active site. (b) Superposition of SCP3 (green) with a dimer of the tetrameric *Haemophilus influenzae* deoxy-D-mannose-oculosonate 8-phosphatase (red and gray, PDB ID: 1K1E). Mg^{2+} ions are shown as pink spheres, and conserved phosphate-binding loops are shown in yellow. The capping domain of 1F5S occludes the active site entrance. In 1K1E, the second subunit of the dimer plays a similar role. Figure also appears in the Color Figure section.

dephosphorylates phosphotyrosine (Peisach et al., 2004), the catalytic site also appears to be highly solvent accessible. Recently published work suggests that MDP-1 might recognize posttranslationally encoded protein sugar phosphates, which would also require a substantially more solvent accessible catalytic site (Fortpied et al., 2006). This architectural type is also present in the phosphatase domain of T4 polynucleotide kinase, a HAD superfamily member that utilizes polynucleotides substrates (Galburt et al., 2002). It is remarkable that all of these HAD phosphatases recognize their polymeric substrates and perform catalysis at sites present at termini or linker regions. It, therefore, appears that a combination of catalytic site accessibility and substrate dynamics is required to support the biological activity of these polymer substrate specific phosphatases.

CHRONOPHIN

Chronophin is an example of the burgeoning family of "moonlighting" proteins (Jeffery, 1999; Jeffery, 2003), as it plays a central role in vitamin B6 metabolism and serves as a major regulator of the actin cytoskeleton (Wiggan, Bernstein, and Bamburg, 2005). This protein was first identified as pyridoxal phosphatase, which specifically dephosphorylates pyridoxal 5′-phopshate (PLP), the coenzymatically active form of vitamin B6 that participates in a remarkable range of enzymatic transformations (Fonda, 1992). PLP synthesis requires a flavin mononucleotide-dependent pyridoxine 5′-phosphate (PNP) oxidase and an ATP-dependent pyridoxal kinase. Degradative pathways for PLP include the action of one or more pyridoxal phosphatases. The overall mechanism by which PLP levels are regulated is complex to say the least and includes PLP biosynthetic and degradative pathways, PLP binding proteins, and proteins that regulate availability and/or transport of synthetic precursors. Given its central role in PLP metabolism, it is not surprising that the PLP phosphatase is expressed in all human tissues examined and is particularly abundant in brain, suggesting a specialized role in the central nervous system (CNS) (Jang et al., 2003).

More recently, Bokoch and colleagues demonstrated that chronophin plays a direct role in regulating the actin cytoskeleton (Gohla, Birkenfield, and Bokoch, 2004) (Figure 40.9). Under physiologic conditions, monomeric actin (G-actin) spontaneously polymerizes to form actin filaments (F-actin) with chemically and structurally distinct ends, termed barbed and pointed (Pollard and Cooper, 1986, Schafer and Cooper, 1995). Further assembly of F-actin into an array of higher order assemblies underpins all actin-based dynamic processes, including cell motility, vesicle movement, and cytokinesis. *In vivo*, actin polymerization is dominated by monomer addition to barbed ends, making barbed end generation a central focus in studies of dynamic actin-based processes (Schafer and Cooper, 1995; Zigmond, 1996). Cofilin is a 15 kD protein that severs the actin filament by disrupting the noncovalent bonds between monomers comprising the filament (Ono, 2007). Filament severing increases the number of polymerisation-competent barbed ends, and results in increased rates of actin polymerization. Cofilin is also thought to be involved in depolymerization, because under certain conditions cofilin not only increases the number of filaments, but can also increase the rate of monomer dissociation from the newly created pointed ends (Ono, 2007). Cofilin itself is regulated by a phosphorylation–dephosphorylation cycle: LIM kinase-mediated phosphorylation of Ser3 inactivates cofilin (Scott and Olson, 2007) while the action of the slingshot phosphatases return cofilin to the active state (Huang, DerMardirossian, and Bokoch, 2006 Ono, 2007). Bokoch's work exploited a wide range of techniques to demonstrate chronophin specificity. siRNA knockdowns of chronophin activity (reduced phosphatase activity) increases phosphocofilin levels, while overexpression of chronophin decreases phosphocofilin levels. Decreased chronophin activity also results in stabilization of F-actin structures *in vivo* and causes massive defects in cell division (Gohla, Birkenfield, and Bokoch, 2004).

The NYSGXRC structure of human chronophin (PDB ID: 2OYC) confirmed that this protein is indeed a member of the HAD phosphatase family. It possesses a typical core domain that contains the catalytic signature sequence DXDX. There is also a very substantial

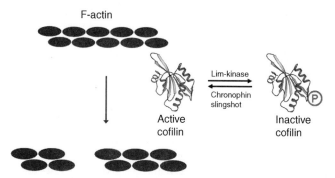

Figure 40.9. Cofilin-mediated F-actin severing. F-actin severing activity of cofilin is regulated by a phosphorylation cycle involving the LIM kinase and the slingshot and chronophin phosphatases. Actin monomers are represented by ellipses.

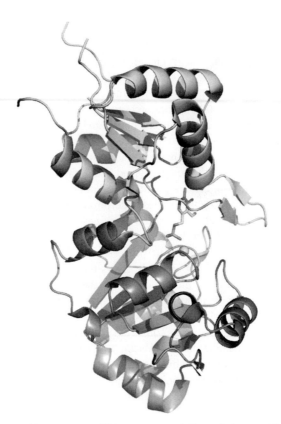

Figure 40.10. Chronophin structure. Ribbon representation of chronophin/PLP phosphorylase with bound PLP and Ca^{2+}. The core domain is the lower half, the capping domain is the upper half, PLP is shown with stick representation, and the Ca^{2+} is shown as a sphere. The catalytic site lies at the interface between the core and capping domains.

capping domain that abuts the catalytic site (Figure 40.10). The structure of the PLP-bound form of the enzyme was obtained by substituting Mg^{2+} (PDB ID: 2P27) with the catalytically inert Ca^{2+} (PDB ID: 2P69). This substitution results in a change of metal ligation from six to seven coordinate via bidentate coordination of the catalytic aspartic acid (Asp25), which results in the near-complete loss of activity (Figure 40.11). Our structure demonstrates that bound PLP is largely buried, with the phosphate being completely shielded from solvent (Figures 40.10 and 40.12). Immediately after our first structure was deposited, three chronophin structures from Kang and coworkers were released from the PDB (PDB IDs: 2CFR, 2CFS, and 2CFT).

Demonstration of cofilin phosphatase activity is remarkable given the structural and functional features associated with all previously characterized polymer-specific HAD family phosphatases. As noted above, two members of the SCP family, MDP-1 and T4 polynucleotide kinase, all lack a capping domain, resulting in a relatively open and solvent-accessible active site. In contrast, chronophin appears unique among the polymer-directed HAD phosphatases examined to date, because the presence of a substantial capping domain largely occludes the catalytic center (Figures 40.12 and 40.13). Analysis of our chronophin structure suggests that conformational reorganization of the core and capping domains is required to facilitate binding of the phosphorylated N-terminus of cofilin. The N-terminus of

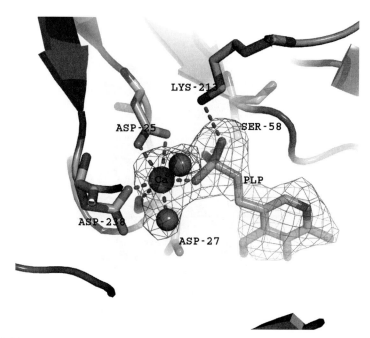

Figure 40.11. Chronophin catalytic site. The active site of chronophin with its ligand PLP and inhibitory Ca²⁺. The Ca²⁺ (green sphere) is hepta-coordinated and participates in a bidentate interaction with the active site nucleophile Asp-25. Figure also appears in the Color Figure section.

Figure 40.12. Chronophin capping domain. Superposition of chronophin and SCP (CTD phosphatase) (PDB ID: 2HHL), which lacks a capping domain. The core domains share 11% sequence identity and superimpose with a DALI Z score of 6.6 and an RMSD of ~2.9Å-for 116 structurally equivalent Cα atoms.

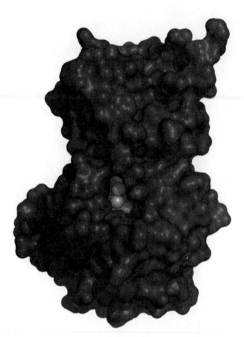

Figure 40.13. Inaccessibility of PLP in the chronophin catalytic site. Surface representation of chronophin (dark grey) with bound ligand PLP (light grey) in the same orientation as shown in Figure 40.12. The PLP is viewed on the edge and the phosphoryl group is completely buried from solvent.

cofilin is unstructured in solution (Blanchoin et al., 2000; Pope et al., 2004) and, like the dynamic properties described above for other HAD-associated polymeric substrates, this behavior is almost certainly essential for it to gain access to the chronophin catalytic site. It, therefore, appears that the HAD family has evolved various mechanisms by which to perform chemistry on polymeric substrates. Availability of our structures and protein reagents provides the necessary foundation for a detailed study of chronophin function. These results will be particularly relevant to our fundamental understanding of cancer, as the phosphorylation status of cofilin is directly implicated in the actin-based mechanisms underlying invasion, intravasation, and metastasis of mammary tumors (Wang et al., 2006).

OTHER PHOSPHATASES

In addition to the human phosphatase structures described above, the NYSGXRC has determined and made publicly available X-ray structures from *T. gondii* (PDB IDs: 2I44 and 2ISN), mosquito (TAB1, PDB ID: 2IRM; and PPM1G, PDB ID: 2I0O), and *T. brucei* (PDB ID: 2QJC), which will be described elsewhere.

BIOMEDICAL THEME TARGETS: POTENTIAL IMPACT ON DRUG DISCOVERY

As highlighted by some of the examples described in detail above, systematic structural characterization of the protein phosphatases provides significant mechanistic and functional insights. Notwithstanding these and other successes, we believe that full exploitation of the growing structural phosphatome database must include efforts to discover new therapeutic phosphatase inhibitors and generate reagents that allow specific inhibition of signaling pathways in cell culture and whole animal model systems. There is much interest in the pharmaceutical and biotechnology industries in protein phosphatases as drug discovery targets, as evidenced by recent publications on PTP1B and related targets (reviewed in Pei et al., 2004). We briefly highlight in the following sections two structure-based approaches to the problem of discovering potent phosphatase inhibitors from academe.

FRAGMENT CONDENSATION LEAD DISCOVERY STRATEGY APPLIED TO PTP1B

The concept of inhibitor (or agonist) design via condensation of individual small-molecule fragments known to bind proximally in the vicinity of a target protein active site has received growing attention over the past decade (Hajduk and Greer, 2007). The simple rationale is that the geometrically appropriate covalent linkage of multiple low affinity fragments/functionalities will generate high-affinity species. A rigorous thermodynamic treatment of this phenomenon was published by Jencks as early as 1981 (Jencks, 1981) and demonstrated that the enhanced affinities of the final species is the consequence of both the additivity of the intrinsic binding energies (ΔG) of the individual fragments.

Fragment libraries are typically composed of 1000–10,000 low molecular weight compounds representing a breadth of chemically diverse substructures that possess various chemical functional groups, which can facilitate either target binding or further chemical elaboration. Selection for solubility, lipophilicity, H-bonding donors/acceptors, and known toxicity or ADME (absorption, distribution, metabolism, and excretion) properties are typical considerations in building the library.

Various approaches to fragment binding detection have been adopted, among them are mass spectrometry (Erlanson, Wells, and Braisted, 2004), NMR (Shuker et al., 1996; Baurin et al., 2004), crystallography (Blaney, Nienaber, and Burley, 2006; Mooij et al., 2006) and combinations therefrom (Moy et al., 2001; Liu et al., 2003). Special considerations for fragment-based discovery using X-ray crystallography include the necessity for obtaining a well-characterized crystal form of the target protein that diffracts well ($d_{min} < 2.5$ Å), possesses a lattice amenable to soaking experiments (i.e., without occlusion of the target site), and is able to withstand exposure to modest quantities of organic solvents (e.g., ethanol and DMSO are popular solvents for chemical fragments). Initial crystal soaking experiments can be conducted on mixtures of the library to speed the initial screening with follow-up soaks using individual fragments, where an individual component cannot be conclusively identified from the resulting electron density maps.

There are numerous examples in the literature of the success of this method, many coming from small biotechnology companies. Hartshorn et al. (2005) have published a summary of their initial findings on five very different targets (p38 MAP kinase, CDK2, thrombin, ribonuclease A, and PTP1B). A detailed summary of a similar approach applied to spleen tyrosine kinase has been published by Blaney, Nienaber, and Burley (2006). The power of the method is highlighted by the following example for PTP1B, which represents the simplest possible approach to fragment screening with a screening library composed solely of phosphotyrosine.

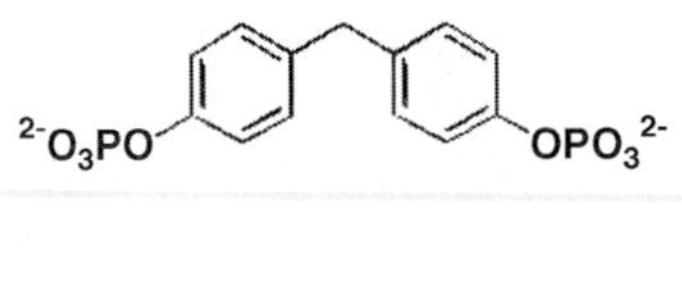

Figure 40.14. Compounds that bind PTP1B. The low K_m, low k_{cat} substrate, compound **1**, compound **2**; and a highly selective, nonhydrolyzable bidentate inhibitor, compound **3**.

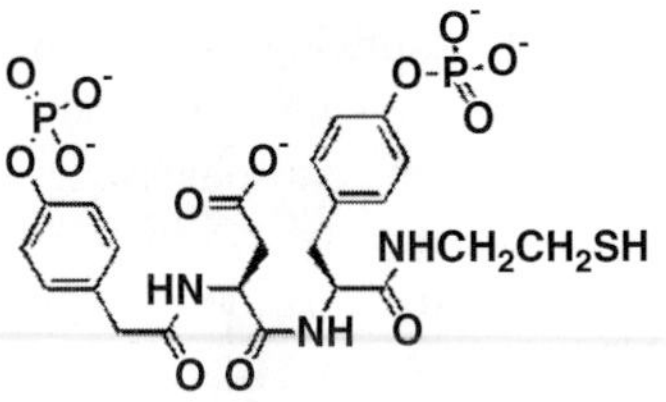

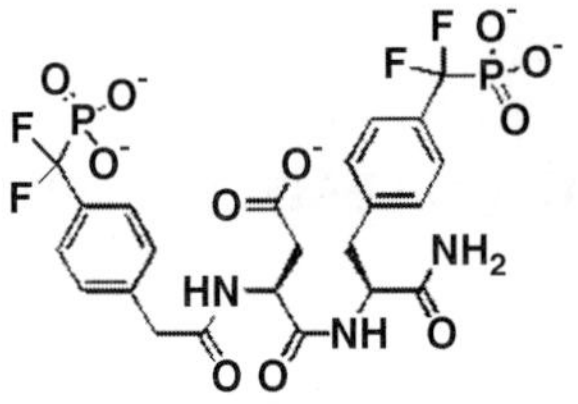

An analysis of the substrate specificity of PTP1 (the rat ortholog of human PTP1B) revealed that this enzyme catalyzes hydrolysis of a wide variety of low molecular weight phosphate monoesters. One of these substrates (compound **1**; Figure 40.14) exhibits poor turnover ($6.9/s$) coupled with an extraordinarily low K_m ($16 \mu M$) value for a nonpeptidic species. This latter property suggested that substrate may bind tightly to the enzyme in a less than optimal fashion with respect to catalysis. Crystallographic analysis of compound **1** bound to a catalytically incompetent C215S mutant form of PTP1B revealed that the substrate can occupy one of the two mutually exclusive binding modes: (1) a catalytically competent active site-bound form and (2) a nonproductive peripheral site-bound form (Figure 40.15a) (Puius et al., 1997). Furthermore, the less sterically demanding phospho-tyrosine (library of one) was observed to simultaneously bind to both sites (Figure 40.15b). Although the active site is highly conserved among all PTPase family members, the peripheral binding site is not. This latter observation suggested that a bidentate ligand, capable of occupying both positions on the surface of the phosphatase, would not only display enhanced affinity but also enhanced selectivity for PTP1B.

Based on these observations, a combinatorial library of 184 compounds was designed containing (1) a fixed phosphotyrosine moiety, (2) 8 structurally diverse aromatic acids (terminal elements) to target the unique peripheral site, and (3) 23 structurally diverse "linkers" to tether the phosphotyrosine with the array of terminal elements (Shen et al., 2001). Compound **2** was identified as the lead species from the library and the corresponding nonhydrolyzable difluorophosphonate (compound **3**) was synthesized. Compound **3** exhibits high affinity ($K_i = 2.4 \, nM$) and extraordinary selectivity (1000–10,000-fold versus an array of phosphatases) for PTP1B. Analogues of this compound have recently been shown to serve as insulin sensitizers and mimetics in cell culture and as appetite suppressors in animal models. These results are consistent with the role of PTP1B as a negative regulator of the insulin and leptin signaling pathways (Taylor and Hill, 2004).

(a)

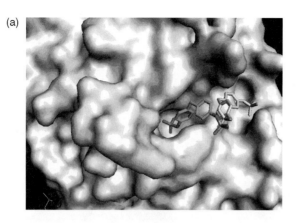

(b)

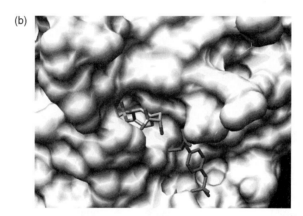

Figure 40.15. Surface representation of the PTP1B active site. (a) Compound **1** binds to a catalytically incompetent form of PTP1B via one of the two mutually incompatible binding modes, which encompass either the active site (multicolor) or a secondary peripheral site (black and white). (b) Phosphotyrosine simultaneously binds to the active site (multicolor) and a secondary peripheral site (red). Figure also appears in the Color Figure section.

The structural biology finding that served as the source of inspiration to prepare the library of bidentate phosphatase inhibitors was based on the surprising finding that PTP1B can simultaneously bind two phosphotyrosine amino acid residues. However, Nature had an additional surprise in store with the bidentate ligand compound **3**, designed to coordinate both the active and peripheral sites. Subsequent crystallographic analysis revealed that although compound **3** binds to the active site in the expected fashion, it does not coordinate to the predicted peripheral site (Figure 40.16). Instead, the bisphosphonate compound **3** coordinates in an unanticipated fashion to a completely different secondary site on the surface of the enzyme (Sun et al., 2003). This secondary site, however, like the initially identified peripheral site, is not conserved among PTPase family members. The latter appears to provide the structural basis for the extraordinary selectivity displayed by compound **3** and its congeners for PTP1B. This serendipitous finding was the product of structure-guided library-screening that relied on access to purified protein

(a)

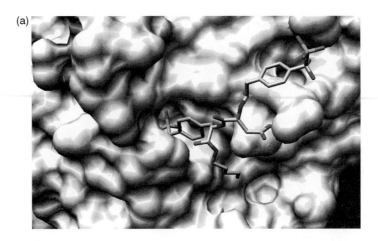

(b)

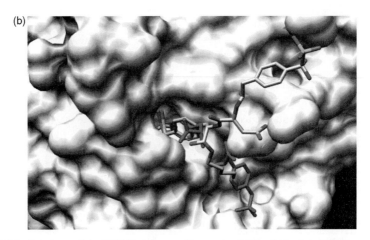

Figure 40.16. Structure of the PTP1B/compound **3** complex (a) The bidentate ligand, compound **3**, is bound to the active site and a secondary peripheral site different from that observed with compound **1** and phosphotyrosine. (b) Overlay of the double binding mode of phosphotyrosine (red) and the bidentate ligand compound **3** (multicolor). Figure also appears in the Color Figure section.

samples, and underscores the potential value of the PSI Materials Repository and its contents.

VIRTUAL SCREENING STRATEGY APPLIED TO PP2Cα

In silico virtual ligand screening (VS) uses computational approaches to identify small-molecule ligands of target macromolecules. The process can be usefully divided into two stages, including docking and scoring. Docking utilizes a structure of the macromolecular target to calculate whether or not a particular compound can fit within a putative binding cleft (e.g., enzyme active site). Scoring methods estimate the free energy of binding for a particular ligand bound to the target in a particular pose, thereby permitting prioritization of predicted target–ligand complexes according to calculated binding energy.

Compound library selection plays an important role in VS. In principle, vast compound sets of diverse properties (including size, lipophilicity, and chemical substructure) are accessible to VS due to the availability of powerful computational resources. For example, Irwin and Shoichet have compiled a freely available database of ~4.6 million commercially available compounds with multiple defined subsets, useful search functions, literature references, chemical similarity cross-references, vendor information, and atomic coordinates (http://blaster.docking.org/zinc/; Irwin and Shoichet, 2005). Moreover, the National Cancer Institute maintains a repository of over 140,000 compounds together with associated literature and structural information; of these 1900 are arrayed on multiwell plates intended for use in ligand discovery (http://dtp.nci.nih.gov/index.html). For a review of the growing availability of chemical databases, see Baker (2006).

While there certainly exists a benefit to having a source of the actual compound for follow-up *in vitro* studies, hypothetical small-molecule compounds can also be generated *in silico* and serve as inputs for VS calculations. In practice, electronic compound libraries are processed (Cummings, Gibbs, and DesJarlais, 2007) and filtered for a variety of properties (e.g., Lipinski's rules) either prior to VS (forward filtering) to generate a smaller test set on which more thorough calculations can be undertaken or after VS (backward filtering) to eliminate excessive downstream chemistry on candidates with poor chemical properties (Klebe, 2006). Moreover, in contrast to fragment approaches (see above) for which very small molecules are observed to bind specifically, albeit sometimes quite weakly, to the target protein, VS methods encounter difficulty placing fragment molecules unless the binding landscape is limited to a manageably small area of the target protein surface. Thus, for VS, library components are usually larger and somewhat more elaborated compounds.

Selection of an appropriate "druggable" macromoleular target is also of critical concern. The availability of high-resolution structural data not only support VS in general, but also allow the objective assessment, selection, and accurate boundary definition of the protein surface site to be interrogated (An, Totrov, and Abagyan, 2005; Vajda and Guarnieri, 2006).

A VS campaign against a phosphatase target for which no nonphosphate-based inhibitor was known has been published recently (Rogers et al., 2006). The authors selected PP2Cα as a target due to its importance in cell cycle and stress response pathways and the availability of a crystal structure (PDB ID: 1A6Q; Das et al., 1996). While inspection of the enzyme binding site revealed three putative binding pockets, PP2Cα represents a more difficult case for VS due to the presence of two bound metal ions coordinated by six waters leading to ambiguity concerning the relevant target structure. Using AutoDock 3.0 (Morris et al., 1998), the authors ran docking and scoring calculations against control compounds (pSer, pThr, and pNPP docked readily with their phosphate groups overlaying a bound phosphate ion observed in the X-ray structure) and the NCI Diversity Set (1900 compounds). This initial screening exercise was followed by a chemical similarity search of the Open NCI Database (currently > 260,000 small-molecule structures) resulting in a second generation library, which was also docked against PP2Cα and the resulting hits scored. Additional docking calculations were run on the most attractive-looking hits against the apo form of the protein (deleting metals and waters) and all 64 possible permutations of water deletions (while retaining the active site metals). Perhaps in an indication of binding to pockets surrounding the metals, there was some agreement among the runs with metal/water deletions when compared to the original run. Postscoring filtering for solubility (compounds with clog P < 6 were retained) yielded some leads, which were then requested from the NCI for inhibition assays against PP2Cα and a panel of three additional Ser/Thr phosphatases. Additional assays of nonspecific inhibition due to aggregation by varying enzyme concentration or adding detergent were conducted. Remarkably, at compound concentrations of 100 μM, many of VS hits demonstrated robust inhibitory activity; compound 109268 showed ~80% inhibition of PP2Cα and reasonable selectivity for PP2Cα and PP1 (80% inhibition) versus PP2A and PP2B (40% inhibition) with no appreciable aggregation effects. No further elaboration of these leads was described.

In light of this result, it is of considerable interest that the NYSGXRC has recently determined the structure of a human PP2Cβ (PPM1B) fragment (PDB ID: 2P8E) and the *T. gondii* ortholog of PPM2C (PDB ID: 2I44; herein referred to as PP2Ctg). The structures of PP2Cα and PP2Cβ are nearly identical (RMSD ~ 0.8 Å; 80% sequence identity), although the PP2Cβ expression construct did not include ~90 C-terminal residues, which form a small α-helical subdomain distal to the active site in PP2Cα. The structure of Ser/Thr phosphatase PP2Ctg (Figure 40.17) resembles those of human PP2Cα (Das et al., 1996) and human PP2Cβ structures. Pair-wise RMSDs comparing the structure of PP2Ctg to hPP2Cα and hPP2Cβ are ~1.9 Å (23% sequence identity) and ~1.8 Å (25% sequence identity), respectively (Figure 40.18). The four-layered αββα architecture, the number of β-strands, and the location and coordination of the dimetal center are conserved among these structures. Although the core catalytic domain structure is conserved, there are significant differences between the human and the *T. gondii* enzymes. The α-helices in PP2Ctg are longer than those found in the human enzyme structures in one of the two α-helical layers and PP2Ctg lacks the

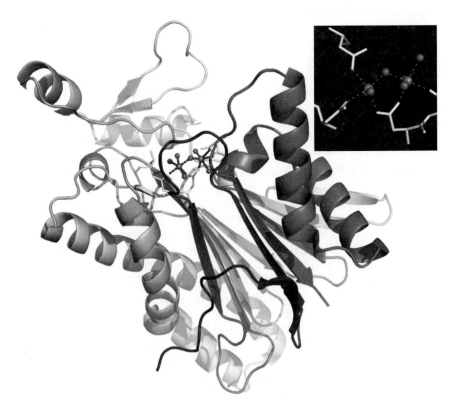

Figure 40.17. X-ray structure of *T. gondii* PP2C (PDB ID: 2i44). *Inset*: two calcium ions supported by conserved aspartates and coordinated waters.

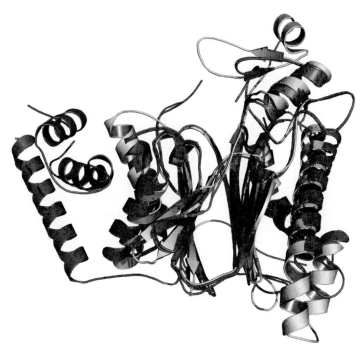

Figure 40.18. Structural alignment and overlay of PP2Cs. PP2Ctg, PP2Cα, and PP2Cβ.

C-terminal 3 α-helices present in the human PP2C structures. Remarkably, in the vicinity of the active site, PP2Ctg possesses two insertions (PP2Ctg residues 207–213 and 218–232) forming an α-helix and a β-sheet, which are not found in the human PP2Cs (Figures 40.18 and 40.19a). Another significant difference affecting the nature of the active site results from the disposition of the N-terminus. In both human PP2C structures, the N-terminus is close to the active site and forms an additional β-strand along one side of the β-sandwich (Figure 40.19b), whereas in PP2Ctg, the N-terminus folds down away from the metals and makes contacts on both sides of the β-sandwich. Several other residues found in the vicinity of the active site and identified as potentially important for ligand binding to human PP2Cα (Rogers et al., 2006) are not conserved: V34 → K, E35 → H, H62 → T, A63 → V, and R186 → F (using PP2Cα residue numbering). This sequence/structure divergence gives rise to significant variation in the active sites of human versus *T. gondii* PP2Cs (Figure 40.19a and b) and may provide the basis for discovery and development of parasite selective phosphatase inhibitors.

BIOMEDICAL THEME TARGETS: CONCLUSION

The impact of various structural genomics efforts on our structural and functional understanding of the human protein phosphatases and protein phosphatases from a wide range of biomedically relevant pathogens is already apparent. Current coverage of the human protein phosphatome is ~45%, with the promise of many more structures to come within the next few years. These data will provide insights into both normal and pathophysiologic processes,

(a)

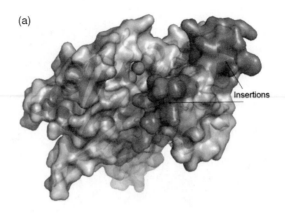

(b)

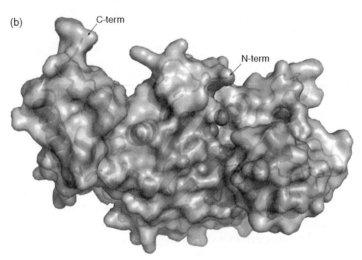

Figure 40.19. Surface representations of PP2Cs. (a) PP2Ctg: Ca^{2+} shown as green spheres; surface corresponding to amino acid insertions 207–213 and 218–232 colored orange. (b) PP2Cα in similar orientation as PP2Ctg: Mn^{2+} shown as green spheres. Figure also appears in the Color Figure section.

including transcription regulation, signal transduction, neural development, and type 1 diabetes. They will also help to stimulate and support discovery of specific small-molecule inhibitors of phosphatases, for use both in biomedical research and in various clinical settings.

COMMUNITY-NOMINATED TARGETS: MOTIVATION

Our initial strategy for community-nominated targets was to identify multidisciplinary collaborative research programs for which determination of a large number of novel protein structures could make a material difference, as opposed to the incremental progress that might be anticipated from individual targets proposed by disparate research groups. Two

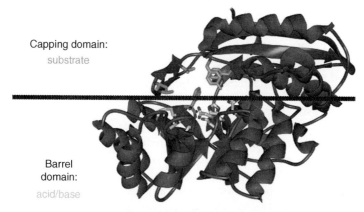

Capping domain:
substrate

Barrel
domain:
acid/base

Figure 40.20. Ribbon representation of the EN/AH superfamily $\alpha_8\beta_8$ barrel fold.

members of the NYSGXRC (Almo and Sali) are also involved in a Project Program Grant (PPG) to examine the structural and chemical basis for functional evolution of the enolase and amidohydrolase metalloenzyme superfamily, in collaboration with John A. Gerlt (University of Illinois, Urbana-Champaign), Frank M. Raushel (Texas A&M University), Patricia C. Babbitt (UCSF), Matthew Jacobson (UCSF), and Brian Shoichet (UCSF). This highly successful program is developing an integrated sequence–structure-based strategy for functional assignment of unknown proteins by predicting the substrate specificities of unknown members of the functionally diverse AH and EN superfamilies that share the ubiquitous $\alpha_8\beta_8$ barrel fold (Figure 40.20, reviewed in Gerlt and Ruashel, 2003). The PPG integrates extensive expertise in functional, structural, and computational enzymology. These diverse, complementary approaches to studies of enzyme substrate specificity should aid in deciphering the ligand binding of other functionally uncharacterized proteins. Their choice of the EN/AH superfamily was motivated by the marked segregation of amino acids contributing to acid/base catalysis (largely derived from the barrel domain) from those dictating enzyme substrate specificity (largely derived from the overlying cap domain).

COMMUNITY-NOMINATED TARGETS: SELECTION AND PROGRESS

Target selection for the EN/AH superfamily was a collaborative process overseen by Babbit (PPG) and Sali (NYSGXRC/PPG). The later stages of target selection were also extensively guided by mechanistic insights from Gerlt (PPG) and Raushel (PPG). Our joint objective was to select structure determination targets from among both known enolases and amidohydrolases and newly identified ENs and AHs (particularly from organisms for which genomic DNA is readily available). Known and putative EN and AH sequences can be found within the PPG-curated structure–function linkage database SFLD (http://sfld.rbvi.ucsf.edu). In addition to this original set of target sequences, new EN and AH sequences were identified from the GenBank nonredundant sequence database, and 11 EN sequences with low similarity to those of known ENs were added to the NYSGXRC community target list in November 2006. Genes for these have all been synthesized and two have already yielded structures (PDB ID: 2PCE and 2PMQ).

To date, we have targeted a total of 279 amidohydrolases and 214 enolases. Expression clones for 370 AH/EN targets have been generated (212 AH, 148 EN), 71% through gene synthesis. One hundred and eight amidohydrolases and sixty enolase targets have been successfully purified (58% from synthetic clones). More than 82 AH/EN targets have yielded crystals (44 AH, 38 EN), and over 850 crystals have been harvested and screened for crystal quality. Diffraction data sets for 59 distinct AH/EN targets have been collected (23 AH, 36 EN). As of October 2007, structures for 34 distinct ENs and 18 distinct AHs have been deposited into the PDB. They have been observed with Mg, Zn, Fe, or Zn bound (and, in some cases, no metal). A structure summary is provided in Table 40.3.

Success rates with the EN superfamily (57% of successfully purified proteins progressed to structures) are significantly better than those observed for the AHs (17%). Our relatively poor success with the AHs reflects our continuing challenge with a subset of the AHs for which unusually broad peaks are observed in electrospray ionization mass spectra following purification to apparent homogeneity as judged by gel electrophoresis. Denaturation of the protein and metal removal (using EDTA or 1,10-phenanthroline) had no effect on appearance of the ESI–MS peaks. Roughly 5–10% of the purified AHs exhibited ESI–MS peak broadening, which appears to be due to covalent modification of solvent-accessible residues. Five AHs were analyzed by proteolytic digestion and peptide mapping. In all cases, correct protein identity was confirmed (~80% polypeptide chain coverage). We have not yet overcome this problem, but will continue to characterize this target family with other biophysical methods and to evaluate their performance in crystallization. Only one of the purified AHs with a broad ESI–MS peak led to successful structure determination. We are also evaluating whether or not *E. coli* fermentation in different media or expression in insect cells may eliminate this unwanted heterogeneity.

COMMUNITY-NOMINATED TARGETS: FUNCTIONAL CHARACTERIZATION

NYSGXRC structures have already had considerable impact on the PPG. All members of the mandelate racemase subfamily of the EN superfamily that have been characterized to date contain a KXK sequence at the end of the second β-strand that is responsible for substrate recognition and chemistry. The NYSGXRC determined the first structure of a target that contains KXD at this position (PDB ID: 2GL5). This structure is now being analyzed computationally and experimentally for novel function. Moreover, our efforts have highlighted the importance of frequent communication between the NYSGXRC and the community-target nominator(s). For example, our work on a particular EN yielded a structure of the apoenzyme (PDB ID: 2GDQ), which was only of partial use to the PPG, as all members of the EN/AH superfamilies are metal-dependent enzymes. Additional NYSGXRC efforts were required to obtain the structure of the enzyme with $Mg2^+$ ion bound (PDB ID: 2GGE). We believe that such special efforts are both justified and essential for prosecution of community-solicited targets. It does not advance the science of the community nominator should the resulting structure fail to provide the desired information (in this case, proper metal placement and coordination to support computational ligand identification).

Gerlt (PPG) and Raushel (PPG) are currently using small-molecule compound library screening methods to provide direct functional annotations with NYSGXRC-provided protein samples.

To date, Gerlt has identified biochemical function for seven ENs, including four o-succinylbenzoate synthases, two L-rhamnonate dehydratases, and a D-galacturonate

TABLE 40.3 NYSGXRC Structures of Community-Nominated Enolases and Amidohydrolases

Enolases			Amidohydrolases		
PDB	TargetID	Deposition	PDB	TargetID	Deposition
2gdq	9293a	3/16/2006	2gok	9252b	4/13/2006
2gge	9293a	3/23/2006	2i5g	9311a	8/24/2006
2gl5	9270a	4/4/2006	2i9u	9246a	9/6/2006
2gsh	9265a	4/25/2006	2ics	9295a	9/13/2006
2hne	9435a	7/12/2006	2imr	9256a	10/4/2006
2hzg	9278a	8/9/2006	2ogj	9244b	1/5/2007
2i5q	9265b	8/25/2006	2ood	9231a	1/25/2007
2nql	9279a	10/31/2006	2oof	9252h	1/25/2007
2o56	9270b	12/5/2006	2p9b	9350a	3/25/2007
2og9	9382a	1/5/2007	2paj	9339a	3/27/2007
2okt	9307b	1/17/2007	2puz	9252b	5/9/2007
2ola	9307a	1/19/2007	2q01	9229a	5/18/2007
2oo6	9228c	1/25/2007	2q08	9247a	5/21/2007
2opj	9312b	1/29/2007	2q09	9252h	5/21/2007
2oqh	9291a	1/31/2007	2q6e	9247a	6/5/2007
2oqy	9375a	2/2/2007	2qah	10053d	6/15/2007
2ovl	9301a	2/14/2007	2qs8	9355e	7/30/2007
2ox4	9270c	2/19/2007	2qt3	9364b	8/1/2007
2oz3	9265m	2/23/2007	2plm	455f	8/2/2007
2oz8	9287a	2/25/2007	2qvh	9312b	8/8/2007
2p0i	9265j	2/28/2007	2r8c	9260b	9/10/2007
2ozt	9306e	2/28/2007	2rag	9257a	9/17/2007
2p3z	9265a	3/10/2007	3b40	9240b	10/23/2007
2pce	9436e	3/29/2007			
2pge	9393a	4/9/2007			
2pgw	9387a	4/10/2007			
2pmq	9437a	4/23/2007			
2pod	9253e	4/26/2007			
2poz	9283a	4/27/2007			
2ppg	9279b	4/30/2007			
2ps2	9440a	5/4/2007			
2qde	9379a	6/20/2007			
2qdd	9436c	6/20/2007			
2qgy	9384a	6/29/2007			
2qq6	9367a	7/26/2007			
2qye	9436d	8/14/2007			

dehydratase. Structural analyses themselves can also provide information regarding putative function. For example, the structure of the unliganded form of target 9265a demonstrated well-ordered "20s and 50s" loops, which are typically disordered in the absence of bound ligand. These features result in an active site that appears to be considerably "more open" than most unliganded ENs, suggesting that the natural ligand/substrate might be polymeric. This hypothesis is currently being tested by Gerlt and coworkers.

To date, Raushel has screened small-molecule compound libraries appropriate for the AHs and produced five tentative biochemical annotations for three putative dipeptidases and two putative N-acyl-D-amino acid deacetylases (DAAs). These studies are now being complemented by direct metal analysis. NYSGXRC efforts on the AHs also enabled a remarkable example of structure-guided functional annotation. The structure of Tm0936, a putative member of the AH family, was determined during PSI-I (by both NYSGXRC and JCSG). Shoichet (PPG), Raushel (PPG), Almo (NYSGXRC and PPG) and coworkers performed extensive computational docking to successfully identify substrates of the enzyme (Hermann et al., 2007). A unique feature of this computational effort was the use of high-energy conformations that mimic reaction intermediates; such high-energy inter-mediates may be better suited to capturing substrate–protein interactions than are ground states. Shoichet's docking hit list was dominated by 6-amino-purine analogues. Raushel then demonstrated that three of these compounds, including 5-methylthioadenosine and s-adenosylhomocysteine, exhibited substantial turnover in the presence of the enzyme (105/Ms). Definitive confirmation of this related group of substrates followed with Almo's X-ray crystal structure determination of Tm0936 bound to the product of s-adenosylho-mocysteine deamination, s-inosylhomocysteine (PDB ID: 2PLM). The experimental struc-ture is remarkably similar to that predicted by Shoichet with computational docking. Further work documented that these deamination products are further metabolized by *Thermotoga maritima* via a previously uncharacterized s-adenosylhomocysteine degradation pathway. Our collaborative efforts with the Gerlt/Raushel PPG clearly demonstrate the power and potential of structure-guided computational approaches for ligand/substrate discovery and functional annotation, provided both protein reagents and X-ray structures are readily available.

COMMUNITY-NOMINATED TARGETS: CONCLUSION

The NYSGXRC community-nominated target structure determination program is proving highly effective as (1) it is servicing a large programmatic effort addressing functional assignment, a major issue in the postgenomic era; (2) it leverages the expertise/platform of the NYSGXRC; (3) it maximizes the value of each purified protein, as, in addition to our crystallographic efforts, these materials support a wide range of mechanistic solution-based studies, and (4) it has already provided novel structural information that can be translated into function/mechanistic insight.

OVERALL CONCLUSIONS

Structural genomics studies of protein superfamilies by the NYSGXRC during the first two years of the second phase of the NIH-funded Protein Structure Initiative have already yielded a considerable wealth of new experimental X-ray crystal structures. In addition, our efforts have produced reagents (i.e., expression clones, protocols for *E. coli* fermentation and protein purification and crystallization, and purified proteins) that are proving valuable to researchers seeking to understand protein structure/function relationships. During the remainder of PSI-II, the NYSGXRC plans to continue its biomedical theme studies of the protein phosphatase family and its community-nominated collaborative studies of the enolase and amidohydrolase superfamily with Gerlt/Raushel and coworkers.

ACKNOWLEDGMENTS

The NYSGXRC is supported by NIH Grant U54 GM074945 (principal investigator: S.K. Burley). We gratefully acknowledge the efforts of all members of the NYSGXRC, past and present.

REFERENCES

Abbott KL, Archambault J, Xiao H, Nguyen BD, Roeder RG, Greenblatt J, Omichinski JG, Legault P (2005): *Biochemistry* 44:2716–2731.

Ahn SH, Kim M, Buratowski S (2004): *Mol Cell* 13:67–76.

An J, Totrov M, Abagyan R (2005): *Mol Cell Proteomics* 4(6):752–761.

Baker M (2006): *Nat Rev Drug Discov* 5:707–708.

Baurin N, Aboul-Ela F, Barril X, Davis B, Drysdale M, Dymock B, Finch H, Fromont C, Richardson C, Simmonite H, Hubbard RE (2004): *J Chem Inf Comput Sci* 44(6):2157–2166.

Blanchetot C, den Hertog J (2000): *J Biol Chem* 275(17):12446–12452.

Blanchetot C, Tertoolen LG, Overvoorde J, den Hertog J (2002): *J Biol Chem* 277(49):47263–47269.

Blanchoin L, Robinson RC, Choe S, Pollard TD (2000): *J Mol Biol* 295(2):203–211.

Blaney J, Nienaber V, Burley SK (2006): In: Jahnke W, Erlanson DA,editors. *Fragment-Based Approaches in Drug Discovery*. KGaA, Weinheim: Wiley-VCH Verlag GmbH & Co. pp 215–248.

Buist A, Zhang YL, Keng YF, Wu L, Zhang ZY, den Hertog J (1999): *Biochemistry* 38:914–922.

Chagnon MJ, Uetani N, Tremblay ML (2004): *Biochem Cell Biol* 82:664–675.

Corden JL, Cadena DL, Ahearn JM Jr, Dahmus ME (1985): *Proc Natl Acad Sci USA* 82:7934–7938.

Cummings MD, Gibbs AC, DesJarlais RL (2007): *Med Chem* 3(1):107–113.

Dahmus ME (1996): *J Biol Chem* 271(32):19009–19012.

Das AK, Helps NR, Cohen PT, Barford D (1996): *EMBO J* 15:6798–6809.

Davidson HW, Watts C (1989): *J Cell Biol* 109:85–92.

DiLorenzo TP, Peakman M, Roep BO (2007): *Clin Exp Immunol* 148:1–16.

Erlanson DA, Wells JA, Braisted AC (2004): *Annu Rev Biophys Biomol Struct* 33:199–223.

Fonda ML (1992): *J Biol Chem* 267:15978–15983.

Fortpied J, Maliekal P, Vertommen D, Van Schaftingen E (2006): *J Biol Chem* 281(27):18378–18385.

Galburt EA, Pelletier J, Wilson G, Stoddard BL (2002): *Structure* 10:1249–1260.

Gerlt JA, Ruashel FM (2003): *Curr Opin Chem Biol* 7:252–264.

Gohla A, Birkenfield J, Bokoch GM (2004): *Nat Cell Biol* 7:21–29.

Goodrich JA, Tjian R (1994): *Cell* 77:145–156.

Gross S, Blanchetot C, Schepens J, Albet S, Lammers R, den Hertog J, Hendriks WJ (2002): *Biol Chem* 277(50):48139–48145.

Hajduk PJ, Greer J (2007): *Nat Rev Drug Discov* 6(3):211–219.

Hartshorn MJ, Murray CW, Cleasby A, Frederickson M, Tickle IJ, Jhoti H (2005): *J Med Chem* 48 (2):403–413.

Hengartner CJ, Myer VE, Liao SM, Wilson CJ, Koh SS, Young RA (1998): *Mol Cell* 2:43–53.

Hermann JC, Marti-Aborna R, Federov AA, Federov E, Almo SC, Shoichet BK, Raushel FM (2007): *Nature* 448:775–779.

Ho CK, Shuman S (1999): *Mol Cell* 3:405–411.

Huang TY, DerMardirossian C, Bokoch GM (2006): *Curr Opin Cell Biol* 18(1):26–31.

Irwin JJ, Shoichet BK (2005): *J Chem Inf Model* 45(1):177–182.

Jang YM, Kim DW, Kang TC, Won MH, Baek NI, Moon BJ, Choi SY, Kwon OS (2003): *J Biol Chem* 278(50):50040–50046.

Jeffery CJ (1999): *Trends Biochem Sci* 24(1):8–11.

Jeffery CJ (2003): *Ann Med* 35(1):28–35.

Jemc J, Rebay I (2007): The eyes absent family of phosphotyrosine phosphatases: properties and roles in developmental regulation of transcription. *Ann Rev Biochem* 76:513–538.

Jencks WP (1981): *Proc Natl Acad Sci USA* 78:4046–4050.

Jiang G, den Hertog J, Su J, Noel J, Sap J, Hunter T (1999): *Nature* 401:606–610.

Kamenski T, Heilmeier S, Meinhart A, Cramer P (2004): *Mol Cell* 15:399–407.

Kashuba VI, Li J, Wang F, Senchenko VN, Protopopov A, Malyukova A, Kutsenko AS, Kadyrova E, Zabarovska VI, Muravenko OV, Zelenin AV, Kisselev LL, Kuzmin I, Minna JD, Winberg G, Ernberg I, Braga E, Lerman MI, Klein G, Zabarovsky ER (2004): *Proc Natl Acad Sci USA* 101:4906–4911.

Ke Y, Kapp JA (1996): *J Exp Med* 184:1179–1184.

Kim HY, Heo YS, Kim JH, Park MH, Moon J, Kim E, Kwon D, Yoon J, Shin D, Jeong EJ, Park SY, Lee TG, Jeon YH, Ro S, Cho JM, Hwang KY, J Biol Chem 277:(2002): 46651–46658.

Klebe G (2006): *Drug Discov Today* 11(13/14):580–594.

Knockaert M, Sapkota G, Alarcon C, Massague J, Brivanlou AH (2006): *Proc Natl Acad Sci USA* 103:11940–11945.

Lampasona V, Bearzatto M, Genovese S, Bosi E, Ferrari M, Bonifacio E (1996): *J Immunol* 157:2707–2711.

Lan MS, Lu J, Goto Y, Notkins AL (1994): *DNA Cell Biol* 13:505–514.

Lan MS, Wasserfall C, Maclaren NK, Notkins AL (1996): *Proc Natl Acad Sci USA* 93:6367–6370.

Lee S, Faux C, Nixon J, Alete D, Chilton J, Hawadle M, Stoker AW (2007): *Mol Cell Biol* 27:1795–1808.

Liu G, Xin Z, Pei Z, Hajduk PJ, Abad-Zapatero C, Hutchins CW, Zhao H, Lubben TH, Ballaron SJ, Haasch DL, Kaszubska W, Rondinone CM, Trevillyan JM, Jirousek MR (2003): *J Med Chem* 46(20):4232–4235.

Magistrelli G, Toma S, Isacchi A (1996): *Biochem Biophys Res Commun* 227:581–588.

McCluskey A, Sim AT, Sakoff JA (2002): *J Med Chem* 45(6):1151–1175.

Meinhart A, Kamenski T, Hoeppner S, Baumli S, Cramer P (2005): *Genes Dev* 19(12):1401–1415.

Mooij WT, Hartshorn MJ, Tickle IJ, Sharff AJ, Verdonk ML, Jhoti H (2006): *Chem Med Chem* 1(8):827–838.

Morris GM, Goodsell DS, Halliday RS, Huey R, Hart WE, Belew RK, Olson AJ (1998): *J Comput Chem* 19(14):1639–1662.

Moy FJ, Haraki K, Mobilio D, Walker G, Powers R, Tabei K, Tong H, Siegel MM (2001): *Anal Chem* 73(3):571–581.

Nam HJ, Poy F, Krueger NX, Saito H, Frederick CA (1999): *Cell* 97:449–457.

Nam HJ, Poy F, Saito H, Frederick CA (2005): *J Exp Med* 201:441–452.

Ono S (2007): *Int Rev Cytol* 258:1–82.

Ouyang Q, Standifer NE, Qin H, Gottlieb P, Verchere CB, Nepom GT, Tan R, Panagiotopoulos C (2006): *Diabetes* 55:3068–3074.

Pei Z, Liu G, Lubben TH, Szczepankiewicz BG (2004): *Curr Pharm Des* 10(28):3481–3504.

Peisach E, Selengut JD, Dunaway-Mariano D, Allen KN (2004): *Biochemistry* 43(40):12770–12779.

Pollard TD, Cooper JA (1986): *Annu Rev Biochem* 55:987–1035.

Pope BJ, Zierler-Gould KM, Kuhne R, Weeds AG, Ball LJ (2004): *J Biol Chem* 279(6):4840–4848.

Puius YA, Zhao Y, Sullivan M, Lawrence DS, Almo SC, Zhang ZY (1997): *Proc Natl Acad Sci USA* 94 (25):13420–13425.

Pulido R, Serra-Pages C, Tang M, Streuli M (1995): *Proc Natl Acad Sci USA* 92:11686–11690.

Rodriguez CR, Cho EJ, Keogh MC, Moore CL, Greenleaf AL, Buratowski S (2000): *Mol Cell Biol* 20:104–112.

Rogers JP, Beuscher AE IV, Flajolet M, McAvoy T, Nairn AC, Olson AJ, Greengard P (2006): *J Med Chem* 49:1658–1667.

Saeki K, Zhu M, Kubosaki A, Xie J, Lan MS, Notkins AL (2002): *Diabetes* 51:1842–1850.

Sapieha PS, Duplan L, Uetani N, Joly S, Tremblay ML, Kennedy TE, Di Polo A (2005): *Mol Cell Neurosci* 28:625–635.

Sapkota G, Knockaert M, Alarcon C, Montalvo E, Brivanlou AH, Massague J (2006): *J Biol Chem* 281 (52):40412–40419.

Sayre MH, Tschochner H, Kornberg RD (1992): *J Biol Chem* 267:23376–23382.

Schafer DA, Cooper JA (1995): *Annu Rev Cell Biol* 11:497–518.

Schneider TD, Stephens RM (1990): *Nucleic Acids Res* 18:6097–6100.

Scott RW, Olson MF (2007): LIM kinases: function, regulation and association with human disease. *J Mol Med* 85:555–568.

Shatkin AJ, Manley JL (2000): *Nat Struct Biol* 7:838–842.

Shen K, Keng YF, Wu L, Guo XL, Lawrence DS, Zhang ZY (2001): *J Biol Chem* 276 (50):47311–47319.

Shuker SB, Hajduk PJ, Meadows RP, Fesik SW (1996): *Science* 274(5292):1531–1534.

Simitsek PD, Campbell DG, Lanzavecchia A, Fairweather N, Watts C (1995): *J Exp Med* 181:1957–1963.

Siu R, Fladd C, Rotin D (2007): *Mol Cell Biol* 27:208–219.

Solimena M, Dirkx R Jr, Hermel JM, Pleasic-Williams S, Shapiro JA, Caron L, Rabin DU (1996): *EMBO J* 15:2102–2114.

Su YA, Lee MM, Hutter CM, Meltzer PS (1997): *Oncogene* 15:1289–1294.

Sun JP, Fedorov AA, Lee SY, Guo XL, Shen K, Lawrence DS, Almo SC, Zhang ZY (2003): *J Biol Chem* 278(14):12406–12414.

Taylor SD, Hill B (2004): *Expert Opin Investig Drugs* 13(3):199–214.

Tertoolen LG, Blanchetot C, Jiang G, Overvoorde J, Gadella T.W. Jr, Hunter T., den Hertog J. (2001): *BMC Cell Biol* 2:8.

Thompson J, Lepikhova T, Teixido-Travesa N, Whitehead MA, Palvimo JJ, Janne OA (2006): *EMBO J* 25:2757–2767.

Tonks NK (2006): *Nat Rev Mol Cell Biol* 7(11):833–846.

Vajda S, Guarnieri F (2006): *Curr Opin Drug Discov Devel* 9(3):354–362.

Verge CF, Gianani R, Kawasaki E, Yu L, Pietropaolo M, Jackson RA, Chase HP, Eisenbarth GS (1996): *Diabetes* 45:926–933.

Villinger F, Mayne AE, Bostik P, Mori K, Jensen PE, Ahmed R, Ansari AA (2003): *J Virol* 77:10–24.

Wallace MJ, Fladd C, Batt J, Rotin D (1998): *Mol Cell Biol* 18:2608–2616.

Wang W, Kim R, Jancarik J, Yokota H, Kim SH (2001): *Structure* 9:65–71.

Wang W, Mouneimne G, Sidani M, Wyckoff J, Chen X, Makris A, Goswami S, Bresnick AR, Condeelis JS (2006): *J Cell Biol* 173:395–404.

Watts C, Lanzavecchia A (1993): *J Exp Med* 178:1459–1463.

Wiggan O, Bernstein BW, Bamburg JR (2005): *Nat Cell Biol* 7(1):8–9.

Xie H, Zhang B, Matsumoto Y, Li Q, Notkins AL, Lan MS (1997): *J Immunol* 159:3662–3667.

Xie H, Deng YJ, Notkins AL, Lan MS (1998): *Clin Exp Immunol* 113:367–372.

Xu Z, Weiss A (2002): *Nat Immunol* 3:764–771.

Yeo M, Lee SK, Lee B, Ruiz EC, Pfaff SL, Gill GN (2005): *Science* 307:596–600.

Zhang B, Lan MS, Notkins AL (1997): *Diabetes* 46:40–43.

Zhang Y, Kim Y, Genoud N, Gao J, Kelly JW, Pfaff SL, Gill GN, Dixon JE, Noel JP (2006): *Mol Cell* 24:759–770.

Zhou M, Halanski MA, Radonovich MF, Kashanchi F, Peng J, Price DH, Brady JN (2000): *Mol Cell Biol* 20:5077–5086.

Zigmond S (1996): *Curr Opin Cell Biol* 8:66–73.

INDEX

Ab initio structure prediction, 506, 665

Active-site enzymes structures, architectures, 420

Acyl-CoA thioesterase, 200
  cross-links connecting, 200

Algebraic system, 583

All-atom contact analysis, methodology of, 379

Alzheimer's disease, 479

Alzheimer's tangles, 947

Amide hydrogens, 174
  exchange, 175
  features affect, 175

Amino acid deacetylases, 1014

Amino acid interactions, 975

Amino acid partners-wrong interactions, 747

Amino acid residue, types, 364

Amino acid sequence, 16, 17, 300, 468
  types/structure, 16–19
  Z-values of, 974

Analogy-based function analysis, 750

Analytical gel filtration chromatography, 992

Angle-distance hydrogen bond, 461–462
  assignment, 465–467

Anisotropic network model (ANM) potential, 914

ANOLEA program, 364

Anthrax lethal toxin, 184

Antibody
  conformational changes in, 851
  residues, 852
  structure function, 849–850

Antigen, 852

Application program interface (API), 271, 287–289
  technologies, 287

Aromatic interactions, 256

Arp2/3 complex, 148
  principle of, 3D reconstruction, 148

Artificial neural network (ANN), 862

ASTRAL compendium, 326, 428
  rapid access format (RAF), 326
  sequence/structure analysis, 326
  stereochemical check score (SCS), 327

Atlas pages, 309–312

Atom-labeled amino acids, 100

ATOM records, 272

Atomic distribution, 146
  Fourier components of, 146
  rapid access format (RAF), 326

Atomic model, parameterization of, 84

Atomic positions, 358
  typical uncertainties, 358

Atomic resolution structures, 144

Atomic scale descriptors, 316

Automated comparative modeling, 333
  swiss-model server, 333

Automated domain classification, 545

Automated NMR analysis, 115
  errors in modeling approximations, 116

Automatic homology, 561, 719

Average positional error, 349

β-barrel structures, 889

B-cell epitope prediction, 849, 850, 857
  objective of, 850

B-DNA structures, 56

Backbone-backbone interaction energy, 463–465

Backbone-dependent libraries, 215, 216

Base pair covariation, 793–794
  phylogenetic analysis of, 794

Bayes' rule, 118

Structural Bioinformatics, Second Edition   Edited by Jenny Gu and Philip E. Bourne
Copyright © 2009 John Wiley & Sons, Inc.

Bayesian inference, 209
   approach, 209
Bayesian statistics, *see* Bayesian inference,
   209–211
Benchmarking experiments, 670
   evaluation, 891
   launcher, 689
Beta-barrel membrane proteins (TMB), 688
Bioinformatics, 3–4, 93
   applications, 3–4
   approaches, 93
   challenges, 7
   definition of, 3
   domains, 6
   epitope prediction, 855, 858
   methods, 858
   tools, 888
Biological assay, 635–636
Biological macromolecule crystallization
      database (BMCD), 324
   NASA protein crystal growth archive, 324
   X-ray crystallography, 324
Biological macromolecule, *see* Protein
Biological membranes, 882–884
   glycerol-based phospholipids, 884
Biological Sciences Research Council
      (BBSRC), 888
Biomolecular binding partners, 577
Biomolecular electrostatics, 577
   methods, 577
Biomolecular NMR, 94–95,129
   database for, 129
   structures, 112
   validation of, 112
Biomolecules, 587
   electrostatic properties of, 587
      characterization of, 587
Biopolymers, DNA, 576
Biotin, 635
Blue-Gene team, 908
Blue Gene computer, 926
BMRB entry, 121–122, 131
   potential information content of, 131
Boltzmann constant, 121, 346
Boltzmann equation, 582–584
   equation coefficients, 578
Bond-containing fold, 565
Bond-dependent domains, 565
Boundary element methods, 584
Bovine rhodopsin, 890
Brownian dynamics, 926–927
   diffusion-controlled reactions, 927

CATH protocol, 440
   schematic presentation, 440
C-terminal domain phosphatases, 996–998
   family, 998
C-terminal globular domain, 950
Calcium-bound bovine calmodulin, 950
Calmodulin, conformational variations, 951
Cambridge structural database (CSD), 325, 347
   Cambridge crystallographic data center,
      347–348
   crystal structure information, 325–326
      types of, 325
cAMP-dependent protein kinase, 946
Cancer-associated proteins, 940
Carbohydrate modifications, 31
Carcinogenic adducts, 59
Cardiovascular disease, 947
   role, 947
Cartesian meshes, 582
CASP experiment, 665, 674, 776
   boundaries, 672
   division, 665
   evaluators, 776
   goal, 666
Catalytic cycle, 946
   stage of, 946
CATH-based Web resource, 444
CATH database, 435, 436
   phylogenetic/phonetic relationships, 435–436
   schematic representation, 436
CATH domain, 433, 498, 542, 545, 548
   bar graph, 545
   boundaries, 437
   database, 433–434
   distribution of, 548
   historical development, 434–435
CATH Web site, 446–447
   hierarchy, 435, 451
   molscript representations, 437
   multidomain architectures, 543
   plot, 449
   superfamilies, 445–446
   wheel, 447
   workflow, 543
Cathedral server, 447
Cathode ray tube (CRT), 244
Cell-based methods, 100–101
Cell-free biological systems, 100
Cell-substrate interactions, 985
Cellular localization, 516
Cellular membrane, 889
   schematic representation, 889

Cellular retinoic acid-binding protein type II (CRABP), 362
  protein of, 362
Center for biologics evaluation and research (CBER), 814
Centre for eukaryotic structural genomics, 888
Checking structure coordinates, 366
  programs for, 366
Chemical shifts, 114, 120, 122
  linear analysis of, 114, 122
  measurement of, 114
Chiral molecules, 17
Chronophin capping domain, 1001
Chronophin catalytic site, 1001–1002
Chronophin phosphatases, 999
Chronophin/PLP phosphorylase, 1000
Circular dichroism driven assignment, 471–472
Circular dichroism spectroscopy, 478
Classical molecular mechanics, 916–922
  Langevin/Brownian dynamics, 926–927
    Newton's equation of motion, 926
  Langevin equation, 927
  QM-MM methods, 927–928
  stochastic equation, 926
Coarse-grained models, 912–916
Cofilin phosphatase activity, 1000
  demonstration of, 1000
Coiled coil proteins, 424
  α-helices, 424
Collision-induced dissociation (CID), 174
  fragments molecular ions, 174
Combining virtual screening methods, 828
Common object request broker architecture, 287
Community-driven molecular biology data collections, 293
Community-nominated enolases, 1013
  NYSGXRC structuresm, 1013
Community-nominated targets, 1012
  motivation, 1010
  selection/progress, 1011–1012
Community-wide assessments, 665
  measurement, 665–666
Community-wide benchmarking services, 667
  history/findings, 667–673
Community-wide blind test, 775
Complementarity determining regions (CDRs), 849
Computational methods, 518, 888
  identification, 888–889
Computational scientists, 4
Computer hardware, 243–244

Computer software, 244–245
  environment, 86
Computer visualization, 245
Conditional log-linear methods (CLLM), 798
Conservative expansion, 550
Constrained geometric simulation (CGS), 821–822
Constraint-based force fields, 977
Continuous epitope prediction, 855
Continuum electrostatics methods, 587
  investigate macromolecular, 587
  software, 588
Contrast transfer function (CTF), 146
  graph of, 146
Cooper–Dryden model, 946
CORA templates, 445
Correct partners-wrong interactions, 746, 748
Coulomb energy, 462
Coulomb hydrogen bond energy, 462
Coulomb law, 919
Coulomb/van der Waals interaction energy calculations, 461, 463
Covalent bonds, 565
^{13}C relaxation parameters, 99
Critical assessment of structure prediction (CASP), 334, 665
  prediction methods, 334
Cross-link identification, 194
  techniques for, 194
    affinity tagged cross-linking reagents, 194
    cleavable cross-linking reagents, 194
    fragmentation of DTSSP cross-linker, 194
    heavy cross-linking reagents, 194
Cross-linked protein complexes, 198
  formation of, 198
  screen proteolytic mass spectra maps of, 198
Cross-linker insertions, 192
  types of, 192
Cryoelectron microscopy, 241, 284
Crystallographic structures, 220
CSP algorithm, 800
Current methods, 756
CYANA, 107
  cis/trans peptide bond isomers, 110
  NOEASSIGN module of, 107
  Ramachandran statistics, 112
  Root mean squared deviation (RMSD), 108
CYANA calculations, 106
CYANA supports, 106
Cytoplasmic polyhedrosis virus, 151
Cytoskeleton filaments, 149

DACA method, 364
Database of converted restraints (DOCR), 131
Databases of protein structures (PDB), 113
2D dynamic programming, 742, 745
2D exchange spectroscopy, 99
de novo design, 828–829
de novo structure prediction, 766–767
DEE criteria, 219
   protein conformations, 219
Density functional theory (DFT), 928
Density modification, 80
Deposited structures errors, 356
   obsolete structures, 356
   serious errors, 356
     PDB entries, 356
     X-ray/NMR structures, 356
   typical errors, 358
Derived secondary structure of proteins (DSSP), 328
   PDB files, 328–329
Deuterium-labeled cross-linker isomers, 199
Deuterium-labeled cross-linking reagents, 198
Deuterium exchange mass spectrometry (DXMS), 172
Dictionary-based representations, 271
Dictionary of homologous superfamilies (DHS), 445–446
Dictionary of secondary structure of proteins (DSSP), 459
   concepts, 460–461
   history, 460
Dielectric constant, 918, 922
Diffraction methods, 84
Dihedral-angle analysis, 389
Dihydrofolate reductase, 823
Dimeric proteins, 106
Dirichlet boundary condition, 579
Distributed annotation system (DAS), 554
Disulfide bond-containing domains, 565
   emergence of, 567
Disulfide bond-containing fold, 565
Disulfide bond-dependent domains, 567
Disulfide bond, 565
DNA, 41, 738
   binding activity, 59
   conformation of, 49
   discovery, 41
   double helices, forms, 61
   five-member ring, 49
   pseudorotation parameters, 50
   pseudorotation wheel, 49, 50
   replication, 736

   single-crystal structure, 57
   structure, 41
DNA binding domain, 962
   Zn-finger motif, 962
DNA interaction with drug molecules, 59
DNA manipulation techniques, 961
   polymerase chain reaction (PCR), 961
DNA polymer strand, 285
DNA recognition, 604
   structure-based prediction of, 604
DNA/RNA oligonucleotides, 305
DNA/RNA polymerases, 390
2D NMR pulse schemes, 99
DOCK search algorithm, 637
Docking algorithm, 196, 197
   distance constraints, 196
   flow diagram of, 197
   performance of, 644–645
Docking methods, 644, 645, 669
Docking molecules, 618
   proteins, 618
Docking tools, 196
Domain assignment methods, 503–504, 508
   abundance, 563
   combination, 567
   duplication, 561
   graphical performance, 503
   identification, 441
   transfers, 567
DomainFinder works, 543
Domain-level properties, 529
DomainParser algorithm, 500, 504
   schematic representation, 500
Double-stranded RNA, 60
3D protein structure, 487, 679
   hypothesis, 679
Drug-DNA complex, 282, 285
Drug-DNA hydrogen bonding interactions, 600
Drug design, 819
   protein flexibility, 819
   web sites, 238
Drug discovery, 811, 816, 1002–1003
   cycle, 635–637
   historic development of, 809–810
   macromolecular target, 810
   potential impact, 1002–1003
   process, 811
   structure-activity relationships (SAR), 810
   target-based, 814
Drug targets, 813–814
   macromolecules, 813
     types of, 813

Druggability, 816–817
  principles of, 816–817
  quantitative assessment of, 818
DSSP method, 465
  assignment, 465–467
  explanation, 466
  hydrogen bond conformations, 470
DTSSP cross-linked fragments, 195
  unique fragmentation of, 193, 195
DXMS methodology, 176
  automated data analysis, 179
  digestion optimization, 177
  H/D exchange MS data, 178
  H/D exchange reaction, 176
  HPLC separation, 177
  mass analysis, 177
  protein fragmentation, 177
Dynamic programming algorithm, 744

Electron cryomicroscopy, 144, 145, 147,
    157
  classification levels, 516
  signal-to-noise ratio in, 144
  structure prediction methodology, 159
  three-dimensional electron density, 145
  three-dimensional structure, 147
Electron crystallographic reconstructions, 159
  transmembrane proteins, 159
Electron crystallography, 149, 154
  combines elements of, 154
Electron density, 349, 350
  effect of resolution on the quality, 350
  single side chain, 349
Electron diffraction, 147
Electron micrograph, 147
Electron microscope, 145
  image formation, 145
    schematic diagram, 145
Electron microscopy, 143, 144, 147, 153, 155,
    344
  X-ray crystallography, 144, 156, 343
  electron optical resolution of, 147
  imaging macromolecules, 155
  imaging protein dynamics, 153–155
Electron tomography, 160
  pattern recognition, 160
Electron wave functions, 928
  computation of, 928
Electrostatic equations, 577
Electrostatic interactions, 576, 580
  history, 576–577
Electrostatic models, 576

Electrostatic potential, 578, 580
Electrostatics calculations, 581
Electrostatics methods, 577, 584
  application of, 584
  efficient/scalable, 577
Empirical hydrogen bond, 463
  angles/distances, 463
Endocyclic torsion angles, 49
Energetic interactions, 575
Energy-based fold recognition, 739
Energy-based scoring function, 223
Energy parameterization, 745
ENM-related methods, 913
Enzymatic catalysis, 420–421
  evolution of, 420–421
Enzyme active-site residues, 526
Enzyme activity, active site, 976
Enzyme Commission numbers, 516
Ergodic hypothesis, 211, 212, 908
ERRAT program, 363
Escherichia coli, 823
  cells, 100
Estimated standard deviation (ESD), 352
Eukaryotic cells, 100
Eukaryotic root, 564
European Bioinformatics Institute (EBI), 328,
    540
European Macro-molecular Structure Database
    (EMSD), 434
Evolutionary biology, 559
Evolutionary events, 567
  domain combination, 567
Evolutionary units, 566
Extensible markup language (XML), 284
Extra Precision Glide (GlideXP), 644
Extracytoplasmic region, 687

Familial amyotrophic lateral sclerosis (FALS),
    964
Family-dependent conserved positions, 621
FDA-approved drug, 636
Fibroblast growth factor, 2 (FGF2) protein,
    199
Fibrous proteins, 8
  discovery, 41
Filtered restraints database (FRED), 131
Fit docking protocol, 647
Flexible structure alignment, 407–408
Fluid dioleoylphosphocoline (DOPC), 883
  structure of, 883
Fluid lipid bilayers, 883
Fold recognition algorithms, comparing, 748

Fold recognition method, 733
  approach, 733
  perspectives, 735
Folded protein, 176
  free energy, 176
  H/D exchange rates, 176
Folding process of membrane proteins, 884
Foldspin plot, 450
Four-axis space, 213
Fourier space, 148
Fourier transform, 8, 149
  ion cyclotron resonance (FTICR) mass
    spectrometry, 174
Fourier transform ion cyclotron resonance mass
    spectrometers (FTICR-MS), 189
Fourier transform mass spectrometry (FTMS),
    199
Free energy expressions, 580
Frequently occurring domains, 453
FTDock rigid-body method, 602
FTICR mass spectrometry, 177
Full-sized counterparts, 563
Functional epitopes, 851

G-protein-coupled receptors (GPCRs), 887
Gaussian charge densities, 919
Gaussian distribution, 123
GenBank sequence database, 434
Gene-content methods, 561
Gene ontology, classifications, 517
Gene silencing, 64
Generating protein structures, 812–813
  structure-based drug design (SBDD) program,
    812
Genetic algorithm approaches, 797–798
Genome annotation, 541
  availability, 540
  resources, 554
  structural, 554–555
Genome International Sequencing
    Consortium, 4
Genome-sequencing projects, 515
Genome-wide scale, 561
Genomics networks, 888
Genomics threading database (GTD), 544, 554
Geochemical environment, 565–566
Geochemical properties, 566
Geometric hashing algorithm, TESS, 526
Gerstein/Levitt algorithm, 332
Gibbs free energy, 952
Global minimum energy conformation
    (GMEC), 218

Globular protein, 32, 424, 882, 895
  structures, 424
Globular regions, 687
Glucocorticoid receptor, 947–949
  anatomy of, 948
  binding activity, 947
  boundaries, 947
Glucose-6-phosphate dehydrogenases, 426
Glycerophospholipids, 882
Glycosidic torsion angles, 51
GMEC incompatible member, 219
GMEC solutions, 220
Goldstein criteria, 220
  DEE claims, 220
GPCR database, 898
GPCR subfamily classifier, 900
Gram-negative bacteria, 883
Graphical user interface (GUI), 245
Graphics visualization pipeline, 249
Guanine-rich DNA sequences, 60
Guanosine-containing nucleotides, 51
Guanosine nucleoside, 52
  structure, 52

H-bond pattern-based assignment, 465
H/D exchange MS data, 173
  diagram of, 173
H/D exchange reaction, 176
  FTICR mass spectrometry, 174
  quadrupole ion-trap (QIT), 174
HAD-associated polymeric substrates, 1002
  family members, 1004
HAD phosphatase family, 999
HADDOCK method, 602
Halobacterium NRC-1, 769
  genome of, 769
Hammersly-clifford theorem, 116
Harmonic bond, 917
Harmonic potential, 913
Heavy atoms, 79
Helical crystallization, 150
Helical membrane proteins, 884
  folding of, 884
  structures, 886
  topology of, 687
α-helices, 21–23
  diagram presentation, 22
Helicoidal system, 472–473
Heme-like structures, 566
Heterocyclic molecules, 42
Hidden Markov models (HMM), 856
Hierarchical fashion, 919

High-energy barriers, 224
High-quality prediction methods, 468
High-resolution protein structure, 772
High-resolution iterative frequency
        identification (HIFI), 117
High-resolution methods, 775
High-resolution protein folding, 772
High-resolution structure, 887, 948
    prediction, 757
High-speed CCD data collection devices, 78
High-throughput crystallographic macro
        molecular structure, 77
    density modification, 80
    determination, 77
    map interpretation, 81
    molecular replacement, 80
    refinement, 83
    validation, 84
High-throughput proteomics, 584
High-throughput structural genomics, 284
HIV protease database, 330
    inhibitors, 6
HMM-based genome assignments, 428
HMM-based methods, 891
Homo sapiens, 213
Homobifunctional cross-linkers, 189
Homodimeric SCAN domain, 108
Homologous superfamily, 445
Homology-based comparative modeling, 756
Homology detection, approach, 540
Homology modeling, 345
    3D structures, 345
    steps, diagrammatically, 717
HTS liquid samples, 636
Human immunodeficiency virus, 330
Human phosphatase structures, 1002
Human phosphatome phylogenetic tree, 987
    structure of, 976
Human receptor PTPs, expression of, 992
Human T-cell leukemia virus type, 849
Hybrid methods, 619
Hybrid threading/sequence fold, 750
Hydro-phobic-hydrophobic interactions, 945
Hydrogen atoms, 256, 349
    colors, covalent and noncovalent interactions,
        256
    electron density, 349
Hydrogen bond, 460, 466
    calculation, 461
    energy, 461
Hydrogen bond energy calculation, 462
Hydrolytic enzyme cutinase, 967

Hydrophilic probes, 644
Hydrophobic factor, 894
Hydrophobic interactions, 11, 256
    protein-protein interactions, 256
Hydrophobic membrane-exposed surface, 894
Hydrophobic side chains, 379

IgG mouse antibody, 850
    intact anticanine lymphoma monoclonal, 850
        structure of, 850
Image shifts, 148
Immunoglobulin domain, 421
Intrinsically unstructured proteins, *see* Protein
        disorder
in silico hypothetic cross-linking analysis, 195
in silico virtual ligand screening, 1006
Inferential structure determination, 126
Influenza A virus, structure of, 852
Influenza virus neuraminidase, 851
    B-cell epitopes, 852
        structural/functional, 853
Insulinoma-associated protein, 994
    family, 996
Integral membrane proteins, 889
    β-barrels, 889
Interaction partners, 615
Interactive graphics systems, 82
Interface definition language (IDL), 287
International union of crystallography, 325
    cambridge crystallographic data center, 325
Internet resources, 399
Intra-backbone hydrogen bonds, 460
Intrinsic advantages, 563
Investigational new drugs, 811
Ion-accessible volume, 578
Ion cyclotron resonance (ICR), 189
Ionic strength, 576
Isotope-edited NOESY spectra, 106

KAKSI helix, 473
Knobs-into-holes, 682
Knowledge-based scoring, 756

Labeling techniques, 79
    selenomethionyl incorporation, 79
Langevin/Brownian dynamics, 926
Large-scale computational analysis, 891
Large-scale genome-sequencing projects, 530
Large-scale sequencing technology, 559
Last universal common ancestor, 563
Lattice models, 758
    energy evaluation, 761

Lead identification, 822
Lead molecule, 811
  process of optimizing, 811
Leontis-Westhof nomenclature, 48
Leucine residues, 215
  superposition, 215
LFN, 185
  deuterium uptake for, 186
  DXMS Analysis of, 185
  H/D exchange and mutagenesis, 188
  H/D exchange behavior, 185
    pepsin fragmentation, 185
  mutaganesis analysis, 187
LGIC database, 899
Ligand-based perspectives, 641
Ligand-based virtual screen (LBVS) technique,
    825
Ligand-gated ion channels database (LGIC),
    898
Ligand-induced disorder, 946
Ligand binding domains, 948
  sites, 575
Linear analysis of chemical shift (LACS),
    114
Liquid chromatography-mass spectrometry
    (LC-MS), 172
Liquid crystal display (LCD), 244
Low-complexity regions, 544
Low-energy domains, 51
Low-frequency components, 583
Low-quality noninformative data, 212
Low-resolution scoring functions, 765
Low-resolution structure, 757
LUCA's domain, domain content, 564
LUCA fold set, 564
Luzzati plot, 349
Lysozyme, 524
  enzymatic action, 985
  structure, 568

Macromolecular complexes, 585
Macromolecular cystallographic information
    file, 114
Macromolecular dynamics, 926
Macromolecular motions, 331
  database, 331
Macromolecular structures, 321
  access format, 326
  database, 326
  flow of, 324
  popular resources, 322
  sequence/structure analysis, 326

Macromolecular structure database at the
    European bioinformatics institute
    (MSD-EBI), 293
Macromolecular systems, 284
MALDI-TOF/TOF analysis, 195
Manganese transport regulator (MntR), 179
  DXMS analysis of, 179
    deuterated samples, 179
    manganese Transport Regulator, 179
    metal-mediated DNA binding mechanism,
      188
    N-terminal DNA binding domain, 180
Map interpretation, 86
Markov models, 891
Mass spectrometers, 173
  basic components, 173
Mass spectrometry techniques, 193
  limitations, 193
    electrospray ionization mass spectrometers,
      193
    Fourier transform ion cyclotron resonance
      mass spectrometers, 194
    MALDI-TOF mass spectrometers, 193
Matthews' correlation coefficient (MCC), 962
MBT toolkit, 259
MC protocol, 224
  convergence of, 224
MC scoring function, 221
MC simulations, 224
MC variations, 226
MD methods, 926
  macromolecular motion, 926
MD protein simulations, 908
MD simulations, 908, 912, 928
MDB dictionary, 284
Medical subject headings, 335
Membrane-bound proteins, 8
Membrane-bound transporter proteins, see zinc
    transporter
Membrane-spanning residues, 898
Membrane axis, 882
Membrane helix, 687
Membrane proteins, 32, 201, 424, 882, 883, 884,
    888, 897, 898
  3D structures, 886, 896, 900
  folding process, 884
  histograms, 598
  PDB, 884
    variation of, 886
  classification of, 897
  identification/classification, 888, 897
  web collections of, 896

X-ray crystallography, 201
  resources for, 892
  web resources for, 896
  prediction of, 895
Membrane protein data bank (MPDB), 896
  classification of, 898
  determination, 887
  explorer program, 890
  schematic diagram of, 885
  simulations, 908
  synthesis/folding, 885
Membranous cell, 882
  schematic representation, 882
Memory storage, 245
MEMSAT method, 891
Metabolic networks, 551
  evolution of, 551
Metal-containing domains, 566
Metal-containing proteins, 566
Metal ions, 566
Metal/water deletions, 1007
Metalloprotein Database and Browser (MDB),
      331
  extensible markup language (XML), 331
  hypertext transfer protocol, 331
  SQL queries, 331
  XML-RPC-based interface, 331
Microscopic information, 908
Microsecond-to-millisecond time, 99
Microtubule electrostatics, 585
Microtubule protofilament assembly, 585
Midwest Center for Structural Genomics
      (MCSC), 529
Minimalist models, 913
Misguided analysis, 734
mmCIF categories, 283, 285
  schematic diagram, 283
mmCIF data files, 275
mmCIF dictionary, 275, 279, 286, 288
mmCIF methodology, 284
mmCIF molecular entities, 280
mmCIF syntax, 276
Model proteins, 213
Modeling methods, 895
Modern drug discovery, cycle, 635
Modern structure repertoire, 565
Modular design, 962
Molecular biology databases, 323
  nucleic acids research, 323
Molecular docking, 200
Molecular dynamics simulations, 908, 938
Molecular graphics, 237

Molecular modelling database (MMDB), 327
Molecular models, 250
  computer-generated, 253
  implementing algorithms for, 251
  visualization of, 250
  wireframe, stick, CPK, ball stick, 254
Molecular replacement, 80
Molecular surface, 578
Molecular systems, 273
  ribosomal subunit structures, 273
Molecular targets, 814
  assessment, 815
  selecting, 816
Molecular visualization, 237, 238, 243, 260
  advantages/disadvantages, 259
  animation, 257
  applications of, 260
  computer-generated models, 243
  vs. molecular modeling, 258
Molecular visualization program, 246, 256
  graphical user interface, 257
  types of, 258
MolMovDB, 332
  PDB identifier, 332
  XML files and MYSQL, 332
Moloney murine leukemia virus reverse
      transcriptase, 56
Molten globule, 974
Monovalent ion distribution, 579
Monte Carlo algorithms, 119
Monte Carlo–based docking method (MONTY),
      602
Monte Carlo simulations, 223
Monte Carlo transition probability, 926
Moonlighting proteins, chronophin, 998
Morph server software, 332
Mouse Par-3, 112
  structure statistics, 112
MPtopo database, 898
  membrane protein topologies, 898
Multi-domain proteins, 424
Multidimensional phase space, 212
Multilevel hierarchy, 582
Multiline strings, 276
Multiple alignment construction, 740
Multiple alignments proteins, 739
Multiple polypeptide chains, 35
Multiple sequence alignments (MSA), 621, 668
  prediction of, 621
Multiple structure alignment, 406
Multiwavelength anomalous diffraction
      (MAD), 9

N-terminal domain, 948, 950
N-terminal residues, 687
NAMD-based simulation, 926
National Center for Biotechnology Information (NCBI), 327
National Institutes of Health (NIH), 984
Native states, normal mode analysis (NMA) methods, 913
NCBI method, 503
NDB Atlas pages, graphical views, 311
NDB database, 307–309
  applications, 312–316
  processing system, 306
  relational database, 309
  use, 309
  web interface, 309
Neural network algorithm, 942
  RONN, 942
Neural network system, 683
  diagramatical presentation, 683
New York SGX Research Center for Structural Genomics (NYSGXRC), 983
Newtonian equations, 926
Newton's equation of motion, 924
Newtonian molecular mechanics, 908
NIGMS protein structure centers, 983
Nitrogen-containing heterocyclic bases, 42
NLM DTD-based XML document, 334
NMA simulations, 909
NMR analogue, 108
NMR analysis, 115
  front-end of, 115
NMR chemical shifts, 121
NMR data analysis, 123
NMR data classes, 129
  structural/functional genomics, 130
NMR experiments, 9, 295, 950
NMR methods, 96, 887
  sequence-specific assignments, 96
NMR models, 468, 469, 896
NMR signals, 99
NMR spectroscopy, 5, 95, 96, 127, 343, 354
  BMRB, 129
  data analysis, 127
  schematic illustrating, 95
  types of model, 343
  crystallographic factor, 354
  error estimates in, 353
    global parameters, 353
  quality of, 354
NMR strategies, 94

NMR structure, 108, 110, 114
  determinations, 295
  average constraint density, 110
  back-end process of, 125
  chemical shifts, 114
  deposition of, 114
  high-resolution iterative frequency identification (HIFI), 117
NMR structures, 355
  rules of thumb, 355
NMR/X-ray experiment, 717
NMRSTAR dictionary, 284
NOE assignments, 108
NOEASSIGN algorithm, 108
NOEASSIGN module, 107
  automated structure calculations, 107
  NMR molecular replacement, 108
NOE-based refinement, 108
NOESY spectrum, 107
Noncrystallographic symmetry (NCS), 80
Nonredundant database, 214
Nonredundant protein database, 542
Nonredundant structural databases, 214
Nonzero ionic strength, 576
NP-complete problem, 744
Nuclear magnetic resonance (NMR), 241, 907, 940
Nuclear overhauser effect spectroscopy, 343
Nucleic acid binding activity, 527
  binding sites, 528
  helix-turn-helix, 527
Nucleic acid crystallography, 57
Nucleic acid database (NDB), 113, 321
  history of database, 305
Nucleic acids, 42, 341, 366
  3D structures of, 341
  chemical structure, 42
  composition and nomenclature, 43
  structures of nucleic acids, 54–67
  types, 42
  DNA duplexes, 54
  DNA quadruplexes, 60
  RNA duplexes, 60
  structural forms, 60
Nucleophile/phoshoryl acceptor, 997
Nucleotide monomers, 285
Nucleotide polymorphisms, 965
Null hypothesis, 209, 210
NYSGXRC biomedical theme project, 990–994
NYSGXRC protein phosphatase structures, 991

o-glycosyl hydrolase, 524
Oligomeric protective antigen, 184
  binding of Anthrax lethal factor, 184
  LFN, 185
    DXMS analysis of, 185
Oligosaccharide binding (OB)-fold, 357
Optical tweezers, 926
Orientations of Proteins in Membranes database
      (OPM), 898
Overall phylogeny, 562

P-Curve assignment, 472–473, 475
P-Curve parameters, 473
Pair-wise alignments, 400
Pair-wise interaction, 943
Pair-wise sequence alignment, 4
Pancreatic serine proteases, 938
Pancreatic trypsin inhibitor, 908
Parameterization process, 918
Parsimony methods, 564
Particle-mesh Ewald (PME), 919
PAS domain, 181
  DXMS analysis of, 181
Pattern recognition, 158
Patterson-based techniques, 79
Patterson maps, 79
PDBML schema, 286
PECAN software, use of, 121
Peptide amide hydrogens, 175
  exchange rate of, 175
Peptide bond, 19
  graphical structure, 20
Peptide H/D exchange mass spectrometry, 173
  basic principles of, 173
  collision-induced dissociation (CID), 174
  fragments molecular ions, 174
  free induction decay (FID), 174
  H/D exchange chemistry, 174–175
  H/D exchange thermodynamics, 175–176
  mass analyzers, 174
Peripheral site, 1004
PHD key players, 682
PHENIX integrated crystallographic software,
      387
PHENIX system, 83
Phospholipid bilayer, 883
Photoactive yellow protein (PYP), 172, 181, 928
  active-site hydrogen bonding network, 181
  amide H/D exchange for, 182
  crystallographic structures, 182
  dark state, 182
    DXMS analysis of, 182

H/D exchange rates, 182
  integrating structural changes, 184
  signal transduction pathways for, 184
Photoreactive cross-linkers, 192
  UV, 192
Phylogenetic methods, 559, 561
Phylogenetic profiling, 623
Phylogenetic purposes, 563
Physical interaction energies, 459
  usage, 459
Physical techniques, *see* X-ray crystallography
Picosecond timescale, 99
PISTACHIO assignment algorithm, 119
  software package, 127
Platinum-containing drugs, 59
Plekstrin homolgy domain, 453
PLP binding proteins, 999
  degradative pathways, 998
  metabolism, 999
Poisson-Boltzmann theory, 579–580
  calculations, 580
  equation, 582
Polygons, building atoms, 250
Polymer chain identifier, 272
PoRiN Database Server (PRNDS), 898
Position-specific mutation rules, 739
Positron emission tomography (PET), 241
Possible special coordinates, 758
Post-translational modification, 100
Posttranscriptional inhibition of gene
      expression, 64
Prediction methods, 464
Probabilistic data collection, 117
  protein NMR spectroscopy, 117
Probabilistic models, 4
Probabilistic steps, 123
  integration of, 123
PROCHECK program, 363
Profile-based methods, 441
Profile-profile alignment program, 738
Progressive expansion, 550
PROMOTIF program, 328
Proteins, 15, 191, 238, 341, 353, 359, 419–421
  2D drawings of, 238
  3D structures of, 341
  catalysts, 15
  cross-linkers, 191
    chemical reactions, 191
  evolution of, 419
  peptide geometry in, 353
  quaternary structure, 35–37
  Ramachandran Plot, 359

Proteins (*Continued*)
  secondary structure, 20–25
  solvent-accessible surface, 255
  tertiary structure, 25–30
  three-dimensional structures, 560
Protein alphabet, 971
  hypothetical diagram, 972
  structure-sequence relationship, 971
Protein binding, 176
  free energy, 176
Protein chain backward, 362
  electron density, 362
Protein classes, 449
Protein coding sequence, 739
Protein complex, 913
  chemical cross-linking and massspectrometry, 190
  determination, 190
Protein core elements, 461
Protein crystal structures, 108, 937
  molecular replacement methods, 108
Protein curvature-based assignment, 472
Protein Data Bank (PDB), 4, 9, 32, 78, 114, 293, 321, 427, 433, 448, 517, 670
  BMRB, 125
  Brookhaven National Laboratory, 293
  content of, 295
  contents guide, 272
  cryo-electron microscopy, 293
  electron tomography, 345
  exchange dictionary (PDBx), 271, 283, 284, 294
  exchange, 283
  file, 215, 273
  fold space, 213
  growth chart of, 297
  header records of, 348
  history, 272
  information on, 323
  macromolecular structure, 323
  neutron diffraction, 345
  PDB_extract software, 114
  primary information, 323
  structures of, 855
  validation approach, 122
  X-ray crystallography, 293
  X-ray structures, 349–350
Protein Data Bank Data access, 298
  FTP, 298
  Web sites, 299
  Data acquisition, 294
  data deposition sites, 294
  validates structures, 294

Data category, 276
Data collection information, 295
Data deposition sites, 294
Data dictionaries, 283
Data processing statistics, 297
Protein Data Bank (PDB) format, 271, 272, 273
  chart, 297
  data, 276, 286
  identifiers, 554
  Markup language, 271
  representation of, 286
Protein-DNA complexes, 305, 314
  interfaces of, 314
Protein/DNA structures, 387
Protein-DNA interactions, 600
Protein-DNA interface, 316
  hydration, 316
Protein-DNA recognition, 593
Protein-DNA systems, 596
  ROSETTA potential function, 596
Protein-ligand binding (PLB) index, 648
  model, 825
Protein-ligand complexes, 78
Protein-ligand interactions, 329
Protein/nucleic acid complexes, 307, 600
  flexibility in, 600
  histone proteins, 307
  ribosomal, 307
Protein-nucleic acid interactions, 594, 602, 606
  AMBER package, 595
  applications, 602
  CHARMM package, 595
  classification of, 596
  DNA deformation, 595
  hidden Markov model (HMM), 595
  statistical potential functions, 596–599
    molecular docking, 602
  modeling of, 606
    predictive tool, 606
  structure-based prediction of, 600
Protein-protein interactions, 434, 585, 615, 948, 962, 978
  binding sites, 978
  flexibility, 821
  hot spots, 820
  interfaces, 227
  targets, 820
Protein-RNA complex, 607
Protein-RNA interactions, 596, 603, 607
Protein-RNA recognition, 593
Protein-RNA structures, 603
  computational analysis of, 603

Protein design, 962, 976
  energetics, 977
  experimental/computational approaches, 976
Protein design scheme, 221
Protein disorder, 937–940, 943–945, 949–951
  biological consequences, 945–947
  biophysical explanation, 950–954
  computational approaches, 940–945
  definition, 941
  disorder predictors, 939
  ensemble-based description, 951
  importance of, 938
Protein domains, 97, 423, 429, 540, 560,
      561,566–568, 962
  α-helices, 424
  β-sheet, 424
  content, 560
  distribution, 567
  evolutionary history, 566
  history of, 566
  phylogeny by, 561
  position of, 561
Protein domain evolution, 966–971
  determinants, 964
  fragmental rearrangements, 964
  mechanisms, 966
Protein domain loss, identification of, 566
Protein domains/motifs, 27
Protein domain universe graph (PDUG), 967
Protein energetic conformational analysis from
      NMR chemical shifts (PECAN), 120
  energy function, 120
Protein engineering, 961
Protein evolution, 31–32
Protein expression systems, 100
Protein families, 330, 429
Protein fold, 420, 423–424, 460
  architecture, 420
  hierarchical levels, 423
  identification of, 424
  SCOP classification, 423
Protein function server, 519
Protein homology modeling, 716
Protein interactions, 322
  resources of, 322
Protein kinases, 825
  ATP cleft of, 825
  like superfamily, 566
  role of informatics, 811
Protein model, 913
Protein modelling, 715, 824
  goal, 715

Protein modifications, 30
Protein motion, 332, 477, 909, 911
  characteristic division, 909
  classified, 332
  simulation, 926
  timescales/size/simulation, 922
Protein NMR spectroscopy, 117, 124
  modeled relationships linking, 124
  multidimensional data collection, 117
Protein phosphatases, 983–987
  bargraph, 989–990
  role, 992
  significance, 986
  structural studies, 989
Protein phosphorylation, 985
Protein purification, 887
Protein quaternary structure (PQS), 328
  PDB formatted files, 329
Protein secondary structure, 120
Protein sequence, 433, 567, 737, 743
  analysis, 737–739
Protein side-chain packing, 601
  dead-end elimination (DEE) algorithm, 601
Protein sphere, 576
Protein stability, 973
  lessons, 973
  stability, 973
Protein structure, 15, 35, 459, 679
  analysis, 489, 831
  determination, 94
  domains, 566
  evolution, 970
  four-tier hierarchy, 35
  history, 576
  introduction, 575
  manipulation, 961
  mechanism for, 968
  prediction, 679, 776
  primary structure, 16–20
  quaternary, 459
  representation, 402
  resources classifying, 322
  significance, 15
  tertiary, 459
Protein structure validation, 322
  software and resources for, 322
Protein tyrosine, 987
  dendrogram of, 987
Protein tyrosine phosphatase, 990–994
Protein ubiquitin, 127
  representation of automated assignments
      determined, 127

Pseudo-rotational parameters, 308
PTP/DSP tree, active site motif, 988
PTP1B active site, 1005
  surface representation of, 1005
Public library of science (PLoS), 334
Purine/pyminidine bases, 799
  interacting edges, 799

QM-MM methods, 910, 916–917
  simulations, 927, 928
    electronic structure theory, 927
Quality checks software, 367
Quantitative structure-activity relationship
    (QSAR), 221
Quantum mechanical, 634
  calculations, 910
  receptor/ligand, 634
Quaternion dynamics for rigid fragments, 925

Ramachandran plot, 214, 360, 668
  D-amino acids, 360
  differences in, 360
  protein structure, 360–361
  schematic representation, 21
Ramachandran/geometry evaluations, 385
Raman spectroscopy, 460
Random errors, 341, 342
Rapid sampling techniques, 977
Rational drug design, 6
Receiver-operating characteristic (ROC),
    823
Receptor-based drug design cycle, 636
Recognition algorithms, 735
Recurring domains, 549
Reduced complexity models, 758, 761
  scoring functions, 761
Reductive evolution, 623
Refinement methods, 84, 109
  final stages of, 109
    NOE assignments, 109
Refine NMR structures, 113
  software packages, 113
Regulatory networks, 552
Relaxation dispersion methods, 99
REMARK records, 273
Replica exchange methods, 775
Research Collaboratory for Structural
    Bioinformatics (RCSB), 114
Resolving symmetric dimers, 106
Restrictive models, 953
Ribbon models, 254
Ribbons algorithm, 254

Ribosomal RNAs (rRNA), 791
Ribosome, 63, 65–67
  50S subunit, 65
  70S subunit, 65
Rigid-body based approaches, 938
Root-mean-square distance, 774
  values, 386
RNA crystal structures, 798
RNA duplexes, 60, 791
RNA enzymes, *see* Ribozymes
RNA expression data, 5
RNA interference (RNAi), 64
RNA molecule, forms, 11, 62, 791
  role in posttranscriptional gene control, 64
RNA ontology consortium (ROC), 316
RNA processing events, 996
  capping, 996
  processing, 996
RNA structure models, 792, 800
  history of, 792
  predicting secondary structures, 793
    methods for, 793
  motifs, 800
Robust database, 305
Root-mean-squared deviation (RMSD), 354,
    668, 973
Rosetta method, 766
  design algorithm, 601
  function, 762
Rosetta structure prediction protocol, 768
  schematic presentation, 768
Rotamer description, 215
Rotamer frequencies, 225
Rotational/translational parameters, 45–46
  pictorial definitions, 47
  structural illustration, 47
PTPs tandem phosphatase domains, 993
  structural comparison, 993

Saccharomyces cerevisiae, 891
SAR investigations, 636
Scanning tunneling microscopy (STM), 241
  molecular imaging, 241
  schematic representation, 970
SCMF, 222
  applications of, 222
  approach, 225
    SCMF-biased MC method, 225
SCMF methods, 225
SCMF, Mean field theory, 221
SCP catalytic sites, 997
Sec61 complexes, 886

Secondary structure elements, 962
   α-helices, 964
   applications, 478
   methods, 464
   schemes, 475
   β-strands, 962
   definition, 471
SecY/Sec61 translocon complex, 886
   structure of, 886
SEG, 941
Self-consistent field methods
Sequence-based disorder predictors, 939
   DISOPRED, 939
   PONDR, 939
Sequence-based methods, 436, 505, 619, 858
   categories, 505
   homology recognition, 733
Sequence-independent terms/secondary
     structure arrangement, 766
Sequence-only fold recognition/distant
     homology, 741
   overview, 741
Sequence alignments, 619
   specific assignments, 105
Sequential structure alignment program (Ssap),
     443
Ser-His-Asp catalytic triad, 523
Serine/threonine protein phosphatases, 986
Side-chain conformations, 215
Side-chain flipping, 215
Signal peptides, 688
Signal recognition particle (SRP), 884
Signal transduction pathways, 8, 985
   effects, 985–988
Simian immunodeficiency virus (SIV), 330
Simple object access protocol (SOAP)
     technology, 429
   web services, 529
Simulation methods, 911
Single nucleotide polymorphisms (SNPs), 975
Site-directed labelling, 946
Small molecules, 239
   representations of, 239
SMILES string format, 237
Solute-solvent interactions, 926
Solvation-based scores, 761
Solvent-screened interactions, 576
Spectroscopic techniques, 940
   Fourier transform infrared, 940
   infrared circular dichroism, 940
SPHGEN algorithm, 644
SRP complex, 885

SSAP algorithm, 444
SSEP-domain, 506
Stable isotopes, 100
Statistical potential function, 597
   creation of, 597
   distance-dependent, 598
   orientation-dependent, 598
Stereo-array isotope labeling (SAIL) *see*
     Designer labeling scheme, 97
Stereochemical check score (SCS), 327
Stereochemical parameters, 359
   bad contacts, 363
   side chain torsion angles, 363
Stick model, 252
Stimulus-sensing domain, 962
Stochastic methods, 226
Stochastic *vs.* Deterministic search methods,
     217
Straw-clutching measures, 345
STRIDE residue number, 470
Structural bioinformatics, 4, 5, 11, 207, 815
   databases, 78, 831
Structural classification of proteins (SCOP), 419
   data set, 499
   database, 419, 426–430
   domains, 428, 545
   hierarchy, 421, 427, 428, 429
   perspective, 430
   usage, 430
Structural databases, 334
   impact of structural genomics, 335
   integration over multiple resources, 334
   targetDB, 335
Structural domains, 486
   algorithms, 488–493
   definition, 486–487
   features, 567
Structural genomics, 336, 428
   pipelines, 887
Structure-activity relationships and homology
     (SARAH), 825
Structure alignment algorithms, 745
Structure-based drug design (SBDD), 815
Structure-based drug discovery cycle, 634
Structure-based genome annotation, 554, 603
   resources, 553
Structure-based method, 530
Structure-based pharmacophore (SBP), 827
Structure-based virtual screening, 822
Structure activity relationship and homology
     (SARAH), 815
Structure predictions, evaluation of, 775–776

Structure/energy-based fold recognition
    methods, 733
Sulfhydryl-reactive reagents, 192
Supramolecular assemblages, 575
Swiss-model, 333, 345
    PDB format, 333
        ProModII, 333
    steps, 333
PDBViewer, 333
server, 333
Symmetry-related molecules, 296
NOESY distance, 107
Systematic errors, 341
Systems biology, 3

T. gondii, 1008
    X-ray structure of, 1008
Tandem affinity purification (TAP), 622
Target-ligand complexes, 1006
Targeting protein-protein interactions, 819–820
TATA binding protein-DNA complex, 57
Tentative domains, 494
Text strings, 276
Thermodynamic hypothesis, 743
Thimet metallo endo-peptidase, 478
Thin film, 144
Threading algorithm, 196, 743
    distance constraints, 196
    overview, 742
Threading approximations, 744–746
Threading force field, 743–744
Threading methods, 345
Threading using cross-link derived constraints,
    196
    postfiltering of, 196
Three-Dimensional Reconstruction, 147–148
    2D Fourier transform, 148
    crystalline arrays, 149
        application for, 152
        charge-coupled device (CCD), 152
        of eukaryotic cell, 153
        single-particle image-processing methods,
            153
    helical assemblies, 149–150
    imaging protein dynamics, 153–155
        actin- and microtubule-based motor
            proteins, 154
        electron microscopic reconstructions, 155
    motor proteins, 150
    single-particle analysis, 150–152
Three-domain combinations, 567
TIM barrel, 443, 449, 498

fold, 453, 735
Time-dependent classical macromolecular
    motion simulation, 916
Time-dependent mechanism, 907
Time-dependent simulation, 908
Time-resolved crystallography, 910
TMDET algorithm, 894, 896
TMDET score, 898
TMH predictions, 688
Top-scoring docked structure, 200
Top-scoring population, 226–227
Torsion angles, 469
    distributions, 296
Toxic effects, 887
Transcription factors, 552
Transfer RNAs (tRNA), 791
Translation-libration-screw (TLS) parameters,
    86
Translocons, *see* Sec 61 complexes
Transport proteins, 899
Transport systems, 899
Transporter classification database (TCDB), 898
    system, 899
Transverse relaxation-optimized spectroscopy
    (TROSY), 96
tRNA anticodon base pairs, 65
tRNA family, 791
    multiple sequence alignment of, 793
TROSY methodology, 97
Truncation-based approximation schemes, 919
Tsukuba Workshop on nucleic acid structure and
    interactions, 307
Two-domain proteins, 549
Tyrosine kinases, 515

Validate protein structures, 365
van der Waals interaction energy, calculations,
    463
van der Waals radii, 256
van der Waals sphere, 495
Vector alignment search tool (VAST), 327
Versatile technique, 623
Virtual ligand screening, 1006
Visual images, 237
VoTAP helices, 472

Water-catalyzed reaction, 174
    pH values, 174
Water-soluble proteins, 887, 890, 893, 895
Water refinement, 110
    CYANA calculation, 110
    validation of structural models, 111

Watson-Crick/Hoogsteen edges, 48
Watson-Crick base pairs, 48
  arrangements, 45
Web-based resources, 898, 900, 901
Web browser, 554
Web resources, 955
Well-represented proteins, 213
WHAT_CHECK program, 365
Wheat germ cell-free system, 100
Whole genome domain, 561
Wireframe model, 252
  molecular models, 252
Worldwide PDB, 293, 341
  biological magnetic resonance bank (BMRB),
    294
WPD loop, 995
wwPDB member sites, 294

X-DEE/extended-DEE application, 221
X-ray crystallography, 10, 77, 94, 127, 144,
    157, 248, 341, 342, 347, 348, 539,
    851, 937
  B-factor model, 342
  crystallographic data, 5
  data, 242
  error estimates, 347

high-resolution structural information, 157
protein model, 342
rules of Thumb, 353
schematic diagram, 348
types of model, 343
  standard uncertainties, 347
X-ray diffraction, 347
  data, 351
  pattern, 8
X-ray models, 344
X-ray structures, 348, 350, 355
  global parameters, 348
X/Y displacements base pair, 46
XML attribute, 288
XML data, 286
XML-encoded definitions, 287

Yeast phenylalanine tRNA, 63
  crystal structure, 63

Z-DNA binding protein, 58
Zinc-finger motif, 31
Zinc-rich environment, 566
Zinc transporters, 8
ZNF24 SCAN homodimer, automated structure
    calculation of, 109